Basic Physics

Basic

Physics

By

Prof. (Dr.) G.C. Trigunayat
(Former Head, Department of Physics and Astrophysics, University of Delhi)

BPB PUBLICATIONS
B-14, CONNAUGHT PLACE, NEW DELHI-110001

FIRST EDITION 2012

ISBN : 978-81-8333-461-7

LIMITS OF LIABILITY AND DISCLAIMER OF WARRANTY

The information contained in this book is true and correct and the best of Author's & Publisher's knowledge. We are committed to serve students with best of our knowledge and resources. We have taken utmost care and attention while editing and printing of this book but Author and Publisher should not be held responsible for errors, omissions and any other unintentional mistake that might have crept in. However, errors brought to our notice shall be gratefully acknowledged and attended to.

Distributors:

COMPUTER BOOK CENTRE
12, Shrungar Shopping Centre,
M.G.Road, BENGALURU–560001
Ph: 25587923/25584641

MICRO BOOKS
Shanti Niketan Building,
8, Camac Street, KOLKATA-700017
Ph: 22826518/22826519

BPB PUBLICATIONS
B-14, Connaught Place,
NEW DELHI-110001
Ph: 23325760/43526249

MICRO MEDIA
Shop No. 5, Mahendra Chambers,
150 DN Rd. Next to Capital Cinema,
V.T. (C.S.T.) Station, MUMBAI-400 001
Ph: 22078296/2207829

BPB BOOK CENTRE
376 Old Lajpat Rai Market,
DELHI-110006
Ph: 23861747

DECCAN AGENCIES
4-3-329, Bank Street,
HYDERABAD-500195
Ph: 24756967/24756400

INFOTECH
G-2, Sidhartha Building,
96 Nehru Place,
NEW DELHI-110019
Ph: 26438245

BUSINESS PROMOTION BUREAU
8/1 Ritchie Street, Mount Road,
CHENNAI-600002
Ph: 28410796/28550491

Published by Manish Jain for BPB Publications, B-14, Connaught Place, New Delhi-110001 and Printed by him at Balaji Printer, New Delhi.

DEDICATED TO THE MEMORY OF

|

MY LATE PARENTS

Preface

The presently available physics books usually have a frightening size (running into 1500 pages or more). My attempt has been to reduce the size as much as possible, without losing on the necessary details, whether conceptual or otherwise. Accordingly, all descriptions in the book will be found brief and to the point. The subject matter has been presented in a simple language, in an easy to understand manner. Clarity of expression has been specially aimed at. The emphasis has been on quality, while not doing away with necessary quantity.

Apart from the solved examples in the text, each chapter is followed by a set of variety of problems, viz. conceptual, solved numericals, numerical exercises, and three types of questions. They include **HOTS** (High Order Thinking Skills) questions. To avoid 'spoon-feeding', no readymade answers have been provided to the questions. The students can find the answers themselves by searching in the text, which will help them to gain some more knowledge beyond what they may be seeking initially. Alongwith, the students are further advised to look into the **SOLVED CONCEPTUAL PROBLEMS,** which contain due answers to several important questions. All kinds of problems provided in the book are of a representative nature, covering various difficulty levels. However, problems of very high difficulty level have been excluded, in harmony with the latest education policy.

The book should be found equally useful to the engineering and medical streams of the students, as also to the candidates for the entrance examinations for these streams.

A **SUBJECT INDEX** has been provided at the end of the book, for a ready reference. The readers will find it extremely useful in quickly locating the description of a desired topic.

The author will feel duly rewarded, if the readers, the students in particular, find the book beneficial.

I am pleased to acknowledge with gratitude the necessary help received at various stages from profs. Gulzari Lal Bhalla, S. Annapoorni, Binay Kumar, Dr. Kulvinder Sing and Sh. M. M. Sharma. The help received in the process of publication and printing from Dr. G. C. Jain, Sh. Manish Jain, Sh. Mohd. Shafique Siddiqi and Sh. Mohd. Shakeel is also thankfully acknowledged.

(G.C. Trigunayat)

For the Kind Attention of the Readers

1. **A SUBJECT INDEX** has been provided at the end of the book, for ready reference. As an example of its utility, if a reader is looking for the description of **Full Wave Rectifier**, he simply has to look under either 'F' or 'R' to locate the relevant page number (P.538).

2. For a ready use by the students, as many as 7 **APPENDICES** have been provided towards the end of the book. They include the Greek alphabet and the prefixes used for various measurements, various systems of units, some useful conversion factors and mathematical signs & relations, some fundamental physical constants and important astronomical values, and important properties of some selected compounds. The Appendix A includes Hindi pronunciations, too, of the Greek alphabet.

3. Special care has been taken to make the diagrams simple, yet fully expressive. It will help the students not only to understand the subject matter conveniently, but also to reproduce them, wherever necessary.

4. Boxes, with the heading **Additional knowledge** have been provided in the various chapters. They are intended to supplement the students' knowledge, particularly in relation to uses in common life.

5. The book consists of 10 Units. Each Unit consists of two Chapters, except for the small Units 5, 7 and 10. Thus, there are a total of 17 Chapters. The Figure Nos. and the Equation Nos. find continuation between the two chapters belonging to a particular unit.

6. Some terms of common usage in both the British form and the American form have been used in the book, e.g. centre, metre, colour, travelled, analyser, polarization (British usage) and center, meter, color, traveled, analyzer, polarization (American usage).

CONTENTS

UNIT 1
(Electrostatics)

CHAPTER 1 (A)
(Electrostatic Charges, Force & Field)

CHAPTER 1(B)
(Electric Potential, Flux and Capacitance)

UNIT 2
(Current Electricity)

CHAPTER 2(A)
(Electric current, voltage, resistance and their measurement)

CHAPTER 2(B)
(Electric Energy and Power; Electrolytes)

UNIT 3
(Magnetic Effects of current and magnetism)

CHAPTER 3 (A)
(Magnetic Effects of Current)

CHAPTER 3(B)
(Modern Magnetism)

UNIT 4
(Electromagnetic Induction & Alternating Current)

CHAPTER 4(A)
(Electromagnetic Induction; Self & Mutual Inductance)

CHAPTER 4(B)
(Alternating Current, AC & DC Generators and Transformer)

UNIT 5
(Electromagnetic Waves)

CHAPTER 5
(Electromagnetic Waves)

UNIT 6
(Optics)

CHAPTER 6(A)
(Ray optics; Reflection & Refraction; Optical Instruments)

CHAPTER 6(B)

(Wave Optics; Huygens' Construction; Interference; Diffraction; Polarization)

UNIT 7
(Dual Nature of Radiation and Matter)

CHAPTER 7
(Dual Nature of Radiation and Matter)

UNIT 8
(Atomic Nucleus)

CHAPTER 8(A)
(Composition & Properties of Nucleus; Radioactivity)

CHAPTER – 8(B)
(Nuclear Forces, Nuclear Fission & Fusion)

UNIT 9
(Solids & Semiconductor Devices)

CHAPTER 9(A)
(Energy Bands in Solids; Semiconductors)

CHAPTER 9(B)
(Semiconductor Devices)

UNIT 10

(COMMUNICATION SYSTEMS)

CHAPTER 10
(COMMUNICATION SYSTEMS)

Appendices

NOTE

Each chapter of the book is followed by the following items:

(I) Solved Conceptual Problems (Including HOTS Questions) (HOTS: High Order Thinking Skill)

(II) Solved Numerical Problems

(III) Very Short Answer Questions and Short Answer Questions (Including HOTS Questions) + Long Answer Questions

(IV) Numerical Exercises

UNIT 1
ELECTROSTATICS

CHAPTER 1 (A)
Electrostatic Charges, Force & Field

CHAPTER 1 (B)
Electric Potential, Flux & Capacitance

CHAPTER 1 (A)

ELECTROSTATIC CHARGES, FORCE & FIELD

Electrostatics is the branch of physics which deals with the study of electricity at rest. Since the charges are static, i.e. they do not flow from one body to another, such electricity is also termed as **'static electricity'.** The localized static charges can be easily produced by rubbing two suitable bodies together, e.g. glass and wool; so it is also called **'frictional electricity'.**

1A. 01. Historical

About 2500 years back the Greek philosopher Thales observed that when amber, a yellowish resinous substance, was rubbed against fur, it acquired the property of attracting light bodies like pieces of paper or straw. That is how the concept of **'electrification'** was perceived, which we now better understand in terms of the so-called **'electric charges',** to be explained in the next section.

However, this finding went largely unnoticed until 1600A.D. when the British scientist William Gilbert discovered that several other materials, like glass and ebonite, when rubbed with suitable materials developed a similar property of attraction. He pointed out that the rubbing produced two kinds of electric charges, viz. **'resinous'** on the resins like amber and **'vitreous'** on the substances like glass. Later, towards the middle eighteenth century, the American scientist Benjamin Franklin assigned these charges as negative and positive, respectively. These terms continue to be used till today. Two bodies with like charge (both negative or both positive) repel each other and those with unlike charge (one positive and the other negative) attract each other.

The Latin name for amber is **'electrum'** and the Greek name is **'electron',** which gave birth to the present familiar terms like **'electric'** and **'electricity'.**

1A. 02. Explanation of 'Electrification' by Friction; Origin of Electric 'Charge'

The **'electrification'** produced by rubbing together of two bodies is said to create **'electricity'** in the bodies. We still do not know what it exactly is. But broadly we may describe it as an unknown agent which makes its existence felt by some effects that it produces, e.g. the aforesaid observed effect of attraction of small bits of paper, straw, etc., towards the **'electrified'** (or **'charged'**) bodies. The quantity of **'electricity'** possessed by a body can be expressed in term of the so-called **'electric charge'** carried by it. Thus, in a qualitative way, we can **define 'electric charge'** on a body as a measure of the degree of electrification of the body. Conventionally, the quantity of charge on an electron is assigned as one unit of negative charge, which is also called **'elementary charge',** expressed as **'e'.** Similarly, the charge on a proton is assigned as one unit of positive charge. As further described in Section 1A.06, the unit employed for the measurement of electric charge is coulomb, denoted as C. In absolute terms the conventional 'elementary charge (e)' on an electron is -1.6×10^{-19} C.

Although the process of 'electrification' produced by rubbing together of two bodies remained a mystery for the early scientists for more than two thousand years, it is quite easy to understand for a present day student who is fairly familiar with the concept of atoms and atomic structure. According to the classical picture of an atom, the nucleus of the atom consists of positively charged protons and electrically neutral neutrons. Negatively charged electrons revolve around the nucleus in fixed orbits (or shells). The number of electrons is equal to the number of protons and so normally an atom is always electrically neutral. The electrons in the various shells (and sub-shells) are held in position by a balance between their strong net force of electrostatic attraction towards the nuclear protons and the large centrifugal force resulting from their fast orbital motion (which is often upto about one tenth the velocity of light), in the opposite direction.

The force of attraction progressively decreases for higher orbits. Therefore, compared to the electrons of inner shells the electrons in the outermost shells are very weakly bound to the nucleus. An additional factor is that if opportunity is available all elements tend to acquire a stable inert gas configuration of electrons in the various shells, viz. with the number of electrons as 2, 8, 18, 32, etc., in the K, L, M, N, etc., shells. For example, sodium (Na) has one electron in its outermost shell, which it tends to give away in order to attain the stable electronic structure of neon (Ne), the inert gas preceding sodium in the periodic table. Chlorine (Cl), on the other hand, has one less electron than the inert gas argon in its outermost shell and so it always tends to receive an extra electron. Thus, when Na and Cl atoms come together, the 'extra' electron of the former transfers to the latter, thus satisfying the electronic requirement of both. (This transfer turns the Na atom into a positive ion and the Cl atom into a negative ion. Because of their opposite electric charges the two ions are pulled together by the electrostatic force between them, forming a stable NaCl molecule. Consequently, the compounds like NaCl are termed as 'ionic compounds').

The process of rubbing two bodies together provides necessary energy to the outermost electrons of some surface atoms to be detached from one body, to become free and settle down on the other body. When a glass rod is rubbed with silk, the negatively charged electrons are detached from the glass rod. So the glass rod is left with a net positive charge, which renders it 'positively electrified,' or better say 'positively charged'. If all the detached electrons are transferred to the silk, both the glass rod and the silk will have equal amount of charge, but of opposite nature. On the other hand, when an ebonite rod is rubbed with a piece of flannel, the electrons from the flannel are stripped off and settle upon the ebonite rod. Thus the flannel becomes positively charged and the ebonite negatively charged.

The rubbing of any substance against another substance under suitable conditions results in detachment and subsequent transfer of electrons, thus making them electrically charged. But the extent of charging may vary considerably from one substance to another. The basic factor to consider here is work function, which is defined as the energy required to remove an electron from a solid at absolute zero of temperature (–273 °C). Although laboratory experiments are usually conducted at room temperature, which is nearly 300 °C higher than absolute zero, the work function of a material can still be taken as a rough guide for the charging produced by rubbing the material. It is obvious that when two solids are rubbed together, the one with lower work function will lose more electrons. Thus subsequently it is found as positively charged, and the other solid is found as negatively charged.

1A. 03. Repulsion as a Sure Test for Determining the Nature of Charge; Charging by Induction & Conduction

The kind of charge on a given body can only be tested by the repulsion produced when another body with similar charge is brought near to it, because attraction will be produced not only when a body with dissimilar charge is brought near to it but also when an uncharged body is brought near to it. For the same reason, to know if a body is charged or not, both a positively charged body (e.g. a rubbed glass rod) and a negatively charged body should be brought near to it turn by turn. If the body is uncharged, attraction will be experienced in both the cases. Otherwise, attraction will result in one case and repulsion in the other case. Thus we find that repulsion, not attraction, is the sure test for determining the nature of charge (positive, negative or neutral) on a given body. [A positively charged glass rod attracts an uncharged object like a piece of paper because the loosely bound electrons at and near the edge of the paper close to the rod get detached and stick to the edge, making it negatively charged at the edge. Similarly, a negatively charged ebonite rod attracts the paper piece because it repels the electrons at the edge, thus leaving it with a localized net positive charge there. This process of charging of a neutral object like paper, in which a localized charge of opposite sign develops on a neighbouring object, is called **'induction'.** A given body can also be charged by connecting it to an already charged body, e.g. by a wire. Then the sign of charge on both the bodies is same. This process of charging a body is called **'conduction'.**].

1A. 04. Quantization of Charge

In the eighteenth century, the electric charge was visualized as a continuous fluid which, in quantitative terms, could have all possible values . However, our later experimental knowledge, acquired through many important experiments (e.g. the famous Milikan oil drop experiment), has revealed that the charge on a body is always a whole number multiple of the elementary charge e. It may be expressed as $q = ne$ (where $n = \pm 1, \pm 2, \pm 3, \pm 4,$ ....). The + and – signs refer to positive and negative charge, respectively. n is always an integer and can never have a fractional value, i.e. we can never have a fractional charge like 0.5e or 25.7e on a body . In other words, the charge cannot be continuous or, say, it cannot be divided indefinitely.

This property of charge is called **'quantization of charge'** (sometimes also as 'atomicity of charge'). We may **define** it as "the property by virtue of which any charge can only exist as a discrete value, which is an integral multiple of the elementary charge e".

It follows that the change in charge on a body can only be in steps, not continuous. The smallest step is e. Since the value of e is extremely small, viz. -1.6×10^{-19} C, normally such changes go undetected. The smallness of this value may be judged from the fact that the little charge which is produced by rubbing a glass rod with silk is of the order of one microcoulomb, i.e. 1μC ($=1\times10^{-6}$ C). The number of elementary charge e that it contains is,

$$N = \frac{1\times10^{-6}\text{ C}}{1.6\times10^{-19}\text{ C}} \approx 6\times10^{12}$$

Thus, even this little charge developed on the glass rod consists of nearly six thousand billion elementary charges. Clearly, it will be impossible to detect the change in charge by just one step, viz. 1.6×10^{-19} C. Thus in daily life and in laboratory experiments any changes in the magnitude of flow of current, which consists of moving charges, appear continuous and not stepwise. They can be detected only through very sensitive special instruments. That is why the earlier scientists were led to think of electrical charges as 'electric fluid'.

Although the concept of 'quantization of charge' is purely an experimental fact, it perfectly matches with the other known concepts of quantization of matter (which consists of discrete atoms and molecules), quantization of energy, and quantization of angular momentum. The concept is not difficult to grasp for a present day student who now knows that an electric current consists of moving charges, which are usually electrons, that can only be in whole numbers and never in fractions. The same applies to static charge residing on a body, e.g. the positive charge produced on a glass rod by detachment of electrons, when the rod is rubbed against a silk cloth.

Neutrons (no charge) and protons (positively charged; of the same charge as for electrons, viz. 1.6×10^{-19} C each) are now known to consist of 'quarks' which carry fractional charges, $\pm e/3$, $\pm 2e/3$,............ etc. But the 'quarks' cannot be detected individually. They are not reckoned as elementary charge. Thus their discovery does not mean that the charge is not quantized, i.e. it is not an integral multiple of the elementary charge e.

1A. 05. Conservation of Electric Charge

As mentioned in Section 1A. 02, when an ebonite rod is rubbed against flannel, the flannel becomes positively charged and the ebonite becomes negatively charged. When the quantity of charge on both is measured, it is found that (if no charge is lost in the process of rubbing) the positive charge on the flannel is exactly equal to the negative charge on the ebonite. The same is found to be true in any other such process, e.g. glass rubbed with silk. It follows that the process of rubbing neither creates nor destroys any charge. It simply transfers the charge from one body to another. The total charge remains the same before the rubbing or after the rubbing, i.e. the total charge is conserved during the process of rubbing. This fact is found to be true for all the processes in nature that involve a transfer of charge within a given system, i.e. no charge enters or goes out of the system. A few more examples will be given in the following. Thus, we are led to the following **principle of conservation of charge:**

"The total amount of electric charge in an isolated system remains constant. The charge can neither be created nor destroyed, but can only be transferred from one part of the system to another".

The student has earlier learnt about the laws of conservation of energy, linear momentum, and angular momentum. The conservation of charge is a further addition to this list.

Other examples

(i) Annihilation processes: When a negatively charged electron (charge -e) and its positively charged antiparticle positron (charge +e) come together, they annihilate each other to produce two uncharged γ-ray photons,

$$e^- + e^+ \rightarrow \gamma + \gamma \qquad \text{(ANNIHILATION)}$$

Note that due consideration has to be given to the sign (+ or –) of the charges, when added up. In the above equation, the net charge is zero on both the sides. So the charge is conserved.

(ii) Pair production process: It is the converse of annihilation process; now a γ-ray photon transforms into an electron and a positron,

$$\gamma \rightarrow e^- + e^+ \qquad \text{(PAIR PRODUCTION)}$$

So, again the charge is conserved. (The γ-ray energies are different in the processes of annihilation and pair production).

(iii) Radioactive decay process: In this process a particular nucleus transforms into another nucleus. For instance, a nucleus of the uranium isotope ${}^{238}_{92}U$ (charge: 92e) decays into a thorium nucleus ${}^{234}_{90}Th$ (charge: 90e) by emitting an α-particle, which is a helium nucleus, ${}^{4}_{2}He$ (charge: 2e)

$${}^{238}_{92}U \rightarrow {}^{234}_{90}Th + {}^{4}_{2}He \qquad \text{(RADIOACTIVE DECAY)}$$

Again, the charge is conserved.

Example 1. A conductor has a positive charge of 9.6×10^{-19} C on it. How many electrons are in excess or short on it ?

Solution. Given : $q = 9.6\times10^{-19}$ C

We know, $q = ne$ ($e = 1.6\times10^{-19}$ C)

$$\therefore \quad n = \frac{q}{e} = \frac{(9.6\times10^{-19})}{(1.6\times10^{-19})} = 6 \quad \textbf{(Answer).}$$

Since the conductor has positive charge, it is short of 6 electrons.

Example 2. A copper nucleus is surrounded by 29 electrons. The atomic weight of copper is 63.5 g/mole. If one electron is removed per 1000 atoms from a copper piece of weight 10 g, what will be the charge on it ? Given that,

$$e = 1.6\times10^{-19}\text{ C}; \quad \frac{1}{4\pi\epsilon_0} = 9\times10^9\ \text{Nm}^2\text{C}^{-2}; \quad \text{Avogadro no.} = 6\times10^{23}/\text{ mole}$$

Solution. As 1 mole (63.5 g of copper) contains 6×10^{23} atoms, the number of atoms in 10g of copper will be

$$N = \frac{(6\times10^{23})\times(10)}{(63.5)} = 9.45\times10^{22}$$

Out of 1000 atoms, 1 electron is removed. ∴ the total number of the electrons removed will be,

$$n = \frac{1}{1000}\times(9.45\times10^{22}) = 9.45\times10^{19}$$

Due to the removal of these electrons, the copper piece will become positively charged, with the magnitude of charge as,

$$q = ne = (9.45\times10^{19})\times(1.6\times10^{-19}) = 15.12\text{ C} \quad \textbf{(Answer).}$$

1A. 06. Coulomb's Law

As stated earlier, unlike charges attract each other and like charges repel each other. In 1785, it was experimentally established by Coulomb that the magnitude of **electrostatic force of interaction (attraction or repulsion)** between two charges q_1 and q_2 separated by a distance r is directly proportional to the product of charge and inversely proportional to the square of distance between them (Fig. 1.1). Thus, assuming the charges to be highly localized so that they could be treated as point charges, one gets,

Fig. 1.1

$$F \propto \frac{q_1 q_2}{r^2}$$

$$\Rightarrow \quad \boxed{F = k\frac{q_1 q_2}{r^2}} \quad \text{(Coulomb's Law)} \qquad \text{..... (1.1)}$$

where **k** is termed as **'electrostatic force constant'** or simply **'electrostatic constant'.** For the charges situated in air/vacuum its value in the CGS system is taken as 1 and in the SI system as $1/4\pi\epsilon_0$, where ϵ_0 is called **'permittivity constant'** or **'absolute electrical permittivity of free space'.** Its value is, $\epsilon_0 = 8.85\times10^{-12}\ C^2\ N^{-1}m^{-2}$.

$$k = \frac{1}{4\pi\epsilon_0} \text{ (SI system)}$$

Then the Coulomb's law can also be expressed as,

$$F_{vac} = \frac{1}{4\pi\epsilon_0}.\frac{q_1 q_2}{r^2} \quad \text{(Coulomb's Law; another form)} \qquad \text{..... (1.2)}$$

When the medium in which the charges exist is other than air or vacuum, the electrostatic force is given by,

$$F_{med} = \frac{1}{4\pi\epsilon_0}.\frac{q_1 q_2}{r^2} \qquad \text{..... (1.3)}$$

ϵ is called **absolute permittivity** or simply **permittivity,** of the medium. It may be written as,

$$\epsilon = \epsilon_r \times \epsilon_0 \text{ (or, } K\times\epsilon_0)$$

Then the Coulomb's law may also be expressed as,

$$F_{med} = \frac{1}{4\pi\epsilon_0\epsilon_r}.\frac{q_1 q_2}{r^2} \quad \text{(Coulomb's Law; another form)} \qquad \text{..... (1.4)}$$

ϵ_r (or **K**) is a dimensionless constant, called **'dielectric constant'** or **'relative permittivity'** or **'specific inductive capacity'** of the medium. The medium is referred to as a 'dielectric'. For all dielectrics, the value of ϵ_r is always greater than 1 ($\epsilon_r > 1$). Consequently, whenever the medium between the charges is other that air or vacuum, i.e. the medium is a dielectric with $\epsilon_r > 1$, the electrostatic force between the charges is reduced.

From Eqs. 1.2 and 1.4, we have,

$$\epsilon_r = \frac{F_{vac}}{F_{med}}$$

Thus, the dielectric constant ϵ_r (or **K**) can be **defined** as the ratio of electrostatic force between two point charges in vacuum to the electrostatic force between them in another medium.

Units and Dimensions of ϵ_0 and ϵ. From Eq. 1.2,

$$\epsilon_0 = \frac{1}{4\pi}.\frac{q_1 q_2}{Fr^2}$$

As SI units of force and charge are newton (N) and coulomb (C), respectively, the unit of ϵ_0 is $C^2N^{-1}m^{-2}$. Similarly, the dimension of ϵ_0 is, $[AT]\,[AT] / [MLT^{-2}]\,[L^2] = [M^{-1}L^{-3}T^4A^2]$. The same applies to ϵ. Further, we have,

$$\frac{1}{4\pi\epsilon_0} = k = 9\times10^9\ Nm^2C^{-2}$$

Coulomb's law in vector form. The Coulomb's law as expressed by Eq. 1.2 gives the magnitude of the electrostatic force between two point charges, However, since force is a vector quantity, it is useful to express the law in a vector form, too. In Fig. 1.2, the position vectors of the charges q_1 and q_2 with respect to the point of observation P are r_1 and r_2, respectively. Let the force exerted by q_1 on q_2 be $\vec{F}_{12}$ and the force exerted by q_2 on q_1 be $\vec{F}_{21}$. The vector leading from charge 1 to charge 2 is,

$$\vec{r}_{21} = \vec{r}_2 - \vec{r}_1$$

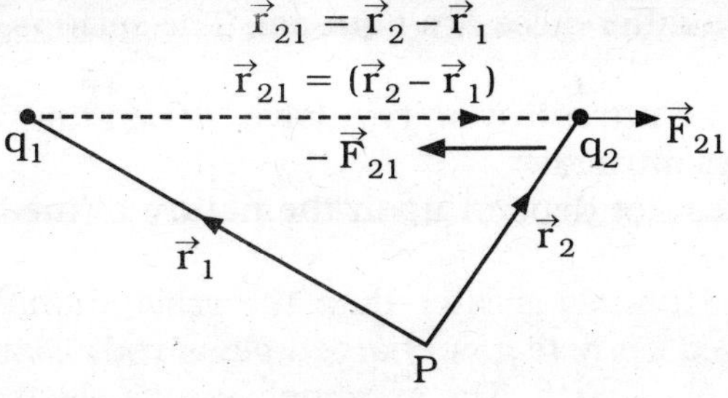

Fig. 1.2

Similarly, the vector leading from charge 2 to charge 1 is,

$$\vec{r}_{12} = \vec{r}_1 - \vec{r}_2$$

The magnitudes of the two vectors $\vec{r}_{21}$ and $\vec{r}_{12}$ are equal,

$$|\vec{r}_{21}| = |\vec{r}_{12}| = r_{21} = r_{12}$$

For the sake of convenience, we specify the direction of the vectors $\vec{r}_{21}$ and $\vec{r}_{12}$ by unit vectors $\hat{r}_{21}$ $(= \vec{r}_{21}/r_{21})$ and $\hat{r}_{12}$ $(= \vec{r}_{12}/r_{12})$, respectively. Then the coulomb's law in vector form is expressed as,

$$\vec{F}_{21} = \frac{1}{4\pi\epsilon_0} \cdot \frac{q_1 q_2}{r_{21}^2} \hat{r}_{21} \quad \text{OR} \quad \frac{1}{4\pi\epsilon_0} \frac{q_1 q_2}{|\vec{r}_{21}|^3} \vec{r}_{21} \qquad \text{..... (1.5)}$$

(Coulomb's Law; vector form)

This equation holds for all possible combinations of charges. For like charges, i.e. for q_1, q_2 being either both positive or both negative, $\vec{F}_{21}$ and $\hat{r}_{21}$ are positive (q_1 and $q_2 > 0$), which means that the force exerted by q_1 on q_2 is in the direction of the arrow shown in the figure, i.e. it is directed away from q_1. Thus q_1 repels q_2, as expected for like charges. For unlike charge, one positive and the other negative, $\vec{F}_{21}$ and $\hat{r}_{21}$ turn out to be negative ($q_1 q_2 < 0$). So the direction of the arrow in the figure is reversed, which implies that q_2 is drawn towards q_1, i.e. q_1 attracts q_2, which is again as expected. Thus the validity of Eq. 1.5, expressing the Coulomb's law in vector form, is well verified.

The same results are obtaind if the Coulomb's law is written in the equivalent form,

$$\vec{F}_{12} = \frac{1}{4\pi\epsilon_0} \cdot \frac{q_1 q_2}{r_{12}^2} \cdot \hat{r}_{12} \qquad \text{..... (1.6)}$$

Since $\hat{r}_{21} = -\hat{r}_{12}$, we have,

$$\vec{F}_{21} = -\vec{F}_{12}$$

It follows that the forces $\vec{F}_{21}$ and $\vec{F}_{12}$ are equal and opposite to each other, i.e. the charge q_1 exerts the same force on the charge q_2 as q_2 exerts upon q_1. This is in perfect agreement with the Newton's third law of motion, as it should be.

Point charge. The Coulomb's law only holds for point charges, not for charges of all sizes. A point charge is one which has a spread of dimensions far less than its distance from the point of observation ($x \ll r$, where x is the spread and r is the said distance).

Unit of charge [coulomb (C)]. In Eq. 1.2, if $q_1 = q_2 = 1$ and $r = 1$, then $F_{vac} = (1/4\pi\epsilon_0)$. Thus we can **define one coulomb of charge** in terms of electrostatic force as "the charge which when placed in vacuum at a distance of one metre from an equal and similar charge, repels it with a force of 9×10^9 N ". A smaller unit is 1mC ($= 10^{-6}$C).

Coulomb's Law and Newton's Law of Gravitation. The Coulomb's law [$F = k(q_1q_2)/r^2$; Eq. 1.1] resembles the inverse square law of gravitation,

$$F = G\frac{m_1m_2}{r^2}$$

The charges q_1, q_2 are replaced by the masses m_1, m_2 and the constant of proportionality k is replaced by the gravitational constant G. These differences are nominal. The serious difference lies in the nature of the two forces, as follows:

(i) The electrical forces may be attractive or repulsive, depending upon the sign of charges, but the gravitational force is always attractive.

(ii) The gravitational force does not depend upon the nature of medium, but the electrical force does (Eq. 1.3).

(iii) The electrical forces are extremely greater than the gravitational forces. An obvious consequence of this fact is that a charged body (e.g. a charged glass rod) easily picks up small pieces of paper against the gravitational pull of earth. The extremely strong electrostatic forces within an atom keep the moving electrons bound to the nucleus.

Example 3. Calculate the ratio between the magnitudes of electrostatic and gravitational force acting between two electrons separated by a finite distance.

Solution. Electrostatic force: $F_e = \frac{1}{4\pi\epsilon_0}.\frac{e.e}{r^2}$

Gravitational force: $F_g = G\frac{m_e.m_e}{r^2}$

$$\therefore \frac{Fe}{Fg} = \frac{(1/4\pi\epsilon_0) \times e^2}{G \times (m_e)^2} = \frac{(9\times10^9) \times (1.6\times10^{-19})^2}{(6.67\times10^{-11}) \times (9.1\times10^{-31})^2}$$

$$= 4.2\times10^{42}$$

Thus the ratio $F_e : F_g$ is found to be $\sim 10^{42}$. **(Answer).**

1.A. 07. Principle of Superposition for Charges

When more than two charges are present, the total force exerted on a particular charge is determined by the **principle of superposition for charges.** The principle **states** that "when multiple charges are present, the net force on a given charge is the vector sum of the individual forces exerted on the charge due to all the other charges".

In Fig. 1.3, three charges q_1, q_2, q_3 with position vectors as $\vec{r}_1$, $\vec{r}_2$, $\vec{r}_3$ are shown distributed about the test charge q_0, which has position vector as $\vec{r}_0$. The system can be imagined to have more charges like q_1, q_2, q_3, with total number as n. The net force on the charge q_0 due to all the n charges will be,

$$\vec{F} = \vec{F}_{01} + \vec{F}_{02} + \vec{F}_{03} + \text{..........} + \vec{F}_{0n}.$$

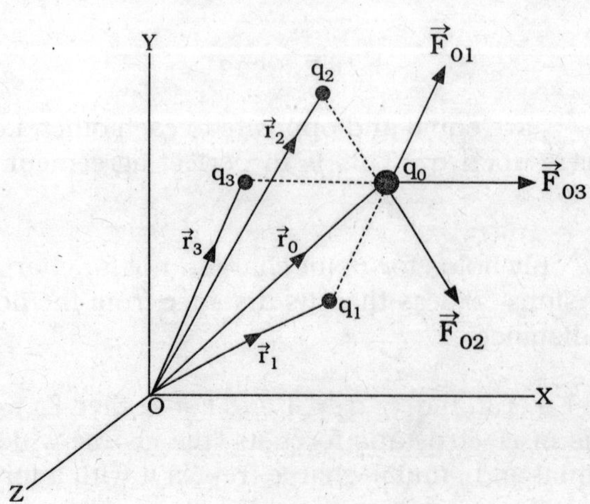

Fig. 1.3

$\vec{F}_{01}, \vec{F}_{02}, \vec{F}_{03}, \ldots\ldots\ldots, \vec{F}_{0n}$ are the individual forces exerted on q_0 due to the charges $q_1, q_2, q_3 \ldots\ldots\ldots, q_n$, respectively. Substituting for the $\vec{F}$ values from Eq. 1.5, we get (r_{01}, r_{02}, r_{03}, etc., are the distances measured from the charges q_1, q_2, q_3, etc., towards q_0),

$$\vec{F} = \frac{1}{4\pi\epsilon_0}\left[\frac{q_0 q_1}{r_{01}^2}\hat{r}_{01} + \frac{q_0 q_2}{r_{02}^2}\hat{r}_{02} + \ldots\ldots\ldots\ldots\ldots\ldots + \frac{q_0 q_n}{r_{0n}^2}\vec{r}_{0n}\right]$$

$$= \frac{1}{4\pi\epsilon_0}\sum_{i=1}^{i=n}\frac{q_0 q_i}{r_{0i}^2}\hat{r}_{01}$$

$$= \frac{q_0}{4\pi\epsilon_0}\sum_{i=1}^{i=n}\frac{q_i}{r_{0i}^3}\vec{r}_{0i}$$

$$= \frac{q_0}{4\pi\epsilon_0}\sum_{i=1}^{i=n}\frac{q_i}{|\vec{r}_0 - \vec{r}_i|^3}(\vec{r}_0 - \vec{r}_i) \qquad \ldots\ldots (1.7)$$

1A. 08. Continuous Charge Distribution

In section 1A.04 we had learnt that the electric charge is discrete. However, if the charges in a given system are distributed extremely close to each other, the system is said to form a continuous charge distribution.

We can have three types of continuous chage distributions. For all of them, the force on a test charge can be determind by the pinciple of superposition for charges. We shall treat them one by one.

(i) Linear charge distribution. In this case the charge is continuously (or, say, uniformly) distributed along a line, which can be straight or curved or both (Fig. 1.4a). Let the **linear density of charge,** i.e. the charge per unit length, be λ. It can also be expressed as $\lambda = (dq/d\ell)$, where $d\ell$ is a small element of length along the line, of total lenght L. The amount of charge on $d\ell$ is,

$$dq = \lambda\, d\ell$$

The small force exerted by this element of charge on a test charge q_0 located at a distance $\vec{r}$ from $d\ell$ will be,

$$\vec{dF} = \left[\frac{q_0}{4\pi\epsilon_0}\cdot\frac{dq}{r^2}\right]\vec{r} = \left[\frac{q_0 \lambda d\ell}{4\pi\epsilon_0\, r^2}\right]\vec{r}$$

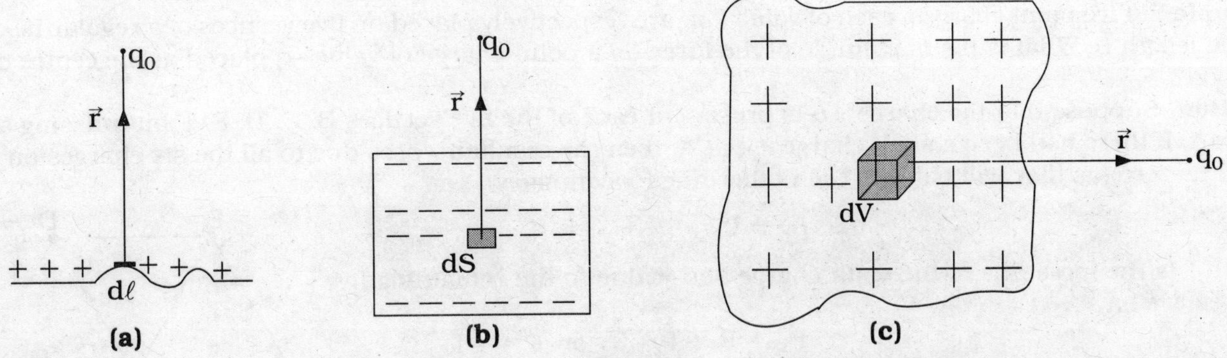

Fig. 1.4

The total force on q_0 due to the entire length L is found by vectorially adding up all the contributions like $\vec{dF}$. Since the charge is continuously distributed along the line, the addition amounts to integration over the length L,

$$\vec{F} = \frac{q_0}{4\pi\epsilon_0}\int_L \frac{\lambda}{r^2}\hat{r}\,d\ell \qquad \text{..... (1.8)}$$

(ii) Surface charge distribution. Here the charge is distributed over some area, e.g. a small metal sheet (Fig. 1.4b). Let the **Surface charge density,** i.e. the charge per unit area, be $\sigma = (dq/dS)$, where dS is a small element of the total surface area S.

Proceeding in the same way as before, the total force due to S on a test charge q_0 will be,

$$\vec{F} = \frac{q_0}{4\pi\epsilon_0}\int_L \frac{\lambda}{r^2}\hat{r}\,dS \qquad \text{..... (1.9)}$$

(iii) Volume charge distribution. The charge is now distributed over a 3-dimensional volume (Fig. 1.4c). Let the **volume charge density,** i.e. charge per unit volume, be $\rho = (dq/dV)$, where dV is a small element of the total volume V. Proceeding as before, the total force on q_0 is,

$$\vec{F} = \frac{q_0}{4\pi\epsilon_0}\int_S \frac{\rho}{r^2}\hat{r}\,dV \qquad \text{..... (1.10)}$$

Example 4. Two equally charged identical metal spheres A and B repel each other with a force of 2×10^{-5} N. Another identical charged sphere C is first touched with B and then placed at the mid-point between A and B. What is the net electric force on C ?

A C B

Solution. Let initially the charge on each sphere be q and the separation between their centres be r. Given that,

$$F = 2\times10^{-5}\text{ N}$$

From Coulomb's law,

$$F = \frac{q^2}{4\pi\epsilon_0 r^2} = 2\times10^{-5}$$

When the sphere C touches B, q will distribute equally on B & C, as the spheres are identical conductors.

$$\therefore \quad q_B = q_C = q/2$$

So the sphere C will experience a force,

$$F_{CA} = \frac{1}{4\pi\epsilon_0}\frac{q\,(q/2)}{(r/2)} = 2F, \text{ along } \overrightarrow{AB}, \text{ due to the charge on A, and a force,}$$

$$F_{CB} = \frac{1}{4\pi\epsilon_0}\frac{(q/2)\,(q/2)}{(r/2)} = F, \text{ along } \overrightarrow{BA}, \text{ due to the charge on B.}$$

∴ The net force on C will be,

$$F_R = F_{CA} - F_{CB} = 2F - F = F = 2\times10^{-5}\text{ N (along } \overrightarrow{AB}\text{)}$$ **(Answer).**

Example 5. Five point charges, each of value +q, are respectively placed on five vertices of a regular hexagon of side length L. What is the magnitude of the force on a point charge of value –q placed at the centre of the hexagon?

Solution. Suppose that the charge +q is present at each of the five vertices B, C, D, E, F but missing at the vertex A. If there had been a sixth charge +q at A, then the resultant force due to all the six charges on -q at O will be zero (as they will balance each other due to symmetry), i.e.,

$$\vec{F}_R = 0$$

Now, if $\vec{f}$ is the force due to the sixth charge and $\vec{F}$ due to the remaining five charges, we have,

$$\vec{F}_R = \vec{F} + \vec{f} = 0 \quad \text{or} \quad \vec{F} = -\vec{f}$$

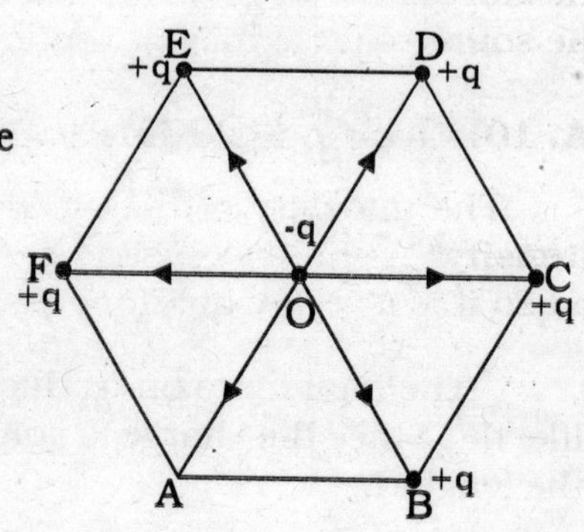

The negative sign of $\vec{f}$ is of no importance, since we are only interested in the magnitude of $\vec{f}$. From the geometry of the figure, OA = EF = L. Therefore,

$$F = f = \frac{1}{4\pi\epsilon_0}\frac{q\times q}{L^2} = \frac{1}{4\pi\epsilon_0}\left(\frac{q}{L}\right)^2$$ **(Answer).**

Example 6. Two opposite corners of a square carry charge Q each and the other two opposite corners of the same square carry charge q each. If the resultant force on either charge q is zero, how are Q and q related ?

Solution. Consider the charge q at the corner D. We have,

$$\vec{F}_{DA} = \frac{1}{4\pi\epsilon_0}\frac{Qq}{a^2}(\hat{i}),\ \vec{F}_{DC} = \frac{1}{4\pi\epsilon_0}\frac{Qq}{a^2}(-\hat{i})\ \text{(a : side length of the square)}$$

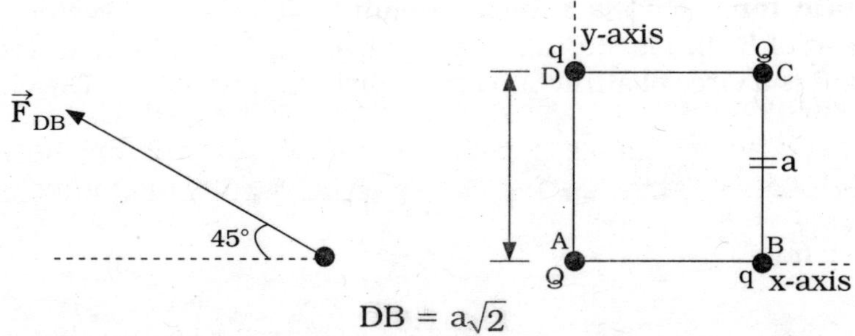

$$DB = a\sqrt{2}$$

$\vec{F}_{DB}$ is the sum of its components along DA and DC,

$$\vec{F}_{DB} = \frac{1}{4\pi\epsilon_0}\frac{Q^2}{\left(a\sqrt{2}\right)^2}\left[\cos 45°(-\hat{i}) + \sin 45°(\hat{j})\right] = \frac{1}{4\pi\epsilon_0}\frac{Q}{a^2 2\sqrt{2}}(-\hat{i}+\hat{j})$$

Now, $\vec{F}_{DA} + \vec{F}_{DC} + \vec{F}_{DB} = 0$

$$\therefore \quad \frac{1}{4\pi\epsilon_0 a^2}\left\{Q\hat{j} + Q(-\hat{i}) + \frac{q}{2\sqrt{2}}(-\hat{i}+\hat{j})\right\} = 0$$

$$\Rightarrow \quad \left\{Q(-\hat{i}+\hat{j}) + \frac{q}{2\sqrt{2}}(-\hat{i}+\hat{j})\right\} = 0$$

$\therefore \quad Q2\sqrt{2} + q = 0 \quad$ or $\quad q = -2\sqrt{2}Q$ **(Answer).**

1A. 09. Action at a Distance; Invisible Forces; Electric Field

As stated earlier, when a point charge q_0 is brought into the vicinity of another localised charge q, it experiences a force (attraction or repulsion), whose magnitude and direction can be determined by the Coulomb's law. The force is invisible, but real. If q is appreciably large, it can actually make the charge q_0 move towards it or away from it, depending upon whether the two are unlike charges or like charges. Such 'action at a distance', as it is called in electrostatics, can be understood by realising that the charge q produces some kind of invisible electrical influence in the space around it, which is felt by any other charge present in this space. We refer to such an 'electrical influence' as an **elecric field.** It may be **broadly defined** as the "space around a charge in which another charge experiences an electrostatic force of attraction or repulsion". It can also be defined specifically in terms of electrostatic force, as described in the next section. The aforesaid charge **q** is called **'source charge'** and the charge $\mathbf{q_0}$ is called **'test charge'** or **'probe charge'**. The source charge may be a point charge, a group of point charges, or a continuous distribution of charge.

1A. 10. Electric Field Intensity $\vec{E}$ (or, simply, Electric Field $\vec{E}$); Unit and Dimensions of $\vec{E}$.

The intensity of the electric field at a given point around a source charge is called **'electric field intensity'** or simply **'electric field'** at that point. As we shall see shortly, it has both magnitude and direction and so it is a vector quantity. It is denoted as $\vec{E}$.

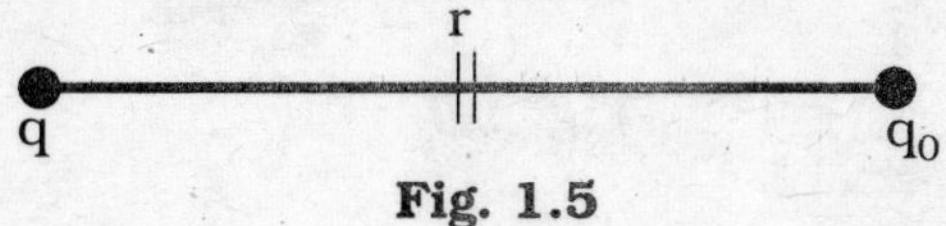

Fig. 1.5

Suppose a test charge $+q_0$ is placed at a distance r from a source charge q and experiences a foce $\vec{F}$ (r) due to q (Fig. 1.5). Then the electric intensity at the position of q_0 is given by,

$$\vec{E}\,(r) = \frac{\vec{F}(r)}{q_0} \qquad \text{..... (1.11)}$$

Here it is essential that the magnitude of the test charge is assumed to be vanishingly small, otherwise it will affect the actual electric field $\vec{E}$ due to q and modify it to give a different value.

Since the **electrostatic force $\vec{F}$(r)** is a vector quantity, $\vec{E}$ is also a vector quanitty. Obviously, its direction is the same as that of $\vec{F}$. In Fig. 1.6(a), q is positive. So $\vec{F}$ is a force of repulsion on q_0, directed away from q. In Fig. 1.6(b), $\vec{F}$ is a force of attraction on q_0, directed towards q. The electric field $\vec{E}$ is directed accordingly, as shown.

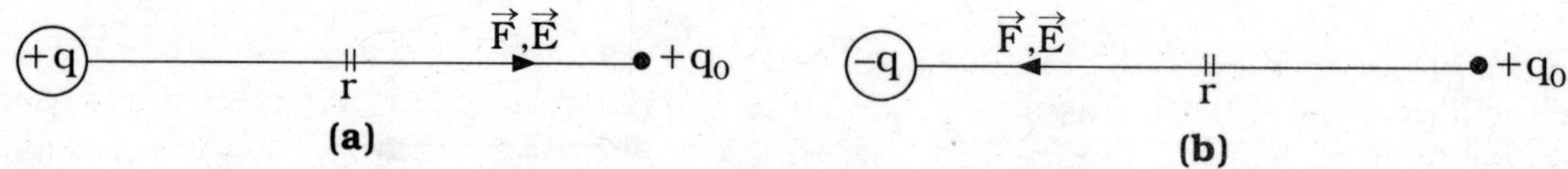

Fig. 1.6

We can use Eq. 1.11 to **define the electric field** in another (concrete) way as "the electric field at a given point is the electrostatic force $\vec{F}$ experienced by a positive test charge q_0 placed at that point ($\vec{E} = \vec{F}/q_0$)".

The SI unit of the electric field intensity is newton/coulomb (N/C). Its dimensions are,

$$\begin{aligned} \text{newton/coulomb} &= (\text{kg}\times\text{metre} - \text{sec}^{-2})/(\text{amp}\times\text{sec}) \\ &= \text{kg}\times\text{metre}\times\text{sec}^{-3}\times\text{amp}^{-1} \\ &= [\text{MLT}^{-3}\text{A}^{-1}] \end{aligned}$$

Physical significance of electric field. (i) From the knowledge of the electric field at a given position we can readily calculate the force experienced by a charge placed at that position. (This point will become more clear in the following sections, where we shall derive expressions for $\vec{E}$ for different systems of charges).

(ii) The electric field produced by a given system of charges is characteristic of that system. It has its own independent existence, irrespecive of the presence of the test charge q_0. This will become more clear after the student has acquired the knowledge of electromagnetic waves. All electromagnetic waves (radio waves, light waves, X-rays, γ-rays, etc.) consist of oscillations of electric and magnetic fields. They all travel with a finite velocity, namely the velocity of light, c. All waves carry energy and thus make their presence felt wherever they arrive. An accelerated moving charge becomes a source of electromagnetic waves. Thus, when a source charge q is brought within a distance r of the test charge q_0, the e.m. waves produced by it take a finite time, t = r/c, to reach the position of q_0. So it is clear that the influence of q is not instantaneously felt by q_0, as may initially appear and give a false impression that the existence of q_0 is a must to the identity of $\vec{E}$. Thus, the independent existence of the electric field in space is confirmed.

1A. 11. Electric Field Intensity due to a Point Charge

Suppose a highly localised charge +q exists at a point O in vacuum (or air) at a distance r from a test charge $+q_0$ placed at P, as shown in Fig. 1.7. Applying the Coulomb's law (vide Eq. 1.2), the force experienced by the test charge q_0 at P is,

$$\vec{F} = [(1/4\pi\epsilon_0)\,(qq_0/r^2)]\,\hat{r}$$

where r is the unit vector directed from q to q_0 along $\vec{r}$. Then the electric field intensity at P is (vide Eq. 1.11),

$$\vec{E} = \vec{F}/q_0$$

O +q r P $+q_0$ $\vec{F},\vec{E}$

Fig. 1.7

$$\Rightarrow \vec{E} = \frac{1}{4\pi\epsilon_0} \cdot \frac{q}{r^2} \hat{r} \qquad \text{..... (1.12)}$$

The direction of $\vec{E}$ is oriented away from q if q is positive (as shown in the figure) and oriented towards it if q is negative.

If q is placed in a medium other than vacuum or air, having dielectric constant K, then,

$$\vec{E} = \frac{1}{4\pi\epsilon_0 K} \cdot \frac{q}{r^2} \hat{r} \qquad \text{.....(1.13)}$$

In scalar form, giving magnitude of the electric field, we have,

$$E = |\vec{E}| = \frac{1}{4\pi\epsilon_0 K} \cdot \frac{q}{r^2} \qquad \text{.....(1.14)}$$

If q is assumed to be located at the origin O of a cartesian system of coordinates (Fig. 1.8), the electric field $\vec{E}$ around it will possess spherical symmetry, i.e. in all the directions around O, $\vec{E}$ will have same magnitude at a distance r from q. But if q is located at a point A, other than the origin O, the spherical symmetry is lost. The value of $\vec{E}$ is also changed. We shall now evaluate the changed value of $\vec{E}$ at the position P. Let the coordinates of A and P be (a, b, c) and (x, y, z), respectively, We have,

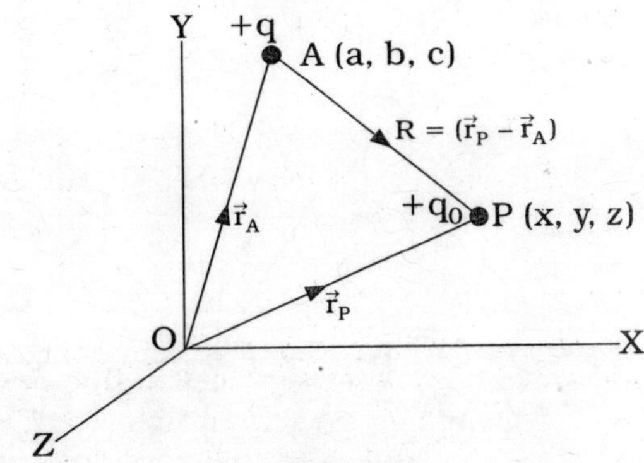

Fig. 1.8

$$\vec{r}_A = a\hat{i} + b\hat{j} + c\hat{k}$$

$$\vec{r}_P = x\hat{i} + y\hat{j} + z\hat{k}$$

In Fig. 1.8, $\vec{R} = \vec{r}_P - \vec{r}_A$

$$\Rightarrow \quad \vec{R} = (x-a)\,\hat{i} + (y-b)\,\hat{j} + (z-c)\,\hat{k} \qquad \text{..... (A)}$$

$$\therefore \quad |\vec{R}| = [(x-a)^2 + (y-b)^2 + (z-c)^2]^{1/2} \qquad \text{..... (B)}$$

The electric field intensity $\vec{E}$ at P can be written as (vide Eq. 1.12),

$$\vec{E} = \frac{1}{4\pi\epsilon_0 K} \frac{R}{|\vec{R}|^2} \hat{R} = \frac{1}{4\pi\epsilon_0 K} \frac{q\vec{R}}{|\vec{R}|^3} \qquad \text{..... (1.15)}$$

For vacuum or air (K = 1), we have,

$$\vec{E} = \frac{1}{4\pi\epsilon_0} \frac{q\vec{R}}{|\vec{R}|^3} \qquad \text{..... (1.16)}$$

This is the required expression for the changed value of $\vec{E}$. We can transform it to another simpler expression without any loss of generality. Substituting for $\vec{R}$ and $|\vec{R}|$ from Eqs. (A) and (B), we have,

$$\vec{E} = \frac{q}{4\pi\epsilon_0} \frac{[(x-a)\hat{i} + (y-b)\hat{j} + (z-c)\hat{k}]}{\left[(x-a)^2 + (y-b)^2 + (z-c)^2\right]^{3/2}}$$

Now, let the origin of our coordinate system be shifted to A, the position of the source charge +q. Then the coordinates of A become (0, 0, 0) and, therefore, a = b = c = 0. Then,

$$\vec{E} = \frac{q}{4\pi\epsilon_0} \frac{[(x\hat{i}+y\hat{j}+z\hat{k})]}{\left[x^2+y^2+z^2\right]^{3/2}} \quad \text{..... (C)}$$

If $\vec{E}_x$, $\vec{E}_y$, $\vec{E}_z$ are the components of $\vec{E}$ along the three coordinate axes x, y, z, then,

$$\vec{E} = \vec{E}_x + \vec{E}_y + \vec{E}_z \quad \text{..... (D)}$$

Separating and equating the x, y, z components in Eqs. (C) and (D), we have,

$$\left.\begin{aligned} E_x &= (qx/4\pi\epsilon_0)\,[1/(x^2 + y^2 + z^2)^{3/2}] \\ E_y &= (qy/4\pi\epsilon_0)\,[1/(x^2 + y^2 + z^2)^{3/2}] \\ E_z &= (qz/4\pi\epsilon_0)\,[1/(x^2 + y^2 + z^2)^{3/2}] \end{aligned}\right\} \quad \text{..... (1.17)}$$

These equations express the changed value of $\vec{E}$ in another form. However, it should be kept in mind that with the shift of the origin from O to A, the values of x, y, z for the point P will also change accordingly.

1A. 12. Electric Field intensity due to a Group of Charges

If a large number of charges q_1, q_2, q_3, etc., are present, then according to the principle of superposition for forces the resultant electric field produced by them at a given point P is given by the vector sum of the individual electric fields produced by them at P (Fig. 1.9).

$$\vec{E} = \vec{E}_1 + \vec{E}_2 + \vec{E}_3 + \text{..................} + \vec{E}_n \quad (n = 1, 2, 3 \text{ ....})$$

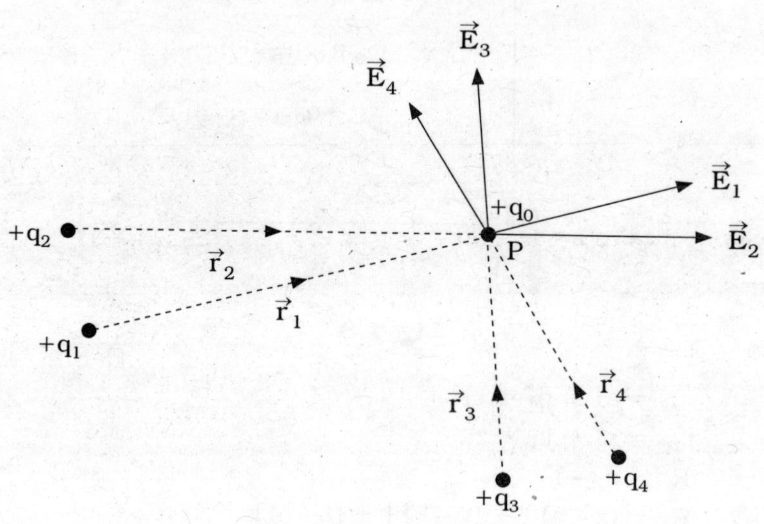

Fig. 1.9

From Eq. 1.12, the electric field intensity due to a point charge, say q_n, in vacuum (or air) is given by,

$$\vec{E}_n = \frac{1}{4\pi\epsilon_0} \frac{q_n}{r_n^2} \hat{r}_n$$

Then, the resultant field becomes,

$$\vec{E} = \frac{1}{4\pi\epsilon_0} \left[\frac{q_1}{r_1^2}\hat{r}_1 + \frac{q_2}{r_2^2}\hat{r}_2 + \text{..............} + \frac{q_n}{r_n^2} r_n \right]$$

$$\Rightarrow \quad \vec{E} = \frac{1}{4\pi\epsilon_0} \sum_{i=1}^{n} \frac{q_i}{r_i^2} \hat{r}_i \quad \text{..... (1.18)}$$

Here r_i is the distance between the ith charge q_i and P and $\vec{r}_i$ is the unit vector directed from q_i to P.

In the above treatment it is assumed that any one charge neither affects nor is affected by the electric fields of other charges, i.e. it ignores the presence of other charges. Accordingly, the value of $\vec{E}$ for each charge is written as if it was located at the origin of a cartesian system of coordinate. If a common origin is chosen for all the charges, then we have the situation as earlier shown in Fig. 1.8. Then the position vectors of the various charges are given by $\vec{r}_1$, $\vec{r}_2$, $\vec{r}_3$, etc., and the position vector of P is given by $\vec{r}_p$. The resultant field is given by (vide Eq. 1.16),

$$\vec{E} = \frac{1}{4\pi\epsilon_0} \sum_{i=1}^{n} \frac{q_i(\vec{r}_p - \vec{r}_i)}{|(\vec{r}_p - \vec{r}_i)|^3} \quad \text{..... (1.19)}$$

Continuous distribution of charges. If the various charges q_1, q_2, q_3, etc, are continuously disributed in space, then the summation in the Eq. 1.19 is replaced by integration. We have three possible cases:
(i) Linear charge distribution. The charges are continuously aligned along a line of length L. If λ be the charge distribution density and $d\ell$ a small element of length along the line, $\vec{E}$ is given by,

$$\vec{E} = \frac{1}{4\pi\epsilon_0} \int_L \frac{\lambda(\vec{r}_p - \vec{r}_i)}{|(\vec{r}_p - \vec{r}_i)|^3} d\ell \quad \text{..... (1.20)}$$

(ii) Surface charge distribution. If the charge distribution density be σ and a surface area element be dS, we have,

$$\vec{E} = \frac{1}{4\pi\epsilon_0} \int_S \frac{\sigma(\vec{r}_p - \vec{r}_i)}{|(\vec{r}_p - \vec{r}_i)|^3} dS \quad \text{..... (1.21)}$$

(iii) Volume charge distribution. If ρ is the density of charge distribution and dV is a volume element, we have,

$$\vec{E} = \frac{1}{4\pi\epsilon_0} \int_V \frac{\rho(\vec{r}_p - \vec{r}_i)}{|(\vec{r}_p - \vec{r}_i)|^3} dV \quad \text{..... (1.22)}$$

1A. 13. Electric Field Intensity due to a Circular Loop of Charge (Charged Ring)

The Fig. 1.10 shows an extremely thin wire bent into the shape of a cicular ring AB, held perpendiular to the plane of paper. Its radius is R and it carries a charge +q uniformly distributed along its circumference. We proceed to find the electric field intesity at a point P on the axis OX of the ring, such that OP = x.

We imagine the ring as consisting of a large number of infinitesimal elements, each of length $d\ell$. Then each such element may be treated as a point charge. One such element is shown at A in Fig. 1.10. Since the charge +q is uniformly distributed along the circumference of the ring, the charge per unit length is $q/2\pi R$ and so the charge dq on the element $d\ell$ at A is,

$$dq = \frac{q}{2\pi R} d\ell$$

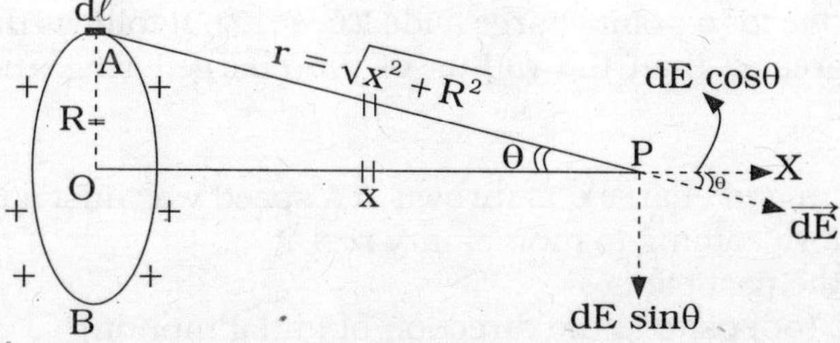

Fig. 1.10

The magnitude of the electric field intensity at P due to this charge will be (vide Eq. 1.12),

$$dE = |\vec{dE}| = \frac{1}{4\pi\epsilon_0} \frac{dq}{r^2}$$

$$\Rightarrow \qquad dE = \frac{q.d\ell}{(4\pi\epsilon_0)(2\pi R)(x^2+R^2)} \qquad \text{..... (G)}$$

$$(\because r^2 = x^2+R^2)$$

The direction of $\vec{dE}$ is along AP, as shown in the figure. $\vec{dE}$ can be resolved into two components, one of magnitude dEcosθ along the axis of the ring and the other of magnitude dEsinθ perpendicular to the axis. For a pair of diametrically opposite elements on the loop, like those at A and B, the components of $\vec{dE}$ perpendicular to OX have opposite sign and therefore they mutually cancel. On the other hand, the components along OX are similarly aligned and so they add up. The ring can be imagined as made up of a large number of such diametrically opposite elements. Then the resultant field intensity $\vec{E}$ solely consists of the components dEcosθ. Thus, we have,

$$\int_0^{2\pi R} dE \sin\theta = 0$$

$$\therefore \qquad |\vec{E}| = \int_0^{2\pi R} dE \cos\theta = \int_0^{2\pi R} \frac{x}{(x^2+R^2)^{1/2}} dE$$

Substituting for dE from Eq. (G),

$$|\vec{E}| = \frac{q.x}{(4\pi\epsilon_0)(2\pi R)(x^2+R^2)^{3/2}} \int_0^{2\pi R} d\ell$$

$$= \frac{q.x(2\pi R)}{(4\pi\epsilon_0)(2\pi R)(x^2+R^2)^{3/2}}$$

$$\Rightarrow \qquad |\vec{E}| = \frac{qx}{4\pi\epsilon_0\,(x^2+R^2)^{3/2}} \qquad \text{..... (1.23)}$$

The direction of $\vec{E}$ lies along OX, the axis of the ring. Two special cases arise.

Special Cases

(i) If P lies at the centre of ring, then x = 0.
Hence, from Eq. 1.23, $|\vec{E}| = 0$.
Thus, no electric field exists at the centre of the ring.
(ii) If P lies very far away from the ring, x>>R.
Then, in Eq. 1.23, R^2 can be ignored compared to x^2. So we have,

$$|\vec{E}| = \frac{1}{4\pi\epsilon_0} \cdot \frac{q}{x^2}$$

This equation is similar to the one for a point charge (vide Eq. 1.12). It follows that if the point of observation is situated at a distance far greater than the radius of the charged ring, the ring behaves like a point charge.

Example 7. A particle of mass m and charge q is thrown at a speed v against a uniform electric field E. How much distance will it travel before coming to momentary rest ?
Soultion. The force acting on the particle is,

Fe = qE (opposite to the direction of initial motion)

$$\therefore \quad \text{Retardation,} \qquad a = \frac{E}{m} = \frac{qE}{m}$$

Now, $v_f^2 = u^2 - 2aS$ (S: distance; v_f and u: final and initial speed)

$$\therefore \qquad 0 = v^2 - 2\frac{qE}{m}S \quad \text{or} \quad \frac{2qE}{m}S = v^2 \text{ (Given : } u = v \;\&\; v_f = 0)$$

$$\Rightarrow \qquad S = \frac{v^2 m}{2\,qE} \qquad \textbf{(Answer).}$$

Example 8. Two charges of value 1μC each are respectively placed at two corners of an equilateral triangle, with side length 1 m. Find the electric field intensity at the third corner.

Solution. The two charges are placed at the corners A and B of the equilateral triangle ABC.

Each charge has value 1μC (= 1×10^{-6} C).

The electric field intensity $\vec{E}$ at C is due to the charges at A and B, i.e.

$$\vec{E} = \vec{E}_A + \vec{E}_B$$

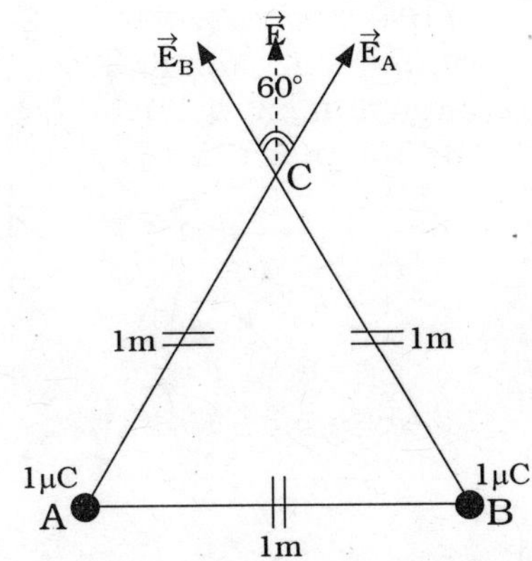

Now,

$$E_A = \frac{1}{4\pi\epsilon_0}\cdot\frac{q_A}{(AC)^2}$$

$$= (9\times10^9)\times\frac{(1\times10^{-6})}{(1)^2}$$

$$= 9\times10^3 \text{ N/C}$$

Similarly,

$$E_B = \frac{1}{4\pi\epsilon_0}\cdot\frac{q_B}{(BC)^2} = 9\times10^3 \text{ N/C}$$

From the vector law of addition,

$$E = \sqrt{E_A^2 + E_B^2 + 2E_AE_B\cos 60°}$$

$$= (9\times10^3)\sqrt{3} \qquad \left(\cos 60° = \frac{1}{2}\right)$$

$$= 1.56\times10^4 \text{ N/C} \quad \textbf{(Answer).}$$

By symmetry, the direction of $\vec{E}$ is perpendicular to AB and away from C, as shown in the figure.

1A. 14. Electric Field Lines (or Electric Lines of Force)

The electric field lines are a means of pictorially representing the electric field intensity around a given configuration of charges. As mentioned earlier, a free positive test charge when placed in an electric field moves along a linear or curved path, by virtue of the electrostatic force experienced by it in the field. So Faraday (19th century) originally termed such paths as **'electric lines of force'.** We now better refer to them as **electric field lines,** because it becomes difficult to seek an exact parallel for them in magnetism, as a magnetic monopole does not exist. If the field line is straight, it directly represents the direction of the dielectric field. If it is curved, the tangent drawn to it at a given point represents the direction of the electric field at that point.

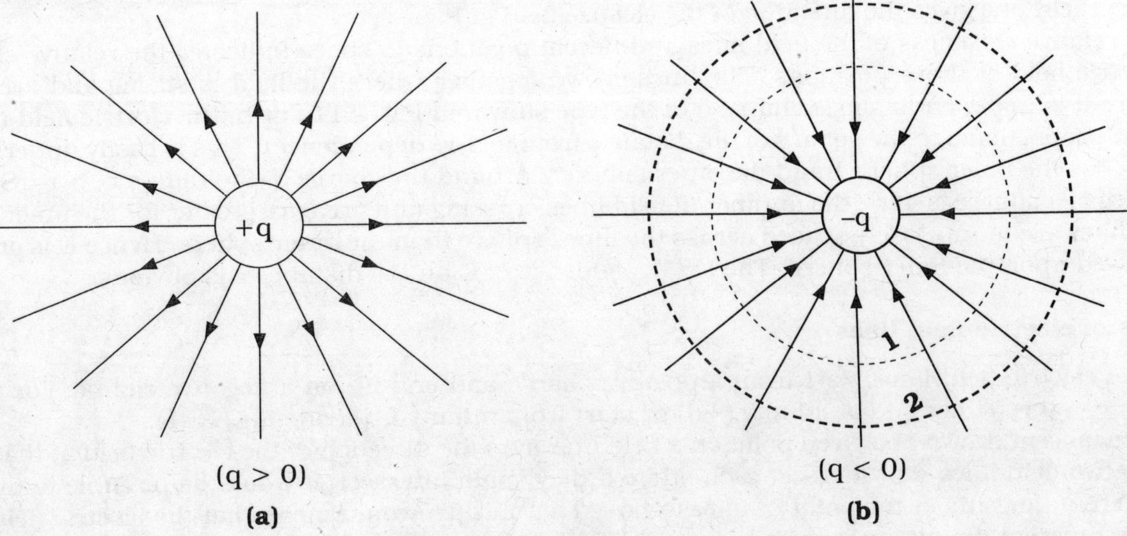

Fig. 1.11

Thus, we can **define** an **electric field line** as "an imaginary linear or curved path along which a free isolated test charge moves and the tangent drawn at a given point on the curve gives the direction of electric field at that point".

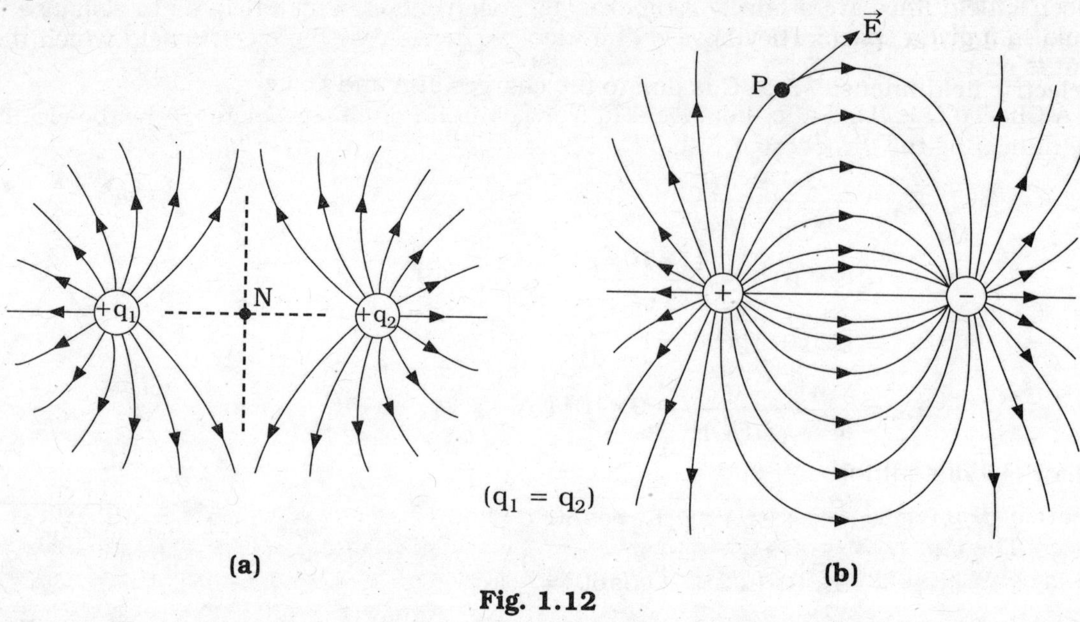

Fig. 1.12

The Figs. 1.11 and 1.12 show the electric field lines around some simple charge configurations. It should be kept in mind that the lines actually occupy the entire 3-dimensional space around the charges.

The field lines of a single positive charge (+q) are directed radially outwards, because a positive test charge placed anywhere in the space around q will experience a repulsive force (Fig.1.11a). On the other hand, it will experience a force of attraction around a negative charge –q. So the field lines are directed radially towards q (Fig.1.11b). Two like charges, both positive or both negative, repel each other. The field lines for such a configuration of two equal positive charges are shown in Fig. 1.12a ($q_1=q_2$, the product q_1q_2 is positive). At the midpoint N between q_1 and q_2, a test charge experiences no net electrostatic force, as the forces exerted upon it separately due to q_1 and q_2 are equal and opposite and so they balance each other. N is called **'neutral point'.** If q_1 and q_2 are unequal, N will not be centrally located between q_1 and q_2. For example, if $q_1 > q_2$, N will be shifted towards q_2. If the distances of q_1 and q_2 from N are r_1 and r_2, respectively, then $q_1/r_1^2 = q_2/r_2^2$. Two unlike charges attract each other. The field lines for a configuration of two equal and opposite charges, referred to as an electric dipole (the product q_1q_2 is negative), are depicted in Fig. 1.12b. Now N still lies along the line joining the two charges, but outside. The tangent drawn at a given point P on a field line gives the direction of the electric field at P.

The relative closeness of the field lines at different positions in space indicates the relative strengths of the electric field at those positions. The lines crowd together where the field is strong and are spaced apart where it is weak. For a single charge q of the type shown in Fig. 1.11a or b, the electric field intensity is inversely proportional to the square of the distance from q ($1/r^2$ dependence). This is easily understood in Fig. 1.11b. For the inner sphere 1 and the outer sphere 2 around the charge q, the radius $r_2 > r_1$. Since the surface area of a sphere is $4\pi r^2$, the number of field lines crossing unit area are greater for the inner sphere. Thus, the lines are more closely spaced across the inner sphere than the outer sphere. Hence $\vec{E}$ is greater at points located upon the inner sphere. The $1/r^2$ dependence of $\vec{E}$ on the distance r is obvious.

Properties of electric field lines

(i) The electric field lines start from a positive charge and end up on a negative charge. For a single charge, they either end at infinity (+q) or start from infinity (-q) (vide Fig. 1.11).

(ii) The tangent drawn at a given point on a field line gives the direction of the electric field at that point.

(iii) No two field lines can intersect each other. If they could intersect, it would be possible to draw two diferent tangents at the point of intersection. That in turn would mean that the electric field $\vec{E}$ had two different directions at the same point, which is impossible.

(iv) The magnitude of the electric field $\vec{E}$ in a given small region of space is proportional to the density of the field lines, i.e. the number of lines crossing unit area, in the region.

(v) No field lines exist inside a charged conductor.

The electric field lines are a purely geometrically construction, which help us to visualise the nature of electric field in a given space. They have no physical existence, but the electric field which they occupy and represent is real.

Example 9. A Charge Q is fixed at a distance d in front of an infinite metal plate. Draw the electric lines of force, clearly indicating the directions.

Solution.

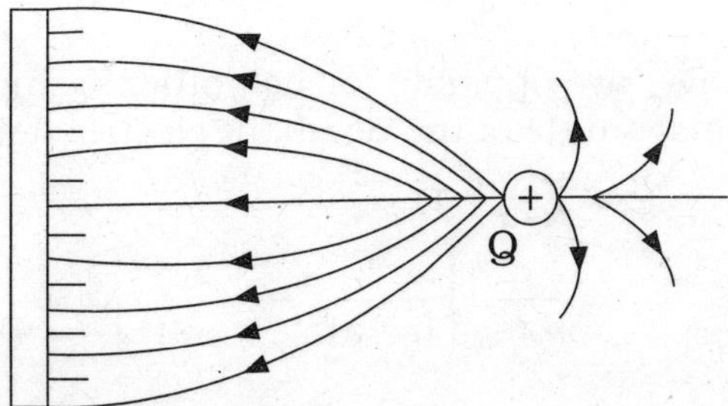

1A. 15. Electric Dipole; Dipole Moment

An **electric dipole** is "a combination of two equal and opposite point charges separated by a very small distance". The Fig. 1.13 shows two such charges separated by a distance 2a. The direction of the axis of the dipole is conventionally taken to run from -q to +q.

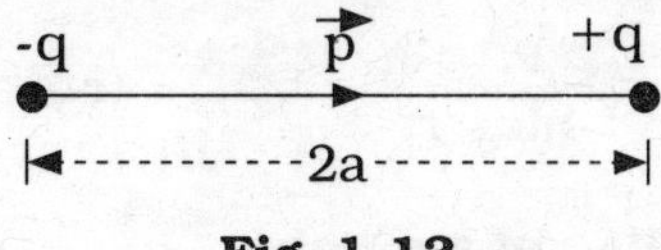

Fig. 1.13

Dipole Moment. The **dipole moment** $\vec{p}$ of an electric dipole is a vector quantity whose magnitude is **defined** as "the product of either of the two charges of the dipole and the distance between them" (Fig. 1.13).

$$\vec{p} = q.2\vec{a} \qquad \text{..... (1.24)}$$

$$|\vec{p}| = 2aq \qquad \text{..... (1.25)}$$

The direction of $\vec{p}$ lies along the axis of the dipole, pointing from the negative charge to the positive charge. The dipole moment is a measure of the strength of the dipole. Its SI **unit** is coulomb metre (C-m).

1A. 16. Dipole Field

The electric field produced by an electric dipole is called **dipole field.** It constitutes the space around the dipole in which any electrical effect due to the dipole is felt. Initially it may appear that the dipole field should be zero, because the dipole is made up of two equal and opposite charges +q and –q whose numerical sum is zero. It is not so because the electric field $\vec{E}$ is a vector quantity and therefore the field separately produced by +q and -q have to be added up vectorially, not numerically.The addition follows the principle of superposition (vide Section 1A.07).We shall evaluate $\vec{E}$ for two cases, viz. along the axis of the dipole and perpendicular to the axis of the dipole.

(a) Electric field along the axial line of a dipole

Consider a dipole AB with separation 2a between its two charges -q and +q. We shall evaluate the electric field intensity $\vec{E}$ at a point P along the axial line of the dipole, distant r from the mid-point O of the dipole (Fig. 1.14).

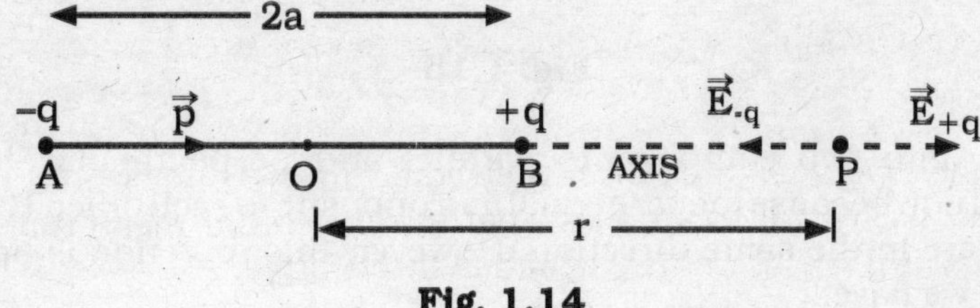

Fig. 1.14

The electric field $\vec{E}_{-q}$ due to the charge –q at P is,

$$\vec{E} = \frac{1}{4\pi\epsilon_0}\cdot\frac{(-q)}{(AP)^2}\hat{p} = \frac{(-q)}{4\pi\epsilon_0\,(r+a)^2}\hat{p}$$

Here $\hat{p}$ is a unit vector along the dipole axis, in the direction pointing from –q to +q. The electric field at P due to the charge +q is,

$$\vec{E}_{+q} = \frac{1}{4\pi\epsilon_0}\cdot\frac{(+q)}{(BP)^2}\hat{p} = \frac{(+q)}{4\pi\epsilon_0\,(r-a)^2}\hat{p}$$

Note that the directions of $\vec{E}_{-q}$ and $\vec{E}_{+q}$ are opposite to each other. Using the principle of superposition for electric fields (Section 1A.12), the magnitude of the resultant electric field at P will be,

$$\vec{E} = \vec{E}_{+q} + \vec{E}_{-q}$$

$$= \frac{q}{4\pi\epsilon_0}\cdot\left[\frac{1}{(r-a)^2} - \frac{1}{(r+a)^2}\right]\hat{p} = \frac{q.4ar.\hat{p}}{4\pi\epsilon_0\,(r^2-a^2)^2}$$

Now, (q.2a) $\hat{p}$ = q.2$\vec{a}$ ($\because \hat{p}$ is a unit vector). But q.2$\vec{a} = \vec{p}$ (Eq. 1.24). Therefore, we have,

$$\vec{E} = \frac{\vec{p}}{4\pi\epsilon_0}\cdot\frac{2r}{(r^2-a^2)^2} \qquad \text{..... (1.26)}$$

Since the separation 2a for a dipole is a very small value, we usually have $r \gg a$. Then, a^2 can be neglected in comparison to r^2. So we have,

$$\vec{E} = \frac{\vec{p}}{4\pi\epsilon_0}\cdot\frac{2}{r^3} \qquad (r \gg a) \qquad \text{..... (1.27)}$$

(b) Electric field along the equatorial line of a dipole

We now evaluate $\vec{E}$ at a point P situated along the equatorial line of the dipole, i.e. in a direction perpendicular to the axis of the dipole. The line bisects the axis at the mid-point O of the dipole. OP = r (Fig. 1.15). Again using the superposition principle, the resultant intensity is,

$$\vec{E} = \vec{E}_{+q} + \vec{E}_{-q}$$

Fig. 1.15

On resolving E_{+q} and E_{-q} into two components parallel and perpendicular to AB, the perpendicular components $\vec{E}_{+q}\sin\theta$ cancel out, because they are equal and opposite to each other. But the parallel components $\vec{E}_{+q}\cos\theta$ add up, as they are in the same direction. However, this direction is opposite to the direction of $\vec{p}$, as seen in the figure. So we have,

$$\vec{E} = -[E_{+q}\cos\theta + E_{-q}\cos\theta]\,\hat{p}$$
$$= -[E_{+q} + E_{-q}]\cos\theta.\hat{p}$$

Here, again, $\hat{p}$ is a unit vector in the direction of the dipole moment $\vec{p}$. Now, the magnitude of the electric field at P due to the point charge +q is,

$$\vec{E}_{+q} = \frac{q}{4\pi\epsilon_0}\cdot\frac{1}{(BP)^2} = \frac{q}{4\pi\epsilon_0}\cdot\frac{1}{(r^2+a^2)}$$

Similarly,
$$E_{-q} = \frac{q}{4\pi\epsilon_0}\cdot\frac{1}{(AP)^2} = \frac{q}{4\pi\epsilon_0}\cdot\frac{1}{(r^2+a^2)}$$

$$\therefore \quad \vec{E} = -\frac{2q}{4\pi\epsilon_0}\cdot\frac{1}{(r^2+a^2)}.\cos\theta.\hat{p}$$

Now,
$$\cos\theta = \frac{OB}{BP} = \frac{a}{(r^2+a^2)^{1/2}}$$

$$\therefore \quad \vec{E} = -\frac{(q.2a)\hat{p}}{(4\pi\epsilon_0)(r^2+a^2)^{3/2}}$$

Now, $(q.2a)\,\hat{p} = q.2\vec{a}$ ($\because \hat{p}$ is a unit vector). But $(q.2\vec{a}) = \vec{p}$, the dipole moment (vide Eq. 1.24). Hence,

$$\vec{E} = -\frac{\vec{p}}{4\pi\epsilon_0\,(r^2+a^2)^{3/2}} \qquad \text{..... (1.28)}$$

Again, because the separation 2a for a dipole is a very small value, we usually have $r \gg a$. Then, a^2 can be neglected in comparison to r^2. Hence, we have,

$$\vec{E} = -\frac{\vec{p}}{(4\pi\epsilon_0)}\cdot\frac{1}{r^3} \qquad (r \gg a) \qquad \text{..... (1.29)}$$

From Eqs. 1.28 and 1.29, we note the following:

(i) $E \propto 1/r^3$. Therefore for both the axial points and the equatorial points, the electric field intensity falls off as $1/r^3$ (not as $1/r^2$, as for a single point charge).

(ii) For the same value of r, the electric field intensity at an axial point is twice that for the corresponding equatorial point (vide Eq. 1.27).

(iii) For the axial points, the electric field $\vec{E}$ points in the same direction as the dipole moment $\vec{p}$. But for the equatorial points, the direction of $\vec{E}$ is opposite to $\vec{p}$.

1A. 17. Dipole in a Uniform Electric Field; Torque on the Dipole

Consider a dipole AB of moment $|\vec{p}|$ placed in a uniform electric field $\vec{E}$, such that the angle between $\vec{p}$ and $\vec{E}$ is θ (Fig. 1.16a). The points charges –q and +q of the dipole are separated by a small distance 2a. The electric field will exert force on both +q and -q. The force on +q will be,

$$|\vec{F}_{+q}| = q\,|\vec{E}| \qquad \text{(along } \vec{E}\text{)}$$

The force on -q will be,
$$|\vec{F}_{-q}| = q\,|\vec{E}| \qquad \text{(opposite to } \vec{E}\text{)}$$

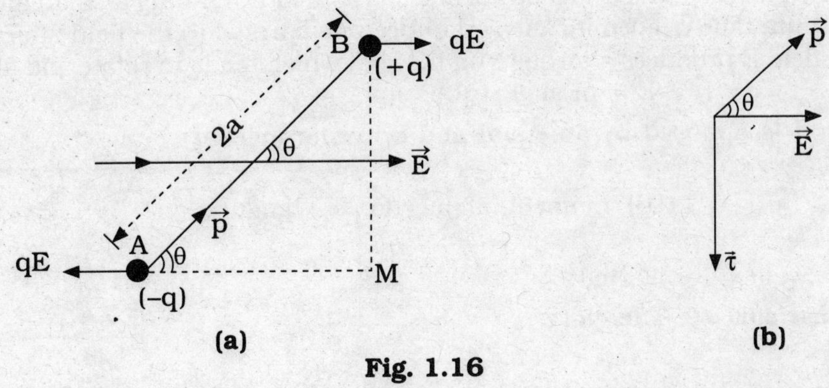

Fig. 1.16

Thus, the two forces are equal in magnitude. Hence, the net translatory force on the dipole is zero. But since the two forces are mutually opposite, they form a couple, which tends to rotate the dipole in the clockwise direction, so that $\vec{p}$ becomes aligned along $\vec{E}$. The torque (moment of the couple), $\vec{\tau}$, produced by $\vec{E}$ will be,

$$\vec{\tau} = (\text{Force}) \times (\perp \text{ distance between the two forces})$$
$$= (F) \times (BM) = (qE) \times 2a \sin\theta$$
$$= (2aq) \times (E) \sin\theta$$

Since the dipole moment p = 2aq, we have,

$$\vec{\tau} = pE \sin\theta \quad \text{..... (1.30)}$$

In vector form, we can write this equation as,

$$\vec{\tau} = \vec{p} \times \vec{E} \quad \text{(Torque on a Dipole)} \quad \text{..... (1.31)}$$

From this equation it follows that $\vec{\tau}$ is perpendicular to the plane containing $\vec{p}$ and $\vec{E}$. Its direction can be determined by the right-handed screw rule, depicted in Fig. 1.16b.

The unit of $\vec{\tau}$ is N-m. Its dimensions are $[ML^2T^{-2}]$.

Special cases. Two cases arise, (i) $\vec{p} /\!/ \vec{E}$ and (ii) $\vec{p} \perp \vec{E}$.

(i) $\vec{p} /\!/ \vec{E}$: There are two possibilities, viz. $\vec{p}$ along $\vec{E}$ (θ=0) and $\vec{p}$ opposite to $\vec{E}$ (θ=180°).

In both the cases, $\vec{\tau}$ = PE sinθ = 0. However, when $\vec{p}$ is directed opposite to $\vec{E}$, the torque tends to rotate the dipole through 180°, so that its dipole moment becomes oriented along the electric field. Thus, the position θ = 180° is said to be the position of 'unstable equilibrium' for the dipole. In comparison, the position θ = 0 is said to be the position of 'stable equilibrium', because then there is no torque acting on the dipole.

(ii) $\vec{p} \perp \vec{E}$: Now, τ = pE sin90° = pE. So in this case the dipole experiences maximum torque,

$$\tau_{max} = pE.$$

Example 10. Three charges are arranged on the vertices of an equilateral triangle, as shown in the figure. Find the dipole moment of the combination.

Solution. D is the middle point of AC. The combination of charges at A and C (–q each) is equivalent to a charge –2q concentrated at D. Now,

$$BD = \sqrt{(BC)^2 - (CD)^2} = \sqrt{d^2 - (d/2)^2} = \sqrt{\frac{3d^2}{4}} = \frac{d\sqrt{3}}{2}$$

Now, effectively the dipole consists of the charge +2q at B and the charge –2q at D.

$$\therefore \quad |\vec{p}| = 2q.\frac{d\sqrt{3}}{2} = qd\sqrt{3} \quad \text{(directed from D to B)} \quad \textbf{(Answer)}.$$

Example 11. A particle of mass M is attached to one end of a massless rigid non-conducting rod of length L. Another particle of the same mass is attached to the other end of the rod. The two particles carry charges +q and –q, respectively. The rod is held in a uniform electric field $\vec{E}$, making a small angle θ (a few degrees) with $\vec{E}$. Find the minimum time needed for the rod to become parallel to the field after it is set free.

Solution. The electric field $\vec{E}$ produces a torque τ on the rod, which tends to rotate and align the rod along $\vec{E}$,

$$\tau = \vec{p} \times \vec{E} = pE\sin\theta = (qL)E\sin\theta$$

The rotation of the rod is opposed by an equal and opposite inertial torque,

$$\tau = I\alpha \text{ (I: moment of inertia; } \alpha \text{ : angular acceleration)}$$

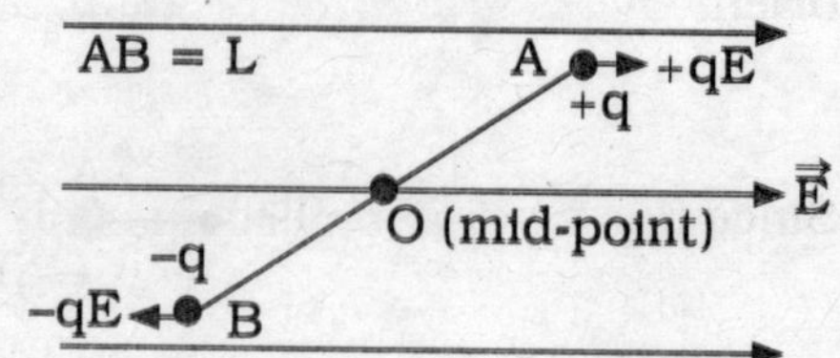

$$\therefore \quad I\alpha = (qL)E \sin\theta$$

Since θ is small, we have sin$\theta \approx \theta$. Therefore,

$$I\alpha = (qL)E\theta$$

$$\Rightarrow \quad \alpha = \left(\frac{qEL}{I}\right)\theta$$

This equation has the form, $\alpha = \omega^2\theta$, for angular simple harmonic motion. Thus, the motion of the rod is an angular S.H.M., with

$$\omega^2 = \frac{qeL}{I}$$

Now, the moment of inertia of the rod about O is,

$$I = M\left(\frac{L}{2}\right)^2 + M\left(\frac{L}{2}\right)^2 = \frac{ML^2}{2}$$

$$\therefore \quad \omega^2 = \frac{2qeL}{ML^2} = \frac{2qe}{ML}$$

$$\Rightarrow \quad \omega = \sqrt{\frac{2Qe}{ML}}$$

The minimum time needed for the rod to become parallel to the field is T/4 ($\theta = 0$ to $\theta = 90^0$), where T is the time period of rotation, i.e.

$$t = \frac{T}{4} = \frac{2\pi/\omega}{4} \quad (\because T = 2\pi/\omega)$$

$$= \frac{2\pi}{4\omega} = \frac{2\pi}{4\sqrt{2Qe/ML}}$$

$$= \frac{\pi}{2}\sqrt{\frac{ML}{2Qe}}$$ **(Answer).**

1A. 18. Work Done in Rotating a Dipole; Potential Energy of the Dipole

In the last section we have seen that for a dipole of moment $\vec{p}$ placed in a uniform electric field $\vec{E}$ at an angle θ (Fig. 1.16), the torque acting on the dipole due to $\vec{E}$ is (Eq. 1.30),

$$\tau = pE\sin\theta$$

If the dipole is now rotated through a small angle $d\theta$ against the torque, the small work dW done in the process will be,

$$dW = \tau d0 = pE \sin 0 \, d0$$

The total work W done in rotating the dipole from a given position θ_1 to another position θ_2 is found by integrating dW in the limits θ_1 to θ_2,

$$W = \int_{\theta_1}^{\theta_2} pE\sin\theta\, d\theta = pE\left[-\cos\theta\right]_{\theta_1}^{\theta_2}$$

$$\Rightarrow \quad W = -pE(\cos\theta_2 - \cos\theta_1) \qquad \text{..... (1.32)}$$

This work done in rotating the dipole through the angular range $(\theta_2 - \theta_1)$ is stored in the dipole as its electrical **potential energy, U.** Thus, the Eq. 1.32 also represents the electrical potential energy, U, of the dipole (as we know, the potential energy of a body is the energy possessed by the body by virtue of its position in a given force field; more about it will be explained in the next chapter in Section 1B.01). However, we can derive a more general and simple expression for U as follows. We imagine that initially the dipole is oriented $\perp$ to $\vec{E}$ and then rotates through an angle θ. Then, $\theta_1 = 90°$ and $\theta_2 = \theta$. So, from Eq. 1.32,

$$W = -pE(\cos\theta - \cos 90°)$$

$$= -pE\cos\theta$$

Since this work done becomes the p.e., U, possessed by the dipole, we can write,

$$W = U = -pE\cos\theta \qquad \text{..... (1.33)}$$

In vector form,

$$W = U = -(\vec{p}.\vec{E}) \quad ..... (1.34)$$

Since the scalar product of two vectors is a scalar quantity, U is a scalar quantity. Being a measure of the work done, it is measured in joule in the SI system of units.

In the next chapter (Section 1B.11), we shall derive the Eq. 1.34 once again by a different approach, viz. by calculating the work done in bringing the dipole charges from infinity to the electric field, $\vec{E}$.

Example 12. An electric dipole experiences maximum torque of 5×10^{-6} Nm, when placed in a uniform field. Calculate its potential energy when it is oriented at an angle of 60° with the field.

Solution. The torque experienced by a dipole is given by,

$$\tau = pE\sin\theta$$

The maximum torque is experienced at $\theta = 90°$,

$$\therefore \quad \tau_{max} = pE \sin 90° = pE$$

It is given that $\tau_{max} = 5\times10^{-6}$ Nm.

$$\therefore \quad pE = 5\times10^{-6}$$

The potential energy of a dipole in a uniform field is given by,

$$U = -pE\cos\theta$$

$$= -(5\times10^{-6}) \times \cos 60° \qquad (\cos 60° = 0.5)$$

$$= -2.5\times10^{-6}\ J$$ **(Answer).**

Example 13. An electric dipole consists of two opposite charges of 2 μC each, separated by 1cm. It is placed in an external uniform field of magnitude 10^4 NC^{-1}. Find (i) the maximum torque exerted by the field on the dipole and (ii) the work done in rotating the dipole through 180°, starting from the position $\theta = 0°$.

Solution. Given $q = 2\ \mu C = 2\times10^{-6}$ C; $2a = 1$ cm $= 0.01$ m; $E = 10^4$ NC^{-1}

(i) The torque is given by,

$$\vec{\tau} = \vec{p} \times \vec{E} = pE\sin\theta$$

For maximum torque, $\theta = 90°$. $\therefore \sin\theta = 1;\ p = 2a = (2\times10^{-6})\times0.01 = 2\times10^{-8}$

$$\therefore \quad \vec{\tau}_{max} = pE$$

$$= (2\times10^{-8})\times10^4\ Nm$$

$$= 2\times10^{-4}\ Nm$$ **(Answer).**

(ii) $\theta_1 = 0$ and $\theta_2 = 180°$. The work done is given by,

$$W = pE(\cos\theta_1 - \cos\theta_2)$$

$$= (2\times10^{-4})(1+1)$$

$$= 4\times10^{-4}\ J$$ **(Answer).**

Solved Conceptual Problems

[Including HOTS (Higher Order Thinking Skills) Questions]

Problem 1. What is electrostatics ?

Solution. The branch of physics which deals with charges at rest is called electrostatics.

Problem 2. What is a charge ?

Solution. Charge is a property possessed by a body due to which it experiences a force in the presence of another charged body. It is a measure of the degree of electrification of the body.

Problem 3. Briefly explain the main methods of charging a body.

Solution. There are mainly two methods:

(i) Frictional electricity: In this process, two bodies, e.g. a glass rod and silk cloth, are rubbed together. It provides energy to the outermost electrons of some surface atoms of the body with lower work function to get detached from the body and settle on the other body. Thus the first body becomes positively charged and the second body becomes negatively charged.

(ii) Induction Process: In this process a localized charge of opposite sign develops on an electrically neutral body, when a charged body is brought close to it. If the initially neutral body is non-conducting, the charge developed on it will remain localized. If it is conducting, it will spread all over the body.

Problem 4. How will you demonstrate that there are two types of charges ?

Solution. Take two glass rods rubbed with silk cloth and an ebonite rod rubbed with fur. When the glass rods are brought close to each other, they repel each other but when a glass rod is brought close to the ebonite rod, the two rods attract each other. This shows that the charges on the glass rod and the ebonite rod are different in nature.

Problem 5. A negatively charged ebonite rod is brought near a neutral conducting sphere, which is then connected to earth but later insulated from earth. Finally, the rod is also removed. Show with diagrams the stepwise charging of the sphere.

Solution.

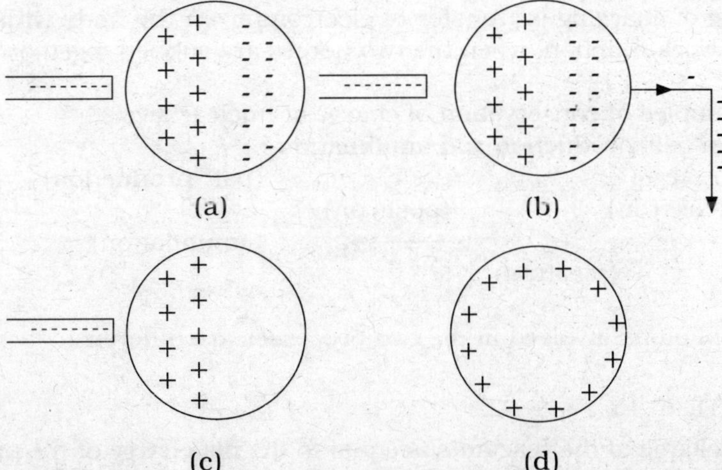

(a) The charge on the neutral metallic sphere is redistributed when the ebonite rod is brought near to it. The part of the sphere near the rod has positive charge on it.

(b) The sphere is then earthed. The excess electrons leave the conductor and pass on to the earth.

(c) The connecting wire to the earth is removed. The sphere now contains an excess of induced positive charge.

(d) When the ebonite rod is removed, the positive charge is uniformly distributed all over it, because of repulsion among the charges.

Problem 6. How many types of electric charges exist in nature ? What are vitreous and resinous charges ?
Solution. Two types of electric charges exist in nature, viz. positive and negative. The charges developed on a vitreous substance, e.g. glass (when rubbed with silk) are called vitreous. The charges developed on a resin like amber or ebonite (when rubbed with wool or fur) are called resinous.

Problem 7. Give important properties of electric charge.
Solution. (i) Two kinds of charges exist in nature: positive and negative. Like charges repel and unlike charges attract each other.
(ii) Charge is quantized: It means that all charges are an integral multiple of an elementary charge, ($q = \pm ne$, where n is an integer and e is the charge on an electron).
(iii) Additive nature of charge: The additivity means that the total charge on a body is the algebraic sum of all the charges present over different parts of it.
(iv) Charge conservation: The total charge on an isolated system does not change, i.e. it remains constant. In other words the charge can neither be created nor destroyed; it can merely be transferred from one body to another.

Problem 8. Give important points of distinction between charge and mass.
Solution.

Charge

(1) Force due to charges can be attractive or repulsive. It is so because the charges can be both positive or negative.
(2) The charge on a body does not depend on its velocity.
(3) It is well established that charge is quantized.

Mass

(1) Force due to mass is always attractive, because the mass is never negative.
(2) Mass of a body depends upon its velocity $\left(m = \frac{m_0}{\sqrt{1 - v^2/c^2}}\right)$
(3) The quantization of mass is yet to be established.

Problem 9. What is the cause of charging in frictional electricity ?
Solution. The basic cause of charging is transfer of electrons from the body with lower work function to another body with higher work function, when the two bodies are rubbed together.

Problem 10. Give two examples of conservation of charge at nuclear levels.
Solution. (i) Phenomena of pair production and annihilation :

$$\gamma \longrightarrow \underset{\text{(electron)}}{e^-} + \underset{\text{(positron)}}{e^+} \quad \text{(pair production)}$$

$$\underset{\text{(electron)}}{e^-} + \underset{\text{(positron)}}{e^+} \longrightarrow 2\gamma \quad \text{(annihilation)}$$

(The energies of the γ-ray photons involved in the two processes are different)
(ii) Decay of ${}^{238}_{92}U$:

$${}^{238}_{92}U \rightarrow {}^{234}_{90}Th + {}^{4}_{2}He$$

In both the cases, the net charge of the reactants is equal to the net charge of the product.

Problem 11. Tabulate the given materials under the headings 'positive charge' and 'negative charge', according to the nature of charge acquired by them on mutual rubbing in their respective pairs.
(1) Glass rod - Silk cloth (2) Ebonite - Fur
(3) Woollen cloth-Plastic material (4) Dry hair - Comb.

Solution.

Positive Charge	Negative Charge
Glass rod	Silk cloth
Fur	Ebonite
Woollen cloth	Plastic material
Dry Hair	Comb

Problem 12. It is observed that when a metal rod is rubbed while holding it in hand, it does not get charged. What is the reason ?

Solution. The human body is a fairly good conductor of electricity. Therefore, when the metal rod is held in one's hand, whatever charge develops upon it is immediately transferred to the earth through the body. Hence no static charge is retained on the metal rod, i.e. it does not get charged.

Problem 13. What is the order of the ratio between electrostatic and gravitational force for two electrons separated by a finite distance.

Solution. $$\frac{F_e}{F_g} \sim 10^{39}$$

Problem 14. How a force of attraction develops between a charged body and an uncharged body ?

Solution. When the uncharged body is brought near to the charged body, then by the process of induction opposite charge develops on the end of the uncharged body nearest to the charged body.

Problem 15. The force of attraction between two point charges placed at a distance r from each other is F. What distance apart should they be kept in the same medium so that the force between them becomes F/3 ?

Solution.

$$F = \frac{1}{4\pi\epsilon_0}\frac{q_1q_2}{r^2} \qquad \text{..... (i)}$$

$$\frac{F}{3} = \frac{1}{4\pi\epsilon_0}\frac{q_1q_2}{(r'^2)^2} \qquad \text{(r' : new distance)} \qquad \text{..... (ii)}$$

from (i) and (ii),

$$r'^2 = 3r^2$$

$$\therefore \quad r' = \sqrt{3}r \quad \textbf{(Answer).}$$

Problem 16. What is the charge on a photon ? Give reason for your answer.

Solution. The charge on a photon is zero, because the rest mass of a photon is zero and no charge can exist without mass.

Problem 17. An observer finds two charged bodies at rest. If he starts moving away from the bodies, what change will he observe in respect of the force between the bodies ?.

Solution. In the frame of reference where all charges are at rest, the forces are purely electrical. So initially he will observe only electrostatic force between the two bodies. But when he starts moving, the bodies appear to be interacting in terms of both electrical and magnetic forces (electromagnetic forces).

Problem 18. What are quarks ?

Solution. Recent discoveries in high energy physics have indicated that the elementary constituents of protons and neutrons are quarks. They are mainly of two types:

(i) Up-quark (charge $= +\frac{2}{3}e$).

(ii) Down-quark (charge $= -\frac{1}{3}e$).

Solved Numerical Problems

Electric Charges

Problem 1. One milli-mole of electrons are removed from an electrically neutral body. Calculate the net electric charge on the body.

Solution. Let n be the number of electrons removed,

$$n = (6.023\times10^{23})\times10^{-3} = 6.023\times10^{20}$$

Let Q be the total charge removed,

$$Q = ne = (6.023\times10^{20})\times(1.6\times10^{-19})$$
$$= 1.6\times6.023 \times 10$$
$$= 96.37$$

Since the electrons are negatively charged, their removal leaves the body with a net positive charge of +96.37 C. **(Answer).**

Problem 2. Calculate how many electrons should be removed from an electrically neutral body, so that it is left with a net positive charge of +1 coulomb.

Solution. Charge needed on the body, Q = 1 C.

Let n be the number of electrons required to be removed. We know,

$$Q = ne$$

$$\Rightarrow \quad n = \frac{Q}{e} = \frac{(1C)}{(1.6\times10^{-19}C)}$$
$$= 6.25\times10^{18}$$

Thus, 6.25×10^{18} electrons are required to be removed from the body. **(Answer).**

Problem 3. A body has a charge of 1.2×10^{-6} C on it. Suggest how the charge can be raised to 2.0×10^{-6} C, by a suitable transfer of electrons.

Solution. Initial charge, $Q_1 = 1.2\times10^{-6}$ C

Final charge, $Q_2 = 2.0\times10^{-6}$ C

The electrons are negatively charged particles. As the final charge is more, the electrons need to be removed from the body. Let the number to be removed be n. We have,

$$\Delta Q = Q_2 - Q_1 = ne$$

$$\Rightarrow \quad n = \frac{Q_2 - Q_1}{e}$$

$$= \frac{(2.0\times10^{-6}) - (1.2\times10^{-6})}{(2.0\times10^{-19})} = 5\times10^{12}$$

Thus, 5×10^{12} electrons should be removed from the body. **(Answer).**

Problem 4. A conducting cube of side length 5 cm has a charge of 1 milli- coulomb on it. Calculate the surface charge density.

Solution. Charge, $Q = 1\times10^{-3}$ C

Total surface area of the cube, $S = 6a^2$, where a = 5 cm = 5×10^{-2} m. (A cube has 6 faces of same side length).

Now, $$Q = \sigma S \quad (\sigma\text{: surface charge density})$$

$$\Rightarrow \quad \sigma = \frac{Q}{S}$$

$$= \frac{(1\times10^{-3})}{6 \times (5\times10^{-2})^2} = 6.67\times10^{-2}\ C/m^2$$ **(Answer).**

Problem 5. Two uniformly charged spheres of radii 4 cm and 5 cm, respectively, have equal charge. Find the ratio of their volume charge densities.

Solution. Let A and B be the two spheres. Their radii are,

$$r_A = 4\times10^{-2}\ m$$
$$r_B = 5\times10^{-2}\ m$$

The electric charge on each of them is Q.
The volume charge density,

$$\rho = \frac{Q}{V} \text{ (V : volume)}$$

$$\therefore \quad \frac{\rho_A}{\rho_B} = \frac{Q/\left[\left(\frac{4}{3}\right)\pi r_A^3\right]}{Q/\left[\left(\frac{4}{3}\right)\pi r_B^3\right]} = \frac{r_B^3}{r_A^3}$$

$$= \frac{(5\times10^{-2})^3}{(4\times10^{-2})^3} = \frac{(5)^3}{(4)^3} = \frac{125}{164} = 1.95$$

Thus the required ratio is 125 : 64 or 1.95. **(Answer).**

Problem 6. A metal wire is bent in the shape of a semi-circle of radius 10 cm. If the total charge on it is 2 μC, calculate its linear charge density.

Solution. Radius, r = 10 cm = 0.1 m.

Length, $\ell = \pi r$; chargc, $q = 2\times10^{-6}$ C

The linear charge density is,

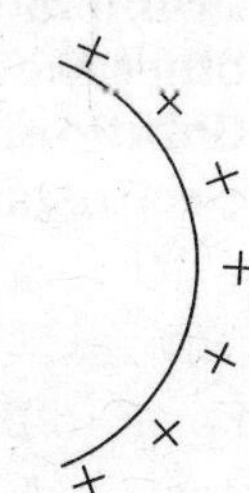

$$\lambda = \frac{q}{\ell} = \frac{(2\times10^{-6})}{(3.14\times0.1)} = 6.37\times10^{-6} \text{ C/m}$$ **(Answer).**

Coulomb's Law

Problem 7. Calculate the electrostatic force between two point charges of value 10 μC each and separated by a distance of 1 mm in vacuum.

Solution. Charges, $q_1 = q_2 = 10\times10^{-6} = 10^{-5}$ C

Separation, $r = 1\text{mm} = 10^{-3}$ m.

$$\text{Electrostatic force, } F = \frac{1}{4\pi\epsilon_0}\cdot\frac{q_1q_2}{r^2}$$

$$= \frac{(9\times10^9)\times(10^{-5})\times(10^{-5})}{(10^{-3})^2}$$

$$= 9\times10^5 \text{ N}$$ **(Answer).**

The force is repulsive, because both the charges are positive. **(Answer).**

Problem 8. Two α-particles are separated by a distance of 10 Å. Find the electrostatic force between them and compare it with the gravitational force between them.

Solution. Charge on an α–particle, $q = 2e = 2(1.6\times10^{-19})$ C

Separation, $r = 10 \text{ Å} = 10^{-9}$ m. ($1 \text{ Å} = 10^{-10}$ m)

$$\text{Electrostatic force, } F_e = \frac{1}{4\pi\epsilon_0}\frac{q_1q_2}{r^2}$$

$$= \frac{1}{4\pi\epsilon_0}\frac{(2e)(2e)}{r^2}$$

$$= \frac{(9\times10^9)\times4\times(1.6\times10^{-19})^2}{(10^{-9})^2}$$

$$= 9.22\times10^{-10} \text{ N}$$ **(Answer).**

The force is repulsive, because both the particles have positive charge. **(Answer).**

Mass of an α–particle, $m_\alpha = 4u$ (u : atomic mass unit)

$= 4(1.66\times10^{-27}\text{ kg})$

$= 6.64\times10^{-27}\text{ kg}$

Gravitational force, $$F_g = \frac{m_\alpha m_\alpha}{r^2} G \quad (G = 6.67\times10^{-11}\text{ Nm}^2/\text{kg}^2)$$

$$= \frac{(6.64\times10^{-27})^2}{(10^{-9})^2} \times (6.67\times10^{-11}) = 2.94\times10^{-45}\text{ N}$$

The ratio of the electrostatic force and the gravitational force is,

$$\frac{Fe}{Fg} = \frac{9.22\times10^{-10}}{2.94\times10^{-45}}$$

$$= 3.14\times10^{35} \quad \textbf{(Answer).}$$

Thus, the electrostatic force is very much stronger than the gravitational force. The electrostatic force is repulsive, whereas the gravitational force is attractive. **(Answer).**

Problem 9. A charge Q is to be divided into two parts, so that when kept at a certain distance from each other the force between them is maximum. Find out how the charge Q should be suitably divided.

Solution. Let the charge Q be divided into two parts q and (Q - q), which are kept at a distance r apart. The electrostatic force between them is,

$$F = \frac{1}{4\pi\epsilon_0}\frac{q(Q-q)}{r^2}$$

For F to be maximum,

$$\frac{dF}{dq} = 0$$

$$\Rightarrow \quad \frac{d}{dq}\left[\frac{1}{4\pi\epsilon_0}\frac{q(Q-q)}{r^2}\right] = \frac{1}{4\pi\epsilon_0 r^2}\cdot\frac{d}{dq}(qQ - q^2) = 0$$

$$\Rightarrow \quad \frac{1}{4\pi\epsilon_0 r^2}(Q - 2q) = 0$$

$$\Rightarrow \quad q = \frac{Q}{2}$$

$$\therefore \quad (Q - q) = \frac{Q}{2}$$

Thus, the charge Q should be divided into two equal parts of value Q/2 each. **(Answer).**

Problem 10. The force between two point charges in vacuum, when separated by a distance of 1m, is 4 newton. What will be the force between them if the distance between them is 1/2 m and they are placed in a medium of relative permittivity 4 ?

Solution. Let the charges be q_1 and q_2.

Initial separation, $r_1 = 1$ m

Initial force, $$F_1 = \frac{1}{4\pi\epsilon_0}\cdot\frac{q_1q_2}{r_1^2} = 4\text{ N} \quad \text{..... (1)}$$

Relative permittivity, $\epsilon_r = 4$

New separation, $r_2 = \frac{1}{2}$ m

New force, $$F_2 = \frac{1}{4\pi\epsilon_0\epsilon_r}\cdot\frac{q_1q_2}{r_2^2} \quad \text{..... (2)}$$

Dividing Eq. 2 by Eq. 1, $$\frac{F_2}{F_1} = \frac{r_1^2}{\epsilon_r r_2^2} = \frac{1}{4\times\frac{1}{4}} = 1 = 1$$

Thus, the new force $F_2 = F_1 = 4$ N. **(Answer).**

Problem 11. If the separation between two point charges of equal value is increased by 0.1% and their values are increased by 0.2% each, calculate the percentage change in the electrostatic force between them.

Solution. From Coulomb's law, the electrostatic force between two equal charges of value q each, at a separation r, is,

$$F = \frac{1}{4\pi\epsilon_0} \cdot \frac{q^2}{r^2}$$

Taking logarithm on both sides, we get,

$$\log_e F = \log_e\left(\frac{1}{4\pi\epsilon_0}\right) + 2\log_e(q) - 2\log_e(r)$$

Differentiating, we get,

$$\frac{dF}{F} = 0 + 2\frac{dq}{q} + 2\frac{dr}{r}$$

$$\Rightarrow \quad \frac{dF}{F} \times 100\% = 2\left(\frac{dq}{q}\times 100\% - \frac{dr}{r}\times 100\%\right)$$

Given that,

$$\frac{dq}{q} \times 100\% = 0.1\%$$

and

$$\frac{dr}{r} \times 100\% = 0.2\%$$

$$\therefore \quad \frac{dF}{F} \times 100\% = 2(0.1 - 0.2) = -0.2\%$$

Thus, there is a reduction in the electrostatic force by 0.2%. **(Answer).**

Problem 12. A small pith ball of mass 0.05 g carries a charge of 1 μC. Another pith ball of the same mass is placed vertically above it at a distance of 1 cm. What should be the charge on the second ball, so that it is held stationary, in spite of the force of gravity acting upon it ?

Solution. For holding the second ball stationary, its weight has to be balanced by the electrostatic force of repulsion between the two balls. Let the charge on the second ball be q_2. The charge on the first ball is, $q_1 = 1\ \mu C = 10^{-6}$ C . The separation between the two balls is,

$$r = 1\text{cm} = 10^{-2}\text{ m}$$

In the condition of balance,

$$mg = \frac{1}{4\pi\epsilon_0} \cdot \frac{q_1 q_2}{r^2}$$

$$\Rightarrow \quad q_2 = \frac{mg(4\pi\epsilon_0)r^2}{q_1}$$

Now, $m = 0.05$ g $= 5 \times 10^{-5}$ kg; $g = 9.81$ ms^{-2}; $\frac{1}{4\pi\epsilon_0} = 9\times 10^9$ Nm^2/C^2

$$\therefore \quad q_2 = \frac{(5\times10^{-5})\times(9.81)\times(10^{-2})^2}{(9\times10^9)\times(10^{-6})}$$

$$= 5.45\times10^{-12}\text{ C}$$

$\vec{F}$

F = mg

q (Second ball)

mg

1μC (First ball)

Thus, the charge on the second ball should be + 5.45×10^{-12} C. **(Answer).**

Problem 13. Two small balls having equal charge are embedded in ice at a mutual separation of 20 cm. To maintain the same electrostatic mutual force of repulsion after the ice has melted into water, the separation has to be reduced to 4.2 cm. Taking the dielectric constant of water as 81, calculate the dielectric constant of ice.

Solution. Let the charges be of value q each. Separation in ice, r_1 = 20 cm = 0.2 m.
The electrostatic force between the charges in ice is,

$$F_i = \frac{1}{4\pi\epsilon_0\epsilon_i} \cdot \frac{q^2}{(0.2)^2} \qquad \text{..... (1)}$$

$\in_i$ is the dielectric constant of ice. The electrostatic force in water is, (r_2 = 4.2 cm = 0.042 m),

$$F_w = \frac{1}{4\pi\epsilon_0\epsilon_w}\cdot\frac{q^2}{(0.042)^2} \quad ..... (2)$$

($\in_w$ = 81, is the dielectric constant of water)
Dividing Eq. 2 by Eq. 1,

$$\frac{F_w}{F_i} = \frac{\epsilon_i}{\epsilon_w}\cdot\frac{(0.2)^2}{(0.042)^2} = 1 \quad (\because \text{Given that } F_w = F_i)$$

$$\therefore \quad \in_i = \in_w\left(\frac{0.042}{0.20}\right)^2$$

$$= 81 \times \left(\frac{0.021}{0.100}\right)^2$$

$$= 81\times (0.21)^2$$

$$= 3.57$$

Thus the dielectric constant of ice is found to be 3.57. **(Answer).**

Problem 14. The force of attraction between two point charges placed at a distance d apart in a medium is 'F'. What should be the distance apart in the same medium so that the force of attraction between them becomes $\frac{F}{3}$?

Solution. Let the charges be q_1 and q_2.
Initial separation, r_1 = d.

Initial force, $$F = \frac{1}{4\pi\epsilon_0}\cdot\frac{q_1q_2}{d^2} \quad ..... (1)$$

Let the final separation be r_2. Then the force becomes F/3,

$$\frac{F}{3} = \frac{1}{4\pi\,\epsilon_0}\cdot\frac{q_1q_2}{r_2^2} \quad ..... (2)$$

Dividing Eq. 2 by Eq. 1,

$$\frac{1}{3} = \frac{d^2}{r_2^2}$$

$$\Rightarrow \quad r_2{}^2 = 3d^2$$

$$\Rightarrow \quad r_2 = d\sqrt{3}$$

Thus, the required distance is found to be $d\sqrt{3}$ **(Answer).**

Problem 15. Two point charges q_1 = +1 μC and q_2 = 2 μC are placed at points P_1 and P_2, respectively. The position vectors of P_1 and P_2 are $(2\hat{i} + 3\hat{j})$ and $(7\hat{i} + 8\hat{j})$, respectively. Find the electrostatic force on q_2 due to q_1 in vector form. Take the positions as expressed in metre.
Solution. The vector form of the Coulomb's law is (vide Eq. 1.5),

$$\vec{F}_{21} = \frac{1}{4\pi\epsilon_0}\cdot\frac{q_1q_2}{|\vec{r}_2 - \vec{r}_1|^3}.(\hat{r}_2 - \hat{r}_1)$$

$$q_1 = +1\times10^{-6}\text{ C};\ q_2 = -2\times10^{-6}\text{ C}$$

$$\hat{r}_1 = (2\hat{i} + 3\hat{j});\ \hat{r}_2 = 7\hat{i} + 8\hat{j};\ \frac{1}{4\pi\epsilon_0} = 9\times10^9\,\frac{\text{Nm}^2}{\text{C}^2}$$

$$\therefore \quad (\hat{r}_2 - \hat{r}_1) = (7\hat{i} + 8\hat{j}) - (2\hat{i} + 3\hat{j})$$

$$= 5\hat{i} + 5\hat{j}$$

Also (see Fig.), $(\vec{r}_2 - \vec{r}_1) = \sqrt{(5)^2 + (5)^2}\ 5\sqrt{2}$ metre

$\therefore$
$$F_{21} = \frac{(9\times10^9)\times(1\times10^{-6})\times(-2\times10^{-6})}{(5\sqrt{2})^3}(5\hat{i} + 5\hat{j})$$

$$= -\frac{18\times10^{-3}}{125\times2\sqrt{2}}(5\hat{i} + 5\hat{j})$$

$$= -\frac{18}{50\sqrt{2}}\times10^{-3}(\hat{i} + \hat{j})$$

$$= -2.55\times10^{-4}(\hat{i} + \hat{j})$$ **(Answer).**

$\vec{r}_1$, 5m, $(\vec{r}_2 - \vec{r}_1)$, 5m, $\vec{r}_2$

The negative sign means that the electrostatic force between the charges is a force of attraction. **(Answer).**

Superposition Principle

Problem 16. Two point charges +4q and +9q are kept at a distance d apart in air. A third charge $+q_0$ is to be kept along the line joining the two charges, in such a way that the net force on q_0 is zero. Calculate the distance between the charges $+q_0$ and +4q.

Solution. Suppose q_0 is placed at a distance x from 4q. The net force on q_0 will be zero if the force $\vec{F}_{4q}$ on it due to 4q is equal and opposite to the force $\vec{F}_{9q}$ on it due to 9q. As all the charges are positive, the forces $\vec{F}_{4q}$ and $\vec{F}_{9q}$ are mutually opposite in direction.

For the net force $\vec{F}$ on q_0 to be zero, we have,

$\vec{F}_{4q}$, $+q_0$, $\vec{F}_{9q}$, $+q_0$, +4q, +9q, x, d-x, d

$$|\vec{F}| = |\vec{F}_{4q}| - |\vec{F}_{9q}| = 0$$

$$\Rightarrow \quad |\vec{F}_{4q}| = |\vec{F}_{9q}|$$

$$\Rightarrow \quad \frac{1}{4\pi\epsilon_0}\cdot\frac{q_0.4q}{x^2} = \frac{1}{4\pi\epsilon_0}\cdot\frac{q_0.9q}{(d-x)^2}$$

$$\Rightarrow \quad \frac{4}{x^2} = \frac{9}{(d-x)^2}$$

$$\Rightarrow \quad \frac{2}{x} = \frac{3}{(d-x)} \quad \text{(taking square roots)}$$

$$\Rightarrow \quad 2(d-x) = 3x \quad \text{Or} \quad 2d = 5x$$

$$\Rightarrow \quad x = \frac{2}{5}d$$

Thus, q_0 should be placed at a distance of (2/5)d from the charge 4q. **(Answer).**

Problem 17. Two point charges of value +q and +2q are kept at a distance d apart from each other in air. A third charge +Q is to be kept along the same line in such a way that the net force acting on either q or 2q is zero. Calculate the position of charge Q in terms of the separation d.

Solution. Suppose the distances of Q from q and 2q are x_1 and x_2, respectively. For the net force on q to be zero,

$$\vec{F}_{qQ} + \vec{F}_{(2q)q} = 0$$

$$\Rightarrow \quad \frac{1}{4\pi\epsilon_0}\cdot\frac{qQ}{(x_1)^2} = \frac{-1}{4\pi\epsilon_0}\cdot\frac{q.(2q)}{(x_1 + x_2)^2}$$

$$\Rightarrow \quad \frac{Q}{x_1^2} = \frac{-(2q)}{(x_1 + x_2)^2} \quad \text{..... (1)}$$

Similarly, for the net force on 2q to be zero,

$$\vec{F}_{(2q)Q} + \vec{F}_{(2q)q} = 0$$

$$\Rightarrow \quad \frac{1}{4\pi\epsilon_0}\cdot\frac{(2q).Q}{x_2^2} = \frac{-1}{4\pi\epsilon_0}\cdot\frac{q.(2q)}{(x_1+x_2)^2}$$

$$\Rightarrow \quad \frac{Q}{x_2^2} = \frac{-q}{(x_1+x_2)^2}$$

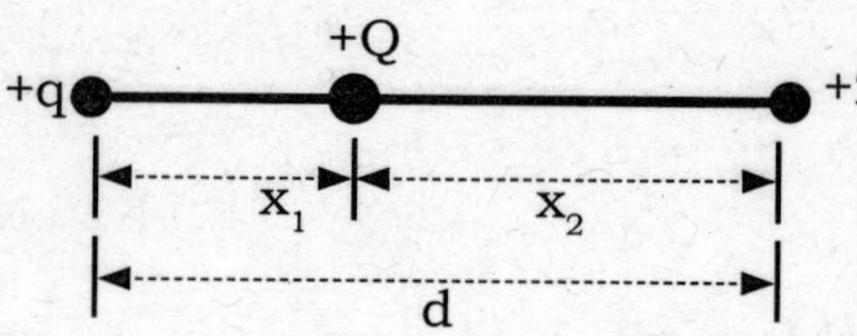

+Q

+q ● ● +2q

x_1 x_2

d

Multiplying on both the sides by 2,

$$\frac{2Q}{x_2^2} = \frac{-(2q)}{(x_1+x_2)^2} \qquad \text{..... (2)}$$

From the Eqs. 1 and 2 we have,

$$\frac{Q}{x_1^2} = \frac{2Q}{x_2^2}$$

$$\Rightarrow \quad \frac{x_2}{x_1} = \sqrt{2} \quad \text{Or} \quad x_2 = x_1\sqrt{2}$$

Now, $x_1 + x_2 = d$

$$\Rightarrow \quad x_1 + x_1\sqrt{2} = d$$

$$\Rightarrow \quad x_1 = \frac{d}{\left(1+\sqrt{2}\right)}$$

and $x_2 = x_1\sqrt{2} = \dfrac{d\sqrt{2}}{\left(1+\sqrt{2}\right)}$

These values of x_1 and x_2 give the position of Q in terms of d. **(Answer).**

Problem 18. Four point charges $q_A = 1\ \mu C$, $q_B = -1\ \mu_C$, $q_C = 2\ \mu C$ and $q_D = -2\ \mu C$ are placed along x - axis at $x_A = 0$, $x_B = 2$ cm, $x_C = 4$ cm and $x_D = 6$ cm, respectively. Find the magnitude and direction of the electrostatic force on the charge q_B.

Solution. The charge q_B is attracted towards q_A and q_C and repelled by q_D. The total electrostatic force on q_B is given by,

$$\vec{F}_B = \vec{F}_{BA} + \vec{F}_{BC} + \vec{F}_{BD}$$

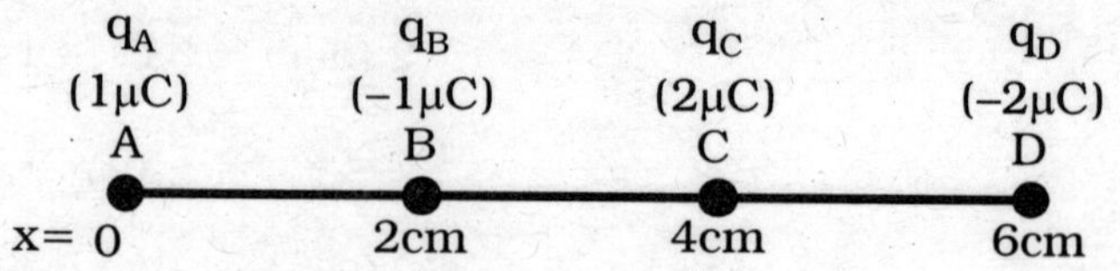

q_A	q_B	q_C	q_D
(1μC)	(−1μC)	(2μC)	(−2μC)
A	B	C	D
x= 0	2cm	4cm	6cm

Now, $|\vec{F}_{BA}| = \dfrac{1}{4\pi\epsilon_0}\cdot\dfrac{q_B.q_A}{(AB)^2}$

$(q_A = q_B = 1\ \mu C = 1\times10^{-6}\ C;\ AB = 2\times10^{-2}\ m)$

$$= \frac{(9\times10^9)\times(1+10^{-6})\times(1\times10^{-6})}{(2\times10^{-2})^2}$$

$= 22.5$ N (directed towards A)

$$|\vec{F}_{BC}| = \frac{1}{4\pi\epsilon_0}\cdot\frac{q_B.q_C}{(BC)^2}$$

$$= \frac{(9\times10^9)\times(1\times10^{-6})\times(1\times10^{-6})}{(2\times10^{-2})^2}$$

$= 45$ N (towards C)

$$|\vec{F}_{BD}| = \frac{1}{4\pi\epsilon_0}\cdot\frac{q_B.q_D}{(BD)^2}$$

$$= \frac{(9\times10^9)\times(1\times10^{-6})\times(2\times10^{-6})}{(4\times10^{-2})^2}$$

$= 11.25\text{N}$ (towards B)

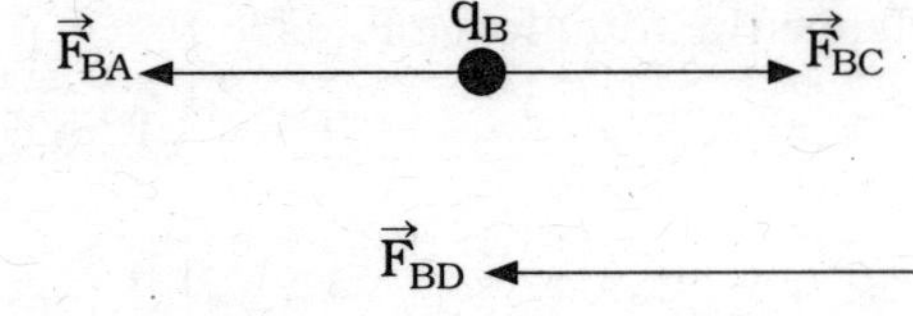

∴ The total force on q_B is,

$$F_B = F_{BC} - (F_{BA} + F_{BD})$$
$$= 45 - (22.5 + 11.25)$$
$$= 11.25 \text{ N (towards C)}$$ **(Answer).**

Problem 19. Four equal point charges of value 1 μC each are respectively placed at the four corners of a square of side length $\sqrt{2}$ m. Calculate the magnitude of electrostatic force on any one charge.

Solution. Let the charges be placed at the corners A, B, C, D of the square. We shall evaluate the electrostatic force on the charge at C. We have

$$CB = CD = \sqrt{2} \text{ m}; \; 1\,\mu C = 10^{-6}\,C$$

$$AC = \sqrt{\left(\sqrt{2}\right)^2 + \left(\sqrt{2}\right)^2} = 2 \text{ m}$$

The electrostatic force at C is,

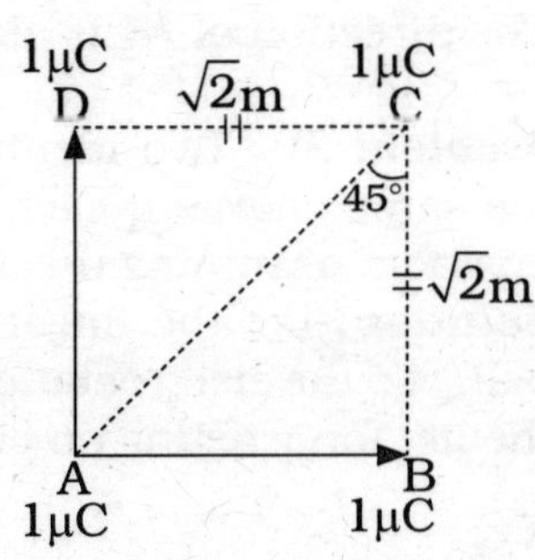

$$\vec{F}_C = \vec{F}_{CA} + \vec{F}_{CB} + \vec{F}_{CD}$$

We have,

$$|\vec{F}_{CA}| = \frac{1}{4\pi\epsilon_0} \cdot \frac{q_C q_A}{(AC)^2}$$

$$= \frac{(9\times10^9)\times(10^{-6})\times(10^{-6})}{(2)^2} \quad (1\,\mu C = 10^{-6}\,C)$$

$$= \frac{9}{4}\times10^{-3}\,\text{N}$$

$$|\vec{F}_{CB}| = \frac{1}{4\pi\epsilon_0} \cdot \frac{q_C q_B}{(CB)^2}$$

$$= \frac{(9\times10^9)\times(10^{-6})\times(10^{-6})}{\left(\sqrt{2}\right)^2}$$

$$= \frac{9}{2}\times10^{-3}\,\text{N}$$

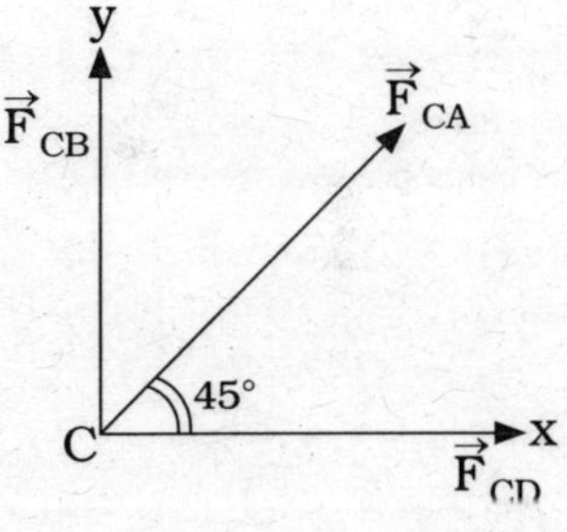

$$|\vec{F}_{CD}| = \frac{1}{4\pi\epsilon_0} \cdot \frac{q_C q_D}{(CD)^2}$$

$$= \frac{(9\times10^9)\times(10^{-6})\times(10^{-6})}{\left(\sqrt{2}\right)^2}\,\text{N}$$

$$= \frac{9}{2}\times10^{-3}\,\text{N}$$

for adding the above forces vectorially, we take CD along +x-axis and CB along +y-axis. The total x-component of the force $\vec{F}_C$ is,

$$F_x = F_{CD} + F_{CA}\cos 45°$$

$$= \left(\frac{9}{2} + \frac{9}{4}\times\frac{1}{\sqrt{2}}\right)10^{-3}\,\text{N}$$

Similarly, the total y-component of F_c is,

$$F_y = F_{CB} + F_{CA}\sin 45°$$

$$= \left(\frac{9}{2} + \frac{9}{4\sqrt{2}}\right)\times 10^{-3}\,\text{N}$$

Hence the total force F_C is,

$$F_C = \sqrt{F_x^2 + F_y^2}$$

$$= \left[\left\{\left(\frac{9}{2}+\frac{9}{4\sqrt{2}}\right)10^{-3}\right\}^2 + \left\{\left(\frac{9}{2}+\frac{9}{4\sqrt{2}}\right)10^{-3}\right\}^2\right]^{1/2}$$

$$= \left[\left\{\left(\frac{9}{2}+\frac{9}{4\sqrt{2}}\right)10^{-3}\right\}^2 (1+1)\right]^{1/2}$$

$$= \left(\frac{9}{2}+\frac{9}{4\sqrt{2}}\right)\sqrt{2}\times 10^{-3} = 9\left(\frac{1}{\sqrt{2}}+\frac{1}{4}\right)10^{-3}$$

$$= 8.61\times 10^{-3}\text{ N} \quad \textbf{(Answer).}$$

The direction of $\vec{F}_C$ is along CA.

Problem 20. Two identical balls of mass m each are hung from silk threads of length ℓ each and carry the same charge q each. Obtain an expression for the linear separation between the balls in equilibrium position, assuming the angular separation between the balls to be small.

Solution. Let the linear separation between the balls (A and B) be d. Three forces act together on each ball, viz. the electrostatic force $\vec{F}$, the weight $\overrightarrow{mg}$, and the tension $\vec{T}$ in the thread. In the equilibrium position, the net force acting on either ball should be zero, i.e.

$$\vec{F} + \overrightarrow{mg} + \vec{T} = 0$$

Choosing the +x axis as the direction of $\vec{F}$ and +y axis as the direction of $\overrightarrow{mg}$, and resolving these forces along x and y, we have, in the equilibrium condition,

$$T\sin\theta = F$$

$$T\cos\theta = mg$$

$$\therefore \quad \tan\theta = \frac{F}{mg}$$

Now,
$$F = \frac{1}{4\pi\epsilon_0}\cdot\frac{q^2}{d^2}$$

For small θ–values,
$$\tan\theta \approx \theta \approx \frac{d/2}{\ell} = \frac{d}{2\ell}$$

$$\therefore \quad \frac{d}{2\ell} = \frac{1}{4\pi\epsilon_0}\cdot\frac{q^2}{d^2.mg}$$

$$\Rightarrow \quad d^3 = \frac{q^2\ell}{2\pi\epsilon_0\, mg}$$

$$\Rightarrow \quad d = \left[\frac{q^2\ell}{2\pi\epsilon_0\, mg}\right]^{1/3} \quad \textbf{(Answer).}$$

Problem 21. Three point charges $q_A = +1\ \mu C$, $q_B = -1\ \mu C$ and $q_C = +1\ \mu C$ are placed at the corners of an equilateral triangle of side length 1 m. Calculate the electrostatic force on the charge q_C.

Solution. The given charges q_A, q_B, q_C lie at the corners A, B, C of the equilateral triangle ABC. The electrostatic force on q_c is given by,

$$\vec{F}_C = \vec{F}_{CA} + \vec{F}_{CB}$$

$\vec{F}_{CA}$ is a force of repulsion, directed away from C,

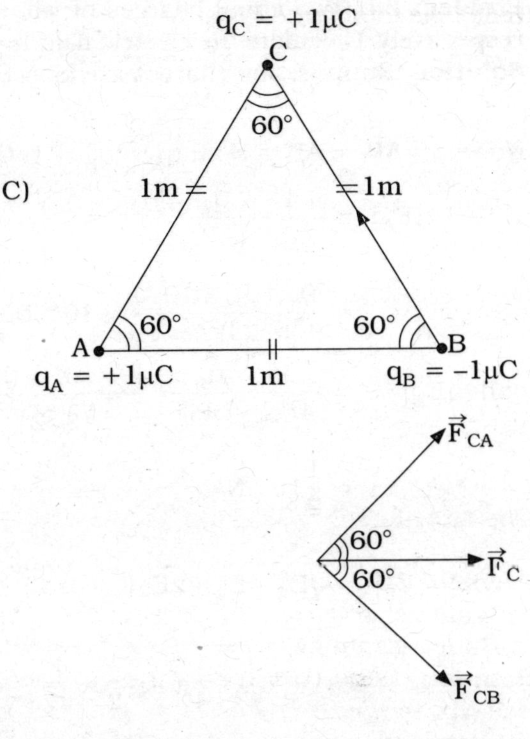

$$|\vec{F}_{CA}| = \frac{1}{4\pi\epsilon_0}\cdot\frac{|q_C q_A|}{(CA)^2}$$

$$= \frac{(9\times10^9)\times(10^{-6})\times(10^{-6})}{(1)^2} \quad (1\mu C = 10^{-6} C)$$

$$= 9\times10^{-3} \text{ N}$$

$\vec{F}_{CB}$ is a force of attraction, directed towards B,

$$|\vec{F}_{CB}| = \frac{1}{4\pi\epsilon_0}\cdot\frac{q_C q_B}{(CB)^2}$$

$$= \frac{(9\times10^9)\times(10^{-6})\times(10^{-6})}{(1)^2}$$

$$= 9\times10^{-3} \text{ N}$$

The angle between $\vec{F}_{CA}$ and $\vec{F}_{CB}$ is 120°,

$$\therefore \quad |\vec{F}_C| = \sqrt{F_{CA}^2 + F_{CB}^2 + 2F_{CA}F_{CB}\cos 120^\circ}$$

$$= 9\times10^{-3}\sqrt{1+1-2\times\frac{1}{2}}$$

$$= 9\times10^{-3} \text{ N} \quad \textbf{(Answer).}$$

By symmetry, $\vec{F}_C$ lies parallel to AB, in the direction A to B. **(Answer).**

Electric Field

Problem 22. Calculate the electric field intensity at a point 2 m away from a point charge of 1 mC, placed in air.

Solution. Charge, q = 1mC = 1×10^{-3} C

Distance of the given point from the charge, r = 2 m.

∴ Electric field intensity is given by,

$$E = \frac{1}{4\pi\epsilon_0}\cdot\frac{q}{r^2}$$

$$= \frac{(9\times10^9)\times(10^{-3})}{(2)^2} = 2.25 \times 10^6 \text{ N/C} \quad \textbf{(Answer).}$$

Problem 23. Two charges 4 μC and 16 μC are placed at a distance of 36 cm from each other. Find the electric field intensity at a distance of 12 cm from the charge 4 μC, in the direction of the charge 16 μC.

Solution. Suppose the charges are placed at A and B. A point P is located at a distance of 12 cm from A. The electric field intensity at P is jointly due to A and B.

$$\vec{E}_P = \vec{E}_A + \vec{E}_B$$

AP = 12 cm = 12×10^{-2} m; BP = 24 cm = 24×10^{-2} m;

q_A = 4 μC = 4×10^{-6} C; q_B = 16 μC = 16×10^{-6} C

$$|\vec{E}_A| = \frac{1}{4\pi\epsilon_0}\cdot\frac{q_A}{(AP)^2} = \frac{(9\times10^9)\times(4\times10^{-6})}{(12\times10^{-2})^2} = 2.5\times10^6 \text{ N/C}$$

4μC 16μC
$\vec{E}_A$ $\vec{E}_B$
A P B
12cm 24cm

$$|\vec{E}_B| = \frac{1}{4\pi\epsilon_0}\cdot\frac{q_B}{(BP)^2} = \frac{(9\times10^9)\times(16\times10^{-6})}{(24\times10^{-2})^2} = 2.5\times10^6 \text{ N/C}$$

As the two fields $\vec{E}_A$ and $\vec{E}_B$ are equal and opposite, the total field $\vec{E}_P$ at P is zero. **(Answer).**

Problem 24. Two equal charges of value 1 μC each are located at the coordinates (0,0) m and (0,3) m, respectively. Calculate the electric field intensity at a point located at x = 3 m along the x-axis.
Solution. Suppose the charges are placed at A and B in XAY plane. The electric field intensity at P is,

$$\vec{E}_P = \vec{E}_A + \vec{E}_B$$

Now, AP = AB = 3 m; $q_A = q_B = 1\mu C = 10^{-6}$ C; BP = $3\sqrt{2}$ m.

$$\therefore \quad |\vec{E}_A| = \frac{1}{4\pi\epsilon_0}\cdot\frac{q_A}{(AP)^2}$$

$$= \frac{(9\times10^9)\times(10^{-6})}{(3)^2} = 10^{-3} \text{ N/C}$$

$$\text{and } |\vec{E}_B| = \frac{1}{4\pi\epsilon_0}\cdot\frac{q_B}{(BP)^2} = \frac{(9\times10^9)\times(10^{-6})}{\left(3\sqrt{2}\right)^2}$$

$$= \frac{1}{2}10^{-3} \text{ N/C}$$

$$\therefore \quad |\vec{E}_P| = \sqrt{E_A^2 + E_B^2 + 2E_AE_B\cos 45°}$$

$$= 10^{-3}\sqrt{1+\frac{1}{4}+2\times\frac{1}{2}\times\frac{1}{\sqrt{2}}}$$

$$= 1.39 \times 10^{-3} \text{ N/C} \quad \textbf{(Answer).}$$

To find the direction of $\vec{E}_P$, suppose $\vec{E}_P$ makes an angle α with the x-axis. Then,

$$\tan\alpha = \frac{(\vec{E}_P)_{y\text{-component}}}{(\vec{E}_P)_{x\text{-component}}} = \frac{E_B \sin 45°}{E_A + E_B \cos 45°}$$

$$= \frac{\left(\frac{1}{2}10^{-3}\right)\frac{1}{\sqrt{2}}}{\left(1+\frac{1}{2}\cdot\frac{1}{2}\right)10^{-3}} = \frac{1/2\sqrt{2}}{\left(1+2\sqrt{2}\right)/2\sqrt{2}}$$

$$= \frac{1}{1+2\sqrt{2}} = 0.261$$

$$\therefore \quad \alpha = \tan^{-1}(0.261) = 14.62°$$

Thus, $\vec{E}_P$ makes an angle of 14.62° with the x-axis. **(Answer).**

Problem 25. A charged particle of mass 0.5 mg is held stationary in a uniform vertical electric field of strength 2×10^2 N/C, directed upwards. Calculate the charge on the particle. What is the nature of the charge ?
Solution. For the particle P to remain stationary, the upwards electrostatic force on it must be equal to the weight of the particle, i.e.

$$qE = mg \quad \text{or} \quad q = \frac{mg}{E}$$

m = 0.5 mg = 5×10^{-7} kg; E = 2×10^2 N/C; g = 9.81 ms^{-2}

$$\therefore \quad q = \frac{mg}{E} = \frac{(5\times10^{-7})\times(9.81)}{(2\times10^2)} \text{ C}$$

$$= 2.45\times10^{-8} \text{ C} \quad \textbf{(Answer).}$$

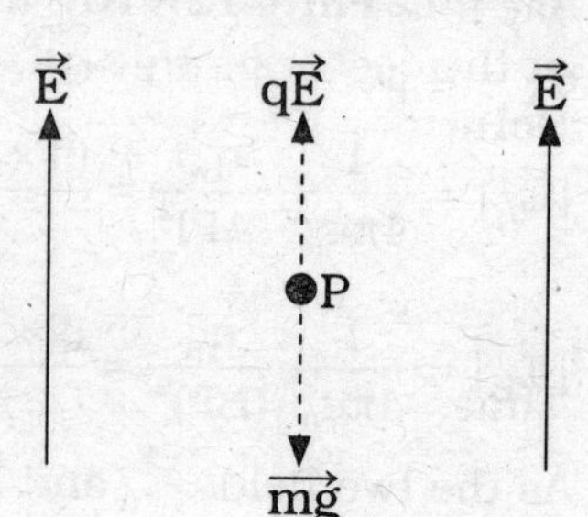

The nature of the charge must be positive (only then the electrostatic force on it due to the field will be directed upwards). **(Answer).**

Problem 26. An electron enters a uniform electric field of strength 10^3 N/C in the direction of the field, with an initial velocity of 4×10^6 m/s. Calculate the distance travelled by it, before it comes to a momentary rest.

Solution. The electron, being negatively charged, experiences retardation (i.e. negative acceleration) upon entering the field. The retardation force (= mass × acceleration) is given by,

$$F = ma = eE \qquad \text{or} \qquad a = \frac{eE}{m} \qquad \text{(a : acceleration)}$$

Given that the initial velocity, $u = 4\times10^6$ m/s and the final velocity, $v = 0$. To find the required distance, S, travelled by the electron, we use the formula,

$$v^2 = u^2 + 2as$$

$$\Rightarrow \quad 0 = (4\times10^6)^2 - 2\times\left(\frac{eE}{m}\right)S$$

$$\Rightarrow \quad S = \frac{(16\times10^{12})\,m}{2\,eE}$$

Now, $m = 9.1\times10^{-31}$ kg; $e = 1.6\times10^{-19}$ C; $E = 10^3$ N/C

$$\therefore \quad S = \frac{(16\times10^{12})\times(9.1\times10^{-31})}{2\times(1.6\times10^{-19})\times(10^3)} = 4.55\times10^{-2}\text{ m}$$ **(Answer).**

Problem 27. A pendulum is at rest in a horizontal uniform electric field of strength 2×10^4 V/C. Its bob has mass $\sqrt{3}\times10^{-2}$ g and carries a charge 5×10^{-9} C. Find the tension in the suspending string of the pendulum. Take $g = 10\text{ ms}^{-2}$.

Solution. Mass of bob $= \sqrt{3}\times10^{-2}\text{ g} = \sqrt{3}\times10^{-5}$ kg; $g = 10\text{ ms}^{-2}$; $q = 5\times10^{-9}$ C; $E = 2\times10^4$ V/C.

Let the angle made by the string with the vertical be θ. In the equilibrium state (when the pendulum is at rest), we have,

$$T\sin\theta = qE$$

$$T\cos\theta = mg$$

$$\therefore \quad \tan\theta = \frac{qE}{mg}$$

$$= \frac{(5\times10^{-9})\times(2\times10^4)}{(\sqrt{3}\times10^{-5})\times10} = \frac{1}{\sqrt{3}}$$

$$\therefore \quad \theta = \tan^{-1}(1\sqrt{3}) = 30^\circ$$

$$\therefore \quad T = \frac{qE}{\sin 30^\circ} = \frac{(5\times10^{-9})\times(2\times10^4)}{(1/2)}$$

$$= 2\times10^{-4}\text{ N}$$ **(Answer).**

Problem 28. Four point charges $q_A = 1\ \mu C$, $q_B = 1\ \mu C$, $q_C = -1\ \mu C$ and $q_D = -1\ \mu C$ are placed respectively at the four corners of a square ABCD. The side length of the square is 1m. Calculate the electric field intensity at the centre of the square.

Solution. The electric field $\vec{E}$ at the centre O is due to all the four charges. Thus, we have,

$$\frac{OP}{OA} = \sin 45^\circ = \frac{1}{\sqrt{2}}$$

$$\Rightarrow \quad OA = OP\sqrt{2} = (1/2)\sqrt{2} = (1/\sqrt{2})\text{ m}$$

Thus, $OA = OB = OC = OD = 1/\sqrt{2}$

By symmetry,

$$|\vec{E}_A| = |\vec{E}_B| = |\vec{E}_C| = |\vec{E}_D| = \frac{1}{4\pi\epsilon_0} \cdot \frac{q}{r^2}$$

$$= \frac{(9\times10^9) \times (1\times10^{-6})}{(1/\sqrt{2})^2}$$

$$= 1.8\times10^4 \text{ N/C}$$

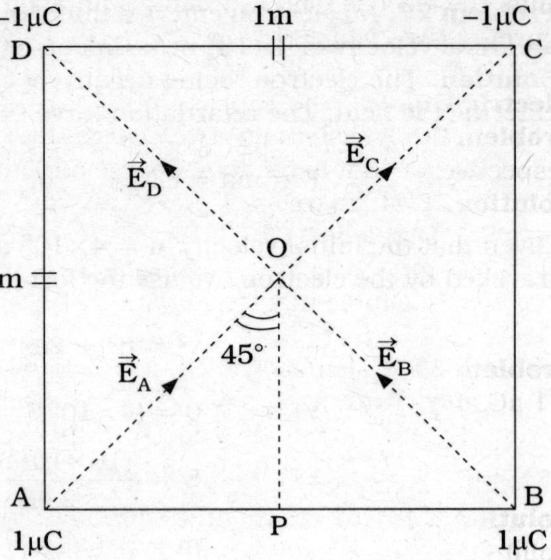

The directions of $\vec{E}_A$, $\vec{E}_B$, $\vec{E}_C$, $\vec{E}_D$ are marked in the figure. The total field along OC is,

$$E_{OC} = E_A + E_C = (1.8\times10^4 + 1.8\times10^4)$$
$$= 3.6\times10^4 \text{ N/C}$$

Similarly,

$$E_{OD} = E_B + E_D = 3.6\times10^4 \text{ N/C}$$

$$\therefore \quad E = \sqrt{E_{OC}^2 + E_{OD}^2} = 3.6\sqrt{2}\times10^4 \text{ N/C} \quad \textbf{(Answer).}$$

The direction of $\vec{E}$ is $\perp$ to DC, along PO (It follows from the definition of electric field). **(Answer).**

Problem 29. Three point charges $q_A = +28.8\ \mu C$, $q_B = -8.1\ \mu C$, $q_C = +25.6\ \mu C$ are placed at the corners A, B, C, respectively, of a right-angled triangle ABC ($\angle A = 90°$), with AB = 3 m and AC = 4 m. Calculate the electric field at the foot of the perpendicular drawn from A upon BC.

Solution. P is the foot of the perpendicular drawn from A on BC. The electric field at P is due to all the three charges at A, B, C. Thus,

$$\vec{E}_P = \vec{E}_A + \vec{E}_B + \vec{E}_C$$

$$BC = \sqrt{3^2 + 4^2} = 5\text{m}$$

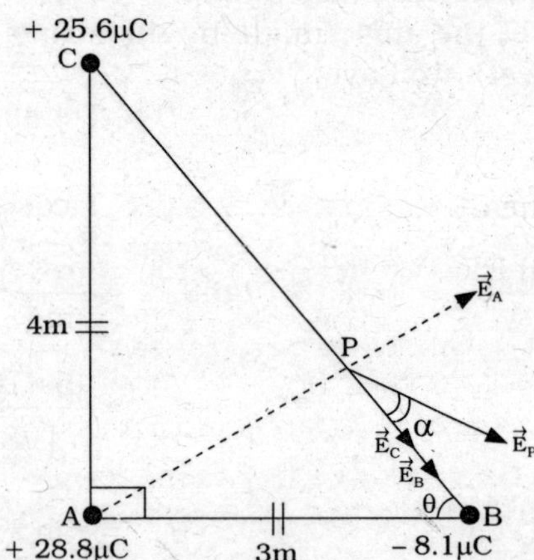

Let $\angle B = \theta$. We have,

$$\sin\theta = \frac{AC}{BC} = \frac{AP}{AB}$$

$$\Rightarrow \quad \frac{4}{5} = \frac{AP}{3}$$

$$\Rightarrow \quad AP = \frac{12}{5} \text{ m}$$

and

$$\cos\theta = \frac{AB}{BC} = \frac{3}{5} = \frac{BP}{AB} = \frac{BP}{3}$$

$$\Rightarrow \quad BP = \frac{9}{5} \text{ m}$$

$$\therefore \quad PC = BC - BP = 5 - \frac{9}{5} = \frac{16}{5} \text{ m}$$

$$|\vec{E}_A| = \frac{1}{4\pi\epsilon_0} \cdot \frac{q_A}{(AP)^2} = \frac{(9\times10^9)\times25\times(28.8\times10^{-6})}{144} = 4.5\times10^4 \text{ N/C} \qquad \left[(AP)^2 = \left(\frac{12}{5}\right)^2\right]$$

$$|\vec{E}_B| = \frac{1}{4\pi\ \epsilon_0} \cdot \frac{q_B}{(BP)^2} = \frac{(9\times10^9)\times25\times(8.1\times10^{-6})}{81} = 2.25\times10^4 \text{ N/C} \qquad \left[(BP)^2 = \left(\frac{9}{5}\right)^2\right]$$

$$|\vec{E}_C| = \frac{1}{4\pi\epsilon_0} \cdot \frac{q_C}{(PC)^2} = \frac{(9\times10^9)\times25\times(25.6\times10^{-6})}{256} = 2.25\times10^4 \text{ N/C} \qquad \left[(PC)^2 = \left(\frac{16}{5}\right)^2\right]$$

The directions of $\vec{E}_A$, $\vec{E}_B$, $\vec{E}_C$, $\vec{E}_P$ are marked on the figure. We have,

$$E_P = \sqrt{(E_B + E_C)^2 + (E_A)^2} = 4.5\times10^4\sqrt{2} = 6.36\times10^4 \text{ N/C} \qquad \textbf{(Answer).}$$

Let α be the angle made by $\vec{E}_P$ with the side BC. We have,

$$\alpha = \tan^{-1}\left(\frac{E_A}{E_B}\right) = \tan^{-1}\left(\frac{4.5\times10^4}{4.5\times10^4}\right) = \tan^{-1}(1) = 45°$$

Thus, the electric field at P makes an angle of 45° with BC and it is directed away from P, as shown in the figure. **(Answer).**

Electric Dipole

Problem 30. A system has two charges of +1 μC and –1 μC located at A (0, 0, –10 cm) and B (0, 0, +10 cm), respectively. What is the total charge and electric dipole moment of the system ?

Solution. Total charge = 1 μC – 1 μC = 0 **(Answer).**

Dipole length, $2a = 10 - (-10) = 20\text{ cm} = 0.2\text{m}$

Charge $q = 1\ \mu C = 1\times10^{-6}$ C

∴ Dipole moment $= 2a.q = (0.2) \times (1\times10^{-6}) = 2\times10^{-7}$ C.m **(Answer).**

Problem 31. Calculate the electric field due to an electric dipole of length 6 cm, with end charges of value +1 μC and –1 μC, at

(i) 4 cm from the centre of the dipole along axial line,

(ii) 3 cm from the centre of the dipole along equatorial line.

Solution. O is the centre of the dipole. The point P is situated at 4 cm from O along the axial line and the point Q is situated at 3 cm from O along the equatorial line.

Dipole length, $2a = 6\text{ cm} = 6\times10^{-2}$ m

∴ $a = 3\text{ cm} = 3\times10^{-2}$ m; $q = 1\ \mu C = 1\times10^{-6}$ C

∴ Dipole moment, $p = q.2a$

$= (1\times10^{-6}) \times (6\times10^{-2}) = 6\times10^{-8}$ C.m

(i) Electric field along the axial line ($r = 4\text{ cm} = 4 \times 10^{-2}$ m)

Q, 3cm, –1μC, +1μC, A, O, B, P, 6cm, 4cm

$$E_P = \frac{1}{4\pi\epsilon_0}\cdot\frac{2pr}{(r^2-a^2)^2} = \frac{(9\times10^9)\times2\times(6\times10^{-8})\times(4\times10^{-2})}{(16-9)^2\times10^{-8}}$$

$= 8.82\times10^7$ N/C **(Answer).**

The direction of $\vec{E}_P$ is along the dipole moment $\vec{p}$, i.e. from A to B. **(Answer).**

(ii) Electric field along the equatorial line ($r = 3\text{ cm} = 3 \times 10^{-2}$ m)

$$E_Q = \frac{1}{4\pi\epsilon_0}\cdot\frac{P}{(r^2+a^2)^{3/2}}$$

$$= \frac{(9\times10^9)\times(6\times10^{-8})}{(9+9)^{3/2}\times10^{-6}}$$

$= 7.07\times10^6$ N/C **(Answer).**

The direction of $\vec{E}_Q$ is opposite to $\vec{p}$, i.e. from B to A. **(Answer).**

Problem 32. An electric dipole of moment 5×10^{-3} cm is kept in a uniform electric field of strength 2×10^4 N/C. If 50 N-m of torque is required to keep the dipole in position, find the angle between the dipole moment and the electric field.

Solution. Dipole moment, $p = 5\times10^{-3}$ cm

Electric field strength, $E = 2\times10^4$ N/C

Torque on the dipole, $\tau = pE\sin\theta$

$\Rightarrow \quad 50 = (5\times10^{-3}) \times (2\times10^4)\sin\theta$

$\Rightarrow \quad \sin\theta = \frac{1}{2}$

∴ $\quad \theta = \sin^{-1}\left(\frac{1}{2}\right) = 30°$ **(Answer).**

Very Short Answer Questions
[Including HOTS (Higher Order Thinking Skills) Questions]
(1 mark each)

1. Give SI units of permittivity and electric field intensity.

2. State two basic properties of electric charge.

3. Define dielectric constant of a medium in terms of force between two point charges.

4. A dry plastic rod rubbed with a piece of silk can lift small pieces of paper. Is there any transfer of mass between rod and silk ?

5. An electric dipole is placed in a uniform electrostatic field. Will the net translatory force experienced by it be zero when the torque experienced by it is maximum ?

6. Why don't two electric field lines intersect each other ?

7. Draw field lines to show electrostatic field pattern in a plane containing
(a) a point charge $q > 0$ **(b)** a point charge $q < 0$

8. Is the Coulomb's law valid when the two charges are in motion ?

9. There is a point charge q each at the four corners of a square. What will be the ratio of force between two charges at adjacent corners and the force between two charges at opposite corners of the square ?

10. There is a uniform electrostatic field $\vec{E}$ along x-axis. An electron is released in the field, such that it is free from the effect of gravity. What is the direction of its motion ? [Hint: Along $\vec{E}$].

11. Two point charges q and -2q lie a fixed distance apart. If $\vec{E}$ is the field strength at the location of q, then what is the field strength at the location of –2q ?

12. An electron and a proton are released from rest in a uniform electrostatic field. Which of them will have larger acceleration ?

13. An electric dipole has its dipole moment vector directed along x-axis. What will be the direction of the field strength at a point on its
(a) axial line ? **(b)** equatorial line ?

14. State whether the electric field is symmetrical or asymmetrical in the following cases:
(a) Around a point charge **(b)** Around an electric dipole

15. Two conducting spheres of equal size carry charges of value 60 μC and –2 μC, respectively. What will be the charges on them after they have been brought in contact and then separated ?

16. How is 1 μC of charge related to one coulomb of charge ?

17. A glass rod is rubbed with a piece of silk cloth. As per the sign convention, what is the sign of charge acquired by the glass rod ?

18. Two point charges kept at a finite distance apart in vacuum experience a mutual electrostatic force of magnitude F. What will be the magnitude of force when they are immersed together in a non-conducting liquid of dielectric constant 5 ?

19. Define linear charge density.

20. What is the force experienced by a charge q in the presence of an electric field $\vec{E}$?

21. Can electric field lines exist inside a conductor ?

22. If the closeness of electric field lines is increased, will the electric field become stronger or weaker ?

23. What is pair production ? How does it show conservation of charge ?

24. How does an electric field line help to find the direction of electric field at a given point ?

25. Is the Coulomb's law in agreement with the Newton's third law of motion ?

26. The protons and neutrons have now been found to consist of quarks, having charges $\pm\frac{1e}{3}$ and $\pm\frac{2e}{3}$. Does it mean that the quantisation of charge is no longer valid ?

27. Name the process of charging a body by another body, in which both the interacting bodies have same type of charges.

28. What is the value of electric field at the centre of a charged ring ?

29. Two identical metal spheres A and B of equal masses are charged, such that A has a positive charge and B an equal negative charge. Which of the two will have greater mass ?

Short Answer Questions

[Including HOTS Questions]

(2 or 3 marks each)

1. Write any two characteristics of the force between a pair of point charges, as given by the Coulomb's law.

2. Can any set of parallel field lines represent a uniform electrostatic field ? Justify your answer.

3. Obtain an expression for electric field strength at a point lying at a distance r from a point charge q in vacuum.

4. An oil drop of mass m carrying charge –q is to be held stationary in the gravitational field of the earth. What is the magnitude and direction of the electrostatic field required for this purpose ?

5. A point charge of value 10 μC is placed at each corner of a square of side length 10 cm. What will be the field strength at the centre of the square if one of the charges is removed ?

6. Write any two properties of electric charge and illustrate each with an example.

7. Vehicles carrying inflammable materials usually have metallic ropes touching the ground during motion. What is the reason for it ? [Hint : To make any charge developed on the vehicle's body to be conducted away to earth and thus prevent any sparking on the vehicle].

8. Why does an electric dipole of dipole moment $\vec{p}$ tend to align itself parallel to the uniform electrostatic field $\vec{E}$ in which it is initially placed with the angle $\theta \neq 0$ (θ: angle between $\vec{p}$ and $\vec{E}$) ?

9. The rubber tyres of aircrafts are made slightly conducting. Why is it necessary ? [Hint : To make any charge developed on the tyres due to friction with ground, during take-off or landing, to conduct away to earth].

10. Calculate the Coulomb force between two alpha-particles separated by a distance of 3.2×10^{-15} m in air.

11. Two point charges 4e and e are separated by a distance x. Where should a third charge q be placed between them so that the electrostatic force experienced by it is zero ? Does your answer depend on the magnitude and/or sign of q ?

12. Two like point charges of value q each and separated by a distance r in air exert an electrostatic force F on each other. If they are immersed in a non-conducting oil of dielectric constant 4, what should the new separation between them so that the force F remains unchanged ?

13. Why does a comb run through one's dry hair attract small bits of paper ? What happens if the hair are wet or if it is a rainy day ? [Hint (2nd part): Wetting or excessive moisture in air leads to a drastic reduction in friction and so in the development of charge, which further tends to leak to earth (through body) and to surroundings (through moisture). Hence the comb fails to attract paper bits].

14. Draw field lines to show the electrostatic field pattern in a plane containing

(a) a pair of point charges q and -q lying a small distance apart,

(b) a pair of point charges q and q lying a small distance apart.

15. Why is it necessary to have the test charge infinitely small, while obtaining the electric field intensity at a particular point in a charge distribution ?

16. Briefly explain the various methods of charging a body.

17. What is meant by continuous volume charge distribution ? Obtain an expression for the electric field intensity due to a continuous volume distribution of charge.

18. Can charge exist without mass ? Can mass exist without charge ? Give reason for your answer.

19. "Neutrino is a massless but charged particle". Is this statement true? Give reason for your answer.

20. Repulsion is regarded as the true test of electrification of two bodies. Explain why.

21. Define electric field intensity. Give its unit and dimensions. Is it a scalar or a vector quantity ?

22. An electron is orbiting the nucleus of a singly ionised helium atom. Which one of the two, the electron or the nucleus, exert larger force on the other? Give reason for your answer. [Hint : The mutual force of attraction is essentially the same for the two; $F = k.(q_1q_2)/r^2$].

23. Can we say that a wire carrying a steady current is charged ? Give reason.

24. "The equilibrium of a charged particle under the action of electrostatic forces alone is always unstable". Comment on this statement.

25. Define an electric field line. State four properties of electric field lines.

26. Explain how frictional electrification is consistent with the law of conservation of charge.

27. Explain how does the electric field due to a point charge vary with distance r from the charge.

28. A glass rod rubbed with silk is brought close to two uncharged metallic spheres in contact with each other, inducing charges on them, as shown in the figure below. Describe what happens when

(i) the spheres are slightly separated, and

(ii) the glass rod is subsequently removed, and finally

(iii) the spheres are separated far apart.

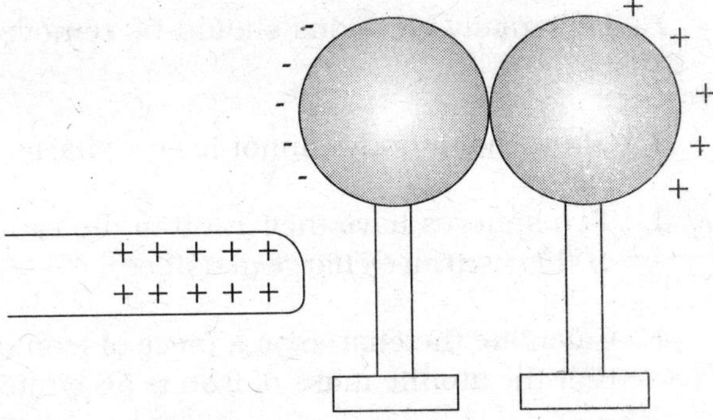

Long Answer Questions

(5 marks each)

1. An electric dipole consists of a pair of point charges of magnitude q each, separated by a distance 2r. Obtain an expression for the field strength at a point

(a) on its axial line **(b)** on its equatorial line.

Take the distance of the point from the dipole as x.

2. An electric dipole of dipole moment $\vec{p}$ is placed in a uniform electric field of strength $\vec{E}$, such that the angle between $\vec{p}$ and $\vec{E}$ is θ. Obtain an expression for the torque experienced by the dipole.

3. Two like point charges of value q each are placed at two corners of an equilateral triangle of side length x. Find the net electric field strength at the third corner.

4. State the principle of superposition for electric charges. Obtain an expression for the total force on a test charge q_0 in a collection of N point charges q_1, q_2, q_3,............., q_N, distributed around it.

5. What is continuous charge distribution ? Derive expressions for the force experienced by a test charge for linear, surface, and volume charge distributions.

6. Define electric field intensity. Obtain an expression for the electric field intensity due to a point charge q, at a point distant r from it.

7. Derive an expression for the electric field E at a point located on the axis of a ring, over which charge is uniformly distributed. Hence show that for a point located at a large distance from the ring, the ring behaves like a point charge.

8. Derive an expression for the electric field at a point located along the equatorial line of an electric dipole. Also derive an expression for the electric field at a point located along the axial line of the dipole.

Numerical Exercises

Charge and its Properties

1. How many electrons should be removed from an uncharged body to give it a charge of $+ 9.6\times10^{-8}$ C ?
[60]

2. Show that a body cannot have a charge of 2×10^{-18} C.

3. Two spheres have their radii in the ratio 1 : 2 and their charges in the ratio 1 : 3. Calculate the ratio of their surface charge densities.
[9:2]

4. Calculate the charge on a piece of iron of mass 1 mg, if 0.01% of eletrons are removed from it. Given that the atomic mass of iron is 56 g/mol.
[1.71 × 10^{-4} C]

5. A polythene piece rubbed with wool is found to have a negative charge of 3.2×10^{-7} C. Estimate the number of electrons transferred (from wool or polythene ?)
[2 × 10^{12} (approx.), from wool to polythene]

Coulomb's Law

6. What is the electrostatic force between two charges of value 1μC each, kept 1m apart in vacuum? Is the force attractive or repulsive ?
[9×10^{-3} N (repulsive)]

7. What is the magnitude and nature of the electrostatic force between two small charged spheres having charges of 2×10^{-7} C and 3×10^{-7} C, respectively, and placed 30 cm apart in air?
[6×10^{-3} N (repulsive)]

8. The electrostatic force on a small sphere of charge 0.4 μC due to another small sphere of charge –0.8 μC in air is 0.2 N. (a) What is the distance between the two spheres ? (b) What is the force on the second sphere due to the first ?
[12 cm; 0.2 N (attractive)]

9. Four point charges q_A = 2μC, q_B = –5μC, q_C = 2μC, q_D = –5μC are respectively located at the corners of a square ABCD of side length 10 cm. What is the force on a charge of 1μC placed at the centre of the square ? (NCERT).
[Zero]

10. Calculate Coulomb's force between two α-particles separated by a distance of 320 fermi in air. (1 fermi = 10^{-15} m).
[9×10^{-3}N]

11. Two charged particles of charge q each, held a small distance apart in isolated space, are released from rest. The initial acceleration of the first particle due to the electrostatic force of repulsion is twice that of the second. Find the ratio of electrostatic force and gravitational force between them. Take the mass of the first particle as m.
$$\left[\frac{1}{8\pi\epsilon_0 G}\frac{q^2}{m^2}\right]$$

F_e m F_e

q q

12. Two equal and similar charges are placed 0.03 m apart in air. They repel each other with a force of 45N. Find the charge on the bodies.
[2.12μC]

13. Four point charges 2μC, –5μC, 2μC and –5μC are respectively located in air at the corners A, B, C, D of a square ABCD of side length 10 cm. What is the electrostatic force experienced by the charge placed at A ?
[1.86 N; towards C]

14. Three charges $+5\ \mu C$, $+10\ \mu C$ and $-10\ \mu C$, respectively, are placed in air at the corners of an equilateral triangle ABC of side length 5×10^{-2} m. Determine the electrostatic force experienced by the charge at A. **[180 N; parallel to BC]**

15. Two charges A & B of values $+1\mu C$ and $+4\mu C$, respectively, are separated from each other by a distance of 2 cm. Find the position of the neutral point between them. **[0.67 cm from A]**

16. Three equal and similar charges, of value q each, are placed at three respective corners of a square of side length 'a'. Show that the electric field intensity at the fourth corner is

$$\frac{q(2\sqrt{2}+1)}{8\pi\epsilon_0 a^2},$$

along the diagonal of the square, directed away from the centre.

17. Two small spheres of mass m and charge q each are hung from a point by insulating threads, each of length ℓ but of negligible mass. If θ is the angle which each thread makes with the vertical, when equilibrium has been attained, show that

$$q^2 = 4mg\ell^2 \sin^2\theta \tan\theta\,(4\pi\varepsilon_0)$$

18. If an oil drop of weight 3.2×10^{-23} N is balanced in an electric field of strength 5×10^{-5} N/C directed vertically upwards, find the charge on the oil drop. **[6.4×10^{-19} C]**

19. Two identically charged spheres are suspended by strings of equal length. The strings make an angle of 30° with each other. When suspended in an insulating liquid of density 0.8 g/c.c., the angle remains the same. Calculate the dielectric constant of the liquid. The density of the material of the spheres is 1.6 g/c.c. **[2]**

20. Two positive charges q_1 and q_2 are placed at a distance r apart. A third charge q is placed between them such that all the three charges are in equilibrium. Find the magnitude, sign and position of the charge q. **[q = –4e/9; at distance a/3 from charge e]**

21. The force between two point charges when separated by a distance of 0.5 m in air is 50 N. What will be the force between them if the distance becomes 0.25 m and system is put in a liquid of dielectric constant 2 ? **[100 N]**

22. If the distance between two point charges is increased by 0.1%, calculate the percentage change in the electrostatic force between them.

[reduction by 0.2%]

23. Four point charges of value q each are placed equidistant from each other on a circle of radius r. Find the magnitude of electrostatic force on any one of the charges. $\left[\dfrac{0.957q^2}{4\pi\epsilon_0 r^2}\right]$

24. Two identical conducting spheres having charges 4q and -2q, respectively, are kept a small distance apart. Their electrostatic force of attraction has magnitude F. If they are made to touch each other and then removed to the previous positions, calculate the magnitude of the new force of attraction between them. What is the nature of the force? **[F/8; repulsive]**

25. What charge must be given to a small sphere of mass M placed directly above a pith ball of mass m and charge q, in order to hold it in equilibrium ? $\left[\dfrac{4\pi\epsilon_0 GMm}{q}\right]$

26. Three point charges Q_1, Q_2 and Q_3 in that order are equally spaced along a straight line. Q_2 and Q_3 are equal in magnitude but opposite in sign. If the net force on Q_3 is zero, show that $Q_1 = -4Q_2 = 4Q_3$.

27. How can a charge Q be divided into two parts Q_1 and Q_2, so that the repulsive force between them is maximum when separated by a small distance ?

$$\left[Q_1 = Q_2 = \frac{Q}{2}\right]$$

28. Two point charges are separated by a small distance r in vacuum. If they are immersed in a non-conducting medium of dielectric constant K, what change should be made in r so that the electrostatic force between them remains the same ?

$$\left[r\left(1-\frac{1}{\sqrt{K}}\right)\right]$$

Electric Field

29. Find the electric field strength due to a point charge of 5 μC at a distance of 80 cm from the charge.

[7×10^3 N/C]

30. A hollow conducting sphere of radius 8 cm is given a charge 16 μC. What is the electric field intensity

(i) at the centre of the sphere and

(ii) on the outer surface of the sphere ? **[0; 2.25×10^7 N/C]**

31. Two point charges q_A = 3 μC and q_B = –3 μC are located at points A and B, respectively, 20 cm apart in vacuum.

(a) What is the electric field at the midpoint O of the line AB joining the two charges ?

(b) If a negative test charge of magnitude 1.5×10^{-9} C is placed at this point, what are the magnitude and direction of the electrostatic force experienced by it ?

[5.4×10^6 N/C along OB; 8.1×10^{-3} N along OA]

32. A point charge $+10^{-9}$ C is located at the origin and another point charge Q at (2, 0, 0), in free space. If the x-component of the electric field at (3, 0, 0) is zero, calculate the value of Q. What is the sign of Q ?

$$\left[-\frac{1}{9}10^{-9}C\right]$$

33. An electron enters a uniform electric field of strength 10^3 N/C with an initial velocity of 1600 km/s parallel to the field. Calculate the distance travelled by the electron before it momentarily comes to rest.[Hint: Use the formula $v^2–u^2=2as$]. **[7.28 mm]**

34. Calculate the electric field intensity just sufficient to balance the earth's gravitational attraction on an electron (mass and charge of the electron are 9.1×10^{-31} kg and 1.6×10^{-19} C, respectively).

[5.6×10^{-11} N/C]

35. Two charges q and –4q are placed at a distance d apart. Find the location of the neutral point. [Hint: Let N be distant x from q. Use $q_1/r_1^2 = q_2/r_2^2$, leading to $q/x^2 = 4q/(d+x)^2$ (disregard –ve sign of –4q). q cancels out, take square root, and solve to get x = d].

[d from q, on the line joining the two charges]

Electric Dipole

36. A system has two charges $q_A = 2.5\times10^{-7}$ C and $q_B = -2.5\times10^{-7}$ C located at points A (0, 0, –15 cm) and B (0, 0, +15 cm), respectively. What are the total charge and the electric dipole moment of the system ? **[0;7.5×10^{-8} C–m]**

37. An electric dipole of moment 4×10^{-9} C-m is aligned at 30° with respect to the direction of a uniform electric field of magnitude 5×10^4 NC^{-1}. Calculate the magnitude of the torque acting on the dipole.

[10^{-8} N–m]

38. The electric field in a region is directed radially outwards. Its magnitude is, E = kr. Find the charge contained in a sphere of radius 20 cm lying in the region. Take k = 100 N/cm. **[8.89×10^{-11} C]**

39. An electric dipole of length 2 m is placed in a uniform electric field of strength 10^5 N/C, with its axis making an angle of 30° with the field. It experiences a torque of $10\sqrt{3}$ N.m. Calculate the magnitude of the charge on the dipole. **[$\sqrt{3}\times10^{-2}$ C]**

40. Calculate the electric field intensity at a point located 10 cm from one end along the axis of a short electric dipole of moment 1 C–m. **[1.8×10^{14}N /C]**

41. Calculate the electric field intensity at a point located 20 cm along the equatorial line of a short electric dipole of moment 2 C–m. **[2.5×10^{12} N/C]**

42. A point charge placed at a distance r from the mid-point of an electric dipole, along the axis of the dipole, experiences a force F. Find the force acting on the charge when its distance from the dipole is doubled. **[F/8]**

CHAPTER 1 (B)

ELECTRIC POTENTIAL, FLUX & CAPACITANCE

1B. 01. Electric (or Electrostatic) Potential & Potential Energy

The electric (or electrostatic) potential of a body is a measure of the amount of charge residing on the body or, say, electrification of the body. We know that flow of electricity in a system essentially means a flow of electric charges in the system. The concept of electric potential is easily grasped by comparison with flow of water or flow of heat in a body. Just as water naturally flows from a higher level to a lower level and heat in a body naturally flows from a position at higher temperature to a position at lower temperature, electricity flows from a body at higher potential to another body at a lower potential, when the two bodies are connected together. The flow stops when the potentials of the two bodies become equal, i.e. when the bodies acquire equal electric charge.

The concept of electric potential can be understood more specifically in terms of electrostatic (or electric) potential energy of charges in an electrostatic field, $\vec{E}$. In the last chapter (Section 1A.06), we have seen that the electrostatic Coulombian forces mathematically resemble with the Newtonian gravitational forces, as both have inverse square dependence on the distance, r. Accordingly, just as a body possesses mechanical potential energy by virtue of its particular position in a gravitational field of force, a charge possesses electric potential energy by virtue of its position in an electric field of force. Consider a positive charge Q in space (Fig. 1.18a). It gives rise to an electrostatic field $\vec{E}$ around it. Suppose a tiny positive test charge q initially exists at the position A. It will experience a repulsive force due to the charge Q, since both are positive charges. Now if q is slowly moved to another position B nearer to Q, as shown in Fig. 1.18a, work will be done against the repulsive force and the potential energy of q will be raised. If U_A and U_B are the potential energies of q at A and B, respectively, and W is the work done, we have,

$$W = -(U_B - U_A)$$

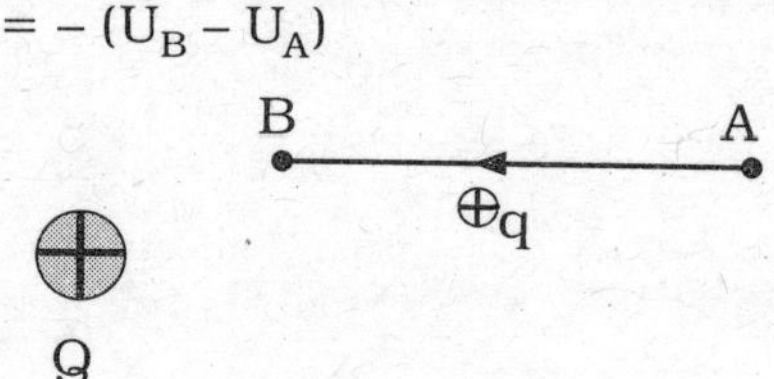

Fig. 1.18 (a)

The negative sign comes because the work is done against the electrostatic force. So we finally have,

$$W = (U_A - U_B) \qquad \text{..... (1.35)}$$

Note that the value of W depends only on the positions of A and B. It does not matter at all what path is chosen for the test charge for arriving at B from the initial position A. In other words, the work done is independent of the path between A and B.

If the test charge be initially taken to exist at infinity, its potential energy due to Q will be zero, i.e.

$$U_A = U_\infty = 0.$$

Then,

$$U_B = U = -W_\infty \qquad \text{..... (1.36)}$$

It follows that the potential energy of the test charge at any given position in an electric field is equal to the work done in bringing q from infinity to that position (the negative sign only means that work is done against the repulsive force). U_B has been put equal to the general term U, because the point can be chosen anywhere in the electric field. Thus, we arrive at the following **definition of electric (or electrostatic) potential energy:**

"The electric (or electrostatic) potential energy of a charge at a given position in an electric field is the work done in bringing the charge from infinity to that position."

The SI unit of electric potential energy is joule (J).

The electrostatic force $\vec{F}$ experienced by a charge q existing at a particular position in an electric field $\vec{E}$ is equal to the product of q and the value of $\vec{E}$ at that position ($\vec{F} = q\vec{E}$; Chapter 1A, Section 1A.10;

Eq. 1.11). Since Work = (Force × Distance), it follows from Eq. 1.36 that the potential energy U of a charge q existing in an electric field is proportional to the magnitude of the charge q, i.e. $U \propto q$. However, the potential energy per unit charge, i.e. U/q, is independent of q, because if the value of q is increased, the value of $\vec{F}$ and therefore U is proportionately increased. It has a unique value at any given position in an electric field. It is characteristic only of the electric field existing at that position. It is called **electric (or electrostatic) potential, V,** at that position,

$$V = U/q \qquad \text{..... (1.37)}$$

The value of V varies from one point to another in the field, but at any given point it has a specific value. Thus we find that the **electric potential is a property of an electric field.** It does not matter whether any charge is present in the field or not. Note that **it is a scalar quantity.**

Now, the potential difference between the points B and A is,

$$\Delta V = V_B - V_A = (U_B/q - U_A/q) = (U_B - U_A)/q$$

Or,

$$\Delta V = \Delta U/q \qquad \text{..... (1.38)}$$

From Eq. 1.35, $\Delta U = (U_B - U_A) = -W$. So we have,

$$V_B - V_A = \Delta U/q = -(W/q) \qquad \text{..... (1.39)}$$

From Eqs. 1.38 and 1.39, we find that the potential difference between any two points in an electric field can be taken either as the (i) potential energy difference per unit charge between the two points or (ii) work done by the electrostatic force to move a unit charge from one point to the other.

We have seen that if the charge q is initially taken to exist at infinity, its potential energy in the electric field is zero ($U_A = U_\infty = 0$). Then, from Eq. 1.37, the electric potential at infinity (U_∞) is also zero. From Eqs. 1.36 and 1.37 we have,

$$V = -(W_\infty/q) \qquad \text{..... (1.40)}$$

Thus, we arrive at the following **definition of electric potential:**

"The electric potential at any point in an electric field is the amount of work done in moving a unit charge along any path from infinity to that point, against the electrostatic force."

Depending on the values of W_∞ and q, the potential V can be positive, negative or zero.

Unit of potential (volt); electron-volt (eV)

The SI units of work and charge are joule (J) and coulomb (C), respectively. So from Eq. 1.40, the SI unit of potential and potential difference (p.d.) is joule per coulomb (JC^{-1}). However, these terms are so frequently used in physics and other branches of science and technology, and even in common life, that a special simple unit, **volt (V),** is employed for them, instead of 'joule per coulomb':

$$1 \text{ volt (V)} = 1 \text{ joule per coulomb } (JC^{-1}) \qquad \text{..... (1.41)}$$

This new unit for potential and p.d. has further enabled to evolve a new convenient unit for the electric field $\vec{E}$. In the last chapter (Section 1A.06), the unit for $\vec{E}$ was mentioned as newton per coulomb (NC^{-1}). From Eq. 1.41, ($1VCJ^{-1}$) =1. Similarly, since Work = Force × Distance and the units for work, force and length are J (joule), N (newton) and m (metre), respectively, we have ($1JN^{-1}m^{-1}$) =1.

So we can write,

$$\begin{aligned} 1NC^{-1} &= (1NC^{-1}).(1VCJ^{-1}).(1JN^{-1}m^{-1}) \\ &= 1\ (Vm^{-1}) \end{aligned}$$

Thus, we now have an alternative convenient unit, volt per metre (V/m or Vm^{-1}), for electric field $\vec{E}$.

A related unit of energy which is commonly used in the description of atomic and subatomic phenomena is **electron-volt (eV).** It is **defined** as the "work required to move an elementary charge e through a potential difference of one volt." From Eq. 1.39, the magnitude of this work is $q.\Delta V = e(1V.)$ Therefore we get, from Eq. 1.41, the value of one electron-volt (eV) as,

$$\begin{aligned} 1 \text{ eV} &= e(1V) = (1.60\times10^{-19}C)\ (1JC^{-1}) \\ &= 1.60\times10^{-19}\ J \end{aligned}$$

Example 1. A point charge is taken from one point to another over an equipotential surface. Calculate the work done in this process.

Solution. The work done in moving a point charge of value q from one point to another in an electrostatic field is given as, $W = q\Delta V$, where ΔV is the potential difference between the two points. Since the potential difference between any two points on an equipotential surface is zero (vide Section 1B.06), the work done in taking a point charge from one point to another over the surface is zero. **(Answer).**

Example 2. Two joule (2 J) of work is required to be done for moving an electric charge of value –2 mC from a point A to another point B. The potential at A is 200 V. Find the potential at B.

Solution. Given: $W = 2$ J; $q = -2$ mC $= -2\times10^{-3}$C; $V_A = 200$ V

The work done is given by,

$$W = -q\,(V_B - V_A)$$

$$\Rightarrow \quad 2 = -(-2\times10^{-3})\,(V_B - 200)$$

$$\Rightarrow \quad (V_B - 200) = 10^3 = 1000$$

$$\Rightarrow \quad V_B = 1200 \text{ V}$$ **(Answer).**

1B. 02. Electric Potential due to a Point Charge

Consider a positive point charge Q, as shown in Fig. 1.18 (b). We shall evaluate the electric potential due to Q at a point P distant r from Q. By definition, it is equal to the work done in bringing a unit positive charge from infinity to P along any arbitrary path. For the sake of convenience, we choose a straight path, shown by dashed lines in the figure. Now consider an arbitrary point S on the path, distant x from Q. The repulsive electrostatic force experienced by the unit charge at S due to Q will be (Chapter 1A; Section 1A.06; Eq. 1.6),

$$\vec{F} = \frac{Q\times1}{4\pi\epsilon_0\, x^2}\hat{x} = \frac{Q}{4\pi\epsilon_0\, x^2}\hat{x}$$

Q ⊕ —— r —— P - - - - - x - - - - S - - - ∞

Fig. 1.18(b)

$\hat{x}$ is a unit vector along the path. If the unit charge at S now moves an infinitesimal distance $\vec{dx}$ towards Q, the small amount of work done against the repulsive force will be,

$$dW = -(\vec{F}.\vec{dx}) = -\left[\frac{Q}{4\pi\epsilon_0\, x^2}(\hat{x}.\vec{dx})\right]$$

$$= -\frac{Q}{4\pi\epsilon_0\, x^2}dx$$

$\hat{x}.\vec{dx} = |\hat{x}|.dx.\cos 0 = dx$. The negative sign comes because the work is done against the repulsive force. The work done in bringing the unit charge from infinity to the point P is obtained by integrating dW in the limits $x = \infty$ to $x = r$,

$$W = -\int_{\infty}^{r}\frac{Q}{4\pi\epsilon_0\, x^2}dx = \frac{Q}{4\pi\epsilon_0}\int_{\infty}^{r}\frac{1}{x^2}dx = \left[\frac{Q}{4\pi\epsilon_0\, x}\right]_{\infty}^{r}$$

$$= \frac{Q}{4\pi\epsilon_0}\left[\frac{1}{r} - \frac{1}{\infty}\right] = \frac{Q}{4\pi\epsilon_0\, r}$$

By definition, this work is equal to the potential V at the point P due to Q,

$$V = \frac{Q}{4\pi\epsilon_0\, r} \qquad \text{..... (1.42)}$$

If Q happens to be a negative charge, the same result is obtained, but with the difference that Q becomes –Q on the right hand side. The following points should be noted:

(i) A positive charge produces a positive electric potential and a negative charge produces a negative potential.

(ii) At an equal distance r from Q in all the directions, V has the same value. This is true for all values of r. Thus, we find that the electric potential due to a single charge is **spherically symmetric.**

(iii) The electrostatic force and electric field have an inverse square relationship with distance ($F \propto 1/r^2$ and $E \propto 1/r^2$; Chapter 1A, Eqs. 1.2 and 1.12), but the electric potential has an inverse linear relationship with distance ($v \propto 1/r$).

Example 3. Calculate the electric potential at the surface of a copper (Z=29) nucleus, whose radius is 3.4×10^{-15} m.

Solution. Potential, $V = \dfrac{q}{4\pi\epsilon_0 r}$

Given: $q = 29e = 29\times(1.6\times10^{-19}\text{ C})$; $r = 3.4\times10^{-15}$ m; $\dfrac{1}{4\pi\epsilon_0} = 9\times10^9\ \text{Nm}^2\text{C}^{-2}$

$$\therefore \quad V = \frac{(9\times10^9)\times[29\times(1.6\times10^{-19})]}{(3.4\times10^{-15})}$$

$$= 1.23\times10^7 \text{ volt} \quad \textbf{(Answer).}$$

1B. 03. Electric Potential due to a Group of Point Charges

When multiple charges are present, the net potential due to them at a given point can be determined with the help of the principle of superposition for charges (Chapter 1A; Section 1A.07). Suppose a total number of n charges are present, at distances $r_1, r_2, \ldots\ldots, r_i, \ldots\ldots, r_n$ from the point P, at which the net potential, V, is to be found. Then, according to the superposition principle, V is the algebraic sum of the potentials due to the individual charges $Q_1, Q_2, \ldots\ldots, Q_i, \ldots\ldots, Q_n$.

$$V = V_1 + V_2 + \ldots\ldots + V_i + \ldots\ldots + V_n$$

$$= \sum_{i=1}^{i=n} V_i$$

From Eq. 1.42, $$V_i = \frac{Q_i}{4\pi\epsilon_0 r_i}$$

$$\therefore \quad V = \frac{1}{4\pi\epsilon_0}\sum_{i=1}^{n}\frac{Q_i}{r_i} \quad \text{(n point charges)} \qquad \ldots\ldots (1.43)$$

An important point to be noted here is that the addition involved in the above equation is an **algebraic addition,** not a vectorial addition, because potential is a scalar quantity. The vectorial addition can be very cumbersome, particularly when n is a large value. This is a great advantage of adding up potentials, as compared to the addition of electric fields, which are vector quantities. The various terms in Eq. 1.43 can be positive as well as negative, depending on whether the concerned charge Q_i is positive or negative.

Example 4. Calculate the electric potential at the centre of a square ABCD of side length 2 m due to point charges of 1 μC each kept at its corners.

Solution. Let O be the centre of the square ABCD.

$$AO = \frac{1}{2}(AC) = \frac{1}{2}\sqrt{(2)^2+(2)^2} = \sqrt{2}\text{m}$$

From the principle of superposition, the potential at O is due to all the charges at the corners of the square,

$$V_O = V_A + V_B + V_C + V_D$$

$$= \frac{1}{4\pi\epsilon_0}\cdot\left[\frac{q_A}{(AO)}+\frac{q_B}{(BO)}+\frac{q_C}{(CO)}+\frac{q_D}{(DO)}\right]$$

Now, AO (=BO=CO=DO)=$\sqrt{2}$ m and $q_A=(q_B=q_C=q_D)=1\times10^{-6}$ C

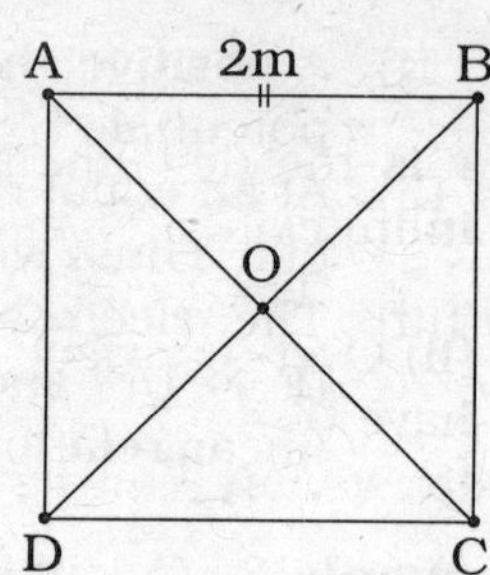

$$\therefore \quad V_0=\frac{1}{4\pi\epsilon_0}\frac{q_A}{(OA)}[1+1+1+1]$$

$$=4\times(9\times10^9)\times\frac{1\times10^{-6}}{\sqrt{2}}$$

$$=2.54\times10^4\text{ V}\quad \textbf{(Answer).}$$

Example 5. Two point charges $q_A=+6\ \mu C$ and $q_B=-2\ \mu C$ are located 16 cm apart. At what point on the line joining the two charges is the electric potential zero (take the potential at infinity to be zero).

Solution. $q_A=+6\ \mu C=6\times10^{-6}$ C; $q_B=-2\ \mu C=-2\times10^{-6}$ C; AB = 16 cm. There are two possible points P_1 and P_2 where the electric potential can be zero, with P_1 lying between A and B and P_2 lying beyond B.

Let $AP_1=x_1$ cm and $AP_2=x_2$ cm.

$\therefore\ BP_1=(16-x_1)$ cm

(i) For the electric potential to be zero at P_1, we have,

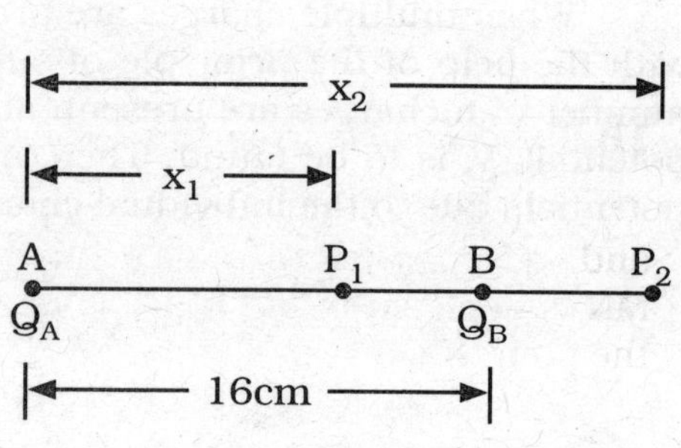

$$V_1=0=\frac{1}{4\pi\epsilon_0}\left[\frac{q_A}{x_1}+\frac{q_B}{(16-x_1)}\right]$$

$$\Rightarrow \frac{q_A}{x_1}=\frac{-q_B}{(16-x_1)}$$

$$\Rightarrow \frac{(6\times10^{-6})}{x_1}=\frac{(2\times10^{-6})}{(16-x_1)}$$

$$\Rightarrow \frac{3}{x_1}=\frac{1}{(16-x_1)}$$

$$\Rightarrow x_1=12\text{ cm}$$

(ii) For the potential to be zero at P_2, we have,

$$\Rightarrow V_2=0=\frac{1}{4\pi\epsilon_0}\left[\frac{q_A}{x_2}+\frac{q_B}{(x_2-16)}\right]$$

$$\Rightarrow \frac{q_A}{x_2}=\frac{-q_B}{(x_2-16)}$$

$$\therefore \frac{(6\times10^{-6})}{x_2}=\frac{(2\times10^{-6})}{(x_2-16)}$$

$$\Rightarrow \frac{3}{x_2}=\frac{1}{(x_2-16)}$$

$$\Rightarrow x_2=24\text{ cm}$$

Thus, the required zero potential points P_1 and P_2 lie at distances 12 cm and 24 cm, respectively, from A, along the line **(Answer).**

1B. 04. Electric potential due to a Continuous Distribution of charge

The Eq. 1.43 cannot be used if the electric charge is continuously distributed over a body, e.g. over a metal sheet or along a wire. Then we select an infinitesimal charge element dQ of the body, find the small potential dV produced by it at the point of observation P, and then integrate dV over the whole body to obtain the net potential due to the body at P,

$$V = \int dV = \frac{1}{4\pi\epsilon_0}\int \frac{dQ}{r} \quad \text{(continuous charge distribution)} \quad \ldots\ldots (1.44)$$

r is the distance between dQ and P. Here the charge element dQ is treated as a point charge, as it is infinitesimally small.

There can be three possible cases of continuous charge distribution: (i) along a line L (e.g. a thin wire), (ii) Over a surface S (e.g. a metal sheet) and (iii) over a volume V´ (e.g. a solid conductor). Accordingly, we have (i) $dQ = \lambda d\ell$ (λ: linear charge density), (ii) $dQ = \sigma\, ds$ (σ : surface charge density) and (iii) $dQ = \rho dv'$ (ρ : volume charge density), in the three respective cases. The three respective values of the net potential are:

$$V_\ell = \frac{1}{4\pi\epsilon_0}\int_L \frac{\lambda d\ell}{r} \quad \ldots\ldots (1.45)$$

$$V_S = \frac{1}{4\pi\epsilon_0}\int_S \frac{\sigma ds}{r} \quad \ldots\ldots (1.46)$$

$$V_{V'} = \frac{1}{4\pi\epsilon_0}\int_{V'} \frac{\rho dv'}{r} \quad \ldots\ldots (1.47)$$

If all the four types of charge distribution are present, viz. multiple discrete charges and the above three types of continuous charge distributions, the net potential is given by the sum of Eqs. 1.43, 1.45, 1.46 and 1.47:

$$V = \left[\sum_{i=1}^{n}\frac{Q_i}{r_i} + \int_L \frac{\lambda d\ell}{r} + \int_S \frac{\sigma ds}{r} + \int_{V'} \frac{\rho dv'}{r}\right] \quad \ldots\ldots (1.48)$$

1B. 05. Potential due to an Electric Dipole

In the last chapter, we had learnt that an electric dipole consists of two equal but unlike charges +q and –q, separated through a small distance 2a. The magnitude of the dipole moment p = 2aq. Such a dipole MN is depicted in Fig. 1.19. O is the centre of the dipole. We shall calculate the potential produced by it at the point P. OP = r, NP = r_1, MP = r_2.

We draw MB and NA ⊥ to OP. In Δ OAN, we have,

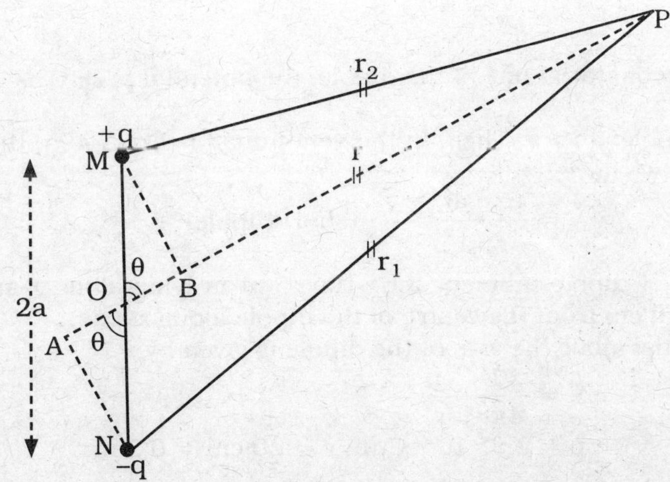

Fig. 1.19

$$\frac{OA}{ON} = \frac{OA}{a} = \cos\theta$$

Or $$OA = a\cos\theta$$

Similarly, $$OB = a\cos\theta$$

While the length 2a of a dipole is essentially very small, the distance of the point of interest P from the dipole is usually large. Therefore, r, r_i, r_2 >>a. Hence,

$$r_1 = NP \approx AP = OP + OA$$
$$= r + a\cos\theta$$

and
$$r_2 = MP \approx BP = OP - OB$$
$$= r - a\cos\theta$$

The net potential at P due to the dipole is the sum of the potentials due to +q and –q,

$$V = \frac{(+q)}{4\pi\epsilon_0 r_2} + \frac{(-q)}{4\pi\epsilon_0 r_1} = \frac{q}{4\pi\epsilon_0}\left[\frac{1}{r_2} - \frac{1}{r_1}\right]$$

$$= \frac{q}{4\pi\epsilon_0}\left[\frac{1}{r - a\cos\theta} - \frac{1}{r + a\cos\theta}\right]$$

$$= \frac{q}{4\pi\,\epsilon_0}\left[\frac{2a\cos\theta}{r^2 - a^2\cos^2\theta}\right]$$

Since 2aq=p, the dipole moment, we finally have,

$$V = \frac{p\cos\theta}{4\pi\epsilon_0\,(r^2 - a^2\cos^2\theta)} \qquad \text{..... (1.49a)}$$

There are two cases of particular interest:

(i) The point **P lies along the axial line of the dipole,** i.e. along the length of the dipole: Then,

$$\theta = 0°;\ \cos\theta = 1$$

$$\therefore \quad V = \frac{P}{4\pi\epsilon_0\,(r^2 - a^2)}$$

Since r>>a, a^2 can be neglected in comparison to r^2. Hence,

$$V = \frac{p}{4\pi\epsilon_0}\frac{1}{r^2} \qquad \text{..... (1.50)}$$

Comparing with Eq. 1.42 for a single charge $[V = Q/(4\pi\epsilon_0 r)]$, we find that for a dipole the potential has an inverse square relationship with distance, instead of an inverse linear relationship for a single charge.

(ii) The point **P lies along the equatorial line** of the dipole, i.e. ⊥ to the length of the dipole: Then,

$$\theta = 90°;\ \cos\theta = 0$$

$$\therefore \quad V = 0$$

Thus, at all points along the equatorial line of the dipole, the potential is zero.

Short dipole: For a short dipole, r >> a. Then, in the denominator (Eq. 1.49a) the term $a^2\cos^2\theta \ll r^2$ and so it can be neglected. Thus, we have,

$$V = \frac{p\cos\theta}{4\pi\epsilon_0 r^2} \quad \text{(short dipole)} \qquad \text{..... (1.49 b)}$$

Example 6. A short dipole of dipole moment 2.6×10^{-3} C-m is lying along x-axis. Calculate the electric potential at a point located 20 cm from the centre of the dipole along x-axis.

Solution. The electric potential along the axis of the dipole is given by,

$$V = \frac{1}{4\pi\epsilon_0}\cdot\frac{p}{r^2}$$

Given: $p = 2.6\times10^{-3}$ C-m; r = 20 cm = 0.2 m

$$\therefore \quad V = \frac{(9\times10^9)\times(2.6\times10^{-3})}{(0.2)^2}$$

$= 5.85\times10^8$ V **(Answer).**

1B. 06. Equipotential Surfaces

An **equipotential surface** is **defined** as "the surface over which the electric potential has the same value at all points".

The following two properties of equipotential surfaces are notable:

(i) We select two arbitrary points M and N over an equipotential surface. From Eq. 1.39, the work done in moving a charge q from M to N is related to the potential difference between M and N as,

$$\Delta V = V_M - V_N = -(W_{MN}/q)$$

By the definition of an equipotential surface,

$$V_M = V_N. \text{ So, } V_M - V_N = 0.$$

$$\therefore \quad W_{MN} = 0$$

It follows that no work is done in moving a charge from one point to another along an equipotential surface.

(ii) Suppose we move a unit charge through an infinitesimal distance $\vec{d\ell}$ along an equipotential surface. The small work done against the electrostatic force $\vec{F}$ will be (work = force × distance),

$$dW = \vec{F} . \vec{d\ell}$$

Since $\vec{F} = q\vec{E}$ and we have taken q as a unit charge, $\vec{F} = \vec{E}$. So we have,

$$dW = \vec{E} . \vec{d\ell} = E. d\ell. \cos\theta$$

But, by the definition of an equipotential surface,

$$dW = 0.$$

So,

$$E. d\ell \cos\theta = 0$$

$$\Rightarrow \quad \cos\theta = 0$$

$$\Rightarrow \quad \theta = 90°, \qquad \text{i.e. } E \perp d\ell$$

If follows that at every point of an equipotential surface, the electric field at that point is perpendicular to the surface.

[This property can also be proved in a purely qualitative way. If $\vec{E}$ was not ⊥ to the surface, it will have a non-zero finite component along the surface. Then, in moving a charge along the surface from one point to another, a finite amount of work will be done. But that is not possible according to the definition of an equipotential surface. Hence, $\vec{E}$ has to be ⊥ to the surface everywhere].

Equipotential surfaces and electric field lines. We have just seen that the electric field $\vec{E}$ is ⊥ to the equipotential surface. In the last chapter, we had seen that $\vec{E}$ is tangential to the electric field lines (e.g. Fig. 1.13b). It follows that the equipotential surfaces are always ⊥ to the electric field lines. Three cases are depicted in Fig. 1.20: (a) a uniform electric field, (b) electric field of a positive point charge and (c) electric field of a dipole. Some electric field lines and some cross-sections of the equipotential surfaces are shown. They may be usefully compared with the electric field lines shown earlier in the last chapter (Figs. 1.12 and 1.13). For a uniform field the equipotential surfaces are parallel planes and for a point charge they are concentric spheres.

Like the electric field lines, the spacing between the equipotential surfaces provides a qualitative picture of the magnitude and direction of the electric field $\vec{E}$. Near the charges, where $\vec{E}$ is strong, the surfaces are close together. Away from the charges, where $\vec{E}$ is weak, the surfaces are far apart. The direction of the electric field is in the direction of decreasing potential. These points will become more clear in the next section, where relevant mathematical relationships will be derived.

1B. 07. Relation between Electric Potential (V) and Electric Field ($\vec{E}$)

Qualitatively, it is obvious that electric potential and electric field are related with each other. We can also develop a mathematical relationship between them. It can be done in two different ways, to serve two different purposes: (i) calculation of the potential from the knowledge of the field and (ii) calculation of the field from the knowledge of the potential.

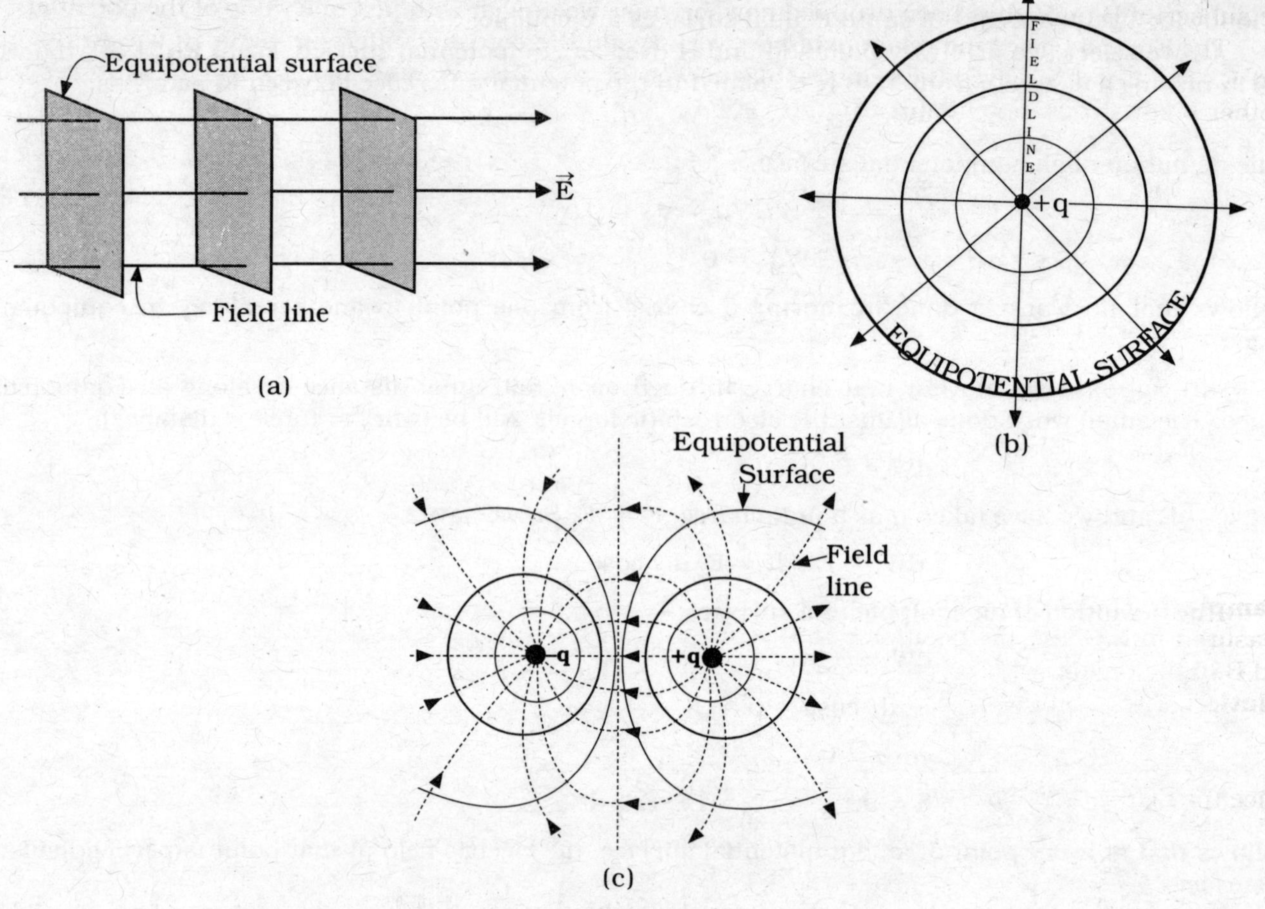

Fig. 1.20

(i) Calculation of the potential (V) from the field ($\vec{E}$)

Suppose we move a positive test charge q_0 along an arbitrary path, from an arbitrary point A to another arbitrary point B in an electric field $\vec{E}$, through an infinitesimal distance $\vec{d\ell}$. The amount of work done in this process will be (work = force × distance),

$$dW = (\vec{F}.\, \vec{d\ell}) \qquad ..... (1.51)$$

$\vec{F}$ is the electrostatic force due to $\vec{E}$ on the charge q_0. Since $\vec{F} = q_0\vec{E}$, we have,

$$dW = q_0\, (\vec{E}.\, \vec{d\ell}) \qquad ..... (1.52)$$

The total work done in moving q_0 from A to B is found by integrating the above as,

$$W = q_0 \int_A^B \vec{E}.\vec{d\ell} \qquad ..... (1.53)$$

[In writing Eqs. 1.51, 1.52, 1.53, we have disregarded the sign (positive or negative) for the amount of work done].

From Eq. 1.39, we have,

$$V_B - V_A = \Delta V = (-W/q_0)$$

Substituting for W from this equation into the Eq. 1.53, we get,

$$V_B - V_A = -\int_A^B \vec{E}.\vec{d\ell} \qquad ..... (1.54)$$

Thus, we find that the p.d. $\Delta V = (V_B - V_A)$ between any two points A and B in the field is equal to the negative of the line integral of $\vec{E}.\, \vec{d\ell}$ from A to B (the word 'line' here means any arbitrary path between A and B; it can be curved or linear or a combination of the two).

If we choose the potential at A to be zero, e.g. if we take A to be at infinity, the Eq. 1.54 becomes,

$$V = -\int_A^B \vec{E}.\vec{d\ell} \qquad ..... (1.55)$$

The subscript B on V_B has been dropped now, because we are left with just one value of the potential.

The Eq. 1.54 gives the relationship between $\vec{E}$ and the potential difference between two points in the field $\vec{E}$. The Eq. 1.55 gives the relationship between $\vec{E}$ and the potential at a given point with respect to another point at zero potential.

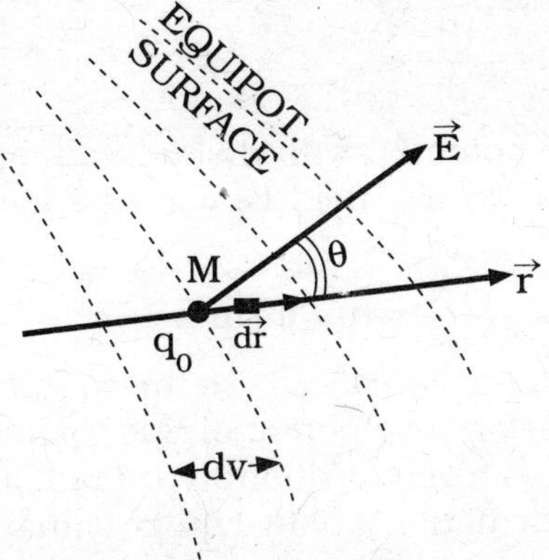

Fig. 1.21

Example 7. The electric field in a region of space lies along x-axis, given by, $\vec{E} = 3x^2 - 2x + 1$. The field is measured in N/C and the position x in metre (m). Calculate the potential difference between two points A and B in the region, with $x_A = 2$ m and $x_B = 4$ m.

Solution. The potential difference between A and B is given by,

$$V_B - V_A = -\int_A^B \vec{E}.\vec{dx} = -\int_2^4 E dx \cos\theta$$

Since the field E lies along the x-axis, $\theta = 0$. $\therefore \cos\theta = 1$

$$\therefore \quad V_B - V_A = -\int_2^4 (3x^2 - 2x + 1)\, dx$$

$$= \left| -x^3 + x^2 - x \right|_2^4 = (-52) - (-6)$$

$$= -46 \text{ volt} \quad \textbf{(Answer).}$$

(ii) Calculation of the field ($\vec{E}$) from the potential (V)

A family of closely spaced equipotential surfaces is shown in Fig. 1.21. The potential difference between any two adjacent surfaces is dV. We draw an arbitrary path $\vec{r}$ across the surfaces and choose an arbitrary point M on a surface. We know that the electric field $\vec{E}$ at M is ⊥ to this surface, as discussed in the last section.

Let a positive test charge q_0 move a distance $\vec{dr}$ along $\vec{r}$ from this surface to the next adjacent surface. Since the surfaces are closely spaced, $\vec{dr}$ is an infinitesimal distance. The p.d. through which the charge moves is dV. From Eq. 1.39, the small amount of work done in this movement is,

$$dW = -q_0\, dV$$

But, from Eq. 1.52, we also have

$$dW = q_0\, (\vec{E}.\vec{dr}) = q_0\, (E\cos\theta)\, dr$$

Or

$$E\cos\theta = -\frac{dV}{dr} \qquad \text{..... (1.56)}$$

As seen in the figure, $E\cos\theta$ is the component (E_r) of $\vec{E}$ along $\vec{r}$. So Eq. 1.56 can be written as,

$$E_r = -\frac{\partial V}{\partial r} \qquad \text{..... (1.57)}$$

The use of partial derivative symbols simply means that the Eq. 1.57 only involves the variation of V along a specified direction. Here this direction is along $\vec{r}$. But in general it can be any one of the directions x, y, or z for a cartesian system of coordinates in space. So, in general, we can write,

$$E_x = -\frac{\partial V}{\partial x},\ E_y = -\frac{\partial V}{\partial y},\ E_z = -\frac{\partial V}{\partial z} \qquad \text{..... (1.58)}$$

If follows from Eqs. 1.57 and 1.58 that the component of the field $\vec{E}$ in any direction is the negative of the rate of change of the potential with distance in that direction.

For a particular direction, the partial derivative symbol may be dropped. Thus for the direction $\vec{r}$, the Eq. 1.57 may be written as,

$$E_r = -\frac{dV}{dr} \qquad \text{..... (1.59)}$$

dV/dr, which represents the change of potential with distance along $\vec{r}$, is called **potential gradient** along $\vec{r}$.

In the simple case of a uniform electric field $\vec{E}$, e.g. as shown in Fig. 1.20a, the subscript r can be dropped. Then the Eq. 1.59 becomes,

$$E = -\frac{dV}{dr} \quad \text{(uniform field)} \qquad \text{..... (1.60)}$$

Here both $\vec{r}$ and $\bar{E}$ are ⊥ to the equipotential surfaces. So the component of $\vec{E}$ in any direction parallel to the surfaces is zero. Note that although the electric potential V is a scalar quantity, the electric field $\vec{E}$ which is its derivative with respect to distance, is a vector quantity.

The Eq. 1.60 also contains information about equipotential surfaces. It implies that greater is the value of E, the lesser is the rate of change of V with distance (note the negative sign of dV/dr). Hence the equipotential surfaces are spaced closely where the electric field is strong and they are spaced widely where the field is weak. Looking at the Fig. 1.21, the Eq. 1.60 also means that $\vec{E}$ points in the direction of decreasing V.

Example 8. The electric potential in a region of space is given by,

$$V = -3x^4 + 4x^2 - 3x$$

x represents the position coordinate in metre (m). Find the electric field at a point x = 2 m.

Solution. The relation between electric potential and electric field is given by,

$$E = -\frac{dV}{dr}$$

In our case, the variation of potential is along the x-axis alone. Hence,

$$E = -\frac{d}{dx}[-3x^4 + 4x^2 - 3x]$$

$$= 12x^3 - 8x + 3$$

$$\therefore \quad E_{x=2} = (12\times8) - (8\times2) + 3$$

$$= 83 \text{ N/C} \quad \textbf{(Answer).}$$

The positive value of E means that the electric field is directed towards positive x.

1B. 08. Conservative Forces; Conservative Nature of Electrostatic Force ($\vec{F}$) and Electric Intensity ($\vec{E}$)

If a ball is tossed up from the ground, the kinetic energy (k.e.) imparted to it is gradually converted into potential energy (p.e.), which is stored up in the ball. The p.e. comes from the work done on the ball by the gravitational force. It is maximum at the highest position of the ball. The work done when the ball rises is negative (–W), because it is done against the gravitational force, which transfers p.e. to the ball from the k.e. of the ball. From the highest position, the ball starts falling back and the same amount of work is done in reaching back to the ground (work = force × distance). But now the whole process is reversed. The work done is positive (+W), because it is performed in the direction of the gravitational force, which now transfers k.e. to the ball from the p.e. of the ball. Hence in a round trip of the ball the net work done by the gravitational force and therefore the net transfer of potential energy is zero. In spite of all its up and down movement the net potential energy of the ball remains unchanged, i.e. it is conserved. In this sense, the gravitational force is regarded as a conservative force. Note that our arrangement does not depend on how high the ball is tossed up or what is the shape of the path followed by it. The situation can be generalised for any **closed path.** In fact, we can generally **define a conservative force** as "that force by which the net work done on a body moving around any closed path is zero". The closed path may be short or long, or may be of any shape, e.g. circular, zigzag, elliptical, etc.

Like the gravitational force, the electrostatic force ($\vec{F}$) and the electric field ($\vec{E}$) are also conservative in nature [remember that the electric field intensity (also simply called electric field), at any point in the field is defined as the force experienced by a unit charge placed at that point; vide Section 1A. 11 of the last chapter]. We can simply prove it as follows. Suppose a point charge q_0 moves clockwise along an arbitrary closed path, from the position A to B to A (Fig. 1.22). From Eqs. 1.40 and 1.54, the work done per unit charge (W/q_0) in moving from A to B is,

$$(W_{AB}/q_0) = -\int_A^B \vec{E}.d\ell = (V_B - V_A)$$

The work done in moving from B to A is,

$$(W_{AB}/q_0) = -\int_B^A \vec{E}.\vec{d\ell} = (V_A - V_B)$$

Therefore, the net work done in moving around the entire closed path is,

$$(V_B - V_A) + (V_A - V_B) = \text{Zero}$$

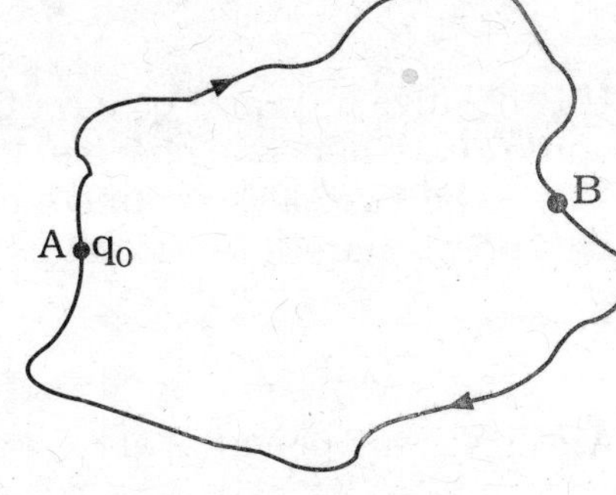

Fig.1.22

Thus, no net work is done in moving the charge around the closed path. So the electric field represents a conservative force. Since $\vec{F} = q\vec{E}$, the same argument applies to $\vec{F}$. Thus, the electrostatic force is also a conservative force. In a mathematical form, the said conservative nature can be expressed through closed line integrals as,

$$\oint \vec{E}.\vec{d\ell} = 0 \quad ..... (1.61)$$

$$\oint \vec{F}.\vec{d\ell} = 0 \quad ..... (1.62)$$

(The word 'line' here includes any curved or zigzag path, not merely a linear path)

The work done on a given charge by an electrostatic force (or by the force of an electric field) is stored up in the charge as its electric potential energy.

1B. 09. Electric Potential Energy of a System of Two Point Charges (With No Initial Electric Field)

The Fig. 1.23 shows a simple system of two point charges $+q_1$ and $+q_2$, placed at a mutual distance r. We proceed to evaluate its electric potential energy. For this purpose, we first make out how this system came into existence. We imagine that initially the two charges lie far away at infinity at rest. Now we first bring the charge $+q_1$ slowly to the position A. In doing so, no amount of work is done because no electric field and therefore no electrostatic force is present (work = force × distance). Next, we slowly bring the charge $+q_2$ from ∞ to the position B. Now some work, W_∞, is done because the charge $+q_1$ has its electric field around it and therefore an electrostatic force acts on the charge $+q_2$ during its movement. From the definition of potential (or from Eq. 1.40, disregarding the negative sign, as we are only concerned with the quantity here), we have,

$$W_\infty = W \text{ (say)} = q_2 V$$

V is the potential created by q_1 at the position B, where the charge q_2 has been brought. From Eq. 1.42 for a single point charge,

$$V = \frac{1}{4\pi\epsilon_0} . \frac{q_1}{r}$$

Fig. 1.23

From the definition of electric potential energy, the work done, $W = q_2V$, is stored up as potential energy, U, of our pair of charges q_1 and q_2. Substituting for V, we get,

$$U = W = q_2V = \frac{1}{4\pi\epsilon_0} \cdot \frac{q_1q_2}{r} \qquad \text{..... (1.63)}$$

U is positive if q_1 and q_2 have the same sign, i.e. a repulsive force exists between them. It is negative if q_1 and q_2 have opposite charges, i.e. they have a mutual force of attraction.

The foregoing treatment can be extended to a system of more than two charges. The p.e. of a system of N point charges located at positions r_1, r_2, r_3, etc., with respect to an origin is found to be,

$$U_N = \frac{1}{2} \sum_{m=1}^{m=N} q_m V_m \qquad \text{..... (1.64)}$$

where V_m, the potential at r_m due to all the other charges, is given by,

$$V_m = \sum_{\substack{n=1 \\ n \neq m.}}^{n=N} \frac{1}{4\pi \epsilon_0} \cdot \frac{q_n}{r_{mn}} \qquad \text{..... (1.65)}$$

The factor 1/2 occurs in Eq. 1.64 because each pair of charges is counted twice in the summation.

1B. 10. Electric Potential Energy of a System of Two Point Charges in an Electric Field

Suppose the two point charges q_1 and q_2 shown in Fig. 1.23 are situated in an external electric field $\vec{E}$. Let their distance from an origin be $\vec{r}_1$ and $\vec{r}_2$, respectively. Let the potentials at the positions A and B due to the field $\vec{E}$ be $V(\vec{r}_1)$ and $V(\vec{r}_2)$, respectively.

Now, when the charge q_1 is brought from infinity to A, work is done on it against the field $\vec{E}$. The work done is,

$$W_1 = q_1V(\vec{r}_1)$$

When the charge q_2 is brought from infinity to B, an amount of work, W_2, is done on it against two fields: (i) $\vec{E}$ and (ii) the field due to q_1. Therefore,

$$W_2 = q_2V(\vec{r}_2) + \frac{q_1q_2}{4\pi\epsilon_0 r}$$

Hence the total work done in bringing q_1 and q_2 into the field is,

$$W = W_1 + W_2 = q_1V(\vec{r}_1) + q_2V(\vec{r}_2) + \frac{q_1q_2}{4\pi\epsilon_0 r}$$

This work is stored up as electric potential energy, U, of the system,

$$U = q_1V(\vec{r}_1) + q_2V(\vec{r}_2) + \frac{q_1q_2}{4\pi\epsilon_0 r} \qquad \text{..... (1.66)}$$

1B. 11. Electric Potential Energy of a Dipole in an Electric Field

An electric dipole consists of two point charges +q and –q, separated through a small distance 2a. We assume that the external electric $\vec{E}$ in which it is placed is a uniform field (Fig. 1.24). This case is similar to the one discussed in the last section, with $q_1 = +q$ and $q_2 = -q$. Therefore, from Eq. 1.66 the potential energy of the dipole is,

$$U_{\text{Dipole}} = q[V(\vec{r}_1) - V(\vec{r}_2)] - \frac{q^2}{4\pi\epsilon_0 . 2a}$$

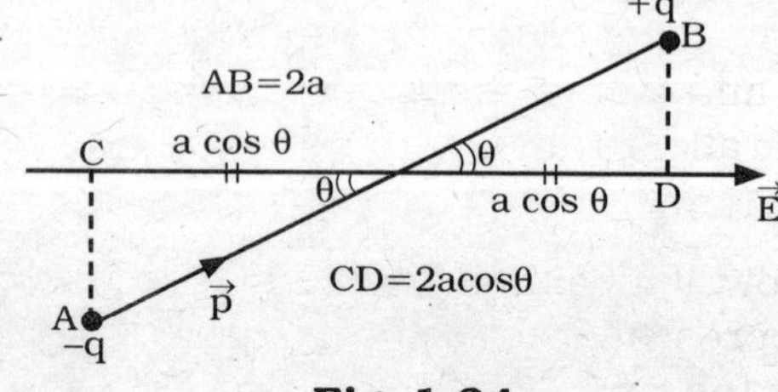

Fig. 1.24

The potential difference between the positions A and B of the two charges is equal the work done in moving a unit positive charge from B to A against the field $\vec{E}$. From Fig. 1.24, the component of the distance moved in the direction of $\vec{E}$ is CD = 2acosθ. Hence (work = force × distance),

$$V(\vec{r}_1) - V(\vec{r}_2) = -E.2a\cos\theta$$

The negative sign appears because the work is done against the field $\vec{E}$. Then we have,

$$U_{Dipole} = -qE.2a\cos\theta - \frac{q^2}{4\pi\epsilon_0.2a}$$

For a given dipole, the second term is a constant value. Since the first term involves the angular value θ, any value of U_{Dipole} has to be measured with respect to a reference value of θ (usually taken as θ = 90°, i.e. the dipole oriented ⊥ to the field $\vec{E}$). Thus, any value of U_{Dipole} is a difference between two values of potential energy, both of which include the constant 2nd term, which therefore cancels out. Hence the 2nd term is of no significance to us and may be left out. So we can write,

$$U_{Dipole} = -(2aq).E\cos\theta$$

But 2aq = p, the dipole moment (Chapter 1A; Section 1A.16). So we have,

$$U_{Dipole} = -pE\cos\theta$$

Or $$U_{Dipole} = -(\vec{P}.\vec{E}) \qquad \text{..... (1.67)}$$

The same expression was derived earlier in the last chapter (Section 1A.19) in a different way, viz. by calculating the work done in rotating a dipole.

Example 9. An electric dipole consists of two point charges of 2 μC and –2 μC, respectively, separated by a distance of 2 cm. The dipole is placed in an external uniform electric field of strength 10^6 N/C . If θ is the angle between electric field and dipole moment, calculate the work done in rotating the dipole from an angle θ_1= 30° to the angle θ_2= 60°.

Solution. The work done is equal to the difference in potential energy,

$$W = U_2 - U_1$$

The potential energy of a dipole in an electric field is given by,

$$U = -pE\cos\theta$$

Dipole moment, $p = q(2a) = (2\times10^{-6})\times(2\times10^{-2}) = 4\times10^{-8}$ C.m $(2a = 2cm = 2\times10^{-2}m)$

$$\therefore \quad W = (U_2 - U_1) = -pE(\cos\theta_2 - \cos\theta_1)$$

$$= -(4\times10^{-8})\times(10^6)[\cos60° - \cos30°] \quad (E = 10^6 \text{ N/C})$$

$$= -4\times10^{-2}\left(\frac{1}{2} - \frac{\sqrt{3}}{2}\right)$$

$$= -4\times10^{-2}\times\frac{1}{2}(-0.732)$$

$$= 1.46\times10^{-2} \text{ J}$$ **(Answer).**

1B. 12. Vectorial Representation of Area

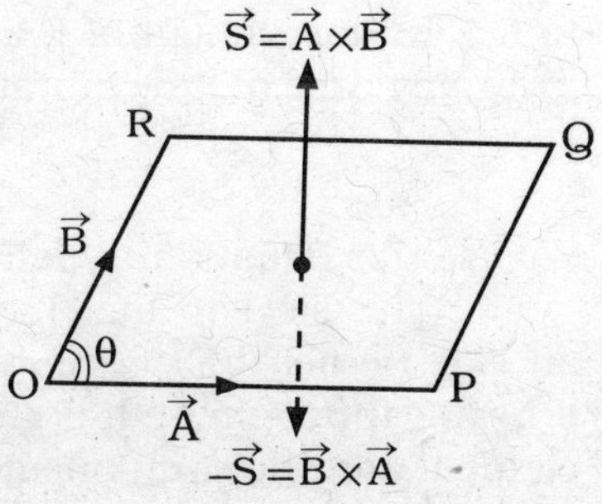

Fig. 1.25

The Fig. 1.25 shows a parallelogram with length of its two sides as A and B and the smaller angle between the two sides as θ. The area of the parallelogram is,

$$S = A \times B \times \sin\theta.$$

Now, if we assign a sense of direction to A and B, as seen in the figure, they are represented as vectors $\vec{A}$ and $\vec{B}$. Then the right hand side above can be written as the cross product of $\vec{A}$ and $\vec{B}$,

$$\vec{A} \times \vec{B} = A \times B \times \sin\theta$$

Since a cross product of two vectors is also a vector quantity, the area S is also to be represented as a vector $\vec{S}$,

$$\vec{S} = \vec{A} \times \vec{B} \qquad \text{..... (1.68)}$$

The direction of $\vec{S}$ is the direction of advancement of a right-handed screw rotated from the first vector $\vec{A}$ to the second vector $\vec{B}$ through the smaller angle θ between their positive directions. As seen in the figure, this direction points normally outwards from the plane of $\vec{A}$ and $\vec{B}$, i.e. the plane of the paper. Thus the area S is now conveniently represented as a vector $\vec{S}$, lying in the direction of the normal to the plane of $\vec{A}$ and $\vec{B}$.

In the scalar form, it does not matter if we interchange A and B, since

$$S = A \times B \times \sin\theta = B \times A \times \sin\theta$$

But in the vector form, if we exchange $\vec{A}$ and $\vec{B}$, the sense of rotation changes, because in the vector product $\vec{B} \times \vec{A}$ the rotation has to be made from the vector $\vec{B}$ to $\vec{A}$. Then the right-handed screw moves in the opposite direction, i.e. the area vector $\vec{S}$ now points downwards from the plane of paper, although its magnitude remains the same.

$$(\vec{A} \times \vec{B}) = -(\vec{B} \times \vec{A})$$

Thus, we have a simple convenient way of representing a surface area as a vector quantity.

1B. 13. Electric Flux

The word 'flux' is derived from the latin word 'fluxus', which means 'to flow'. Suppose we have water flowing in a stream or a river. Now we hold a plane metal sheet with a large hole in the middle against the flow of water. Maximum amount of water will flow through the hole when the sheet is held $\perp$ to the direction of flow. If the sheet is gradually tilted, progressively less amount of water will flow through it as the angle of tilt increases. In the extreme position when the sheet becomes parallel to the direction of flow, i.e. the angle of tilt is 90°, minimum (zero) amount of water will flow through the hole. Thus, it is noted that the amount of water flowing through the sheet, which we may call as the flux of water, depends on the orientation of the sheet with respect to the direction of flow. Also, this amount depends on the size of the hole and the velocity of flow of water. The greater the hole size and the velocity, the more will be the amount. Here we have assumed that the velocity of flow is uniform throughout the volume of water. Now suppose it is not so and the velocity is different for different sections of the water, (which does happen in a stream or a river). Then the amount flowing through different sections of the hole will be different. Since velocity is a vector quantity, we can assign a separate velocity vector to each section. The collection of these vectors may be said to form a **velocity field** across the hole. Then the flux of water through the sheet is jointly determined by the hole size, the orientation of the sheet, and the velocity field.

A similar situation exists for electric flow. Here the flow of electric charge becomes the equivalent of flow of water. The electric field becomes the equivalent of the velocity field. In the last chapter (Section 1A.14) we had described that electric field lines pictorially represent the electric field intensity $\vec{E}$ around a given configuration of charges. The lines are more closely spaced where the field is strong. The magnitude of $\vec{E}$ in a small region of space is proportional to the density of the field lines, i.e. the number of lines crossing unit area, in the region. We may call this density as electric flux density. Consider a small element of area ΔS in an electric field $\vec{E}$ (Fig. 1.26). It is equivalent to the aforesaid middle hole in a metal sheet. As mentioned in the last section, the area ΔS is vectorially represented by the normal $\overrightarrow{\Delta S}$ drawn perpendicular to it. The component of $\vec{E}$ along $\overrightarrow{\Delta S}$ is E cosθ, where θ is the angle between $\vec{E}$ and $\overrightarrow{\Delta S}$. In analogy with the foregoing case of flow of water, the electric flux $\Delta\phi$ through the area element ΔS is given by,

$$\Delta\phi = \Delta S \times E\cos\theta = \vec{E}.\overrightarrow{\Delta S} \quad \text{..... (1.69a)}$$

We use this equation to **define electric flux** as follows:

"The electric flux ($\Delta\phi$) across a small element (ΔS) of space in an electric field ($\vec{E}$) is represented by the number of electric field lines crossing the element. It is given by the scalar product of the area $\overrightarrow{\Delta S}$ of the element and the component of the electric field $\vec{E}$ normal to the element ($\Delta\phi = \vec{E}.\overrightarrow{\Delta S}$)."

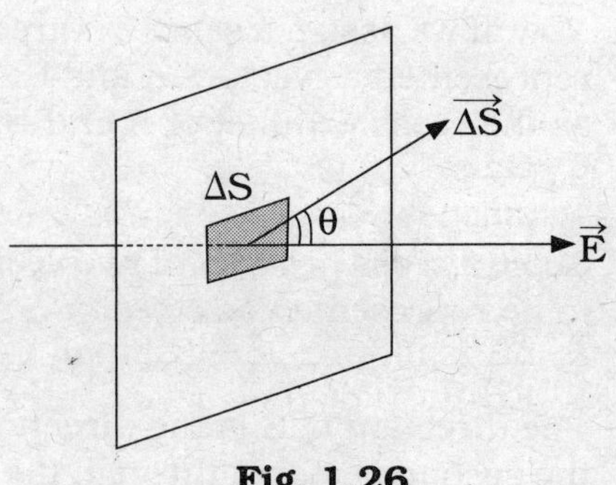

Fig. 1.26

Note that the electric flux is a scalar quantity, because the scalar product of two vectors is a scalar. Its **SI unit is Nm^2C^{-1}** (The SI unit of $\vec{E}$ is NC^{-1} and that of area ΔS is m^2). For a plane surface (e.g. a plane circular surface) of area S, the total flux ϕ is given by,

$$\phi = \vec{E} \cdot \vec{S} = ES\cos\theta \qquad \text{..... (1.69b)}$$

Example 10. A plane circular surface of radius 1 cm is kept in YZ-plane in a uniform electric field of strength 10^3 N/C , directed along x-axis. Calculate the flux associated with the surface.

Solution. Flux, $\phi = ES \cdot \cos\theta$

Given: $\theta = 0;\ E = 10^3$ N/C

$S = \pi r^2 = 3.14 \times (1\times10^{-2})^2 \qquad (r = 1\text{cm} = 1\times10^{-2}\text{m})$

$= 3.14 \times 10^{-4}\text{m}^2$

$\therefore \quad \phi = E.S.\cos\theta$

$= (10^3)\times(3.14\times10^{-4})\times(1)$

$= 0.31\ \text{Nm}^2\text{C}^{-1}$ **(Answer).**

Electric flux through a Gaussian surface

For a full description of electric flux, it is necessary to consider a so-called **Gaussian surface.** A Gaussian surface is an imaginary closed surface. It can have any shape, although for convenience usually symmetrical shapes (e.g. sphere, oval shape, cylinder, etc.) are taken into consideration. A part of a Gaussian surface is shown in Fig. 1.27. It lies in a non-uniform electric field, i.e. the value of $\vec{E}$ varies from one position to another across the surface. Now we divide the surface into a large number of small elements of area ΔS. Four such elements; marked as 1, 2, 3, 4, are shown in the figure. Each element is represented by an area vector $\overrightarrow{\Delta S}$ of the type shown earlier in Fig. 1.26. Each vector $\overrightarrow{\Delta S}$ is $\perp$ to the surface at the position of the area element ΔS and it is directed outwards from the interior of the surface. The angle between $\overrightarrow{\Delta S}$ and the electric field $\vec{E}$ at its position is θ. Because of different curvatures of the surface at different positions, the orientation of the elements differs from one position to another, as seen in the figure for the elements 1, 2, 3, 4.

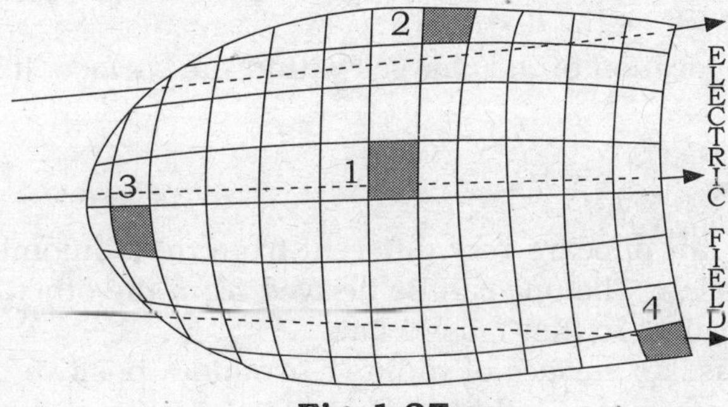

Fig. 1.27

As already mentioned, the electric field $\vec{E}$ differs from one position to another over the surface. However, because of the small area ΔS of the elements, $\vec{E}$ can be regarded as constant over a particular element. From Eq. 1.69a, the flux of the electric field through the surface for this area element is $\vec{E}.\overrightarrow{\Delta S}$ = (E. ΔS. cosθ). The flux for the other area elements is also given by the same expression. The total flux ϕ for the entire Gaussian surface is given by the algebraic sum of all these contributions,

$$\phi = \sum_N \vec{E}.\overrightarrow{\Delta S} \qquad \text{..... (1.70)}$$

N is the total number of the area elements ΔS. Since $\vec{E}$ and cosθ vary for different positions on the closed Gaussian surface, the value of the scalar product $\vec{E}.\ \overrightarrow{\Delta S}$ also varies from one place to another across the surface. At the positions θ=90°, it becomes zero. Its sign also varies from place to place. So it can be both positive and negative. These factors are taken into consideration while carrying out the summation.

The approximation of regarding $\vec{E}$ as constant for an area element ΔS becomes really true only when ΔS is infinitesimally small ($\Delta S \rightarrow 0$). Then we replace ΔS by the differential value dS and the summation in Eq. 1.70 by the surface integral over the closed Gaussian surface,

$$\phi = \oint_S \vec{E}.\overrightarrow{dS} \qquad \text{..... (1.71)}$$

S is the total surface area of the Gaussian surface. The small circle on the integral sign denotes the closed nature of the surface.

1B. 14. Gauss's Theorem (or Gauss's Law)

The Gauss's theorem (or Gauss's law) connects the net charge Q within a closed surface (Gaussian surface) with the flux ϕ of the electric field through the surface:

$$\phi = \frac{1}{\epsilon_0} Q \qquad \text{..... (1.72)}$$

(Gauss's theorem)

Using Eq. 1.71, we may write,

$$\phi = \oint_S \vec{E}.\, d\vec{S} = \frac{1}{\epsilon_0} Q \qquad \text{..... (1.73)}$$

(Gauss's theorem; expanded from)

Thus, the **Gauss's theorem** states that "the electric flux through a closed surface S, given by the surface integral of the electric field $\vec{E}$, is $1/\epsilon_0$ times the net charge enclosed by the surface".

The net charge Q includes all possible charges (positive, negative, and continuous charge distributions). It is the algebraic sum of all these charges. The Eqs. 1.72 and 1.73 hold for charges located in a vacuum (or air). If the medium is other than air, e.g. an oil, the equations are suitably modified. The following points are noteworthy:

(i) If Q is positive, the flux ϕ is directed outwards from the surface. If it is negative, ϕ is directed inwards from the surface.

(ii) Any charges lying outside the Gaussian surface affect the value of $\vec{E}$, the electric field, but do not affect the value of ϕ at all. The value of ϕ is solely governed by the net charge Q enclosed by the surface.

(iii) ϕ is not affected by the location of the charges within the surface. It is solely determined by the net value Q.

Gauss's law vs. Coulomb's law

Prima facie, the Gauss's law appears very different from the Coulomb's law. But, in reality, it is just another form of the Coulomb's law. The one can be derived from the other. In the following we shall show how the Coulomb's law is equivalent to the Gauss's law.

We draw a spherical Gaussian surface of radius r around a positive point charge +q (Fig. 1.28). We divide the surface into a large number of infinitesimal area elements dS each. At any point of the surface, the area vector $\overrightarrow{dS}$ is ⊥ to the surface (directed outwards) (vide Section 1B.12). At this point the electric field $\vec{E}$ is also ⊥ to the surface (and directed outwards; see Fig. 1.12, Section 1A.14, Chapter 1A). Thus the angle θ between $\vec{E}$ and $\overrightarrow{dS}$ is zero. So from Eq. 1.73 for the Gauss's law we have ($\vec{E}.\,\overrightarrow{dS}$ = E. dS. cos θ = E. dS),

$$\oint \vec{E}.d\vec{S} = \oint E dS = q/\epsilon_0$$

The value of $\vec{E}$ decreases as one goes away from the point +q, i.e. as r increases, but it is a constant value all over the spherical surface, since r is same all over the surface. Hence in the above surface integral, E is to be treated as constant. So we have,

$$E \oint dS = q/\epsilon_0$$

The integral $\oint$ dS taken over the entire spherical surface equals the area $4\pi r^2$ of the surface. Therefore,

$$E.\, 4\pi r^2 = q/\epsilon_0$$

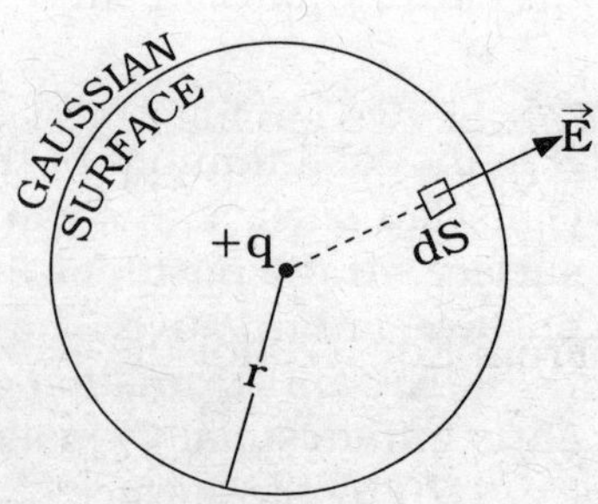

Fig. 1.28

Or
$$E = = \frac{q}{4\pi\epsilon_0 . r^2}$$

We had derived the same value of $\vec{E}$ due to a point charge, using Coulomb's law (Chapter 1A; Section 1A.11; Eq. 1.12). Thus we find that the Gauss's law is equivalent to the Coulomb's law.

Applications of Gauss's theorem

In the following sections, we shall apply the Gauss's theorem to determine electric field due to (i) an infinitely long straight wire, (ii) a uniformly charged infinite plane sheet and (iii) a uniformly charged thin spherical shell.

1B. 15. Electric Field due to a Charged Infinitely Long Straight Wire (Linear Charge)

The Fig. 1.29 depicts a section of an infinitely long straight wire. Suppose it carries a positive charge. Let the linear charge density (i.e. charge per unit length) be λ. We shall apply Gauss's law to determine the magnitude of the electric field intensity $\vec{E}$ at a point P. Let the perpendicular distance of P from the wire be r. We draw our Gaussian surface as a circular cylinder of radius r and length ℓ, with the wire as the axis of the cylinder. In addition to the round curved surface, the total Gaussian surface (which has to be a closed surface) is capped by two circular end surfaces at A and B, shown shaded in the figure.

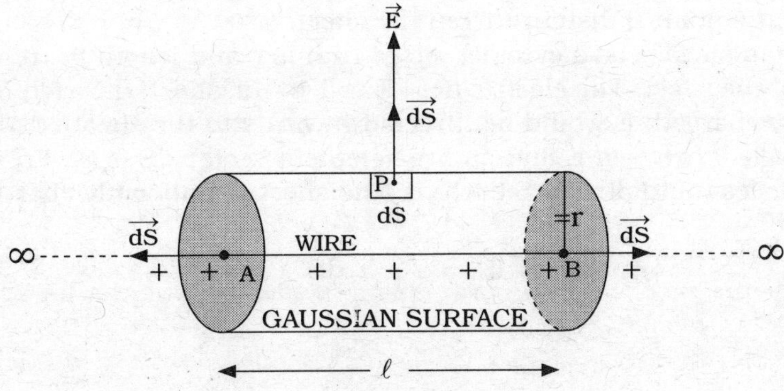

Fig. 1.29

The area S of the curved surface is $2\pi r\ell$. All points like P lying on this surface are situated at an equal perpendicular distance r from the wire. Hence for all these points the electric field $\vec{E}$ has the same magnitude E and it is directed radially outwards from the wire (if the wire had negative charge, the direction of $\vec{E}$ would be radially inwards to the wire). Further, the area element $\vec{dS}$ at P will be $\perp$ to the surface. The same will be the case for all other points like P on the surface. So the angle θ between $\vec{E}$ and $\vec{dS}$ is zero. From Eq. 1.71, the electric flux through the curved surface is,

$$\phi = \oint_S \vec{E}.\vec{dS} = \oint_S E.dS.\cos\theta$$
$$= E\oint_S dS \cos 0^0 = E\oint_S dS$$
$$= E\,(2\pi r\ell)$$

For the two circular end surfaces at A and B, the angle θ between $\vec{E}$ and $\vec{dS}$ is 90°. So, $\phi = E\oint_S dS \cos 90° = 0$, for them. Thus the total flux through the Gaussian surface is,

$$\phi = E\,(2\pi r\ell)$$

From Eq. 1.73 for the Gauss's law,

$$\phi = \frac{1}{\epsilon_0} Q$$

$$\therefore \quad E\,(2\pi r\ell) = (1/\epsilon_0)\,Q$$

The charge Q enclosed by our Gaussian surface is $\lambda\ell$. So we have,

$$E\,(2\pi r\ell) = \frac{1}{\epsilon_0}\,(\lambda\ell)$$

Or
$$E = \frac{\lambda}{2\pi\epsilon_0\, r} \quad \text{(Linear charge)} \qquad ..... (1.74)$$

Thus, for a linear charge like a charged straight wire, we find that the electric field intensity is inversely proportional to r,

$$E \propto (1/r)$$

A plot of E as a function of r is shown in Fig. 1.30.

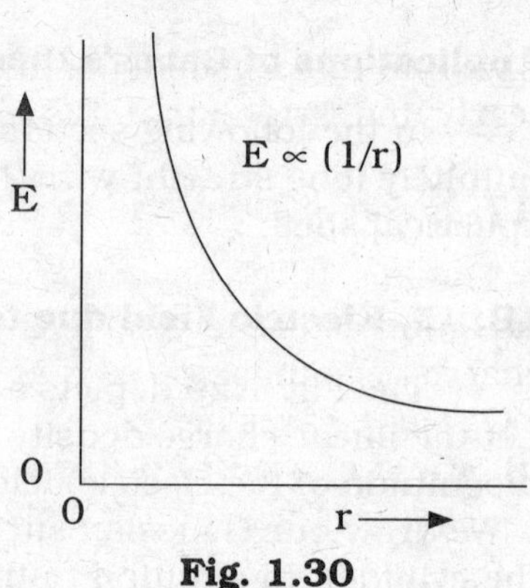

Fig. 1.30

The Eq. 1.74 also holds good for a straight wire of finite length, except for the portions near to the two ends of the wire, where the value of E is modified.

1B. 16. Electric Field due to a Uniformly Charged Infinite Plane Sheet

The Fig. 1.31 shows a part KLMN of a uniformly charged plane sheet extending upto infinity. It is taken as positively charged. Let the surface charge density (i.e. charge per unit area) be σ. We shall apply the Gauss's law to find the electric field intensity at a point P, distant ℓ from the sheet.

We draw our Gaussian surface as a cylinder AB of radius r and length 2ℓ. It symmetrically extends a distance ℓ on either side of the sheet. The electric field $\vec{E}$ is $\perp$ to the sheet, directed outwards from the sheet (if the sheet was negatively charged, $\vec{E}$ would be directed inwards to the sheet). At the two end faces of the surface at A and B, the angle θ between $\vec{E}$ and an area element vector $\overrightarrow{dS}$ is 0°. From Eq. 1.71, the electric flux through the two end faces (of total area S each) is (the sheet is uniformly charged, so E is constant for all points on the faces),

$$\phi = \oint_S \vec{E}.\overrightarrow{dS} = 2\oint_S E.dS.\ \cos\theta$$

$$= 2E.S$$

Fig. 1.31

For all points lying on the curved surface of the cylinder, $\vec{E}$ is $\perp$ to $\overrightarrow{dS}$. $\therefore$ $\oint_S \vec{E}.dS = \oint_S E.dS.\cos 90° = 0$, for them. Hence the curved surface does not contribute to the flux at all and the total flux from the entire Gaussian surface is contributed by the end faces alone. So the total flux is,

$$\phi = 2E.S$$

From Eq. 1.73 for the Gauss's law,

$$\phi = (1/\epsilon_0)\ Q$$

$$\therefore \quad 2E.S = (1/\epsilon_0)\ Q$$

The charge Q enclosed by the Gaussian surface is σS ($\because$ S is the cross-sectional area of the cylinder; $S = \pi r^2$). So we have,

$$2E.S = (1/\epsilon_0)\,\sigma.S$$

Or $$E = \frac{\sigma}{2\epsilon_0} \qquad \text{(planar charge)} \qquad \text{..... (1.75)}$$

Thus we find that E is independent of ℓ, the distance of the point of observation from the sheet. But it should be clearly understood that it is only true for a sheet extending upto infinity on all sides.

1B. 17. Electric Field due to a Uniformly Charged Thin Spherical Shell

The Fig. 1.32 shows the section in the plane of paper of a uniformly charged thin spherical shell of radius R and centre located at O. The shell is taken as positively charged. Let the surface charge density (i.e. the charge per unit area) be σ. We shall apply Gauss's law to evaluate the electric field at a point P distant r from O. Three cases arise:

(i) Field at a point outside the shell. We draw our Gaussian surface as a concentric sphere of radius r ($r > R$). By symmetry, the value of $\vec{E}$ is same at all points of the surface, directed outwards from the surface (if the shell was negatively charged, it would be directed inwards to the surface). The area element vector $\vec{dS}$ at any point also has the same direction as $\vec{E}$ at that point. So the angle θ between $\vec{E}$ and $\vec{dS}$ is zero. From Eq. 1.71, the flux through the surface is (area of the surface $= 4\pi r^2$),

$$\phi = \oint_S \vec{E}.\vec{dS} = \oint_S E.dS.\cos 0$$
$$= \oint_S E.dS = E.S = E.\,4\pi r^2$$

From Eq. 1.73 for the Gauss's law,

$$\phi = (1/\epsilon_0)\,Q$$

$$\therefore \quad E.\,(4\pi r^2) = (1/\epsilon_0)\,Q$$

Or $$E = \frac{1}{4\pi\epsilon_0}\cdot\frac{Q}{r^2} \qquad \text{(spherical shell; } r \geq R) \qquad \text{..... (1.76)}$$

This is the same expression for E as obtained earlier for a point charge (Chapter 1A; Section 1A.11; Eq. 1.12). Thus, for the points outside it, the spherical shell behaves as if its entire charge were concentrated at the centre of the shell.

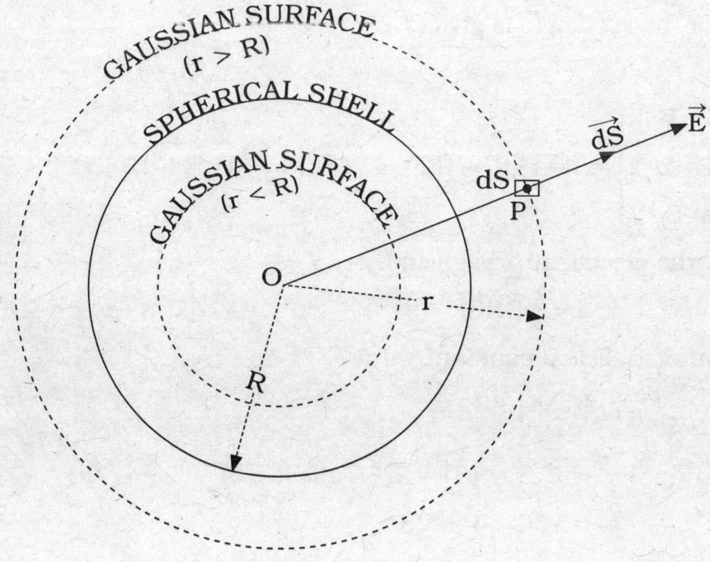

Fig. 1.32

(ii) Field at a point located upon the shell. The Eq. 1.76 applies to this case, too, with r = R. Since the charge $Q = 4\pi R^2.\sigma$, we have

$$E = \frac{1}{4\pi\epsilon_0} \cdot \frac{4\pi R^2.\sigma}{R^2}$$

Or

$$E = \frac{\sigma}{\epsilon_0} \quad \text{(Sph. shell; r = R)} \qquad \text{..... (1.77)}$$

(iii) Field at a point inside the shell. We draw our Gaussian surface inside the shell (r < R). Since all the charge of the shell is located on the surface of the shell, the charge Q enclosed by this Gaussian surface is zero. Hence, from Eq. 1.73,

$$E = 0 \quad \text{(Sph. shell; r < R)} \qquad \text{..... (1.78)}$$

Thus, no electric field exists inside a uniformly charged spherical shell. Hence, if a charged particle happens to be inside the shell, it experiences no electrostatic force due to the shell.

A plot of E vs. r is shown in Fig. 1.33

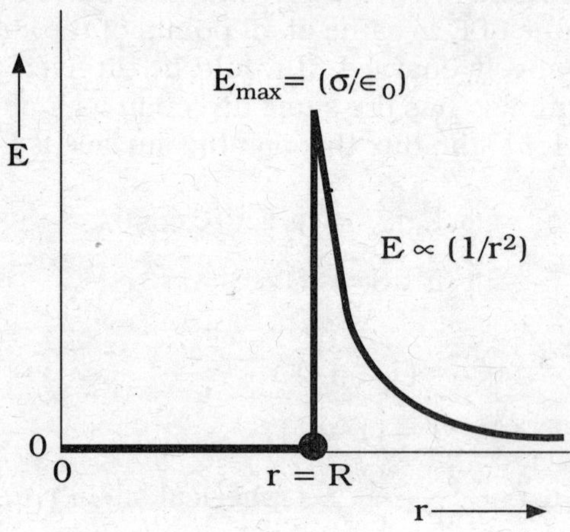

Fig. 1.33

Example 11. In a uniform electric field of value $(3\hat{i} - 2\hat{j} + 4\hat{k})$ N/C, a rectangular surface of side lengths $\ell_1 = (\hat{i} - 2\hat{j} + 2\hat{k})$ m and $\ell_2 = (2\hat{i} + 3\hat{j} + 2\hat{k})$ m is formed. Calculate the electric flux associated with the surface.

Solution. The area vector of the rectangle is given by,

$$S = \vec{\ell}_1 \times \vec{\ell}_2$$

$$= \begin{vmatrix} \hat{i} & \hat{j} & \hat{k} \\ 1 & -2 & 2 \\ 2 & 3 & 2 \end{vmatrix} = \hat{i}(-4-6) - \hat{j}(2-4) + \hat{k}(3+4) = (-10\hat{i} + 2\hat{j} + 7\hat{k})$$

From the Gauss's theorem, the electric flux is given by,

$$\phi = \int \vec{E}.\vec{dS}$$

Since the electric field is uniform, $\vec{E}$ is a constant value.

$$\therefore \quad \phi = \vec{E}\int_S dS$$

$$= \vec{E}.\vec{S} = (3\hat{i} - 2\hat{j} + 4\hat{k}).(-10\hat{i} + 2\hat{j} + 7\hat{k})$$

$$= -30 - 4 + 28$$

$$= -6 \text{ Nm}^2\text{C}^{-1}$$ **(Answer).**

Example 12. Find the total charge enclosed in a cube of side length 10 cm, which has a net outward electric flux of 4×10^3 Nm^2 C^{-1}.

Solution. From the Gauss's law,

$$\phi = \frac{\Sigma q}{\epsilon_0}$$

$$\Rightarrow \quad \Sigma q = \phi\, \epsilon_0 = (4\times10^3)\times(8.85\times10^{-12})$$

$$= 3.54\times10^{-8}\text{C}$$

Thus, the total charge enclosed within the cube is 3.54×10^{-8} C. **(Answer).**

1B. 18. Conductors and Insulators (or Non-conductors)

The conduction (i.e. transmission or flow) of electricity in materials takes place by the movement of charge carriers. For instance, in metals the carriers of electric charge are the negatively charged electrons. As explained earlier in Chapter 1A (Section 1A.02), a huge number of such **free electrons** in a metal are produced by their easy detachment from the outermost orbits of the atoms of the metal. When an electric field is applied (e.g. when a metallic wire is connected to a battery), the electrons rush towards the positive end of the field, thus making an electric current to flow through the metal. **The materials in which the charge carriers can move freely are called CONDUCTORS.** Some efficient conductors are gold, silver, copper and mercury. Earth is a good conductor and so is the human body.

In many other materials, the number of the said free electrons is negligibly small. So they conduct no or little electricity, i.e. they are bad (or poor) conductors of electricity. **The materials which cannot freely conduct electricity are called INSULATORS (or NON-CONDUCTORS).** Some common examples are wood, paper, glass, plastics, etc.

In a metal, when electrons from the outermost orbits are detached to become 'free', positively charged atoms (positive ions) are left behind. They do not move and are said to form **bound charges** in the conductor. The atoms in an insulator also constitute bound charges.

The electrons are not the only charge carriers. The motion of any charged particles, e.g. positive or negative ions, constitutes the flow of current. In the electrolytes (liquid conductors), the conduction of electricity is through the movement of such ions. We shall discuss more about them in the next chapter.

Intermediate between the conductors and insulators are the materials called **SEMICONDUCTORS.** On the order hand, there are materials which permit an absolutely free flow of current in them. They are called **SUPERCONDUCTORS** (the aforesaid 'conductors' offer some resistance to the flow of electrons in them, to be discussed in the next chapter). More about the semiconductors and the superconductors will be described in the next chapter.

1B. 19. Some Important Electrostatic Properties of Conductors

From our knowledge of electrostatics gained so far, a set of important properties of conductors follow, both in the absence and presence of an external electric field. For the sake of convenience, we shall discuss them with reference to metallic conductors.

(i) Net electric charge and electric field inside a conductor are zero

Two cases arise here, viz. (a) when no external electric field is present and (b) when an external field is present:

(a) When no external electric field is present: As already mentioned, in a metal the negatively charged free electrons are produced by detachment of outermost electrons from the atoms of the metal. The detachment renders the neutral atoms into positive ions. So, naturally, the total negative charge (on the free electrons) is exactly equal to the total positive charge (on the ions left behind) in the metal. Thus, the net electric charge is zero. Mathematically, we express it by writing $\rho = 0$, where ρ is the volume charge density, i.e. amount of charge per unit volume.

From the Gauss's theorem (Eq. 1.73) we have,

$$\oint \vec{E}.\vec{dS} = Q/\epsilon_0$$

Since the net charge Q enclosed by the surface of the conductor is zero, $\vec{E} = 0$. Thus, there is no electric field inside a conductor.

(b) When an external electric field is present: Suppose we have a metallic conductor KLMN placed between two plates PQ and RS, as shown in Fig. 1.34. As discussed above, initially there is no electric field present inside the conductor. Now let an electric field $\vec{E}$ be set up across PQ and RS (e.g. by connecting them with a battery) so that PQ is at a positive potential and RS at a negative potential. As soon as the field is set up, the free electrons in the conductor rush towards PQ. Hence the end MN of the conductor is depleted of the negative charge initially present there and so it acquires a positive charge. The free electrons move upto the end KL and make it negatively charged. Thus, finally an equal and opposite field $\vec{E}_2$ is set up inside the conductor (this happens almost instantaneously). The result is that the net electric field inside the conductor is zero ($|\vec{E}_2| = |-\vec{E}_1|$).

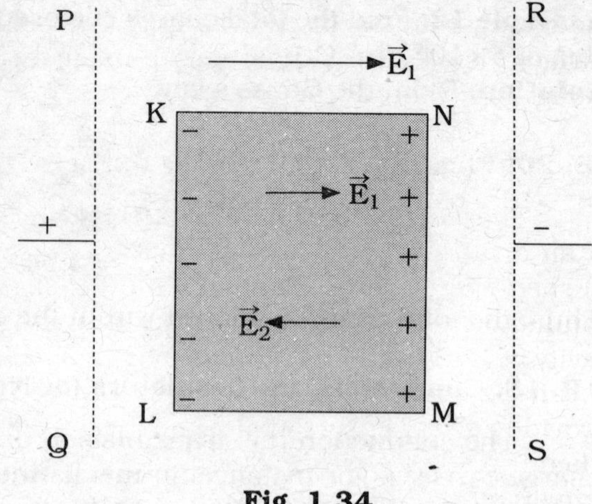

Fig. 1.34

Again, from the Gauss's theorem, if the net field $\vec{E}$ is zero, the net charge ρ inside the conductor is also zero.

The situation is pictorially depicted in Fig. 1.35, for a conductor of an arbitrary shape.

(ii) All the electric charge resides on the surface of the conductor

We have just seen that the electric charge inside a conductor is always zero. Hence if any charge is imparted to the conductor from outside, it can only reside on the surface of the conductor.

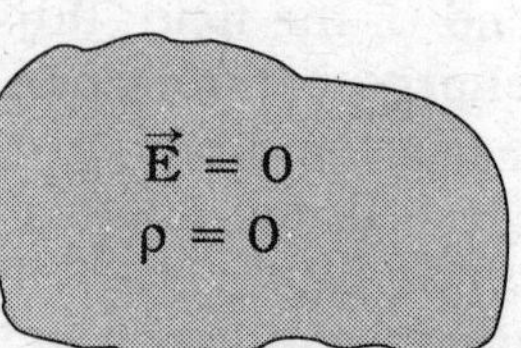

Fig. 1.35

(iii) Electric field at the surface of a charged conductor is normal to the surface at every point

Consider a charged conductor, i.e. a conductor to which electric charge has been imparted from outside. We know that all this charge resides on the surface of the conductor. Let $\vec{E}$ be the electric field due to the charge, just near to the surface. If $\vec{E}$ is not normal to the surface, it will have a finite tangential component along the surface. This tangential component will make the charges residing on the surface to move, which is against observation. Hence it follows that $\vec{E}$ is normal to the surface at every point. If the conductor is positively charged, $\vec{E}$ is directed away from the surface. If it is negatively charged, $\vec{E}$ is directed inwards to the surface. This is diagrammatically shown in Figs. 1.36 (a) and (b).

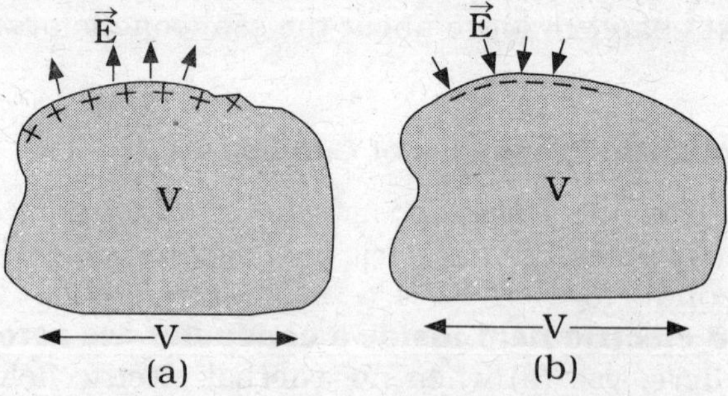

Fig. 1.36

(iv) Electric potential inside and at the surface of a conductor has the same value

We have known that the electric field E inside a conductor is zero. Therefore, from Eq. 1.60 we have,

$$E = -\frac{dV}{dr} = 0$$

Or

$$V = \text{constant}$$

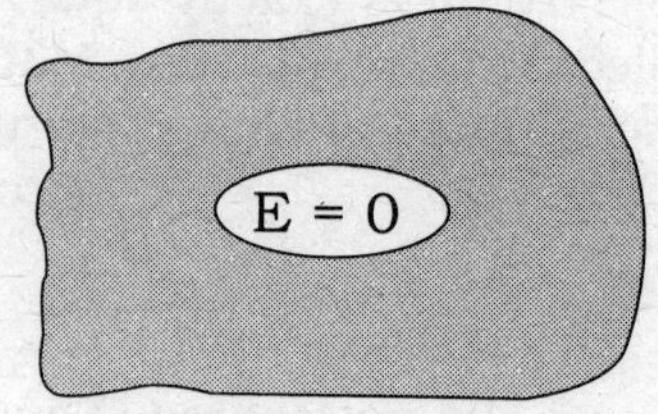

Fig. 1.37

Hence the electric potential inside (throughout the volume) and at the surface of the conductor is a constant value. Further, it follows that the surface of a conductor is an equipotential surface. The equality of potential inside and at the surface is shown for a conductor of an arbitrary shape in Fig. 1.36 (both Figs. (a) and (b)).

1B. 20. Electrostatic Shielding (or Screening)

The fact that the electric field inside a conductor is zero is remarkable. The Fig. 1.37 shows a conductor of an arbitrary shape with a cavity. The electric field inside the cavity has to be zero, irrespective of the (i) shape and size of the conductor, (ii) shape, size and location of the cavity, (iii) amount of charge on the conductor, and (iv) presence of any external field around the conductor. Hence any body or material inside the cavity is totally protected from all outside electrical influences. It is termed as electric shielding (or screening). It can be made use of in several practical applications. The coaxial cables used for telecommunication provide effective protection against external electrical (or electronic) disturbances. During a thunderstorm, when there is a great risk of being struck by lightning from the sky during an outing, it is absolutely safe to remain seated in a car (or bus) with metallic body.

1B. 21. Dielectrics (Polar and Non-polar); Dielectric Polarization; Dielectric Constant; Dielectric Strength

The dielectrics are essentially insulators, i.e. they are non-conductors of electricity, as they contain no or little amount of free charge carriers. Still they possess the very important property that they can transmit electrical effects through them (the word 'di' is derived from the Greek word 'dia', meaning 'through'). They do so by a suitable alignment of their constituent molecules under the influence of an applied electric field. This process is known as electric polarization, which we shall describe in some detail. There are two categories of dielectrics, polar and non-polar.

(i) Polar dielectrics

In an atom, the positive charge in the nucleus is symmetrically surrounded by the negatively charged electrons. So the centre of the +ve charge distribution coincides with the centre of the -ve charge distribution. This is the case with all monoatomic materials, e.g. sulphur, phosphorus, argon, etc. In polyatomic materials, which consist of molecules, the two centres in a molecule may or may not coincide (a molecule is made up of two or more atoms). If they do not coincide, i.e. the centre of the positive charge distribution is separated from the centre of the negative charge distribution, **each molecule possesses a permanent electric dipole moment even in the absence of an external electric field.** Such dielectric materials are known as **polar dielectrics.** Examples are water (H_2O), hydrochloric acid ($HC\ell$), alcohol, etc. (Fig. 1.38). Their molecules have **symmetric shape,** which gives rise to the said separation between the two centres.

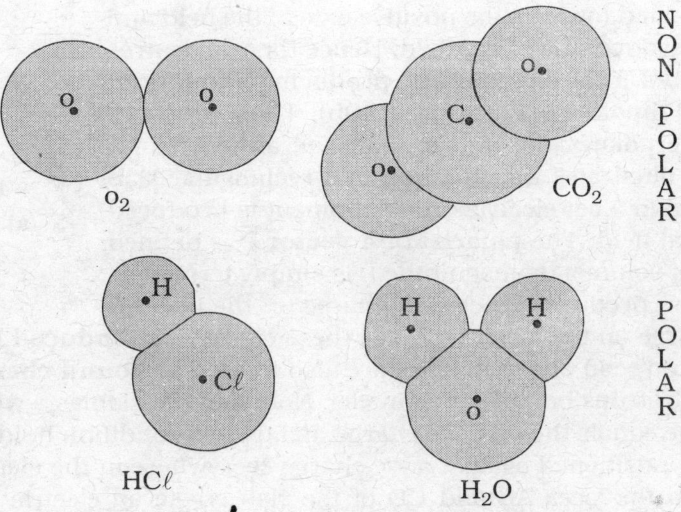

Fig. 1.38

(ii) Non-polar dielectrics

The molecules of the materials like oxygen (O_2), Carbon dioxide (CO_2), nitrogen (N_2), etc. have **symmetric shape.** The centre of positive charge distribution concides with the centre of negative charge distribution. Therefore, in the absence of an external electric field, the net dipole moment is zero. Such dielectric materials are known as **non-polar dielectrics.** Two examples are depicted in Fig. 1.38.

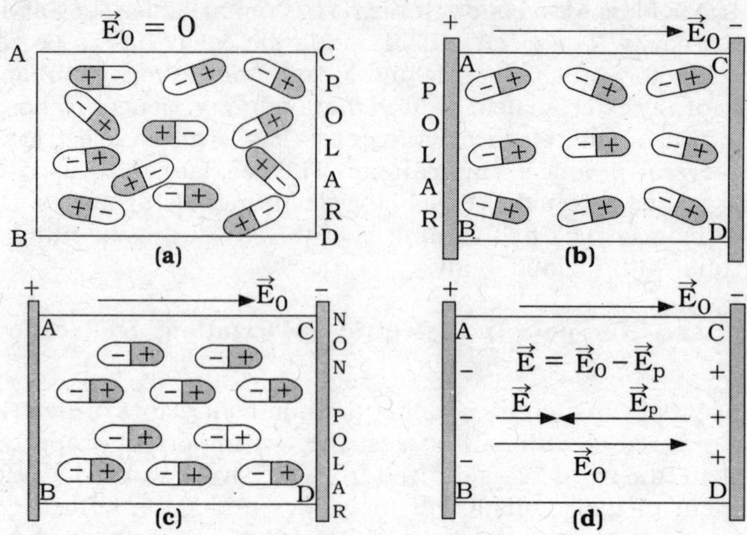

Fig. 1.39

Dielectric polarization

In the absence of an external electric field, the various dipoles in a polar material are randomly oriented, as shown in Fig. 1.39a, which depicts a rectangular slab ABCD of the material. On an average, as many dipoles are oriented in a particular direction as in the opposite direction. Hence, in spite of each dipole possessing a finite dipole moment, the net dipole moment is zero. When an external electric field $\vec{E}_0$ is applied, the field exercises a torque on the dipoles, which tends to align them parallel to the direction of the field. Although this tendency is opposed by the thermal agitation (which always tries to randomise the dipoles), the dipoles are able to align themselves wholly or partially along the field direction (Fig. 1.39b). The process is termed as **dielectric polarization** (or **electric polarization).** It naturally produces a net dipole moment in the material. A partially aligned dipole has a component of its dipole moment along the applied field.

In a non-polar material, in the absence of an external field the centres of +ve and -ve charges in a molecule coincide at a common point P (Fig. 1.40a). When a field $\vec{E}_0$ is applied, the -ve charge is attracted towards the positive side of the field and the +ve charge towards the negative side of the field. Hence the two centres P_1 and P_2 get slightly separated from one another, producing an induced dipole and hence an induced dipole moment (Fig. 1.40b). Thus, again we have the process of dielectric polarization in the material, although in a different way. The Fig. 1.39c illustrates the situation in a rectangular slab of the material. Thus, once again a net electric dipole moment is produced in the presence of an external field. The **polarization vector $\vec{P}$** is defined as the dipole moment per unit volume. Its magnitude P is simply termed as **polarization.** The polarization produces a negative charge on the face AB of the slab and a positive charge on the face CD. Since these charges are **induced** by the application of the electric field, they are called **induced charges.** They are also termed as **bound charges,** as they result from a slight displacement of the electrons bound to the nuclei. Note that the slab as a whole remains electrically neutral, as the total +ve charge equals the total -ve charge. In fact this condition holds in any volume element of the slab, with a +ve charge positioned against a -ve charge everywhere in the element.

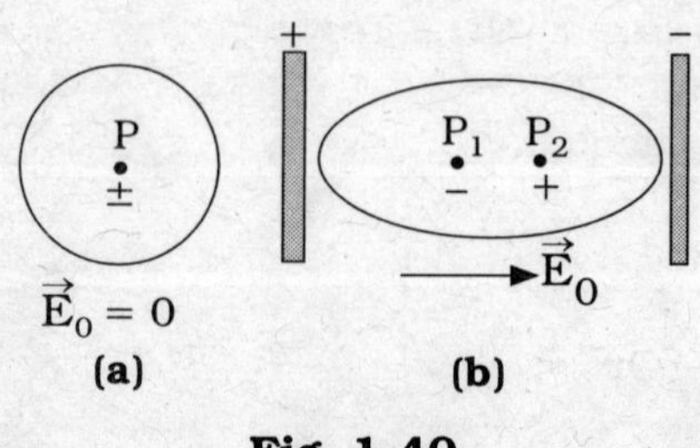

Fig. 1.40

The induced charges on the faces AB and CD of the slab create an electric field $\vec{E}_p$ in the direction opposite to that of the applied field $\vec{E}_0$ (Fig. 1.39d). Hence, inside the dielectric the resultant field $\vec{E}$ ($= \vec{E}_0 - \vec{E}_p$)

has he same direction as $\vec{E}_0$, but it has a smaller magnitude. The permanent dipoles produced in a polar material by the application of an external electric field $\vec{E}_0$ also behave in the same way as the induced dipoles in a non-polar material. Hence in a non-polar material, too, $\vec{E}\ (= \vec{E}_0 - \vec{E}_P) < \vec{E}_0$.Thus, for both the polar and non-polar dielectric materials, an applied electric field is weakened within the material. The opposing field $\vec{E}_P$ produced by polarization, is often referred to as **polarizing field.**

Dielectric constant

Since the magnitude E_0 of the applied electric field is reduced to a value $E\ (= E_0 - E_p)$ inside a dielectric, we can write,

$$E = E_0/K$$

Or
$$E_0/E = K \quad \text{..... (1.79)}$$

K, which is a constant value for a given dielectric, is called **dielectric constant.** It may be **defined** as "the ratio of the electric field applied to a dielectric to the reduced value of the field inside the dielectric". Note that K cannot be less than 1. The values of K for some common materials are given in Table 1. For vacuum, there is no polarization. Hence $E = E_0$ and $K = 1$. The value of K for air is also practically the same, viz. unity.

Table 1

Material	Dielectic Constant	Dielectric Strength (kV/mm)
Vacuum	1	∞
Air (1 atmosphere pressure)	1.00054	3
Paper	3.5	16
Water (20 °C)	80.4	-
Transformer oil	4.5	-
Ruby Mica	5.4	-
Polystyrene	2.6	-

Dielectric strength

If a very high electric field be applied to a dielectric, the outer electrons in the atoms of the dielectric may get detached from their positions and become free electrons. Then the dielectric will begin to conduct electricity, i.e. it will become a conductor. This phenomenon is known as **dielectric breakdown.** The minimum value of the applied electric field at which the breakdown occurs in a dielectric material is called the **dielectric strength** of the material. Accordingly, we may **define** the dielectric strength of a material as "the maximum value of the applied electric field that the dielectric material can withstand before its electric breakdown".

The values of dielectric strength of a few common materials are also given in Table 1.

1B. 22. Capacitor and Capacitance

(i) Capacitance of a conductor

If some electric charge Q is given to an isolated conductor, say an isolated metallic ball, it acquires an electric potential V. Higher the value of Q, the greater is the value of V. It is found that,

$$V \propto Q \qquad \text{or} \qquad Q \propto V$$

Or
$$\boxed{\begin{array}{c} Q = CV \\ \text{[conductor]} \end{array}} \quad \text{..... (1.80)}$$

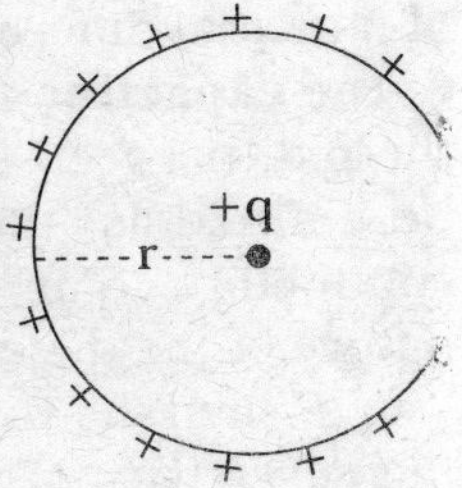

Fig. 1.41

The constant of proportionality C is called **capacitance** (earlier called capacity) of the conductor, because for a given value of V, the larger can be the charge stored on the conductor the larger is the value of C. The value of C depends on the shape and size of the conductor, but not on the nature of its material.

If V = 1, we have C = Q. Thus the **capacitance of a conductor** is **defined** as "the amount of charge required to raise the potential of the conductor by unity". Its SI unit is coulomb/volt, which is written as **farad (F).** One farad is a very large capacitance. Therefore, smaller units of one **microfarad (1 μF = 10^{-6} F)** or **one picofarad (1 pF or 1 μμF = 10^{-12} F)** are generally used.

We know that for any conductor, whether solid or hollow, the entire charge uniformly resides over its surface. Thus, for a spherical conductor (e.g. our earth), every point on its symmetrical outer surface is at the same potential. So the entire charge q can be taken as concentrated at the centre of the sphere (Fig. 1.41). Then, from Eq. 1.42 (Section 1B. 02), the potential at any point on the surface of the sphere is given by,

$$V = \frac{q}{(4\pi\epsilon_0)r}$$

r is the radius of the sphere. Substituting q = CV from Eq. 1.80, we get

$$C = 4\pi\epsilon_0 r \qquad \text{..... (1.81)}$$

Thus, we find that the capacitance of an isolated spherical conductor is directly proportional to the radius of the conductor. That is, the radius of such a conductor, or say the size of the conductor, is a direct measure of its capacitance. If C is in farad, r is in metre.

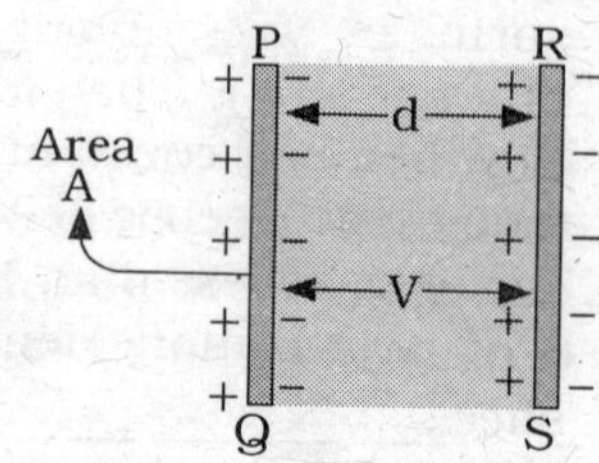

Fig. 1.42

Depending on its value of C, there is a limit to which a conductor can be charged. Since $V \propto Q$, the potential of the conductor rises as Q is increased. Beyond a limiting maximum value, the charge begins to leak out to the surroundings. The surrounding air is ionised, with -ve ions attracted towards a positively charged conductor, which partially neutralise the charge on the conductor.

(ii) Capacitance of a capacitor

The capacitance of a conductor is substantially increased when another conductor is brought near to it and the space between the two is filled with a suitable dielectric material. "A combination of two conductors placed close to each other with a dielectric between them" is called a **capacitor.** One of them is imparted a +ve charge and the other an equal -ve charge. The conductors can have any shape and size. However, for the sake of simplicity we consider a parallel-plate capacitor, which consists of two parallel conducting plates separated by a small distance d. The medium between the two plates is a dielectric material (which can be air, too). The Fig. 1.42 shows such a capacitor. Initially the two plates PQ and RS of the capacitor may be imagined as carrying the same charge on their respective two faces, i.e. +ve charge on both the faces of PQ and -ve charge on both the faces of RS (this can happen as soon as the two plates are connected to the terminals of a battery). Immediately afterwards, polarization will be set up in the dielectric, resulting into an induced -ve charge on the inner face of PQ and an induced +ve charge on the inner face of RS. The situation is similar to the one earlier depicted in Fig. 1.39d. The polarization of the molecules of the dielectric produces an opposite electric field $\vec{E}_p$, which reduces the original electric field $\vec{E}_0$ between the plates to a value $\vec{E} = (\vec{E}_0 - \vec{E}_p)$. The value of any electric field is directly related to the value of the potential difference in the field (vide Eq. 1.54). So a reduced value of the electric field means that the potential difference between the plates is reduced. [From Eq. 1.54, the p.d. $\Delta V = \int \vec{E}.d\ell = \vec{E}\int d\ell$, since $\vec{E}$ is very nearly uniform in our case. If d is the spacing between the plates, $\int d\ell = d$, and so $\Delta V = Ed$. Thus the p.d. is directly proportional to the magnitude of $\vec{E}$].

The potential difference V between the plates of a capacitor is found to be proportional to the charge Q of the capacitor (+Q on the positive plate and –Q on the negative plate),

$$V \propto Q \qquad \text{or} \qquad Q \propto V$$

Or

$$\boxed{\begin{matrix} Q = CV \\ \text{[Capacitor]} \end{matrix}} \qquad \text{..... (1.82)}$$

The Eq. 1.82 is the same as the Eq. 1.80 for the capacitance of a conductor, but now V represents the value of potential difference, not just potential. The constant of proportionality C now represents the **Capacitance of the capacitor.** If V = 1, C = Q. So as before, C can be **defined** as "the amount of charge required to raise the p.d. between the two conductors of the capacitor by unity". Alternatively, since C = (Q/V), it **can also be defined** as "the ratio of the charge Q on the two conductors of the capacitor to the p.d. V between the conductors". The value of C depends on the shape, size and spacing between its two conductors and on the nature of the dielectric material between the conductors.

When the value of V is reduced by the polarization of the dielectric between the plates, the capacitance increases (C = Q/V). Hence, for a given value of the potential difference V, more charge can be stored in a capacitor. If desired, C can be further increased by choosing a dielectric of higher dielectric constant and by a suitable adjustment of the shape, size and spacing between its two conductors. Thus, a capacitor can hold a large amount of charge even for a small p.d. between its conductors. This fact is of great practical importance.

Note that although Q is called the charge of the capacitor, actually it is the charge on any one of its two conductors. The total charge of the capacitor (+Q – Q) is zero.

Although the aforesaid treatment has been made for a parallel-plate capacitor, the considerations made therein generally apply to all types of capacitors. The symbol for a capacitor is ⊣ ⊢. As for a conductor, the SI unit of capacitance is **farad (coulomb volt^{-1})**, but the smaller units **μF (microfarad)** and **μμF or pF (picofarad)** are generally used.

Just as potential energy can be mechanically stored in a given object, e.g. in a ball by lifting it, in a spring by stretching it, etc., electrical energy can be stored in a capacitor. This is one of the important uses of a capacitor. The energy so stored can be released later at will, e.g. in the photoflash unit of a camera. Besides, capacitors of various sizes and shapes are extensively used in a multitude of electrical circuits, including fans, motors, ignition system of engines, radio, television, computers (which have microscopic size capacitors in their memory banks), etc. A capacitor in plane shape can be simply made by enclosing a piece of paper or mica between two thin metallic sheets (or foils) or in a cylindrical shape by rolling the sheets (or foils).

In early days, a capacitor was called condenser, in the sense that it was believed to condense 'electric fluid' (as electricity was believed to exist in early times) within it. The existing modern terms capacitor and capacitance (in place of 'condenser' and 'capacity') are in uniformity with the other similar electrical terms resistor (or resistance) and inductors (or inductance).

The p.d. V for a given capacitor should not exceed a specified maximum value, otherwise dielectric breakdown will take place, making the electricity to conduct through the capacitor (Section 1B. 21).

Additional knowledge

In Fig. 1.42, one face of the plate PQ (or RS) has +ve charge and the other face has -ve charge. Initially this may appear contradictory, because the electric charge on a conductor is supposed to distribute itself uniformly all over the surface of a symmetrically shaped conductor. However, from Figs. 1.39 b, c it is seen that while on one side the -ve charge on a molecule touches the inner face of PQ, on the other side it is bound with the +ve charge on the molecule. Therefore it cannot move from its position to any other point on the plate. In fact this molecule makes a chain – + – + – + ... with further neighbouring charges and so becomes further strongly bound at its position. Thus, the negative charge on the inner face of AB becomes 'locked up' at its position. Same is the case for the plate RS.

1B. 23. Relation between Capacitance and dielectric Constant

From Eq. 1.79, the dielectric constant is given by the ratio of the electric field in the absence of the dielectric to the field in the presence of the dielectric,

$$K = E_0/E = E_0/(E_0 - E_p) \qquad \text{..... [A]}$$

By definition, for vacuum (and very nearly for air) the dielectric constant is unity, i.e. $K_{air/vac.} = 1$. So, for air/ vacuum we have,

$$1 = E_0/(E_0 - E_p)$$

$$\Rightarrow \quad E_0 - E_p = E_0$$

$$\Rightarrow \quad E_p = 0$$

(Physically it means that no polarizing field is created when vacuum or air exists between the two conductors of a capacitor, i.e. there is no polarization in vacuum/air).

$$\therefore \; E_{air/vac} = E_0 \qquad \text{..... [B]}$$

From the relations C = Q/V and V = E.d, we have for air/vacuum,

$$C_{air/vac} = Q/V_{air/vac} = Q/(E_{air/vac}d) = Q(E_0.d) \qquad \text{(from Eq.B)}$$

For the dielectric, we have,

$$C_{diel} = Q/V_{diel} = Q(E_{diel}d) = Q/[(E_0 - E_p).d]$$

$$\therefore \quad \frac{C_{diel}}{C_{air/vac}} = \frac{Q/[(E_0 - E_p).d]}{Q/(E_0.d)} = \frac{E_0}{E_0 - E_p} = K \qquad \text{(from Eq. A)}$$

Or

$$C_{diel} = K\,C_{air/vac.} \qquad \text{..... (1.83)}$$

Thus, it is seen that the capacitance of a capacitor increases by the factor K when a dielectric material is inserted between the conductors of the capacitor. Although we have arrived at this result for a parallel-plate capacitor, it applies generally to all types of capacitors. (This important relation was first experimentally established around 1840 by Michael Faraday. In fact, the whole concept of capacitance was largely given by him. The unit of capacitance, farad (F), is named after him).

1B. 24. Parallel-Plate Capacitor

As mentioned earlier, a parallel-plate capacitor consists of two thin conducting plates slightly separated from each other. The charge on each plate is Q and the potential difference between them is V. The Fig. 1.42 shows such a capacitor. For clarity, the separation d has been shown quite large, but actually it is a very small value. Let the area of each plate be A and let σ be the charge density of each plate. We shall treat two cases: (i) when the medium between the plates is air ($K \approx 1$) or vacuum ($K = 1$) and (ii) when the medium is a dielectric of larger K value ($K > 1$), e.g. paper, mica, etc.

(i) Air (or vacuum) as medium between the plates (K = 1 or ≈ 1)

Since the separation d between the plates is very small, we can treat each plate as an infinite charged plane sheet. From Eq. 1.75 (Section 1B. 16), the electric field intensity due to such a sheet is $\sigma/2\epsilon_0$. Thus the field intensity due to the two plates in our case will be

$$E = \sigma/2\epsilon_0 + \sigma/2\epsilon_0 = \sigma/\epsilon_0 \qquad \text{..... (1.84a)}$$

The electric field is uniform all along the plates, except for a slight distortion at the edges (known as 'fringing of field'). Therefore, we can write,

$$V = Ed = (\sigma/\epsilon_0)d$$

The charge density (charge per unit area), $\sigma = Q/A$. Therefore we have,

$$V = (Q.d)/(A.\epsilon_0)$$

The capacitance C of the capacitor is then given by,

$$C = Q/V = \frac{Q}{(Q.d)/(A.\epsilon_0)}$$

Or

$$\boxed{C_{air/vac} = \frac{\epsilon_0 A}{d} \quad [K=1 \text{ or } \approx 1]} \qquad \text{..... (1.84b)}$$

Thus we find that C is directly proportional to A and inversely proportional to d. Hence, to achieve a large capacitance, the area of the plates should be as large as possible and their separation should be as small as possible.

(ii) Dielectric as medium between the plates (K>1)

In the last section, we have seen that the capacitance of a capacitor increases by the factor K (dielectric constant), when a dielectric material is inserted between the two conductors of the capacitor (Eq. 1.83),

$$C_{diel} = KC_{air/vac}$$

Accordingly, the capacitance of a parallel plate capacitor with a dielectric (K>1) as the medium between the plates is,

$$C_{diel} = K\frac{\epsilon_0 A}{d} \quad [K>1] \qquad \text{..... (1.85)}$$

Thus, apart from the dependence of C on A and d, the value of C further increases when a material of higher dielectric constant is inserted between the plates.

Example 13. The capacitance of a parallel-plate capacitor, with a separation of 0.4 cm between the plates and having vacuum between the plates, is 2 μF. The separation is reduced by half and a material of dielectric constant 2.8 is introduced between the plates. Find the new capacitance of the capacitor.

Solution. Capacitance, $C = \frac{A\epsilon_0\epsilon_r}{d}$ (ϵ_r: Dielectric constant)

With vacuum between the plates, $\epsilon_r = 1$ Therefore,

$$C_{vac} = \frac{A\epsilon_0}{d} = 2\ \mu F \qquad \text{..... (1)}$$

With the new separation $d' = d/2$ and with the dielectric material having $\epsilon_r = 2.8$, the new capacitance is,

$$C = \frac{A\epsilon_0\epsilon_r}{(d/2)} = \frac{A\epsilon_0}{d}(2\ \epsilon_r)$$

Substituting for $(A\epsilon_0)/d$ from Eq. 1, we have,

$$C = 2 \times (2\times2.8)$$
$$= 11.2\ \mu F$$ **(Answer).**

Example 14. What is the area of the plates of a parallel-plate capacitor with plate separation as 1 mm and capacitance as 15 μF, when vacuum exists between the plates ?

Solution. The capacitance of a parallel-plate capacitor is given by,

$$C = \frac{A\epsilon_0}{d}$$

$$\Rightarrow A = \frac{C.d}{\epsilon_0}$$

Given: $C = 15\ \mu F = 15 \times 10^{-6}$ F; $d = 1$ mm $= 10^{-3}$ m

$$\therefore \quad A = \frac{(15\times10^{-6})\times(10^{-3})}{(8.85\times10^{-12})}$$

$$= 1.70\times10^{3}\ m^2$$ **(Answer).**

1B. 25. Combination of Capacitors

For use in electric circuits, many times a number of capacitors need to be connected together. There are two common ways of connecting them: (i) in series with each other and (ii) in parallel with each other. In both cases, we can calculate the equivalent capacitance of the combination, as follows.

(i) Capacitors in series

The Fig. 1.43 shows three capacitors C_1, C_2, C_3, connected in series to a battery. The series connection means that the capacitors are connected together one after another. The connection to the battery produces and maintains a voltage difference V across the two ends of the series.

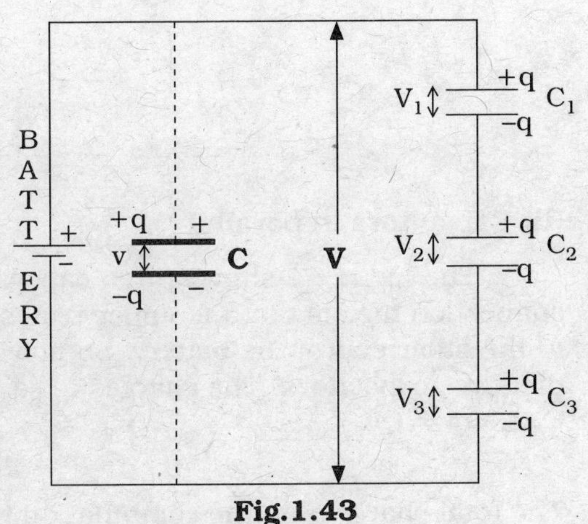

Fig. 1.43

As soon as the battery is connected, the negatively charged electrons flow towards the lower plate of C_3, producing a charge $-q$ on it. This charge repels the free electrons from the upper plate of C_3 leaving it positively charged with charge $+q$. The repelled electrons move to the lower plate of C_2, and impart a charge $-q$ to it. The chain process continues until the upper plate of C_1 gets a charge $+q$ on it, as seen in the figure. Finally, each of the three capacitors has an identical charge $+q$ on it. But, depending on the value of the capacitance, they will have different potential differences V_1, V_2, V_3 across them. From the relation $Q = CV$, we have,

$$V_1 = q/C_1, \quad V_2 = q/C_2, \quad V_3 = q/C_3$$

The sum of the three p.d. values V_1, V_2, V_3 should be equal to the total p.d. V due to the battery,

$$V = V_1 + V_2 + V_3$$

$$= q(1/C_1 + 1/C_2 + 1/C_3)$$

Let C be the capacitance equivalent to the series combination of C_1, C_2, C_3. If we imagine it to be connected to the battery, as shown in the figure by broken lines, the p.d. across it will be V and the charge on its plates will be q. So we shall have,

$$C = q/V = \frac{q}{q(1/C_1 + 1/C_2 + 1/C_3)}$$

$$\Rightarrow \quad 1/C = 1/C_1 + 1/C_2 + 1/C_3$$

For a combination of n capacitors in series, we have the general relation,

$$\frac{1}{C} = \sum_{i=1}^{n} \frac{1}{C_i} \qquad \text{..... (1.86)}$$

[n capacitors in series]

The student can work out that a series combination results in a reduced value of C. This value is less than any individual capacitance in the combination. However, the combination serves the same purpose as increasing the thickness of the dielectric. It offers greater dielectric strength and so it can withstand higher potential difference across its ends without breakdown.

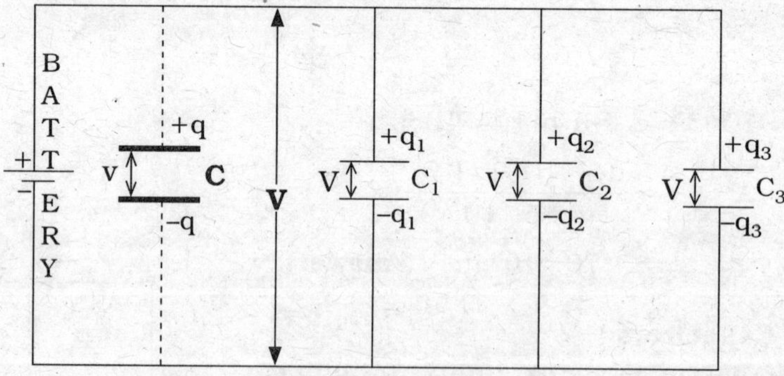

Fig. 1.44

(ii) Capacitors in parallel

The Fig. 1.44 shows three capacitors in a parallel combination connected to a battery. The parallel connection means that the upper plates of all the capacitors are connected to one end and the lower plates to the other end of the battery. So now the battery produces and maintain the same p.d. V across each of the three capacitors. The charges q_1, q_2, q_3 produced on the three capacitors depend on the values C_1, C_2, C_3 as ($Q = CV$),

$$q_1 = C_1V, \quad q_2 = C_2V, \quad q_3 = C_3V$$

The total charge q on the combination is,

$$q = q_1 + q_2 + q_3$$
$$= (C_1 + C_2 + C_3)V$$

As shown by the broken lines in the figure, if an equivalent capacitance C is imagined to be connected to the battery, it will have the same p.d. V across it as for C_1, C_2 or C_3 and the total charge as q on it. So we have,

$$C = \frac{(C_1 + C_2 + C_3)V}{V}$$
$$= C_1 + C_2 + C_3$$

Thus, for a parallel combination, the equivalent capacitance is simply found by adding up the individual capacitances in the combination. In general, for a parallel combination of n capacitances, we have,

$$C = \sum_{i=1}^{n} C_i \qquad \text{..... (1.87)}$$

[n capacitors in parallel]

It is clearly seen that a parallel combination results in an enhanced value of C. This value is higher than any individual capacitance in the combination.

Example 15. Three capacitors, $C_1 = 5\ \mu F$, $C_2 = 3\ \mu F$ and $C_3 = 2\ \mu F$, are connected as shown in the figure. Find the equivalent capacitance of the combination.

Solution. C_1 and C_2 are in series. Therefore their equivalent capacitance C_{12} is given by,

$$\frac{1}{C_{12}} = \frac{1}{C_1} + \frac{1}{C_2}$$

$$\Rightarrow \qquad C_{12} = \frac{C_1 C_2}{C_1 + C_2} = \frac{(5)\times(3)}{(5+3)} = \frac{15}{8}\ \mu F$$

Next, C_{12} and C_3 are in parallel. Hence the capacitance of the entire combination is given by,

$$C_{123} = C_{12} + C_3 = \left(\frac{15}{8} + 2\right) = \frac{31}{8}\ \mu F$$
$$= 3.88\ \mu F \quad \textbf{(Answer).}$$

Example 16. Find the equivalent capacitance of the given combination. The capacitance of each capacitor is 2 μF.

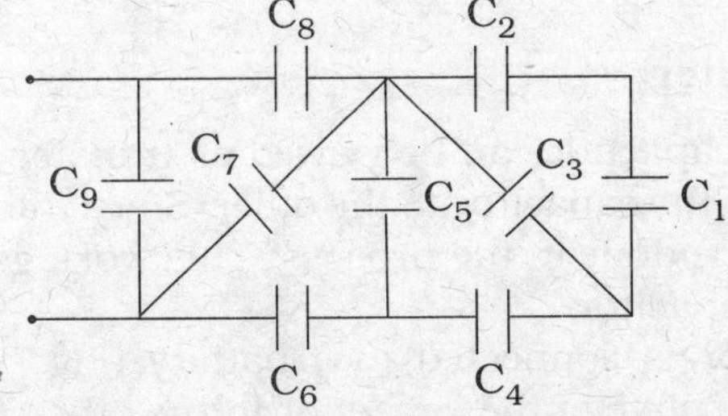

Solution. C_1 and C_2 are in series. Therefore, their equivalent capacitance is,

$$C_{12} = \frac{C_1 C_2}{C_1 + C_2}$$
$$= \frac{(2)\times(2)}{(2+2)} = 1\ \mu F$$

Next, $\qquad C_{12}$ and C_3 are in parallel.

$$\therefore \qquad C_{123} = C_{12} + C_3 = (1+2) = 3\ \mu F$$

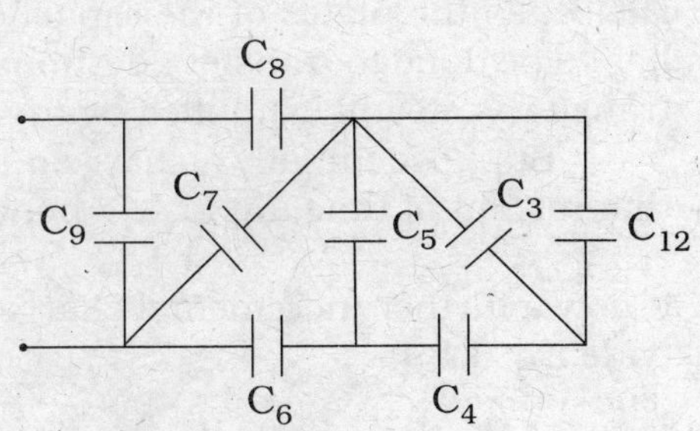

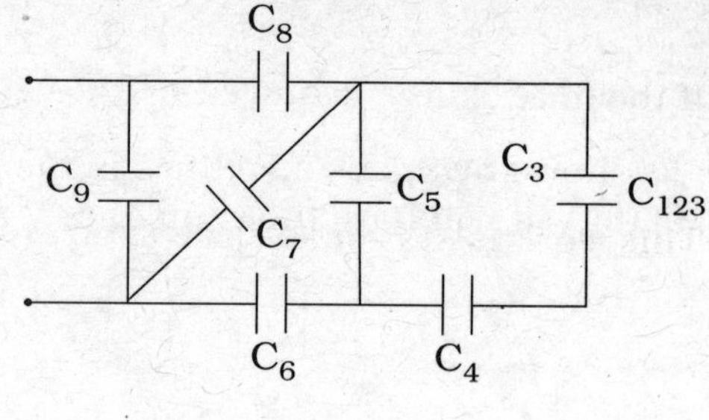

Next C_{123} & C_4 are in series.

$$\therefore \quad C_{1234} = \frac{C_{123} \times C_4}{C_{123} + C_4} = \frac{(3)\times(2)}{(3+2)} = \frac{6}{5}\ \mu F$$

Next, C_{1234} & C_5 are in parallel.

$$\therefore \quad C_{12345} = C_{1234} + C_5 = \left(\frac{6}{5}+1\right) = \frac{11}{5}\ \mu F$$

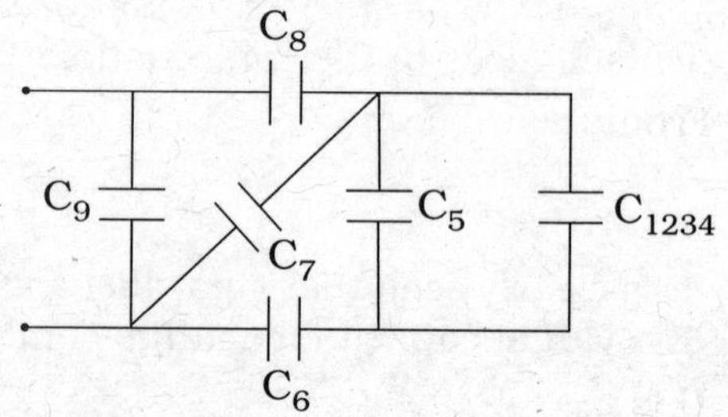

Next, C_{12345} & C_6 are in series.

$$\therefore \quad C_{123456} = \frac{C_{12345} \times C_6}{C_{12345} + C_6} = \frac{(11/5)\times(2)}{(11/5+2)} = \frac{22}{21}\ \mu F$$

Next, $C_{1\text{to }6}$ & C_7 are in parallel.

$$\therefore \quad C_{1\text{ to }7} = C_{1\text{to}6} + C_7 = \frac{22}{21} + 2 = \frac{64}{21}\ \mu F$$

Next, $C_{1\text{ to }7}$ and C_8 are in series.

$$\therefore \quad C_{1\text{ to }8} = \frac{C_{1\text{to}7} \times C_8}{C_{1\text{to}7} + C_8} = \frac{(64/21)\times(2)}{(64/21+2)} = \frac{128}{106}\ \mu F$$

Next, $C_{1\text{to }8}$ and C_9 are in parallel.

$$\therefore \quad C_{\text{eqiv.}} = C_{1\text{to }8} + C_9$$

$$= \left(\frac{128}{106} + 2\right)$$

$$= \frac{340}{106} = 3.21\ \mu F \quad \textbf{(Answer).}$$

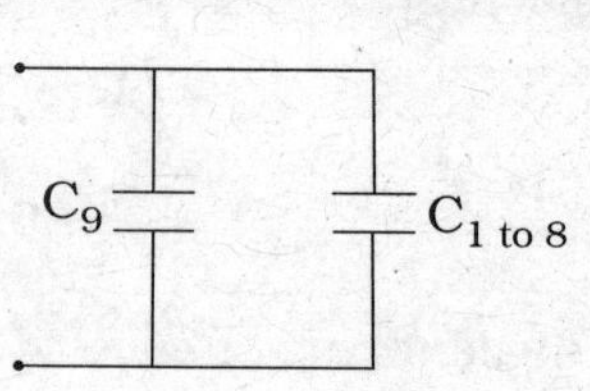

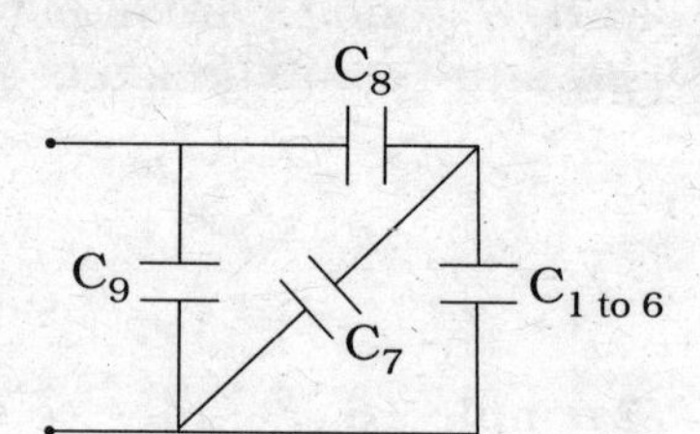

1B. 26. Energy Stored in a Capacitor

The process of charging of a capacitor can be viewed as transfer of charge (viz. free electrons) from one plate of the capacitor to the other. Since the electrons are material particles, work has to be done in the process. This work gets stored as electric potential energy in the capacitor.

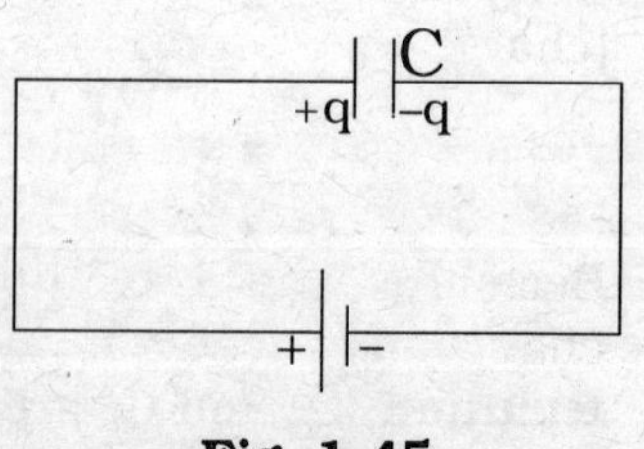

Fig. 1.45

When an uncharged capacitor is connected to a battery (Fig. 1.45), the electrons begin to transfer from one plate of the capacitor to the other. The plate losing the electrons begins to get positively charged and the one receiving the electrons begins to get negatively charged. Thus an electric field begins to build up between the plates such that it opposes any further transfer of the charge. Hence, as the plates of the capacitor gradually become more charged, increasingly larger amount of work has to be done to transfer additional electrons between the plates. (The necessary energy for the continuous transfer is steadily supplied by the battery, which has chemical energy stored in it).

Suppose initially a charge q is transferred from one plate of the capacitor to the other plate. Then, at that instant of time, the p.d. V between the plates is,

$$V = q/C$$

If now a further incremental charge dq is transferred, the incremental work dW done in this process will be (vide Eq. 1.39),

$$dW = V.dq = (q/C)dq$$

If the final charge acquired by the capacitor is Q, the total work done in the process will be,

$$W = \int_0^Q dW = \int_0^Q (q/C)dq = (1/C)\int_0^Q q.dq = Q^2/2C$$

This work is stored as electric potential energy U in capacitor,

$$U_{electric} = \frac{Q^2}{2C} \quad \text{[Electric potential energy]} \qquad ..... (1.88)$$

From the relation Q = CV, the Eq. 1.78 can also be written as,

$$U_{electric} = \frac{1}{2}CV^2 \quad \text{[Electric potential energy]} \qquad ..... (1.89)$$

U is essentially the energy of the electric field between the two plates of the capacitor. Thus, it resides in the dielectric medium between the plates. Its SI unit is joule (J), if the charge Q is in coulomb (C), the p.d. V is in volt (V) and the capacitance C is in farad (F).

Example 17. A parallel-plate capacitor has plate area as 80 cm^2 and plate separation as 2 mm. The capacitor is charged to a potential of 200 V. Calculate the electrostatic energy stored in the capacitor.

Solution. The energy stored,

$$U = \frac{1}{2}CV^2$$

$$= \frac{1}{2}\frac{A\epsilon_0}{d}V^2 \qquad \left(\because C = \frac{A\epsilon_0}{d}\right)$$

Given: $A = 80\ cm^2 = 8\times10^{-3}\ m^2$; $V = 200$ volt; $d = 2$ mm $= 2\times10^{-3}$ m

$$\therefore \quad U = \frac{1}{2}\times\frac{(8\times10^{-3})\times(8.85\times10^{-12})}{(2\times10^{-3})}(200)^2$$

$$= 7.08\times10^{-7}\ J \quad \textbf{(Answer).}$$

1B. 27. Corona Discharge (Spray of Charge at Sharp Ends)

If we have a spherical conductor of radius R, carrying a charge Q on it, its surface density of charge (charge per unit area) is (surface area of a sphere = $4\pi R^2$).

$$\sigma = \frac{Q}{4\pi R^2}$$

Because of the inverse square proportionality of σ with R, the charge density changes very rapidly with change in R. If the radius of curvature becomes nearly one tenth, which means that the curvature becomes ten times, the charge density becomes one hundred times!

Now consider a charged conductor PQRS of an arbitrary shape, shown in Fig. 1.46. It has different curvature (and therefore different R) at different positions on its surface. Accordingly, the charge density varies along its surface. It is highest at its narrow end P, where the curvature is very small. It can be easily imagined that if this end becomes pointed like the end of a needle, the charge density will become extremely high. Correspondingly, the electric field intensity becomes so high that the surrounding air gets ionized, which causes the charge on the conductor to jump into the air around the sharp end. Thus the charge on the conductor at the sharp end starts 'spraying' around the end. This phenomenon is called **Corona discharge.** Often this discharge is

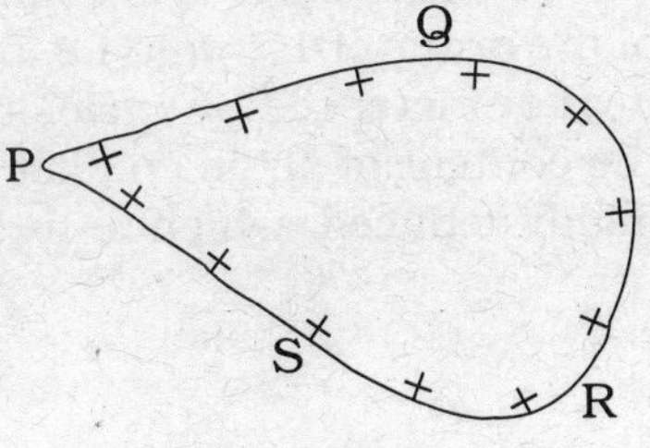

Fig. 1.46

accompanied with a visible glow. Then the end acquires a visible halo around it (the Latin word 'corona' means 'crown').

1B. 28. Van de Graaf Generator

The Van de Graaf generator is a high voltage electric machine, which can generate voltages of the order of ten million volts (10^7 V). It was designed by the American scientist Robert J. Van de Graaf around the year 1930. Such high voltages are mostly needed in nuclear physics experiments, conducted to study the ultimate structure of matter.

Principle. It is based on the phenomenon of corona discharge, i.e. spray of charge at the pointed ends of a conductor. The charge is finally received by a large spherical conductor, where it becomes uniformly distributed over the outer surface.

Construction. A large hollow metallic sphere is mounted on pillars made of an insulating material, e.g. glass. A pulley P_1 is mounted at the lower end of the pillars and another pulley P_2 at the centre of the sphere. P_1 is connected both to a motor and to the pulley P_2 through belts made of an insulating material, e.g. rubber (Fig. 1.47). The motor continuously drives the pulley P_1, which in turn drives the upper pulley P_2. Two comb-shaped conductors C_1 and C_2 are mounted near the pulleys. The teeth of the combs have sharp needle-shaped ends projecting towards the belt. The comb C_1 is maintained at a high positive potential of the order of 10^4 volts by connecting it to the positive terminal of a high voltage supply (the negative end of the supply is conveniently earthed). The comb C_2 is connected to the metal sphere.

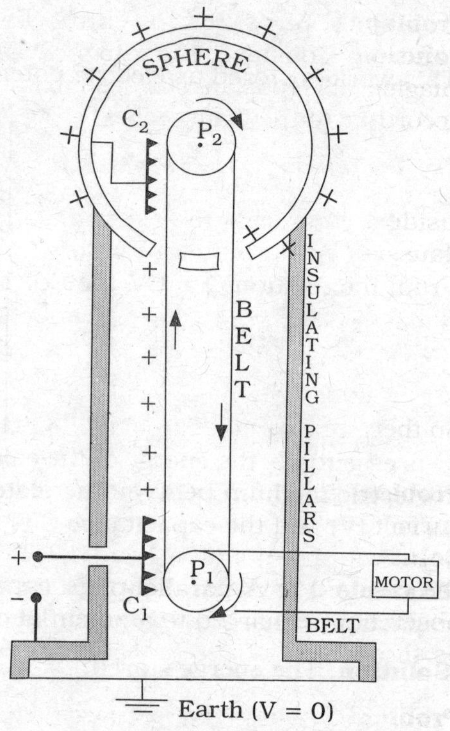

Fig.1.47

Working. The high electric field near the sharply pointed ends of C_1 causes corona discharge there. The air surrounding the ends becomes ionized. The negative charge (negatively charged ions) is attracted towards the teeth of the comb and neutralises some of its positive charge. However, the constant supply of charge from the power supply immediately compensates for this loss of charge and maintains the constant level of potential at C_1. The positive charge (the left out positively charged ions) scatters towards the belt and sticks to it. In this sense, the comb C_1 acts like a 'spray comb', spraying positive charge towards the belt. The continuous rotation of the belt results in accumulation of a large amount of positive charge on the belt. This huge charge again raises the electric field near the belt and causes corona discharge and ionization in the surrounding air. In the neighbourhood of the comb C_2, the negative ions move towards the positively charged belt and the positive ions move to the teeth of the comb. Hence the comb acquires a positive charge, which is instantly transferred to the metallic sphere and becomes uniformly distributed over the surface of the sphere.

As more and more charge is transferred to the sphere, the potential of the sphere keeps on increasing ($Q = CV$, where C is the capacitance of the sphere). A limit is reached when the potential becomes so high that the electric field due to it in the vicinity of the sphere causes dielectric breakdown of air. The air gets ionized. Then the charge on the sphere starts leaking to the surroundings and so no more charge accumulates upon it. The capacitance can be increased by increasing the radius of the sphere, which reduces the surface charge density, i.e. charge per unit area (the surface area of the sphere is $4\pi R^2$) and therefore more charge can be put on the sphere. It is estimated that a spherical conductor of radius 1 metre can be raised to a potential of the order of 10^6 volt, i.e. one million volt. The charge on the conductor, and therefore its potential, can be further increased by enclosing the conductor in a highly evacuated chamber. The evacuation greatly reduces the content of air in contact with the conductor. Consequently, the ionization of the atoms present in air is highly reduced, which in turn reduces the chance of leakage of charge from the conductor to air.

Solved Conceptual Problems

(Including HOTS Questions)

Problem 1. Show that the entire charge on a conductor resides on its surface.

Solution. Consider a conductor carrying a charge q, as shown in the figure. Imagine a Gaussian surface inside the conductor just near its boundary. According to the Gauss's law,

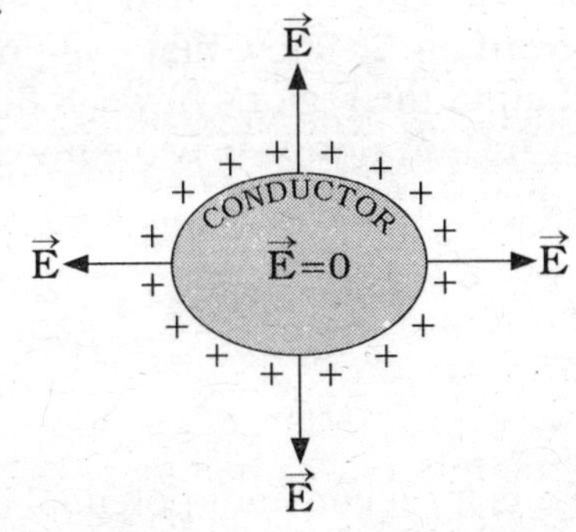

$$\phi = \oint \vec{E}.d\vec{S} = \frac{q}{\epsilon_0}$$

Inside a conductor the electric field $\vec{E} = 0$. Therefore the flux ϕ through the Gaussian surface must be zero,

$$\phi = \oint \vec{E}.d\vec{S} = 0$$

$$\Rightarrow \quad \frac{q}{\epsilon_0} = 0$$

$$\Rightarrow \quad q = 0$$

So there can be no charge inside the conductor. The entire charge must lie on its surface.

Problem 2. Show that the electric field at any point just outside the surface of a conductor is normal to its surface.

Solution. If it is not so, the electric field will have a tangential component on the surface, which will set the charged particles of conductor in motion, causing a surface current. But no such current is actually observed to exist. Therefore, the electric field must be normal to the surface at every point of the conductor, as shown in the figure in Problem 1.

Problem 3. Show that a charge +q suspended in the cavity of a hollow conductor induces a charge -q on the surface of the cavity.

Solution. Let us consider a hollow conductor with S_1 as its outer surface and S_2 as its inner surface, as shown in the figure. A positive charge +q is suspended inside the cavity. Imagine a Gaussian surface, shown by dash marks, inside the conductor. We know that the electric field inside the conductor is zero ($\vec{E} = 0$). Therefore, according to the Gauss's theorem,

$$\text{Flux, } \phi = \int \vec{E}.d\vec{S} = \frac{\text{Total charge enclosed}}{\epsilon_0} = 0$$

Or $\qquad$ Total charge enclosed = 0

Thus, the total charge enclosed by the Gaussian surface is zero. This can happen only if a charge -q is induced on the inner surface of the conductor, i.e. on the surface of the cavity.

If initially the conductor is neutral, the charge conservation will require that it will now carry a charge +q on its outer surface.

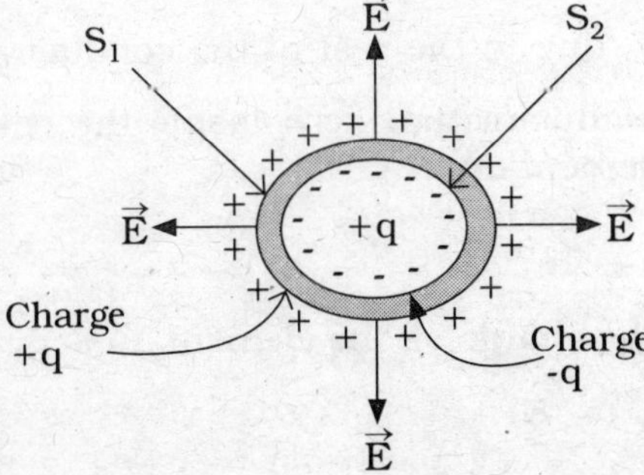

Problem 4. Explain electrostatic shielding. Give its two applications.

Solution. The electric field inside a cavity within a conductor is zero. So any region of space can be protected from an electric field by surrounding it with a conducting sheet. This phenomenon of protecting a region of space from any electric field, by surrounding it with a conducting sheet, is called electrostatic shielding.

Applications

1. During a lightning, it is safer to sit inside a car than under a tree. The metallic body of the car provides electrostatic shielding against the lightning.
2. The connecting wires of sensitive electrical instruments are covered with metal sheaths, which protect the instruments from any electrical disturbances.

Problem 5. In defining electric potential, why infinity is taken as a reference point ?
Solution. Infinity is taken as a reference point, because the potential of a charge is supposed to be zero at infinity.

Problem 6. Show that equipotential surfaces are more closely spaced in the regions of strong electric field than in the regions of weak field.
Solution. Suppose two equipotential surfaces, separated by a distance dr, differ by a potential dV from each other. We know,

$$E = \frac{dV}{dr}$$

$$\Rightarrow \qquad dr = \frac{dV}{E}$$

Thus for a constant potential difference dV between the two surfaces, we have,

$$dr \propto \frac{1}{E}$$

It follows that for large E, dr is small and vice versa. In other words, the equipotential surfaces in the regions of strong field are closer than in the regions of weak field.

Problem 7. Explain why the charge density along the surface of a conductor of arbitrary shape is not uniform.
Solution. For a spherical conductor of radius R, carrying a charge Q on it, the surface density of charge (i.e. charge per unit area) is (surface area of a sphere = $4\pi R^2$),

$$\sigma = \frac{Q}{4\pi R^2}$$

Because of the inverse square proportionality of σ with R, the surface charge density σ changes very rapidly with change in R. Smaller is the value of R, larger is the value of σ. Since the curvature of a surface is inversely proportional to the radius of curvature of the surface, it follows that larger is the curvature of the surface, larger is the value of σ [if the surface becomes pointed (i.e. nearly infinite curvature), σ tends to become a very high value]. A conductor of arbitrary shape has different curvatures at different positions along its surface. Hence the surface charge density σ varies along the surface.

Problem 8. Find an expression for the mechanical force per unit area on a charged conductor.
Solution. Let us consider a small area element ΔA on a conductor. Let E_1 be electric field due to ΔA and E_2 due to the rest of the conductor. The electric field due to the whole conductor is $E = \frac{\sigma}{\epsilon_0}$ outside the conductor and zero inside the conductor. Therefore the average electric field E_{av} surrounding the area element ΔA is,

$$E_{av} = \frac{(\sigma/\epsilon_0)+0}{2} = \frac{\sigma}{2\epsilon_0}$$

The charge on the element ΔA is σ.ΔA. Therefore the average force on it is (F = qE),

$$(\sigma\Delta A).(\sigma/2\epsilon_0) = \frac{\sigma^2\Delta A}{2\epsilon_0}$$

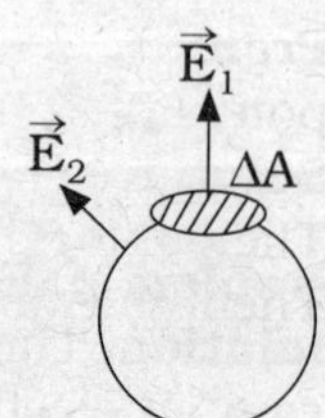

∴ The average force per unit area is,

$$\frac{(\sigma^2\Delta A)/2\epsilon_0}{\Delta A} = \frac{\sigma^2}{2\epsilon_0}$$

Since the electric field outside the conductor is $E = \sigma/\epsilon_0$ (∴ $\sigma = E\epsilon_0$), the force per unit area outside the conductor is,

$$\frac{\sigma^2}{2\epsilon_0} = \frac{(E\,\epsilon_0)^2}{2\epsilon_0} = \frac{1}{2}E^2\epsilon_0 \text{ (outside the conductor)}$$

Problem 9. Mathematically the capacitance of a capacitor is defined as the ratio of charge (Q) to potential difference (V), viz. C = Q/V. Does it mean that the capacitance arises only when the capacitor is charged ? Explain.
Solution. The equation C = Q/V is valid if the capacitor is connected to a source of e.m.f. But even if the capacitor is not connected to such a source, its capacitance value remains the same. For example, the capacitance of a parallel-plate capacitor is $A\epsilon_0/d$, where A is plate area and d is spacing between the plates. Thus, the capacitance basically depends on the dimensions of the capacitor and not on the charge or the potential difference.

Problem 10. A parallel-plate capacitor with vacuum between the plates is charged by a source of emf and then the source is removed. Then a dielectric is inserted between the plates. What will happen to the electrostatic energy stored in the capacitor ?
Solution. On inserting a dielectric between the plates, the capacitance value will increase, as $C = A\epsilon_0 K/d$ (K is dielectric constant; K>1). However, in the absence of the source of emf the charge on the capacitor will remain same. Hence the electrostatic potential energy, $U = \frac{1}{2}\frac{Q^2}{C}$, will decrease.

Problem 11. A parallel-plate capacitor is charged by a source of emf, after which the source is not removed. Then a dielectric is inserted between the plates. What will happen to the electrostatic energy stored in the capacitor ?
Solution. On inserting a dielectric between the plates, the capacitance value will increase, as $C = A\epsilon_0 K/d$ (K is dielectric constant; K>1). However, in the presence of the source of emf the voltage on the capacitor will remain the same. Hence the electrostatic potential energy, $U = \frac{1}{2}CV^2$, will increase.

Solved Numerical Problems

Electric Potential & Electric Field

Problem 1. Calculate the electric potential due to a point charge 1μC at a point 9 cm away from it.
Solution. Given: Charge, $q = 1\mu C = 10^{-6}$ C; Distance, $r = 9 \times 10^{-2}$ m

1μC —— 9 cm —— P

$$\therefore \quad \text{Potential,} \quad V = \frac{1}{4\pi\epsilon_0}\cdot\frac{q}{r}$$

$$= \frac{(9\times10^9)\times(10^{-6})}{(9\times10^{-2})} = 10^5 \text{ volt} \quad \textbf{(Answer).}$$

Problem 2. Two tiny spheres carrying charges 1 μC and – 2 μC are located 40 cm apart. Find the electric potential at the midpoint of the system of charges.
Solution. Given: $q_1 = 1\ \mu C = 10^{-6}$ C; $q_2 = -2\ \mu C = -2\times10^{-6}$ C; d = 40 cm
The point P lies midway, i.e. 20 cm from each charge, i.e. $r_1 = r_2 = 2\times10^{-2}$m.
Then, the electric potential at P is,

q_1 —— P —— q_2 (40 cm)

$$V_P = V_1 + V_2$$

$$= \frac{1}{4\pi\epsilon_0}\cdot\frac{q_1}{r_1} + \frac{1}{4\pi\epsilon_0}\cdot\frac{q_2}{r_2}$$

$$= \frac{(9\times10^9)\times(1\times10^{-6})}{(0.2)} - \frac{(9\times10^9)\times(2\times10^{-6})}{(0.2)}$$

$$= -4.5\times10^4 \text{ V} \quad \textbf{(Answer).}$$

Problem 3. Four point charges of values 1 μC, 2 μC, -3 μC and 4 μC, respectively, lie at the corners of a square of side length 2 m. Calculate the electric potential at the centre of the square.

Solution. The distance, r, of the centre O from each corner is,

$$r = \sqrt{(1)^2 + (1)^2} = \sqrt{2} \text{ m}$$

The electric potential, V_O, at O is contributed by all the charges at A, B, C & D, respectively,

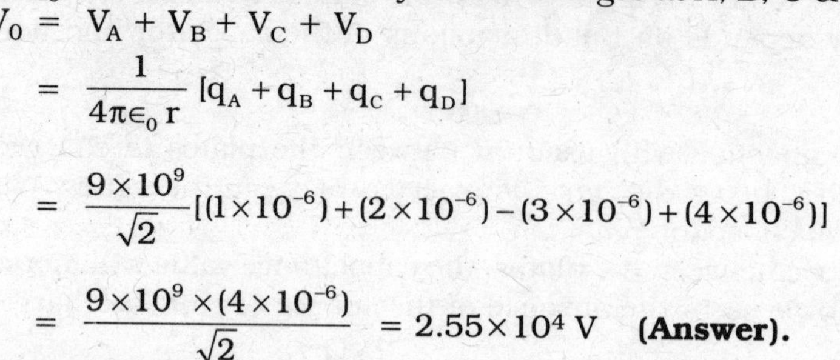

$$V_O = V_A + V_B + V_C + V_D$$

$$= \frac{1}{4\pi\epsilon_0 r}[q_A + q_B + q_C + q_D]$$

$$= \frac{9\times10^9}{\sqrt{2}}[(1\times10^{-6}) + (2\times10^{-6}) - (3\times10^{-6}) + (4\times10^{-6})]$$

$$= \frac{9\times10^9 \times (4\times10^{-6})}{\sqrt{2}} = 2.55\times10^4 \text{ V} \quad \textbf{(Answer).}$$

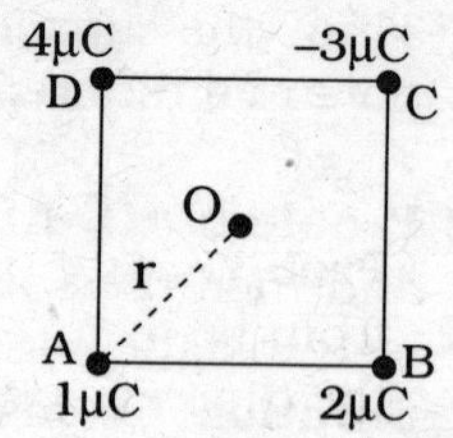

Problem 4. Point charges of value q each are located at the corners of a regular hexagon of side length d metre. Calculate the electric potential at

(i) the centre of the hexagon

(ii) one of the corners.

Solution. A charge q is present at each corner of the regular hexagon ABCDEF. The distance from each corner to the centre, OA = OB = OC = OD = OE = OF = d.

(i) The electric potential, V_O, at O is due to all the charges, i.e.,

$$V_O = V_A + V_B + V_C + V_D + V_E + V_F$$

$$= 6\times\left[\frac{1}{4\pi\epsilon_0}.\frac{q}{d}\right]$$

$$= \frac{3q}{2\pi\epsilon_0 d} \text{ volt} \quad \textbf{(Answer).}$$

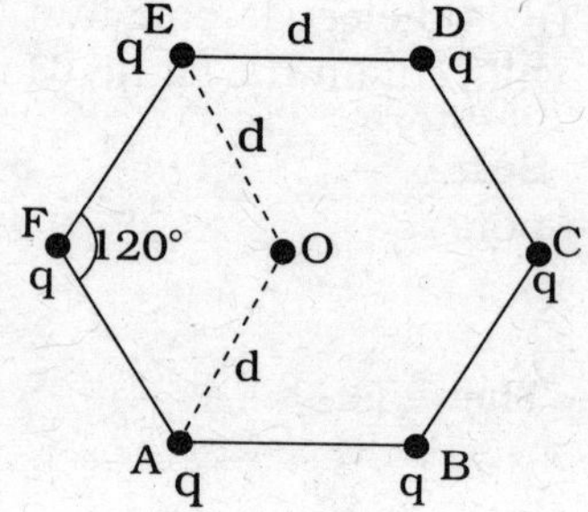

(ii) Let us calculate the electric potential, V_F at the corner F. All the corner charges, except the one at F, will contribute to it. If V_A, V_B, ... are the contributions due to the charges at A, B, ..., we have

$$V_F = V_A + V_B + V_C + V_D + V_E$$

Now, AF = EF = d and CF = 2d. Further,

$$BF = DF = \sqrt{(AF)^2 + (AB)^2 - 2(AF).(AB)\cos 120°} \quad \text{(from cosine law)}$$

$$= \sqrt{d^2 + d^2 - 2d.d.(-1/2)}$$

$$= \sqrt{3} \text{ d metre}$$

$$\therefore \quad V_F = \frac{q}{4\pi\epsilon_0}\left[\frac{1}{d} + \frac{1}{\sqrt{3}d} + \frac{1}{2d} + \frac{1}{\sqrt{3}d} + \frac{1}{d}\right]$$

$$= \frac{q}{4\pi\epsilon_0 d}\left[2 + \frac{1}{2} + \frac{2}{\sqrt{3}}\right] = \frac{3.65}{4\pi\epsilon_0}.\frac{q}{d}$$

$$= \frac{0.91}{\pi\epsilon_0}.\frac{q}{d} \text{ volt} \quad \textbf{(Answer).}$$

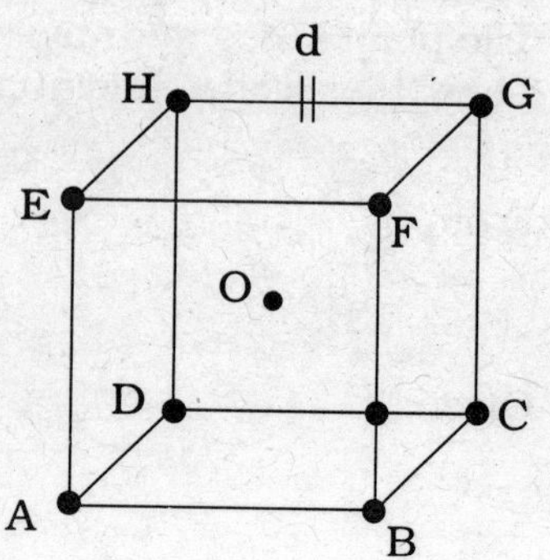

Problem 5. A cube of side length d metre has a charge q located at each of its corners. Determine the electric potential of the system at the centre O of the cube.

Solution. The length of the long diagonal AG is,

$$AG = \sqrt{d^2 + d^2 + d^2} = \sqrt{3}d$$

The distance of the centre O from each corner is,

$$\frac{1}{2}AG = \frac{\sqrt{3}}{2}d$$

All the eight charges at the corners of the cube contribute to the electric potential, V_0, at O.

$$\therefore \quad V_0 = 8\left[\frac{q}{4\pi\epsilon_0 \, .(\sqrt{3}/2)d}\right]$$

$$= \frac{4q}{\pi\epsilon_0 d\sqrt{3}} \text{ volt} \quad \textbf{(Answer).}$$

Problem 6. Calculate the electric potential due to a short dipole of moment 3×10^{-6} C–m at a point 2 m away from the centre of the dipole, such that the line joining the point and the centre makes an angle of 60° with the dipole axis.

Solution. The potential at a point due to a short dipole is,

$$V = \frac{1}{4\pi\epsilon_0}.\frac{p\cos\theta}{r^2}$$

$$= \frac{(9\times10^9)\times(3\times10^{-6})\times(\cos 60°)}{(2)^2}$$

$$= 3.38\times10^3 \text{ volt} \quad \textbf{(Answer).}$$

P
2m
D 60°
+q –q

Problem 7. Two point charges 2 μC and -4 μC are separated by a distance of 50 cm in air. Find the location of a point on the line joining the charges where the electric potential is zero.

Solution. Let the charges be located at A and B, as shown in the figure. Let the point at which the electric potential is zero be located at P, along the line joining A and B. Let PA = x metre.

$$\therefore \quad PB = (x + 0.5) \text{ metre.}$$

The potential, V_p, at P is contributed by both the charges, i.e.

P x A 50 cm B
2μC –4μC

$$V_P = V_A + V_B = 0$$

$$\Rightarrow \quad \frac{(2\times10^{-6})}{4\pi\epsilon_0 (x)} - \frac{(4\times10^{-6})}{4\pi\epsilon_0 (x+0.5)} = 0$$

$$\Rightarrow \quad \frac{1}{x} = \frac{2}{(x+0.5)}$$

$$\Rightarrow \quad x = 0.5 \text{ m}$$

Thus, the point P lies at a distance of 0.5 m from A. **(Answer).**

Problem 8. Two conducting spheres of radii 1cm and 2 cm, respectively, are carrying equal charges of 2 μC each. What will be the respective charges on them if they are connected to each other ?

Solution. When the two spheres are connected together, they will transfer charges between them, such that finally they are at an equal electric potential. Suppose finally the charges on them are q_1 and q_2, respectively. From the principle of charge conservation,

$$q_1 + q_2 = 2 \times (2\times10^{-6} \text{ C}) \quad [1\mu C = 10^{-6} C]$$

$$= 4\times10^{-6} \text{ C} \quad \text{..... (i)}$$

From the equality of electric potential,

$$\frac{q_1}{4\pi\epsilon_0 r_1} = \frac{q}{4\pi\epsilon_0 r_2} \quad \text{Or} \quad \frac{q_1}{r_1} = \frac{q_2}{r_2}$$

Given: $r_1 = 1\times10^{-2}$ m; $r_2 = 2\times10^{-2}$ m

$$\therefore \quad \frac{q_1}{1} = \frac{q_2}{2}$$

$$\Rightarrow \quad q_2 = 2q_1$$

Substituting in Eq. (i) we get,

$$3q_1 = 4\times10^{-6}\ C$$
$$\Rightarrow \quad q_1 = 1.33\times10^{-6}\ C \quad \textbf{(Answer).}$$
$$\therefore \quad q_2 = 2q_1 = 2.66\times10^{-6}\ C \quad \textbf{(Answer).}$$

Problem 9. N equally charged water droplets, each of radius r and charge q, coalesce to form a single drop of radius R. Calculate the electric potential at the drop.

Solution. When N small drops of radius r each coalesce to form a large single drop of radius R, the total volume of the N small drops is very nearly the same as that of the single large drop, i.e.,

$$N\cdot\frac{4}{3}\pi r^3 = \frac{4}{3}\pi R^3$$
$$\Rightarrow \quad R = N^{1/3}r \qquad \text{..... (1)}$$

From the principle of charge conservation, the total charge Q on the large drop is,

$$Q = Nq \qquad \text{..... (2)}$$

∴ The electric potential at the drop is,

$$V = \frac{Q}{4\pi\epsilon_0 R} = \frac{Nq}{4\pi\epsilon_0 (N^{1/3})r}$$
$$= \frac{(N)^{2/3}q}{4\pi\epsilon_0 r} \quad \textbf{(Answer).}$$

Problem 10. Two point charges of value 1μC and –2 μC, respectively, are separated by a distance of 45 cm in air. Calculate the electrostatic potential energy of the system.

Solution. Given: $q_1 = 1\times10^{-6}$ C; $q_2 = -2\times10^{-6}$ C; $r = 45$ cm $= 0.45$ m

Potential energy,
$$U = \frac{q_1q_2}{4\pi\epsilon_0 r}$$
$$= \frac{(9\times10^9)\times(1\times10^{-6})\times(-2\times10^{-6})}{(0.45)}$$
$$= -(4\times10^{-2})\ J \quad \textbf{(Answer).}$$

Problem 11. An electron is accelerated through a potential difference of 400 V. What is the final velocity of the electron ?

Solution. The kinetic energy acquired by the electron is equal to the difference in its electrostatic potential energy, i.e.

$$\frac{1}{2}mv^2 = q.\Delta V \quad \text{Or} \quad v = \sqrt{\frac{2q\Delta V}{m}}$$

We have, $\Delta V = 400$ V (Given); $q = e = 1.6\times10^{-19}$ C; $m = 9.1\times10^{-31}$ kg

$$\therefore \quad v = \sqrt{\frac{2\times(1.6\times10^{-19})\times(400)}{(9.1\times10^{-31})}} = \sqrt{\frac{12.8}{9.1}\times10^{14}}$$
$$= 1.19\times10^7 ms^{-1} \quad \textbf{(Answer).}$$

Problem 12. An alpha particle is projected towards an atom of gold (atomic number = 79), with an initial speed of 2×10^6 m/s. Find the distance of closest approach of the α-particle.

α Au r_0

Solution. Let the distance of closest approach be r_0. At this distance of the α-particle from the gold atom, its entire kinetic energy is converted into electrostatic potential energy. i.e.,

$$\frac{1}{2}mv^2 = U = \frac{1}{4\pi\epsilon_0}\cdot\frac{q_\alpha \cdot q_{Au}}{r_0}$$

$$\Rightarrow \qquad r_0 = \frac{2}{4\pi\epsilon_0} \cdot \frac{q_\alpha . q_{Au}}{mv^2}$$

Now, $q_\alpha = 2e = 2\ (1.6\times10^{-19})$ C; $\quad q_{Au} = 79(1.6\times10^{-19})$ C; $\quad v = 2\times10^6\ ms^{-1}$(Given); $m=9.1\times10^{-31}$ kg ; $1/4\pi\epsilon_0 = 9\times10^9$

$$\therefore \qquad r_0 = \frac{2\times(9\times10^9)\times(3.2\times10^{-19})\times(79\times1.6\times10^{-19})}{(9.1\times10^{-31})\times(2\times10^6)^2}$$

$= 2.0\times10^{-8}$ m **(Answer).**

Problem 13. Three point charges of value 2 μC each are placed at the corners of an equilateral triangle of side length 1 m. What is their mutual potential energy ?

Solution. The total potential energy of the system of charges is equal to the sum of the energies of each pair of charge, i.e.

$$U_{total} = U_{AB} + U_{AC} + U_{BC}$$

Now, $q_A = q_B = q_C = 2\times10^{-6}$ C; AB = AC = BC = 1 m

$$\therefore \qquad U_{total} = \frac{1}{4\pi\epsilon_0}\left[\frac{q_A q_B}{(AB)} + \frac{q_A q_C}{(AC)} + \frac{q_B q_C}{(BC)}\right]$$

$$= \frac{3}{4\pi\epsilon_0}\left[\frac{(2\times10^{-6})\times(2\times10^{-6})}{1}\right] J$$

$= 3 \times (9\times10^9)\times(4\times10^{-12})$ J $= 1.08\times10^{-1}$ J **(Answer).**

2μC C

A 2μC   1m   B 2μC

Problem 14. The work done in rotating a dipole from an angle $\theta_1 = 30°$ to $\theta_2 = 60°$ in a uniform electric field of 10^3 N/C is 4×10^{-2} J. Calculate the dipole moment.

Solution. The work done in rotating a dipole is given by,

$$W = pE\ (\cos\theta_1 - \cos\theta_2)$$

Given: $\theta_1 = 30°$; $\theta_2 = 60°$; $E = 10^3$ N/C; $W = 4\times10^{-2}$ J.

$$\therefore \qquad p = \frac{W}{E(\cos\theta_1 - \cos\theta_2)} = \frac{(4\times10^{-2})}{10^3\left(\frac{\sqrt{3}}{2} - \frac{1}{2}\right)} = \frac{8\times10^{-5}}{(\sqrt{3}-1)}$$

$= 1.09\times10^{-4}$ C-m **(Answer).**

Problem 15. An electric dipole of length 40 cm is placed with its axis making an angle of 60° to a uniform electric field of strength 10^4 N/C. If its potential energy is -10^{-3} J, calculate the magnitude of charge on the dipole.

Solution. $U = -pE\cos\theta$

$= -q\ (2a).\ E\cos\theta$

$$\therefore \qquad q = -\frac{U}{(2a)E\cos\theta}$$

Given: $\quad U = -10^{-3}$J; $\quad 2a = 0.4$m; $\quad E = 10^4$N/C; $\quad \theta = 60°$

$$\therefore \qquad q = \frac{(10^{-3})}{(0.4)\times(10^4)\times(\cos 60°)} = \frac{10^{-3}}{0.4\times10^4\times(1/2)}$$

$= 5\times10^{-7}$ C **(Answer).**

Problem 16. Four charges of value 1 μC each are to be placed at the corners of a square of side length 1m. Calculate the

(i) total work required to put together this arrangement and

(ii) the work done in placing a charge of –1 μC at the centre of the square.

Solution. (i) The total work W required to put together this arrangement is equal to the sum of potential energies of each pair of charges of the system,

$$W = \frac{1}{4\pi\epsilon_0}\left(\frac{q_A q_B}{(AB)} + \frac{q_A q_C}{(AC)} + \frac{q_A q_D}{(AD)} + \frac{q_B q_C}{(BC)} + \frac{q_B q_D}{(BD)} + \frac{q_C q_D}{(CD)}\right)$$

Now, $AB = AD = BC = CD = 1$ m; $AC = BD = \sqrt{2}$ m;
$q_A = q_B = q_C = q_D = 10^{-6}$ C.

$$\therefore \quad W = (9\times10^9)\left(\frac{10^{-12}}{1} + \frac{10^{-12}}{\sqrt{2}} + \frac{10^{-12}}{1} + \frac{10^{-12}}{1} + \frac{10^{-12}}{\sqrt{2}} + \frac{10^{-12}}{1}\right)$$

$$= 9\times10^{-3}\left[4+\sqrt{2}\right]$$

$= 4.87\times10^{-2}$ J **(Answer).**

(ii) The potential at the centre O of the square is

$$V = V_A + V_B + V_C + V_D$$

$$= 4(V_A) = 4\left[\frac{(10^{-6})}{4\pi\,\epsilon_0\,.(AO)}\right] \quad \text{(By symmetry, } V_A = V_B = V_C = V_D\text{)}$$

$$AO = (1/2)AC = (1/2)\sqrt{(1)^2(1)^2} = 1/\sqrt{2}$$

$$\therefore \quad V = \frac{4\times(9\times(10^9)\times10^{-6}}{\left(\frac{1}{\sqrt{2}}\right)} = 5.09\times10^4 \text{ volt}$$

The work done, $W = qV$

$$= (5.09\times10^4)\times(-1\times10^{-6})$$

$= -(5.09\times10^{-2})$ J **(Answer).**

Electric Flux & Gauss's Theorem

Problem 17. Calculate the electric flux over the surface of a sphere of radius 2 m, enclosing an α-particle.

Solution. From the Gauss's theorem, the electric flux, $\phi = \frac{Q}{\epsilon_0}$

Enclosed charge, $Q = 2e = 2(1.6\times10^{-19}\text{ C}) = 3.2\times10^{-19}$ C

$$\therefore \quad \phi = \frac{(3.2\times10^{-19})}{(8.85\times10^{-12})}$$

$= 3.62\times10^{-8}\ \text{Nm}^2\text{C}^{-1}$ **(Answer).**

Problem 18. The electric flux associated with a spherical surface of radius 1m is $\pi\times10^{-3}\ \text{Nm}^2\text{C}^{-1}$. The surface surrounds a point charge lying at its centre. Find the electric field intensity on the surface.

Solution. The electric flux through a Gaussian surface is given by,

$$\phi = \oint \vec{E}.d\vec{S} = \oint E.dS\cos\theta$$

For a spherical surface, E is constant over the surface and $\theta=0$ ($\because$ $\vec{E}$ has the same direction as $\vec{S}$ at all points on the surface).

$$\therefore \quad \cos\theta = 1.$$

$$\therefore \quad \phi = \oint E.dS = E\oint dS = E.4\pi r^2$$

$$\Rightarrow \quad E = \frac{\phi}{4\pi r^2} = \frac{(\pi\times10^{-3})}{4\pi\times(1)^2} = 2.5\times10^{-4}\ \text{NC}^{-1}$$ **(Answer).**

Problem 19. A thin ring of radius R is uniformly charged to a linear charge density λ. A spherical surface of radius R is constructed with its centre at a point on the periphery of the ring. The sphere encloses one-third of the periphery. Obtain an expression for the electric flux through the spherical surface.

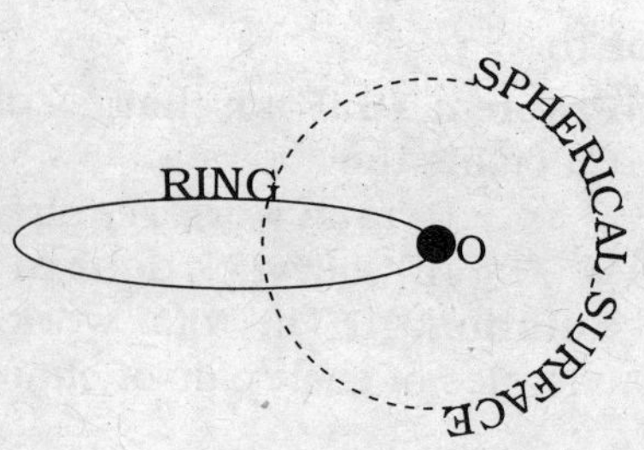

Solution. 1/3rd periphery of the ring is inside the spherical surface. Therefore the net charge inside the sphere is (total periphery=$2\pi R$),

$$Q = \lambda\left(\frac{1}{3}\times 2\pi R\right)$$

$$= \frac{2}{3}\lambda\pi R$$

From the Gauss's law, the electric flux through the surface is given by,

$$\phi = \frac{Q}{\epsilon_0} = \frac{2}{3}\cdot\frac{\lambda\pi R}{\epsilon_0} \quad \textbf{(Answer).}$$

Problem 20. A uniformly charged rod is lying along the x-axis, with one end of the rod lying at the origin. Multiple spherical surfaces are constructed with the origin as the centre. Draw a graph between the electric flux through the surfaces and the radii of the surfaces.

Solution. As the radius of the spherical surface increases, the length of the part of the rod inside the sphere and therefore the charge inside the sphere keeps on increasing. The flux is given by

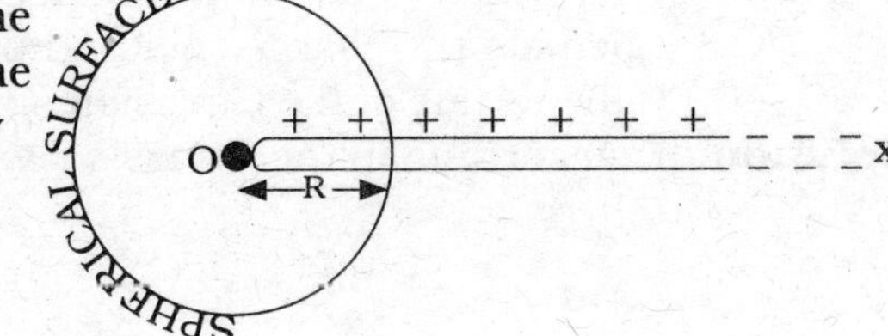

$$\phi = \frac{Q}{\epsilon_0}$$

Since ϵ_0 is a constant value, we have,

$$\phi \propto Q$$

As the radius R and therefore the charge Q increases, the flux linearly increases, as shown in the figure.

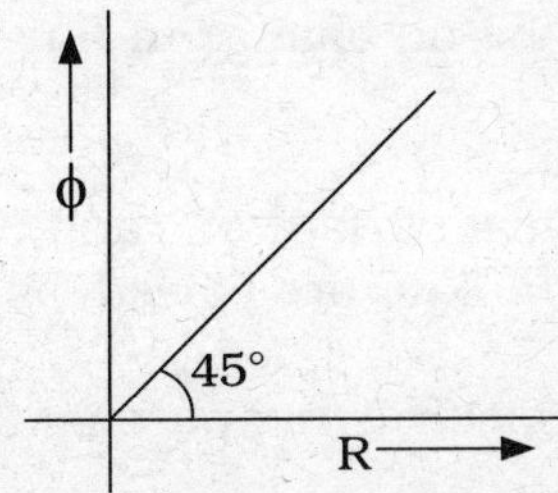

Problem 21. An electric field $\vec{E} = 2x^{1/2}\ \hat{i}\ NC^{-1}$ exists in a region of space. A cube of side length 10 cm is constructed in the region along the x-axis at a distance of 10 cm from the origin, as shown in the figure. Calculate the electric flux through the cube and the charge enclosed by the cube.

Solution. The electric field on the face ABCD is, $\vec{E}_1 = 2(0.1)^{1/2}\ \hat{i}\ NC^{-1}$ and the field on the face EFGH is, $\vec{E}_2 = 2(0.2)^{1/2}\hat{i}\ NC^{-1}$ (10cm=0.1m).

The surface vector corresponding to ABCD points towards the -ve x-axis, whereas for EFGH it points towards the +ve x-axis,

$$\vec{S}_1 = -[(0.1\times 0.1)\hat{i}] = -[(0.01)\hat{i}]m^2;\ \vec{S}_2 = +[(0.01)\hat{i}]m^2$$

Therefore the total flux through the two surfaces is,

$$\phi = \vec{E}_1.\vec{S}_1 + \vec{E}_2.\vec{S}_2 = E_1S_1\cos 180^\circ + E_2S_2\cos 0^\circ$$

$$= -E_1S_1 + E_2S_2$$

$$= -2(0.1)^{1/2}(0.01) + 2\times(0.02)^{1/2}\times(0.01)$$

$$= 0.02[(0.2)^{1/2} - (0.1)^{1/2}]$$

$$= 0.02[0.4472 - 0.3162] = 2.62\times 10^{-3}\ Nm^2C^{-1} \quad \textbf{(Answer).}$$

For the other four surfaces of the cube, $\vec{E}$ and surface vectors are perpendicular to each other. Hence $\vec{E}.\vec{S}$ =ES cos 90° = 0 for them. Therefore these surfaces do not contribute towards electric flux.

From the Gauss's law,

$$\phi = \frac{Q}{\epsilon_0}$$

$$\Rightarrow \quad Q = \phi.\epsilon_0 \quad \text{(charge enclosed by the surface)}$$

$$= (2.62\times 10^{-3})\times(8.85\times 10^{-12})$$

$$= 2.32\times 10^{-14}\ C. \quad \textbf{(Answer).}$$

Problem 22. In a region of space the electric field is given by $\vec{E} = (4\hat{i} - 3\hat{j} + 5\hat{k})\ NC^{-1}$. Calculate the electric flux associated with a surface of area $2m^2$ in the yz-plane.

Solution. The yz-plane is perpendicular to the x-axis. Therefore the surface vector for the given surface (of area $2\ m^2$) is given by,

$$\vec{S} = 2\hat{i}$$

Therefore the flux associated with the surface is,

$$\begin{aligned}\phi &= \vec{E}.\vec{S}\\ &= (4\hat{i} - 3\hat{j} + 5\hat{k}).(2\hat{i}) = (4\hat{i}.2\hat{i}) + 0 + 0\\ &= 8\ Nm^2C^{-1}\ \textbf{(Answer).}\end{aligned}$$

Problem 23. By placing a charge at the centre of a spherical Gaussian surface of radius 10 cm, an electric flux of $-3.2\times10^2\ Nm^2C^{-1}$ enters the surface.

(i) Calculate the charge enclosed by the Gaussian surface.

(ii) If the radius of the Gaussian surface is doubled, how much flux would pass through the surface ?

Solution. (i) According to the Gauss's law,

$$\phi = \frac{Q}{\epsilon_0}$$

$$\begin{aligned}\therefore\quad Q &= \phi\epsilon_0\\ &= (-3.2\times10^2) \times (8.85\times10^{-12})\\ &= 2.83\times10^{-9}\ C\ \textbf{(Answer).}\end{aligned}$$

(ii) The electric flux passing through a Gaussian surface of any size only depends on the charge enclosed by the surface. Therefore there will be no change in flux upon doubling the radius of the surface. **(Answer).**

Problem 24. We have two concentric spherical surfaces S_1 and S_2 of radii 10 cm and 40 cm, respectively. Three charges $q_1 = 1\ \mu C$, $q_2 = -3\ \mu C$ and $q_3 = 4\ \mu C$ are placed, as shown in the figure. Calculate the electric flux through S_1 and S_2.

Solution. According to the Gauss's law, the electric flux passing through a closed surface is,

$$\phi = \frac{Q}{\epsilon_0}$$

Q is the total charge inside the closed surface. Therefore the flux through S_1 is,

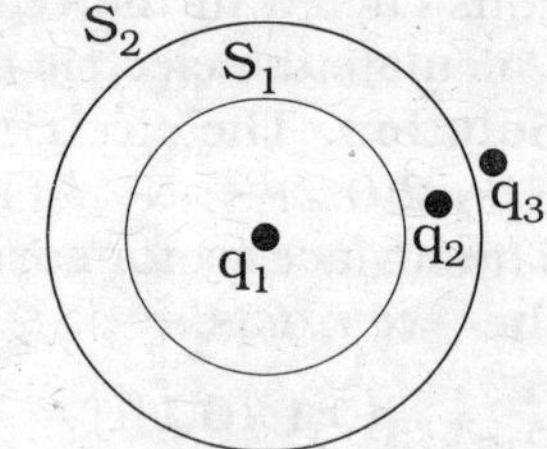

$$\begin{aligned}\phi_1 &= \frac{q_1}{\epsilon_0}\\ &= \frac{(1\times10^{-6})}{(8.85\times10^{-12})} = 1.13\times10^5\ Nm^2C^{-1}\ \textbf{(Answer).}\end{aligned}$$

For the surface S_2, the charge enclosed is $(q_1 + q_2)$. Therefore,

$$\begin{aligned}\phi_1 &= \frac{q_1+q_2}{\epsilon_0} = \frac{(-3\times10^{-6})+(1\times10^{-6})}{(8.85\times10^{-12})}\\ &= -\frac{2\times10^{-6}}{8.85\times10^{-12}} = -2.26\times10^5\ Nm^2C^{-1}\ \textbf{(Answer).}\end{aligned}$$

The charge q_3 lies outside both S_1 and S_2. Therefore, it does not contribute to the electric flux at all.

Problem 25. A uniformly charged sphere of radius 2 m has a volume charge density, $\rho = 1\ \mu C/m^3$. It is made of a material of dielectric constant, $\epsilon_r = 4$. Find the electric flux associated with concentric spherical surfaces of radius (i) 1 m and (ii) 3 m.

Solution. As shown in the figure, S_2 is the surface of the given sphere. S_1 and S_3 are the other given concentric spherical surfaces. The radii of S_1, S_2 and S_3 are 1 m, 2 m and 3 m, respectively. Let the volumes enclosed by the surfaces S_1, S_2, S_3 be V_1, V_2, V_3, respectively.

For a medium other than vacuum (or air), by Gauss's law the electric flux associated with a surface is given by,

$$\phi = \frac{Q}{\epsilon_0\epsilon_r}$$

(i) For the surface S_1, the charge Q_1 enclosed by the surface is [volume = $(4/3)\ \pi R^3{}_1$],

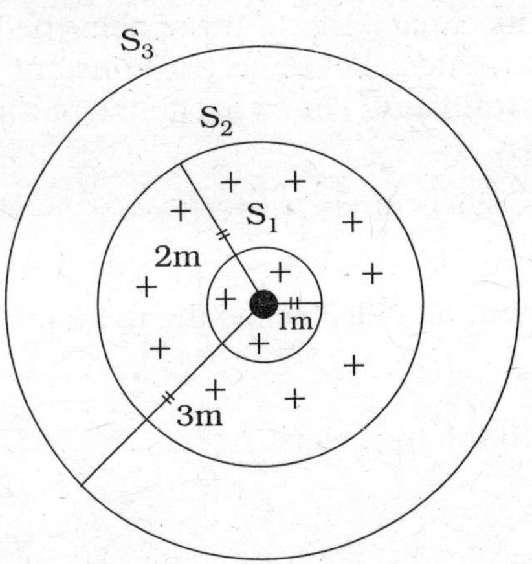

$$Q_1 = \rho V_1 = (1\times10^{-6})\times\frac{4}{3}\pi\times(1^3) = \frac{4\pi}{3}\times 10^{-6} \quad (1\mu C = 10^{-6}C)$$

$$\phi_1 = \frac{Q_1}{\epsilon_0\epsilon_r} = \frac{4\pi\times10^{-6}}{3\times(8.85\times10^{-12})\times(4)} = 1.18\times10^5\ Nm^2C^{-1}$$ **(Answer).**

(ii) For the surface S_2, which lies in air ($\epsilon_r = 1$), the flux is given by,

$$\phi_2 = \frac{Q_2}{\epsilon_0} = \frac{(1\times10^{-6})\times(4/3)\pi(2)^3}{(8.85\times10^{-12})} = \frac{32\times10^{-6}\times\pi}{3\times(8.85\times10^{-12})}$$

$$= 3.78\times10^6\ Nm^2C^{-1}$$ **(Answer).**

Capacitors and Capacitance

Problem 26. On giving 8 μC of charge to a conductor, its electric potential is raised by 2 volt. Calculate its capacitance.

Solution. For a single conductor,

$$C = \frac{Q}{V} \qquad (Q = 8\times10^{-6}\ C)$$

$$= \frac{8\times10^{-6}}{2} = 4\times10^{-6}\ F = 4\ \mu F$$ **(Answer).**

Problem 27. Assuming the earth to be a spherical conductor of radius 6400 km, calculate its capacitance.

Solution. The capacitance of a spherical conductor is given by,

$$C = (4\pi\epsilon_0)R$$

Given: $R = 6400\ km = 6.4\times10^6\ m$; $\frac{1}{4\pi\epsilon_0} = 9\times10^9\ Nm^2C^{-2}$

$$\therefore \quad C = \frac{(6.4\times10^6)}{(9\times10^9)}$$

$$= 7.11\times10^{-4}\ F\ (\text{or } 711\ \mu F)$$ **(Answer).**

Problem 28. Calculate the capacitance of a parallel-plate capacitor of plate area $10^{-4}\ m^2$ and with spacing between the plates as 1 mm, when

(i) the space between the plates is evacuated, and

(ii) the space between the plates is filled by a material of dielectric constant 4.

Solution.(i) Plate area, $A = 10^{-4}\ m^2$; spacing, $d = 1mm = 1\times10^{-3}\ m$.

With vacuum (or air) between the plates, the capacitance is given by,

$$C_{vac} = \frac{A\epsilon_0}{d}$$

$$= \frac{(10^{-4})\times(8.85\times10^{-12})}{(10^{-3})} = 0.89\times10^{-12}\ F\ (\text{or } 0.89\ pF)$$ **(Answer).**

(ii) With a dielectric as the medium between the plates, the capacitance is given by,

$$C_{diel} = K\frac{A\epsilon_0}{d}$$

$$= 4\times C_{vac} = 4\times0.89\times10^{-12}$$

$$= 3.56\times10^{-12}\ F\ (\text{or } 3.56\ pF)$$ **(Answer).**

Problem 29. A parallel-plate capacitor having air between the plates has a capacitance of 6 μF. What will be its capacitance if the spacing between the plates is reduced to half and the space between them is filled with a material of dielectric constant 4 ?

Solution. The capacitance of a parallel-plate capacitor having vacuum (or air) between the plates is given by,

$$C_{vac} = \frac{A\epsilon_0}{d}$$

With a dielectric as the medium between the plates, the capacitance is given by,

$$C_{diel} = K\frac{A\epsilon_0}{d}$$

In the first case,

$$C_{vac} = \frac{A\epsilon_0}{d} = 6\ \mu F$$

In the second case,

$$C_{diel} = K\frac{A\epsilon_0}{(d/2)} = 2K\left(\frac{A\epsilon_0}{d}\right)$$

$$= 2KC_{vac} = 2\times4\times6$$

$$= 48\ \mu F \quad \textbf{(Answer).}$$

Problem 30. Calculate the force on a plate of a parallel-plate capacitor, having plates charged equally but oppositely, with surface charge densities as $+\sigma$ and $-\sigma$, respectively.

Solution. The charge on each plate is $q = \sigma A$ (A = plate area). The electric field produced by either plate is,

$$E = \frac{\sigma}{2\epsilon_0}$$

Hence the force on the plate is,

$$F = qE$$

$$= \sigma A \times \frac{\sigma}{2\epsilon_0}$$

$$= \frac{1}{2}\frac{\sigma^2 A}{\epsilon_0} \quad \textbf{(Answer).}$$

(Note: the electric field between the plates is $\frac{\sigma^2}{\epsilon_0}$, but the electric field experienced by a plate is only due to the charge on the other plate and not because of its own charge).

Problem 31. Calculate the equivalent capacitance of three capacitors of capacitance 2μF each, when they are connected (i) in series and (ii) in parallel.

Solution. (i) For a series combination of capacitors, the equivalent capacitance is given by,

$$\frac{1}{C_{eq}} = \frac{1}{C_1} + \frac{1}{C_2} + \frac{1}{C_3}$$

$$= \frac{1}{(2\mu F)} + \frac{1}{(2\mu F)} + \frac{1}{(2\mu F)} = \frac{3}{2\mu F}$$

$$\Rightarrow \quad C_{eq} = \frac{2}{3}\ \mu F = 0.67\ \mu F \quad \textbf{(Answer).}$$

(ii) For a parallel combination, the equivalent capacitance is,

$$C_{eq} = C_1 + C_2 + C_3$$

$$= 2\ \mu F + 2\ \mu F + 2\ \mu F$$

$$= 6\ \mu F \quad \textbf{(Answer).}$$

Problem 32. You are given three capacitances of value 1μF each. How will you connect them to get
(i) minimum total capacitance and
(ii) maximum total capacitance ?
Also calculate the ratio of maximum to minimum total capacitance.

Solution. $C_1 = C_2 = C_3 = 1\mu F$

(i) For minimum total capacitance, all the capacitors are to be connected in series. Then the total capacitance is given by,

$$\frac{1}{C_{eq}} = \frac{1}{C_1} + \frac{1}{C_2} + \frac{1}{C_3} = \frac{1}{1} + \frac{1}{1} + \frac{1}{1} = 3$$

$$\Rightarrow \quad C_{eq} = C_{min} = \frac{1}{3}\,\mu F \quad \textbf{(Answer).}$$

(ii) For maximum total capacitance, all the capacitors are to be connected in parallel. Then the total capacitance is,

$$C_{eq} = C_{max} = C_1 + C_2 + C_3 = (1+1+1) = 3\,\mu F \quad \textbf{(Answer).}$$

The ratio, $C_{max} : C_{min} = (3) : \left(\frac{1}{3}\right) = 9 : 1$ **(Answer).**

Problem 33. Obtain the total capacitance of the combination shown in the first figure below.

Solution. The capacitors C_2 and C_3 are in a parallel combination. So their equivalent capacitance is,

$$C_{23} = C_2 + C_3 = (2 + 4) = 6\,\mu F$$

The circuit now simplifies, as shown in the second figure. Now the capacitances, C_1 and C_3 are in a series combination. Their equivalent capacitance C_{123} is found as,

$$\frac{1}{C_{123}} = \frac{1}{C_1} + \frac{1}{C_{23}} = \frac{1}{(4)} + \frac{1}{(6)} = \frac{5}{12}$$

$$\Rightarrow \quad C_{123} = \frac{12}{5}\,\mu F$$

Now the circuit is further simplified. The capacitors C_4 and C_{123} are in a parallel combination. Their equivalent capacitance is,

$$C_{1234} = C_4 + C_{123}$$

$$= (1) + \frac{12}{5} = \frac{17}{5}\,\mu F$$

$$= 3.4\,\mu F \quad \textbf{(Answer).}$$

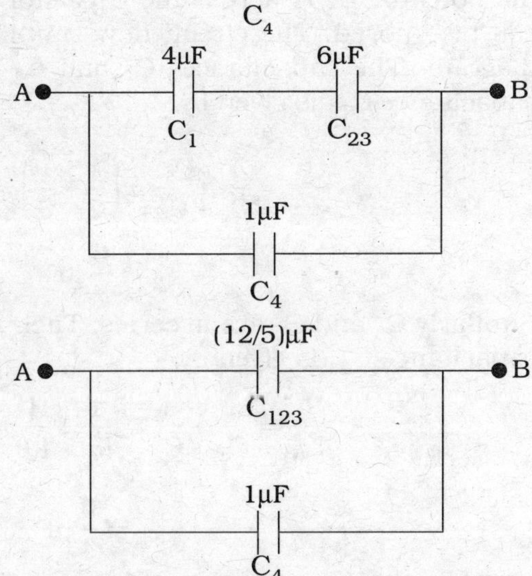

Problem 34. Obtain the equivalent capacitance of the given mixed combination of five capacitances across AB.

Solution. $C_1 = C_2 = C_3 = C_4 = 3\,\mu F$; $C_5 = 2\,\mu F$.
A close look at the given circuit reveals that it is the same as the circuit shown in the next figure. Now we find that C_1, C_2, C_3 form a parallel combination, whose equivalent capacitance is,

$$C_{123} = C_1 + C_2 + C_3 = (3 + 3 + 3)\,\mu F = 9\,\mu F$$

Thus the circuit is simplified. The capacitances C_{123} and C_4 are in a series combination. Their equivalent capacitance C_{1234} is given by,

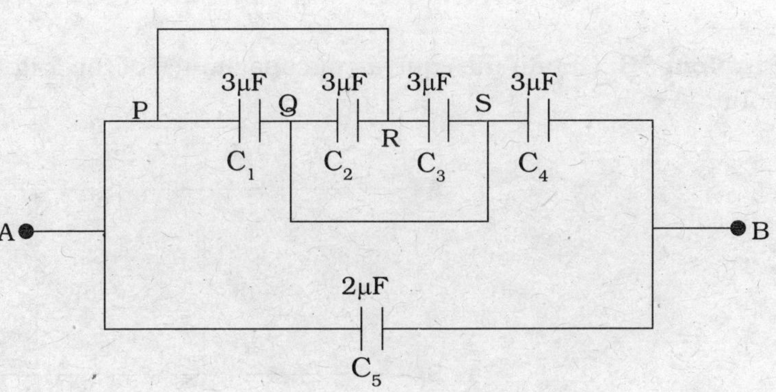

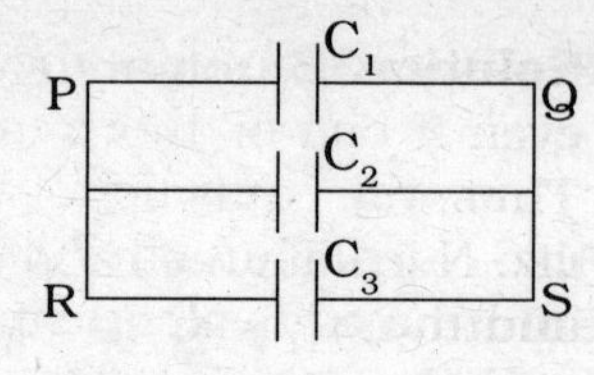

$$\frac{1}{C_{1234}} = \frac{1}{C_{123}} + \frac{1}{C_4}$$

$$= \frac{1}{9} + \frac{1}{3} = \frac{4}{9}$$

$$\Rightarrow \quad C_{1234} = \frac{9}{4}\ \mu F$$

The circuit is now very much simplified. The capacitances C_{1234} and C_5 are in a parallel combination. Their equivalent capacitance C_{12345} (which is the total capacitance across AB) is,

$$C_{12345} = C_{1234} + C_5$$

$$= \left(\frac{9}{4} + 2\right)$$

$$= 4.25\ \mu F \quad \textbf{(Answer).}$$

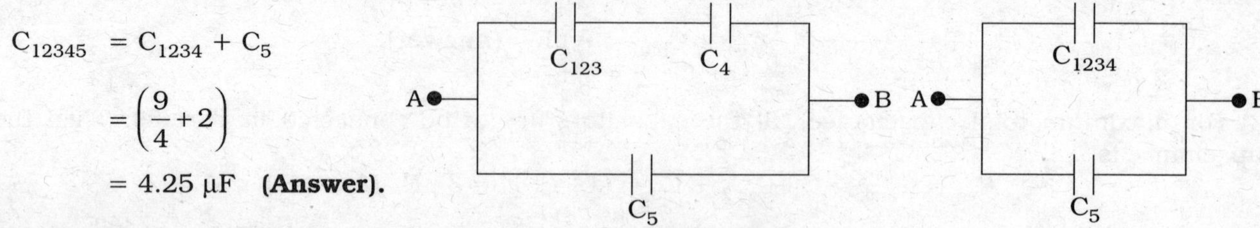

Problem 35. Five capacitors of capacitances $C_1 = 2\ \mu F$, $C_2 = 6\ \mu F$, $C_3 = 3\ \mu F$, $C_4 = 9\ \mu F$ and $C_5 = 4\ \mu F$ are joined together across A and B, as shown in the diagram. Obtain their equivalent capacitance.

Solution. We find that,

$$\frac{C_1}{C_2} = \frac{C_3}{C_4} = \frac{1}{3}$$

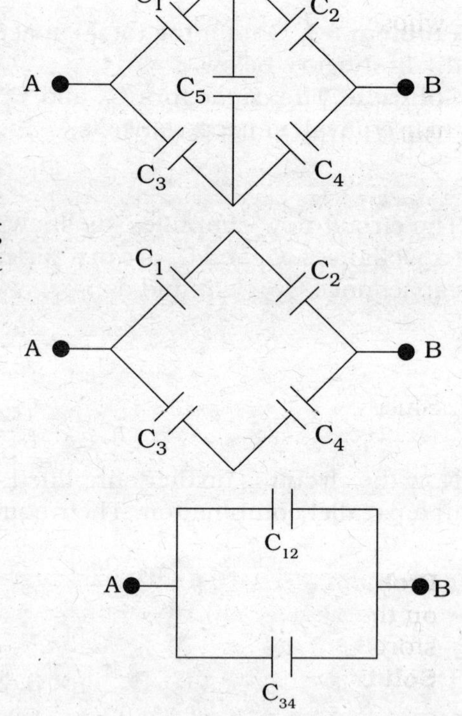

Hence the combination forms a balanced Wheatstone bridge. Therefore, the potential drop across the capacitor C_5 is zero. Hence its existence can be ignored. The circuit now simplifies, as shown in the following diagram. The capacitances C_1 and C_2 are in series. Their equivalent capacitance C_{12} is given by,

$$\frac{1}{C_{12}} = \frac{1}{C_1} + \frac{1}{C_2} = \left(\frac{1}{2} + \frac{1}{6}\right) = \frac{2}{3}$$

$$\Rightarrow \quad C_{12} = \frac{3}{2}\ \mu F$$

Similarly C_3 and C_4 are in series. Their equivalent capacitance C_{34} is given by,

$$\frac{1}{C_{34}} = \frac{1}{C_3} + \frac{1}{C_4} = \frac{1}{(3)} + \frac{1}{(9)} = \frac{4}{9}$$

$$\Rightarrow \quad C_{34} = \frac{9}{4}\ \mu F$$

Now C_{12} and C_{34} are in parallel. So their equivalent capacitance (which is the total equivalent capacitance across AB) is,

$$C_{total} = C_{12} + C_{34} = \left(\frac{3}{2} + \frac{9}{4}\right) = \frac{15}{4}\ \mu F = 3.75\ \mu F \quad \textbf{(Answer).}$$

Problem 36. Obtain the equivalent capacitance of the following infinite symmetrical combination across the points A & B.

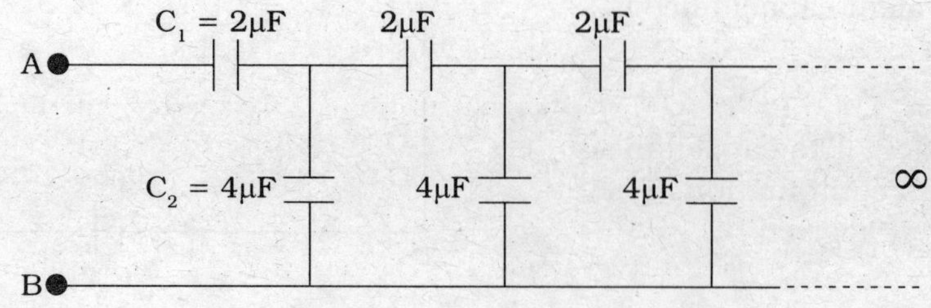

Solution. Suppose the equivalent capacitance across A and B is N μF. In an infinite symmetric combination, even if one or more units are removed, the equivalent capacitance will remain unchanged $[\because (\infty - n) = \infty]$. Thus, the equivalent capacitance across the set of points P and Q is the same as about the points A and B, viz. N μF. Thus, the circuit becomes simplified as shown in the next figure. Now the capacitors C_2 = 4 μF and the equivalent capacitor C_x = N μF form a parallel combination, whose equivalent capacitance C_{2x} is given by,

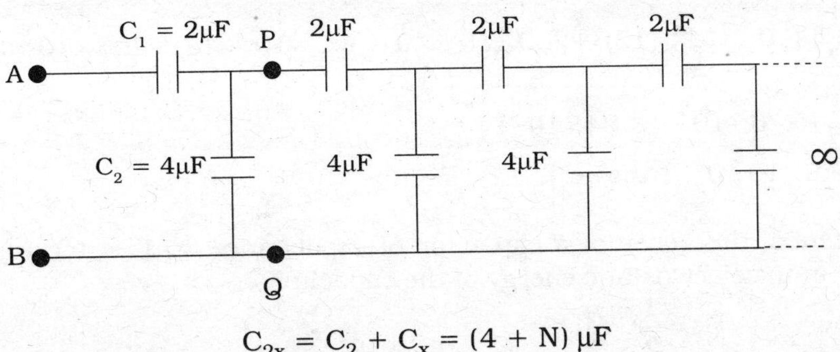

$$C_{2x} = C_2 + C_x = (4 + N)\ \mu F$$

Now the circuit is further simplified, as shown in the next figure. Now C_1 and C_{2x} form a series combination, whose equivalent capacitance C_{12x} is given by,

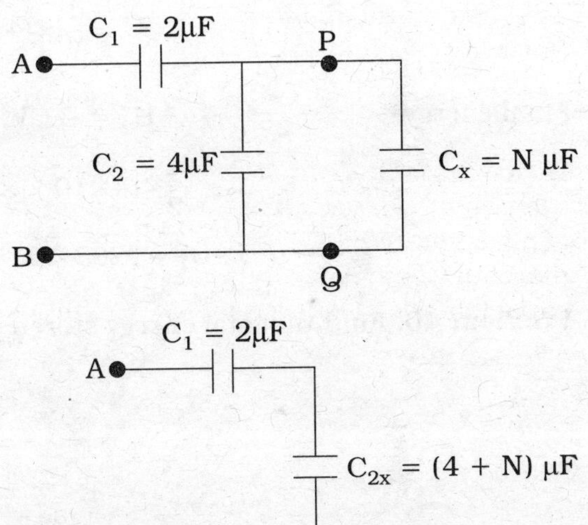

$$\frac{1}{C_{12x}} = \frac{1}{C_1} + \frac{1}{C_{2x}} = \frac{1}{(2)} + \frac{1}{(4+N)} = \frac{(6+N)}{2(4+N)}$$

But, C_{12x} = N μF. Therefore we have,

$$\frac{1}{N} = \frac{(6+N)}{2(4+N)}$$

$$\Rightarrow \quad 8 + 2N = 6N + N^2$$

$$\Rightarrow \quad N^2 + 4N - 8 = 0$$

$$\therefore \quad N = \frac{-4 \pm \sqrt{16+32}}{2}$$

Since the capacitance cannot be negative, we have,

$$N = -2 + \frac{\sqrt{48}}{2}$$

$$= 1.46\ \mu F \quad \textbf{(Answer).}$$

Problem 37. A 10 μF parallel-plate capacitor is connected to a 12 V battery. (i) What is the charge developed on the plates ? (ii) How much electrostatic energy is stored in the capacitor ? (iii) What is the form of this stored energy ?

Solution. (i) Charge, $q = CV = (10\times10^{-6}) \times (12) = 1.2\times10^{-4}$ C $[10\ \mu C = 10\times10^{-6}\ C]$. **(Answer).**

(ii) Energy, $U = \frac{1}{2}CV^2 = \frac{1}{2}\times(10\times10^{-6})\times(12)^2 = 7.2\times10^{-4}$ J. **(Answer).**

(iii) The energy is stored in the form of electric potential energy created between the plates. **(Answer).**

Problem 38. A 400 μF capacitor is charged by a 12 volt electric supply. It is then disconnected from the supply and connected to another uncharged capacitor of capacitance 200 μF. What is the common potential acquired by the two capacitors and how much energy is lost in the process ?

C_1

C_2

Solution. Common potential, $V_C = \dfrac{C_1V_1 + C_2V_2}{C_1 + C_2}$

[Initially, the total charge of the two capacitors is $Q = (C_1V_1+C_2V_2)$. After they are connected together the total charge becomes, $Q = C_1V_C+C_2V_C$. $\therefore V_C(C_1+C_2) = (C_1V_1+C_2V_2)$].

Given: C_1=400 μF; C_2 = 200 μF; V_1=12 V; V_2= 0

$$\therefore \quad V_C = \frac{(400 \times 12) + 0}{(400 + 200)} = 8 \text{ volt} \quad \textbf{(Answer).}$$

The loss of energy =(Initial stored energy - Final stored energy)

$$= \frac{1}{2} C_1 V_1^2 + \frac{1}{2} C_2 V_2^2 - \frac{1}{2}(C_1 V_C^2 + C_2 V_C^2)$$

$$= \frac{1}{2} [400 \times 10^{-6} \times (12)^2 + 200 \times 10^{-6} \times 0 - 400 \times 10^{-6} \times (8)^2 - 200 \times 10^{-6} \times (8)^2]$$

$$= \frac{1}{2} (5.76 \times 10^{-2} - 3.84 \times 10^{-2})$$

$$= 9.6 \times 10^{-3} \text{ J} \quad \textbf{(Answer).}$$

Problem 39. The potential difference of a capacitor of capacitance 4 μF is increased from 6 V to 12 V. Calculate the increase in the electrostatic energy of the capacitor.

Solution. Initial energy, $U_1 = \frac{1}{2} C V_1^2$

Given: $C = 4\ \mu F = 4 \times 10^{-6}$ F; $V_1 = 6$ V; $V_2 = 12$ V

$$\therefore \quad U_1 = \frac{1}{2} \times (4 \times 10^{-6}) \times (6)^2 = 72 \times 10^{-6} \text{ J}$$

Final energy, $U_2 = \frac{1}{2} C V_2^2$

$$\therefore \quad U_2 = \frac{1}{2} \times (4 \times 10^{-6}) \times (12)^2 = 288 \times 10^{-6} \text{ J}$$

$$\therefore \quad U_2 - U_1 = (288 - 72) \times 10^{-6} \text{ J} = 216 \times 10^{-6} \text{ J} \quad \textbf{(Answer).}$$

Problem 40. Find the total energy stored in the capacitors in the given network.

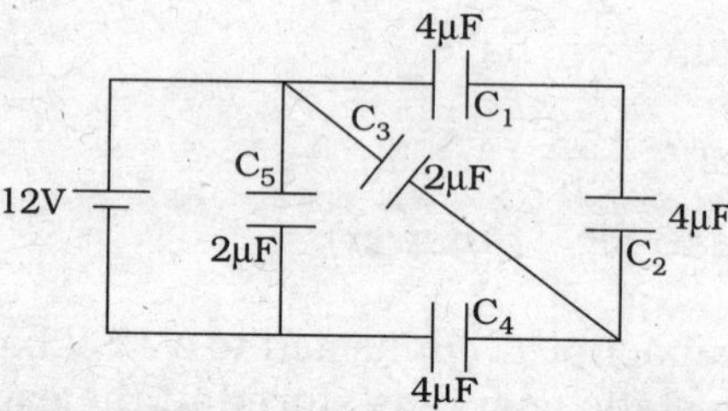

Solution. The capacitors C_1 and C_2 are in series. Their equivalent capacitance C_{12} is given by,

$$\frac{1}{C_{12}} = \frac{1}{C_1} + \frac{1}{C_2} = \frac{1}{4\mu F} + \frac{1}{4\mu F} = \frac{1}{2\mu F}$$

$$\therefore \quad C_{12} = 2\ \mu F$$

C_{12} and C_3 are in parallel. Their equivalent capacitance C_{123} is,

$$C_{123} = (2+2)\ \mu F = 4\ \mu F$$

C_{123} and C_4 are in series. Their equivalent capacitance C_{1234} is given by,

$$\frac{1}{C_{1234}} = \frac{1}{C_{123}} + \frac{1}{C_4} = \frac{1}{4\mu F} + \frac{1}{4\mu F} = \frac{1}{2\mu F}$$

$$\therefore \quad C_{1234} = 2\ \mu F$$

C_{1234} and C_5 are in parallel. Their equivalent capacitance (which is the net capacitance of the whole circuit) is,

$$C_{12345} = C_{1234} + C_5 = 2\ \mu F + 2\ \mu F = 4\ \mu F = 4\times 10^{-6}\ F$$

$\therefore$ The net energy stored is,

$$U = \frac{1}{2}CV^2 = \frac{1}{2}\times(4\times10^{-6})\times(12)^2 = 2.88\times10^{-4}\ J \quad \textbf{(Answer).}$$

Problem 41. In a Van de Graaff generator, the potential of a spherical metal shell has to be raised to 2000 kV. The strength of the electric field surrounding the metal is $4\times10^7\ Vm^{-1}$. What should be the minimum radius of the spherical shell ?

Solution. Electric field, $$E_{max} = \frac{Q}{4\pi\epsilon_0 R_{min}^2} = 4\times10^7\ Vm^{-1}$$

Potential, $$V_{max} = \frac{Q}{4\pi\epsilon_0 R_{min}} = 2000\ kV = 2\times10^6\ V$$

$$\therefore \quad R_{min} = \frac{V_{max}}{E_{max}} = \frac{2\times10^6}{4\times10^7} = 0.05\ m = 5cm \quad \textbf{(Answer).}$$

Problem 42. A battery of voltage 12 V is connected across a combination of capacitors $C_1 = 1\mu F$, $C_2 = C_3 = 2\ \mu F$, $C_4 = 1\mu F$, as shown in the diagram. Calculate the charge and the potential across each capacitor.

Solution. The circuit can be redrawn, as shown in the next diagram. C_2 & C_3 are seen in a series combination. Their equivalent capacitance C_{23} is given by,

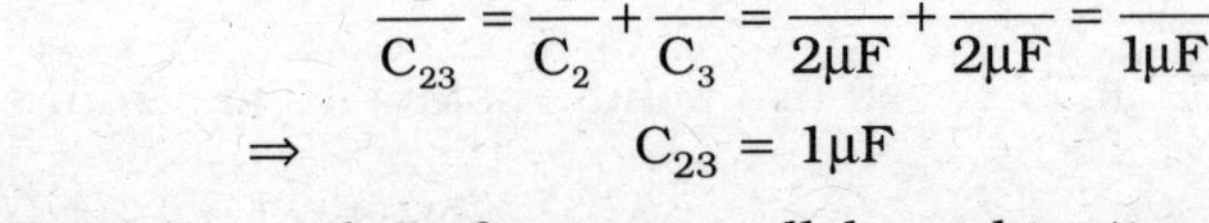

$$\frac{1}{C_{23}} = \frac{1}{C_2} + \frac{1}{C_3} = \frac{1}{2\mu F} + \frac{1}{2\mu F} = \frac{1}{1\mu F}$$

$$\Rightarrow \quad C_{23} = 1\mu F$$

Further, C_{23} and C_1 form a parallel combination, with their equivalent capacitance C_{123} as,

$$C_{123} = C_1 + C_{23} = (1+1)\ \mu F = 2\ \mu F$$

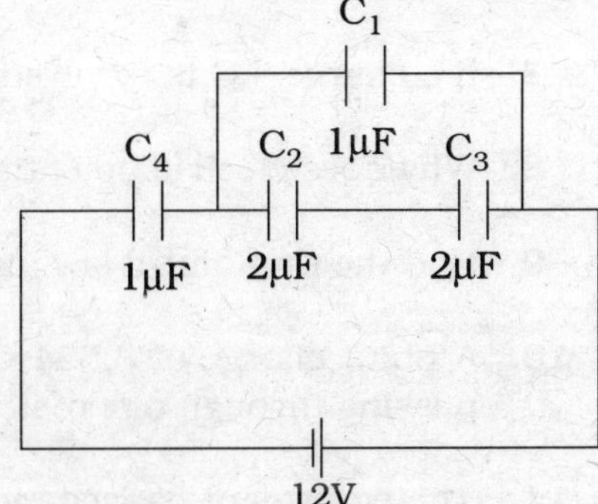

Next, C_{123} & C_4 are in a series combination. In a series combination, the charge on the various capacitors is the same. Let the potential drop across C_4 be V_4 and across C_{123} be V_{123}. Then the equality of charge gives ($Q = CV$),

$$C_4V_4 = C_{123}V_{123} \quad \text{Or} \quad 1\times V_4 = 2\times V_{123}$$

$$\Rightarrow \quad V_4 = 2(V_{123})$$

Also, $$V_4 + V_{123} = 12\ V$$

$$\Rightarrow \quad 2(V_{123}) + (V_{123}) = 12\ V$$

$$\Rightarrow \quad V_{123} = 4\ V$$

$$\therefore \quad V_4 = 12\ V - 4\ V = 8\ V$$

The potential difference across C_1 and the combination (C_2+C_3) is the same, viz. V_{123}.

$$\therefore \quad V_1 = V_{123} = 4\ V$$

Hence the charge on the capacitor C_1 is,

$$q_1 = C_1V_1 = (1\times10^{-6})\times 4 = 4\times10^{-6}\ C$$

Next, C_2 and C_3 form a series combination. $\therefore$ The charge on them is the same, i.e.,

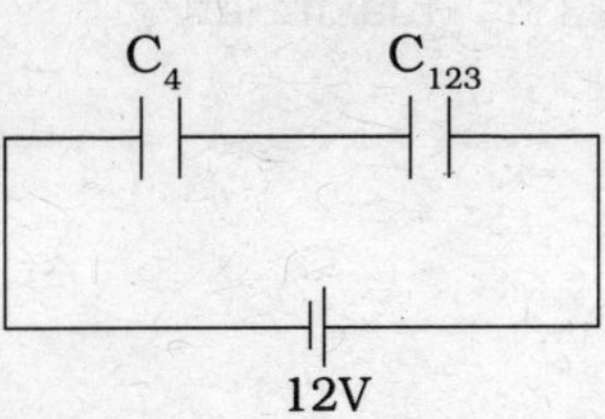

$$C_2V_2 = C_3V_3$$

$$\Rightarrow \quad (2\times10^{-6})V_2 = (2\times10^{-6})V_3$$

$$\Rightarrow \quad V_2 = V_3$$

Also, $$V_2 + V_3 = V_{123} = 4\ V$$

But since $V_2 = V_3$, we get,

$$V_2 = V_3 = 2\ V$$

∴ The charges on C_2 and C_3 are,

$$q_2 = C_2V_2 = (2\times10^{-6}) \times 2 = 4\times10^{-6}\ C\ (\text{or } 4\ \mu C)$$
$$q_3 = C_3V_3 = (2\times10^{-6}) \times 2 = 4\times10^{-6}\ C\ (\text{or } 4\ \mu C)$$

The charge on C_4 is,

$$q_4 = C_4V_4 = (1\times10^{-6}) \times 8 = 8\times10^{-6}\ C\ (\text{or } 8\ \mu C)$$

Thus, the respective potentials and charges are,

$V_1 = 4\ V;\ V_2 = V_3 = 4\ V;\ \ V_4 = 8\ V$ **(Answer).**

$q_1 = q_2 = q_3 = 4\ \mu C;\ \ q_4 = 8\mu C$ **(Answer).**

Very Short Answer Questions

(Including HOTS Questions)

(1 mark each)

1. Define electric potential at a point.

2. Define capacitance of a capacitor.

3. How is dielectric constant expressed in terms of capacitance of a capacitor ?

4. Define the term electric moment.

5. How much work is done in moving a charge of 200 μC between two points distant 10 cm from each other on an equipotential surface ?

6. Is electrostatic potential necessarily zero at a point where electric field strength is zero ?

7. If a charge +Q is revolved round another charge +q in a circle of radius r, how much work is done ?

8. Why does electric potential decrease in the direction of electrostatic field ?

9. Had the Coulomb's law involved $1/r^4$ dependence, will Gauss's law still hold ?

10. A point charge of 8.854 C is placed at the centre of a cube of side length 12 cm. Calculate the flux passing through one of its faces.

11. The equivalent capacitance of the combination of two capacitors between the points X and Y in the given figure is 12 μF. Calculate the capacitance of the capacitor C.

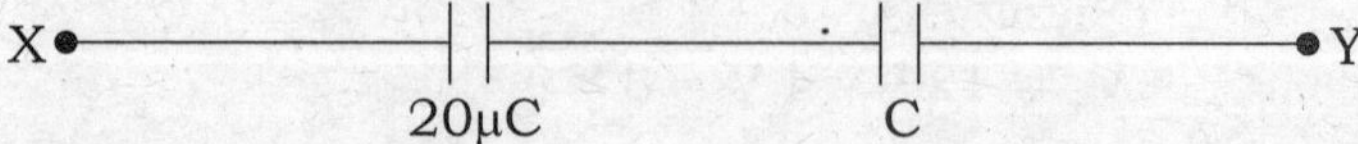

12. Name the physical quantity whose SI unit is Joule/Coulomb. Is it a scalar or a vector ?

13. A charge q_0 is taken turn by turn from P to A, P to B and P to C. In which case the work done is maximum ?

14. Define dielectric strength of a medium.

15. Distances of three points, at A, B and C from a point charge Q are x, 2x and 3x, respectively. If V_A, V_B and V_C are the electric potentials at A, B and C, respectively, what is the ratio between (V_A-V_B) and (V_B-V_C) ?

16. A point charge –q goes round another point charge Q in a semi-circular path of radius R. What will be the change in the electrostatic potential energy of the system of charges ?

17. From the relation Q=CV, explain why the capacitance of a parallel-plate capacitor decreases upon increasing the separation between the plates.

18. What information about the variation in electric field can be gained from the observed spacing between the equipotential surfaces in a region of space ?

19. The electric field strength at a distance x from a charged infinite flat sheet has magnitude E. What will be the field strength at a distance 2x from the sheet ?

20. Three points A, B, C lie along the line joining two point charges Q and -Q, as shown in the figure. What is the sign of the potential difference

(a) $(V_A - V_B)$? **(b)** $(V_B - V_C)$?

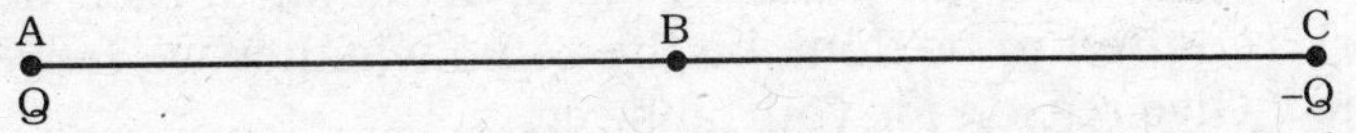

A B C
Q –Q

21. Name the physical quantity which has its unit as coulomb/volt. Is it a scalar or a vector ?

22. Where and in what form is energy stored in a charged capacitor ?

23. A spherical Gaussian surface encloses a charge of 8.85×10^{-8} C. What is the electric flux passing through it ?

Short Answer Questions

(Including HOTS Questions)

(2 marks each)

1. You are given three identical capacitors. When two of them are connected in series, the net capacitance is 9 μF. What will be the net capacitance when all three are connected in parallel ?

2. The net capacitance of three identical capacitors connected in parallel is 12 μF. What will be the net capacitance when two of them are connected in series ?

3. What should be the charge on a metal sphere of radius 4 cm so that when it is brought in contact with another metal sphere of radius 3 cm carrying a charge of 20 μC, there is no net transfer of charge between the two spheres ?

4. A uniform electrostatic field exists in a region of space along x-axis. Does the electric potential decrease or remain unchanged along (a) x-axis and (b) y-axis. Justify your answer.

5. A charge of 12 μC is lying at the corner of a cube of side length 12 cm. What is the electric flux through one of the faces of the cube ?

6. At which of the points A, B and C in a uniform electrostatic field, as shown, will the electric potential be (a) minimum, (b) maximum ?

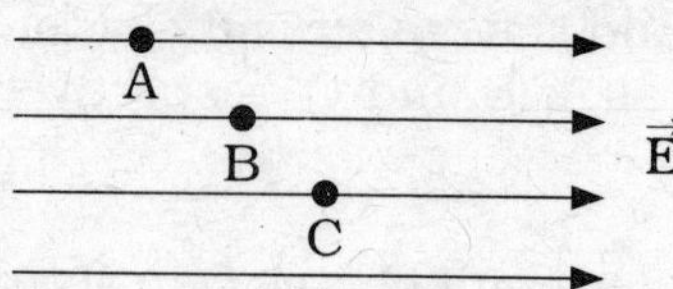

7. You are given an isolated parallel capacitor of capacitance C charged to a potential difference V. What will happen to the following when the separation between the plates is doubled ?
   (a) charge on the plates
   (b) potential difference across the plates
   (c) field between the plates
   (d) energy stored in the capacitor

8. There is an upper limit to the charge that can be given to a metal sphere, without leakage to the surroundings. What factors decide this limit and how ?

9. Does the maximum charge that can be given to a metal sphere of radius R depend on whether the sphere is hollow or solid ? Give reason for your answer.

10. Two point charges 10 μC and 5 μC are placed 4 cm apart in air. If their separation is to be decreased to 1 cm, will the work required to be done be the same whether we move the charge 10 μC alone or the charge 5 μC alone ? Give reason for your answer.

11. Illustrate with an example that there can be an electric potential at a point where no electrostatic field exists. Give justification for the same.

12. How does the electrostatic field near an infinite flat charge sheet differ from that of a point charge ?

13. A point charge Q exists in free space. Using the concept of equipotential surface, show that the work done in moving another charge q in the electrostatic field of Q is independent of the path along which q is moved.

14. The adjoining figure shows two point charges 4q and -q. Also in the figure are two curves and a line in the plane of the charges, marked 1, 2 and 3. Which of them can possibly be equipotential? Justify your answer.

15. Use the Gauss's theorem to show that the electrostatic force between two point charges is inversely proportional to the square of the distance between them.

16. Why is the electrostatic field around a charged metal sphere always normal to the surface of the sphere ?

17. Deduce the expression for the energy stored in a charged parallel-plate capacitor.

18. Three capacitors of equal capacitance when connected in series have a net capacitance C_1 and when connected in parallel, they have a net capacitance C_2. What is the ratio of C_1 to C_2 ?

19. On what factors does the capacitance of a parallel-plate capacitor depend ? Give the formula used.

20. A charge of 12 μC is given to a hollow metal sphere of radius 0.4 m. Find the potential at (i) the surface of the sphere and (ii) the centre of the sphere.

21. How much energy will be stored in a capacitor of capacitance 80 μF, when charged by a battery of 10 V ?

22. State the Gauss's theorem in electrostatics. Give its mathematical form. Does the electric flux through the surface depend on the position of the charge enclosed by the surface ?

23. What is corona discharge ? Explain why a hollow spherical conductor is preferred for the accumulation of charge in a Van de Graaff generator ?

24. Two point charges of unknown magnitude and sign are placed at a distance r apart. The electric field intensity is zero at a point not midway between the charges, along the line joining the charges. Write two essential conditions for this to happen.

25. (i) Calculate the electric potential at a point P due to a charge of 8 μC distant 12 cm from it.
(ii) Also calculate the work done in bringing a charge of 5×10^{-9} C from infinity to the point P.

26. Show that the electric field is related to the potential as $E = -\frac{dV}{dr}$.

27. Two point charges 4×10^{-8} C and -2×10^{-8} C are located 12 cm apart in air. Find at what point along the line joining these charges the electric potential is zero.

28. S_1 and S_2 are two hollow concentric spheres in air, enclosing charges q and 2q, respectively, as shown in the figure.

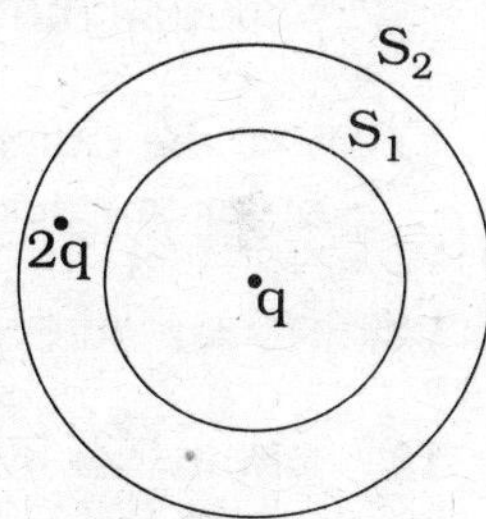

(i) What is the ratio of the electric flux values passing through S_1 and S_2, respectively ?
(ii) How will the electric flux through the sphere S_1 change, if a medium of dielectric constant 5 is introduced in the space inside S_1, in place of air ?

29. Explain the process of charging of a capacitor.

30. If the dielectric strength of air is 2 MV/m, can a metal sphere of radius 1 cm hold a charge of 1 coulomb ?

Short Answer Questions

(Including HOTS Questions)

(3 marks each)

1. Derive an expression for the capacitance of a parallel-plate capacitor having plates of area A each and separated by a distance d, when the space between the plates is filled by a medium of dielectric constant K.

2. State two uses of a capacitor. What is the area of the plates of a parallel-plate capacitor, of capacitance 4 F, in which the separation between the plates is 2 cm ?

3. Briefly explain the phenomenon of dielectric polarization. What is meant by dielectric breakdown and dielectric strength ?

4. Explain how does the applied electric field decrease inside a polar dielectric. A parallel-plate capacitor with air between the plates has a capacitance of 12 pF. What will be the capacitance if the distance between the plates is reduced by half and the space between them is filled with a material of dielectric constant 4.

5. Use the Gauss's theorem to show that no electrostatic field can exist inside a metallic conductor.

6. Define the term electric flux. What is its SI unit ? An electric flux of -8.854×10^{3} Nm^2C^{-1} passes normally through a spherical Gaussian surface of radius 10 cm. If the radius of the Gaussian surface is doubled, how much flux would pass through the surface ?

7. Explain the concept of electric potential energy. Derive an expression for the potential energy of a system of two point charges in an electric field.

8. How far apart are two equipotential surfaces, whose potential difference is 5 volt in a uniform electric field of strength 1000 V/m ?

9. A conducting slab of thickness t is introduced between the plates of a parallel-plate capacitor, separated by a distance d ($t<d$). Derive an expression for the capacitance of the capacitor. What will be its capacitance, when $t = d$?

10. A capacitor of capacitance C is charged to a potential difference V_1. Then, it is further charged to a potential difference $V_2(V_2>V_1)$. Obtain an expression for the additional energy stored in the capacitor.

11. Use the Gauss's theorem to show that any charge given to a metallic conductor always resides on the outer surface of the conductor.

12. Obtain an expression for the electric potential V at a distance R from a point charge Q.

13. Obtain an expression for electrostatic potential energy of a pair of point charges of value Q each, separated by a distance R in free space.

14. How does the introduction of a dielectric slab between the plates of a charged parallel-plate capacitor, which initially had air between the plates, change its
(i) capacitance, (ii) electric field, and (iii) electric energy ?
(Assume that the charging source is removed after charging the capacitor)

15. Four point charges of value Q each are lying on a circle of radius R, such that they are equidistant from each other. What is the electric field strength and electric potential at the centre of the circle ?

16. What will be the work done in the field of a nucleus for moving an electron in a circular orbit ?

17. A charge of 4 μC is located at the origin. Calculate the work done in taking a small charge of -2×10^{-12} C from a point A (0,0,4cm) to a point B (0, 6cm, 0) via a point C (0,4cm, 5cm).

18. Calculate the capacitance of a parallel-plate capacitor, which is filled by a conducting slab of same area as that of the plate and with thickness two-third of the spacing between the plates.

19. A metal sphere X of radius 54 cm is carrying a charge of 10 μC. Another metal sphere Y of radius 80 cm is carrying a charge of 10 μC. The two spheres are kept a large distance apart. When brought in contact through a conducting wire, in which direction would the charge flow ? Find the final charges acquired by the two spheres.

20. A metal wire is bent in the shape of a circle of radius 8 cm. It is given a charge of 32 μC, which is spread uniformly over it. Calculate the electric potential at the centre of the circle.

21. A charge of 10 μC is given to a hollow metallic sphere of radius 10 cm. Find the electric potential at
(a) the surface of the sphere
(b) the centre of the sphere.

Long Answer Questions

(5 marks each)

1. A dielectric slab of thickness t is introduced between the plates of a parallel-plate capacitor, separated by a distance d ($t<d$). Derive an expression for the capacitance of the capacitor. A capacitor of capacitance 10 μF and charged to 400 V is connected in parallel to another capacitor of capacitance 10 μF charged to 200 V. Find their common potential.

2. Show that when charged conductors having different capacitances and different potentials are joined together, there is always a loss of energy.

3. Two charges of value -32×10^{-9} C each are placed 4 cm apart. Find the electric potential at a point located at 9 cm from the centre of the electric dipole
   (i) along the axial line and
   (ii) along the equatorial line.

4. What is an equipotential surface ? Discuss its properties. Sketch equipotential surfaces for
   (i) a uniform electric field and
   (ii) an isolated point charge (both positive and negative).

5. Find the equivalent capacitance across A and B in the following circuit. The capacitance of each capacitor is 2 μF.

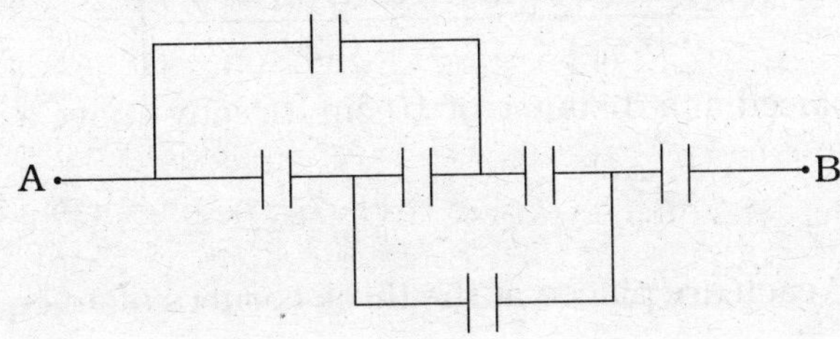

Also determine the charge and voltage across each capacitor, when the combination is connected across a 32 V electric supply.

6. A charge is uniformly distributed over a ring of radius a. Obtain an expression for the electric potential at a point along the axis of the ring.

7. A dipole consists of two charges $+5\times10^{-9}$ C and -5×10^{-9} C, separated by a distance of 1.25×10^{-8} m. If it is placed in a uniform electric field of 2.0×10^{5} V/m, find
   (i) the potential energy of the dipole in the equilibrium position and
   (ii) the work necessary to rotate the dipole by 180° from the equilibrium position.

8. Show that the energy stored per unit volume in a capacitor is equal to $(1/2)\epsilon_0 E^2$. A capacitor of capacitance 2 μF is charged to 30 μC. Calculate the electrostatic energy stored in it.

9. Briefly discuss the principle, construction and working of a Van de Graaff generator.

Numerical Exercises

1. Determine the electric potential at the surface of a charged sphere of radius 1 cm and with charge density of 8.85×10^{-10} C/cm^2. **[1 volt]**

2. Three charges of value q each are placed along a straight line, with a separation r between successive charges. What is the potential difference between the middle and an end charge ? $\left[\dfrac{q}{8\pi\epsilon_0 r}\right]$

3. An infinite number of charges of value q each are placed along the x-axis at x=1m, x=2m, x=3m,............ Find the potential at the point x = 0.

$$\left[\text{Hint: } V=\frac{q}{4\pi\epsilon_0}\left(\frac{1}{1}+\frac{1}{2}+\frac{1}{4}+\ldots\ldots\ldots\right)=\frac{q}{4\pi\epsilon_0}\left(\frac{1}{1-1/2}\right)\right].\qquad \left[\frac{q}{2\pi\epsilon_0}\right]$$

4. A spherical liquid drop having electric potential of V volt, splits up into twenty seven identical drops. Calculate the potential on each drop. [Hint: If R and r are radii of large and small drop, respectively, then, because the total volume will remain unchanged, $(4/3)\,\pi R^3 = 27\times(4/3)\pi r^3$, or R = 3r.

$$\therefore \text{Potential on a small drop, } v=\frac{(Q/27)}{4\pi\epsilon_0 r}=\frac{(Q/27)}{4\pi\epsilon_0 (R/3)}=\frac{Q}{4\pi\epsilon_0 R}\cdot\frac{1}{9}=\frac{1}{9}V\Bigg].\qquad \left[\frac{1}{9}V\right]$$

5. A regular hexagon of side 20 cm has charges of 1 μC each at its vertices. Calculate the potential difference at the centre of the hexagon. [Hint: 10 cm = 0.2 m.

$$V=\frac{6\times q}{4\pi\epsilon_0 r}=\frac{6\times(10^{-6})\times(9\times10^{9})}{0.2}\Bigg].\qquad [2.7\times10^5\ V]$$

6. A point charge of 1 μC is placed at a distance of 10 cm directly above a square of side 20 cm. What is the flux through the square ?

[1.88×10^4 Nm^{2}C^{-1}]

7. Three charges of value 1μC each are placed at the three corners of an equilateral triangle of side 1 cm.

Calculate he electric potential at the centroid of the triangle. $\left[\text{Hint: } C=\dfrac{3\sqrt{3}q}{4\pi\epsilon_0 a}=\dfrac{3\times\sqrt{3}\times(10^{-6})\times(9\times10^{9})}{0.2}\right].$

[4.68×10^6V]

8. A charge q of -2.3×10^{-8} C is placed at a point. Calculate the work done to carry 1C of charge around q in a semicircle of radius 10 cm. **[Zero]**

9. Find the area of a parallel-plate capacitor of capacitance 1μF, having spacing of 1μm and the medium as air between the plates. **[0.113 m^2]**

10. Find the charge on each capacitor in the following figure.

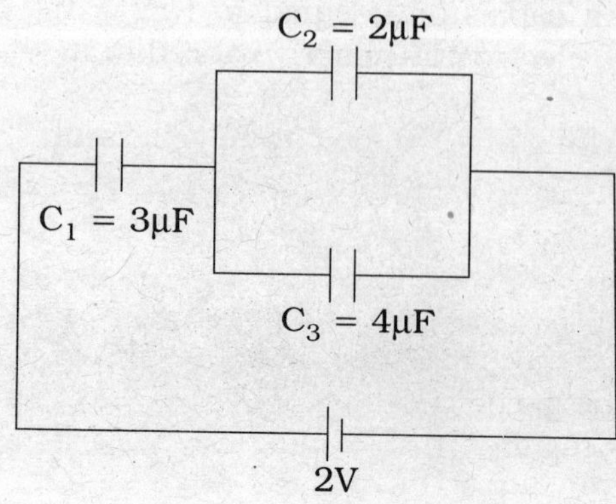

[q_1 = 4 μC ; q_2 = 1.33 μC; q_3 = 2.66 μC]

11. It is found that for moving a charge of +2 C from a place with electric potential of –10 V to another place with potential of V_P volt, a work of 40 J needs to be done. Determine the value of V_P. **[–30V]**

12. Determine the magnitude and sign of charge at a point located 0.1 m away from the charge and having an electric potential of 45 volt.
[0.5×10^{-9} C; +ve]

13. A wire is shaped as a circle of radius 5 cm and given a uniform charge of 250 μC. Calculate the electric potential at its centre. **[4.5×10^{7} V]**

14. The electric field in a region is given by $\vec{E}= \frac{4}{5} E_0\hat{i} + \frac{3}{5}E_0\hat{j}$, with $E_0 = 4.0\times10^3$ N/C. Calculate the flux due to it, passing through a rectangular cross-section of area 0.4 m^2, situated parallel to the yz-plane. **[1280 Nm^2/C]**

15. A charge of -3.8×10^{-7} C is uniformly spread over the surface of a glass rod of length 4.4 m and radius 1.8 mm. Calculate the electric field on the surface around the midpoint of the rod.
[-8.6×10^{5} NC^{-1}]

16. In a given region of space, an electric field is directed radially outwards from the origin. It varies with distance r as E = 500r Vm^{-2}. Find the charge located within a sphere of radius 0.1 m, with its centre at the origin. **[0.555×10^{-10} C]**

17. What is the work done in moving a 2 μC point charge from the corner A to the corner B of a square ABCD, when a 10 μC charge exists at the centre of square. (CBSE; 1998) **[Zero]**

18. In a particular region the electric potential V(x) varies with distance x (in metre) as $V = 5x^2$. If a charge of 1 μC is located at x = 1m, find the force experienced by it. **[-10×10^{-6} N]**

19. Two electrons approach each other, each with a velocity of 2×10^6 ms^{-1}. Calculate the closest distance reached between them. **[0.64×10^{-10}m]**

20. Two identical particles with charge 4.0×10^{-4} C each and of mass 20 g each are first held 10 cm apart and then released. Calculate the speed of the particles, after they are appreciably separated from one another. **[6×10^{2} ms^{-1}]**

21. Two particles of mass 5 g each and with charge 2.0×10^{-5} C and -2.0×10^{-5} C, respectively, are initially held 2.0 m apart, before being released. Calculate their speed, when their separation becomes 1.0 m. **[18.97 ms^{-1}]**

22. Three point charges –q, Q and –q are located along a straight line at equal distance from each other. Assuming that the potential energy of this system of charges is zero, calculate the ratio Q:q. **[1 : 4]**

23. An electric field is expressed as $\vec{E} = 6\hat{i} + 3\hat{j} + 4\hat{k}$. Find the electric flux passing through a region of area 20 units and lying in the yz – plane. **[120 units]**

24. An infinite thin plane sheet of charge density 10^{-8} $C.m^{-2}$ is held in air. How far apart are two equipotential surfaces, whose p.d. is 5 volt ? (CBSE; 1999). **[8.85×10^{-3}m]**

25. A particle of mass 10×10^{-6} g is held over a uniformly charged sheet, having charge density of 8×10^{-6} $C.m^{-2}$. What charge should the particle have, so that if released it does not separate from the sheet ? How many electrons need to be removed from the particle to impart this charge ?
[Hint: Use (i) $F_e = F_G$ or qE = mg or q $(\sigma/2\epsilon_0)$ = mg and (ii) q = ne]. **[2.16×10^{-13}C; 1.36×10^{6}]**

26. A bullet of mass 2 g has a charge of 2 μC. Through what potential difference must it be accelerated, starting from rest, to acquire a speed of 10 m/s ? (CBSE PMT; 2004). **[5×10^4 V]**

27. The electric potential at the surface of a hollow metal sphere is 20 volt. Find the potential at a distance of 5 cm from the centre of the sphere. **[20 volt]**

28. Two metallic spheres of radii 1cm and 2 cm are given charges of 10^{-2} C and 3×10^{-2} C, respectively. If they are connected by a conducting wire, what will be the final charge on the smaller sphere ? (CBSE; 1995). **[2×10^{-2} C]**

29. A cube of side length 15 cm has six dipoles with end charges of +e and -e each located within it. What is the total electric flux across it ? **[Zero]**

30. Two point charges A and B of value 15 μC and 9 μC are kept 18 cm apart in air. Calculate the work done when B is moved by 3 cm towards A. (CBSE; 2000). **[1.35 J]**

31. A transfer of 2.0×10^{12} electrons from one conductor to another results in a potential difference of 20 volt between them. Calculate the capacitance of this two-conductor system. **[1.6×10^{-8} F]**

32. An 80 μF capacitor is charged by a 50 V battery. The capacitor is disconnected from the battery and then connected across another uncharged 320 μF capacitor. Calculate the charge on the second capacitor. (CBSE; 1994). **[3.2×10^{-3}C]**

33. An air capacitor has plates of area 0.3×0.2 m^2 each. If it has to have the same capacitance as that of a spherical conductor of radius 1.5 m, what should be the separation between the plates ? **[3.18 mm]**

34. Three capacitors are mutually connected, as shown. If the p.d. across C_2 is 50 V, find the p.d. across C_1 and C_3.

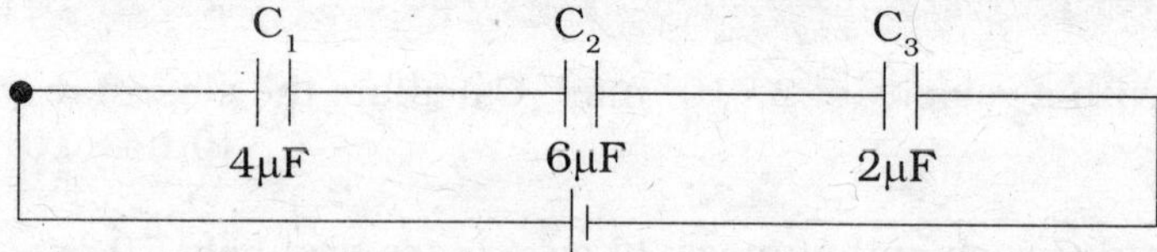

[75 V; 150 V]

35. Find the equivalent capacitance between A and B in the following circuit.

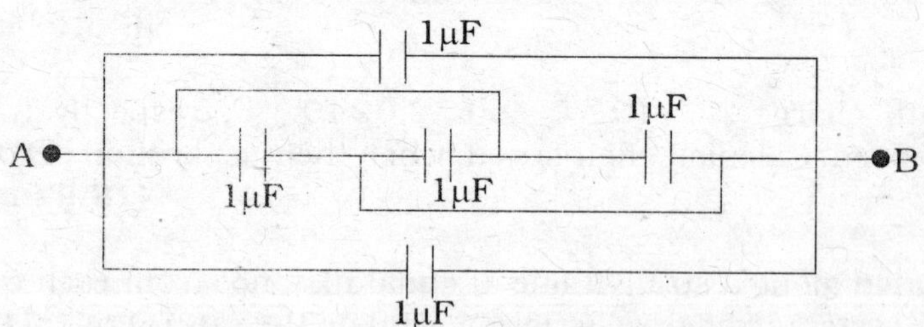

[5 μF]

36. Find the equivalent capacitance between A and B in the following circuit.

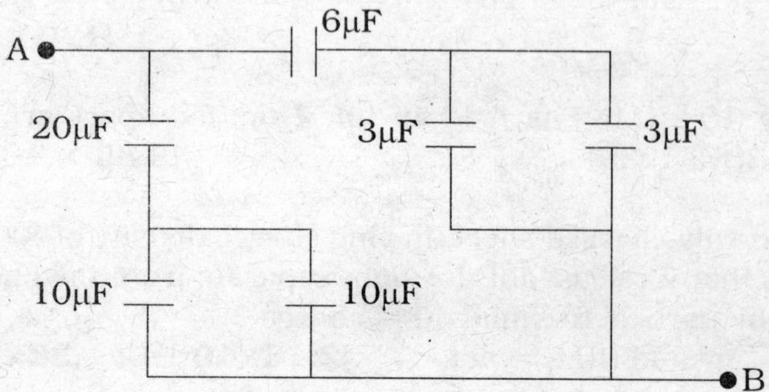

[13 μF]

37. Find the equivalent capacitance between A and B in the following circuit.

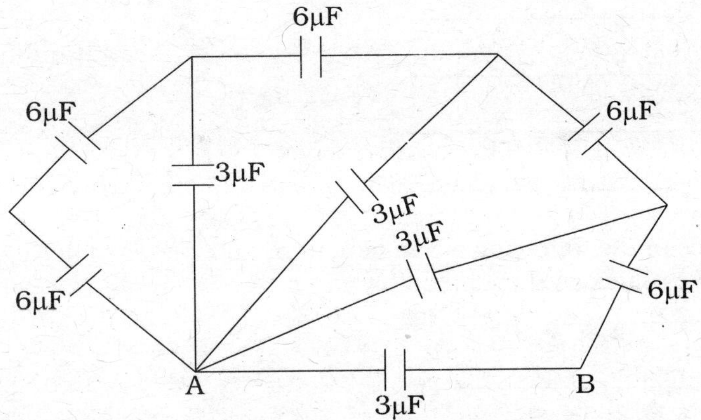

[6 μF]

38. Find the equivalent capacitance between. A an B in the following circuit.

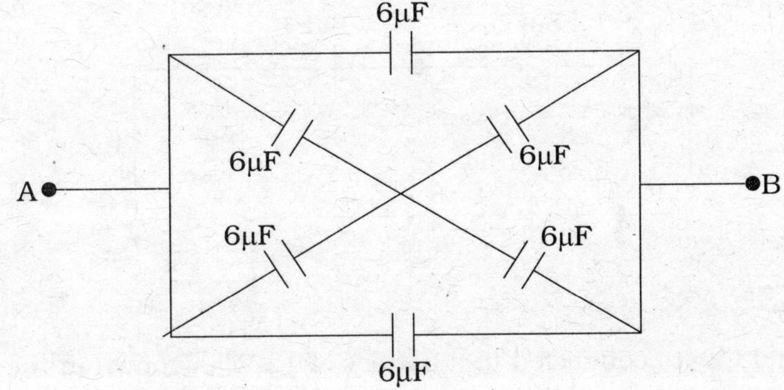

[18 μF]

39. Find the equivalent capacitance between A and B in the following circuit.

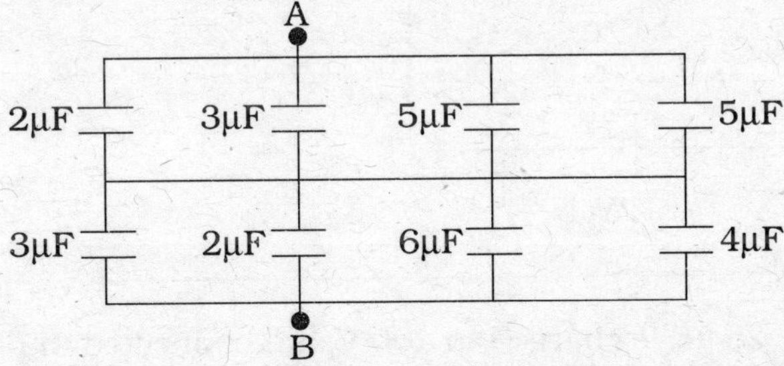

[7.5 μF]

40. Four plates of area A each are arranged as shown, with a separation d each between the neighbouring plates. What is the capacitance of the arrangement, between B and C ?

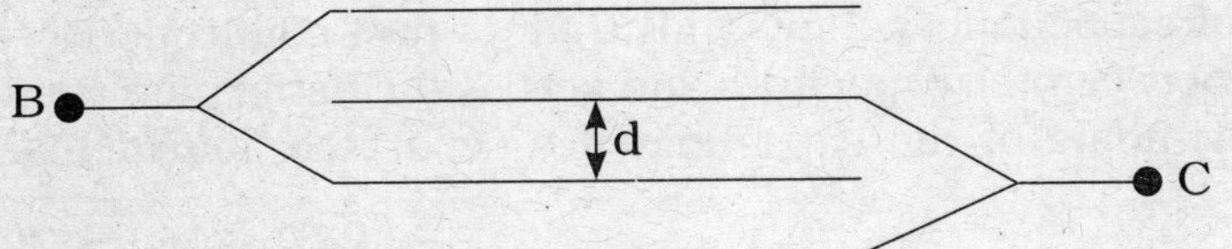

$\left[\frac{3\epsilon_0 A}{d}\right]$

41. What is the capacitance of the following arrangement of four plates of area A each and with separation between two successive plates as d each ?

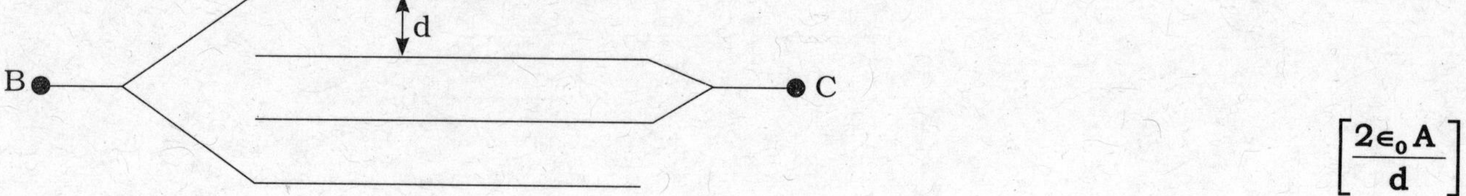

$\left[\frac{2\epsilon_0 A}{d}\right]$

42. What is the capacitance of the following arrangement of four plates of area A each and with separation between two successive plates as d each.

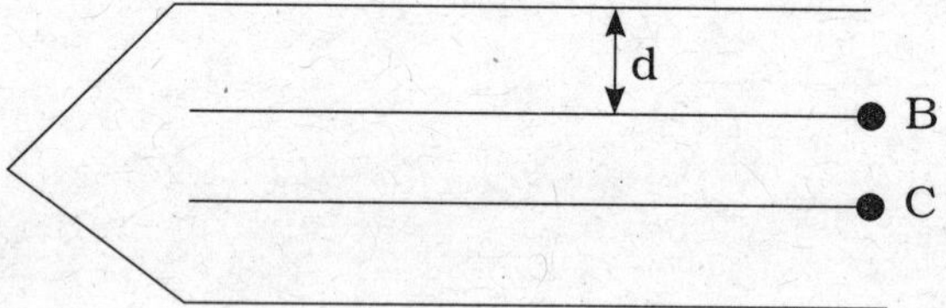

$\left[\frac{\epsilon_0 A}{2d}\right]$

43. Calculate the equivalent capacitance between the points A and B in the combination shown below.

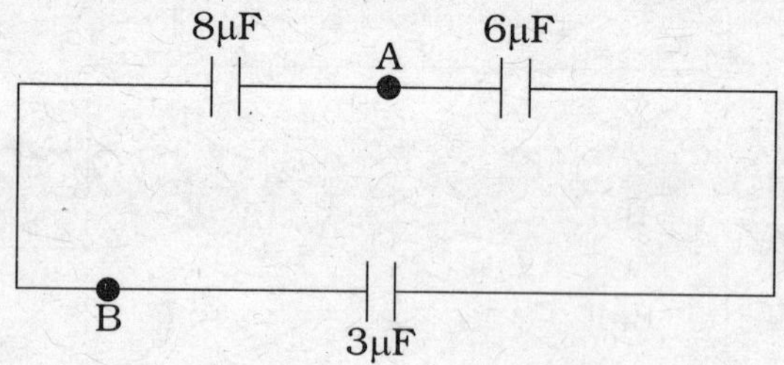

[10 μF]

44. Two capacitors C_1 and C_2 are connected to a battery of 12 V, as shown in the figure. Find the charge on each capacitor.

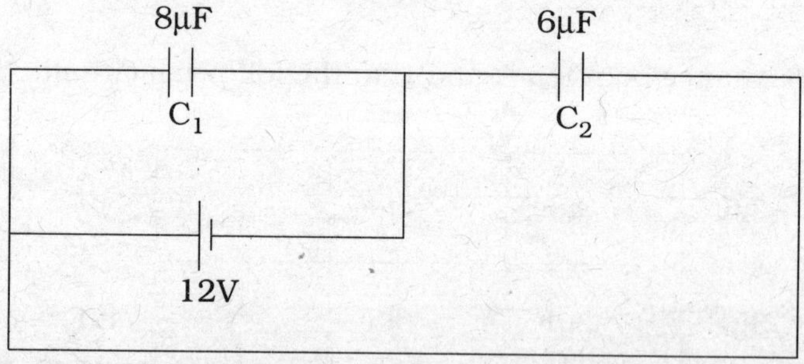

[96 μC; 72 μC]

45. A capacitor of capacitance 25 μF is charged to 400 V and connected in parallel with another capacitor of capacitance 75 μF charged to 200 V. Find their common potential. **[250 V]**

46. The separation between the plates of a parallel-plate capacitor is 2 cm. A slab of dielectric constant 5 and having a thickness of 1 cm is inserted between the plates. The separation between the plates is so adjusted that the capacitance of the capacitor remains unchanged. Calculate the value of the new separation between the plates. **[2.8 cm]**

47. A series combination of three capacitors of capacitances 2 μF, 3 μF, 6 μF is first connected across a 12V battery. Then all the connecting wires between the capacitors and across the battery are removed. Next, the three +ve terminals and -ve terminals of the capacitors are separately joined together. Calculate the charge on each capacitor.

$\left[\frac{24}{11}\,\mu C; \frac{36}{11}\,\mu C; \frac{72}{11}\,\mu C\right]$

48. In the circuit sketched below, find the charge residing on the 4 μF capacitor.

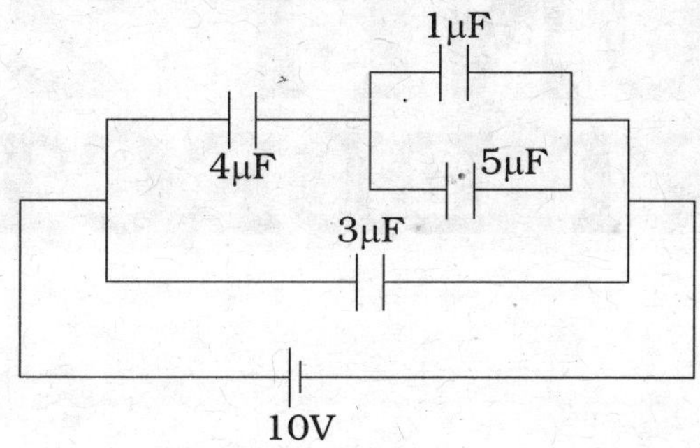

[24 μC]

49. Initially, the capacitance of a parallel-plate capacitor is measured as 10 μF. Find the increase in its capacitance, when a conducting slab of thickness equal to one-fourth of the separation between the plates is inserted between the plates. **[3.33 μF]**

50. A capacitor of capacitance 300 μF is charged by connecting to a 400 V supply. Then the supply is disconnected and the capacitor is connected to an uncharged 300 μF capacitor. Determine the electrostatic energy lost in the process. **[12 joule]**

UNIT 2
CURRENT ELECTRICITY

CHAPTER 2 (A)
Electric Current, Voltage, Resistance and Their Measurement

CHAPTER 2 (B)
Electric Energy and Power; Electrolytes

CHAPTER 2 (A)

ELECTRIC CURRENT, VOLTAGE, RESISTANCE AND THEIR MEASUREMENT

In the previous chapter on electrostatics we had dealt with the phenomena related to charges at rest, i.e. with 'static' electricity. When the charges are set into motion, they constitute an electric current. Then a new exciting set of phenomena related to 'current electricity' become observable. We shall study some of them in the present and the next chapter, alongwith the laws on which they are based and the basic quantities and characteristics that are associated with them.

The 'static electricity' is of little practical use. The reason is that electric charge as such is not energy. In contrast, the electric charge in motion, i.e. the 'current electricity', is one of the best forms of useful energy known to man. Hence it is of great practical value.

2A. 01. Electric Current and Charge Carriers; Current Density

The flow of electric current, e.g. in a wire, is similar to the flow of water in a pipe. The fundamental carriers in the 'water current' are the H_2O molecules. The flow of an electric current is basically a flow of electric charges. But since the charges have no independent existence, they need suitable carriers to be transported from one place to another. [It is somewhat like numerous children seated in floating boats, with each child riding his own boat. The floating of all the boats together in one direction, e.g. along a river stream, will constitute the 'current' of the children]. In solids, usually the charge carriers (or, say current carriers) are electrons. In liquids and gases, the charge carriers are positive and negative ions, in addition to the electrons.

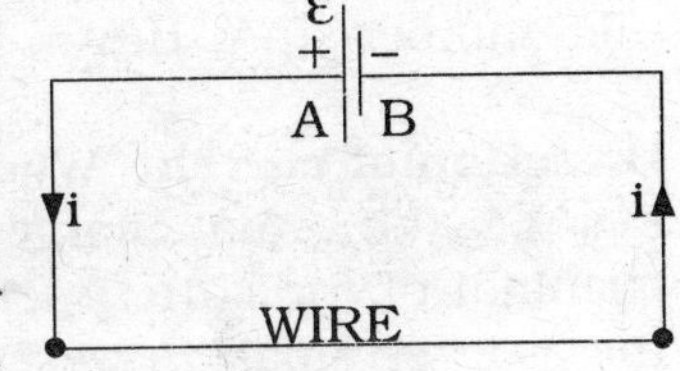

Fig. 2.1

The flow of electric current may also be visualized as similar to the flow of heat in a metallic rod from its hot end to the cold end. To maintain a steady flow of heat energy, the hot end needs to be constantly heated.

Consider the simple electric circuit shown in Fig. 2.1, consisting of a battery connected to a piece of wire. The negatively charged free electrons in the wire move towards the positive terminal A of the battery, which maintains a constant potential difference across its two terminals through chemical reaction. Thus a constantly flowing electric current is set up in the circuit. Let us examine the process ab initio, i.e. from the very beginning. As soon as the battery is connected to the wire, current starts building up in the wire, starting from a zero value. Very soon, within a tiny fraction of a second, it builds up to a steady value i, which then remains constant. So **i** is called **steady-state value of the current.** During the building-up period, a **transient current** flows through the circuit, whose value keeps on changing, until it becomes equal to i. We shall mostly restrict our study to the phenomena related to the steady state flow of electric current.

In simple terms, an electric current is a current of electricity, just like a current of water in a stream or a river.

Definition and units of electric current. The Fig. 2.2 depicts a steady-state current i flowing through a bat-shaped metallic conductor. For any imaginary plane across the conductor, like AB, CD, or EF, the current is the same, irrespective of the position or orientation of the plane. It is so because the charge has to be conserved (vide Chapter 1A, Section 1A.05). Any electron that passes through the plane EF has also to pass through any other plane like AB and CD. (The same applies to the water running through a pipe. Every drop of water entering at one end of the pipe has to leave at the other end. So the water is the quantity conserved here).

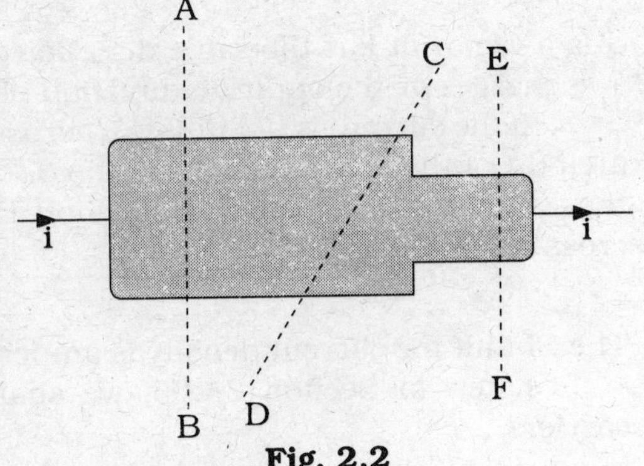

Fig. 2.2

If a small charge dq passes through a plane like AB or CD in time dt, then the current i passing through the plane is given by,

$$i = \frac{dq}{dt} \quad \text{[Electric current; instantaneous value]} \qquad \text{..... (2.1)}$$

In the steady-state, i is a constant value. Then we can write,

$$i = \frac{q}{t} \quad \text{[Electric current; steady-state value]} \qquad \text{..... (2.2)}$$

If t = 1, i = q. So we can **define steady-state electric current** as "the amount of charge flowing through any cross-section of a conductor in unit time".

From Eq. 2.1, in general, the amount of charge that passes through any cross-section of a conductor during a time interval zero to t is found by integration over the dq values,

$$q = \int dq = \int_0^t i dt \qquad \text{.....(2.3)}$$

From Eq. 2.2, the **SI unit for current** is expressed as coulomb per second (C/s), which is popularly termed as **ampere (A)**,

1A (1 ampere) = $1 Cs^{-1}$ (1 coulomb per second)

The definition of ampere follows from magnetic effects of current, which will be discussed in the next chapter (Section 3A.14). Smaller units of current are frequently used. These are miliampere (mA; 1mA = 10^{-3} A), microampere (μA; 1μA = 10^{-6}A), nanoampere (nA; 1nA= 10^{-9}A), etc.

Direction of current. When an electric current flows through a solid conductor, e.g. in a wire, as shown in Fig. 2.1, the actual charge carriers (electrons, which are negatively charged) move away from the negative terminal of the battery to the positive terminal. So actually the current flows from the –ve terminal to the +ve terminal. However, for historical reasons, conventionally the direction of current is taken as the opposite direction, i.e., the direction in which imaginary positive charge carriers will move through the conductor, as shown by the arrow marks in the figure.

Although we use an arrow mark to indicate the direction of flow of an electric current, note that the **current is a scalar quantity.** In Eq. 2.2 (or Eq. 2.1), both the charge q and the time t are scalars. So their ratio (i =q/t) has to be a scalar. The arrow only indicates the direction of movement of charges in the conductor, not a direction in open space.

Current density. If a uniform current i flows through a conductor, the amount of current flowing per unit area of a perpendicular cross-section of the conductor is called current density, $\vec{J}$. If the total cross-sectional area is A, then we have,

$$J = \frac{i}{A} \qquad \text{.....(2.4)}$$

$\vec{J}$ is a vector. It has the same direction as the electric current i, i.e. the imagined direction of movement of +ve charges or the opposite direction of actual movement of –ve charges.

If the current is not uniform across the conductor, the amount of current flowing through an elemental area $\overrightarrow{dA}$ of the cross-section will be $\vec{J}.\overrightarrow{dA}$ (note that $\overrightarrow{dA}$ is the area vector of the element, perpendicular to the element; vide Chapter 1B, Section 1B.12). Then the total current i is found by integrating over the total cross-sectional area,

$$i = \int \vec{J}.\overrightarrow{dA} \qquad \text{.....(2.5)}$$

The SI unit for current density is ampere-metre^{-2} (Am^{-2}).

Later, in Section 2A.05, we shall express the current $\vec{J}$ in terms of drift velocity of the charge carriers.

Example 1. How many electrons pass through a cross-section of a conductor in 1 minute, when the current flowing through it is 1mA ?

Solution. We know, i = q/t and q = eN (N: No. of electrons)

$$\therefore \quad N = \frac{it}{e}$$

Given: i = 1mA = 1×10^{-3} A; t = 60s; (e = 1.6×10^{-19} C)

$$\therefore \quad N = \frac{(10^{-3})\times(60)}{(1.6\times10^{-19})} = 3.75\times10^{17}$$ **(Answer).**

Example 2. An electric current of 5 mA flows through a wire of radius 1mm. Calculate the current density.

Solution. Current density, $J = \frac{i}{A} = \frac{i}{\pi r^2}$

Given: i = 5 mA = 5×10^{-3} A; r = 1 mm = 10^{-3} m

$$\therefore \quad J = \frac{(5\times10^{-3})}{3.14\times(10^{-3})^2} = 1.59\times10^3\ \text{Am}^{-2}$$ **(Answer).**

2A. 02. emf (Electromotive Force)

The Fig. 2.1 shows flow of electric current in a metallic wire connected to a battery. The current begins to flow because the negatively charged free electrons present in the wire immediately move towards the positive terminal A of the battery, as soon as it is connected to the wire. They neutralise the positive charge at A. At the same time, the depletion of electrons at the negative terminal reduces the negative charge at B. Thus, the electric potentials at A and B tend to become equal. Then how the flow of charge, i.e. the flow of electrons, is maintained ? Obviously, a mechanism of constant supply of +ve and –ve charges to the terminals A and B, respectively, is provided by the battery, so that a constant potential difference between them is maintained. As we know, this mechanism is provided by suitable chemical reactions in the battery. In this way, the battery becomes a source of constant supply of charges to the two terminals. Thus, a 'driving force' becomes available for a constant movement of the electrons in the circuit. So we say that in a literal sense the battery acts as a source of **'electromotive force' (emf).**

However, a careful consideration reveals that in the technical sense the term 'electromotive force' is a misnomer, i.e. a wrongly used term. In reality no force is involved in the supply of charges. Only a necessary amount of work is done in maintaining a constant p.d. between A and B. As explained earlier in Chapter 1B (Section 1B. 01), the potential difference between any two points in an electric field is a measure of the work done to move a unit charge from one point to the other. To maintain the necessary p.d. between A and B, +ve charge carriers present in the electrolyte in the battery move from a region of low electric potential at B to a region of higher potential at A. So a required amount of work has to be done. (Note that this movement of charges is just opposite to what the electric field between the terminals is causing to do inside the battery, viz. to move the charge carriers (electrons) in the circuit from the +ve terminal A to the –ve terminal B). Work is also done in bringing necessary –ve charge carriers present in the electrolyte to the negative terminal B of the battery and in maintaining a constant flow of electrons in the circuit.

Thus, we have a source of energy or, say, a source of emf, existing within the battery. As we know, in a battery this source is of chemical nature. But we have emf sources of other nature, too. The most familiar example is an 'electric generator', which can be of small size but also so huge that it is capable of supplying electricity to a whole city or more than one city ! It is of a mechanical nature. The energy derived from fall of water in a dam (hydroelectric power) or by burning oil, gas, coal, etc., drives a set of turbines to generate electricity. The 'solar cells' derive their energy from the sun, and so on.

In a closed circuit, e.g. when the battery has been connected to the wire in Fig. 2.1, the total work done can be divided in two parts: (i) work done by the moving charges (the electrons) in negotiating the potential difference V between the two terminals of the battery and thus keep on circulating (here the charges 'fall' through the p.d.). From the definition of p.d., V is the work done by the electric field per unit charge. (ii) Work done (dW) within the battery itself in transporting +ve and –ve charges to A and B, respectively. Since the total work done is the **emf, $\mathcal{E}$,** of the battery, we have,

$$\mathcal{E} = V + dW \quad \text{(closed circuit)} \qquad .....(2.6)$$

In an open circuit, e.g. when the battery is disconnected from the wire in Fig. 2.1, there is no flow of current and so there is no work to be done inside the battery, i.e. dW = 0. Then we have,

$$\mathcal{E} = V \quad \text{(open circuit)} \qquad .....(2.7)$$

Thus we find that the emf of a battery is equal to the potential difference between its two terminals in an open circuit. The same holds for all other sources of emf, e.g. an electric generator.

Since the emf is a potential difference value, its **unit** is the same as for the potential, viz, volt (V).

In section 2A. 12 we shall find how dW is related to the internal resistance of the cell.

2A. 03. Resistance, Resistor, Resistivity, and Conductivity

Suppose we take identical pieces of wire, i.e. of same length and area of cross-section, of different materials. We connect them to a common source of emf V and separately measure the current flowing through each of them. Then it is found that the values of current i_1, i_2, i_3, etc., flowing through the various wires are all different from one another. It follows that something inside each wire opposes, or say resists, the flow of current through it and that the degree of resistance offered varies from material to material.

Next, if we focus our attention to a particular piece of wire, i.e. to a particular material, it is found that as the voltage V is increased, the resulting current i proportionately increases,

$$V \propto i \qquad .....(2.8\text{ a})$$

$$\Rightarrow \quad V = iR \qquad .....(2.8\text{ b})$$

The constant of proportionality R is called the electrical **resistance** of the particular conductor (the particular piece of wire in this case). It may be **defined** as "the property of opposing the flow of electric current through a conductor". The greater the resistance, the smaller is the current. From Eq. 2.8b, it may **also be defined** as "the ratio of the potential difference V applied across a conductor to the resulting current i flowing through the conductor (R = V/i)". The resistance is a very important fundamental property of the conductors.

From Eq. 2.8b, the **SI unit for resistance** is found to be volt/ampere (V/A). it is popularly known as **ohm** (Ω),

$$1\text{ohm } (\Omega) = 1 \text{ volt-ampere}^{-1} \text{ (V/A)}.$$

Further, from the same equation, if V=1 volt and i = 1 ampere, we have R = 1 ohm. So we can **define one ohm** as " the resistance of a conductor through which a current of one ampere flows, when a potential difference of one volt is applied across it".

Resistor: So far we have been using the term 'conductor' for a body which conducts electricity through it. All these bodies offer resistance, high or low, to the passage of electricity through them. So they are called **resistors.** Thus all conductors are also resistors (an exception are 'superconductors'; Section 2A. 09).

We can make resistors of specified values, like 0.1 Ω, 3Ω, 700 Ω, 2000 Ω, etc., for use in various electric circuits, as will be described further in Section 2A. 06. Accordingly, we usually reserve the term resistor for a conductor which has a specified value of resistance.

A resistance or resistor is represented by the **symbol** -WW-.

Resistivity (or specific resistance). When we measure the resistance R (= V/i) of wires of different length ℓ and thickness (i.e. area of cross-section A) of a given material, we find that,

(i) $R \propto \ell$ when A is same (i.e. different lengths, but same thickness)

(ii) $R \propto (1/A)$, when ℓ is same (i.e. different thickness but same length)

Therefore, when both ℓ and A vary, we have,

$$R \propto (\ell/A)$$

$$\Rightarrow \quad R = \rho\frac{\ell}{A} \qquad .....(2.9)$$

The constant of proportionality **ρ** is called the **resistivity (or specific resistance)** of the material. Unlike the resistance, it depends only on the nature of the material, not on its dimensions (length or thickness). From Eq. 2.9, if ℓ =1 and A = 1, we have ℓ = R. So we can **define** the resistivity (or specific resistance) of a material as "the resistance offered by a conductor of that material of unit length and unit area of cross-section".

Using the Eq. 2.9, the value of resistance of a conductor can be easily calculated from the knowledge of resistivity, The values of resistivities of some common materials are listed in Table 2.1.

The **SI unit of resistivity is ohm-metre (Ω.m),** as easily seen from Eq. 2.9

The Eq. 2.9 can also be expressed in a more general way, in terms of the electric field strength $\vec{E}$ at a point in a conductor (instead of the p.d. V) and the current ensity $\vec{J}$ (instead of the current i),

$$\vec{E} = \rho \vec{J} \quad .....(2.10)$$

Later, in Section 2A.05, we shall examine the atomic level mechanisms that give rise to the resistance offered by a material. The resistance and resistivity will be expressed in terms of some ultimate quantities like e, m, drift velocity, etc.

Table 2.1. Resistivities (ρ) and Temperature Coefficients of Resistivity (α) of some common materials at room temperature (20 °C).

Material	ρ (Ωm)	α (K^{-1} or $°C^{-1}$)
	CONDUCTORS (METALS/ALLOYS)	
Silver	1.6×10^{-8}	4.1×10^{-3}
Copper	1.7×10^{-8}	4.3×10^{-3}
Aluminium	2.8×10^{-8}	4.4×10^{-3}
Iron	1.0×10^{-9}	6.5×10^{-3}
Platinum	1.1×10^{-9}	3.9×10^{-3}
Manganin (alloy)	48×10^{-8}	2×10^{-6}
Nichrome (alloy)	1×10^{-6}	4×10^{-4}
	SEMICONDUCTORS	
Silicon	2.5×10^{3}	-7×10^{-2}
Germanium	5×10^{-1}	-5×10^{-2}
Carbon (graphite)	4×10^{5}	-5×10^{4}
	INSULATORS	
Water	3×10^{5}	—
Glass	$10^{10}-10^{14}$	—
Hard Rubber	$10^{13}-10^{16}$	—
Mica	$\sim10^{15}$	—

Conductivity. The reciprocal of resistance is called **Conductance.** Its unit is **Ohm^{-1} [(Ω)$^{-1}$],** which is also expressed as **mho**, i.e. 'ohm' written in reverse order.

The reciprocal of resistivity (or specific resistance) is termed **conductivity (or specific conductance).** Its unit is **(ohm-metre) $^{-1}$ [(Ω) $^{-1}$] or mho-(metre)$^{-1}$.** A high value of resistivity means a low value of conductivity and vice-versa. From Eq. 2.9, we have,

$$\sigma = \frac{1}{\rho} = \frac{\ell}{RA}$$

If $\ell = 1$ and $A = 1$, we have $\sigma = 1/R$. Accordingly, the electrical conductivity of a material may be **defined** as "the reciprocal of the resistance offered by a conductor of that material of unit length and unit area of cross-section".

Among the metals, copper has a very low resistivity and therefore a very high conductivity. Also, it is not very expensive. So it is commonly used in all types of connecting wires and coils.

Example 3. What is the resistivity of the material of a wire of length 10 cm and area of cross-section of 10^{-6} m^2? The resistance of the wire is 6 Ω.

Solution. Given: R = 6 Ω; A = 10^{-6} m^2; ℓ = 10 cm = 0.1 m

We know,

$$R = \rho \frac{\ell}{A} \text{ (ρ: Resistivity)}$$

$$\therefore \quad \rho = \frac{RA}{\ell} = \frac{(6)\times(10^{-6})}{(0.1)}$$

$$= 6\times10^{-5}\ \Omega m \quad \textbf{(Answer).}$$

Example 4. A wire of resistance 2 Ω is drawn out so that its length is doubled. Calculate the resistance of the elongated wire.

Solution. Let ℓ_1, A_1 be the initial length and cross-sectional area of the wire and let ℓ_2, A_2 be the final values. Upon drawing the wire, its volume (= $A \times \ell$) remains unchanged.

$$\therefore \qquad \ell_1 A_1 = \ell_2 A_2$$

Since $\ell_2 = 2\ell_1$, we have, $A_2 = \dfrac{A_1}{2}$

Now, Initial resistance, $R_1 = \rho \dfrac{\ell_1}{A_1}$

Final resistance, $R_2 = \rho \dfrac{\ell_2}{A_2}$

$$\therefore \qquad \frac{R_1}{R_2} = \frac{\ell_1 A_2}{\ell_2 A_1} = \frac{\ell_1}{2\ell_1} \times \frac{A_1}{2A_1} = \frac{1}{4}$$

$$\therefore \qquad R_2 = 4R_1 = 4 \times 2 \qquad \text{(Given: } R_1 = 2\ \Omega)$$

$$= 8\ \Omega \quad \textbf{(Answer).}$$

2A. 04. Ohm's Law and Its Limitations

In the last section, it was described that if the voltage V across a metallic conductor increases, the resulting current i proportionately increases (Eqs. 2.8 a, b),

$$V \propto i$$

$$\Rightarrow \qquad V = iR$$

This important result was experimentally established in the year 1826 by the German physicist George Simon Ohm. Named after him, it is known as **Ohm's law.** It may be **stated** as follows:

"The current flowing through a conductor between any two points is directly proportional to the potential difference between them, if the temperature and all other physical conditions remain unchanged". (However, now we know of many departures from the law, as follows).

The ohm's law can also be expressed in terms of current density and electric field as,

$$\vec{E} = p\vec{J} \qquad \text{..... (2.10)}$$

A simple experimental arrangement to verify the ohm's law is shown in Fig. 2.3. A battery is connected to the two fixed ends of a rheostat (variable resistance). A fixed end and the sliding end of the rheostat are connected to any two points A and B of the conductor, respectively, thus providing a source of variable voltage V across A and B. The resulting variable current i is measured by an ammeter connected in series with the conductor. If we change the polarity of the battery, i.e. we interchange the +ve and -ve terminals of the battery, the direction of the current is reversed, but its magnitude remains unchanged. Thus, if initially the value of the current is +i, then after the reversal of polarity it becomes –i.

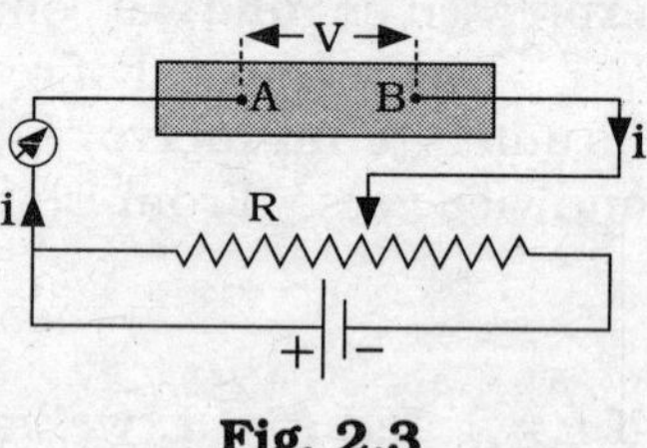

Fig. 2.3

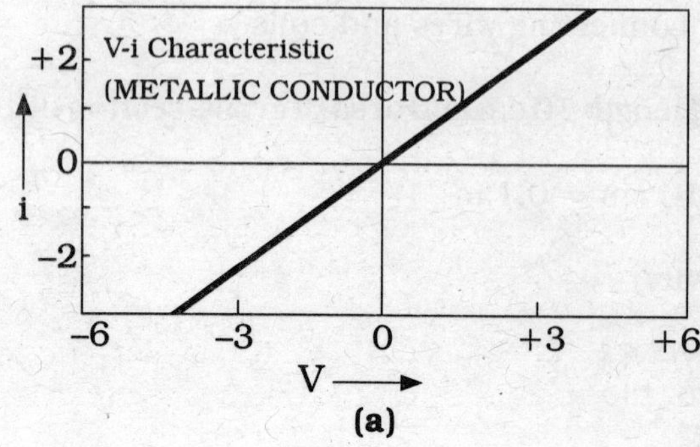

(a) **(b)**

Fig. 2.4

The V-i characteristic, which is a plot of voltage vs. current, for a typical conductor (of the type listed in Table 1) is shown in Fig. 2.4a. As expected from the relation V = iR, it is **linear,** i.e. the plot is a straight line. It follows that the Ohm's law is obeyed in this case. Both +ve and –ve values of i have been included in the plot.

However, when a similar plot is made for a solid state device called p-n junction diode, which also conducts electricity (i.e. it can also be called a conductor), the V-i characteristic is found to be non-linear (Fig. 2.4b). It means that the Ohm's law is not obeyed in this case. For all -ve values (i.e. for opposite polarity of V) and for the relatively small +ve values of V, the resulting current i is zero, i.e. the diode does not conduct electricity at all or, in other words, it behaves like an insulator for these values of V. Even for the relatively higher values of V, the plot is not a straight line, as seen in the figure. Similar or even more serious breakdown of the Ohm's law is observed in many more cases. One such example is another solid state device called thyristor, for which the V-i characteristic is not only found to be non-linear but also displays the same current value i for two different values of the voltage V. For some part of the curve, i is found to increase with decreasing V. A similar behaviour is observed for the semiconducting material gallium arsenide (GaAs).

Further, it is found that there is a departure from the Ohm's law in all cases if the electric field is very strong (i.e. for very high values of $\vec{E}$ in the relation $\vec{E} = \rho\vec{J}$ or for very high values of the p.d. V in the relation V =iR). Then the linear part of the V–i characteristic is found to bend for high values of V. The law also does not hold for certain alloys at extremely low temperatures, nearing absolute zero.

Thus, we find that the Ohm's law is not universally valid. Prominent departures from the law are observed in the following cases:

(i) Semiconducting materials and solid state devices.
(ii) In all the cases, when the applied electric field is too strong.
(iii) For certain alloys, at extremely low temperatures, nearing absolute zero.

Example 5. A potential difference of 12 V is maintained across a conductor of resistance 20 Ω. If the area of cross-section of the conductor is 10^{-6} m^2, calculate the current density in the conductor.

Solution. According to the Ohm's law, V= iR

$$\therefore \quad i = \frac{V}{R}$$

$$\therefore \quad \text{Current density, } J = \frac{i}{A} = \frac{V}{RA}$$

Given: V = 12 volt; R = 20 Ω; A = 10^{-6} m^2

$$\therefore \quad J = \frac{(12)}{(20)\times(10^{-6})} = 6\times10^{5}\ \text{Am}^{-2}$$ **(Answer).**

Additional knowledge

Although we find that the Ohm's law is not universally valid, this fact does not conflict with the validity of the relation V = iR. This relation can be used for any material or device under all physical conditions, to calculate the value of R(=V/i) for a particular value of V. The real test for the Ohm's law to be valid is that the V-i characteristic should be linear. Thus, it should be clearly understood that, strictly speaking, the relation V=iR cannot always be taken to represent the Ohm's law. But the relation itself is always valid for all conductors.

2A. 05. Origin of Resistivity (Atomic Level mechanisms); Drift Velocity; Mobility

Let us consider electric conduction, i.e. flow of electric current, in a metallic conductor. We know that a conductor offers some resistance to the flow of current. Since the current consists of flow of electrons, some mechanisms should be existing within the conductor which are responsible for producing resistance to a free flow of the electrons. We shall now examine these mechanisms. We shall do so for copper, which is the most commonly used conductor in everyday life, but the considerations made here will apply equally well to other conductors.

As we know, innumerable electrons are obtained in a conductor by detachment from outermost orbits of the atoms. We shall base our considerations on the **free-electron model,** according to which the electrons exist in the conductor as an electron gas (or electron cloud), such that they are free to move inside the conductor like the gas molecules in a chamber. We shall assume that during their movements the electrons only collide with the atoms of the conductor, i.e. we shall neglect their mutual collisions (this assumption is well justified, because the electrons are very light as compared to the atoms). (The number of atoms is enormous and so is the number of electrons. Even a volume of 1mm × 1mm × 1mm has about 10^{18} atoms i.e. a billion billion atoms!) The electron movement arises from thermal agitation at all temperatures above absolute zero.

Consider a rectangular conductor of length ℓ, as shown in Fig. 2.5. First consider the situation when no electric field is applied, i.e. the battery in the figure is disconnected from the conductor. The heavy atoms remain fixed at their respective positions, while the light electrons repeatedly keep colliding with them and keep getting **randomly scattered** in all possible directions, as shown in the figure. The velocities acquired by the electrons after collisions have widely distributed values. But an estimate of the average velocity between two successive collisions can be made, which is found to be $v_{av} \approx 10^6$ metre per second for copper. There is no net movement of the electrons along the length of the conductor, because on an average at any given time as many electrons move in one direction as in the opposite direction.

We express it by saying that the electrons have no net drift velocity v_d in the absence of electric field,

$$v_d = 0 \text{ (no electric field present)}$$

We can **define** the **drift velocity** as "the average velocity with which the free charge carriers inside a conductor move towards one end of the conductor".

The situation changes when an electric field is applied across the conductor, e.g. by connecting it to a battery, as shown in Fig. 2.5. Then two opposite mechanisms operate together inside the conductor. The force due to the field $\vec{E}$ accelerates the electrons in the direction opposite to $\vec{E}$ and so tends to move them towards the +ve end of the conductor (remember that conventionally the direction of the field $\vec{E}$ and the current i is taken from the +ve end to the -ve end of $\vec{E}$, while the negatively charged electrons actually travel from the -ve end to the +ve end). But their repeated random collisions with the atoms tend to keep them confined between the atoms and so strongly retard their movement along the length of the conductor. Under the combined influence of these two opposite tendencies, finally the electrons only succeed to drift very slowly towards the +ve end of the conductor with a very small average velocity,

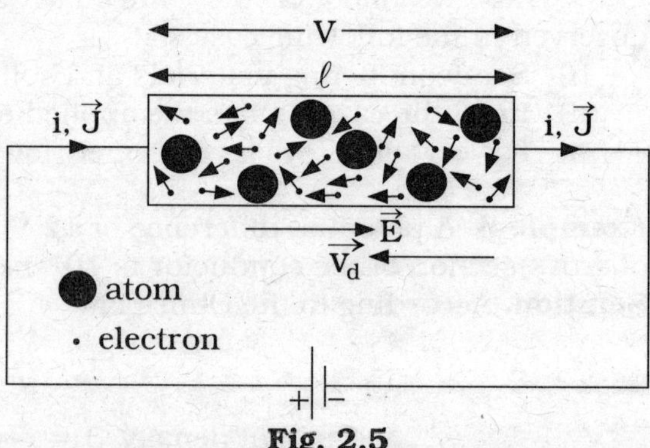

Fig. 2.5

$$v_d > 0 \text{ (when electric field is present)}$$

For copper, $v_d \approx 10^{-6}$ m/s, i.e. just a few mm per hour. So $v_d << v_{av}$.

We proceed to establish relations between (i) v_d and the current i and (ii) v_d and the field $\vec{E}$.

(i) Relation between v_d and i. Let the number density of free electrons, i.e. the number of electrons per unit volume, be n. Let ℓ be the length of the conductor and A be its cross-sectional area. Then the volume of the conductor is $A\ell$ and so the number of free electrons contained in it is $nA\ell$. Each electron has the charge e, disregarding the negative sign on e. Hence the total charge of the free electrons in the conductor is,

$$q = (nA\ell)e$$

The electrons move along the length ℓ of the conductor with the average velocity v_d. Hence the time taken by the total charge to move through any cross-section of the conductor is,

$$t = \frac{\ell}{v_d}$$

From Eq. 2.2, for a steady-state flow of current, we have,

$$i = \frac{q}{t}$$

$$= \frac{nA\ell e}{(\ell / v_d)} = nAev_d$$

$$\Rightarrow \quad v_d = \frac{i}{nAe} \quad ..... (2.11)$$

Since the current density $\vec{J}$ is current per unit area, i.e. $\vec{J} = i/\vec{A}$, the Eq. 2.11 can also be written in terms of the current density $\vec{J}$ as,

$$\vec{J} = (ne)\vec{v_d} \quad ..... (2.12)$$

(ii) Relation between v_d and E; Relaxation time. When the electric field is applied, each electron experiences a force eE (actually it is –eE, but we disregard the negative sign on e). Since force = mass×acceleration, the acceleration experienced by an electron is,

$$a = \frac{eE}{m} \quad ..... (A)$$

The acceleration is momentary, since each electron continuously suffers collisions with the atoms. After each collision, an electron makes a fresh start. Let the average time spent between two successive collisions be τ. Since v_d is average drift velocity of the electrons, we have (acceleration = velocity ÷ time),

$$a = \frac{v_d}{\tau} \quad ..... (B)$$

From Eqs. (A) and (B), we get,

$$v_d = \frac{eE\tau}{m} \quad ..... (2.13)$$

τ is called **relaxation time** (as if an electron gets some time to 'relax' between two consecutive collisions). If λ be the average distance covered between two successive collisions, we have

$$\tau = \frac{\lambda}{v_d} \quad ..... (2.14)$$

Accordingly, **τ is also known as 'mean free time'**. λ is called **'mean free path'** between two collisions.

The relaxation time may be **defined** as "the average time spent by a free electron between two successive collisions with the atoms in a conductor".

Mobility. The electric current flows in a conductor because the charge carriers can move, i.e. they are mobile. The **mobility (μ)** of the charge carriers is given by,

$$\mu = \frac{v_d}{E} \quad ..... (2.15)$$

If E = 1, $\mu = v_d$. So the mobility can be **defined** as "the magnitude of the drift velocity of the charge carriers per unit electric field".

Substituting for v_d from Eq. 2.13 into Eq. 2.15, we get.

$$\mu = \frac{e\tau}{m} \quad ..... (2.16)$$

Effect of heating on drift velocity and relaxation time

This has been discussed at the end of Section 2A. 10, where the effect of temperature on resistance and resistivity has been described.

Example 6. Calculate the drift velocity of electrons in a conducting wire, in which the value of current density is 2×10^2 A.cm^{-2}. The number density of the electrons in the wire is 8.4×10^{28} m^{-3}.

Solution. Drift velocity, $v_d = \frac{J}{ne}$ (J: current density)

Given: $J = 2\times10^2$ A.cm^{-2} = 2×10^6 A.m^{-2}; $n = 8.4\times10^{28}$ m^{-3}; ($e = 1.6\times10^{-19}$ C)

$$\therefore \quad v_d = \frac{(2\times10^6)}{(8.4\times10^{28})\times(1.6\times10^{-19})}$$

$$= 1.49\times10^{-4} \text{ ms}^{-1} \quad \textbf{(Answer).}$$

Example 7. A potential difference of 10 V is applied across a conductor of length 5 cm. If the average velocity of the charge carriers is 2×10^{-4} ms^{-1}, calculate the mobility of the carriers.

Solution. Mobility,
$$\mu = \frac{V_d}{E} = \frac{V_d\ell}{V} \left(\text{Electric field}, E = \frac{V}{\ell}\right)$$

Given: $V_d = 2\times10^{-4}$ ms^{-1}; $\ell = 5$ cm $= 5\times10^{-2}$ m; $V = 10$ volt

$$\therefore \quad \mu = \frac{(2\times10^{-4})\times(5\times10^{-2})}{(10)} = 10^{-6}\ \text{m}^2\text{s}^{-1}\text{V}^{-1}$$ **(Answer).**

Additional knowledge

An electric lamp in a room lights up as soon as its switch is turned on. Prima facie, this appears incosistent with the fact that the drift velocity of the electrons in copper wiring is extremely small, viz. just a few mm per hour. However, a deeper consideration reveals that the flow of current in the circuit, which includes the wiring and the lamp, actually depends on the establishment of the necessary potential difference, or say the eletric field, in the circuit. In other words, the driving force for the electrons is provided by the electric field. These necessary changes in the field occur almost with the speed of light ($c = 3\times10^8$ metre per second). The free electrons do drift at a very slow speed, but the process of drifting starts everywhere in the circuit immediately after the switch is turned on. The current in a wire is a measure of the quantity of charge and the rate of movement of charge through a cross-section of the wire. For instance, a current of 5 ampere means a charge of 5 coulomb passing per second. The speed at which the charge carriers pass is not really significant; a 5 ampere current can be generated both by a few carriers moving very fast or by numerous carriers moving slowly. The signal to the carriers to move travels almost at the speed of light; it is issued by the source of emf, which is usually the electric supply company's generator. So it is received by the charge carriers almost simultaneously.

2A. 06. Derivation of Ohm's Law on the Concept of Drift Velocity; Relation between R, ρ and τ

From Eq. 2.13, we have,
$$v_d = \frac{eE\tau}{m}$$

If the p.d. across the ends of the conductor is V and ℓ is the length of the conductor (Fig.2.5), then the field strength E is given by,
$$E = \frac{V}{\ell}$$

Hence,
$$v_d = \frac{eV\tau}{m\ell}$$

From Eq. 2.11, we have,
$$i = nAev_d$$
$$= \frac{nAe.eV\tau}{m\ell}$$
$$= \left(\frac{nAe^2\tau}{m\ell}\right)V$$
$$\Rightarrow \quad V = \left(\frac{m\ell}{nAe^2\tau}\right)i \quad \text{.... (2.17)}$$

The quantity within the brackets is a constant value for a given conductor. Therefore,
$$V = (\text{constant})\ i$$
$$\Rightarrow \quad V \propto i$$
$$\Rightarrow \quad V = iR$$

As we know, this relation represents the Ohm's law. R represents the resistance of the conductor,

$$R = \frac{m}{ne^2\tau} \cdot \frac{\ell}{A} \qquad \text{..... (2.18)}$$

From Eq. 2.9, we have,

$$R = \rho\frac{\ell}{A},$$

where ρ is the resistivity of the material of the conductor. Comparing with Eq. 2.18, we have,

$$\rho = \frac{m}{ne^2\tau} \qquad \text{..... (2.19)}$$

The Eqs. 2.18 and 2.19 express the relationship between the relaxation time (τ), the resistance (R), and the resistivity (ρ).

2A. 07. Carbon Resistors and Their Colour Code

Resistors of widely different values are commonly required for use in all types of electric circuits. They are commercially available. Two types of resistors are mainly produced:

(i) Wire-wound resistors

(ii) Carbon resistors

The wire-wound resistors are made by winding wire of a high resistivity material on an insulating base. Some suitable materials are nichrome (alloy of Ni, Fe, Cr), manganin (alloy of Cu, Ni, Mn), constantan (alloy of Cu, Ni), etc. (vide Table 1, Section 2A.03). These materials also have the advantage of having low values of temperature coefficient of resistance. But these resistors are relatively expensive and heavier and also occupy more space, as compared to the carbon resistors, to be described in the following.

The carbon resistors are made by mixing powdered carbon with a suitable binding material and then moulding the mixture into a cylindrical shape and enclosing the resistor within a ceramic or plastic jacket. Wire leads are connected to the two ends of the resistor. A typical carbon resistor is shown in Fig. 2.6. These resistors are compact, light and cheap. So wherever possible, they are preferred over wire-wound resistors. Carbon also has the advantage of possessing high resistivity and low temperature coefficient of resistance (Table 1; Section 2A.03).

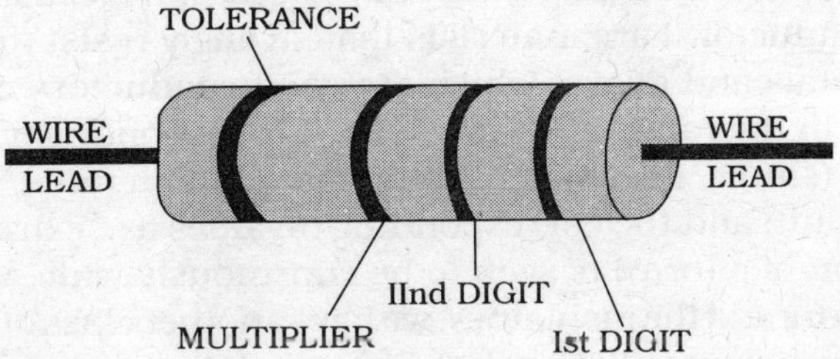

Fig. 2.6

Colour code for carbon resistors. A colour code is used to indicate the value of the resistance (Ω) of a resistor and its percentage accuracy. Each colour is assigned a digit, as listed in Table 2.2. A set of four coaxial coloured bands are marked on the surface of the resistor, as shown in Fig. 2.6. The colours of the first and second bands give the first two digits of the resistance value R. The colour of the third band gives the factor (multiplier) by which these digits are to be multiplied. The colour of the fourth band only indicates the tolerance, i.e. possible percentage variation in the value of R. If the fourth band is missing, it means that the tolerance is 20%. The following examples are illustrative:

(i)	Colour :	Brown	Green	Red	Gold
	Indication :	1	5	10^2	5%
	$R = (15 \times 10^2 \pm 5\%)\ \Omega$		or	$(1500 \pm 75)\ \Omega$	
(ii)	Colour :	Violet	Brown	Blue	Silver
	Indication :	7	1	10^6	10%
	$R = (71 \times 10^6 \pm 10\%)\ \Omega$		or	$R = (71 \pm 7.1)\,M\ \Omega$	

A helpful way to remember the colour code is to memorise the following sentence:

Big **B**reezy **R**ain **O**r **Y**our **G**reetings **B**rings **V**ery **G**ood **W**eather

Table 2.2. Colour code for resistors.

Colour	Digit	Multiplier	Tolerance
Black	0	10^0 (=1)	-
Brown	1	10^1	-
Red	2	10^2	-
Orange	3	10^3	-
Yellow	4	10^4	-
Green	5	10^5	-
Blue	6	10^6	-
Violet	7	10^7	-
Grey	8	10^8	-
White	9	10^9	-
Gold	-	10^{-1}	5%
Silver	-	10^{-2}	10%
No colour	-	-	20%

Example 8. A carbon resistor has the following sequence of colour bands marked upon it: green, violet, red, silver. What is its resistance ?

Solution. The colours green, violet and red represent the digits 5,7 and 2, respectively. The silver band represents 10% tolerance. Hence the value of the resistance is,

$$R = 57 \times 10^2 \pm 10\% = (5700 \pm 10\%)\ \Omega \quad \textbf{(Answer).}$$

2A. 08. Classification of Materials on the Basis of Conductivity; Conductors, Semiconductors, and Insulators

The good conductors are materials that permit easy passage of electricity through them. On the other hand, the insulators (or non-conductors) are materials that strongly resist the passage of electricity through them. For some well-known metals and alloys, which are good conductors of electricity and listed in Table 2.1, the resistivity (ρ) values lie in the range 10^{-6} – 10^{-9} Ωm. Correspondingly, the conductivity (σ) values are very high, in the range 10^6 – 10^9 $(\Omega m)^{-1}$, i.e. about 1 million to 1 billion units. For insulators, the ρ-values are very high, viz. about 10^5 – 10^{16} unit, and the corresponding σ-values are extremely low, viz 10^{-5} – 10^{-16} units. Thus, the total range of variation of ρ (or σ) is seen to be enormousiy wide, viz. about 10^{25} units !

Between the good conductors and the insulators, we have another class of materials, the semiconductors. As the name suggests, they have intermediate values of resistivity (or conductivity), as seen in Table 2.1 for two well-known semiconductors, silicon and germanium. A very important characteristic of these materials is that by **doping,** which consists of adding a very small amount of an impurity to the material, the resistivity can be highly reduced (or, say the conductivity can be highly increased). By controlling the doping, the conductivity and other electrical properties can be controlled. That is why the semiconductors, e.g. transistors, are so important for practical applications. In silicon, the resistivity comes down from $\sim 10^3$ units to $\sim 10^{-3}$ units, i.e. about one million times reduction in value (correspondingly, there is about one million times increase in conductivity).

The difference in the conductivity of conductors, semiconductors, and insulators can be understood in terms of the quantity of charge carriers available in them. Even at room temperature, the good conductors like metals and alloys have plenty of charge carriers (the free electrons, detached from the outer orbits of the atoms). When an electric field is applied across the conductor, the extra energy that becomes available frees some more electrons from the atoms. Similarly, if the temperature is raised, the thermal energy that becomes available also has the same effect of freeing more electrons. But at the same time the frequency of collision of the electrons with the atoms also increases which, in turn, reduces the value of the relaxation time τ in the Eq. 2.19,

$$\rho = \frac{m}{ne^2\tau}$$

The reduction in τ outweighs the increase in the value of n. The net effect is that the resistivity of the conductor increases, and so the conductivity reduces, as the temperature is increased. In the insulators, the electrons are strongly bound to the nuclei in the atoms and so a very small number of charge carriers exist in them at the room temperature. A rise in temperature or the application of electric field has little effect. Hence the insulators have a high resistivity, and therefore a low conductivity, both at the room temperature and higher temperatures.

The semiconductors have an appreciable number of charge carriers available in them even at the room temperature, but much less than the conductors. So they have an intermediate range of resistivity and conductivity. However, in a semiconductor the number of charge carriers increases very rapidly as the temperature increases. So in the Eq. $\rho = (m/ne^2\tau)$, the resistivity ρ decreases very rapidly. Hence the conductivity rises very rapidly (although the collisions also increase, but their effect is negligibly small). Thus, the effect of rise in temperature not only has the opposite nature as compared to the conductors, it is also of a far greater magnitude. The opposite nature is expressed by assigning a negative sign to the temperature coefficient of resistivity for semiconductors (Table 2.1).

A brief comparison of the characteristics of conductors, semiconductors, and insulators is provided in Table 2.3.

Table 2.3. Comparison of characteristics of conductors, semiconductors, and insulators.

Characteristic	Conductors	Semiconductors	Insulators
Resistivity (ρ)	very low	Intermediate	very high
Conductivity (σ)	very high	Intermediate	very low
Effect of rise in temperature on conductivity (σ)	σ decreases	σ increases very rapidly	negligible effect

2A. 09. Superconductivity; Superconductors

An ideal conductor should have infinite conductivity. An ideal insulator should have infinite resistivity. No such ideal conductors and insulators are commonly available in nature.

However, in 1911, Onnes discovered an ideal conductor under special laboratory conditions. He found that at a very low temperature of 4.2 K (i.e. nearly –269 °C) the resistance of mercury almost suddenly dropped to zero (Fig. 2.7). The phenomenon came to be known as **superconductivity** and the corresponding materials as **superconductors.** The temperature at which the sudden drop in the resistivity is observed is called **critical temperature,** or **transition temperature, T_c.** Later work has led to the discovery of many more materials that show superconductivity at relatively much higher T_C–values. Some of them are listed in Table 2.4. The earlier ones included metals, alloys, organic compounds, oxides, etc., but they were all non-cuprates, i.e. they did not contain copper. T_C–values upto 30K were observed. A breakthrough occurred in 1987 when an unusually high value of T_C = 90K was observed in the cuprate YBaCuO, which is a ceramic mateiral. Since then even higher T_C values have been observed in more ceramic cuprates. The highest observed is 153 K in HgBaCaCuO. These latter compounds have come to be known as **high-T_C superconductors.** It is noteworthy that at room temperature these ceramic materials are insulators, but when sufficiently cooled they become superconductors. (YBaCuO and Hg BaCaCuO are not the chemical formulae. They only indicate the various elements present in them, not their proportions).

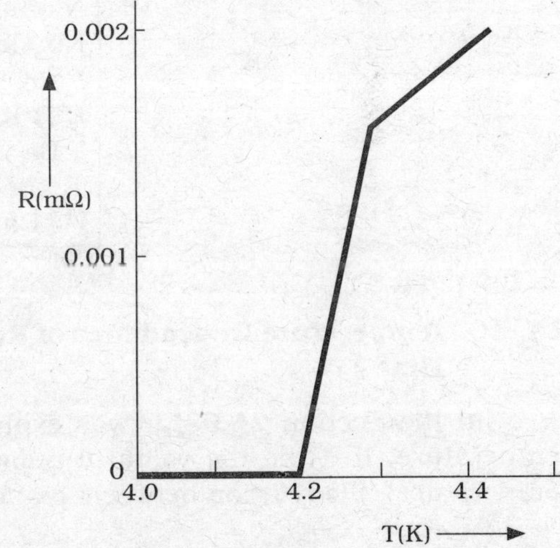

Fig. 2.7

Because of the zero resistivity of a superconductor, once an electric current begins to flow in it, it keeps on flowing afterwards without any reduction, even after the source of current is withdrawn. Such currents have been observed to flow years together in rings of superconducting materials, until the experimenters tired of observation and closed the experiment themselves !

Normally, when an electric current i flows in a conductor of resistance R, heat energy equal to i^2R is produced (vide Chapter 2B, Section 2B.01). But since R = 0 for a superconductor, no heat is produced at all in it. This fact is already being used in building very powerful compact **supermagnets,** which will otherwise require huge space and a flow of hundreds of litres of chilled water per minute for cooling the current-carrying coils. Other applications include several diverse fields like medicine, industry, energy, electronics, computers, information processing, non-destructive testing, high energy physics, geophysics, etc.

Another remarkable phenomenon displayed by the superconductors is **Meissner effect,** observed as follows. When a superconducting material is placed in a magnetic field and is gradually cooled, the magnetic field lines are completely ejected out of the material as soon as the transition temperature T_C is reached, i.e. as soon as the material becomes a superconductor. It means that the material then behaves like a perfect diamagnet, with no magnetic field existing inside it. The property of diamagnetism will be described in Chapter 3B (Section 3B. 13; Fig. 3.54). This has led to the possibility of making magnetically levitated superfast trains, which could run at speeds of several hundred kilometres per hour (vide 'Additional knowledge', at the end of Section 3B.13, Chapter 3B).

The phenomenon of superconductivity, first observed in mercury by Onnes and shortly afterwards in a number of other materials, could be explained on the basis of formation of pairs of electrons in the material, known as **'Cooper pairs',** which behave in a special way. A full explanation is out of the scope of this book. However, a satisfactory theory for explaining superconductivity of the later high-T_C ceramic materials is still not available. Search also continues for a stable high-T_C superconductor which will be superconducting at room temperature. When that is achieved, we shall be entering a new technological revolution in our world ! The necessary research efforts are being made all over the world.

Table 2.4. T_C-values of some superconducting materials.

Material	T_C(K)
NON-CUPRATES	
Mercury (Hg)	4.2
Lead (Pb)	7.20
Niobium (Nb)	9.25
Nb_3Ge(Alloy)	23.2
Ba K BiO	30
CUPRATES	
Y Ba Cu O	90
T Ca Ba Cu O	125
Hg Ba Ca Cu O	153

2A. 10. Temperature Dependence of Resistance and Resistivity; Effect on Drift Velocity and Relaxation Time

In the Section 2A.08, it was explained that the resistivity ρ of a conductor increases with rise in temperature. If ρ_0 be the value of resistivity at a reference temperature T_0 (which is usually the room temperature), the relation between ρ and the temperature T is approximately given by,

$$\rho_T = \rho_0[1 + \alpha(T - T_0)] \qquad \text{.... (2.20)}$$

$$\Rightarrow \quad \alpha = \frac{(\rho_T - \rho_0)}{\rho_0(T - T_0)} \qquad \text{..... (2.21)}$$

α is called **temperature coefficient of resistivity.** If $(T - T_0) = 1$, we have $\alpha = (\rho - \rho_0)/\rho_0$. So α can be **defined** as "the fractional change in the value of resistivity when the temperature is raised by unity". The unit of α is (K^{-1}) or (°C^{-1}).

The relationship between α and T for a conductor is non-linear, as shown in Fig. 2.8 for a typical case. However, the departure from linearity for a limited range of temperature, selected anywhere along the curve, is not large.

As mentioned in Section 2A.07, certain alloys are specially suited for making wire-wound resistors, as they have low values of α. These are (Table 1, Section 2A.03):

(i) Nichrome: Alloy of Ni, Fe, Cr
(ii) Manganin: Alloy of Cu, Ni, Mn
(iii) Constantan: Alloy of Cu, Ni

Carbon, used for making carbon resistors, also has a low value of α (Table 1), but it is negative.

The semiconductors have negative values of α (Table 1), i.e. their resistivity decreases with rise in temperature. The explanation for this opposite behaviour, as compared to the conductors, has already been provided earlier in Section 2A.08.

The relation between resistance (R) and α can be similarly expressed as the Eqs. 2.20 and 2.21 for the resistivity ρ,

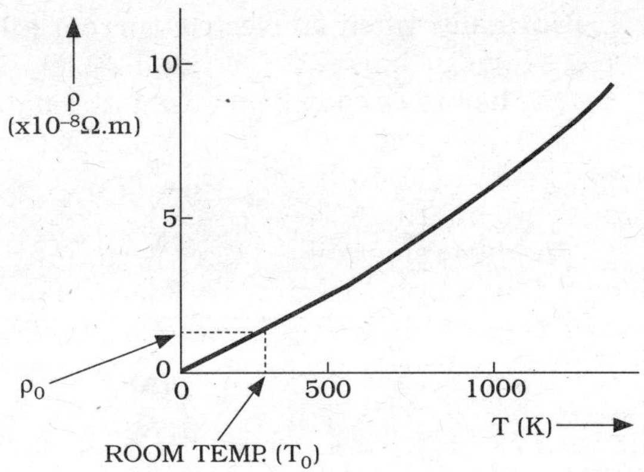

Fig. 2.8

$$R_T = R_0[1 + \alpha(T - T_0)] \quad \text{..... (2.22)}$$

$$\Rightarrow \quad \alpha = \frac{(R_T - R_0)}{R_0(T - T_0)} \quad \text{..... (2.23)}$$

Effect of temperature on drift velocity (v_d) and relaxation time (τ). As the temperature of a conductor increases, the thermal agitation of the atoms also increases. Hence the rate of collision of the free electrons, i.e. the number of collisions made per unit time, increases. The result is that the value of the relaxation time (τ), which is the average time spent by an electron between two successive collisions, is reduced.

From Eq. 2.14, $\tau = \lambda/v_d$, where λ is the mean free path, viz. the average distance covered between two successive collisions. The value of λ does not change with change in temperature. Therefore, as τ is reduced, v_d $(=\lambda/\tau)$ increases. Thus, the drift velocity v_d increases when the temperature rises.

Example 9. A conducting wire has resistance 1.8 Ω at 27° C and 2.1 Ω at 100° C. Determine the temperature coefficient of resistance, averaged over the temperature range 27° C to 100° C. Also find the resistance at 0° C, assuming the temperature coefficient of resistance to remain the same between 0° C and 27° C.

Solution. We know,
$$\alpha = \frac{(R_2 - R_1)}{R_1(t_2 - t_1)}$$

Given: $R_1 = 1.8\ \Omega$; $R_2 = 2.1\ \Omega$; $t_1 = 27°\ C$; $t_2 = 100°\ C$

$$\therefore \quad \alpha = \frac{(2.1 - 1.8)}{(1.8) \times (100 - 27)}$$

$$= 2.28 \times 10^{-3}\ °C^{-1} \quad \textbf{(Answer).}$$

Since, $R_1 = R_0(1 + \alpha t)$, we have,

$$R_0 = \frac{R_1}{1 + \alpha t_1} = \frac{(1.8)}{1 + (2.28 \times 10^{-3}) \times (27)}$$

$$= 1.69\ \Omega \quad \textbf{(Answer).}$$

2A. 11. Combination of Resistors (in Series and in Parallel)

If we are required to connect a large number of resistors together between two given points, across which a potential difference V is applied, it can be done in several different ways. However, there are two simplest ways to connect them:

(i) In series combination (Fig. 2.9a): Here the same current i flows through all the resistors but the potential drops across the various resistors are different, if their resistance values are different (V = iR).

(ii) In parallel combination (Fig. 2.9b): Here the potential drop across each resistor in the combination is the same as the applied potential drop V, but the current flowing through the various resistors is different, if their resistance values are different (i = V/R).

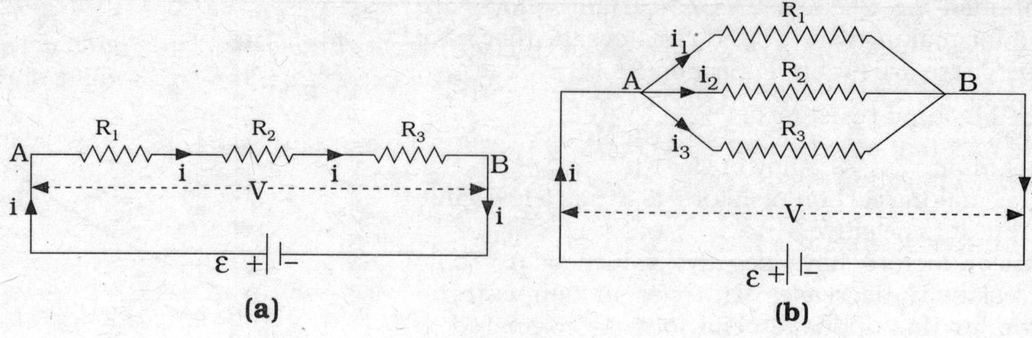

Fig. 2.9

In each case, we can calculate the **resultant resistance (or equivalent resistance)** of the combination, as follows. This resistance, when put in the circuit in place of the combination of several resistances, gives the same current in the circuit as with the combination.

(i) Series combination

Three resistors R_1, R_2, R_3 are connected one after another between the points A and B, which are connected to a source of emf $\mathcal{E}$ (Fig. 2.9a). Let the p.d. between A and B be V.

Since the same current i flows through all the resistors, from Ohm's law (V = iR) the p.d. values across different resistors will be,

$$V_1 = iR_1$$
$$V_2 = iR_2$$
$$V_3 = iR_3$$

But $V = V_1+V_2+V_3$. If the equivalent resistance is R, we have,

$$iR = iR_1+iR_2+iR_3$$
$$= i(R_1+R_2+R_3)$$
$$\Rightarrow \quad R = R_1+R_2+R_3$$

In general, for n resistors,

$$R = \sum_{m=1}^{n} R_m \qquad \text{..... (2.24)}$$

Thus for a series combination of resistors, the equivalent resistance is equal to the sum of the individual resistances in the combination. Clearly, the equivalent resistance is always larger than any individual resistance in the combination.

(ii) Parallel combination

A parallel combination of three resistors R_1, R_2, R_3, between the points A and B, is shown in Fig. 2.9b. The p.d. across all the resistors is the same as the applied p.d. V across the points A and B. The separate values of the current flowing through them are (i=V/R),

$$i_1 = V/R_1$$
$$i_2 = V/R_2$$
$$i_3 = V/R_3$$

The sum of i_1, i_2, i_3 has to be equal to the total current i in the circuit,

$$i = i_1+i_2+i_3$$
$$\Rightarrow \quad (V/R) = V/R_1+V/R_2+V/R_3$$
$$= V\,(1/R_1+1/R_2+1/R_3)$$
$$\Rightarrow \quad \frac{1}{R} = \frac{1}{R_1}+\frac{1}{R_2}+\frac{1}{R_3}$$

In general, for n resistors,

$$\frac{1}{R} = \sum_{m=1}^{n} \frac{1}{R_m} \qquad \text{..... (2.25)}$$

Thus, for a parallel combination of resistors, the reciprocal of the equivalent resistance R is equal to the sum of the reciprocals of the individual resistances in the combination. Hence the value of R is less than any one of the individual resistances.

Thus, we find that by connecting together a number of resistors in series their equivalent resistance increases, but by connecting them together in parallel it decreases.

2A. 12. Internal Resistance of a Cell

In the Section 2A.02, we had seen that when electric current flows in a circuit, work has to be done within the cell to maintain the current. Some positive charge has to flow constantly to the positive electrode of the cell and similarly some negative charge to the negative electrode. These charges encounter resistance to their movement by other atoms and ions within the electrolyte of the cell and by the atoms of the two electrodes. The combined resistance offered to the flow of electric current in a circuit by the electrolyte and the electrodes of a cell is called its **internal resistance.** It depends upon:

(i) Material of the electrolyte.
(ii) Material of the electrodes.
(iii) Spacing between the electrodes (more is the spacing, more is the internal resistance).
(iv) Area of the electrodes (more is the area, lesser is the internal resistance, since then the charges can move more freely inside the electrodes).

When a current is drawn in a circuit containing a cell of emf $\mathcal{E}$ (e.g. in Fig. 2.1), we have from Eq. 2.6,

$$\mathcal{E} = V + dW$$

The emf $\mathcal{E}$ represents the total work done in the circuit. V represents the work done in the circuit outside the cell and dW represents the work done inside the cell. V is also a measure of the potential difference between the two terminals (or, say electrodes) of the cell. Let R be the resistance of the circuit outside the cell and r the internal resistance of the cell. Let the current flowing in the circuit be i. Then by Ohm's law we have,

$$V = iR$$

Since both V and dW are measures of work done, it follows that,

$$dW = ir$$

So the Eq. 2.6 becomes,

$$\mathcal{E} = iR + ir$$
$$= i\,(R+r)$$

$$\Rightarrow \qquad i = \frac{\mathcal{E}}{(R+r)} \qquad \text{..... (2.26)}$$

Thus, the current i flowing in the circuit is determined by the external resistance and the internal resistance together. The Eq. 2.6 is now written as,

$$\mathcal{E} = V + ir \qquad \text{..... (2.27)}$$

When the cell is on the open circuit, i.e. it is not connected with any external circuit, no current is drawn from the cell. So i = 0 and we have,

$$\mathcal{E} = V \quad \text{(open circuit)}$$

We had concluded the same earlier in Section 2A.02 (Eq. 2.7). So in an open circuit, the terminal potential difference V of a cell is equal to $\mathcal{E}$, the emf of the cell. In a closed circuit, $V<\mathcal{E}$.

From Eq. 2.27, we have,

$$r = \frac{(\mathcal{E} - V)}{i}$$

But V = iR and so i = V/R. Therefore,

$$r = \left(\frac{\mathcal{E} - V}{V}\right)R \qquad \text{..... (2.28)}$$

2A. 13. Combination of Electric Cells

If we have more than one electric cells to be connected together between two given points, it can be done in several ways. But once again, as for the resistors (Section 2A.11), there are two simplest ways to do it: (i) in series combination and (ii) in parallel combination.

(i) Series combination

In this combination the cells are connected end to end, with the -ve terminal of the first cell joined to the +ve terminal of the next and so on. Suppose the combination has n identical cells of emf $\mathcal{E}$ each and external resistance in the circuit is R. The Fig. 2.10a depicts such a combination for three cells. We have,

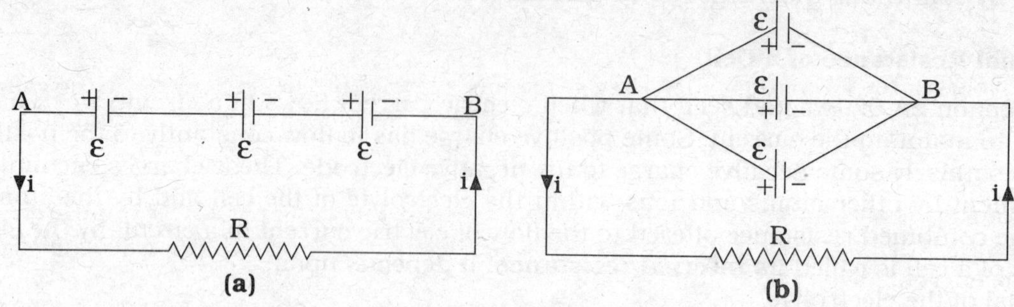

Fig. 2.10

Total internal resistance of the cells $= nr$

$\therefore$ Total resistance in the circuit $= (R+nr)$

Total emf of the cells $= n\mathcal{E}$

Therefore, the current in the circuit will be,

$$i = \frac{n\mathcal{E}}{(R+nr)} \qquad \text{..... (2.29)}$$

Two cases are of particular interest:

Case (a). $R \gg nr$. So in the denominator nr can be neglected as compared to R and we have,

$$i = n(\mathcal{E}/R)$$

i.e., the effective current in the circuit is n times the current due to a single cell.

Case (b). $R \ll nr$. now R is negligible in comparison to nr and we have,

$$i = \frac{n\mathcal{E}}{nr} = \frac{\mathcal{E}}{r}$$

i.e. the effective current in the circuit due to the combination is the same as that due to a single cell.

(ii) Parallel combination

Now the +ve terminals of all the cells are connected together at one end A and similarly the -ve terminals at the other end B (Fig. 2.10b). Since now the internal resistances of the n cells in the combination are connected together in parallel, the equivalent internal resistance r_n of the combination is given by (vide Eq. 2.25),

$$\frac{1}{r_n} = \sum_n \frac{1}{r_n} = \frac{n}{r}$$

$$\Rightarrow \qquad r_n = \frac{r}{n}$$

$\therefore$ Total resistance in the circuit $= R + \dfrac{r}{n}$

The potential at either end of each cell is the same as that for the combination. Therefore, the effective emf of the combination of cells is the same as the emf $\mathcal{E}$ of a single cell. Hence the current in the circuit is,

$$i = \frac{\mathcal{E}}{(R + r/n)}$$

$$\Rightarrow \qquad i = \frac{n\mathcal{E}}{(nR + r)} \qquad \text{..... (2.30)}$$

Again, we have two cases of particular interest:

Case (a). R>>r. Then in the denominator r can be neglected in comparison to nR and we have,

$$i = \frac{n\mathcal{E}}{nR} = \frac{\mathcal{E}}{R}$$

i.e. the current in the circuit is the same as will result due to a single cell.

Case (b). R<<r. Now, if the number of cells (n) is not extraordinarily large, nR can be neglected as compared to r and we have,

$$i = \frac{n\mathcal{E}}{R}$$

i.e. now the current in the circuit is n times the current due to a single cell.

2A. 14. Kirchhoff's Laws

In relatively simple electric circuits, the use of Ohm's law is sufficient to calculate the values of electrical parameters like voltage and current. But for complicated circuits, the use of Kirchhoff's laws becomes necessary.

1. Kirchhoff's first law (also known as Kirchhoff's current law or Kirchhoff's junction rule)

It **states** that "the algebraic sum of the currents entering any junction in an electric circuit is equal to the sum of the currents leaving the junction".

This law follows from the law of conservation of charge. There has to be a steady flow of charge through any given point in the circuit. The charge can neither accumulate at any point nor get depleted at that point. As an example, in Fig. 2.9b, at the junction A (or B) of four currents i, i_1, i_2, i_3 we must have,

$$i = i_1 + i_2 + i_3$$

The law also amounts to saying that the algebraic sum of all the currents meeting at a junction is zero. Thus, for the above example,

$$i - i_1 - i_2 - i_3 = 0$$

or, in general,

$$\Sigma i = 0$$

In writing the above equations, we have chosen the **sign convention** that the direction of a current entering the junction is taken as +ve and the direction of the current leaving the junction as –ve.

2. Kirchhoff's second law (also known as Kirchhoff's voltage law or Kirchhoff's loop rule).

It **states** that "the algebraic sum of all the potential differences along a closed loop in a circuit is zero".

$$\Sigma dV = 0$$

Starting from any point, the loop may be traversed either clockwise or anticlockwise, to reach back to the same point. For deciding the sign of a particular p.d., we adopt the following **sign conventions:**

(a) Resistance. The p.d. iR is to be taken as -ve if we move through the resistance R in the direction of the current i; otherwise +ve.

(b) emf. The emf $\mathcal{E}$ is to be taken as -ve if we move through the emf source from +ve terminal to the –ve terminal; otherwise +ve.

For example, in Fig. 2.9a there is only one loop, containing a battery of emf $\mathcal{E}$ and three resistances R_1, R_2, R_3 in series. Suppose we traverse the loop clockwise, starting from the point A and returning to it. The p.d.'s iR_1, iR_2, iR_3 are to be taken as –ve. The emf $\mathcal{E}$ is to be taken as +ve. Then, the Kirchhoff's second law for this loop gives,

$$\mathcal{E} - iR_1 - iR_2 - iR_3 = 0$$

The law is based on the conservation of energy. If it was not valid, any amount of energy could be generated by a repeated movement of charge in a closed loop.

The usefulness of the Kirchhoff's laws will become more clear later from the sample problems to follow.

2A. 15. Wheatstone Bridge

The simplest way to find the value R of a resistance is to measure the voltage V across it with a voltmeter and measure the current i in the circuit with an ammeter. Then use Ohm's law to calculate R (=V/i). However, since errors in two different measurements are involved, the accuracy of measurement is relatively low.

The Wheatstone bridge does away with the measurement of voltage and current altogether. It consists of a simple network of four resistances and a galvanometer connected to a sourse of emf, with the keys K_1

and K_2 suitably inserted in the circuit, as shown in Fig. 2.11. Three resistances (P,Q,S) are of known adjustable values and the fourth (R) is unknown. It is based on the principle of balancing the bridge resistances, such that the potential difference between two junctions of the resistances becomes zero. Then the galvanometer, connected between the two junctions, shows no deflection and the value of the unknown resistance R can be simply calculated from the knowledge of the values of P, Q, S as follows. Note that the galvanometer is used only as an indicator, viz. the indicator of **null-point** (no deflection). It does not measure any quantity. So its accuracy of measurement does not matter. Only the value of the resistances P,Q,S should be known accurately. For balancing the bridge, first the key K_1 is closed. Then the key K_2 is intermittently closed, while adjusting the value of S.

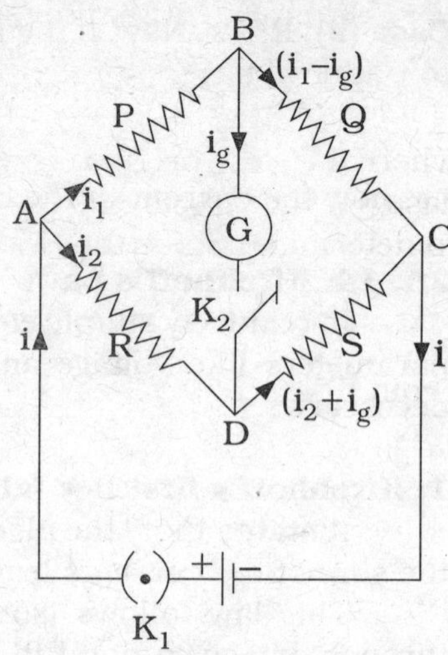

Fig. 2.11

The total current flowing in the circuit is i. At the point A, it is divided in the arms AB and CD of the bridge as i_1 and i_2. At B it is further divided into i_g and i_1-i_g, assuming that the point B is at a higher potential than D. When the bridge is balanced and the galvanometer shows no deflection, the points B and D are at the same potential (V_B=V_D) and i_g = 0. Then the current flowing through the arm BC is also i_1, like the arm AB. Similarly, the current flowing through the arms AD and DC is the same, viz. i_2. Using Ohm's law (potential difference V = iR),

$$(V_A-V_B) = i_1P \quad \& \quad (V_B-V_C) = i_1Q$$

$$(V_A-V_D) = i_2R \quad \& \quad (V_D-V_C) = i_2S$$

But under the condition of balance, V_B=V_D. Therefore,

$$(V_A-V_B) = (V_A-V_D) \quad \text{or} \quad i_1P = i_2R \qquad \text{..... (1)}$$

$$(V_B-V_C) = (V_D-V_C) \quad \text{or} \quad i_1Q = i_2S \qquad \text{..... (2)}$$

Dividing Eq. 1 by Eq. 2, we get

$$\frac{P}{Q} = \frac{R}{S} \qquad \text{..... (2.31 a)}$$

$$\Rightarrow \quad R = \frac{P}{Q}S. \qquad \text{..... (2.31 b)}$$

Since the values P, Q, S are known, R is calculated. Usually the ratio (P/Q) is chosen to be equal to 1, so that we simply get R = S, without any calculation. The bridge is most sensitive when the values of the four resistances P, Q, R, S are equal. Then the accuracy of determination of R is highest.

2A. 16. Wheatstone Bridge Application for Temperature Measurement

A suitable resistor is made by winding platinum wire of nearly half metre (50 cm) length on an insulating frame and inserted into the arm AD of the Wheatstone bridge (Fig. 2.11). The ratio (P/Q) is conveniently chosen to be unity, i.e. P = Q, but any other value can also be chosen. The experiment has the following three steps:

(i) First the platinum resistor is kept in ice, i.e. at zero temperature, and the bridge is balanced. Let R_o be the resistance of the wire at this temperature and let S_1 be the resistance value in the arm CD. Then,

$$\frac{P}{Q} = \frac{R_o}{S_1} \quad \text{or} \quad R_o = \left(\frac{P}{Q}\right)S_1 \qquad \text{..... (3)}$$

(ii) Next, the resistor is kept in steam, i.e. at 100°C, and the bridge is balanced. If R_{100} is the resistance of the wire at this temperature and S_2 the value of the resistance in the arm CD, then, as before, we have,

$$R_{100} = \left(\frac{P}{Q}\right)S_2 \qquad \text{..... (4)}$$

(iii) Finally, the resistor is kept in the bath whose temperature T is to be determined and the bridge is balanced. If R_T be the resistance of the wire at temperature T and S_3 the value of the resistance in the arm CD, we have,

$$R_T = \left(\frac{P}{Q}\right)S_3 \qquad \text{..... (5)}$$

From Eq. 2.23, the temperature coefficient of resistance α is given by,

$$\alpha = \frac{R_T - R_0}{R_0 T}, \qquad ..... (6)$$

where R_0 is the original resistance of a resistor and R_T is its increased resistance at a temperature T. In our case, R_0 is the resistance of the platinum wire at 0°C and R_T the resistance at the temperature T that we have to determine. Since the Eq. 6 is a general expression, we apply it to the steam temp. (100°C) and get,

$$\alpha = \frac{R_{100} - R_0}{R_0.100} \qquad ..... (7)$$

From Eqs. 6 and 7 we get,

$$\frac{R_T - R_0}{R_0 T} = \frac{R_{100} - R_0}{R_0.100}$$

$$\Rightarrow \quad T = \frac{R_T - R_0}{R_{100} - R_0} \times 100$$

$$= \frac{S_3 - S_1}{S_2 - S_1} \times 100 \qquad \text{(from Eqs. 3,4,5)}$$

Since the values of S_1, S_2, S_3 are known, the value of the unknown temperature T can be calculated.

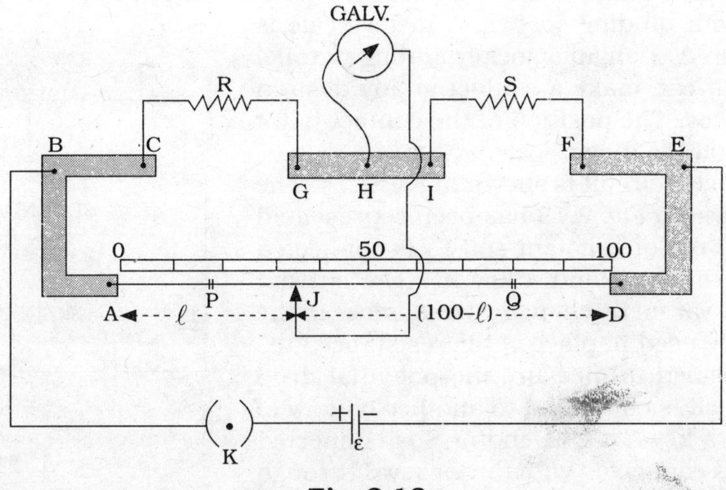

Fig. 2.12

2A. 17. Meter Bridge (or Slide-wire Bridge)

A meter bridge (or slide-wire bridge) is a practical form of the Wheatstone bridge. It consists of a thin wire AD of 1 metre length stretched over a wooden board by the side of a meter scale (Fig. 2.12). It is assumed to be of uniform cross-section. The ends of the wire are soldered to two L-shaped copper strips ABC and DEF fixed to the board. Between these two strips, another straight strip GHI is fixed to the board, thus providing two gaps CG and IF on the board. All the three strips have 3 binding screws each on them, as shown in the figure. A sliding jockey J moves in a groove on the board, making a sharp knife edge contact with the wire. The position of the jockey is read on the meter scale.

The terminals of a cell are connected through a key K to the binding screws B and E. A resistance box S is connected in the right gap and the unknown resistance R in the left gap. The ends of a galvanometer are connected to the binding screw H and the jockey J. Since the broad thick copper strips have negligible resistance, this network is identical with the Wheatstone bridge (Fig. 2.11). On sliding the jockey along the wire AD, a position is found out where the galvanometer shows no deflection. Then the bridge is balanced. This position is adjusted to be near the middle of the wire, by using a suitable resistance S from the resistance box, because then the sensitivity of the bridge is maximum.

In the balanced condition, if the length AJ of the wire is ℓ, the length of the remaining part JD of the wire is $(100-\ell)$. Let P and Q be the resistances of these two parts of the wire. If r be the resistance per unit length of the wire, we have,

$$P = \ell r \quad \text{and} \quad Q = (100-\ell)r$$

From Eq. 2.31a, in the balanced condition,

$$\frac{P}{Q} = \frac{R}{S} \quad \text{or} \quad R = \left(\frac{P}{Q}\right)S$$

$$\Rightarrow \quad R = \frac{\ell}{(100-\ell)}S \qquad \text{..... (2.32)}$$

Since the values of ℓ and S are known, the value of the unknown resistance can be calculated.

2A. 18. Potentiometer and its Applications

The potentiometer is one of the most accurate instruments used for electrical measurements. As its name implies, it primarily measures potential difference. However, it can also be employed for accurate measurement of a resistance or a current.

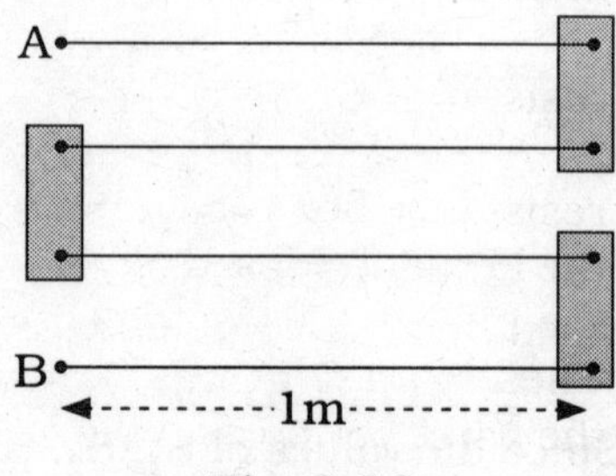

Fig. 2.13

It consists of a wood board on which a number (say N) of wires are stretched parallel to each other. The wires have uniform cross-section and each wire is 1 metre long. They are joined end to end by thick copper strips. So they **behave like a single wire** of length N metre. A series of four such wires is shown in Fig. 2.13. So the effective length of the wire is 4 m. The end terminals are provided with binding screws. A meter scale is fixed parallel to the wires . A triangular jockey sliding parallel to the wires is provided. It can make a contact at any desired point of any one of the wires. The position of the contact point can be conveniently read on the meter scale.

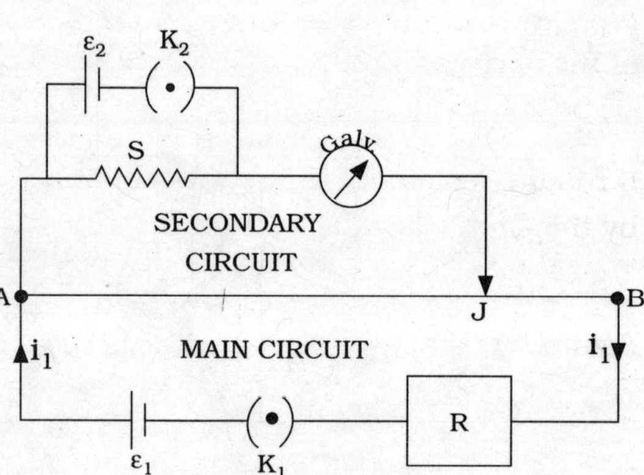

Fig. 2.14

An actual potentiometer circuit is shown in Fig. 2.14. The combination of wires shown in Fig. 2.13 has been represented by a single wire AB. A battery of constant emf $\mathcal{E}_1$ is connected to AB through a resistance box R and a key K_1. The current flowing through AB can be varied by changing resistances in the box. A rheostat can also be used in place of the resistance box. The circuit has been designed to measure the potential drop across a resistance S, which is connected to another battery of constant emf $\mathcal{E}_2$, through a key K_2. One end of S is connected with A and the other end is connected with the jockey J through a galvanometer.

Principle of potentiometer

The working of the potentiometer is based on the following two principles:

(i) When the potentiometer wire (of uniform cross-section) is connected to a source of constant emf, the potential drop V across a particular length ℓ of the wire is proportional to ℓ,

$$V \propto \ell$$

(ii) When another source of emf is connected to the wire such that it produces current in the opposite direction in the wire, the variable length ℓ can be so adjusted that no net current flows in the wire,

$$i = 0 \text{ (Two opposing emf's balanced)}$$

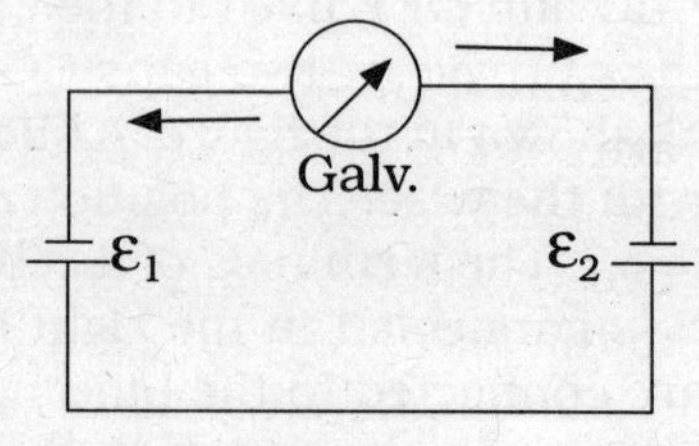

Fig. 2.15

The principle of balancing of emf's is easily understood from Fig. 2.15. The +ve terminals of two batteries of emf $\mathcal{E}_1$ and $\mathcal{E}_2$, respectively, are connected together. Their –ve terminals are separately connected to the two terminals of a shunted galvanometer. The two batteries tend to send current in opposite directions through the galvanometer. Therefore the galvanometer shows a deflection one way or the other, depending upon whether $\mathcal{E}_1 > \mathcal{E}_2$ or $\mathcal{E}_2 > \mathcal{E}_1$. If one of them, say $\mathcal{E}_2$, is adjustable, we can make $\mathcal{E}_1 = \mathcal{E}_2$. Then no current flows through the galvanometer, i.e. the galvanometer indicates a balance point. On either side of this balance point the galvanometer will show a deflection to the left or to the right, depending upon whether $\mathcal{E}_1 > \mathcal{E}_2$ or $\mathcal{E}_2 > \mathcal{E}_1$.

Example 10. Find the potential gradient along a potentiometer wire of length 5 m, which has a resistance of 5 Ω and a current of 40 mA flowing through it.

Solution. Given: $i = 40\text{ mA} = 4\times10^{-2}$ A; $R = 5\ \Omega$; $L = 5$ m

The total potential drop across the potentiometer is given by,

$$V = iR = (4\times10^{-2}) \times (5) = 0.2 \text{ volt}$$

$$\therefore \text{ Potential gradient } = \frac{V}{L} = \frac{(0.2)}{5}$$

$$= 4\times10^{-2}\ \text{Vm}^{-1} \quad \textbf{(Answer).}$$

Measurement of potential difference

The circuit is made (Fig. 2.14), as already described. It is intended to measure the p.d. across the resistance S.

First the key K_1 is closed so that a current i_1 flows through the wire AB. The resistance R in the resistance box is so adjusted that the potential drop across AB is greater than the p.d. to be measured. Next the key K_2 is closed, so that a current flows through the resistance S and a p.d. develops across S. By sliding the jockey J along the wire AB, the balance position is found, when the galvanometer shows no deflection. Then the p.d. across S, which we want to measure, is equal to the potential drop V across the length AJ of the wire. Let $AJ = \ell$. We know that,

$$V \propto \ell$$

$$\Rightarrow \quad V = P\ell$$

P is the constant of proportionality. If $\ell = 1$, then P = V. So P represents the potential drop per unit length of the wire. If AB = L (total length of the wire) and V_1 is the p.d. across AB, we have

$$P = \frac{V_1}{L}$$

If r is the resistance of the wire AB and R is the resistance from the resistance box at the balance point, then by the Ohm's law the current i_1 in the wire due to $\mathcal{E}_1$ is,

$$i_1 = \frac{\mathcal{E}_1}{(r+R)}$$

Again, by Ohm's law, the p.d. V_1 across the wire AB is,

$$V_1 = i_1 r = \frac{\mathcal{E}_1}{(r+R)}\cdot r$$

∴ The p.d. per unit length along AB is,

$$P = \frac{V_1}{L} = \frac{\mathcal{E}_1}{(r+R)}\cdot\frac{r}{L}$$

$$\therefore \quad V = P\ell = \frac{\mathcal{E}_1}{(r+R)}\cdot\frac{r\ell}{L}$$

Since $\mathcal{E}_1$, r, R, ℓ and L are known, V can be calculated.

Comparison of emf's of two cells

The circuit is made as shown in Fig. 2.16. The +ve terminals of the two cells whose emf's are to be compared are connected to the end A of the potentiometer wire. Their -ve terminals are connected to the terminals 1 and 2 of a two-way key, the third common terminal of which is joined to the jockey J through a shunted galvanometer.

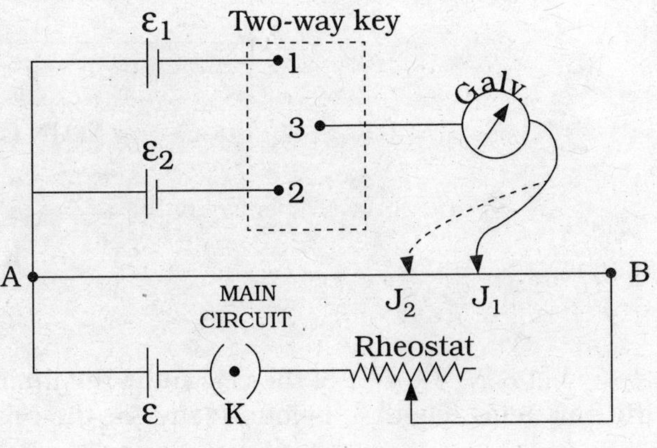

Fig. 2.16

The key K is closed and the rheostat is so adjusted that the p.d. across the potentiometer wire AB is greater than either $\mathcal{E}_1$ or $\mathcal{E}_2$. First the cell $\mathcal{E}_1$ is brought in the circuit by connecting the terminals 1 and 3 of the two-way key and the balance point J_1 is obtained by a suitable movement of the jockey along the wire. Next, keeping the current in the wire unchanged, the cell $\mathcal{E}_2$ is brought in the circuit by connecting the terminals 2 and 3 of the 2-way key and the new balance point J_2 is obtained.

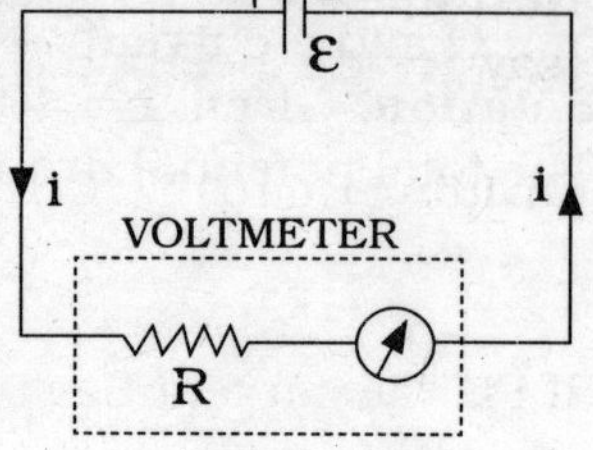

Fig. 2.17

If $AJ_1 = \ell_1$ and $AJ_2 = \ell_2$, then according to the principle of potentiometer,

$$\varepsilon_1 \propto \ell_1 \quad \text{and} \quad \varepsilon_2 \propto \ell_2$$

$$\Rightarrow \quad \frac{\varepsilon_1}{\varepsilon_2} = \frac{\ell_1}{\ell_2} \qquad \text{..... (2.33)}$$

Thus, the emf's of the two cells are compared.

If one of the two emf values, say ε_2 is known, the other is simply calculated as,

$$\varepsilon_1 = \frac{\ell_1}{\ell_2}.\varepsilon_2$$

Sensitivity of potentiometer

The **sensitivity** of a potentiometer is "the smallest potential difference that can be measured with the potentiometer".

The smaller is the potential gradient P, i.e. the potential drop per unit length, the more sensitive is the potentiometer. P can be reduced by (i) increasing the length of the potentiometer wire and (ii) reducing the current in the wire, by increasing the resistance connected in series with the wire.

Superiority of potentiometer over voltmeter

A potentiometer is superior to a voltmeter for measuring a given emf for the following reasons:

(i) It draws no current from the source of the emf. In this sense, it offers infinite resistance to the flow of current. The voltmeter, which has a high resistance connected in series with it, does draw a small finite current from the source (Fig. 2.17). So there is a small potential drop (V = iR) across it. Hence the value measured by it is not the true value of the emf. If the true value is ε, the measured value is $(\varepsilon - V)$.

(ii) The sensitivity of a potentiometer is relatively much higher.

Determination of Internal Resistance of a Cell

The circuit employed for determining the internal resistance of a cell is shown in Fig. 2.18. The method essentially consists of first balancing the potentiometer with the cell alone and next with the cell in combination with a resistance R.

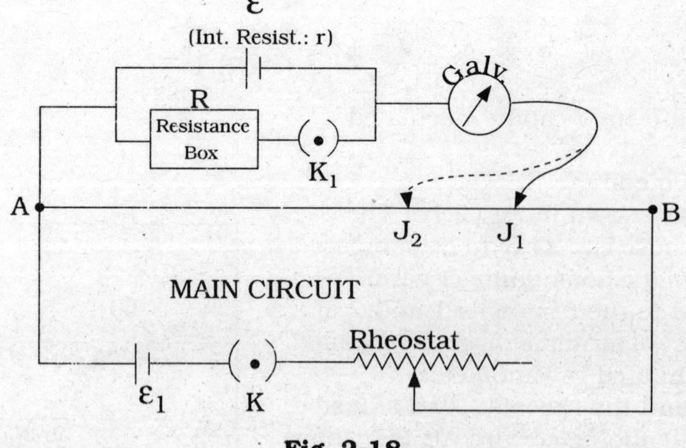

Fig. 2.18

Let ε be the emf of the cell and r the internal resistance of the cell. First, the key K_1 is kept open and the balancing length ℓ_1 is found (say, for the position J_1 of the jockey). Then,

$$\varepsilon = k\ell_1 \qquad \text{..... (i)}$$

Next the key k_1 is closed and the new balancing length ℓ_2 is found with a resistance R in the resistance box (say, for the position J_2 of the jockey). If V is the terminal voltage of the cell, we have,

$$V = k\ell_2 \qquad \text{..... (ii)}$$

Dividing Eq. (i) by Eq. (ii), we get,

$$\frac{\mathcal{E}}{V} = \frac{\ell_1}{\ell_2} \qquad \text{..... (iii)}$$

If i is the current flowing in the cell's circuit, we have,

$$\mathcal{E} = i\,(R+r) \qquad \text{..... (iv)}$$

and

$$V = iR \qquad \text{..... (v)}$$

Dividing Eq. (iv) by Eq. (v), we get,

$$\frac{\mathcal{E}}{V} = \frac{(R+r)}{R} = \left(1+\frac{r}{R}\right) \qquad \text{..... (vi)}$$

From the Eqs. (iii) and (vi), we get,

$$1 + \frac{r}{R} = \frac{\ell_1}{\ell_2}$$

$$\Rightarrow \qquad r = \left(\frac{\ell_1}{\ell_2} - 1\right)R \qquad \text{..... (vii)}$$

Thus, r is calculated from the known values of ℓ_1, ℓ_2, and R.

Example 11. The balance point for a cell on open circuit is found at 220 cm length of a potentiometer wire. When the cell is short circuited through a resistance of 20 Ω, the balance point is found at 200 cm of the wire. Calculate the internal resistance of the cell.

Solution. From the Eq. (vii), the internal resistance is given by,

$$r = \left(\frac{\ell_1}{\ell_2} - 1\right)R$$

Here, $\ell_1 = 220$ cm; $\ell_2 = 200$ cm; $R = 20\ \Omega$

$$\therefore \qquad r = \left(\frac{220}{200} - 1\right)\times(20)$$

$$= 2\ \Omega \quad \textbf{(Answer).}$$

Solved Conceptual Problems

(Including HOTS Questions)

Problem 1. At room temperature, the free electrons in a metallic conductor have velocities of the order of a few hundred kilometer per second. But no electric current is detectable in the conductor when directly connected to an ammeter. Why ?

Solution. Although at room temperature the free electrons in a metallic conductor move with fairly high velocities, their directions of motion are randomly distributed. Hence their average velocity is zero. Therefore, no electric current is detected.

Problem 2. A current i passes through a conductor for time t. What is the charge on the conductor ?

Solution. The charge on the conductor is zero. During the current flow through a conductor, equal amount of charge enters and leaves the conductor. Hence the net charge passing through the conductor is zero.

Problem 3. Under normal conditions, the drift velocity of free electrons in a conductor is of the order of 10^{-5} ms^{-1}. Then how a large current (e.g. ~ 10 A) is able to flow through the conductor ?
Solution. The current flowing through a conductor is given by,

$$i = neAv_d$$

n is the number density of free electrons, i.e. it is the number of free electrons per unit volume. In a good conductor, the number of free electrons per cubic metre is of the order of 10^{29}. Hence, in spite of a small value of the drift velocity v_d, the flow of a large current through the conductor becomes possible.

Problem 4. An electric lamp is located at a distance of 10 metre from its switch. The velocity of free electrons in the connecting wires is 8 mm/s. Approximately, after how much time will the lamp light up, when the switch is closed ?
Solution. The lamp will light up almost instantaneously. When the switch is closed, an electric field is instantly established in the wire all along its length with the speed of light ($c = 3\times10^8$ m/s). Hence all the free electrons in the wires start moving instantly.

Problem 5. When a conductor is subjected to an electric field, the free electrons inside it are accelerated. Then, why does their drift velocity not increase with time ?
Solution. The accelerated electrons collide with each other and with the positive ions, which randomly changes their paths. In this process, they keep on losing energy and hence cannot increase their drift velocity with time.

Problem 6. Why metals have lower resistivity and temperature coefficient of resistance, as compared to alloys ?
Solution. In metals, the atomic arrangement is more streamlined as compared to the alloys, because only one type of atoms are involved. Hence the motion of free electrons between the atoms is less random as compared to the alloys, resulting in a higher drift velocity and therefore lower resistivity and temperature coefficient of resistance.

Problem 7. Does the bending of a wire change its electrical resistance ? Justify your answer.
Solution. The bending of a wire does not change the mean free path of the electrons, because their collisions between themselves or with the +ve ions remain unaffected. Hence their drift velocity and therefore the electrical resistance of the wire remains unaffected.

Problem 8. Why is a potentiometer preferred to a voltmeter for measuring the emf of a cell ?
Solution. When a voltmeter is connected across the cell for measuring its emf, a small current passes through the cell and the voltmeter, causing a potential drop ($V = iR$, where R is the internal resistance of the cell) across the cell. Hence the measured emf is not the true value $\mathcal{E}$, but the value ($\mathcal{E} - V$). On the other hand, the potentiometer measures the emf in the null position of the galvanometer, i.e. when no current passes through the measuring instrument, viz. the galvanometer. Hence the true value of $\mathcal{E}$ is measured by the galvanometer.

Problem 9. Two potentiometers A and B have their wire lengths as 5 m and 10 m, respectively. Explain which of them will give a more accurate measurement of the internal resistance of a cell.
Solution. The least count of the potentiometer B is smaller, viz. half that of A (because the same potential drop is distributed over double the length). Hence it will give more accurate measurement of the internal resistance of a cell.

Problem 10. Why is it preferable to obtain balance point in the middle of the wire of a potentiometer or a meter bridge ?
Solution. When the balance point is obtained in the middle of the wire, the effect of the end error on the measurement is minimum. Further, in this condition the sensitivity of the galvanometer or the meter bridge is maximum.

Solved Numerical Problems

Charge carriers; Ohm's law; Drift velocity/mobility

Problem 1. One billion electrons are found passing through a cross-section of a wire in one microsecond. Find the current flowing through the wire.

Solution. The charge on one electron is, $e = 1.6\times10^{-19}$ C. Therefore, the charge on one billion (1×10^9) electrons is,

$$q = 10^9\times(1.6\times10^{-19}) = 1.6\times10^{-10}\text{ C}$$

Given: $t = 1$ microsecond $(1\mu s) = 1\times10^{-6}$ s

$\therefore$ the current, $$i = \frac{q}{t} = \frac{(1.6\times10^{-10})}{(1\times10^{-6})}$$

$$= 1.6\times10^{-4}\text{ A (or 0.16 mA)}$$ **(Answer).**

Problem 2. A current of 800μA flows through an electric bulb. How many electrons pass through a cross-section of the filament of the bulb in 1 second?

Solution. Given: $i = 800\ \mu A = 800\times10^{-6}\text{ A} = 8\times10^{-4}\text{ A};\ t = 1s$

Let the required number of electrons be N. Then the total charge residing on them is,

$$q = N\times(1.6\times10^{-19})\text{ C}$$

We know, $$q = it$$

$$\Rightarrow \quad N\times(1.6\times10^{-19}) = (8\times10^{-4})\times(1)$$

$$\Rightarrow \quad N = \frac{8\times10^{-4}}{1.6\times10^{-19}} = 5\times10^{15}$$ **(Answer).**

Problem 3. 6.4×10^{17} electrons pass through a cross-section of a wire in 2 second. Calculate the current flowing through the wire.

Solution. Given: $n = 6.4\times10^{17};\ t = 2$ s

We know, $$i = \frac{q}{t} = \frac{ne}{t}$$

$$= \frac{(6.4\times10^{17})\times(1.6\times10^{-19})}{(2)}$$

$$= 5.12\times10^{-2}\text{ A (or 51.2 mA)}$$ **(Answer).**

Problem 4. As suggested in the Bohr's model of hydrogen atom, an electron moves around the nucleus in a circular orbit of radius r_0, with a frequency ν. What is the current associated with this motion ?

Solution. Let the electron take time T to complete one revolution, i.e. the time period of motion of the electron is T. We know,

$$i = \frac{q}{t}$$

$$= \frac{e}{T} \quad (\because \text{ in our case, } q = e\ \&\ t = T)$$

$$= e\nu \quad (\because \nu = 1/T)$$ **(Answer).**

e

Nucleus

r_0

Problem 5. The number of hydrogen ions (i.e. protons) passing per second through a cross-section of a discharge tube is 1×10^{16}, while the number of electrons drifting per second in the opposite direction is 2×10^{16}. Calculate the magnitude of current flowing through the tube.

Solution. The protons are positively charged and the electrons are negatively charged. Therefore, they both contribute current in the same direction.

Let N_e and N_p be the number of electrons and protons, respectively, passing through the cross-section in time t. We are given that,

$$N_e = 2\times10^{16};\ N_p = 1\times10^{16};\ t = 1s;\ (e = 1.6\times10^{-19}\text{ C})$$

The current is given by,

$$i = \frac{q}{t} = \frac{(N_e + N_p)e}{t}$$

$$= \frac{(2+1)10^{16} \times (1.6\times10^{-19})}{(1)}$$

$= 4.8\times10^{-3}$ A (or 4.8 mA) **(Answer).**

Problem 6. The charge flowing through a conductor varies with time as q = $(2t^3 + 4t + 6)$. Find the current flowing through the conductor at time t = 2 second.

Solution. $i = \frac{dq}{dt} = \frac{d}{dt}(2t^3 + 4t + 6)$

$= 6t^2 + 4$

$\therefore$ $i(2) = 6\times(2)^2 + 4$

$= 28A$ **(Answer).**

Problem 7. If the electric current flowing through a wire varies as i = 10 sin (100πt), calculate the charge passing through a cross-section of the wire in the time interval t = 0 to 1/100 second.

Solution. $i = \frac{dq}{dt}$

$$\therefore \quad q = \int_{t_1}^{t_2} dq = \int_{t_1}^{t_2} i.dt = \int_{t_1}^{t_2} [10\sin(100\pi t)].dt$$

$$= \left|-\frac{10}{100\pi}\cos(100\pi t)\right|_0^{1/100} = \frac{1}{10\pi}[1-(-1)]$$

$$= \frac{2}{10\pi} = 6.37\times10^{-2}\text{ C}$$ **(Answer).**

Problem 8. Consistent with the Bohr's model of hydrogen atom, an electron is moving with a speed of 2.2×10^6 ms^{-1} in a circular orbit of radius 5.1×10^{-11} m. Calculate the current associated with its motion.

Solution. $i = \frac{q}{t} = \frac{e}{T}$ (T: time period of electron's revolution)

$$T = \frac{\text{Distance traversed in 1 revolution}}{\text{Speed}} = \frac{2\pi r_0}{v}$$

$$\therefore \quad i = \frac{ev}{2\pi r_0}$$

Given: $r_0 = 5.1\times10^{-11}$ m; $v = 2.2\times10^6$ ms^{-1}; $(e = 1.6\times10^{-19}$ C)

$$\therefore \quad i = \frac{(1.6\times10^{-19})\times(2.2\times10^6)}{2\times3.14\times(5.1\times10^{-11})} = (1.1\times10^{-3})\text{ A}$$ **(Answer).**

Problem 9. The number densities of electrons and holes in an extrinsic semiconductor of cylindrical shape are 7×10^{13} cm^{-3} and 4×10^{13} cm^{-3}, respectively. The cross-sectional area of the semiconductor is 10^{-6} m^2 and its length is 5 mm. If the electrons and holes take 10 second and 25 second, respectively, to pass through a cross-section of the semiconductor, calculate the current flowing through the semiconductor.

Solution. $$i = \frac{q}{t} = \left(\frac{q_e}{t_e} + \frac{q_h}{t_h}\right) = eA\ell\left(\frac{n_e}{t_e}\right) + eA\ell\left(\frac{n_h}{t_h}\right) = Ae\ell\left(\frac{n_e}{t_e} + \frac{n_h}{t_h}\right)$$

Given: A $= 10^{-6}$ m^2; $\ell = 5\times10^{-3}$ m; $(e = 1.6\times10^{-19})$; $n_e = 7\times10^{13}$ $cm^{-3} = 7\times10^{19}$ m^{-3}; $t_e = 10$ s;

$n_h = 4\text{x}10^{13}$ $cm^{-3} = 4\times10^{19}$ m^{-3}; $t_h = 25$ s

$$\therefore \quad i = (10^{-6})\times(1.6\times10^{-19})\times(5\times10^{-3})\left[\frac{7}{10} + \frac{4}{25}\right]\times10^{19}$$

$= 6.88\times10^{-9}$ A **(Answer).**

Problem 10. The variation of charge flowing through a conductor is given by,

$$q = 6t - \frac{3}{2}t^2 + \frac{5}{3}t^3$$

Find (a) the initial value of current, (b) the time after which the current becomes minimum, and (c) the minimum value of current.

Solution. (a) Current, $i = \frac{dq}{dt} = \frac{d}{dt}\left(6t - \frac{3}{2}t^2 + \frac{5}{3}t^3\right)$

$$= 6 - 3t + 5t^2$$

For initial current, t = 0.

$$\therefore \quad i_{initial} = 6 - 3\times(0) + 5\times(0)$$

$$= 6 - 0 + 0 = 6 \text{ A} \quad \textbf{(Answer).}$$

(b) $\frac{di}{dt} = -3 + 10t$

For i to be minimum,

$$\frac{di}{dt} = 0$$

$$\Rightarrow \quad -3 + 10t = 0$$

$$\Rightarrow \quad t = 0.33 \text{ s} \quad \textbf{(Answer).}$$

(c) for minimum value of current, t = 0.33 s

$$\therefore \quad i_{min} = 6 - 3\times(0.33) + 5\times(0.33)^2$$

$$= 5.56 \text{ A} \quad \textbf{(Answer).}$$

Problem 11. A metal wire of uniform cross-section and having a resistance of 16 Ω is melted and cast into a new wire of different uniform cross-section. The diameter of the new wire is double that of the original wire. Calculate the resistance of the new wire.

Solution. Let R, A, l and R′, A′, ℓ′ be the resistance, area of cross-section, length of the original and the new wire, respectively. We are given that R = 16 Ω.

When the original wire is cast into new wire, its volume (= Area × length) does not change. Further, when the diameter becomes double, the area of cross-section ($A = \pi r^2$) becomes 4 times.

$$\therefore \quad A' \times \ell' = A \times \ell$$

$$\Rightarrow \quad \ell' = \left(\frac{A}{A'}\right)\times \ell = \frac{1}{4}\ell = \frac{\ell}{4}$$

The resistance of the old wire is given by (ρ: resistivity).

$$R = 16\Omega = \frac{\rho\ell}{A}$$

The resistance of the new wire is given by,

$$R' = \frac{\rho\ell'}{A'} = \frac{\rho(\ell/4)}{A'}$$

$$= \frac{\rho(\ell/4)}{(4A)} = \frac{1}{16}\left(\frac{\rho\ell}{A}\right)$$

$$= \frac{1}{16}(R) = \frac{1}{16}\times(16)$$

$$= 1\ \Omega$$

Thus, the resistance of the new wire is found to be 1 Ω. **(Answer).**

Problem 12. The number density of free electrons in a conductor of cross-sectional area 10^{-4} m^2 is 4×10^{28} m^{-3}. A current of 3.2 A is maintained in the conductor. Determine the drift velocity of electrons.

Solution. Given: $i = 3.2$ A; $A = 10^{-4}$ m^2; $n = 4\times10^{28}$ m^{-3}

Drift velocity, $$v = \frac{i}{nAe}$$

$$= \frac{(3.2)}{(4\times10^{28})\times(10^{-4})\times(1.6\times10^{-19})}$$

$$= 5\times10^{-6}\ ms^{-1}$$ **(Answer).**

Problem13. A copper wire of length 0.4 m and having area of cross-section 2 mm^2 is connected to a battery of 4 volt. Calculate the current flowing through the wire (number density of electrons in copper = 8.5×10^{28} m^{-3}; electron mobility, $\mu = 6\times10^{-6}$ $m^2V^{-1}s^{-1}$).

Solution. Given: $\ell = 0.4$ m; $A = 2$ $mm^2 = 10^{-6}$ m^2; V=4 volt; $n = 8.5\times10^{28}$ m^{-3}; $\mu = 6\times10^{-6}$ $m^2V^{-1}s^{-1}$

The electric field set up across the conductor is given by,

$$E = \frac{V}{\ell} = \frac{(4)}{(0.4)} = 10\ Vm^{-1}$$

The current through the wire is given by,

$$i = n.A.e.v_d = n.A.e.(\mu E)$$

$$= (8.5\times10^{28}) \times (2\times10^{-6}) \times (1.6\times10^{-19}) \times (6\times10^{-6}) \times (10)$$

$$= 1.63\ A$$ **(Answer).**

Problem 14. A copper wire of uniform cross-section has length 1m, diameter 0.7 mm, and resistance 0.08 Ω. Calculate the relaxation time of the electrons flowing through the wire (number density of electrons = 10^{28} m^{-3}; drift velocity of electrons = 10^{-1} $cm.s^{-1}$).

Solution. Given: ℓ = 1m; d = 0.7 mm = 0.7×10^{-3} m; R = 0.08 Ω; $v_d = 10^{-1}cm.s^{-1} = 10^{-3}ms^{-1}$; $n = 10^{28}m^{-3}$

Area of cross-section of the wire is,

$$A = \pi r^2 = \frac{\pi d^2}{4} = \frac{3.14\times(0.7\times10^{-3})^2}{4} = 38.5\times10^{-8}\ m^2$$

We know, $$R = \frac{m\ell}{ne^2\tau A} \quad (\tau\text{: relaxation time})$$

$$\therefore \quad \tau = \frac{m\ell}{ne^2RA} = \frac{(9\times10^{-31})\times(1)}{(10)^{28}\times(1.6\times10^{-19})^2\times(0.08)\times(38.5\times10^{-8})}$$

$$= 1.14\times10^{-13}\ s$$ **(Answer).**

Drift velocity

Problem 15. A charge of 16 C takes 1 minute to pass through a wire of cross-section 10^{-2} cm^2. If the number density of the charge carriers is 10^{28} m^{-3}, calculate the drift velocity of the carriers.

Solution. $$i = neAv_d \quad \text{or} \quad v_d = \frac{i}{nea}$$

Also, $$i = \frac{q}{t}\ ; \quad \therefore\ v_d = \frac{q}{neat}$$

Given: $q = 16$ C; $t = 60$ s; $n = 10^{28}$ m^{-3}; $A = 10^{-2}$ $cm^2 = 10^{-6}$ m^2

$$\therefore \quad v_d = \frac{(16)}{(10^{28})\times(1.6\times10^{-19})\times(10^{-6})\times(60)}$$

$$= 1.67\times10^{-4}\ ms^{-1}$$ **(Answer).**

Problem 16. A copper wire of cross-section 1 mm² carries a current of 0.1 A. Assuming that each copper atom contributes one free electron, calculate the average drift velocity of the electrons (density of copper = 8.9 g cm^{-3}; atomic mass no. of copper = 63.5 amu).
Solution. Let n be the number density of free electrons. Each copper atom contributes one electron.
We know, 1 mole of copper gives: 6.023×10^{23} electrons (Avogadro no.)

i.e. 63.5 g of Cu gives: 6.023×10^{23} electrons

Now, volume of 1 mole of Cu $= \dfrac{\text{Mass}}{\text{Density}} = \dfrac{63.5\,\text{g}}{8.9\,\text{g.cm}^{-3}} = 7.13\ \text{cm}^3 = 7.13 \times 10^{-6}\ \text{m}^{-3}$

∴ The number density is, $n = \dfrac{(6.023 \times 10^{23})}{(7.13 \times 10^{-6})} = 8.45 \times 10^{28}\ \text{m}^{-3}$

Now, i = 0.1 ampere; Area, A = 1 mm² = 1×10^{-6} m³

∴ Drift velocity, $v_d = \dfrac{i}{neA} = \dfrac{0.1}{(8.45 \times 10^{28}) \times (1.6 \times 10^{-19}) \times (10^{-6})}$

$= 7.4 \times 10^{-6}\,\text{ms}^{-1}$ **(Answer).**

Problem 17. Two copper wires A and B of radius r and 2r, respectively, are connected in series. If the drift velocity of electrons in the wire A is 2.5×10^{-3} ms^{-1}, what is the drift velocity in wire B ?
Solution. The radii of the wires are,

$r_A = r$; $r_B = 2r$

Their areas of cross-section are,

$A_A = \pi r^2$; $A_B = 4\pi r^2$

Since in a series combination the electric current has to be the same in the two wires, we have,

$i = neA_Av_A = neA_Bv_B$

$\Rightarrow A_Av_A = A_Bv_B$

$\Rightarrow v_B = \dfrac{A_Av_A}{A_B} = \dfrac{\pi r^2 \times v_A}{4\pi r^2} = \dfrac{v_A}{4}$

$= \dfrac{(2.5 \times 10^{-3})}{4} = 6.25 \times 10^{-4}\ \text{ms}^{-1}$ **(Answer).**

Colour Code; Temperature Dependence of Resistance

Problem 18. Express a resistance of 3700 Ω ± 10% by its colour code.
Solution. The value of the resistance is,

$3700 \pm 10\% = 37 \times 10^2 \pm 10\%\ \Omega$

The colours representing the digits 3,7,2 are orange, violet, red, respectively. The tolerance of 10% is represented by silver shade. Thus, the sequence of colour bands marked on the given carbon resistor will be,

ORANGE, VIOLET, RED & SILVER **(Answer).**

Problem 19. A voltage of 30 V is applied across a carbon resistor, with first, second and third rings of blue, black and yellow colours, respectively. Find the value of current through the resistor. (CBSE; 2005).
Solution. The digits represented by the blue, black, and yellow colours are 6, 0, and 4, respectively. ∴ The resistance of the carbon resistor is,

$R = 60 \times 10^4\ \Omega$

∴ The current through the resistor is,

$i = \dfrac{V}{R} = \dfrac{(30)}{(60 \times 10^4)} = 5 \times 10^{-5}\text{A}$ **(Answer).**

Problem 20. The resistance of a platinum wire is measured to be 10 Ω at 0 °C and 20 Ω at 273 °C. Calculate the temperature coefficient of resistance of the wire.
Solution. Given: $R_0 = 10\ \Omega$; $R_T = 20\ \Omega$; $T_0 = 0$ °C; T = 273 °C
The temperature coeff. of resistance is given by,

$$\alpha = \frac{R_T - R_0}{R_0(T - T_0)} = \frac{(20)-(10)}{(10)\times(273-0)}$$
$$= 3.66\times10^{-3}\ °C^{-1} \quad \textbf{(Answer).}$$

Combination of Cells; Internal Resistance of a Cell

Problem 21. Three identical cells of emf 4 V each and internal resistance 'r' each are connected in series to a 6 Ω resistor. If the current flowing in the circuit is 1.5 A, calculate (i) the internal resistance of each cell and (ii) the terminal voltage across the cells (CBSE.; 2002).

Solution.

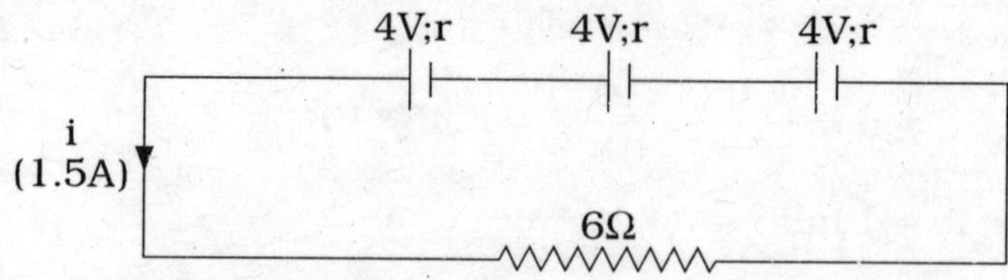

Total emf of the cells = 4+4+4 = 12 V
Total internal resistance = r+r+r = 3r
External resistance = 6 Ω
Current flowing in the circuit (i) = 1.5 A

We know, $$i = \frac{\text{Total emf}}{\text{Total int. resistance + External resistance}}$$

$$\Rightarrow \quad 1.5 = \frac{(12)}{(3r)+(6)}$$

$$\Rightarrow \quad 4.5r + 9 = 12$$

$$\Rightarrow \quad r = \frac{2}{3}\ \Omega \quad \textbf{(Answer).}$$

The terminal voltage across each cell is given by,

$$V = E - ir = (4)-(1.5)\times(2/3)$$
$$= 3 \text{ volt}$$

∴ The total terminal voltage across all the cells is,

$$(V)_{total} = 3+3+3 = 9 \text{ volt} \quad \textbf{(Answer).}$$

Problem 22. Two identical cells are connected to an external resistance of 1 Ω. They are found to give the same current in the circuit, whether connected together in parallel or in series. Calculate the internal resistance of either cell.

Solution. Let E and r be the emf and internal resistance, respectively, of each cell. Given: External resistance, R = 1 Ω.

When the two cells are connected in series, we have,

Total emf of the cells = E + E = 2E

Total resistance in the circuit = R + r + r = 1 + 2r

$$\therefore \quad \text{Current in the circuit, } i_1 = \frac{2E}{1+2r}$$

When the two cells are connected in parallel, we have,

Total emf of the cells = E

$$\text{Total int. resistance of the cells} = \frac{r}{2} \quad \left(\frac{1}{r_{total}} = \frac{1}{r} + \frac{1}{r}\right)$$

$$\text{Total resistance in the circuit} = R + \frac{r}{2} = 1 + \frac{r}{2}$$

$$\therefore \quad \text{Current in the circuit, } i_2 = \frac{E}{1+(r/2)}$$

We are given that,

$$i_1 = i_2$$

$$\Rightarrow \quad \frac{2E}{1+2r} = \frac{E}{1+(r/2)} = \frac{2E}{2+r}$$

$$\Rightarrow \quad r = 1\ \Omega \quad \textbf{(Answer).}$$

Combination of Resistances

Problem 23. Three resistances of 12 Ω each are first combined in series and then in parallel. Calculate the equivalent resistance in each case.

Solution. $R_1 = R_2 = R_3 = 12\ \Omega$

(i) For the series combination, the equivalent resistance is given by,

$$(R)_{series} = R_1+R_2+R_3 = (12)+(12)+(12)$$

$$= 36\ \Omega \quad \textbf{(Answer).}$$

(ii) For the parallel combination, the equivalent resistance is given by,

$$\frac{1}{R_{parallel}} = \frac{1}{R_1}+\frac{1}{R_2}+\frac{1}{R_3}$$

$$= \frac{1}{(12)}+\frac{1}{(12)}+\frac{1}{(12)} = \frac{1}{4}$$

$$\Rightarrow \quad (R)_{parallel} = 4\ \Omega \quad \textbf{(Answer).}$$

Problem 24. Obtain the equivalent resistance across A and B for the following combination,

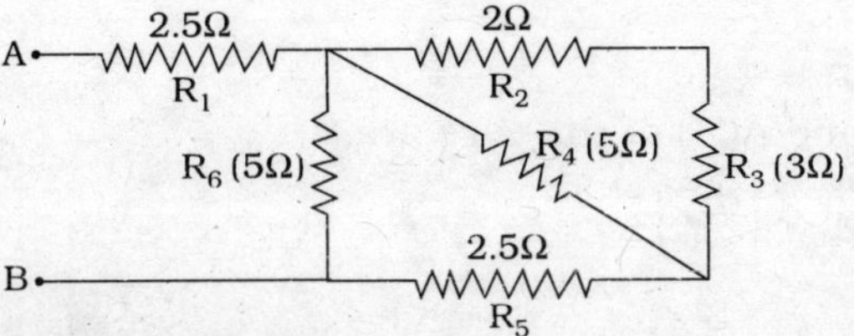

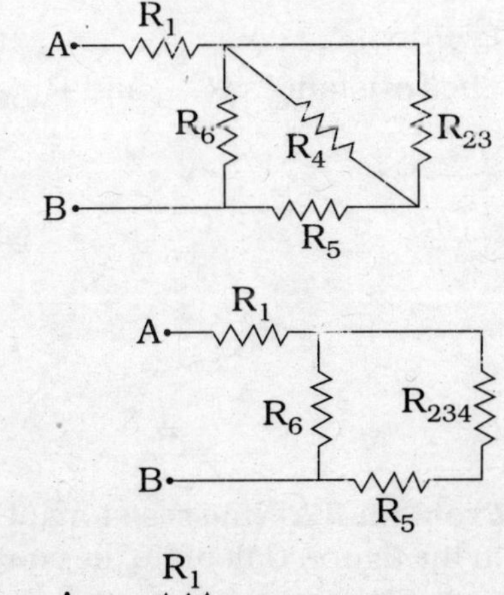

Solution. R_2 and R_3 are in series. Their equivalent resistance is,

$$R_{23} = R_2+R_3 = (2)+(3) = 5\ \Omega$$

Now R_{23} and R_4 are in parallel. Their equivalent resistance is given by,

$$\frac{1}{R_{234}} = \frac{1}{R_{23}}+\frac{1}{R_4} = \frac{1}{(5)}+\frac{1}{(5)} = \frac{2}{5}$$

$$\Rightarrow \quad R_{234} = 2.5\ \Omega$$

Now R_{234} and R_5 are in series. Their equivalent resistance is,

$$R_{2345} = R_{234} + R_5 = (2.5)+(2.5) = 5\ \Omega$$

Now R_{2345} and R_6 are in parallel. Their equivalent resistance is given by,

$$\frac{1}{R_{23456}} = \frac{1}{R_6}+\frac{1}{R_{2345}} = \frac{1}{(5)}+\frac{1}{(5)} = \frac{2}{5}$$

$$\Rightarrow \quad R_{23456} = 2.5\ \Omega$$

Now R_{23456} and R_1 are in series. Their equivalent resistance, which is also the total resistance of the network, is given by,

A R1 R23456 B

$$R_{total} = (2.5)+(2.5) = 5\ \Omega \quad \textbf{(Answer).}$$

Problem 25. Four resistances of 10 Ω each are connected as shown in the figure. Calculate the equivalent resistance between the points A and B.

Solution. We rearrange the resistances in the form of a Wheatstone bridge, as shown in the adjoining figure ($R_1 = R_2 = R_3 = R_4 = R_5 = 10\ \Omega$). We find,

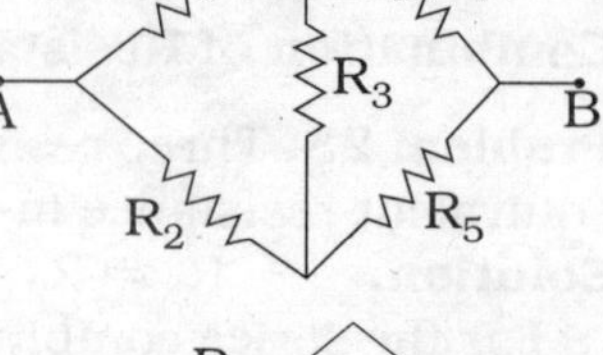

$$\frac{R_1}{R_4} = \frac{R_2}{R_5}$$

∴ The bridge is balanced. Hence no current will pass through the resistance R_3, as if it was non-existent. Thus the network now simplifies as shown alongside. It is seen that the pairs R_1, R_4 and R_2, R_5, respectively, are in series combination. Their equivalent resistances are,

$$R_{14} = R_1 + R_4 = (10) + (10) = 20\ \Omega$$

$$R_{25} = R_2 + R_5 = (10) + (10) = 20\ \Omega$$

Now R_{14} and R_{25} are in parallel combination. Their equivalent resistance, which is also the total resistance of the network, is given by,

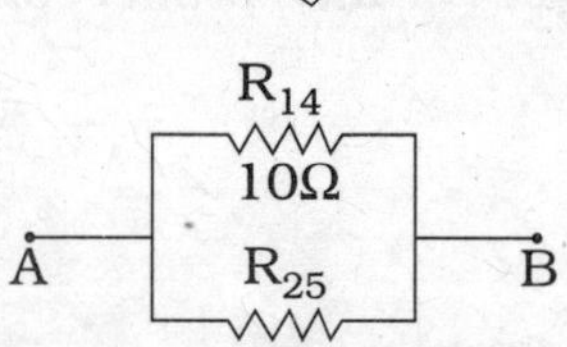

$$\frac{1}{R_{total}} = \frac{1}{R_{14}} + \frac{1}{R_{25}} = \frac{1}{(20)} + \frac{1}{(20)} = \frac{1}{10}$$

$$\Rightarrow \quad R_{total} = 10\ \Omega \quad \textbf{(Answer).}$$

Problem 26. A uniform wire of length ℓ and resistance R is bent into a circular shape, as shown in the figure. What is the resistance between the points A and B ?

Solution. Let λ be the resistance per unit length of the wire. Then the resistance of the wire is given by,

$$R = \lambda \ell$$

The resistances of the parts APB and AQB of the wire are,

$$R_{APB} = \left(\frac{1}{4}\ell\right)\lambda$$

$$R_{AQB} = \left(\frac{3}{4}\ell\right)\lambda$$

The resistances R_{APB} and R_{AQB} are in parallel. Their equivalent resistance is given by,

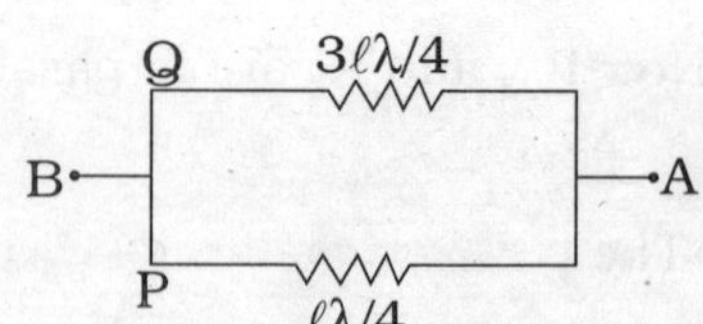

$$\frac{1}{R_{eq}} = \frac{1}{R_{APB}} + \frac{1}{R_{AQB}} = \frac{1}{(\ell\lambda)/4} + \frac{1}{(3\ell\lambda)/4}$$

$$= \frac{4}{\ell\lambda}\left(1 + \frac{1}{3}\right) = \frac{16}{3\ell\lambda}$$

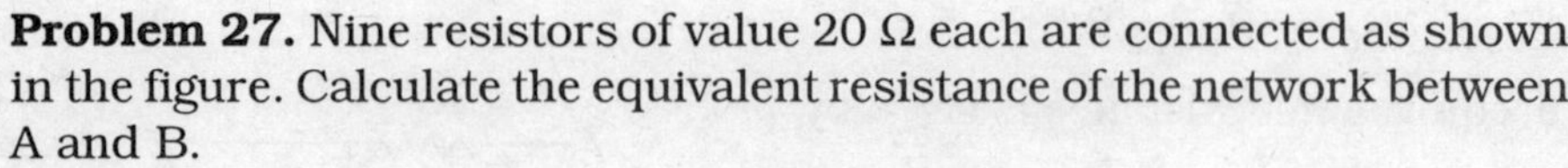

$$\Rightarrow \quad R_{eq} = \frac{3}{16}\ell\lambda = \frac{3}{16}R \quad \textbf{(Answer).}$$

Problem 27. Nine resistors of value 20 Ω each are connected as shown in the figure. Calculate the equivalent resistance of the network between A and B.

Solution. We rearrange the resistors as follows,

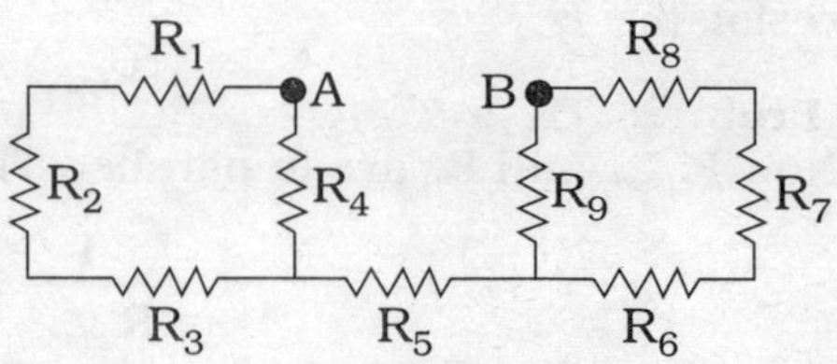

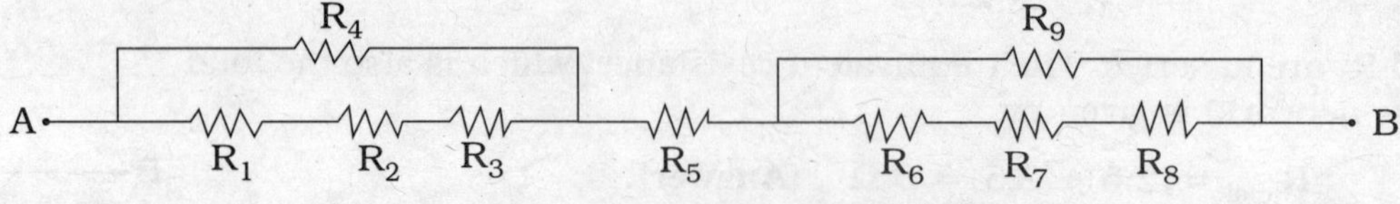

The groups of resistors R_1, R_2, R_3 and R_6, R_7, R_8, respectively, are in series combination. Their equivalent resistances are,

$$R_{123} = R_1+R_2+R_3 = (20)+(20)+(20) = 60\ \Omega$$
$$R_{678} = R_6+R_7+R_8 = (20)+(20)+(20) = 60\ \Omega$$

Now the circuit simplifies as shown. The pairs R_{123}, R_4 and R_{678}, R_9, respectively, are in parallel combination. Their equivalent resistances are given by,

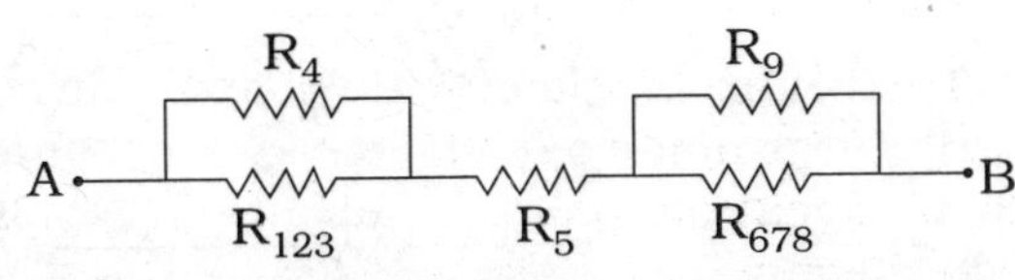

$$\frac{1}{R_{1234}} = \frac{1}{R_{123}} + \frac{1}{R_4} = \frac{1}{(60)} + \frac{1}{(20)} = \frac{4}{(60)} = \frac{1}{15}$$

$$\frac{1}{R_{6789}} = \frac{1}{R_{678}} + \frac{1}{R_9} = \frac{1}{(60)} + \frac{1}{(20)} = \frac{1}{15}$$

$$\Rightarrow \qquad R_{1234} = R_{6789} = 15\ \Omega$$

Now the circuit further simplifies, as shown. The resistors R_{1234}, R_5, R_{6789} are in series. Their equivalent resistance, which is also the total resistance of the network, is given by,

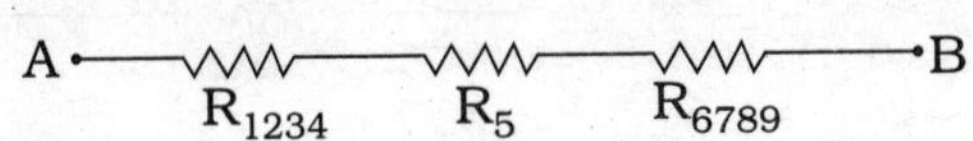

$$R_{total} = R_{1234}+R_5+R_{6789}$$
$$= (15)+(20)+(15)$$
$$= 50\ \Omega$$ **(Answer).**

Problem 28. In the given circuit, calculate the readings shown by the ammeter and voltmeter.

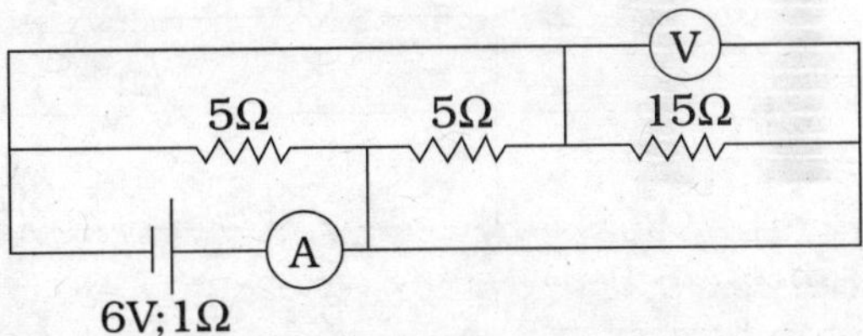

Solution. The given circuit can be redrawn as shown. Since the three resistors are in parallel, their equivalent resistance is given by,

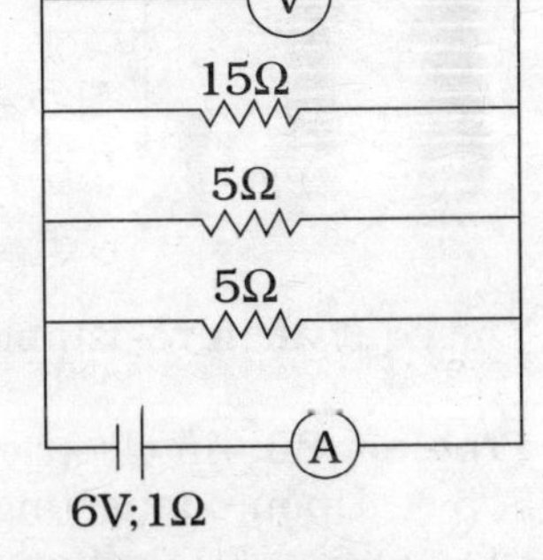

$$\frac{1}{R_{eq}} = \frac{1}{(5)} + \frac{1}{(5)} + \frac{1}{(15)} = \frac{7}{15}$$

$$\Rightarrow \qquad R_{eq} = \frac{15}{7}\ \Omega$$

∴ The current in the circuit, as shown by the reading in the ammeter, is given by

$$i = \frac{E}{R_{eq}+r} \quad \text{(r : Int. resistance of the cell)}$$

$$= \frac{(6)}{(15/7)+(1)} = \frac{42}{22} = 1.91\ \text{A}$$ **(Answer).**

The p.d. across the 15 Ω resistor, which is the reading shown by the voltmeter, is given by,

$$V = iR_{eq} = \left(\frac{42}{22}\right) \times \left(\frac{15}{7}\right) = \frac{45}{11}$$
$$= 4.1\ \text{volt}$$ **(Answer).**

Problem 29. In the given circuit, determine the equivalent resistance between A and B.

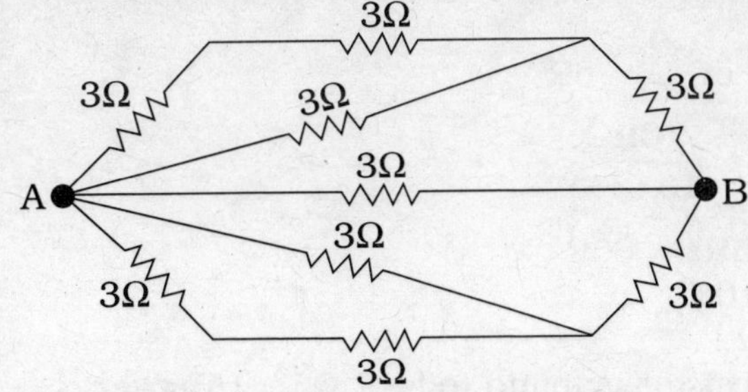

Solution.

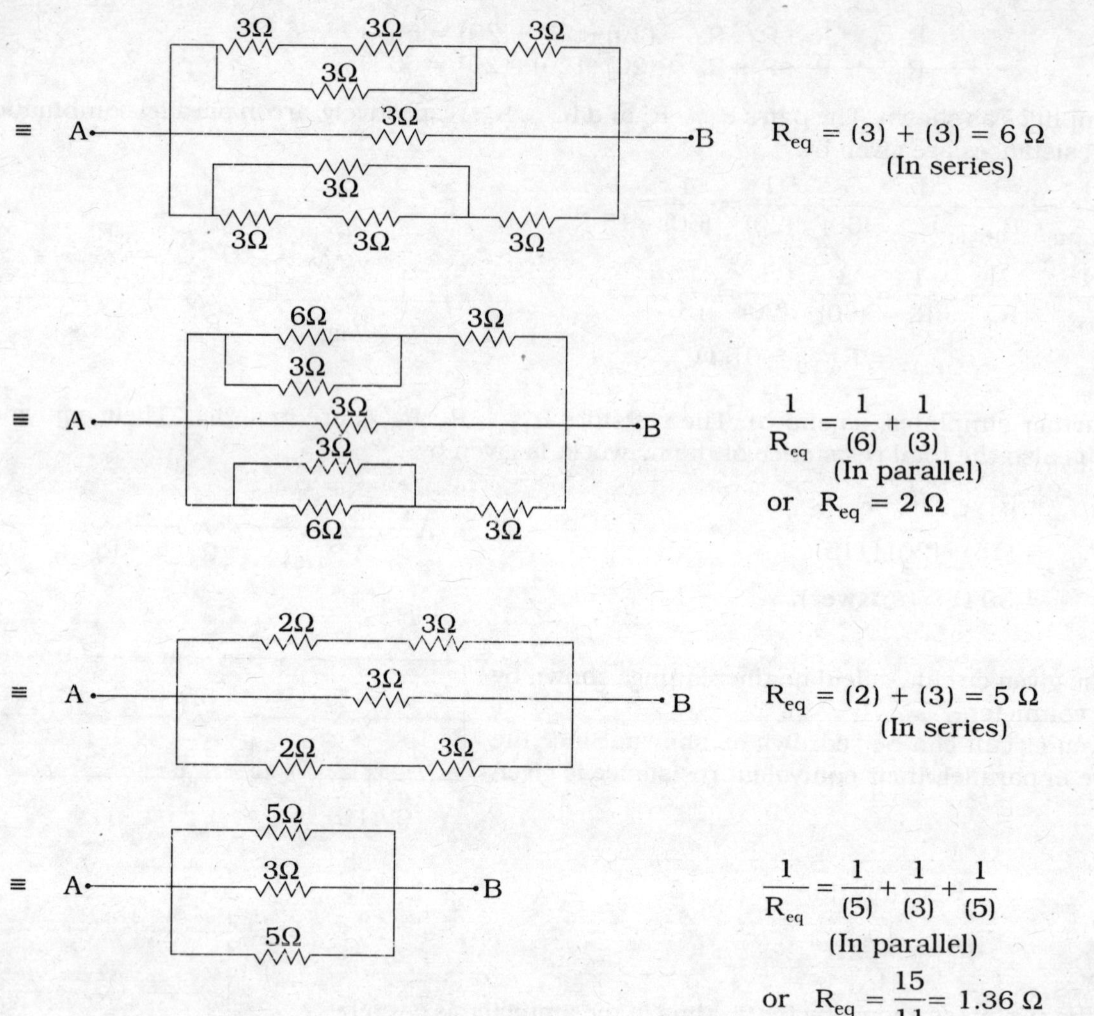

$R_{eq} = (3) + (3) = 6\ \Omega$
(In series)

$$\frac{1}{R_{eq}} = \frac{1}{(6)} + \frac{1}{(3)}$$
(In parallel)

or $R_{eq} = 2\ \Omega$

$R_{eq} = (2) + (3) = 5\ \Omega$
(In series)

$$\frac{1}{R_{eq}} = \frac{1}{(5)} + \frac{1}{(3)} + \frac{1}{(5)}$$
(In parallel)

or $R_{eq} = \dfrac{15}{11} = 1.36\ \Omega$

The equivalent resistance of the network, between A and B, is found to be 1.36 Ω. **(Answer).**

Problem 30. An electric circuit consists of an ideal battery and a resistor. The current flowing in the circuit is 5 A. Upon connecting another resistor of resistance 10 Ω in parallel with the first resistor, the current rises to 6 A. Determine the resistance of the first resistor.

Solution. Let the emf of the battery be E and the resistance of the first resistor be R ohm. Then,

$$i = 5 = \frac{E}{R} \text{ or } E = 5R$$

When the second resistor of 10 Ω is connected in parallel, the total resistance R′ in the circuit is given by,

$$\frac{1}{R'} = \frac{1}{R} + \frac{1}{10} = \frac{10+R}{10R}$$

Now, the current in the circuit becomes 6 A. So, we have,

$$\frac{E}{R'} = 6 \text{ or } E = 6R'$$

$$\Rightarrow \quad 5R = 6 \times \frac{10R}{10+R}$$

$$\Rightarrow \quad 50R + 5R^2 = 60R$$

$$\Rightarrow \quad 5R^2 = 10R$$

$$\Rightarrow \quad R = 2\ \Omega$$

Thus, the resistance of the first resistor is found to be 2 Ω. **(Answer).**

Problem 31. Calculate the value of the current read in the ammeter in the following circuit.

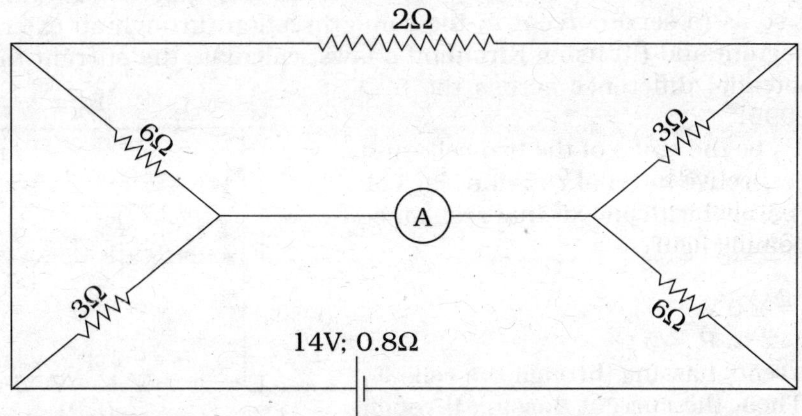

Solution. In the given circuit, the 3 Ω and 6 Ω resistors are connected in parallel. Therefore, the circuit can be redrawn as follows.

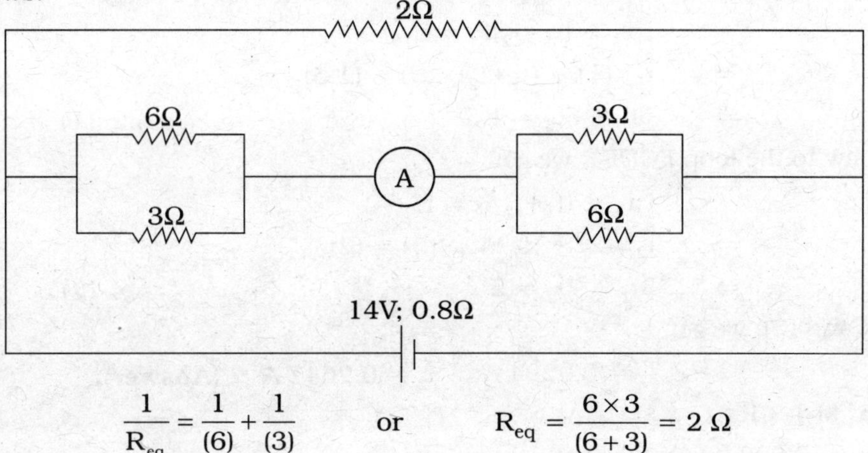

Now, $$\frac{1}{R_{eq}} = \frac{1}{(6)} + \frac{1}{(3)} \quad \text{or} \quad R_{eq} = \frac{6 \times 3}{(6+3)} = 2\ \Omega$$

Thus, the circuit is redrawn as follows.

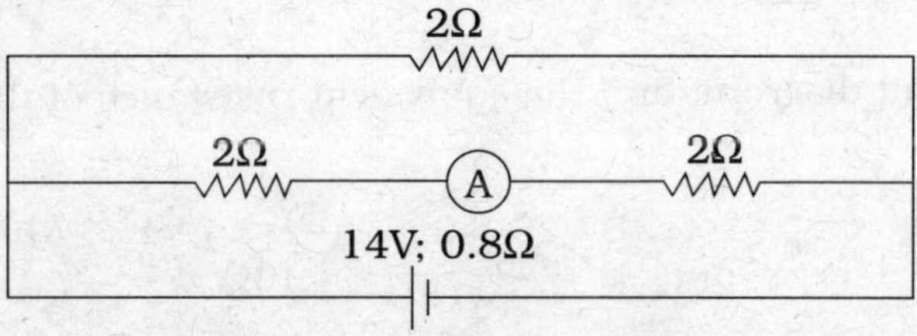

Now the resistors of value 2 Ω each are in series. ∴ Their equivalent resistance is (2+2) = 4 Ω. Thus, now the circuit is redrawn as shown. Now the resistors 2 Ω and 4 Ω are in parallel. Their equivalent resistance is given by,

$$\frac{1}{R_{total}} = \frac{1}{(2)} + \frac{1}{(4)}$$

$$R_{total} = \frac{4}{3}\ \Omega$$

2Ω
4Ω
A
14V; 0.8Ω

∴ The current in the circuit is given by,

$$i_{total} = \frac{E}{R_{total} + r} = \frac{14}{(4/3) + (0.8)} = \frac{42}{6.4} = 6.56\ \text{A}$$

The current through the 4 Ω resistor, which is the reading shown by the ammeter, is given by,

$$(i)_{4\Omega} = (i)_{total} \times \frac{R_{total}}{(4)} = (6.56) \times \frac{(4/3)}{(4)}$$

$$= 2.19\ \text{A} \quad \textbf{(Answer).}$$

Kirchhoff's Laws

Problem 32. Two cells of emf's 1.5 V and 2.0 V and internal resistances 1 Ω and 2 Ω, respectively, are connected in parallel, so as to send current in the same direction through an external resistance of 5 Ω. (i) Draw the circuit diagram and (ii) using Kirchhoff's laws, calculate the current through each branch of the circuit and the potential difference across the 5 Ω resistance. (AISSCE; 1995).

Solution. Let E_1 and E_2 be the emf's of the two cells and let r_1 and r_2 be their respective internal resistances. The cells are connected in parallel with an external resistance R, as shown in the adjoining figure.

Given: $E_1 = 1.5$ V; $E_2 = 2.0$ V;

$r_1 = 1\ \Omega$; $r_2 = 2\ \Omega$; $R = 5\ \Omega$

Let i_1 and i_2 be the currents passing through the cells E_1 and E_2, respectively. Then, the current passing through the resistance R is (i_1+i_2).

Applying the Kirchhoff's second law to the loop ABDEA, we have,

$$i_1r_1 + (i_1+i_2)R = E_1$$

$$\Rightarrow \quad i_1\times(1) + (i_1+i_2)\times(5) = (1.5)$$

$$\Rightarrow \quad 6i_1 + 5i_2 = 1.5 \qquad \text{..... (1)}$$

Applying the same law to the loop FCDEF, we get,

$$i_2r_2 + (i_1+i_2)R = E_2$$

$$\Rightarrow \quad i_2\times(2) + (i_1+i_2)\times(5) = (2)$$

$$\Rightarrow \quad 5i_1 + 7i_2 = 2 \qquad \text{..... (2)}$$

Solving the Eqs. (1) and (2), we get,

$i_1 = 0.0294$ A; $i_2 = 0.2647$ A **(Answer).**

∴ P.D. across R $= (i_1+i_2)\times R$

$= (0.0294 + 0.2647)\times 5$

$= 1.47$ volt **(Answer).**

Problem 33. In the given circuit diagram, find the equivalent resistance of the network between the points P and Q.

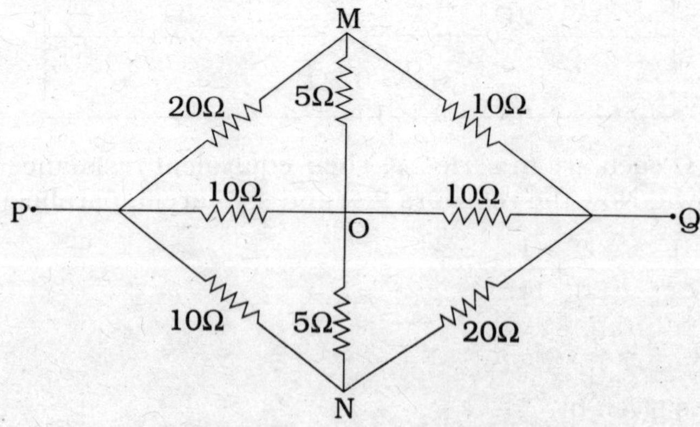

Solution. We connect a battery of emf E between P and Q (see next figure). The total current entering the network at P and leaving at Q is $(i_1+i_2+i_3)$, where i_1, i_2, i_3 are the values of current through PM, PQ, PN, respectively. Because of the symmetry of the network, the current distribution is as shown in the figure. Now we apply the Kirchhoff's second law to various loops in the network. For the loop POQCDP, we have,

$$10i_2 + 10i_2 = E$$

$$\Rightarrow \quad E = 20\ i_2 \qquad \text{..... (i)}$$

Let R be the equivalent resistance of the network between P and Q. Then,

$$(i_1 + i_2 + i_3)R = E \qquad \text{..... (ii)}$$

From Eqs. (i) and (ii),

$$(i_1 + i_2 + i_3)R = 20i_2 \qquad \text{..... (iii)}$$

For the loop POMP,

$$20i_1 + 5(i_1 - i_3) - 10i_2 = 0$$

$$\Rightarrow \quad 25i_1 - 10i_2 - 5i_3 = 0 \qquad \text{..... (iv)}$$

For the loop MQOM,

$$5(i_1 - i_3) + 10i_2 - 10i_3 = 0$$

$$\Rightarrow \quad 5i_1 + 10i_2 - 15i_3 = 0 \qquad \text{..... (v)}$$

Adding Eqs. (iv) and (v),

$$30i_1 - 20i_3 = 0$$

$$\Rightarrow \quad i_1 = \frac{2}{3}i_3 \qquad \text{..... (vi)}$$

For the loop PNOP,

$$10i_2 + 5(i_1 - i_3) - 10i_3 = 0$$

$$\Rightarrow \quad 5i_1 + 10i_2 - 15i_3 = 0$$

$$\Rightarrow \quad i_1 + 2i_2 - 3i_3 = 0$$

$$\Rightarrow \quad \frac{2}{3}i_3 + 2i_2 - 3i_3 = 0 \quad \text{(From Eq. vi)}$$

$$\Rightarrow \quad i_3 = \frac{6}{7}i_2$$

From Eq. (vi),

$$i_1 = \frac{2}{3}i_3 = \frac{2}{3}\times\frac{6}{7}i_2 = \frac{4}{7}i_2$$

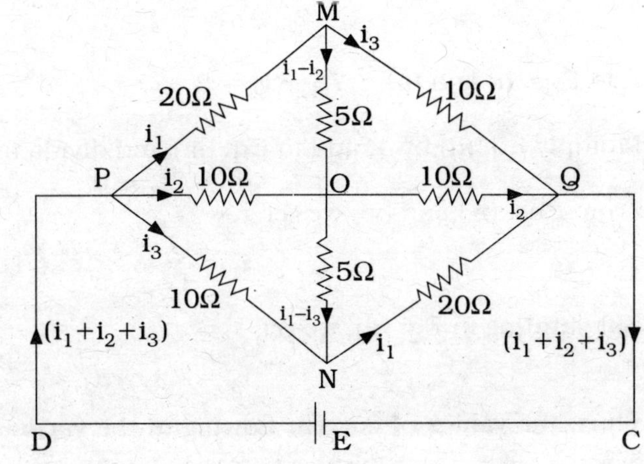

Substituting the values of i_1, i_3 in Eq. (iii),

$$\left(\frac{4}{7}i_2 + i_2 + \frac{6}{7}i_2\right)R = 20i_2$$

$$\Rightarrow \quad \left(\frac{17}{7}i_2\right)R = 20i_2$$

$$\Rightarrow \quad R = \frac{20\times 7}{17} = 8.24\ \Omega \quad \textbf{(Answer).}$$

Problem 34. Using Kirchhoff's laws, calculate the values of electric currents in the given electric network. (CBSE; 1994).

Solution. Applying Kirchhoff's first law at the junction A, we get,

$$i_1 + i_3 = i_2$$

$$\Rightarrow \quad i_3 = i_2 - i_1 \qquad \text{..... (i)}$$

Applying Kirchhoff's second law to mesh 1 and mesh 2, we get,

Mesh 1:

$$5i_1 + 2i_2 - 12 = 0$$

$$\Rightarrow \quad 5i_1 + 2i_2 = 12 \qquad \text{..... (ii)}$$

Mesh 2:

$$3i_3 + 2i_2 - 6 = 0$$

$$\Rightarrow \quad 2i_2 + 3i_3 = 6$$

$$\Rightarrow \quad 2i_2 + 3(i_2 - i_1) = 6 \quad \text{(From Eq. (i)}$$

$$\Rightarrow \quad 5i_2 - 3i_1 = 6 \qquad \text{..... (iii)}$$

A, i_3, i_2, 5Ω, 3Ω, i_1, 1, 2, 12V, 6V, B

Solving the Eqs. (ii) and (iii), we get,

$$i_1 = \frac{48}{31} = 1.55\ \text{A} \quad \textbf{(Answer).}$$

$$\therefore \quad i_2 = \frac{66}{31} = 2.13\ \text{A} \quad \textbf{(Answer).}$$

Substituting in Eq. (i), we get,

$$i_3 = 0.58\ \text{A} \quad \textbf{(Answer).}$$

Problem 35. Determine the current in each branch of the following network. (NCERT).

Solution. The currents flowing through the various arms of the circuit are marked in the adjoining figure.

We apply Kirchhoff's second law to the loops ECBAE, ADBA and BDCB.

Loop ECBAE:

$$10\,(i_1+i_2) + 10i_1 + 5(i_1-i_3) - 10 = 0$$

$$\Rightarrow \quad 25i_1 + 10i_2 - 5i_3 = 10$$

$$\Rightarrow \quad 5i_1 + 2i_2 - i_3 = 2 \quad \text{..... (i)}$$

Loop ADBA: $10i_1 - 5i_2 + 5i_3 = 0$

$$\Rightarrow \quad 2i_1 - i_2 + i_3 = 0 \quad \text{..... (ii)}$$

Loop BDCB: $5(i_1-i_3) - 10(i_2+i_3) - 5i_3 = 0$

$$\Rightarrow \quad 5i_1 - 10i_2 - 20i_3 = 0$$

$$\Rightarrow \quad i_1 - 2i_2 - 4i_3 = 0 \quad \text{..... (iii)}$$

Add Eqs. (i) and (ii): $7i_1 + i_2 = 2$ ..... (iv)

Multiply Eq. (ii) by 4, add to Eq. (iii) and divide by 3. We get,

$$3i_1 - 2i_2 = 0 \quad \text{..... (v)}$$

From Eqs. (iv) and (v), we get,

$$i_1 = \frac{4}{17}\text{A} \quad ; \quad i_2 = -\frac{6}{17}\text{A}$$

Substituting in Eq. (ii), we get,

$$i_3 = -2i_2 + i_2 = -\frac{2}{17}\text{A}$$

Thus, the values of current flowing in the various branches of the network are,

$$i_1 = \frac{4}{17}\text{A} \;;\quad i_2 = \frac{6}{17}\text{A} \;;\quad i_3 = -\frac{2}{17}\text{A}$$

$$(i_1 - i_3) = \frac{6}{17}\text{A}; \quad (i_2 + i_3) = \frac{4}{17}\text{A} \quad \textbf{(Answer)}.$$

Metre Bridge & Potentiometer

Problem 36. (a) In a metre bridge, the balance point is found to be at 39.5 cm from the end A, when the resistor Y is of 12.5 Ω. Determine the resistance of X. Why are the connections between resistors in a metre bridge made of thick copper strips ?

(b) Determine the balance point of the given bridge, if X and Y are interchanged.

(c) What happens if the galvanometer and cell are interchanged at the balance point of the bridge ? Would the galvanometer show any current ? (NCERT)

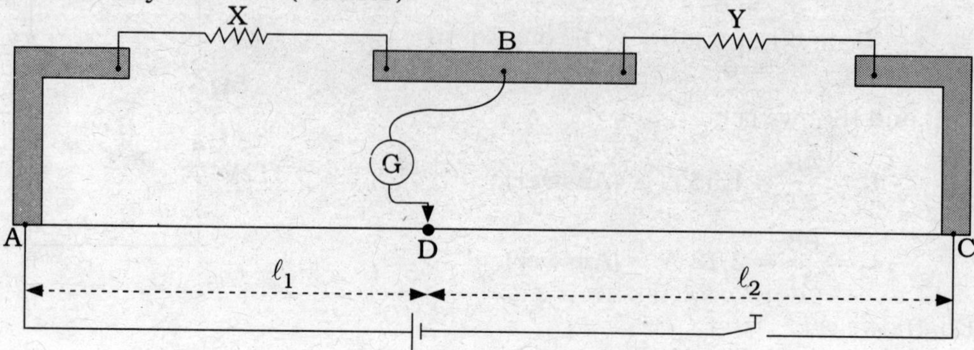

Solution. (a) In a Wheatstone bridge, at the balance point we have,

$$\frac{P}{Q} = \frac{R}{S}$$

∴ For our case,

$$\frac{X}{Y} = \frac{\ell_1}{\ell_2}$$

Given: Y = 12.5 Ω; ℓ_1 = 39.5 cm. ∴ ℓ_2 = 100–39.5 = 60.5 cm

$$\therefore \quad \frac{X}{12.5} = \frac{39.5}{60.5}$$

$$\Rightarrow \quad X = \frac{39.5 \times 12.5}{60.5} = 8.16\ \Omega \quad \textbf{(Answer).}$$

Thick copper strips are used to minimise the resistance at the joints in the circuit, because this resistance is assumed to be zero when we use the bridge formula P/Q = R/S. **(Answer).**

(b) If X and Y are interchanged, ℓ_1 and ℓ_2 will also get interchanged. Hence now ℓ will become 60.5 cm, i.e. the new balance point will be obtained at 60.5 cm from the end A. **(Answer).**

(c) If the galvanometer and the cell are interchanged at the balance point of the bridge, the condition of balance will still hold good. Hence the galvanometer will continue to show zero deflection. **(Answer).**

Problem 37. The length of a potentiometer wire is 600 cm and it carries a current of 40 mA. For a cell of emf 2 V and internal resistance 10 ohm, the null point is found to be at 500 cm. If a voltmeter is connected across the cell, the balancing length is decreased by 10 cm. Find (i) the resistance of the whole wire, (ii) reading of the voltmeter and (iii) resistance of the voltmeter. (CBSE;2000).

Solution. (i) Let R be the total resistance of the potentiometer wire and i be the current flowing through it. Then,

Potential drop across the wire = iR

The potential gradient (i.e. p.d. per unit length) across the wire is given by,

$$k = \frac{iR}{\ell} = \frac{iR}{(600\text{cm})} \quad (\text{Given} : \ell = 600 \text{ cm})$$

Next, emf of the cell, ε = k × Balancing length

$$\Rightarrow \quad 2 = \frac{iR}{(600\text{cm})} \times (500\text{cm}) \quad (\text{Given} : \varepsilon = 2\text{V; Bal. length} = 500 \text{ cm})$$

$$\Rightarrow \quad R = \frac{6 \times 2}{5i} = \frac{12}{5 \times (40 \times 10^{-3})} \quad (\text{Given: } i = 40 \text{ mA} = 40 \times 10^{-3} \text{ A})$$

$$\Rightarrow \quad R = 60\ \Omega \quad \textbf{(Answer).}$$

(ii) Voltmeter reading = V = k × Decreased Bal. length

$$= \frac{iR}{\ell} \times (500\text{cm} - 10\text{cm})$$

$$= \frac{(40 \times 10^{-3}) \times (60) \times (490\text{cm})}{(600\text{cm})}$$

$$= 1.96 \text{ volt} \quad \textbf{(Answer).}$$

(iii) Let i′ be the current flowing through the voltmeter and the cell. Then the voltage drop across the cell is given by,

$$\varepsilon - V = i'r \quad (r\text{: Int. resistance of cell})$$

$$\Rightarrow \quad i' = \frac{\varepsilon - V}{r} = \frac{(2) - (1.96)}{(10)} \quad (\text{Given: } r = 10\ \Omega)$$

$$= 4 \times 10^{-3} \text{ A}$$

$$\therefore \quad \text{Voltmeter resistance} = \frac{V}{i'} = \frac{(1.96)}{(4 \times 10^{-3})}$$

$$= 490\ \Omega \quad \textbf{(Answer).}$$

Problem 38. A metre bridge with 1m long constantan wire of uniform cross-section has resistances X and Y, respectively, in its gaps, as shown in Fig.(a). The null point is found for the value of AD = 33.7cm. When the resistance Y is shunted with a resistance of 12 Ω (Fig. b), the null point is found to be shifted by a distance of 18.2 cm. Find the values of X and Y.

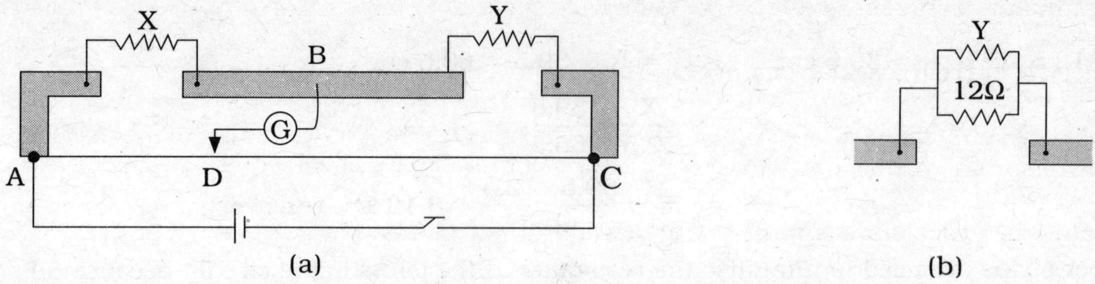

(a) (b)

Solution. Since the wire has a uniform cross-section, the ratio of resistances of its segments AD and DC is the same as the ratio of the lengths AD and DC. According to the condition of balance of a Wheatstone bridge, we have,

$$\frac{X}{Y} = \frac{\ell_1}{\ell_2} \quad \text{(Given: } \ell_1 = 33.7 \text{ cm)}$$

$$\Rightarrow \frac{X}{Y} = \frac{(33.7)}{(100-33.7)} = \frac{33.7}{66.3} \qquad \text{..... (i)}$$

Let Y′ be the resultant resistance of the parallel combination Y and 12 Ω (Fig.b). We have,

$$\frac{1}{Y'} = \frac{1}{Y} + \frac{1}{12} = \frac{12+Y}{12Y} \qquad \text{..... (ii)}$$

Since Y′<Y, the new ratio $\frac{X}{Y'} > \frac{X}{Y}$. Therefore, now the null point will be shifted towards C (so that the new ℓ – values have $\ell_1' > \ell_2'$). The shift is given to be 18.2 cm. Hence, we now have,

$$\frac{X}{Y'} = \frac{(33.7+18.2)}{(66.3-18.2)} = \frac{51.9}{48.1}$$

$$\Rightarrow \frac{X(12+Y)}{12Y} = \frac{51.9}{48.1} \quad \text{(From Eq. ii)}$$

$$\Rightarrow (12+Y) = \frac{51.9}{48.1} \times 12 \times \frac{Y}{X}$$

$$= \frac{51.9}{48.1} \times 12 \times \frac{66.3}{33.7} \quad \text{(From Eq. i)}$$

$$= 25.47$$

$$\Rightarrow \quad Y = (25.47-12) = 13.47\ \Omega \quad \textbf{(Answer).}$$

Substituting in Eq. (i), we get,

$$X = \frac{33.7}{66.3} \times Y$$

$$= \frac{33.7}{66.3} \times 13.47 = 6.85\ \Omega \quad \textbf{(Answer).}$$

Very Short Answer Questions

(Including HOTS Questions)

(1 mark each)

1. What makes a current to flow in an electric circuit ?

2. Is electric current a scalar or a vector quantity ? Give reason for your answer.

3. An electric bulb lights almost mistantaneously when switched on, although the drift velocity of the free electrons in the connecting wires is not very high. State the reason.

4. What is the order of magnitude of the number of free electrons in metals ?

5. Does a rise in temperature increase or reduce the relaxation time of electrons in a metal ?

6. What is the effect of rise in temperature on the resistivity of metals ?

7. Does the resistivity of a semiconductor increase or decrease with rise in temperature ?

8. If the radius of a copper wire is doubled, what will be the effect on the resistivity of the material of the wire ?

9. What is the cause of internal resistance of an electric cell ?

10. What is the conductance of a wire of resistance 2 Ω ?

11. A wire of resistivity ρ is stretched to three times its original length. What will be its new resistivity ?

12. How three resistors of resistance 2 Ω each can be joined together, so that their combined resistance is 3 Ω ?

13. A metal wire of resistance 4 Ω is bent upon itself at its mid-point and the two halves are twisted together, which doubles the thickness of the wire. What is its new resistance value ? [Hint: 0.5 Ω].

14. Can the terminal potential difference of an electric cell in a circuit be greater than its emf ?

15. What should be the sequence of colour bands to be marked on a carbon resistor of 45 kΩ ?

16. Will the drift velocity of free electrons in a metal rise or fall with increase in temperature ?

17. A uniform wire of resistance 4 Ω is bent into circular shape. Imagine that a diameter has been drawn across the circle. What is the resistance between the end points of the diameter ? [Hint: 1 Ω].

18. Name two elements which do not obey Ohm's law.

19. Two wires, one made of copper and the other of manganin, have equal length and same resistance. Which of the two has greater thickness ?

20. In an electric circuit, two copper wires A and B of same length have a current i flowing through them. If the thickness of B is double that of A, how are the drift velocities of free electrons in the two wires related to each other ?

21. Are the Kirchhoff's laws applicable to both AC and DC circuits ?

22. In a potentiometer, copper strips are used instead of copper wires for connections between various components of the circuit. Why ?

23. What are the two fundamental laws of physics on which the Kirchhoff's laws are based ?

24. State the principle of working of a potentiometer.

25. Why is the measurement of the emf of a cell by a potentiometer considered more accurate than the measurement by a voltmeter ?

26. What is the principle of working of a metre bridge ?

27. What happens to the balance point if the positions of the electric cell and the galvanometer are interchanged in a metre bridge ?

28. What is potential gradient ? Give its SI unit.

29. Why sometimes one fails to obtain the balance point all along the potentiometer wire ? [Hint: When the p.d. to be measured > the total p.d. across the potentiometer wire].

30. Should the resistivity of a potentiometer wire be a low value or a high value ?

Short Answer Questions

(Including HOTS Questions)

(2 or 3 mark each)

1. Define ampere, volt and ohm.

2. State the units of conductance and conductivity. Give their dimensional formulae.

3. State the factors on which the resistance of a conductor depends. Give the relation between these factors and the resistance.

4. For making standard resistance coils, manganin or eureka wires are used. Explain why.

5. Draw the voltage vs. current graph for an ohmic and a non-ohmic conductor. Also give one example of each.

6. Discuss the relative merits of carbon resistors and wire-wound resistors.

7. A voltage of 200 V is applied across a colour-coded carbon resistor, with first, second and third ring of blue, black and yellow colours. What is the current flowing through the resistor ? (HBSE; 1999). [Hint: 3.33×10^{-4} A].

8. Find the condition under which the value of current flowing through a resistor remains the same, when it is connected to a combination of n identical cells, either in series or in parallel.

9. The following figure represents a part of a closed circuit. What is the potential difference between the points A and B. [Hint: Apply Kirchhoff's law; 15V].

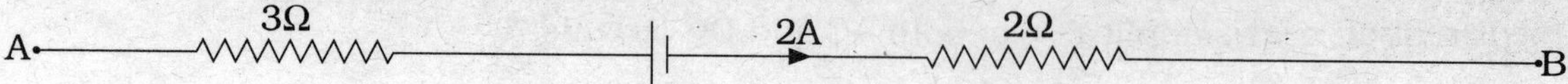

10. Calculate the conductivity of a wire of length 4 m, area of cross-section 2 cm^2 and resistance 10^{-3} Ω. [Hint: 2×10^7 $\Omega^{-1}m^{-1}$].

11. How many electrons flow per second through a section of a conductor, when a current of 1A flows through the conductor. [Hint: 6.25×10^{18}].

12. If a metallic wire is stretched to increase its length by 1%, what will be the percentage change in its resistance ? [Hint: Since volume ($=A \times \ell$) remains constant, 1% elongation means 1% less area of cross section; 1%].

13. A bend in a water-carrying pipe slows down the flow of water. Explain if a similar bend in a current-carrying wire will enhance the resistance of the wire.

14. Define the term temperature coefficient of resistivity. Show graphically the variation of resistivity with temperature for nichrome. (CBSE; 2005).

15. Briefly describe how does the resistance of metals, semiconductors and insulators change with increase in temperature.

16. Explain how the emf of a cell is different from the potential difference across its terminals measured by a voltmeter.

17. Mention the factors on which the internal resistance of a cell depends. Is the electromotive force (emf) of a cell really a force ?

18. Explain in terms of relaxation time τ that how the drift velocity of free electrons in a conductor increases with increase in the temperature of the conductor.

19. Define the term resistivity of a conductor. Give its SI unit. Show that the resistance R of a conductor is given by $m\ell/(Ane^2\tau)$, where the symbols have their usual meanings. (CBSE; 2002).

20. Explain how the drift velocity of free electrons in a conductor is zero in the absence of electric field, but has a small finite value in the presence of the field.

21. Explain how does the resistivity of a conductor depend upon (i) number density n of free electrons and (ii) relaxation time τ. (CBSE; 2004).

22. Keeping the potential difference unchanged at the ends of a wire, if the length of the wire is doubled, explain what will be the effect on (i) the resistance of the wire and (ii) the drift velocity of the electrons.

23. A steady electric current flows through a copper wire of radius r. If the same current is made to flow through a copper wire of radius 2r, explain what will be the effect on the drift velocity of free electrons. Further, if the same current flows through an iron wire of radius r, explain if the drift velocity will increase or decrease.

24. In a potentiometer measurement of emf, if the emf of the experimental cell is reduced, explain how will it effect the position of zero deflection along the potentiometer wire.

25. Explain how the position of zero deflection along a potentiometer wire will be affected, when the length of the wire is (i) doubled and (ii) halved.

26. Explain how the position of zero deflection along a potentiometer wire will be affected, when the current flowing through the wire is reduced.

27. Explain why a long potentiometer wire is preferred for accurate measurements.

28. How a potentiometer of a given length of wire can be made more sensitive by using a resistance box with it ?

29. Explain how the Kirchhoff's first law obeys the law of conservation of charge.

30. Why should the area of cross-section of a metre bridge wire be uniform ?

31. Briefly describe, with necessary circuit diagram, how a metre bridge is employed for the measurement of an unknown resistance.

32. What is meant by the end error in a metre bridge ? How is it removed ?

33. Write the equation expressing the principle of a Wheatstone bridge. Deduce it using Kirchhoff's laws.

34. Explain under what condition a Wheatstone bridge is most sensitive.

Long Answer Questions

(5 mark each)

1. What is meant by drift velocity and relaxation time of free electrons in a current-carrying metallic conductor ? Establish a relation between them. Also, derive Ohm's law on the concept of drift velocity.

2. Define resistivity and conductivity of a material. State their SI units. Discuss their variation with temperature for metals, semiconductors and insulators.

3. What is the essential difference between a series combination and a parallel combination of a set of resistors ? Derive expressions for equivalent resistance separately for the two combinations.

4. What is meant by the emf of a cell ? State the factors on which its value depends. Derive a relation between the emf $\mathcal{E}$, potential difference V, internal resistance r, and external resistance R. Prove that the emf is always greater than the potential difference.

5. State the principle of a potentiometer. How is it of advantage to use a long wire in a potentiometer ? Describe with necessary diagram the comparison of emf's of two cells with a potentiometer.

6. What is meant by the internal resistance of a cell ? On what factors does it depend ? Describe with necessary circuit diagram how the internal resistance of a cell is determined using a potentiometer.

7. Give the principle of Wheatstone bridge. Describe with necessary theory and circuit diagram how it can be used for measurement of temperature.

Numerical Exercises

1. A current of value 10 A flows for 4 minute through a conductor of resistance 10 Ω. Find the total number of electrons and the total charge passing through any cross-section of the conductor during this time. **[15×10^{21};2400 C]**

2. Upon drawing a wire of resistance 10 Ω, its length becomes four times the original length, without any change in the density and resistivity of the material of the wire. Calculate the resistance of the elongated wire. **[160 Ω]**

3. An electric circuit consists of an external resistance of 3.8 Ω and a battery of emf 6.0 V and internal resistance 0.2 Ω. Calculate the potential difference between the two terminals of the battery. **[5.7 V]**

4. A series electric circuit has a current of 10 A flowing through it. Upon introducing an additional resistance of 4 Ω in the circuit, the current reduces to 8 A. Find the total resistance in the original circuit. **[16 Ω]**

5. An electric circuit consists of a parallel combination of three resistances 10 Ω, 20 Ω and 30 Ω, with a current of 6 A passing through the first resistance. Find the values of current passing through (a) 20 Ω resistance, (b) 30 Ω resistance, and (c) combination of the three resistances. **[3 A; 2 A; 11 A]**

6. The values of resistance of a silver wire are measured to be 2.1 Ω and 2.7 Ω at temperatures 27.5 °C and 100 °C, respectively. Calculate the temperature coefficient of resistance of silver. **[3.9 × 10^{-3} °C^{-1}]**

7. Calculate the equivalent resistance between the points A and B for the combination of resistances shown in the figure. **[6 Ω]**

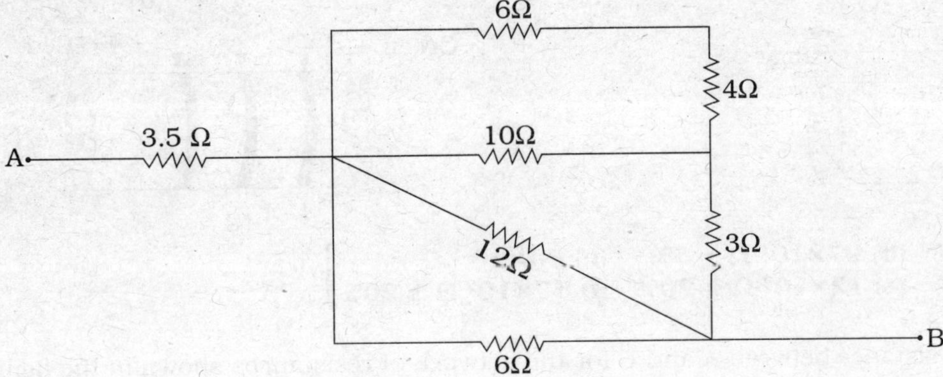

8. Determine the voltage drop across the resistor R_1 in the circuit given below, with E = 6.5 V, R_1 = 50 Ω, R_2 = 100 Ω, R_3 = 100 Ω, R_4 = 300 Ω. (CBSE; 1994) **[12 V]**

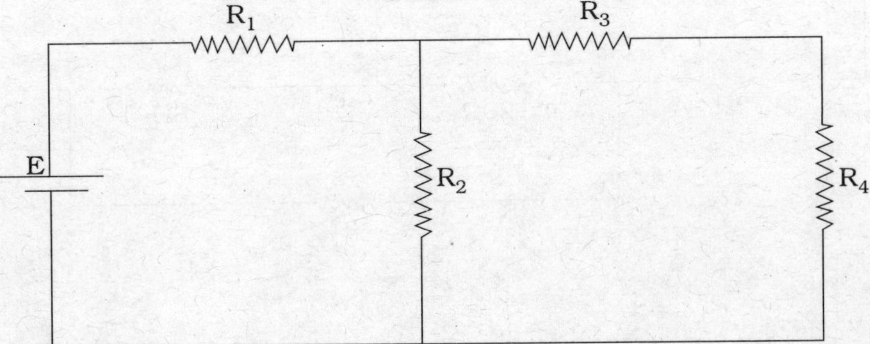

9. An electric supply of voltage 12 V and internal resistance 0.5 Ω is connected to an infinite network of resistances, as shown in the figure. The value of each resistance is 1 Ω. Calculate the current drawn from the supply. **[3.71 A]**

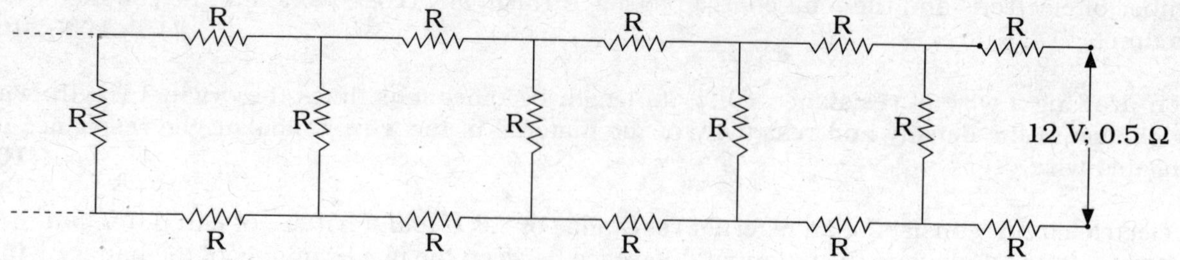

10. Calculate the value of each of the resistors shown in the figure.

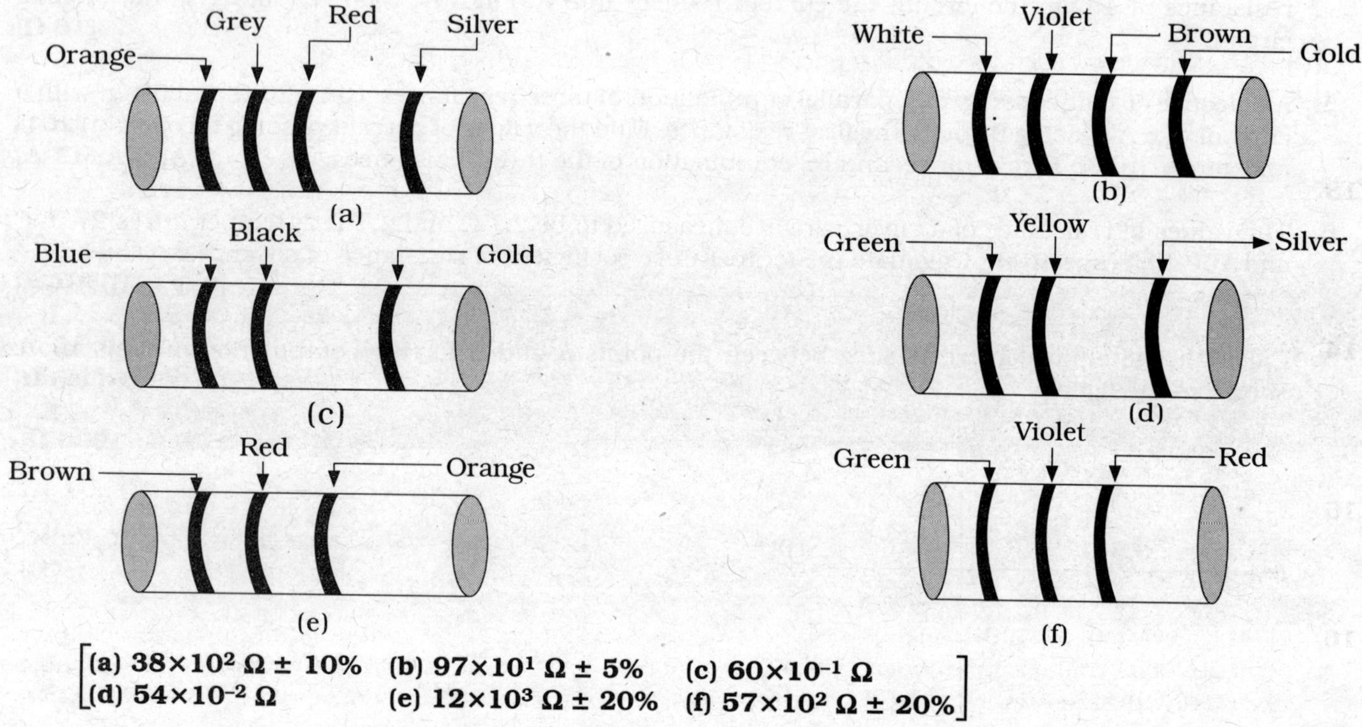

[(a) 38×10^2 Ω ± 10% (b) 97×10^1 Ω ± 5% (c) 60×10^{-1} Ω (d) 54×10^{-2} Ω (e) 12×10^3 Ω ± 20% (f) 57×10^2 Ω ± 20%]

11. Find the equivalent resistance between A and B for the network of resistances shown in the figure. **[4 Ω]**

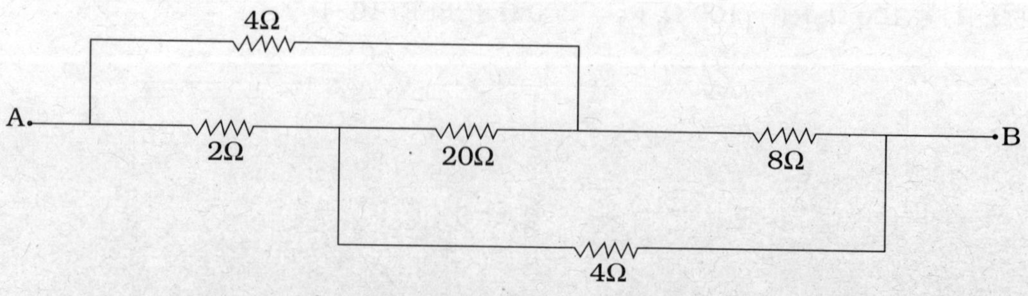

12. In the network of resistances shown in the figure, a potential difference of 2 V is applied between the points A and B. Calculate,

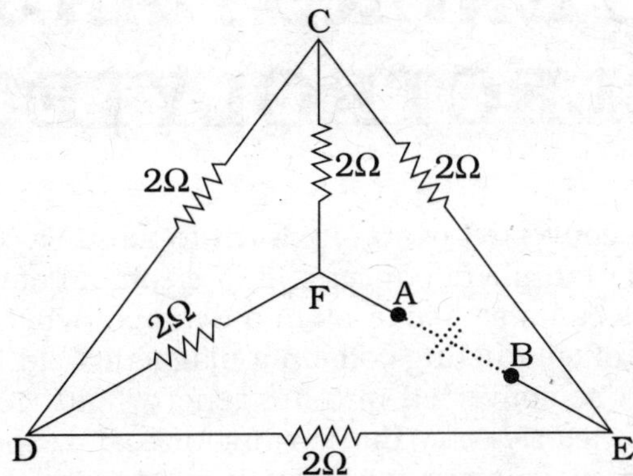

(i) The equivalent resistance of the network across the point A and B, and
(ii) the magnitudes of currents flowing in the arms AFCEB and AFDEB. (CBSE; 1998).

[2 Ω; 0.5 A each]

13. A potentiometer wire of length 100 cm has a resistance of 10 Ω. It is connected in series with a resistance and a cell of emf 2 V and of negligible internal resistance. A source of emf 10 mV is balanced by a length of 40 cm of the potentiometer wire. What is the value of the external resistance ? (IIT). **[790 Ω]**

14. A cell of emf 2 V and a resistance box are connected in series with a potentiometer wire of length 10 m and having resistance as 4 Ω/m. Determine the values of resistance which should be introduced in the resistance box to obtain potential gradients of (i) 0.05 V/m and (ii) 0.2 mV/m, respectively.

[120 Ω; 3996 Ω]

15. In the determination of internal resistance of a cell by potentiometer, the null point is found at 200 cm when the cell is shunted with a resistance of 5 Ω. Next, the null point is located at 300 cm upon shunting the cell with a resistance of 20 Ω. Calculate the internal resistances of the cell. **[4 Ω]**

16. In an experiment for the determination of internal resistance of a cell with potentiometer, the balance point is located at 120 cm when the cell is on open circuit and at 100 cm when the cell is short circuited with a resistance of 12 Ω. Find the internal resistance of the cell. **[2 Ω]**

CHAPTER 2 (B)

ELECTRIC ENERGY AND POWER; ELECTROLYTES

We know that energy can be converted from one form to another. The law of conservation of energy holds for all such conversions. An example of conversion of electrical energy into mechanical energy is an electric motor. The reverse process can also occur, as in a dynamo, where mechanical energy is converted into electric energy by the rotation of conducting coils in a magnetic field.

Similarly, electric energy can be converted into heat energy and vice versa. Familiar examples are a room heater and the generation of electricity by the burning of coal, respectively. The chemical energy can also be converted into electric energy, as when we draw current from a battery. Then the battery is said to be getting gradually 'discharged', i.e. it loses charge, through some chemical reactions in the battery. The reverse process of 'charging' takes place, i.e. the charge is gradually regained through another set of reactions in the battery, when the battery is connected to a source of electricity which sends current through it in the opposite direction. Then the electric energy is converted into chemical energy.

The basic factors behind the thermal and chemical effects of electric current and some of their applications will be described in this chapter.

2B. 01. Electric Energy & Power; Heating Effect; Joule's Law

As already explained in the last chapter (Section 2A.05), when a metallic conductor is connected with a source of emf, an electric field is set up across the conductor, which accelerates the free electrons in the conductor. A current begins to flow in the conductor. For example, as shown in Fig. 2.19, when a battery of emf $\mathcal{E}$ is connected with a resistor of resistance value R, a p.d. V develops across the resistor and a current i starts flowing in the circuit. But the electrons are unable to move freely along the field, because they suffer repeated collisions with the atoms of the conductor and so they largely remain confined between the atoms. Hence the kinetic energy gained by the electrons due to their acceleration is converted into heat energy and the conductor gets heated. The heat is finally lost to the surroundings. Thus the electric energy is said to be dissipated (i.e. lost or wasted), as it cannot be regained and so the process is irreversible.

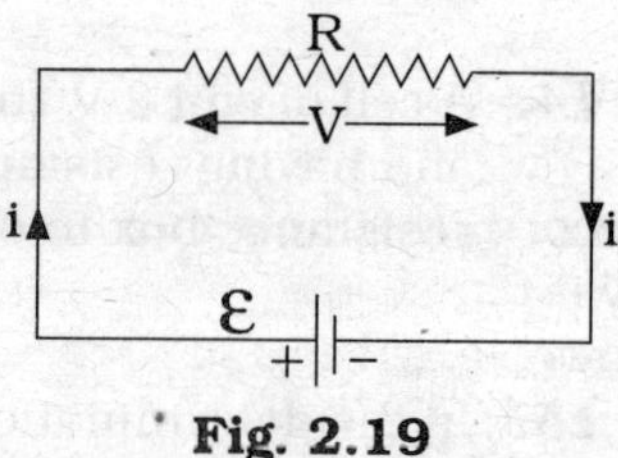

Fig. 2.19

Joule's law

We can calculate the heat produced in the conductor in a time t. From Eq. 2.2 of the last chapter, the total amount of steady state charge flowing across the conductor in time t is

$$q = it$$

The p.d., V = work done in carrying a unit charge from one end of the conductor to the other.

∴ the total work done in similarly carrying a charge q is,

$$W = V.q = Vit \quad \text{..... (2.34)}$$

Since V = iR (Ohm's law), we get (we assume that the internal resistance of the battery is negligible as compared to R),

$$W = i^2Rt \quad \text{..... (2.35)}$$

We assume that this work is entirely converted into heat. Then the heat produced in the conductor is given by

$$H = i^2Rt \quad \text{..... (2.36a)}$$

$$\Rightarrow \quad H = \frac{V^2}{R}t \quad (\because V = iR) \quad \text{..... (2.36b)}$$

The Eqs. 2.36a,b express the so called **Joule's law.** From this law it follows that when current flows through a conductor, the amount of heat produced in the conductor is

(i) directly proportional to the square of the current, for a given conductor and a given time (so the heat is independent of the direction of flow of the current),
(ii) directly proportional to the resistance of the conductor, for a given current and a given time,
(iii) directly proportional to the time of flow of the current, for a given conductor and a given current,
(iv) inversely proportional to the resistance of the conductor, for a given p.d. across the conductor and a given time.

The heating produced in a conductor when an electric current passes through it is popularly known as **Joule heating.**

Electric power & Electric energy

In general, power is defined as the rate of doing work or the rate of production (or dissipation) of energy. Accordingly, the **electric power** of an electric circuit is **defined** as " the rate at which work is done by the emf source in maintaining a steady current in the circuit", or as "the rate of production of energy when a steady current flows in the circuit". (It may also be defined as the rate of dissipation of energy, provided the entire energy that is produced is dissipated in the circuit).

We also know that generally the energy is defined as the capacity to do work, i.e. the amount of work that a system can do. Accordingly, the **electric energy** of a circuit is **defined** as "the total work done by the emf source in maintaining a steady current in the circuit for a given time".

From Eq. 2.34, the work done in the circuit when a steady current i flows through a conductor of resistance R and with p.d. V across it is given by,

$$W = Vit$$

In the SI system, V is expressed in volt, i in ampere and t in seconds. Then W is expressed in joule (J),

$$W = Vit \text{ joule}$$

Then, the electric power P, which is the rate of doing work, is expressed in watt (W),

$$P = W/t = Vi \text{ watt (or } Js^{-1})$$

If V = 1 volt and i = 1 ampere, we have P = 1 watt,

1 watt = 1 ampere × 1 volt (or 1 ampere - volt)

Accordingly, **one watt (1W)** is **defined** as "the power produced by a source of electric energy (i.e. work done by it per second), when it drives a current of one ampere under a potential difference of one volt" or as "the power of an electric circuit when a steady one ampere current flows in it against a potential difference of one volt".

Bigger units of power are commonly used. These are **kilowatt (kW)** and **megawatt (MW),**

$$1 \text{ kW} = 1000 \text{ W (or } 10^3 \text{ W)}; \quad 1 \text{ MW} = 10^6 \text{ W}$$

In terms of horse power (H.P.), which is the commonly employed unit of mechanical power, the electric power in watt is expressed as,

$$1 \text{ H.P.} = 746 \text{ W}$$

From Ohm's law (V = iR), the power P can also be expressed as,

$$P = Vi = i^2R = \frac{V^2}{R} \qquad ..... (2.37)$$

The electric energy is determined by the total work done in the system, which from Eqs. 2.34 and 2.37 is given by,

$$W = Vit = P.t$$

Thus, **energy = power × time**

∴ **1 joule = 1 watt-second**

A bigger measure of energy is **watt-hour** (when a current of 1A flows against a potential difference of 1V for 1 hour),

$$1 \text{ watt-hour} = 1 \times 60 \times 60 \text{ joule} = 3600 \text{ joule}$$

For commercial purposes, the Board of Trade uses a still larger unit of electric energy, called **kilowatt-hour (kWh)** or Board of Trade Unit (**B.O.T.U.** or simply **one unit of electricity),**

$$\begin{aligned} 1 \text{ kWh (or B.O.T.U.)} &= 1 \text{ kilowatt} \times 1 \text{ hour} \\ &= 1000 \text{ watt} \times 1 \text{ hour} \\ &= 1000 \times 3600 \text{ joule} \\ &= 36 \times 10^5 \text{ J} \end{aligned}$$

Example 1. A 1000 W room heater is designed to operate from a 200 V mains supply. If the supply drops to 160 V, by what percentage will its heat output be reduced ?

Solution. Given: P = 1000 W; V = 200 volt

$$\therefore \text{Resistance of the heater, } R = \frac{V^2}{P} = \frac{(200)^2}{(1000)} = 40\ \Omega$$

When the supply drops to 160 volt, the heat produced is given by,

$$\frac{V^2}{R} = \frac{(160)^2}{(40)} = 640\ \text{W}$$

$$\therefore \text{Drop in heat output} = 1000\text{–}640 = 360\ \text{W}$$

$$\therefore \text{\%age drop in heat output} = \frac{360}{1000} \times 100$$

$$= 36\%$$ **(Answer).**

Example 2. An AC source of 110 volt is connected to an electric lamp with the marking 220 V, 110 W. What is the power consumed by the lamp ?

Solution. Given: P = 100 W; V = 220 volt

$$\therefore \text{Resistance of the bulb, } R = \frac{V^2}{P} = \frac{(220)^2}{(100)} = 484\ \Omega$$

Source voltage, V′ = 110 volt

$$\therefore \text{Power consumed by the bulb} = \frac{(V')^2}{R} = \frac{(110)^2}{(484)}$$

$$= 25\ \text{W}$$ **(Answer).**

Example 3. If a 20 W, 6 V lamp has to be operated at a higher voltage of 200 V, should an additional resistance be inserted in series or in parallel with it ? Calculate the value of this resistance.

Solution. Given: P = 20 W; V = 6 volt

$$\text{Resistance of the lamp} = \frac{V^2}{P} = \frac{(6)^2}{(20)} = 1.8\ \Omega$$

$$\text{Current capacity of the lamp,} \quad i_{max} = \frac{P}{V} = \frac{(20)}{(6)} = \frac{10}{3}\ \text{A}$$

New voltage, V′ = 200 V (given)

Since the new voltage is a higher value, the circuit resistance will have to be increased, so that the current in the circuit does not exceed i_{max} ($V = iR$; $\therefore i = V/R$). It means that the additional resistance will have to be inserted in series with the lamp. **(Answer).**

Let R′ be the resistance to be connected in series with the lamp, to have a maximum current of (10/3) A in the circuit. Then,

$$i_{max} = \frac{V'}{(R'+1.8)}$$

$$\Rightarrow \left(\frac{10}{3}\right) = \frac{(200)}{R'+1.8}$$

$$\Rightarrow R' = \frac{(200\times 3)-18}{10} = 58.2\ \Omega$$ **(Answer).**

Example 4. Calculate the electric power consumed by a lamp of filament resistance 1936 Ω, when connected to a 220 V mains supply.

Solution. Given: V = 220 volt; R = 1936 Ω

$$\therefore \text{Electric Power, } P = \frac{V^2}{R} = \frac{(220)^2}{(1936)}$$

$$= 25\ \text{W}$$ **(Answer).**

Example 5. Calculate the current drawn by a motor of 1.5 H.P., when connected to a 220 V supply.

Solution. Given: $P = 1.5 \text{ H.P.} = 1.5\times746 \text{ W}; \quad V = 220 \text{ volt}$

$$\therefore \quad \text{Current, } i = \frac{P}{V} = \frac{(1.5\times746)}{(220)}$$

$$= 5.1 \text{ A} \quad \textbf{(Answer).}$$

Example 6. Calculate the number of units of energy consumed by an electric lamp in 30 days, when operating for 6 hours in a day.

Solution. Given: $P = 200$ W; $t = 6$ hours per day$\times$30 days $= 180$ hours

$$\therefore \text{ Energy consumed} = \text{Power}\times\text{Time}$$

$$= (200)\times(180) \text{ Wh}$$

$$= \frac{200\times180}{1000} \text{ kWh}$$

$$= 36 \text{ units} \quad \textbf{(Answer).}$$

2B. 02. Some Common Applications of the Heating Effect of Current

(i) Electric bulb (or Incandescent electric lamp). It is the most common application of the heating effect of current, in which artificial white light is produced by strongly heating a filament. When a body is strongly heated, it begins to emit a dull red light (~500°C), which turns to bright orange and finally to white (~1500°C). (Incandescence means emission of light by heating, particularly when it becomes white hot).

The ordinary bulb consists of a fine metal filament enclosed in a sealed glass bulb and heated by passing a current through it. The filament is usually coiled, as it raises the efficiency of the lamp. The bulb is filled with some inert (i.e. chemically inactive) gas, so that the filament does not gradually burn out. Usually it contains argon, with some nitrogen, at a pressure of about 3/4th of an atmosphere. The material of the filament is usually tungsten, which has a very high melting point. The filament has a high resistance, because it is very thin, i.e. it has a very small area of cross-section. So the heating effect ($P = i^2R$) is very strong. The thick copper wires in the circuit, which carry the same current as the filament, do not get much heated as their resistance value is comparatively very low (the resistivity of copper is only about 1/3rd of tungsten).

The wattage and operating voltage are marked on the bulb, e.g. (40 W, 220 V), (100 W, 220 V), etc. From these values, the values of current and resistance for a given lamp can be calculated ($P = Vi = V^2/R$; so $i = P/V$ and $R = P/V^2$). The resistance of a 40 W lamp is 2.5 times that of a 100 W lamp. The 40 W lamp has thinner wire.

(ii) Electric heating devices (heater, heating rod, geyser, toaster, press, oven, etc.). These devices need large amounts of electric power, as they provide strong heating over extended areas. The current in the circuit is a large value ($P = Vi$) and so even the connecting copper wires tend to get heated, in spite of their low resistance ($H = i^2R$). The material of the heating coils must have a high melting point. The wire used to make the coils should also be thick enough to avoid melting. The material should have a high resistivity, so that in spite of the large thickness of the wire, the required resistance value is achieved for the available limited length of the wire ($P = V^2/R$ or $R = V^2/P$).

The alloy nichrome (60% nickel, 25% iron,15% chromium) is found to be a suitable material. It has both a high melting point and a high resistivity (about 60 times that of copper). It is also highly resistant to oxidisation in air. The heating coils are wound on mica or porceilane, which both are strong electric insulators and can also withstand prolonged heating. The values of the wattage and supply voltage are labelled on the device, e.g. (1 kW, 220 V) on a room heater, which means 1 kW of power consumption, when operated for 1 hour at 220 V. As before, the values of the current and resistance can be easily calculated ($i = P/V$; $R = V^2/P$). The higher is the wattage of the device, the higher is the current and so thicker has to be the wire of the heating coil. Also, higher is the wattage, lower is the resistance of the coil.

(iii) Fuse. If by mistake or by accident an electric installation is 'short circuited', i.e. the circuit gets completed by connection through a low resistance, very large current, and therefore excessive heating, is caused in the installation. Not only it can damage all the electric devices (e.g. household appliances, machines, high voltage transformer, etc.), but can also cause fire in buildings. A fuse provides a cheap but very effective protection against such possible damages. Usually it is a small wire or strip connected in series with the circuit. When the current exceeds a specified value (e.g. 5 ampere, for a 5 A fuse), it melts down and breaks the circuit and thus saves the installation. The wire is supported on a fire-proof material, like glazed porceilane. The material of the fuse wire should have a low melting point. A common material for the wire is a tin-lead alloy (63% tin and 37% lead). Other kinds of fuses are also available, e.g. cartridge fuse, which has combustible conducting powder enclosed within porceilane. They are safer but expensive. Nowadays, the use of M.C.B's (miniature circuit breakers) is on increase, because they break the circuit without any damage to themselves and so need not be replaced again and again. However, they are relatively very expensive.

Efficiency of an electric device

The **efficiency** η of an electrical device is **defined** as "the ratio of the power (or energy) delivered by it (called its **output**) to the power (or energy) supplied to it (called its **input**),

$$\eta = \frac{\text{Output power}}{\text{Input power}}$$

Generally this ratio is multiplied by 100, so that the efficiency is conveniently expressed as a percentage. For instance, η = 22% means that 22% of the power supplied to the device is actually available for use.

Example 7. An electric motor operates on a 50 V supply and draws a current of 12 A. If the motor yields a mechanical power of 150 W, what is the percentage efficiency of the motor ? (NCERT).

Solution. Given: V = 50 volt; i = 12 A; P = 150 W

Input Electric Power = Vi = (50)×(12) W

Output mechanical Power = 150 W

$$\therefore \quad \text{\% age Efficiency of the Motor} = \frac{\text{Output}}{\text{Input}} \times 100$$

$$= \frac{(150)}{(50 \times 12)} \times 100$$

$$= 25\% \quad \textbf{(Answer).}$$

2B. 03. Electrolytes; Anions & Cations

So far our study of the flow of electricity has been concerned with solids, mainly metallic conductors. The liquid metal mercury also conducts electricity in the same way. In all these cases, Ohm's law is obeyed and no chemical changes occur in the conductor due to the passage of current. The carriers of electricity are the free electrons, which are abundantly present inside the conductor. The atoms or molecules of the conductor play no direct role in the passage of current.

Most of the liquids in their pure form, e.g. distilled water, are bad conductors of electricity. However, solutions of salts, acids and bases in water are very good conductors of electricity (e.g. tap water, which generally contains some dissolved salts, is a good conductor). In these cases, chemical decompositions takes place in the liquid conductor when the electric current passes through it. The material of the conductor dissociates into +ve and -ve ions.

"The process of dissociation of a conducting solution into +ve and –ve ions when an electric current passes through it" is called **electrolysis.** "A conducting solution which dissociates into +ve and -ve ions on passing an electric current through it" is called an **electrolyte.** However, although electrolytes are generally liquids, some solid electrolytes, e.g. caustic soda and silver iodide, also exist.

The experimental arrangement for studying the phenomenon of electrolysis consists of two metallic conductors (usually metal plates) dipped into an electrolyte and connected to the terminals of a source of emf, e.g. a battery. They are called **electrodes.** The electrode connected to the +ve terminal of the emf source is called **anode** and the one connected to the -ve terminal is called **cathode.** Thus the anode is at a higher potential and the cathode at a lower potential. Since conventionally we take electric current to flow from the +ve end of the emf source to its –ve end throught the circuit, inside the electrolyte the current flows from the anode to the cathode.

The vessel in which the electrolysis takcs place is called an **electrolytic cell** or a **voltameter.** It essentially contains an electrolyte in which two electrodes are dipped. The electricity produced by a chemical action is called 'voltaic'. Hence the name 'voltameter' (not 'voltmeter').

The process of conduction of electricity through electrolytes is fundamentally different from that in the solid conductors. Ohm's law needs to be modified for application to the electrolytes. Besides, the conduction in the electrolytes is not due to the transport of the very light free electrons, but due to the physical transport of the relatively very heavy **ions,** which are charged atoms or groups of atoms. An atom is said to be ionized when it loses some of its electrons or gains some extra electrons. A **positive ion** is created in the former case and a **negative ion** in the latter case. "The ions which migrate towards the anode during electrolysis" are called **anions.** They are negatively charged. "The ions which migrate towards the cathode during electrolysis" are called **cations.** They are positively charged.

Hydrogen and all metallic radicals produce +ve ions (cations), as they easily lose their loosely bound outer electrons. They always move toward the cathode. Oxygen and the basic radicals capture the free electrons, producing -ve ions (anions), which move towards the anode.

The chemical properties of an ion are very different from the properties of the corresponding neutral atom or molecule (or just a group of atoms). An ion is inert, i.e. chemically inactive. For instance, the sodium ion (Na^+) is stable in an aqueous solution. But the neutral sodium atom (Na) vigorously reacts with water (in warfare, sodium bombs are made to explode in water). However, when the ions reach their respective electrodes, the charges on them are neutralised by the opposite charges on the electrodes. Then they are converted into neutral atoms or radicals. Whenever possible, these neutralised atoms and radicals chemically react with the electrolyte or the materials of the electrodes.

The conductivity of the electrolytes is very much smaller (down by a factor of $10^{-5} - 10^{-6}$) as compared to the metals. The main reasons are: much larger mass of the ions leading to lower drift velocities, greater viscosity of the medium, smaller number density of the ions as compared to the free electrons, etc.

Solved Conceptual Problems
[Including HOTS Questions]

Problem 1. If the current flowing in an electric circuit is tripled, without any change in the resistance in the circuit, explain how will it affect the power dissipation in the circuit.
Solution. The power dissipated in an electric circuit is, $P = i^2R$. Thus, $P \propto i^2$, when R is constant. When i becomes 3 times, P becomes $(3)^2 = 9$ times, i.e. the power dissipation will become 9 times.

Problem 2. A wire connected to an electric bulb does not glow, when the bulb is switched on, but the filament of the bulb glows, although the same current flows through both of them. Explain why.
Solution. From the relation $H = i^2Rt$, the amounts of heat produced in a time t in different components of an electric circuit are proportional to the respective resistance values of the components.

The connecting wires have very small resistance as compared to the resistance of the filament of the bulb. Hence, while the filament gets heated to a high (glowing) temperature, the heating of a connecting wire is comparatively very small. So the wire does not glow.

Problem 3. By what percentage will the illumination of an electric lamp be reduced, if the current in the circuit drops by 20%. Assume that the illumination is directly proportional to the heat produced in the lamp.
Solution. Let R be the resistance of the filament of the lamp and i be the current flowing through it for time t. Then the heat produced in time t is given by,

$$H = i^2Rt$$

When the current drops by 20%, the reduced current i′ (say) in the circuit becomes 80% of the initial value i, i.e.,

$$i' = \frac{80}{100}\times i = \frac{4}{5}i$$

$$\therefore \text{Reduced heat produced, } H' = \left(\frac{4}{5}i\right)^2 Rt = \frac{16}{25}i^2Rt$$

$$\therefore \text{ \% age Heat Reduction} = \left(\frac{H-H'}{H}\right)\times 100 = \left(1-\frac{H'}{H}\right)\times 100$$

$$= \left[1-\frac{(16/25)i^2RT}{i^2RT}\right]\times 100$$

$$= \frac{9}{25}\times 100$$

$$= 36\% \quad \textbf{(Answer).}$$

Problem 4. Long distance power transmission is carried at high voltages. Explain why ?
Solution. Electricity can be produced in different ways (e.g. hydroelectricity, nuclear reactor, burning of coal or gas, etc.). It is transmitted from the source to the consumer end by long connecting cables, through a system of transformers. A substantial amount of the electric power is lost to the atmosphere due to Joule heating in the cables, given by i^2R, where R is the resistance of the cables and i is the current flowing through the cables.

Since the total power produced, $P = V\times i$, we have, $i = P/V$. Therefore the power loss is,

$$\text{Power loss} = i^2R = \frac{P^2\times R}{V^2}$$

Since P and R are constant, we find that the power loss is inversely proportional to V^2. Hence it can be substantially reduced by transmitting the electric power at a high voltage, e.g. 11000 volt.

Problem 5. Why is the conductivity of an electrolyte very low as compared to that of a metal, at room temperature ?
Solution. The conductivity of the electrolytes is relatively very low, being just about 10^{-5} to 10^{-6} times that of the metals, at room temperature. The main reasons are,

(i) the number density (i.e. no. per unit volume) of the free ions in an electrolyte is very small, as compared to the number of the charge carriers, viz. free electrons, in a metal.

(ii) due to the large mass of an ion as compared to that of a free electron, the ions drift much more slowly than the electrons, in an electric field.

(iii) the motion of the ions slows down in a solution, due to the viscocity of the solution, i.e. the electrolytes have a naturally high resistance.

Solved Numerical Problems

Problem 1. A heating element is marked 210 V, 630 W. What is the current drawn by the element when connected to a 210 V DC mains ? What is the resistance of the element ? (NCERT).
Solution. Given: V = 210 volt; P = 630 W
The power consumed by the element is given by,

$$P = \frac{V^2}{R}$$

$$\Rightarrow \quad R = \frac{V^2}{P} = \frac{(210)^2}{(630)} = 70\ \Omega \quad \textbf{(Answer).}$$

Next, we know, $P = Vi$
∴ The current drawn by the element is given by

$$i = \frac{P}{V} = \frac{(630)}{(210)} = 3\text{ A} \quad \textbf{(Answer).}$$

Problem 2. A source of emf is connected across a series combination of three resistors of equal resistance, which together dissipate power of P watt. How many times more power will be dissipated for a parallel combination of these resistors, when connected across the source ?

Solution. For a series combination of 3 resistances of value R each, the total resistance is given by,

$$R_s = R + R + R = 3R$$

If V is the emf of the source, the power dissipated for this combination is,

$$P_S = \frac{V^2}{3R}$$

For a parallel combination, the total resistance is given by,

$$\frac{1}{R_p} = \frac{1}{R} + \frac{1}{R} + \frac{1}{R}$$

$$\Rightarrow \quad R_p = \frac{R}{3}$$

$\therefore$ The power dissipated for the parallel combination is,

$$P_p = \frac{V^2}{(R/3)} = \frac{3V^2}{R}$$

Hence,
$$\frac{P_p}{P_S} = \frac{(3V^2/R)}{(V^2/3R)} = 9$$

$\therefore$ The power dissipated will be 9 times more for the parallel combination. **(Answer).**

Problem 3. An electric bulb is marked 100 W, 230 V. If the voltage drops to 115 V, what heat and light energy is produced by the bulb in 20 minutes ? (NCERT).

Solution. Given: P = 100 W; V = 230 volt; t = 20 min = 20×60 = 1200 second

Let R be the resistance of the bulb. The expected rate of production of energy (i.e. power produced; taking into account both heat energy and light energy) is given by,

$$P = \frac{V^2}{R}$$

Therefore the resistance of the bulb is,

$$R = \frac{V^2}{P} = \frac{(230)^2}{(100)} = 529\ \Omega$$

Next, the voltage drops to, V′ = 115 V

$\therefore$ Energy produced is given by,

$$H = i^2Rt = \frac{V^2}{R}t \quad (\because V = iR)$$

$$= \frac{(115)^2 \times (1200)}{(529)} = 30{,}000\ \text{J (or 30 kJ)}$$ **(Answer).**

Problem 4. A standard 50 W electric bulb in series with a room heater is connected across the mains. If the 50 W bulb is replaced by a 100 W bulb, will the heater output be larger, smaller or remain the same ? (IIT; 1973).

Solution.

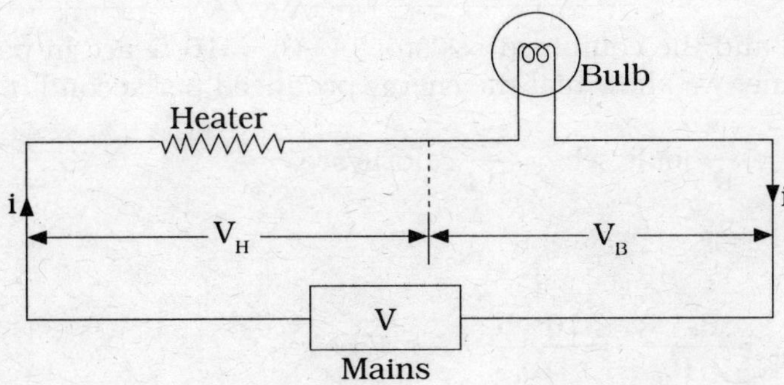

Let,
V = Applied p.d. across heater and bulb
i = current in the circuit
V_H = p.d. across heater
V_B = p.d. across bulb
R_H = Resistance of heater

The power output of the heater is given by ($V_H = iR_H$),

$$P = \frac{V_H^2}{R_H} = i^2R_H \quad \text{..... (i)}$$

(Since the heater and the bulb are in series, same current i flows through them).

For a given voltage supply, the wattage is inversely proportional to the resistance R (wattage = V^2/R). It means that the resistance of a 100 W bulb is less then that of a 50 W bulb,

$$(R)_{100\,W} < (R)_{50\,W}$$

When the 50 W bulb is replaced by a 100 W bulb, the total resistance ($R_H + R_{Bulb}$) in the circuit decreases. Then the current i in the circuit is increased [V = iR (Ohm's law); ∴ i = V/R]. Hence, from Eq.(i), the heater output will increase when the 50 W bulb is replaced by a 100 W bulb. **(Answer).**

Problem 5. The water in an electric kettle begins to boil in 15 minutes after being switched on. Using the same mains supply, should the length of the wire used for the heating element be increased or decreased, if the water is to boil in 10 minutes ? Why ? Neglect the heat loss to the surroundings. (IIT; 1977).

Solution. If V is the voltage of the mains supply and R the resistance of the heating wire, then the heat developed in the wire in a time t is given by,

$$H = P \times t = \frac{V^2}{R} \times t \quad \text{joule}$$

If the water is to boil in lesser time with the same mains supply (i.e. V = constant) and the same amount of heat (i.e. H = constant), then the value of R should be such that,

$$\frac{t}{R} = \text{constant, i.e. } R \propto t$$

$$\Rightarrow \quad \frac{\rho L}{A} \propto t \left(\because R = \frac{\rho L}{A}\right)$$

(ρ: resistivity ; L: length of wire; A: cross-sectional area of wire)

$$\Rightarrow \quad L \propto t \quad (\because \rho \text{ and } A \text{ are constant values})$$

If follows that to boil water in lesser time (i.e. smaller t), the length of the wire should be decreased. It has to be reduced to (10/15), i.e. 2/3rd of the initial value. **(Answer).**

Problem 6. In the circuit shown below, the heat produced in 5 ohm resistor due to current flowing through it is 10 calorie per sec. What is the heat dissipated in the 4 ohm resistor ? (IIT; 1981).

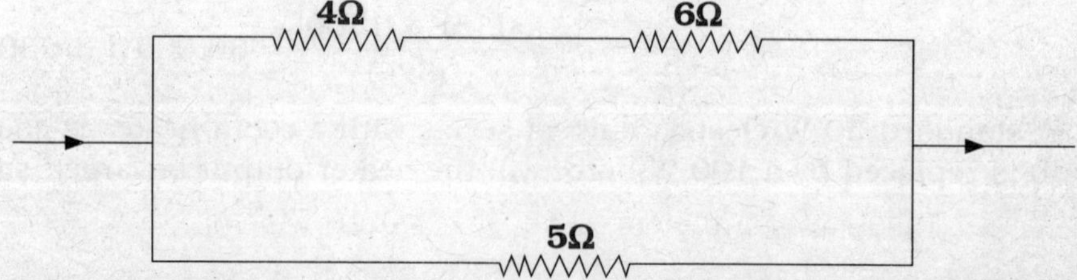

Solution. The resistor 5Ω and the combined resistor (4+6) = 10 Ω are in parallel. Therefore the p.d. V across them will be the same. We know that the energy produced per second (i.e. the power) is given by,

$$P = \frac{V^2}{R} \text{ joule/sec.} = \frac{V^2}{RJ} \text{ calorie/sec.} \quad \text{..... (i)}$$

$$\therefore \quad \frac{P_{10}}{P_5} = \frac{R_5}{R_{10}}$$

$$\Rightarrow \quad P_{10} = \frac{P_5 \times R_5}{R_{10}} = \frac{10 \times 5}{10} = 5 \text{ cal/sec.}$$

Next, since the resistors 4 Ω and 6 Ω are in series, the current flowing through them is the same. Hence the power produced in them is given by ($P = V^2/RJ = i^2R/J$ cal./sec.),

$$P_{10} = P_4 + P_6 = 5 \text{ cal./sec.}$$

Again, from Eq.(i),
$$\frac{P_4}{P_6} = \frac{R_6}{R_4} = \frac{6}{4} = \frac{3}{2}$$

$\Rightarrow$ $P_6 = (2/3)\ P_4$

$\therefore$ $P_4 + (2/3)\ P_4 = 5$

$\Rightarrow$ $(5/3)\ P_4 = 5$

$\Rightarrow$ $P_4 = 3$ calorie/second **(Answer).**

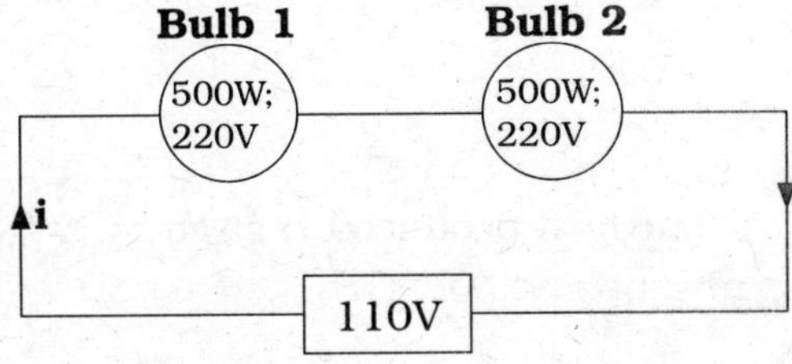

Problem 7. If two electric bulbs, each designed to operate with a power of 500 W in a 220 V line, are put in series in a 110 V line, what will be the power generated by each bulb ? (IIT; 1977).

Solution. Let the resistance of each bulb be R. Then the total resistance in the circuit is (R+R)=2R. Since the two bulbs are in series, the voltage drop across each bulb is,

$$V_B = \frac{R}{2R}\times(110) = 55 \text{ V}$$

The power generated by each bulb is given by,

$$P = \frac{V_B^2}{R} = \frac{(55)^2}{R}$$

But, $$R = \frac{(220)^2}{(500)} \quad \text{(in general, } P = V^2/R. \therefore R = \frac{V^2}{P}\text{)}$$

$\therefore$ $$P = \frac{(55)^2 \times 500}{(200)^2} = 31.25 \text{ W}$$

Thus, each bulb will generate 31.25 W of power. **(Answer).**

Problem 8. An electric kettle has two coils. When one coil is switched on, it takes 15 minutes to boil water and when the second coil is switched on it takes 30 minutes. How long will it take to boil water when both the coils are used in (a) series (b) parallel. (Roorkee; 1992).

Solution. The heat produced in a resistance R in time t is given by,

$$H = Pt = \frac{V^2}{R}t$$

$\therefore$ For individual coil (1 minute = 60 second),

$$H_1 = \frac{V^2}{R_1}(15\times60)$$

$$H_2 = \frac{V^2}{R_2}(30\times60)$$

In both the cases, the heat is sufficient to boil the water.

$$\therefore H_1 = H_2$$

$\Rightarrow$ $$\frac{V^2\times15\times60}{R_1} = \frac{V^2\times30\times60}{R_2}$$

$\Rightarrow$ $$\frac{15}{R_1} = \frac{30}{R_2} \Rightarrow R_2 = 2R_1$$

(a) When the coils are used in series, the heat produced is,

$$H_s = \frac{V^2}{R_1+R_2}\times t_s$$

$$= \frac{V^2}{3R_1} \times t_s \quad (R_2 = 2R_1)$$

But, $H_s = H_1 \ (= H_2)$

$$\Rightarrow \quad \frac{V^2}{3R_1} \times t_s = \frac{V^2}{R_1} \times (15 \times 60)$$

$$\Rightarrow \quad t_s = (45 \times 60) \text{ second}$$

$$= 45 \text{ minute} \quad \textbf{(Answer).}$$

(b) When the coils are used in parallel, their combined resistance R_p is given by,

$$\frac{1}{R_p} = \frac{1}{R_1} + \frac{1}{R_2} = \frac{1}{R_1} + \frac{1}{2R_1} = \frac{3}{2R_1}$$

$$\Rightarrow \quad R_p = \frac{2}{3} R_1$$

∴ The heat produced is given by,

$$H_p = \frac{V^2}{(2/3)R_1} \times t_p = \frac{3V^2}{2R_1} \times t_p$$

But, $H_p = H_1 \ (= H_2)$

$$\Rightarrow \quad \frac{3V^2}{2R_1} \times t_p = \frac{V_2}{R_1} \times (15 \times 60)$$

$$\Rightarrow \quad t_p = (10 \times 60) \text{ second}$$

$$= 10 \text{ minute} \quad \textbf{(Answer).}$$

Problem 9. A house has electricity supplied at 220 V and priced as Rs. 2.50 per kWh. It is fitted with 15 lamps marked 100 W each, 5 ceiling fans consuming 0.5 A each, and an electric kettle of resistance 200 Ω. If all the fittings are daily used for 6 hour, calculate the charges to be paid for the electric consumption for 30 days.

Solution. Daily consumption by lamps $= (15) \times (100 \times 10^{-3}) \times 6$ h

$= 9$ kWh

Daily consumption by fans $(P = Vi)$ $= (5) \times (220 \times 0.5 \times 10^{-3}) \times 6$

$= 3.3$ kWh

Daily consumption by kettle $\left(P = \frac{V^2}{R}\right) = \frac{(220)^2}{(200)} \times 10^{-3} \times 6$

$= 1.45$ kWh

∴ Total daily consumption of electric energy $= 9 + 3.3 + 1.45$

$= 13.75$ kWh

∴ Total consumption in 30 days $= 13.75 \times 30$

$= 412.5$ kWh

∴ Total cost of the energy consumed $=$ Rs. (412.5×2.5)

$=$ Rs. 1031.25 **(Answer).**

Problem 10. A fuse wire of radius 0.15mm is found to blow up when a current of value 15 A or above passes through it. If a fuse wire of the same material is to blow up when a current of 30 A or above passes through it, what should be its radius ?

Solution. When a current i flows through a wire of a material of resistivity ρ and having length and radius of cross-section as ℓ and r, respectively, the rate of heat production in the wire is given by,

$$H = i^2R = i^2\left(\rho \frac{\ell}{\pi r^2}\right)$$

The wire simultaneously loses heat to the atmosphere through its surface area ($= 2\pi r. \ell$). Suppose the heat lost per unit area per second is h. Then, the rate of heat loss in the wire is,

$$H' = (2\pi r\ell)h$$

A fuse wire blows up when the rate of heat production in it exceeds its rate of heat loss. In the steady state, the rate of heat production is just equal to the rate of heat loss, i.e.

$$H = H'$$

$$\Rightarrow \quad i^2 \frac{\rho\ell}{\pi r^2} = 2\pi r\ell h$$

$$\Rightarrow \quad i^2 = \left(\frac{2\pi^2 h}{\rho}\right) r^3$$

$$\text{i.e.} \quad i^2 \propto r^3$$

For two fuse wires made of the same material, having radii of cross-section as r_1 and r_2, and with currents i_1 and i_2 passing through them, respectively, we have at the steady state,

$$\frac{i_1^2}{i_2^2} = \frac{r_1^3}{r_2^3}$$

$$\Rightarrow \quad r_2^3 = \left(\frac{i_2}{i_1}\right)^2 \times r_1^3$$

$$\Rightarrow \quad r_2 = \left(\frac{i_2}{i_1}\right)^{2/3} \times r_1$$

We are given that $i_1 = 15$ A, $i_2 = 30$ A, and $r_1 = 0.15$ mm $= 0.15\times10^{-3}$ m. Therefore, we have,

$$r_2 = \left(\frac{30}{15}\right)^{2/3} \times (0.15\times10^{-3})$$

$$= 1.587\times0.15\times10^{-3}$$

$$= 0.24\times10^{-3} \text{ m (or 0.24mm)}$$ **(Answer).**

Very Short Answer Questions

[Including HOTS Questions]

(1 mark each)

1. Name the material of the element commonly used in electric heaters.

2. Mention two characteristics of the material of the element commonly used in electric heaters.

3. What do the following specifications on two different electric appliances mean: (i) 5 A; 220 V and (ii) 40 W; 220 V ?

4. Which of the two bulbs having the following specifications has higher resistance: (i) 100 W; 220 V and (ii) 60 W; 220 V ?

5. The coil of an electric heater is first cut into three equal parts and then one of the parts is used to make a new heater. What is the ratio of the heat produced by the original heater to that produced by the new heater ? [Hint: 3:1].

6. What is the characteristic of the material used in a fuse wire ?

7. Two electric heaters have specifications marked as 220 V; 1000 W each. If they are connected together in series, what will be the electric power of the combination ?

8. When a coil is connected across a 20 V source, it develops heat of 800 calorie per second. What is the resistance of the coil ? [Hint: 0.12 Ω].

9. How much electric charge passes through a 60 W lamp, when a current of 0.5 A flows through it for one hour ? [Hint: 1800 C].

10. If one unit of electric energy is said to be consumed in a domestic use, what does it physically mean ?

11. The current flowing through an electric bulb changes by 1%. What is the percentage change in the power consumed by it ?

12. Name the physical quantity which has its units as joule coulomb^{-1}. Is it a scalar or a vector quantity ?

13. What are the charge carriers in an electrolyte ? Distinguish between anode and cathode.

14. Distinguish between a voltmeter and an ammeter, in terms of the value and mode of connection of the resistance in their construction.

15. What do you understand by anions and cations ?

16. Distinguish between a voltmeter and a voltameter.

17. What do you mean by the capacity of a cell ?

18. Define the efficiency of an electric device.

Short Answer Questions

[Including HOTS Questions]

(2 or 3 marks each)

1. Explain the working of a fuse wire in an electric circuit. How is it different from the other connecting wires in the circuit?

2. One electric lamp with a metal filament and another with carbon filament are connected together in series to the mains. The two lamps have the same wattage. Explain which lamp will glow brighter and why.

3. A heater and a 50 W lamp are connected together in series to the mains. If the lamp is replaced by another 100 W lamp, explain with due reasoning if now the heater will produce same, more, or less heat.

4. The cross-sectional areas of two wires X and Y of equal length and of same material bear the ratio 1:4. Find the ratio of the heat produced in the wires, when same voltage is applied across each of them. [Hint: 1:4].

5. A resistor of value R dissipates power P_1 when connected to a voltage source. Assuming the resistance of the source to be zero, find the power P_2 dissipated by the resistor when another resistor of value R is connected in series with it. [Hint: $P_1/4$].

6. Calculate the time taken by an immersion heater of rating 836W to heat 2 litre of water from room temperature (20°C) to 40°C. (1calorie = 4.18 joule).

7. In an electric circuit, two resistors of resistance R each are first connected in series and then in parallel. Calculate the ratio of the heat produced by the combination in the two cases. [Hint: 1:4].

8. Calculate the units of electric energy consumed in a month of 30 days, when twelve 100 W lamps are operated for 10 hours per day. [Hint: 360 kWh].

9. If the current in an electric circuit drops by 20%, determine the percentage reduction in the illumination of a lamp in the circuit, assuming that the light produced by the lamp is directly proportional to the heating of the lamp filament. [Hint: 36%].

10. Explain how the conduction of electricity in an electrolyte is different from that in a solid conductor.

11. Briefly describe three applications of heating effect of current, to (i) produce light, (ii) produce large amount of heat and (iii) protect against flow of large current in an electric circuit.

Long Answer Questions

(5 marks each)

1. Explain the atomic process that leads to the heating of a conductor when current flows through it. Derive the equation for the Joule heating. Define electric power, electric energy, and watt. What is meant by "one unit of electricity" ?

Numerical Exercises

1. An electric heater consists of 20 m length of manganin wire of cross-sectional area 0.23 m^2. Calculate the wattage of the heater when the p.d. across it is 200 V (resistivity of manganin = 46×10^{-7} Ω–m). (DBSE; 1980). **[10^9 W]**

2. A fuse wire of circular cross-sectional radius 0.2 mm is found to blow off at 20 A current. If another fuse wire made of the same material blows off at 40 A current, determine its circular cross-sectional radius. **[0.32 mm]**

3. The markings on an electric lamp read as 200 W, 230 V. Calculate the heat and light energy produced by the lamp in 20 minutes, if the supply voltage drops to 115 V. **[30 kJ]**

4. A current of 5.0 A flows through an electric press of resistance 44 Ω. Calculate the energy consumed by the press in 5 minutes. (CBSE; 1993). **[3.3×10^5 J]**

5. Three identical resistors of resistance R each when connected in series with a DC source dissipate power X. If the resistors are connected in parallel to the same source, how much power will be dissipated. (CBSE; 1999). **[9X watt]**

6. A student has two wires of iron and copper, respectively, having same length and same diameter. He first joins the two wires in series and passes a gradually increasing current through the combination. Next, he joins the two wires in parallel and repeats the process of passing the current. Which wire will glow more in each case and why? (CBSE Sample Paper). **[Series: Iron wire; parallel: Copper wire]**

UNIT 3
MAGNETIC EFFECTS OF CURRENT AND MAGNETISM

CHAPTER 3 (A)
Magnetic Effects of Current

CHAPTER 3 (B)
Modern Magnetism

CHAPTER 3 (A)

MAGNETIC EFFECTS OF CURRENT

3A. 01. Introduction

In Chapter 1, we had learnt that a stationary electric charge produces an electric field around it. If any other electric charge is brought in the field, it experiences an electrostatic force. We essentially assumed the charges to be in rest position. The situation changes when the charges are not at rest, e.g. in an electric current (which consists of moving charges) flowing in a wire, electrons moving around the nucleus in an atom, a beam of moving electrons in a cathode ray tube, etc. It is natural to ask asto what additional effects appear when the charges are in motion. The basic question and some related consequences will be addressed to in the present chapter. The chapter has three main sections. The

(i) first deals with determining the additional field (called magnetic field, $\vec{B}$) existing in the surroundings of moving charge (s),

(ii) second is devoted to the interaction of this field with other moving charges, and

(iii) third describes some related applications.

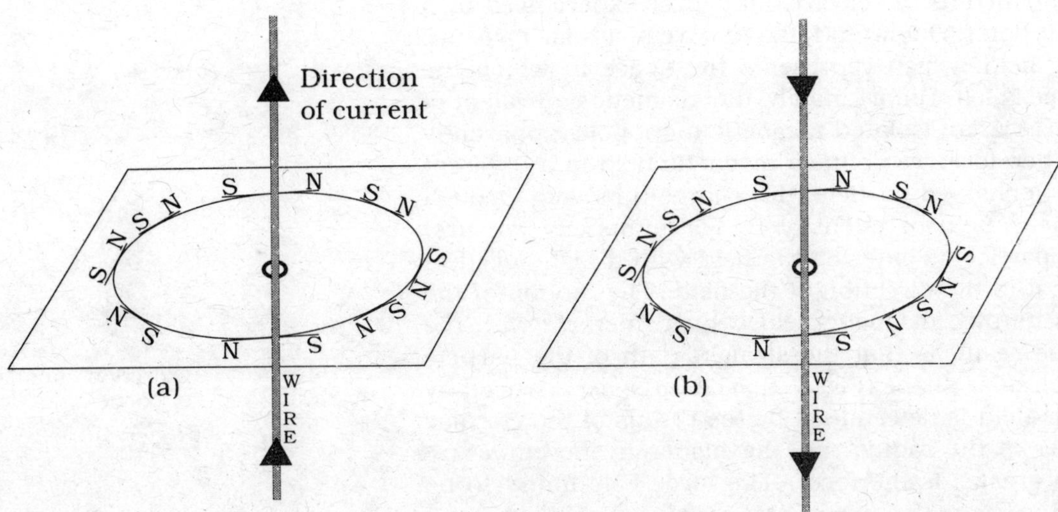

Fig. 3.1

3A. 02. Oersted's Experiment; Electromagnetism

In 1820, the Danish physicist Hans Christian Oersted observed that a current-carrying wire influenced the alignment of a magnetic compass needle placed in its neighborhood. When the needle was placed at different positions on a circular path around the wire, it always aligned itself tangentially to the curvature of the circle. (Fig. 3.1a). On reversing the direction of current in the wire, the orientation of the needle also reversed (Fig. 3.1b), Further, the influence on the needle reduced, i.e. the needle aligned itself less quickly, on decreasing the current or on increasing the radius of the circle. Before the experiment was performed by Oersted, such effects on a magnetic needle were only expected from the presence of a magnet around it. Thus, the experiment led to the conclusion that moving charges (electrons in the case of a current-carrying wire) produce magnetic effects.

Until the Oersted's experiment was performed in 1820, electricity and magnetism remained as two distinct branches of physics, primarily due to different methods employed to study them. The Oersted's experiment established a direct link between them, leading to the new branch of **ELECTROMAGNETISM** in physics. It includes a wide variety of phenomena and applications, some of which will be described in the present chapter and the next two chapters.

Additional knowledge

We now understand that an electric current produces a magnetic field around it. Then, how does a bar magnet produce its magnetic field without any current passing through it ? The answer lies in the atomic structure. The bar magnet consists of a vast number of atoms (billions of them). In each atom the electrons revolving around the nucleus constitute tiny current loops, each of which becomes a source of magnetic field. Collectively, the atoms of the magnet give rise to an appreciable magnetic field. One may argue then why not every material should be a magnet ? The answer lies in the fact that in general as many loops have current flowing clockwise in them as have the current flowing anticlockwise. The result is that there are as many tiny magnets pointing N-S as pointinhg S-N. So the net magnetic field is zero. For a magnet (or a magnetic material, in general), there is a dominance of one kind of tiny magnets, either N-S or S-N. So the net magnetic field has a non-zero finite value in magnets. The electrons are also believed to spin about their own axes, which produces additional magnetic field, More about it will be explained later in chapter 3B (Section 3B. 08). (Howerver, even here it may be cautioned that actually the electrons neither orbit the nucleus nor spin about an axis. Only the equivalent effets are observed).

3A. 03. Magnetic Field (or Magnetic Field Intensity) and Lorentz Force

In Chapter 1 (Section 1A. 10), the electric field intensity at a given point was defined as the electrostatic force experienced by a test charge at that point. One would like to have a similar measure for the magnetic field, which represents the space in which the magnetic influence is felt. Unfortunately, the magnetic equivalent of an electric charge, viz. an isolated magnetic monopole, apparently does not exist. Therefore, we define a magnetic field in a different way. We consider a charged particle with charge q moving through a uniform magnetic field of intensity $\vec{B}$. For simplicity, we first assume that the particle is entering the field with a high speed in a direction normal to the direction of the field. The motion of the charge produces its own magnetic field, which interacts with the existing field. The result is that the straight path of the particle bends to take a circular shape (Fig. 3.2). The force experienced by the particle in the field is determined by the radius of the circular path. The smaller is the radius, i.e. the higher is the curvature of the circle, the greater is the force. The force F is found to be proportional to the

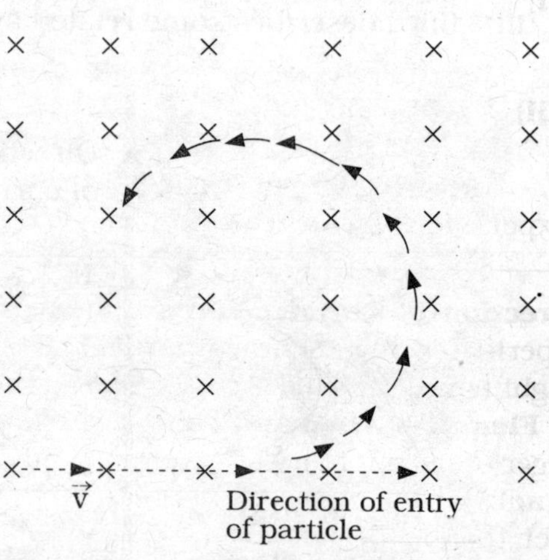

Fig. 3.2

(i) magnitude of charge, q
(ii) speed of the particle, v
(iii) magnetic field intensity, B

i.e. $F \propto qvB$

$\Rightarrow$ $F = k\,qvB$

where k is a constant of proportionality. In SI system, k = 1, so,

$$F = qvB \text{ (SI system)} \qquad \text{..... (3.1)}$$

If q and v each are taken to be unity, F = B. Thus, the **magnetic field intensity or magnetic field** can be **defined** as "the amount of force experienced by a unit charge entering normally in the field with unit speed."

The magnetic field is also referred to as **magnetic induction or magnetic flux density,** which will be further explained in Chapter 3B (Section 3B. 04). If the particle enters the field in a direction making an angle θ with the field direction, the force experienced by it depends upon the component of its velocity in the direction perpendicular to the field (Fig. 3.3),

$$F = q\,(v \sin\theta)\,B = qv\,B \sin\theta \qquad \text{..... (3.2)}$$

Fig. 3.3

In vector form it can be expressed as,

$$\vec{F} = q(\vec{v} \times \vec{B}) \quad \text{..... (3.3a)}$$

The above equation is known as **magnetic Lorentz force equation.** More about it will be described in Section 3A. 13.

If an electric field $\vec{E}$ is also present, the charged particle will also experience an electric force $q\vec{E}$ in the direction of the field. Thus the **complete Lorentz Force equation** is given as,

$$\vec{F} = q(\vec{v} \times \vec{B}) + q\vec{E} \quad \text{..... (3.3b)}$$

Additional knowledge

From Eq. 3.2,

$$F = qvB \sin\theta$$

Special Cases

(i) If $\theta = 0°$ or 180°,

$$F = 0$$

i.e. if a charged particle enters the magnetic field in a direction parallel to the field, the force experienced by it is zero.

(ii) If $v = 0$,

$$F = 0$$

i.e. there is no force on a charged particle at rest in the magnetic field.

(iii) If $\theta = 90°$,

$$F = qvB$$

i.e. if a charged particle enters the magnetic field in a direction perpendicular to the field, the force experienced by it is maximum.

Direction of Lorentz Force. Since the Eq. 3.3 involves vector multiplication, the direction of the force experienced by a charged particle in the magnetic field can be given either by 'Fleming's left hand rule' or 'Right hand grip rule':

(i) Fleming's left hand rule. If the first finger and second finger of the left hand are stretched out at an angle θ and the thumb is stretched out perpendicular to both the fingers, such that the first and second finger point in the direction of magnetic field and current (motion of the charged particle), respectively, the thumb represents the direction of force experienced by the particle (Fig. 3.4).

(ii) Right hand grip rule. If one grips something in his right hand such that the fingers bend from the direction of current to the direction of magnetic field and the thumb is held perpendicular to the plane of the grip, then the thumb represents the direction of magnetic force (Fig. 3.5).

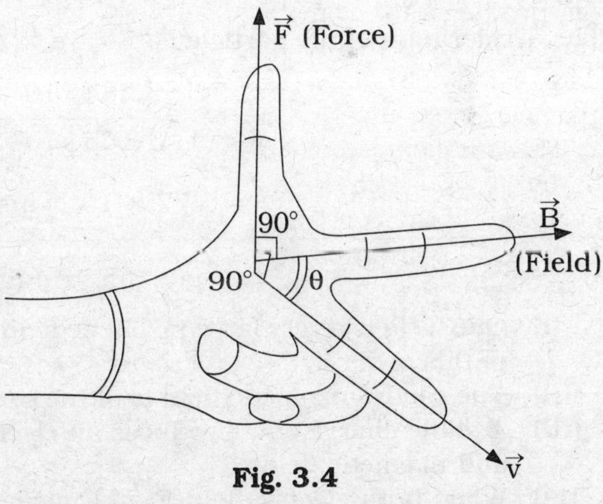

Fig. 3.4

Work done in the magnetic field (Energy of the particle). The magnetic force $\vec{F}$ acts on the particle in a direction perpendicular to the direction of motion (along $\vec{v}$) of the particle. Therefore, work done (= force × displacement) = $\vec{F}.\vec{d} = F \times d \times \cos 90°) = 0$. Consequently, the total energy of the particle remains unchanged [since it includes kinetic energy ($= 1/2\, mv^2$), it follows that the velocity of the particle remains constant in the field].

Additional knowledge

(i) The magnetic Lorentz force is always perpendicular to the direcions of current and the magnetic field (or, we can as well say that it is always perpendicular to the plane containing the directions of $\vec{v}$ and $\vec{B}$).

(ii) The Fleming's left hand rule and the right hand grip rule are valid even when the angle between $\vec{v}$ and $\vec{B}$ is not 90°, but this angle should not exceed 180°.

Units and dimensions of magnetic field. The SI unit of magnetic field, B, is tesla (T). From Eq. 3.1,

$$B = \frac{F}{qv}$$

If q = 1 coulomb, v = 1m/s and F = 1N, then B is called 1 tesla (T). Thus **one tesla (T)** is the "magnetic field in a region where a particle of charge one coulomb, moving with a velocity $1 ms^{-1}$ perpendicular to the field, experiences a force of one newton". One tesla is also equal to $1(weber)/(meter)^2$ (or Wb/m^2).
The C.G.S. unit of magnetic field is gauss (G). It is related to tesla as,

$$1G = 10^{-4}T$$

From the relation B = F/qv, the dimensions of the magnetic field are easily worked out to be $[MA^{-1}T^{-2}]$ (M: mass, A: current, T: time).

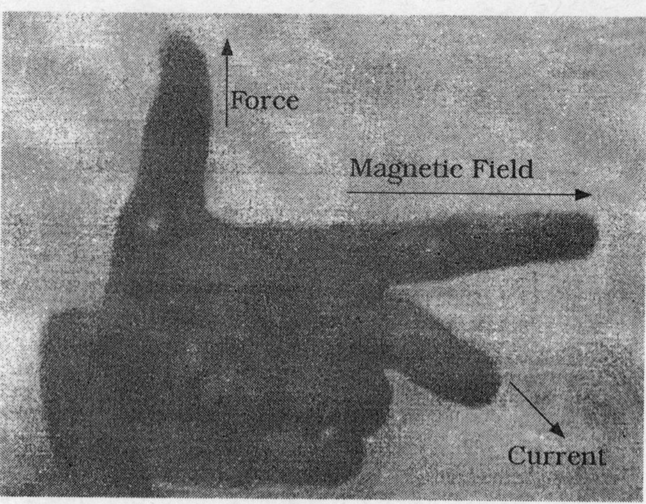

Fig. 3.5

Example 1. An α-particle enters a magnetic field of strength 0.3T with a velocity of 4×10^6 ms^{-1}, in a direction perpendicular to the field. Find the force experienced by the particle and its acceleration in the field (mass of alpha-particle = 6.65×10^{-27}kg).

Solution. Given: m = 6.65×10^{-27} kg; q = 2d = 2 × (1.6×10^{-19}) C; v = 4×10^6 ms^{-1}; B = 0.3T; θ = 90°

The force on the particle is given by,

$$F = qv B \sin\theta = (2\times1.6\times10^{-19}) \times (4\times10^6) \times (0.3) \times \sin 90°$$

$$= (2\times1.6\times4\times0.3\times1) \times 10^{-13}$$

$$= 3.84\times10^{-13} \text{ N}$$

∴ The acceleration of the particle is (F = m × a),

$$a = \frac{F}{m} = \frac{(3.84\times10^{-13})}{(6.65\times10^{-27})}$$

$$= 5.77 \times 10^{-13} \text{ ms}^{-2}$$

Additional knowledge

(i) One tesla is a very large magnetic field. A very large size electromagnet produces a magnetic field of this order.
(ii) The earth's magnetic field is of the order of 10^{-4}T, i.e. of the order of one gauss.
(iii) A static charge only produces an electric field, whereas a moving charge produces both electric and magnetic fields.
(iv) When both electric field $\vec{E}$ and magnetic field $\vec{B}$ exist perpendicular to each other in the same space, they are said to be 'crossed fields', e.g. in a cathode ray tube.
(v) As mentioned earlier, unlike for the electric charges, an isolated magnetic pole does not exist. Therefore, the source of a magnetic field is always other than a magnetic monepole.

3A. 04. Biot-Savart Law

In the preceding sections, while considering the behavior of a moving charged particle in an applied magnetic field, we had ignored the magnetic field produced by the moving charge itself. It was done for the sake of convenience. Such a simplification is justified only when the number of moving charges is too small to produce a net magnetic field of appreciable magnitude. In reality, e.g. current flowing through a wire, the number of moving charges is so vast that the combined magnetic field produced by them effectively interacts with the applied magnetic field. So it cannot be ignored. Consequently, it becomes necessary to evaluate the **magnetic field produced in the space around a moving charge.**

Fig. 3.6 shows a wire AB of arbitrary shape, carrying a current i. We assume it to be made up of a succession of tiny elements of length $d\ell$ each. One such element CD is marked in the figure. Its direction continuously changes according to the shape of the wire. So it is a vector quantity. We now define a current-length element $i\,\vec{d\ell}$. Obviously, it is also a vector quantity with its direction given by the tangent drawn to the length $\vec{d\ell}$.

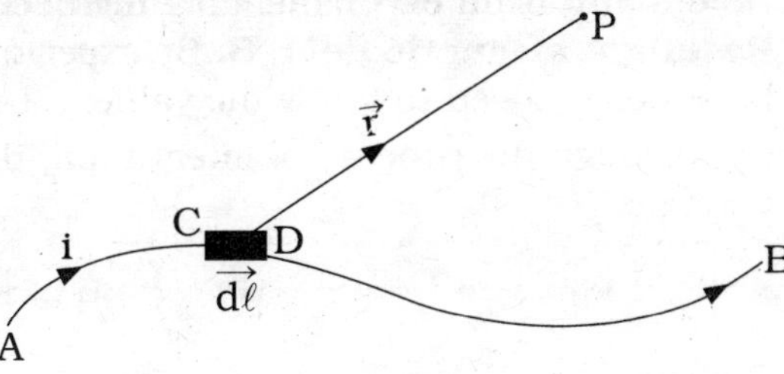

Fig. 3.6

The experiments carried out by Savart and his coworkers revealed that the magnetic flux intensity $\vec{dB}$ (also called magnetic field induction or simply magnetic field), produced by a current-length element $i\,\vec{d\ell}$ at a point P near the wire depends on four factors, as follows.

(i) Directly proportional to the current i flowing in the wire,

$$dB \propto i$$

(ii) Directly proportional to the length dl of the current-length element $i\,\vec{d\ell}$,

$$dB \propto d\ell$$

(iii) Directly proportional to the sine of the angle θ between $\vec{d\ell}$ and $\vec{r}$, the position vector of P,

$$dB \propto \sin\theta$$

Since both $\vec{d\ell}$ and $\vec{r}$ are vectors, we can combine (ii) and (iii) to write in vector form as,

$$dB \propto |\vec{d\ell} \times \vec{r}| \quad \text{..... (3.4)}$$

(iv) Inversely proportional to the square of r,

$$dB \propto (1/r^2)$$

Combining all the factors together, we may write,

$$dB \propto (id\ell \sin\theta)/r^2$$

$$\Rightarrow \quad dB = C\,(id\ell \sin\theta/r^2)$$

Where C is a constant of proportionality. Its value is found to be $(\mu_0/4\pi)$ in SI units, where μ_0 is called absolute permeability (or permeability constant or permeability of free space)*. Its value is defined to be exactly $4\pi \times 10^{-7}$ TA^{-1}m $\approx 1.26 \times 10^{-6}$ TA^{-1}m.

Thus,

$$dB = \left(\frac{\mu_0}{4\pi}\right)\cdot\frac{id\ell\sin\theta}{r^2} \quad \text{..... (3.5)}$$

[BIOT-SAVART LAW; scalar form]

In vector form, $\vec{dB}$ is given as (vide Eq. 3.4),

$$\vec{dB} = \frac{\mu_0}{4\pi}\cdot\frac{i\left(\vec{d\ell}\times\vec{r}\right)}{r^3} \quad \text{..... (3.6)}$$

[BIOT-SAVART LAW; vector form]

The Eqs. 3.5 and 3.6, in scalar and vector form, respectively, express the Biot-Savart law. It is an inverse square law in terms of r. The factor r^3 in Eq. 3.6 appears in the denominator only because of the presence of $\vec{r}$ in vector form in the numerator. If we express the position vector as unit vector $\hat{r}$, the Biot-Savart law in the vector form can also be expressed as,

$$\vec{dB} = \frac{\mu_0}{4\pi}\cdot\frac{i\left(\vec{d\ell}\times\hat{r}\right)}{r^2} \quad \text{..... (3.7)}$$

[BIOT-SAVART LAW; vector inverse square form]

*In general, the value of C depends upon the medium between the current-length element $i\vec{d\ell}$ and the point of observation P. Then it is expressed as $C = \mu/4\pi = (\mu_0\mu_r)/4\mu$, where $\boldsymbol{\mu}$ is called **permeability** of the medium and $\boldsymbol{\mu_r}$ is called **relative permeability** of the medium. For free space, $\mu_r = 1$.

Note that it is an experimentally deduced law.

Resultant magnetic field, $\vec{B}$. By experimentation it is found that magnetic fields, like the electric fields, can be superimposed to know the value of the resultant field. Thus, the resultant field $\vec{B}$ is evaluated by adding up, through the process of integration, the tiny contributions from all the current-length elements $i\vec{d\ell}$ of the wire, as,

$$\vec{B} = \int_{line} d\vec{B} = \frac{\mu_0}{4\pi}\int_{line}\frac{i(\vec{d\ell}\times\vec{r})}{r^3} = \frac{\mu_0}{4\pi}\int_{line}\frac{i(\vec{d\ell}\times\hat{r})}{r^2} \qquad \text{..... (3.8)}$$

3A. 05. Magnetic field due to a straight current-carrying conductor

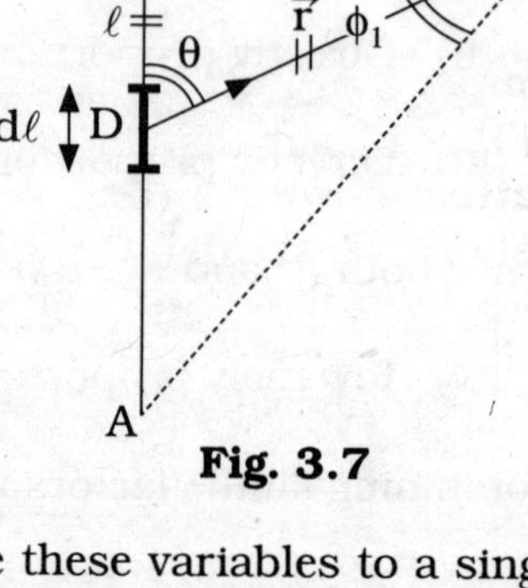

Fig. 3.7

Consider a straight conductor AB, carrying current i and lying in the plane of the paper (Fig. 3.7). A nearby point P is situated at a perpendicular distance OP = a from the conductor. A small current-length element $i\vec{d\ell}$ of length $d\ell$ is situated at D, at a distance ℓ from O and a distance r from P. The angle between DO and DP is θ. Let the angle subtended by DP with OP be ϕ. Since we are free to choose our current-length element anywhere along the conductor, ϕ can possibly range from the value ϕ_1 (on the negative side of OP) to the value ϕ_2 (see figure).

According to the Biot - Savart law, the magnetic field contribution by the current - length element $i\vec{d\ell}$ at P is (vide Eq. 3.6),

$$d\vec{B} = (\mu_0/4\pi)(i\vec{d\ell} \times \vec{r})/r^3$$

$$\Rightarrow \quad |d\vec{B}| = dB = \frac{\mu_0}{4\pi}\cdot\frac{i d\ell \sin\theta}{r^2} \qquad \text{..... (1)}$$

We cannot integrate the above equation to obtain the net magnetic field $\vec{B}$ at P, because it involves as many as three variables $\vec{\ell}$, θ and ϕ. So we first reduce these variables to a single variable. From the triangle DOP,

$$\theta = (\pi/2 - \phi) \qquad \text{..... (2)}$$

Also, $\tan\phi = \ell/a$ or $\ell = a\tan\phi$

On differentiating, $d\ell = a\sec^2\phi.d\phi$ ..... (3)

Further, $r = a/\cos\phi = a\sec\phi$ ..... (4)

Substituting from Eqs. 2, 3, 4 into Eq. 1, we get,

$$dB = (\mu_0/4\pi)\, i\,(a\sec^2\phi.d\phi)\cos\theta/(a\sec\phi)^2$$

$$= \frac{\mu_0}{4\pi}\cdot\frac{i\cos\phi d\phi}{a}$$

Integrating, we get,

$$B = \int_{-\phi_1}^{\phi_2} dB \int_{-\phi_1}^{\phi_2}\frac{\mu_0 i\cos\phi.d\phi}{4\pi a}$$

$$= \frac{\mu_0 i}{4\pi a}|\sin\phi|_{-\phi_1}^{\phi_2}$$

$$= \frac{\mu_0 i}{4\pi a}[\sin\phi_2 - \sin(-\phi_1)]$$

Or

$$B = \frac{\mu_0 i}{4\pi a}(\sin\phi_1 + \sin\phi_2) \qquad \text{..... (3.9)}$$

(Straight current-carrying conductor)

From the above relation, the following statements can be readily made:

(i) B is directly proportional to the current i flowing in the conductor [Fig. 3.8a].

(ii) B is inversely proportional to the perpendicular distance (a) of the conductor from the point of observation [Fig. 3.8b].

(iii) B is directly proportional to ($\sin\phi_1 + \sin\phi_2$).

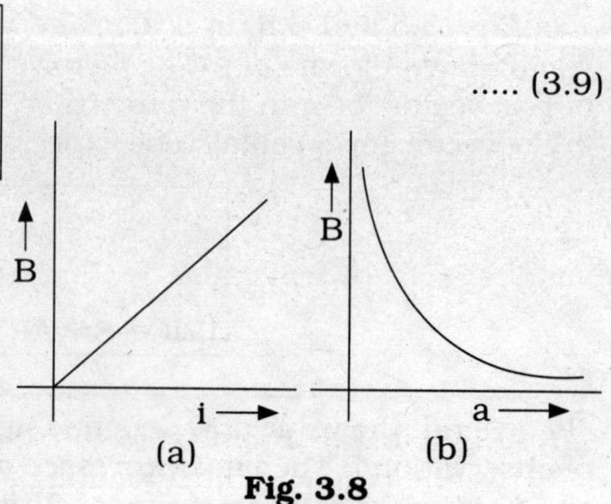

Fig. 3.8

Special Cases; Infinitely long conductor (e.g. a very long wire)

$$\phi_1 = \phi_2 \approx 90°$$

(i) If the point of observation, P, lies near the centre of the wire,

$$\phi_1 = \phi_2 \approx 90°$$

$$\therefore \quad B = \frac{\mu_0 i}{4\pi a}(1+1) = \frac{\mu_0 i}{2\pi a} \quad \text{..... (3.10a)}$$

(ii) If P lies near to one end of the conductor, say B,

$$\phi_1 = 90°, \phi_2 \approx 0$$

$$\therefore \quad B = \frac{\mu_0 i}{4\pi a}(1+0) = \frac{\mu_0 i}{4\pi a} \quad \text{..... (3.10b)}$$

The same expression will be obtained if P is taken to lie near the other end A of the conductor.

Example 2. A current of 8A flows through a long straight wire. Determine the magnitude of the magnetic field $\vec{B}$ at a point distant 15 cm from the wire.

Solution. Given: current (i) = 8 A; distance (a) = 15 cm = 0.15 m (unless stated otherwise, in such questions we assume the wire to be infinitely long and the point of observation to he situated near to the centre of the wire).

The field is given by,

$$B = \frac{\mu_0 i}{2\pi a}$$

$$= \frac{(4\pi \times 10^{-7}) \times (8)}{2\pi \times (0.15)}$$

$$= 10.67 \times 10^{-6} \text{ T (or } 10.67 \ \mu\text{T)}$$ **(Answer).**

Direction of the magnetic field $\vec{B}$. The Eq. 3.6 (or Eq. 3.7) for $d\vec{B}$ involves the cross product $(\vec{d\ell} \times \vec{r})$. It follows that the direction of $d\vec{B}$, and therefore the direction of the magnetic field $\vec{B}$, is perpendicular to the plane containing $\vec{d\ell}$ and $\vec{r}$, i.e. it is perpendicular to both (i) the current-length element $\vec{d\ell}$ and (ii) the position vector $\vec{r}$. In Fig. 3.7, $\vec{d\ell}$ and $\vec{r}$ lie in the plane of the paper. Therefore $\vec{B}$ lies in the direction of normal to the plane of the paper. For the direction of $i\,\vec{d\ell}$ and $\vec{r}$ shown in Fig. 3.7, the vector $(\vec{d\ell} \times \vec{r})$, and therefore the magnetic field $\vec{B}$, is directed into the plane of the paper. If the direction of current is reversed, $\vec{B}$ will be directed outwards from the plane of the paper. The direction of the magnetic field can also be conveniently obtained by any one of the following two rules:

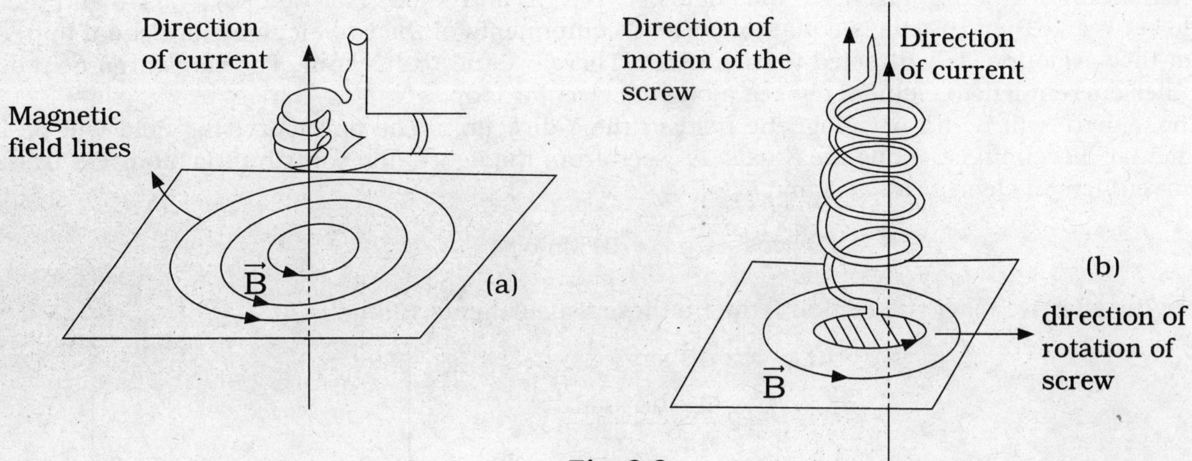

Fig. 3.9

(i) Right hand thumb rule. If one grips the conductor in the palm of the right hand, such that the thumb points in the direction of flow of current, then the curls of the finger give the direction of the magnetic field, taking nails as arrow heads (Fig. 3.9a).

(ii) Maxwell's cork screw rule. If a right-handed cork screw is rotated so that it moves in the direction of flow of current in the conductor, then the direction of the magnetic field lines is given by the direction of rotation of the screw (Fig. 3.9 b).

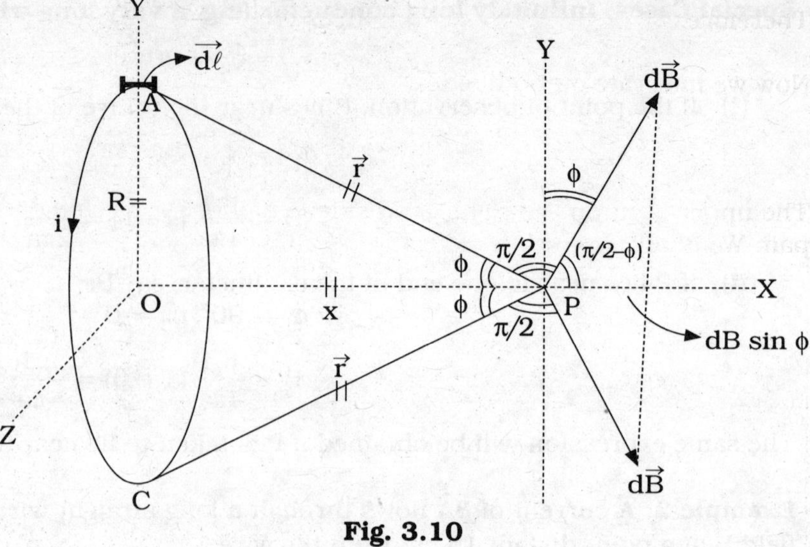

Fig. 3.10

3A. 06. Magnetic Field due to a Current-Carrying Circular Loop at an Axial Point

Consider a circular loop of radius R, carrying a current i and lying in the plane YOZ perpendicular to the plane of paper. Out point of observation P lies along the axis of the loop, at a distance x from the centre, O, of the loop (Fig. 3.10). We select a small element of length $d\ell$ of the loop at the point A. According to the Biot-Savart law, the magnetic field at P due to the current-length element $i\vec{d\ell}$ is (Eq. 3.6),

$$\vec{dB} = \frac{\mu_0}{4\pi}\cdot\frac{i(\vec{d\ell}\times\vec{r})}{r^3}$$

The element of length $\vec{d\ell}$ lies in the plane YOZ (perpendicular to the plane of the paper) and $\vec{r}$ lies in the plane YOX (in the plane of the paper). So they are perpendicular to each other, giving $(\vec{d\ell}\times\vec{r}) = d\ell\times r.\sin(\pi/2) = d\ell.r.$

Hence,
$$dB = \frac{\mu_0}{4\pi}\cdot\frac{(i\vec{d\ell})}{r^2}$$

As stated earlier, $\vec{dB}$ is perpendicular to both $\vec{d\ell}$ and $\vec{r}$. Since $\vec{d\ell}$ lies in the plane YOZ and $\vec{r}$ in the perpendicular plane YOX, it implies that the direction of $d\vec{B}$ lies perpendicular to $\vec{d\ell}$ in the plane YOZ (shown as directed upwards in the upper half of Fig. 3.10). If we now consider the magnetic field contribution from a similar current-length element $i\vec{d\ell}$ situated at the point C diametrically opposite to A, its magnitude will be the same, viz. $(\mu_0/4\pi).(id\ell)/r^2$, but its direction will be different (shown as directed downwards in the lower half of Fig. 3.10, because the direction of flow of current in the loop at the element $i\vec{d\ell}$ at C is opposite to that in the element $i\vec{d\ell}$ at A. From Fig. 3.10, the angle between $d\vec{B}$ and X-axis is $\pi-(\pi/2+\phi) = \pi/2-\phi$ and therefore the angle between $d\vec{B}$ and Y-axis is equal to ϕ. The Y-components of the two elements are equal to $d\vec{B}\cos\phi$ each, but they are oppositely directed to each other. Therefore they cancel out. The same argument applies to any other current-length element chosen along the circular loop.

Thus, there will be no net magnetic field in the Y-direction. The only surviving field will be in the perpendicular direction, i.e. along the X-axis. As seen from the figure, the contribution along OX from each of the current-length element idl at A and C is,

$$dB\left(\cos\frac{\pi}{2}-\phi\right) = dB\sin\phi$$

Therefore the effective magnetic field at P due to these two elements will be,

$$dB_{eff} = 2\,dB\sin\phi$$
$$= 2\cdot\frac{\mu_0}{4\pi}\cdot\frac{(id\ell)\sin\phi}{r^2}$$

From Fig. 3.10, $r = (R^2 + x^2)^{1/2}$ and $\sin\phi = \frac{AP}{OP} = \frac{P}{(R^2+x^2)^{1/2}}$

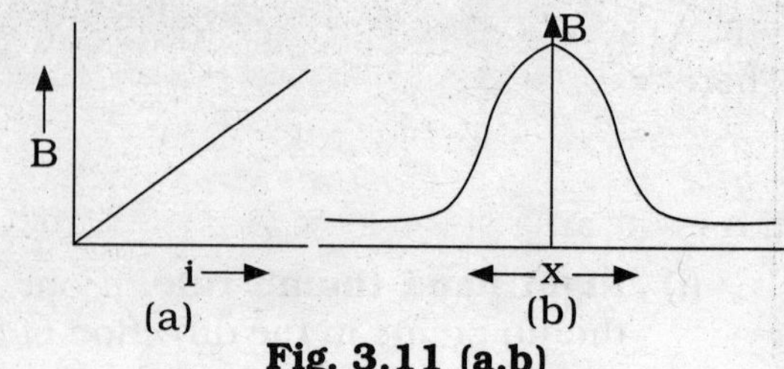

Fig. 3.11 (a,b)

Therefore, $$dB_{eff} = \frac{\mu_0}{2\pi} \cdot \frac{(id\ell).R}{(R^2 + x^2)^{3/2}}$$

Now we integrate on both sides,

$$\int_0^B dB_{eff} = \int_0^{\pi R} \left[\frac{\mu_0}{2\pi} \cdot \frac{iR}{(R^2 + x^2)} \right] d\ell$$

The upper limit on the right hand side is half of the circumference, because the contributions are taken in pair. We finally get,

$$\int_0^B dB_{eff} = \left[\frac{\mu_0}{2\pi} \cdot \frac{iR}{(R^2 + x^2)^{3/2}} \right] \int_0^{\pi R} d\ell$$

$$= \left[\frac{\mu_0}{2\pi} \cdot \frac{iR}{(R^2 + x^2)^{3/2}} \right] \pi R$$

Or,

$$B = \frac{\mu_0 iR^2}{2(R^2 + x^2)^{3/2}} \qquad \text{..... (3.11)}$$

(Circular Loop; Axial Point)

Two statements follow from Eq. 3.11: (a) B is directly proportional to the current i in the circular loop (Fig. 3.11a) and (b) B is inversely proportional to $(R^2 + x^2)^{3/2}$ (Fig. 3.11b).

Direction of the magnetic field. Since only the x-component survives in the process of summing up of contributions from the various current-length elements $i\vec{d\ell}$ along the loop, the magnetic field can also be conveniently obtained by the following **Right Hand Thumb Rule:** If the loop is held in the palm of the right hand such that the fingers bend in the direction of the current, the direction of the magnetic field lies along the thumb (Fig. 3.11c).

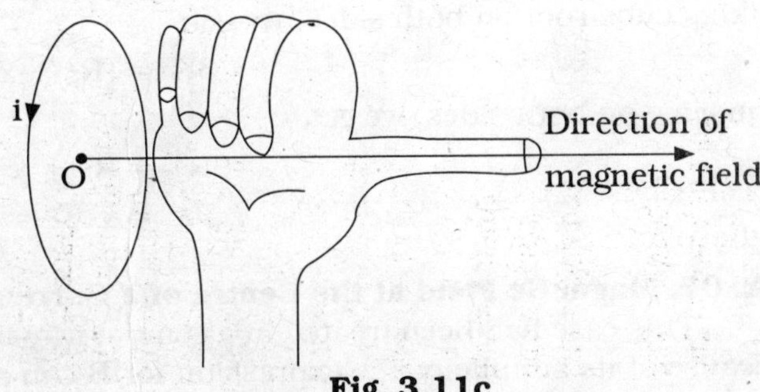

Fig. 3.11c

Special cases

(i) Magnetic field at an axial point due to a circular coil of N turns. Assuming that the radii of all the turns are same and the current flows through each in the same sense (clockwise or anticlockwise), the total magnetic field is N times the field due to one turn, i.e.

$$B = \frac{N.\mu_0 iR^2}{2(R^2 + x^2)^{3/2}} \qquad \text{..... (3.12)}$$

(ii) Magnetic field due to a circular loop at its centre. At the centre of the loop, x = 0. So,

$$B = \frac{\mu_0 i}{2R} \qquad \text{..... (3.13)}$$

Direction of the field lies along the axis.

This important case is also treated separately in the next section.

(iii) Magnetic field due to a circular loop at a point far away from centre of the loop. In this case, $x \gg R$. Hence in the term $(R^2 + x^2)^{3/2}$, R^2 can be neglected.

Therefore,

$$B = \frac{\mu_0 iR^2}{2x^3} \qquad \text{..... (3.14)}$$

If A is the area of the loop ($A = \pi R^2$), the magnetic field can also be expressed as

$$B = \frac{\mu_0 iA}{2\pi x^3}$$

Example 3. A circular coil with number of turns as n and radius R carries a current i. Determine the distance from the centre along the axis of the coil at which the magnitude of the magnetic field due the coil is reduced to 1/27th of its value at the centre.

Solution. Let B_1 and B_2 be the magnitudes of magnetic field at a distance x from the centre of the coil along the axis and at the centre of the coil, respectively. We have,

$$B_1 = \frac{\mu_0 iNR^2}{2(R^2 + x^2)^{3/2}}$$

$$B_2 = \frac{\mu_0 iN}{2R} \quad (\because x = 0)$$

We are given that,

$$B_1 = \frac{1}{27} B_2$$

$$\Rightarrow \quad \frac{\mu_0 iNR^2}{2(R^2 + x^2)^{3/2}} = \frac{1}{27} \times \frac{\mu_0 iN}{2R}$$

$$\Rightarrow \quad \frac{R^2}{(R^2 + x^2)^{3/2}} = \frac{1}{27} \times \frac{1}{R}$$

$$\Rightarrow \quad 27 R^3 = (R^2 + x^2)^{3/2}$$

Taking cube root on both sides, we get,

$$3R = (R^2 + x^2)^{1/2}$$

Squaring on both sides, we get,

$$9R^2 = R^2 + x^2$$

$$\therefore \quad x = \sqrt{8r^2} = 2\sqrt{2}\, R \quad \textbf{(Answer).}$$

3A. 07. Magnetic Field at the Centre of a Current-Carrying Coil

This case had been treated in a general way in the last section. However, because of its simplicity, the expression for B can also be obtained in a direct and simpler manner, as follows.

Consider a circular coil with centre O and radius r in the plane of the paper. A current i flows in the coil in clockwise direction (Fig. 3.12). The coil may be divided into a large number of current elements of length $d\ell$ each. One such element is shown in the figure.

According to the Biot-Savart law (Ex. 3.6), the magnetic field at the centre of the coil due to the current-length element $id\ell$ will be,

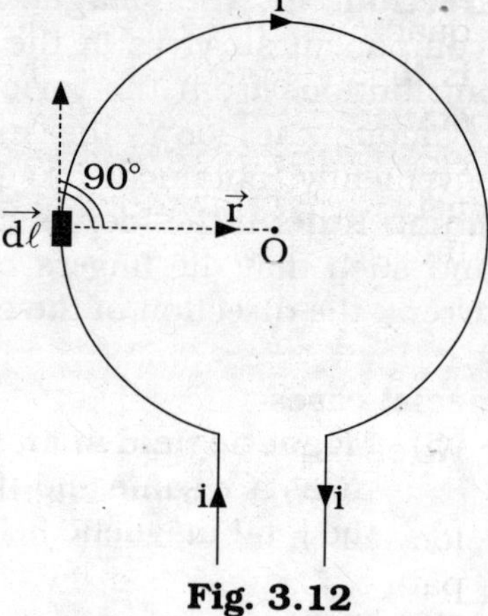

Fig. 3.12

$$d\vec{B} = \frac{\mu_0}{4\pi} \cdot \frac{i(\vec{d\ell} \times \vec{r})}{r^3}$$

$$\Rightarrow \quad dB = \frac{\mu_0}{4\pi} \cdot \frac{id\ell . r \sin\theta}{r^3} = \frac{\mu_0}{4\pi} \cdot \frac{id\ell \sin\theta}{r^2}$$

As seen from Fig. 3.12, for all positions of $\vec{d\ell}$ along the coil, the angle θ between $\vec{d\ell}$ and $\vec{r}$ is 90°. Therefore,

$$dB = \frac{\mu_0}{4\pi} \cdot \frac{id\ell}{r^2}$$

The resultant magnetic field at O for the whole coil is found by integrating the above equation.

$$B = \int dB = \frac{\mu_0}{4\pi} \cdot \frac{i}{r^2} \int d\ell$$

Now, $\int d\ell$ = circumference of the coil = $2\pi r$.

$$\therefore \quad B = \frac{\mu_0 i}{2r} \quad \text{(single turn)} \qquad \text{..... (3.13a)}$$

Which is the same as the Eq. 3.13 derived earlier. If the coil has n turns, the magnetic field is,

$$B = \frac{\mu_0 in}{2r} \quad \text{(n turns)} \quad ..... (3.13b)$$

The direction of the field $\vec{B}$ is conveniently found by either the right hand thumb rule or the Maxwell's cork screw rule, stated earlier. In this case it is found to be ⊥ to the plane of the paper (which is the same as the plane of the coil) and directed inwards.

Example 4. A current of 5 A is found flowing clockwise in a horizontally placed circular coil, when viewed from top. The coil has 20 turns of radius 12 cm each. What is the magnitude of the magnetic field produced at the centre of the coil ?

Solution. Given : n = 20; i = 5 A; r = 12 cm = 12×10^{-2} m

The magnetic field at the centre of the coil is given by,

$$B = \frac{\mu_0 in}{2r} = \frac{(4\pi \times 10^{-7}) \times (5) \times (20)}{2 \times (12 \times 10^{-2})}$$

$$= 5.23 \times 10^{-4} \text{ T} \quad \textbf{(Answer).}$$

3A. 08. Ampere's Circuit Law

In the previous sections we have known how Biot-Savart law can be applied to obtain magnetic field at a given point due to a current-carrying conductor. In dealing with cases where a number of current-carrying conductors are present, the law needs to be applied separately for each such conductor and then the various contributions have to be summed up, e.g. for a solenoid. This exercise involves complicated mathematical steps, mainly because vector quantities are involved. However, in some cases the problem is highly simplified by the **Ampere's circuit law.** The law **states** that "the line integral of the magnetic field $\vec{B}$ over a closed circuit (or path) in free space is equal to μ_0 times the net current (i_e) threading the circuit" (it can be derived from the Biot-Savart law).

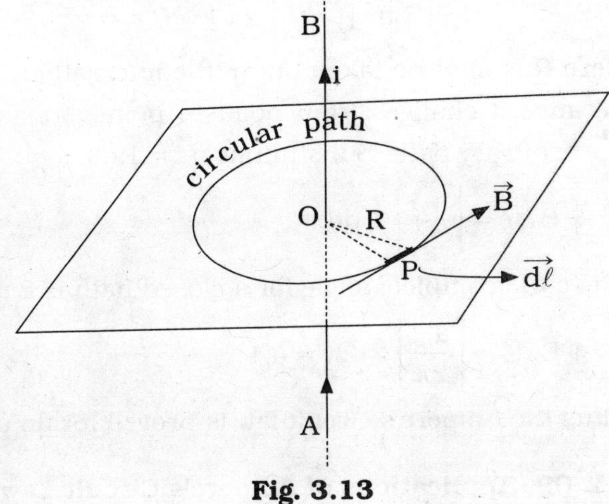

Fig. 3.13

$$\oint \vec{B}.\vec{d\ell} = \mu_0 i_e \quad ..... (3.15)$$

(Ampere's circuit law)

We shall prove the law for two representative cases:

(a) A Circular path with a straight long conductor passing through its centre. Fig. 3.13 shows a long extended straight conductor AB perpendicular to the paper and carrying current i through it. A circular path of radius R is drawn around the conductor with the point O in the plane of paper as the centre. A small element of length $d\vec{\ell}$ is chosen along it at the point P. From Eq. 3.19, the magnitude of the field $\vec{B}$ at P is,

$$B = \frac{\mu_0 i}{2\pi R}$$

According to the right hand thumb rule, the direction of $\vec{B}$ lies along the tangent to the circular path at P, which means that $\vec{d\ell}$ and $\vec{B}$ are parallel to each other i.e. the angle between them is zero.

Hence the line integral of the magnetic field over the circular path is given as,

$$\oint \vec{B}.\vec{d\ell} = \oint B.d\ell.\cos 0 = \oint B.d\ell$$

Substituting the value of B,

$$\oint B.d\ell = \oint \left[\frac{\mu_0 i}{2\pi R}\right].d\ell = \left[\frac{\mu_0 i}{2\pi R}\right] \oint d\ell$$

$$= \left[\frac{\mu_0 i}{2\pi R}\right] (2\pi R)$$

$$= \mu_0 i$$

Thus the Ampere's circuit law is proved.

(b) An arbitrary path with a straight long conductor passing through its centre. Consider a long straight conductor AB, carrying current i through it, as shown in Fig. 3.14. We draw an arbitrary closed path around O and take a small element of length $d\ell$ along it at the point P (OP = R).]

The angular width of $d\ell$ is,

$$d\theta = \frac{d\ell}{R}$$

$$\Rightarrow \quad d\ell = r d\theta$$

Again, the magnetic field at the point P is,

$$B = \frac{\mu_0 i}{2\pi R}$$

The direction of $\vec{B}$ at the point P is along the tangent to the path, i.e. along $\vec{d\ell}$. Therefore, the line integral of $\vec{B}$ over the entire path is,

$$\oint \vec{B}.\vec{d\ell} = \oint B.d\ell.\cos 0 = \oint B d\ell$$

$$= \oint \left[\frac{\mu_0 i}{2\pi R}\right] d\ell = \left[\frac{\mu_0 i}{2\pi}\right] \oint \frac{d\ell}{R}$$

Fig. 3.14

Here R cannot be taken out of the integration, because it changes from point to point along the arbitrary path. Substituting $\vec{d\ell} = R d\theta$,

$$\oint \vec{B}.\vec{d\ell} = \left[\frac{\mu_0 i}{2\pi}\right] \oint d\theta$$

Since the complete angle for a closed path is 2π, we have,

$$\oint \vec{B}.\vec{d\ell} = \left[\frac{\mu_0 i}{2\pi}\right] \times 2\pi = \mu_0 i$$

Thus the Ampere's circuit law is proved for an arbitrary circuit, too.

3A. 09. Applications of Ampere's Circuit Law

(a) Magnetic field due to a long straight current-carrying conductor. The Fig. 3.13 is valid for this case. AB is a long extended straight conductor carrying current i. A circular path of radius R is drawn perpendicular to it in the plane of the paper, with O as centre, and a small element of length $\vec{d\ell}$ is chosen along it at P. Using the right hand thumb rule, the magnetic field $\vec{B}$ is found to lie in the direction of the tangent drawn at P. So $\vec{B}$ is parallel to $\vec{d\ell}$ and hence the angle between them is zero. Taking the line integral of the magnetic field over the closed path,

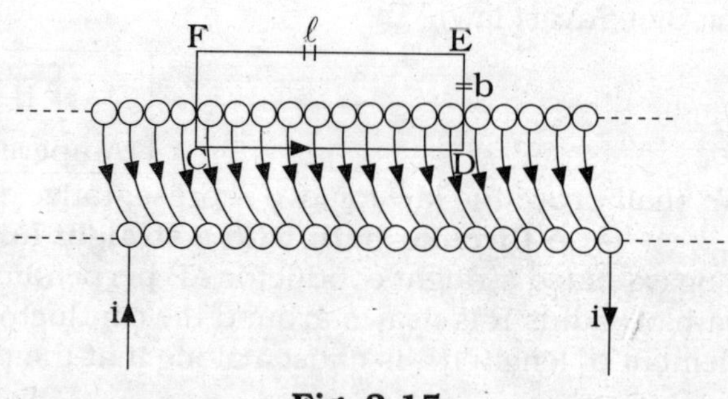

Fig. 3.15

$$\oint \vec{B}.\vec{d\ell} = \oint B.d\ell.\cos 0 = \oint B d\ell$$

Due to circular symmetry, the magnitude of the magnetic field $\vec{B}$ is same everywhere along the closed path. So it may be taken as a constant value in the integral and we have,

$$\oint \vec{B}.\vec{d\ell} = B \oint d\ell = B.2\pi R$$

Also, according to the Ampere's circuit law,

$$\oint \vec{B}.\vec{d\ell} = \mu_0 i$$

$$\therefore \quad B.2\pi R = \mu_0 i$$

$$\Rightarrow \quad B = \frac{\mu_0 i}{2\pi R}$$

Thus we get the same expression as Eq. 3.10a, as it should be.

(b) Magnetic field due to a long straight solenoid. A straight solenoid consists of a cylindrical coil of wire. The wire is wound around a cylinder of non-magnetic material. The Fig. 3.15 shows a long straight and tightly wound solenoid. So the length of the solenoid is much larger than its diameter. We select a rectangular loop CDEF of size ($\ell \times$ b) across the solenoid, as our Amperian loop. The line integral of the magnetic field $\vec{B}$ along the loop can be divided as,

$$\oint_{CDEF} \vec{B}.\vec{d\ell} = \int_C^D \vec{B}.\vec{d\ell} + \int_D^E \vec{B}.\vec{d\ell} + \int_E^F \vec{B}.\vec{d\ell}$$

Experimentally, it is known that the magnetic field inside a solenoid is directed along the axis and the field just outside the solenoid is zero. The magnetic field lines of a solenoid are shown in Fig. 3.16. For the sides CF and DE of the loop, the field $\vec{B}$ is oriented perpendicular to them, i.e. $\vec{B} \perp \vec{d\ell}$.

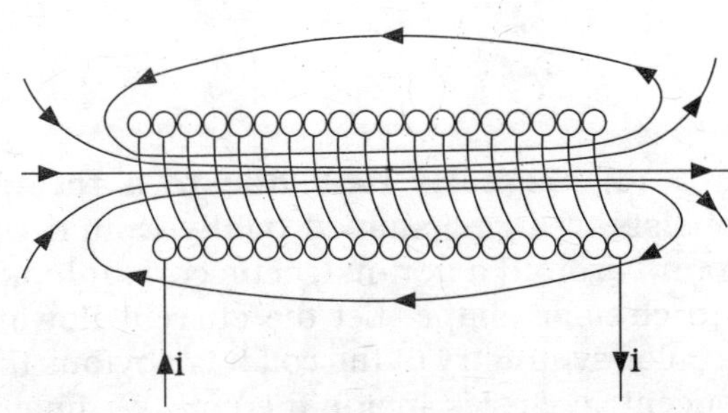

Fig. 3.16

Hence,

$$\int_C^F \vec{B}.\vec{d\ell} = 0 \text{ and } \int_D^E \vec{B}.\vec{d\ell} = 0$$

Further, the field just outside the solenoid is zero. Hence,

$$\int_E^F \vec{B}.\vec{d\ell} = 0$$

Thus we are left with,

$$\oint \vec{B}.\vec{d\ell} = \int_C^D \vec{B}.\vec{d\ell}$$

For the side CD, $\vec{d\ell}$ lies along the axis of the solenoid, i.e. $\vec{d\ell}$ is parallel to $\vec{B}$. So the angle between $\vec{d\ell}$ and $\vec{B}$ is zero. Further, since the solenoid is long, the magnetic field $\vec{B}$ remains practically uniform in the middle region of the solenoid i.e. B remains practically constant in the region. Consequently,

$$\int_{CDEF} \vec{B}.\vec{d\ell} = \int_C^D \vec{B}.\vec{d\ell} = \int_C^D B.d\ell \cos 0 = B\int_C^D d\ell = B\ell \quad \text{..... (A)}$$

Applying the Ampere's circuit law,

$$\int_{CDEF} \vec{B}.\vec{d\ell} = \mu_0 \Sigma i \quad \text{..... (B)}$$

Since the turns of the coil of the solenoid are closely spaced, each turn may be treated as a circular loop. The net magnetic field is the sum of the fields due to all the turns. Let the number of turns of the solenoid per unit length be n. Then the total number of turns threading the loop CDEF is $n\ell$. Therefore the net current enclosed within the loop is,

$$i_e = \Sigma i = i\,(n\ell) \quad \text{..... (C)}$$

From the relations (A), (B) and (C), we have,

$$B\ell = \mu_0 n i \ell$$

Or

$$\boxed{B = \mu_0 n i} \quad \text{..... (3.16)}$$

(Long straight solenoid)

Special cases

(i) The magnetic field lines widen as one moves from the centre outwards to the ends of the solenoid (Fig. 3.16). Consequently, the magnitude of the magnetic field falls as either end of the solenoid is approached. At points **near to the ends,** the magnetic field has a value nearly half of that at the centre, i.e.

$$B_{end} \approx \frac{1}{2}\mu_0 n i$$

(ii) If the coil of the solenoid is wound around a cylindrical core of a material having relative permeability μ_r, then,

$$B = \mu_0 \mu_r n i = \mu n i \quad \text{..... (3.17)}$$

The current in adjacent circular turns of a solenoid flows in the same direction. Therefore **when a current begins to flow in a solenoid,** the neighboring turns attract one another and **the solenoid contracts.**

Example 5. A current of 2 ampere flows through a solenoid of length 60 cm and having 180 turns. Calculate the magnetic field inside the solenoid.

Solution. Given : N = 180; ℓ = 60 cm = 0.6 m; i = 2 A

$$\therefore \quad \text{No. of turns per unit length, } n = \frac{N}{\ell} = \frac{180}{0.6} = 300 \text{ m}^{-1}$$

The magnetic field inside a solenoid is given by,

$$B = \mu_0 ni$$
$$= (4\pi\times10^{-7}) \times (300) \times (2)$$
$$= 24\pi\times10^{-5} \text{ T (or } 75.4\times10^{-5} \text{ T)} \qquad \textbf{(Answer).}$$

(c) Magnetic field due to a toroidal solenoid. A toroidal solenoid consists of a ring-shaped tightly wound coil of wire (Fig. 3.17). The wire is wound around a non-magnetic core. It looks like a long straight solenoid bent into circular shape. Let the current flowing through the coil be i. From the circular symmetry of the coil it is obvious that the magnetic field lines lie along concentric circles inside the coil. We imagine a concentric circle of radius r traversed clockwise and take it as our Amperian loop. We evaluate the field $\vec{B}$ at a point P along the loop. Again, due to circular symmetry, the magnitude of $\vec{B}$ is same all along the circumference of the loop and $\vec{B}$ is parallel to $\vec{d\ell}$ at all points on the circumference.

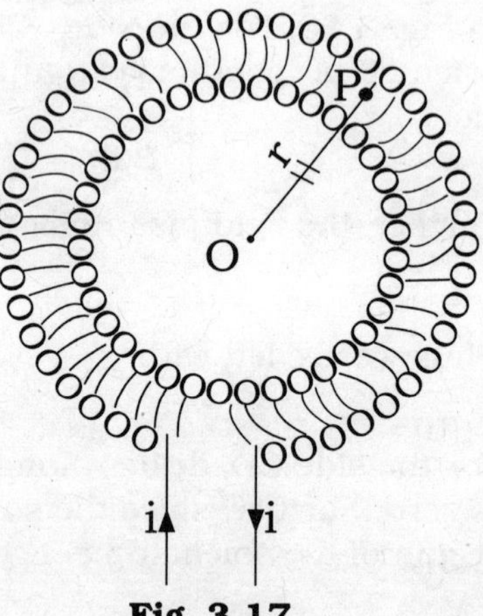

Fig. 3.17

Taking the line integral of $\vec{B}$ along the circumference,

$$\oint \vec{B}.\vec{d\ell} = \oint B.d\ell \cos 0 = B\oint d\ell = B.2\pi r$$

According to the Ampere's circuit law,

$$\oint \vec{B}.\vec{d\ell} = \mu_0 \Sigma i$$

If N be the total number of turns of the coil, the net current enclosed by them is,

$$i_e = \Sigma i = Ni$$

So,
$$\oint \vec{B}.\vec{d\ell} = \mu_0 iN$$

$$\Rightarrow \quad B.2\pi R = \mu_0 iN$$

$$\therefore \quad B = \frac{\mu_0 iN}{2\pi r} \qquad \text{..... (3.18 a)}$$

(Long straight solenoid)

However, if we assume r to be the average radius of the toroid coil (which will be nearly true when r is much larger than the width of the coil) and if the coil has n turns per unit length,

$$N = n.2\pi r$$

Then the Eq. 3.18a becomes,

$$B = \mu_0 ni \qquad \text{.....(3.18 b)}$$

It is of the same form as the Eq. 3.16 for a long straight solenoid. Since i = 0 outside the toroid, no magnetic field exists outside the toroid.

Example 6. In a toroid, 4000 turns of a wire are wound around a non-magnetic core, having inner radius of 30 cm and outer radius of 32 cm. A current of 12 A is made to flow in the wire. Give the values of the resulting magnetic field (a) inside the core, (b) just outside the toroid, and (c) in the space around the toroid.

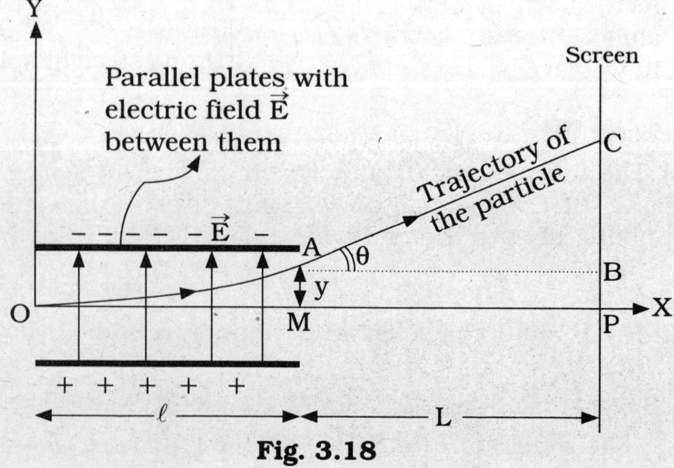

Fig. 3.18

Solution. Inner radius, $r_1 = 30$ cm

Outer radius, $r_2 = 32$ cm

$\therefore$ Mean radius, $r = \frac{r_1 + r_2}{2} = \frac{30+32}{2}$ cm

$= 31$ cm $= 0.31$ m

Total length of toroid $= 2\pi r = 2\pi \times (0.31)$

$= 0.62\pi$ metre

Total number of turns $= 4000$

$\therefore$ No. of turns per unit length, $n = \frac{4000}{0.62\pi}$

Current, $i = 12$ A

(a) Magnetic field inside the core is given by,

$$B = \mu_0 n i = (4\pi \times 10^{-7}) \times \left(\frac{4000}{0.62\pi}\right) \times (12)$$

$= 3.1 \times 10^{-2}$ T **(Answer).**

(b) Since $i = 0$ outside the toroid, no magnetic field exists just outside the toroid. **(Answer).**

(c) For the same reason as in (b), no field exists in the space around the toroid. **(Answer).**

3A. 10. Motion of a Charged Particle in a Uniform Transverse Electric Field

The Fig. 3.18 shows the trajectory of a charged particle moving through a uniform electric field $\vec{E}$ existing between two parallel plates. $\vec{E}$ is directed along y-axis and the particle, of mass m and charge q, enters the field in the perpendicular direction along x-axis with an initial velocity v. If V be he potential difference between the plates and d the distance between them, the electric field along the y-axis is,

$$E_y = \frac{V}{d}$$

The electric force on the particle is qE_y. There is no field in the x-direction and therefore when the particle is just entering the field, the electric force along the x-axis is zero ($F_x = 0$). Accordingly, the components of the acceleration, a, of the charged particle are (F = ma),

$$a_x = 0; \qquad a_y = \frac{qE_y}{m}.$$

The components of the initial velocity are (the particle enters the field along the x-axis),

$$v_x = v; \qquad v_y = 0$$

As the particle moves through the field, its velocity $v = (v_x)$ remains unchanged, since there is no acceleration along x ($a_x = 0$). If ℓ be the length of the region of electric field and t the time taken by the particle to traverse it, we have,

$$\ell = vt$$

Using the equation of kinematics,

$$s = ut + \frac{1}{2} at^2$$

We have along the x-axis and y-axis, respectively,

$$x = \ell = vt \qquad \text{..... (F)}$$

$$y = 0 + \frac{1}{2}\left(\frac{qE_y}{m}\right)t^2 \qquad \text{..... (G)}$$

Substituting for t from Eq. (F) into Eq. (G),

$$y = \frac{qE_y}{2m} \cdot \frac{x^2}{v^2}$$

$$\Rightarrow \qquad x^2 = \left(\frac{2mv^2}{qE_y}\right)y$$

The quantity $(2mv/qE_y)$ is a constant value, say k. So we may write,

$$x^2 = ky,$$

which represents a parabola. Thus, the **trajectory of a charged particle in a transverse uniform electric field is a parabola.**

When the charged particle just emerges from the electric field, its vertical displacement is AM = y. Then it moves along a straight path. As seen in the figure, this path is located at an angle θ to the x-axis, such that,

$$\tan\theta = \left.\frac{dy}{dx}\right|_{x=\ell} = \frac{qE_y\ell}{mv^2} \qquad \text{..... (H)}$$

The displacement of the particle along the y-axis is usually measured along a screen, held at a distance L from the farther end of the plates. From the figure,

$$\tan\theta = \frac{BC}{L}$$

$$\Rightarrow \qquad BC = L\tan\theta$$

$$= \frac{LqE_y\ell}{mv^2} \qquad \text{(from Eq. H)}$$

The total displacement Y, on the screen is,

$$Y = BP + BC = Y + BC$$

$$= \frac{qE_y\ell^2}{2mv^2} + \frac{LqE_y\ell}{mv^2}$$

$$\Rightarrow \qquad Y = \frac{qE_y\ell}{mv^2}\left[\frac{\ell}{2} + L\right] \qquad \text{..... (3.19)}$$

3A.11. Motion of a Charged Particle in a Uniform Magnetic Field; Circular and Helical (Spiral) Paths

As discussed earlier in Section 3.03, when a charged particle with initial velocity $\vec{v}$ moves through a uniform magnetic field, it is acted upon by the so-called Lorentz force. We have two cases:

(1) If the particle enters the field in a perpendicular direction, the Lorentz force is given by

$$F = qvB \qquad \text{(vide Eq. 3.1)}$$

(2) If the particle enters the field in an arbitrary direction, the Lorentz force is given by

$$\vec{F} = q(\vec{v} \times \vec{B}) \qquad \text{(vide Eq. 3.3)}$$

As explained in Section 3.03, in this case $\vec{v}$ has a component v cos θ parallel to $\vec{B}$ and another component v sin θ, perpendicular to $\vec{B}$ (Fig. 3.19),

$$v_\perp = v\sin\theta, \qquad v_{||} = v\cos\theta$$

Fig. 3.19

Now we shall treat the two cases in detail.

Case 1

If the particle enters in a direction perpendicular to $\vec{B}$, the Lorentz force makes the particle to move along a circular path in a plane perpendicular to $\vec{B}$. The situation is pictorially represented in Fig. 3.20. The direction of the magnetic field $\vec{B}$ lies ⊥ to the plane of paper. The Lorentz force $\vec{F}$ is ⊥ to the plane containing $\vec{v}$ and $\vec{B}$ [$\vec{F} = (\vec{v} \times \vec{B})$], such that it is always directed towards the centre of the circular path. From our knowledge of the mechanics of a particle in uniform circular motion, we know that the centripetal force, that maintains the particle in circular motion, is directed towards the centre of the circle and its magnitude is given by (force = mass × acceleration),

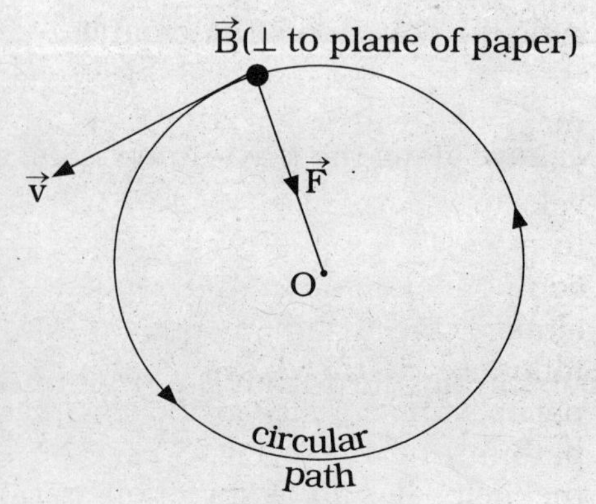

Fig. 3.20

$$F = m \cdot \frac{v^2}{r}$$

Clearly, it is akin to our case of circular motion of a charged particle in a magnetic field. The equivalent of the said centripetal force is the Lorentz force F = qvB. Therefore, we have,

$$qvB = \frac{mv^2}{r}$$

$$\Rightarrow \quad r = \frac{mv}{qB} \qquad ..... (3.20)$$

The time period, T, of the particle (time taken to complete one revolution) is equal to the circumference divided by the velocity,

$$T = \frac{2\pi r}{v} = \frac{2\pi m}{qB} \qquad ..... (3.21)$$

The frequency of revolution (no. of revolutions made in unit time, i.e. in one second) will be,

$$f = \frac{1}{T} = \frac{qB}{2\pi m} \qquad ..... (3.22)$$

Accordingly, the angular frequency (total angular value traversed per unit second) is,

$$\omega = f = \frac{qB}{m} \qquad ..... (3.23)$$

f is called **cyclotron frequency.**

Example 7. An electron being accelerated through 100 V enters a uniform magnetic field of strength 0.004 T, placed perpendicular to the direction of motion of the electron. Calculate the radius of path traversed by the electron. (AI; 1992).

Solution. Given : V = 100 volt; B = 0.004 T; θ = 90°

Kinetic energy of electron, $E = eV = (1.6\times10^{-19}) \times (100)$

$$= 1.6\times10^{-17}\ J$$

The radius of the circular path traversed by a charged particle in a magnetic field is given by,

$$r = \frac{mv}{qB} = \frac{\sqrt{2mE}}{qB} = \frac{\sqrt{2m_eE}}{eB}$$

$$= \frac{\sqrt{2\times(9.1\times10^{-31})\times(1.6\times10^{-17})}}{(1.6\times10^{-19})\times(0.004)}$$

$$= \frac{5.396\times10^{-24}}{0.0064\times10^{-19}}$$

$$= 8.43\times10^{-3}\ m \text{ (or 8.43 mm)}$$ **(Answer).**

Case 2

When the charged particle enters the magnetic field in an arbitrary direction, say with $\vec{v}$ making an angle θ with $\vec{B}$, the velocity component $v_\perp$ = v sin θ tends to move it along a circular path and the component $v_{||}$ (= v cos θ) tends to move it along a linear path. Under the combined influence of these two components the particle finally follows a helical (or spiral) path, as shown in Fig. 3.21. The axis of the helix lies in a direction parallel to the direction of the field $\vec{B}$. The distance

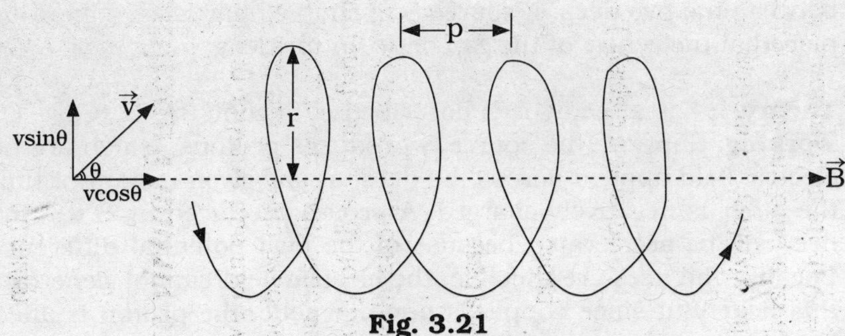

Fig. 3.21

between two neighbouring turns of the helix is the pitch p of the helix. The radius of any one of the turns is the radius r of the helix. From Eq. 3.20, r = (mv)/(qB). The pitch represents the linear distance traversed by the particle during the time taken to complete one revolution, i.e. (from Eq. 3.21),

$$p = v\cos\theta . T = v\cos\theta . \left(\frac{2\pi m}{qB}\right) = \frac{2\pi mv\cos\theta}{qB}$$

Example 8. An electron emitted by a heated filament is first accelerated by passing between two plates, with a potential difference of 3.0 kV between them, and then enters a uniform magnetic field of strength 0.10 T, in a direction making an angle of 30° with the direction of the field. Determine its trajectory in the field.

Solution. Given : V = 3.0 kV = 3000 volt; B = 0.10 T; θ = 30°

k.e. of electron, $E = eV = (1.6\times10^{-19})\times3000 = 48\times10^{-17}$ J

Since the direction of motion of the electron at the time of entry into the magnetic field is inclined at 30° to the field, the electron will traverse a helical path, with the radius given by,

$$r = \frac{mv\sin\theta}{qB} = \frac{\sqrt{2m_eE}\,\sin\theta}{eB} \quad \text{(for an electron)}$$

$$= \frac{\sqrt{2\times(9.1\times10^{-31})\times(48\times10^{-17})}\times(1/2)}{(1.6\times10^{-19})\times(0.1)}$$

$$= \frac{29.56\times10^{-24}}{32\times10^{-21}}$$

$= 0.92\times10^{-3}$ m (or 0.92 mm) **(Answer).**

3A. 12. Cyclotron

A cyclotron is a device which can accelerate charged particles to extremely high energies. For this purpose, it uses both electric field and magnetic field in its construction.

Working Principle. The working of a cyclotron is based on the principle that (i) a charged particle moving in a uniform magnetic field adopts a circular path under the action of the Lorentz force acting upon it and (ii) it can be accelerated step by step to extremely high energies by subjecting it to a suitable oscillating electric field.

Construction. It mainly consists of two D-shaped hollow flat half-cylinders, enclosed in an evacuated chamber (Fig. 3.22). Because of their shape, the half-cylinders are called 'dees'. They are metallic (made of copper) and kept in an evacuated chamber. Their diametric edges are held horizontal, parallel and slightly separated from one another. They are kept centrally between the poles of a vertical uniform magnetic field ⊥ to their planes. They are connected with a ratio frequency oscillator, which provides a rapidly alternating electric field (and therefore an alternating potential difference) across the gap between the two dees. A source S of charged particles, e.g. protons, is placed at the centre of the gap between the dees.

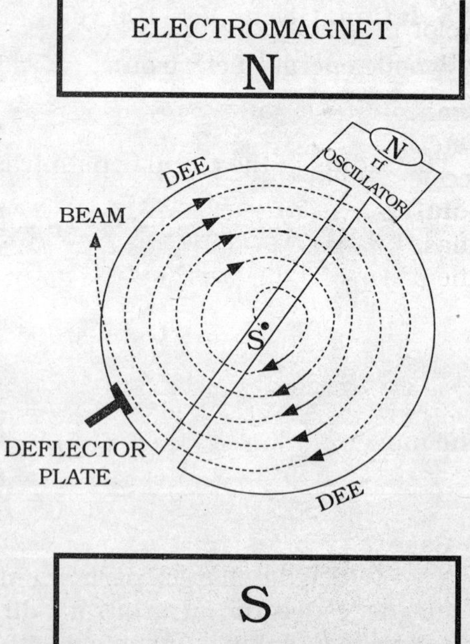

Fig. 3.22

Theory. It has already been described in Section 3A.11 (Case 1).

Working. Suppose the source S produces protons, which are positively charged particles. By virtue of the electric field applied across the dees, at any given instant of time one of the dees is positively charged and the other is negatively charged. A proton produced by S will intially move towards the negatively charged dee, will be accelerated because of the high potential difference between the dees, and so enter the dee. Because the dees are metallic, the electric field cannot penetrate them and so the proton is shielded from this field. But since copper is non-magnetic, the proton is affected by the magnetic field, which makes it to move in a circular path by virtue of the Lorentz force exerted upon it. The force accelerates the proton.

Consequently, the velocity $\vec{v}$ of the proton and hence the radius of the circular path increases ($r = mv/qB$; Eq. 3.20). The frequency of oscillation of the r.f. oscillator is so adjusted that when the proton comes out of the dee after traversing its semi-circle, the electric field is reversed. Then the opposite dee becomes negatively charged, attracting the proton to enter it. There the proton is further accelerated and the radius of circular path further increases. The process continues, with the circulating proton always in 'resonance' with the oscillating electric field. When the radius of the path equals the radius R of a dee, the proton emerges with the maximum velocity (vide Eq. 3.20),

$$v_{max} = \frac{qBR}{m}$$

and with maximum kinetic energy,

$$K_{max} = \frac{1}{2} mv_{max}^2 = \frac{q^2B^2R^2}{2m}$$

Thus, to get a highly energetic beam of charged particles from a cyclotron, one needs,

(i) Dees of large radius R

(ii) High magnetic field B

To facilitate a smooth emergence of the proton beam at the edge of the dee, a deflector plate with positive potential is provided at the edge.

In the 'resonance condition', the frequency of revolution of the proton is equal to the frequency of the oscillator,

$$f = f_{osc}$$

Or, from Eq. 3.22,

$$\boxed{\begin{array}{c} qB = 2\pi m f_{osc} \\ \text{(Resonance condition)} \end{array}} \qquad \text{..... (3.24)}$$

Only those protons which have their initial velocities in horizontal planes $\perp$ to $\vec{B}$ finally emerge from the cyclotron. Other protons adopt helical paths after entering the dee and so they are soon lost by collisions with the walls of the dees.

Example 9. A proton beam is to be accelerated by a cyclotron, having oscillator frequency as 12 MHz. Determine the magnitude of the necessary operating magnetic field. Also find the kinetic energy of the accelerated protons in MeV, if the radius of the dees of the cyclotron is 50 cm.

Solution. Given : $f = 12 \text{ MH}_z = 12 \times 10^6 \text{ H}_z$; $R = 50 \text{ cm} = 0.5 \text{ m}$

(mass of a proton $= 1.67 \times 10^{-27}$ kg; charge of a proton $= 1.6 \times 10^{-19}$ C)

The necessary magnetic field is given by,

$$B = \frac{f \times 2\pi m}{q} = \frac{(12 \times 10^6) \times (2 \times 3.14) \times (1.67 \times 10^{-27})}{(1.6 \times 10^{-19})}$$

$$= 0.787 \text{ T}$$

The maximum kinetic energy of the protons is given by,

$$K_{max} = \frac{q^2B^2R^2}{2m}$$

$$= \frac{(1.6 \times 10^{-19})^2 \times (0.787)^2 \times (0.5)^2}{2 \times (1.67 \times 10^{-27})}$$

$$= 0.11868 \times 10^{-11} \text{ J}$$

$$= \frac{0.11868 \times 10^{-11}}{(1.602 \times 10^{-19})}$$

$$= 7.4 \times 10^6 \text{ eV} = 7.4 \text{ MeV} \quad \textbf{(Answer).}$$

Limitations (Proton Synchrotron). (i) From Eq. 3.22, the frequency of revolution is given by,

$$f = \frac{qB}{2\pi m}$$

Thus, prima facie it appears that f is independent of v, the velocity of the particle. However, this is true only when $v \ll c$, the velocity of light. For high velocities ($v \geq$ nearly c/10), the mass changes according to the relation,

$$m = m_0\sqrt{1-\frac{v^2}{c^2}}$$

m_0 is the normal 'rest' mass. m is referred to as 'relativistic' mass. Then f changes and stops matching with f_{osc}, the frequency of the r.f. oscillator, leading to a breakdown of the essential 'resonance' condition. The problem starts appearing for proton energies greater than nearly 50 MeV (million electron-volts).

There is another problem. From Eq. 3.20 (r = mv/qB) and the relation k.e. = $(1/2)mv^2$, the radius of the circular path for a 500 GeV proton in a magnetic field of 1.5 T works out to be 1.1 km. Then the electromagnet needs to have its pole faces of area nearly 2 km × 2 km, which is prohibitively expensive to arrange!

[Both these problems are overcome in the so-called **proton synchrotron.** Both the magnetic field B and the oscillation frequency f_{osc} are made to vary suitably with time, instead of being fixed values. Then the resonance condition is maintained and the electromagnet also has a manageable size. However, r (= mv/qB) remains a large value. In the Fermilab in U. S. A., the proton synchrotron is capable of producing protons of energies of about one thousand billion electron-volt (10^{12} eV = 1TeV, i.e. one trillion electron-volt). It has $r \approx 1$ km (circumference > 6 km)].

(ii) Light particles like electrons gather speed very fast (v = qBr/m). Hence the resonance condition starts breaking down for them for relatively much lower particle energies. For this reason, the cyclotron is not a suitable instrument to produce high energy light particles like electrons.

(iii) For obvious reasons, the cyclotron cannot be used to accelerate uncharged particles, e.g. neutrons. Being electrically neutral, they remain unaffected both by electric and magnetic fields.

3A. 13. Force on a Current-Carrying Conductor Placed in a Uniform Magnetic Field.

Consider a metallic straight conductor (e.g. a wire) of length L and uniform cross-section A, placed in a uniform magnetic field $\vec{B}$ and carrying current i (Fig. 3.23). The field is ⊥ to the plane of paper, directed downwards into the plane. Let n be the free electron concentration, i.e. the number of free electrons per unit volume, of the material. Let $\vec{v}_d$ be the average drift velocity of a free electron. The Lorentz force $\vec{f}$ experienced by such an electron is (vide Eq. 3.3; the charge on the electron is –e),

Fig. 3.23

$$\vec{f} = -e(\vec{v}_d \times \vec{B})$$

The total number of free electrons in the sample is,

$$N = n(\text{volume}) = nAL$$

Hence, the net Lorentz force on all the free electrons together in the conductor is,

$$\vec{F} = N\vec{f}$$

$$= -neAL\,(\vec{v}_d \times \vec{B})$$

$$\Rightarrow \quad F = (-neA).L.v_d.B \quad (\because \vec{v}_d \perp \vec{B})$$

The direction of flow of electrons is assumed to be opposite to the direction of current. The current $i = (-nea).v_d$ (charge on an electron × number of electrons flowing across A per second). Then we have,

$$F = iLB \qquad \text{..... (3.24)}$$

If we assume L to be a 'length vector' having the same direction as the current i, then in the general case when the magnetic field $\vec{B}$ is not perpendicular to the length of the conductor but makes an angle ϕ with it, the Lorentz force experienced by the electrons is,

$$F = iLB\sin\theta$$

$$\Rightarrow \quad \vec{F} = i[\vec{L} \times \vec{B}] \qquad \text{..... (3.25)}$$

(Force on a current or a conductor)

Since the electrons remain contained within the conductor under the action of the magnetic field, the conductor also experiences the same force as the electrons. The Eq. 3.25 is known as Ampere's formula for magnetic force.

From Eq. 3.25 it follows that the Lorentz force on a conductor is directly proportional to (i) the current flowing through it, (ii) its length, and (iii) the magnitude of the magnetic field $\vec{B}$.

Special cases. (i) $\theta = 0°$ or $180°$, $F = 0$.
i.e. if the current-carrying conductor is **placed parallel to the field** direction, the magnetic force experienced by it is **zero.**
(ii) If $\theta = 90°$, $F = iLB$
i.e. if the conductor is **placed perpendicular to the field,** the force experienced by it is **maximum.**
(iii) The Eq. 3.25 is valid only if the magnetic field is uniform and the conductor is straight. If either the field is not uniform or the conductor is not straight, the force is given by,

$$\vec{F} = \int_L i(d\vec{\ell} \times \vec{B})$$

Here $d\vec{\ell}$ is an infinitesimal element of length of the conductor, which experiences a magnetic field $\vec{B}$.
(iv) The **direction of force** is conveniently given by the **Fleming's left hand rule.** If the forefinger and the second finger of left hand are stretched out at a mutual angle ϕ and the thumb is stretched out perpendicular to the plane of the two fingers, such that the forefinger and the second finger point in the direction of magnetic field and current, respectively, the thumb represents the direction of force on the conductor (Fig. 3.24).

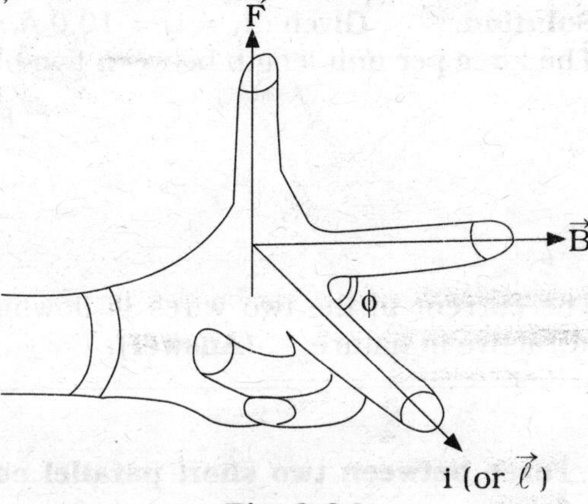

Fig. 3.24

3A. 14. Force between Two Parallel Infinitely Long Straight Current-Carrying Conductors

In Section 3A. 04, we had learnt that a magnetic field is produced around a current-carrying conductor. If another current-carrying conductor is placed near to it, then the second conductor will experience a force due to the magnetic field produced by the first conductor and vice-versa. Consider two parallel straight conductors AB and CD placed at a distance r from each other and carrying current i_1 and i_2 respectively. The conductors are long, approaching infinity in the limiting case. First we consider the case when i_1 and i_2 are in the same direction (Fig. 3.25a). The magnetic field produced by AB at a point lying on CD is given by (vide Eq. 3.9),

$$B_1 = \frac{\mu_0 i_1}{2\pi r}$$

According to the Fleming's left hand rule, the direction of B_1 is perpendicular to the plane of paper, pointed downwards. Due to the field B_1, a unit length of the conductor CD will experience a magnetic force (vide Eq. 3.25),

$$F_{21} = i_2.1.B_1 \sin 90° = \frac{\mu_0 i_1 i_2}{2\pi r}$$

The force F_{21} is directed towards AB. Similarly, a unit length of the conductor AB will experience a force F_{12} of the same magnitude due to the conductor CD,

$$F_{12} = \frac{\mu_0 i_1 i_2}{2\pi r}$$

The force F_{12} is directed towards CD. Thus both the conductors attract each other.

(a) (b)
Fig. 3.25

Similarly, it can be shown that if currents in the conductors are in opposite directions, they repel each repel each other (Fig. 3.25b).

In either case, the mutual force per unit length acting between the two conductors is given by,

$$F = \frac{\mu_0 i_1 i_2}{2\pi r} \quad \text{..... (3.26)}$$

(Force between two conductors)

Also note that,

Parallel currents mutually attract and antiparallel currents mutually repel.

This behavior is opposite to that in electrostatics, where we had found that unlike charges attract and like charges repel each other.

Definition of ampere. The Eq. 3.26 is employed to define one ampere of current. If r = 1m and $i_1 = i_2$ = 1A, then $(\mu_o/4\pi = 10^{-7})$.

$$F = 2\times10^{-7}\ Nm^{-1}$$

Thus, "one **ampere** is that value of electric current which when flowing in each of two parallel long straight conductors placed one metre apart in free space, produces a force of $2\times10^{-7}\ Nm^{-1}$ on each conductor".

Example 10. Two long wires 5.0 cm apart carry a current of 10.0 A in each wire. Find the magnitude and nature of the force acting on a unit length of each wire.

Solution. Given : $i_1 = i_2 = 10.0$ A; $r = 5.0\text{ cm} = 5\times10^{-2}$ m

The force per unit length between two parallel conducting wires is given by,

$$F = \frac{\mu_o i_1 i_2}{2\pi r}$$

$$= \frac{(4\pi\times10^{-7})\times(10)\times(10)}{2\pi\times(5\times10^{-2})}$$

$$= 4\times10^{-4}\ Nm^{-1}$$ **(Answer).**

The current in the two wires is flowing in the same direction. ∴ The mutual force on each of them is attractive in nature. **(Answer).**

Additional knowledge

Force between two short parallel current-carrying conductors; Magnetic force vs. Electrostatics force

Consider two short conductors A and B of lengths $d\ell_1$ and $d\ell_2$ and carrying current i_1 and i_2, respectively, placed parallel to each other at a distance r apart (Fig. 3.26). If $\vec{r}_{12}$ is a unit vector directed from A to B, the magnetic field at B due to the conductor A will be (Eq. 3.7),

$$d\vec{B}_{21} = \frac{\mu_o}{4\pi} i_1 \frac{d\vec{\ell}_1 \times \hat{r}_{12}}{r^2}$$

The direction of the field is downwards perpendicular to the plane of the paper and has magnitude

$$dB_{21} = \frac{\mu_o}{4\pi} i_1 \frac{d\ell_1}{r^2}$$

Fig. 3.26

The force due to this field experienced by the conductor B is directed towards A and from Eq. 3.25 it is given as ($\hat{r}_{21}$ is a unit vector directed from B to A),

$$\vec{F}_{21} = i_2\,(d\vec{\ell}_2 \times d\vec{B}_{21}) = i_2 d\ell_2 dB_{21} \sin 90^\circ\, \hat{r}_{21}$$

$$\Rightarrow \quad \vec{F}_{21} = \frac{\mu_o}{4\pi}\cdot\frac{(id\ell_1).(id\ell_2)}{r^2}\hat{r}_{21} \qquad \text{..... (i)}$$

Similarly, force on A due to B is direted towards B and is given as,

$$\vec{F}_{12} = \frac{\mu_o}{4\pi}\cdot\frac{(i_1 d\ell_1).(i_2 d\ell_2)}{r^2}\hat{r}_{12} \qquad \text{..... (ii)}$$

The mutual force $\vec{F}_{21}$ (or $\vec{F}_{12}$) is attractive in nature. If any one of the conductors carries current in opposite direction, the mutual force experienced by them will be repulsive in nature.

The current elements $i_1 d\ell_1$ and $i_2 d\ell_2$ can also be expressed as charges q_1 and q_2 in motion with velocities v_1 and v_2, respectively, such that,

$$i_1 d\ell_1 = q_1 v_1 \text{ and } i_2 d\ell_2 = q_2 v_2$$

In terms of charges the force is given as,

$$\vec{F}_{12} = \frac{\mu_o}{4\pi}\frac{(q_1 v_1)(q_2 v_2)}{r^2}\hat{r}_{12} \qquad \text{..... (iii)}$$

and

$$\vec{F}_{21} = \frac{\mu_0}{4\pi}\frac{(q_1v_1)(q_2v_2)}{r^2}\hat{r}_{21} \quad \text{..... (iv)}$$

From Eq. (iii), the magnetic force between two charges q_1 and q_2 moving with velocities v_1 and v_2, respectively, parallel to each other is given as,

$$F_m = \frac{\mu_0}{4\pi}\frac{q_1q_2}{r^2}.v_1v_2 \quad \text{..... (v)}$$

From Coulomb's law, the correspoding electrostatic force between the charges is,

$$F_e = \frac{1}{4\pi\epsilon_0}\cdot\frac{q_1q_2}{r^2} \quad \text{..... (vi)}$$

Dividing (v) by (vi) ($\mu_0\epsilon_0 = 1/c^2$, where c (= 3×10^8 ms^{-1}) is the velocity of light),

$$\frac{F_m}{F_e} = (\mu_0\epsilon_0)\,v_1v_2 = \frac{v_1v_2}{c^2}$$

The drift velocity of electrons in a conductor is of the order of 10^{-5}m/s. So,

$$\frac{F_m}{F_e} \approx \frac{10^{-10}}{(3\times10^8)^2} \approx 10^{-27}$$

It follows that in comparison to the electrostatic force existing between two parallel current-carrying conductors, the associated magnetic force existing between them is extremely weak.

3A. 15. Torque on a Current-Carrying loop (or Coil) in a Uniform Magnetic Field; Principle of Electric Motors

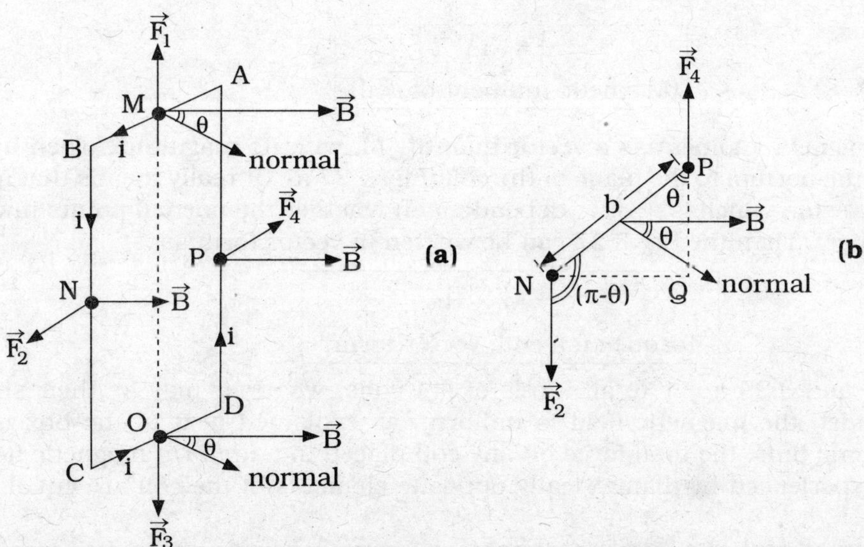

Fig. 3.27

Consider a metallic rectangular loop ABCD suspended in a uniform magnetic field $\vec{B}$, such that the normal to the plane of the loop makes an angle θ with $\vec{B}$ (Fig. 3.27 a). Then the angle made by $\vec{B}$ with AB or CD is (90° + θ) and with BC or DA it is 90°. Let AB = CD = b and BC = DA = ℓ. Let M, N, O, P be the midpoints of the sides. AB, BC, CD and DA, respectively. Let i be the current flowing through the loop in anticlockwise direction. We shall initially treat the sides AB, BC, CD and DA of the loop as separate conductors, each carrying a current i. The magnetic force exerted by the field $\vec{B}$ on AB is (vide Eq. 3.25),

$$\vec{F}_1 = i(\overrightarrow{AB} \times \vec{B})$$

F_1 lies in the direction of OM, which is perpendicular to both $\overrightarrow{AB}$ and $\vec{B}$. Its magnitude is,

$$F_1 = i.b.B\sin(90°+\theta) = ibB\cos\theta \qquad \text{..... (H)}$$

Similarly, the force experienced by CD is,

$$\vec{F}_3 = i(\overrightarrow{CD} \times \vec{B}) = i.b.B\sin(90+\theta) = ibB\cos\theta$$

Thus $\vec{F}_3$ has the same magnitude as $\vec{F}_1$. But since the direction of flow of current in CD is opposite to that in AB, the direction of $\vec{F}_3$ is opposite to $\vec{F}_1$, i.e. along MO. So $\vec{F}_1$ & $\vec{F}_3$ are collinear and have the same magnitude but opposite orientation. Hence they cancel out, with one force exactly balancing the other. In a similar way, the magnitude of either $\vec{F}_2$ and $\vec{F}_4$ is,

$$i.\ell.B \sin 90° = i\ell B$$

Thus $\vec{F}_2$ and $\vec{F}_4$ have the same magnitude. However, since the directions of flow of current in BC and DA are mutually opposite, the directions of $\vec{F}_2$ and $\vec{F}_4$ are also mutually opposite. But $\vec{F}_2$ and $\vec{F}_4$ are not collinear. Hence they together form a couple, which produces a torque on the loop. The torque tends to rotate the loop to make its normal coincide with $\vec{B}$, in which position $\theta = 0$ and so $\vec{F}_2$ and $\vec{F}_4 = 0$. The torque has the value,

$$\tau(\text{Torque}) = (\text{Magnitude of } \vec{F}_2 \text{ and } \vec{F}) \times (\perp \text{ distance between } \vec{F}_2 \text{ and } \vec{F}_4)$$

The $\perp$ distance between $\vec{F}_2$ and $\vec{F}_4$ is NQ = PN $\sin\theta$ = b$\sin\theta$ (Fig. 3.27 b). So we have,

$$\tau = i\ell B.b\sin\theta$$
$$= i(\ell \times b)B\sin\theta = (iA)B\sin\theta,$$

where $(\ell \times b) = A$ is the area enclosed by the loop ABCD. If the rectangular loop is replaced by a flat rectangular coil with N turns, the same considerations will apply to each turn of the coil. Then the net torque acting on the coil will be,

$$\boxed{\begin{array}{l}\tau = (NiA)B\sin\theta \\ \text{(Torque on a coil)}\end{array}} \qquad \text{..... (3.27)}$$

The quantities N (no. of turns), i (current in the coil), and A (area of the coil) are all characteristic of the coil. We call the product (NiA) as the magnetic moment of the coil,

$$\boxed{\begin{array}{c}M = NiA \\ \text{(Magnetic moment of coil)}\end{array}} \qquad \text{..... (3.28)}$$

We can specify the magnetic moment as a vector quantity $\vec{M}$, with its magnitude given by Eq. 3.28 and its direction lying along the normal to the plane of the coil (Fig. 3.27 a). (It really means that now we treat A, the area of the coil, as a vector, which is + or –, depending on whether the normal points inwards or outwards from the plane of paper). Then the Eq. 3.27 can be written in vector form as,

$$\boxed{\begin{array}{c}\vec{\tau} = \vec{M} \times \vec{B} \\ \text{(Torque on a coil; vector form)}\end{array}} \qquad \text{..... (3.29)}$$

The Eqs. 3.27 and 3.29 apply to all kinds of flat coils, whatever may be their shape (e.g. circular, elliptical, etc.), provided the magnetic field is uniform, as explained next in the box under 'Additionnal knowledge'. At the same time, the total force on any coil placed in a uniform magnetic field is always zero, because the forces experienced by diametrically opposite elements of the coil are equal and opposite and hence cancel out.

The foregoing treatment of a current-carrying coil, experiencing a magnetic force and so tending to rotate in a uniform magnetic field, forms the basis of construction of all kinds of **electric motors** that are so widely used in daily life. It is necessary to reverse the direction of flow of current in the coil after every half-revolution, because then the torque starts acting in the opposite direction. This task is accomplished with the help of a suitable commutator. Then the torque always acts in the same direction, ensuring a continuers rotation of the coil.

For $\theta = 0$ and 180°, the torque is zero. Then the coil is said to be in zero equilibrium condition. It is 'stable equilibrium' for the position $\theta = 0$, because a slight disturbance of the coil from this position makes the coil to tend to regain its original position. For $\theta = 180°$, it is 'unstable equilibrium', because a slight disturbance now makes the coil to 'fly away' from this position, instead of tending to return to the original position.

Example 11. A current of 6 A flows in a circular coil having 24 turns and radius 12 cm. It is placed in a uniform magnetic field $\vec{B}$ of strength 0.20 T. Find the torque on the coil when the normal to its plane (i) is perpendicular to $\vec{B}$, (ii) makes an angle of 30° with $\vec{B}$, and (iii) lies along $\vec{B}$. What is the total (net) force experienced by the coil in these cases?

Solution. Given : N = 24; i = 6.0 A; r = 12 cm = 0.12 m; B = 0.20 T

$\therefore$ Area of coil, $A = \pi r^2 = 3.14 \times (0.12)^2 = 0.045 \text{ m}^2$

We shall use the following formula for finding the torque in the three cases,

$$\tau = N B i A \sin\theta$$
$$= (24) \times (0.20) \times (6) \times (0.045) \sin\theta$$
$$= 1.296 \sin\theta$$

(i) In this case, $\theta = 90°$. $\therefore \sin\theta = 1$

$\therefore$ $\tau = 1.296 \times 1 \approx 1.3$ N.m **(Answer).**

(ii) Here, $\theta = 30°$. $\therefore$ $\sin\theta = 0.5$

$\therefore$ $\tau = 1.296 \times 0.5 = 0.65$ N.m **(Answer).**

(iii) Here $\theta = 0$. $\therefore$ $\sin\theta = 0$

$\therefore$ $\tau = 0$ **(Answer).**

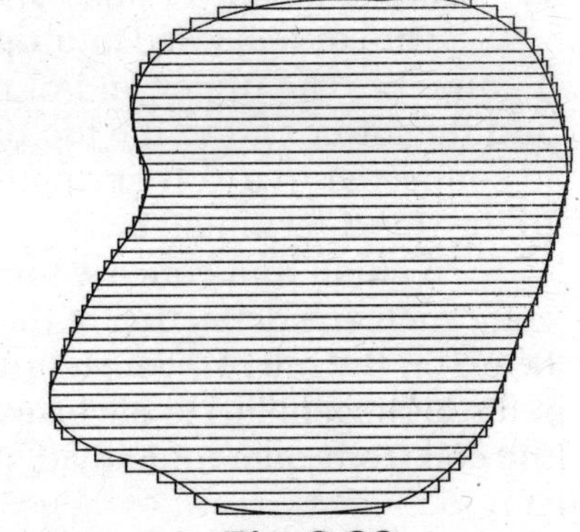

Fig. 3.28

In all the cases, the total force (i.e. net force) is zero, because the forces experienced by diametrically opposite elements of the coil are equal and opposite and hence mutually cancel out. **(Answer).**

Additional knowledge

The equation $\vec{\tau} = \vec{m} \times \vec{B}$ is appliable to all shapes of loops or coil. Let us consider a loop of arbitrary shape, as shown in Fig. 3.28. It can be visualized as made up of a large number of successive rectangular strips. Each strip will contribute a small torque,

$$\overrightarrow{\Delta\tau} = i(\overrightarrow{\Delta A} \times \vec{B})$$

Since $\vec{B}$ is a uniform field and the same current i flows through all the strips, we have,

$$\vec{\tau} = \Sigma \overrightarrow{\Delta\tau} = i[(\Sigma \overrightarrow{\Delta A}) \times \vec{B}]$$
$$= i(\vec{A} \times \vec{B})$$

Or $$\vec{\tau} = \vec{M} \times \vec{B}$$

3A. 16. Moving-Coil Galvanometer

It is an instrument used for the detection of current in a circuit. When suitably modified, it can also be employed to measure current and voltage, as we shall find in the next sections 3A. 17 and 3A. 18.

Principle. It works on the principle that when a current-carrying coil is placed in a uniform magnetic field, it experiences a torque, which tends to rotate the coil [Eq. 3.27].

Construction. It consists of a rectangular coil made by winding thin insulated copper wire on a rectangular frame. The frame is made of a non-magnetic material. The frame is suspended by a thin strip of gold (or phosphor-bronze), G, from a torsion head at the top (Fig. 3.29 a). The coil lies between the pole pieces N, S of a strong permanent magnet. It is actually a U-shaped magnet, as separately shown in Fig. 3.29 b. The pole faces are cylindrical in shape. A cylindrical soft iron rod R is placed coaxially with the pole faces. It is rigidly attached to the body of the instrument. It forms the core of the coil. The axis of suspension of the coil is also coaxial with this cylinder R. The coil lies just outside R, such that it can freely rotate in the air gap between R and the pole faces.

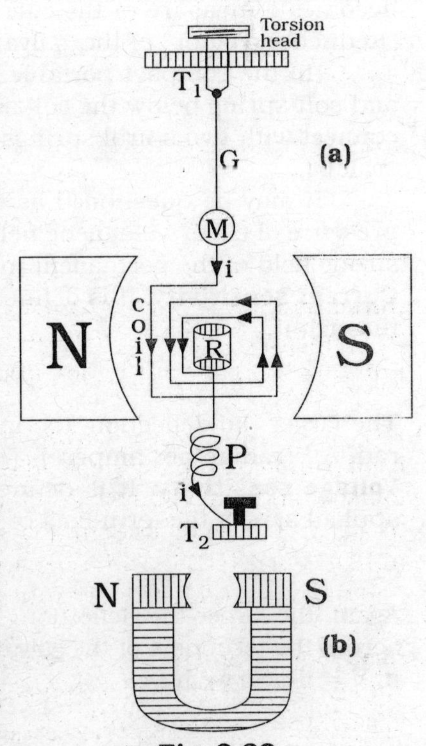

Fig. 3.29

The purpose of having the pole faces cylindrically concave is to generate a radial field, i.e. a field in which the magnetic lines of force lying in a vertical cross-section are directed along the radii of a circle whose centre lies on the axis of suspension (Fig. 3.30). Then the plane of the coil is in the direction of magnetic field for all positions of the coil. The soft iron cylinder R is meant to improve the radial nature of the field, as well as to increase the strength of the field (due to crowding of magnetic lines of force through the soft iron core).

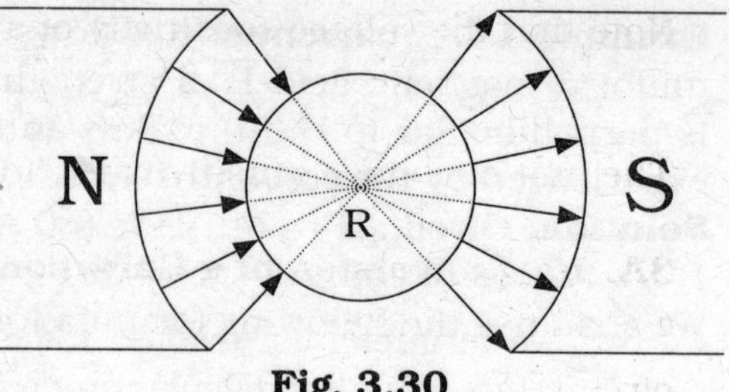

Fig. 3.30

The current is led into and out of the coil through the terminals T_1 and T_2. The upper end of the suspension G is connected to T_1 and the lower end to the top of the coil. The bottom end of the coil is connected to a soft spring P, the free end of which is connected to the fixed terminal T_2.

A small thin concave mirror M is attached to the frame of the coil. Then any deflection of the coil can be read by a lamp and scale arrangement. The lamp focusses a light beam on the mirror, which reflects it back on a translucent plastic scale. The whole set-up (except lamp and scale arrangement) is enclosed in a metal casing having a glass window in front of the mirror M. It protects the coil from dust and any disturbance from wind. The casing is provided with three levelling screws.

Working. When the galvanometer is put in an electric circuit, the current i flowing through the coil gives rise to a torque, which rotates the coil. Because of the radial nature of the magnetic field, the plane of the coil is parallel to the field for all positions of the coil. Or, we may say that the angle between the magnetic field $\vec{B}$ and the normal to the plane of the coil is always 90°. If the coil has N turns of area A each, the torque experienced by the coil is (vide Eq. 3.27),

$$\tau = (NiA)\,B\sin 90° = (NiA)\,B$$

As the coil rotates, the suspension wire G (and the spring P) are twisted. This produces a restoring torque, which acts opposite to the existing torque τ and tends to bring the coil back to its original rest position. When the two torques become equal, the coil comes to an equilibrium position. If the coil twists through an angle ϕ and the restoring torque per unit angle of twist is C, then the net restoring torque is $C\phi$. At equilibrium,

$$C\phi = \tau = (NiA)B$$

$$\Rightarrow \quad i = \left(\frac{C}{NAB}\right)\phi \qquad \text{..... (3.30)}$$

Thus, the current i is found to be directly proportional to the angle of deflection ϕ, or we can say that ϕ becomes a measure of the current i. The quantity C/NAB, which is constant for a galvanometer, is called "Reduction Factor" of the galvanometer. C is called spring constant or torsional constant.

[In the compact portable galvanometers, commonly used in class laboratories, the suspension strip and soft spring below the coil are replaced by sturdy spindles mounted on jewelled bearings. The spindles connect with two spiral springs, which control the movement of an aluminium pointer across a graduated scale].

It may be questioned as to why the working of a moving-coil galvanometer is not influenced by the presence of earth's magnetic field. The influence is there, but it can be safely neglected in comparison to the strong field of the permanent magnet provided in the galvanometer.

Current sensitivity. It is defined as the deflection produced in the galvanometer when a unit current flows through it,

$$\text{current sensitivity} = i_s = \frac{\phi}{i} = \frac{NAB}{C}$$

The larger the deflection, the more current sensitive is the galvanometer. The unit of current sensitivity is rad. A^{-1} (radian per ampere).

Voltage sensitivity. It is defined as the deflection produced in the galvanometer when a unit voltage is applied across the terminals of the galvanometer,

$$\text{voltage sensitivity} = v_s = \frac{\phi}{V}$$

Again, the larger the deflection, the more voltage-sensitive is the galvanometer. Here V is the voltage applied across the terminals of the galvanometer. If R is the resistance of the coil and i is the current passing through it, V = iR. So we have,

$$v_s = \frac{\phi}{iR} = \frac{NAB}{CR}$$

Note that the voltage sensitivity of a galvanometer is related to its current sensitivity as,

$$V_s = \frac{i_s}{R}$$

The unit of voltage sensitivity is rad. V^{-1} (radian per volt).

3A. 17. Conversion of a Galvanometer into an Ammeter; Shunt Resistance

An ammeter is used to measure the current flowing in an electric circuit. It is a modified galvanometer. The modification consists of inserting a low resistance in parallel with the galvanometer. This resistance is called shunt resistance or simply shunt. It has two-fold purpose:

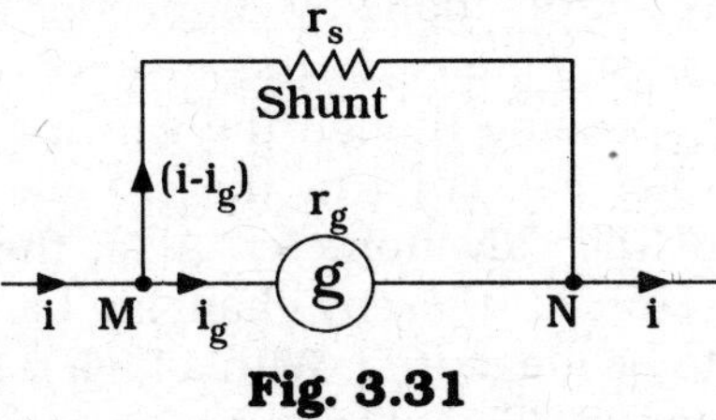

Fig. 3.31

(i) If the current has a high value, the deflection of the galvanometer coil may be so large that the supporting suspension or spring will be damaged. If the galvanometer is provided with a shunt, the bulk of the current passes through the shunt because of its low value and only a small fraction passes through the galvanometer (Fig. 3.31).

(ii) The galvanometer has its own resistance (resistance of the coil and other associated parts). When put into the electric circuit, it alters the total resistance of the circuit and therefore the current flowing through the circuit is also altered. So, ideally, a galvanometer should have zero resistance, which is impossible to achieve. However, the insertion of the shunt resistance highly reduces the total galvanometer resistance, as we shall find in the following. So the current passing through the circuit is negligibly altered.

The required value of a shunt resistance is easily calculated as follows. Let i_g be the current required to be passed through the galvanometer for full scale deflection. Let r_g be the resistance of the galvanometer and r_s the resistance of the shunt, connected in parallel with the galvanometer (Fig. 3.31). When the galvanometer G is inserted in the electric circuit at MN, the current i flowing in the circuit is divided at M into a small part i flowing through G and a large part $(i - i_g)$ flowing through r_s. By Ohm's law, the potential difference V_{mn} across the points M and N is,

$$V_{mn} = i_g r_g = (i - i_g)\, r_s$$

$$\Rightarrow \qquad r_s = \frac{i_g}{(i - i_g)} \times r_g \qquad \text{..... (3.31)}$$

which is the required value of the shunt resistance. Further, since r_g and r_s are connected in parallel, the combined 'effective' resistance R_{eff} of the galvanometer is given by,

$$\frac{1}{R_{eff}} = \frac{1}{r_g} + \frac{1}{r_s}$$

$$\therefore \qquad R_{eff} = \frac{r_g r_s}{r_g + r_s} \qquad \text{..... (3.32)}$$

If $r_g \gg r_s$, which is always the case, then $R_{eff} \cong r_s$ ($\because$ in the denominator, r_s can be neglected). Thus the effective resistance of a galvanometer is highly reduced when it is converted into an ammeter, as required.

Note that an ammeter has always to be connected in series with the electric circuit in which the current is to be measured. Since the current flowing though the circuit is a uniform value, the ammeter can be inserted anywhere in the circuit to measure it.

Example 12. A galvanometer having a coil resistance of 100 Ω gives full scale deflection when a current of 1mA is passed though it. Calculate he value of resistance required to convert it into an ammeter of the range of 1A. (AI ; 1993).

Solution. Given : $r_g = 100\ \Omega$; $i = 1A$; $i_g = 1mA = 10^{-3}$ A

The shunt resistance needed to convert a galvanometer into an ammeter is given by,

$$r_s = \frac{i_g}{(i - i_g)} \times r_g$$

$$= \frac{(10^{-3})}{(1) - (10^{-3})} \times 100 = \frac{0.1}{1 - 0.001}$$

$$= \frac{0.1}{0.999} \approx 0.1\ \Omega \qquad \textbf{(Answer).}$$

3A. 18. Conversion of a Galvanometer into a Voltmeter

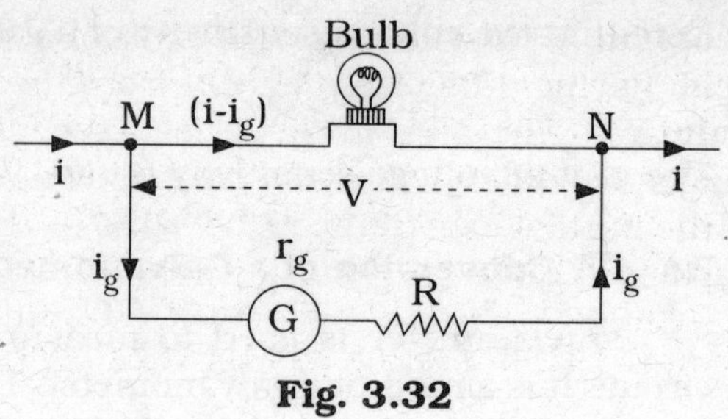

Fig. 3.32

A voltmeter is used to measure the potential drop across a given component in an electric circuit. Therefore, unlike an ammeter, the voltmeter has to be connected in parallel with the component (an electric bulb in Fig. 3.32) across which the potential difference is to be measured. It is a modified galvanometer. The modification is simply made by connecting a high resistance R in series with the galvanometer. In the absence of such a high resistance, the current i_g passing though the galvanometer will be a large fraction of the total current i in the circuit, which may damage sensitive parts of the galvanometer. Also, the combined resistance ($R_{comp.}$ and $(r_g + R)$ in parallel) will be appreciably reduced, thereby changing i and so giving a wrong value of the potential difference (= current×resistance) to be measured. With a high resistance R connected in series with he galvanometer, both these problems are solved. Then the combined resistance remains nearly equal to R_{comp}. Ideally, R should be infinite.

The value of R is found as follows. Let i_g be the current required to pass through the galvanometer G for full scale deflection and let r_g be the resistance of the galvanometer. R is the resistance required to convert the galvanometer into a voltmeter of range V (maximum voltage for full scale deflection). According to the Ohm's law,

$$V = i_g (r_g + R)$$

$$\Rightarrow \quad R = \frac{V}{i_g} - r_g \qquad \text{..... (3.33)}$$

The effective resistance of the voltmeter is,

$$R_{eff} = (r_g + R) = \frac{V}{i_g} \qquad \text{..... (3.34)}$$

Example 13. Calculate the value of the resistance needed to convert a galvanometer of resistance 100 Ω, which gives a full scale deflection for a current of 5 mA, into a voltmeter of 0–10 V range. (AI ; 1994).

Solution. Given : $r_g = 100\ \Omega$; $V = 10$ volt ; $i_g = 5\text{mA} = 5\times10^{-3}$ A

The series resistance needed to convert a galvanometer into a voltmeter is given by,

$$R = \left(\frac{V}{i_g}\right) - r_g$$

$$= \frac{(10)}{(5\times10^{-3})} - 100 = 2000 - 100$$

$$= 1900\ \Omega \quad \textbf{(Answer).}$$

Solved Conceptual Problems

Problem 1. How the Oersted's experiment led to a linkage between electricity and magnetism ?

Solution. Oersted observed that a current-carrying wire influenced the alignment of a magnetic compass needle placed in its neighbourhood. When placed at different positions along a circular path around the wire, it always aligned tangentially to the path. The orientation of the needle reversed upon reversing the current. The alignment became slower upon reducing the current or increasing the radius of the path. The experiment led to the conclusion that an electric current produces magnetic effect.

Problem 2. Why the straight path of a moving charged particle becomes curved after entering a magnetic field, having direction different from the particle's straight path ?

Solution. The motion of the charged particle produces its own magnetic field, which interacts with the magnetic field in which the particle is moving. Under the combined action of the two fields, the straight path of the particle bends (except when it is moving in the direction of the field). The magnetic force acting on the particle is given by,

$$F = q\,v\,B \sin\theta$$

(q : charge ; $\vec{v}$: velocity ; $\vec{B}$: magnetic field ; θ : angle between $\vec{v}$ and $\vec{B}$)

Problem 3. How can the direction of force experienced by a charged particle moving in a magnetic field be determined ?

Solution. The said direction can be determined by any one of the following rules:

(i) Fleming's left hand rule : If the first finger and the second finger of the left hand are stretched out at an angle θ and the thumb is stretched out perpendicular to both the fingers, such that the first and second finger point in the direction of magnetic field and current (motion of charged particle), respectively, the thumb represents the direction of force experienced by the particle.

(ii) Right hand grip rule : If one grips something in his right hand such that the fingers bend from the direction of current to the direction of field and the thumb is held perpendicular to the plane of the grip, then the thumb represents the direction of force.

Problem 4. Does a static electric charge experience a force in a magnetic field ? Justify your answer.

Solution. No. The magnetic force acting on a charged particle moving with a velocity $\vec{v}$ in a magnetic field $\vec{B}$, with the angle between $\vec{v}$ and $\vec{B}$ equal to θ, is given by,

$$F = q\,v\,B \sin\theta$$

For a static charge, $v = 0$. $\therefore F = 0$

Problem 5. A current-carrying wire has net electric charge equal to zero. Then how does it experience a force in a magnetic field ?

Solution. The conductor has two types of charge, produced by ionization of atoms: free electrons (–vely charged) and the resulting +ve ions (the unionized atoms contained in the conductor are electrically neutral). Out of them, only the electrons move when a potential difference exists across the wire and so an electric current flows through it (they move towards the +ve end of the wire). Only they experience a magnetic force, because such a force only acts on a moving charge ($F = qvB \sin\theta$). Hence in turn the conductor as a whole experiences the same force. The +ve ions, being comparatively too heavy, do not move and therefore have no such force acting on them.

Problem 6. If the path of a charged particle does not change in moving through a certain region of space, does it necessarily mean that no magnetic field exists in the region ? Justify your answer.

Solution. No. The magnetic force experienced by a charged particle is given by,

$$F = q\,vB \sin\theta$$

Therefore, if the particle moves in the direction of the field $\vec{B}$, $\theta = 0$ ($\vec{v}$ parallel to $\vec{B}$) or $180°$ ($\vec{v}$ antiparallel to $\vec{B}$).
$\therefore F = 0$, in spite of the presence of the magnetic field.

Problem 7. The free electrons present in a conductor are constantly in vigorous motion at room temperature. Yet they do not seem to experience any force in a magnetic field, unless an electric current passes through the conductor. Explain why.

Solution. In the absence of an electric current, the free electrons are in random motion, with as many electrons moving in one direction as in the opposite direction. Hence their net velocity is zero ($v = 0$) and no magnetic force acts on them ($F = qvB \sin\theta$). On the other hand, when a current flows through the conductor, the electrons acquire a definite drift velocity parallel to the direction of current. Hence they experience a magnetic force.

Problem 8. For a charged particle moving in a uniform magnetic field, under what condition the magnetic force acting upon it is (i) maximum and (ii) minimum ?

Solution. The force acting on a charged particle in a magnetic field is given by

$$F = qvB \sin \theta.$$

(i) When $\theta = 90°$, F = qvB (maximum value). Thus, if the particle enters the field in a direction perpendicular to the field, the force experienced by it is maximum.
(ii) When $\theta = 0$ or $180°$, F = 0 (minimum), i.e. if the particle enters the field in a direction parallel or antiparallel to the field, the force acting on it is minimum.

Problem 9. For a charged particle moving in a direction perpendicular to a uniform magnetic field, explain why work done is zero (and so its energy remains unchanged).
Solution. The magnetic force $\vec{F}$ acts on the particle in a direction perpendicular to the direction of motion (along $\vec{v}$) of the particle. ∴ Work done (= force × displacement) = $\vec{F} \cdot \vec{d} = F \times d \times \cos 90° = 0$. [Hence the total energy of the particle remains unchanged. Since it includes kinetic energy (= $1/2\ mv^2$), it follows that the velocity of the particle remains constant in the field].

Problem 10. A moving charged particle is deflected by a magnetic field as well as by an electric field. Explain how the (i) nature of deflection and (ii) change in k.e. of the particle are different in the two cases.
Solution. (i) In a magnetic field, the Lorentz force [$\vec{F} = q(\vec{v} \times \vec{B})$] and hence the deflection of the particle is in a direction perpendicular to the field. In an electric field, the Lorentz force ($\vec{F} = q\vec{E}$) and hence the deflection is in the plane of the field (the direction depends upon the angle between the velocity vector ($\vec{v}$) and $\vec{E}$).

(ii) As already explained, for a magnetic field the displacement of the particle is normal to the plane of field and so the particle's kinetic energy remains unchanged. For an electric field, the displacement of the particle is in the plane of the field. So a finite amount of work is done and hence the energy of the particle is changed.

Problem 11. Is it possible that a charged particle passing through (a) magnetic field or (b) an electric field experiences no force ? If so, explain how.
Solution. (a) In a magnetic field, the force acting on the particle ($F = qvB \sin \theta$) can be zero in spite of the field, if the particle is moving parallel to the field ($\theta = 0$) or antiparallel to the field ($\theta = 180°$). (b) In an electric field, the force ($F = qE$) cannot be zero unless $E = 0$. Therefore, in the presence of a finite electric field, the particle always experiences a force.

Problem 12. The path of an electron changes upon entering a certain field. How can the nature of field be found, i.e. whether it is a magnetic field or an electric field ?
Solution. The nature of field can be determined by knowing the trajectory of the electron and its kinetic energy. In a magnetic field, the path of an electron beam entering perpendicular to the field becomes circular and the kinetic energy (and therefore the speed) of the beam does not change. In an electric field, the path of an electron beam entering perpendicular to the field becomes a parabola and the k.e. (and so the speed) of the beam changes.

Problem 13. Is a cyclotron suitable for accelerating electrons ? If not, explain why.
Solution. Because of its tiny mass, an electron gathers speed very fast in a cyclotron ($v = qBr/m$). Therefore its mass changes relativistically, leading to a breakdown of the resonance condition for relatively much lower particle energies. Thus a cyclotron is not suited to produce high energy electrons.

Problem 14. Can neutrons be accelerated in a cyclotron ? If not, explain why.
Solution. The neutrons are electrically neutral. They carry no charge. Therefore they remain unaffected by either an electric field or a magnitude field. Hence they cannot be accelerated in a cyclotron.

Problem 15. How does a torque arise in a rectangular coil suspended in a magnetic field, in spite of the net force acting on the coil being zero ?
Solution. The magnetic forces acting on the two parallel horizontal arms of the coil are equal, opposite and collinear. Hence they mutually cancel out. On the other hand, the magnetic forces acting upon the two parallel vertical arms of the coil are also equal and opposite, but non-collinear. Hence they together form a couple, which produces a torque on the coil.

Problem 16. What is meant by a radial magnetic field ? How one arranges to obtain such a field in a moving - coil galvanometer ?
Solution. A radial magnetic field is one in which the magnetic lines of force lying in a vertical cross-section are directed along the radii of a circle drawn with the centre lying on the axis of suspensions of the coil. Then the plane of the coil lies in the direction of magnetic field for all positions of the coil.

It is achieved by (i) making the pole faces of the magnet cylindrically concave and (ii) using a soft iron cylindrical core. The soft iron core improves the radial nature of the field, as well as increases the strength of the field, due to crowding of magnetic lines of force through the core (this, in turn, enhances the sensitivity of the galvanometer).

Problem 17. The working of a moving-coil galvanometer is supposed not to be affected by the earth's magnetic field. Explain why.
Solution. The influence of the earth's magnetic field exists, but it is negligible in comparison to the influence of the field of the strong permanent magnet provided in the galvanometer, because the former is very weak compared to the latter.

Problem 18. Foe converting a galvanometer into an ammeter, a small resistance is connected with the galvanometer in parallel. Why not a high resistance is connected instead of the small shunt resistance ?
Solution. The purpose of inserting a small shunt resistance in parallel is two-fold: (i) to make bulk of the current to pass through it instead of the galvanometer and thus protect sensitive parts of the galvanometer from any damage and (ii) to highly reduce the total resistance of the galvanometer, so that the current flowing through the circuit is negligibly altered. If a high resistance is connected in place of the small shunt resistance, both these purposes will be defeated.

Problem 19. For converting a galvanometer into a voltmeter, a high resistance is connected in series with the galvanometer. Why not a low resistance is connected instead of a high resistance ?
Solution. The insertion of a high resistance in series with the galvanometer (i) reduces the current passing through the galvanometer, thus saving its sensitive parts from possible damage and (ii) makes the combined resistance in the circuit to remain nearly the same as the resistance of the component across which potential difference is to be measured. The use of a low resistance in place of the high resistance R will have opposite effects. So its use is not permissible.

Problem 20. In an electric circuit, an ammeter is connected in series and a voltmeter in parallel. Why is it not the other way round ?
Solution. An ammeter is meant to measure the current in the circuit. Therefore its insertion in the circuit has to be such that (i) the same current flows through it as in the other components of the circuit and (ii) its resistance is low, so that the current in the circuit does not change appreciably. These purposes are served by (i) connecting a shunt resistance in parallel with a galvanometer (which converts the galvanometer into an ammeter) and (ii) connecting the ammeter in series with the circuit. If the ammeter is connected in parallel with some component in the circuit, the effects will be opposite.

A voltmeter is meant to measure voltage (or, say, potential difference) across a component in the circuit. Its insertion has to be such that the current flowing through the component remains nearly the same as before. This purpose is served by first connecting a high resistance in series with a galvanometer (which converts the galvanometer into a voltmeter) and then connecting the voltmeter in parallel with the component. If the voltmeter is connected in series with the circuit, the effect will be opposite.

Solved Numerical Problems

Problem 1. A current of 120 A flows in an overhead power line running from east to west direction. Find the magnitude and direction of the magnetic field produced by it at a position 2 metre below the line.
Solution. Given : i = 120 A; a = 2 m
The magnetic field due to an infinitely long straight conductor is given by,

$$B = \frac{\mu_0 i}{2\pi a} = \frac{(4\pi \times 10^{-7}) \times (120)}{2\pi \times (2)}$$

$$= 1.2 \times 10^{-5}\ \text{T} \quad \textbf{(Answer).}$$

According to the right hand thumb rule, the direction of the field should be directed as north to south. **(Answer).**

Problem 2. A wire of length 56 cm is bent into a circular shape and another wire of the same length is bent into a square shape. Both the wires have a current of 12 A flowing them. Calculate the magnetic field at the centre in the two cases.
Solution. We take up the two cases one by one.
(i) Circular shape : if the radius of the circle is r, we have,

$$2\pi r = 56\ \text{cm}$$

$$\therefore \quad r = \frac{56}{2\pi} = \frac{28}{\pi}\ \text{cm} = \frac{0.28}{3.14}\ \text{m} = 0.089\ \text{m}$$

The magnetic field at the centre of a circular loop is given by,

$$B = \frac{\mu_0 i}{2r} = \frac{(4\pi \times 10^{-7}) \times (12)}{2 \times (0.089)}$$

$$= 8.47 \times 10^{-5}\ \text{T} \quad \textbf{(Answer).}$$

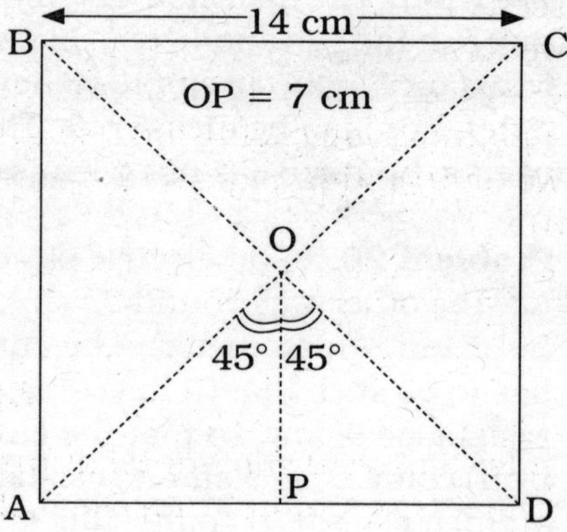

(ii) Square shape : As seen from the figure, when the wire acquires a square shape, each side has a length equal to (56/4) = 14 cm. Each side equally contributes to the magnetic field at the centre 0. The contribution from one side is given by,

$$B = \frac{\mu_0 i}{4\pi a}(\sin\phi_1 + \sin\phi_2)$$

$$\therefore \quad B_{Total} = 4 \times \frac{\mu_0 i}{4\pi a}(\sin\phi_1 + \sin\phi_2)$$

We have (see figure) $\phi_1 = \phi_2 = 45°$ and a = OP = 7 cm = 0.07 m

$$\therefore \quad B_{Total} = \frac{4 \times (4\pi \times 10^{-7}) \times (12)}{4\pi \times (0.07)} \times (\sin 45° + \sin 45°)$$

$$= (6.86 \times 10^{-5}) \times (1.414)$$

$$= 9.7 \times 10^{-5}\ \text{T} \quad \textbf{(Answer).}$$

[Note that for the same length of wire the square shape provides higher magnetic field than the circular shape].

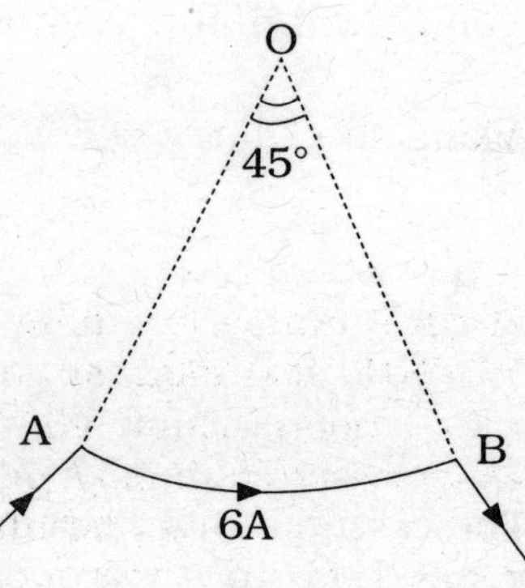

Problem 3. A segment of a circular coil of radius 12 cm subtends an angle of 45° at its centre. If it carries a current of 6 ampere, determine the magnitude and direction of the magnetic field produced at its centre.
Solution. Given : r = 12 cm = 0.12 m;
$\theta = 45° = (\pi/4)$ radian; i = 6 A
The segment AB, with its centre at 0, is shown in the figure. We know that the magnetic field produced by a circular coil at its centre is given by,

$$B_{coil} = \frac{\mu_0 i}{2r}$$

This field corresponds to the entire angle of 2π radian subtended by the coil at its centre. ∴ The field corresponding to the segment AB is given by,

$$B_{segment} = \frac{\mu_0 i}{2r} \times \frac{\theta}{2\pi} = \frac{\mu_0 i\theta}{4\pi r}$$

$$= \frac{(4\pi \times 10^{-7}) \times (6) \times (\pi / 4)}{4\pi \times (0.12)}$$

$$= \frac{6 \times 3.14 \times 10^{-7}}{4 \times 0.12}$$

$$= 39.25 \times 10^{-7}\ \text{T}$$ **(Answer).**

If we assume the segment AB as lying in the plane of paper, then application of the right hand thumb rule gives the direction of $\vec{B}$ as downwards and normal to the plane of paper. **(Answer).**

Problem 4. A uniform current i flows in a thick straight wire, having circular cross-section of radius R (see figure). Determine the magnitudes of magnetic field in the regions (i) x>R and (ii) x<R, respectively. What is the essential difference between these two values ?

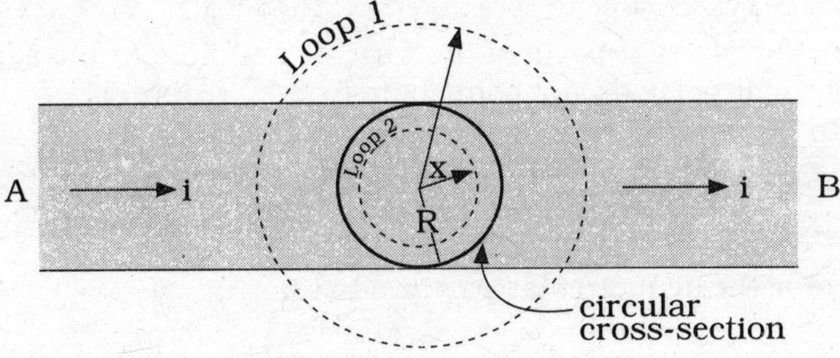

Solution. Both the loop 1 (x>R) and loop 2 (x<R) are circular and concentric with the circular cross-section of the wire (of radius R). For both of them, we have,

$$\oint_\ell d\ell = 2\pi x$$

We consider the two cases one by one.

(i) x > R (Loop 1):

The current enclosed by this loop is i. ∴ according to the Ampere's circuit law, we have,

$$\oint_\ell \vec{B}.\vec{d\ell} = \mu_0 i$$

$$\Rightarrow \quad B \times 2\pi x = \mu_0 i$$

$$\Rightarrow \quad B = \frac{\mu_0 i}{2\pi x} = \frac{\mu_0 i}{2\pi}\left(\frac{1}{x}\right)$$ **(Answer).**

(ii) x < R (Loop 2):

The current enclosed by this loop is a smaller value, say i′. It is related to i as,

$$\frac{i'}{i} = \frac{\pi x^2}{\pi R^2}$$

$$\Rightarrow \quad i' = \frac{ix^2}{R^2}$$

Again, according to the Ampere's law,

$$\oint_\ell \vec{B}.\vec{d\ell} = \mu_0 i'$$

$$\Rightarrow \quad B \times 2\pi x = \frac{\mu_0 ix^2}{R^2}$$

$$\Rightarrow \quad B = \frac{\mu_0 i}{2\pi R^2} x$$ **(Answer).**

The essential difference between the two cases is that for x > R, B is inversely proportional to x and for x < R, B is directly proportional to x. **(Answer).**

Problem 5. A part of a straight copper wire is bent into semicircular shape of radius 3 cm [Fig. (i)]. A current of 15A flows through it. Determine the magnetic field $\vec{B}$ at the centre of the semicircle due to (a) straight parts of the wire, (b) semicircular part, and (c) oppositely bent semicircular part [Fig. (ii)].

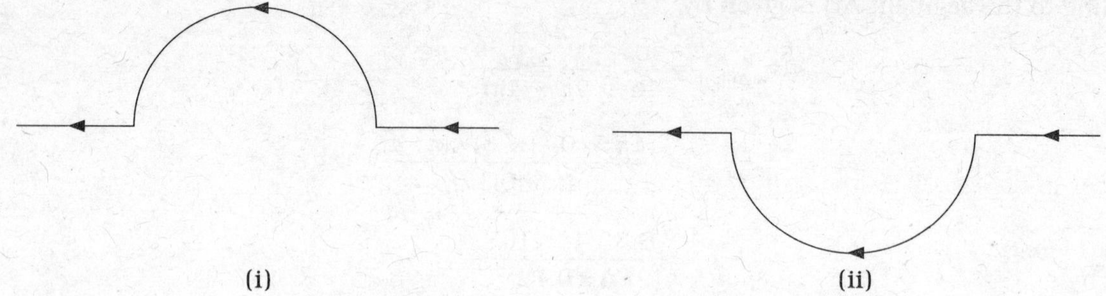

Solution. We consider the three cases one by one. (a) In the general expression,

$$\vec{dB} = \frac{\mu_0}{4\pi} \cdot \frac{i(\vec{d\ell} \times \vec{r})}{r^3},$$

for the straight parts of the wire, $\vec{d\ell}$ and $\vec{r}$ lie in the same direction. ∴ $(\vec{d\ell} \times \vec{r}) = 0$. Hence $\vec{B} = 0$. Thus, the straight parts do not contribute to $\vec{B}$. **(Answer).**

(b) The magnetic field at the centre of a circular loop is given by,

$$(B)_{loop} = \frac{\mu_0 i}{2r}$$

∴ For the semicircular part, we have,

$$(B)_{semicircle} = \frac{\mu_0 i}{4r}$$

$$= \frac{(4\pi \times 10^{-7}) \times (15)}{4 \times (3 \times 10^{-2})} \quad (3 \text{ cm} = 3 \times 10^{-2} \text{ m})$$

$$= 1.57 \times 10^{-4} \text{ T} \quad \textbf{(Answer).}$$

If the wire lies in the plane of paper, then according to the right hand thumb rule $\vec{B}$ is directed normally into the plane of paper. **(Answer).**

(c) $\vec{B}$ has same magnitude as in (b), viz. 1.57×10^{-4} T, but now it is directed normally outwards from the plane of paper. **(Answer).**

Problem 6. A current of 2.8 A flows through a circular coil of radius 12 cm and having 120 turns. Calculate the magnetic moment of the coil and the magnetic field at the centre of the coil. If the coil is suspended vertically in an external horizontal magnetic field and rotates through 90° under the influence of this field, find the torque experienced by the coil.

Solution. Given : $N = 120$; $i = 2.8$ A; $r = 12$ cm $= 0.12$ m; $\theta = 90°$

(i) The magnetic field at the centre of the coil is given by,

$$B = \frac{\mu_0 N i}{2r} = \frac{(4\pi \times 10^{-7}) \times (120) \times (2.8)}{2 \times (0.12)}$$

$$= 1.76 \times 10^{-3} \text{ T (or 1.76 mT)} \quad \textbf{(Answer).}$$

(ii) The magnetic moment is given by,

$$M = NiA = Ni(\pi r^2)$$

$$= (120) \times (2.8) \times (3.14) \times (0.12)^2$$

$$= 15.2 \text{ A.m}^2 \quad \textbf{(Answer).}$$

(iii) The torque on the coil is given by,

$$\tau = MB \sin\theta = (15.2) \times (1.76 \times 10^{-3}) \times \sin 90°$$

$$= 26.75 \times 10^{-3} \text{ N.m} \quad \textbf{(Answer).}$$

Problem 7. A current of 20 A flows through a solenoid of length 1.5 m and diameter 3 cm. The magnetic field within the solenoid is 25 mT. Calculate the length of wire used up in making the solenoid.

Solution. Given : ℓ = 1.5 m; i = 20 A; d(diameter) = 3 cm = 3×10^{-2} m; B = 25 mT = 25×10^{-3} T

The magnetic field inside a solenoid is given by,

$$B = \mu_0 ni$$

$$\therefore \quad n = \frac{B}{\mu_0 i} = \frac{(25\times10^{-3})}{(4\pi\times10^{-7})\times(20)}$$

$$\approx 995$$

n is the number of turns per unit length. ∴ the total number of turns is,

$$N = n\times\ell = 995\times1.5 \approx 1493$$

∴ The required length of wire is,

$$\text{Length} = (2\pi r)\times N = \pi d\times N \quad (\because d = 2r)$$
$$= 3.14\times(3\times10^{-2})\times(1493)$$
$$= 140.64 \text{ m} \quad \textbf{(Answer).}$$

Problem 8. A current of 0.4 A flows through a solenoid of length 0.5 m and diameter 12 cm. The solenoid has 500 turns. Find the magnetic field inside it.

Solution. Given : N = 500; ℓ = 0.5 m; d(diameter) = 12 cm = 0.12 m; i = 0.4 A

The no. of turns per unit length, $n = \frac{N}{\ell} = \frac{500}{0.5} = 1000$

The required magnetic field within the solenoid is given by,

$$B = \mu_0 n i$$
$$= (4\pi\times10^{-7})\times(1000)\times(0.4)$$
$$= 0.5\times10^{-3} \text{ T (or 0.5 mT)} \quad \textbf{(Answer).}$$

Problem 9. A positron moves with a velocity of 9×10^6 ms^{-1} in a magnetic field of strength 78 mT. Find (i) the maximum and the minimum force experienced by the positron in the field and (ii) the angle between the velocity vector and the field at the instant when the acceleration of the positron has magnitude 4.8×10^{14} ms^{-2}.

Solution. Given : v = 9×10^6 ms^{-1}; B = 78 mT = 78×10^{-3} T; a = 4.8×10^{14} ms^{-2}.

(i) The force on a charged particle moving in a magnetic field is given by,

$$F = qvB \sin\theta = evB \sin\theta \text{ (for a positron)}$$
$$\therefore \quad F_{max} = evB \sin 90° = evB$$
$$= (1.6\times10^{-19})\times(9\times10^6)\times(78\times10^{-3})$$
$$= 1.12\times10^{-13} \text{ N} \quad \textbf{(Answer).}$$
$$F_{min} = evB \sin 0 = 0 \quad \textbf{(Answer).}$$

(ii) Force on the positron is,

$$F = m_e a = (9.1\times10^{-31})\times(4.8\times10^{14})$$
$$= 43.68\times10^{-17} \text{ N}$$

$$\therefore \quad \sin\theta = \frac{F}{evB} = \frac{(43.68\times10^{-17})}{(1.12\times10^{-13})} = 0.0039$$

$$\Rightarrow \quad \theta = \sin^{-1}(0.0039)$$
$$\approx 13 \text{ minutes (or } 0.22°) \quad \textbf{(Answer).}$$

Problem 10. A current of 32 A flows through a wire placed horizontally in a magnetic field. If the linear density (mass per unit length) of the wire is 0.05 kg per meter, find the magnitude and direction of the minimum value of $\vec{B}$ required to suspend the wire in air against the gravitational pull of earth.

Solution. Given : i = 32 A, linear density, m/L = 0.05 kgm^{-1}

To suspend the wire in space, the force $\vec{F}_B$ on it due to the magnetic field $\vec{B}$ has to act upwards, to be able to balance the downwards gravitational force $\vec{F}_g$ (=mg) due to earth, as seen in the figure. If the direction of flow of current in the wire is normal to and projecting outwards from the plane of paper, then according to the Fleming's left hand rule, the magnetic field should be directed to right (see figure). Thus,

$$F_B = F_g$$

$$\Rightarrow \quad i\,L\,B \sin\theta = mg$$

If B has to be minimum, as required, then sin θ = 1 or θ = 90° (thus $\vec{B}$ has to be perpendicular to the length of wire). So we have,

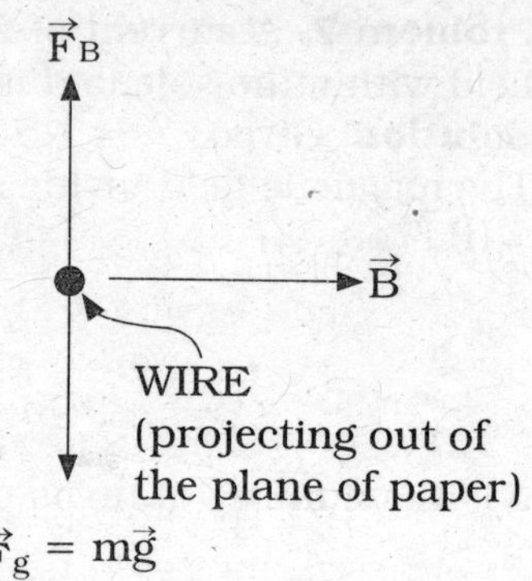

$$i\,L\,B = mg$$

$$\Rightarrow \quad B = \frac{mg}{iL} = \left(\frac{m}{L}\right) \times \frac{g}{i}$$

$$= (0.05) \times \frac{(9.8)}{(32)}$$

$$= 0.0153 \text{ T}$$

$$= 15.3 \text{ mT}$$ **(Answer).**

(Note that B is ~ 150 times the strength of earth's magnetic field). Further, as mentioned above, $\vec{B}$ has to be perpendicular to the wire. **(Answer).**

Problem 11. An α - particle traverses a circular path of radius 5 cm in a uniform magnetic field of strength 1.4T. Calculate its (i) velocity, (ii) period of revolution, and (iii) kinetic energy. To attain the calculated energy, what potential difference should it be accelerated through ? (mass of α - particle = 4 a.m.u.).

Solution. Given : B = 1.4T; r = 5 cm = 5×10^{-2} m; m = 4 a.m.u. = $4\times(1.66\times10^{-27}$ kg) = 6.64×10^{-27} kg

Also, we know that the charge on an α - particle is 2e.

$$\therefore q = 2e = 2\times(1.6\times10^{-19}\text{C}) = 3.2\times10^{-19}\text{C}$$

(i) The radius of the circular path is given by,

$$r = \frac{mv}{qB}$$

$$\therefore \quad v = \frac{rqB}{m} = \frac{(5\times10^{-2})\times(3.2\times10^{-19})\times(1.4)}{(6.64\times10^{-27})}$$

$$= 3.37\times10^{6} \text{ ms}^{-1}$$ **(Answer).**

(ii) The time period of revolution is given by,

$$T = \frac{2\pi m}{qB} = \frac{2\times3.14\times(6.64\times10^{-27})}{(3.2\times10^{-19})\times(1.4)}$$

$$= 9.31\times10^{-8} \text{ s}$$ **(Answer).**

(iii)
$$\text{k.e.} = \frac{1}{2}mv^2 = \frac{(6.64\times10^{-27})\times(3.37\times10^{6})^2}{2}$$

$$= 37.7\times10^{-15} \text{ J} = \frac{37.7\times10^{-15}}{(1.6\times10^{-19})} \text{ eV}$$

$$= 23.57\times10^{4} \text{ eV}$$ **(Answer).**

To attain an energy of 23.57×10^4 eV, the required potential difference for the electron is simply 23.57×10^4 volt. **(Answer).**

Problem 12. A deuteron, an α - particle, and a proton are required to move in a common circular path in a uniform magnetic field. The kinetic energy of the proton is found to be 2.4 meV. What are the corresponding energies of the deuteron and the α - particle, respectively ? (The deuteron and the α - particle have charges as e and 2e, respectively, and their masses are 2 a.m.u and 4 a.m.u., respectively).

Solution. Given : k.e. of proton, E_1 = 2.4 MeV;

(charge of a proton = q_1 = e; its mass = m_1 = 1 a.m.u);
charge and mass of deuteron : q_2 = e, m_2 = 2 a.m.u;
charge and mass of α - particle : q_3 = 2e, m_3 = 4 a.m.u.

Let the k.e.'s of deuteron and α - particle be E_2 and E_3, respectively. The radius of circular path of a charged particle in a magnetic field is given by,

$$r = \frac{mv}{qB} = \frac{\sqrt{2mE}}{qB} \quad (\because E = \frac{1}{2}mv^2)$$

Since the three particles have to traverse a common circular path, the r-value is the same for all of them.

$$\therefore \quad \frac{\sqrt{2m_1E_1}}{q_1B} = \frac{\sqrt{2m_2E_2}}{q_2B} = \frac{\sqrt{2m_3E_3}}{q_3B}$$

(i) From the first equality we get,

$$\frac{E_2}{E_1} = \left(\frac{q_2}{q_1}\right)^2 \times \left(\frac{m_1}{m_2}\right)$$

∴ The k.e. of the deuteron is,

$$E_2 = E_1 \times \left(\frac{q_2}{q_1}\right)^2 \times \left(\frac{m_1}{m_2}\right) = 2.4 \times \left(\frac{e}{e}\right)^2 \times \left(\frac{u}{2u}\right)$$

$$= 1.2 \text{ MeV} \quad \textbf{(Answer).}$$

(ii) Similarly, the k.e. of the α - particle is,

$$E_3 = E_1 \times \left(\frac{q_3}{q_1}\right)^2 \times \left(\frac{m_1}{m_3}\right) = 2.4 \times \left(\frac{2e}{e}\right)^2 \times \left(\frac{u}{4u}\right)$$

$$= 2.4 \text{ MeV} \quad \textbf{(Answer).}$$

Problem 13. The kinetic energies of a proton (q_1 = e, m_1 = 1 a.m.u.), a deuteron (q_2 = e, m_2 = 2 a.m.u.), and an α - particle (q_3 = 2e ; m_3 = 4 a.m.u.) have the same magnitude. Compare the radii of the circular paths traversed by them after being projected into a uniform magnetic field $\vec{B}$ in a direction normal to $\vec{B}$.

Solution. Given : q_1 = e, q_2 = e, q_3 = 2e;
m_1 = 1u, m_2 = 2 u, m_3 = 4 u

Let the radii of the paths traversed by the proton, the deuteron and the α - particle be r_1, r_2 and r_3, respectively. The radius of the circular path of a charged particle is given by,

$$r = \frac{mv}{qB} = \frac{\sqrt{2mE}}{qB} \quad (\because E = \frac{1}{2}mv^2)$$

The three particles have the same kinetic energy, E. ∴ The ratio of their radii is given by,

$$r_1 : r_2 : r_3 = \frac{\sqrt{2m_1E}}{q_1B} : \frac{\sqrt{2m_2E}}{q_2B} : \frac{\sqrt{2m_3E}}{q_3B}$$

$$= \frac{\sqrt{2 \times u \times E}}{e \times B} : \frac{\sqrt{2 \times 2u \times E}}{e \times B} : \frac{\sqrt{2 \times 4u \times E}}{2e \times B}$$

$$= \sqrt{2} : 2 : \frac{2\sqrt{2}}{2} \quad \left(\frac{\sqrt{uE}}{eB} \text{ is common}\right)$$

$$= 1 : \sqrt{2} : 1 \quad \textbf{(Answer).}$$

Problem 14. An electron with kinetic energy of 3 keV traverses a helical path after entering a uniform magnetic field $\vec{B}$, with its initial direction of motion inclined at an angle of 78° to $\vec{B}$. If the field strength B be 0.15 T, determine the (i) radius of the path, (ii) time period, and (iii) pitch of the path.

Solution. Given : B = 0.15 T; θ = 78°; k.e. (E) = 3 keV = $3 \times (1.6 \times 10^{-16}$J) = 4.8×10^{-16}J
($q = e = 1.6 \times 10^{-19}$ C, $m = m_e = 9.1 \times 10^{-31}$ kg).

(i) The radius of the helical path is given by,

$$r = \frac{mv \sin\theta}{qB} = \frac{\sqrt{2mE} \sin\theta}{qB} \quad (\because E = \frac{1}{2}mv^2)$$

$$= \frac{\sqrt{2 \times (9.1 \times 10^{-31}) \times (4.8 \times 10^{-16})} \times (0.9781)}{(1.6 \times 10^{-19}) \times (0.15)} \quad (\sin 78° = 0.9781)$$

$$\approx 1.2 \times 10^{-3} \text{ m (or 1.2 mm)} \quad \textbf{(Answer).}$$

(ii) The time period is given by,

$$T = \frac{2\pi m}{qB} = \frac{2\times 3.14\times(9.1\times10^{-31})}{(1.6\times10^{-19})\times(0.15)}$$

$= 0.24\times10^{-9}$ s (or 0.24 ns) **(Answer)**

(iii) The pitch of the helical path is given by,

$$p = \frac{2\pi(mv)\cos\theta}{qB} = \frac{2\pi\left(\sqrt{2mE}\right)\cos\theta}{qB}$$

$$= \frac{2\times3.14\times\sqrt{2\times(9.1\times10^{-31})\times(4.8\times10^{-16})\times(0.2079)}}{(1.6\times10^{-19})\times0.15} \quad (\cos 78^\circ = 0.2079)$$

$= 1.58\times10^{-3}$ m (or 1.58 mm) **(Answer)**

Problem 15. A proton with kinetic energy 6.0 MeV, moving horizontally from south to north, enters a uniform magnetic field $\vec{B}$ of strength 1.5 mT, directed vertically upwards. Determine the magnitude and direction of the deflecting force experienced by the proton in the field.

Solution. Given : B = 1.5 mT = 1.5×10^{-3} T; k.e. (E) = 6.0 MeV = $6\times(1.6\times10^{-13})$J = 9.6×10^{-13}J ($m_p = 1.67\times10^{-27}$ kg)

The velocity of the proton is given by,

$$v = \sqrt{\frac{2E}{m_p}} \quad (\because E = \frac{1}{2}mv^2)$$

$$= \sqrt{\frac{2\times(9.6\times10^{-13})}{(1.67\times10^{-27})}} = \sqrt{11.5}\times10^7$$

$= 3.39\times10^7$ ms^{-1}

The magnitude of the deflecting Lorentz force on the proton is given by,

$F = qvB \sin\theta = qvB$ ($\because \theta = 90^\circ$)

$= (1.6\times10^{-19})\times(3.39\times10^7)\times(1.5\times10^{-3})$

$= 8.13\times10^{-15}$ N **(Answer).**

The direction of $\vec{F}$ is found by applying the Fleming's left hand rule, as directed west-east in the horizontal plane. **(Answer).**

Problem 16. The radius of the dees of a cyclotron is 60 cm each. If the oscillator frequency of the cyclotron is 15 MHz, determine the magnitude of the magnetic field to be applied for accelerating deuterons in the cyclotron and the resulting kinetic energy of the deuterons (mass of a deuteron = 3.34×10^{-27}kg).

Solution. Given : f_{osc} = 15 MHz = 15×10^6 Hz; R = 60 cm = 0.6 m; m = 3.34×10^{-27} kg (q = e = 1.6×10^{-19} C)

The magnetic field is given by,

$$B = \frac{2\pi m f_{osc}}{q} = \frac{2\times3.14\times(3.34\times10^{-27})\times(15\times10^6)}{(1.6\times10^{-19})}$$

$= 1.97$ T **(Answer).**

For determining the k.e., we first find v as,

$$v = \frac{qBR}{m} = \frac{(1.6\times10^{-19})\times(1.97)\times(0.6)}{(3.34\times10^{-27})}$$

$= 5.66\times10^7$ ms^{-1}

$$\therefore \quad \text{k.e.} = \frac{1}{2}mv^2 = \frac{(3.34\times10^{-27})\times(5.66\times10^7)^2}{2}$$

$$= 5.35\times10^{-12} \text{ J} = \frac{5.35\times10^{-12}}{(1.6\times10^{-13})} \text{ MeV}$$

$= 33.4$ MeV **(Answer).**

Problem 17. The magnetic field employed in a cyclotron is 1.1 T. If a proton moves in the cyclotron in a circle of radius 0.6 m, find the oscillator frequency of the cyclotron and the kinetic energy of the proton.
Solution. Given : B = 1.1T; R = 60 cm = 0.6 m ($q = e = 1.6\times10^{-19}$ C; $m = m_p = 1.67\times10^{-27}$ kg)
The oscillator frequency is given by,

$$f_{osc} = \frac{qB}{2\pi m} = \frac{(1.6\times10^{-19})\times(1.1)}{2\times3.14\times(1.67\times10^{-27})}$$

$$= 1.678\times10^{7}\ \text{Hz}$$

$$= 16.78\ \text{MHz} \quad \textbf{(Answer)}.$$

The maximum kinetic energy of the proton is given by,

$$(k.e.)_{max} = \frac{q^2B^2R^2}{2m} = \frac{(1.6\times10^{-19})^2\times(1.1)^2\times(0.6)^2}{2\times(1.67\times10^{-27})}$$

$$= 33.39\ \ 10^{-13}\ \text{J} = \frac{33.39\times10^{-13}}{(1.6\times10^{-13})}\ \text{MeV}$$

$$= 20.9\ \text{MeV} \quad \textbf{(Answer)}.$$

Problem 18. Deuterons are accelerated in a cyclotron, having accelerating potential between its dees as 75 kV. If they are finally ejected with a kinetic energy of 33.4 MeV and the operating magnetic field has magnitude of 1.97 T, calculate the total path length traversed by a deuteron in the cyclotron (charge and mass of a deuteron are 1.6×10^{-19} C and 3.34×10^{-27} kg, respectively).
Solution. Given : B = 1.97 T; accelerating potential = 75 kV = 75×10^{3}V; $q = 1.6\times10^{-19}$C;
$m = 3.34\times10^{-27}$ kg; final k.e. = 33.4 MeV

The total distance (D) traversed by a deuteron is equal to the total number of revolutions (n) multiplied by the circumference of the orbit ($2\pi r_{av}$), corresponding to the average energy of a deuteron,

$$\text{Total distance} = D = n\times2\pi r_{av}$$

The deuteron accelerates twice in each cycle, receiving an energy of 75×10^{3} eV each time. Since its final energy is 33.4 MeV (= 33.4×10^{6} eV), the total number of revolutions made by it is,

$$n = \frac{(33.4\times10^{6})\text{eV}}{2(75\times10^{3}\text{eV})} \approx 223$$

Its average energy (E_{av}) during the accelerating process is $(33.4\div2)=16.7$ MeV=(16.7×10^{6}) eV = $(16.7\times10^{6})\times(1.6\times10^{-19})$J. The corresponding average radius of the orbit is given by,

$$r_{av} = \frac{m\,v_{av}}{qB}$$

$$= \frac{\sqrt{2E_{av}m}}{qB} \quad (\because E = \frac{1}{2}mv^2)$$

$$= \frac{\sqrt{2\times(16.7\times10^{6})\times(1.6\times10^{-19})\times3.34\times10^{-27}}}{(1.6\times10^{-19})\times(1.97)}$$

$$= 0.424\ \text{m}$$

∴ The total distance traversed by the deuteron is,

$$D = n\times2\pi r_{av}$$

$$= (223)\times(2\times3.14\times0.424)$$

$$= 593.8\ \text{m} \quad \textbf{(Answer)}.$$

Problem 19. A circular coil having 40 turns and a radius of 10 cm is suspended in a uniform magnetic field $\vec{B}$ of strength 1.2 T. A current of 8 A flows through the coil. The magnetic field is horizontal, with the vector $\vec{B}$ making an angle of 30° with the normal to the plane of the coil. If it is intended to prevent the coil from turning in the field, determine the counter-torque needed to apply to the coil. Will its value change if the coil is replaced by a plane but irregularly-shaped coil ?
Solution. Given : N = 40; i = 8 A ; B = 1.2 T; r = 10 cm = 0.10 m; θ = 30°

Area of the coil, $A = \pi r^2 = 3.14\times(0.1)^2 = 314\times10^{-4}\ m^2$

The torque produced in the coil is given by,

$$\tau = NiAB \sin\theta = NiAB \sin 30^\circ$$
$$= (40)\times(8)\times(314\times10^{-4})\times(1.2)\times0.5$$
$$= 6.03 \text{ N.m}$$

∴ To prevent the coil from turning in the magnetic field, it will be necessary to apply a counter-torque of 6.03 N.m in the opposite direction. **(Answer).**

The equation, $\tau = NiAB \sin\theta$, is independent of the shape of the coil. It only involves the area, A, of the coil. ∴ There will be no change in the value of counter-torque upon replacing the coil with another planar irregularly - shaped coil. **(Answer).**

Problem 20. Two moving coil galvanometers G_1 and G_2 are identical in all respects, except for the values of resistance (12 Ω and 16 Ω, respectively), number of turns of coil (40 and 60, resp.), area of the coil ($4.0\times10^{-3}\ m^2$ and $2.0\times10^{-3} m^2$, resp.), and magnitude of magnetic field (0.3 T and 0.6 T, resp.). Calculate the ratio of their current sensitivities and voltage sensitivities.

Solution. Given : $N_1 = 40$; $R_1 = 12\ \Omega$; $A_1 = 4\times10^{-3} m^2$; $B_1 = 0.3$ T

$N_2 = 60$; $R_2 = 16\ \Omega$; $A_1 = 2\times10^{-3} m^2$; $B_2 = 0.6$ T

The current sensitivity of a moving coil galvanometer is given by,

$$i_s = \frac{NAB}{C}$$

Since the galvanometers are identical, the value of C (restoring torque per unit angle of twist; also called spring constant or torsional constant) is same for G_1 and G_2.

$$\therefore \quad \frac{(i_s)_1}{(i_s)_2} = \frac{N_1A_1B_1}{N_2A_2B_2} = \frac{(40)\times(4\times10^{-3})\times(0.3)}{(60)\times(2\times10^{-3})\times(0.6)}$$
$$= \frac{2}{3}$$
$$\therefore \quad (i_s)_1 : (i_s)_2 = 2 : 3 \quad \textbf{(Answer).}$$

The voltage sensitivity is given by,

$$V_s = \frac{NAB}{CR} = \frac{i_s}{R}$$
$$\therefore \quad \frac{(V_s)_1}{(V_s)_2} = \frac{(i_s)_1}{(i_s)_2}\times\frac{R_2}{R_1} = \frac{2}{3}\times\frac{(16)}{(12)} = \frac{8}{9}$$

Thus, $(V_s)_1 : (V_s)_2 = 8 : 9$ **(Answer).**

Problem 21. In the electric circuit shown in the accompanying figure, calculate the values of current when the measuring meter happens to be (i) a galvanometer having resistance as $r_g = 80\ \Omega$, (ii) the same galvanometer with a shunt resistance $r_s = 0.01\ \Omega$ and (iii) an ideal ammeter having zero resistance.

Solution. (i) The effective resistance in the circuit is,

$$R_{eff} = r_g + 5 = 80 + 5 = 85\ \Omega$$

∴ The current in the circuit (as measured by the meter) is,

$$i = \frac{\text{voltage}}{R_{eff}} = \frac{(6)}{(85)} = 0.07 \text{ A } \textbf{(Answer).}$$

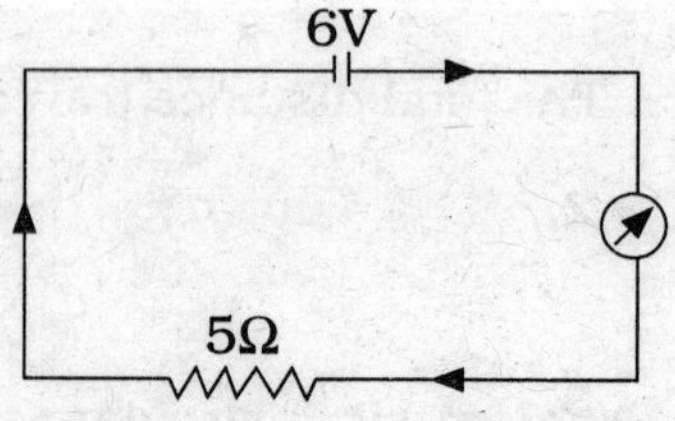

6V

5Ω

(ii) In this case,

$$R_{eff} = \frac{r_g r_s}{r_g + r_s} = \frac{(80)\times(0.01)}{(80)+(0.01)} \approx 0.01\ \Omega$$
$$\therefore \quad i = \frac{6}{(5)\times(0.01)} = \frac{6}{5.01}$$
$$= 1.198 \text{ A} \quad \textbf{(Answer).}$$

(iii) An ideal ammeter has zero resistance. ∴ In this case, the effective resistance is 5 Ω.

$$\therefore \quad i = \frac{6}{5} = 1.2 \text{ A} \quad \textbf{(Answer).}$$

Problem 22. By increasing the resistance of a moving coil galvanometer two times, the current sensitivity of the galvanometer is raised by 50%. Calculate the percentage change in the voltage sensitivity of the galvanometer.
Solution. Suppose the initial and final values of the galvanometer resistance, current sensitivity and voltage sensitivity are (R_1, R_2), (i_{s1}, i_{s2}), and (V_{s1}, V_{s2}) respectively. We are given that,

$$R_2 = 2\,R_1 \qquad \text{and} \qquad i_{s2} = 1.5\,i_{s1}$$

We know that the voltage sensitivity is given by,

$$V_s = \frac{NAB}{CR} = \frac{i_s}{R}$$

$$\therefore \qquad \frac{V_{s2}}{V_{s1}} = \frac{i_{s2}}{i_{s1}} \times \frac{R_1}{R_2} = 1.5 \times \frac{1}{2} = \frac{3}{4}$$

$$= 0.75$$

Thus the voltage sensitivity is found to be reduced by a factor of 0.25, i.e. by $(0.25 \times 100) = 25\%$. **(Answer).**

Problem 23. The rectangular coil of a moving coil galvanometer has 300 turns and measures 3.0 cm in height and 1.5 cm in width. It is suspended between the poles of a permanent magnet of field strength 0.3 T. If a current of 120 μA has to produce an angular deflection of 30° of the coil, calculate the necessary torsional constant of the spring of the galvanometer.
Solution. Given : $N = 300$; $B = 0.3T$; $i = 120\,\mu A = 120 \times 10^{-6} A$; $A = (3.0 \times 1.5) cm^2 = 4.5 \times 10^{-4}\, m^2$; $\phi = 30°$

The torsional constant C is the restoring torque per unit twist of the coil (it is measured as per degree, not as per radian). It is given by,

$$NiAB = C\phi$$

$$\Rightarrow \qquad C = \frac{NiAB}{\phi} = \frac{(300) \times (120 \times 10^{-6}) \times (4.5 \times 10^{-4}) \times (0.3)}{(30)}$$

$$= 16.2 \times 10^{-8} \text{ N.m. degree}^{-1}$$ **(Answer).**

Very Short Answer Questions

[Including HOTS Questions]

(1 mark each)

1. When an electric current flows in a wire, what is the source of magnetic field created around the wire ?

2. Both a magnetic field and an electric field are associated with electric charges. Then how their sources differ from each other ? [Hint: The former depends on charge velocity, but the latter does not].

3. Write Biot-Savart relation for magnetic field due to a current element.

4. State a rule relating the direction of current with the associated magnetic field.

5. Write the equation for the magnetic force acting on a charge q moving with velocity $\vec{v}$ in a magnetic field $\vec{B}$.

6. Which vectors are always perpendicular to each other in the magnetic Lorentz force equation, $\vec{F} = q(\vec{v} \times \vec{B})$?

7. What is the force experienced by a stationary charge in a magnetic field ? Give reason for your answer.

8. Will the south pole of a magnet situated near a stationary proton experience any force ? Give reason for your answer.

9. State Fleming's left hand rule.

10. What kind of path a moving charged particle attains after entering a uniform magnetic field in a direction perpendicular to the field ?

11. What is the nature of path of a charged particle moving in a uniform magnetic field, in a direction along the field ?

12. What is the nature of path of a moving charged particle after entering a uniform magnetic field at an arbitrary angle to the field ?

13. Under what condition no magnetic force acts on a charged particle moving in a uniform magnetic field ?

14. Give the condition under which maximum force will act on a proton moving in a uniform magnetic field.

15. Write the equation for the radius of circular path followed by an electron with speed $\vec{v}$, after entering a uniform magnetic field, at right angles to the field.

16. For an electron and proton with same speed v entering a uniform magnetic field at right angles to the field, which one will have smaller radius of circular path ?

17. Give the direction of magnetic force experienced by an electron moving along x-axis, after entering a uniform magnetic field aligned along y-axis.

18. Does the energy of a charged particle change in moving through a uniform magnetic field ?

19. Does the velocity of a charged particle change in moving through a uniform magnetic field ?

20. If a proton is not deflected in passing through a given region of space, does that necessarily mean an absence of magnetic field in the region ? Give reason for your answer.

21. If an electron is found to deflect in a region of space, how can we decide whether an electric field or a magnetic field is the cause of deflection ?

22. Does a magnetic field exist (i) inside and (ii) outside a long metal pipe carrying an electric current ?

23. The force acting upon an electron in a certain field is found to depend on the velocity of electron, becoming zero when the electron comes to rest. What is the nature of field ?

24. Write the equation for Lorentz force acting on an electron moving in a region of space, where both an electric field and a magnetic field exist.

25. Out of an α-particle and a β-particle, which one will experience greater force when injected with same velocity into a magnetic field, in a direction normal to the field ?

26. Can a cyclotron accelerate neutrons ? Give reasons for your answer.

27. Which one out of an α-particle and a β-particle will have lesser frequency of revolution when injected with same velocity into a magnetic field, in a direction normal to the field ?

28. If the current flowing through a circular coil is doubled and the coil's radius is halved, how will it change the magnetic field at the centre of the coil ?

29. Why a force acts on a current - carrying conductor placed in a magnetic field ?

30. Write an equation for the force experienced by a current element $\vec{d\ell}$ of a current-carrying conductor placed in a magnetic field.

31. In which position of a current-carrying conductor kept in a magnetic field the force acting on the conductor has maximum value ?

32. Name and state a rule which enables to find the direction of magnetic force acting upon a current-carrying conductor placed in a magnetic field.

33. Write an equation for the force existing between two parallel short wires carrying current.

34. Electric currents of equal value flow in the same direction through two very long parallel wires. Give an equation for the magnetic field at mid-point of the wires.

35. Give the direction of force between two parallel wires carrying currents in opposite direction.

36. A force of magnitude F exists between two parallel current-carrying conductors. How will it change if the current in each conductor is doubled ?

37. Two neighboring electron beams moving in the same direction repel each other, while two neighboring parallel wires carrying current in the same direction attract each other. Why this opposite behavior ?

38. Give an expression for the magnitude of magnetic field created at a point well inside a solenoid, through which a current flows.

39. A solenoid contracts a bit when a current starts flowing through it. Why ?

40. What is the magnitude of total (net) force acting on a current-carrying rectangular or circular loop, placed in a magnetic field ?

41. Write an equation for the torque experienced by a current-carrying coil placed in a uniform magnetic field.

42. A current-carrying coil can freely turn in a magnetic field. State the conditions in which the torque experienced by it will be (a) maximum and (b) minimum.

43. If an ammeter and a millimeter (measuring small currents of the order of a few milliamperes) are converted from the same galvanometer, which of the two has higher resistance ?

44. Both an ammeter and a voltmeter are made by converting a galvanometer. What is the essential difference between the construction of the two ?

45. Define the current sensitivity of a galvanometer.

46. Define the voltage sensitivity of a galvanometer.

47. Mention two factors on which the sensitivity of moving-coil galvanometer depends.

48. What is the nature of magnetic field existing in a moving-coil galvanometer ?

49. What is the purpose of the soft iron cylindrical core in a moving-coil galvanometer ?

50. Why is the shape of the pole pieces of the magnet in a moving-coil galvanometer made concave ?

51. Why is the effect of earth's magnetic field on the working of a galvanometer not considered significant ?

52. When a shunt resistance r_p is put across the terminals of a galvanometer of resistance r_g, what is the effective resistance of the resulting ammeter ?

Short Answer Questions

[Including HOTS Questions]

(2 or 3 marks each)

1. State the main observations made in the Oersted's experiment.

2. Deduce an equation for the force experienced by a moving charge in a magnetic field.

3. What is the net charge on a conductor when a current flows through it ? Explain how a force acts on it when it is placed in a magnetic field.

4. Although free electrons in a conductor are always in motion, yet they do not experience any magnetic force unless a current flows through the conductor. Explain why.

5. When an irregularly shaped flexible wire loop carrying current is placed in a magnetic field, it acquires a circular shape. Explain how it happens.

6. You are provided with a high resistance r_1,a low resistance r_2, and a moving-coil galvanometer. Explain how would you use them to make an instrument for measuring current in a circuit and another for measuring voltage across a component in the circuit.

7. The velocity of a charged particle circulating at right angles to a uniform magnetic field remains unchanged and its frequency of revolution is independent of its velocity. Explain how.

8. Equal currents flow through two very long parallel wires. Explain with due reason if a magnetic field will exist at a position midway between the wires when the current flows in (a) the same direction and (b) opposite directions in the two wires.

9. Briefly explain the functions of the electric field and the magnetic field in a cyclotron.

10. A current flows through a straight wire of infinite length. Derive an expression for the magnetic field produced at a point near the wire.

11. State Biot-Savart law. Use it to derive an expression for the magnetic field produced at the centre of a current-carrying circular loop.

12. State Ampere's circuit law. Use it to derive an expression for the magnetic field existing inside a toroidal solenoid.

13. Currents of different values flow through two infinitely long straight wires placed parallel to each other. Derive an expression for the force per unit length between them and use it to define one ampere of current.

14. A current flows through a rectangular coil suspended in a uniform magnetic field. Derive an expression for the torque experienced by the coil due to the field.

Long Answer Questions

(5 marks each)

1. State Biot-Savart law. Employ it to derive an expression for the magnetic field produced by a current-carrying circular loop at a position along the axis of the loop.

2. Use Biot-Savart law for deriving an expression for the magnetic field produced inside a straight solenoid. What is the relative value of the field near the ends of solenoid ? Sketch the magnetic lines of force for the solenoid.

3. Describe with a labelled diagram the principle, construction, and working of a moving-coil galvanometer. Show that the deflection of the coil is directly proportional to the current flowing through the coil. Explain why a radial magnetic field is needed in the galvanometer and how is it achieved.

4. Describe with a labelled diagram the principle, construction, and working of a cyclotron.

Numerical Exercises

1. A current of 60A flows in a horizontally placed long straight wire in north-south direction. Find the magnitude and direction of the resulting magnetic field $\vec{B}$ at a position located 2 m east of the wire.

[6×10^{-6}T; vertically upwards]

2. A current of 12A flows through a long straight wire placed in a uniform magnetic field $\vec{B}$ of strength 0.2 T. Determine the magnitude of magnetic force experienced by it, if it is inclined at 30° to $\vec{B}$.

[1.2 Nm^{-1}]

3. A 2.5 cm long piece of wire is inserted inside a solenoid in a direction normal to the axis of solenoid. If the magnetic field within the solenoid is 0.3 T and a current of 12A flows through the wire, find the magnitude of the magnetic force experienced by the wire.

[9×10^{-2} N]

4. Electric currents of magnitude 6A and 4A flow in the same direction through two long and straight parallel wires W_1 and W_2, respectively. If the wires are spaced 3 cm apart, find the force experienced by a 8 cm long segment of the wire W_1.

[1.28×10^{-5} N; attractive; $\perp$ to W_1, directed towards W_2]

5. Two long and straight parallel wires W_1 and W_2 are separated by a distance of 4 cm and have currents of magnitudes 6 A and 8 A, respectively, flowing through them. Determine the force on W_1 due to W_2.

[2.4×10 Nm^{-1}]

6. A long wire of the shape shown in the figure carries a current of 10A. The parts AB and DE are straight and the part BCD is a semicircle of radius 12 cm. Find the magnetic field $\vec{B}$ at the centre O of the semicircle. **[428 mT; directed vertically upwards]**

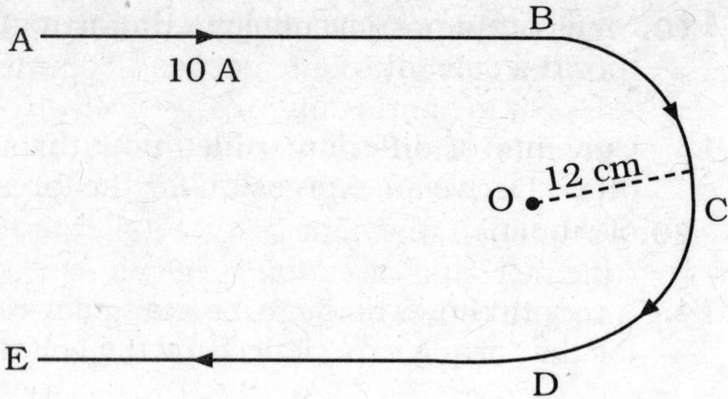

7. An electric current of 2 A flows through a circular coil of radius 12 cm and having 120 turns. Find the magnitude of magnetic field at the centre of the coil. **[12.56×10^{-4} T]**

8. An external magnetic field exists at the centre of a circular coil with 800 turns and of radius 60 cm. Determine the magnitude of current to be flown in the coil which could just counterbalance the field. **[5.9×10^{-8} A]**

9. A long wire W_1 is bent to form a plane coil of single turn. Another identical wire W_2 of same length is bent to form a plane coil with three turns. Find the ratio of the magnetic fields produced at the centres of the two wires. **[$B_{w1} : B_{w2} = 1: 9$]**

10. A solenoid consisting of 250 turns and having a length of 30 cm and a radius of 6 cm carries a current of 0.4 A. Find the magnitude of the magnetic field produced within the solenoid. (Hint: radius does not matter). **[0.42mT]**

11. A current of 6 A flows through a solenoid having a length of 1.32 m and outer diameter of 3.8 cm. It comprises three close-packed layers of 880 turns each. Find the magnetic field at its centre. (Hint: radius does not matter). **[15.1 mT]**

12. A current of 3 A flows through a long wire of length 2 m and of mass 240 g. A uniform horizontal magnetic field $\vec{B}$ helps to suspend the wire horizontally in air. Determine the magnitude of $\vec{B}$. **[0.39 T]**

13 An electron with a kinetic energy of 25 eV enters a uniform magnetic field $\vec{B}$ of strength 5×10^{-4} T, with its velocity vector making an angle of 60° with $\vec{B}$. Find the pitch of the helical path traversed by the electron. **[10.6 cm]**

14. An α-particle enters a uniform magnetic field $\vec{B}$ of strength 0.05 T with a velocity $\vec{v}$ = 600 ms^{-1}. The angle between $\vec{B}$ and $\vec{v}$ is 60°. Calculate the magnetic force on the particle and the resulting acceleration of the particle (charge and mass of the α - particle are 3.2×10^{-19} C and 6.6×10^{-27} kg, respectively). Does the speed of the particle change upon entering the field ? **[8.32×10^{-18} N; 12.61×10^{8} ms^{-2}; No]**

15. A proton enters a uniform magnetic field $\vec{B}$ of magnitude 3 mT, with its velocity making an angle of 45° with $\vec{B}$. If the force acting on the proton due to $\vec{B}$ is 7×10^{-17} N, find the speed and kinetic energy of the proton. **[2.1×10^{5} ms^{-1}; 222 MeV]**

16. A beam of electrons is travelling with a speed of 1.5×10^{6} ms^{-1}. A magnetic field $\vec{B}$ applied in a direction normal to the beam makes the electrons to move along an arc of radius 0.4 m. Calculate the magnitude of $\vec{B}$. **[21.3 μT]**

17. An initially stationary electron is accelerated by moving through a potential difference of 320 V. Then it enters a region of a uniform magnetic field $\vec{B}$ of strength 180 mT, with its velocity vector normal to $\vec{B}$. Determine the speed of the electron and the radius of the circular path followed by the electron in the field. $\left(\text{Hint} : v = \sqrt{2eV/m}\right)$. **[$10.6\times10^{6}$ ms^{-1}; 3.35×10^{-4} m]**

18. An electron traverses a circular path of radius 24 cm in a plane normal to the direction of a magnetic field $\vec{B}$. If it possesses a kinetic energy of 1.10 keV, find its speed, frequency, and time period and the magnitude of $\vec{B}$. **[1.97×10^{7} ms^{-1}; 7.65 MHz; 131 ns; 467 μT]**

19. A charged particle enters a uniform magnetic field $\vec{B}$ of strength 20 mT with velocity $\vec{v}$ equal to 8 ms^{-1} and circles in a plane perpendicular to $\vec{B}$. $\vec{v}$ is perpendicular to $\vec{B}$. The charge and mass of the particle are 80 μC and 8 mg, respectively. How much time will pass for the particle to return to its original position after completing the first circle ? **[31.4 s]**

20. A uniform magnetic field of magnitude 0.8 T is set up in south to north direction. An α-particle enters the field in a direction vertically upwards with a velocity of 2.5×10^4 km s^{-1}. Determine the magnitude and direction of the force acting on the particle due to the field (charge of α-particle = 3.2×10^{-19} C). **[6.4×10^{-12} N; directed towards west]**

21. A current of 7.5 A flows in a wire of radius 6 cm. The wire lies in a plane normal to the direction of a uniform magnetic field $\vec{B}$ of strength 0.7 T. Calculate the magnitude of the force acting on the wire due to the field. **[0.63 N]**

22. A current of 2.4 A flows through a horizontally held straight wire of length 1.2 m and mass 12 mg. Determine the minimum value of a magnetic field to be set up around the wire, which produces enough force on the wire to counterbalance its weight. **[4.1×10^{-5}T]**

23. An electron circles along a path of radius 8 cm in a uniform magnetic field $\vec{B}$. Its kinetic energy is 80 eV. Calculate the magnitude of $\vec{B}$ and the number of revolutions made by the electron in one second. **[3.77×10^{-4} T; 10.54×10^6]**

24. A charged particle attains a speed of 1.4×10^6 ms^{-1} after being accelerated through a potential difference of 14000 volt. Then it is projected into a uniform magnetic field of magnitude 0.3 T in a direction normal to the field. Calculate the radius of the circular path traced by the electron in the field. **[6.67 cm]**

25. An electron describes a circular path of radius 0.8 m in a plane perpendicular to a uniform magnetic field of magnitude 0.4 T. Calculate its speed. Also determine the speed of a proton circulating under identical conditions in the field. **[5.63×10^{10} ms^{-1}; 3.06×10^7 ms^{-1}]**

26. A uniform magnetic field pointing east to west is located in a region of space. Its strength is 0.35 T. Determine the speed with which a particle of charge 1.5×10^{-5} C and with mass 0.02 g should be injected into the field, which enables the particle to circulate in the field with a uniform velocity. **[37.33 ms^{-1}]**

27. It is intended to accelerate protons in a cyclotron with oscillator frequency of 12 MHz and the radius of its dees as 0.5 m each. Calculate the magnitude of the necessary magnetic field to be applied and the resulting kinetic energy of the protons. **[0.79 T; 7.46 MeV]**

28. Protons are to be accelerated in a cyclotron, having radius of the dees as 60 cm each and the oscillator frequency as 15 MHz. Find (a) the magnitude of the necessary magnetic field and (b) the resulting kinetic energy of the protons. If the magnetic field is doubled, what will be the (c) required oscillator frequency of the cyclotron and (d) the new resulting k.e. of the protons ? **[0.98 T; 16.5 MeV; 30 MHz; 66 MeV]**

29. A square-shape coil comprising 16 turns is vertically suspended in a horizontal uniform magnetic field $\vec{B}$ of strength 0.75 T. The length of the sides of the coil is 8 cm each and the current flowing through it is 10 ampere. If the normal to the plane of the coil is inclined at 45° to $\vec{B}$, find the torque acting on the coil in the field. **[0.54 N.m]**

30. A rectangular coil measuring 7 cm×5 cm is located in a uniform magnetic field, with its plane parallel to the field. It consists of 130 turns and a current of 3 A flows through it. The torque experienced by the coil in the field is found to be 0.3 N.m. Determine the strength of the field $\vec{B}$. **[0.22 T]**

31. A galvanometer has to be converted into a voltmeter of range 0–20 V. It shows a full-scale deflection for a current of 4 A and the resistance of its coil is 14 Ω, Find the value of the necessary resistance to be connected with the galvanometer. Should it be joined is series or in parallel ?

[4986 Ω; in series]

32. A galvanometer has to be converted into an ammeter of range 0–10 A. It shows a full scale deflection for a current of 5 mA and the resistance of its coil is 20 Ω. Determine the value of the necessary resistance to be connected with the galvanometer. Should it be joined in series or in parallel ?

[10 mΩ; in parallel]

33. The resistance of moving-coil galvanometer goes up 1.4 times, when its current sensitivity is raised by 15%. Determine the percentage change in the voltage sensitivity of the galvanometer. **[82 %]**

34. In a galvanometer there is a deflection of 10 divisions per mA. The internal resistance of the galvanometer is 78 Ω. If a shunt of 2 Ω is connected to the galvanometer and there are 75 divisions in all on the scale of the galvanometer, calculate the maximum current which the galvanometer can read. (AI; 2001).

[0.3 A]

CHAPTER 3 (B)
MODERN MAGNETISM

3B. 01. Introduction

About two and a half thousand years back, certain stones were found that could 'magically' attract small bits of some metals like iron, nickel and cobalt towards them. They were 'naturally occurring' magnets discovered by man. They were found in the district of magnesia in Greece, from which the word 'magnet' was derived afterwards (however, the Chinese are believed to have used them more than 4000 year back for magic tricks and determination of direction; the ore 'magnetite' (Fe_3O_4) is now known to be widely distributed over earth, including India). Later, it was found that such a stone when freely suspended with a thread always aligned itself in a definite orientation, which led to the concept of some huge magnet existing inside the earth. Much later, it also led to the use of such stones in navigation by ships, i.e. for locating direction, and thus 'leading' the course of the ships. The ancient word 'lode' (or load) is related to the English word 'lead', so they were called 'lodestones' (or loadstones).

The loadstones consist of oxides of iron, as we now know. The process of acquisition of magnetism by them will be described later in Section 3B. 14. When a piece of iron is repeatedly rubbed along its length with such a natural magnet, it also acquires magnetism. That is how an 'artificial magnet' was first produced, although now we have more superior processes available, for producing such magnets. The artificial magnets can be made in any shape and of great magnetic strength and so they are superior to the 'natural' magnets. A **'compass needle'**, which is a light piece of steel with pointed ends and horizontally pivoted (enclosed in a small case with glass top), is a familiar simple example of an artificial magnet, frequently used in class laboratories. The experiments conducted with artificial magnets led to our elementary knowledge of several magnetic properties, e.g. mutual attraction and repulsion. However, in the last few decades our knowledge of magnetism and magnetic materials has extensively enhanced, deep down to atomic and sub-atomic levels. This in turn has led to the emergence of a very wide variety of magnetic materials and magnetic devices. Actually, now they form an integral part of our daily life, being employed in audio and video cassettes, telephones, instruments like galvanometers (vide Section 3A. 16) and huge instruments like a cyclotron (vide Section 3A. 12). A few basic aspects of magnetism, including earth's magnetism, will be presented in this chapter.

3B. 02. Classical Approach of Magnetism; Magnetic Dipole

S $2\vec{a}$ N

(–m)   (+m)

Fig. 3.33

In the previous chapter (Section 3A. 02; Oersted's experiment), it was seen that an electric current produces a magnetic field around it. Since much before that the magnets were already known to possess magnetic fields around them, this observation led to the linking of fields of electricity and magnetism together, giving birth to the new exciting field of **ELECTROMAGNETISM.** However, in electricity we have isolated charges which can exist independently, e.g. an electron, which has a unit negative charge. Two equal and opposite point charges, of magnitude q each, separated by a small distance 2a constitute an electric dipole, with its dipole moment as $\vec{p} = 2\vec{a}q$ (vide Chapter 1, Section 1A. 16). A small bar magnet may be considered as equivalent to an electric dipole, i.e. it may be considered as a 'magnetic dipole'. However, unlike an isolated electric charge, so far a magnetic 'monopole' is not known to exist. Still, without any loss of generality, one can imagine that in the said bar magnet two monopoles of equal but opposite strength +m (taken as north pole) and –m (taken as south pole) exist at the two respective ends, separated by a distance 2a (Fig. 3.33). Then it becomes a 'magnetic dipole' of dipole moment $\vec{p}_m$, given by,

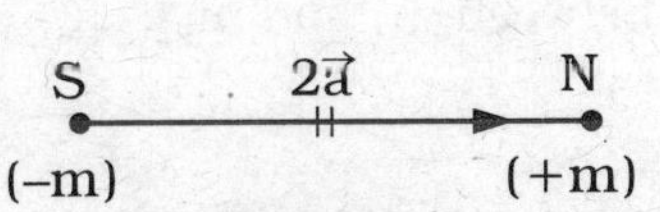

$$\vec{p}_m = 2\vec{a}m \qquad \text{..... (3.35)}$$

The vector $2\vec{a}$ is taken as pointing from south pole to north pole. The SI unit of pole strength is ampere-meter (Am).

Thus, in the classical approach of magnetism we imagine a monopole each as existing at the two poles of a magnet, although in reality a monopole is not known to exist. The poles are taken as located slightly inwards of the edges of the magnet, such that the distance between them is nearly 6/7 times the actual length of the magnet. The classical approach is both useful and necessary for deriving relevant expressions in magnetism.

Magnetic dipole and dipole moment

As already discussed, a **magnetic dipole** is made up of two unlike poles of equal strength, separated by a small distance. From Eq. 3.35 ($\vec{p}_m = 2\vec{a}m$), the **magnetic dipole moment** is defined as the product of pole strength and the distance of separation between the two poles. Another definition of magnetic dipole moment in terms of torque experienced by it in a magnetic field will be provided in the next section (Eq. 3.49). Since $\vec{a}$ is taken as directed from the south pole to the north pole, the magnetic dipole moment $\vec{p}_m$ is also directed likewise (cf. electric dipole moment, which is taken as directed from negative charge to positive charge). Obviously, it is a vector quantity. Its SI unit is ampere-metre² (Am^2) or joule/tesla (JT^{-1}) (cf. next section, which gives two more SI units of magnetic dipole moment).

A small bar magnet and a compass needle are the simplest examples of a magnetic dipole. As discussed in Section 3A. 06, a current-carrying loop has a magnetic field around it. So it is also equivalent to a magnetic dipole. For the same reason, an electron revolving around the nucleus in an atom also constitutes a magnetic dipole. A detailed treatment of these will be made in Sections 3B. 05 and 3B. 08.

Properties of magnetic poles and dipoles

Some important properties of magnetic poles and dipoles are as follows.

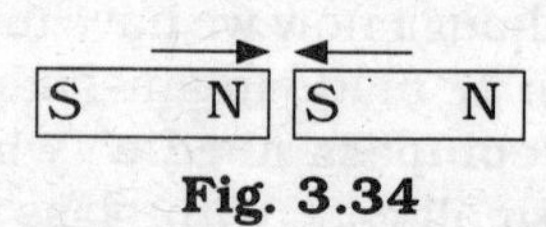

Fig. 3.34

(i) Like poles repel each other and unlike poles attract each other. It can be readily demonstrated by bringing together two bar magnets, once with like poles facing each other and another time with unlike poles facing each other (Fig. 3.34).

(ii) The force of interaction (attraction or repulsion) between two magnetic poles is (a) directly proportional to the product of their pole strengths and (b) inversely proportional to the square of the distance of separation between them, as shown in the following.

Force of interaction between two magnetic poles. If two poles of strength m_1 and m_2 are located at distance $\vec{r}_1$ and $\vec{r}_2$, respectively, from the origin O (Fig. 3.35), then in scalar form the force is given by (r = distance between m_1 and m_2),

$$F \propto \frac{m_1 m_2}{r^2}$$

Or $$F = K\frac{m_1 m_2}{r^2} \text{ (scalar form)} \qquad \text{..... (3.36)}$$

Fig. 3.35

K is a constant of proportionally, with value $\frac{\mu_0}{4\pi}$ in SI system. $\boldsymbol{\mu_0}$ is called **permeability of free space.** Its value is equal to $4\pi \times 10^{-7}$ $WbA^{-1}m^{-1}$.

In vector form, the Force on m_1 due to m_2 is given by,

$$\vec{F}_{12} = \frac{\mu_0}{4\pi}\frac{m_1 m_2}{|\vec{r}_1 - \vec{r}_2|^2}\hat{r}_{21} \text{ (vector form)} \qquad \text{..... (3.37)}$$

where $\hat{r}_{21}$ is a unit vector in the direction from pole 2 to pole 1.Similarly the force on m_2 due to m_1 is given by,

$$\vec{F}_{21} = \frac{\mu_0}{4\pi}\frac{m_1 m_2}{|\vec{r}_1 - \vec{r}_2|^2}\hat{r}_{12} \text{ (vector form)} \qquad \text{..... (3.38)}$$

Example 1. Two bar magnets of length 16 cm each are placed such that their north poles face each other at a distance of 6 cm apart. The poles of each magnet have a strength of 10 Am each. Calculate the force exerted by one north pole on the other.

Solution. Given : $m_1 = m_2 = 10$ Am; r = 6 cm = 0.06 m

The required force is given by,

$$F = \frac{\mu_0}{4\pi} \cdot \frac{m_1 m_2}{r^2} = \frac{(4\pi \times 10^{-7})}{4\pi} \times \frac{(10) \times (10)}{(0.06)^2}$$

$$= 2.78 \times 10^{-3} \text{ N} \quad \textbf{(Answer).}$$

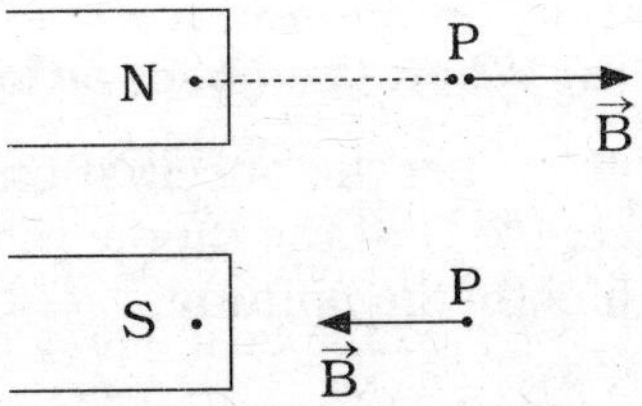

N • P $\vec{B}$

S • P $\vec{B}$

Fig. 3.36

(iii) Just as an electric charge is surrounded by an electric field, a magnetic monopole is surrounded by a magnetic field.

Magnetic field due to a monopole. As for the electric field, the magnetic field at a given point due to a monopole of strength m is defined as the force experienced by a monopole of unit strength at that point. Mathematically,

$$\vec{B} = \frac{\vec{F}}{m} \qquad \text{..... (3.39)}$$

Conventionally, the direction of the magnetic field due to the north pole is always taken as directed away from the pole and that due to the south pole is taken to be directed towards the pole (Fig. 3.36). The magnitude of the magnetic field (also called 'magnetic field strength') due to a monopole of strength m at a point distant r from it is given by,

$$B = \frac{\mu_0}{4\pi} \frac{m}{r^2} \qquad \text{..... (3.40)}$$

(iv) Just as an electric dipole is surrounded by an electric field, a magnetic dipole is surrounded by a magnetic field.

Magnetic field due to a dipole. We shall evaluate the magnetic field associated with a magnetic dipole for two possible cases.

(a) When the observation point lies along the axial line of the dipole

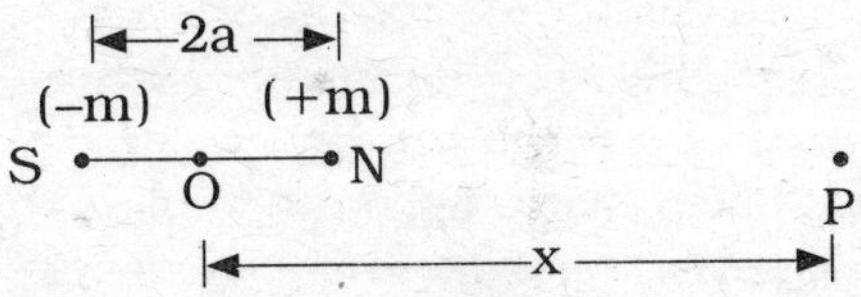

Consider a dipole SN of magnetic moment p_m, consisting of two monopoles of strength –m and +m, separated by a distance 2a (Fig. 3.37). Let the observation point P be situated at a distance x from the centre O of the dipole. The magnetic field due to the north pole N at P is (conventionally) directed from N to P and has the magnitude,

2a; (–m) (+m); S • O • N • P; x

Fig. 3.37

$$B_N = \frac{\mu_0}{4\pi} \frac{m}{(x-a)^2}$$

The magnetic field due to the south pole S at P is directed from P to S and has the magnitude,

$$B_S = \frac{\mu_0}{4\pi} \frac{m}{(x+a)^2}$$

The total magnetic field, taken as directed from N to P, is

$$B_{ax} = B_N - B_S \text{ (towards PN)}$$

$$= \frac{\mu_0}{4\pi} \cdot \frac{m}{(x-a)^2} - \frac{\mu_0}{4\pi} \cdot \frac{m}{(x+a)^2}$$

$$= \frac{\mu_0 m}{4\pi} \left[\frac{1}{(x-a)^2} - \frac{1}{(x+a)^2} \right]$$

$$= \frac{\mu_0 m}{4\pi} \frac{(4ax)}{(x^2-a^2)^2} = \frac{2\mu_0 (2am)x}{4\pi (x^2-a^2)^2}$$

$$= \frac{2\mu_0 p_m x}{4\pi (x^2-a^2)^2} \qquad (p_m = 2am; \text{ vide Eq. 3.35})$$

In vector form, we have,

$$\vec{B}_{ax} = \frac{\mu_0}{4\pi} \cdot \frac{2\vec{p}_m x}{(x^2 - a^2)^2} \quad \text{..... (3.41)}$$

For a small dipole, a ≪ x. Then,

$$\vec{B}_{ax} \approx \frac{2\mu_0 \vec{p}_m}{4\pi x^3} \quad \text{..... (3.42)}$$

(b) When the observation point lies along the equatorial line of the dipole

Let the observation point P be located along the equatorial line of the dipole, at a distance x from the centre O of the dipole (Fig. 3.38). The magnetic field at P due to the north pole is directed from N to P and has the magnitude,

$$B_N = \frac{\mu_0}{4\pi} \cdot \frac{m}{(x^2 + a^2)}$$

Similarly, the magnetic field at P due to the south pole is directed from P to S and has the magnitude,

$$B_S = \frac{\mu_0}{4\pi} \cdot \frac{m}{(x^2 + a^2)}$$

Note that $B_N = B_S = \frac{\mu_0}{4\pi} \cdot \frac{m}{(x^2 + a^2)}$ = B (say). The components of B_N and B_S along the equatorial line OP are opposite in sign and hence cancel out. Therefore, the magnetic field along OP is zero,

$$B_{OP} = B \sin\theta - B \sin\theta = 0$$

On the other hand, the components in the perpendicular direction PR (which is parallel to NS) add up to give the net magnetic field at P as,

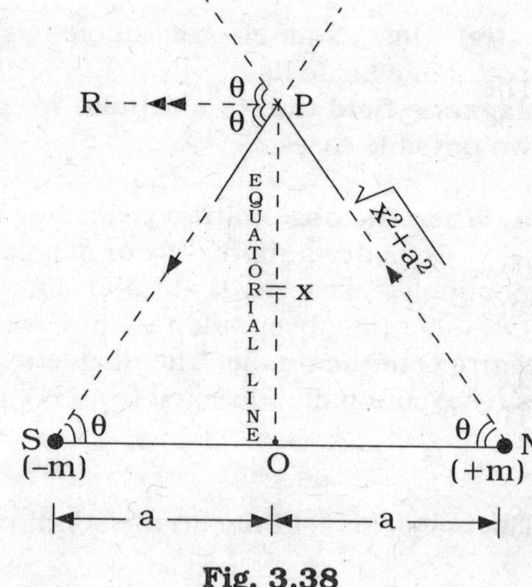

Fig. 3.38

$$B_{eq} = 2B\cos\theta = \frac{2\mu_0 m}{4\pi(x^2 + a^2)} \cos\theta$$

$$= \frac{2\mu_0 m}{4\pi(x^2 + a^2)} \frac{a}{(x^2 + a^2)^{1/2}}$$

$$= \frac{\mu_0 (2am)}{4\pi(x^2 + a^2)^{3/2}} = \frac{\mu_0 p_m}{4\pi(x^2 + a^2)^{3/2}}$$

The direction of the field is along NS.
In vector form it is given as,

$$\vec{B}_{eq} = \frac{\mu_0 \vec{p}_m}{4\pi (x^2 + a^2)^{3/2}} = \quad \text{..... (3.43)}$$

For a small dipole, a ≪ x. Then,

$$\vec{B}_{ax} \cong \frac{\mu_0 \vec{p}_m}{4\pi x^3} \textbf{ (Small dipole)} \quad \text{..... (3.44)}$$

From Eqs. 3.42 and 3.44, one finds that for the same value of x, $B_{ax} = 2B_{eq}$. Thus, for the same distance of the observation point from the centre of a small bar magnet, the magnetic field along the axial line is two times the field along the equatorial line.

(v) A magnetic dipole experiences a torque when placed in a uniform magnetic field.

Torque on a dipole. We shall examine how the torque arises and derive an expression for it in terms of magnetic moment and magnetic field strength.

Consider a dipole NS of magnetic moment $\vec{p}_m$ placed in a uniform magnetic field $\vec{B}$, such that the angle between $\vec{p}_m$ and $\vec{B}$, is θ (Fig. 3.39).

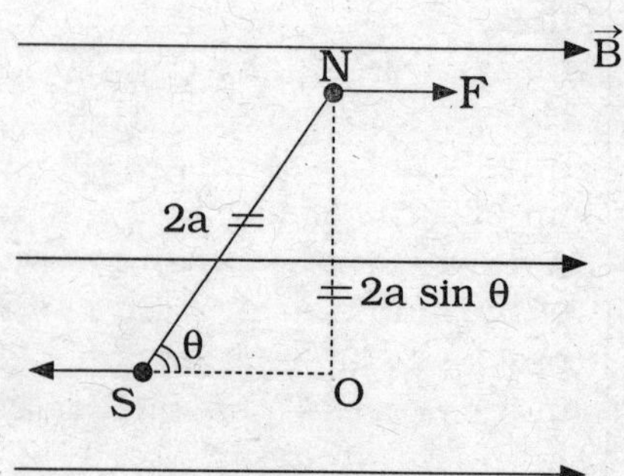

Fig. 3.39

The force F exerted by the magnetic field on the north pole is equal to mB. It is directed along the direction of $\vec{B}$. The force exerted on the south pole is also equal to m$\vec{B}$, but it is directed opposite to $\vec{B}$. Thus the net force experienced by the dipole is zero. However, since the two forces are non-collinear, they constitute a couple, due to which the dipole experiences a torque of the magnitude (force × perpendicular distance between the two forces),

$$\tau = (mB)\,(2a\sin\theta)$$

$$= (2am)\,B\sin\theta$$

Or

$$\tau = P_m B\sin\theta \qquad .....(3.45)$$

In vector form,

$$\vec{\tau} = \vec{P}_m \times \vec{B} \qquad .....(3.46)$$

The torque tends to rotate the dipole such that it becomes aligned along the magnetic field $\vec{B}$, i.e. θ tends to become zero.

(vi) When a magnetic dipole is placed in a uniform magnetic field, it has a potential energy associated with it.

Potential energy of a dipole. We shall examine how the potential energy arises and derive an expression for it in terms of magnetic moment and magnetic field strength. As stated above, for any given position θ of the dipole, the torque acting upon the dipole tends to rotate it, in the attempt to align it along the magnetic field $\vec{B}$. When the dipole is aligned parallel to $\vec{B}$ and we want to bring it to a different position θ, work has to be done against the torque. This work is stored as potential energy of the dipole. Thus, as expected, the potential energy arises by virtue of the position θ occupied by the dipole in the magnetic field. the work done in rotating the dipole through a small angle dθ will be,

$$dW = \tau d\theta = p_m \sin\theta\, d\theta \qquad \text{(vide Eq. 3.45)}$$

The work done in rotating the dipole from a reference position θ_1 to another position θ_2 will be,

$$W = \int_{\theta_1}^{\theta_2} p_m B\sin\theta d\theta = p_m B\left[-\cos\theta\right]_{\theta_1}^{\theta_2}$$

$$= -p_m B\,(\cos\theta_2 - \cos\theta_1)$$

If we choose $\theta_1 = 90°$ as the reference position for the dipole (when it is oriented ⊥ to the magnetic field $\vec{B}$ and has maximum value of $\tau = p_m B$) and put $\theta_2 = \theta$, we have ($\cos 90° = 0$),

$$W = -p_m B\cos\theta$$

This work is stored as the potential energy U of the dipole,

$$U = -p_m B\cos\theta \qquad .....(3.47)$$

The minus sign indicates that work has to be done against the torque for rotating a dipole. In vector form,

$$U = (-\vec{P}_m \cdot \vec{B}) \qquad .....(3.48)$$

Additional knowledge

Origin of magnetism. The earlier understanding of magnetism was based on the concept of each molecule of a magnetic material treated as a tiny magnet, called 'weber element' (Weber's molecular theory of magnetism). The non-magnetic materials were believed to contain self-neutralising 'closed chains' of such elements, in which as many elements pointed in a given direction as in the opposite direction. In the magnetic materials, during the process of magnetization the chains supposedly broke up and the elements aligned themselves in a particular direction to produce a net magnetic moment. Although this theory provided a rough picture of the origin of magnetism, with our present understanding of the materials at an atomic level and even sub-atomic level, a correct picture has now emerged. In the previous chapter (Section 3A. 02; 'Additional knowledge'), the fundamental role played by the electrons in the process of magnetization was outlined. In Section 3B. 13, more about it will be learnt in terms of magnetic 'domains'.

3B. 03. Properties of Magnets; Bar Magnet

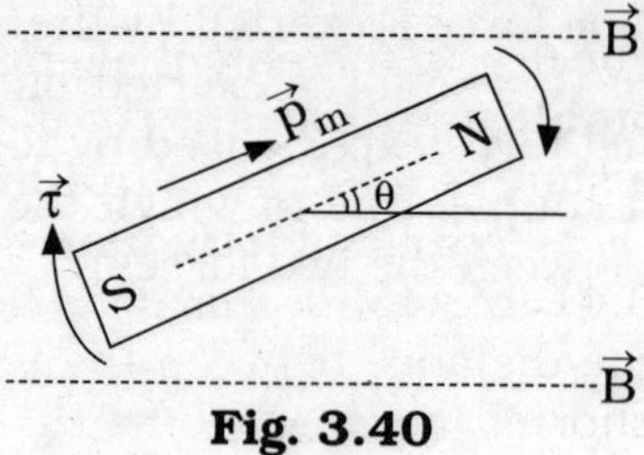

Fig. 3.40

We shall briefly describe fundamental properties of magnets. For the sake of convenience, we shall do so for bar magnets. As the name implies, a bar magnet is a magnet of rectangular shape. The two ends of the bar are termed as the poles of the magnet. The two poles are always of equal strength. When such a magnet is freely suspended, one of the poles always points towards the earth's geographical north. It is referred to as 'North' pole (N). The other one, which points towards south, is called 'South' pole (S). In other words, a bar magnet is a **magnetic dipole,** analogous to the electric dipole described earlier in Chapter 1 (vide Section 1A. 15). One can associate a magnetic dipole moment, $\vec{p}_m$, with a bar magnet, on similar lines as for an electric dipole. We shall enumerate some important properties of bar magnets.

(i) If a bar magnet is freely suspended in a uniform magnetic field, it experiences a torque $\vec{\tau}$, which tries to align the magnet parallel to the field (Fig. 3.40),

$$\vec{\tau} = \vec{p}_m \times \vec{B} \qquad \text{..... (Earlier Eq. 3.46)}$$

$$\Rightarrow \quad \tau = p_m B \sin\theta \qquad \text{..... (Earlier Eq. 3.45)}$$

If $\theta = 90°$, τ has maximum value,

$$\tau = p_m . B \qquad \text{..... (3.49)}$$

The Eq. 3.49 enables to define the **magnetic dipole moment** as "the torque experienced by the dipole per unit magnetic field, when it is oriented at an angle of 90° to the field". (cf. the definition given earlier in the last section in terms of pole-strength).

The direction of $\vec{p}_m$ is taken from south pole to north pole. The SI unit of the magnetic dipole moment is (Newton–metre)/(tesla) [$N\text{-}mT^{-1}$] (cf. last section, in which two other SI units were given).

When a bar magnet is freely suspended in the earth's magnetic field, it experiences a torque due to the field, which aligns it in the direction north-south.

(ii) If unlike poles of two bar magnets are brought close to each other, they attract each other. But when like poles are similarly brought close, they repel each other, as already described in the last section (Property (i)).

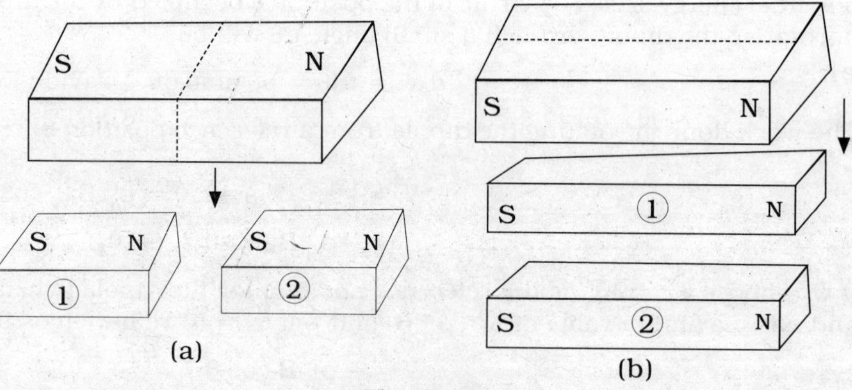

Fig. 3.41

(iii) If a bar magnet is cut (or broken) into two parts in any manner, new poles appear, such that each part becomes a new smaller bar magnet, possessing its own north and south poles. The Fig. 3.41 depicts two cases: (a) cut perpendicular to the length of the magnet and (b) cut along the length of the magnet. In the first case, each new magnet has the same pole strength as before, but the magnetic moment is halved due to the length becoming half. In the second case, the magnetic moment is again halved due to the pole strength being halved, although the length remains the same.

If the cutting (or breaking) is further continued, one keeps on getting more and more (but increasingly smaller) magnets, each of which is a magnetic dipole with its own north pole and south pole. The process continues down to molecular, atomic and subatomic dimensions.

(iv) Unlike the known existence of single electric charges, e.g. an electron, a proton, etc., a magnetic monopole is not known to exist. Nevertheless, it is the simplest magnetic system assumed to exist.

(v) When a piece of a material like soft iron, nickel, cobalt, etc., is placed close to a bar magnet, it gets magnetized for so long as it remains there. If it is cyclically rubbed with a bar magnet, it is magnetized such that the magnetism is retained after removing the bar magnet. In both the cases the piece is said to have acquired 'induced magnetism'. In the latter case, it is said to have become an artificial permanent magnet.

(vi) As described in the last section (Eqs. 3.47 and 3.48), a bar magnet of dipole moment $\vec{p}_m$ when placed in a uniform magnetic field of strength $\vec{B}$ acquires a potential energy,

$$U = -\vec{p}_m . \vec{B} = -p_m . B \cos\theta \qquad \text{..... (3.50)}$$

For $\theta = 0$, the potential energy of the dipole is minimum ($-p_mB$) and the dipole is said to be in the state of **stable equilibrium.** For $\theta = 180°$, the potential energy is maximum ($+p_mB$) and the dipole is said to be in a state of **unstable equilibrium.**

(vii) Magnetic field due to a bar magnet. Since a bar magnet is essentially a magnetic dipole, the Eqs. 3.41 to 3.45 for magnetic field due to a dipole also apply for a bar magnet. The length 2a has to be taken as the distance between the two poles of the magnet. It is called **magnetic length** of the magnet. It is slightly shorter than the geometrical length of the magnet.

Example 2. (a) Find the magnetic field due to a bar magnet of magnetic length 12 cm at a point distant 10 cm from the centre along the axis towards south pole. The pole strength of the magnet is 4.5 Am. (b) Also find the field for the same magnet at an equal distance from the centre at a point located along the equatorial line.

Solution. Given : 2a = 12 cm = 0.12 m; m = 4.5 A.m; x = 10 cm = 0.10 m; (∴ p_m = 2a.m = (0.12)×(4.5) = 0.54 A.m); a = (0.06 m)

(a) The required field is given by,

$$B_{ax} = \frac{\mu_0}{4\pi} \cdot \frac{2p_m x}{(x^2 - a^2)^2} = \frac{(4\pi \times 10^{-7})}{4\pi} \times \frac{2 \times (0.54) \times (0.1)}{\left[(0.1)^2 - (0.06)^2\right]^2}$$

$$= 10^{-7} \times \frac{1.08 \times 10^{-1}}{[0.0064]^2} = 10^{-8} \times \frac{1.08}{4096 \times 10^{-8}}$$

$$= 2.64 \times 10^{-4} \text{ T} \quad \textbf{(Answer).}$$

(b) The required field is given by,

$$B_{eq} = \frac{\mu_0}{4\pi} \cdot \frac{p_m}{(x^2 + a^2)^{3/2}} = \frac{(4\pi \times 10^{-7})}{4\pi} \times \frac{(0.54)}{\left[(0.1)^2 + (0.06)^2\right]^{3/2}}$$

$$= 10^{-7} \times \frac{0.54}{(0.0136)^{3/2}} = 10^{-7} \times \frac{54 \times 10^{-2}}{1585 \times 10^{-6}}$$

$$= 3.41 \times 10^{-5} \text{ T} \quad \textbf{(Answer).}$$

Example 3. A magnet is freely suspended in a uniform magnetic field of strength 6×10^{-5} T. Its magnetic moment is 2.5×10^4 JT^{-1}. Determine the potential energy gained by it, when it is slowly rotated from a position parallel to the field to a position oriented at 30° to the field.

Solution. Given : $p_m = 2.5 \times 10^4$ JT^{-1} ; B = 6×10^{-5} T; $\theta_1 = 0$; $\theta_2 = 30°$

The potential energy is given by,

$$U = -(\vec{p}_m \cdot \vec{B})$$

∴ The change in potential energy is given by (it is the same as the work done in moving from position 1 to position 2; it is stored in the magnet as p.e.),

$$\Delta U = U_1 - U_2 = -p_m B(\cos 0° - \cos 30°)$$

$$= -(2.5 \times 10^4) \times (6 \times 10^{-5}) \times (1 - 0.866)$$

$$= -0.20 \text{ J} \quad \textbf{(Answer).}$$

3B. 04. Magnetic Field Lines (or Magnetic Lines of Force) ; Magnetic Flux Density (or Magnetic Induction).

In Chapter 1 (Section 1A. 14), we had described how electric field lines provide a means for visualizing the nature of electric field $\vec{E}$ in a given space. The magnetic field lines similarly provide an effective means of pictorially representing the variation of magnetic field strength $\vec{B}$ in a given space. But there is an important point of difference. An electric field line represents the path along which a free positive test charge is supposed to move in the electric field, by virtue of the electrostatic force experienced by it in the field. Since an isolated magnetic pole (a monopole) is not known to exist, a tiny compass needle, which is a magnetic dipole, has to be employed for tracing a magnetic field line. A force now acts on the compass needle by virtue of the torque experienced by it in the magnetic field (vide Eq. 3.46). As the value of $\vec{B}$ varies from point to point in the field, the force changes accordingly, making the compass needle to orient itself differently at different points of the field. In this sense, we can also call the magnetic field line as 'magnetic lines of force'. At most positions in the field the magnetic lines are curved. The tangent drawn to the curve at a given point represents he direction of magnetic field at that point.

Thus, we can **define** a **magnetic field line** as "an imaginary linear or curved path along which a tiny magnetic dipole is differently oriented at different points and the tangent drawn at a given point on the curved path gives the direction of magnetic field at that point." Unlike the electric field lines along which the movement of a free charge needs to be imagined, the magnetic field lines for a given magnetic field can actually be traced with the help of a compass needle, as done by the students in their laboratory experiments. A dot is marked at the south tip for successive positions of the needle and later all the dots are joined to draw the curve. The magnetic field lines drawn for a few different configurations of bar magnets are shown in Fig. 3.42 [(a) isolated bar magnet shielded from the earth's field, (b) bar magnet with north pole facing geographical north, (c) bar magnet with south pole facing geographical north].

If iron filings are sprinkled around a bar magnet, they align with the magnetic field of the magnet (Fig. 3.43). Their pattern of distribution closely resembles the magnetic field lines traced for the bar magnet (Fig. 3.42 a).

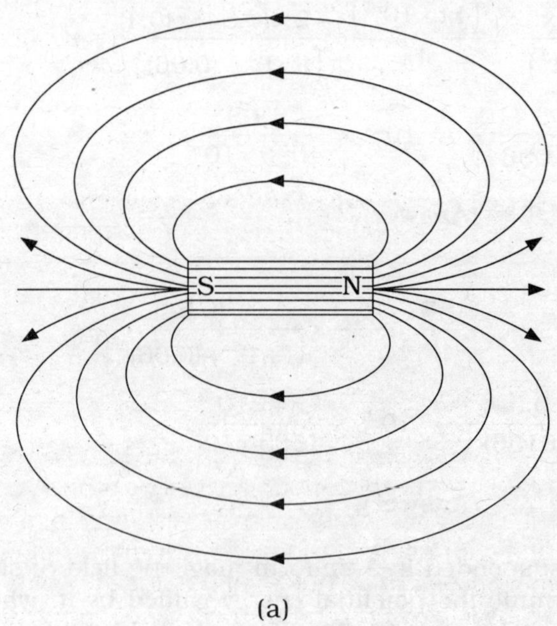

(a)

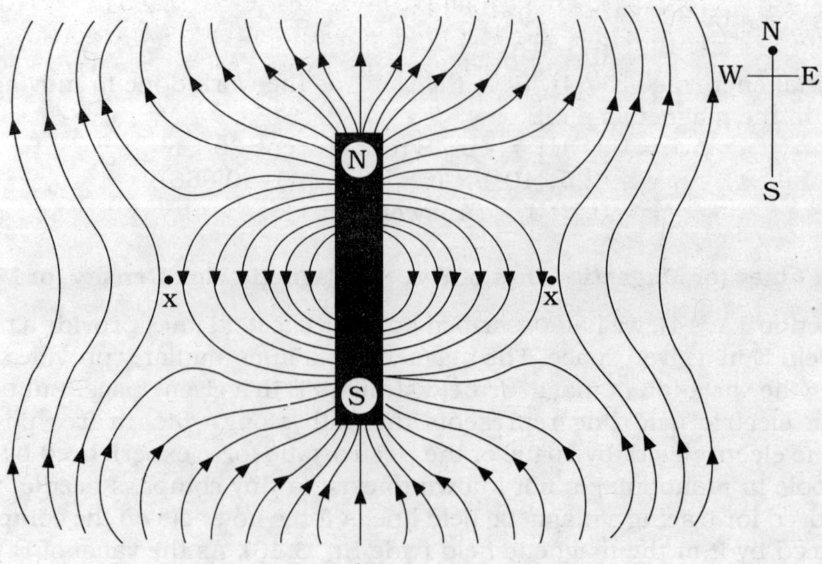

(b)

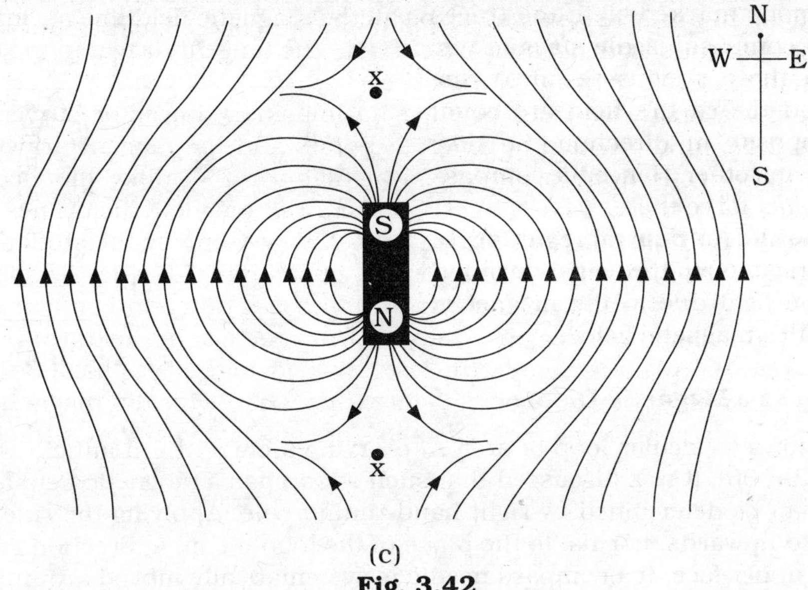

(c)
Fig. 3.42

Properties of magnetic field lines

(i) The magnetic field lines form closed loops, extending through the body of the magnet. Thus, they are continuous curves. This is unlike the electric field lines which are discontinuous curves, because no field lines can exist inside a charged conductor.

(ii) Outside a magnet, the lines are taken as directed from the north pole to the south pole of the magnet.

(iii) The tangent drawn at a given point on a field line gives the direction of magnetic field at that point.

(iv) No two field lines can intersect each other (the reason is same as for the electric field lines (vide Section 1A. 14, Chapter 1A, viz. it will mean existence of two different field values at the point of intersection, which is not possible).

Fig. 3.43

(v) The greater is the clustering of field lines in a region of space around a magnet, the greater is the magnitude of magnetic field $\vec{B}$ in the region. Thus the degree of crowding of field lines in a given region is a measure of the magnetic field strength in that region. In fact, the **magnetic flux density,** i.e. the magnetic field lines passing per unit area perpendicular to the lines at a given position in the field, becomes a measure of the magnetic field intensity at that position. It is also termed as **magnetic induction** (Chapter 4A; Section 4A. 01).

Magnetic field lines for various configurations of bar magnets; neutral points. In section 3B. 11, we shall explain how earth behaves as a natural magnet, producing a uniform magnetic field, i.e. a field that has the same magnitude and direction in a given region of space. This field is always superposed on the magnetic field of any magnet on the earth. While tracing the magnetic field lines of a magnet on earth, the horizontal component H of the earth's field is to be taken into consideration in its interaction with the field of the magnet.

For a powerful bar magnet, the magnitude of the earth's magnetic field is comparatively small. Hence, in regions near to the magnet the magnetic field lines have very nearly the same shape as for an isolated bar magnet shielded from the effect of the earth's field. This is seen by comparing the trace (a) with the traces (b) and (c) in Fig. 3.42, which shows the traces of magnetic field lines for a few representative configurations of bar magnets.

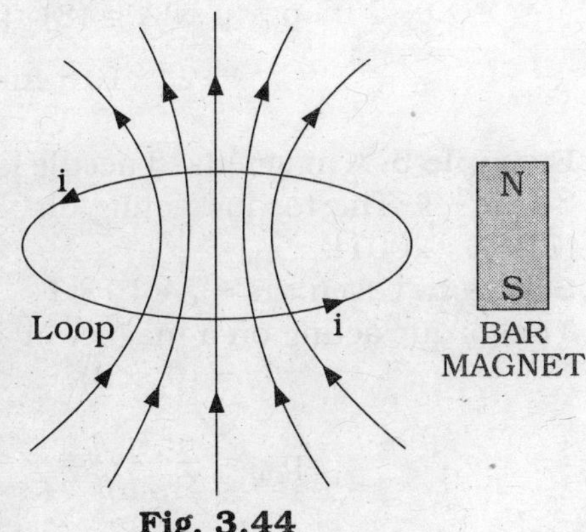

Fig. 3.44

In the small regions marked 'x' in the traces of Figs. 3.42 b, c, the compass needle stays in any arbitrary position. In these regions the field due to the bar magnet and the earth's field are equal in magnitude but opposite in direction. So they neutralise (or cancel) each other. Hence the compass needle experiences no net force there.

These **Neutral points (or neutral regions)** are defined as "points (or regions) in the neighbourhood of a magnet where the field due to the magnet is neutralized by the earth's magnetic field".

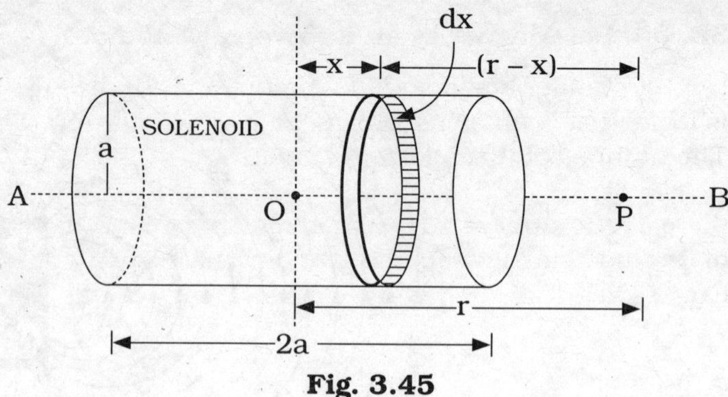

Fig. 3.45

3B. 05. Current Loop as a Magnetic Dipole

The Fig. 3.44 shows a circular loop of area A, carrying a current i in anticlockwise direction. In the last Chapter (Section 3A. 06), it was discussed that such a loop has a magnetic field associated with it. The direction of the field can be determined by right hand thumb rule. Applying the rule, we find for our case that the field is directed upwards, normal to the plane of the loop, i.e. in a direction running from the lower face of the loop to the upper face. If a compass needle is systematically moved around the loop, it will trace magnetic field lines as shown in the figure. **Thus the loop behaves like a small bar magnet** (see Fig. 3.44, right hand side), **with the upper face of the loop acting as north pole and the lower face as south pole. In other words, the loop behaves like a magnetic dipole.**

We define a magnetic dipole moment $\vec{p}_m$ of the loop as having magnitude iA, i.e. product of current and area of the loop,

$$|\vec{p}_m| = iA \qquad \text{..... (3.51a)}$$

If we regard the area as a vector quantity $\vec{A}$, with direction perpendicular to the plane of the loop, the above equation can be written in the vector form as,

$$\vec{p}_m = i\vec{A} \qquad \text{..... (3.52a)}$$

Thus, $\vec{p}_m$ has the same direction as that of the magnetic field, as it should be. The SI unit of $\vec{p}_m$ is ampere – metre^2 (Am^2).

For a coil with N number of turns, the Eqs. 3.51a and 3.52a become,

$$p_m = NiA \qquad \text{..... (3.51b)}$$

$$\vec{p}_m = Ni\vec{A} \qquad \text{..... (3.52b)}$$

Example 4. A current of 4 A flows through a coil of area of cross-section as 2×10^{-4} m^2. If the coil consists of 3 turns, find the magnetic moment associated with it.

Solution. Given : N = 3; i = 4 A; A = 2×10^{-4} m^2

The magnetic moment associated with the coil is given by,

$$p_m = NiA = (3)\times(4)\times(2\times10^{-4})$$

$$= 24\times10^{-4} \text{ Am}^2 \quad \textbf{(Answer).}$$

Example 5. A magnetized needle is placed at 30° with the direction of a uniform magnetic field of magnitude 3×10^{-2} T. The torque acting on the needle is 7.2×10^{-4} J. Calculate the magnetic moment of the needle. (CBSE ; 2001).

Solution. Given : B = 3×10^{-2} T; $\tau = 7.2\times10^{-4}$ J; $\theta = 30°$

The torque acting on a magnet in a magnetic field is given by,

$$\vec{\tau} = (\vec{p}_m \times \vec{B}) \quad \text{or} \quad \tau = p_m B \sin\theta$$

$$\therefore \quad p_m = \frac{\tau}{B\sin\theta} = \frac{(7.2\times10^{-4})}{(3\times10^{-2})\times(\sin 30°)}$$

$$= 4.8\times10^{-2} \text{ JT}^{-1} \quad \textbf{(Answer).}$$

3B. 06. Bar Magnet as an Equivalent Solenoid

Consider a solenoid of length 2a and radius R, consisting of n turns per unit length. Its central point is located at O along its axis AB (Fig. 3.45). We choose a circular element of width dx at a distance x from O. The number of turns in the element are,

$$dN = n.dx$$

Let a current i flow through the solenoid. We know that the magnetic field due to a circular loop of radius R at a distance d from the centre of the loop along the axial line is (vide Section 3A. 06; Eq. 3.11),

$$B = \frac{\mu_0 i R^2}{2\left(d^2 + R^2\right)^{3/2}}$$

Accordingly, the magnetic field due to the element at x at a point P, distant r from O along the axis AB, will be,

$$dB = \frac{\mu_0 i(dN)R^2}{2\left[(r-x)^2 + R^2\right]^{3/2}} = \frac{\mu_0 i(n.dx)R^2}{2\left[(r-x)^2 + R^2\right]^{3/2}}$$

The total magnetic field at P due to the solenoid is found by integrating dB (taking the half length a of the solenoid on the right of O as +ve and on the left as –ve),

$$B = \int_{-a}^{a} dB \frac{\mu_0 inR^2}{2} \int_{-a}^{a} \frac{dx}{\left[(r-x)^2 + R^2\right]^{3/2}}$$

We consider the case when the point of observation P is situated far off from the solenoid, so that r >> a and also r >> R. Then both x and R in the denominator can be neglected in comparison to r and we have,

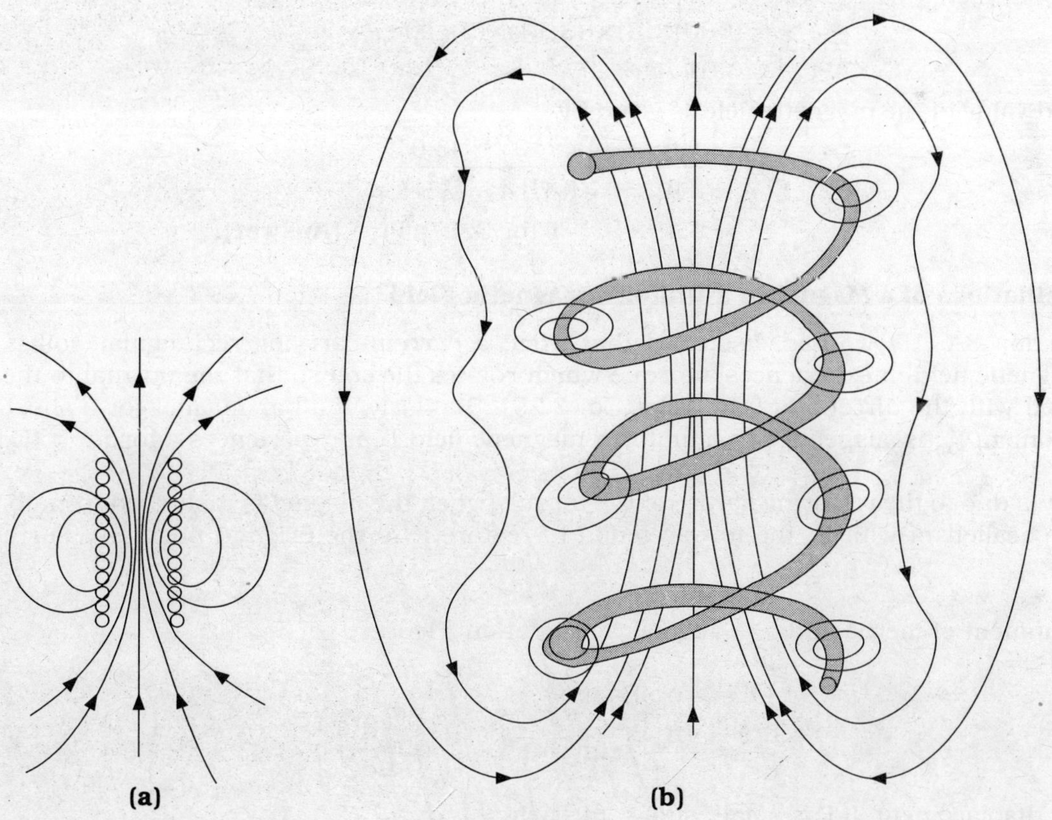

Fig. 3.46

$$B \approx \frac{\mu_0 i n R^2}{2} \int_{-a}^{a} \frac{dx}{r^3}$$

$$\Rightarrow \quad B \approx \frac{\mu_0 i n R^2}{2r^3} \cdot 2a$$

$$\Rightarrow \quad B = \frac{\mu_0}{2\pi r^3}(2\pi a\, niR^2) = \frac{\mu_0}{2\pi r^3}[n\,(2a)\, i\,(\pi R^2)]$$

The quantity i (πR^2) is the dipole moment of a single turn of the solenoid (vide Eq. 3.51a, Section 3B. 05). The total dipole moment is, therefore (total number of turns = n.2a),

$$p_m = n\,(2a)\, i\pi R^2$$

Hence, $$B \approx \frac{\mu_0\, p_m}{2\pi\, r^3} \quad \text{..... (3.53)}$$

If the magnetic field of a bar magnet of dipole moment p_m is measured at a long distance r along its axis, it turns out to be the same value as above. Thus, a bar magnet is found to be equivalent to a solenoid. This is further confirmed by comparing the magnetic field lines of (i) a short tightly wound solenoid (Fig. 3.46a) and (ii) a long loosely wound solenoid (Fig. 3.46b) with those of an isolated bar magnet (Fig. 3.42a). They have strong similarity with each other.

Example 6. A current of 8 A flows through a solenoid of length 15 cm and radius 2 cm. If the solenoid consists of 100 turns, determine the magnetic field at an axial point distant 12 cm from the centre.

Solution. Given : N = 300; i = 8 A; r = 12 cm = 12×10^{-2} m; radius of solenoid = r_s = 2 cm = 2×10^{-2} m

(The given length of the solenoid is of no relevance to this question)

The solenoid is equivalent to a bar magnet of magnetic moment given by,

$$p_m = N\,i\,A = N\,i\,(\pi\, r_s^2)$$

$$= (300)\times(8)\times[(3.14)\times(2\times10^{-2})^2]$$

$$= 3.01\ \text{Am}^2$$

The required value of the magnetic field is given by,

$$B = \frac{\mu_0 p_m}{2\pi r^3} = \frac{(4\pi\times10^{-7})\times(3.01)}{2\pi\times(12\times10^{-2})^3}$$

$$= 3.48\times10^{-6}\ \text{T (or } 3.48\ \mu\text{T)}$$ **(Answer).**

3B. 07. Oscillations of a Magnet in a Uniform Magnetic Field

In Section 3A. 15, we had discussed that when a current-carrying rectangular coil is placed in a uniform magnetic field, it experiences a torque which rotates the coil so that the normal to the plane of the coil is aligned with the direction of the field (Eq. 3.29). Similarly, if a bar magnet (or a short solenoid) of magnetic moment $\vec{p}_m$ is suspended in a uniform magnetic field $\vec{B}$, it experiences a torque $\vec{\tau}$ (Eq. 3.46),

$$\vec{\tau} = -(\vec{p}_m\times\vec{B})$$

The –ve sign is due to the restoring nature of the torque (when the magnet is disturbed from its equilibrium position of parallelism with $\vec{B}$, the torque tends to restore it to the original position). Further, we know that

$$\vec{\tau} = I\vec{\alpha}$$

where I is moment of inertia and $\vec{\alpha}$ is angular acceleration. Hence,

$$I\vec{\alpha} = -\vec{p}_m\times\vec{B}$$

$$\Rightarrow \quad I\alpha = -p_m\, B\sin\theta$$

$$\Rightarrow \quad \frac{d^2\theta}{dt^2} = -\frac{p_m B}{I}\sin\theta \qquad \left(\alpha = \frac{d^2\theta}{dt^2}\right)$$

θ is angular displacement. If θ is small, $\sin\theta \approx \theta$. Then,

$$\frac{d^2\theta}{dt^2} = \left(-\frac{p_m B}{I}\right)\theta$$

This relation shows that the acceleration is proportional to the angular displacement. So it represents a simple harmonic motion of the magnet in the magnetic field, with the angular frequency ω and the time period T given by,

$$\omega = \sqrt{\frac{p_m B}{I}}; \quad T = 2\pi\sqrt{\frac{I}{p_m B}} \qquad \text{..... (3.54)}$$

By experimentally determining the time period of oscillation of the magnet, the above relation for T can be used to determine the magnetic moment p_m of the magnet (I, the moment of inertia, can be calculated from the mass and dimensions of the magnet).

3B. 08. Magnetic Dipole Moment of a Revolving Electron

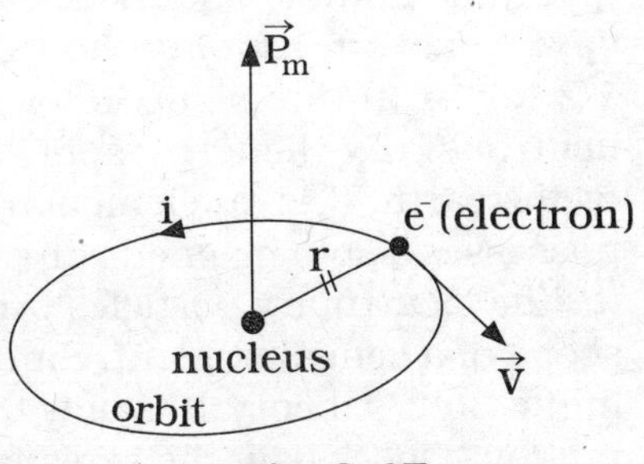

Fig. 3.47

According to the classical Bohr's model of an atom, electrons revolve around the nucleus in various orbits. In Fig. 3.47, an electron of charge –e is shown revolving around the nucleus with velocity $\vec{v}$ in a circular orbit of radius r. Such a moving charge obviously constitutes an electric current. Since the electron is moving clockwise, the direction of current flow is conventionally taken as anticlockwise. As an electric current represents the rate of flow of charge, its magnitude in our case is (T = time period of revolution of the electron),

$$i = e/T$$

Now, $T = 2\pi r/v$ (circumference ÷ velocity)

Therefore, $i = ev/2\pi r$

The magnetic moment associated with this current is (vide Eq. 3.52),

$$p_m = i\,(\pi r^2) = (ev/2\pi r).\,\pi r^2$$

$$\Rightarrow \quad p_m = evr/2 \qquad \text{..... (3.55)}$$

According to the right hand thumb rule (vide Section 3A. 06), the direction of p_m is upwards from the plane of the orbit, as shown in Fig. 3.47.

The Eq. 3.55 can be expressed in terms of the angular momentum ℓ of the electron about the nucleus ($\ell = mvr$)

$$p_m = (e/2m)\,\ell \qquad \text{..... (3.56)}$$

In vector form,

$$\vec{p}_m = -\,(e/2m)\,\vec{\ell} \qquad \text{..... (3.57)}$$

The negative sign appears because the angular momentum of the electron is opposite in direction to its magnetic moment ($\vec{v}$ is clockwise, i is anticlockwise). Thus we find that magnetic dipole moments exist even at an atomic level.

Additional knowledge

According to the classical Bohr's atomic model, the angular momentum ℓ can only have a set of discrete values, given by

$$\ell = n\,(h/2\pi), \qquad n = 1, 2, 3,\ldots.. \qquad \text{..... (3.56)}$$

Such discrete values are said to be **quantised** (Bohr quantisation condition). [h is called Planck's constant. It is a very important fundamental constant of nature, as it relates energy to frequency through the famous relation, E = n (hv). Thus it figures in all energy transfers (which, again, are quantised) in the universe. Its value is 6.626×10^{-34} Js]. Then, the possible 'discrete' values of the magnetic moment of a revolving electron become,

$$p_m = n\left(\frac{eh}{4\pi m}\right) \qquad \text{..... (3.58)}$$

The smallest value of p_m corresponds to n = 1. If one substitutes n = 1 and the known values of e, h, m in Eq. 3.57, p_m is calculated to be,

$$(p_m)_{\text{minimum}} = 9.27\times10^{-24}\ \text{Am} \qquad \text{..... (3.58)}$$

This value (= eh/4πm) is known as **Bohr magneton.**

The magnetic dipole moment $\vec{P}_m$ expressed by Eq. 3.57 is termed **Orbital magnetic dipole moment** and ℓ is termed **Orbital magnetic quantum number.** Besides this dipole moment, an electron also possesses an **intrinsic** magnetic dipole moment, called **spin magnetic dipole moment,** which has the same numerical value as p_m. The corresponding quantum number is labelled as **s,** called **spin quantum number.** Its value is always 1/2 (or –1/2) (the same is for protons and neutrons). By the term 'spin', the student should not be misled to think that an electron spins like a top or the earth. It is an elementary particle and does not have a well-defined axis like earth to rotate about. It possesses a spin magnetic dipole moment, just as it possesses a mass (m_e) and a charge (-e). The same holds for the elementary particles protons and neutrons. The causes lie in quantum physics, which is beyond the scope of a school student. Similarly, actually the electrons do not orbit the nucleus of an atom, as assumed in the Bohr's theory. It is only that the observed magnetic effects are equivalent to such a hypothetical orbital motion. Once again the explanation lies in quantum physics.

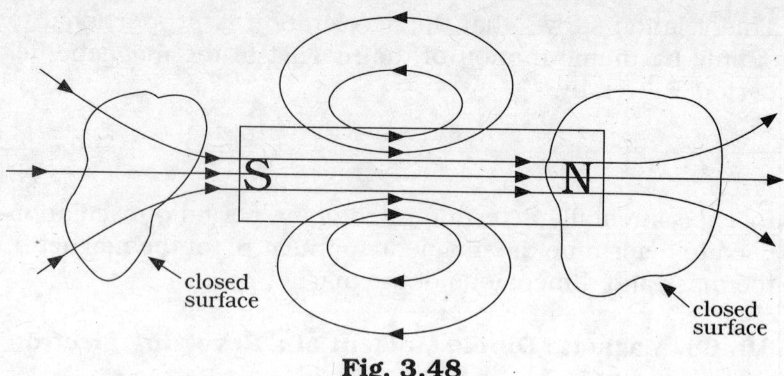

Fig. 3.48

Both the orbital magnetic dipole moment and the spin magnetic dipole moment of the electrons play a fundamental role in determining the magnetism of all materials, about which we had partly discussed in the last chapter (Section 3A. 02) and will discuss further in Section 3B. 13. However, as emphasized above, for a full understanding of such a role one needs the advanced knowledge of quantum physics.

3B. 09. Gauss's Law (or Theorem) for Magnetic Fields

In our study of electrostatics, we had described that in general the Gauss's law for electric fields states that the surface integral of the electric field $\vec{E}$ over a closed surface (which represents the net electric flux through the surface) is $1/\epsilon_0$ times the total electric charge enclosed by the surface (Chapter 1B, Section 1B. 14). If the closed surface does not enclose any net charge (e.g. when it has both a positive charge and an equal and opposite negative charge within it) but has an electric field associated with it, the net electric flux is zero. It is so, because then the number of lines entering the surface is equal to the number of lines leaving the surface. In the case of magnetic fields, since the magnetic field lines always form closed loops, the number of field lines entering the closed surface is always equal to the number of field lines going out of it (Fig. 3.48). Hence, the net magnetic flux is always zero. In mathematical form, the Gauss's law is expressed as,

$$\phi_B = \int \vec{B}.\vec{dS} = 0 \qquad \text{..... (3.59)}$$

ϕ_B is the net magnetic flux through the surface, $\vec{B}$ is magnetic field and $\vec{dS}$ is a small element of the surface area. The **law** can be **stated** as 'the net magnetic flux through a closed surface is always zero'.

The law is an expression of the fact that a magnetic monopole does not exist. The simplest real magnetic entity is a magnetic dipole, for which, in classical terms, the net magnetic pole strength is essentially zero.

3B. 10. Electrostatic Analog

In our study so far we have found that the electrostatic and magnetic phenomena are analogous to each other in several ways. Some salient magnetic quantities and expressions along with their electrostatic analogs are listed in Table A for ready reference.

3B. 11. Earth's Magnetic Field; Terrestrial Magnetism

Earth is a natural magnet and has its own magnetic field. The alignment of a compass needle in a particular direction at a given place is a simple proof of the earth's magnetism (also referred to as terrestrial magnetism). Another proof is the existence of neutral points (Section 3B. 04). This magnetism can be imagined as arising from the presence of an extremely large and powerful magnet (a huge magnetic dipole) running across the centre of the earth (however, the student should be clear in his mind that actually no such magnet exists inside the earth. Its existence is entirely fictitious). Before describing the characteristics

Table A

Serial No.	Electrostatics	Magnetism
1.	$1/\in_0$ ($\in_0$: Absolute electrical permittivity of free space)	μ_0 (Permeability of free space)
2.	$\vec{E}$ (Electric field)	$\vec{B}$ (Magnetic field)
3.	$\vec{p}$ (Electric dipole moment)	$\vec{p}_m$ (Magnetic dipole moment)
4.	$\vec{E} = \frac{2\vec{p}}{4\pi\in_0 r^3}$ (Electric field due to an electric dipole along axial line)	$\vec{B} = \frac{\mu_0 2\vec{p}_m}{4\pi r^3}$ (Magnetic field due to a magnetic dipole along axial line)
5.	$\vec{E} = \frac{-\vec{p}}{4\pi\in_0 r^3}$ (Electric field due to an electric dipole along equatorial line)	$\vec{B} = \frac{-\mu_0\vec{p}_m}{4\pi r^3}$ (Magnetic field due to a magnetic dipole along equatorial line)
6.	$\vec{\tau} = \vec{p} \times \vec{E}$ (Torque on an electric dipole in uniform electric field)	$\vec{\tau} = \vec{p}_m \times \vec{B}$ (Torque on a magentic dipole in uniform magnetic field)
7.	$U = -\vec{p}.\vec{E}$ (Potential energy of an electric dipole in uniform electric field)	$U = -\vec{p}_m.\vec{B}$ (Potential energy of a magentic dipole in uniform magnetic field)
8.	$\phi = \int \vec{E}.ds = \frac{1}{\in_0}\Sigma q$ (Electric flux; Gauss's law for electric fields)	$\phi_B = \int \vec{B}.d\vec{S} = 0$ (Magnetic flux; Gauss's law for magnetic fields)

of the earth's magnetic field, it will be useful to briefly revise one's knowledge of some related salient geographical elements.

Geographical elements (The students are strongly advised to use a globe or a round object, like a ball, for following this section). Assuming earth to be a perfect sphere, its circular cross-section in the plane of the paper is shown in Fig. 3.49. RR is the axis about which the earth rotates. Its points of intersection with the earth's surface are the **geographical north pole (N_G)** and **geographical south pole (S_G)**. The regions of earth around N_G and S_G are labelled as **arctic region and antarctic region,** respectively. The great circle (i.e. a diametrical circle) passing horizontally through the earth is its equatorial section, marked as EE in the figure. It is called **geographical equator.** The position of any place on the earth's surface is generally indicated by two angles, termed as the **latitude** and **longitude** of the place. The **latitude** of a point on the earth's surface is its angular distance (expressed in degrees) from the equator, either north or south. The latitude of the equator itself is zero and that of N_G or S_G is 90°. The other lines of latitude (30° and 60° lines are drawn in Fig. 3.49) are actually parallel small circles running east-west, joining points of equal latitude on the earth's surface. Delhi has a latitude of 28°40′ (Miami in U.S.A., Cairo in Egypt, Brisbane in Australia, and Chungking in China have nearly the same value, around 30°; the value of latitude largely determines the annual weather pattern of a place). The **longitudes** are the lines formed where vertical great circles (which all contain the axis of rotation RR) run north-south to join at the poles and meet the surface of the earth. The great circles, which all are vertical planes passing through various places on the earth, are called **geographic meridians.** The one passing through Greenwich in England, called **prime meridian,** is taken as the reference position, at 0° longitude. The **longitude** at a place is measured as the angular distance from this position, extending upto 180°, both east or west. In Fig. 3.49, RR is the longitude corresponding to the **prime meridian.** Two other longitudes are shown at 60° and 120°. Delhi has the longitude 77°14′.

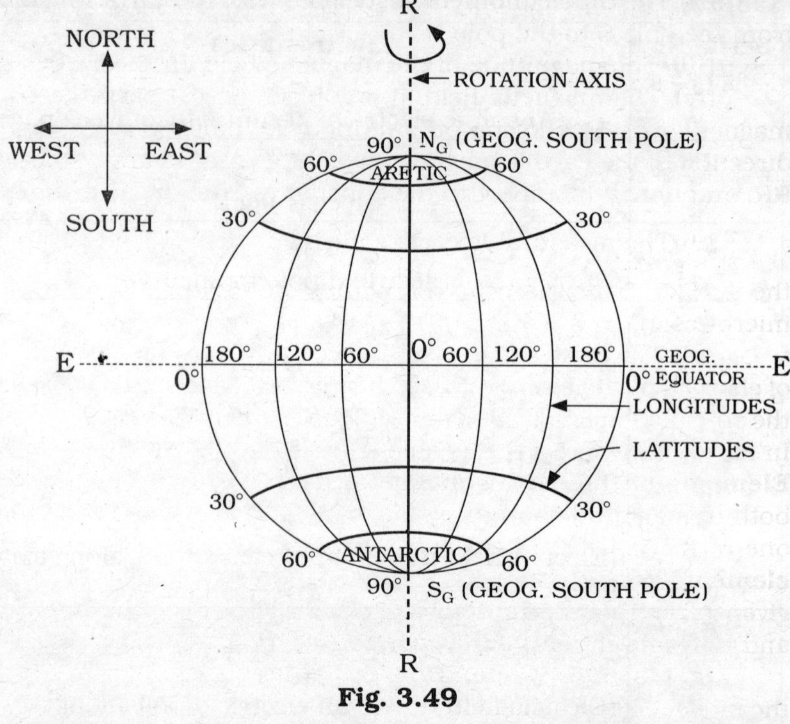

Fig. 3.49

Some salient features of the earth's magnetic field

(i) The axis MM of the earth's magnetic dipole makes an angle of nearly 12° with the earth's rotation axis RR (Fig. 3.50). As we shall find later, the value of this angle is subject to change with time. The south pole S of the dipole is situated in the northern hemisphere and the north pole N in the southern hemisphere (that is why the north pole of a magnetic needle always points towards earth's northern region). The points N_M and S_M, where the axis MM intersects the earth's surface in the northern and southern regions, are known as **geomagnetic north pole** (or **north magnetic pole**) and **geomagnetic south pole** (or **south magnetic pole**), respectively. Thus, note that what we call as 'north magnetic pole' N_G is actually the south pole S of the earth's magnet. The magnetic field lines run from the north pole N to the south pole S of the dipole. [The distance N_MN_G (or S_MS_G) between magnetic and geographic poles, measured along the surface of the earth, is nearly 2000 km].

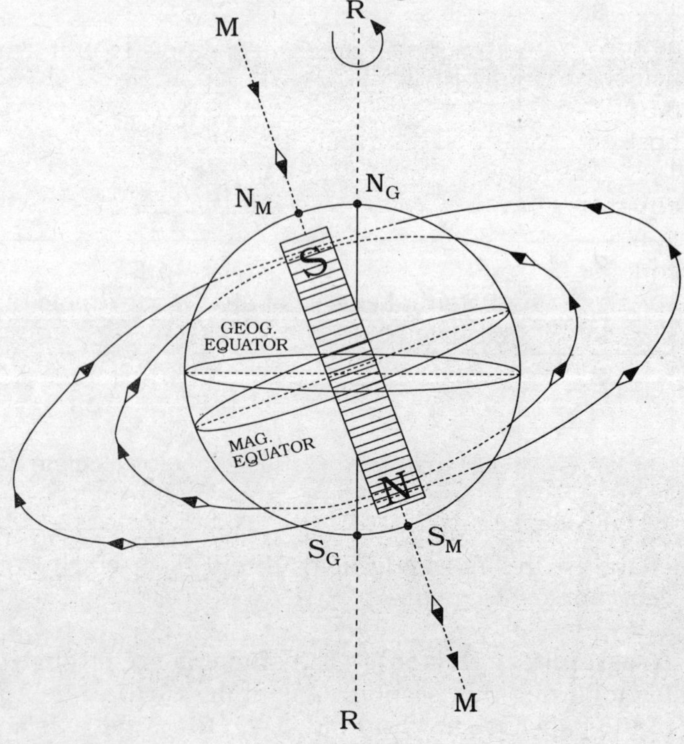

Fig. 3.50

(ii) The dipole moment associated with the earth's field has the magnitude 8.0×10^{22} J/T. It is directed from the pole S to the pole N.

(iii) The magnitude of the magnetic field on the earth's surface is less than 10^{-4} tesla.

(iv) The magnetic field at any place on the earth keeps on changing, i.e. the direction of the earth's magnetic axis MM keeps on shifting with time. The reasons for variation are not known. Therefore, the direction of the earth's magnetic field at a place is always specified with reference to the fixed rotation axis, RR, and not with respect to the constantly changing direction MM.

(v) The measurable earth's magnetic field extends upto the height of approximately 30,000 km above the surface, which is nearly $2\frac{1}{2}$ times the earth's diameter. At this height the field falls below 10^{-6} T (one micro-tesla).

(vi) The earth's field is influenced by the frequently occurring solar flares, in which enormous amounts of charged particles are released from sun (solar wind). They distort the magnetic lines of the field. They cause the so-called magnetic storms, which seriously disturb radio, telegraphic, telephonic, etc., communications. In Fig. 3.50, where a few magnetic lines have been shown, it is assumed that no solar flares exist.

Elements of the earth's magnetic field: The magnetic field of the earth varies from one place to another both in magnitude and direction. This is easily visualized by the movement of a magnetic needle along any one of the magnetic field lines shown in Fig. 3.50. There are **three elements** which completely specify the earth's magnetic field at a given place: **magnetic declination, magnetic inclination (or dip),** and **horizontal component of the earth's magnetic field.**

(a) Magnetic declination, θ: The vertical plane passing through the magnetic north and south poles (N_M and S_M in Fig. 3.50) at a given place is called **magnetic meridian** at that place. Since a magnetic needle orients itself in the direction $N_M - S_M$ at any place, the **magnetic meridian** can **also be defined** as the vertical plane passing through a magnetic needle at that place. The **declination, θ,** at a given place is defined as the angle subtended between the geographic meridian and the magnetic meridian at that place (Fig. 3.51). It arises from the fact that the axis of rotation of earth does not coincide with the axis of the earth's magnetic dipole.

Fig. 3.51

(b) Magnetic inclination (or dip), δ, at a given place is **defined** as the angle subtended between the direction of the earth's magnetic field at that place and a horizontal line in the magnetic meridian (Fig. 3.51). A freely suspended magnetic needle sets itself in a position inclined at the angle δ with respect to the horizontal direction. Hence the name 'inclination' or 'dip'.

The locus of those points where the freely suspended magnetic needle shows no dip is called **magnetic equator** (Fig. 3.50). In 1825, a ship actually traced the magnetic equator that circles the earth. Its position is not very different from the geographic equator. The southernmost region of India lies on the magnetic equator.

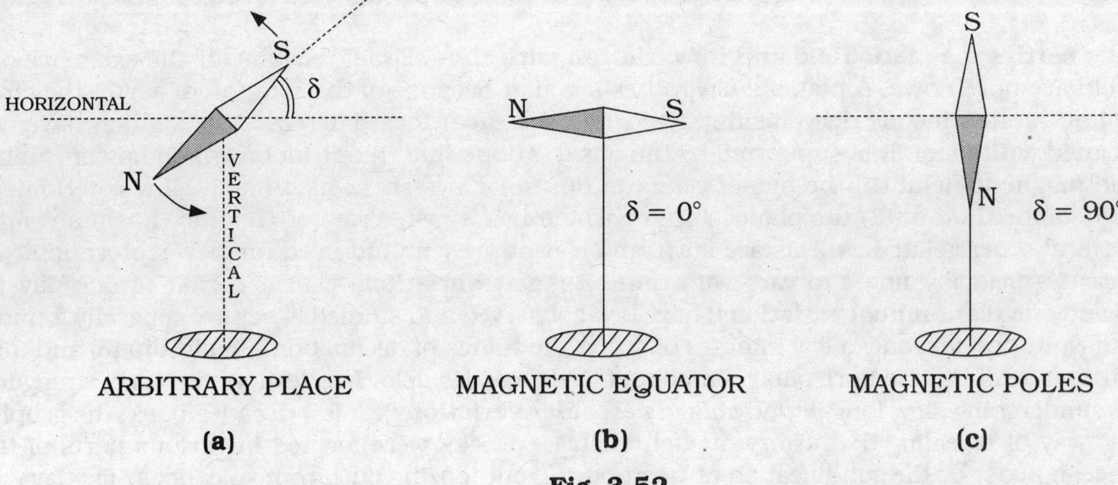

Fig. 3.52

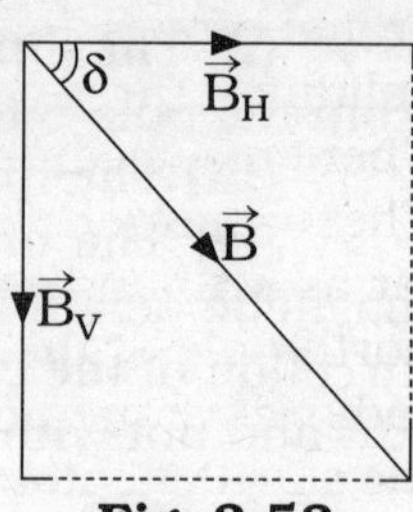

Fig. 3.53

The angle of dip is measured with the help of the instrument called **dip circle,** which essentially has a magnetic needle rotating about a horizontal axis in a vertical plane, with graduations (in degree) marked on a circular dial; the plane is made to coincide with the magnetic meridian. An outline is shown in Fig. 3.52a, depicting the position of the needle at an arbitrary place on the earth's surface. The needle becomes horizontal at all points along the magnetic equator. i.e. the angle of dip, $\delta = 0$ at the magnetic equator (Fig. 3.52b). The needle becomes vertical at the magnetic poles (either north or south), i.e. $\delta = 90°$ at the magnetic poles (Fig. 3.52c). (The needle lies in the vertical plane of the magnetic meridian, which coincides with plane of the paper in the figure).

(c) Horizontal component of the earth's magnetic field (B_H or H). It is the horizontal component of the magnetic field of earth in the magnetic meridian (Fig. 3.51). As seen from Fig. 3.51 and Fig. 3.53, the total magnetic field $\vec{B}$ can be resolved into two components, $\vec{B}_H$ and $\vec{B}_V$ as,

$$\left.\begin{aligned} B_H &= B\cos\delta \\ B_v &= B\sin\delta \end{aligned}\right\} \qquad \text{..... (3.60)}$$

Also
$$B = \sqrt{B_H^2 + B_v^2} \qquad \text{..... (3.61)}$$

and
$$\tan\delta = B_v/B_H \qquad \text{..... (3.62)}$$

Three values of the angle of dip, δ, are of particular interest:

(i) At the magnetic equator, $\delta = 0°$ (Fig. 3.52b).

$$\therefore \qquad B_H = B$$

(ii) At the magnetic poles (either north or south), $\delta = 0°$ (Fig. 3.52c).

$$\therefore \qquad B_H = 0$$

(iii) For $\delta = 45°$, $B_H = B\cos 45° = B\sin 45° = B_V$.

$$\therefore \qquad B_H = B_V$$

For practical purposes, knowledge of the value of horizontal component B_H is usually more important than the value of the total earth's magnetic field B. It can be accurately determined with the help of the instruments called magnetometers (either a **deflection magnetometer** or a **vibration magnetometer**). If necessary, the magnitudes of B_V (= $B_H \tan\delta$) (Eq. 3.62) and B = ($B_H \sec\delta$) (Eq. 3.60) can be calculated. The angle of dip, δ, can be measured with the help of a dip circle. When the declination, θ, is also known, the direction of B can also be expressed, with reference to the true north-south (geographic) direction. Thus, a knowledge of the three quantities θ, δ and B_H at a given place is sufficient to specify the intensity of the earth's magnetic field at that place, in both magnitude and direction. Hence, these three quantities, viz. declination (θ), dip (δ) and horizontal component (B_H), are termed as **elements of the earth's magnetic field, $\vec{B}$,** at a given place on the surface of earth. The horizontal component B_H is also simply denoted as **H.**

Origin of the earth's magnetic field and its variation with time. Exact reasons for the existence of earth's magnetic field are not known. A plausible hypothesis is that because of the rotation of earth, the circulating ions of the hot, molten metals deep inside the core of the earth form current loops, which have magnetic fields associated with them. It is supported by the observations that (i) our moon which has no such molten core, has no magnetic field, (ii) the planet venus in our solar system, which has a slow rotation rate, has a weak magnetic field, and (iii) the planet Jupiter which has a faster rate of rotation, has a stronger field. However, several other related aspects are far from clear at present and need further exploration.

The earth's field is known to vary with time. **Regular variations** with a period of one day **(diurnal variations)** and one year **(annual variations)** have been observed and studied. They are generally found to have very small magnitudes, viz. only a few minutes of arc in the values of declination (θ) and dip (δ) and just about 0.001% in the value of B_H, the horizontal component of the earth's field. In addition, the earth's magnetic field is constantly undergoing slow long period changes **(secular variations).** A few decades back, the geophysicists found a new way of revealing the history of such changes. Rocks were formed in various parts of the earth (including ocean beds) by the solidification of lava pouring out (oozing out) from volcanoes. The lava contains grains of iron that get magnetized by the earth's field and act like floating compasses. As the lava cools and solidifies, these grains are 'frozen' to become permanent records of the direction of the earth's field at that time. There may be hundreds of such frozen lava layers in a rock, one on top of the other, formed at different times. These 'magnetic maps' preserved in each layer provide a picture of the earth's magnetic field, going back as far as six hundred million years. They seem to indicate that during this enormously long period the magnetic north pole of the earth had wandered haphazardly around the geographic poles, like a drunkard! The latest radioactive methods of dating have permitted to determine the age of the lava layers. The studies have led to the startling information that the earth's field completely reverses (viz. north pole becomes south pole and vice versa) over a period of the order of one million years!

Example 7. At a certain place, the values of horizontal and vertical components of the earth's magnetic field are found to be 0.24 gauss and 0.42 gauss, respectively. Determine the angle of dip and resultant intensity of the field at the place.

Solution. Given : $B_H = 0.24$ gauss; $B_V = 0.42$ gauss

The angle of dip is given by,

$$\tan\delta = \frac{B_V}{B_H} = \frac{0.42}{0.24} = 1.75$$

$$\Rightarrow \quad \delta = \tan^{-1}(1.75) = 60°15' \quad \textbf{(Answer).}$$

The resultant intensity of the earth's field is given by,

$$B = \sqrt{B_H^2 + B_V^2} = \sqrt{(0.24)^2 + (0.42)^2}$$

$$= \sqrt{0.234}$$

$$= 0.48 \text{ gauss} \quad \textbf{(Answer).}$$

3B.12. Intensity of Magnetization (or simply Magnetization, $\vec{M}$), Magnetic Intensity ($\vec{H}$) & Magnetic Susceptibility (χ)

We have seen that when a magnetic field is applied across a material, the atomic dipoles (whether already present or created by the applied field) tend to get aligned parallel to the direction of the field. Thus, a net magnetic moment is created in the material parallel to the field or, say, the material becomes 'magnetized'. (The net moment can be opposite to the field, too, as we shall find in the next section for diamagnetic materials).

The intensity of magnetization (or, simply, **magnctization**) is **defined** as 'the magnetic moment per unit volume produced in a material by an applied magnetic field'. It is usually denoted by $\vec{M}$ and has SI unit as Am^{-1} (ampere/metre). If the total dipole moment produced by the atomic dipoles be $\vec{p}_m$ and the volume of the material be V, we have,

$$\vec{M} = -\frac{\vec{p}_m}{V} \qquad \text{..... (3.63)}$$

Thus, we find that on applying an external magnetic field across a material, an additional magnetic field is created inside the material. This additional field is found to be equal to $\mu_0\vec{M}$, where μ_0 is he permeability of free space, i.e. vacuum. If the applied field be $\vec{B}_0$, then the total magnetic field $\vec{B}$ inside the material becomes,

$$\vec{B} = \vec{B}_0 + \mu_0\vec{M} \qquad \text{..... (3.64)}$$

We have earlier seen in the last chapter that the magnetic field produced by any object is a product of μ_0 and a quantity (say, $\vec{H}$) which depends upon certain physical factors related to the object. For instance, the magnetic field due to a current-carrying circular loop is $\mu_0[iR^2/2\ (R^2+x^2)^{3/2}]$ (i: current; R: radius of the loop; x: distance of the point at which the field is measured from the centre of the loop; vide Section 3A. 06, Eq. 3.11). Similarly, the field due to a long straight solenoid is $\mu_0[ni]$ (i: current; n: no. of turns in the solenoid; vide Section 3A. 09, Eq. 3.16). Thus, in general, we can write,

$$\vec{B}_0 = \mu_0\vec{H} \qquad \text{..... (3.65)}$$

$\vec{H}$ has the same units as $\vec{M}$, viz. Am^{-1}. It is called **magnetic intensity or magnetic field strength.**

The magnetization $\vec{M}$ results from the application of the external field $\vec{B}_0$ to the material. Since $\vec{B}_0 = \mu_0 H$ and μ_0 is a constant value for a particular material, it means that $\vec{M}$ should be directly related to the magnetic intensity $\vec{H}$. The relationship is mathematically expressed as,

$$\vec{M} = \chi\vec{H} \qquad \text{..... (3.66)}$$

χ is called **magnetic susceptibility.** Obviously, it is a dimensionless quantity ($\vec{M}$ and $\vec{H}$ have same units). It can have both +ve and –ve values, e.g. $\chi_{silver} = -2.6\times10^{-5}$ and $\chi_{aluminium} = 2.3\times10^{-5}$. From Eq. 3.65, a negative value of χ implies that the resultant magnetic field inside the material is less than the applied field, i.e. the atomic dipoles get so oriented by the applied field that the net magnetic field produced by them opposes the applied field. As we shall see in the next section, this happens for diamagnetic materials. For paramagnetic and ferromagnetic materials, χ is positive, but appreciably large (χ>>1) for the latter. Thus, χ is a measure of the nature of the material. From Eq. 3.66, χ can be **defined** as 'the ratio of the intensity of magnetization $\vec{M}$ produced in a material to the magnetic intensity $\vec{H}$ of the applied field that produces the magnetization'.

Substituting for $\vec{M}$ from Eq. 3.66 into the Eq. 3.65, we get

$$\vec{B} = \mu_0 (1+\chi)\vec{H} \qquad \text{..... (3.67)}$$

We now introduce two new quantities, μ_r and μ as follows,

$$\mu_r = (1+\chi) \qquad \text{..... (3.67)}$$

$$\mu = \mu_0 \mu_r = \mu_0 (1+\chi) \qquad \text{..... (3.69)}$$

Then the Eq. 3.67 becomes,

$$\vec{B} = \mu_0\mu_r \vec{H} \qquad \text{..... (3.70)}$$

Or,

$$\vec{B} = \mu\vec{H} \qquad \text{..... (3.71)}$$

Further, from Eqs. 3.65, 3.71, and 3.69, we have,

$$\frac{\vec{B}}{\vec{B}_0} = \frac{\mu}{\mu_0} = \mu_r$$

μ_r is called **relative magnetic permeability** of the material. It is a dimensionless quantity ($\mu_r = 1+\chi$ and χ is dimensionless). μ is called **magnetic permeability** of the material. It has the same dimensions and unit as μ_0 [$\mu = \mu_0(1+\chi)$]. Since μ, μ_r and χ are interrelated through Eqs. 3.68 and 3.69, the knowledge of any one of them enables to calculate the rest two.

From Eq. 3.71 ($\vec{B} = \mu\vec{H}$), the **magnetic permeability** (μ_r) of a material may be **defined** as 'the ratio of the magnetic permeability (μ) of the material to the magnetic permeability (μ_0) of free space'. Further, from Eq. 3.72 ($\vec{B}/\vec{B}_0 = \mu_r$), it may **also be defined** as 'the factor by which the total magnetic field $\vec{B}$ increases or decreases when the material is brought into an external field $\vec{B}_0$.

3B. 13. Diamagnetic, Paramagnetic, and Ferromagnetic Materials

Initially, the magnetic properties were believed to be confined to a few substances like iron, nickel, cobalt and some of their compounds and alloys. These substances are strongly attracted by a magnet. They were termed as 'magnetic substances' and all others as 'non-magnetic substances'.

However, in 1845, Faraday convincingly demonstrated that magnetism was not confined to these few substances alone. By suspending small needle-like samples of different materials between the poles of a very strong electromagnet, he found that most of them aligned themselves in north pole to south pole direction, like a compass needle. A few substances like bismuth, sulphur, graphite, etc., set themselves at right angles to the field, which meant that they acquired reverse polarities in the field of the electromagnet. Thus he established that all the materials in nature are 'magnetic' in one way or the other. **Based on their visible physical behaviour in a magnetic field,** he was led to the following **classification of the materials in three categories:**

(i) Ferromagnetic materials. They are strongly attracted by magnets and align themselves along the magnetic field in which they are placed. Examples are iron, nickel, cobalt, gadolinium, etc. (also, their compounds and alloys). Even today, many people are inclined to refer to these materials alone as 'magnetic materials' (which is obviously not justified).

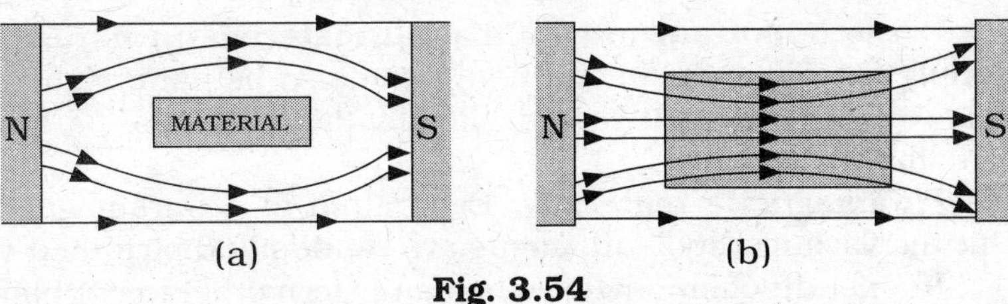

Fig. 3.54

(ii) Paramagnetic materials. They are weakly attracted by magnets of normal strength. Like the ferromagnetic materials, they also align along the magnetic field in which they are placed. Examples are: aluminium, calcium, sodium, manganese, oxygen (at STP), etc.

(iii) Diamagnetic materials. They are repelled by magnets. However, this repulsion is so weak that it cannot be detected unless a very powerful magnet is used. **All materials in nature are diamagnetic.** In addition, they may also be either paramagnetic or ferromagnetic (which masks their diamagnetic character). They align themselves perpendicular to the direction of the magnetic field in which they are placed. Prominent examples are : bismuth, copper, silicon, nitrogen (at STP), water, etc.

With our present day knowledge of the atomic structure down to the electronic level, we now have a much deeper understanding of the aforesaid classification of materials. In Section 3B. 08, it was explained that an electron possesses both an orbital magnetic dipole moment and a spin magnetic dipole moment. The resultant moments of all the electrons in an atom further combine vectorially to produce a total magnetic moment for the atom as a whole. Then all such atomic moments vectorially combine together to yield a net magnetic moment for the material as a whole. In the absence of an external magnetic field, the various atomic moments are randomly oriented inside the material. So, statistically, as many moments are oriented in a particular direction as in the opposite direction. Hence the net magnetic moment, and therefore the net magnetic field, inside the material is zero. The situation changes when an external magnetic field is applied across the material (or, say, the material is placed in an external magnetic field). Then a net magnetic moment, and therefore a net magnetic field, is found to develop inside the material. The materials are found to behave in three different ways. Accordingly, they are classified in three categories, as before: **diamagnetic paramagnetic, and ferromagnetic materials.** A brief description follows, which now includes **a further understanding of the materials on the basis of the behaviour of their atomic dipole moments.**

(i) Diamagnetic materials. A purely diamagnetic material possesses no net magnetic dipole moment. A weak magnetic moment is produced when the material is placed in an external magnetic field. The weak magnetic field so produced has its direction opposite to that of the external field. The explanation for this phenomenon requires knowledge of quantum physics, which is behond the scope of this book. Thus the magnitude of the external field is slightly reduced inside the material. Hence the total field intensity $\vec{B}$ becomes less than the magnetizing intensity $\vec{H}$. Therefore, from the relation $\vec{B} = \mu\vec{H}$ (Eq. 3.71), it follows that the magnetic permeability, $\mu<1$. Further, from Eq. 3.71, ($\vec{B} = \mu\vec{H}$) Eq. 3.65 ($\vec{B}_0 = \mu_0\vec{H}$), and Eq. 3.69 ($\mu = \mu_0\mu_r$), we have,

$$\vec{B}/\vec{B}_0 = \mu/\mu_0 = \mu_r$$

Since the value of the applied field $\vec{B}_0$ is less than the total field $\vec{B}$ inside the material, we find that the relative permeability $\mu_r<1$ and so $\mu<\mu_0$. Then, since $\mu_r = 1+\chi$ (Eq. 3.68), we further have that the magnetic susceptibility, $\chi<1$, i.e. a negative value of susceptibility. In summary, we have for a diamagnetic material,

$$\chi < 1; \qquad \mu, \mu_r < 1; \qquad \mu < \mu_0$$

The magnetic field lines of the applied field are found to be 'repelled' by a diamagnetic material, or, say, 'expelled' from the interior of the material (Fig. 3.54a). If the applied field is removed (which is easily done by simply switching off the current, if the field is that of an electromagnet), the net dipole moment produced by the field completely disappears, i.e. the material is restored to its original condition.

If the material is made in the shape of a needle, the needle tends to orient itself perpendicular to the direction of the applied field.

The diamagnetism is not affected by change of temperature, i.e. it is independent of temperature.

It is **important** to note that **all materials in nature are diamagnetic.** However, a material can be doubly magnetic, i.e. in addition it can also be paramagnetic or ferromagnetic. Then, since diamagnetism is always very weak, it gets masked by the additional paramagnetism or ferromagnetism. So, very often, it remains undetected.

(ii) Paramagnetic materials. Each atom of a paramagnetic material has a permanent dipole moment. But the various dipole moments are randomly distributed within the material and so the material has no or little net dipole moment. When an external field is applied, the atomic moments get partially aligned in the direction of the field and thus produce a net magnetic field of their own. Hence the value of the total magnetic field $\vec{B}$ inside the field is increased and we have $\vec{B} > \vec{H}$ and $\vec{B} < \vec{B}_0$. Then, from the relations $\vec{B} = \mu\vec{H}$, $\vec{B}/\vec{B}_0 = \mu_r$, and $\mu_r = 1+\chi$, we have, in summary,

$$\chi > 1; \qquad \mu, \mu_r > 1; \qquad \mu > \mu_0$$

The magnetic field lines of the external field now prefer to pass through the material than the medium (air) between the poles of the magnet (Fig. 3.54b). However, the flux density is small, because the additional field is generally weak. If the applied field is removed, the net dipole moment inside the material completely disappears, i.e. the material returns to its original condition.

When in needle shape, the material tends to orient itself along the direction of the applied field.

The magnetization of the material in the external field $\vec{B}_0$ at a temperature T is found to be governed by the **Curie law,**

$$\vec{M} = C(\vec{B}_0 / T) \qquad \text{..... (3.73)}$$

where **C** is called **Curie constant.** Using Eq. 3.66 ($\vec{M} = \chi\vec{H}$), the Curie's law is also expressed as,

$$\chi = C\,(\mu_0/T) \qquad \text{..... (3.74)}$$

From Eqs. 3.73 and 3.74 it follows that the magnetization, and therefore the paramagnetic susceptibility of a material, decreases with increase in temperature. (Also note that the magnetization within the material increases with rising value of the applied field. However, the Curie's law is only approximately valid. It holds only for values of B_0/T not very large).

(iii) ferromagnetic materials. A ferromagnetic material contains numerous small pockets, called domains, in which the atomic dipole moments are aligned along a common direction and therefore each domain possesses a strong permanent dipole moment. The common direction differs from one domain to other. When an external field $\vec{B}_0$ is applied, these dipole moments become aligned along the field, producing a strong additional magnetic field inside the material in the same direction as the applied field. Thus a large value of the total field $\vec{B}$ results. As before, in summary, we now have,

$$\chi \gg 1; \qquad \mu, \mu_r \gg 1; \qquad \mu \gg \mu_0$$

(For iron, $\chi > 1000$).

The magnetic field lines again prefer to pass through the material, as for the paramagnetic substances, but now the flux density is very high, i.e. the lines are heavily concentrated inside the material.

When needle-shaped, a ferromagnetic material orients itself along the direction of the external field.

The ferromagnetism, like paramagnetism, also depends on temperature. As the temperature rises, the domain structure continuously breaks down and so the ferromagnetism is gradually lost. At a **critical temperature, T_c, known as Curie temperature,** the ferromagnet becomes a paramagnet. In the paramagnetic phase ($T > T_c$), the susceptibility is given by,

$$\chi = \frac{C}{(T - T_0)} \; (T > T_c) \qquad \text{..... (3.75)}$$

The value of T_c for iron is 770 °C (or 1043 K). At temperatures higher than T_c and approaching the melting point of the material of a magnet, the magnetization is completely lost (the melting point of iron is nearly 1000 °C).

When the external field is removed, the various ferromagnetic materials behave differently. In one group of materials, called 'hard' magnetic materials or 'hard' ferromagnets, the magnetization persists after the removal of the field (one such material is **alnico**, which is an alloy of iron, nickel, cobalt, copper, and aluminium). In another group of ferromagnetic materials, the magnetization disappears on the removal of the applied field. Such materials are called 'soft' materials (an example is soft iron). More about it will be described in Section 3B. 15.

Prominent differences in the characteristics of diamagnetic, paramagnetic and ferromagnetic materials are listed in Table 1.

Table 1. Differences in characteristics of diamagnetic, paramagnetic and ferromagnetic materials.

Serial No.	Characteristic	Diamagnetic	Paramagnetic	Ferromagnetic
1.	χ	< 1	> 1	$\gg 1$
2.	μ, μ_r	< 1	> 1	$\gg 1$
3.	μ vs. μ_0	$\mu < \mu_0$	$\mu > \mu_0$	$\mu \gg \mu_0$
4.	Behaviour near a magnet	weakly repelled	weakly attracted	strongly attracted
5.	Alignment in an external field (needle-shaped material)	$\perp$ to the field	along the field	along the field
6.	Direction of the additional field produced inside the material	opposite to the external field	along the external field	along the external field
7.	Behaviour on removal of the external field	returns to the original condition	returns to the original condition	magnetization persists in some materials and disappears in others
8.	Magnetic field lines	repelled by the material	pass through the material	pass through the material
9.	Flux density	zero	small	large
10.	Variation with temperature	no	yes	yes
11.	Examples	Bi, Cu, Si, N_2 (at STP), H_2O, etc.	Al, Ca, Na, Mn, O_2 (at STP), etc.	Fe, Ni, Co, Gd, etc.

Additional knowledge

(i) For the diamagnetic materials, usually the susceptibility χ is a small negative value. For bismuth, $\chi = -1.7 \times 10^{-5}$. Similarly, the relative permeability μ_r is a small negative value. From the difinitions of χ and μ_r, it follows that usually there is only a slight reduction in the value of the applied field when a diamagnetic materials is placed in the field. However, for type - I superconductors, which we had briefly described in Chapter 2A (Section 2A.09), $\chi = -1$ and $\mu_r = 0$. It means that these superconductors completely expel the magnetic field lines from their interiors when placed in a magnetic field, so that no magnetic field can exist inside such a superconductor (this remarkable phenomenon is known as **Meissner effect).** In other words, such a superconductor acts as a perfect diamagnetic material. It repels a magnet and so it is also repelled by magnets. This fact is very important from application point of view. As an example, attempts are afoot to make superfast magnetically levitated trains, which will run at enormous speeds, because the slight raise from the railway track will render their running almost frictionless.

(ii) The strong dipole moment in a ferromagnetic domain is created by the interaction of electron spins of one atom with those of its neighbouring atoms. The interaction gives rise to the so-called **exchange forces** in the material, which help to align the various atomic dipole moments in a common direction and so produce a large net dipole moment in the domain. Once again, an adequate explanation of these so-called 'cooperative' phenomena requires the knowledge of quantum physics, which is beyound the scope of this book.

(iii) Ferrimagnetic and antiferromagnetic materials. In addition to the three categories of magnetic materials discussed by us, viz. diamagnetic, paramagnetic and ferromagnetic, we have a few materials belonging to special categories: (i) ferrimagnetic materials and (ii) antiferromagnetic materials. They are characterized by special spin orders, a detailed discussion of which is beyond the scope of this book.

3B. 14. Hysteresis

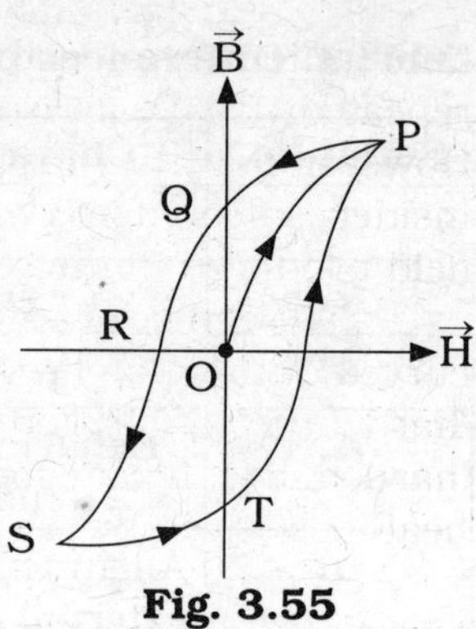

Fig. 3.55

Let us consider change in the magnetization ($\vec{B}$) of a ferromagnetic material with change in the magnetizing field intensity ($\vec{H}$). It is depicted in Fig. 3.55. $\vec{H}$ can be easily varied by using an electromagnet and varying the current in the coils of the magnet. As $\vec{H}$ is raised from zero value (point O), $\vec{B}$ gradually rises to a saturation value (point P). (Saturation means that $\vec{B}$ does not increase further even if $\vec{H}$ is further increased). When $\vec{H}$ is gradually reduced back to zero, $\vec{B}$ does not fall to zero (point Q). Thus, some magnetism is retained in the material in spite of the magnetizing force becoming zero. OQ, which represents the value of magnetization $\vec{B}$ after the magnetizing intensity $\vec{H}$ returns to zero, is called **retentivity** or **residual magnetism.** If now $\vec{H}$ is reversed (by reversing the current in the coil) and gradually increased, $\vec{B}$ gradually decreases to zero (pt. R). It follows that to bring down the magnetization of the material to zero, a reverse magnetizing force represented by OR has to be applied. Hence OR is called **Coercivity** of the material (the word coerce means 'to persuade by using force or threat'). Further increase of $\vec{H}$ makes us to arrive at another saturation point S. Then a gradual reduction of $\vec{H}$ brings us to point T, after which a further reversal and gradual increase of $\vec{H}$ makes us to reach back the point P. Thus the cycle of magnetization is completed.

We find that the magnetization curve OP is not retraced when the magnetizing field returns to zero (points Q and T). This **phenomenon of lack of retraceability of magnetization of a ferromagnetic material** is called **hysteresis** (it is derived from the Greek word 'Hussteresis', meaning 'deficiency' or 'coming late'). The cyclic curve PQRSTP is called **hysteresis loop.** The phenomenon is well understood in terms of the concept of magnetic domains described for ferromagnetic materials in the last section. As the applied field rises to the saturation point P, the magnetic dipole moments in the domains align themselves along the field. When the field gradually reduces back to zero (point Q), the magnetic dipoles do not return completely to their original configuration. They do so only when a 'reverse force' is applied, by reversing the direction of the applied field (point R). Then the dipoles begin to flip, i.e. reverse their orientation. In other words, the domains retain some 'memory' of their existing alignment even after the applied field is switched off. Such retention of 'memory' of ferromagnetic materials is of immense practical value. An example is its utility for storage of information in computers, cassette tapes, credit cards, etc. The naturally occurring lodestones (vide Section 3B. 01) also owe their magnetism to the existence of magnetic domains in them. When a lightning strikes, it sends strong currents through numerous vertical zigzag paths to earth. The strong magnetic field associated with these currents magnetizes those parts of the rocks on earth which contain ferromagnetic materials like Fe, Ni, Co, etc. These parts then become the so-called lodestones by retaining the 'memory' in their domains.

The linear $\vec{B} - \vec{H}$ relationship ($\vec{B} = \mu\vec{H}$) is not valid for ferromagnetic materials. Otherwise in Fig. 3.55 we would have got a straight line plot instead of the hysteresis loop. The value of the relative permeability μ_r varies for different relative values of $\vec{B}$ and $\vec{H}$ along the hysteresis loop. At the saturation point P in the figure, μ_r can be as high as nearly 10,000!

During the cycle of magnetization of a ferromagnetic material, represented by the hysteresis loop in Fig. 3.55, the magnetic dipoles are continuously oriented and reoriented, i.e. we have a continuous molecular motion. So work is done, which manifests as heat and the material gets heated. The heat so produced, e.g. in the iron core of a transformer where the current (AC) is continuously reversed and so the hysteresis cycle is continuously repeated, is dissipated to the surroundings. It can be shown that energy per unit volume of the material per cycle of magnetization is equal to the area of the $\vec{B} - \vec{H}$ hysteresis loop.

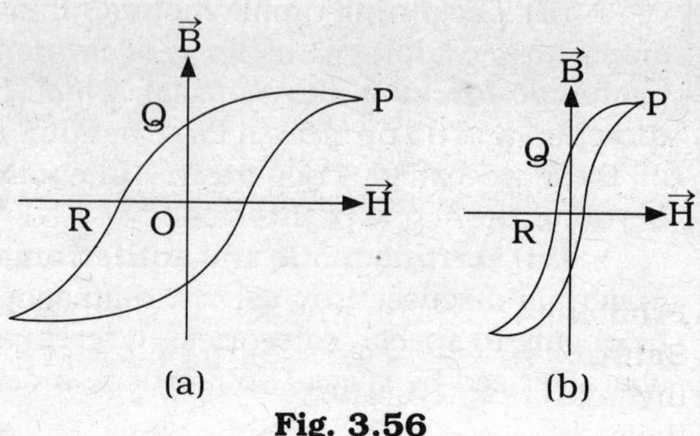

Fig. 3.56

3B. 15. Permanent Magnets and Electromagnets

Any magnet in use slowly tends to lose its magnetism, because temperature effect, handling, proximity to other magnetic fields, etc., constantly keep on disturbing the alignment of its magnetic dipole moments. The **permanent magnets** are those which can retain their magnetism for long periods of time (upto several years) at room temperature. The earlier methods of making a permanent magnet were to repeatedly strike an iron rod held in north-south direction, to cyclically move a magnet over an iron bar, etc. However, now the most convenient and efficient way is to place a ferromagnetic material rod (e.g. an iron rod) inside a solenoid and pass a current through the coils of the solenoid for a sufficiently long time. The magnetic field produced in the solenoid due to the flow of electric current in the coils magnetizes the rod.

The choice of a right material for making a permanent magnet is important. For this purpose, hysteresis curves of the materials are very useful guides. In the previous Section 3B. 13, we had mentioned abut the 'hard' and 'soft' ferromagnetic materials (e.g. steel and iron, respectively). Typical hysteresis curves for the 'hard' and 'soft' categories are shown in Fig. 3.56a, b, respectively. The retentivity (OQ), coercively (OR) and heat loss (given by the area of the loop) are higher for the 'hard' materials.

For making a permanent magnet, the material should have a high retentivity (OQ), so that it continues to possess strong magnetism after the process of magnetization. It should also have a high permeability, so that a magnetic field of moderate strength can magnetize it ($\vec{B} = \mu\vec{H}$). It should also possess a high coercivity (OR), so that when in use its magnetism does not easily erase by handling, rise in temperature, presence of other magnetic fields, etc. A common choice is steel. Some iron alloys, e.g. alnico (an alloy of Fe, Co, Ni & Cu), are also available. Depending upon the requirement, the permanent magnets are made in several shapes and sizes.

An **electromagnet,** as the name suggests, is made by passing electric current through a coil wound around a suitable ferromagnetic material. Electromagnets are also made in different shapes and sizes, depending on their use. Here the main requirement for the selection of material is a low retentivity, so that little magnetism is retained after the current is switched off. In other words, the material remains a strong magnet only so long as the current flows in the coil. A high permeability is also needed, so that a strong magnetism is produced when the current is switched on. A low value of coercively is useful. Thus, a soft ferromagnetic material is the right choice for an electromagnet (Fig. 3.56b). Soft iron is a common choice. Some iron alloys are also available, e.g. mu metal, radio metal, etc. The use of electromagnets ranges from small electric bells to huge cranes employed for lifting machinery and other large articles.

A list of characteristics (requirements) for making permanent magnets and electromagnets is provided in Table 2. Many applications involve the flow of alternating current (AC) through the coils enclosing a ferromagnetic material, e.g. transformer cores, telephone diaphragms, etc. Then the material continuously undergoes repeated cycles of magnetization and so the consideration of heat loss becomes very important. This calls for the area enclosed by the hysteresis loop to be small, for which again a 'soft' ferromagnetic material like soft iron is suitable.

Table 2. Characteristics (requirements) for making permanent magnets and electromagnets

Permanent magnets	Electromagnets
High retentivity	Low retentivity
High coercively	Low coercively
High permeability	High permeability
Require 'hard' ferromagnetic materials,	Require 'soft' ferromagnetic materials e.g. soft iron, some iron alloys, etc.

Solved Conceptual Problems

[Including HOTS Questions]

Problem 1. Explain why two magnetic field lines cannot cross each other.

Solution. If two magnetic field lines could intersect, it would be possible to draw two different tangents at the point of intersection. That in turn would mean that the magnetic field $\vec{B}$ had two different directions at the same point, which is impossible. Hence no two field lines can cross each other.

Problem 2. Explain how the magnetic field lines are continuous curves, but the electric field lines are discontinuous curves, forming loops.
Solution. The magnetic field lines due to a magnet extend through the body of the magnet. Thus, they are continuous curves, forming closed loops. On the other hand, no electric field lines can exist inside a charged conductor. Hence they are discontinuous curves.

Problem 3. Is the earth's magnetic field produced by a huge magnet existing in the core of earth ? If not, what else can be the most likely source of earth's magnetism ?
Solution. No magnet exists inside the earth. The most possible source of the earth's magnetism are the ions of hot molten metals deep inside the core of earth. Because of the rotation of earth, the ions also circulate continuously, forming current loops, which have magnetic fields associated with them.

Problem 4. If a magnetic field exists at a place, how can one decide if it is due to the earth's magnetism alone or also due to a nearby current-carrying conductor ?
Solution. It can be decided with the help of a compass needle. If the needle always points north-south at different positions in the place, then the magnetic field at the place is solely due to the earth's magnetism. If its direction changes at different positions, then it means the presence of an additional magnetic field due to a nearby current-carrying conductor. It can be confirmed by switching off the current flowing in the conductor, when the needle will point north-south.

Problem 5. What are the pole strengths and magnetic moments of two pieces of magnets obtained by cutting a long bar magnet (a) along its length and (b) perpendicular to its length ?
Solution. When a bar magnet is cut or broken into two parts, new poles appear, such that each part becomes a new smaller bar magnet, possessing its own north and south pole.
(a) The pole strength becomes half. The magnetic moment also becomes half ($p_m = m \times 2a$; m is halved, but the length 2a remains unchanged).
(b) The pole strength remains unchanged. The magnetic moment becomes half ($\because$ the length is halved).

Problem 6. You are given two soft iron bars, identical in size and shape, of which only one is a magnet. Using nothing else except the bars themselves, how can you decide which of the two is a bar magnet ?
Solution. Let us call the magnetized bar and the unmagnetized bar as A and B, respectively. Place one of them horizontally on the table and touch its ends and mid-region with one end of the other bar. If the bar placed on the table is A, then the end of B will experience attraction at each end of A, but not in the mid-region. If the bar placed on the table is B, then the end of A will experience equal attraction all along the length of B.

Problem 7. If a compass needle is placed exactly at the geomagnetic north pole or south pole, in which direction will it point ? Give reason for your answer.
Solution. The earth's magnetic field is exactly vertical at the geomagnetic poles (angle of dip, $\delta = 90°$). Therefore its component in the horizontal plane is zero. Since the compass needle is free to rotate in the horizontal plane, it may point in any arbitrary direction.

Problem 8. If a bar magnet is heated upto the melting point of its material, will it retain its magnetism ? Give reason for your answer.
Solution. A bar magnet can be taken as consisting of an extremely large number of tiny magnetic dipoles, all of which are aligned in the same direction. At the high temperatures approaching the melting point, the severe molecular agitation inside the magnet makes these dipoles to be strongly disturbed and hence makes them to point in random directions. Hence the magnet loses its magnetism.

Problem 9. If a magnet is brought near iron filings, the filings not only tend to translate towards the magnet, but also tend to rotate. Explain how.
Solution. The field produced by a magnet is, in general, non-uniform. Hence it exerts both a net force and a torque on a filing. Therefore the filing both translates towards the magnet and rotates, before sticking to the magnet.

Problem 10. (i) It is believed that the earth's magnetic field has reversed a number of times in the last nearly 5 billion years. What is the basis of this belief ?

(ii) Does the earth's magnetic field regularly change with time in more that one way ? If so, briefly indicate how ?

Solution. (i) This information is based on the study of a large number of solidified lava layers contained in rocks. The lava flowing out of volcanoes from time to time, contains grains of iron that get magnetized by the earth's field and so act like floating compasses. As the lava cools and solidifies, these grains get 'frozen', to become permanent records of the direction of the earth's field at that time. Different lava layers are formed at different times.

(ii) Regular variations in the earth's magnetic field, ranging from a period of one day (diurnal variations) to one year (annual variations), have been observed. In addition, constant long period variations (secular variations) have also been observed.

Problem 11. The earth's core contains huge quantities of iron, yet it is not deemed as the source of earth's magnetism. Explain why.

Solution. The iron contained in the earth's core is in molten form and therefore it is non-magnetic. Hence it cannot be related to the magnetic field of earth.

Problem 12. Why magnetic lines of force preferentially pass through ferromagnetic materials ?

Solution. In the presence of a magnetic field, the innumerable tiny dipoles inside a ferromagnetic material readily tend to align themselves along the direction of the field. In other words, the ferromagnetic materials have high values of magnetic permeability. Hence the magnetic lines of force preferentially tend to pass through them.

Problem 13. How can a selected region of space be shielded from the influence of external magnetic fields ?

Solution. The region can be shielded by surrounding it with a wide soft iron ring. The magnetic field lines due to any external fields are drawn into the ring, thus freeing the region from the influence of the external fields.

Problem 14. If we have two identical samples, one paramagnetic and the other ferromagnetic, can they be magnetized to the same maximum extent ? Give reason for your answer.

Solution. Yes, they can be magnetized to the same maximum extent, which we can term as saturation point. But, because paramagnetic materials have very small susceptibility (χ), the paramagnetic sample will require application of extremely high magnetic fields to acquire the saturation ($\vec{M} = \chi \vec{H}$; M : magnetization; $\vec{H}$: magnetic intensity of the applied magnetic field).

Problem 15. What are the guiding considerations for the choice of materials for making a permanent magnet ?

Solution. For making a permanent magnet, the material should have a high retentivity, so that it continues to possess strong magnetism after the process of its magnetization is completed. It should also have a high permeability, so that a magnetic field of moderate strength can magnetize it ($\vec{B} = \mu\vec{H}$). It should also possess a high coercively, so that when in use its magnetism does not easily erase by handling, rise in temperature, presence of other magnetic fields, etc.

Problem 16. What are the guiding considerations for the choice of material for making an electromagnet ?

Solution. The main requirement is a low retentivity, so that little magnetism is retained after the current passing through the coil (wrapped around the magnet) is switched off. A high permeability is also needed, so that a strong magnetism is produced when the current is switched on. A low value of coercively is desirable.

Problem 17. Explain why a ferromagnetic material gets heated when undergoing repeated cycles of magnetization and demagnetization.

Solution. During the repeated cycles of magnetization and demagnetization of a ferromagnetic material, the innumerable tiny magnetic dipoles inside are continuously oriented and disoriented, i.e. we have a continuous molecular motion. So work is done, which manifests as heat energy. Thus the material gets heated.

Solved Numerical Problems

Problem 1. The magnetic moment of a small bar magnet is 0.52 JT^{-1}. Determine the strength and direction of the magnetic field produced by the magnet at a distance of 12 cm from the centre of the magnet, along the (i) axis of the magnet and (ii) normal bisector of the magnet.

Solution. Given : $p_m = 0.52\ JT^{-1}$; $x = 12\text{ cm} = 12\times10^{-2}\text{ m}$

(i) For a small bar magnet, the magnetic field at an axial point is given by,

$$B_{ax} = \frac{2\mu_0 p_m}{4\pi x^3} = \frac{2\times(4\pi\times10^{-7})\times(0.52)}{4\pi\times(12\times10^{-2})^3}$$

$$= 6.02\times10^{-5}\text{ T (or 60.2 }\mu\text{T)} \quad \textbf{(Answer).}$$

The direction of the field is south-north. **(Answer).**

(ii) For a small bar magnet, the magnetic field at an equatorial point is given by,

$$B_{eq} = \frac{\mu_0 p_m}{4\pi x^3} = \frac{B_{ax}}{2}$$

$$= \frac{6.02\times10^{-5}}{2}$$

$$= 3.01\times10^{-5}\text{ T (or 30.1 }\mu\text{T)} \quad \textbf{(Answer).}$$

The direction of the field is north-south **(Answer).**

Problem 2. A torque of 0.02 Nm acts on a small bar magnet located in a uniform magnetic field of strength 1000 gauss, with its axis inclined at 45° to the field. (i) Calculate the magnetic moment of the magnet and (ii) the work done in moving the magnet from its most stable position to the most unstable position. (iii) If the magnet is substituted by an equivalent solenoid consisting of 800 turns and having area of cross-section as $3\times10^{-4}\text{ m}^2$, find the magnitude of current flowing through the solenoid.

Solution. Given : $B = 1000\text{ gauss} = 1000\times10^{-4}\text{ T} = 10^{-1}\text{ T}$; $\tau = 0.02$ Nm; $\theta = 45°$; $N = 800$ turns; $A = 3\times10^{-4}\text{ m}^2$

(i) In terns of torque, the magnetic moment is given by,

$$\tau = p_m B \sin\theta$$

$$\Rightarrow \quad p_m = \frac{\tau}{B\sin\theta} = \frac{(0.02)}{(10^{-1})\times(\sin 45°)} = \frac{0.02}{0.1\times0.7071}$$

$$= 0.28\text{ Am}^2 \quad \textbf{(Answer).}$$

(ii) The most stable position of the magnet corresponds to $\theta_1 = 0°$ and the most unstable position to $\theta_2 = 180°$. The work done is given by,

$$W = p_m B(\cos\theta_1 - \cos\theta_2) = p_m B(\cos 0° - \cos 180°)$$

$$= p_m B[1-(-1)] = 2\,p_m B = 2\times(0.28)\times(10^{-1})$$

$$= 0.056\text{ J} \quad \textbf{(Answer).}$$

(iii) The magnetic moment of the equivalent solenoid is given by,

$$p_m = NiA$$

$$\therefore \quad i = \frac{p_m}{NA} = \frac{(0.28)}{(800)\times(3\times10^{-4})}$$

$$= 1.17\text{ A} \quad \textbf{(Answer).}$$

Problem 3. A bar magnet is placed in a uniform magnetic field of magnitude 0.24 T, with its axis aligned along the field. Its magnetic moment is 1.8 JT^{-1}. (i) Calculate the work needed to align its axis (a) normal to the field and (b) opposite to the field. (ii) Also evaluate the torque acting on the magnet in these cases.

Solution. Given : $p_m = 1.8\ JT^{-1}$; $B = 0.24$ T

We know that the work done in turning the magnet from a position θ_1 to another position θ_2 is given by,

$$W = p_m B(\cos\theta_1 - \cos\theta_2)$$

(i) Case (a) : Here $\theta_1 = 0°$ and $\theta_2 = 90°$

∴ $W = p_m B (\cos 0° - \cos 90°)$

$= (1.8) \times (0.24)(1-0)$

$= 0.43$ J **(Answer).**

Case (b) : Here $\theta_1 = 0°$ and $\theta_2 = 180°$

∴ $W = p_m B (\cos 0° - \cos 180°)$

$= (1.8) \times (0.24) [1-(1)] = 1.8 \times 0.24 \times 2$

$= 0.86$ J **(Answer).**

(ii) The torque acting on the magnet is given by,

$\tau = p_m B \sin\theta$

Case (a) : Here $\theta = (90° - 0°) = 90°$

∴ $\tau = p_m B \sin 90° = (1.8) \times (0.24) \times 1$

$= 0.43$ Nm **(Answer).**

Case (b) : Here $\theta = (180° - 0°) = 180°$

∴ $\tau = p_m B \sin\theta = p_m B (\sin 180°)$

$= p_m B \times$ (zero)

= Zero **(Answer).**

Problem 4. A current of 3 A flows through a closely wound solenoid, which is suspended such that it can freely turn about its centre in a horizontal plane. It consists of 1800 turns and has cross-sectional area of 1.5×10^{-4} m^2. (i) Calculate the magnetic moment associated with the solenoid. (ii) If the solenoid is surrounded by a uniform magnetic field of strength 6×10^{-2} T, making on angle of 45° with the solenoid axis, evaluate the (a) torque and (b) magnetic force, acting on it.

Solution. Given : N = 1800; i = 3A; $A = 1.5 \times 10^{-4}$ m^2; $B = 6 \times 10^{-2}$ T; $\theta = 45°$

(i) The magnetic moment associated with a solenoid is given by,

$p_m = NiA = (1800) \times (3) \times (1.5 \times 10^{-4})$

$= 0.81$ Am^2 **(Answer).**

$\vec{p}_m$ lies along the axis of the solenoid, with its direction related to the direction of flow of current through the right hand screw rule. **(Answer).**

(ii) (a) The torque acting on the solenoid is given by,

$\tau = p_m B \sin\theta$

$= (0.81) \times (6 \times 10^{-2}) \times \sin 45°$

$= 0.81 \times 6 \times 0.7071 \times 10^{-2}$

$= 3.44 \times 10^{-2}$ Nm^{-1} **(Answer).**

$\vec{\tau}$ will act in a direction such that the torque will tend to align the solenoid axis along $\vec{B}$. **(Answer).**

(b) The force acting on a bar magnet (the solenoid is equivalent to a bar magnet) in a uniform field is zero. **(Answer).**

Problem 5. The magnitude of the earth's magnetic field at the equator is nearly 0.4 gauss. Determine the approximate value of the dipole moment of earth (radius of earth ≈ 6.4×10^6 m).

Solution. Given : $B_{eq} = 0.42$ G $= 0.42 \times 10^{-4}$ T; r (earth) $= 6.4 \times 10^6$ m

The (imaginary) magnet responsible for producing the earth's magnetic field is believed to be located in the central region of earth. Therefore its length (2a) << radius (r) of earth, and so the equatorial field is given by,

$$B_{eq} = \frac{\mu_0 p_m}{4\pi x^3} \qquad (x = r >> a)$$

$$\Rightarrow \qquad p_m = \frac{4\pi r^3 B_{eq}}{\mu_0}$$

$$= \frac{4\pi \times (6.4 \times 10^6)^3 \times (0.42 \times 10^{-4})}{(4\pi \times 10^{-7})}$$

$$= 1.10 \times 10^{23} \text{ Am}^2$$ **(Answer).**

Problem 6. In a city located in the region of magnetic meridian, the angle of dip and the horizontal component of the earth's magnetic field are measured to be 45° and 0.32 gauss, respectively. Evaluate the earth's field at the city's location.

Solution. Given : $\delta = 45°$; $B_H = 0.32$ G

The relation between the earth's field B and its horizontal component B_H is given by,

$$B_H = B \cos \delta$$

$$\therefore \quad B = \frac{B_H}{\cos \delta} = \frac{(0.32)}{(\cos 45°)}$$

$$= \frac{0.32}{0.7071} = 0.45 \text{ gauss} \quad \textbf{(Answer).}$$

Problem 7. The horizontal component of the earth's magnetic field at a certain location is found to be 0.36 G. For a short magnet, having its north pole pointing towards north direction, the neutral point is found to be located 10 cm away from the centre of the magnet. Find the magnetic moment of the magnet.

Solution. Given : $B_H = 0.36 \text{ G} = 0.36 \times 10^{-4}$ T; $x = 9 \text{ cm} = 9 \times 10^{-2}$ m

For a small magnet, the field at an axial point is given by,

$$B_{ax} = \frac{2\mu_0 p_m}{4\pi x^3}$$

At the neutral point (located at a point along the axis of the magnet, in this case),

$$B_{ax} = B_H$$

$$\Rightarrow \quad \frac{2\mu_0 p_m}{4\pi x^3} = B_H$$

$$\Rightarrow \quad p_m = \frac{4\pi x^3 B_H}{2\mu_0} = \frac{4\pi \times (9\times10^{-2})^3 \times (0.36\times10^{-4})}{2\times(4\pi\times10^{-7})}$$

$$= 0.13 \text{ Am}^2 \quad \textbf{(Answer).}$$

Problem 8. In a hydrogen atom, the electron revolving in the orbit of radius 0.05 nm makes 10^{16} revolutions per second. Determine the magnetic moment associated with this orbital motion of the electron.

Solution. Given : $r = 0.05 \text{ nm} = 0.05\times10^{-9}$ m; $v = 10^{16}$ r.p.s.

The required magnetic moment is given by,

$$p_m = iA = (ev) \times (\pi r^2)$$

$$= [(1.6\times10^{-19}) \times (10^{16})] \times [(3.14) \times (0.05\times10^{-9})^2]$$

$$= 1.26\times10^{-23} \text{ Am}^2 \quad \textbf{(Answer).}$$

Very Short Answer Questions

[Including HOTS Questions]

(1 mark each)

1. What is a natural magnet ?
2. Is it possible to have a monopole (a single north pole or a single south pole) ?
3. Do the two poles of a magnet have same strength ?
4. Why repulsion, not attraction, is considered to be the sure test for the magnetism of a material ?
5. What constitutes a magnetic dipole ? Is a small bar magnet a dipole ?
6. Define magnetic moment. What is its unit ?
7. Is magnetic moment a scalar quantity or a vector quantity ? If vector, what is its direction ?

8. In what position a magnetic dipole held in a uniform magnetic field has (a) minimum potential energy and (b) maximum potential energy ?

9. For a magnet held in a uniform magnetic field, in which position the torque acting on it is (a) maximum and (b) minimum ?

10. An iron filing on being attracted by a magnet moves towards it and sticks to it. What is the source of its kinetic energy? [Hint : a part of the magnetic energy is converted into kinetic energy].

11. What is the approximate distance from earth upto which the earth's magnetic field can be detected and measured ?

12. What is the approximate value of the horizontal component of the earth's magnetic field in India ?

13. Why does a freely suspended magnet or a pivoted compass needle always point north-south ?

14. What is magnetic meridian ?

15. What is geographic meridian ?

16. What is geomagnetic equator ?

17. Name the elements of the magnetic field of earth.

18. What is meant by angle of dip at a place ?

19. What is meant by angle of declination at a place ?

20. What is the value of angle of dip at (a) equator and (b) south pole ?

21. What is the magnitude of horizontal component of the earth's magnetic field at the geomagnetic north pole or south pole ?

22. At a certain place on earth, the magnitudes of the horizontal component and the vertical component of the earth's magnetic field are found to be equal. What is the angle of dip at the place ?

23. At a certain place the ratio of the horizontal component of earth's magnetic field to the total value of the field is found to be equal to 1. What is the angle of dip at the place ?

24. The values of angle of dip at two places on earth are measured to be 0° and 90°, respectively. What are the locations of these places ?

25. An iron box is kept in a room. Is the magnitude of earth's magnetic field inside it less or more than that existing outside the box ?

26. What is meant by a neutral point ?

27. A magnet is placed with its north pole pointing towards the geographic north. Where will the neutral point be located ?

28. If the neutral point is found to be located along the axis of a magnet, how is the magnet oriented with respect to the earth's magnetic field ?

29. What is meant by magnetizing field intensity ? Give its SI unit.

30. What is meant by intensity of magnetization ? Give its SI unit.

31. What is meant by magnetic induction ? Give its SI unit.

32. What is meant by magnetic permeability ? Give its SI unit.

33. What is meant by magnetic susceptibility ? Give its SI unit.

34. What is meant by relative permeability ? How is it related to magnetic susceptibility ?

35. What is meant by diamagnetism ?

36. What is meant by paramagnetism ?

37. What is meant by ferromagnetism ?

38. The magnetic susceptibility of a material is found to be a small positive value. To which class of magnetic materials does it belong ?

39. The magnetic susceptibility of a material is found to be a small negative value. To which class of magnetic materials does it belong ?

40. The magnetic permeability of a material is 0.97. To which class of magnetic materials does it belong ?

41. What is meant by Curie point of a magnetic material ?

42. What is meant by hysteresis of a magnetic material ?

43. What does the area of hysteresis loop indicate physically ?

44. Distinguish between a permanent magnet and an electromagnet.

45. Name two substances commonly employed for making permanent magnets.

46. Name a material commonly used for making electromagnets.

Short Answer Questions

[Including HOTS Questions]

(2 or 3 marks each)

1. Define dipole moment of a magnet in terns of the torque acting on it when located in a uniform magnetic field. Give its SI unit. Is it a vector quantity ?

2. What do you mean by a uniform magnetic field ? Sketch the magnetic field lines to represent it.

3. What are magnetic field lines ? Sketch them for a bar magnet with its north pole pointing towards geographic north.

4. Compare the magnetic behaviours of a straight solenoid and a bar magnet.

5. Distinguish between diamagnetic, paramagnetic and ferromagnetic materials. Give one example of each.

6. What is meant by retentivity and coercively of a magnetic material ?

7. A magnetic dipole is located in a uniform magnetic field. Deduce an expression for the torque experienced by it in the field. Hence define magnetic moment.

8. A magnetic dipole is placed in a uniform magnetic field, making an angle θ with the direction of the field. Derive an expression for its potential energy in the field.

9. What are magnetic field lines ? State their important characteristics.

10. Name the elements of the earth's magnetic field and briefly explain the meaning of each of them.

11. What is a neutral point ? Through line diagrams, indicate its positions for a bar magnet having its north pole pointing towards (a) geographic north and (b) geographic south.

12. Using a labelled diagram, indicate the positions of the elements of earth's magnetic field and derive relations between them.

13. Briefly describe the phenomenon of hysteresis in magnetic substances.

14. Explain how an atom is considered equivalent to a magnetic dipole and derive a relation for the orbital magnetic moment of an electron. Use it to define Bohr magneton.

15. Explain the origin of the phenomenon of paramagnetism.

16. Explain the origin of the phenomenon of ferromagnetism on the basis of magnetic domains.

Numerical Exercises

1. Determine the strength of the magnetic field produced by a 9 cm long magnet at an axial point distant 15 cm from the centre of the magnet. The pole strength of the magnet is 100 CGS units (1 CGS unit of pole strength = 0.1 A.m). **[0.19 μT]**

2. Find the magnitude of the magnetic field due to a bar magnet of length 8 cm and magnetic moment 0.5 Am^2, at a point distant 30 cm from its centre, both along its axis and along the bisector drawn at the centre. **[18.52×10^{-7} T; 37.04×10^{-7} T]**

3. A small bar magnet is located in a uniform magnetic field of magnitude 0.3 T, with its axis inclined at 60° to the direction of the field. If a torque of magnitude 5×10^{-2} N.m acts upon it, calculate its magnetic moment. **[0.19 JT^{-1} (or Am^2)]**

4. A small bar magnet of magnetic moment 0.45 JT^{-1} is suspended in a uniform magnetic field of strength 0.18 T. Find its potential energy, both in the position of stable equilibrium and unstable equilibrium. **[-8.1×10^{-2} J; + 8.1×10^{-2} J]**

5. A current of 1.5 A flows through a solenoid, with number of turns as 800 per metre length. If the relative permeability of the core of the solenoid is 250, determine the (i) magnetizing intensity, H, (ii) magnetic field, B, and (iii) intensity of magnetization, M. **[1.2×10^3 Am^{-1}; 0.38 T; 3×10^5 Am^{-1}]**

6. A dipole possesses a magnetic moment of 1.8 Am^2. Find the magnitude and direction of the magnetic field produced by it at a point distant 60 cm from its centre, along a radial direction inclined at 30° to its axis. **[15×10^{-7} T; $\phi = \tan^{-1}(1/2\sqrt{3})$, with the radial direction]**

7. A torque of magnitude 0.6 JT^{-1} acts on a small bar magnet located in a uniform magnetic field, with its axis making an angle of 60° with the field. If the magnetic moment of the magnet is 0.04 N.m, determine the magnitude of the magnetic field. **[0.08 T]**

8. For a small bar magnet having its north pole pointing towards north direction, the neutral point is found to be located 12 cm away from its centre. If the magnitude of the horizontal component of the earth's magnetic field at the location of the magnet is 0.5 G, find the magnetic moment of the magnet. **[0.86 Am^2]**

9. A torque of magnitude 1.0×10^{-3} N.m acts on a magnetic needle of magnetic moment 50 Am^2, pointing towards the geographical north. If the horizontal component of the earth's magnetic field at the location of the needle is 35×10^{-6} $Wb.m^{-2}$, find the magnetic declination at the location. **[34° 51′]**

10. The angle of dip in a region on earth is measured to be 45°. The horizontal component of the earth's magnetic field in the region is 0.5×10^{-4} T. Determine the vertical component and the resultant magnetic field of the earth in the region. **[0.50×10^{-4} T; 0.71×10^{-4} T]**

UNIT 4
ELECTROMAGNETIC INDUCTION AND ALTERNATING CURRENT

CHAPTER 4 (A)
Electromagnetic Induction; Self and Mutual Inductance

CHAPTER 4 (B)
Alternating Current; AC Generator and Transformer

CHAPTER 4 (A)

ELECTROMAGNETIC INDUCTION; SELF & MUTUAL INDUCTANCE

In the Chapter 3A (Section 3A. 15), we had seen that when a coil was placed in a uniform magnetic field and a current was passed through the coil, the coil experienced a torque, which made it to rotate. In other words, the passage of current through the coil placed in an uniform magnetic field **induced** the coil to move ('induce' means 'to cause to happen'). This is the principle of construction of all electric motors (Section 3A. 15) as well as the moving coil galvanometers (Section 3A. 16). The electric current in the coil produces a magnetic field, which interacts with the magnetic field of the magnet. So, essentially, we have the phenomenon of electricity producing magnetism.

As may be expected, the opposite phenomenon also takes place, i.e. if a magnet is moved through a coil, an electric current is **induced** in the coil. The phenomenon is known as electromagnetic induction. Here magnetism produces electricity. It was discovered by the British scientist Michael Faraday and even earlier by the American scientist J. Henry around the same time in 1831. However, the results were first published by Faraday. So the discovery goes by his name. The phenomenon has led to applications of great value to the mankind. In the next chapter (Sections 4B.09 and 4B.10), as examples, we shall see how it is employed for the construction of electric generators which produce electric energy, and the transformers which efficiently transmit this energy across long distances.

4A. 01. Magnetic Flux; Magnetic Induction (or Magnetic Flux Density) Electromagnetic Induction; Faraday's Laws

Magnetic flux

For a clear understanding of the Faraday's experiments and the consequent Faraday's laws of electromagnetic induction, it is useful to have the knowledge of magnetic flux. Like the electric flux, discussed earlier in Chapter 1B (Section 1B.13), the **magnetic flux** can be **defined** as follows:

"The magnetic flux $\Delta\phi_B$ across a small element ΔS of space in a magnetic field $\vec{B}$ is represented by the number of magnetic field lines crossing the element. It is given by the scalar product of the area $\overrightarrow{\Delta S}$ of the element and the component of the magnetic field $\vec{B}$ normal to the element".

$$\Delta\phi_B = \vec{B}.\overrightarrow{\Delta S} \qquad \text{..... (4.01)}$$

(Remember that earlier the magnetic field lines were also called 'magnetic lines of force'. The Eq. 4.01 is the magnetic equivalent of Eq. 1.59 of Section 1B.13).

Further, as before, for a Gaussian surface (which is an imaginary closed surface of arbitrary shape), the magnetic flux is given by,

$$\phi_B = \oint_S \vec{B}.\overrightarrow{ds} \qquad \text{..... (4.02)}$$

$\overrightarrow{ds}$ is an infinitesimal element of the surface (the Eq. 4.02 is the magnetic equivalent of Eq. 1.61 of Section 1B.13). S is the total surface area. It is represented as a vector $\vec{S}$ normal to the plane of the area (vide Chapter 1B, Section 1B.12).

Now, suppose a coil of surface area $\vec{A}$ is placed in a magnetic field $\vec{B}$. Then the magnetic flux through the coil is given by,

$$\phi_B = \int_A \vec{B}.\overrightarrow{dA} \qquad \text{..... (4.03)}$$

$\overrightarrow{dA}$ is an infinitesimal element of the area A. If $\vec{B}$ is a uniform magnetic field, we have

$$\phi_B = \vec{B}.\int_A \overrightarrow{dA}$$

Or

$$\boxed{\phi_B = \vec{B}.\vec{A}} \qquad \text{..... (4.04)}$$

(Mag. flux; uniform mag. field)

Note that the magnetic flux is a scalar quantity, because the scalar product of two vectors is a scalar.

Next, if the magnetic field be perpendicular to the plane of the coil, we have,

$$\phi_B = B.A \cos\theta = B.A \cos 0°$$

Or

$$\boxed{\phi_B = B.A} \quad \text{(Mag. flux; B uniform and} \perp \text{A)} \qquad ..... (4.05)$$

(Remember that the vector $\vec{A}$ is $\perp$ to the plane of the coil. $\therefore \theta = 0$). From Eqs. 4.04 and 4.05, it is seen that the **SI unit of magnetic flux** is **tesla-square metre ($T.m^2$)**, which is conveniently termed as **weber** (Wb):

$$1 \text{ weber (Wb)} = 1 \text{ tesla. square metre } (T.m^2)$$

Magnetic Induction (or **magnetic Flux Density**). "The magnetic field lines passing per unit area perpendicular to the lines" is called **magnetic flux density** or **magnetic induction.** It is a measure of the magnetic field intensity and represented as $\vec{B}$. We had earlier referred to it in chapter 3B (Section 3B.04).

Note that the terms magnetic field (or magnetic field intensity), magnetic field strength (or magnitude), or magnetic flux density (or magnetic induction) physically mean the same. The unit of magnetic field is tesla (T) or weber per sq.metre ($Wb.m^{-2}$), as earlier given in chapter 3A (Section 3A.03).

Electromagnetic Induction

The Fig. 4.1(a) shows a simple circuit consisting of a metallic coil connected to a sensitive galvanometer. If a magnet is quickly moved towards the coil, the galvanometer shows a deflection, which means that an electric current appears in the circuit. The current disappears when the magnet stops. It is obvious that the current is produced by the change of magnetic flux linked with the circuit. It is called **induced current.** The emf that causes the current, i.e. makes the free electrons to move in the circuit, is called **induced emf.** Quantitatively, it is equal to the work done per unit charge to produce the current. "The phenomenon of appearance of an induced current (and an induced emf) in a circuit by a change of magnetic flux in the circuit" is called **electromagnetic induction.** Obviously, the induced current has a very short duration.

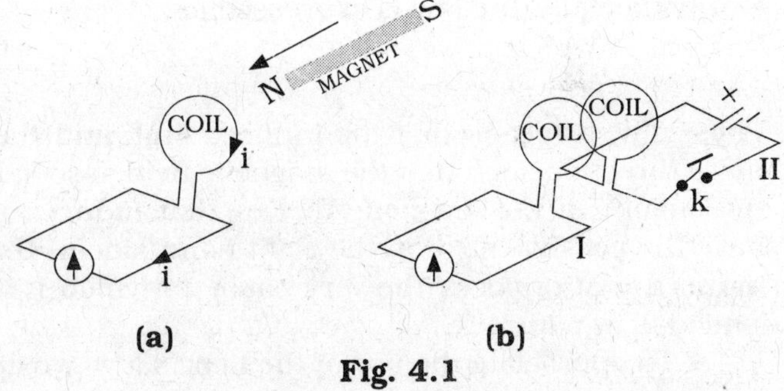

Fig. 4.1

If the magnet is moved away from the coil, again an induced current is produced, but its direction is found to be reversed (the galvanometer needle deflects in the opposite direction). Similarly, if instead of the north pole the south pole of the magnet moves towards the coil, the direction of the current is reversed. The faster is the motion, the greater is the current. Further, the same effects are produced if the magnet is held stationary and the coil is quickly moved towards it or away from it. Thus it is concluded that the induced current and emf are produced by a relative motion between the coil and the magnet which, in turn, causes a change in magnetic flux.

Similar effects are observed if the coil is placed close to another coil, which has a battery and a spring-loaded key in its circuit (Fig. 4.1b). As soon as the key K is closed, so that a current starts building up in the circuit II, an induced current is produced in the circuit I. The induced current stops when finally a steady current is set up in the circuit II. When the key K is released, again an instantaneous induced current is produced in the circuit I, but in the opposite direction. How an induced current/emf is produced in this experiment ? In Chapter 3A (Section 3A.06) we had known that a current-carrying coil has a magnetic field associated with it. As soon as the key K is closed, the current and its associated magnetic field start building up from a zero value to a steady maximum value in the second circuit. Since the two coils are situated close to each other, the changing magnetic flux passes through the first coil too and so induces a current in it. The same happens when he key K is released. Thus, once again, the origin of induced current/emf lies in the changing magnetic flux linked with the coil.

Faraday's laws

A series of experiments of the type described in the foregoing were conducted by Faraday. He was able to recognize the fundamental role piayed by a changing magnetic flux in the production of the observed induced current/emf. His findings have led to the formulation of the following two **Faraday's laws of electromagnetic induction:**

Law I. Whenever there is a change in the magnetic flux linked with a circuit, an induced emf is produced in the circuit, which lasts only as long as the flux in changing.

Law II. The magnitude of the induced emf is equal to the rate of change of the flux ϕ_B with time,

$$\varepsilon = -\frac{d\phi_B}{dt} \quad \text{..... (4.06 a)}$$

(Thus, in Fig. 4.1a, the faster the magnet moves relative to the coil, the greater is the magnitude of the induced emf)

The first law defines the phenomenon of electromagnetic induction. The second law gives the magnitude of the emf that is induced. The direction of the emf is given by the Lenz's law, to be discussed in the next section (it will also explain the –ve sign of the emf in Eq. 4.06a).

4A. 02. Lenz's Law

Shorty after the Faraday's laws were formulated, the German scientist Lenz deduced a convenient rule for determining the direction of the induced current/emf. It is known as **Lenz's law :**

"The induced emf, and therefore the induced current, in a circuit is always in such a direction that it opposes the very cause which produced it".

Mathematically, the law is expressed as,

$$\varepsilon = -\frac{d\phi_B}{dt}$$

Physically, it means that the induced emf, and therefore the induced current, has a direction such that the magnetic field associated with the current opposes the change in the magnetic flux ϕ_B that induces the current. That is why we have the negative sign on the right hand side of the above equation. (Since the induced emf opposes the very cause to which it is due, it is often said to be 'suicidal' in nature !).

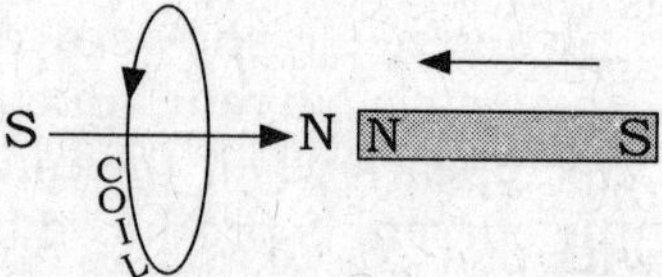

Fig. 4.2

To appreciate the use of the Lenz's law, we apply it to the situation depicted in Fig. 4.1a, where a bar magnet moves towards a conducting coil, with its north pole N facing the coil. For the sake of clarity, it is again shown in Fig. 4.2, with the detecting galvanometer removed from the circuit. The changing magnetic flux produces an induced current in the coil. As discussed in the last chapter (Section 3B. 05; Fig 3.44), then the coil behaves like a magnetic dipole, with its dipole moment $\vec{p}_m$ directed from its south pole to the north pole. The Lenz's law tells us that the direction of the induced current has to be such that the north pole of the dipole faces the north pole of the magnet, as shown in Fig. 4.2, so that the two repel each other. Then the direction of the induced current can be found by applying the right hand thumb rule, stated and illustrated earlier in Chapter 3A (Section 3A. 06; Fig. 3.11c). In this case it is found that, looking towards the north pole of the dipole, the direction of the induced current must be anticlockwise, as shown in the Figs. 4.1a and 4.2.

If the bar magnet is moved away from the coil, the induced current has to flow such that the south pole of the dipole is presented towards the north pole of the bar magnet, so that the two poles attract each other and thus the receding motion of the bar magnet is opposed. Application of the right hand thumb rule will now give clockwise flow of the induced current in the coil.

Note that the induced current/emf has no direction of its own. Its direction depends on a number of external factors, e.g. whether the N-pole or the S-pole of the magnet faces the coil, whether the magnet moves towards the coil or away from the coil, or whether the coil moves or the magnet moves while the other is stationary.

Lenz's law and the law of conservation of energy

The induced current generated in the coil's circuit (Fig. 4.1a) represents an amount of electrical energy (we know that it manifests as Joule's heat in the coil, given by i^2R, where i is the current and R is resistance of the coil). According to the principle of conservation of energy, it cannot be created out of nothing. Then where does it come from ? As already discussed, in Fig. 4.1a (or Fig. 4.2), when the N-pole of the magnet approaches the coil, N-polarity is induced in the face of the coil towards the magnet. Therefore, in pushing

Additional knowledge

(i) The aforesaid situation can alternatively be interpreted in terms of change of magnetic flux. The bar magnet's motion towards the coil increases the magnetic flux through the coil. Then, according to the Lenz's law the induced current has to set up its associated magnetic field such that its magnetic flux opposes the original flux. Once again, the direction of the induced current can be found by applying the right hand thumb rule.

(ii) As an alternative to the right hand thumb rule, the direction of the induced current can also be found by one of the following :

(a) Fleming's right hand rule. Stretch the thumb, forefinger and middle finger of the right hand perpendicular to each other. Imagine that the magnet is stationary and the conducting coil is moving relative to it. If the forefinger is in the direction of the magnetic field and the thumb in the direction of motion of the conductor, then the middle finger indicates the direction of the induced current.

**(b) Lorentz force equation [$\vec{F} = q(\vec{v} \times \vec{B})$]; vide Chapter 3A; Section 3A. 03, Eq. 3.3]. Again assume that the magnet is stationary and the coil is moving relative to it. Focus attention on a free electron in the coil moving in the direction of motion of the coil (alongwith the coil). The direction of the Lorentz force on it can be found by applying the Flemings left hand rule (vide Section 3A.03; Fig. 3.4). The direction of the induced current is opposite to it.

the magnet nearer to the coil work is done against the mutual repulsion between the two N-poles. **It is this mechanical work that is transformed into electrical energy.** Similarly, when the N-pole of the magnet is withdrawn, S-polarity is induced in the face of the coil towards the receding magnet. Work is now done against the mutual attraction between the two opposite poles, as they are removed farther from each other. Again, this work is converted into electrical energy. Similarly, in Fig. 4.2b, where no movement is involved, the necessary energy is transferred from the battery. Thus, it is found that the Lenz's law is **in agreement with the law of conservation of energy.**

4A. 03. Methods of Producing Induced emf

We now know that the basic cause for producing an induced emf in a circuit is the change in the magnetic flux linked with the circuit. From Eq. 4.04, the magnetic flux ϕ_B linked with a plane conductor of surface area A placed in a uniform magnetic field $\vec{B}$ is given by,

$$\phi_B = \vec{B}.\vec{A}$$
$$= B.A \cos\theta$$

θ is the angle between the direction of $\vec{B}$ and the normal to the conductor. Thus, we find that the magnitude of ϕ_B depends on three different factors, viz. the magnetic field B, surface area A of the conductor and the angle θ. Accordingly, the magnetic flux can be varied to produce an induced emf in three possible ways:

(i) By changing the magnetic field $\vec{B}$. We have already discussed two examples, presented in Figs. 4.1 (a) and (b). In one case $\vec{B}$ changes due to a relative motion between the magnet and the conductor and in the other case due to a gradual build-up of electric current in the second circuit.

(ii) By changing the surface area A of the conductor. Suppose the conductor is vertically placed in the gap between the poles of a large horse-shoe magnet. If the conductor is laterally moved across the gap, its effective area within the magnetic field will change (decrease) giving rise to an induced emf. The same will happen if the conductor is laterally inserted inside the gap (increase in area).

(iii) By changing the angle θ between the magnetic field $\vec{B}$ and the conductor. If instead of laterally moving the conductor across the gap, it is rotated in the gap, then θ will change, again giving rise to an induced emf. The same happens when the coil of a moving-coil galvanometer rotates. As we shall see later, it forms the principle of construction of an a.c. generator (vide Chapter 4B; Section 4B. 10).

Example 1. A rapid motion of a copper coil in a uniform magnetic field changes the magnetic flux linked with the coil from 18×10^{-3} weber to 9×10^{-3} weber in 0.005 second. Find the emf induced in the coil.

Solution. Given : $\phi_{B1} = 18\times10^{-3}$ Wb; $\phi_{B2} = 9\times10^{-3}$ Wb; dt = 0.005 s

The induced emf is given by,

$$\varepsilon = -\frac{d\phi_B}{dt} = -\frac{\phi_{B2} - \phi_{B1}}{dt}$$

$$= -\frac{(9\times10^{-3})-(18\times10^{-3})}{(0.005)}$$

$$= 1.8 \text{ volt} \quad \textbf{(Answer).}$$

emf induced in a straight conductor. Suppose a straight conductor of length ℓ moves a distance x in a uniform magnetic field, in a direction perpendicular to both its length and the field. It sweeps through an area ($\ell \times x$) in doing so. Thus it crosses a magnetic flux $\phi_B = B.(\ell.x)$ (B represents magnetic flux density, i.e. flux per unit area). Hence the emf induced in the conductor is given by,

$$\mathcal{E} = -\frac{d\phi_B}{dt} = -\frac{d}{dt}(B.\ell.x)$$

$$= B\ell\frac{dx}{dt}$$

dx/dt represents the velocity of the conductor. Therefore we have,

$$\mathcal{E} = B\ell v \quad \text{..... (4.06b)}$$

(conductor perp. to field)

It is easily seen that if the conductor moves at an angle θ to the field, the induced emf will be,

$$\mathcal{E} = B\ell v \sin\theta \quad \text{..... (4.06c)}$$

(conductor inclined to field)

Example 2. An 8 cm long aluminium wire moves across a uniform magnetic field of strength 1.5 Wbm^{-2} in a direction perpendicular to the field, with a velocity of 12 ms^{-1}. Calculate the emf inducted in the wire.

Solution. Given : B = 1.5 $Wb.m^{-2}$; ℓ = 8 cm = 0.08m; v = 12 ms^{-1}

The emf induced in the wire is given by,

$$\mathcal{E} = B\ell v$$

$$= (1.5) \times (0.08) \times (12)$$

$$= 1.44 \text{ volt} \quad \textbf{(Answer).}$$

4A. 04. Eddy Currents

The **eddy currents** are the 'electric currents induced in the body of a conductor, when the magnetic flux linked with the conductor changes". (They are also knows as Focault currents, after the name of the scientist who discovered them).

The eddy currents are essentially induced currents. Suppose instead of a coil we move a solid conducting plate in a magnetic field. The relative motion of the conductor and the field induces currents in the conductor. But now unlike the case of the coil where just one path was followed by the conducting electrons, numerous closed paths are available to the electrons in the plate. (The direction of the induced current along any path is given by the Lenz's law). Thus the electrons swirl about within the plate as if they had been trapped in a whirlpool. ('eddy' is a popular word for a whirlpool of water. A whirlpool is created when a water current runs against the mainstream of water). Hence the name 'eddy current'.

Physical effects of eddy currents

(i) Heating effect. We know that the flow of current in a conductor produces Joule heating in the conductor. It is given by i2R, where i is the current and R is the resistance of the conductor. Thus eddy currents heat up the conductor. The quicker the magnetic flux ϕ_B changes, greater is the induced emf [$\mathcal{E} = -(d\phi_B/dt)$]. So, the greater is the induced current and the resulting heating.

(ii) Magnetic effect. We know that any current-carrying conductor has a magnetic field associated with it. So a conductor in which eddy currents are produced has a magnetic field, and therefore a magnetic flux, associated with it. According to the Lenz's law, this magnetic flux opposes the original flux due to which it is produced.

The existence of both these effects together is well illustrated by a simple experiment. A solid conducting plate is suspended from the pivot P between the two poles of a magnet, such that it is parallel to the pole faces (Fig. 4.3a). When the plate is stationary, no current exists in it, in spite of magnetic flux passing through it. Now if it is made to oscillate like a pendulum in its own plane, i.e. parallel to the pole faces, the magnetic flux passing through it continuously changes, setting up eddy currents to flow through it. So the plate starts getting heated. Each time the plate enters or leaves the field of the magnet, a part of its mechanical energy converts into thermal energy. The heat tends to dissipate into the surroundings. At the same time, the magnetic flux associated with the eddy currents opposes the original flux ϕ_B of the magnet and hence slows down the movement of the plate. This is called **electromagnetic damping.** The faster are the initial oscillations, the larger is the induced emf and the induced current $[\mathcal{E} = -(d\phi_B/dt)]$. Hence the greater is the opposing magnetic flux and the resulting damping. After several swings of the plate, all mechanical energy supplied to the plate for starting the oscillations is spent up. Then the warmed up plate hangs still between the poles of the magnet.

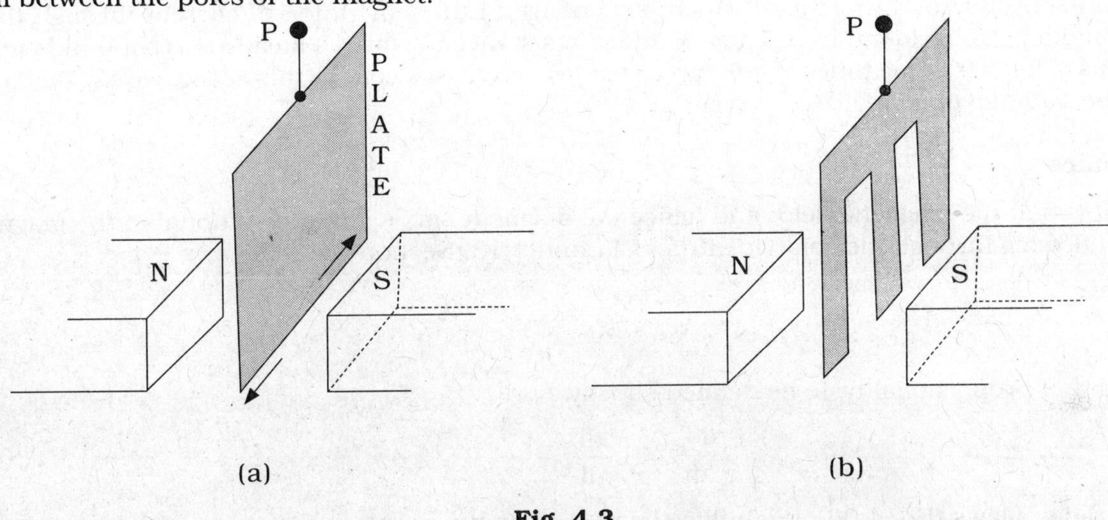

Fig. 4.3

The eddy currents can be reduced by cutting slots in the plate (Fig. 4.3b) or by punching holes in the plate. Then the number of possible paths for the eddy currents are reduced. Hence both the heating and the electromagnetic damping are reduced.

Some applications of eddy currents

(i) Induction furnace. High temperatures ranging from a few hundred °C to ~ 2000 °C (sufficient to melt most metals) can be attained by using the heating effect of the eddy currents. The material to be heated is placed in a constantly changing high frequency magnetic field and put within a thick-walled ceramic material enclosure, which provides necessary thermal insulation. The arrangement is called **induction furnace.**

(ii) Electromagnetic brakes. It utilises the electromagnetic damping. In electric trains, a strong magnetic field is applied to a metallic drum mounted on the same axle that connects the rotating wheels. The strong eddy currents set up in the drum oppose the motion of the drum and so immediately slow down the motion of the train. Thus an effective electromagnetic braking system is obtained. The greater is the speed of the train, the greater is the braking effect, since the induced emf is greater $[\mathcal{E} = -(d\phi_B/dt)]$. For the same reason, the braking effect is smooth, too.

(iii) Dead beat galvanometers. In a moving-coil galvanometer, if the coil is wound on a non-conducting frame, e.g. wood or plastic, it keeps on oscillating for a long time before coming to rest, when a current is passed through it. Thus a great deal of time is wasted in taking readings during an experiment. The problem is overcome by winding the coil on a metallic frame, which is well insulated from the coil. The electromagnetic damping produced in the frame brings the coil to a quick stop. Such a galvanometer is therefore called a **dead beat galvanometer.**

(iv) Induction motor (or AC motor). An induction motor (also called AC motor) consists of two parts, a 'stator' and a 'rotor'. The stator is a suitable arrangement for producing a rotating magnetic field of constant value. The rotor has a shaft passing through a cylindrical laminated core of metal bars of copper, aluminium and suitable alloys. It is placed in the magnetic field of the stator. The eddy currents produced in the rotor interact with the magnetic flux of the stator in such a way that the rotor starts revolving in the same direction as the rotating flux.

4A. 05. Self-Induction; Self-Inductance; Inductor; Back-emf

"The phenomenon of production of induced emf in a coil due to change of current in the coil itself" is called **self-induction.**

We know that when a current flows in a coil, it produced a magnetic field (vide Section 3A. 06; Chapter 3A). If the current changes with time, the magnetic flux through the coil changes and hence an emf is produced in the coil because of change in current/magnetic flux within the coil itself. It has been discussed and illustrated (Fig. 4.1a) earlier in Section 4A. 01. Hence the name self-induction. From the Lenz's law, we know that the induced emf opposes the cause to which it is due. Therefore, the self-induction is said to be the property of a coil by virtue of which any change in the magnitude of current through the coil is opposed by inducing a back-emf in it. Thus, unlike a resistance in which a steady current builds up almost instantaneously, it takes some time to build up a steady current in a coil. In this sense, we say that a coil has 'inertia' to the building up of a steady current in it.

Self-inductance

We know that the magnetic field, and hence the magnetic flux ϕ_B, is proportional to the magnitude of the current i flowing through the coil (vide Eqs. 3.11 and 3.13, Section 3A. 06),

$$\phi_B \propto i$$

Or

$$\phi_B = Li \qquad \text{..... (4.07)}$$

L is a constant of proportionality, to be defined shortly.

$$\therefore \quad \frac{d\phi_B}{dt} = L\frac{di}{dt}$$

From the Faraday's laws (Eq. 4.06), the induced emf is given by,

$$\mathcal{E} = -\frac{d\phi_B}{dt}$$

$$\therefore \quad \mathcal{E} = -L\frac{di}{dt} \qquad \text{..... (4.08)}$$

The constant of proportionality L is called **self-inductance** or **coefficient of self-induction.** From Eq. 4.07, if i = 1, we have $L = \phi_B$. Similarly, from Eq. 4.08, if (di/dt) = 1,we have $L = \mathcal{E}$ (disregarding the negative sign). Accordingly, the self-inductance can be **defined** in either of the following two ways :

"The self-inductance of a coil is equal to the magnetic flux linked with the coil, when unit current flows through the coil".

OR

"The self-inductance of a coil is numerically equal to the induced emf produced in the coil, when the rate of change of current through the coil is unity".

The value of the self-inductance L depends upon the following factors:

(i) Number of turns of the coil
(ii) Area of cross-section of the coil
(iii) Nature of material of the core of the coil.

The SI unit of self-inductance is henry (H).

From Eq. 4.07, $L = (\phi_B/i)$. Hence

1 henry = (1 weber /1 ampere)

From Eq. 4.08, $L = [\mathcal{E}/(d\phi_B/dt)]$. Hence,

1 henry = [1 volt/(1 ampere sec^{-1})]

Therefore,

1 henry (H) = 1 weber/ampere ($Wb.A^{-1}$)
= 1 volt-sec./ampere ($V.s.A^{-1}$)

Thus, the self-inductance of a coil is said to be 1 henry if "one ampere current flow through it produces one weber magnetic flux linked with it" OR "a change of current at the rate of one ampere/sec. through it induces an emf of one volt".

Inductor; Ideal inductor

Since a coil exhibits the phenomenon of self-induction, it is called an **inductor.** It is represented by the following **symbol:** —ꝏꝏ—
A coil possesses both self-induction and resistance. An **'ideal inductor'** is one which possesses only an inductance, i.e. its resistance is assumed to be zero.

Example 3. The current flowing in a copper coil is changed from 7A in one direction to 7A in the opposite direction, in a time duration of 0.10 second, which induces an emf of 0.35 volt in the coil. Calculate the self-inductance of the coil.
Solution. Given : di = 7 – (–7) = 14 A; dt = 0.10 s; $\mathcal{E}$ = 0.35 V
We know that the emf induced in a coil is given by,

$$\mathcal{E} = -L\frac{di}{dt}$$

∴ The self-indutance (L) of the coil is,

$$L = -\mathcal{E}\frac{di}{dt} = -(0.35) \times \frac{(0.10)}{14}$$

$$= -2.5\times10^{-3}\text{ H (or 2.5 mH)}$$ **(Answer).**

(The –ve sign may be disregarded, as here we are only interested in the magnitude of L).

4A. 06. Energy Stored in a Current-Carrying Inductor

Suppose an inductor of self-inductance L is connected to a battery through a key. When the key is closed an induced emf starts building up in the inductor, until a steady current i is set up in the inductor. According to the Laplace's laws, the induced emf $\mathcal{E}$ opposes the emf of the battery. Hence the battery has to do some work against the induced emf until the current i is established in the circuit. This work is stored up as energy in the inductor.

If W is the work done, the resulting power (which is the rate of work done) is given by (vide Eq. 2.37; Chapter 2B),

$$\frac{dW}{dt} = \mathcal{E}i$$

From Eq. 4.08, ignoring the –ve sign (as we are only interested in the magnitude or $\mathcal{E}$), we have,

$$\frac{dW}{dt} = Li\frac{di}{dt}$$

Now,
$$\frac{di^2}{dt} = 2i\frac{di}{dt}$$

∴
$$\frac{dW}{dt} = \frac{1}{2}L\frac{di^2}{dt}$$

Integrating, we get,

$$W = U_B = \frac{1}{2}Li^2 \qquad \text{..... (4.09)}$$

U_B is the energy stored in the inductor. (The constant of integration = 0, since W = 0 for i = 0). The subscript B in U_B represents the magnetic nature of the energy, since the induced emf results from a changing magnetic flux linked with the coil.

4A. 07. Self-Inductance of a Long Straight Solenoid

A long straight solenoid has its length much larger than its diameter. In Chapter 3A (Section 3A. 09; Eq. 3.16) we had found that the magnetic field inside a long solenoid with air core is given by,

$$B = \mu_0 ni$$

i is the current flowing through the solenoid, n is the number of turns per unit length of the solenoid, and μ_0 is the absolute permeability (also called 'permeability constant' or 'permeability of free space').

The magnetic flux linked with a single turn of area A is B.A. If the coil has total N turns and its length is ℓ, the total flux ϕ_B linked with the coil is,

$$\phi_B = B.A.N = (\mu_0 ni)\,NA$$
$$= (\mu_0 ni)\,(n\ell)\,A = (\mu_0\, n^2\ell A)\,i$$

But from Eq. 4.07, we have,

$$\phi_B = Li$$

$$\therefore \quad Li = (\mu_0 n^2\ell A)\,i$$

Or

$$L = \mu_0 n^2\ell\, A \qquad \text{..... (4.10a)}$$

Since $N = n\ell$, we can also write,

$$L = \frac{\mu_0 N^2 A}{\ell} \qquad \text{..... (4.10b)}$$

If the air core of the solenoid is replaced by a material of permeability μ, then μ_0 is replaced by μ in the Eqs. 4.10(a) and (b) ($\mu = \mu_0\mu_r$, where μ_r is called the 'relative permeability' of the material).

Additional knowledge

(i) Note that when the current is switched on in an electric circuit consisting of an inductance, the induced emf in the inductance opposes the original emf that causes the current to flow in the circuit. But when the current gradually falls on switching off the circuit, the induced emf assists the original emf.

(ii) All the three electrical entities, resistance, capacitance and self-inductance depend on geometrical factors and an intrinsic material property. (a) Resistance depends on length and cross-sectional area of the resistor, (b) capacitance, for a parallel plate capacitor, depends on the plate area and plates' spacing, plus the dielectric constant K of the material between the plates, and (c) self-inductance depends on the cross-sectional area of the coil and the number of turns of the coil, plus the nature of material of the core on which the coil is wound.

(iii) Note the similarity in the expressions for energy stored in an inductor $[(1/2)\,Li^2]$ and a capacitor $[(1/2)\,CV^2]$ with the kinetic energy of a body of mass m $[(1/2)mv^2]$.

(iv) As already stated, physically the self-inductance plays the role of inertia in an electric circuit, just as the role played by mass in mechanics (vide Newton's laws of motion). That is why a tubelight, which only contains an inductance (choke) in its circuit, takes a bit of time to light up when the current is switched on in the circuit. But in an electric circuit consisting of a bulb alone, the bulb lights up almost instantaneously when the circuit is switched on. If an inductor (choke) is included in the bulb's circuit, then the bulb will also take a little longer time to light up.

4A. 08. Mutual Induction; Mutual Inductance

"The phenomenon of production of induced emf in a coil due to change in current in a neighboring coil" is called **mutual induction.**

The phenomenon of self-induction involves only one coil. When the current rises or falls in the coil, the magnetic flux linked with the coil changes, which causes an induced emf in the coil (Fig. 4.1a; Section 4A.01). Mutual induction involves two interacting coils. The production of induced emf in this case was discussed and illustrated earlier in Section 4A.01 (Fig. 4.1b). For the sake of clarity, it is further illustrated in Fig. 4.4, for two coils with multiple turns. When the current rises in circuit 2 upon closing the key, a changing magnetic flux is produced, which is linked both with the coil 2 and the coil 1. Thus, in addition to an induced emf in coil 2 (self-induction), an induced emf is caused in the coil 1, too (mutual induction). The same will happen in coil 2 if the two circuits are interchanged, i.e. the battery and key are inserted in circuit 1 and the galvanometer is inserted in circuit 2.

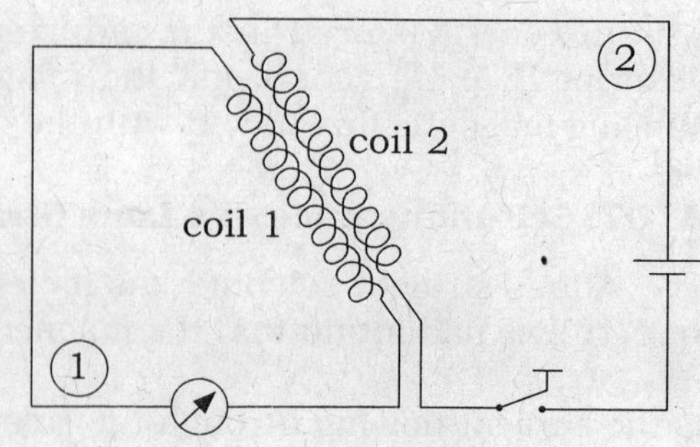

Fig. 4.4

According to the Lenz's law, the induced emf opposes the emf of the battery to which it is due. Thus, like self-induction, the mutual induction can be regarded as the property of a pair of coils, by virtue of which each coil opposes the emf in the other coil by generating an opposite induced emf.

Further, as for the case of self-induction, when the current i is building up in the circuit 2, the magnetic flux ϕ_{B1} linked with the coil 1 is proportional to i,

$$\phi_{B1} \propto i$$

Or $$\phi_{B1} = M_{12}\, i, \qquad \text{..... (1)}$$

where the constant of proportionally M_{12} is called the mutual inductance or coefficient of mutual induction of coil 1 w.r.t. coil 2. Now, let the battery be placed in circuit 1 and the galvanometer in circuit 2. If the same current i builds up in circuit 1, then the magnetic flux ϕ_{B2} in circuit 2 will be

$$\phi_{B2} = M_{21}\, i, \qquad \text{..... (2)}$$

where M21 is the mutual inductance of coil 2 w.r.t coil 1. It can be proved that

$$M_{12} = M_{21} \qquad \text{..... (4.11)}$$

[This equality holds even if the current values in the two cases are different, e.g. i_1 and i_2]. Then we have $\phi_{B1} = \phi_{B2}$ (say) and so the Eqs. 1 and 2 can be written in a general way as,

$$\phi_B = Mi, \qquad \text{..... (4.12)}$$

where M is the mutual inductance of the pair of coils 1 and 2. The induced emf in the circuit 1 due to the current flow in circuit 2 (or in circuit 2 due to current flow in circuit 1) is

$$\mathcal{E} = -\frac{d\phi_B}{dt} = -M\frac{di}{dt} \qquad \text{..... (4.13)}$$

From Eq. 4.12, if i = 1 we have $M = \phi_B$. Similarly, from Eq. 4.13, if (di/dt) = 1, we have $M = \mathcal{E}$ (disregarding the negative sign, as we are only interested in the magnitude of M). Accordingly, the **mutual inductance** (also called coefficient of mutual induction) can be **defined** in either of the following two ways :

"The mutual inductance of a pair of coils is equal to the magnetic flux linked with either of the coils, when unit currents flows in the other coil".

OR

"The mutual inductance of a pair of coils is numerically equal to the induced emf produced in either of the coils, when the rate of change of current through the other coil is unity".

Like the self-inductance, the mutual inductance depends upon the following factors :

(i) Number of turns of the coils.

(ii) Area of cross-section of the coils.

(iii) Separation between the coils and their mutual orientation.

(iv) Nature of material of the core of the coils.

This dependence will become more clear in the next section, where we shall derive an expression for the mutual inductance of two coaxial solenoids.

Further, like the self-inductance, the SI unit of mutual inductance is also henry (H). Hence **1 henry (H)** can **also be defined** as "the mutual inductance of a pair of coils when either one ampere current through one coil produces one weber magnetic flux linked with the other coil **or** a change of current at the rate of one ampere/second through one of the coils induces an emf of one volt in the other coil".

$$1 \text{ henry (H)} = 1 \text{ weber/ampere } (\text{Wb.A}^{-1})$$
$$= 1 \text{ volt–second/ampere } (\text{V.s.A}^{-1})$$

Example 4. In a pair of coils, a change of current of 5 A causes the magnetic flux in the second coil to change by 14×10^{-4} Wb per turn. If the second coil consists of 1500 turns, calculate the mutual inductance of the pair.

Solution. Given : N = 1500; i = 5 A; magnetic flux per turn = 14×10^{-4} Wb

Total magnetic flux, $\phi = (1500) \times (14\times10^{-4})$
$= 2.1$ Wb

∴ The required mutual inductance is given by,

$$M = \frac{\phi}{i} = \frac{(2.1)}{(5)}$$

$$= 0.42 \text{ H} \quad \textbf{(Answer).}$$

Example 5. In a pair of coils, when the current in one coil changes from 14 A to 7 A in 0.05 second, an emf of 70 mV is induced in the neighbouring coil. Calculate the mutual inductance of the two coils.

Solution. Given : $\mathcal{E} = 70$ mV $= 70\times10^{-3}$V; dt = 0.05 s; di = 7–14 = –7 A

We know that the mutually induced emf is given by,

$$\mathcal{E} = -M\frac{di}{dt}$$

∴ The mutual inductance is given by,

$$M = -\mathcal{E}\frac{dt}{di} = -(70 \times 10^{-3}) \times \frac{(0.05)}{(-7)}$$

$$= 0.5\times10^{-3} \text{ H (or 0.5 mH)}$$ **(Answer).**

4A. 09 Mutual Inductance of Two Long Coaxial Solenoids

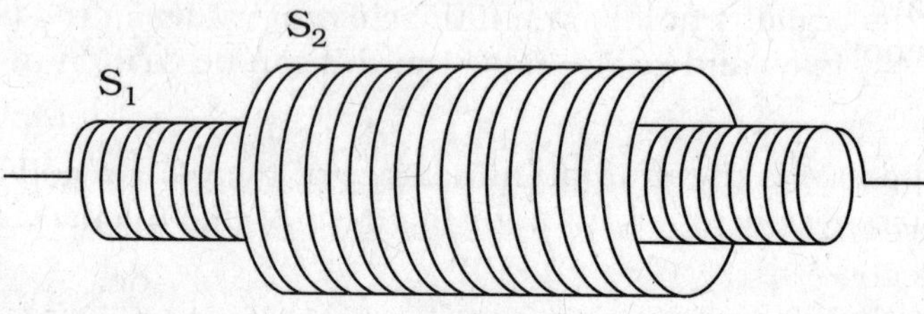

Fig. 4.5

Suppose we have two coaxial solenoids S_1 and S_2. The smaller solenoid S_1 is placed inside the larger solenoid S_2. Let their numbers of turns per unit length be n_1 and n_2, respectively. Let the area of cross-section of the inner solenoid S_1 be A.

If a current i is passed through the solenoid S_1, it will produce a magnetic field $B = \mu_0 n_1 i$. The resulting magnetic flux linked with each turn of the outer solenoid S_2 will be,

$$B.A = \mu_0 n_1 iA$$

The total flux ϕ_B linked with all the turns in the total length ℓ of the solenoid will be,

$$\phi_B = (\mu_0 n_1 iA)\, n_2 \ell = (\mu_0 n_1 n_2 iA\ell)i$$

But from Eq. 4.12,

$$\phi_B = Mi$$

Comparing the above two equations, we get,

$$M = \mu_0 n_1 n_2 A\ell \qquad \text{..... (4.14)}$$

If the total number of turns in the length ℓ for the solenoids S_1 and S_2 be N_1 and N_2, respectively, we have,

$$N_1 = n_1\ell \quad \text{and} \quad N_2 = n_2\ell$$

Then the Eq. 4.14(a) can also be written as,

$$M = \frac{\mu_0 N_1 N_2 A}{\ell} \qquad \text{..... (4.14b)}$$

The Eqs. 4.14(a) and (b) are the expressions for mutual inductance for coils with air core. If the core material is different, with permeability μ, then μ_0 in the two equations is replaced by μ.

4A. 10. Displacement Current

An induced emf $\mathcal{E}$ produces an induced current i in a coil. This induced current must have an associated induced electric field $\vec{E}$ present around the coil. By considering the work done on a charge q_0 moving along a closed path, it can be shown that $\mathcal{E}$ is related to $\vec{E}$ as,

$$\mathcal{E} = \oint \vec{E}.\vec{d\ell} \qquad \text{..... (4.15)}$$

$\vec{d\ell}$ is a small element of length along the closed path. Thus the Faraday's second law of induction (Eq. 4.06) can also be written as,

$$\oint \vec{E}.\vec{d\ell} = -\frac{d\phi_B}{dt} \qquad \text{..... (4.16)}$$

The Eq. 4.16 implies that a changing magnetic flux induces an electric field. By analogy, a changing electric flux should induce a magnetic field. It is really found to be so, which is mathematically expressed by the following so-called **Maxwell's law of induction** (ϕ_E : electric flux),

$$\oint \vec{B}.\vec{d\ell} = \mu_0\epsilon_0\frac{d\phi_E}{dt} \qquad \text{..... (4.17)}$$

We now recall that a current i_e flowing in a closed circuit, e.g. a circular loop, also has a magnetic field associated with it, which is given by the Ampere's law as (Chapter 3A; Section 3A.08; Eq. 3.15),

$$\oint \vec{B}.\vec{d\ell} = \mu_0 i_e \qquad \text{..... (4.18)}$$

Since the left hand side in the Eqs. 4.17 and 4.18, representing magnetic field, is the same, the two equations can be combined into the following single equation (known as **Ampere-Maxwell law),**

$$\oint \vec{B}.\vec{d\ell} = \mu_0 \epsilon_0 \frac{d\phi_E}{dt} + \mu_0 i \qquad \text{..... (4.19)}$$

The first term on the right hand side represents the contribution to the magnetic field due to change in electric flux linked with the circuit and the second term represents contribution due to the flow of current in the circuit. The comparison of these two terms immediately suggests that the product ϵ_0 $(d\phi_E/dt)$ must have the dimension of the electric current i. This imaginary current is called **displacement current.** It is denoted by the symbol i_d,

$$i_d = \epsilon_0 \frac{d\phi_E}{dt} \qquad \text{..... (4.20)}$$

The choice of the name 'displacement' is unfortunate, because there are no physical displacements in the circuit, but somehow this term has become popular (the change in electric field is loosely conceptualized as electric 'displacement').

The **displacement current** may be **defined** as "an imaginary current, which arises due to change in electric field with time in the circuit [$i_d = \epsilon_0(d\phi_E/dt)$]".

The usefulness of the concept of displacement current, in spite of its imaginary existence, is best illustrated for the case of a parallel-plate capacitor. Let the plates be circular in shape, with area A each (Fig. 4.6). We consider the situation when the capacitor is charging. The current starts rising from its zero value until it attains its maximum (real) value i, in a very short time. The changing current changes the electric field $\vec{E}$ between the plates, which produces the (imaginary) displacement current i_d. Let us examine how the two currents i and i_d are related to each other. At any instant of time during the charging, the charge q on the plates is related to E as [vide Chapter 1B; Section 1B.24; Eq. 1.84a; σ = q/A (charge density)],

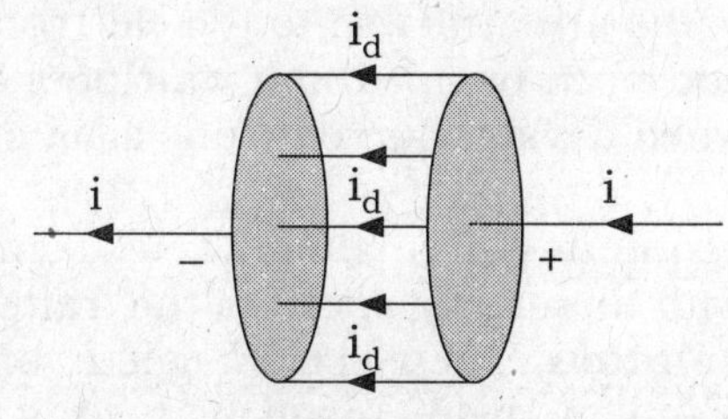

Fig. 4.6

$$q = \epsilon_0 AE$$

Therefore the real current i is given by,

$$i = \frac{dq}{dt} = \epsilon_0 A \frac{dE}{dt} \qquad \text{..... (4.21)}$$

Assuming that the electric field $\vec{E}$ between the plates of the capacitor is uniform, we have for the electric flux,

$$\phi_E = EA$$

Then, from Eq. 4.20 the imaginary current i_d is found as,

$$i_d = \epsilon_0 \frac{d\phi_E}{dt} = \epsilon_0 \frac{d(EA)}{dt} = \epsilon_0 A \frac{dE}{dt} \qquad \text{..... (4.22)}$$

The Eqs. 4.21 and 4.22 tell us that the real current i flowing in the circuit has the same magnitude as the imaginary displacement current i_d between the plates of the capacitor. It means that the imaginary current i_d across the capacitor gap finds continuity with the real conduction current i flowing in the circuit (Fig. 4.6). That is how the concept of displacement current is necessary and important. Otherwise, the insulation between the plates would provide nearly infinite resistance to the flow of the real current i.

In spite of the fact that no electric charge really flows across the capacitor gap, the concept of the displacement current i_d can be employed to find the direction and magnitude of the induced magnetic field. The direction can be found with the help of right-hand thumb rule (vide Chapter 3A; Section 3A.05), just as for the real current i. The magnitude can be found by imagining the space between the circular plates to be an imaginary circular wire (vide Chapter 3A; Section 3A.06).

Solved Conceptual Problems

[Including HOTS Questions]

Problem 1. A current is induced in a stationary copper coil when a magnet moves through it. Will also a current be induced in the coil if the magnet is held stationary and the coil is moved ? Justify your answer.
Solution. Yes. The basic cause for the current to be induced is change in magnetic flux associated with the coil. Any relative motion between the magnet and the coil produces a change in the magnetic flux associated with the coil, resulting in an induced current.

Problem 2. When a magnet moves towards a metal coil, the current induced in the coil has an electrical energy (i^2R) associated with it. Explain how this energy is created.
Solution. Suppose the north pole of the magnet is approaching the coil. Then the direction of induced current in the coil is such that north polarity is induced in the face of the coil towards the magnet. Hence in moving the magnet further towards the coil, work is done against the mutual repulsion between the two north poles. It is this mechanical work that is transformed into electrical energy (in accordance with the law of conservation of energy).

Problem 3. Explain why the **eddy** currents are named so.
Solution. The eddy currents are essentially induced currents. Suppose a solid conducting plate is moved in a magnetic field. The relative motion between the plate and the field induces current in the conductor. But unlike the case of a coil where a single path is followed by the conducting electrons, now numerous closed paths are available to the electrons in the plate. Hence the electrons swirl about within the plate as if they had been trapped in a whirlpool ('eddy' is a popular word for a whirlpool of water. A whirlpool is created when a water current runs against the mainstream of water). Hence the name 'eddy' currents.

Problem 4. Explain how electromagnetic braking works in electric trains. How the braking effect increases with increasing speed of the train ?
Solution. The electromagnetic braking is based on the principle of electromagnetic damping. In electric trains, a strong magnetic field is applied to a metallic drum mounted on the axle connecting the rotating wheels. The strong eddy currents set up in the drum oppose the motion of the drum and so immediately slow down the motion of the train. The greater is the speed of the train, the greater is the braking effect, since the induced emf is greater [$\mathcal{E} = (- d\phi_B/dt)$].

Problem 5. Explain why the phenomenon of self-induction is named so.
Solution. When a current flows in a coil, a magnetic field is produced. If the current changes with time, the magnetic flux through the coil changes and hence an emf is induced in the coil. Thus, an emf is induced in the coil because of change in current/magnetic flux within the coil itself. Hence the name self-induction.

Problem 6. It is known that some work is done when a coil is removed from a magnetic field. When will the work done be more : in case the removal is rapid or in case it is slow ? Justify your answer.
Solution. The work done is more when the removal is rapid. The faster is the removal, the faster is the change is magnetic flux passing through the coil. Therefore greater is the opposing emf induced in the coil. Hence the faster is the removal, the greater is the work done.
Problem 7. A metal plate held in a uniform magnetic field is rapidly moved in a direction parallel to the field. Will an electric current be induced in the plate ? Give reason for your answer.
Solution. No current will be induced. When the plate moves parallel to the field, the magnetic flux passing through it remains unchanged. Hence no emf (and so no current) is induced in the plate.

Problem 8. Two copper coils A and B are mutually positioned as shown in the figure. What will be the direction of induced current in B, just after (i) the circuit containing A is switched on and (ii) this circuit is switched off ?

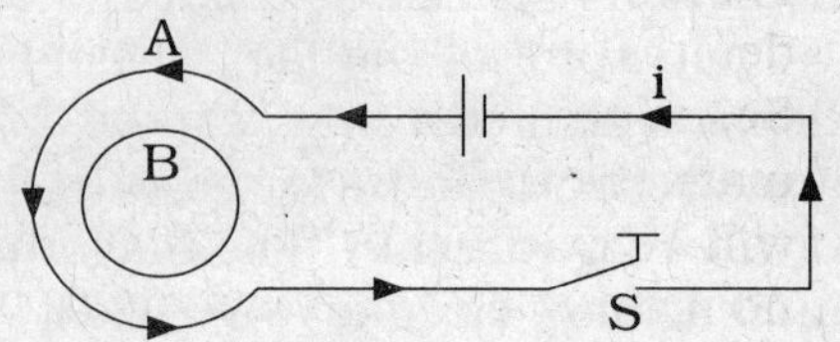

Solution. When the circuit is switched on (by closing S), the current in A starts growing, with its direction of flow being anticlockwise. This will induce a current in B, such that it opposes the growth in A (according to the Lenz's law). Hence the flow of current in B will be clockwise.

When the circuit is switched off (by opening S), the current in A reduces from a maximum to zero. The induced current in B will oppose this reduction. Hence now the flow of current in B will be anticlockwise.

Problem 9. A magnet is made to fall through a horizontally held metal loop. Will its acceleration during the fall be the same as the value of **g** at that place ? Give reason for your answer.
Solution. No. As the magnet falls, the magnetic flux linked with the coil changes, thus inducing a current in the coil. According to the Lenz's law, the direction of the induced current is such that the downward motion of the magnet is opposed, both when the magnet approaches the coil or moves away from it. Hence the acceleration of the magnet during the fall becomes less than **g**.

Problem 10. In the foregoing problem, how will the value of acceleration be affected if a small cut is made somewhere in the loop ?
Solution. If a cut is made in the loop, induced emf will develop in the loop, but no induced current will flow because of the circuit remaining incomplete. Hence the loop will acquire no magnetism and so it will provide no opposing force to the downward motion of the magnet. Therefore, the acceleration of the magnet during its fall will be equal to **g**.

Problem 11. A circular metal loop is freely suspended in a vertical plane. A horizontal bar magnet, with its south pole facing the loop, is rapidly moved towards the loop. Will the position of the loop be disturbed ? If so, explain how.
Solution. According to the Lenz's law, an induced current will be set up in the loop, with its associated magnetic field, such that the face of the loop towards the magnet will develop south polarity. The repulsion between the two south poles will make the loop to move away from the magnet.

Problem 12. An electric circuit comprises a 25 W lamp, a battery, a copper coil with a large number of turns, and a switch. When the circuit is switched on, it takes a small finite time for the lamp to glow with full intensity. Explain why.
Solution. When the circuit is switched on, same current flows through the lamp and the coil. Since the lamp is purely resistive in nature, maximum current tends to flow through it almost instantaneously. But an induced emf is set up in the coil, which opposes the growth of current. Thus a small finite time passes before maximum current is built up in the circuit. Only then the lamp glows with maximum intensity.

Problem 13. A 40 W lamp is switched on, once only with a resistance of 100 Ω in the circuit and next only with an inductance of the same value of resistance in the circuit. (i) In which case the lamp will begin glowing earlier and (ii) in due course of time, will it finally glow with the same intensity in the two cases ? Justify your answer.
Solution. (i) In the first case the lamp will light up to its full intensity almost instantaneously, but in the second case it will begin glowing weakly and a little late because the induced emf set up in the coil will be such that it will oppose the growth of current in the circuit.

(ii) After a small time passes, the same maximum amount of current will flow in the circuit in the two cases. Hence finally the lamp will glow with the same maximum intensity in both the cases.

Problem 14. A vertically held iron pole is dropped from a house top to fall freely in the plane of magnetic meridian. Will any electric current be induced in it ?
Solution. During its fall in the magnetic meridian, the pole will neither cross the horizontal component $\vec{H}$ nor the vertical component $\vec{V}$ of the earth's magnetic field (both $\vec{H}$ and $\vec{V}$ lie in the magnetic meridian). Hence no emf, and therefore no current, will be induced in the pole.

Problem 15. If an astronaut descends upon moon with necessary scientific equipment, how can he readily detect if a magnetic field exists on the moon ?
Solution. He should connect a coil to a sensitive galvanometer (or ammeter) and rotate the coil. If a magnetic field exists, the magnetic flux through the coil will change, inducing a current in the coil, which will be detected by the galvanometer (or ammeter). If the galvanometer shows no deflection, it will mean that no magnetic field exists on moon.

Problem 16. A bird sitting on a high voltage wire flies away when current is switched on in the wire. Explain why.

Solution. The blood circulating in the body of the bird is electrically conducting (just as in the human body). When current starts growing in the wire, induced current starts building up in the body of the bird. The bird gets shocked and frightened. So it flies away.

Problem 17. Two identical coils, one made of copper and the other made of iron, move with the same speed across a uniform magnetic field. Will the (i) induced emf and (ii) induced current be the same in the two cases ? Give reason for your answer.

Solution. (i) The induced emf in both the coils will be the same, because it does not depend on the nature of material of the coils.

(ii) The induced current will be more in the copper coil, as the resistivity of copper is less than that of iron.

Problem 18. When a high wattage appliance, e.g. a room heater, is switch off, sparking is seen to take place in the switch. Explain why.

Solution. At the instant of switching off, the break in the circuit is very sudden, leading to an extremely rapid change in the current in the circuit. Hence the associated magnetic flux also changes very rapidly, giving rise to a high magnitude of induced emf across the gap in the switch, leading to the observed sparking $(\mathcal{E} = -\frac{d\phi_B}{dt})$.

Problem 19. If in an electric circuit including an AC source and a capacitor, the frequency of the source is reduced, will the displacement current also reduce or increase ?

Solution. The displacement current (i_d) has the same magnitude as the normal conduction current (i). A reduction in the frequency of the source will lead to an increase in the rectance of the capacitor ($X_c = 1/\omega c$) and therefore to a reduction of i. Hence i_d (= i) will also be reduced.

Solved Numerical Problems

Problem 1. A rectangular coil of N turns and area of cross-section A, is held in a time-varying magnetic field given by $B = B_0 \sin \omega t$, with the plane of the coil normal to the magnetic field. Deduce an expression for the emf induced in the coil. [CBSE ; 2000].

Solution. Given : No. of turns = N; Area of cross-section = A; Magnetic field, $B = B_0 \sin \omega t$; $\theta = 0$

The magnetic flux linked with the coil is given by,

$$\phi_B = NBA \cos \theta$$
$$= N (B_0 \sin \omega t) A (\cos 0)$$
$$= N B_0 A \sin \omega t$$

The emf induced in the coil is given by,

$$\mathcal{E} = -\frac{d\phi_B}{dt} = -\frac{d}{dt}(N B_0 A \sin \omega t)$$

$$= - N B_0 A\omega \cos \omega t \quad \textbf{(Answer).}$$

Problem 2. A 70 cm long straight conductor moves in a uniform magnetic field in a direction normal to both the field and its length, with a velocity 3 ms^{-1}. If the magnetic field intensity is 1.2×10^4 gauss, determine the emf induced in the conductor.

Solution. Given: $B = 1.2 \times 10^4$ G = 1.2 T; ℓ = 70 cm = 0.7 m; v = 3 ms^{-1}; $\theta = 90°$

The emf induced in a straight conductor is given by,

$$\mathcal{E} = B\ell v \sin \theta$$
$$= (1.2) \times (0.7) \times (3) \times \sin 90° \; (\sin 90° = 1)$$
$$= 2.52 \text{ volt} \quad \textbf{(Answer).}$$

Problem 3. A large copper ring is located in a uniform magnetic field of strength 0.03 T. If the ring is made to shrink rapidly at a rate of 1.5 mm per second, what will be the emf induced in it at the moment when its radius has been reduced to just 1 cm ?

Solution. Given : B = 0.03 T; r = 1 cm = 10^{-2}m; $\frac{dr}{dt} = 1.5 \text{ mm.s}^{-1} = (1.5 \times 10^{-3}) \text{ m.s}^{-1}$

Let the radius of ring be r at an instant t. Since the area of the ring is πr^2, the magnetic flux crossing the ring is given by,

$$\phi_B = (\pi r^2)\, B$$

Hence the induced emf is given by,

$$\mathcal{E} = \frac{d\phi_B}{dt} = \frac{d}{dt}(\pi r^2\, B)$$

$$= 2\,\pi r\, B \frac{dr}{dt}$$

$$= 2 \times 3.14 \times (10^{-2}) \times (0.03) \times (1.5 \times 10^{-3})$$

$$= 2.83 \times 10^{-6} \text{ V (or } 2.83\ \mu\text{V)}$$ **(Answer).**

Problem 4. A square-shaped loop made of aluminium is held vertically in a uniform magnetic field of strength 0.15 T. The plane of the loop points east-west and the field points north-east. Each side of the loop measures 9 cm and its total resistance is 0.6 Ω. If the field strength is rapidly brought down to zero in just 0.5 s, calculate the emf and current induced in the loop.

Solution. Given : B = 0.15 T; A = $(9 \times 10^{-2}\text{m})^2 = 81 \times 10^{-4}\ \text{m}^2$; θ = 45°; Δt = 0.5 s; R = 0.6 Ω

The initial value of the magnetic flux is given by,

$$\phi_1 = BA \cos\theta = (0.15) \times (81 \times 10^{-4}) \times \cos 45°$$

$$= 8.9 \times 10^{-4} \text{ Wb } (\cos 45° = 0.707)$$

The final value of the flux is,

$$\phi_2 = 0$$

∴ Change in flux, $\Delta\phi_B = 0 - (8.59 \times 10^{-4}) = -8.59 \times 10^{-4}$

The induced emf is given by,

$$\mathcal{E} = -\frac{\Delta\phi_B}{\Delta t} = -\frac{(-8.59 \times 10^{-4})}{0.5}$$

$$= 1.72 \times 10^{-3} \text{ V (or 1.72 mV)}$$ **(Answer).**

The induced current is given by,

$$i = \frac{\varepsilon}{R} = \frac{(1.72 \times 10^{-3})}{(0.6)}$$

$$= 2.87 \times 10^{-3} \text{ A (or 2.87 mA)}$$ **(Answer).**

Problem 5. A coil of radius 3.6 cm and having 120 turns is wrapped around a long solenoid of radius 1.8 cm and having 80 turns per cm. The resistance of the coil is 18 Ω and it is short-circuited through a small resistance of negligible value. The current flowing through the solenoid is rapidly switched over from 6 A in one direction to 6 A in the opposite direction. Calculate the resulting electric charge flowing through the coil.

Solution. Given : r (solenoid) = 1.8 cm = 1.8×10^{-2} m; n (solenoid) = 80 turns cm^{-1} = 8000 turns m^{-1}; N (coil) = 120 turns; R (coil) = 18 Ω [radius of the coil (= 3.6 cm) is immaterial here]

Let the current in the solenoid be i_1 and the induced current in the coil be i_2. We know that no magnetic field will exist outside the solenoid. The field inside the solenoid (parallel to its axis) is given by,

$$B = \mu_0 n i_1$$

The resulting magnetic flux linked with the coil is given by (A = area of cross-section)

$$\phi_B = B(\pi r^2)\, N$$

∴ The emf induced in the coil is given by,

$$\mathcal{E} = \frac{d\phi_B}{dt} = \pi r^2 . N . \frac{dB}{dt}$$

$$= \pi r^2.N.\mu_0 n \frac{di_1}{dt}$$

The induced current in the coil is given by,

$$i_2 = \frac{\mathcal{E}}{R} = \frac{\pi r^2.N.\mu_0 n}{R} \cdot \frac{di_1}{dt}$$

$$\Rightarrow \qquad i_2 dt = \frac{\pi r^2 N \mu_0 n}{R} di_1$$

i_2dt is the charge dq passing through the coil in time dt, for a change di in the solenoid current. ∴ The total charge passing through it for a change in solenoid current equal to Δi_1 is given by,

$$\Delta Q = \int i_2 dt = \frac{\pi r^2 N \mu_0 n}{R} \int di_1$$

$$= \frac{\pi r^2 N \mu_0 n}{R} \Delta i_1$$

Now, it is given that the solenoid current changes from 6 A in one direction to 6 A in the opposite direction. Therefore,

$$\Delta i_1 = 6A - (-6A) = 12A$$

$$\therefore \Delta Q = \frac{3.14 \times (1.8 \times 10^{-2})^2 \times (120) \times (4 \times 3.14 \times 10^{-7}) \times (8000) \times (12)}{(18)}$$

$$= 818 \times 10^{-6} \text{ C (or 818 } \mu\text{C)} \qquad \textbf{(Answer).}$$

Problem 6. A square loop of area 3.6×10^{-3} m^2 and having 130 turns is placed in a plane perpendicular to a magnetic field of magnitude 0.5 T, such that its side CD coincides with the edge of the field. If it has a resistance of 140 Ω, calculate the work done to uniformly pull it out of the field in 1.2 second, in a direction normal to the field.

Solution. Given : N = 130 turns; Area = 3.6×10^{-3} m^2; B = 0.5 T; R = 140 Ω; t = 1.2 s

The length of each side of the loop is,

$$\ell = \sqrt{3.6 \times 10^{-3}} = \sqrt{36 \times 10^{-4}} = 0.06 \text{ m}$$

For pulling the loop out of the field, the side AB has to move a distance BC (= 0.06 m) in 1.2 second. ∴ the velocity of the loop is,

$$v = \frac{0.06}{1.2} = 0.05 \text{ ms}^{-1}$$

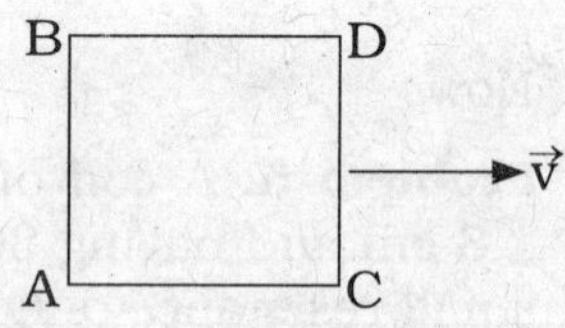

The emf induced in the side AB of the loop is given by,

$$\mathcal{E} = N\,(Bv\ell) = (130) \times (0.5) \times (0.05) \times (0.06)$$

$$= 0.195 \text{ V}$$

∴ The current induced in the loop is,

$$i = \frac{\mathcal{E}}{R} = \frac{(0.195)}{(140)} = 139 \times 10^{-5} \text{ A}$$

The force on AB due to the magnetic field is given by,

$$F = i\ell B = (139 \times 10^{-5}) \times (0.06) \times (0.5)$$

$$= 4.17 \times 10^{-5} \text{ N}$$

This force is directed towards left side. To pull the loop out of the field, a force of the same magnitude has to be applied to the right side, which moves it through a distance BC. ∴ The work done is given by (work = force×distance),

$$W = (4.17 \times 10^{-5}) \times (0.06)$$

$$= 2.5 \times 10^{-6} \text{ J (or 2.5 } \mu\text{J)} \qquad \textbf{(Answer).}$$

Problem 7. A circular coil of radius 12 cm and having 600 turns has its plane normal to the horizontal component (B_H) of the magnetic field of earth. If the resistance of the coil is 3Ω, find the emf and current induced in it, when it is uniformly rotated about its vertical diameter for 0.3 second (take $B_H = 3.6 \times 10^{-5}$ T).

Solution. Given : r = 12 cm = 0.12 m; N = 600 turns; R = 3Ω; B = 3.6×10^{-5} T;
$\Delta t = 0.3$ s; $\theta_1 = 0°$ and $\theta_2 = 180°$

$$\text{Area of the coil, } A = \pi r^2 = 3.14 \times (0.12)^2 = 0.0452$$

The magnetic flux through the coil before rotation is,

$$\phi_{B1} = B\,A\cos\theta_1 = (3.6\times10^{-5}) \times (0.0452) \times \cos 0$$
$$= -\,0.163\times10^{-5}\text{ Wb} \qquad (\cos 0 = 1)$$

The flux through the coil after 180° rotation is,

$$\phi_{B2} = B\,A\cos\theta_2 = (0.163\times10^{-5}) \times \cos 180°$$
$$= -\,0.163\times10^{-5}\text{ Wb} \qquad (\cos 180° = -1)$$

∴ change in flux, $\Delta\phi_B = \phi_{B1} - \phi_{B2} = [0.163 - (-\,0.163)] \times 10^{-5}$

$$= 2\times0.163\times10^{-5}\text{ Wb}$$

∴ The induced emf is given by,

$$\mathcal{E} = -\,N\frac{\Delta\phi}{\Delta t} = -\frac{(600)\times(2\times0.163\times10^{-5})}{(0.3)}$$
$$= 6.52\times10^{-3}\text{ V (or 6.52 mV)} \quad \textbf{(Answer).}$$

The induced current is,

$$i = \frac{\mathcal{E}}{R} = \frac{(6.52\times10^{-3})}{(3)}$$
$$= 2.17\times10^{-3}\text{ A (or 2.17 μA)} \quad \textbf{(Answer).}$$

Problem 8. A wheel revolves in a plane perpendicular to the direction of the horizontal component (B_H) of the magnetic field of earth. It has 12 steel spokes of length 60 cm each and makes 180 revolutions per minute. Calculate the emf induced between its axle and the rim (take B_H = 0.36 G).
Solution. Given : Radius of wheel, R = 60 cm = 0.6 m; angular speed, w = 180 r.p.m.;
No. of spokes = 12; $B_H = 0.36\text{ G} = 0.36\times10^{-4}$ T
The emf generated across each element of length dr of the wheel is,

$$d\mathcal{E} = B.v.dr = B.r\omega.dr \text{ (v: linear speed)}$$

∴ The emf generated across the entire length R of the spoke becomes,

$$\mathcal{E} = \int dE = \int_0^R B\omega.rdr = \frac{B\omega R^2}{2}$$

Now, $\omega = 180\text{ r.p.m.} = 180 \times 2\pi\text{ radian/minute}$

$$= \frac{180\times2\pi}{60}\text{ rad/second} = 6\pi\text{ rad s}^{-1}$$

$$\therefore \quad \mathcal{E} = \frac{B\omega R^2}{2} = \frac{(0.36\times10^{-4})\times(6\pi)\times(0.6)^2}{2}$$
$$= 122\times10^{-6}\text{ volt (or 122 μV)} \quad \textbf{(Answer).}$$

The emf's generated across all the spokes are in parallel with each other. Hence the number of spokes does not matter.

Problem 9. Derive an expression for the mutual inductance of two concentric coils of radii r_1 (inner coil) and r_2 (outer coil), such that $r_2 \gg r_1$.
Solution. Suppose a current i flows through the outer coil. It will produce a magnetic field B_2 at the centre of the coil.

$$B_2 = \frac{\mu_0 i}{2r_2}$$

Since the smaller inner coil has a comparatively very small radius ($r_1 \ll r_2$), the magnitude of the magnetic field B_2 over its area ($\pi r_1{}^2$) may be assumed to be constant. Hence the magnetic flux linked with it is given by,

$$\phi = (\pi r_1^2) B_2 = \frac{\mu_0 \pi r_1^2}{2r_2} i \qquad \text{..... (i)}$$

But we also have, $\phi = Mi$ ..... (ii)

Comparing Eqs. (i) and (ii) we get,

$$M = \frac{\mu_0 \pi r_1^2}{2r_2}$$ **(Answer).**

Problem 10. A horizontal 90 cm long thin rod is dropped from the top of a tower perpendicular to the horizontal component (B_H) of the magnetic field of earth, such that it remains horizontal throughout the fall, pointing east-west. Determine the magnitude and direction of the emf induced in it at the instant its speed is 6 ms^{-1}. Which one of its two ends is at higher potential ? Take $B_H = 0.36 \times 10^{-4}$ Wb.

Solution. Given : $\ell = 90$ cm $= 0.9$ m; $v = 6\ ms^{-1}$; $B = 0.36 \times 10^{-4}$ Wb

The magnitude of the instantaneous value of induced emf is given by,

$$\mathcal{E} = Bv\ell = (0.36 \times 10^{-4}) \times (6) \times (0.9)$$
$$= 1.94 \times 10^{-4}\ V$$ **(Answer).**

Assuming the circuit to be closed and applying Fleming's right hand rule, it is found that the direction of induced current in the rod is east-west. Since conventionally the direction of electric current is taken to be opposite to the direction of flow of electrons, it follows that the electrons flow from west to east, which in turn means that the east end of the rod is at a higher potential (i.e. the east end is +ve and the west end is –ve). The direction of the induced emf is the same as that of the flow of electrons, viz. west-east. **(Answer).**

Problem 11. The magnitude of the earth's magnetic field at a place is 4 gauss and the angle of dip is 45°. An aircraft is flying westward with a velocity of 1200 km per hour. It measures 30 metre end to end, along its wings. Calculate the potential difference developed between the ends of the wings.

Solution. Given : $B_E = 4\ G = 4 \times 10^{-4}\ T$; $\theta = 45°$; $\ell = 30$ m; $v = 1200\ km.h^{-1} = 1200 \times \frac{1000}{3600}\ ms^{-1} = \frac{1000}{3}\ ms^{-1}$

Since normally an aircraft remains horizontal during its flight, its wings only intercept the vertical component (B_V) of the earth's field. Now,

$$B_V = B_E \sin\theta = (4 \times 10^{-4}) \times \sin 45°$$
$$= 4 \times 10^{-4} \times 0.707 = 2.83 \times 10^{-4}\ T$$

The potential difference developed between the ends of the wings is the emf induced between the ends, given by,

$$P.D. = \mathcal{E} = B_V v \ell = (2.83 \times 10^{-4}) \times \left(\frac{1000}{3}\right) \times (30)$$
$$= 2.83\ \text{volt}$$ **(Answer).**

Very Short Answer Questions

[Including HOTS Questions]

(1 mark each)

1. Define magnetic flux and give its SI unit.
2. Define magnetic flux density and give its SI unit.
3. What causes an emf to be induced in a metal coil ?
4. If a metal plate is moved in a uniform magnetic field parallel to the field, will an emf be induced in it ?
5. In an open circuit, does the change in magnetic flux through a metal coil induce an emf or a current or both in the coil ?

6. A small straight wire falls freely from a height, such that it points north-south throughout the fall. Will an emf be induced in it ?

7. A small straight wire falls freely from a height, such that it points east-west throughout the fall. Will an emf be induced in it ?

8. A copper coil is held stationary normal to a magnetic field existing between the poles of a very strong permanent magnet. Will a current be induced in the coil ?

9. An aluminium coil is held stationary in a non-uniform magnetic field. Will an emf be induced in it ? [Hint: No, since the magnetic flux does not change].

10. A current-carrying straight wire P rapidly approaches another parallel straight wire Q. Will an induced current be set up in Q ? If yes, what will be its direction with respect to the direction of current in P ? [Hint: Opposite to the direction in P].

11. In the previous question, what will be the direction of current in Q, if it rapidly moves away from P ? [Hint: same direction as in P].

12. What is meant by back emf ?

13. A copper plate and a plastic plate of the same size are simultaneously dropped from the top of a tower. Will they reach the ground simultaneously ? If not, which one will reach earlier ?

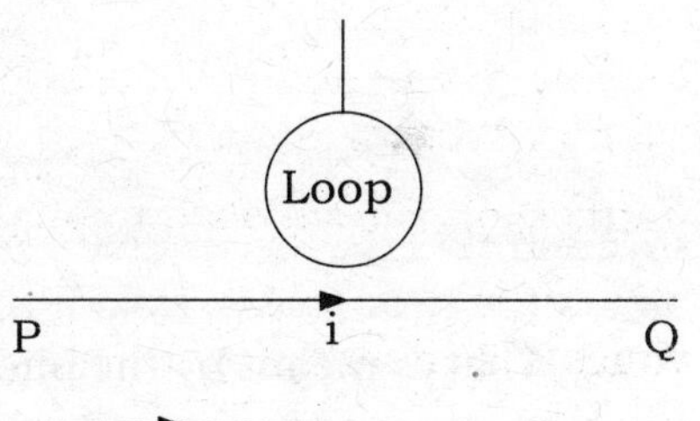

14. If the current i flowing in the wire PQ in the figure is made to fall rapidly, will the direction of the current induced in the metal loop suspended above the wire be clockwise or anticlockwise ?

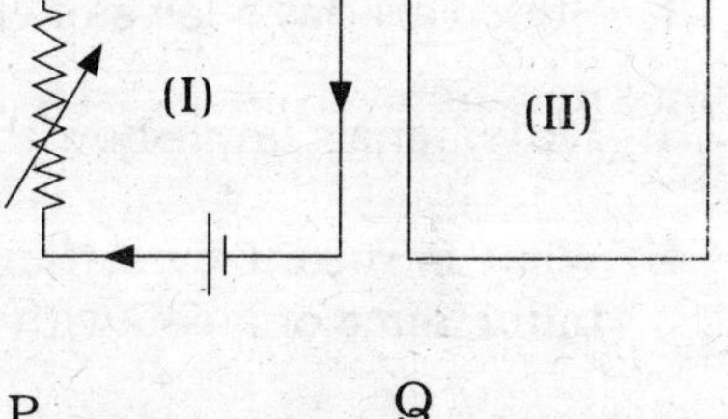

15. In the figure shown, the resistance in circuit (I) is rapidly reduced. Will the direction of current induced in circuit (II) be clockwise or anticlockwise ?

16. The figure shows a horizontal solenoid PQ connected to a battery B and a switch S. A copper ring R is placed on a frictionless track, with the axis of the ring along the axis of the solenoid. What would happen to the ring as the switch S is closed ? [Hint: The ring moves away, because the current induced in it opposes the growth of current in PQ]. (AI; 2001).

17. A magnet is moved in the direction indicated by an arrow between two coils AB and CD, as shown in the figure. Suggest the direction of current in each coil. (AI; 2001).

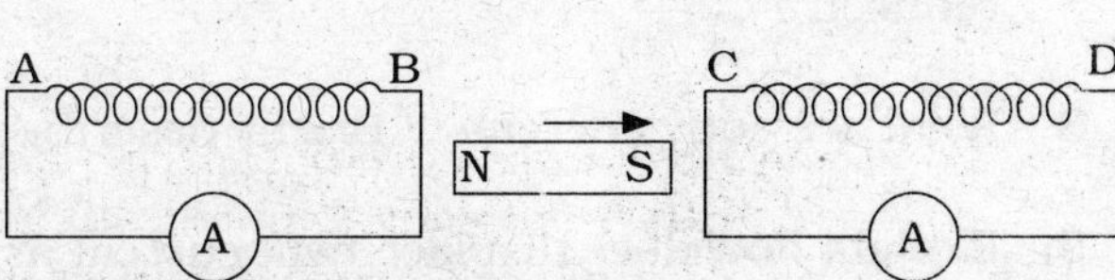

18. The inductance coils are usually made of copper. What is the underlying reason ? [Hint: To obtain a large induced current, as copper has small resistivity].

19. The length of a solenoid and its number of turns are increased two-fold, without changing its area of cross-section. How will the value of inductance be affected ? [Hint: Doubled; $L = \mu^2 N^2 A/\ell$].

20. The use of soft iron core leads to increase in the self-inductance of an inductor. How ? [Hint: The large relative permeability, μ_r, of soft iron raises the magnetic flux μ_r times, leading to a proportionate rise in inductance].

21. A copper coil of area of cross-section A is located in a uniform magnetic field, with its plane lying perpendicular to the direction of field. What will be the change in magnetic flux across the coil when it is rotated through 180° ? [Hint: 2BA].

22. A 10 cm long glass rod is moved perpendicular to a uniform magnetic field with velocity 50 ms^{-1}. What will be the value of emf induced in the rod ? [Hint: zero, since glass is non-conducting].

23. What is the cause of strong damping of oscillations of a copper sheet in a magnetic field ?

24. Define displacement current. What is its SI unit ?

25. If the charging current for a parallel-plate capacitor is 0.5 A, what is the displacement current across its plates ? [Hint: Same].

Short Answer Questions

[Including HOTS Questions]

(2 or 3 marks each)

1. What is meant by the phenomenon of electromagnetic induction. Illustrate it with one example.

2. State Faraday's laws of electromagnetic induction, along with mathematical expressions.

3. State Lenz's law. Show that it is consistent with the principle of conservation of energy.

4. What is meant by self-inductance ? Give its SI unit. Mention two factors which determine the self-inductance of a coil with air core.

5. What is meant by mutual inductance ? Give its SI unit. Mention two factors which determine the mutual inductance of a pair of coils.

6. What is meant by electromagnetic damping. How it helps to make a galvanometer dead beat ?

7. How can the heating and electromagnetic damping produced by eddy currents in a conductor be reduced ?

8. What are eddy currents ? Briefly describe their two applications.

9. Explain how does displacement current arise. How is it different from the usual conduction current ?

Long Answer Questions

(5 marks each)

1. Distinguish between self-induction and mutual induction. Define henry. Derive expressions for the (i) self-inductance of a long solenoid and (ii) energy stored in a current-carrying inductor.

Numerical Exercises

1. A circular coil is located in a uniform magnetic field of magnitude 3000 tesla, with its plane normal to the field. When the field is rapidly reduced to 600 tesla in 2 second, an emf of 12 V is induced in the coil. If the average diameter of the coil is 3 cm, find the number of turns in the coil. **[14 turns]**

2. A uniform magnetic field of strength 9 T acts normal to a circular coil, having 60 turns and occupying an area of 80 cm^2. Determine the emf induced in the coil, when it is quickly removed from the field in just 0.12 second. **[36 volt]**

3. A horizontal copper wire of length 14 cm moves with a speed of 12 ms^{-1} perpendicular to a uniform magnetic field of magnetic induction 1.5 T. Determine the emf induced in the wire. **[2.52 volt]**

4. When the current in a circuit is quickly brought down from 8A to zero, an average emf of 150 V is induced in the circuit. Find the self-inductance of the circuit. **[3.75 H]**

5. A circular copper coil of area 1.6 cm^2 is kept inside a solenoid, at right angles to the axis of the solenoid. If the solenoid has 20 turns per cm and the current flowing in it steadily rises from 3 A to 6 A in just 0.05 s, find the emf induced in the coil. **[24.1 μV]**

6. At a certain place, a horizontally held 30 cm long aluminium rod pointing north-south is steadily moved towards east at a speed of 12 ms^{-1}. If the horizontal component of the earth's magnetic field and the angle of dip at the place are 4×10^{-4} T and 48°, respectively, determine the emf induced in the rod. [Hint: The moving rod will intercept the vertical component, B_V, of the earth's field]. **[16.0 μV]**

7. A square-shaped loop is made from copper wire of diameter 1 mm and length 60 cm. It is placed in a uniform magnetic field, with its plane held normal to the direction of field. If the magnetic field is varied at a steady rate of 0.03 tesla per second, calculate the current induced in the loop. **[5.2×10^{-2} A]**

8. An electric current of 4 ampere flows through an inductor of inductance 80 mH. Find the energy stored in the inductor. **[0.64 J]**

9. An 80 cm long solenoid has radius 3 cm and consists of 1200 turns. A copper wire is wound over its central region, such that it has 120 turns. Calculate the mutual inductance of the combination. **[0.64 mH]**

10. Two neighboring coils have a mutual inductance of 1 henry. If the current in one of them steadily rises from zero to 15 A in 0.3 second, calculate the change in magnetic flux linked with the other coil. **[15 Wb]**

11. A circular copper disc 10 cm in radius rotates at 20π rad/s about an axis passing through its centre and perpendicular to the disc.
(i) Calculate the potential difference developed between the axis of the disc and the rim.
(ii) What is the induced current, if the resistance of the disc is 2 Ω ? (AI; 2001).

[0.02 V; 0.01 A]

12. A wheel consists of 40 steel spokes, each of length 40 cm. It is made to revolve in a plane at right angles to the magnetic field of earth, of magnitude 0.6×10^{-4} tesla. If the angular speed of revolution of the wheel is 15 radian per second, find the emf induced between the axle and the rim of the wheel.

[72 μV]

13. A coil of inductance 12 H and a resistor of 8 Ω are connected in series to a battery of 16 V, through a switch, so that a steady current flows in the circuit. When the switch is opened, the current falls to zero in 0.5 ms. Determine the average emf induced across the coil. **[48000 V]**

CHAPTER 4 (B)

ALTERNATING CURRENT; AC GENERATOR AND TRANSFORMER

So far we have considered electric circuits in which the source of supply of current made the current to continuously flow in one direction. Such a current source, e.g. a battery, is termed as a **direct current (DC) source** and the resulting current as **direct current (DC).** However, in our daily life we are mostly served by current sources which supply current that rapidly reverses its direction in a periodic manner. In India, the frequency of current supply is 50 hertz, i.e. 50 cycles per second. The current reaches its maximum value up and down twice in one cycle. It means that the current reverses its direction $50 \times 2 = 100$ times in one second. Such a current is termed as **alternating current (AC)** and the current source as an **AC source.** [In a loose way, the AC is called 'AC current', which means 'alternating current current'! Such wrong usage is unfortunate but cannot be helped, because now it has become universally acceptable].

The use of AC voltage offers several advantages over the DC voltage. A major advantage is that one voltage value can be readily converted into another desired value through a transformer. Another is that electric power can be economically transmitted over long distances, without a heavy loss. In the present chapter, we shall study some basic features of the AC circuits, besides describing the construction and working of transformers and AC generators.

4B. 01. Wave Motion,

As already mentioned, the AC current and voltages involve rapid periodic changes with time. Therefore, before embarking on a study of characteristics of the AC circuits, the student should usefully revise his knowledge of wave motion and simple harmonic motion, studied earlier. In the following, we shall very briefly revise some essential features of the wave motion.

Any motion that repeats itself at regular intervals is termed **periodic motion** or **harmonic motion.** Two familiar examples are shown in Figs. 4.7 and 4.8. In Fig. 4.7, a particle rotates along a circle with centre O and radius y_m. Its circular motion is a periodic motion, with period 2π. The projection of the particle P along the vertical diameter AB of the circle is Q. As P revolves around the circle, its projection point Q undergoes an up and down vibratory (or, say, oscillatory) motion along AB. This vibratory (or oscillatory) motion of Q is also a periodic motion, repeating itself after every total displacement $(y_m + y_m + y_m + y_m) = 4y_m$. The same result is obtained if the projection of P along any other diameter, e.g. CD, is considered. If the displacement y of the projection point Q is plotted as a function of time, the **wave motion** depicted in Fig. 4.8 is obtained, which is also a periodic motion. Three positions Q, Q_1, Q_2 marked in Fig. 4.8 correspond with the similarly marked positions in Fig. 4.7.

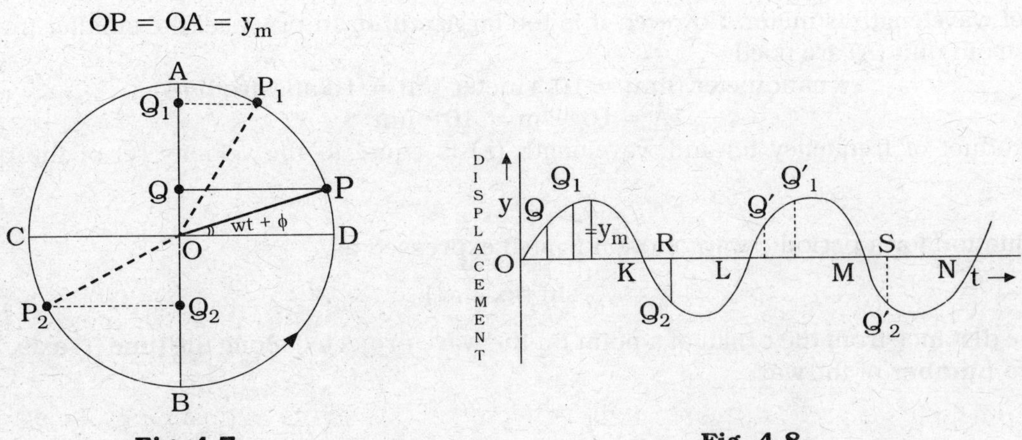

Fig. 4.7 **Fig. 4.8**

In both the Figs. 4.7 and 4.8, the displacement y of the projection particle from the origin O is represented as a function of time by the equation,

$$y = y_m \sin(\omega t + \phi), \quad \text{..... (4.23)}$$

where y_m, ω, and ϕ are constant values. Such a periodic motion, which is a sinusoidal function of time, is called **simple harmonic motion (SHM).** It can be equally well represented as a cosine function, viz. $y = y_m \cos(\omega t + \phi)$, but it has become customary to represent it as a sinusoidal variation (the cosine variation corresponds with the motion of the projection of P along the diameter CD is Fig. 4.7). The sine and cosine waves look alike, having the appearance shown in Fig. 4.8.

Since a sine (or cosine) function varies between O and ±1, the maximum value of the displacement is y_m. It is called **amplitude** of the periodic motion.

The time taken by the particle to complete one vibration (or oscillation), i.e. time taken to return to its starting position, is called **time period (T)** of the motion. The number of vibrations (or oscillations or cycles) completed in one second represent the **frequency (v)** of vibration (or oscillation). We clearly have the following relationship between T and ν,

$$\nu = \frac{1}{T} \quad \text{..... (4.24)}$$

The SI unit of frequency is hertz (Hz). 1 hertz (Hz) = 1 cycle per second ($1 s^{-1}$).

A since function repeats after every angular change of 2π. Also, from the definition of the time period, it should repeat after every time change T. Accordingly, the Eq. 4.23 can be written in either of the following two equivalent ways,

$$y = y_m \sin[(\omega t + \phi) + 2\pi]$$

and,

$$y = y_m \sin[\omega(t + T) + \phi]$$

The equality of the argument of the sine function in the two equations (within square brackets) readily yields,

$$(\omega t + \phi) + 2\pi = \omega(t + T) + \phi$$

Or

$$\omega = (2\pi/T) = 2\pi\nu \quad \text{..... (4.25)}$$

ω is called **angular frequency** of the motion. Its SI unit is radian per second.

At any given time t, the angular position of the particle in Fig. 4.7 is given by (ωt + ϕ). The constant value ϕ is its angular position at t = 0. It is called the **phase constant** (or **phase angle).** The time-varying factor (ωt + ϕ), which is included in Eq. 4.23, is called the **phase** of the periodic motion. It represents the **state of vibration (or oscillation)** of the particle. In Fig. 4.8, the positions Q and Q′ of equal upwards displacement are the positions of same phase; similarly for the positions Q_1 and Q'_1. The positions Q_2 and Q'_2 of equal downwards displacement are also same phase positions. Note that the positions K and L, for which the displacement is same (viz. zero), are not the same phase positions. At K the particle is tending to go down from the mean position, while at L it is tending to go up. So the states of vibration are different at K and L. But K and M are same phase positions. The same is true for the positions O, L, N.

The distance between any two successive positions on the wave corresponding to the same phase is called **wavelength** (λ). Thus in Fig. 4.8, we have,

$$OL = KM = LN = RS = \lambda$$

The SI unit of wavelength is meter. However, it is too large a unit. In practice, the smaller units nanometer (nm) or ängstrom unit (Å) are used

$$1 \text{ nanometer (nm)} = 10^{-9} \text{ meter (m)} = 10 \text{ ängstrom (Å)}$$

$$1\text{Å} = 10^{-10} \text{ m} = 10^{-1} \text{ nm}$$

The product of frequency (ν) and wavelength (λ) is equal to the velocity (v) of the propagation of wave,

$$\nu\lambda = v \quad \text{..... (4.26)}$$

The equation for a periodic wave motion is also expressed as,

$$y = y_m \sin(kx - \omega t), \quad \text{..... (4.27)}$$

where x is the distance from the origin of a point on the wave projected along the time (t) axis. **k** is called the **angular wave number** of the wave,

$$k = \frac{2\pi}{\lambda} \quad \text{..... (4.28)}$$

4B. 02. Instantaneous Peak, Average and rms Values of Current/Voltage; AC Circuit Containing a Resistor Alone

As already expressed, usually the electricity produced at the generating stations and then supplied for use in daily life is AC. It is sinusoidal in nature (vide Fig. 4.8). The supply voltage and the resulting current in the electric circuit are expressed as,

$$V = V_0 \sin \omega t \quad \text{..... (4.29)}$$

$$i = i_0 \sin \omega t \quad \text{..... (4.30)}$$

V and i are the values of voltage and current, respectively, at a given time t. So they are called **instantaneous values** of the voltage/current. They can be **defined** as "the values of voltage/current at a given instant of time t". Since the maximum value of a sine function is 1, the corresponding maximum values of V and i are V_0 and i_0, respectively. They are called **peak values** of the voltage/current. They can be **defined** as "maximum values of AC voltage/current". For a given value of V, V_0 is a constant value. Similarly for i and i_0.

ω is the **angular frequency** of voltage/current. It is related to the **frequency** ν as (vide Eq. 4.25),

$$\omega = 2\pi\nu$$

In India, ν = 50 Hz, i.e. 50 cycles per second. The AC voltage in an electric circuit is represented by the symbol (~).

Comparing the Eqs. 4.29 and 4.30 with the general equation (Eq. 4.23) for a periodic motion, it is found that the term ϕ (phase angle) has been dropped. This has been done for the sake of convenience. It does not lead to loss of generality.

Because of the sinusoidal variation, the current continuously changes with time, alternately becoming positive and negative. However, for an infinitesimal interval of time dt it may be treated as a constant value. Then we can write the average value $\bar{i}$ of the current over a time interval 0 to t as,

$$\bar{i} = \left(\int_0^t i\,dt\right)\Big/\left(\int_0^t dt\right) = (1/t)\int_0^t i\,dt$$

$$= (1/t)\int_0^t [i_0 \sin\omega t]\,dt \quad \text{(from Eq. 4.30)}$$

$$= (-i_0/t)\,(1/\omega)\,[\cos\omega t]_0^t \quad (\because i_0 \text{ and } \omega \text{ are constant})$$

$$= -(i_0/t\omega)\,[\cos\omega t - \cos 0°]$$

(We have taken the integration constant = 0, because at t = 0 there is no current in the circuit and so both $\bar{i}$ and i_0 = 0)

For one complete cycle of variation of the current, t = T. Further, from Eqs. 4.24 and 4.25, $\omega T = 2\pi\nu.(1/\nu) = 2\pi$. Therefore we have,

$$\bar{i} = (-i_0/t\omega)[\cos 2\pi - \cos 0°]$$

$$= (-i_0/t\omega)[1-1]$$

$$= 0$$

Thus the **average value of an AC current over a complete cycle** is zero. Obviously, the same is true for the average value taken over a long period of time. The same argument applies to the AC voltage, i.e. the **average value of an AC voltage over a complete cycle** or over a long time is also zero.

$$[\bar{i}]_{\text{complete cycle}} = 0 \quad \text{..... (4.31a)}$$

$$[\bar{V}]_{\text{complete cycle}} = 0 \quad \text{..... (4.32a)}$$

However, if calculated over a half-cycle, the average values are > 0. For a half-cycle, the limits of integration become o to T/2 and we have as before,

$$\bar{i} = -\frac{i_0}{t\omega}[\cos\omega t - \cos 0°]$$

$$= -\frac{i_0}{(T/2)\omega}[\cos\omega(T/2) - 1]$$

$$= -\frac{2i_0}{\omega T}\left[\cos\frac{\omega T}{2} - 1\right]$$

$$= -\frac{i_0}{\pi}[\cos\pi - 1] \quad (\because \omega T = 2\pi)$$

$$= -\frac{i_0}{\pi}[-1-1] = \frac{2i_0}{\pi}$$

Similarly the value of $\overline{V}$ for a half-cycle can be calculated as $(2V_0/\pi)$. Thus the **average values of an AC current and an AC voltage over a half-cycle** are given by,

$$[\bar{i}]_{\text{half-cycle}} = \frac{2i_0}{\pi} = 0.637\ i_0 \qquad \text{..... (4.31b)}$$

$$[\overline{V}]_{\text{half-cycle}} = \frac{2V_0}{\pi} = 0.637\ V_0 \qquad \text{..... (4.32b)}$$

The instantaneous current i can be +ve or –ve at a given time. But i^2 is always a positive value. Hence its average value is also always positive. As before for $\bar{i}$, the average value $\overline{i^2}$ over a complete cycle is given by,

$$\overline{i^2} = \left(\int_0^T i^2 dt\right)\Big/\left(\int_0^T dt\right) = (1/T)\int_0^T \left[i_0^2 \sin^2 \omega t\right] dt$$

$$= \left(i_0^2/2T\right)\int_0^T \left[1-\cos 2\omega t\right] dt$$

$$= \left(i_0^2/2T\right)\left[t-(\sin 2\omega t/2\omega)\right]_0^T$$

$$= \left(i_0^2/2T\right)\left[T-(\sin 4\pi - \sin 0)/2\omega\right] \qquad (\because \omega T = 2\pi)$$

$$= \left(i_0^2/2T\right)\left[T-(0-0)\right] = \left(i_0^2/2T\right).T$$

$$= \frac{i_0^2}{2}$$

(As before, the integration constant = 0, because at t = 0 both $\bar{i}$ and i_0 = 0). $\overline{i^2}$ is called **mean square current.** Its square root is known as **root mean square current** or **rms current** (also called **effective current** or **virtual current).** Thus we have,

$$i_{rms} = \sqrt{\overline{i^2}} = \frac{i_0}{\sqrt{2}}$$

$$= 0.707\ i_0 \qquad \text{..... (4.31c)}$$

In the same way, the **root mean square voltage** or **rms voltage** (also called **effective voltage** or **virtual voltage)** is given by,

$$V_{rms} = \sqrt{\overline{V^2}} = \frac{V_0}{\sqrt{2}}$$

$$= 0.707\ V_0 \qquad \text{..... (4.32c)}$$

Example 1. The household electric mains supply in India is 220 V, 50 Hz. Give the corresponding equation for instantaneous voltage.

Solution. Given : V_{rms} = 220 V; ν = 50 Hz

The instantaneous voltage is given by,

$$V = V_0 \sin\omega t$$
$$= V_0 \sin(2\pi\nu)t$$
$$= (\sqrt{2}\times 220)\sin(2\times 3.14\times 50)t$$
$$= 311 \sin 314t \text{ volt}$$

Thus the required equation is,

$V = 311 \sin 314t$ **(Answer).**

The significance of the rms values of voltage/current is realised when considering an AC circuit containing a resistor alone (Fig. 4.9). The current i and the voltage V in the circuit are given by

$$i = i_0 \sin\omega t$$
$$V = V_0 \sin\omega t$$

The thermal energy (Joule heating) generated in the resistor during an infinitesimal time interval dt is,

$$(i^2R)dt = [i_0^2 R \sin^2 \omega t]dt$$

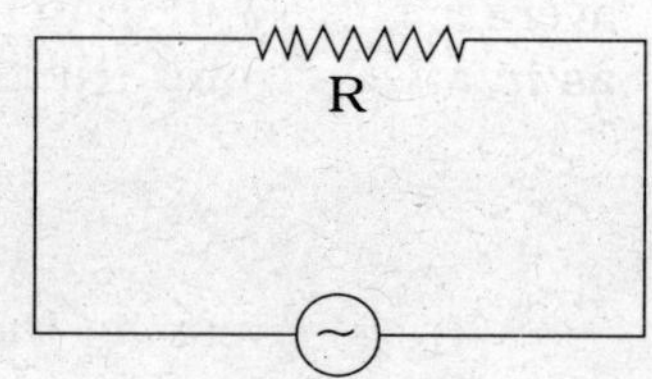

Fig. 4.9

Therefore the heat energy generated during one complete cycle (t = T) is,

$$H = \int_0^T [i^2R]dt = R\int_0^T [i_0^2 \sin^2 \omega t]dt$$
$$= RT\,[1/T]\int_0^T [i_0^2 \sin^2 \omega t]dt$$
$$= RT\left(i_0^2/2\right) \quad \text{(vide the foregoing derivation of } \overline{i^2};\ \omega T = 2\pi)$$
$$= \left(i_{rms}^2 R\right) T$$

(The integration constant = 0, because at t = 0 there is no heat produced and so H = 0). Thus we find that the thermal energy produced in a resistor during one cycle, on passing an AC current i through it, is equivalent to the thermal energy produced by passing a constant current i_{rms} through the resistor. Similarly, the thermal energy produced by the voltage $V = V_0 \sin\omega t$ is the same as that produced by applying a constant voltage V_{rms} across the resistor.

Accordingly, we can **define** the rms values of AC voltage and current as follows:

(a) **rms value of AC current.** "It is that value of steady current which when flowing through a resistor produces the same amount of heat as produced by the AC current flowing through the same resistor in the same time".

$$i_{rms} = \frac{i_0}{\sqrt{2}} = 0.707\, i_0$$

(b) **rms value of AC voltage (or emf).** "It is that value of voltage which when applied across a resistor produces the same amount of heat as produced by an AC voltage (or emf), when applied across the same resistor for the same time".

$$V_{rms} = \frac{V_0}{\sqrt{2}} = 0.707\, V_0$$

In practice, all AC voltages and current are supposed to be rms values. All AC meters give rms values as their readings. Hence it has become customary to drop the subscript 'rms' from the values of AC voltage and current.

Two more points of interest are to be noted in a purely resistive AC circuit, i.e. an AC circuit containing a resistor alone:

(i) In Fig. 4.9, the form of the AC voltage applied across the resistor of resistance R is,

$$V = V_0 \sin\omega t \qquad \text{..... (A)}$$

This should be equal to the voltage drop iR across the resistor,

$$iR = V_0 \sin\omega t$$

This gives the value of the current i in the circuit as,

$$i = \frac{V_0}{R} \sin\omega t$$
$$\Rightarrow \quad i = i_0 \sin\omega t \qquad \text{..... (B)}$$

V_0 is the maximum value of the applied voltage V. Therefore i_0 is the maximum value of the current i in the circuit,

$$i_0 = \frac{V_0}{R}$$
$$\Rightarrow \quad V_0 = i_0 R$$

If we compare this with the known expression V = iR for the Ohm's law for DC circuit, we find that the Ohm's law holds for the peak values of voltage/current in the AC circuits as well (it holds for the instantaneous and average values, too). In other words, a resistor **in a purely resistive AC circuit behaves in the same way as in a DC circuit.** Hence in such a circuit, the average power (P_{rms} or $\overline{P}$) produced is given by,

$$P_{rms}\ (\text{or } \overline{P}) = V_{rms}\, i_{rms} \left(= i_{rms}^2 R = \frac{V_{rms}^2}{R}\right)$$

(ii) Comparing the Eqs. (A) and (B), we find that the **voltage and current in a purely resistive circuit are in phase with each other.** This is diagrammatically illustrated in Fig. 4.10, in which plots of V and i as functions of ωt have been made. Note that the maxima and minima for V and i occur at the same positions.

Example 2. A heating rod of resistance 150 Ω is connected to an AC source of 220 V, 50 Hz. Find the average power consumed by the rod.
Solution. Given : R = 150 Ω; V_{rms} = 220 V; the given value of ν = 50 Hz is immaterial here
The average power is given by (with a resistance alone in the circuit),

$$\bar{P} = V_{rms}\, i_{rms} = \frac{V_{rms}^2}{R}$$

$$= \frac{(220)^2}{150} = 322.7 \text{ W} \quad \textbf{(Answer).}$$

Example 3. An electric lamp is marked as 60 W, 220 V. Calculate (i) peak voltage of the supply, (ii) resistance offerred by the lamp, and (iii) rms current passing through the filament of the lamp.
Solution. Given : $\bar{P}$ = 60 W, V_{rms} = 220 V

(i) The peak voltage is given by,

$$V_0 = \sqrt{2}\, V_{rms} = (1.414)\times(220)$$

$$= 311 \text{ V}$$

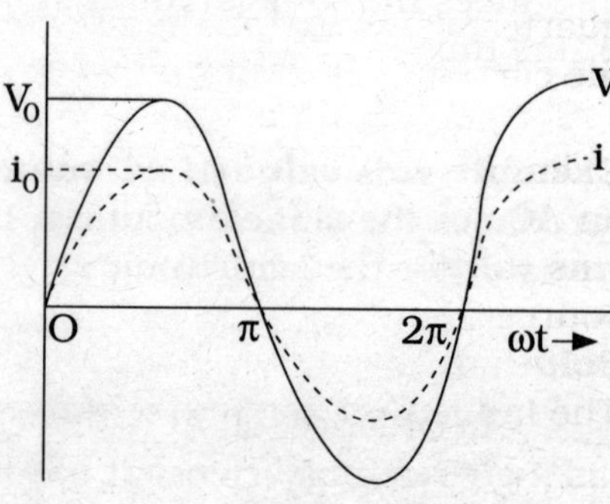

Fig. 4.10

(ii) The resistance of the filament is given by (P = Vi = V²/R),

$$R = \frac{V^2}{P_{rms}} = \frac{(220)^2}{(60)}$$

$$= 806.7\ \Omega \quad \textbf{(Answer).}$$

(iii) The rms current through the filament is given by,

$$i_{rms} = \frac{\bar{P}}{V_{rms}} = \frac{(60)}{(220)}$$

$$= 0.27 \text{ A} \quad \textbf{(Answer).}$$

4B. 03. AC Circuit Containing an Inductor Alone

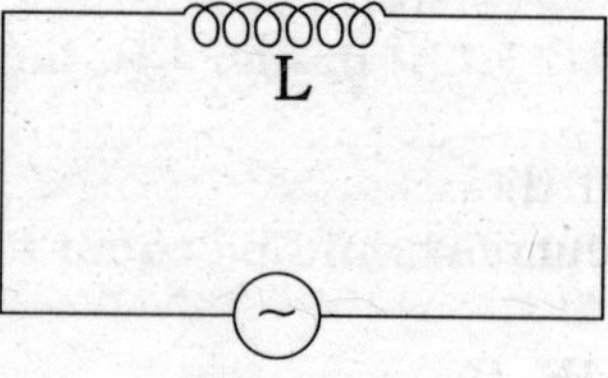

Fig. 4.11

The circuit shown in Fig. 4.11 comprises an AC source connected to an inductor of inductance L. We assume the inductor to be a pure inductor, i.e. its resistance is zero. (However, in practice it can never be so. Nevertheless, usually the coiled wire has a large thickness and therefore a low resistance). As before, the applied voltage is represented as,

$$V = V_0 \sin\omega t \qquad \text{..... (4.29)}$$

When the voltage is applied, it gradually builds up a current i in the circuit. The rate of change of current is (di/dt). According to the Faraday's laws of induction, it gives rise to a self-induced emf –L(di/dt) across the inductor. Applying Kirehoff's loop law (Chapter 2A; Section 2A. 14), we have

$$V - L\frac{di}{dt} = 0$$

$$\Rightarrow \quad \frac{di}{dt} = \frac{1}{L}V = \frac{1}{L}V_0 \sin\omega t$$

Integrating, we get the current as,

$$\int \frac{di}{dt} = \frac{V_0}{L}\int \sin\omega t\, dt$$

$$\Rightarrow \quad i = -\frac{V_0}{\omega L}\cos\omega t + \text{constt.}$$

Since our applied voltage V is sinusoidal, the resulting current i should also be sinusoidal. So its average over one cycle should be zero. On the right hand side, the average of cosωt over one cycle will be zero. Hence the integration constant should also be zero. So we have,

$$i = -\frac{V_0}{\omega L}\cos\omega t$$

$$= \frac{V_0}{\omega L} \sin(\omega t - \pi/2)$$

$$\Rightarrow \qquad i = i_0 \sin(\omega t - \pi/2), \qquad \text{..... (4.33)}$$

where, $$i_0 = \frac{1}{\omega L} V_0 \qquad \text{..... (4.34)}$$

In Eq. 4.33, i_0 is the amplitude (max. value) of the current. If we compare Eq. 4.34 with the Ohm's law [$V = iR$; $\therefore i = (1/R)V$], we find that the quantity **ωL** is equivalent to the resistance R. It is called **inductive reactance.** Just as a resistance R limits the current in a purely resistive circuit, the inductance L limits the current in a purely inductive circuit. The inductive reactance is denoted by $\mathbf{X_L}$,

$$X_L = \omega L \qquad \text{..... (4.35)}$$

X_L has the same dimensions as R. Its SI unit is Ohm(Ω). It is directly proportional to the frequency ω of the voltage/current, as well as to the value of L. Greater the value of ω, greater is X_L.

If we compare the Eqs. 4.29 and 4.33, it is seen that the current lags the voltage by ω/2, i.e. by one-quarter of a cycle (cf. the previous case of a purely resistive circuit, in which the voltage was in phase with the current). The relevant wave diagram is shown in Fig. 4.12.

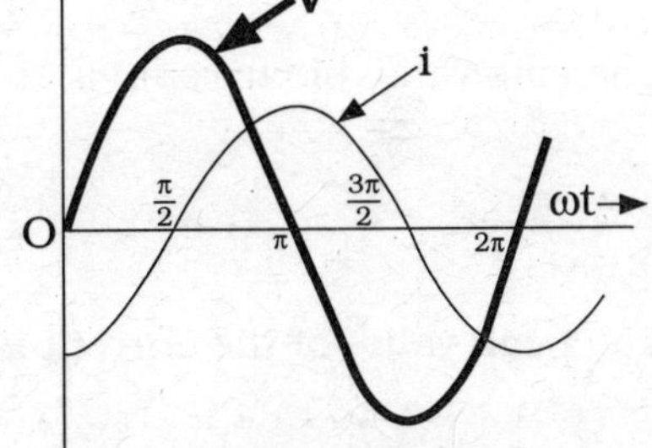

Fig. 4.12

Example 4. An electric circuit only consists of an inductor of 100 mH and an AC source of 220 V, 50 Hz. Determine the inductive reactance and the rms value of current in the circuit. How these values will be affected if the source frequency is halved ?

Solution. Given : V_{rms} = 220 V; L = 100 mH = 0.1 H; ν = 50 Hz

The inductive reactance is given by,

$$X_L = 2\pi\nu L = 2\times3.14\times(50)\times(0.1)$$
$$= 31.4\ \Omega \qquad \textbf{(Answer).}$$

The rms value of the current is,

$$i_{rms} = \frac{V_{rms}}{X_L} = \frac{(220)}{(31.4)}$$

$$= 7.0\ \text{A} \qquad \textbf{(Answer).}$$

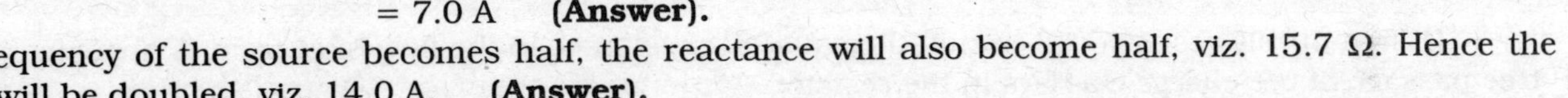

If the frequency of the source becomes half, the reactance will also become half, viz. 15.7 Ω. Hence the current will be doubled, viz. 14.0 A. **(Answer).**

4B. 04. AC Circuit Containing a Capacitance Alone

The Fig. 4.13 shows an AC source connected to a capacitor of capacitance C. As before, the applied voltage is represented as,

$$V = V_0 \sin\omega t \qquad \text{..... (4.29)}$$

Let q be the charge on the capacitor at a given instant t. Then the voltage across the capacitor at the instant t is,

$$V = \frac{q}{C}$$

Applying the Kirchhoff's loop rule to the circuit in Fig. 4.13, the AC supply voltage has to be equal to the voltage across C at all times,

$$V_0 \sin\omega t = \frac{q}{C}$$

$$\Rightarrow \qquad q = CV_0 \sin\omega t$$

Fig. 4.13

The current in the circuit is given by,

$$i = \frac{dq}{dt} = \frac{d}{dt}[CV_0 \sin\omega t]$$
$$= \omega CV_0 \cos\omega t$$
$$= (\omega C)V_0 \sin(\omega t + \pi/2)$$

$$\Rightarrow \qquad i = i_0 \sin(\omega t + \pi/2) \qquad \text{..... (4.36)}$$

where, $$i_0 = (\omega C)V_0 \qquad \text{..... (4.37)}$$

i_0 is the amplitude of the current i, which oscillates with the same angular frequency ω as the applied voltage V. Comparing the Eq. 4.37 with the expression for Ohm's law [V = iR or i = (V/R)], we find that the quantity (1/ωC) in a purely capacitive AC circuit is equivalent to a resistance R in a DC circuit. It is called **capacitive reactance.** It is denoted by $\mathbf{X_C}$,

$$X_C = \frac{1}{\omega C} \qquad \text{..... (4.38)}$$

The dimensions of X_C are the same as that of the resistance. Its SI unit is ohm (Ω). It is inversely proportional to ω and C. The greater the angular frequency ω, the smaller is the value of X_C.

If we compare the Eqs. 4.29 and 4.36, it is found that in this case the current leads the applied voltage by (π/2), i.e. by one-quarter of a cycle. The corresponding wave diagram is shown in Fig. 4.14.

Example 5. An electric circuit has only a capacitor of 20 μF and an AC source of 220 V, 50 Hz. Calculate the values of capacitive reactance, rms current, and peak current.

Solution. Given : V_{rms} = 220 V; ν = 50 Hz; $C = 20\ \mu F = 20 \times 10^{-6}$ F

The capacitive reactance is,

$$X_C = \frac{1}{2\pi\nu c} = \frac{1}{2 \times 3.14 \times (50) \times (20 \times 10^{-6})}$$

$$= 159.2\ \Omega \quad \textbf{(Answer).}$$

The rms value of current is,

$$i_{rms} = \frac{V_{rms}}{X_C} = \frac{(220)}{(159.2)}$$

$$= 1.38\ A \quad \textbf{(Answer).}$$

The peak value of the current is,

$$i_0 = \sqrt{2}\, i_{rms} = 1.414 \times (1.38)$$

$$= 1.95\ A \quad \textbf{(Answer).}$$

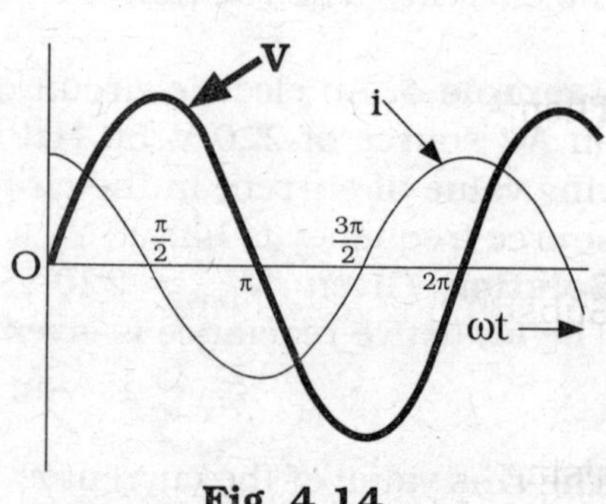

Fig. 4.14

Additional knowledge

(i) In a purely resistive AC circuit, the cause of resistance to the flow of current is obstruction to a free passage of the charge carriers in the resistor.

(ii) In a purely inductive AC circuit, the cause of resistance to the flow of current is the self-induced emf, which builds up in the inductor and opposes the emf of the AC source.

(iii) In a purely capacitive AC circuit, the capacitor is alternately charged and discharged, as the supply AC voltage reverses each half cycle. When it is charging, it opposes the flow of current through it (produced by the AC source). In a DC circuit, the current in the circuit falls to zero when the capacitor becomes fully charged. But in the AC circuit, the flow of charge is not completely prevented because of the continuous quick reversal of the direction of flow of charge. Thus the capacitor only limits the current flowing in the circuit.

(iv) Since the average value of an AC current or voltage over a complete cycle is zero, DC ammeters and voltmeters cannot be used for AC measuremetns. They show zero value for all AC currents and voltages. Hence the voltmeters and ammeters employed for AC measurements are based on heating effect. They can be used for DC measurements, too. They are termed as **hot wire ammeters and voltmeters.**

4B. 05. Series LCR Circuit; Impedance

The Fig. 4.15 shows an A.C. source connected to a resistor (R), an inductor (L), and a capacitor (C), in series. As before, the applied voltage is represented as,

$$V = V_0 \sin \omega t \qquad \text{..... (4.29)}$$

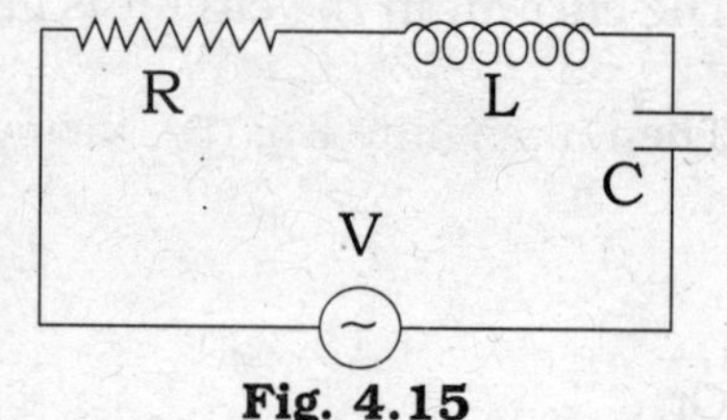

Fig. 4.15

Since this is a series circuit, the same current i flows through all the components R, L, and C. However, since the circuit contains both L and C, in general it may be expected that the current i will differ in phase with the applied voltage V. Let the phase difference be ϕ (in special cases ϕ may be zero, as we shall see later). Then the current i may be expressed as,

$$i = i_0 \sin(\omega t - \phi) \qquad \text{..... (4.39)}$$

To know the value of the current i in the circuit, we have to find the values of the current amplitude i_0 and the phase constant ϕ, as described in the following.

By applying the Kirchhoff's second law to the circuit shown in Fig. 4.15, we get,

$$V_L + V_R + V_C = V$$

Or, from Eq. 4.29,

$$L\frac{di}{dt} + iR + \frac{q}{C} = V_0 \sin\omega t$$

Since i = (dq/dt), we have,

$$L\frac{d^2q}{dt^2} + R\frac{dq}{dt} + \frac{q}{C} = V_0 \sin\omega t \qquad \text{..... (A)}$$

Let a solution of this equation be,

$$q = q_0 \sin(\omega t + \theta) \qquad \text{..... (B)}$$

Then,

$$\frac{dq}{dt} = q_0\omega \cos(\omega t + \theta)$$

$$\frac{d^2q}{dt^2} = -q_0\omega^2 \sin(\omega t + \theta)$$

Substituting these values in Eq. (A), we get,

$$q_0\omega\left[-L\omega \sin(\omega t + \theta) + R\cos(\omega t + \theta) + \frac{1}{\omega C}\sin(\omega t + \theta)\right] = V_0 \sin\omega t$$

Using the relations $X_L = \omega L$ and $X_C = (1/\omega C)$, we get,

$$q_0\omega\,[R\cos(\omega t + \theta) + (X_C - X_L)\sin(\omega t + \theta)] = V_0 \sin\omega t$$

Now we multiply and divide throughout by a factor Z,

$$Z = \sqrt{R^2 + (X_L - X_C)^2} \qquad \text{..... (4.40)}$$

Then,

$$q_0\omega Z\left[\frac{R}{Z}\cos(\omega t + \theta) + \frac{(X_C - X_L)}{Z}\sin(\omega t + \theta)\right] = V_0 \sin\omega t$$

We put,

$$\frac{R}{Z} = \cos\phi$$

$$\frac{X_C - X_L}{Z} = \sin\phi$$

Then,

$$q_0\omega Z\,[\cos\phi\cos(\omega t + \theta) + \sin\phi\sin(\omega t + \theta)] = V_0 \sin\omega t$$

Or,

$$q_0\omega Z\cos(\omega t + \theta - \phi) = V_0 \sin\omega t = V_0\cos\left(\omega t - \frac{\pi}{2}\right)$$

Comparing the two sides of this equations and putting $i_0 = q_0\omega$, we get,

$$V_0 = q_0\,\omega Z = i_0 Z$$

$$\theta - \phi = -\frac{\pi}{2} \quad \text{or} \quad \theta = -\frac{\pi}{2} + \phi$$

Then, using the Eq. (B), the current in the circuit is given by,

$$i = \frac{dq}{dt} = q_0\omega\cos(\omega t + \theta)$$

Or

$$i = i_0\cos(\omega t + \theta) = i_0\cos\left[\omega t + \left(\phi - \frac{\pi}{2}\right)\right]$$

$$= i_0 \cos\left[\frac{\pi}{2} - (\omega t + \phi)\right] \quad [\because \cos(-\theta) = \cos\theta]$$

Or
$$i = i_0 \sin(\omega t + \phi) \quad \text{..... (4.41)}$$

Where,
$$i_0 = \frac{V_0}{Z} = \frac{V_0}{\sqrt{R^2 + (X_L - X_C)^2}} \quad \text{..... (4.42)}$$

and
$$\tan\phi = \frac{X_L - X_C}{R} \quad \text{..... (4.43)}$$

$$\left[\text{Note that } (X_C - X_L)^2 = (X_L - X_C)^2 \text{ and } \tan\phi = \frac{\sin\phi}{\cos\phi} = \frac{\sin(-\phi)}{\cos(-\phi)} = \frac{-\sin\phi}{\cos\phi} = \frac{(X_L - X_C)/Z}{R/Z} = \frac{(X_L - X_C)}{R}\right].$$

Comparing the Eq. 4.42 with the Ohm's law expression V = iR (or i = V/R), we find that Z is equivalent to the total effective resistance in the circuit. It is called **impedance** of the circuit. As for reactance, its unit is ohm. The value of Z, and therefore the value of i, depends on R and on the difference (ωL – $1/\omega c$). It does not matter whether $\omega L > (1/\omega c)$ or $\omega L < (1/\omega c)$, because the difference is squared and therefore it is always a positive quantity.

The relation between the impedance Z, the ohmic resistance R and the net reactance (ωL – $1/\omega c$) is well appreciated through the graphical construction shown in Fig. 4.16, called **impedance triangle.** From the right-angled triangle ODA, we have,

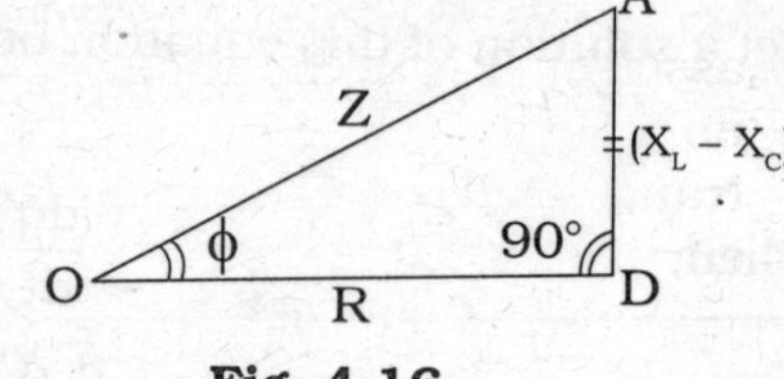

Fig. 4.16

$$Z = \sqrt{R^2 + (X_L - X_C)^2} \quad \text{..... (4.40)}$$

$$\tan\phi = \frac{X_L - X_C}{R} \quad \text{..... (4.43)}$$

The Eqs. 4.42 and 4.43, for the current amplitude i_0 and the phase constant ϕ, enable us to find the current i in the circuit (Eq. 4.39).

Different cases. Depending on the relative values of X_L and X_C, we have there different cases :

(i) $X_L > X_C$ (Dominantly inductive circuit). ϕ is a +ve value (Eq. 4.43), i.e. the current lags the applied voltage (Fig. 4.12).

(ii) $X_C > X_L$ (Dominantly capacitive circuit). ϕ is –ve. The current leads the applied voltage (Fig. 4.14).

(iii) $X_C = X_L$ (Resonant circuit). $\phi = 0°$. The current and the applied voltage are in phase (Figs. 4.9 and 4.10). We shall consider this case in detail in the next to next section.

Difference between Resistance, Reactance, and Impedance. The resistance (R) in a circuit is a direct measure of the obstruction caused to the flow of current in the circuit.

The reactance of an inductor (X_L) or a capacitor (X_C) is the effective resistance offered by the inductor or the capacitor to the flow of current in an AC circuit ($X_L = \omega L$; $X_C = 1/\omega C$).

The **impedance** (Z) of an AC series LCR circuit is the effective resistance offered by the three components L,C, and R together to the flow of current in the circuit $\left[Z = \sqrt{R^2 + (X_L - X_C)^2}\right]$.

4B. 06. LC Circuit; LC Oscillations

When a capacitor is connected to an AC source, charge starts building up in the capacitor, resulting in a rise in its electrical energy [=(1/2) CV^2 = (1/2) (q^2/C)]. The energy rises to a maximum value at the end of the quarter cycle (0 to $\pi/2$) and then gradually goes down to zero in the next quarter cycle ($\pi/2$ to π; Fig 4.14). The process keeps on repeating every half cycle (o to π). Thus we have oscillation of electrical energy alone, such that every quarter cycle the energy is stored in the capacitor, which is returned to the source in the next quarter cycle.

Similarly, when an ideal inductor is connected to an AC source, the magnetic flux and therefore the related magnetic energy [=(1/2) Li^2, where i is the current resulting from the change of flux] rises and falls

Additional knowledge

The foregoing treatment of the series LCR AC circuit assumes the existence of a steady current in the circuit. However, when the applied voltage is turned on, then starting from the initial value V = 0, it takes a small finite time before the steady state current i begins to flow in the circuit. The brief current that flows during this build-up period is called **transient current** (vide Chapter 4A; Section 4A.11). The same happens when the circuit is switched off. So, strictly speaking, the solution of our problem of finding the value of the current i in the circuit (Eq. 4.39) really consists of two parts, viz. transient-state solution and the steady-state solution.

Sometimes the magnitude of the transient current can be very large. In a high power AC motor it can be so large that unless the motor circuit is properly designed, it can seriously damage the motor. For the same reason, the mains switch for the motor is always a specially designed switch (called 'motor starter'), which can withstand the initial large transient current. An ordinary power switch readily burns out.

The transient currents are not caused by any part of the applied voltage. They are totally associated with the changes in stored energy in the inductors and capacitors in the circuit. There are no transients in a purely resistive circuit, because a resistor has no stored energy. The duration of a transient depends upon the so-called inductive time constant $\tau_L = L/R$ and the capacitative time constant $\tau_C = RC$.

every half cycle (Fig. 4.12). Thus in this case we have oscillation of magnetic energy alone, which gets stored in the inductor every quarter cycle, and then returned to the source in the next quarter cycle.

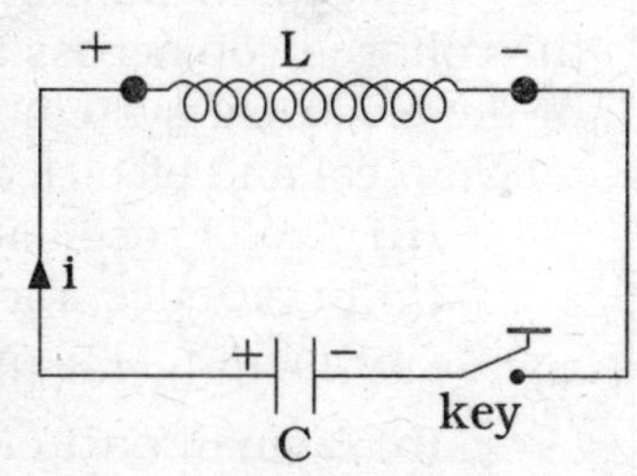

Fig. 4.17

An interesting phenomenon of oscillation of energy between a capacitor and an inductor occurs when a charged capacitor is connected across an inductor (Fig. 4.17). We assume that the inductor has no resistance, i.e. it is an ideal inductor. Now the total energy of the circuit alternates because the energy of the electric field of the capacitor and the energy of the magnetic field of the inductor alternate together in a characteristic manner. Thus we have the important case of the LC circuit becoming a **source of electromagnetic oscillations.** The angular frequency of oscillation is found to be,

$$\omega_0 = \frac{1}{\sqrt{LC}} \qquad ..... (4.44)$$

(ω_0 is related to the linear frequency ν_0 as, $\omega_0 = 2\pi\nu_0$).

The whole process keeps on repeating in the further cycles, resulting into a constant oscillation of the electromagnetic energy in the circuit. It leads to the propagation of **electromagnetic waves** all around the circuit. The waves are sinusoidal in nature, of the type shown in Fig. 4.12 (or Fig. 4.14). We shall study more about them in the next chapter. The transmitters, which broadcast radio and TV programmes all over the globe, are really powerful sources of propagation of electromagnetic waves of different frequencies. The frequency depends on the values of L and C.

Example 6. An inductor of 25 mH is connected to a charged capacitor of 40 μF. Determine the frequency of free oscillations of the circuit.

Solution. Given : $L = 25 \text{ mH} = 25\times10^{-3}$ H; $C = 40\ \mu F = 40\times10^{-6}$ F

The frequency of oscillation is given by,

$$\nu = \frac{1}{2\pi\sqrt{LC}} = \frac{1}{2\times3.14\sqrt{(25\times10^{-3})\times(40\times10^{-6})}}$$

$$= 159.2 \text{ Hz}$$ **(Answer).**

4B. 07. Resonance in the Series LCR Circuit; Quality Factor (Q)

In our study of the series LCR circuit in Section 4B.05, the current amplitude was found to be (Eq. 4.42),

Additional knowledge

(i) The LC oscillation is analogous to the mechamical oscillaition of a body attached to a rigid support with a spring. There is a constant exchange between the potential and kinetic energy of the body during the oscillations. If x is the displacement of the body at an instant t, its equation of motion is given by,

$$(d^2x/dt^2) + \omega_0^2 x = 0,$$

where the natural frequency ω_0 of the oscillation is given by (m: mass of the body; k: spring constant),

$$\omega_0 = \sqrt{k/m}$$

The corresponding equations for the electrical case are found to be,

$$(d^2q/dt^2) + (1/LC)q = 0$$

$$\omega_0 = \sqrt{1/LC}$$

Comparing the two sets of equations, we find that the charge q is the electrical equivalent of the displacement x and $\sqrt{k/m}$ corresponds to $\sqrt{1/LC}$. Further, for the mechanical system, force = mass × acceleration = m (dv/dt) = m(d^2x/dt^2), and for the electrical system, the emf $\mathcal{E}$ = – L (di/dt) = – L (d^2q/dt^2). Comparing once again, we find that the inductance corresponds to the mass m. Just as mass is a measure of the inertia of the body, L is a measure of the 'electric inertia', in the sense that it resists any change in the current in the circuit. In the expressions for ω_0, 1/C corresponds to the force constant k. k (= F/x) represents the force required to produce a unit displacement, whereas 1/C (= V/q) expresses the voltage drop across the capacitor required to store a unit charge. Thus, L and C play the same role in the electrical system as m and the spring do in the mechanical system. The analogies between the mechanical and electrical systems are listed in Tabel 4B.01.

(ii) The LC oscillations do not continue indefinitely for the following reasons.

(a) In practice, the inductor always has a finite resistance, which causes a grădual dissipation of energy at the inductor (thermal loss).

(b) A small part of the energy of the system is constantly radiated away as electromagnetic waves (radiative loss).

For these reasons, the LC oscillations become damped and so gradually die out. In the mechanical system, too, the oscillations are similarly damped and die away, because of the friction caused by the medium (air) to the movement of the body.

Table 4B. 01. Analogies between mechanical and electrical (LC circuit) systems

Serial No.	Electrical System	Mechanical System
1.	Electromagnetic energy (electrical + magnetic) [(1/2)(q^2/C) + (1/2)Li^2]	Mechanical energy (p.e. + k.e.) [(1/2) kx^2 + (1/2) mv^2]
2.	Mass (m)	Inductance (L)
3.	Displacement (x)	Charge (q)
4.	Force constant (k)	1/C
5.	Velocity, v = (dx/dt)	Current, i = (dq/dt)

$$i_0 = \frac{V_0}{Z} = \frac{V_0}{\sqrt{R^2 + (X_L - X_C)^2}} \qquad \text{..... (4.42)}$$

Now, X_C = (1/ωC) and X_L = ωL. If the angular frequency ω is varied, then at a certain value of ω = ω_0 (say), we shall get $X_L = X_C$. Then the impedance Z is minimum,

$$Z = \sqrt{R^2 + (X_L - X_C)^2} = \sqrt{R^2 + O} = R$$

Hence the current amplitude is maximum. This condition in which the current amplitude in a series LCR circuit is maximum for a particular angular frequency ω_0 is known as the condition of **resonance** in the circuit. $\boldsymbol{\omega_0}$ is termed as the **resonant frequency** of the circuit. Its value is easily found by putting

$$X_L = X_C \quad \text{or} \quad (1/\omega_0 C) = \omega_0 L$$

$$\Rightarrow \qquad \omega_0 = \frac{1}{\sqrt{LC}} \qquad \text{..... (4.45)}$$

The variation of i_0 with ω for a typical case is shown in Fig. 4.18, for two values of R ($R_2 > R_1$). In the region $\omega < \omega_0$, the inductive reactance XL (= ωL) is smaller than the capacitive reactance X_C (=1/ωC). Hence the circuit is said to be capacitive in this low frequency region. In the region $\omega > \omega_0$, we have $X_L > X_C$, and so the circuit is said to be inductive in this high frequency region. The resonance occurs in between ($\omega = \omega_0$) when the impedance is minimum ($X_L = X_C$). Then the current amplitude is maximum.

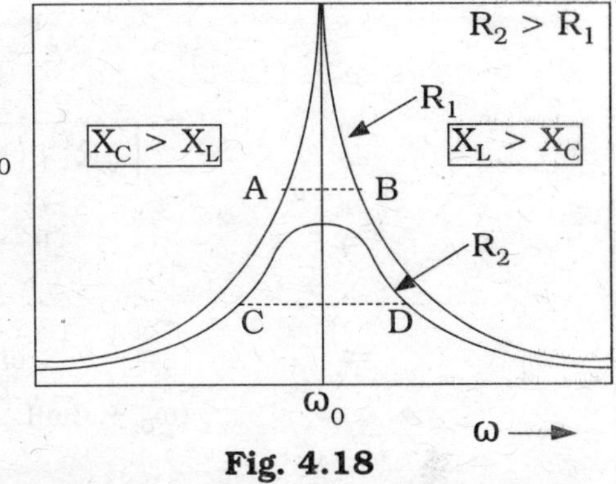

Fig. 4.18

Origin of Resonance; Natural Frequency. Any system (mechanical, electrical, etc.) prefers to oscillate at a particular frequency, which is therefore called its **natural frequency** (ω_0) of oscillation. The value of ω_0 depends on what type of components the system has and what are their respective values. When the system is driven by a force whose frequency is exactly or nearly equal to the natural frequency of oscillation of the system, the amplitude of the oscillation becomes large. Then the system is said to be in the condition of resonance. (This happens in the musical instruments, e.g. in a guitar the resonance occurs when the frequency of vibration of the string matches with the natural frequency of vibration of the air column in the sound box of the guitar). In the last section, we had found that the natural frequency of oscillation of an LC circuit is $\omega_0 = (1/\sqrt{LC})$. It is also the natural frequency of oscillation of the LCR circuit; the presence of the resistance R only acts as a damping factor. When the frequency (ω) of the emf source becomes exactly or nearly equal to ω_0, resonance takes place.

Sharpness of resonance; Quality factor (Q)

Qualitatively, the sharpness of resonance for an LCR circuit is judged from the sharpness of the peaks of the i_0 vs. ω curves (Fig. 4.18). It can be expressed in terms of the so-called half-width of a curve, which is defined as the width of the curve at half of the peak value of i_0. The smaller is the half-width, the sharper (or narrower) is the peak and therefore the sharper is said to be the resonance. Correspondingly, the larger is the value of the current amplitude i_0. In Fig. 4.18, the half-width AB for the curve corresponding to the resistance R_1 is smaller that the half-width CD for the second curve corresponding to R_2 ($R_2 > R_1$). Hence the resonance is sharper for the former, as actually seen in the figure.

Quantitatively, the sharpness of resonance is expressed through the so-called Quality factor (or Q-factor). The **Quality factor (Q)** is a number which determines the sharpness of resonance of a series LCR circuit. It is given by,

$$Q = \frac{\omega_0 L}{R} = \frac{1}{R}\sqrt{\frac{L}{C}} \qquad \text{..... (4.46)}$$

Additional knowledge

The Eq. 4.46 can be derived as follows. A value of ω is selected for which the current awplitude i_0 is $(1/\sqrt{2})$ times the maximum value i_0^{max}. There will be two such possible values, ω_1 and ω_2, such that $\omega_1 > \omega_0$ and $\omega_2 < \omega_0$ (like the points A and B in Fig. 4.18).

Let,

$$\omega_1 = \omega_0 + \Delta\omega$$

$$\omega_2 = \omega_0 - \Delta\omega$$

Then,

$$(\omega_1 - \omega_2) = 2\Delta\omega$$

The ratio $Q = \omega_0/(2\Delta\omega)$ is taken as the measure of the resonance, because for a given circuit the smaller is $\Delta\omega$, the sharper is the resonance peak. We shall evaluate $\Delta\omega$ in terms of L,C,R.($\Delta\omega$ is often referred to as bandwidth of the circuit). For $\omega = \omega_1$ we have,

$$(i_0)\omega_1 = V_0/(Z)_{\omega_1}$$

Also

$$(i_0)\omega_1 = (1/\sqrt{2})\,(i_0)_{max}$$

$$= (1/\sqrt{2})(V_0/R) \quad (\because Z = R, \text{ at resonance})$$

$$\therefore \quad \frac{V_0}{(Z)_{\omega_1}} = \frac{1}{\sqrt{2}}\cdot\frac{V_0}{R}$$

$$\Rightarrow \quad (Z)_{\omega_1} = R\sqrt{2}$$

$$\Rightarrow \quad \left[\sqrt{R^2 + (X_L - X_C)^2}\right]_{\omega_1} = R\sqrt{2}$$

$$\Rightarrow \quad R^2 + (X_L - X_C)^2_{\omega_1} = 2R^2$$

$$\Rightarrow \quad (X_L - X_C)_{\omega_1} = R$$

$$\Rightarrow \quad \omega_1 L - (1/\omega_1 C) = R$$

$$\Rightarrow \quad (\omega_0 + \Delta\omega)L - \frac{1}{(\omega_0 + \Delta\omega)C} = R$$

$$\Rightarrow \quad \omega_0 L\left(1 + \frac{\Delta\omega}{\omega_0}\right) - \frac{1}{\omega_0 C\left(1 + \frac{\Delta\omega}{\omega_0}\right)} = R$$

Now, $\omega_0 = \left(1/\sqrt{LC}\right)$ or $C = \left(1/L\omega_0^2\right)$, which gives,

$$\omega_0 L\left(1 + \frac{\Delta\omega}{\omega_0}\right) - \frac{\omega_0 L}{1 + \frac{\Delta\omega}{\omega_0}} = R$$

Since $\Delta\omega \ll \omega_0$, we have $(\Delta\omega/\omega_0) \ll 1$. Then, we can approximate,

$$\left(1 + \frac{\Delta\omega}{\omega_0}\right)^{-1} \approx \left(1 - \frac{\Delta\omega}{\omega_0}\right)$$

$$\therefore \quad \omega_0 L\left(1 + \frac{\Delta\omega}{\omega_0}\right) - \omega_0 L\left(1 - \frac{\Delta\omega}{\omega_0}\right) = R$$

Or

$$\Delta\omega = \frac{R}{2L}$$

Then, the Quality factor Q is given by,

$$Q = \frac{\omega_0}{2\omega} = \frac{\omega_0}{2(R/2L)}$$

$$\Rightarrow \quad Q = \frac{\omega_0 L}{R} = \frac{1}{\omega_0 CR} = \frac{1}{R}\sqrt{\frac{L}{C}} \quad \left(\because \omega_0^2 = \frac{1}{LC}\right) \qquad \text{..... (4.47)}$$

It can be worked out that Q is a dimensionless quantity. It is just a number. Usually it lies in the range 10 – 100, but higher values are also achieved. Smaller is R, higher is the Q - factor of the circuit and so sharper is the resonance, as evidenced from Fig. 4.18.

The Q - factor is also expressed as the ratio of the p.d. across the inductor (or capacitor) to the applied voltage (which is the same as the p.d. across R). This definition also leads to the Eq. 4.47 for Q.

Example 7. A series LCR circuit, having L = 5 H, C = 200 μF, and R = 50 Ω, is connected to an AC 200 V source of variable frequency. Determine the (i) resonant frequency of the circuit, (ii) impedance of the circuit at resonance, (iii) current amplitude at resonance, (iv) rms value of the potential drop across the inductor, and (v) quality factor of the circuit.

Solution. Given : V_{rms} = 200 V; L = 5 H; C = 200 μF = 2×10^{-4} F; R = 50 Ω

(i) The resonant frequency is,

$$\nu_0 = \frac{1}{2\pi\sqrt{LC}} = \frac{1}{2\times3.14\times\sqrt{(5)\times(2\times10^{-4})}}$$

$$= 5.04 \text{ Hz} \quad \textbf{(Answer).}$$

(ii) At resonance, the impedance is given by,

$$Z = R = 50\ \Omega \quad \textbf{(Answer).}$$

(iii) The amplitude of the current at resonance is,

$$i_0 = \frac{V_0}{Z} = \frac{\sqrt{2}V_{rms}}{Z} = \frac{1.414\times(200)}{(50)}$$

$$= 5.66 \text{ A} \quad \textbf{(Answer).}$$

(iv) The rms voltage across the inductor is,

$$[V_{rms}] = i_{rms} \times X_L = \frac{V_{rms}}{Z} \times 2\pi\nu_0 L$$

$$= \frac{(200)}{(50)} \times 2 \times 3.14 \times (5.04) \times (5)$$

$$= 633 \text{ V} \quad \textbf{(Answer).}$$

(v) The quality factor is,

$$Q = \frac{1}{R}\sqrt{\frac{L}{C}} = \frac{1}{(50)} \times \sqrt{\frac{5}{2\times10^{-4}}}$$

$$= 3.16 \quad \textbf{(Answer).}$$

Tuning of radio and television sets. A common application of the LCR resonant circuit is its use as the tuning circuit of a radio or TV set. The various transmitters situated all over the world, including our country, transmit signals of different frequencies. The antenna of the radio (or TV) set receives these signals and drives current in the tuning circuit. Only the signal which corresponds with the resonant frequency is able to drive appreciable current in the circuit, which is then further amplified and processed. The capacitor in the tuning circuit is of the variable type. So the value of C and therefore the value of the resonance frequency $\omega_0 \left(= 1\sqrt{LC}\right)$ can be varied at will, which facilitates easy 'tuning' of the set to the frequency of the desired programme.

If the resonance is not sharp, then not only the current amplitude i_0 is reduced, the tuning circuit tends to 'catch' more than one signals of different but close frequencies (vide Fig. 4.18). Hence the quality of the selected programme suffers. This underlines the need of a large enough value of the Quality factor Q, which helps to improve the selectivity and quality of the desired programme.

To **sum up,** note the following important points at resonance ($X_L = X_C$):

(i) The impedance Z is minimum and equal to R (Z = R).

(ii) The current amplitude i_0 is maximum.

(iii) The current and the applied voltage are in phase (vide Eqs. 4.29 and 4.39; $\phi = 0$).

(iv) The value of the resonant frequency ω_0 is independent of R, but the value of i_0 depends on R (vide Fig. 4.18).

(v) In the neighbourhood of the position of resonance, the current lags the applied voltage for $X_L > X_C$ (dominantly capacitive circuit).

(vi) The sharpness of resonance depends on the value of R. It is determined by the Quality factor, $Q = (1/R)\left(\sqrt{L/C}\right)$. Larger the Q, the sharper is the resonance.

(vii) The power dissipation in the circuit is maximum (to be discussed in the next section).

4B.08. Power in AC Circuits; Power Factor; Wattless Current

We know that in a series LCR circuit, i.e. a circuit containing all the three elements L, C and R, the applied voltage and the resulting current are given by (vide Eqs. 4.29 and 4.39),

$$V = V_0 \sin\omega t \quad ..... (4.29)$$

$$i = i_0 \sin(\omega t - \phi) \quad ..... (4.39)$$

The instantaneous power dissipated in the circuit (i.e. the rate of dissipation of energy) is

$$\begin{aligned} P &= Vi \\ &= V_0 i_0 \sin\omega t \sin(\omega t - \phi) \\ &= V_0 i_0 \sin\omega t (\sin\omega t \cos\phi - \cos\omega t \sin\phi) \\ &= V_0 i_0 (\sin^2 \omega t \cos\phi - \sin\omega t \cos\omega t \sin\phi) \end{aligned}$$

This is the power dissipated at an instant t in an infinitesimal interval dt. The average power $\overline{P}$, i.e. the power dissipated in a complete cycle (t = 0 to T), will be,

$$\begin{aligned} \overline{P} &= V_0 i_0 \left[\cos\phi \int_0^T \sin^2 \omega t \, dt - \sin\phi \int_0^T \sin\omega t \cos\omega t \, dt\right] \\ &= (1/2) V_0 i_0 \left[\cos\phi \int_0^T (1 - \cos 2\omega t)\, dt - \sin\phi \int_0^T \sin(2\omega t)\, dt\right] \end{aligned}$$

The average of a sine or cosine function over a complete cycle is zero. So we get

$$\begin{aligned} \overline{P} &= \frac{1}{2} V_0 i_0 \cos\phi \\ &= \frac{V_0}{\sqrt{2}} . \frac{i_0}{\sqrt{2}} . \cos\phi \end{aligned}$$

$$\Rightarrow \quad \overline{P} = V_{rms} \cdot i_{rms} \cdot \cos\phi \quad ..... (4.48a)$$

$$\Rightarrow \quad \overline{P} = i_{rms}^2 . Z . \cos\phi \; (\because i_{rms} = V_{rms}/Z) \quad ..... (4.48b)$$

$\overline{P}$ also represents the average power dissipated in a long time (remember that for the household voltage supply in India, at a frequency 50 Hz, even one second is a long time because during this period as many as 50 cycles are completed).

Thus we find that the average power dissipated in an AC circuit not only depends on the values of voltage and current but also on the phase difference ϕ between them. Cosϕ is called **power factor.** From the impedance triangle (Fig. 4.16), we find that cosϕ = (R/Z). Hence the power factor may be **defined** as "the ratio of the resistance in the AC circuit to the impedance of the circuit".

From the Eq. 4.48 (a), $\cos\phi = \overline{P}/(V_{rms} \cdot i_{rms})$. Now, both in a DC and an AC circuit containing resistance alone, the power dissipated is given by the product of voltage and current. So, apparently the product of the measurable quantities V_{rms} and i_{rms} appears to be the power dissipated in the circuit. But we know that the 'true power' dissipated is $\overline{P}$. So we can **also define** the power factor as "the ratio of the 'true power' to the 'apparent power' dissipated in an AC circuit".

$$\text{Power factor} = \cos\phi = \frac{R}{Z} = \frac{\text{True power dissipated}}{\text{Apparent power dissipated}}$$

We shall discuss four representative cases of power dissipation in the various AC circuits:

(i) Purely resistive circuit. We know that in this case there is no phase difference between the supply voltage (V) and the resulting current (i) in the circuit (vide Section 4B.02), i.e. ϕ = 0 and so cos ϕ = 1 (max. value). Hence there is maximum power dissipation in the circuit,

$$\overline{P}_{max} = V_{rms} \cdot i_{rms} \quad ..... (4.49)$$

(ii) Purely inductive or purely capacitive circuit. In both these cases, $\phi = \pi/2$ (either $+\pi/2$ or $-\pi/2$; vide Sections 4B.03 and 4B.04). Hence $\overline{P}$ = 0. Thus, in spite of flow of current in the circuit, there is no dissipation of power.

(iii) Series LCR circuit. From Eq. 4.43, the value of ϕ is given by,

$$\tan\phi = \frac{X_L - X_C}{R}.$$

Therefore, except for $\phi = \pi/2$ (when $\overline{P}$ is zero), there is a finite dissipation of power, given by Eq. 4.48a,

$$\overline{P} = V_{rms}.i_{rms}.\cos\phi \quad \text{..... (4.48a)}$$

However, as discussed earlier in Section 4B.06, as the current flows in the circuit, the electrical energy of the capacitor and the magnetic energy of the inductor simply keep on exchanging with each other. Thus, their collective (total) energy remains constant. Hence the power dissipation takes place only in the resistor.

(iv) At resonance in the series LCR circuit. At resonance in a series LCR circuit (Section 4B. 07),

$$(X_L - X_C) = 0; \quad \phi = 0; \quad Z = R$$

Therefore, $\cos\phi = 1$. Hence maximum power is dissipated at resonance (just as for a purely resistive circuit),

$$\overline{P}_{max} = V_{rms}.i_{rms} \quad \text{..... (4.49)}$$

Wattles (or Idle) current. We have seen that for a purely inductive or purely capacitive circuit, $\phi = \pi/2$, and therefore $\overline{P} = V_{rms}.i_{rms}\cos\phi = 0$. Thus, in spite of a finite current flowing in the circuit, no power is dissipated. Hence this current is termed as **wattless (or idle) current.**

Example 8. A series LCR circuit, with L = 20 mH, C = 600 μF and R = 5 Ω, is connected to a source of sinusoidal voltage. If the peak voltage of the source is 311V and its frequency is 50 Hz,determine the (i) impedance of the circuit, (ii) phase difference between the applied voltage and current, (iii) power factor, and (iv) power dissipated in the circuit.

Solution. Given : $V_0 = 311$ V; $\nu = 50$ Hz; $L = 20$ mH $= 20\times10^{-3}$H; $C = 600\ \mu F = 6\times10^{-4}$ F; $R = 5\Omega$

We first evaluate the values of inductive and capacitive reactances (X_L and X_C).

$$X_L = 2\pi\nu L = 2\times3.14\times(50)\times(20\times10^{-3})$$

$$= 6.28\ \Omega$$

$$X_C = \frac{1}{2\pi\nu C} = \frac{1}{2\times3.14\times(50)\times(6\times10^{-4})}$$

$$= 5.31\ \Omega$$

(i) The impedance is,

$$Z = \sqrt{R^2 + (X_L - X_C)^2} = \sqrt{(5)^2 + (6.28 - 5.31)^2}$$

$$= \sqrt{25 + 0.94} = 5.09\ \Omega \quad \textbf{(Answer).}$$

(ii) The phase difference between the voltage and current is,

$$\phi = \tan^{-1}\frac{X_L - X_C}{R} = \tan^{-1}\frac{(6.28 - 5.31)}{5}$$

$$= \tan^{-1}(0.194) = 11° \quad \textbf{(Answer).}$$

(iii) The power factor is,

$$\text{Power factor} = \cos\phi = \cos 11°$$

$$= 0.98 \quad \textbf{(Answer).}$$

(iv) The power dissipated in the circuit is,

$$\overline{P} = i_{rms}^2 R = \left[\frac{\left(V_0/\sqrt{2}\right)^2}{Z}\right]^2 \times R$$

$$= \left[\frac{(311)}{1.414\times(5.09)}\right]^2 \times (5)$$

$$= 9336\ \text{W} \quad \textbf{(Answer).}$$

Example 9. A series LCR circuit consists of an inductor of 30 mH, a capacitor of 80 μF, a resistor of 40 Ω, and an AC source of 6V, 50 Hz. Calculate the energy dissipated in the circuit in 15 minutes.

Solution. Given : $L = 30 \text{ mH} = 30\times10^{-3}$ H; $C = 80\ \mu F = 80\times10^{-6}$ F; $R = 40\ \Omega$; $V_{rms} = 6$ V; $\nu = 50$ Hz; $t = 15$ minutes $= 15\times60 = 900$ s

We first evaluate the impedance (Z) of the circuit. The inductive reactance is,

$$X_L = 2\pi\nu L = 2\times3.14\times(50)\times(30\times10^{-3})$$
$$= 9.42\ \Omega$$

The capacitive reactance is,

$$X_C = \frac{1}{2\pi\nu C} = \frac{1}{2\times3.14\times(50)\times(80\times10^{-6})} = 39.81\ \Omega$$

$$\therefore \quad Z = \sqrt{R^2+(X_C-X_L)^2} = \sqrt{(40)^2+(39.81-9.42)^2}$$
$$= \sqrt{2523.6}\ \ \Omega$$

The average power dissipated in the circuit is given by,

$$\bar{P} = V_{rms}\, i_{rms} \cos\phi = V_{rms}\cdot\frac{V_{rms}}{Z}\cdot\left(\frac{R}{Z}\right)$$
$$= \frac{V_{rms}^2 . R}{Z^2} = \frac{(6)^2\times(40)}{(2523.6)}$$
$$= 0.57 \text{ W}$$

$\therefore$ The energy dissipated $= \bar{P}.t = (0.57)\times(900)$
$= 513$ J **(Answer).**

4B. 09. Transformer

There are numerous uses of AC voltage and current, including uses in daily life and specialized uses. Depending upon the purpose of the use, sometimes we need relatively high values of the voltage, e.g. to avoid losses in the transmission of power from the generating station to the towns (~ thousands of volt) and sometimes relatively low values, e.g. to ensure personal safety in handling electrical items of household use, like children's toys, chargers for mobile phones and shavers, etc. (~ a few volt). The necessary change from are voltage value to another is achieved through a **transformer.** As the name implies, it 'transforms' one value of the voltage to another.

Principle. It is based on the principle of mutual induction. If two coils lie near to each other, then a change of magnetic flux linked with one coil induces an emf in the second coil.

Construction. It consists of two coils wound around an iron core (Fig. 4.19). The coils have different numbers of turns. They are insulated from the core, e.g. by coating the coils with an insulating material or by enclosing the coil's wire in a sheath of insulating material. One of the coils is connected to the AC source of emf ε. It is called **primary coil.** The other coil, called **secondary coil,** is connected to the circuit in which the transformed output voltage is to be used. In the figure, R represents the total load resistance of the circuit, to which the secondary can be connected or disconnected at will, with the help of a switch. R may include inductive loads as well.

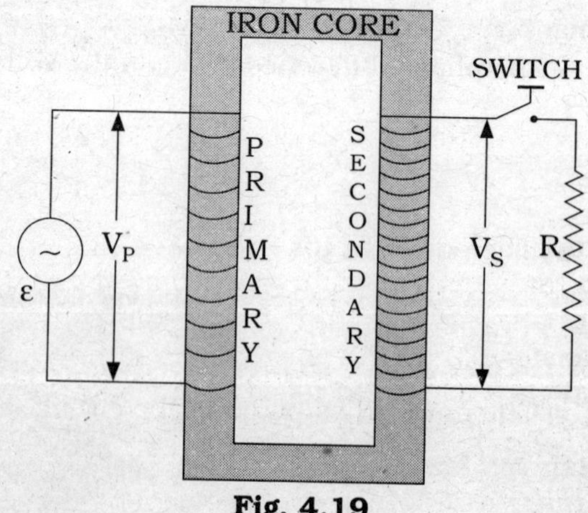

Fig. 4.19

Working. We assume our transformer to be an **ideal transformer,** such that

(i) The coils have negligible (almost zero) resistance.

(ii) No energy losses occur in the transformation of voltage.

Let the numbers of turns in the primary and secondary coils be N_p and N_s, respectively. We first consider the situation when the secondary is on open circuit, i.e. the switch is open and no current flows in the secondary circuit. Initially, a small AC current flows in the primary circuit (because of high impedance of the primary coil), which induces an alternating magnetic flux ϕ_B in the iron core. This flux extends through both the primary coil and the secondary coil. So it induces an emf both in the primary and the secondary coil.

The voltage V_p across the primary is equal to the emf induced in the primary. Similarly, the emf induced in the secondary is equal to the voltage V_S across the secondary. From the Faraday's laws of induction, the induced emf per turn should be the same for the two coils. Its value is,

$$[\mathcal{E}]_{1\ \text{turn}} = \frac{d\phi_B}{dt}$$

Since the number of turns in the primary and secondary are N_p and N_s, respectively, the respective values of the induced emf are,

$$V_p = N_p \frac{d\phi_B}{dt}$$

$$V_s = N_s \frac{d\phi_B}{dt}$$

$$\therefore \qquad \frac{V_s}{V_p} = \frac{N_s}{N_p} \qquad \text{..... (4.50)}$$

$$\Rightarrow \qquad V_s = \left(\frac{N_s}{N_p}\right) V_p \qquad \text{..... (4.51)}$$

Thus, we find that the voltage developed across the secondary is directly proportional to the ratio of the number of turns of the secondary to the number of turns of the primary. If $N_s > N_p$, we have $V_s > V_p$. Such a device is called **step-up transformer.** If $N_s < N_p$, we have $V_s < V_p$. Then the device is termed as **step-down transformer.** The transformer depicted in Fig. 4.19 is a step-up transformer, since it has $N_s > N_p$.

Next we consider the situation when the secondary circuit is closed (by closing the switch). Now an alternating current i_s flows in the secondary circuit and so there is dissipation of energy in the total load R. Further, i_s produces its own alternating magnetic flux in the iron core, which induces an opposing emf (back emf) in the primary (according to the Lenz's law). But this back emf cannot change the value of the voltage V_p, which must always be equal to the generator's emf $\mathcal{E}$. Therefore, to keep V_p unchanged, the generator supplies an additional current i_p to the primary circuit. (The value of i_p is such that the emf induced by it in the primary exactly counteracts the emf induced by i_s).

The power (which is the rate of transfer of energy) in the secondary circuit is $i_s V_s$. According to the principle of conservation of energy, this should be equal to the power $i_p V_p$ in the primary circuit,

$$i_s V_s = i_p V_p$$

Then, from Eq. 4.50, we get,

$$\frac{i_s}{i_p} = \frac{N_p}{N_s} \qquad \text{..... (4.52)}$$

$$\Rightarrow \qquad i_s = \left(\frac{N_p}{N_s}\right) i_p \qquad \text{..... (4.53)}$$

Comparison of Eqs. 4.51 and 4.53 shows that if the voltage is stepped up, the current in the secondary goes down and if it is stepped down, the current goes up. This is as expected, because in the process of transformation of voltage the total power, which is the product of voltage and current, has to remain constant.

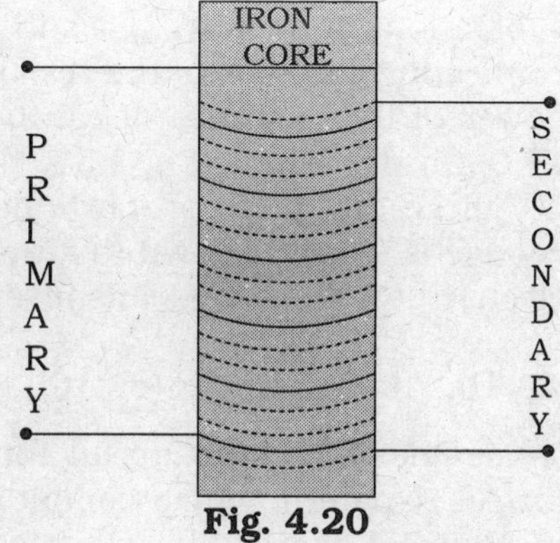

Fig. 4.20

Example 10. The primary and secondary coils of a transformer have 150 turns and 750 turns, respectively. The primary is given a power input of 10 kW at 220 V. Find the current in primary and the output voltage.

Solution. Given : N_p = 150 turns; N_s = 750 turns; P = 10 kW = 10000 W; V_p = 220 V

The current in the primary coil is ($P = V_p i_p$),

$$i_p = \frac{P}{V_p} = \frac{(10000)}{(220)}$$

$$= 45.45 \text{ A} \qquad \textbf{(Answer)}.$$

The output voltage is,

$$V_s = \frac{N_s}{N_p} \times V_p = \frac{(750)}{(150)} \times (220)$$

$$= 1100 \text{ V} \quad \textbf{(Answer).}$$

Energy losses in a transformer

Loss of input energy takes place in a transformer for more than one reason. The various sources of energy loss and their remedies are listed in Table 4B. 02. The suggested remedies do not completely prevent the losses, but only minimise them.

Table 4B.02. Sources of energy loss in a transformer and their remedies (to minimise the loss)

S. N.	Source of energy loss	Remedy
1.	**Flux leakage or incomplete flux linkage:** Not all the flux linked with the primary passes through the secondary.	Using a core shaped to follow the magnetic field lines and by winding the primary and secondary coils one over the other (Fig. 4.20).
2.	**Joule heating:** The wire of the primary or secondary coil has a small finite resistance, resulting in Joule heating (i^2R).	Using thick wires of a material with low resistivity, like copper.
3.	**Hysteresis loss:** Work is done during the repeated reversal of the magnetic dipoles of the iron core, due to the alternating magnetic field.	Using a core material with low hysterisis loss, e.g. soft iron.
4.	**Eddy currents:** The alteranting magnetic flux induces eddy currents in the iron core, which cause heating.	Using a laminated core, which consists of parallel layers of iron, coated with insulating material.

Efficiency of a transformer. The **efficiency** (η) of a transformer is **defined** as the ratio of its power output to the power input.

$$\eta = \frac{\text{Power output}}{\text{Power input}} = \frac{V_s i_s}{V_p i_p} \quad \text{..... (4.54)}$$

An ideal transformer will have η = 1, or in terms of percentage, its efficiency will be 100% (obtained by multiplying by 100). But, because of energy losses, this value is impossible to attain. However, with continuous improvement in design and material, now it has been possible to achieve efficiencies of upto 99%.

Long distance transmission of electric power

When electricity is transmitted from the generating station to the consumer end (viz. the electric substations in the towns), a substantial amount of electric power is lost due to Joule heating (i^2R) in the long connecting cable (R is the resistance of the cable). Since the heat produced is proportional to the second power of the current i, it can be reduced by stepping up the voltage at the generator end, e.g. to a value 11,000 volt (then in Eq. 4.53, $N_s \gg N_p$ and so $i_s \ll i_p$; electric power = Vi and so for a given value of power, if V increases then i proportionately decreases). Then, to ensure personal safety, at the consumer end this voltage is stepped down to a desired low value, e.g. 220 volt, at the various substations. Thus the loss of electric power during transmission is substantially reduced.

4B.10. AC Generator (or AC Dynamo) and DC Generator (or DC Dynamo)

One of the most useful applications of electromagnetic induction is the device known as AC generator (or AC dynamo), which commonly supplies electric power for household and other purposes. It converts mechanical energy into electrical energy, producing an AC voltage. The DC generator (or DC dynamo) is based on the same principle as the AC generator, but finally produces DC voltage. Its construction and working are also almost the same, except that it uses 'split-rings' in place of 'slip-rings' of the AC generator, as described in the following.

Additional knowledge

(i) As already mentioned, when the secondary circuit is open, a small AC current flows in the primary circuit. It is called 'magetizing current' (i_{mag}), because it induces a changing magnetic flux in the iron core. Since the primary circuit is purely inductive, i_{mag} lags the primary voltage V_p by 90° (Section 4B. 03). Therefore the primary's power factor cosϕ, in the Eq. 4.55a for the power $\overline{P}$, is zero. Hence there is no transfer of power from the emf source to the transformer.

However, when the secondary circuit is closed and an additional current i_p flows in the primary circuit, the magnitude and phase constant of i_p has to be such that the emf induced by it in the primary circuit cancels the emf induced by i_s. Thus the phase difference between the current i_p and the voltage V_p is different from 90°. Hence now it becomes possible for i_p to transfer energy to the primary.

(ii) A transformer works on the phenomenon of electromagnetic induction, in which an induced emf and current are produced in a circuit by a changing magnetic flux. Since a transformer is required to give a continuous supply of output voltage, the magnetic flux should also change continuosly. This is only possible when an AC voltage is supplied to the primary. With a DC voltage, only a momentary change of flux occurs, when the current builds up from zero to a steady value upon switching the circuit, and so one gets only a momentary induced emf. Hence a transformer can only transform an AC voltage, not a DC voltage.

(i) AC Generator (or AC Dynamo)

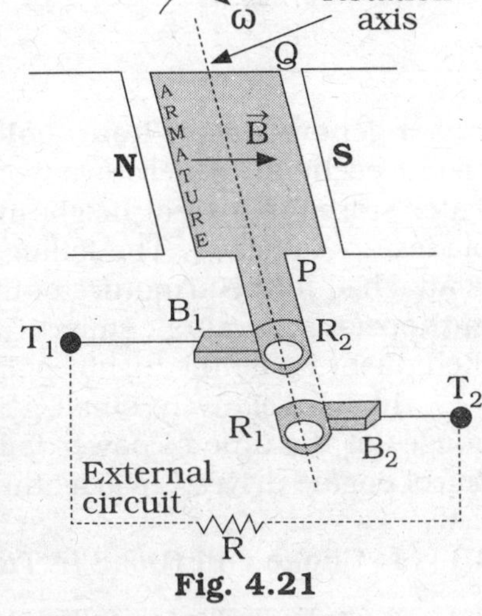

Fig. 4.21

Principle. It is based on the principle of electromagnetic induction, viz. an induced emf is produced in a circuit when the magnetic flux linked with the circuit changes.

Construction. The essential parts of an AC generator are shown in Fig. 4.21

(i) Magnet. A strong uniform magnetic field $\vec{B}$ is produced between the pole pieces of a permanent magnet or an electromagnet. The north and south poles face each other.

(ii) Armature. It is a coil wound over a laminated soft iron core. It is placed between the pole pieces of the magnet. The iron core helps to enhance the magnetic field and hence to enhance the magnetic flux passing through the coil. The coil can rotate about its central axis, lying in the plane of the coil, as shown in the figure. The axis is perpendicular to the magnetic field $\vec{B}$.

(iii) Slip-rings. Two metal slip-rings R_1 and R_2, which are insulated from each other and from the rotating shaft, are attached with the ends of the coil. They rotate alongwith the coil in the magnetic field. During the rotation, they continuously slip against two fixed brushes B_1 and B_2. Hence the name 'slip' rings.

(iv) Brushes. Two small metal (or graphite) plates B_1 and B_2 remain permanently in light contact with the slip-rings R_1 and R_2, respectively. The are called 'brushes', because they brush against the slip rings during the rotation of the coil. They are connected to the terminals T_1 and T_2, respectively, of the external circuit. Their function is to collect +ve and –ve charges, respectively, from the slip-rings B_1 and B_2, respectively, so that an emf is built up across the terminals T_1 and T_2, that can drive a current in an external circuit.

Working. Let,

A = Area of the coil
N = Number of turns in the coil
ω = Angular velocity of rotation of the coil

Suppose initially at time t = 0, the plane of the coil is perpendicular to the magnetic field $\vec{B}$. In this position, the magnetic flux linked with each turn of the coil is B.A. In a time t the coil rotates through an angle ωt. Then the flux per turn of the coil is

$$\phi_B = (B.A)\cos\omega t$$

According to the Faraday's law, the emf induced in each turn of the coil is

$$[\mathcal{E}]_{\text{one turn}} = -\frac{d\phi_B}{dt} = (BA).\omega.\ \sin\omega t$$

$\therefore$ The total emf ($\mathcal{E}$) induced in the coil is,

$$\mathcal{E} = (NBA\omega)\ \sin\omega t$$

Or

$$\mathcal{E} = \mathcal{E}_0\ \sin\omega t \qquad ..... (4.55)$$

where,

$$\mathcal{E}_0 = NBA\omega \qquad ..... (4.56)$$

Thus a periodically varying emf $\mathcal{E}$ with angular frequency ω is produced. $\mathcal{E}_0$ is the maximum value of the emf. Hence it is called 'peak emf'. The AC voltage developed across the terminals T_1 and T_2 drives an AC current in the external circuit. The variation of the voltage and current is of the type shown in Fig. 4.12. Using the Fleming's left hand rule, it can be worked out that during a half-cycle (0 to π) if the current flows in one direction in the coil (e.g. along PQ in Fig. 4.21), then in the next half-cycle (π to 2π) it flows in the opposite direction (along QP). Each cycle corresponds to one revolution of the coil.

Example 11. A circular copper coil is placed perpendicular to a uniform magnetic field of magnitude 5×10^{-2} Wb. m^{-2}. It is rotated at a speed of 400 revolutions per second. If the coil has 130 turns and an area of 300 cm^2, find the maximum emf induced in the coil.

Solution. Given : N = 130 turns; A = 300 cm^{-2} = 3×10^{-2}m^2; ν = 400 rps; B = 5×10^{-2} Wb. m^{-2}

The maximum value of the induced emf is given by,

$$\omega_0 = NBA\omega = NBA\ (2\pi\nu)$$

$$= (130)\times(5\times10^{-2})\times(3\times10^{-2})\times2\times3.14\times(400)$$

$$= 489.8\ \text{V}$$ **(Answer).**

Power generation for household and other uses. Electric power for household and other uses is commonly produced by AC generators working in different but similar ways. In **hydroelectric** generation of power, water stored to a great height in a dam falls on giant size **turbines.** A turbine consists of a rotor, to which blades are attached. The falling water rotates the blades and therefore the rotor of the turbine. The rotor is attached to the armature of the AC generator. So the armature also rotates and AC voltage is generated. In **thermal** generation, superheated steam is produced by boiling water using coal, oil or gas as fuel, which then pushes past the turbines and rotates them.

In the relatively small generator sets (gensets), which are used as stand-by supplies of power in houses at the time of power failure or in marriage and other functions, public meetings, etc., a diesel or petrol engine drives the armature.

4B. 11. Choke coil (Used in Tubelights)

A choke coil (also called 'ballast') is an essential part of a fluorescent mercury tube (popularly known as tubelight), commonly used for lighting in households. It is simply a coil having a large inductance but a small resistance. The tubelight directly works at the mains supply voltage (220 V, 50 Hz; in India).

A tubelight is a form of electric discharge tube, which essentially comprises a glass tube with an electrode at each end. The electrodes are actually metal filaments, heated by the supply voltage (Fig. 4.22). The tube contains a small amount of mercury and a little quantity of argon gas. The inner glass walls are coated with chemicals known as 'phosphors'. Apart from the choke and the filaments, the electric circuit also has a 'starter', which is a small neon-discharge tube with bimetallic strips inside it.

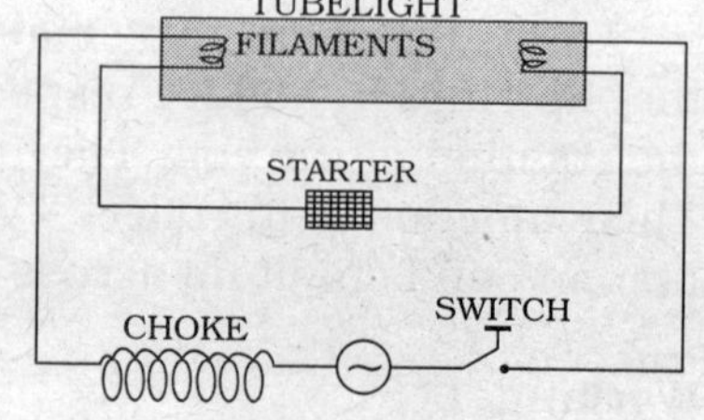

Fig. 4.22

When the switch is turned on, current flows through the two filaments and heats them. So they emit electrons, which ionize argon atoms (argon has a low ionizing potential). The heat also causes mercury to evaporate and the mercury vapour fills the tube. At the same time, neon discharge begins in the starter. As it glows, the bimetallic strips, acting like a thermal switch, break a contact and so the starter is 'released' from the circuit. This opens the circuit, with full mains supply voltage now appearing across the filament electrodes. Then a discharge begins through the gas. The electrodes (filaments) are then maintained 'hot' by bombardment of the heavy +ve ions against them. Thus, they continue to emit electrons and the discharge is maintained.

The light emitted from the discharge is characteristic of mercury atoms' line spectrum. It is very rich in ultraviolet light. This ultraviolet radiation is converted into visible light by the phosphors coated on the glass walls. It excites 'luminescence' (both fluorescence, in which the light is emitted only so long as the radiation lasts, and phosphorescence, in which the light keeps on emitting for some period even after the tubelight is switched off) in the phosphors. The colour of the light depends on the nature of the phosphors used.

If the circuit did not contain the choke coil, a heavy current would flow in the circuit, which would damage the tubelight. The choke reduces the current in the circuit. It is an LR circuit, with impedance $Z = \sqrt{R^2 + \omega^2 L^2}$ (Section 4B.05). Z is a large value, because the inductance L is high. So the current in the circuit is limited. At the same time, the heating produced is much less as compared to the situation in which a pure resistance equivalent to Z had been inserted in the circuit, because the power factor $\cos\phi \ll 1$ (in the expression $\overline{P} = Vi\cos\phi$). [$\tan\phi = (\omega L/R)$; since $\omega L \gg R$, $\tan\phi$ is a large value, which in turn means that $\cos\phi$ is a small value ($\tan\phi = \sin\phi/\cos\phi$)].

Additional knowledge

(i) In India, the household power supply has frequency $\omega = 50$ Hz, i.e. 50 cycles per second. Thus, in one second the direction of flow of current and therefore the direction of movement of conduction electrons in the connecting wires, changes 100 times. Even within a single half-cycle an electron moves hardly about 10 atoms spacing, as can be estimated from the typical value of drift velocity ($\sim 10^{-5}$s) of the electrons in a wire. Then, prima facie it appears highly contradictory that a bulb lights up immediately after its switch is turned on! An in-depth consideration shows that the value of the drift velocity is not really important. The current i flowing through a wire of cross-sectional area A is given by, $i = nAev_d$ (n: no. of electrons per unit volume, v_d : drift velocity; vide Eq. 2.11, chapter 2A,). So a given value of current, say 5A, may result either from a few electrons moving extremely fast or as well from a large number moving slowly. What really matters is how fast the charge carriers receive signal for their movement in either direction. The signal originates in the AC generator from which the household supply of power is derived. It travels nearly at the speed of light. So all the charge carriers receive it instantaneously, i.e. virtually at the same time.

(ii) Some advantages of AC over DC are the following:

(a) An AC voltage can be easily changed from one value to another, using a suitable transformer.

(b) AC can be transmitted over long distances with far less loss of power than DC, resulting in a huge saving of power and money.

(c) AC is easier and cheaper to generate than DC.

(d) The AC in a circuit can be reduced using a choke coil, with little or no dissipation of energy in the circuit.

(e) If necessary, AC can be easily converted into DC.

Solved Conceptual Problems

[Including HOTS Questions]

Problem 1. An AC circuit consists of various components like resistors, inductors and capacitors in series. (i) Will the arithmetic sum of instantaneous voltages across these components be equal to the applied instantaneous voltage ? (ii) Will the same hold for rms voltage, too ? Give reason for your answer.

Solution. (i) Yes. It is true for any instant of time t.

(ii) No. The rms voltages across the different components will differ in phase.

Problem 2. Distinguish between resistance, reactance, and impedance.

Solution. The resistance (R) is a direct measure of the obstruction caused to the flow of current in the circuit.

The reactance (X) of an inductor (X_L) or a capacitor (X_C) is the effective resistance offered by the inductor or the capacitor to the flow of current in an AC circuit ($X_L = \omega L$ and $X_C = 1/\omega C$).

The impedance (Z) of an AC series LCR circuit is the effective resistance offered by the three components L,C,R together to the flow of current in the circuit $\left[Z = \sqrt{R^2 + (X_L - X_C)^2} \right]$.

Problem 3. Graphically show the variation of reactance with frequency for (a) an inductor and (b) a capacitor.

Solution. (i) The reactance of an inductor is given by,

$$X_L = \omega L = 2\pi\nu L$$

$$\therefore \quad X_L \propto \nu$$

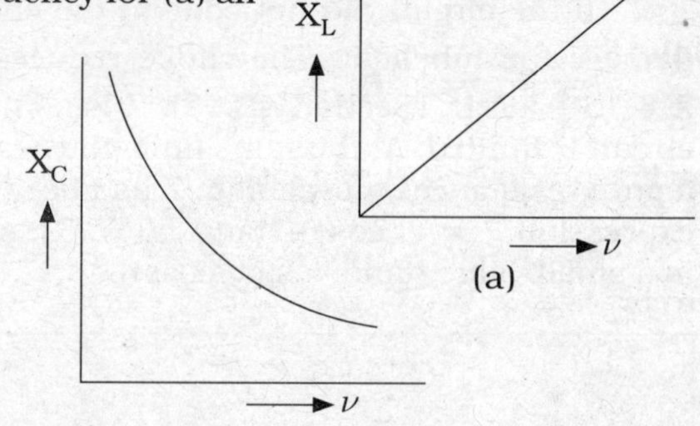

Thus, X_L varies linearly with the frequency ν, as seen in Fig. (a).

(ii) The reactance of a capacitor is given by,

$$X_C = \frac{1}{\omega C} = \frac{1}{2\pi\nu C}$$

$$\therefore \quad X_C \propto \frac{1}{\nu}$$

Thus, the variation of X_C with the frequency ν is hyperbolic, as seen in Fig. (b).

Problem 4. The frequency of AC supply in India is 50 Hz. Can a supply at (a) 15 Hz and (b) 5 Hz be used for AC ?

Solution. (a) Yes. At a 15 Hz (i.e. 15 cycles per second) supply, the current will fluctuate 30 times in a second, which is rapid enough for a bulb to appear as glowing continuously, due to persistence of vision.

(b) No. Now the fluctuations in current are mere 10 times per second, which will make the light from the bulb to appear fluctuating, causing discomfort to the eyes.

Problem 5. A capacitor permits flow of current in an AC circuit but not in a DC circuit. Explain why.

Solution. The capacitive reactance offered by a capacitor in an electric circuit is given by,

$$X_C = \frac{1}{\omega C} = \frac{1}{2\pi\nu C}$$

For an AC circuit, ν has a finite value. ∴ The capacitor does offer a resistance but permits the flow of current in the circuit. On the other hand, for a DC circuit, since the frequency $\nu = 0$, we have,

$$X_C = \frac{1}{0} = \infty$$

Thus, now the resistance offered becomes infinite. ∴ There is no flow of current.

Problem 6. An AC source is connected in series to an electric lamp and a capacitor. If its frequency is raised, will the intensity of the lamp's light change ? If so, explain how. What will happen if the AC source is replaced by a DC source ?

Solution. Yes. The capacitive reactance is given by

$$X_C = \frac{1}{\omega C} = \frac{1}{2\pi\nu C}$$

When the frequency ν increases, X_C decreases. ∴ In turn, the intensity of light increases, i.e. the lamp glows brighter.

If the AC source is replaced by a DC source, the reactance will become infinite. ∴ The lamp will not glow at all.

Problem 7. In an AC circuit, is it possible to use a suitable capacitor in place of a choke coil ? If yes, explain how.

Solution. Yes, we can use a capacitor of equivalent reactance ($X_C = 1/\omega L$). Both have the advantage of reducing current in the circuit without power consumption.

Problem 8. In an AC circuit, why cannot we use a resistor in place of a choke coil ?
Solution. A choke coil, which is essentially an inductor with little resistance, has almost no power consumption in the form of heat (= i^2R). But a resistor does have a considerable power consumption, and so gets heated.

Problem 9. State the main relative advantages and disadvantages of AC and DC supplies of same voltage, e.g. 220 V in India.
Solution. I. Advantages of AC supply (220 V):

(i) With the help of a transformer, an AC voltage can be easily converted from one value to another.

(ii) AC can be transmitted over long distances with far less loss of power than DC, resulting in a huge saving of power and money.

(iii) AC is easier and cheaper to generate than DC.

(iv) The AC in a circuit can be reduced using a choke coil, with little or no dissipation of energy in the circuit.

(v) If necessary, AC can be easily converted into DC.

II. Advantages of DC supply (220 V):

(i) For DC, the peak value is the same as the effective (or instantaneous) value, viz. 220 V. But for AC, for the effective value of 220 V the peak value ($V_0 = \sqrt{2}\, V_{rms} = 1.414 \times 220$ V) is 311 V. Hence 220 V AC gives greater shock than 220 DC.

(ii) The shock of AC is finally attractive (which can be lethal), but that of DC is repulsive.

Problem 10. An AC source is connected in series to a solenoid with air core and an electric lamp. If the air core of the solenoid is replaced by a soft iron core, will the illumination of the lamp increase ? Give reason for your answer. Will the situation be different with a DC source in place of the AC source ?
Solution. The introduction of iron core will increase the inductance of the solenoid, resulting in an increase in the reactance offered by it in the circuit. Hence the current and so the lamp's illumination will go down, i.e. it will glow dimmer. With a DC source, the illumination will remain unchanged, as the resistance of the solenoid will remain unaffected by the iron core.

Problem 11. A DC ammeter can only be used for measuring DC, but a hot wire AC ammeter can be used for measuring both AC and DC. Explain why.
Solution. Both AC and DC produce heat (= i^2R). Therefore they both can be measured by a hot wire ammeter, which is based on heating effect. On the other hand, a DC ammeter is based on the principle of production of magnetic field on passing current through a coil. It gives no deflection with AC, because the average value of AC over a complete cycle is zero. Therefore it can only measure DC.

Problem 12. An AC source is connected turn by turn to (i) a resistor, (ii) an inductor, and (iii) a capacitor. If the frequency of the source is decreased, what will be the effect on the current flowing in the circuit in the three cases ? Justify your answer.
Solution. (i) The value of a resistance is independent of frequency. Hence there will be no effect on the current.

(ii) Since $X_L = \omega L$, the reactance will go down when the frequency is decreased. Hence the current will increase.

(iii) Since $X_C = 1/\omega C$, the reactance will go up upon decreasing the frequency. Hence the current will be reduced.

Problem 13. In an electric circuit, a hot wire ammeter connected in series withe a resistor and a capacitor shows deflection with an AC source in the circuit, but not with a DC source. What is the reason ?
Solution. A capacitor offers infinite resistance ($X_C = 1/\omega C$; $\omega = 0$ for DC) to the flow of DC. Hence there can be no flow of current with a DC source in the circuit.

Problem 14. The value of AC in a circuit can be controlled both by using a resistor or a choke coil (i.e. an inductor). But the latter is preferred. Why ?

Solution. With a resistor, there is considerable wastage of energy in the form of heat (= i^2R). But with a choke coil, which has little resistance, there is little wastage.

Problem 15. How does an LC circuit act as a source of electromagnetic waves ?
Solution. When a charged capacitor is connected across an inductor, the total energy of the circuit alternates, because the energy of the electric field of the capacitor and the energy of the magnetic field of the inductor alternate together in a characteristic manner, with one following the other. Thus the LC circuit becomes a source of electromagnetic oscillations, of frequency $\omega_0 = 1/\sqrt{LC}$.

Problem 16. How does resonance occur in an LCR circuit ? What is the role of the resistance R in the circuit ?
Solution. Any system prefers to oscillate at a particular frequency, called the natural frequency ($\omega_0 = 2\pi\nu_0$) of oscillation. The value of ω_0 depends on the type of components in the system and their values. When the system is driven by a force having frequency exactly or nearly equal to ω_0, the amplitude of oscillation becomes large. Then the system is said to be in the condition of resonance. Thus for an LCR circuit, when the frequency of the emf source is nearly or exactly equal to ω_0, resonance takes place. In this condition, $\omega_0 = 1/\sqrt{LC}$ and $X_C = X_L$.

The sharpness of resonance depends on the value of R. It is determined by the Quality factor, $Q = (1/R)\sqrt{LC}$. Larger the Q, the sharper is the resonance. The resistance R also acts as a damping factor for the oscillations.

Problem 17. In a transformer, the primary and secondary coils have no mutual contact, yet there is a mutual transfer of electrical energy between them. How ?
Solution. The electrical energy transfers from one coil to another by mutual induction. The change in magnetic flux linked with one coil induces an emf in the other coil, for which no material contact is necessary.

Problem 18. An AC voltage can be stepped up or down with the help of a transformer. How the same can be done for a DC voltage ?
Solution. A DC voltage can be increased or decreased by variation of resistance in the circuit, e.g. with the help of a rheostat.

Problem 19. Why the core of a transformer is laminated ?
Solution. The alternating magnetic flux induces eddy currents in the iron core, which cause heating. By using a laminated core, which consists of mutually insulated parallel layers of iron, the generation of eddy currents, and hence the heating, is minimised.

Problem 20. How does a transformer help in an economical transfer of electric power over long distances ?
Solution. During transmission of electricity from the generating station to the consumer end, a large amount of electric power is lost due to Joule heating (i^2R) in the connecting cables. The loss is appreciably reduced by stepping up the voltage to a high value (several thousands of volt) at the generator end, which highly reduces the current flowing in the cables and therefore reduces the heating (heating is proportional to i^2). At the consumer end, the voltage is stepped down with another transformer to a desired low value, e.g. 220 V in India.

Solved Numerical Problems

Problem 1. An electric circuit consists of an AC source of 220 V, 50 Hz and a 200 Ω resistor. Calculate the rms value of current in the circuit and the power consumed in the circuit ove a full cycle.

Solution. Given : V_{rms} = 220 V; ν = 50 Hz; R = 200 Ω

The rms value of the current is given by,

$$i_{rms} = \frac{V_{rms}}{R} = \frac{(220)}{(200)}$$

$$= 1.1 \text{ A} \quad \textbf{(Answer).}$$

The power consumed over a full cycle is given by,

$$\overline{P} = V_{rms} \times i_{rms} = (220)\times(1.1)$$

$$= 242 \text{ W} \quad \textbf{(Answer).}$$

Problem 2. In an AC circuit, the peak voltage of the supply is 360 V and the rms current is 15 ampare. Find the rms voltage and the peak current in the circuit.

Solution. Given : V_0 = 360 V; i_{rms} = 15 A

The rms voltage is,

$$V_{rms} = \frac{V_0}{\sqrt{2}} = \frac{(360)}{1.414}$$

$$= 254.6 \text{ V} \quad \textbf{(Answer).}$$

The peak value of the current is,

$$i_0 = \sqrt{2}\, i_{rms} = 1.414 \times (15)$$

$$= 21.2 \text{ A} \quad \textbf{(Answer).}$$

Problem 3. An inductor of 56 mH is connected to an AC source of 220 V, 50 Hz. Find the (i) rms current in the circuit and (ii) net power absorbed in one cycle.

Solution. Given : V_{rms} = 220 V; ν = 50 Hz; L = 56 mH = 56×10^{-3} H

(i) The inductive reactance in the circuit is,

$$X_L = \omega L = 2\pi\nu L = 2\times3.14\times(50)\times(56\times10^{-3})$$

$$= 17.58\ \Omega$$

$$\therefore \quad i_{rms} = \frac{V_{rms}}{X_L} = \frac{(220)}{(17.58)}$$

$$= 12.5 \text{ A} \quad \textbf{(Answer).}$$

(ii) The net power absorbed in one cycle is,

$$\overline{P} = V_{rms} \times i_{rms} = (220)\times(12.5)$$

$$= 2750 \text{ W} \quad \textbf{(Answer).}$$

Problem 4. A capacitor of 50 μF is connected to an AC source of 200 V, 50 Hz. Find the rms current in the circuit.

Solution. Given : V_{rms} = 200 V; ν = 50 Hz; C = 50 μF = 50×10^{-6} F

The capacitive reactance in the circuit is,

$$X_C = \frac{1}{\omega C} = \frac{1}{2\pi\nu C}$$

$$= \frac{1}{2\times3.14\times(50)\times(50\times10^{-6})}$$

$$= 63.7\ \Omega$$

$$\therefore \quad i_{rms} = \frac{V_{rms}}{X_C} = \frac{(200)}{(63.7)}$$

$$= 3.14 \text{ A} \quad \textbf{(Answer).}$$

Problem 5. In an RC circuit, a 180 Ω resistor and a 20 μF capacitor are connected to an AC supply of 220 V, 50 Hz. Find the current flowing in the circuit. Also find the rms values of the voltage drop across the resistor and the capacitor.

Solution. Given : $R = 180\ \Omega$; $C = 20\ \mu F = 20\times10^{-6}$; $V_{rms} = 220$ V; $\nu = 50$ Hz

We first evaluate the impedance of the circuit,

$$Z = \sqrt{R^2 + X_C^2} = \sqrt{(180)^2 + [1/(2\pi\nu C)]^2}$$

$$= \sqrt{32400 + [1/(2\times3.14\times50\times20\times10^{-6})]^2}$$

$$= \sqrt{32400 + (159.23)^2} = \sqrt{32400 + 25355}$$

$$= 240\ \Omega$$

∴ The current in the circuit is,

$$i_{rms} = \frac{V_{rms}}{Z} = \frac{(220)}{(240)}$$

$$= 0.92 \text{ A} \quad \textbf{(Answer).}$$

Accordingly, the voltage drops across the resistor and the capacitor are

$$V_R = i_{rms}\times R = (0.92)\times(180)$$

$$= 166 \text{ V} \quad \textbf{(Answer).}$$

$$V_C = i_{rms} \times X_C = (0.92)\times(159.23)$$

$$= 147 \text{ V} \quad \textbf{(Answer).}$$

Problem 6. An AC of peak value 8A has frequency 50 Hz. Calculate its rms value. Also determine the time required by it to reach its peak value, starting from zero.

Solution. Given : $i_0 = 8$A; $\nu = 50$ Hz

The rms value of the current is,

$$i_{rms} = \frac{i_0}{\sqrt{2}} = \frac{(8)}{1.414}$$

$$= 5.66 \text{ A} \quad \textbf{(Answer).}$$

The time period of the alternating current (AC) is,

$$T = \frac{1}{\nu} = \frac{1}{(50)} = 0.02 \text{ s}$$

It requires one-fourth of the time period for the current to reach is peak value, starting from zero. Thus, the required time is,

$$t = \frac{1}{4}(T) = \frac{1}{4}(0.02)$$

$$= 0.005 \text{ s (or 5 ms).} \quad \textbf{(Answer).}$$

Problem 7. A 50 Hz AC source of peak emf 220 V is connected to an inductor of 300 mH. (i) Find the peak value of the current in the circuit. (ii) What is the source voltage at the instant when the current has peak value ?

Solution. Given : $L = 300$ mH $= 0.3$ H; $V_0 = 220$ V; $\nu = 50$ Hz

(i) The inductive reactance in the circuit is,

$$X_L = \omega L = 2\pi\nu L = 2\times3.14\times(50)\times(0.3)$$

$$= 94.2\ \Omega$$

∴ The peak current is,

$$i_0 = \frac{V_0}{X_L} = \frac{(220)}{(94.2)}$$

$$= 2.34 \text{ A} \quad \textbf{(Answer).}$$

(ii) We know that in an AC circuit containing an inductor alone, the current lags the voltage by π/2 radian (or 90°). ∴ The voltage is zero at the peak value of the current. **(Answer).**

Problem 8. A capacitor, a resistor and a 40 mH inductor are connected in series to an AC source of frequency 60 Hz. Calcuate the capacitance of the capacitor, if the current is in phase with the voltage. (CBSE; 1996).

Solution. Given: L = 40 mH = 4×10^{-3} H; ν = 60 Hz; $\phi = 0$

Since the voltage and the current are in phase (ϕ = 0), we have,

$$X_C = X_L$$

$$\Rightarrow \quad \frac{1}{\omega C} = \omega L$$

$$\Rightarrow \quad C = \frac{1}{\omega^2 L} = \frac{1}{(2\pi\nu)^2 L}$$

$$= \frac{1}{(2\times3.14\times60)^2\times(40\times10^{-3})} = \frac{1}{5679}$$

$$= 176\times10^{-6} \text{ F (or 176 } \mu\text{F)} \quad \textbf{(Answer).}$$

Problem 9. An AC source of frequency 50 Hz is connected to a 50 mH inductor and a bulb. The bulb glows with some brightness. Calculate the capacitance of the capacitor to be connected in series with the circuit, so that the bulb glows with maximum brightness. (CBSE; 2000).

Solution. Given : ν = 50 Hz; L = 50 mH = 50×10^{-3} H

The bulb will have maximum brightness when the inductive reactance (X_L) will be equal to the capacitive reactance (X_C), because then the impedance in the circuit is minimum.

$$X_C = X_L$$

$$\Rightarrow \quad \frac{1}{\omega C} = \omega L$$

$$\Rightarrow \quad C = \frac{1}{\omega^2 L} = \frac{1}{(2\pi\nu)^2 L}$$

$$= \frac{1}{(2\times3.14\times50)^2\times(50\times10^{-3})}$$

$$= \frac{1}{4930}$$

$$= 203\times10^{-6} \text{ F (or 203 } \mu\text{F)} \quad \textbf{(Answer).}$$

Problem 10. An LC circuit consists of a 70 mH inductor, a switch, and a partially charged 140 μF capacitor, holding a charge of 30 mC. All the components are connected in series. When the switch is closed (t = 0),

(i) Find the natural frequency of the circuit.

(ii) Find the initial total energy stored in the circuit.

(iii) Find at what instants of time this energy is totally electrical (i.e. stored in the capacitor alone) and totally magnetic (i.e. stored in the inductor alone).

(iv) Find at what instants of time this energy is equally shared by the capacitor and the inductor.

(v) How this energy will be affected by the introduction of a resistor of moderate value in series in the circuit ?

Solution. Given : L = 70 mH = 70×10^{-3} H; C = 140 μF = 140×10^{-6}F; q_0 = 30 mC = 30×10^{-3} C

(i) The natural frequency of the circuit is given by,

$$\nu = \frac{1}{2\pi\sqrt{LC}} = \frac{1}{2\times3.14\sqrt{(70\times10^{-3})\times(140\times10^{-6})}}$$

$$= \frac{10^4}{196.6} = 50.9 \text{ Hz} \quad \textbf{(Answer).}$$

(ii) Initially the total energy is stored in the capacitor alone. It is given by,

$$U_E = \frac{q_0^2}{2C} = \frac{(30\times10^{-3})^2}{2\times(140\times10^{-6})}$$

$= 3.21$ J **(Answer).**

(iii) We have, $q = q_0 \cos\omega t$

The share of energy stored in the circuit oscillates between the capacitor and the inductor. It is **totally electrical** (i.e. stored in the capacitor alone) when the charge on the capacitor is maximum ($q = q_0$), i.e.

$$\cos\omega t = \pm 1 \quad \Rightarrow \quad \omega t = 0, \pi, 2\pi, 3\pi, \ldots\ldots,$$

$$\Rightarrow \quad \text{at } t = 0, T/2, T, 3T/2, \ldots\ldots \quad (\because \omega = 2\pi\nu = 2\pi/T)$$ **(Answer).**

The energy is **totally magnetic** (i.e. stored in the inductor alone) when the charge on the capacitor is zero, i.e.

$$\cos\omega t = 0, \quad \Rightarrow \quad \omega t = \pi/2, 3\pi/2, 5\pi/2, \ldots\ldots,$$

$$\Rightarrow \quad \text{at } t = T/4, 3T/4, 5T/4, \ldots\ldots$$ **(Answer).**

(iv) When the total energy is equally shared by the capacitor and the inductor, we have,

$$\text{Energy in the capacitor} = \frac{1}{2}\text{(Total Energy)}$$

$$\Rightarrow \quad \frac{q^2}{2C} = \frac{1}{2}\left(\frac{q_0^2}{2C}\right) = \frac{q_0^2}{4C}$$

$$\Rightarrow \quad q = \frac{q_0}{\sqrt{2}}$$

But, $q = q_0 \cos\omega t$. ∴ We have,

$$q_0 \cos\omega t = \frac{q_0}{\sqrt{2}} \quad \Rightarrow \quad \cos\omega t = \frac{1}{\sqrt{2}}$$

$$\Rightarrow \quad \omega t = \frac{\pi}{4}, \frac{3\pi}{4}, \frac{5\pi}{4}, \ldots\ldots\ldots\ldots$$

$$\Rightarrow \quad \text{at } t = \frac{T}{8}, \frac{3T}{8}, \frac{5T}{8}, \ldots\ldots\ldots\ldots \quad (\because \omega = \frac{2\pi}{T})$$

Thus, the energy is equally shared between the capacitor and the inductor at the above instants of time. **(Answer).**

(v) The effect of the introduction of a resistor of moderate value in the circuit will be to damp out the oscillations in the circuit. Thus, the entire initial energy of 3.21 J will be gradually dissipated as heat. **(Answer).**

Problem 11. An AC source of 110 V, 60 Hz is connected in series to an inductor of 0.4 H and a resistor of 80 Ω. Determine (i) the value of the maximum current in the inductor and (ii) the time lag between the maximum voltage and the maximum current.

Solution. Given : $V_{rms} = 110$ V; $\nu = 60$ Hz; $L = 0.4$ H; $R = 80\ \Omega$

(i) The inductive reactance is,

$$X_L = \omega L = 2\pi\nu L = 2 \times 3.14 \times (60) \times (0.4)$$
$$= 150.7\ \Omega$$

∴ The impedance is,

$$Z = \sqrt{R^2 + X_L^2} = \sqrt{(80)^2 + (150.7)^2}$$
$$= 170.6\ \Omega$$

∴ The maximum current through the inductor is,

$$i_0 = \frac{V_0}{Z} = \frac{V_{rms}\sqrt{2}}{Z} = \frac{(110) \times 1.414}{170.6}$$
$$= 0.91 \text{ A}$$ **(Answer).**

(ii) The phase lag (ϕ) between the voltage and the current is given by,

$$\tan\phi = \frac{X_L}{R} = \frac{2\pi\nu L}{R}$$

$$= \frac{2 \times 3.14 \times (60) \times (0.4)}{80} = 1.884$$

$$\Rightarrow \quad \phi = \tan^{-1}(1.884) = 62° = (62 \times \frac{\pi}{180}) \text{ radian}$$

Now, an angle of 2π corresponds to a time interal T. ∴ An angle ϕ corresponds to a time lag of,

$$\text{Time lag} = \phi \times \frac{T}{2\pi} = \frac{(1/\nu)}{2\pi} = \frac{\phi}{2\pi\nu}$$

$$= \left(\frac{62 \times \pi}{180}\right) \times \frac{1}{2\pi \times (60)}$$

$$= 0.0029 \text{ second (or 2.9 ms)}$$ **(Answer).**

Problem 12. A capacitor of 18 μF is connected to an AC source of frequency 240 Hz. If the rms value of the current in the circuit is 3 ampere, calculate the (i) rms value of the voltage across the capacitor and (ii) average energy stored in the capacitor.

Solution. Given : $C = 18\ \mu F = 18 \times 10^{-6}$ F; $\nu = 240$ Hz; $i_{rms} = 3A$

(i) The capacitive reactance is,

$$X_C = \frac{1}{\omega C} = \frac{1}{2\pi\nu C} = \frac{1}{2 \times 3.14 \times (240) \times (18 \times 10^{-6})}$$

$$= 36.86\ \Omega$$

∴ The rms voltage across the capacitor is,

$$V_{rms} = i_{rms} \times X_C = (3) \times (36.86)$$

$$= 110.6\ V$$ **(Answer).**

(ii) The energy stored in a capacitor is given by

$$U = \frac{1}{2} CV^2$$

∴ The average energy stored in the capacitor in our case is,

$$\bar{U} = \frac{1}{2} C^2_{rms} = \frac{1}{2} \times (18 \times 10^{-6}) \times (110.6)^2$$

$$= 0.11\ J$$ **(Answer).**

Problem 13. The frequency range over which a medium wave (MW) pocket radio set can tune is 600 kHz to 1400 kHz. The value of inductance in the LC circuit of the radio is 300 μH. The capacitance C is variable, to enable the necessary tuning. Calculate the range of variation of C in the set.

Solution. Given : $L = 300\ \mu H = 3 \times 10^{-4}$H; ν (range) = 600 to 1400 kHz = 600×10^{-3} to 1400×10^3 Hz

When the radio has been tuned to a certain frequency of the radio waves, this frequency equals the frequency of free oscillation of the LC circuit, which is given by,

$$\nu = \frac{1}{2\pi\sqrt{LC}}$$

$$\Rightarrow \quad \nu^2 = \frac{1}{4\pi^2 LC}$$

$$\Rightarrow \quad C = \frac{1}{4\pi^2 \nu^2 L}$$

Let ν_1 (= 1400 kHz) and ν_2 (= 600 kHz) be the respective frequencies at the two ends of the frequency range and let C_1 and C_2 be the corresponding values of capacitance.

We have,

$$C_1 = \frac{1}{4\pi^2 \nu_1^2 L} = \frac{1}{4 \times (3.14)^2 \times (1400 \times 10^3) \times (3 \times 10^{-4})}$$

$$= 43 \times 10^{-12} \text{ F (or 43 pF)}$$

$$C_2 = \frac{1}{4\pi^2 \nu_2^2 L} = \frac{1}{4 \times (3.14)^2 \times (600 \times 10^3)^2 \times (3 \times 10^{-4})}$$

$$= 234 \times 10^{-12} \text{ F (or 234 pF)}$$

Thus the range of variation of the capacitance has to be 43 pF to 234 pF. **(Answer).**

Problem 14. A variable frequency AC source of 220 V is connected to a series LCR circuit, consisting of an inductance of 2H, a capacitor of 50 μF, and a resistor of 40 Ω. Calculate the average power transferred to the circuit in one cycle, when the frequency of the source is the same as the natural frequency of oscillation of the circuit.

Solution. Given : $V_{rms} = 220$ V; $L = 2$ H; $C = 50\ \mu F = 50\times10^{-6}$ F; $R = 40\ \Omega$

The frequency of the source is the same as the natural frequency of the circuit in the condition of resonance. Then, $X_L = X_C$ and the impedance of the circuit is,

$$Z = \sqrt{R^2 + (X_C - X_L)^2} = \sqrt{R^2 + 0} = R$$
$$= 40\ \Omega$$

$$\therefore \quad i_{rms} = \frac{V_{rms}}{Z} = \frac{(220)}{(40)}$$
$$= 5.5 \text{ A}$$

∴ The average power transferred to the circuit in one cycle is,

$$\bar{P} = V_{rms} \times i_{rms} \times \cos\phi$$
$$= (220)\times(5.5)\times 1 \quad (\cos\phi = R/Z = 1)$$
$$= 1210 \text{ W.}$$ **(Answer).**

Problem 15. The output voltage of an ideal transformer, connected to a 240 V AC mains, is 24 V. When this transformer is used to light a bulb with rating 24 V, 24 W, calculate the current in the primary coil of the circuit. (AI; 2000).

Solution. Given : $V_p = 240$ V; $V_s = 24$ V; $P = 24$ W

The bulb (of rating 24 V, 24 W) is connected to the secondary. ∴ The current in the secondary is (Power = V × i),

$$i_s = \frac{\text{Power}}{V_s} = \frac{(24)}{(24)} = 1\text{A}$$

Now,
$$\frac{N_s}{N_p} = \frac{V_s}{V_p} = \frac{(240)}{(24)} = 10$$

Further,
$$\frac{i_p}{i_s} = \frac{N_s}{N_p} = 10$$

$$\therefore \quad i_p = 10\times i_s = 10\times(1)$$
$$= 10 \text{ A}$$

Thus, the current in the primary is found to be 10 A. **(Answer).**

Problem 16. A transformer steps down an input 220 V AC emf to 3V AC emf, which is then rectified to 3V DC, for operating a pocket radio set. If the number of turns in the secondary coil of the transformer is 12, what is the number of turns in the primary coil ?

Solution. Given : $V_p = 220$ V; $V_s = 3$V; $N_s = 12$

We know,
$$\frac{V_s}{V_p} = \frac{N_s}{N_p}$$

$$\therefore \quad N_p = \frac{V_p \times N_s}{V_s} = \frac{(220)\times(12)}{(3)}$$
$$= 880 \text{ turns.}$$ **(Answer).**

Very Short Answer Questions

[Including HOTS Questions]

(1 mark each)

1. Write an equation for the instantaneous value of an AC voltage.

2. What is meant by the peak value of an AC voltage supply ?

3. What is the average value of AC over a complete cycle ? Give reason.

4. Give the values of peak voltage and frequency for an AC supply given by, $\mathcal{E} = 300 \sin 648t$.

5. Work out the peak value of voltage for 220 V AC supply.

6. Find the rms value of emf for an AC voltage supply given by, $\mathcal{E} = 200 \sin 314t$.

7. Find the rms value of current, when the instantaneous current for an AC supply is given by, $i = 6 \sin 400t$.

8. A moving coil galvanometer shows no deflection when AC is passed through it. Give reason.

9. What kind of ammeter is used for measuring AC and why ?

10. What is meant by inductive reactance ? Give its SI unit.

11. What is meant by capacitive reactance ? Give its SI unit.

12. If an inductor is included in a DC circuit, what will be its reactance ?

13. A capacitor in an AC circuit behaves almost like a pure conductor for very high values of AC frequency. Give reason. [Hint: Since $X_C = 1/(\omega C)$, the reactance becomes negligibly small for high ω].

14. If the frequency of supply in an AC circuit containing a resistor, an inductor, and a capacitor is increased, how will the values of R, X_L, X_C be affected ?

15. What is the phase relationship between emf and current in an AC circuit containing an inductor alone ?

16. What is the phase relationship between emf and current in an AC circuit containing a capacitor alone ?

17. What is meant by power factor of an AC circuit ?

18. What are the maximum and minimum values of power factor in an AC circuit and under what conditions they result ?

19. What is the phase difference between voltage and current in an AC circuit with power factor of 0.5 ? [Answer : $\pi/3$].

20. What is a wattless current ?

21. In a series LCR circuit at resonance, what is (a) phase difference between the applied voltage and current and (b) impedance ?

22. In a series LCR circuit, out of the three components L, C and R, in which one the power dissipation takes place ?

23. What is quality factor, Q, in a series LCR circuit ? What does it signify ?

24. In an LCR circuit, what factors determine the growth and decay of current : L or R or both ?

25. What is the natural frequency of oscillation in an LC circuit ? What is the reactance at this frequency ?

26. Is the direction of current during the discharge of a capacitor same or opposite to that during charging ? [Hint : Opposite]

27. Where does the energy reside in an inductor when a steady current flows in the circuit ? (Hint : In the magnetic field of the coil).

28. Where does the energy reside in a charged parallel-plate capacitor ? [Hint: In the electric field between the plates].

29. What is the principle of working of a transformer ?

30. A transformer cannot step up or step down a DC voltage. Why ?

31. Does a transformer change the frequency of input voltage, while stepping up or stepping down it ?

32. Why is the core of a transformer made of soft iron ?

33. State two factors responsible for loss of power in a transformer.

34. On which factors does the emf induced in a metal coil rotating in a uniform magnetic field depend ? [Hint : N, B, A, ω ; $\mathcal{E} = NBA\omega \sin\omega t$].

35. What is the principle of working of an AC generator ?

36. On increasing the speed of rotation of the armature of an AC generator, how will the value of the maximum emf produced and the frequency of emf be affected ? [Hint: Both will increase; $\mathcal{E}_0 = NBA\omega$].

Short Answer Questions

[Including HOTS Questions]

(2 or 3 marks each)

1. How does alternating current (AC) differ from direct current (DC) ?

2. Using necessary mathematics, show that the average value of AC over a complete cycle is zero.

3. Derive an expression for the average power ($\bar{P}$) dissipated in an AC circuit and hence show that $\bar{P} = 0$ for a purely inductive or a purely capacitive circuit.

4. If you are provided with a coil with air core, another identical coil with iron core, an electric lamp, and a source of electricity, how can you determine if the source is an AC source or a DC source ? Give reason for your answer. [Hint: Connect the coil, lamp, source in series. With DC, the lamp glows brightly both with air core and iron core ($X_L = \omega L = 0.L = 0$). With AC, the lamp glows dimmer with air core and still dimmer with iron core, due to increase in reactance].

5. Describe the various energy losses that occur in a transformer, along with the remedies employed for minimising them.

Long Answer Questions

(5 marks each)

1. With the help of a labelled diagram, describe the principle, construction, and working of a transformer. Also explain why its core is laminated.

2. With the help of a labelled diagram, describe the principle, construction, and working of an AC generator. Derive an expression for the emf induced in its armature.

Numerical Exercises

1. An AC source of 220 V, 50 Hz is connected to a 25 Ω resistor. Calculate the (i) rms value and maximum instantaneous value of the current in the circuit and (ii) time required for the current to attain its maximum value, starting from its rms value. **[8.8A and 12.44 A; 2.5 ms]**

2. A 20 μF capacitor is connected to an AC source of output voltage $\mathcal{E}$ = (15V) sinωt. Determine the peak values of current in the circuit for ω = 20 s^{-1} and 400 s^{-1}. **[6 mA and 120 mA]**

3. An oscillator with output voltage $\mathcal{E}$ = (15 V) sinωt is connected to a 10 mH inductor. Find the peak values of current in the circuit for ω = 20 s^{-1} and 400 s^{-1}. **[75 A and 3.75 A]**

4. A coil has an inductance of 1 henry. (a) At what frequency will it have a reactance of 3142 Ω ? (b) What should be the capacitance of a capacitor which has the same reactance at that frequency ? **[500 Hz; 0.1 μF]**

5. A coil of inductance 18 mH is connected to an AC source of frequency 50 hertz. Calculate the inductance of the coil. **[5.65 Ω]**

6. A capacitor of 8 μF is connected to an AC source of frequency 2×10^6 Hz. Find its reactance. **[0.01 Ω]**

7. A 10 μF capacitor is connected to an AC source of variable frequency. Find its reactance for the frequencies 20 Hz and 600 Hz. **[79.6 Ω and 2.65 Ω]**

8. An alternating current of 1.5 mA rms and angular frequency ω = 100 rad s^{-1} flows through a 10 kΩ resistor and a 0.50 μF capacitor in series. Calculate the value of rms voltage across the capacitor and the impedance of the circuit (CBSE; 1993). **[30 V; 22.4 kΩ]**

9. When a series combination of a coil of inductance L and a resistor R is connected across a 12 V, 50 Hz supply, a current of 0.5 A flows through the circuit. The current differs in phase from applied voltage by π/3 radian. Calculate the values of L and R. (CBSE; 1997). **[0.07 H; 12 Ω]**

10. A capacitor when connected in series with an inductor of 3 henry causes the circuit to resonate at the frequency of 60 Hz. Determine its capacitance. **[885 μF]**

11. An AC supply of 220 V, 50 Hz is connected in series to a capacitor of 200 μF and a resistor of 60 Ω. Find the maximum current passing through the resistor and the time lag between it and the maximum voltage. [Hint: Find the phase lag φ and then the time lag φ/ω]. **[5.01 A; 8.25 ms]**

12. An AC source of 220 V, 50 Hz is connected in series to 100 mH inductor and an 80 μF capacitor. Find the (i) amplitude of current and rms current, (ii) voltage drop across each component, (iii) average power separately transferred to the inductor and the capacitor, and (iv total average power absorbed in the circuit. **[(i) 37 A and 26.2 A, (ii $(V_C)_{rms}$ = 1041 V; $(V_L)_{rms}$ = 821 V, (iii) zero, zero (iv) zero]**

13. A series LCR circuit consists of a capacitor of reactance 28 Ω, an inductor of inductance 0.3 H, and a resistor of 15 Ω. It is supplied with an AC of 220 V, 50 Hz. Find the current in the circuit and the phase difference between voltage and current. **[i_{rms} = 3.24 A; ϕ = 77.3°]**

14. An AC source of 200 V, 50 Hz is connected to a coil of resistance 40 Ω and inductance 0.6 H. What is the rms current in the circuit ? **[1.04 A]**

15. An AC supply of 220 V, 50 Hz is connected in series with a coil, having resistance 100 Ω and inductance 0.4 H, and a capacitor of capacitance 200 μF. What is the rms current in the circuit ? **[1.48 A]**

16. A series LCR circuit contains an inductance of 120 mH, a capacitor of variable capacitance, a resistor, and an AC supply of frequency 4000 cps. Find the capacitance value that will enable maximum current to flow in the circuit. [Hint: For i_{max}, $\omega L = 1/\omega C$. ∴ $C = 1/\omega^2 L$]. **[13.2 nF]**

17. A variable frequency AC source of 40 V is connected to a series LCR circuit (L = 5H, C = 25 μF, R = 20000 Ω). Find the frequency value for which the current is maximum. Also find the value of maximum current. **[13.9 Hz; 2 mA]**

18. A 200 V variable frequency AC source is connected to a series combination of L = 5H, C = 80 μF and R = 40 Ω. Calculate (i) angular frequency of the source to get maximum current in the circuit, (ii) the current amplitude at resonance and (iii) the power dissipation in the circuit. (CBSE; 2002). **[(i) 50 rad s^{-1}, (ii) 7.07 A, (iii) 1000 W]**

19. A series LCR circuit consists of an inductor of 1H, a capacitor of 20 μF, and a resistor of 20 Ω. Determine the resonant frequency and quality factor of the circuit. **[ω_0 = 35.6 Hz; Q = 11.2]**

20. An AC source of emf $\mathcal{E}$ = (40 V) sin (200 s^{-1})t is connected to a series LCR circuit, consisting of a 160 mH inductor, an 80 μF capacitor, and a 100 Ω resistor. Calculate the impedance, the peak current, and the resonant frequency of the circuit. **[Z = 106 Ω; i_0 = 0.4 A; ν_0 = 44.5 Hz]**

21. A 40 mH inductor is connected across a fully charged 50 μF capacitor. Determine the angular frequency of free oscillations of the circuit. **[ω_0 = 707 rad s^{-1}]**

22. An alternating voltage E = 200 sin 300t is applied across a series combination of R = 10 Ω and an inductor of 800 mH. Calculate :
(i) Impedance of the circuit
(ii) Peak value of current in the circuit
(iii) Power factor of the circuit. (AI; 1994)
[240 Ω; 0.83 A; 0.042]

23. Calculate the current drawn by the primary of a transformer, which steps down 200 V to 20 V to operate a device of resistance 20 Ω. Assume the efficiency of the transformer to be 80%. (CBSE; 2001).
[0.13 A]

24. A step-down transformer with 10000 turns in its primary coil is fed input power at 11000 V through a power transmission cable. If it has to provide output power at 220 V, Find what should be the number of turns in its secondary coil. **[320 turns]**

25. The coil of an AC generator rotates with an angular speed of 50 rad s^{-1} in a uniform magnetic field of magnitude 0.4 T. If it has 100 turns and an area of 4 m^2 and the total resistance of the circuit is 400 Ω, determine the maximum current that can be drawn from the generator. **[20 A]**

UNIT 5
ELECTROMAGNETIC WAVES

CHAPTER 5
Electromagnetic Waves

CHAPTER 5

ELECTROMAGNETIC WAVES

In the last four chapters, we have seen that during the latter half of the eighteenth century and the first half of the nineteenth century the connection between electricity and magnetism had been well established through various experiments and formulations. The Oersted's experiment had demonstrated a close relationship between electricity and magnetism by deflection of a compass needle placed near a current-carrying wire (Chapter 3A). The production of a magnetic field by electric current flowing in a closed loop was formulated through the Ampere's law and the magnetic field due to a current element through the Biot-Savart law (Chapter 3A). The Gauss's law gave the total electric flux through a closed surface enclosing an electric charge (Chapter 1B). The Faraday's laws expressed the electric field induced by a changing magnetic field (Chapter 4A).

In view of such well settled correlation between electricity and magnetism, a brilliant British physicist James Clerk Maxwell was led to argue that if a changing magnetic flux could induce an electric field, the reverse should also be true, i.e. a changing electric flux should also induce a magnetic field. Around 1865, he mathematically unified all the known laws of electricity and magnetism into a set of four general equations, named after him as Maxwell's equations. They form the basis of all known electromagnetic devices known to us, e.g. telephone, radio, television, microwave oven, electric motor, electric generator, transformer, X-ray machine, radar, etc, etc.

The derivation and study of the Maxwell's equations is beyond the scope of a school level student. A very significant prediction of the equations was the existence of electromagnetic waves, consisting of oscillations of electric and magnetic fields, that could propagate in free space (i.e. vacuum). The prediction came true nearly 20 years later, when in 1887 Hertz experimentally demonstrated the existence of the waves. It led to a revolution in the field of communications, that continues till date. Further, according to the Maxwell's equations, the velocity of the waves turned out to be very nearly equal to the velocity of light. It led to the very important conclusion that light should also be a form of electromagnetic waves and thus the fields of electricity, magnetism and light were united.

5.01. Nature and Properties of e.m. waves

An **electromagnetic wave** can be **defined** as "a wave which consists of oscillations of electric and magnetic fields, such that the electric field, the magnetic field, and the direction of propagation of the wave are perpendicular to each other".

Such a wave is diagrammatically shown in Fig. 5.1. The electric field $\vec{E}$ oscillates in the xy-plane along y-direction. The magnetic field $\vec{B}$ oscillates in the xz-plane along z-direction. The wave propagates along x-direction. Thus $\vec{E}$, $\vec{B}$, and the direction of propagation are mutually perpendicular. It is a **sinusoidal wave.** Both $\vec{E}$ and $\vec{B}$ vary sinusoidally and in phase with each other.

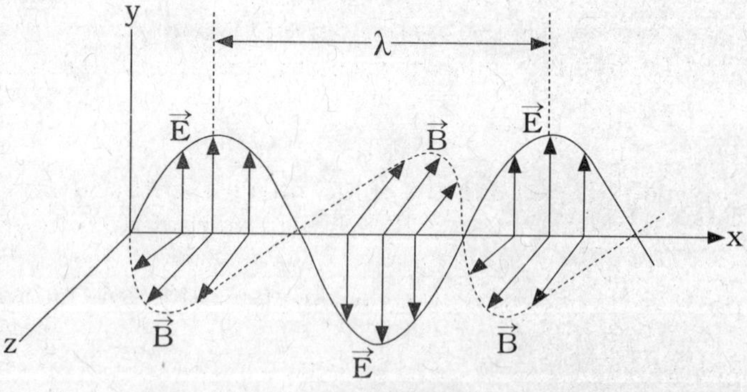

Fig. 5.1

In general, the electric and magnetic fields are vectorially represented as,

$$\vec{E} = E_x \vec{i} + E_y \vec{j} + E_z \vec{k} \quad \text{..... (5.1)}$$

$$\vec{B} = B_x \vec{i} + B_y \vec{j} + B_z \vec{k} \quad \text{..... (5.2)}$$

Since the oscillations of the electric vector $\vec{E}$ and the magnetic vector $\vec{B}$ are restricted to y and z direction, respectively, we have,

$$E = E_y \qquad (E_x, E_y = 0)$$
$$B = B_z \qquad (B_x, B_y = 0)$$

Hence the electric and magnetic fields can be expressed as,

$$E = E_0 \sin (k_x - \omega t) \quad \text{..... (5.3a)}$$

$$B = B_0 \sin (k_x - \omega t) \quad \text{..... (5.4a)}$$

$k = (2\pi/\lambda)$ is the magnitude of the wave vector $\vec{k}$ (physically it means the number of waves contained in a length 2π). E_0 and B_0 are the amplitudes of the electric and magnetic field, respectively. Since $\omega = 2\pi\nu$ and $\nu = (1/T)$, the Eqs. 5.3a and 5.4a can also be expressed as,

$$E = E_0 \sin [2\pi (x/\lambda - t/T)] \quad \text{..... (5.3b)}$$

$$B = B_0 \sin [2\pi (x/\lambda - t/T)] \quad \text{..... (5.4b)}$$

Properties of the e.m. waves

We now enumerate important properties of the e.m. waves. Some of them are obvious. The derivation of the rest of them is beyond the scope of the student.

(i) The electric field $\vec{E}$, the magnetic field $\vec{B}$, and the direction of propagation of the wave are mutually perpendicular.

(ii) $\vec{E}$, and $\vec{B}$ vary sinusoidally. They are always in phase with each other.

(iii) Since the oscillations of both $\vec{E}$ and $\vec{B}$ in the wave are $\perp$ to the direction of propagation of the wave, the e.m. waves are **transverse waves.**

(iv) The magnitudes and amplitudes of $\vec{E}$ and $\vec{B}$ are related as,

$$\frac{E}{B} = \frac{E_0}{B_0} = c \quad \text{..... (5.5)}$$

where c is the velocity of light, a fundamental constant of nature ($c = 2.99792458 \times 10^8$ ms^{-1}, or nearly 3×10^8 ms^{-1}).

(v) The waves can travel not only in a material medium (e.g. air, glass, etc.) but also in free space (vacuum).

(vi) The velocity of the waves in a medium is related to the permeability (μ) and the permittivity (ϵ) of the medium. In free space (vacuum) it is given by,

$$c = 1/\sqrt{\mu_0 \epsilon_0} \quad \text{..... (5.6)}$$

where μ_0 and ϵ_0 are the values of μ and ϵ in free space. (In any other medium, the velocity is given by $v = 1/\sqrt{\mu \epsilon}$).

(vii) Since μ_0 and ϵ_0 do not depend on frequency or wavelength, it follows that all kinds of e.m. waves travel with the same velocity, c.

(viii) The e.m. waves carry energy, which is equally shared by the electric field and the magnetic field. They also carry linear momentum, which is also equally shared.

5.02. Sources of e.m. Waves; General Concepts

The electromagnetic waves are produced by accelerated charges. The acceleration may be either positive or negative (a –ve acceleration is called deceleration or retardation).

An oscillating charge is a continuous source of an e.m. wave. Comparing with the oscillations of a pendulum, when the charge moves from one extreme position to the central position during an oscillation, the velocity changes from zero to the maximum value. So the charge accelerates (+ve acceleration). When it moves from the central position to the other extreme, the velocity falls from the maximum value to zero. So the charge decelerates (–ve acceleration). The process keeps on repeating itself. Thus, we have a continuously accelerated charge, with the acceleration alternately becoming +ve and –ve. Since a charge has an electric field around it, its oscillations produce an oscillating sinusoidal electric field. According to one of the Maxwell's equations, expressing the so-called Maxwell's law of induction, it induces a perpendicular sinusoidal magnetic field, oscillating with the same frequency as the charge. Then, according to another Maxwell's equation, expressing the Faraday's law of induction, such a field induces a perpendicular sinusoidal electric field of the same frequency. The process continues as the charge keeps on oscillating. The two fields continuously keep on creating each other by induction. The resulting mutually perpendicular sinusoidal variations in the field travel as the so-called e.m. wave, both in time and space.

The frequency of the wave is essentially the same as the frequency of oscillation of the charge. Thus, an LC circuit (Chapter 4B; Sections 4B. 06 and 4B .07) can produce e.m. waves of angular frequency.

$$\omega = \frac{1}{\sqrt{LC}}$$

The charge and current in this circuit vary sinusoidally with this frequency. [As we shall see later, the continuous X-rays, which are also a form of electromagnetic waves, solely arise from the deceleration of fast moving electrons that are suddenly stopped by a metal target].

All waves carry energy with them. So the e.m. waves also carry energy with them. The energy is continuously derived from the source which causes and maintains the oscillations of the charge.

A stationary charge cannot be a source of an e.m. wave, because it only produces an electrostatic field around it. Similarly, a charge in uniform motion (e.g. a steady current flowing in a wire), which is accompanied with a magnetic field, cannot produce an e.m. wave, since the field does not vary with time.

A present day source of generation of e.m. waves will be described in Section 5.05.

5.03. Hertz's Experiment

About 20 years after the prediction of the e.m. waves, as expressed by the Maxwell's equations, in 1887 the German physicist Heinrich Hertz actually demonstrated their existence in a laboratory experiment (unfortunately Maxwell himself did not live until then to know about it).

A labelled diagram of the experimental arrangement employed by Hertz is shown in Fig. 5.2. Two large metal spheres S_1, S_2 are connected to two large square-shaped metal plates P_1, P_2. S_1, S_2 are connected to an induction coil I, which produces an oscillating low current high voltage across them. [An induction coil is essentially a step-up transformer, in which the change of magnetic flux is brought about by mechanical means. An automatic mechanical arrangement, called 'interrupter', makes and breaks the primary coil circuit very rapidly]. The high voltage ionizes the air in the gap between S_1, S_2, creating a spark in the gap. The charges (electrons and ions) in the spark oscillate back and forth in the gap, with an angular frequency given by,

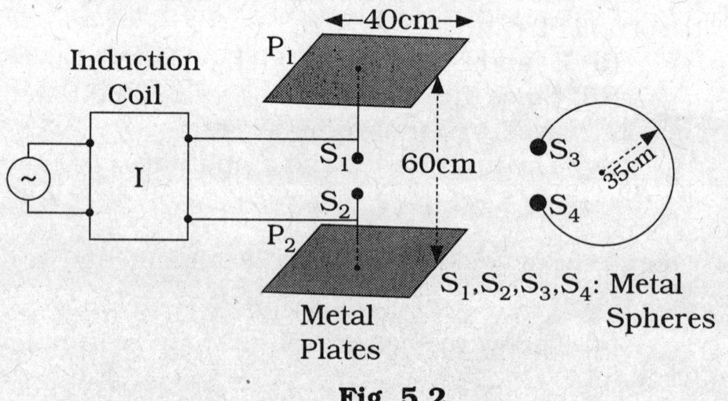

Fig. 5.2

$$\omega = \frac{1}{\sqrt{LC}}$$

L and C are the inductance and capacitance of the circuit, respectively. [The inductance is provided by the secondary of the induction coil and the connecting wires. The capacitance is provided by the plates P_1, P_2. Their large size helps to increase the capacitance substantially, thus enhancing the capacity to hold a large charge between S_1 and S_2. The ionic discharge between S_1, S_2 provides a path to the electric current and thus helps to complete the electric circuit with the induction coil].

The oscillations in the air gap between S_1, S_2 generate e.m. waves of angular frequency ω, which propagate into the space all around the gap. To detect them, a detector consisting of a partially open loop of wire, with the open ends connected to two metal spheres S_3, S_4, was used. The waves reaching the detector were found to produce tiny sparks in the gap S_3 S_4. The sparks arose from the high potential difference created between S_3 and S_4 by the electric field of the e.m. waves. Thus the existence of the e.m. waves was conclusively established.

The intensity of sparking in the detector gap S_3S_4 was found to depend on the position of the gap S_3S_4 with respect to the source gap S_1S_2. It was maximum when the two gaps were parallel and zero when they were at right angles to each other. It followed that the electric field of the wave produced by the source gap S_1S_2 was parallel to S_1S_2, i.e. it was $\perp$ to the direction of propagation of the wave. Thus the transverse nature of the waves was demonstrated.

It was also demonstrated that the waves could be refracted, diffracted and polarised, which further confirmed the wave nature of the radiation produced by the source. Also, stationary e.m. waves could be produced, which enabled to measure their wavelength and hence their velocity ($v = \nu\lambda$, where ν is the frequency of oscillation of the source). The velocity was found to be equal to the velocity of light. Thus, it was concluded that light is also a form of electromagnetic waves.

5.04. Electromagnetic Spectrum

At the time of prediction of electromagnetic waves by Maxwell, the only known e.m. radiation was the light radiation, because it was visible, i.e. seen by the naked eye, e.g. blue sky, green leaves, variously coloured birds and insects, etc. Gradually more kinds of e.m. radiation, all invisible, came to be known to the mankind. By the end of the 19th century, the highly energetic (and so penetrating) X-rays and γ-rays were discovered.

The various e.m. radiations differ in their frequency (and wavelength), mode of production, and properties. Their common features are that they all consist of oscillations of electric and magnetic field of different frequencies (and wavelengths) and travel with the same velocity in free space (vacuum), viz. the velocity of light (c). **The electromagnetic spectrum is a classification of the various e.m. radiations on the basis of their frequencies (and wavelength).** The classification is not based on the properties of the radiations and it is only broadly related to their modes of production or detection.

The e.m. spectrum as we now know is a very wide spectrum. It is depicted in Fig. 5.3. In terms of wavelength, it ranges from a value as small as a fraction of fm (femto meter; 1 fm = 10^{-15}m) for γ-rays to a value as large as thousands of kilometers for long waves. According to the relation $e = h\nu$ (E : energy; h: Planck's constant; ν: frequency), the energy of the radiation increases as the frequency goes up or the wavelength goes down ($\nu\lambda = c$; so $\lambda \propto \frac{1}{\nu}$). Note that **there is no sharp division between two neighboring regions of the spectrum.** In other words, the neighboring regions of the spectrum overlap with each other, e.g. the overlap seen between X-rays and γ-rays (we can only say that γ-rays have shorter wavelength than X-rays).

Production of e.m. radiations. As already stated, the basic source of production of all e.m. radiations is acceleration of charge. The production of radiation of relatively higher energy (infrared, visible, ultraviolet, X-rays, γ-rays) involves atomic level interactions, to understand which the knowledge of quantum physics is needed, which is beyond the scope of a school level student. These interactions usually produce radiations of well defined frequencies (and wavelengths), corresponding to energy differences between well defined electronic and nuclear energy levels (characteristic radiation). The X-rays are also produced when fast moving electrons are suddenly decelerated upon bombarding a metal target. Then the energy of the electrons is partly converted into heat and partly into X-rays of a wide continuous range of frequency or wavelength (continuous or white X-ray radiation). The γ-rays are produced by radioactive materials, like uranium. The infrared radiation is emitted by all hot bodies.

The radio waves are usually produced by an LC-circuit with oscillating current (to be described in the next section). The microwaves, with frequency in the gigahertz (GHz) range, are really short wavelength radio waves. They are produced by special vacuum tubes (klystrons, magnetrons, and Gunn diodes). Because of their short wavelength, they hardly spread or bend around corners of an obstacle during their propagation and so proceed along nearly straight paths like a light beam. The transmission and reception of the radio waves will be described in Chapter 10.

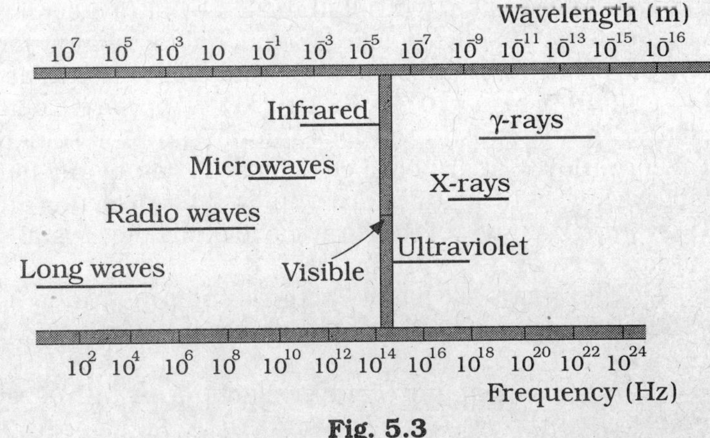

Fig. 5.3

Characteristics and Uses of e.m. Waves

The main characteristics and common uses of various kinds of e.m. radiation will now be described very briefly. The frequency range indicated against each radiation should be taken only as a rough guide.

(i) Long waves (10 Hz – 100 kHz). In India, the electric supply for household and other purposes has the frequency 50 Hz (i.e. 50 cycles per second), which corresponds to a wavelength of the order of 10,000 km.

(ii) Radio waves (10 kHz – 1000 MHz). They are mainly used in radio and television communication. Waves in the frequency range of about 10–500 kHz are used for maritime and aeronautical purposes. The AM (amplitude modulated) medium wave radio band employs waves in the approximate frequency range of 300-1500 kHz. The short wave band uses nearly 1.5-50 MHz. The lower TV broadcast range is nearly 30-300 MHz [VHF (very high frequency]. The FM (frequency modulated) radio band extends over nearly 90-100 MHz. The higher UHF (ultra high frequency) range (300 MHz-3 GHz) is employed for TV broadcast mobile phones, maritime and aeronautical purposes, etc.

The modern MRI (magnetic resonance imaging) technique, being increasingly used for medical diagnosis in hospitals, uses e.m. radiation of frequency around 100 MHz. Unlike the X-rays, it is a harmless non-invasive technique and apart from showing the bone structure, it also reveals the tissu structure in the body.

(iii) Microwaves (10^9 – 5×10^{11} Hz). Since these waves can be transmitted like a light beam because of their short wavelength, they are abundantly used in radar systems, mostly for aircraft navigation. They can locate and judge the distance of a flying or fast moving object. They are also employed to measure speeds of bowling in cricket, tennis serves, running cars, etc. They are also used to transmit radio signals from one point on the earth to another via satellites. Another common use is in microwave ovens for cooking. The water molecules in food articles preferentially absorb these waves and get heated.

(iv) Infrared (10^{11} – 4×10^{14} Hz). They are readily absorbed by H_2O, CO_2, NH_3, etc., molecules which heat up by rise in their thermal motion. So they are also referred to as 'heat waves'. Infrared lamps are commonly employed for physiotherapy of various parts of the body. The infrared radiation helps to maintain the earth's temperature within required limits through the so-called greenhouse effect. The visible light coming from the sun is absorbed by the earth's surface and re-radiated as infrared radiation (of longer wavelength, and so of lesser energy), which is trapped in the earth's atmosphere by greenhouse gases like water vapour and CO_2. The infrared radiation is also being increasingly used in night vision systems and cameras, which enable to see and photograph even in dark. The use extends over several purposes, including warfare.

(v) Visible light (400-750 nm or 4×10^{14}–7.5×10^{14} Hz). It is the most familiar part of the e.m. spectrum, as it is directly detected by the human eye,. The natural sunlight (or moonlight and light from stars) as well as the artificial light from lamps, etc., scattered from all the objects around us enables us to 'see' them. In other words, the world becomes visible to us because of the sensitivity of our eyes to this range of wavelength (or frequency). The range is commonly expressed as **VIBGYOR** (Violet, Indigo, Blue, Green, Yellow, Orange, Red), in increasing order of wavelength. On a rainy day, under favourable conditions all these colours are visible in the sky in a rainbow. The sensitivity of certain other living beings extends even beyond this range, e.g. sensitivity of snakes to infrared and of many insects to ultraviolet.

(vi) Ultraviolet (7.5×10^{14} – 10^{18} Hz). It is used for sterilization, as it instantly kills bacteria. It is also used for forensic tests, e.g. finger prints, as certain materials fluoresce, i.e. emit light of higher wavelength in the visible region, when illuminated by ultraviolet light. It is harmful to the body tissue. The light emitted by sun is rich in ultraviolet radiation, but fortunately for us, most of it is absorbed in the earth's atmosphere by the ozone layer at about 40-50 km above the earth's surface.

(vii) X-rays (10^{17} – 5×10^{19} Hz). They are extensively used for medical diagnosis of all parts of the body. They are also used for radio-therapy, for some forms of cancer and some skin diseases. They are also employed at airports, etc., for detection of any hidden weapons and metallic objects in the passenger luggage. In industry, they are employed for detecting any cracks and other faults in the machinery. Another very important use of the X-rays is to study the crystal structure, i.e. atomic positions, bonding between the atoms, etc., by X-ray diffraction.

The X-rays are highly energetic and penetrating. They cause damage to and even destroy living tissues. Therefore, any prolonged exposure (beyond a few seconds at most) of the body to them has to be scrupulously avoided.

A convenient unit of wavelength for use with X-rays is ängstrom (Å):

$$1 \text{ Å} = 10^{-8} \text{ cm} = 10^{-10} \text{ m} = 10^{-1} \text{ nm}$$

$$\text{or, } 1 \text{ nm} = 10 \text{ Å}$$

(viii) γ-rays (5×10^{18} – 10^{23} Hz). They are extremely energetic and penetrating. Therefore the exposure of the body to them has to be strictly avoided. They are employed for sterilizing food items, medical equipment, etc.

5.05 Source of Radio Waves

In the Section 5.02, we had outlined the general concepts behind the generation of e.m. waves of different kinds. We now describe a source for the generation of radio waves, which involves familiar processes and equipment.

In the Section 4B. 06 (Chapter 4), we had seen that an LC circuit is a source of electromagnetic oscillations. The natural frequency of oscillation of the circuit is,

$$\omega_0 = \frac{1}{\sqrt{LC}}$$

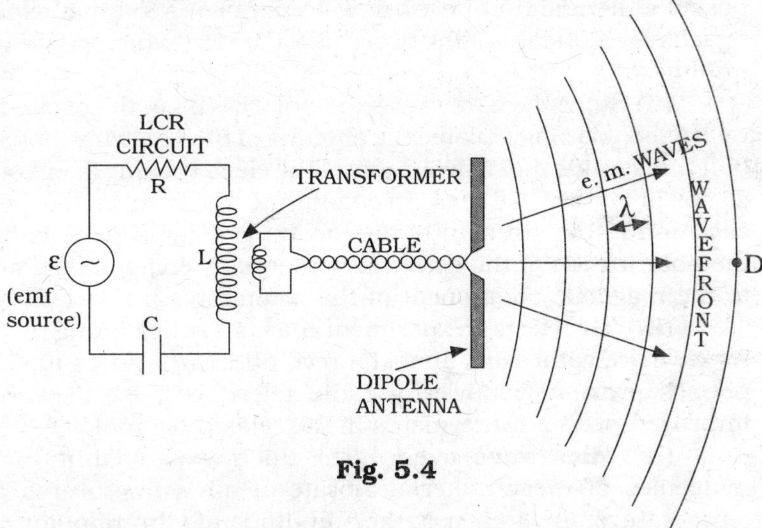

Fig. 5.4

In the Section 4B. 07, we had seen that ω_0 is also the natural frequency of oscillation of a series LCR circuit; the presence of the resistance R only acts as a damping factor for the oscillations. When the angular frequency ω of the emf source becomes equal to ω_0, resonance takes place. Then a current of large amplitude is built up.

The Fig. 5.4 shows the set-up for the generation of the e.m. waves. The voltage of the emf source varies sinusoidally as

$$V = V_0 \sin \omega t$$

At resonance, $\omega = \omega_0 = 1/\sqrt{LC}$. Then the charges and the current in the circuit vary sinusoidally at the frequency ω_0. The circuit is coupled with a transformer and further through a transmission cable to an antenna. The antenna is made up of two thin solid rods of a conducting material. The coupling causes the charge to oscillate sinusoidally along the rods at the frequency ω_0. Accordingly, the associated current in the rods also varies sinusoidally at the same frequency. This effectively makes the antenna as equivalent to an electric dipole, with its dipole moment (a vector quantity) varying along the length of the antenna.

The dipole produces an electric field around it. As the dipole momnent varies in magnetude and direction, the electric field also varies similarly. At the same time, because the current in the rods also varies in magnitude and direction, the magnetic field associated with the current also varies similarly. These combined oscillations of the electric and magnetic fields result in the generation and propagation of electromagnetic waves of angular frequency ω_0. The waves travel away from the antenna at the velocity of light, c. The directions of travel of three such waves are shown by arrows in Fig. 5.4. The points in the same phase on the various waves lie along concentric spheres, called **wavefronts.** They will be discussed in detail in Chapter 6B.

At large distances away from the antenna, like the region around a distant point D in the Fg. 5.4, a part of the wavefront looks almost plane in shape. Since in practice the waves are transmitted over long distances, measuring upto thousands of kilometers and beyond in space, before being recieved by a radio, television, etc., they effectively behave as plane wavefronts. Correspondingly, the waves are said to be **plane e.m. waves.** This makes the theoretical treatment of such waves much simple. In fact, in our description of the e.m. waves in the previous sections, we had tacitly assumed them to be plane electromagnetic waves. However, it should be clearly understood that actually it is the wavefronts that look plane in shape, not the waves themselves.

There is a constant loss of energy in the circuit because of (i) thermal dissipation of energy in the resistance R and (ii) energy radiated away by the travelling e.m. waves. This is constantly compensated for by the emf source, which is usually an electric generator. Thus the e.m. oscillations are sustained in the circuit.

Additional knowledge

(i) The Hertz's experiment to demonstrate the existence of e.m. waves was performed for radio waves, which have relatively low frequency. From that it was concluded that light is also a form of e.m. radiation, because the velocity of the waves was found to be equal to the velocity of light. It was not possible then, and not even now, to perform a similar, experiment directly for the light waves, because they have frequency of the order of $\sim 10^{14}$ Hz, which could not be produced by any electrical circuit even until now.

(ii) Since the e.m. waves are polarized, i.e. the oscillaitons of the electric (and magnetic) vector are restricted to a single plane, the antenna of the receiving instrument, e.g. an AM radio, has to be so oriented that it responds most favourably to the electric component of the incoming waves. Otherwise the incoming signal is greatly reduced, becoming nearly zero in the perpendicular position. In fact, the observed variation in the strength of the incoming signal upon changing the positin of the antenna is a proof of the polarization of the e.m. waves. Some receiving instruments have their antenna designed to respond to the magnetic component of the incoming e.m. waves, which need to be similarly suitably oriented.

(iii) The Hertz's experiment marks the beginning of the field of communication using e.m. waves. Now we are witnessing almost a revolution around us in this field, with the advent of television, mobile phones, extremely powerful radio telescopes, etc. The quality of reception has also tremendously improved, with advancemetns in the related technology.

(iv) Microwave ovens. The microwave cooking has steadily grown popular. The atoms and molecules of every material vibrate at all temperatures above absolute zero (O°K). The higher the temperature, the greater is the amplitude of vibration and hence the greater is the heating produced in the material. Every material possesses a characteristic natural frequency of vibration of its atoms or molecules. For water, it is nearly 3 GHz, which falls within the frequency range of microwaves. Hence, when water is exposed to microwaves of such frequency, it absorbs them by 'resonance' and gets heated. The more is the water content, the more is the heat produced (there is also evaporation). This is the basis of microwave cooking of food items having water content. The heated H_2O molecules pass on their heat to the rest of the food by conduction and thus the food gradually gets cooked. The waves penetrate the food about 2 to 4 cm thickness.

Note that unlike the conventional cooking (with fire or electricity), in which first the utensil containing the food gets heated, which then passes on its heat to the food, in the microwave cooking the food directly receives the necessary energy for cooking from the incident microwaves. Suitable utensils are made of porceilane, glass, or special plastics, as they themselves do not get much heated. Metal utensils have to be avoided, as they get overheated and may also produce electric spark with the walls of the oven, when situated near the walls, as they also tend to get electrically charged.

Solved Conceptual Problems

[Including HOTS Questions]

Problem 1. What does an electromagnetic wave consist of ?

Solution. An e.m. wave consists of oscillations of electric and magnetic field, such that the electric field, the magnetic field, and the direction of propagation of the wave are perpendicular to each other.

Problem 2. Do e.m. waves carry energy and momentum ? Justify your answer.

Solution. Yes. Like any other wave, the e.m. waves also carry energy, which is equalylly shared by the electric field and the magnetic field. They also carry linear momentum, which is also equally shared.

Problem 3. What basically causes the production of e.m. waves ? Briefly explain the process of generation of an e.m. wave.

Solution. The e.m waves are basically produced by accelerated charges. The acceleration can be positive as well as negative (deceleration). An oscillating charge provides a continuous source of an e.m. wave. In each oscillation the velocity of the charge changes from zero to maximum (acceleration) to zero (deceleration). Since a charge has an electric field around it, its oscillations produce an oscillaing sinusoidal electric field, which in turn produces a perpendicular oscillating magnetic field. The oscillating magnetic field again produces a perpendicular oscillating electric field and thus the process continues endlessly (according to the known laws of electromagnetism). Both the electric field and the magentic field oscillate with the same frequency as the frequency of oscillation (say ν) of the charge. Thus, an e.m. wave of frequency ν is produced around the charge, which continues so long as the charge continues to oscillate.

Problem 4. Essentially, how did the Hertz's experiment demonstrate the existence of e.m. waves and establish their trensverse nature ?

Solution. Hertz used an oscillatory LC-circuit for producing e.m. waves, which created a continuous spark across a small linear gap (say, S_1S_2). When a partially open loop of wire, having a similar linear gap (say, S_3S_4), was placed some distance away from S_1S_2, sparking was seen to occur in the gap S_3S_4, too. Thus, the existence of e.m. waves was established.

The sparking in S_3S_4 was found to be maximum when it was parallel to S_1S_2 and zero when it was parpendicular to it. It followed that the electric field of the wave produced by the source gap S_1S_2 was parallel to S_1S_2, i.e. it was perpendicular to the direction of propagation of waves. Thus the transverse character of the waves was established.

Problem 5. How the oscillations in an e.m. wave are sustained, in spite of a constant loss of energy during the propagation of the wave ?

Solution. The e.m. waves are usually produced by an LC circuit. The loss of energy of the waves is constantly compensated for by the emf source connected in the circuit. It is usually a powerful electric generator. Thus the e.m. oscillations are sustained in the circuit.

Problem 6. Sound waves cannot travel in vacuum, but the light waves can. Explain why.

Solution. The light waves are electromagnetic in nature, consisting of oscillations of electric and magnetic field, which do not need a material medium for their propagation. Hence they can travel in free space (vacuum). On the other hand, sound waves are mechanical in nature. They consist of compressions and rarefactions, which essentially need a material medium for their propagation.

Solved Numerical Problems

Problem 1. In a radio transmission, the frequencies of the radio waves fall in the audio range 0.6×10^6 Hz to 1.8×10^6 Hz. Find the shortest and the longest wavelengths in this range.

Solution. Given : $\nu_1 = 0.6 \times 10^6$ Hz; $\nu_2 = 1.8\times10^6$ Hz

The shortest and longest wavelengths correspond to largest and smallest frequencies, respectively.

$$\therefore \quad \lambda_{\text{shortest}} = \frac{c}{\nu_2} = \frac{(3\times10^8)}{(1.8\times10^6)}$$

$$= 1.67\times10^{-2}\text{ m} \quad \textbf{(Answer).}$$

and

$$\lambda_{\text{longest}} = \frac{c}{\nu_1} = \frac{(3\times10^8)}{(0.6\times10^6)}$$

$$= 5\times10^{-2}\text{ m} \quad \textbf{(Answer).}$$

Problem 2. For a harmonic e.m. wave travelling in free space, the amplitude of the magnetic field is 600 nT. Find the amplitue of the electric field in the wave.

Solution. Given : B_0 = 600 nT = 600×10^{-9} T = 6×10^{-7} T

We know, $$\frac{E_0}{B_0} = c$$

$$\Rightarrow \quad E_0 = B_0 c = (6\times10^{-7}) \times (3\times10^{8})$$

$$= 180 \ NC^{-1}$$ **(Answer).**

Problem 3. The amplitude of the electric field in a plane e.m. wave is 180 N.C^{-1}. The frequency of the wave is 60 MHz. Find ω, λ, and the amplitude of the magnetic field in the wave. Also write equations for the electric field and the magnetic field.

Solution. Given : E_0 = 180 N.C^{-1}; ν = 60 MHz = 60×10^6 Hz

(i) $\omega = 2\pi\nu = 2\pi\times(60\times10^6)$

$= 3.71\times10^8$ rad.s^{-1} **(Answer).**

(ii) $\lambda = \frac{c}{\nu} = \frac{(3\times10^8)}{(60\times10^6)}$

$= 5$m **(Answer).**

(iii) $B_0 = \frac{E_0}{c} = \frac{(180)}{(3\times10^8)}$

$= 6\times10^{-7}$ T **(Answer).**

The equations for the electric field and the magnetic field are expressed as,

$$E_y = E_0 \sin (kx - \omega t)$$

$$= 180 \sin\left[\left(\frac{2\pi}{5}\right)x - (3.77\times10^8)t\right] \qquad \left(\because k = \frac{2\pi}{\lambda}\right)$$

$= 180 \sin [(0.4\pi)\, x - (3.77\times10^8)t]$ **(Answer).**

$B_z = (6\times10^{-7}) \sin [(0.4\pi)x - (3.77\times10^8)t]$ **(Answer).**

Very Short Answer Questions

[Including HOTS Questions]

(1 mark each)

1. What do electromagnetic waves consist of ?

2. Are the e.m. waves transverse or longitudinal in nature ? What is their velocity of propagation ?

3. For an e.m. wave propagating in free space, what is the relationship between the amplitudes of its electric and magnetic fields ?

4. From the equation $c = 1/\sqrt{\mu_0 \epsilon_0}$ for the velocity of an e.m. wave, how is it concluded that the velocity of all kinds of e.m. waves in free space is same ?

5. What is the ratio of velocities of light waves of wavelengths 500 nm and 750 nm, respectively ? [Hint: 1].

6. Is the light produced by a tubelight an e.m. wave ? [Hint: Yes. All types of light are so].

7. For the visible part of the e.m. spectrum, what is the approximate range of wavelengths ?

8. What is the name given to that part of electromagnetic spectrum which is used for taking photographs of earth under foggy conditions from great heights ? (CBSE; 2004) [Hint: Infrared].

9. Which of the following has lowest frequency : microwaves, ultraviolet rays, X-rays ? (CBSE; 1996).

10. Arrange the following e.m. radiations in increasing order of wavelength : green light, infrared rays, γ-rays, radio waves, microwaves, X-rays.

11. Name two factors common to all e.m. radiations. [Hint: Transverse nature of wave and velocity of propagation (c)].

12. Name the part of the electromagnetic spectrum of wavelength 10^{-2} m and mention its one application. (CBSE; 2008).

13. What are the directions of electric and magnetic field vectors of an e.m. wave travelling along x-axis ?

14. In the electromagnetic spectrum, what are the orders of magnitude of the frequencies of longest and shortest waves ?

15. Microwaves are used in radar. Why ? (CBSE; 2004).

Short Answer Questions

[Including HOTS Questions]

(2 or 3 marks each)

1. Briefly explain how Maxwell was led to predict the existence of e.m. waves.

2. Give one important use each of the various radiations belonging to the e.m. spectrum.

3. Name different parts of electromagnetic spectrum, alongwith their ranges of frequency (or wavelength).

4. Name the constituent radiation of electromagnetic spectrum which
(a) is used in satellie communication,
(b) is used for studying crystal structure.
(c) is similar to radiations emitted during decay of radioactive nuclei.
(d) has its wavelength range between 390 nm and 770 nm.
(e) is absorbed from sunlight by ozone layer.
(f) produces intense heating effect. (CBSE; 2005).

5. Write any four characteristics of electromagnetic waves. Give two uses each of (i) radio waves and (ii) microwaves. (CBSE; 2007).

6. Draw a labelled diagram of Hertz's experimental set-up to produce electromagnetic waves. Explain the generation of electromagnetic waves using this set-up. (CBSE; 2006).

Numerical Exercises

1. Find the wavelength of electromagnetic waves of frequency 5×10^{19} Hz in free space. (CBSE; 2004).
[0.006 Å]

2. A radio set is able to tune to any transmitting station broadcasting programmes in the frequency band of 8 MHz to 12 MHz. Determine the corresponding wavelength range of tuning by the set.
[37.5 m to 25 m]

3. At an instant of time, a plane e.m. wave travelling in xz-plane along x-axis has the vectorial value of its electric field as $\vec{E} = 5.1\ \vec{j}\ V.m^{-1}$. If the wave frequency is 20 MHz, find the vectorial value of its magnetic field $\vec{B}$ at this instant.
[$\vec{B} = (1.7 \times 10^{-8})\ \vec{k}$ T]

4. The electric field in a plane e.m. wave is oscillating with a frequency of 3×10^{12} Hz. If its amplitude is 60 $V.m^{-1}$, find the amplitude of the oscillating magnetic field in the wave. Also calculate the wavelength of the wave.
[2×10^{-7} T; 1×10^{-4} m]

5. An e.m. wave travelling in free space has its electric field represented as,
$$\vec{E} = (4.2\ N.C^{-1} \cos [(2\ rad.m^{-1})y + (4.8 \times 10^{6}\ rad.s^{-1})\ t]\vec{i}$$
Give (i) the magnitudes of wavelength and frequency of the wave and the direction of wave propagation and (ii) the amplitude of the magnetic field and the equation for the magnetic field.
[λ = 3.14 m; ν = 0.76 MHz; negative y-axis; B_0 = 14 nT; $\vec{B}$ = (14 nT) cos [(2rad m^{-1}) y + 4.8 rad s^{-1})t]

UNIT 6
OPTICS

CHAPTER 6 (A)
Ray Optics; Reflection & Refraction; Optical Instruments

CHAPTER 6 (B)
Wave Optics; Huygens' Construction; Interference; Diffraction; Polarization

CHAPTER 6 (A)

RAY OPTICS; REFLECTION & REFRACTION; OPTICAL INSTRUMENTS

6A. 01. Reflection and Refraction of Light (General Concepts)

In the last chapter, we have learnt that light is a form of electromagnetic waves. If a wave meets a macroscopic obstacle, e.g. the edge of a tin sheet, during its propagation, it tends to bend from its straight path. The larger the wavelength, the larger is the bending. Thus, sound waves can bend through very large angles; they are known to turn around the corners, i.e. through angles upto 180°. However, light waves, which have very small wavelengths [$\lambda \sim 500$ nm, i.e. $\lambda <$ (1/1000) mm], bend through inappreciably small angles. If they meet an obstacle like the edge of a tin sheet on their way, they cast a well defined shadow of the edge. Thus, light is said to travel along a straight path (rectilinear propagation of light). [However, as we shall learn in the study of diffraction of light in the next chapter, this condition breaks down when the light meets a microscopic object, e.g. it passes through a very narrow aperture (size: ~ 1/100 mm) or across a sharp edge].

The straight path followed by a light wave is called a **ray** of light. A small bundle of such rays forms a **beam** of light. The branch of study of properties of light waves under the approximation of rectilinear propagation of light is called **ray optics** or **geometrical optics.**

Earlier, in his junior classes the student has already studied in an elementary way the phenomena of reflection and refraction of light and their application to the construction of optical instruments. In this chapter, these phenomena and the topic of optical instruments will be dealt with in greater detail.

When a light ray from one medium is incident on the surface of separation with another medium, a small part of the light is reflected back into the first medium. The rest of the light enters the second medium (assumed to be transparent to light, like glass or water), as shown in Fig. 6.1. It is said to be **refracted** into the second medium. If the medium 2 is optically denser than the medium 1, the refracted ray is bent towards the normal to the interface, drawn at the point of incidence O. The angle i made between the incident ray and the normal is called **angle of incidence.** The angle r between the normal and the reflected ray is called **plane of incidence.** The angle r between the normal and the reflected ray is called **angle of reflection.** The angle r′ between the normal and the refracted ray is called **angle of refraction.**

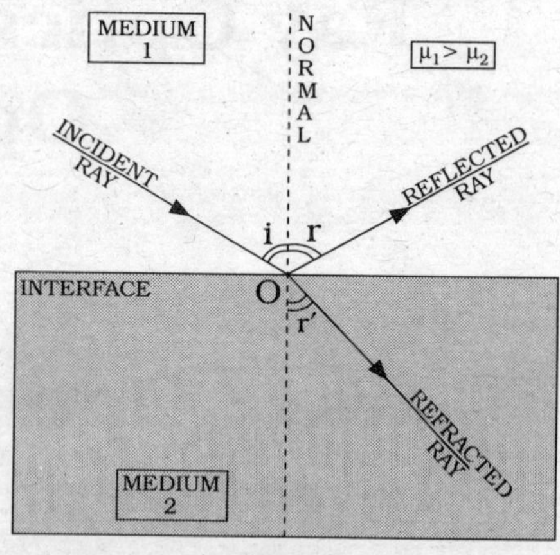

Fig. 6.1

6A.02. Laws of Reflection; Reflection from Plane Mirror and Spherical Mirrors

The students are already familiar with image formation by mirrors. With reference to Fig. 6.1, the **laws of reflection** are stated below:

(i) The reflected ray lies in the plane of incidence (i.e. in the plane containing the incident ray and the normal).

(ii) The angle of incidence is equal to the angle of reflection ($\angle i = \angle r$).

(a) Plane Mirror

The Fig. 6.2 shows the formation of image of a point object A by a plane mirror. The ray AM incident at an angle i is reflected along MB, such that $\angle i = \angle r$. The ray AP is incident along the direction of normal

($\angle i = 0$). Therefore it is reflected back upon itself ($\angle i = \angle r = 0$), as the ray PA. Both the reflected rays MB and PA appear to be coming from the point A′, where they appear to meet when extended backwards. The same holds for any other ray starting from A and reflected by the mirror. Therefore A′ is the image of the point object A, formed by the mirror. It is a **virtual image,** because the light rays from A do not actually meet at A′ after reflection, but only appear to meet there. It cannot be taken on a screen or photographed. We shall shortly discuss more about real and virtual images, when describing the formation of images by spherical mirrors.

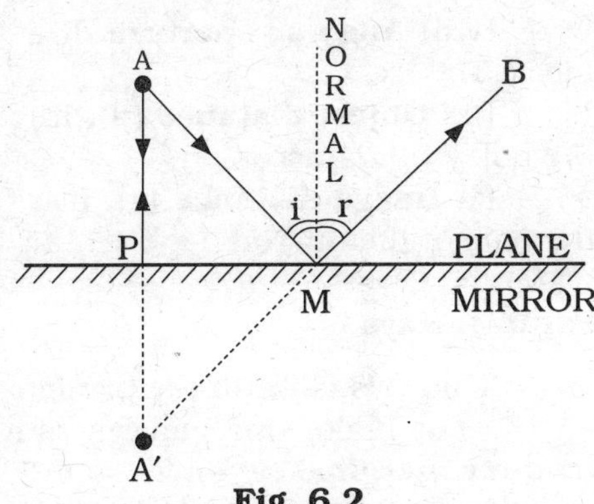

Fig. 6.2

Note that the image formed in a plane mirror lies as much behind the mirror as the object is in front of it (AP = A′P).

If the image of an **extended object,** e.g. a pencil, is formed by the mirror, the size of the image is same as the size of the object and the image is erect. The image is laterally inverted, e.g. the image of the letter "P" looks like "ꟼ". This phenomenon is termed **lateral inversion.** To sum up, the image formed by a plane mirror is erect, virtual, laterally inverted, and of the same size as the object.

(b) Spherical Mirrors

We next consider image formation by mirrors with curved surfaces. Such surfaces may be of various shapes, e.g. spherical, cylindrical, parabolic, etc. For the sake of simplicity, we shall restrict ourselves to the formation of images by spherical mirrors.

A spherical mirror is a part of a hollow sphere. Only one side of the mirror is reflecting. The other side is opaque. Two types of spherical mirrors are possible. In the **concave mirror,** the reflecting surface faces the centre of the sphere (Fig. 6.3; C is the centre of the sphere and $M_1P_1N_1$ is the mirror). In the **convex mirror,** the opaque surface faces the centre of the sphere (the mirror $M_2P_2N_2$ in Fig. 6.3).

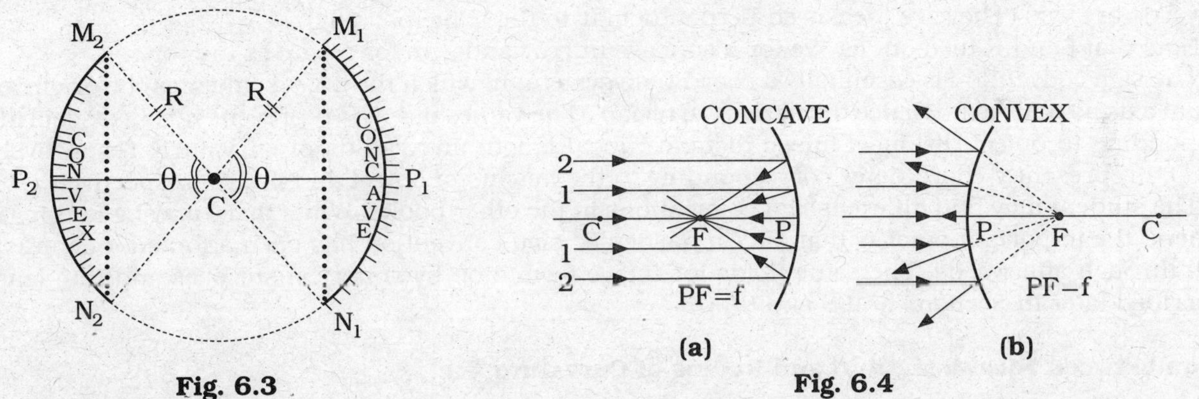

Fig. 6.3 **Fig. 6.4**

In the study of spherical mirrors, it is useful to familiarise with the following terms:

(i) **Centre of curvature.** It is the radius of the sphere of which the mirror is a part (the point C in Fig. 6.3).

(ii) **Radius of curvature (R).** It is the radius of the sphere of which the mirror is a part ($CM_1 = CP_1 = CN_1 = R$; $CM_2 = CP_2 = CN_2 = R$).

(iii) **Pole.** The central point of the mirror (the points P_1 and P_2 in Fig. 6.3).

(iv) **Principal axis.** The extended straight line joining the pole and the centre of curvature (the line PFC in Fig. 6.4).

(v) **Principal focus** (or, simply, **focus;** or **focal point).** It is the point on the principal axis at which the incident rays parallel to the axis actually meet or appear to meet, after reflection from the mirror (the point F in Figs. 6.4a and 6.4b).

(vi) **Focal length (f).** It is the distance between the pole P and the principal focus F (see Fig. 6.4).

(vii) **Aperture (or linear aperture).** It is the diameter of the circular boundary of the mirror (M_1N_1 and M_2N_2, in Fig. 6.3).

(viii) **Angular aperture.** It is the angle subtended at C by the aperture of the mirror (the angle θ in Fig. 6.3).

(ix) **Object distance (u).** It is the distance measured along the principal axis between the object and the pole of the mirror.

(x) **Image distance (v).** It is the distance measured along the principal axis between the image and the pole of the mirror.

Paraxial Rays

The rays close to the principal axis are called **paraxial rays.** They may be parallel to the axis (as in Fig. 6.4) or make small angles with the axis. The rays away from the principal axis are called **peripheral rays (or marginal rays).** Thus in Fig. 6.4a, the pair of rays 1 may be considered as paraxial rays and the pair of rays 2 be considered as peripheral rays.

For the sake of convenience, in our treatment of image formation by mirrors and lenses, we shall assume the light rays to be paraxial in nature (except when considering defects of images).

Sign Convention

For deriving relations between u, v, f, R, h_2 (size of the image), h_1 (size of the object), and m (magnification), which are then used or solving problems regarding image formation, it is necessary to specify a uniform system of ascribing a due sign (+ve or –ve) to each of these quantities. We shall employ the following **new Cartesian sign convention** for this purpose:

(i) The incident rays are taken as travelling from left to right and the reflected rays as travelling from right to left.

(ii) The distances u, v, f, R are measured as starting from the pole of the mirror.

(iii) The distances u, v, f, R measured from left to right (i.e. in the direction of incident rays) are +ve and from right to left (i.e. in the direction of reflected rays) are –ve.

(iv) The object and image heights (h_1, h_2) measured upwards of the principal axis are +ve and measured downwards are –ve. (They are measured perpendicular to the principal axis).

Note that f and R turn out as –ve for a concave mirror and +ve for a convex mirror.

The sign convention is essentially a coordinate system, in which the pole is taken as the origin and the principal axis as the x-axis. Hence the name 'Cartesian'. The values u, v, f, R represent various x-coordinates, corresponding to object distance, image distance, focal length and radius of curvature, respectively. The values O, I represent y-coordinates corresponding to the heights of object and image, respectively.

The student may find different sign conventions in the other books. While there may be nothing wrong with them, the important point is that once a particular sign convention has been adopted, one must stick to it all through afterwards. The same holds for image formation by spherical surfaces and thin lenses, to be described later in Sections 6A.08 and 6A.09.

Relation between Focal Length (f) and Radius of Curvature (R)

The Figs 6.5a and 6.5b show the positions of focus F and centre of curvature C for a concave mirror and a convex mirror, respectively. Any line drawn from C to a point on the mirror represents the direction of normal at that point. Thus CM is the normal at a point M on the mirror. A light ray parallel to the principal axis is incident at M. By definition, it actually passes through F after reflection from the mirror, for the case of concave mirror, and appears to pass though F for the case of convex mirror. The reflection occurs according to the laws of reflection ($\angle i = \angle r$). If ϕ is the angle of incidence and MN is the perpendicular drawn from M on the principal axis, then from the geometry of the figures

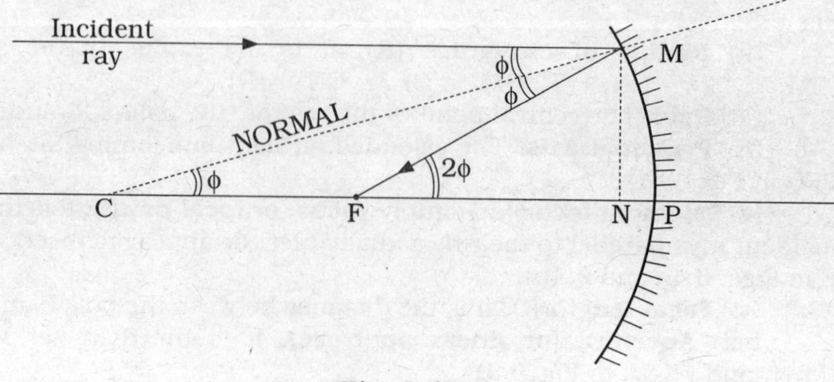

Fig. 6.5 (a)

it is readily seen that,

$\angle CMF = \phi; \angle MFP = 2\phi$

$\tan\phi = \frac{MN}{CN}; \tan 2\phi = \frac{MN}{FN}$

For paraxial rays, ϕ is small. $\therefore \tan\phi \simeq \phi$ and $\tan 2\phi \simeq 2\phi$. Then,

$$\phi \simeq \frac{MN}{CN}; \quad 2\phi \simeq \frac{MN}{FN}$$

$$\Rightarrow \quad FN = \frac{CN}{2}$$

Since the ϕ-values are small for paraxial rays, the point N lies very close to the pole P. $\therefore$ FN = f and CN $\simeq$ R. So we get,

$$f = \frac{R}{2}$$

Fig. 6.5 (b)

Thus for a spherical mirror, the focal length f is half of the radius of curvature R.

Image Formation by Spherical Mirrors; Mirror Formula

Although it is simpler to consider formation of image of a point object, it is preferable to do so for an extended object, because (i) the treatment is more general and (ii) it provides an expression for the magnification, too, in addition to the mirror formula.

To know the position of image of an object, usually it is enough to consider the reflection of just two incident rays from a point on the object. The point where the two reflected rays meet is the image of the chosen point on the object. For the sake of convenience, any two of the following incident rays are chosen:

(i) A ray parallel to the principal axis. After reflection, it passes through the focus F for a concave mirror or appears to pass through F (for a convex mirror).

(ii) A ray passing through C (for a concave mirror) or appearing to pass through C (for a convex mirror). It is reflected back upon itself, because it lies in the direction of a normal to the mirror.

(iii) A ray passing through F (for a concave mirror) or appearing to pass through F (for a convex mirror). It is reflected parallel to the principal axis.

The Figs. 6.6(a) and (b) depict the formation of image (A_2B_2) of a small linear object (A_1B_1) by a concave mirror. If the object is placed beyond the focus F, the image is real and inverted (it may be both diminished or enlarged, depending upon the position of object; In Fig. 6.6a, it is diminished in size). If the object is placed within F, the image is always virtual, erect and enlarged (Fig. 6.6b). The Fig. 6.7 similarly shows the image formation by a convex mirror. The image is virtual, erect and diminished for all positions of the object. In all the figures, one incident ray (A_1M) from A_1 is taken parallel to the principal axis, which after reflection from the mirror (ray MR) either passes through F (Figs. 6.6a,b) or appears to pass through it (Fig. 6.7). Another ray (A_1CN) is taken to pass through C, which reflects back upon itself. A third Ray (A_1P) is incident at P, which is reflected as PQ, with the angle of incidence i equal to the angle of reflection r

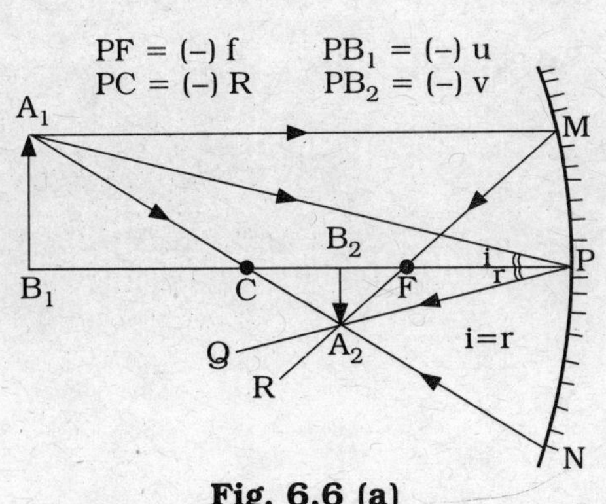

Fig. 6.6 (a)

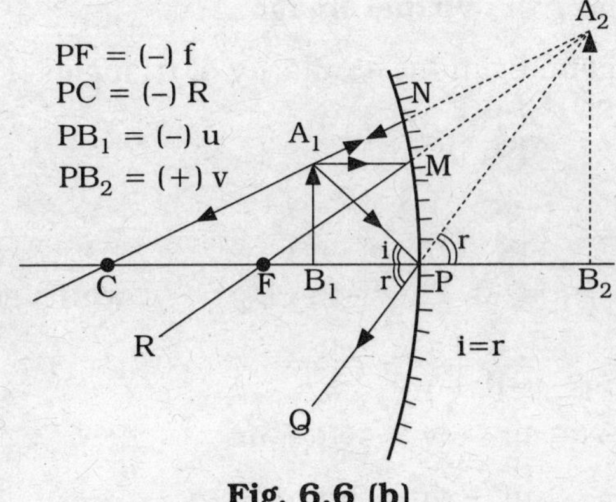

Fig. 6.6 (b)

(B_1P is normal to the mirror at P). All the reflected rays either actually pass through A_2 or appear to pass through it. Hence A_2 is the image of A_1. The perpendicular A_2B_2 drawn from A_2 upon the principal axis is the image of the object A_1B_1.

The relation between object distance (u), image distance (v), and focal length (f) for a spherical mirror is known as the **mirror formula (or mirror equation).** We shall derive it separately for each of the cases depicted in Fig. 6.6a,b and 6.7. It is found to be the same for all the cases.

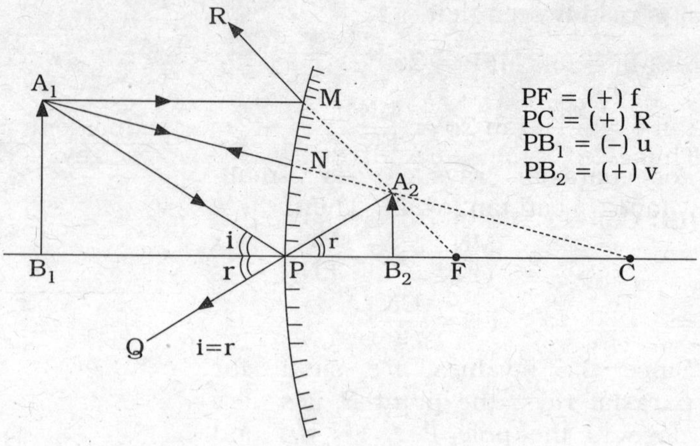

Fig. 6.7

(i) Concave Mirror; Real Image

In Fig. 6.6a, the Δs A_1B_1C and A_2B_2C are similar,

$$\therefore \quad \frac{A_1B_1}{A_2B_2} = \frac{CB_1}{CB_2} \qquad .....(6.1)$$

Further, the Δs A_1B_1P and A_2B_2P are similar, ($\angle i = \angle r$; $\angle B_1 = 90° = B_2$),

$$\therefore \quad \frac{A_1B_1}{A_2B_2} = \frac{PB_1}{PB_2} \qquad .....(6.2)$$

From the Eqs. 6.1 and 6.2, we get,

$$\frac{CB_1}{CB_2} = \frac{PB_1}{PB_2} \qquad .....(6.3)$$

$$\Rightarrow \quad \frac{PB_1 - PC}{PC - PB_2} = \frac{PB_1}{PB_2}$$

According to our sign convention, $PB_1 = -u$, $PC = -R$, $PB_2 = -v$, which gives,

$$\frac{-u + R}{-R + v} = \frac{-u}{-v}$$

$$\Rightarrow \quad ur - uv = uv - vR$$

$$\Rightarrow \quad ur + vR = 2uv$$

Dividing by uvr on both the sides,

$$\frac{1}{v} + \frac{1}{u} = \frac{2}{R}$$

Since f = R/2 (or R = 2f), we get the mirror formula as,

$$\boxed{\frac{1}{u} + \frac{1}{v} = \frac{1}{f}} \qquad .....(6.4)$$

(Mirror formula)

(ii) Concave Mirror; Virtual Image

Once again, from the similarity of triangles (Fig. 6.6b) we get (Eq. 6.3),

$$\frac{CB_1}{CB_2} = \frac{PB_1}{PB_2} \qquad .....(6.3)$$

$$\Rightarrow \quad \frac{PC - PB_1}{PC + PB_2} = \frac{PB_1}{PB_2}$$

In this case, $PB_1 = -u$, $PC = -R$, $PB_2 = +v$, which gives,

$$\frac{-R + u}{-R + v} = \frac{-u}{v}$$

$$\Rightarrow \quad ur - uv = -vR + uv$$

$$\Rightarrow \quad ur + vr = 2uv$$

$$\Rightarrow \quad \frac{1}{u}+\frac{1}{v}=\frac{2}{R}=\frac{2}{2f} \quad \text{(dividing by uvR)}$$

$$\Rightarrow \quad \boxed{\frac{1}{u}+\frac{1}{v}=\frac{1}{f}}$$

Thus, we get the same mirror formula as for the real image formation.

(iii) Convex Mirror (Always a Virtual Image)

Again, from the similarity of triangles (Fig. 6.7) we get (Eq. 6.3),

$$\frac{CB_1}{CB_2}=\frac{PB_1}{PB_2} \qquad .....(6.3)$$

$$\Rightarrow \quad \frac{PB_1+PC}{PC-PB_2}=\frac{PB_1}{PB_2}$$

Here, $PB_1 = -u$, $PC = +R$, $PB_2 = +v$, which gives,

$$\frac{-u+R}{R-v}=\frac{-u}{v}$$

$$\Rightarrow \quad -uv+vR=-uR+uv$$

$$\Rightarrow \quad vR+uR=2uv$$

$$\Rightarrow \quad \frac{1}{u}+\frac{1}{v}=\frac{2}{R}=\frac{2}{2f} \quad \text{(dividing by uvR)}$$

$$\Rightarrow \quad \boxed{\frac{1}{u}+\frac{1}{v}=\frac{1}{f}}$$

Thus, again the same formula is obtained, as for the image formation by a concave mirror.

Linear Magnification (m)

For a spherical mirror, the **linear magnification** (or, simply, **magnification**) is **defined** as the "ratio of the size of image formed by the mirror to the size of the object". It is denoted by **m.**

$$m=\frac{\text{size of image}}{\text{size of object}}=\frac{h_2}{h_1} \qquad ..... (6.5)$$

For **m>1,** the image is enlarged. For **m<1,** the image is diminished in size.

The **sign of m** is determined by the nature of image formed by the mirror. Thus, from the Figs. 6.6a,b and 6.7, it is seen that according to our sign convention (height taken as +ve when measured upwards and taken as –ve when measured downwards), the sign of m ($-h_2/h_1$) is –ve for the formation of real images and +ve for virtual images. The results for the various cases may be summarized as below :

Type of mirror	Nature of image	m	
		Sign	Value
Concave (object beyond F)	Real	–ve	Both m>1 and m<1 possible
Concave (object within F)	Virtual	+ve	m>1
Convex (all positions of object)	Virtual	+ve	m<1

Relation between m, u, v. From Eq. 6.2, we have,

$$\frac{A_2B_2}{A_1B_1}=\frac{PB_2}{PB_1}$$

As worked out below, it leads to the following simple relation between m, v and u for all cases of image formation by spherical mirrors,

$$m=-\frac{v}{u} \qquad .....(6.6a)$$

(i) Concave mirror; object beyond F (real image):

$$\frac{A_2B_2}{A_1B_1} = \frac{PB_2}{PB_1}$$

$$\Rightarrow \quad \frac{-h_2}{h_1} = \frac{-v}{-u}$$

$$\therefore \quad m = \frac{h_2}{h_1} = -\frac{v}{u}$$

(ii) Concave mirror; object within F (virtual image):

In this case, $\frac{h_2}{h_1} = \frac{v}{-u}$

$$\therefore \quad m = \frac{h_2}{h_1} = -\frac{v}{u}$$

(iii) Convex mirror; for all positions of the object (virtual image):

Here, $\frac{h_2}{h_1} = \frac{v}{-u}$

$$\therefore \quad m = \frac{h_2}{h_1} = -\frac{v}{u}$$

Further, using the mirror formula, $\frac{1}{u} + \frac{1}{v} = \frac{1}{f}$, and multiplying throughout once by u and next by v, the following alternative formulae for m (= –v/u) are easily derived,

$$m = \frac{f}{f-u} \qquad \text{.....(6.6b)}$$

$$m = \frac{f-v}{f} \qquad \text{.....(6.6c)}$$

Some Important and Popular Uses of Spherical Mirrors

(i) A concave mirror is commonly used as a shaving mirror or a make-up mirror, as it provides an enlarged and erect image of the face, when held close to the face.

(ii) Large size concave mirrors are employed in the construction of extremely powerful astronomical telescopes (vide Section 6A.17).

(iii) Convex mirrors are commonly used as rear view mirrors in motor vehicles. Since a convex mirror provides diminished (and erect) images of all the objects in front of it, it is able to compress a large field of view into a small one. Hence other vehicles within a wide range at the back of the vehicle being driven are readily seen all together in the mirror, by the driver.

Example 1. A screen is placed in front of a concave mirror of focal length 15 cm, to obtain a clear image of a 4 cm high object, placed at 25 cm before the mirror. Find the distance of the screen from the mirror. What is the nature and size of the image ?

Solution. Given: u = – 25 cm; f = – 15 cm; h_1 (object size) = 4 cm

Now, $$\frac{1}{u} + \frac{1}{v} = \frac{1}{f} \text{ (Mirror formula)}$$

$$\therefore \quad \frac{1}{(-25)} + \frac{1}{v} = \frac{1}{(-15)}$$

$$\Rightarrow \quad \frac{1}{v} = \frac{1}{25} - \frac{1}{15} = -\frac{2}{75}$$

$$\Rightarrow \quad v = -37.5 \text{ cm}$$

Thus, the distance of the screen from the mirror is found to be 37.5 cm. **(Answer).**

Next, Magnification, $m = -\frac{v}{u} = \frac{h_2}{h_1}$ (h_2 : Image size)

$$\Rightarrow \quad h_2 = h_1 \times \left(-\frac{v}{u}\right) = 4 \times \left[-\frac{(-37.5)}{(-25)}\right]$$

$$= -(4 \times 1.5) = -6 \text{ cm}$$

Thus, the image size is 6 cm. The –ve signs of v and h_2 mean that the image is real and inverted **(Answer).**

Example 2. If an image of size double the size of the object is to be obtained from a concave mirror of focal length 40 cm, at what distance from the mirror the object should be placed ? Solve the problem for both the formation of (i) a real image and (ii) a virtual image.
Solution. Given: f = – 40 cm; m = 2 (image size double the object size)

We shall use the formulae,

$$\frac{1}{u}+\frac{1}{v}=\frac{1}{f}\ ;\ m=-\frac{u}{v}$$

Let the object be placed at a distance x in front of the mirror. Then, according to our sign convention, in both the cases we have, u = (–x). ∴ In both the cases,

$$m=-\frac{v}{(-x)} \quad \text{or} \quad v=m\,(x)=2x$$

Case (i): Real (inverted) image (formed on the same side as the object).

$$u=-x;\quad v=-2x;\quad f=-40\text{ cm}$$

$$\therefore \quad \frac{1}{(-x)}+\frac{1}{(-2x)}=\frac{1}{(-40)} \quad \text{or} \quad \frac{3}{2x}=\frac{1}{40}$$

$$\Rightarrow \quad x=60\text{ cm}$$

Thus, in this case, the object distance is 60 cm. **(Answer).**

Case (ii): Virtual (erect) image (formed on the other side of the mirror).

$$u=(-x);\quad v=2x;\quad f=-40\text{ cm}$$

$$\therefore \quad \frac{1}{(-x)}+\frac{1}{2x}=\frac{1}{(-40)} \quad \text{or} \quad -\frac{1}{2x}=-\frac{1}{40}$$

$$\Rightarrow \quad x=20\text{ cm}$$

Thus, in this case, the object distance is 20 cm. **(Answer).**

Example 3. A concave mirror forms a 3 times magnified real image of an object placed before it. Determine the location of the image.
Solution. Magnification is the ratio of image size (h_2) to the object size (h_1), i.e. m = (h_2/h_1). We know that when a concave mirror forms a real and inverted image, m is –ve, i.e.

$$m=-\frac{h_2}{h_1}$$

It is given that the image size is 3 times the object size, i.e.

$$m=-\frac{h_2}{h_1}=-3$$

The magnification is also given as,

$$m=-\frac{v}{u} \quad \text{(given, } u=-10\text{ cm)}$$

$$\Rightarrow \quad v=-(m\times u)=-[(-3)\times(-10)]$$

$$=-30\text{ cm}$$

∴ According to our sign convention, the image is located at 30 cm before the mirror. **(Answer).**

Example 4. The radius of curvature of a convex mirror fitted in a car is 3 m. If another car is located behind at 5 m from the mirror, determine the position, nature and size of the image formed by the mirror.
Solution. Given: R = 3 m; u = – 5 m
The focal length is R/2 and it is +ve for a convex mirror.

i.e. $$f=\frac{R}{2}=\frac{3}{2}=1.5\text{ m}$$

Now, $$\frac{1}{u}+\frac{1}{v}=\frac{1}{f}$$

$$\Rightarrow \frac{1}{(-5)} + \frac{1}{v} = \frac{1}{1.5}$$

$$\Rightarrow \frac{1}{v} = \frac{1}{1.5} - \frac{1}{5} = \frac{13}{15}$$

$$\Rightarrow v = \frac{15}{13} = 1.15 \text{ m}$$

Next, $$m = -\frac{v}{u} = -\frac{1.15}{(-5)} = 0.23$$

The values and signs of v and m mean that a virtual, erect and diminished (0.23 times) image of the car is formed behind the mirror. **(Answer).**

Example 5. A convex mirror of radius of curvature 30 cm has an object of height 5 cm placed at 10 cm in front of it. Determine the nature, position and size of the image formed by it.
Solution. Given: R = 30 cm; u = – 10 cm; h_1 = 5 cm

For a convex mirror, the focal length f (= R/2) is a +ve value,

$$f = \frac{R}{2} = (+)\frac{30}{2} = 15 \text{ cm}$$

Now, $$\frac{1}{u} + \frac{1}{v} = \frac{1}{f}$$

$$\Rightarrow \frac{1}{v} = \frac{1}{f} - \frac{1}{u} = \frac{1}{(15)} - \frac{1}{(-10)}$$

$$= \frac{1}{15} + \frac{1}{10} = \frac{5}{30} = \frac{1}{6}$$

$$\Rightarrow v = 6 \text{ cm}$$

Next, $$m = -\frac{v}{u} = \frac{h_2}{h_1}$$

$$\Rightarrow h_2 = \left(-\frac{v}{u}\right) \times h_1$$

$$= -\frac{(6)}{(-10)} \times (5)$$

$$= 3 \text{ cm}$$

Thus, a virtual, erect and diminished image of height 3 cm is formed at 6 cm behind the mirror. **(Answer).**

6A. 03. Refraction of Light

The refraction of light is "the phenomenon in which the light when passing from one medium to another bends at the surface of separation (interface) of the two media".

In Section 6A.01, we had briefly described how a ray of light bends upon refraction from one medium to another (Fig. 6.1). The phenomenon follows the following laws.

Laws of Refraction; Refractive Index

(i) The incident ray, the refracted ray, and the normal to the interface at the point of incidence lie in the same plane (the plane of incidence).

(ii) The ratio of the sine of angle of incidence to the sine of angle of refraction is constant for a pair of media (Snell's law).

$$\frac{\sin i}{\sin r} = \mu_{21} \text{ (Snell's law; simple form)} \qquad .....(6.7)$$

The constant value μ_{21} is called the **refractive index of medium 2 with respect to medium 1.** It is a dimensionless quantity.

In case the light travels from vacuum (or air) to another medium, the constant is called **absolute index of refraction** or simply **refractive index.** Then we can drop the subscript on μ in Eq. 6.7 and write,

$$\frac{\sin i}{\sin r} = \mu \text{ (w.r.t. vacuum/air)} \qquad .....(6.8)$$

The values of refractive index for some common media are given below:

Medium	**R.I. (μ)**
Vacuum	Unity (1)
Air	1.00029
Water (20°C)	1.33
Sugar solution (80%)	1.49
Common crown glass	1.52
Very heavy flint glass	1.89
Diamond	2.42

It is clear that all values of refractive index are measured with respect to vacuum or air (which very nearly has the same value of R.I. as vacuum, viz. 1). For visible light region (nearly 400-750 nm), all materials have $\mu > 1$ (however, for X-rays, all materials have $\mu < 1$).

In the next chapter, we shall find that the refractive index of a medium can also be expressed as the ratio of two velocities,

$$\mu = \frac{\text{velocity if light in vacuum}}{\text{velocity if light in medium}}$$

Or

$$\mu = \frac{c}{v_{medium}} \qquad .....(6.9)$$

Similarly μ_{21}, the R.I. of medium 2 w.r.t. the R.I. of medium 1, is expressed as the ratio of the velocity of light in medium 1 to that in medium 2,

$$\mu_{21} = \frac{v_1}{v_2} \qquad .....(6.10)$$

The value of μ does not depend on the angle of incidence, but depends on the wavelength (λ) of the light. This will become more clear in Section 6A.05 on dispersion of light by a prism. Further, when light refracts from one medium to another, its frequency (ν) does not change. Only its velocity and wavelength change ($v = \nu\lambda$).

In general, when light refracts from one medium of R.I. μ_1 into another of R.I. μ_2, as shown in Fig. 6.1, the refractive indices of the two media are related as,

$$\mu_1 \sin i = \mu_2 \sin r \text{ (Snell's law; general from)} \qquad .....(6.11)$$

When the light passes from a rare to a dense medium ($\mu_1 < \mu_2$), the refracted ray bends towards the normal, as seen in Fig. 6.1. When the light passes from a dense to a rare medium ($\mu_1 > \mu_2$), it bends away from the normal. It can be visualized in Fig. 6.1 by imagining that the path of the light is reversed, so that the refracted ray becomes the incident ray and vice versa. Or, the same is illustrated in Fig. 6.8, where a light ray passes from air into glass and then further again into air. It is seen that $\angle r_1 < \angle i_1$, but $\angle r_2 > \angle i_2$. Also note that the incident ray AB is parallel to the final emergent ray CD, but CD is laterally displaced w.r.t. AB.

Further, by imagining the reversal of the rays in Fig. 6.1 or Fig. 6.8, it is easily seen that

$$\mu_{21} = \frac{1}{\mu_{12}} \qquad .....(6.12)$$

Fig. 6.8

In the reverse case, the angle of refraction r becomes the angle of incidence i and vice versa.

Thus, if the R.I. of glass w.r.t. air is 3/2, the R.I. of air w.r.t. glass is 2/3.

From Eg. 6.9, we have

$$\mu_1 = \frac{c}{v_1} \text{ and } \mu_2 = \frac{c}{v_2} \qquad .....(6.13)$$

But, from Eq. 6.10 we have, $\mu_{21} = \frac{v_1}{v_2}$.

$$\therefore \quad \mu_{21} = \frac{\mu_2}{\mu_1} \qquad .....(6.14)$$

If a ray of light passes through more than one medium, say three media, it can be easily worked out that the refractive indices of the three media are related as,

$$\mu_{32} = \mu_{31} \times \mu_{12} \qquad .....(6.15)$$

Water-glass interface. We can apply the above equations to obtain the following relations for water-glass refraction:

$$ {}^{w}\mu_g = \frac{\text{velocity of light in glass } (c_g)}{\text{velocity of light in water } (c_w)}$$

$$ {}^{w}\mu_g = \frac{1}{{}^{g}\mu_w} \quad ; \quad {}^{w}\mu_g = \frac{{}^{a}\mu_g}{{}^{a}\mu_w}$$

Observed change in height of objects under water. If a coin lying at the bottom of water in a pond is observed, it looks raised higher. In fact, the entire bottom appears to exist at a higher level. If a long stick is obliquely dipped in water, it appears as if it is broken in two parts at the water surface. Such distortions result from the refraction of light rays at the water surface, where the rays emerge from a dense medium (water) into a rare medium (air). In Fig. 6.9, the light rays emanating from a point object lying at the bottom of water at P_1 appear to be coming from a higher point P_2, to an observer outside water. Thus, to the observer the apparent depth of the object appears to be smaller than the real depth. It can be easily worked out that the apparent and real depths are related to the refractive index of water as,

$$\mu = \frac{\text{Real depth}}{\text{Apparent depth}} \qquad .....(6.16)$$

(Apparent means 'appearing to be real or true').

Example 6. A ray of light travels from air to water, with the angle of incidence at the water surface as 30°. Find (i) the angle of refraction and (ii) speed of light in water. (μ_{water} = 1.33 ; c = 3×10^8 ms^{-1}).

Solution. Given: i = 30°; μ = 1.33; c = 3×10^8 ms^{-1}

(i) We know,

$$\mu = \frac{\sin i}{\sin r}$$

$$\Rightarrow \quad \sin r = \frac{\sin i}{\mu} = \frac{\sin 30°}{1.33}$$

$$= 0.667$$

$$\therefore \quad r = \sin^{-1}(0.667) = 41.8° \quad \textbf{(Answer).}$$

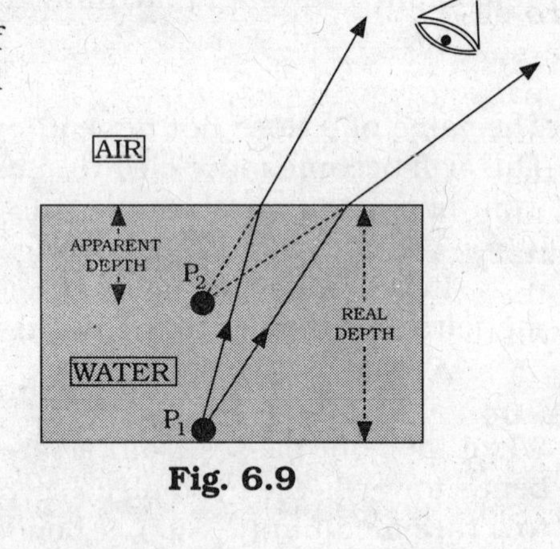

Fig. 6.9

(ii)

$$\mu = \frac{c}{v}$$

$$\Rightarrow \quad v = \frac{c}{\mu} = \frac{(3 \times 10^8)}{(1.33)} = 2.26 \times 10^8 \text{ ms}^{-1} \quad \textbf{(Answer).}$$

Example 7. Refractive indices of water and glass are $\frac{4}{3}$ and $\frac{3}{2}$, respectively. A ray travelling in water is incident on the water-glass interface at 30°. Calculate the angle of refraction. (ICSE; 1996).

Solution. Given: ${}^{a}\mu_w = \frac{4}{3}$; ${}^{a}\mu_g = \frac{3}{2}$; i = 30°

(i) We know,

$$\therefore \quad {}^{w}\mu_g = \frac{{}^{a}\mu_g}{{}^{a}\mu_w} = \frac{3/2}{4/3} = \frac{9}{8}$$

Next,

$$\frac{\sin i}{\sin r} = {}^{w}\mu_g = \frac{9}{8}$$

$$\Rightarrow \quad \sin r = \frac{8}{9} \sin i = \frac{8}{9} \times \frac{1}{2} = 0.4444$$

$$\therefore \quad r = \sin^{-1} 0.4444 = 26.4° \quad \textbf{(Answer).}$$

Example 8. The velocity of light in glass is 2×10^8 ms^{-1} and that in air is 3×10^8 ms^{-1}. By how much would an ink dot appear to be raised when covered by a glass plate 6.0 cm thick ? (PSEB; 1999).

Solution. Given: $v_g = 2\times10^8$ ms^{-1}; $v_{air} = 3\times10^8$ ms^{-1}; glass plate thickness = 6 cm

The refractive index of glass is,

$$\mu_g = \frac{v_{air}}{v_g} = \frac{(3\times10^8)}{(2\times10^8)} = \frac{3}{2}$$

Let, real depth = x and apparent depth = y. Given, x = 6 cm.

We know, $$\mu_g = \frac{\text{Real depth}}{\text{Apparent depth}} = \frac{x}{y}$$

$$\Rightarrow \quad y = \frac{x}{\mu_g} = \frac{(6)}{(3/2)} = 4 \text{ cm}$$

∴ Rise in the position of dot $= x - y = (6) - (4) = 2$ cm **(Answer).**

Observed change in time of sunrise/sunset; atmospheric refraction. The sun becomes visible on earth about 2 minutes before the actual sunrise, i.e. at the time when it actually rises above the horizon. Similarly, it remains visible for about 2 minutes after the actual sunset. These effects result from the refraction of sunlight through the earth's atmosphere. The atmosphere extends upto a few hundred kilometers above the earth's surface. It is most dense near the surface. The denseness gradually diminishes on going up, until it becomes zero on reaching the free space (vacuum). Hence the sun's rays reaching the earth's surface have to pass constantly from rare to dense medium while moving through the atmosphere. The result is that while they move straight in the free space, they continuously bend as they traverse the atmosphere. To an observer on the earth they appear to be coming from a position above the horizon, while the sun is actually under the horizon, as shown in Fig. 6.10. Thus the apparent sunrise is slightly advanced in time. At the sunset, for the same reason the sun continues to be visible for some time, in spite of it having gone down the horizon. So the apparent sunset is delayed in time.

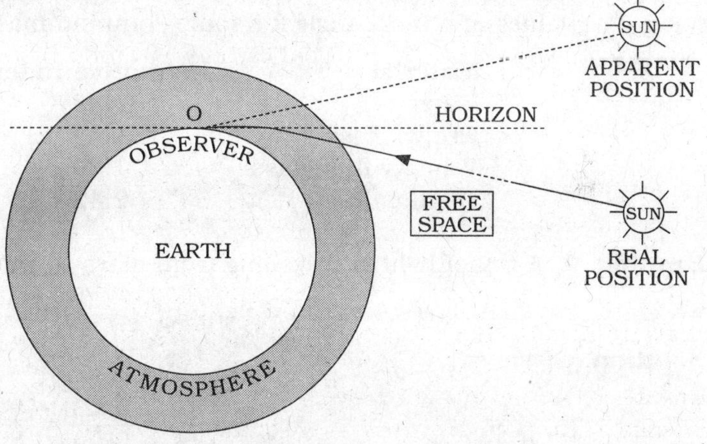

Fig. 6.10

6A. 04. Total Internal Reflection; Critical Angle; Some Applications

The Fig. 6.11 shows refraction of a few rays of light, starting from a point object P, from glass to air. Since the light passes from a dense to a rare medium, the refracted ray bends away from the normal, as seen for the rays marked 2,3. So the angle of refraction r is greater than the angle of incidence i. A small part of the light is reflected back into the dense medium. This happens until the angle $r<90°$. When $r = 90°$, the refracted ray grazes the surface of separation (interface), as seen for the incident ray marked 4. The corresponding angle of incidence i_c is called **critical angle.** When the angle of incidence exceeds i_c ($i>i_c$), as for the ray marked 5, the whole of the incident light is reflected into the dense medium. This phenomenon is called **total internal reflection.** We can **define** it as "the phenomenon in which the light is totally reflected at the interface into the dense medium, when passing towards a rare medium". The **critical angle** may be defined as "that value of the angle of incidence beyond which the light is totally reflected at the interface into the dense medium, when passing towards a rare medium".

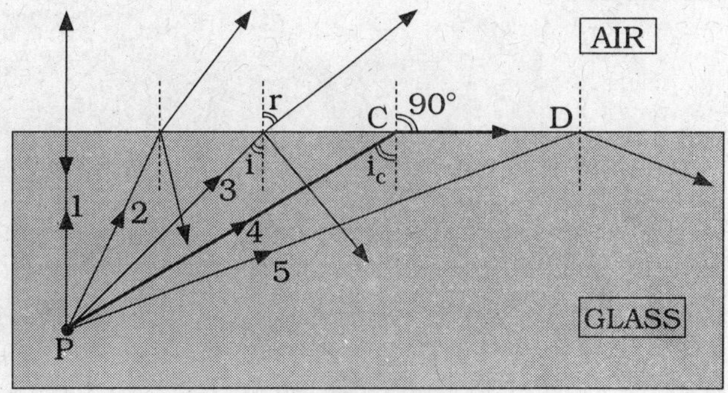

Fig. 6.11

Relation between Refractive Index (μ) and critical angle (i_c). From the Eq. 6.11 for Snell's law, when light passes from a medium of R.I. μ_1 to a medium of R.I. μ_2, we have

$$\mu_1 \sin i = \mu_2 \sin r$$

If i = i_c (critical angle), r = 90°. So we have,

$$\mu_1 \sin i_c = \mu_2 \sin 90° = \mu_2$$

$$\therefore \quad \sin i_c = \frac{\mu_2}{\mu_1} \qquad .....(6.17)$$

If light passes from a medium of R.I. μ to air, we have $\mu_1 = \mu$ and $\mu_2 = 1$. Then

$$\sin i_c = \frac{1}{\mu} \quad \text{(Refraction into air)} \qquad .....(6.18a)$$

Or

$$i_c = \sin^{-1} \frac{1}{\mu} \qquad .....(6.18b)$$

Essential conditions for total internal reflection

(i) Light must travel from a dense medium to a rare medium ($\mu_1 > \mu_2$).

(ii) The angle of incidence must be greater than the value of the critical angle ($i > i_c$).

The values of critical angle for some common materials are given below:

Material	Refractive Index	Critical Angle
Water	1.33	48.8°
Crown glass	1.52	41.1°
V. heavy flint glass	1.89	31.9°
Diamond	2.42	24.4°

Example 9. A ray of light is travelling from glass to water. Determine the minimum angle of incidence for which the ray does not emerge in water. ($\mu_{glass} = \frac{3}{2}$; $\mu_{water} = \frac{4}{3}$).

Solution. Given: ${}^a\mu_g = \frac{3}{2}$; $\quad {}^a\mu_w = \frac{4}{3}$

$$\therefore \quad {}^w\mu_g = \frac{{}^a\mu_g}{{}^a\mu_w} = \frac{(3/2)}{(4/3)} = \frac{9}{8}$$

The minimum angle of incidence for which the incident ray fails to emerge in water is obviously the critical angle i_c for the glass-water interface. We know,

$$\sin i_c = \frac{1}{{}^w\mu_g} = \frac{1}{(9/8)} = \frac{8}{9}$$

$$= 0.8889$$

$$\therefore \quad i_c = \sin^{-1}(0.8889)$$

$$= 62.7°$$ **(Answer).**

Example 10. A small bulb is placed at the bottom of a tank containing water to a depth of 80 cm. What is the area of the surface of water through which light from the bulb can emerge out ? Refractive index of water is 1.33. Consider the bulb to be a point source. (NCERT).

Solution.

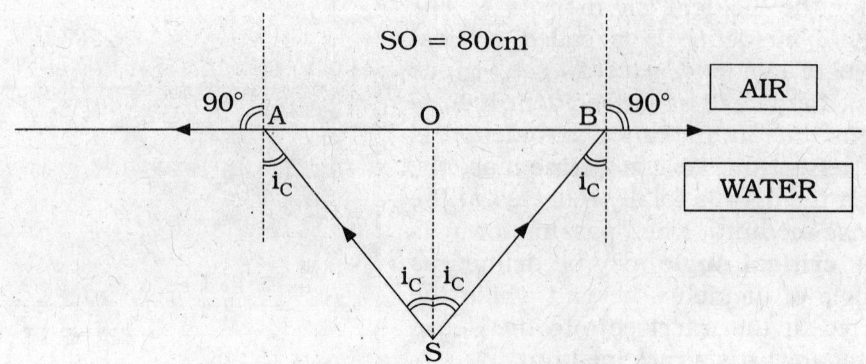

The light source (S) lies 80 cm below the water surface,

$$SO = 80 \text{ cm} = 0.8 \text{ m}$$

The light from the bulb will be able to emerge from water only so long as the angle of incidence is less than the critical angle, i.e. $i<i_c$. When $i \geq i_c$, the light will fail to emerge from water (it will be totally reflected back into the water). As seen in the figure, the area of emergence of light on the water surface is a circle of diameter AB. Its radius is,

$$r = \frac{AB}{2} = OA = OB$$

We know,
$$\mu = \frac{i}{\sin i_c}$$

$$\therefore \quad \sin i_c = \frac{1}{\mu} = \frac{1}{1.33} = 0.75$$

$$\Rightarrow \quad i_c = \sin^{-1}(0.75) = 48.6°$$

In ΔOBS,
$$\tan i_c = \frac{OB}{OS}$$

$$\Rightarrow \quad OB = OS \tan i_c = (0.8) \times \tan(48.6°)$$

$$= 0.8 \times 1.1345 = 0.907 \text{ m}$$

$\therefore$ The required water surface area is,

$$A = \mu r^2$$

$$= 3.14 \times (0.907)^2$$

$$= 2.58 m^2 \quad \textbf{(Answer)}.$$

Additional knowledge

Even the best reflecting surfaces, e.g. a well-polished mirror, does not reflect 100% light incident upon it, because a small part of the incident light is always absorbed at the reflecting surface to produce heat. But in the total internal relflection, the entire amount of the incident light is reflected at the interface of the two media. So the total internal reflection provides the most efficient reflection of light.

Some applications of total internal reflection

(i) Totally reflecting prisms. The critical angle (i_c) for the common crown glass is nearly 41°. So glass prisms in suitable shapes can be made for various purposes, using the phenomenon of total internal reflection, such that the angle of incidence of light at the reflecting surface is always greater than i_c. The glass prisms shown in Figs. 6.12a, b, c achieve the following purposes, using a right-angled prism.

(a) Deviate the light through 90°.
(b) Deviate the light through 180°.
(c) Invert an image without deviation.

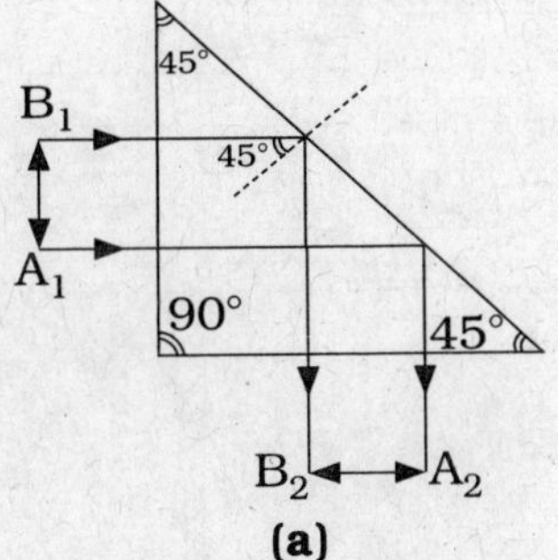

(a)

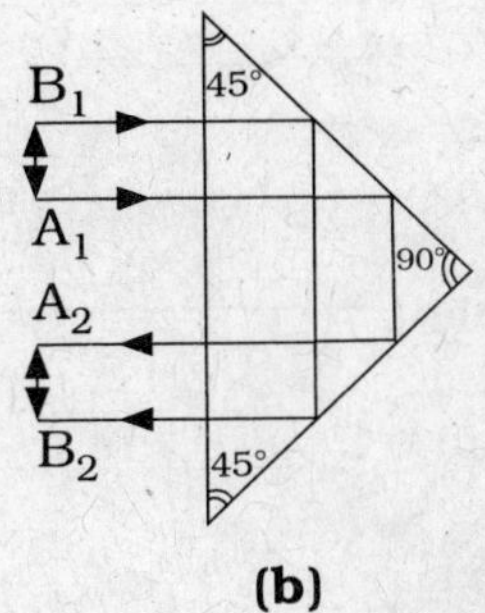

(b)

B1 90° A2
A1 B2
45° 45°
(c)

Fig. 6.12

(ii) Optical fibres; Fibre optics. A modern versatile use of the phenomenon of internal reflection is made through the so-called **optical fires.** The corresponding branch of physics is known as **fibre optics.** An optical fibre is a very thin fibre made of glass, quartz or transparent plastic, with thickness of the order of 10^{-3} mm. It is coated with a thin film of a material of R.I. less than that of the material of the fibre. The principle of transmission of light through such a fibre is illustrated in Fig. 6.13a, which shows a small piece of the fibre. The light incident at one end of the fibre undergoes multiple total internal reflections inside the fibre, before emerging at the other end (although for the sake of clarity the figure shows a large angle of incidence, actually the light is incident almost at a glancing angle, because of the extremely small thickness of the fibre). The total reflection occurs because the angle of incidence is greater than the critical angle ($i>i_c$).

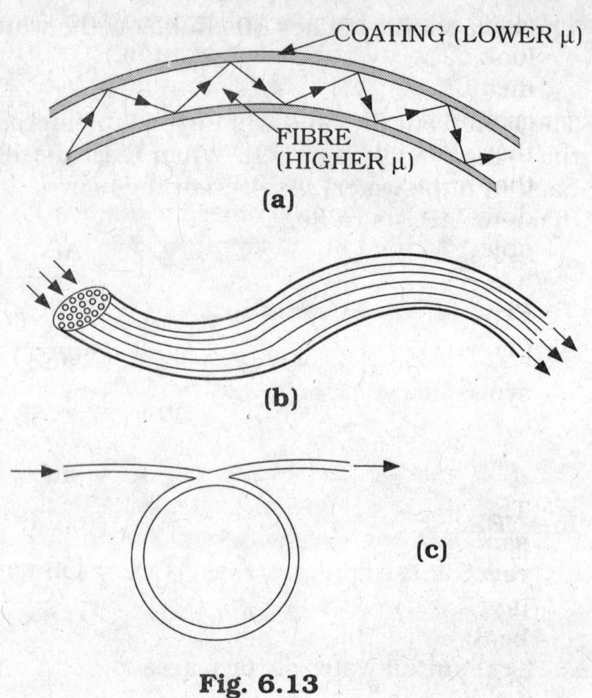

Fig. 6.13

A thin bundle of fibres form the so-called **light pipe** (Fig. 6.13 b). Even if the light pipe is twisted or bent in a complex shape, the light is easily transmitted through it (Fig. 6.13c). There is little loss of intensity of light in travelling through a fibre (over distances of even several kilometers), because of total internal reflections.

The light pipes have proved extremely useful in medical examination (endoscopy). The parts of the human body which are normally inaccessible to seeing, e.g. inside of the stomach, can be viewed by inserting two light pipes in the concerned part. The light beam sent through the outer end of one pipe illuminates the part. A portion of the light scattered by the part travels through the other pipe to give us the image of the part at the outer end of the pipe. The image can be suitably taken on a screen, too, for a convenient and quick examination by the doctor.

Another very important application of the optical fibres is their use in transmitting communication signals, e.g. telephone signals. For this purpose, the electrical signals are first converted into optical signals through suitable transducers. A much greater number of signals (a few thousands) can be transmitted in this way, than what was earliar possible through conventional means. Also, the clarity of the signals is comparatively much better.

(iii) Sparkling of diamonds. Because of the high refractive index (2.42) of diamond, the values of critical angle of diamonds is very low, viz. nearly 24°. Therefore, once a light ray enters a diamond crystal, it is likely to undergo multiple internal reflections at the various cut faces, before coming out. The total reflection of the light at the faces is ensured by suitably cutting the faces of the diamond. Note that all the light rays incident within a range of nearly 25° to 90° will be totally reflected. Normally, when light falls on a crystal face, only a small percentage of it is reflected. Most of it is refracted into the other medium. But when multiple consecutive total internal reflections occur, an incident light beam is virtually trapped inside the diamond. The result is than when it finally emerges from the crystal, it is a strong beam of light (almost as strong as it entered). Thus a diamond crystal sparkles brilliantly, when exposed to light.

The cut glass from which decoration pieces are made also has a high refractive index, achieved by adding a material like lead to ordinary glass.

(iv) Mirage. The mirage is an optical illusion commonly caused in hot deserts, in which an observer on the ground sees a pool of water at some distance in front of him, which really does not exist.

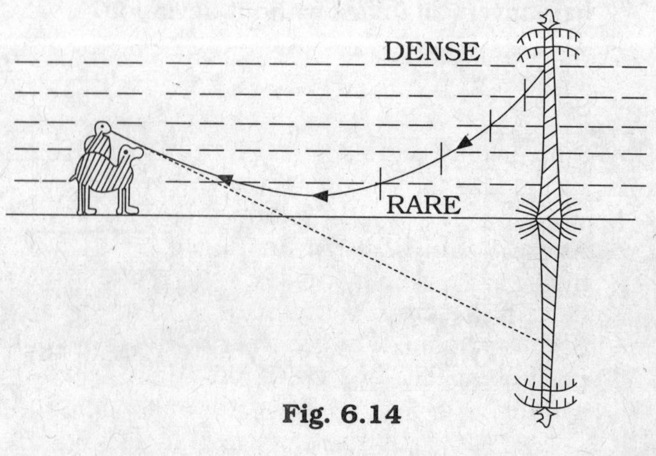

Fig. 6.14

The air near the ground gets heated in summer and becomes hotter than the air upwards. The hotter is the air the less dense it is.. Thus the density of the air is minimum near the ground and gradually increases as one moves up. Accordingly, when an observer looks at a distant tall object (e.g. a tree), the light rays from the object initially travel from dense to rare medium. So they are continuously bent away from the normal, until arriving at such a position above the ground where the angle of incidence exceeds the critical angle ($i>i_c$). Then the rays are totally reflected. Now they start moving from rare to dense medium, and so they are continuously bent towards the normal, until they are received by the observer. To the observer's eyes they appear to be coming from a position vertically downwards from the object, as shown Fig. 6.14. So to the observer it appears to be an inverted image of the object formed in a pool of water.

Principle of Reversibility of Light Rays

It is to be noted that the equations involved in the laws of reflection and refraction are perfectly symmetrical in respect of the values of angles and refractive indices,

$$\angle i = \angle r$$

$$\mu_1 \sin i = \mu_2 \sin r$$

Thus, if the reflected ray becomes the incident ray, the incident ray will become the reflected ray. The same holds for the refracted ray and the incident ray. These considerations readily lead to the **principle of reversibility of light rays,** which states that "if a reflected or refracted ray of light be reversed in direction, it will retrace its original path". The principle holds for all light systems, howsoever complicated they may be, because it applies to each reflecting or refracting surface in the system. Therefore this simple principle is of great value in optics.

6A. 05. Refraction through a Prism; Dispersion of Light; Spectrum; Rainbow

The Fig. 6.15 shows the cross-section of a triangular prism ABC, placed in air. The $\angle BAC$ is the angle of the prism. A ray of light PQ is incident at an angle i_1 on the face AB and refracts into the prism (air to glass) at an angle r_1. It is incident at an angle r_2 on the face AC and refracts out (glass to air) at an angle i_2, called the angle of emergence. The angle through which the incident ray deviates (i.e. bends), i.e. the angle between the incident ray PQ and the emergent ray RS, is called angle of deviation (δ).

Fig. 6.15

From ΔTQR, we have,

$$\delta = \angle TQR + \angle TRQ$$

$$= (i_1 - r_1) + (i_2 - r_2)$$

$$= (i_1 + i_2) - (r_1 + r_2) \quad ..... \text{(i)}$$

In the quadrangle APOR, the sum of its four angles = 360°. But because $\angle AQO = \angle ARO = 90°$, we have,

$$A + \angle QOR = 180°$$

In the triangle QOR, the sum of the three angles = 180°,

$$r_1 + r_2 + \angle QOR = 180°$$

So,

$$r_1 + r_2 = A \quad ..... (6.19)$$

Substituting in Eq.(i), we get,

$$\delta = (i_1 + i_2) - A \quad ..... (6.20)$$

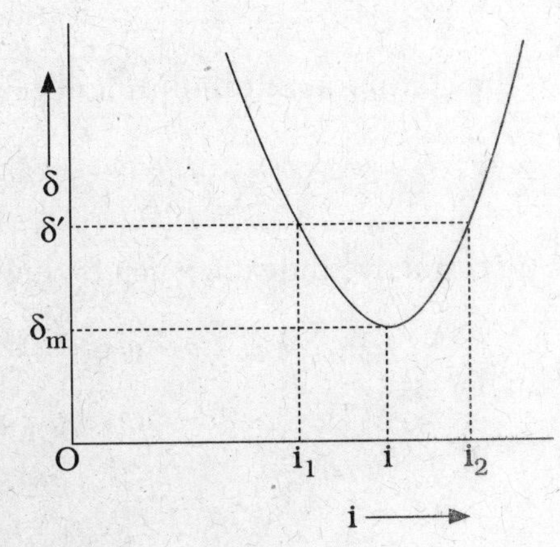

Fig. 6.16

Angle of minimum deviation. In the Eq. 6.20, the angle i_2 is determined by the angle of incidence i_1. It follows that the angle of deviation δ is also determined by i. If a plot of i vs. δ is made, the curve shown in Fig. 6.16 is obtained. In general, for a given value of δ, say δ', there are two possible values of i giving the same value δ'. It is easily seen from the symmetry of Eq. 6.14 that these values are i_1 and i_2, because they are interchangeable. Now, it is seen from Fig. 6.16 that when $i_1 = i_2 = i$ (say), the value of δ is minimum. It is called the **angle of minimum deviation (δ_m)**. Then the refracted ray within the prism passes parallel to the base of the prism, as seen in Fig. 6.15.

Relation between refractive index (μ) and the angle of minimum deviation (δ_m). At the position of minimum deviation, $i_1=i_2=i$. Then the ray QR inside the prism is parallel to the base of the prism, as actually drawn in Fig. 6.15. From the geometry of the figure, then we have, $r_1=r_2=r$ (say). From Eq. 6.20,

$$\delta_m = (i_1 + i_2) - A$$

$$= 2i - A$$

$$\Rightarrow \quad i = \frac{A+\delta_m}{2} \qquad \text{..... (ii)}$$

From Eq. 6.19,

$$r_1 + r_2 = 2r = A$$

Or

$$r = \frac{A}{2} \qquad \text{..... (iii)}$$

From the Eqs. (ii) and (iii), the refractive index of the material of the prism is given by

$$\mu = \frac{\sin i}{\sin r}$$

$$\Rightarrow \quad \mu = \frac{\sin [(A+\delta_m)/2]}{\sin (A/2)} \qquad \text{..... (6.21)}$$

If the angle of prism A is small, the angle δ_m is also small. Then we can write as an approximation ($\sin\theta \approx \theta$, for small θ),

$$\mu = \frac{(A+\delta_m)/2}{A/2}$$

$$\Rightarrow \quad \delta_m = (\mu - 1)\,A \qquad \text{..... (6.22)}$$

Example 11. A prism of angle 7° is made of a material of refractive index 1.64. Calculate the deviation produced by the prism in the path of a light ray passing through it.

Solution. Given: $A = 7°$; $\mu = 1.6$

For small values of A, the angle of deviation is given by,

$$\delta = (\mu - 1)\,A$$

$$= (1.64 - 1)\times 7° = 0.64\times 7°$$

$$= 4.48°$$ **(Answer).**

Example 12. A light ray is incident on a prism of angle 60°, at an angle of 52°. It refracts through the prism with minimum deviation. Determine the angle of minimum deviation and the refractive index of the prism material.

Solution. The angle of prism (A), angle of deviation (δ), angle of incidence (i_1), and angle of refraction (i_2) are related as,

$$\delta = (i_1 + i_2) - A$$

In the condition of minimum deviation ($\delta=\delta_m$), $i_1=i_2$. So we have (Given: $i_1 = 52°$; $A = 60°$),

$$\delta_m = (52° + 52°) - (60°)$$

$$= 104° - 60° = 44°$$ **(Answer).**

The refractive index is given by the prism formula,

$$\mu = \frac{\sin [(A+\delta_m)/2]}{\sin (A/2)} = \frac{\sin [(60+44)/2]}{\sin (60/2)}$$

$$= \frac{\sin 52°}{\sin 30°} = \frac{0.7880}{0.5}$$

$$= 1.58$$ **(Answer).**

Dispersion of light; Spectrum; Rainbow

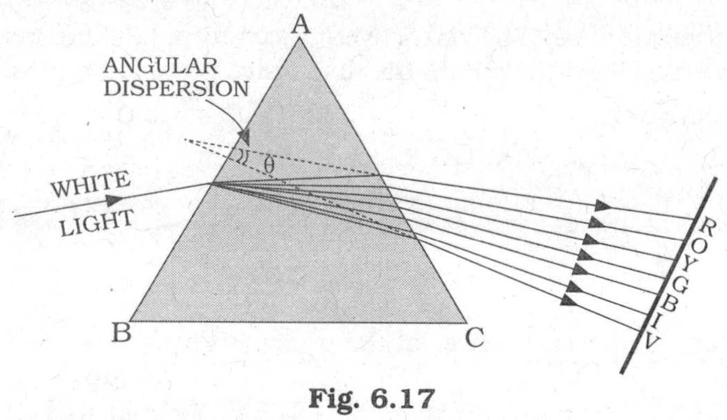

Fig. 6.17

The separation of light into its constituent colours on passing through a medium is called **dispersion** (or chromatic dispersion) of light ('disperse' means to distribute in different directions or over a wide area'; the Greek word KHROMA means 'colour). The medium causing the dispersion is called a **dispersive medium.** The dispersion of light produced by a prism is shown in Fig. 6.17. The pattern of colour produced by the dispersion of light is called **spectrum** of light. It is given by the sequence VIBGYOR (Violet, Indigo, Blue, Green, Yellow, Orange, Red). Other electromagnetic radiations, e.g. infrared, ultraviolet, X-rays, etc., have their own spectra (remember that light forms the visible region of the e.m. radiation).

The dispersion of light is caused by the fact that different wavelengths of light travel with different velocities in a material medium. The frequency of light does not change on passing through the medium. So, according to the relation $\nu\lambda = v$, the velocity becomes different for different λ's. Since the refractive index depends upon velocity as,

$$\mu = \frac{c}{\nu_{medium}},$$

the refractive index of the material of the prism is different for different λ's. The relation between μ and λ is given by the so-called Cauchy's formula (A,B,C are constants),

$$\mu = A + B/\lambda^2 + C/\lambda^4$$

Thus, higher is the wavelength λ, lower is the μ-value. Since $\lambda_{red} \sim 750$ nm and $\lambda_{violet} \sim 350$ nm, we have

$$\lambda_V < \lambda_R$$

and $$\mu_V > \mu_R$$

(For crown glass, $\mu_V \sim 1.53$ and $\mu_R \sim 1.51$).

Thus, on refracting through a prism, the violet light, with lowest μ, bends the most and the red light bends the least, as seen the Fig. 6.17.

The refractive index for all the wavelengths in vacuum is unity (1). So there is no dispersion of light in the vacuum. Accordingly, vacuum is called a non-dispersive medium.

Angular dispersion. The angular separation θ (Fig. 6.17) between the extreme violet and red colours of the spectrum is called **angular dispersion.**

Mathematically, it is given by the difference between the angles of deviation, for the extreme violet and red wavelengths, i.e. as $(\delta_V - \delta_R)$. Since $\delta = (\mu - 1)A$, we have,

$$\text{Ang. Dispersion} = (\delta_V - \delta_R) = (\mu_V - \mu_R)A \quad \text{..... (6.23)}$$

Note that the angular dispersion depends on the angle of the prism.

Example 13. A ray of white light is incident on a prism of angle 7°. The refractive indices of the prism material for violet and red colours are 1.685 and 1.644, respectively. Find the angular dispersion produced by the prism.

Solution. Given: $\mu_v = 1.685$; $\mu_r = 1.644$; $A = 7°$

$$\therefore \text{Angular dispersion} = (\mu_v - \mu_r)\times A$$
$$= (1.685 - 1.644)\times 7°$$
$$= 0.041\times 7°$$
$$= 0.29°$$ **(Answer).**

Dispersive power. The **dispersive power (ω)** of the material of a prism is **defined** as the ratio of angular dispersion to the angle of deviation for the mean wavelength (yellow colour),

$$\omega = \frac{\delta_V - \delta_R}{\delta_Y} = \frac{\mu_V - \mu_R}{\mu_Y - 1} \quad \text{..... (6.24)}$$

The dispersive power is independent of the angle of the prism, but depends upon the material of the prism.

Example 14. The refractive indices for violet and red light for crown glass are $\mu_v = 1.532$ and $\mu_r = 1.515$, respectively. Calculate the dispersive power of a prism made of crown glass.

Solution. Given: $\mu_v = 1.532;\quad \mu_r = 1.515$

∴ The mean refractive index is given by,

$$\mu = \frac{\mu_v + \mu_r}{2} = \frac{(1.532) + (1.515)}{2}$$

$$= \frac{3.047}{2} = 1.524$$

The dispersive power of the prism is given by,

$$\omega = \frac{\mu_v - \mu_r}{\mu - 1} = \frac{(1.532) - (1.515)}{(1.524) - 1}$$

$$= \frac{0.017}{0.524} = 0.032 \quad \textbf{(Answer).}$$

Rainbow. A rainbow is a most common and charming example of the nature's chromatic dispersion of light. It is seen when rain is falling at or near to the place of observation.

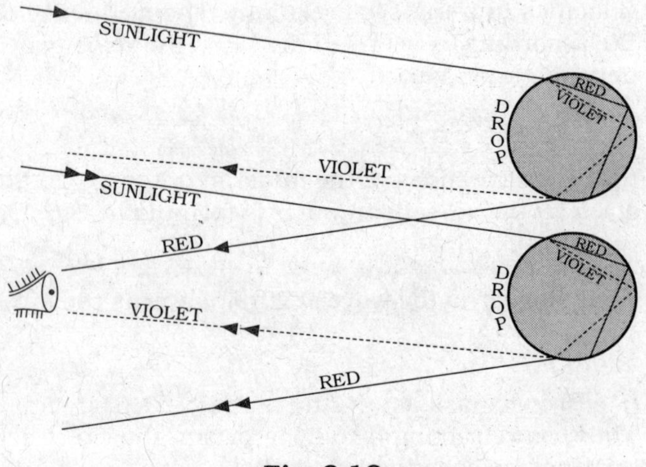

Fig. 6.18

When a falling raindrop passes through the path of white sunlight, a part of the sunlight refracts into the drop, gets reflected from the inner surface of the drop, and then refracts out of the drop. This is schematically shown for two water drops in Fig. 6.18. As for a prism, the first refraction disperses the sunlight, i.e. it separates the sunlight into its constituent colours (VIBGYOR), and the second refraction enhances the dispersion. Since $\mu_{violet} > \mu_{red}$, the refraction (bending) is maximum for the violet light and least for the red light. The intermediate colours bend through intermediate angles. The observer's eye gathers light of different colours from different water drops, e.g. lights of red and violet colour from two different drops in the Fig. 6.18. The collective light coming from the innumuable water drops (millions of them) produces a cone of rays at the observer's eye, as partly seen for two rays in the figure. Hence the rainbow looks like an arc of a circle to the observer on earth. If the rainfall is heavy, the sun is bright and the observer is suitably positioned (at an elevated position), he may see the full circle, extending from one end of the horizon to the next.

The rainbow so formed is called **primary rainbow.** It is spread over about 2°, with the red on the outside. It makes a mean angle of about 41° with the line drawn from the sun through the head of the observer. Note that to be able to see the rainbow, the observer should have his back towards the sun. Each observer gets refracted light from a particular set of water drops (millions of them). Thus each observer's rainbow is a personal one!

Sometimes a **secondary rainbow** is also formed, if the background is dark enough. It has inverted colours w.r.t. the primary rainbow, i.e. it has violet colour on the outside. It is spread over about 3° and makes a mean angle of about 52° with the line drawn from the sun. It is fainter than the primary rainbow, because it is formed by the rays refracting out of the water drops after two internal reflections, unlike the primary rainbow that results from the rays that suffer single internal reflection (Fig. 6.18). Rarely, if the background is very dark and other conditions are favourable a very faint **tertiary rainbow** is also seen.

No rainbow can be seen by an observer at the moon, for the simple reason that unlike the earth the moon has no atmosphere existing around it.

6A. 06. Scattering of Light; Blue Colour of Sky; Red Sun at Sunrise and Sunset

Scattering of light. When light falls on an object (solid, liquid, or gas), each particle of the object deflects it in all possible directions around it. In other words, it 'scatters' the light all around it. This phenomenon is called **scattering** of light. The scattered intensity is not uniform in the various directions. The resultant light in a particular direction is determined by the collective contribution of the light scattered by all the particles together in that direction. All the fundamental optical phenomena, viz. reflection, refraction, diffraction, and polarization, are basically governed by this process. However, it is a very complex process, which is highly mathematical and so beyond the scope of a school student. So we shall not go into its details and present only a very brief rough qualitative picture.

A light wave incident on a transparent medium causes the electrons in the medium to oscillate periodically in response to the oscillating electric vector of the incident wave. Thus, forced oscillations of the same frequency as the incident light are produced and each electron becomes a secondary source of light radiation. Then the wave travelling through the medium is the resultant of the incident wave and the secondary waves from the oscillating electrons. In a solid, the resultant wave has maximum intensity in the forward direction, i.e. in the direction of the incident beam, which falls off vary rapidly on either side. But in a gaseous medium there is an appreciable sideways scattering. The reason is that the oscillating electrons in this case are not bound together in a rigid structure like a solid. They are separated by relatively large distances and so they act independently rather than cooperatively.

The nature of scattering depends on the size of the particles of the medium on which the light falls: (i) very fine particles (size $\ll \lambda$, the wavelength of light; $\lambda \sim 5\times10^{-4}$mm and (ii) fine particles (size $\gg \lambda$). For very fine particles, e.g. atoms and molecules, the so-called Rayleigh scattering governs the process. In the Rayleigh scattering, the scattered intensity is found to be inversely proportional to the fourth power of λ,

$$I_{scattered} \propto \frac{1}{\lambda^4} \qquad \text{..... (6.25)}$$

Thus, the scattered intensity falls of very rapidly with increase in wavelength, e.g. a 2-fold rise in λ brings about 16-fold fall of $I_{scattered}$. If we compare the scattered intensities for blue light ($\lambda \sim 450$ nm) and red light ($\lambda \sim 700$ nm), it works out to roughly six times higher scattering for the blue light.

For particles of relatively larger size (size $\gg \lambda_{light}$), e.g. dust particles, rain droplets and ice particles in clouds, etc., the intensity of the scattered light varies inappreciably with wavelength.

Blue colour of sky. The sunlight (which is white light) travelling towards the earth has to cross a few hundred kilometers of the earth's atmosphere, before reaching the earth. It gets scattered by the innumerable atoms and molecules which it meets on the way. Since the sizes of atoms and molecules $\ll \lambda_{light}$, the process is governed by Rayleigh scattering. As explained above, the scattered intensity is relatively very much higher for the blue light than for the red light. Besides, the sunlight is particularly rich in blue colour. Therefore, if we look above in a direction other than the incident light, i.e. in any direction different than towards the sun, we see a blue sky.*

Red colour of sun at sunrise and sunset. If an observer at O on the surface of earth looks towards the sun around noon time, when it is situated almost overhead, the sunlight just crosses the height OA of the atmosphere, before reaching the observer's eyes (Fig. 6.19). But if he looks towards the sun at the time of sunrise or sunset, when the sun is located in the line of horizon, the sunlight has to cross a few times more of the atmospheric distance, as seen in the figure (OB > OA). Since the earth's atmosphere extends over a height of nearly 500 km, it means that now the sunlight traverses a nearly 1000 km or more longer distance through the atmosphere. The result is that the light radiations of comparatively shorter wavelengths, for which the scattered intensity is high, are effectively removed from the line of view. Only the red height (of longest wavelength) is partly retained in the view. Hence the sun appears red at the time of sunrise and sunset. The reddish tinge is gradually reduced as the sun rises above the horizon and finally after some time the sun looks white, because then scattered light radiation of all wavelengths is received by the eye. The sun continues to look white in appearance, until about the time of sunset.

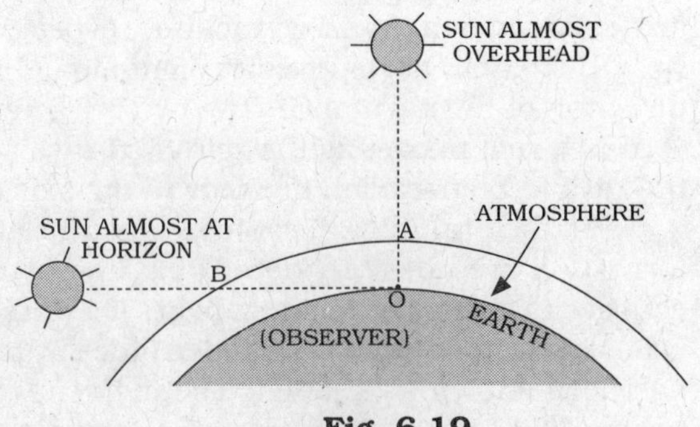

Fig. 6.19

*A careful spectral examination of the skylight reveals that in addition to the strong blue light, it also has other wavelengths. But their proportions are comparatiely so small that to our naked eyes the overall appearance is only blue.

Additional knowledge

(i) If the earth's atmosphere were not present, the sky would appear black. This has been amply verified now by the astronauts going up in the space, who find the sky turning increasingly dark, until it becomes totally black after a few tens of kilometers.

(ii) The scattering process responsible for the blue shade of the sky is also responsible for the generally bluish appearance of smoke (known as 'Tyndall effect'). The effect of particle size also manifests. For example, the smoke emitted from a lighted cigarette usually looks light blue (due to very fine size of particles in air), but looks whitish in the heavy puff from the smoker's mouth.

(iii) The clouds, which have droplets of water or ice particles in them, or the atmosheric dust, for both of which the particle size $\gg \lambda$, generally look whitish, because the scattered intensity varies inappreciably with wavelength. For the same reason, the sky looks deep blue on a clear day, but light blue on a humid day before rains. Similarly, in industrial cities or in cities with heavy atmosphric pollution, the sky has a dull and greyish appearance.

(iv) An electron in an atom or a molecule is bound by strong restoring forces. It has a definite natural frequency of oscillation. When light falls on such an electron, it sets up forced oscillations at the frequency of the incident light. The oscillations have a large amplitude if the external frequency matches with the natural frequency of the electron (resonance condition). The frequency of the blue light is closer to this natural frequency than that of the red light. Hence the blue light is far more effectively scattered.

For the same reason, the full moon when seen near the horizon has a reddish appearance. The red colour of traffic lights is helpful in foggy weather, as less light is lost by scattering, caused by the particles of the fog. Hence it can be seen from much longer distances. Similarly, the yellow light from the streetlight sodium lamps and the cars' headlight lamps is helpful in foggy weather, as it is less scattered as compared to the light from other kinds of strong intensity lamps, e.g. the blue light from a mercury lamp.

6A. 07. Refraction at Spherical Surfaces; Image Formation by Spherical Surfaces

In Section 6A.02, we have seen how images are formed by reflection from spherical mirrors. Now we shall examine how images are formed by refraction through surfaces of transparent materials, like glass. Again, for simplicity, we shall consider only spherical surfaces.

In Fig. 6.20, a convex spherical surface ST separates two transparent media of refractive indices μ_1 and μ_2, with $\mu_2 > \mu_1$ (e.g. air and glass; the shaded area may be imagined as a block of glass, with ST as one of its surfaces). C is the centre of curvature of the surface and R (= CP = CB) is the radius of curvature. Both CP and CB are normals to the surface at the points P and B, respectively. A light ray OB from a point object O incident at B moves along BI upon refraction at the surface (bending towards the normal CB). i and r are the angles of incidence and refraction, respectively, at B. Another ray OP incident at P moves straight along PI, since it is in the direction of the normal CP. The two refracted rays meet at I which, therefore, is the image of the point object O. It is a real image, as the refracted rays actually meet at I. As for the spherical mirrors, we shall assume the rays to be paraxial, i.e. near to the axis and making small angles with the axis.

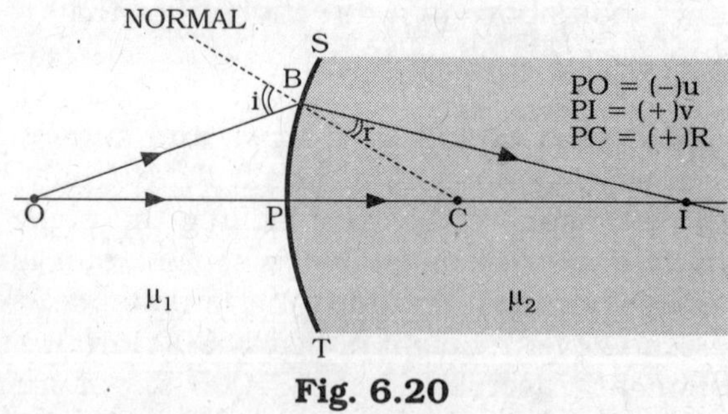

Fig. 6.20

In the Fig. 6.20, the spherical surface is convex towards incident light and $\mu_2 > \mu_1$. More cases of image formation are possible, depending upon whether $\mu_2 > \mu_1$ or $\mu_1 > \mu_2$ and whether the surface is convex or concave towards the incident light. They are depicted in Figs. 6.21 (a - e). In all the six cases, the point object O lies in the medium of refractive index μ_1 to the left of the surface and the medium of higher refractive index has been shown as shaded. The incident ray OB bends towards the normal if $\mu_2 > \mu_1$ and away from the normal if $\mu_1 > \mu_2$. If the refracted rays actually meet after refraction, the image formed is real (Figs. a and b). If they only appear to meet after refraction, the image I is virtual (Figs. c,d,e,f). Note that in Figs. a,b,c,d the image formed is real if the object O lies relatively far from the surface and virtual if O is relatively near to the surface. But in Figs. e,f the image is virtual for all positions of O. Further, the virtual images are formed on the same side of the surface as the object and the real images on the opposite side.

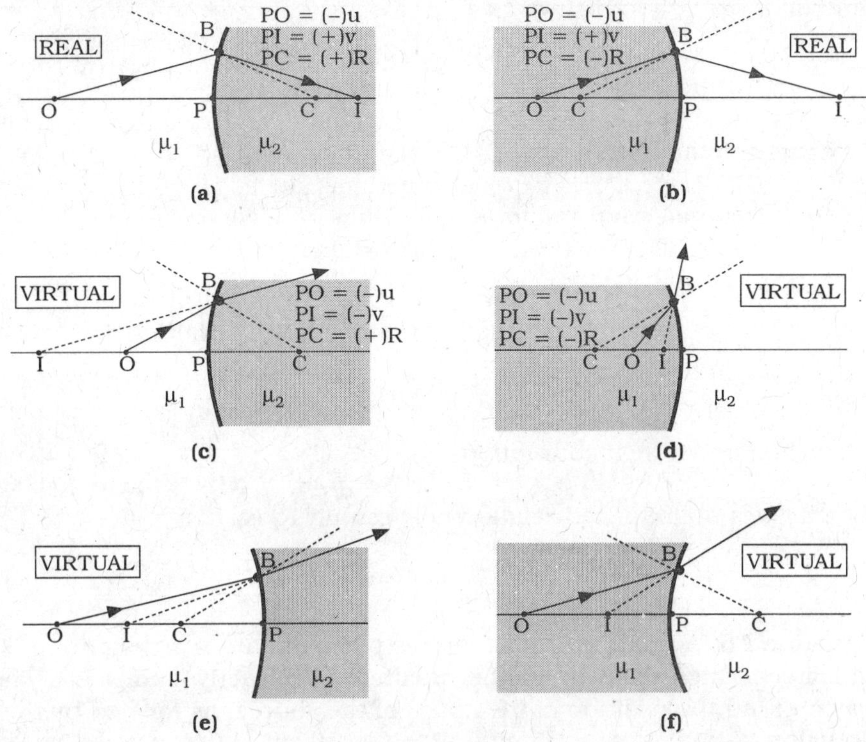

Fig. 6.21

Sign Convention

The same sign convention as employed for the spherical mirrors, viz. the **new Cartesian sign convention,** will be used for the spherical surfaces. The incident rays are taken as travelling from left to right. All the distances are measured starting from the pole of the surface (the point P in Figs, 6.20 and 6.21). The distances u,v,R measured from left to right (i.e. in the direction of the incident ray) are +ve and from right to left are –ve. As before, the radius of curvature (R) turns out as –ve for a concave surface and +ve for a convex surface. The signs of u,v,R for 5 cases for image formation, shown in the Figs. 6.20 and 6.21 (a,b,c,d), have been indicated on the ray diagrams.

Relation between Object Distance (u), Image Distance (v) and Radius of Curvature (R)

The relation between u,v,R for all the cases of image formation by spherical surfaces (Fig. 6.21) is found to be,

$$-\frac{\mu_1}{u}+\frac{\mu_2}{v}=\frac{\mu_2-\mu_1}{R} \quad \text{(Spherical surfaces)} \qquad \text{..... (6.26)}$$

We shall derive the relation for two cases, viz. (i) refraction from a rare to a dense medium for a convex surface (real image) and (ii) refraction from a dense to a rare medium (virtual image). The student can similarly derive the relation for the rest of the four cases of image formation.

(i) Refraction from Rare to Dense; Convex Surface (Real Image)

This case is depicted in Fig. 6.22, showing formation of image I of a point object O. $\mu_2 > \mu_1$.

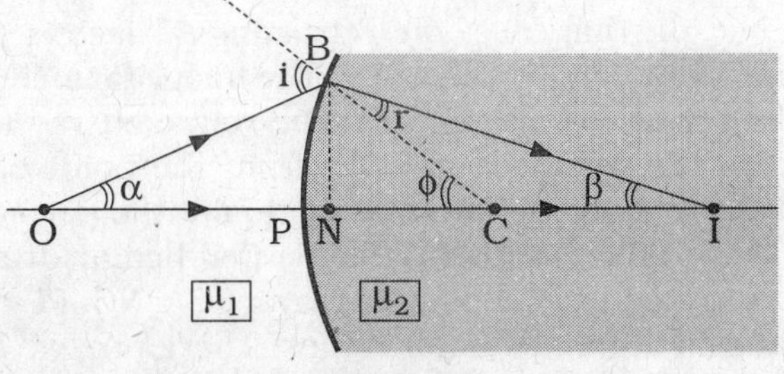

Fig. 6.22

BN is the perpendicular drawn from B upon the axis. For paraxial rays (so that B is near to the axis), PO ≃ NO, PI ≃ NI, and PC ≃ NC. Further, for paraxial rays, the angles α,β,ϕ are small values. Then we have ($\tan\theta \simeq \theta$, for small θ-values),

$$\tan\alpha \simeq \alpha = \frac{BN}{NO}$$

$$\tan\beta \simeq \beta = \frac{BN}{NI}$$

$$\tan\phi \simeq \phi = \frac{BN}{NC}$$

From the ΔBOC, we have (exterior angle i = sum of the interior angles α and ϕ),

$$i = \alpha + \phi = \frac{BN}{NO}+\frac{BN}{NC}$$

Similarly, from the ΔBCI we have,

$$\phi = r + \beta$$

$$\Rightarrow \qquad r = \phi - \beta = \frac{BN}{NC} - \frac{BN}{NI}$$

According to the Snell's law,

$$\mu_1 \sin i = \mu_2 \sin r$$

Since, in general, $\sin\theta \simeq \theta$, for small values of θ, we have,

$$\mu_1 i = \mu_2 r$$

$$\Rightarrow \qquad \mu_1\left(\frac{BN}{NO} + \frac{BN}{NC}\right) = \mu_2\left(\frac{BN}{NC} - \frac{BN}{NI}\right)$$

$$\Rightarrow \qquad \frac{\mu_1}{NO} + \frac{\mu_2}{NI} = \frac{\mu_2 - \mu_1}{NC}$$

According to our sign convention,

$$NO = (-)u, \quad NI = (+)v, \quad NC = (+)R$$

Which gives the required relation between u,v,R as,

$$-\frac{\mu_1}{u} + \frac{\mu_2}{v} = \frac{\mu_2 - \mu_1}{R}$$

Example 15. A point source of light is placed in air at a distance of 1.5m from a convex spherical surface, which separates air from another medium of refractive index 1.4. The radius of curvature of the surface is 30 cm. Find the position of the image of the source, as formed by the surface.

Solution. Given: $u = 1.5$ m; $R = (+)\ 30$ cm $= 0.3$ m; $\mu_1 = 1$; $\mu_2 = 1.4$

We know,

$$-\frac{\mu_1}{u} + \frac{\mu_2}{v} = \frac{\mu_2 - \mu_1}{R}$$

$$\Rightarrow \qquad -\frac{(1)}{(-1.5)} + \frac{(1.4)}{v} = \frac{(1.4)-(1)}{0.3} = \frac{4}{3}$$

$$\Rightarrow \qquad \frac{1.4}{v} = \frac{4}{3} - \frac{2}{3} = \frac{2}{3}$$

$$\Rightarrow \qquad v = \frac{3\times 1.4}{2} = 2.1\text{m}$$

Thus, the image is formed at a distance of 2.1m from the pole of the surface, in the medium of refractive index 1.4. It is a real image. **(Answer).**

(ii) Refraction from Dense to Rare; Concave Surface (Virtual Image)

In this case, the refracting surface is concave towards the incident light and $\mu_1 > \mu_2$. The image formed is virtual (Fig. 6.23), found by extending back the refracted ray to the point I on the axis (the incident ray OB, from the point object O, bends away from the normal after refraction at the surface).

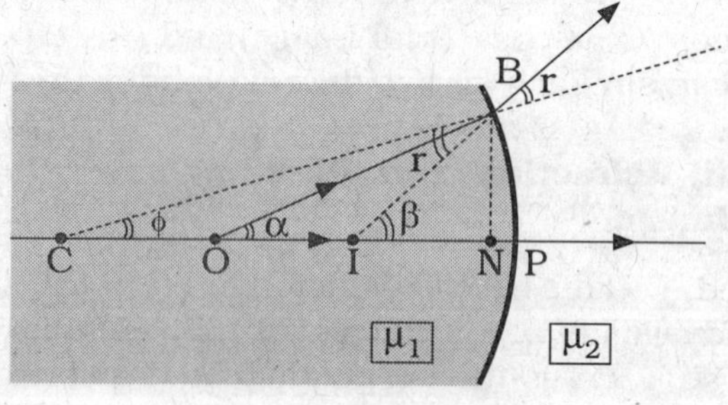

Fig. 6.23

As before, BN is the perpendicular drawn from B upon the axis. Further, for paraxial rays, PO ≃ NO, PI ≃ NI, PC ≃ NC and the angles α,β,φ are small values. Therefore, we have,

$$\tan\alpha \simeq \alpha = BN/NO$$

$$\tan\beta \simeq \beta = BN/NI$$

$$\tan\phi \simeq \phi = BN/NC$$

From the ΔBOC, we have,

$$\alpha = i + \phi \qquad \text{or} \qquad i = \alpha - \phi = \frac{BN}{NO} - \frac{BN}{NC}$$

From the ΔBIC, we have,

$$\beta = \phi + r \quad \text{or} \quad r = \beta - \phi = \frac{BN}{NI} - \frac{BN}{NC}$$

From Snell's law,

$$\mu_1 \sin i = \mu_2 \sin r$$

For small values of angles, in general $\sin\theta \simeq \theta$. Therefore we have,

$$\mu_1 i = \mu_2 r$$

$$\Rightarrow \quad \mu_1\left(\frac{BN}{NO} - \frac{BN}{NC}\right) = \mu_2\left(\frac{BN}{NI} - \frac{BN}{NC}\right)$$

$$\Rightarrow \quad \frac{\mu_1}{NO} - \frac{\mu_2}{NI} = \frac{\mu_1 - \mu_2}{NC}$$

According to our sign convention,

$$NO = (-)u, \quad NI = (-)v, \quad NC = (-)R$$

Which again gives the required relation between u,v,R as,

$$-\frac{\mu_1}{u} + \frac{\mu_2}{v} = \frac{\mu_2 - \mu_1}{R}$$

Example 16. A concave surface bounding a medium of refractive index 1.5 separates the medium from air on the other side. If a virtual image of a point object, placed in the denser medium 50 cm away from the surface, is to be formed at a distance of 40 cm from the surface, what should be the radius of curvature of the surface ?

Solution. Given: $u = -50\text{cm}; \quad v = -40\text{cm}; \quad \mu_1 = 1.5 = 3/2; \quad \mu_2 = 1$

We know, $$-\frac{\mu_1}{u} + \frac{\mu_2}{v} = \frac{\mu_2 - \mu_1}{R}$$

$$\Rightarrow -\frac{3/2}{(-50)} + \frac{1}{(-40)} = \frac{(1) - (3/2)}{R} = -\frac{1}{2R}$$

$$\Rightarrow \frac{1}{2R} = \frac{1}{40} - \frac{3}{100} = \frac{5-6}{200} = -\frac{1}{200}$$

$$\Rightarrow R = -100\text{cm} \quad \textbf{(Answer).}$$

The –ve sign results from the concave shape of the surface.

6A. 08. Refraction through Thin Spherical Lenses; Various Types of Lenses

A **lens** is "a transparent object bounded by two surfaces, of which at least one is a curved surface." In the so-called **spherical lenses,** the curved surfaces are spherical in shape. These lenses are most commonly used. But lenses in other shapes are also made, e.g. cylindrical lenses. Our study will be confined to spherical lenses. They can have six possible shapes, as shown in Fig. 6.24: (a) double convex, (b) plano-convex, (c) concavo-convex, (d) double concave, (e) plano-concave, and (f) convexo-concave.

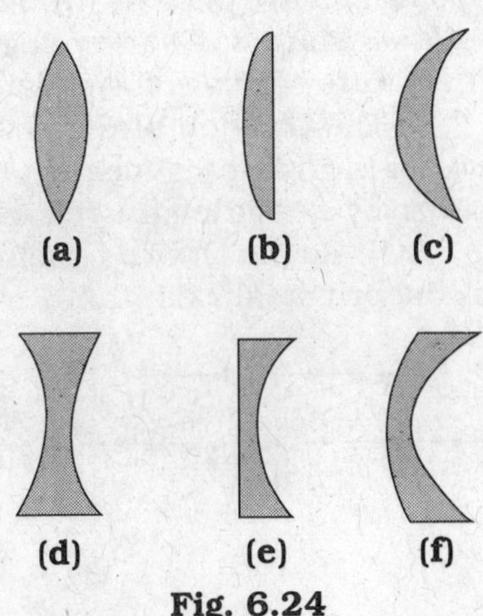

Fig. 6.24

The Fig. 6.25(a) shows how generally a ray of light refracts when incident upon a convex lens placed in air. At the point of incidence A, the ray goes from a rare medium to a dense medium. So it bends towards the normal, drawn at the point A. At the point of emergence B it goes from a dense to a rare medium and so it bends away from the normal. Finally, the ray **converges** towards the point I. The Fig. 6.25(b) shows the refraction through a double concave lens. Here finally the incident ray **diverges** upon emerging from the lens. The emergent ray appears to be coming from a point I.

In general, we have two **radii of curvature R_1 and R_2** and two **centres of curvature C_1 and C_2** for a lens. The line connecting C_1, C_2 and the centre P of the lens is called **principal axis** or **central axis**) of the lens (Fig. 6.25).

A lens that causes incident rays parallel to the principal axis to converge after refraction is called a **converging lens** (Fig. 6.26 a). A lens that causes such rays to diverge is called a **diverging lens** (Fig. 6.26 b). The point F_2 where the rays converge or appear to converge is called **second focus (F_2)** of the lens. The corresponding distance PF_2 measured from the centre of the lens P is called **second focal length (f_2).** The point from which the light rays incident upon a converging lens emerge parallel to the principal axis after refraction is called its **first focus (F_1)** (Fig. 6.26 c). The point towards which light rays incident upon a diverging lens appear to meet after refraction is called its first focus (Fig. 6.26 d). The corresponding distance PF_1 is called **first focal length (f_1)** in each case. The lenses shown in the Figs. 6.24 a,b,c are converging lenses and those shown in the Figs. d,e,f are diverging lenses.

In our further discussion of the spherical lenses, we shall mostly be concerned with the second principal focus. So we shall simply refer to it as **focus (F).**

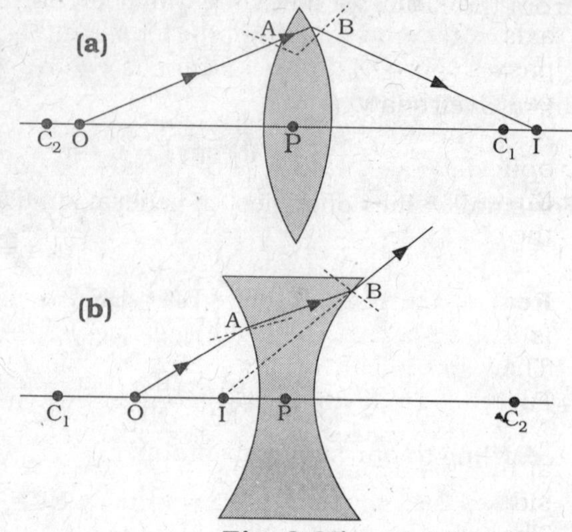

Fig. 6.25

Thin lens. A **thin lens** is one in which the thickness of even the thickest part is small as compared to the radii of curvature (R_1, R_2) and object distance (u) and image distance (v). In our following study of the lenses, we shall assume them to be thin lenses.

Further, we shall assume that the incident rays are paraxial rays, i.e. they lie near to the principal axis and make small angles with the axis. (However, for the sake of clarity, in the diagrams to be presented by us, the rays may apparently be found as inclined at large angles to the axis). Effectively, it means that the aperture of the lens is assumed to be small.

The centre P of a thin lens, which lies on the principal axis (Fig. 6.26), is called the **optical centre** of the lens. (For a thick lens, the position of the optical centre may or may not coincide with the position of the geometrical centre of the lens).

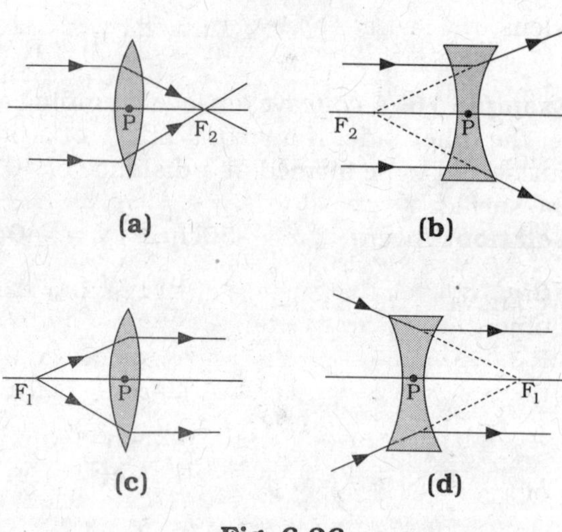

Fig. 6.26

Image formation by thin lenses; real and virtual images

Three representative cases of image formation of an extended object (a pin) are depicted in Fig. 6.27. The positions of the object and the image are marked as O and I, respectively, along the principal axis. The diagrams have been drawn for a double convex lens and a double concave lens, but they apply equally well to other shapes of converging and diverging lenses (Fig. 6.24). Thus the three diagrams generally represent the nature of image formation for all types of converging and diverging lenses.

The position of the image I is located at the intersection of the refracted rays corresponding to any two of the following incident rays starting from the top of the object at O. The point of intersection is found either by extending the refracted rays forward or backward, as the case may be.

(i) Ray 1: Drawn by joining the top of the object to the first focus F_1. It emerges from the lens parallel to the principal axis.

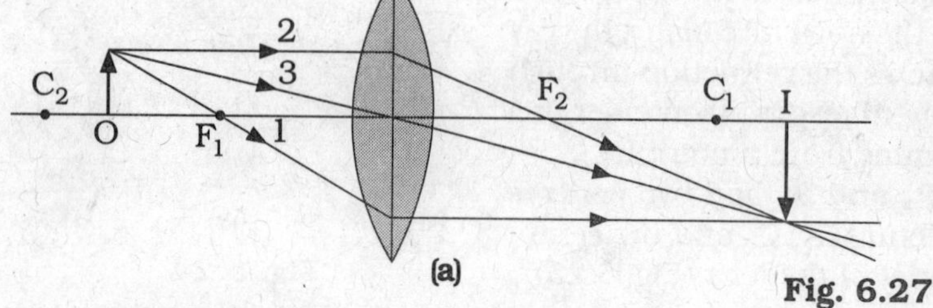

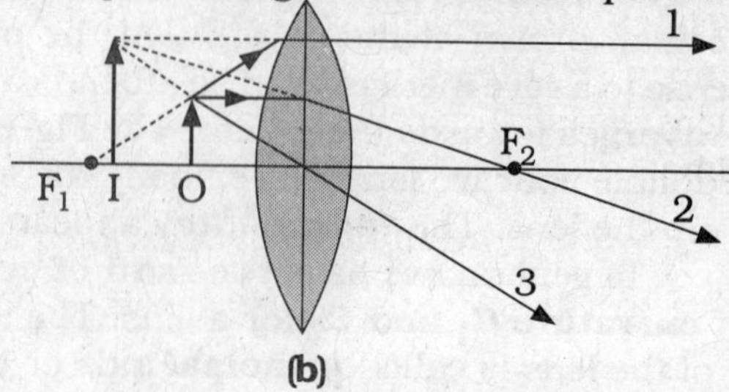

Fig. 6.27

(ii) Ray 2: Drawn parallel to the principal axis. After emerging from the lens, it either actually passes thought the second focus F_2 or appears to pass through F_2.

(iii) Ray 3: Drawn as passing through the optical centre of the lens. It passes undeviated through the lens (the two sides of the lens around the centre are nearly parallel to each other).

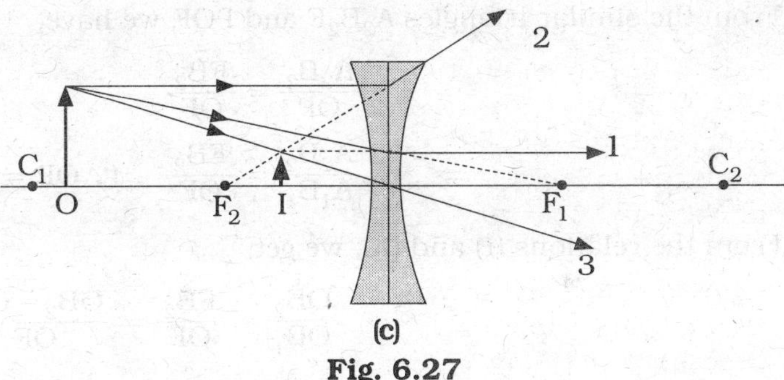

Fig. 6.27

Real and virtual image. A **real image** is one which is formed by actual meeting of the refracted rays. Therefore it can be taken and seen on a screen. Further, it is formed on the opposite side of the object.

A **virtual image** is found by extending the refracted rays backwards. The rays do not actually meet at the position of the image. So it cannot be taken and seen on a screen. Further, it is formed on the same side as the object.

The types of image obtained for the three cases, depicted in Fig. 6.27, are as follows:

(i) Fig. 6.27 (a): The object lies outside the focus F_1. The image is formed on the opposite side of the lens, with respect to the object. It is a real, inverted image.

(ii) Fig. 6.27 (b): The object is placed inside the focus F_1. The image is formed on the same side of the lens as the object. It is an erect (same orientation as the object), virtual image.

(iii) Fig. 6.27 (c): The object is shown as located outside F_2, but it can be anywhere on the left hand side of the lens. The image is formed on the same side as the object. It is an erect, virtual image.

Note that **a converging lens can form both a real image and a virtual image,** depending upon the position of the object. But **a diverging lens always forms a virtual image,** for all positions of the object.

Sign convention. The image formation involves the measurement of the following distances from the lens: object distance (u), image distance (v), focal length (f), and radii of curvature of the two lens surfaces (R_1, R_2). We shall adopt the same **new Cartesian sign convention** for denoting the appropriate signs (+ or –) of these distances as done for image formation by spherical mirrors and spherical surfaces:

(i) The principal axis of the lens is taken as the x-axis, with the optical centre as the origin. All distances are measured from this origin, along the axis. Accordingly, the distances u,v,f,R_1,R_2 represent x-coordinates of the object, image, focus (second focus), and the two centres of curvature, respectively.

(ii) The incident rays are drawn as travelling from left to right. So the +ve direction of the x-axis is along the direction of the incident rays. The distances measured to the right of the optical centre are taken as +ve and those to the left as –ve.

(iii) The vertical distances measured upwards are taken as +ve and those measured downwards are taken as –ve.

As examples, the relevant signs of the distances involved in the Fig. 6.27 are tabulated below.

Figure	u	v	f	R_1	R_2
6.27 (a)	–	+	+	+	–
6.27 (b)	–	–	+	+	–
6.27 (c)	–	–	–	–	+

6A. 09. Thin Lens Formula; Linear (or Lateral) Magnification (m)

Thin Lens Formula. The Fig. 6.28 shows the formation of a real, inverted image A_2B_2 of a vertical object A_1B_1 by a double convex lens. We shall find the relation between the object distance (u), image distance (v), and the focal length (f) of the lens.

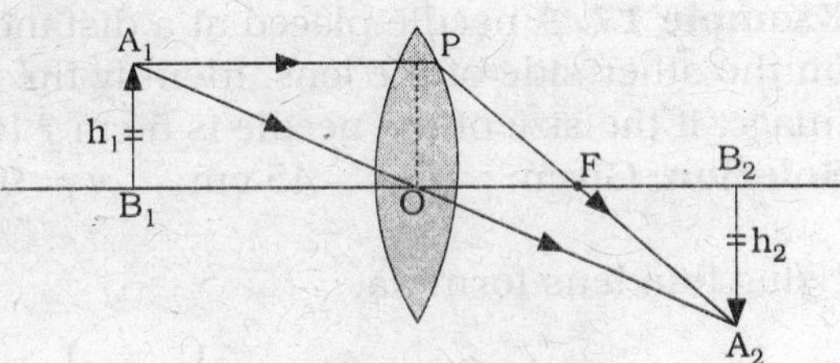

Fig. 6.28

From the similar triangles A_1B_1O and A_2B_2O, we have,

$$\frac{A_2B_2}{A_1B_1} = \frac{OB_2}{OB_1} \quad \text{..... (i)}$$

From the similar triangles A_2B_2F and POF, we have,

$$\frac{A_2B_2}{OP} = \frac{FB_2}{OF}$$

$$\Rightarrow \quad \frac{A_2B_2}{A_1B_1} = \frac{FB_2}{OF} \quad (\because OP = A_1B_1) \qquad \text{..... (ii)}$$

From the relations (i) and (ii), we get,

$$\frac{OB_2}{OB_1} = \frac{FB_2}{OF} = \frac{OB_2 - OF}{OF}$$

According to our sign convention, we have,

$$OB_1 = -u, \quad OB_2 = +v, \quad OF = +f$$

$$\therefore \quad \frac{v}{(-u)} = \frac{v-f}{f}$$

$$\Rightarrow \quad vf = uf - uv \quad \text{or} \quad uf - vf = uv$$

Dividing both sides by uvf, we get,

$$\boxed{\frac{1}{v} - \frac{1}{u} = \frac{1}{f}} \qquad \text{..... (6.27)}$$

(Thin Lens formula)

This relation between u,v,f is known as the **thin lens formula.** Although we have derived it for a (converging) double convex lens, it is valid for all shapes of both converging and diverging lenses and for both real and virtual images. As an exercise, the student may derive it for the earlier described formation of virtual images by a double convex lens and a double concave lens (Figs. 6.27 b,c).

The formula can also be derived by considering refraction of light rays separately from the two surfaces of a lens. We shall do so in the next section on the lens maker's formula.

Linear (or lateral) magnification (m). It is **defined** as "the ratio of the size of the image to the size of the object." In Fig. 6.28,

$$m = \frac{A_2B_2}{A_1B_1} = \frac{h_2}{h_1}$$

From Eq.(i), we get

$$m = \frac{OB_2}{OB_1} = \frac{v}{-u} = -\frac{v}{u}$$

Thus, for a real, inverted image, formed by converging lenses for objects placed outside the first principal focus, the linear magnification is a negative value. It can be easily worked out that for the virtual images formed by the converging lenses (for objects placed within the first principal focus) and by the diverging lenses for all positions of the object, the value of m is positive. In general, we have,

$$m = \frac{v}{u} \qquad \text{..... (6.28)}$$

The Eq. 6.28 is subject to using appropriate sign for v and u, for different cases of image formation.

Example 17. A needle placed at a distance of 45 cm from a lens forms an image on a screen placed 90 cm on the other side of the lens. Identify the type of lens and determine its focal length. What is the size of the image, if the size of the needle is 5 cm ? (CBSE; 1990).

Solution. Given: $u = -45$ cm; $v = 90$ cm; $h_1 = 5$ cm

Using thin lens formula, $\frac{1}{v} - \frac{1}{u} = \frac{1}{f}$

$$\Rightarrow \quad \frac{1}{f} = \frac{1}{(90)} - \frac{1}{(-45)} = \frac{1}{90} + \frac{1}{45} = \frac{1}{30}$$

$$\Rightarrow \quad f = 30 \text{ cm} \quad \textbf{(Answer).}$$

The +ve sign of f means that the lens is a converging lens. **(Answer).**

In this case, u (= 45 cm) > f (30 cm). ∴ A real and inverted image is formed on the other side of the lens. It follows that h_2 is a –ve value. ∴ The magnification is given by,

$$m = -\frac{v}{u} = \frac{-h_2}{h_1}$$

$$\Rightarrow \quad \frac{h_2}{h_1} = \frac{v}{u} = \frac{(90)}{(-45)} = -2$$

$$\Rightarrow \quad h_2 = -2\,h_1 = -2\times(5)$$

$$= -10 \text{ cm}$$

Thus, the image size is 10 cm. Its –ve sign means that the image is inverted. **(Answer).**

Example 18. The refractive index of the material of a diverging lens of focal length 30 cm is 1.5. The lens forms the image of an object at a distance of 20 cm from its centre. Calculate the distance of the object from the lens.

Solution. Given: v = –20 cm; f = –30 cm; μ = 1.5

(v is -ve, because a diverging lens forms a virtual image, on the same side of the lens as the object).

We know,

$$\frac{1}{v} - \frac{1}{u} = \frac{1}{f}$$

$$\Rightarrow \quad \frac{1}{u} = \frac{1}{v} - \frac{1}{f} = \frac{1}{(-20)} - \frac{1}{(-30)} = -\frac{1}{60}$$

$$\Rightarrow \quad u = -60 \text{ cm}$$

Thus, the object is located at a distance of 60 cm from the lens. **(Answer).**

6A. 10 Lens Maker's Formula

Depending upon the requirements of the customers, an optician has to choose different materials (e.g crown glass, flint glass, plastic, etc.) and different shapes (double convex, double concave, plano-convex, concavo-convex, etc.) for the lenses to be used in making spectacles. We shall derive a relation between focal length (f), radii of curvature (R_1, R_2), and refractive index of the material of the lens. It is known as the lens maker's formula. Apart from its use by the opticians, it is useful for many other applications, too.

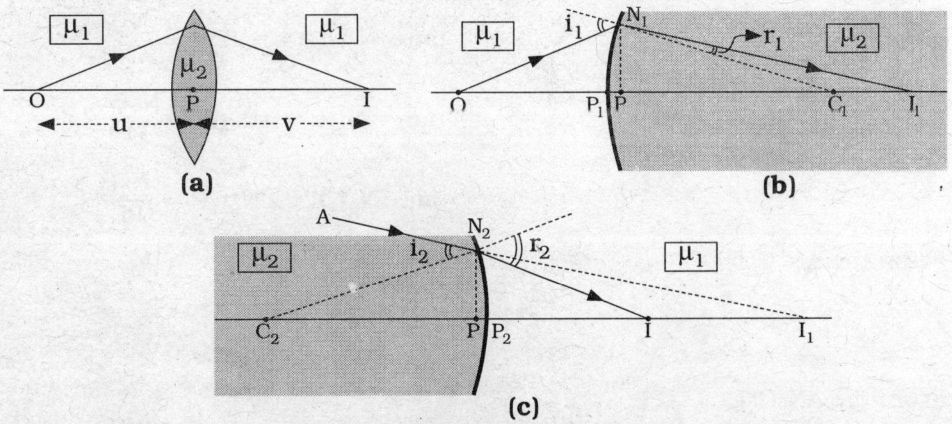

We shall derive the formula by separately considering the refraction of light rays at the two spherical surfaces of a lens. The refractive index of the material of the lens is μ_2 and it is placed in a medium of refractive index μ_1 (Fig. 6.29a). Let $\mu_2 > \mu_1$. The lens forms the image of a point object O at I. We first consider the first surface, shown in Fig. 6.29b. The surface separates two media of refractive index μ_1 and μ_2, respectively. C_1 is the centre of curvature of the surface. Then, the radii of curvature C_1P_1 and C_1N_1 are to be taken as normals drawn from C_1 to the surface at P_1 and N_1, respectively. The incident ray OP_1 from a point object O passes undeviated from the first medium into the second medium, since it goes along the normal C_1P_1. The incident ray ON_1 bends towards the normal C_1N_1 and refracts along N_1I_1 in the second medium. Then, the point of intersection I_1 of the rays N_1I_1 and OI_1 is the image of the point object O. The angles i_1 and r_1 are the angles of incidence and refraction, respectively, at N_1.

Fig. 6.29

N_1P is the perpendicular drawn from N_1 on the principal axis. For a thin lens, $OP_1 \simeq OP$, $C_1P_1 \simeq C_1P$, and $I_1P_1 \simeq I_1P$. Further, for paraxial rays, the angles N_1OP, N_1C_1P and N_1IP have small values. Then we have, ($\tan\theta \simeq \theta$, for small θ-values),

$$\tan \angle N_1OP \simeq \angle N_1OP = \frac{N_1P}{OP}$$

$$\tan \angle N_1I_1P \simeq \angle N_1C_1P = \frac{N_1P}{C_1P}$$

$$\tan \angle N_1I_1P \simeq \angle N_1I_1P = \frac{N_1P}{I_1P}$$

For the ΔN_1OC_1 we have,

$$\angle i_1 = \angle N_1OP + \angle N_1C_1P = \frac{N_1P}{OP} + \frac{N_1P}{C_1P}$$

Similarly, for the $\Delta N_1I_1C_1$ we have,

$$\angle N_1C_1P = \angle r_1 + \angle N_1I_1P$$

$$\Rightarrow \quad \angle r_1 = \angle N_1C_1P - \angle N_1I_1P = \frac{N_1P}{C_1P} - \frac{N_1P}{I_1P}$$

Now, by the Snell's law, we have,

$$\mu_1 \sin i_1 = \mu_2 \sin r_1$$

Since $\sin\theta \simeq \theta$, for small values of θ, we have,

$$\mu_1 i_1 = \mu_2 r_1$$

Substituting for the angles i_1 and r_1 from above, we get,

$$\mu_1\left(\frac{N_1P}{OP} + \frac{N_1P}{C_1P}\right) = \mu_2\left(\frac{N_1P}{C_1P} + \frac{N_1P}{I_1P}\right)$$

$$\Rightarrow \quad \frac{\mu_1}{OP} + \frac{\mu_2}{I_1P} = \frac{\mu_2 - \mu_1}{C_1P} \qquad \text{..... (6.29)}$$

Next, we consider the image formation by the second surface of the lens, with its centre of curvature at C_2. (Fig. 6.29 c). Now the light enters from a medium of R.I. μ_2 into a medium of R.I. μ_1, with $\mu_2 > \mu_1$. The image I_1 formed by the first surface acts as a virtual object for the second surface. It amounts to a light ray being incident at a point N_2 of the surface in the direction AN_2I_1. The ray bends away from the normal C_2N_2 towards I. So I is the image of I_1 formed by the second surface. As before, the ray $C_2P_2I_1$, incident at P_2, goes undeviated. Also, as before, $C_2P_2 \simeq C_2P$ and $P_2I_1 \simeq PI_1$. Further, as before,

$$\tan \angle N_2C_2P \simeq N_2C_2P = \frac{N_2P}{C_2P}$$

$$\tan \angle N_2IP \simeq \angle N_2IP = \frac{N_2P}{PI}$$

$$\tan \angle N_2I_1P \simeq \angle N_2I_1P = \frac{N_2P}{PI_1}$$

From the $\Delta N_2C_2I_1$,

$$\angle i_2 = \angle N_2C_2P + \angle N_2I_1P$$

$$= \frac{N_2P}{C_2P} + \frac{N_2P}{PI_1}$$

From the ΔN_2C_2I,

$$\angle r_2 = \angle N_2C_2P + \angle N_2IP$$

$$= \frac{N_2P}{C_2P} + \frac{N_2P}{PI}$$

By the Snell's law ($\mu_1 \sin i = \mu_2 \sin r$; now the light goes from medium of R.I. μ_2 to the medium of R.I. μ_1),

$$\mu_2 \sin i_2 = \mu_1 \sin r_2$$

$$\Rightarrow \quad \mu_2 i_2 = \mu_1 r_2 \quad \text{(for small angles)}$$

$$\Rightarrow \quad \mu_2\left[\frac{N_2P}{C_2P}+\frac{N_2P}{PI_1}\right]=\mu_1\left[\frac{N_2P}{C_2P}+\frac{N_2P}{PI}\right]$$

$$\Rightarrow \quad \frac{\mu_2}{C_2P}+\frac{\mu_2}{PI_1}=\frac{\mu_1}{C_2P}+\frac{\mu_1}{PI}$$

$$\Rightarrow \quad \frac{\mu_1}{PI}-\frac{\mu_2}{PI_1}=\frac{\mu_2-\mu_1}{C_2P} \qquad \text{..... (6.30)}$$

The Eqs. 6.29 and 6.30 are the respective equations for refraction separately at the two surfaces of a lens. To get the required equation for the lens as a whole, we add up the two equations, to get,

$$\frac{\mu_1}{PI}+\frac{\mu_1}{PO}=(\mu_2-\mu_1)\left(\frac{1}{C_1P}+\frac{1}{C_2P}\right)$$

Applying our sign convention, we have,

$$OP=(-)u; \quad PI=(+)v; \quad C_1P=(+)R_1; \quad C_2P=(-)R_2$$

$$\therefore \quad \frac{\mu_1}{v}-\frac{\mu_1}{u}=(\mu_2-\mu_1)\left(\frac{1}{R_1}-\frac{1}{R_2}\right)$$

$$\Rightarrow \quad \frac{1}{v}-\frac{1}{u}=\left(\frac{\mu_2}{\mu_1}-1\right)\left(\frac{1}{R_1}-\frac{1}{R_2}\right) \qquad \text{..... (6.31)}$$

If the object is situated at infinity, the incident rays are parallel to the principal axis and the image is formed at the focus of the lens, i.e. if $u = \infty$, then $v = f$. Substituting, we get

$$\frac{1}{f}=\left(\frac{\mu_2}{\mu_1}-1\right)\left(\frac{1}{R_1}-\frac{1}{R_2}\right) \qquad \text{..... (6.32)}$$

For a lens situated in air, $\mu_1 = 1$ and $\mu_2 = \mu$ (say). So we get,

$$\boxed{\frac{1}{f}=(\mu-1)\left(\frac{1}{R_1}-\frac{1}{R_2}\right)} \qquad \text{..... (6.33)}$$

(Lens Maker's Formula)

This is the required **lens maker's formula.** Further, the comparison of the Eqs. 6.31 and 6.32 gives,

$$\frac{1}{v}-\frac{1}{u}=\frac{1}{f},$$

Which is the thin **lens formula** that we had derived earlier for a double convex lens (Eq. 6.27), by considering the image formation by the lens as one piece, instead of a combination of two spherical surfaces.

Example 19. The radii of curvature of the bounding surfaces of a double convex lens of focal length 30 cm are 25 cm and 50 cm, respectively. Determine the refractive index of the material of the lens.

Solution. Given: $R_1 = 25$ cm; $R_2 = -50$ cm; $f = 30$ cm

We use the lens maker's formula,

$$\frac{1}{f}=(\mu-1)\left(\frac{1}{R_1}-\frac{1}{R_2}\right)$$

$$\Rightarrow \quad \frac{1}{(30)}=(\mu-1)\left[\frac{1}{(25)}-\frac{1}{(-50)}\right]$$

$$=(\mu-1)\left(\frac{1}{25}+\frac{1}{50}\right)=(\mu-1)\times\frac{3}{50}$$

$$\Rightarrow \quad \mu-1=\frac{50}{90}=\frac{5}{9}$$

$$\Rightarrow \quad \mu=\frac{5}{9}+1=\frac{14}{9}$$

$$= 1.56 \quad \textbf{(Answer).}$$

Example 20. The refractive index of the material of a double concave lens is 3/2. The radii of curvature of its two surfaces are 40 cm each. Find its focal length.

Solution. Given: $R_1 = (-)\ 40$ cm; $R_2 = (+)\ 40$ cm; $\mu = 3/2$

We use the lens maker's formula,

$$\frac{1}{f} = (\mu - 1)\left(\frac{1}{R_1} - \frac{1}{R_2}\right)$$

$$= \left(\frac{3}{2} - 1\right)\left(\frac{1}{-40} - \frac{1}{40}\right)$$

$$= \frac{1}{2} \times \left(-\frac{2}{40}\right) = -\frac{1}{40}$$

$\Rightarrow$ $f = -40$ cm **(Answer).**

The –ve sign of f indicates the diverging nature of the lens.

6A. 11. Power of a Lens (P); Thin Lenses in Contact

Power of a Lens (P)

A lens converges or diverges the light rays falling upon it. The **power of a lens** is "a measure of the degree of convergence or divergence of the light rays incident upon it." The shorter is the focal length of the lens, the greater is the convergence (or divergence) produced by it, as can be readily seen in Fig. 6.26. The power of a lens (P) is **defined** as "the reciprocal of the focal length of the lens",

$$P = \frac{1}{f} \qquad \text{..... (6.34)}$$

The **unit of power of a lens** is **dioptre (D)**, which is **defined** as the reciprocal of focal length in metres. In the Eq. 6.34, f is in meter and P in dioptre. Thus, one dioptre is the power of a lens of focal length one metre ($1D = 1m^{-1}$). P is +ve for a converging lens and –ve for a diverging lens.

Combination of thin lenses in contact. For many applications, it is advantageous to form images by a combination of two or more lenses. The combination helps to eliminate defects in image formation and so provides more clear images. It also gives higher magnification. We first consider a combination of two thin convex lenses of focal lengths f_1 and f_2 in contact with each other (Fig. 6.30). The first lens forms the image I_1 of a point object O. I_1 serves as a virtual object for the second lens, which forms the final image at I. Since the lenses are thin, the point of contact, P, of the lenses may be taken as the optical centre of the combination.

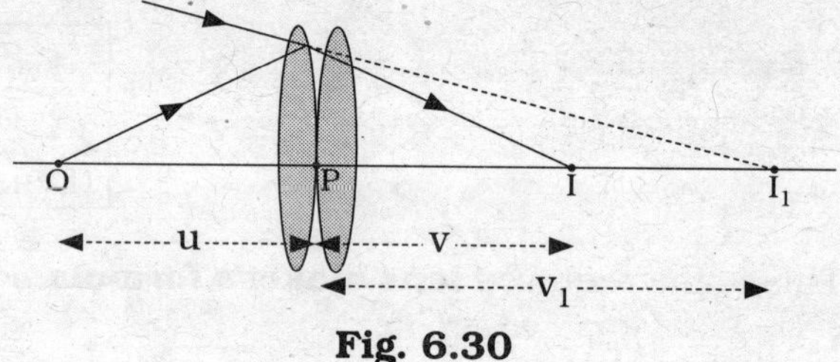

Fig. 6.30

Applying the thin lens formula ($1/v - 1/u = 1/f$) to the formation of the image I_1, we get,

$$\frac{1}{v_1} - \frac{1}{u} = \frac{1}{f_1}$$

Next, applying the formula to the formation for the image I, we get,

$$\frac{1}{v} - \frac{1}{v_1} = \frac{1}{f_2}$$

Adding the two equations, we get,

$$\frac{1}{v} - \frac{1}{u} = \frac{1}{f_1} + \frac{1}{f_2}$$

The combination may be treated as equivalent to a single lens of focal length F, which forms the image of O at I. Applying the lens formula to the combination gives,

$$\frac{1}{v} - \frac{1}{u} = \frac{1}{F}$$

The comparison of the above two equations gives the equivalent focal length (F) of the combination as,

$$\frac{1}{F} = \frac{1}{f_1} + \frac{1}{f_2} \qquad \text{..... (6.35)}$$

Proceeding in the same manner, the equivalent focal length of a combination of more than two lenses of focal lengths f_1, f_2, f_3, f_4,......... will be found as,

$$\frac{1}{F} = \frac{1}{f_1} + \frac{1}{f_2} + \frac{1}{f_3} + \ldots\ldots\ldots \qquad \ldots\ldots (6.36)$$

Since the power of a lens, P = 1/f, we get for the equivalent power, P, of the combination,

$$P = P_1 + P_2 + P_3 + \ldots\ldots\ldots\ldots \qquad \ldots\ldots (6.37)$$

If the combination consists of both converging and diverging lenses, some terms on the right hand side will be –ve and some will be +ve. An interesting case is the combination of a double convex lens and a double concave lens of equal focal length (Fig. 6.31). The power of the combination is zero and its focal length is ∞ [$1/F = 1/f_1 - 1/f_2$ or $F = f_1f_2/(f_2 - f_1)$]. It means that the combination acts like a simple glass plate.

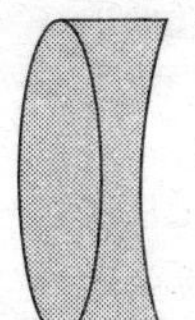

Fig. 6.31

The total magnification m of the combination is a product of the individual magnifications,

$$m = m_1 \times m_2 \times m_3 \times \ldots\ldots\ldots\ldots\ldots\ldots \qquad \ldots\ldots (6.38)$$

Combination of separated lenses. For a combination of two thin lenses of focal lengths f_1 and f_2, separated by a small distance d (Fig. 6.32) it can be worked out that the equivalent focal length F is given by,

$$\frac{1}{F} = \frac{1}{f_1} + \frac{1}{f_2} - \frac{d}{f_1 f_2} \qquad \ldots\ldots (6.39)$$

Such combinations are frequently used in the objectives and eyepieces of the optical instruments, some of which we shall describe in the next section.

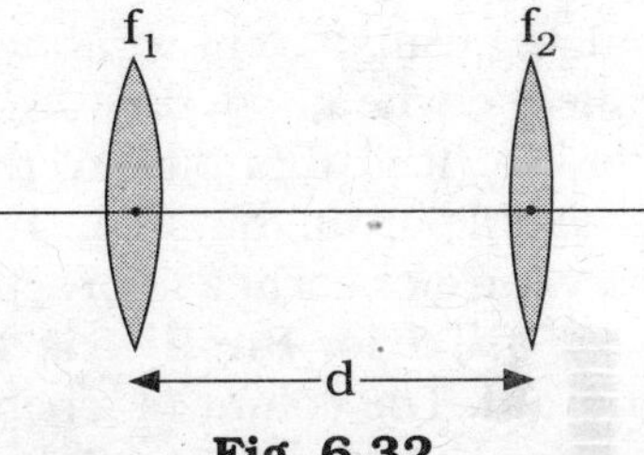

Fig. 6.32

Combination of Lenses and Mirrors

If a mirror is place behind a lens, the image formation consists of the following steps:

(i) With an object placed in front of the lens, the image formed by the lens acts as the object for the mirror.

(ii) Then the image produced by the mirror acts as a new object for the lens, which forms the final image.

Example 21. The focal length of a combination of two thin lenses in contact is 40 cm. If the focal length of one of the lenses is 25 cm, determine the power of the other lens.

Solution. Given: $f_1 = 25$ cm = 0.25 m; F = 40 cm = 0.4 m

The power of the first lens is,

$$P_1 = \frac{1}{f_1} = \frac{1}{(0.25)} = \frac{100}{25} = 4D$$

The power of the combination is,

$$P = \frac{1}{F} = \frac{1}{(0.4)} = \frac{10}{4} = 2.5\ D$$

We know,

$$P = P_1 + P_2$$

$$\Rightarrow \quad P_2 = P - P_1 = (2.5) - (4.0)$$

$$= -1.5\ D$$

Thus, the power of the second lens is –1.5 D. The –ve sign shows that it is a diverging lens. **(Answer).**

6A.12 Optical Instruments

The range of optical instruments is wide. One very efficient optical instrument has been provided to all human beings by nature, viz. the human eye. We shall describe its construction and working. Out of the man-made instruments, the ones of most common use are microscope and telescope. A microscope is employed for viewing small or very small objects that cannot be seen, or cannot be clearly seen, with naked eyes (the name is derived from the Greek words 'mikros' and 'skopein', meaning 'small' and 'look at', respectively). A telescope is employed for viewing distant objects that cannot be seen, or cannot be clearly seen, with naked eyes (Greek word 'tele' means 'far off'). We shall describe them, too.

Human Eye

Construction. The human eye is a small nearly spherical chamber, about an inch (2.5 cm) in diameter. A vertical section of the eye is shown in Fig. 6.33. It has a bulge in the front and a stem behind. It is held

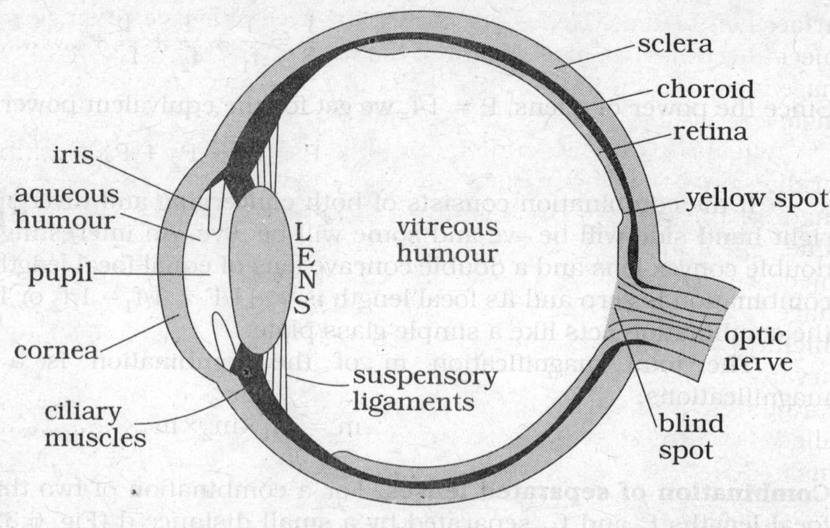

Fig. 6.33

in position by a set of external muscles (not shown in the figure), which help it to rotate so that it can point in different directions. From the optical point of view, i.e. from the point of view of image formation, it consists of the following parts:

(i) Cornea; Sclera; Choroid. The outer shell, **sclera (or sclerotic),** is hard and opaque. It serves to protect the inner delicate parts of the eye from any injury. Beneath it is a coloured layer of blood vessels, called **choroid.** It is dark brown and so prevents the reflection of light entering the eye. The sclera becomes a bulging transparent window in the front, called **cornea,** so that light can enter the eye. It is this part of the eye which is visible from outside. The radius of curvature of cornea is practically the same for all eyes.

(ii) Iris; Pupil. Iris is a coloured adjustable diaphragm, appended to the front thick end of the choroid. The colour of a person's eye is really the colour of the iris. It regulates the amount of light entering the eye. The hole (open space) in the central region of the iris is called **pupil,** which admits light to the interior of the eye. Its size varies according to the amount of light entering the eye, e.g. from about 2 mm in bright light to about 8 mm in dark. It always appears black in colour, because the light entering the eye through it is almost totally absorbed inside the eye, so that no light emerges from it.

(iii) Eye lens; Ciliary muscles; Suspensory ligaments. Behind the pupil is the **eye lens (or crystalline lens),** which is held in position by **suspensory ligaments.** The ligaments are attached to **ciliary muscles,** which form the front thick end of choroid. They can change the focal length of the lens by contracting, which alters the curvature of the two faces of the lens. The refractive index of the material of the lens is not uniform. Its average value is about 1.396.

(iv) Retina; Light sensors (Rods & Cones); Optic nerve. Retina is the innermost layer of the eyeball, under the choroid. It is sensitive to light. It consists of a network of millions of nerve fibres, which are connected with the brain by the so-called **optic nerve** at the back of the eye. Thus the optic nerve consists of a bundle of nerve fibres. The nerve fibre endings have two types of light sensors (or photo receptors), called **rods and cones.** When light falls on either a cone or a rod, an impulse is sent along the corresponding nerve fibre, thus conveying a message to the brain, which in turn makes the person to feel the sensation of light. Further, although an inverted image is formed on the retina by the eye lens, the brain enables to see it as an erect image, through some complex physiological process.

(v) Aqueous humour; Vitreous humour. The lens divides the interior of the eye in two compartments. The space between the cornea and the lens is filled with a transparent dilute salt solution, called **aqueous humour.** The space between the lens and the retina is filled with a transparent watery jelly, called **vitreous humour.** Both aqueous humour and vitreous humour have nearly the same refractive index, about 1.336.

(vi) Yellow spot; Fovea; Blind spot. Yellow spot is a minute depression (about 2 mm across) in the central region of the retina. It is a very light sensitive region. In its middle is a smaller area (about 0.25 mm across), called **fovea.** It is the most light sensitive place on the retina, entirely consisting of cones. The position on the retina at which the optic nerve enters the eye from the brain is called **blind spot,** because it is devoid of either rods or cones and so any light falling upon it remains undetected.

Working (Accommodation; Least distance of distinct vision; Near point; Far point)

It is obvious that the eye forms an optical system, consisting of several coaxial curved surfaces, starting from cornea to vitreous humour. However, the refractive index of cornea is approximately the same (≃ 1.3) as that of the aqueous humour. Therefore the light rays entering the eye from an object undergo refraction mainly at three surfaces, viz. the outer surface of the cornea (air to cornea; μ_{air} = 1.0) and the two surfaces of the lens. Hence, in totality, the eye behaves as a single converging lens, forming the image of the object at retina. The focal length of the lens can be adjusted by the ciliary muscles. Thus, for the purpose of image formation, now onwards the entire optical system of the eye will be deemed by us as equivalent to a single converging lens, like the one shown in Fig. 6.34.

When the eye sees a distance object, e.g. a star, the rays incident upon the said equivalent lens are parallel and focus upon the retina. The distance between the lens and the retina is equal to the focal length of the lens. In this condition, the ciliary muscles are relaxed and the focal length has maximum value. Since the image is formed at the retina, the object is clearly seen by the eye. To clearly see a closer object, e.g. for reading a book, the ciliary muscles have to adjust the focal length in such a way that the image is again formed on the retina. The focal length needs to be reduced. From the lens maker's formula, which relates the focal length of a lens with the radii of curvature of its two faces, it means that the faces' curvatures have to change suitably. Hence the ciliary muscles contract accordingly. "The automatic process of adjustment of focal length (or power) of eye for clearly seeing objects situated at different distances" is called **accommodation.** Thus we find that unlike spectacle lenses, which have fixed focal lengths, the eye lens is flexible.

The muscles cannot be strained beyond a limit. Therefore an object cannot be seen clearly if it is situated closer than a minimum distance from the eye. "The minimum distance at which an object can be clearly seen by the eye" is called **least distance of distinct vision.** Its value varies from person to person and with age. The average value for a normal grown-up person is taken as 25 cm.

"The nearest position at which an object can be clearly seen by an eye" is called **near point** of that eye. Note that the least distance of distinct vision is the distance between the near point and the eye. Thus for a normal eye the near point is located 25 cm away.

"The farthest position upto which an object can be clearly seen by an eye" is called **far point** of that eye. It is infinity for a normal eye. "The distance between the near point and the far point of an eye" is called its **range of accommodation.**

Defects of Eye and Their Correction (Myopia, Hypermetromia, Presbyopia, and Astigmatism)

From optical point of view, a human eye may have one or more of the following defects: myopia, hypermetropia, presbyopia, and astigmatism. We shall describe them in the following, alongwith the common way for their correction by using suitable lenses. Two recent, but as yet uncommon ways of correction, viz. use of contact lenses and laser treatment, will only be mentioned briefly. The medical defects of the eye, like cataract, glaucoma, conjunctivitis, etc., which arise from diseased condition of some part of the eye, will not be discussed, because they are not relevant from optical viewpoint.

Note that the defects of eye which we now proceed to discuss can be corrected by the use of lenses.

(i) Myopia (or Shortsightedness or Nearsightedness)

Myopia is "the defect of eye because of which distant objects cannot be seen clearly by an unaided eye". As compared to the normal eye, for which the near point and the far point are located at ~ 25 cm and ∞, respectively, for a myopic eye the corresponding distances are shorter. For example, in a particular case these distances may be just 15 cm and 50 cm, respectively. Then, unlike a normal eye, this person can clearly see objects located as near as 15 cm to the eye, but cannot clearly see objects located farther than 50 cm from the eye.

For a normal eye, the incident parallel rays from a distant object focus at the retina (Fig. 6.34; at point P), but for a myopic eye they focus short of the retina (Fig. 6.35a; at point Q). In most cases, the **cause of myopia** is that the eyeball loses its spherical shape, such that it becomes elongated parallel to the optic axis of the eye. Hence the retina lies abnormally far from the cornea. Then the rays from a distant point object focus at Q in front of the retina, thus producing a blurred spot image on the retina, instead of a point image. In some cases, the defect may arise from an inherent defect in the crystalline lens, e.g. the lens being too thick.

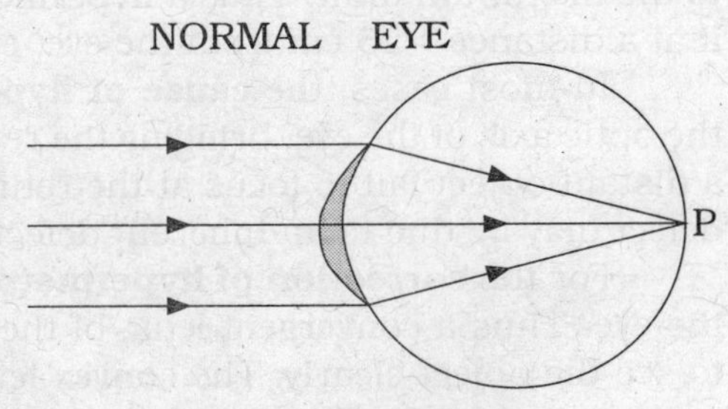

Fig. 6.34

For the **correction of myopia,** the incident rays should be made suitably divergent before entering the eye. Thus a diverging

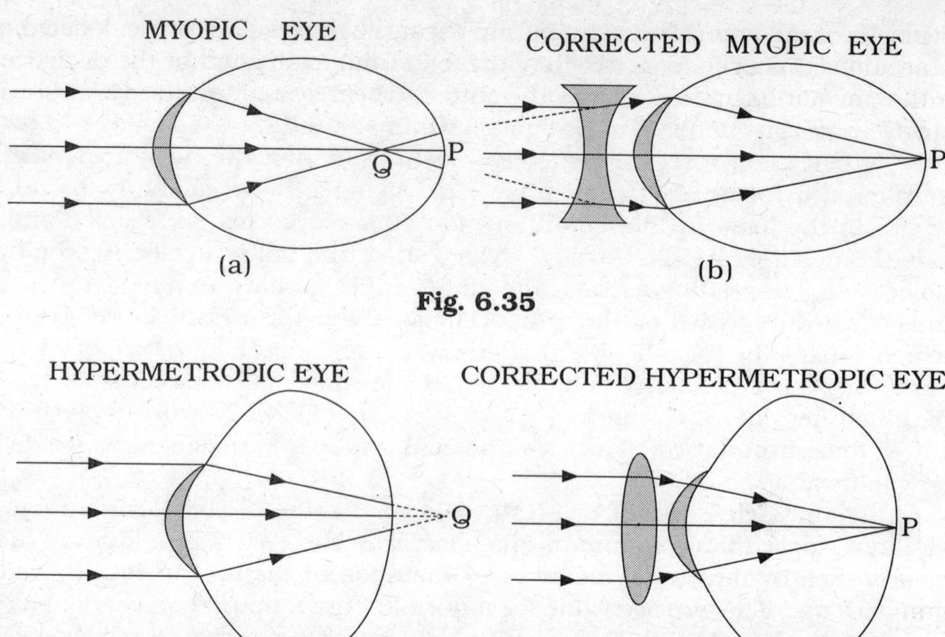

Fig. 6.35

Fig. 6.36

lens, of the type shown in Fig. 6.35b, has to be put in front of the eye, so that the eye can see distant objects clearly. The concave lens enables the eye lens to form a point image of a distant point object at the retina (at P). The virtual image formed by the concave lens acts as the object for the eye lens. Therefore the focal length, and so the power ($P = 1/f$), of the concave lens has to be such that the image of a distant object ($u = -\infty$) is formed at the far point of the eye (the far point is the farthest position at which an object is clearly seen by the eye). If the far point is located at a distance x from the eye, it means that the image distance of the concave lens is nearly x, i.e. $v = -x$ (according to our sign convention). Using the lens formula, we get

$$\frac{1}{v} - \frac{1}{u} = \frac{1}{f}$$

$$\Rightarrow \quad \frac{1}{(-x)} - \frac{1}{(-\infty)} = \frac{1}{f}$$

$$\Rightarrow \quad f = -x$$

The –ve sign appears because the correcting lens is a concave lens. If the far point of a myopic person is 50 cm, the focal length of his concave lens will be – 0.5 m and the power of the lens will be,

$$P = \frac{1}{f} = \frac{1}{(-0.5)} = -2 \text{ dioptre } (-2D)$$

(ii) Hypermetropia (or Longsightedness or Farsightedness)

Hypermetropia is "the defect of eye because of which near objects cannot be seen clearly by an unaided eye". The far point is the same as that for a normal eye, viz. at infinity, but the near point is abnormally far from the eye, say at 40 cm in a particular case.

For a hypermetropic eye, the incident parallel rays from a distant object focus behind the retina (at Q in Fig. 6.36a), when the ciliary muscles of the eye are relaxed. The eye may still focus the rays at the retina, and thus see the object clearly, but only by straining the muscles. (Therefore a person suffering from hypermetropia also tends to suffer from headache, as the eye muscles are constantly in use). But an object lying at a distance of ~ 25 cm cannot be focussed by the eye on the retina, even after straining the muscles to the maximum limit. Thus a hypermetropic person is able to read a book or a newspaper only by keeping it at a distance >25 cm from the eye, e.g. at 40 cm in a particular case.

In most cases, the **cause of hypermetropia** is that the eyeball becomes elongated at right angles to the optic axis of the eye, bringing the retina too close to the cornea (Fig. 6.36a). Then the refracted rays from a distant object fail to focus at the retina, producing a blurred spot image of the object. In some cases, the defect may be due to an inherent defect in the crystalline lens, e.g. the lens being too thin.

For the **correction of hypermetropia,** the incident rays are made suitably convergent before entering the eye. Thus a convergent lens, of the type shown in Fig. 6.36b, has to be put before the eye, to enable it to see the object clearly. The convex lens helps the eye lens to form a point image of a distant point object at the retina (at P). The image produced by this external lens acts as the object for the eye lens. So its focal length, and the power, has to be such that the image of an object located at the near point of normal eye (i.e. at about 25 cm from the eye) is formed at the near point of the defective eye, at which it is clearly seen by the eye. For example, if the near point of the defective eye is located 40 cm away from the eye, we very nearly have u = –25 cm and v = – 40 cm. Then, using the lens formula, we have,

$$\frac{1}{v} - \frac{1}{u} = \frac{1}{f}$$

$$\Rightarrow \quad \frac{1}{f} = \frac{1}{(-40)} - \frac{1}{(-25)} = \frac{3}{200}$$

$$\Rightarrow \quad f = \frac{200}{3}\text{ cm} = \frac{2}{3}\text{ m}$$

$$\therefore \quad P = 1/f = \frac{3}{2}\text{ dioptre} = 1.5\text{ D}$$

Thus, the power of the external lens has to be + 1.5 D. The +ve sign shows that the lens is converging in nature.

(iii) Presbyopia (or Lack of Accommodation)

Presbyopia is "the defect of eye because of which in old age near objects cannot be seen clearly by the an unaided eye". A person with presbyopia cannon see near objects clearly, e.g. he cannot read a newspaper at the least distance of distinct vision (~ 25 cm), while his vision for distant objects remains good. The older he gets, the farther away the objects have to be for being seen clearly, i.e. his near point recedes.

The **causes of presbyopia** are that with advancing age (i) the crystalline lens loses its elasticity (becomes hard) and (ii) the ciliarly muscles become weaker. Thus, the power of accommodation reduces with age. A normal child can focus objects held just a few inches in front of his eyes. This distance gradually increases with age (receding of near point). Many persons suffer total loss of accommodation by about 40 years of age. Distant objects can still be adjusted for them and so the ciliary muscles need not be strained. (Note that shortsighted persons can still be able to read comfortably in old age with unaided eyes, by keeping the reading material at their near point).

It is obvious that in old age the eye lens virtually becomes a fixed-focus lens set at infinity, i.e. set only for distant objects. Therefore, like the case of hypermetropia, for the **correction of presbyopia** the incident rays from the objects are made suitably convergent by putting a convex lens in front of the eye. Note that the spectacles with such lenses have to be used only for viewing the near objects. The distant objects will look blurred if seen through them. Therefore a myopic or a hypermetropic person has to have two pairs of spectacles, one for distant objects and the other for near objects. An alternative is to use a single pair of spectacles with bifocal lenses, which have different curvatures in the lower and upper parts. The lower part is used for viewing near objects and the upper part for distant objects. For a normal eye, i.e. suffering neither from myopia nor from hypermetropia, just one pair of spectacles (for near objects) is enough. (For most people with normal vision, the power of lenses to be used for the correction of presbyopia is about +2.5 D. Therefore spectacles carrying such lenses are cheaply available in the market, as they are mass produced).

(iv) Astigmatism

Astigmatism is "the defect of eye because of which parts of an object lying in different planes are differently focussed by an unaided eye". Thus while some parts of the object are imaged rightly, some others may look distorted or blurred (e.g. to a strongly astigmatic eye, the letter O may look like Q and P as R). In Fig. 6.37, equally thick and equally intense lines have been drawn in four different planes. But to an unaided astigmatic eye, they will appear as having slightly different thickness and intensities, which will alternate if the figure is rotated in the horizontal plane. Astigmatism is the most common defect of eye; nearly everyone suffers from it to a small or large extent.

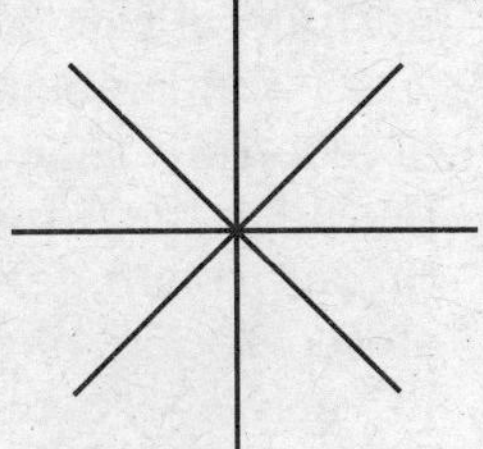

Fig. 6.37

The most common **cause of astigmatism** is asymmetry (i.e. lack of symmetry) of cornea. (Remember that maximum refraction of light rays from the object takes place when they enter the eye from air into cornea). The cornea develops different curvatures in different planes, which means that the focal length of the optical system of the eye becomes different in different planes. Hence the eye cannot see equally well in all the directions. In some cases, the same defect may arise from the asymmetry of the crystalline lens.

We have seen that for correcting myopia, hypermetropia and presbyopia, symmetrical spherical lenses are used. For the **correction of astigmatism,** suitable asymmetrical lenses are used, to compensate for the asymmetry of the cornea of the crystalline lens. These include cylindrical, sphero-cylindrical (one surface spherical and the other cylindrical), and toric (one surface plane and the other curved) lenses.

The images formed by the lenses suffer from several defects (or aberrations). For monochromatic light, there are five such aberrations, one of which is called astigmatism. However, in spite of bearing the same name, its nature is very different from the astigmatism of eye.

Use of contact lenses and laser treatment for the correction of eye defects. Contact lenses made from plastic materials can be fitted under the eyelids in contact with cornea, to correct the defects of eye. The inner surface of the lens exactly conforms with the shape of the cornea, while the outer surface has the required curved shape for correcting the defects.

The laser treatment is essentially a surgical process. A fine laser beam is employed to impart appropriate curved shape to the cornea.

Both the use of contact lenses and the laser treatment have their respective advantages and disadvantages, which have to be carefully looked into by the persons opting for them. We shall not go into their details.

Additional knowledge

(i) The so-called 'crystalline lens' of the eye is not really crystalline. It is composed of layers of transparent, flexible material, with increasing density and hardness towards the centre. The refractive indices of the outermost, middle and innermost layer are 1.405, 1.429 and 1.454, respectively. It is biconvex, with the back surface more curved than the front surface.

(ii) 95% of the retina's light sensors are rods. Each eye has nearly 100 million rods and 5 million cones. The cones have larger size. The cones function in good light and detect colour. The rods need very much less light, but are colour blind.

(iii) The straight line joining the optical centre of the eye lens to the fovea (most sensitive position on the retina) is termed the **visual axis** of the eye (it is inclined at $\sim 5°$ to the **optic axis,** which passes through the optical centre of the eye lens and the centre of cornea). The eye has a natural tendency to look at an object along this line. So the image of the object is automatically formed at the fovea. Hence the name 'visual axis'.

(iv) The image formed by the biconvex lens of the eye on the retina is inverted. The brain has a mechanism to interprct it as an erect image of the object.

(v) The least distance of distinct vision varies with age. In young age, say < 10 years, the ciliary muscles can bear large strain, as they are strong and very flexible. Then the near point may be just about 7 - 8 cm. In old age, the muscles grow weak. Then the near point shifts to large values, e.g. 1 - 2 meter or more.

(vi) After long periods of close work, e.g. reading and writing, repairing, etc., a person feels eye strain, because all the time the muscles are contracted. Therefore, it is desirable to rest the eye intermittently, e.g. just by resting or looking at distant objects.

(vii) The human eye sees objects at different distances by the use of ciliary muscles to change the power of the eye lens. But a fish does so by changing the distance between the lens and the retina, with a bellows-type mechanism.

(viii) The images formed by the eye suffer from spherical aberration and chromatic aberration, as for any other lens. The existence of the former is readily demonstrated by contracting one's eyes, which immediately improves the clarity of the image, because by doing so a large part of the incident peripheral rays are cut off.

(ix) Binocular vision: Why the God has given us two eyes? The two eyes form images of the same object from slightly different angles, which are received by the brain and fused into a single image. This provides the advantages of (a) producing a solid 3-dimensional image and (b) enabling to better judge the approximate distance of the object from the eye. (In a similar way, the existence of two ears provides stereo-sound, as well as a much better estimate of the distance of the source of sound from the person). In early age, it takes several years for the brain to properly coordinate the two images formed by the respective eyes.

(x) Persistence of vision: The sensation produced by an image on the retina does not disappear instantaneously when the light is cut off. It lasts for a small time, which varies from person to person and with the intensity and colour of light. It is of the order of 1/10 second. This effect is termed as persistence of vision. It is made use of in motion pictures, e.g. a movie, by mechanically presenting successive pictures on the screen at intervals ~1/10 second. Since each image lasts for a small finite time before fading, the eye gets impression of continuous action, without any break.

(xi) The angular resolution of the human eye is about 1 minute, i.e. a normal eye can resolve two points subtending an angle of about 1′ at the eye. This limit is determined by the distance apart of the light sensors (rods and cones) of the retina (just as the rsolution of a photographic camera, however powerful be its lens, is limited by the density of grain of the film).

Apparent Size of an Object seen by the Eye

We now proceed to describe two other optical instruments, the microscope and the telescope. However, to understand their functioning, it is useful to examine some basics regarding the size of an object as perceived by the eye, when the object is located at different distances.

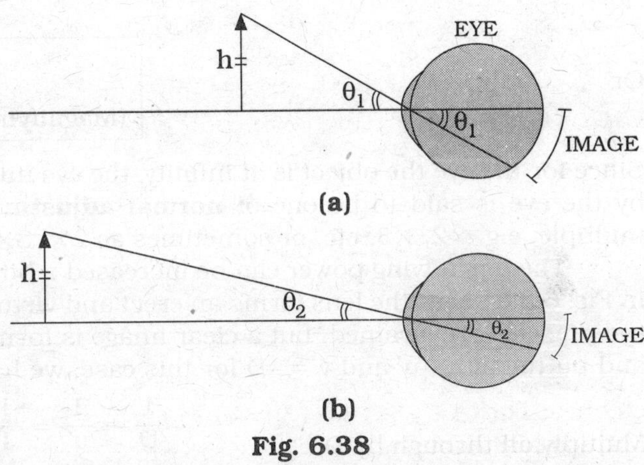

Fig. 6.38

The apparent size of an object as seen by the eye does not depend on the actual size of the object alone. It also depends on the distance of the object from the eye, which effectively means dependence on the **visual angle,** i.e. the angle subtended by the object at the eye. It is easily understood from Fig. 6.38, which shows the image formation by the eye for two different positions of the same object. The image size increases when the object distance becomes smaller and so the visual angle becomes larger ($\theta_1 > \theta_2$). The clarity of the image improves as the image size is increased. The microscope and telescope and other optical instruments help to increase the visual angle θ and thus enable us to see objects clearly, which the naked eyes are not able to do.

6A. 13. Simple Microscope (or Simple Magnifying Lens or Magnifier); Magnifying Power (or Angular Magnification)

To be able to see an object clearly with the naked eye it should lie somewhere between infinity and the near point. For an unaided eye, the visual angle is largest when the object is located at the near point, i.e. at the least distance of distinct vision, D, from the eye (Fig. 6.39a),

$$\theta_D = \frac{h}{D} \qquad \text{..... (6.41)}$$

h is the height of the object. The visual angle can be further increased by inserting a converging lens between the eye and the object which is called a **magnifier** or a **simple microscope** (or a **simple magnifying lens**). It is shown for two different positions of the same object (Figs. 6.39b and c). In the Fig. 6.39b, the object lies at the focus F of the lens. In the Fig. 6.39c, it lies within the focus F. The focal length of the lens is less than D (f < D). When the object is located at F, the incident rays emerge parallel from the lens to enter the eye. So they appear to come from infinity to the eye. Correspondingly, the actual size of the object is infinite. But

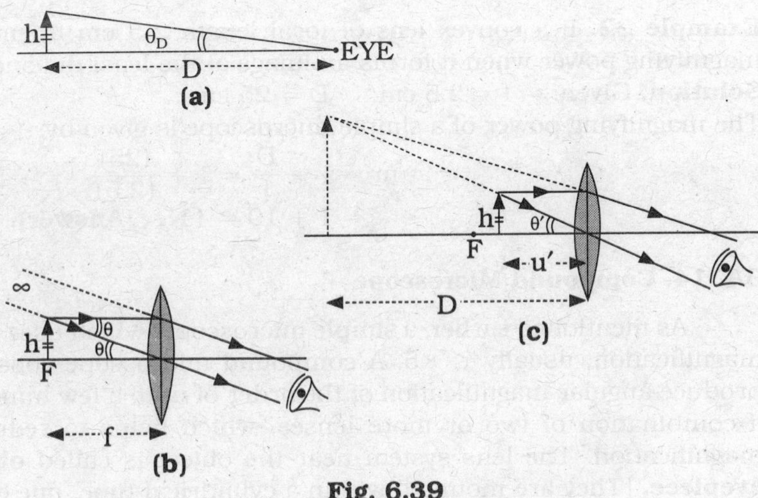

Fig. 6.39

its effective (apparent) size is determined by the angle it subtends at the eye, i.e. by the visual angle θ. In the Fig. 6.39b, the angle subtended on the lens, and therefore on the eye, is,

$$\theta = \frac{h}{f} \quad ..... (6.42)$$

Since f < D, the comparison of Eqs. 6.41 and 6.42 shows that the visual angle is increased by the insertion of the lens ($\theta > \theta_D$). Hence now the size of the image as seen by the eye is bigger.

The **magnifying power** (or **angular magnification**) of a microscope is **defined** as the "ratio of the angle subtended by the final image at the eye to the angle subtended by the object at the eye, when each is located at the least distance of distinct vision form the eye." Like the linear magnification of a lens, it is also denoted by m, but note that the quantity angular magnification is different from the linear magnification Thus it is given by,

$$m = \frac{\theta}{\theta_D} = \frac{h/f}{h/D}$$

Or

$$\boxed{m = \frac{D}{f}} \quad ..... (6.43)$$

(Magnifying Power; Object at F)

Since for the eye the object is at infinity, the eye muscles are relaxed. So in this case the process of focusing by the eye is said to be one of **normal adjustment.** The magnifying power of a lens is expressed as a multiple, e.g. ×2, ×3, etc. or sometimes as 2×, 3×, etc.

The magnifying power can be increased a bit more by bringing the object nearer to the lens, as shown in Fig. 6.39c. Now the lens forms an erect and virtual image of the object, at the same side as the object. The eye muscles are strained, but a clear image is formed at the retina. Using the lens formula (1/v – 1/u = 1/f) and putting u = –u′ and v = –D for this case, we have,

$$-\frac{1}{D} + \frac{1}{u'} = \frac{1}{f}$$

Multiply all through by D,

$$\frac{D}{u'} = 1 + \frac{D}{f}$$

$$\therefore \quad m = \frac{\theta'}{\theta_D} = \frac{h/u'}{h/D} = \frac{D}{u'}$$

Or

$$\boxed{m = 1 + \frac{D}{f}} \quad ..... (6.44)$$

Magnifying Power; Object with in F

The Eqs. 6.43 and 6.44 reveal that smaller is f, larger is the value of m. However, as the focal length is reduced, the defects of the image produced by the lens appreciably increase. Therefore, in practice, the magnifying power is limited to a single digit value, preferably not more than ×5.

Example 22. If a convex lens of focal length 2.5 cm is employed as a simple microscope, calculate its magnifying power when it forms an image at the least distance of distinct vision (D = 25 cm).

Solution. Given: f = 2.5 cm; D = 25 cm

The magnifying power of a simple microscope is given by,

$$m = 1 + \frac{D}{f} = 1 + \frac{(25)}{(2.5)}$$

$$= 1 + 10 = 11 \quad \textbf{(Answer).}$$

6A. 14. Compound Microscope

As mentioned earlier, a simple microscope (which uses a single lens) can produce only a small angular magnification, usually ≤ ×5. A compound microscope consists of two converging lenses, which together produce angular magnification of the order of upto a few hundreds, e.g. ×250. Usually, each lens is actually a combination of two or more lenses, which helps to reduce the defects of image and to increase the magnification. The lens system near the object is called **objective** and the other near the eye is called **eyepiece.** [They are mounted within a cylindrical tube, one each at the two ends of the tube. The objective is fixed but the eyepiece is separately mounted in a small coaxial tube, which can slide within the main tube. Usually the tube is vertically mounted in a metallic stand, with a heavy base. With a simple mechanical arrangement, the tube can be moved up and down to vary the distance between the object and the objective. The object is placed below the objective].

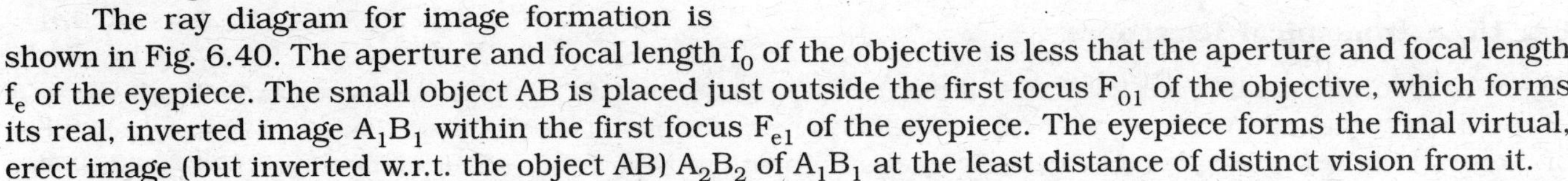

Fig. 6.40

The image formation by a compound microscope is a 2-step process:

(i) The objective forms a magnified (real) image of the tiny object.

(ii) This real image acts as an object for the eyepiece, which forms a further magnified (virtual) image of the object.

The ray diagram for image formation is shown in Fig. 6.40. The aperture and focal length f_0 of the objective is less that the aperture and focal length f_e of the eyepiece. The small object AB is placed just outside the first focus F_{01} of the objective, which forms its real, inverted image A_1B_1 within the first focus F_{e1} of the eyepiece. The eyepiece forms the final virtual, erect image (but inverted w.r.t. the object AB) A_2B_2 of A_1B_1 at the least distance of distinct vision from it.

We now calculate the magnifying power of the microscope. We have,

$$O_1F_{02} = f_0\ ;\quad F_{e1}O_2; = f_e;\qquad B_2O_2 = D$$

The magnification produced by the objective is (vide Eq. 6.28, u_0 and v_0 are object distance and image distance, respectively, for the objective),

$$m_0 = \frac{A_1B_1}{AB} = -\frac{v_0}{u_0}$$

The magnification produced by the eyepiece is (vide Eq. 6.44),

$$m_e = \frac{A_2B_2}{A_1B_1} = 1 + \frac{D}{f_e}$$

By definition, the magnifying power of the microscope is given as,

$$m = \frac{\beta}{\alpha}$$

$$\approx \frac{\tan\beta}{\tan\alpha} \qquad (\because \text{ for small angles, } \tan\theta \approx \theta)$$

$$= \frac{A_2B_2/B_2O_2}{P\,B_2/B_2O_2}$$

$$= \frac{A_2B_2}{P\,B_2} = \frac{A_2B_2}{AB}$$

$$= \frac{A_2B_2}{A_1B_1} \times \frac{A_1B_1}{AB}$$

$$\Rightarrow \qquad m = m_0 \times m_e$$

Or

$$m = -\frac{v_0}{u_0}\left(1 + \frac{D}{f_e}\right) \qquad \text{..... (6.45 a)}$$

Since the tiny object is placed just outside the focus F_{01} of the objective, $u_0 = AO_1 \approx F_{01}O_1 = f_0$. Similarly, B_1 is close to O_2. So, $v_0 = O_1B_1 \approx O_1O_2 = L$ (say). The length L is nearly equal to the length of the microscope tube, since the objective and the eyepiece are situated at the end of the tube. Therefore, we get,

$$m = -\frac{L}{f_0}\left(1 + \frac{D}{f_e}\right), \qquad \text{..... (6.45 b)}$$

which is the required expression for the magnifying power of a compound microscope. The negative sign of m means that the final image is real and inverted.

According to the Eq. 6.45b, the smaller are the focal lengths of objective and eyepiece, the higher is the magnification of the microscope. However, the f-values cannot be reduced beyond a limit, because as f decreases the lens thickness increases, which leads to larger degree of defects of image, which in turn reduces the clarity of the image.

Example 23. A compound microscope has a magnification of ×30. The focal length of its eyepiece is 5 cm. Assuming that the final image is formed at the least distance of distinct vision (25 cm), calculate the magnification produced by the objective. (CBSE; 1991).

Solution. Given: $f_e = 5$ cm; $m = 30$; $D = 25$ cm

The magnification (magnifying power) produced by a compound microscope is given by,

$$m = m_0 \times m_e$$

We know,

$$m_e = 1 + \frac{D}{f_0} = 1 + \frac{(25)}{(5)} = 6$$

$$\therefore \quad m_0 = \frac{m}{m_e} = \frac{30}{6} = 5 \quad \textbf{(Answer).}$$

6A. 15. Astronomical Telescopes

As mentioned earlier, a telescope is meant for viewing distant objects. An astronomical telescope is particularly designed to view very far off heavenly bodies, viz. planets, stars, galaxies, etc. We have two types of them: (i) the **refracting type,** which use a lens to form the first image of the object and (ii) the **reflecting type,** which employ a spherical mirror to form the first image.

(i) Refracting Type Astronomical Telescope

It consists of two converging lenses, the objective (towards the object to be viewed) and the eyepiece (towards the eye). The objective has a large diameter, which has two main advantages: (a) it gathers large amount of light from the distant object like a star, so that even faint stars can be seen and (b) the resolving power of the telescope is increased (to be discussed in the next section). The eyepiece is actually a system of more than one lens, which helps to reduce defects of image and to enhance magnification. [The objective and the eyepiece are mounted at the ends of a cylindrical tube (the eyepiece end is narrow). The tube is mounted on a heavy support, which also enables to tilt the tube in the direction of the heavenly object to be seen. The eyepiece is mounted in a separate small coaxial tube, which can slide within the main tube. This enables to vary its distance from the objective].

The image formation is a 2-step process:

(a) The objective forms a magnified (real) image of the object.

(b) This image serves as the object for the eyepiece, which forms a further magnified (virtual) image.

The ray diagram for the image formation is shown in Fig. 6.41. The parallel light rays from the distant object enter the objective to form a real inverted image A_1B_1, which lies within the (first) focus F_{e1} of the eyepiece. Since the incident rays are parallel, A_1B_1 lies in the (second) focal plane of the objective. A_1B_1 acts as an object for the eyepiece, which then forms its erect (virtual) image A_2B_2 at the least distance of distinct vision (D) from it. So A_2B_2 is the final image seen by the eye (note that it is inverted w.r.t. the object).

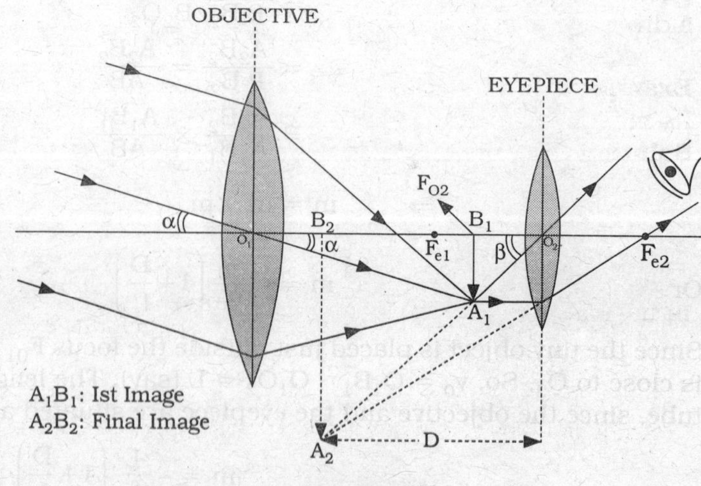

Fig. 6.41

The **magnifying power (or angular magnification) of a telescope** is **defined** as the "ratio of the angle subtended by the final image at the eye to the angle subtended by the object at the eye." As for a microscope, it is denoted by m. Now we shall calculate it in terms of the focal lengths f_0 and f_e of the objective and the eyepiece, as done earlier for the compound microscope. We have,

$$f_0 = O_1B_1; \qquad f_e = O_2F_{e1} \text{ or } O_2F_{e2}; \quad D = O_2B_2$$

By definition,

$$m = \frac{\beta}{\alpha}$$

$$\simeq \frac{\tan\beta}{\tan\alpha} \ (\because \text{ for small angles, } \tan\theta \approx \theta)$$

$$= \frac{A_1B_1/O_2B_1}{A_1B_1/O_1B_1}$$

$$= \frac{O_1B_1}{O_2B_1} = = \frac{f_0}{(-u_e)}$$

u_e denotes the object distance for the eyepiece. The image distance v_e for the eyepiece is $O_2B_2 = -D$ (it is –ve, according to our sign convention). Using the lens formula ($1/v - 1/u = 1/f$) for the eyepiece,

$$\frac{1}{(-D)} - \frac{1}{(-u_e)} = \frac{1}{f_e}$$

$$\Rightarrow \qquad \frac{1}{u_e} = \frac{1}{f_e} + \frac{1}{D} = \frac{1}{f_e}\left(1+\frac{f_e}{D}\right)$$

$$\therefore \qquad m = \frac{f_0}{f_e}\left(1+\frac{f_e}{D}\right) \qquad \text{.....(6.46)}$$

The negative sign of m means that the image is real and inverted.

It follows from the Eq. 6.46 that to have a large value of m, the focal length f_0 of the objective should be large and the focal length f_e of the eyepiece should be small.

If the eyepiece is so adjusted that the objective forms image in the focal plane of the eyepiece, the rays emerging from the eyepiece are parallel. Then the final image is formed at infinity and so the eye is least strained when viewing it (normal adjustment). But the magnification is a bit reduced, given by

$$m = -\frac{f_0}{f_e} \text{ (Normal adjustment)} \qquad \text{..... (6.47)}$$

Since the objective forms image in the focal place of the eyepiece, in this case the distance between the two lenses is equal to $(f_0 + f_e)$.

(ii) Other Refracting Type Telescopes

A **terrestrial telescope** forms an erect image of the distant object. It is achieved by inserting a third converging lens between the objective and the eyepiece. But it substantially increases the length of the telescope tube, which is a major disadvantage. A **Galilean telescope** also forms an erect image, but by using a diverging lens as the eyepiece. But that makes the field of view small, which is a major disadvantage.

Example 24. An astronomical telescope consists of two lenses, placed 50 cm apart, in normal adjustment. Its magnifying power is ×9. Determine the focal lengths of the lenses.

Solution. We are given that,

$$m = \frac{f_0}{f_e} = 9$$

$$\Rightarrow \qquad f_0 = 9\, f_e$$

In normal adjustment, the distance between the lenses is equal to the sum of their focal lengths.

$$\therefore \qquad f_0 + f_e = 50$$

$$\Rightarrow \qquad 9\, f_e + f_e = 50$$

$$\Rightarrow \qquad f_e = \frac{50}{10} = 5 \text{ cm} \quad \textbf{(Answer).}$$

and $\quad f_0 = 9\, f_e = 9\times 5 = 45$ cm **(Answer).**

(iii) Reflecting Type Telescopes

A major requirement of a telescope is to have a high resolving power (to be discussed in the next section), which demands that the objective should have as large a diameter as possible. Besides, the larger is the diameter, the larger is its light-gathering power, which enables to see even faint stars with the telescope. But as the diameter of the objective lens increases, three serious problems arise:

(i) It is difficult to make very large (and so thick) lenses.

(ii) The lens becomes too heavy to be supported on its (thin) edges.

(iii) The defect (or aberrations) of the image increase as the size and thickness of the lens increases. [Two main defects of image are spherical aberration (arising from the spherical shape of the lens surfaces) and chromatic aberration (arising from the existence of light of different wavelengths in the incident white light).

The problems are overcome by using a concave mirror as the objective, in place of the lens, which provides the following advantages:

(i) A mirror has no chromatic aberration (it reflects the light falling upon it).

(ii) The spherical aberration can be eliminated by using a parabolic mirror (instead of a normal spherical mirror).

(iii) A mirror weighs much less than a lens of the same diameter. Besides, unlike a lens it does not need to be supported on its back surface.

(iv) Since the mirror size can be much larger than the maximum possible size of a lens, reflecting type telescopes can have much higher resolving power and light gathering power.

However, a reflecting type telescope has an obvious problem. The parallel light rays from a point on the distant object are incident upon the objective, which reflects and converges them towards a point in its focal plane. The reflected rays are collected by an eyepiece to give the final image of the object. So the observer, looking through the eyepiece, has to sit in a region somewhere in the focal plane of the mirror, i.e. somewhere within the bounds of the mirror [as done for the large 200 inch (~5 m) telescope at Mt. Palomar, California, U.S.A.]. Hence he obstructs a part of the light falling on the mirror. This problem is partly solved by placing a secondary small mirror in the focal plane of the main mirror, as done in the Newtonian and Cassegrainian telescopes, briefly described in the following.

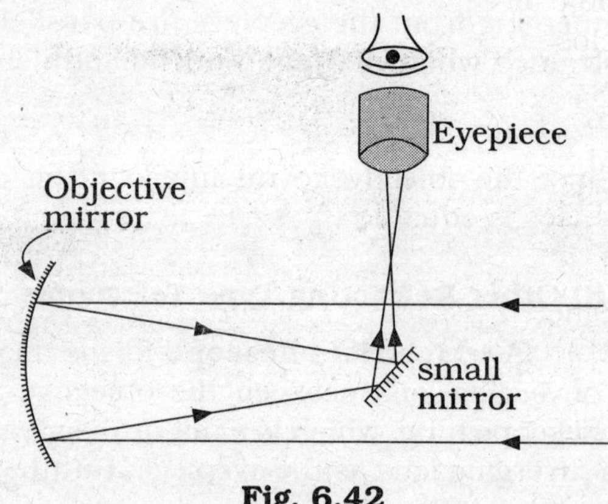

Fig. 6.42

Newtonian Reflecting Type Telescope

The telescope is pointed towards the distant object (Fig. 6.42). Parallel rays from the object are incident upon a parabolic mirror, which reflects them towards the focal plane of the mirror. A small plane mirror centrally placed in the focal plane intercepts the rays and reflects them towards an eyepiece. A real, inverted image of the object is formed within the focus of the eyepiece, which acts as an object for the eyepiece. The eyepiece finally produces a magnified virtual image of this object. Note that the final image is inverted with respect to the main object.

Cassegrainian Reflecting Type Telescope

As before, the telescope is pointed towards the distant object, e.g. a star. The incident parallel rays from the object are reflected by a large parabolic mirror towards the focal plane of the mirror (Fig. 6.43). A small convex mirror centrally placed in the focal plane intercepts the rays and reflects them towards an eyepiece, passing through a central hole provided in the objective mirror. The rays meet in front of the eyepiece, which produces a magnified virtual image of the object.

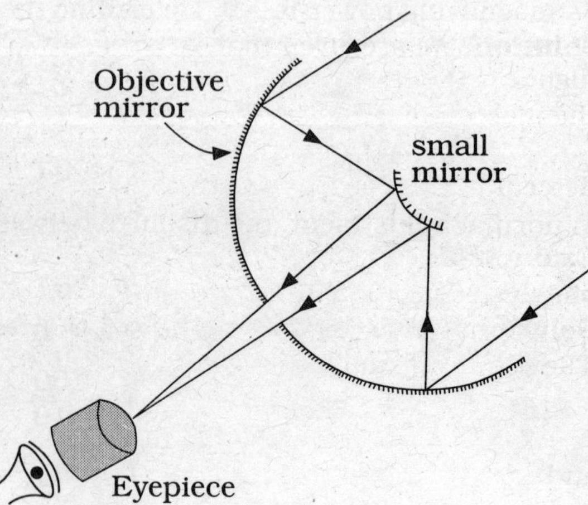

Fig. 6.43

Example 25. A reflecting type telescope has a large mirror for its objective, with radius of curvature equal to 80 cm. What is the magnifying power of the telescope, if the eyepiece used has a focal length of 1.6 cm. (NCERT).

Solution. Given: R = 80 cm; f_e = 1.6 cm

The focal length of the objective mirror is half of its radius of curvature, i.e.

$$f_0 = \frac{R}{2} = \frac{(80)}{2} = 40 \text{ cm}$$

The magnifying power of the telescope is given by,

$$m = \frac{f_0}{f_e} = \frac{(40)}{(1.6)}$$

$$= 25 \quad \textbf{(Answer).}$$

6A. 16. Resolving Power of Optical Instruments

The **resolving power of an optical instrument** is **defined** as the "ability of the instrument to form distinct images of two objects lying close to each other".

For a microscope, the two objects are point objects lying close to the objective. If d is the minimum resolvable distance between the two objects, the resolving power is given by,

$$(\text{R.P.})_{\text{Microscope}} = \frac{1}{d} = \frac{2\mu \sin\theta}{\lambda} \qquad \text{..... (6.48)}$$

[θ: half-angle subtended at the objective by the cone of light rays from any one of the point objects (Fig. 6.44); μ: refractive index of the medium between the object and the objective; λ: wavelength of the light illuminating the object; since the two objects lie very close together, θ is very nearly the same for them].

μ sin θ is called the numerical aperture (n.a.) of the objective. Thus, the larger is the n.a. of the objective, the higher is the resolving power. Also, the lower is the wavelength of light (or higher is the frequency), the higher is the resolving power. Thus the R.P. of a microscope is higher if the objects are illuminated with blue light, than with the red light.

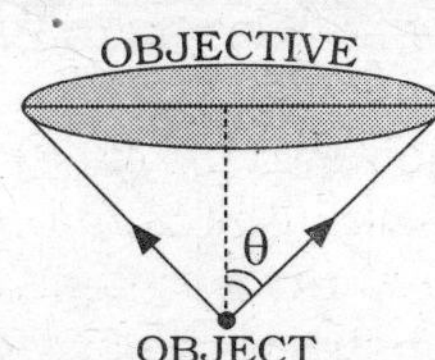

Fig. 6.44

For a telescope, the two objects are distant point objects (e.g. stars) with small angular separation. Since they are very distant, so in spite of their actual huge size they look like point objects to the eye.

If dθ is the angle subtended at the position of the objective by the two objects (Fig. 6.45), the R.P. is given by,

$$(\text{R.P.})_{\text{Telescope}} = \frac{1}{d\theta} = \frac{D}{1.22\,\lambda} \qquad \text{..... (6.49 a)}$$

$$d\theta = \frac{1.22\,\lambda}{D} \qquad \text{..... (6.49 b)}$$

D is the diameter (or aperture) or the objective. The larger is the aperture, the higher is the R.P. Further, the lower is the wavelength (or higher is the frequency) of the incoming light, the higher is the R.P.

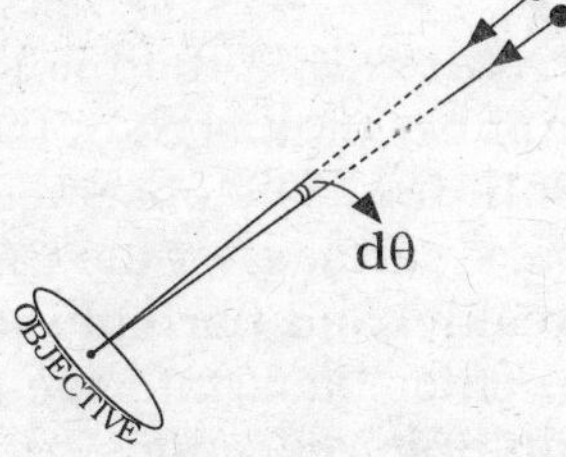

Fig. 6.45

In the next chapter (Section 6B.11) we shall discus how the diffraction of light fundamentally determines the resolving power of an optical instrument.

Example 26. The objective of an astronomical telescope has a diameter of 100 cm. Calculate the smallest angular separation of two neighbouring stars that can be just resolved by the telescope (take λ = 5000 Å).

Solution. Given: D = 100 cm = 1m; λ = 5000 Å = 5×10^{-7} m

The required smallest angular separation of the two stars is given by,

$$d\theta = \frac{1.22\,\lambda}{D}$$

$$= \frac{1.22\times(5\times10^{-7})}{(1)}$$

$$= 6.1\times10^{-7} \text{ radian} \quad \textbf{(Answer).}$$

6A. 17 Prism Binoculars

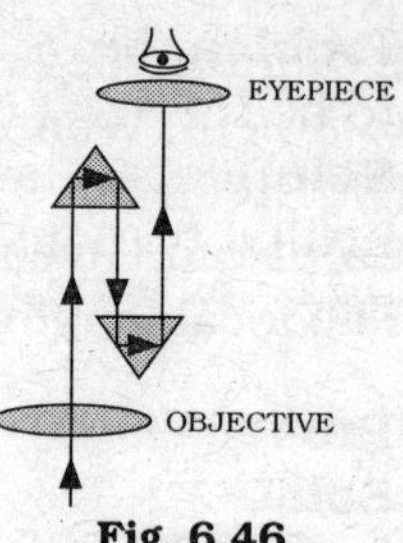

Fig. 6.46

A prism binoculars is a very convenient and compact 2-in-1 telescope. In each of the two arms of the binoculars a pair of totally reflecting prisms are suitably placed, to produce a final erect image of the object (Fig. 6.46). The arrangement also substantially increases the effective length of the telescope, which enhances magnification of the image. Further, the use of both the eyes produces a convenient 3-dimensional image, that not only relaxes the eyes but also provides better perception of depth. [the Latin words 'bini' and 'oculus' mean 'two together' and 'eye', respectively].

Additional knowledge

The world's largest and most powerful optical telescope has been set up on Mount Graham, Arizona state, U.S.A, at a height of 3190 metre above the sea level. It is a binocular telescope, consisting of two similar 8.4 metre wide mirrors, placed at a suitable distance between them. So it can produce 3-dimensional images of the objects. Accordingly, it is known as **Large Binocular Telescope (LBT).**

The telescope has a very high resolving power. Therefore it produces extremely sharp images of the objects. It can view large heavenly objects (and record their images) upto billions of light years away and a small object, like the flame of a candle, upto about 2.5 million kilometres away.

The two mirrors together weigh 16 ton. The telescope took 8 years for its construction, at a cost of $120 million. It is a result of a joint project of three countries, viz U.S.A., Germany and Italy. It started functioning in the year 2005.

Solved Conceptual Problems

[Including HOTS Questions]

Problem 1. An object is placed between two plane mirrors. Why do the distant images get fainter and fainter ? (NCERT).

Solution. The observed large number of multiple images are formed by repeated multiple reflections at the two mirrors. At each reflection, a part of the light energy is absorbed (and so lost). Hence the successive images become increasingly fainter.

Problem 2. A man holding a lighted candle in front of a thick glass mirror and viewing it obliquely sees a number of images of the candle. What is the origin of these multiple images ? (NCERT).

Solution. The observed few multiple images result from repeated reflection of light between the front and back surfaces of the thick glass mirror. At the front surface, the incident light gets partially reflected (only weakly) and partially refracted into the glass. The refracted light is strongly reflected at the back surface, to enter back into the glass and again get partially reflected and refracted at the front surface. Thus the multiple reflections occurring within the thickness of the glass form multiple images. The effect becomes prominent when the light falls obliquely on the mirror and the mirror thickness is large, because then the resulting multiple images are appreciably separated from one another.

Problem 3. For driving in a motor car, what type of mirror is preferred to use for observing the traffic at back side and why ?

Solution. A convex mirror is fitted in a car for conveniently viewing the traffic at the back, for two reasons. (i) For all the objects at the back side, the mirror forms an erect and diminished virtual image and (ii) it provides a much larger field of view than a plane or a concave mirror.

Problem 4. A virtual image cannot be obtained on a screen. Then, how a human eye is able to see it ?

Solution. A virtual image cannot be obtained on a screen, because the rays forming the image do not actually meet but only appear to meet at the position of the image. Our eye is able to see it, because our eye lens focuses the rays at the retina. Thus the virtual image acts as a virtual object for our eye, of which it forms a real image, with the retina as the screen.

Problem 5. We are able to read a newspaper because the light from the surroundings falling on it is believed to be reflected into our eyes. But then why do we not see even a faint image of ourselves in the newspaper ?
Solution. Because of the rough surface of the newspaper the light falling upon it is diffusely scattered by it in all directions. This is called diffuse reflection. The formation of an image requires a regular reflection, i.e. selective scattering of light in one direction.

Problem 6. What makes a diamond to shine brilliantly ?
Solution. The sparkling of diamonds results from total internal reflection of light. Diamond has a high refractive index (2.42). Therefore the value of critical angle i_c for a diamond is low, viz. nearly 24° [$i_c = \sin^{-1}(1/\mu)$]. The faces of the diamond are so cut that when a light beam enters a face, it undergoes multiple repeated total internal reflections at a large number of other faces. Since there is no loss of light intensity upon a total reflection, the intensity of the light beam remains as strong as before, when it emerges from the diamond. Hence the diamond sparkles when exposed to light.

Problem 7. What is the cause of twinkling of stars ? Explain.
Solution. The twinkling of a star is due to variation in refractions of its light in passing through the earth's atmosphere. The atmosphere consists of regions of air of different densities, in which the conditions of temperature, pressure, humidity, etc. are constantly fluctuating. Accordingly, the refractive indices of the regions keep on fluctuating. Hence the apparent position of the star keeps on changing, thus producing the observed twinkle.

Problem 8. Does the apparent depth of a water tank change when viewed obliquely, instead of normally ? If so, does it increase or decrease ?
Solution. When the water tank is viewed obliquely, the angle of refraction in air of the light coming from its bottom increases. Accordingly, the bottom appears as raised, i.e. the apparent depth of water changes, such that it decreases in magnitude.

Problem 9. The two surfaces of a lens have different radii of curvature. The lens forms image of an object lying on its axis. Will the position of the image change if the lens is reversed ? Explain.
Solution. The position of the image will remain unchanged upon reversing the lens. According to the lens maker's formula, we have,

$$\frac{1}{f} = (\mu - 1)\left(\frac{1}{R_1} - \frac{1}{R_2}\right)$$

If R_1 and R_2 are interchanged, the value of f does not change (its sign changes, but that is of no physical significance to us here). Then, according to the relation

$$\frac{1}{v} - \frac{1}{u} = \frac{1}{f}$$

the value of v does not change, if u remains the same.

Problem 10. The refractive index of the material of a concave lens is n. It is immersed in a medium of refractive index n_1. A parallel beam of light is incident on the lens. Trace the path of emergent rays in each of the following cases: (a) $n_1 > n$, (b) $n_1 < n$, (c) $n_1 = n$. (CBSE; 2000, 2001).
Solution.

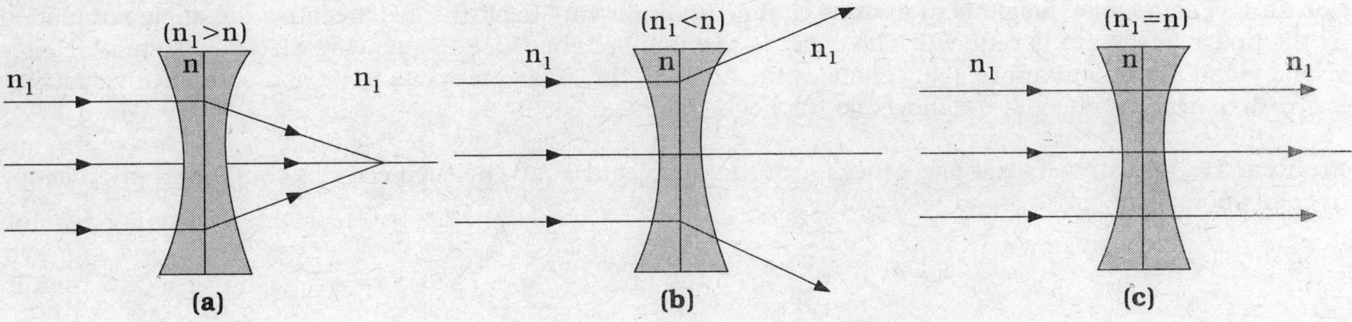

Problem 11. If a concave mirror and a concave lens are immersed in water, will the focal length of either of them change ? Explain why or why not.

Solution. The focal length of the concave mirror will not change, because it is simply equal to half of the radius of curvature (f = R/2), which is independent of the nature of medium in which the mirror exists.

The focal length of the lens will change, because as seen from the lens maker's formula, it depends on the nature of medium surrounding the lens. It will increase upon immersing the lens in water, because,

$$^{water}\mu_{glass} < {}^{air}\mu_{glass}$$

Problem 12. Will the appearance of a blue object change when viewed in sodium lamp light ? If so, how ?

Solution. A sodium lamp gives yellow colour light. If this light falls on a blue object, it will be absorbed by the object. Hence the appearance of the object will change from yellow to black.

Problem 13. If a beam of white light falls on a hollow prism, do we expect to see a spectrum when it emerges on the other side of the prism ? Explain why or why not.

Solution. A hollow prism contains air. Hence the white light passing through it will not be dispersed (the refraction taking place at its thin glass faces will be negligible). Therefore, no spectrum will be formed.

Problem 14. The critical angle for the material of a prism is 45°. The angle of the prism is also 45°. If a light ray falls normally on one of the faces of the prism, trace the course of the refracted ray. Also find the refractive index of the material of the prism.

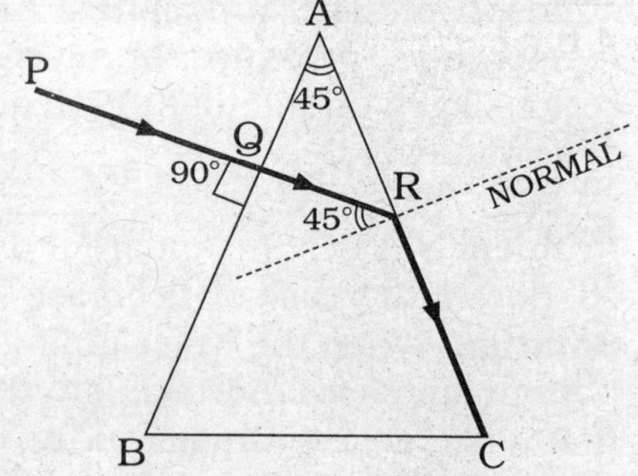

Solution. The ray PQ is normally incident on the face AB of the prism. It moves straight and strikes the face AC at R. The angle of incidence at R is 45°, which is equal to the given value of the critical angle (i_c = 45°). The refraction at R is from dense to rare medium. Therefore, at R the refracted ray QR suffers total internal reflection and moves along RC, as seen in the figure.

The refractive index is found as,

$$\mu = \frac{1}{\sin i_c} = \frac{1}{\sin 45°} = \frac{1}{(1/\sqrt{2})} = \sqrt{2} = 1.41$$

Problem 15. The objective of telescope A has diameter 3 times that of the objective of telescope B. How much greater amount of light is gathered by A as compared to B ? Show that the range of A is three times the range of B. (NCERT).

Solution. Since the objective of A has diameter three times of B, its cross-sectional area is 9 times that of B. Accordingly, the light-gathering capacity of A is 9 times that of B.

The intensity of light arriving from a star on earth is inversely proportional to the square of the distance of the star from earth,

$$\text{Intensity} \propto \frac{1}{(\text{Distance})^2}$$

Accordingly, if the distance of a star increases 3 times, its intensity will reduce 9 times. Then it can only be seen by the telescope A, because its light gathering capacity is 9 times that of A. Thus we find that the viewing range of A is 3 times that of B.

Problem 16. In viewing through a magnifying glass, one usually positions one's eye very close to the lens. Does angular magnification change if the eye is moved back ? (NCERT).

Solution. The angular magnification does change upon moving back the eye, because the angle subtended by the image at the eye is reduced. The effect is pronounced, because the viewing distance is small. Hence even a small movement appreciably changes the angle. If the image happens to be at a comparatively much larger distance, the effect of the movement is negligible.

Problem 17. Explain why is it necessary for both the objective and the eyepiece of a compound microscope to be of short focal length.

Solution. Small values of focal lengths of eyepiece and objective provide high magnifications. The angular magnification of the eyepiece is $\left(1+\frac{D}{f_e}\right)$. Therefore, smaller is the focal length f_e of the eyepiece, the higher is its magnification.

The magnification of the objective is $\frac{v}{u}$. Since in a compound microscope the object is always placed close to the focus, $u \approx f_0$. Therefore the magnification is very nearly equal to $\frac{v}{f_0}$. So, once again, smaller is the focal length f_0 of the objective, higher is its magnification.

Problem 18. How will the resolving power of an astronomical telescope and the intensity of the image produced by it will be affected, if the size of aperture of its objective is doubled.
Solution. Doubling the aperture size means doubling the diameter (D) of the objective. The resolving power (R.P.) is given by,

$$\text{R.P.} = \frac{D}{1.22\,\lambda}$$

Accordingly, when D is doubled, the resolving power is also doubled.

The intensity of the image depends upon the amount of light gathered by the objective. Doubling the diameter increases the cross-sectional area of the objective 4 times. Hence the image intensity also becomes 4 times.

Problem 19. Briefly explain how optical fibres can transmit light over long distances, with little loss of intensity.
Solution. An optical fibre is coated with a thin film of a material of refractive index less than that of the material of the fibre. When light enters at one end of the fibre at near grazing incidence, the angle of incidence is greater than the critical angle. Hence it undergoes repeated multiple total internal reflections within the fibre, thus travelling long distances with little loss of intensity, before emerging at the other end of the fibre.

Problem 20. A rainbow is never observed from the surface of the moon. Why ?
Solution. A rainbow is formed by refraction of sunlight through water drops in the atmosphere. The water drops are initially formed in the clouds by condensation of water vapour around dust or other particles present in the earth's atmosphere. But unlike the earth, the moon has no atmosphere. Hence a rainbow is never formed around it.

Problem 21. If the convex crystalline lens of the eye is replaced by a concave lens, will the eye be able to see objects as before ? If not, explain why.
Solution. The eye sees the objects lying in front of it by virtue of the light sensors (rods and cones) present on the retina, which acts as the screen of the eye. Therefore only a real image can be seen by it, with the light rays actually meeting at the retina. Such an image can only be formed by a convex lens. Since a concave lens always forms a virtual image, for all position of an object, an eye fitted with a concave lens cannot see the objects lying in front of it.

Problem 22. If the position of an object in front of the eye is varied, can a virtual image of the object be formed on the retina.
Solution. The retina acts as the screen of the eye. We know that only a real image can form on a screen, because the refracted rays have to actually meet at the screen. Therefore the formation of a virtual image at the retina is ruled out.

Problem 23. In the process of seeing an object, does only the crystalline lens of the eye participate in the image formation on the retina ? If not, explain why.
Solution. No. The eye forms an optical system, consisting of several coaxial curved surfaces. The rays from the object refract through cornea, aqueous humour, crystalline lens and vitreous humour, before focussing upon the retina. The whole optical system participates in the process of image formation.

Problem 24. Explain why usually a myopic or a hypermetropic person needs two pairs of spectacles (or a single pair with bifocal lenses) in old age.

Solution. In old age, because of loss of accommodation, the crystalline lens of the eye virtually becomes a fixed-focus lens set at infinity, i.e. set only for distant objects. Therefore a myopic or a hypermetropic person has to have two pairs of spectacles, one for seeing distant objects and the other for near objects.

Problem 25. Although the eye lens forms an inverted image of an object on the retina, how is it seen as an erect image by us ?

Solution. When light falls upon the sensors (rods and cones) present on the retina, corresponding impulses are sent to the brain through optic nerve. The brain, through some complex physiological process, enables us to see the inverted image as an erect image.

Solved Numerical Problems

Problem 1. A small candle 2.5 cm in size is placed 27 cm in front of a concave mirror of radius of curvature 36 cm. At what distance from the mirror should a screen be placed in order to receive a sharp image ? Describe the nature and size of the image. If the candle is moved closer to be the mirror, how would the screen have to be moved ? (NCERT).

Solution. $u = -27$ cm, $f = \frac{(-R)}{2} = \frac{(-36)}{2} = -18$ cm, $h_1 = 2.5$ cm.

We use the mirror formula,

$$\frac{1}{v} + \frac{1}{u} = \frac{1}{f}$$

$$\Rightarrow \quad \frac{1}{v} = \frac{1}{f} - \frac{1}{u} = \frac{1}{(-18)} - \frac{1}{(-27)}$$

$$= \frac{1}{27} - \frac{1}{18} = -\frac{1}{54}$$

$$\therefore \quad v = -54 \text{ cm}$$

Thus, the screen should be placed at a distance of 54 cm from the mirror, on the same side as the object. **(Answer).**

Next,

$$m = -\frac{v}{u} = -\frac{(-54)}{(-27)} = -2$$

Since, by definition, the magnification is given by $m = h_2/h_1$, we have,

$$\frac{h_2}{h_1} = -2$$

$$\Rightarrow \quad h_2 = -2h_1 = -2\times(2.5) = -5 \text{ cm}$$

Thus, the image size is found to be 5 cm, i.e. an enlarged image. The –ve signs of v and m mean that the image is real and inverted. **(Answer).**

Next, since $\frac{1}{u} + \frac{1}{v} = \frac{1}{f}$, for a given f any change in the object distance u has to be such that $\left(\frac{1}{u} + \frac{1}{v}\right)$ remains constant. ∴ If the candle is moved closer to the mirror (i.e. u is reduced), the screen has to be moved farther from the mirror (i.e. v has to be increased). However, when $u \leq 18$ cm ($f = 18$ cm), the lens will form a virtual image. Then, the screen has no significance. **(Answer).**

Problem 2. A 4.5 cm long needle is placed 12 cm away from a convex mirror of focal length 15 cm. Give the location of the image and the magnification. Describe what happens as the needle is moved farther from the mirror. (NCERT).

Solution. Given: $u = -12$ cm, $f = +15$ cm, $h_1 = 4.5$ cm.

$$\frac{1}{u} + \frac{1}{v} = \frac{1}{f}$$

$$\Rightarrow \quad \frac{1}{v} = \frac{1}{f} - \frac{1}{u} = \frac{1}{(15)} - \frac{1}{(-12)} = \frac{1}{15} + \frac{1}{12}$$

$$= \frac{4+5}{60} = \frac{9}{60}$$

$$\therefore \quad v = \frac{60}{9} = 6.7 \text{ cm}$$

Thus the image is formed at a distance of 6.7 cm from the mirror, behind the mirror. The +ve sign of v shows that it is virtual and erect. **(Answer).**

$$\text{Magnification, } m = -\frac{v}{u} = -\frac{(6.7)}{(-12)} = 0.56. \quad \textbf{(Answer).}$$

[Since, $m = \frac{h_1}{h_2}$, the image size is given by,

$$\frac{h_1}{h_2} = 0.56$$

$$\Rightarrow \quad h_1 = h_2 \times 0.56 = 4.5 \times 0.56 = 2.5 \text{ cm}$$

The +ve sign of m confirms that the image is virtual and erect. It is diminished in size].

In this casc, any change in the object distance (u) has to be such that $\left(\frac{1}{v}\ \frac{1}{u}\right) = \frac{1}{f}$ remains constant. ∴ if the needle is moved farther from the mirror (i.e. u decreases), the image also moves farther from the mirror (∵ v has to increase). It also continuously reduces in size (the increase in v is slower than the increase in u). **(Answer).**

Problem 3. A concave mirror of radius of curvature 32 cm produces an erect image 4 times the size of the object. Determine the position of the object.

Solution. The magnification m is +ve for a concave mirror forming an erect image. ∴ m = +4. Given, R = –32 cm. ∴ f = –(32/2) = –16 cm. We have,

$$m = 4 = -\frac{v}{u}$$

$$\Rightarrow \quad v = -4u$$

Next,

$$\frac{1}{u} + \frac{1}{v} = \frac{1}{f}$$

$$\Rightarrow \quad \frac{1}{u} + \frac{1}{(-4u)} - \frac{1}{(-16)}$$

$$\Rightarrow \quad \frac{3}{4u} = \frac{1}{16}$$

$$\Rightarrow \quad u = \frac{16 \times 3}{4} = -12 \text{ cm}$$

Thus, the object is situated at a distance of 12 cm from the mirror. The –ve sign of u means that it is placed on the left side in front of the mirror. **(Answer).**

Problem 4. An object is placed 15 cm from a convex mirror of radius of curvature 90 cm. Calculate image position and magnification. (HBSE; 1994).

Solution. Given: $u = -15$ cm, $f = (+90)/2 = 45$ cm

$$\frac{1}{u} + \frac{1}{v} = \frac{1}{f}$$

$$\Rightarrow \quad \frac{1}{v} = \frac{1}{(45)} - \frac{1}{(-15)} = \frac{1}{45} + \frac{1}{15} = \frac{4}{45}$$

$$\therefore \quad v = \frac{45}{4} = 11.25 \text{ cm}$$

Thus, the image is formed at a distance of 11.25 cm from the mirror. The +ve sign of v shows that it is virtual and erect and situated behind the mirror. **(Answer).**

$$\text{Magnification, } m = -\frac{v}{u} = -\frac{(45/4)}{(-15)} = \frac{3}{4} = 0.75 \quad \textbf{(Answer).}$$

Thus, the image is 3/4th the size of the object.

Problem 5. An object is placed in front of a concave mirror of focal length 20 cm. The image formed is three times the size of the object. Calculate the two possible distances of the object from the mirror. (CBSE; 1994).

Solution. Given: $f = -20$ cm

The image may be real or virtual. Accordingly, the magnification, m, may be –ve or +ve, i.e.

$$m = -\frac{v}{u} = \pm 3$$

$\therefore$ $$v = \pm 3u$$

Next, $$\frac{1}{u} + \frac{1}{v} = \frac{1}{f}$$

$$\Rightarrow \quad \frac{1}{u} + \frac{1}{(\pm 3u)} = \frac{1}{(-20)}$$

$$\Rightarrow \quad \frac{3 \pm 1}{3u} = -\frac{1}{20}$$

$$\Rightarrow \quad -3u = (60 + 20) \quad \text{or} \quad (60 - 20)$$

$$= 80 \quad \text{or} \quad 40$$

$\therefore u = -\frac{80}{3}$ or $-\frac{40}{3} = -26.7$ cm or -13.3 cm Thus, the two possible distances of the object from the mirror are –26.7 cm or –13.3 cm, in front of the mirror. For the first position, the image is real ($u > f$) and for the second position it is virtual ($v < f$). **(Answer).**

Problem 6. The magnification obtained upon placing an object at a distance of 80 cm in front of a convex mirror is 1/4. If the magnification needs to be equal to 1/6, find where the object should be placed.

Solution. Given: $u = -80$ cm, $m = \frac{1}{4}$. Thus,

$$m = \frac{1}{4} = -\frac{v}{u} = -\frac{v}{(-80)} = \frac{v}{80}$$

$$\Rightarrow \quad v = \frac{80}{4} = 20 \text{ cm}$$

Next, $$\frac{1}{f} = \frac{1}{u} + \frac{1}{v} = \frac{1}{(-80)} + \frac{1}{20} = \frac{3}{80}$$

$$\Rightarrow \quad f = \frac{80}{3} \text{ cm}$$

In the second case, $$m = \frac{1}{6} = -\frac{v}{u}$$

$\therefore$ $$v = -\frac{u}{6}$$

Again, using the mirror formula, $\frac{1}{u} + \frac{1}{v} = \frac{1}{f}$, we get,

$$\frac{1}{u} + \frac{1}{(-u/6)} = \frac{1}{(80/3)}$$

$$\Rightarrow \quad \frac{1}{u} - \frac{6}{u} = \frac{3}{80}$$

$$\Rightarrow \quad -\frac{5}{u} = \frac{3}{80}$$

$$\Rightarrow \quad u = -\frac{400}{3} \text{ cm} = -133.3 \text{ cm}$$

Thus, to obtain a magnification equal to 1/6, the object should be placed at a distance of 133.3 cm in front of the mirror. **(Answer).**

Problem 7. An object is placed at a distance of 36 cm from a convex mirror. A plane mirror is placed in between so that the two virtual images so formed coincide. If the place mirror is at a distance of 24 cm from the object, find the radius of curvature of convex mirror. (PSEB; 1994).

Solution. Both the plane mirror and the convex mirror form a virtual image each of the object, which coincide.

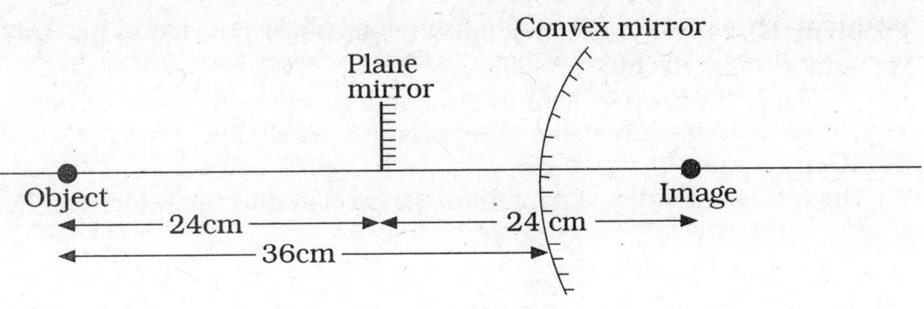

We know that for image formation by the plane mirror, image distance = object distance = 24 cm. It means that virtual image formed by the plane mirror lies 24 cm behind the mirror.

∴ For image formation by the convex mirror, we have,

object distance, u = – 36 cm

Image distance, v = (24 + 24) – 36 = +12 cm

Using the mirror formula,

$$\frac{1}{f} = \frac{1}{u} + \frac{1}{v} = \frac{1}{(-36)} + \frac{1}{(12)} = \frac{2}{36} = \frac{1}{18}$$

Or $\qquad f = 18$ cm

Since f = R/2, we have,

Radius of curvature, R = 2f = 36 cm **(Answer).**

Problem 8. A square wire of side 3.0 cm is placed 25 cm away from a concave mirror of focal length 10 cm. What is the area enclosed by the image of the wire ? The centre of the wire is on the axis of the mirror, with its two sides normal to the axis. (NCERT).

Solution. Given: $\quad u = -25$ cm, $\quad f = -10$cm, $\quad h_1 = 3$ cm

$$\frac{1}{u} + \frac{1}{v} = \frac{1}{f}$$

$$\Rightarrow \quad \frac{1}{v} = \frac{1}{(-10)} - \frac{1}{(-25)} = \frac{1}{25} - \frac{1}{10} = -\frac{3}{50}$$

$$\therefore \quad v = -\frac{50}{3}$$

(The –ve sign of v means that the image is real).

$$m = -\frac{v}{u} = -\frac{(-50/3)}{(-25)} = -\frac{2}{3}$$

Also, $m = (h_2/h_1)$. ∴ we have

$$\Rightarrow \quad h_2 = -\frac{2}{3}h_1 = -\frac{2}{3}\times 3 = -2 \text{ cm}$$

(The –ve sign of the image size, h_2, means that the image is inverted). Since the object is square in shape, the image will also be square in shape. ∴ the area enclosed by it is given by,

Image Area = (side length)^2

= $(-2)^2$ = 4 sq. cm **(Answer).**

Problem 9. The figure below shows the passage of a light ray from air into a liquid of refractive index μ. Find the value of μ.

Solution. The angle of incidence, i = 60°. The incident ray, after refracting into the liquid, deviates through 15° from its original path. ∴ the angle of refraction is,

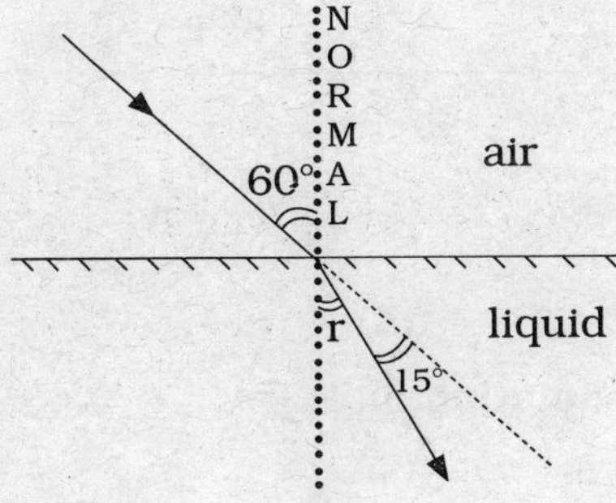

$$r = 60° - 15° = 45°$$

$$\therefore \quad \mu = \frac{\sin i}{\sin r} = \frac{\sin 60°}{\sin 45°} = \frac{\sqrt{3}/2}{1/\sqrt{2}}$$

$$= \frac{\sqrt{3}\times\sqrt{2}}{2} = \frac{\sqrt{3}}{\sqrt{2}} = \sqrt{\frac{3}{2}} = \sqrt{1.5}$$

$$= 1.225.$$ **(Answer).**

Problem 10. (a) The refractive index of glass is 1.5. What is the speed of light in glass ? (speed of light in vacuum, $c = 3\times10^8$ ms^{-1}).

(b) Is the sped of light in glass independent of colour of light ? If not, which of the two colours, red and violet, travels slower in a glass prism ? (NCERT).

Solution. Given: $\mu = 1.5$, $c = 3\times10^8$ ms^{-1}

(a) The refractive index of a medium, μ, is related to the velocity of light in the medium, v_{med}, as

$$\mu = \frac{c}{v_{med}}$$

$$\therefore \quad \mu_{glass} = \frac{c}{v_{glass}}$$

$$\Rightarrow \quad v_{glass} = \frac{c}{\mu_{glass}} = \frac{3\times10^8}{1.5}$$

$$= 2\times10^8 \text{ ms}^{-1} \quad \textbf{(Answer).}$$

(b) The speed of light in glass (v_{glass}) is not independent of the colour of light. In general, v_{med} depends on the wavelength of light, i.e. colour of light. **(Answer).**

We know that $\mu_{violet} > \mu_{red}$. Since μ is inversely proportional to v ($\mu = c/v$), we have,

$$v_{violet} < v_{red}$$

Thus, the violet light travels slower than the red light in a glass prism. **(Answer).**

Problem 11. Monochromatic light of wavelength 600 nm is incident from air on a glass surface. What are the wavelength, frequency and speed of the refracted light ? μ of glass is 1.5. (NCERT).

Solution. Given: $\lambda_{air} = 600 \text{ nm} = 6\times10^{-7}$ m, $\mu_{glass} = 1.5$

(i) The frequency is found from the relation, $c = \nu\lambda$. $\therefore \nu = c/\lambda$.

$$\nu_{air} = \frac{c}{\lambda_{air}} = \frac{3\times10^8}{6\times10^{-7}}$$

$$= 5\times10^{14} \text{ Hz}$$

Upon refraction into glass, the frequency remains unchanged.

$$\therefore \quad \nu_{glass} = \nu_{air} = 5\times10^{14} \text{ Hz} \quad \textbf{(Answer).}$$

(ii) The wavelength changes as ($\lambda' = \lambda/\mu$),

$$\lambda_{glass} = \frac{\lambda_{air}}{\mu} = \frac{600 \text{ nm}}{1.5} = 400 \text{ nm} \quad \textbf{(Answer).}$$

[From the relation $c = \nu\lambda$, we get $c' = \nu'\lambda'$. Since $\nu = \nu'$, we get, on dividing, $\lambda' = \lambda\times(c'/c) = \lambda/(c/c') = \lambda/\mu$].

(iii) The speed changes as ($\mu_{glass} = v_{air}/v_{glass} = c/v_{glass}$),

$$v_{glass} = \frac{c}{\mu_{glass}} = \frac{3\times10^8}{1.5}$$

$$= 2\times10^8 \text{ ms}^{-1} \quad \textbf{(Answer).}$$

Problem 12. The figures A and B show refraction of an incident ray in air at 60° with normal to a glass-air and a water-air interface, respectively. Predict the angle of refraction of incident ray in water at 45° with the normal to a water-glass interface (Fig. C). (NCERT).

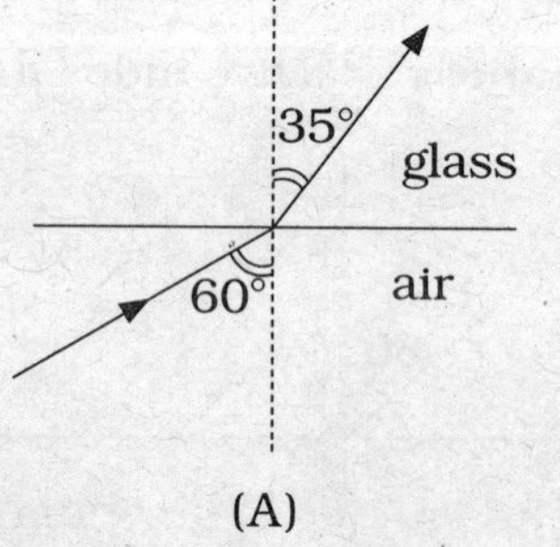

(A)

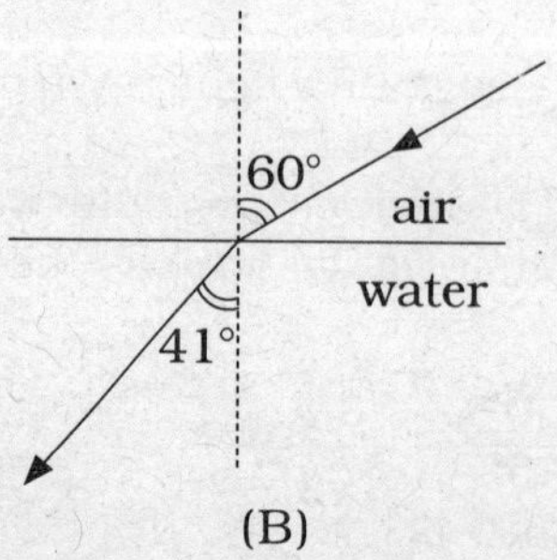

(B)

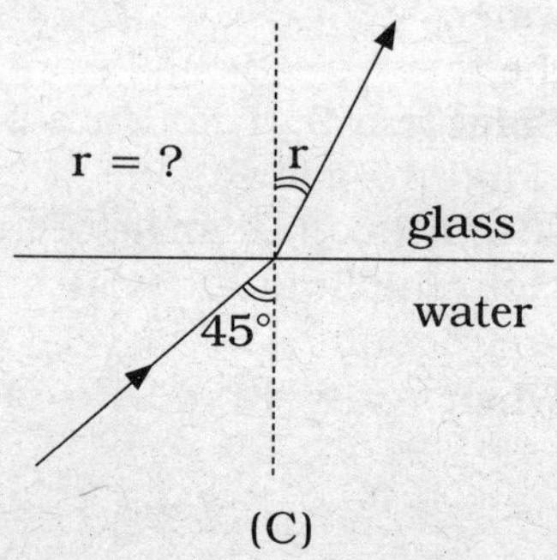

(C)

Solution. In Fig. A, $i = 60°$, $r = 35°$

$$\therefore \quad {}^a\mu_g = \frac{\sin i}{\sin r} = \frac{\sin 60°}{\sin 35°}$$

$$= \frac{0.8660}{0.5736} = 1.51$$

In Fig. B, $i = 60°$, $r = 41°$

$$\therefore \quad {}^a\mu_w = \frac{\sin i}{\sin r} = \frac{\sin 60°}{\sin 44°}$$

$$= \frac{0.8660}{0.6561} = 1.32$$

In Fig. C, $i = 45°$, $r = ?$ We have,

$${}^w\mu_g = \frac{{}^a\mu_g}{{}^a\mu_w} = \frac{\sin i}{\sin r}$$

$$\Rightarrow \quad \frac{1.51}{1.32} = \frac{\sin 45°}{\sin r} = \frac{0.7071}{\sin r}$$

$$\Rightarrow \quad \sin r = \frac{1.32 \times 0.7071}{1.51} = 0.6181$$

$$\therefore \quad r = \sin^{-1}(0.6181)$$

$$= 38.2° \quad \textbf{(Answer).}$$

Problem 13. A tank filled with glycerine ($\mu = 1.47$) upto a depth of 20 m is traversed by a ray of sodium light ($\lambda = 589$ nm) in time t_1. When the tank is filled to the same depth with carbon disulphide ($\mu = 1.63$), the traversal time is found to be t_2. Find the difference $(t_2 - t_1)$.

Solution. Given: $\mu_{gly} = 1.47$, $\mu_{cs_2} = 1.63$, depth = 20 m

$$\mu_{gly} = \frac{c}{v_{gly}}$$

$$\Rightarrow \quad v_{gly} = \frac{c}{\mu_{gly}} = \frac{3 \times 10^8}{1.47}$$

$$= 2.04 \times 10^8 \text{ ms}^{-1}$$

Similarly, $$v_{cs_2} = \frac{c}{\mu_{cs2}} = \frac{3 \times 10^8}{1.63}$$

$$= 1.84 \times 10^8 \text{ ms}^{-1}$$

The respective times taken by the light to traverse the tank are given by,

$$t_2 = \frac{\text{Depth}}{v_{gly}} = \frac{20}{2.04 \times 10^8} = 9.80 \times 10^{-8} \text{ s}$$

$$t_1 = \frac{\text{Depth}}{v_{cs_2}} = \frac{20}{1.84 \times 10^8} = 10.87 \times 10^{-8} \text{ s}$$

$$\therefore \quad (t_2 - t_1) = (10.87 - 9.80) \times 10^{-8} = 1.07 \times 10^8 \text{ s} \quad \textbf{(Answer).}$$

Problem 14. A tank is filled with water to a height of 12.5 cm. The apparent depth of a needle lying at the bottom of the tank is measured by a travelling microscope to be 9.4 cm. What is the refractive index of water ? If water is replaced by a liquid of refractive index 1.63 upto the same height, by what distance would the microscope have to be moved to focus on the needle again ? (NCERT).

Solution. Given: real depth of water = 12.5 cm, apparent depth = 9.4 cm

$$\mu_{water} = \frac{\text{real depth}}{\text{apparent depth}} = \frac{12.5}{9.4}$$

$$= 1.33 \quad \textbf{(Answer).}$$

When water is replaced by another liquid ($\mu = 1.63$), let the apparent depth be y. Then,

$$\mu_{liquid} = 1.63 = \frac{12.5}{y}$$

$$\Rightarrow \quad y = \frac{12.5}{1.63} = 7.7 \text{ cm}$$

Thus the microscope has to be moved up though a distance = (9.4 – 7.7) = 1.7 cm. **(Answer).**

Problem 15. The bottom of a container is 4 cm thick glass (μ = 1.5) slab. The container has two immiscible liquids A and B of depth 6 cm and 8 cm, respectively. What is the apparent position of a scratch on the outer surface of the bottom of the slab, when viewed through the container ? The refractive indices of A and B are 1.4 and 1.3, respectively. (NCERT).

Solution. When a container is filled with a liquid, the image of an object lying at the bottom is seen as displaced upwards. The displacement of the image is given by,

$$\text{Displacement} = \text{Real depth} - \text{apparent depth}$$

$$= \text{Real depth}\left[1-\frac{\text{app. depth}}{\text{real depth}}\right]$$

$$= \text{Real depth}\left[1-\frac{1}{\mu}\right] \quad \left(\because \mu = \frac{\text{real depth}}{\text{app. depth}}\right)$$

$\therefore$ Total displacement of the image of the scratch at the bottom, when seen through 3 media (liquid A + liquid B + glass) is given by,

$$\text{Total Displ.} = 6\left(1-\frac{1}{1.4}\right)+8\left(1-\frac{1}{1.3}\right)+4\left(1-\frac{1}{1.5}\right)$$

$$= \frac{6\times0.4}{1.4}+\frac{8\times0.3}{1.3}+\frac{4\times0.5}{1.5} = 4.89 \text{ cm}$$

Thus the scratch will be seen as raised up by 4.89 cm from its real position. **(Answer).**

Problem 16. A prism ABC has its side AB = BC. If a light ray incident upon it has to travel within it as shown in the figure, calculate the minimum value of the refractive index of the prism material.

Solution. The prism is an isosceles prism, with $\angle B = 90°$. The incident ray suffers total internal reflection at P and Q, such that it turns though 90° at each reflection,

$\therefore$ The angle of incidence at P and Q is,

$$i = \frac{90°}{2} = 45°$$

The total internal reflection will occur when i is equal to or greater than the critical angle i_c ($i \geq i_c$). Since $\mu = 1/\sin i_c$, greater is the refractive index, smaller is the value of i_c. But since under the given condition of refraction, i_c cannot exceed the value of i, the minimum value of μ will correspond to $i = i_c$, i.e. when i_c is just equal to i. Thus,

$$i_c = i = 45°$$

$$\therefore \quad \mu_{min} = \frac{1}{\sin i_c} = \frac{1}{\sin 45°} = \frac{1}{(1/\sqrt{2})}$$

$$= \sqrt{2} = 1.41. \quad \textbf{(Answer).}$$

Problem 17. A cylindrical jar of radius 0.2 m contains glycerine (μ = 1.4) upto a depth of 0.1 m. If a point source of light lies at the centre of the jar's bottom, calculate the surface area of glycerine though which the light is transmitted out.

Solution. As shown in the figure, S is the point source of light. Only those light rays from S which are incident on the glycerine-air interface at angles $i < i_c$ will be able to pass out of glycerine. The rest ($i \geq i_c$) will be totally reflected. The ray with $i = i_c$ passes along the interface, as seen in the figure. Thus only the rays within the circular area of radius OK will pass through.

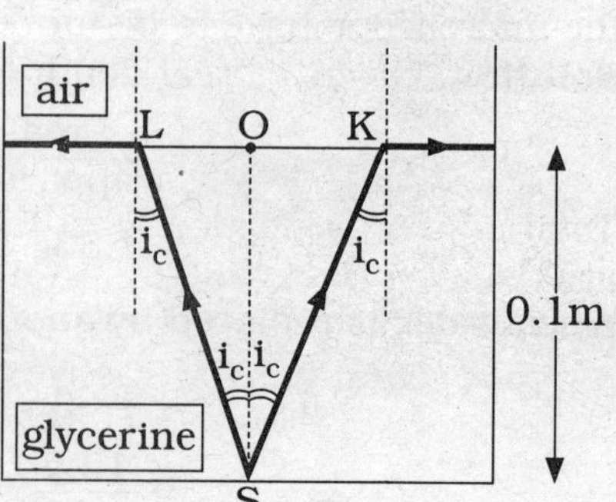

Given: μ = 1.4, OS = 0.1 m

The value of the critical angle i_c is given by,

$$\sin i_c = \frac{1}{\mu} = \frac{1}{1.4} = 0.7143$$

$$\Rightarrow \quad i_c = \sin^{-1}(0.7143) = 45.6°$$

$$\tan i_c = \frac{OK}{OS}$$

$$\Rightarrow \quad OK = OS \tan i_c = 0.1 \times \tan 45.6^\circ$$
$$= 0.1 \times 1.021 = 0.102 \text{ m}$$

$\therefore$ Required surface area $= \pi\,(OK)^2$
$$= 3.14 \times (0.102)^2$$
$$= 0.033 \text{ m}^2 \quad \textbf{(Answer).}$$

Problem 18. If light refracts from a medium into air and the value of the critical angle is 45°, calculate the speed of light in the medium.

Solution. We know that the refractive index is related to the critical angle as,

$$\mu = \frac{1}{\sin i_c} = \frac{1}{\sin 45^\circ} = \frac{1}{(1/\sqrt{2})}$$
$$= \sqrt{2} = 1.41$$

Further, the refractive index of a medium is related to the speed of light as,

$$\mu = \frac{v_{air}}{v_{med}} = \frac{c}{v_{med}}$$

$$\therefore \quad v_{med} = \frac{c}{\mu} = \frac{3 \times 10^8}{1.41}$$
$$= 2.13 \times 10^8 \text{ ms}^{-1} \quad \textbf{(Answer).}$$

Problem 19. The following figure shows a cross-section of a "light pipe" made of a glass fibre of refractive index 1.68. The outer covering of the pipe is made of a material of refractive index 1.44. What is the range of the angles of incident rays with the axis of the pipe, for which total reflection inside the pipe takes place ? (NCERT).

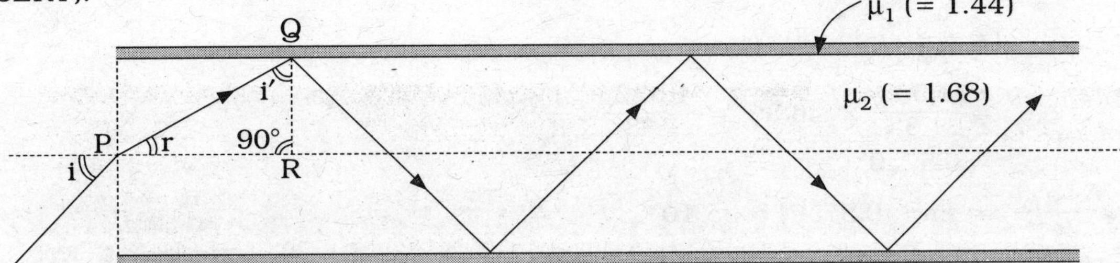

Solution. The light enters the fibre at P, where the angle of incidence is i and the angle of refraction is r. The refracted ray undergoes total internal reflection at Q, where the angle of incidence is i′.

The refractive index of the rare medium (material of the outer covering) is, $\mu_1 = 1.44$, and that of the dense medium (material of the fibre) is, $\mu_2 = 1.68$. Therefore,

$${}^1\mu_2 = \mu \text{ (say)} = \frac{{}^{air}\mu_2}{{}^{air}\mu_2} = \frac{1.68}{1.44}$$

The value of the critical angle i_c is given by the relation,

$$\mu = \frac{1}{\sin i_c}$$

$$\Rightarrow \quad \sin i_c = \frac{1}{\mu} = \frac{1.44}{1.68} = 0.8571$$

$$\Rightarrow \quad i_c = \sin^{-1}(0.8571) = 59^\circ$$

Total internal reflection at Q will occur only when $i' \geq i_c$, i.e. when $i' \geq 59°$. Then, since the angle of refraction at P is given by $r = 90 - i'$, it means that for total reflection at Q, $r \leq 31°$. Thus, we find that the maximum possible value of r is 31°, i.e. $r_{max} = 31°$.

In general, the refractive index of a medium is given by,

$$\mu = \frac{\sin i}{\sin r}$$

We substitute the values of i and r for refraction at P, corresponding to the above condition of total reflection,

$$1.68 = \frac{\sin i_{max}}{\sin r_{max}} = \frac{\sin i_{max}}{\sin 31^\circ}$$

$$\Rightarrow \quad \sin i_{max} = 1.68 \times \sin 31^\circ = 1.68 \times 0.5156 = 0.8662$$

$$\therefore \quad i_{max} = \sin^{-1}(0.8662) = 60^\circ$$

Thus, the light rays with angles of incidence $i \leq 60^\circ$ will undergo total reflection inside the light pipe. **(Answer).**

Problem 20. A prism is made of glass of unknown refractive index. A parallel beam of light is incident on a face of the prism. By rotating the prism, the minimum angle of deviation is measured to be 40°. What is the refractive index of the prism ? If the prism is placed in water (refractive index 1.33), predict the new minimum angle of deviation of a parallel beam of light. The refractive angle of the prism is 60°. (NCERT).

Solution. Given: $A = 60^\circ$, $\delta_m = 40^\circ$

The refractive index of the material of the prism, viz. glass, is given by,

$$^a\mu_g = \frac{\sin[(A+\delta_m)/2]}{\sin(A/2)} = \frac{\sin[(60^\circ+40^\circ)/2]}{\sin(60^\circ/2)}$$

$$= \frac{\sin 50^\circ}{\sin 30^\circ} = \frac{0.766}{0.5} = 1.532$$

Let the new min. angle of deviation (upon placing the prism in water) be δ'_m. When the prism is placed in water, the refractive index of glass with respect to water is given by,

$$^w\mu_g = \frac{\sin[(A+\delta'_m)/2]}{\sin(A/2)}$$

$$\Rightarrow \quad \sin[(A+\delta'_m)/2] = {^w\mu_g} \times \sin(A/2)$$

$$= \frac{^a\mu_g}{^a\mu_w} \times \sin(60^\circ/2)$$

$$= \frac{1.532}{1.33} \times \sin 30^\circ = \frac{1.532}{1.33} \times \frac{1}{2}$$

$$= 0.5759$$

$$\Rightarrow \quad \frac{A+\delta'_m}{2} = \sin^{-1}(0.5759) = 35^\circ 10'$$

$$\Rightarrow \quad A+\delta'_m = 70^\circ 20'$$

$$\therefore \quad \delta'_m = 70^\circ 20' - 60^\circ$$

$$= 10^\circ 20'$$ **(Answer).**

Problem 21. A ray of light is incident at an angle of 60° on one face of a 30° prism. The emergent ray from the prism makes an angle of 30° with the incident ray. Show that the emergent ray is normal to the surface from which it emerges. Calculate the refractive index of the material of the prism. (NCERT).

Solution. Let i_1 and r_1 be the angles of incidence and refraction, respectively, at the point of incidence. Similarly, let i_2 and r_2 be such respective angles at the point of emergence.

Given: $i_1 = 60^\circ$, $A = 30^\circ$, $\delta = 30^\circ$

We know, $i_1 + i_2 = A + \delta$

$$\Rightarrow \quad 60^\circ + i_2 = 30^\circ + 30^\circ = 60^\circ$$

$$\therefore \quad i_2 = 0$$

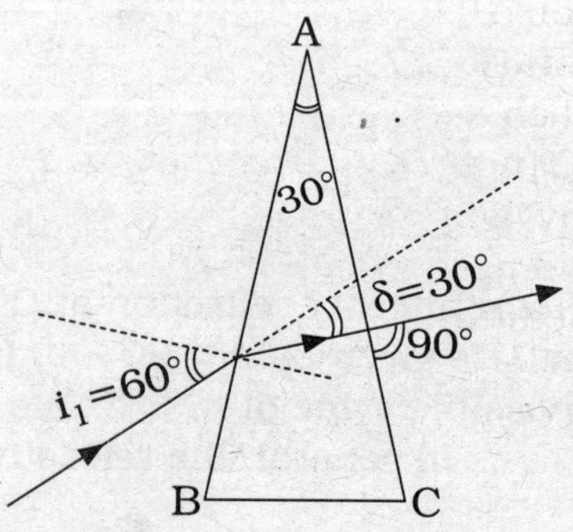

It follows that the emergent ray is normal to the prism surface AC, from which it emerges. **(Answer).**

If the emergent ray is normal to the surface, then the corresponding angle of refraction is also zero, i.e. $r_2 = 0$.

We know, $r_1 + r_2 = A$

$$\Rightarrow \quad r_1 + 0 = 30^\circ$$

$$\Rightarrow \quad r_1 = 30^\circ$$

Then, the refractive index of the prism material is found as,

$$\mu = \frac{\sin i_1}{\sin r_1} = \frac{\sin 60^\circ}{\sin 30^\circ}$$

$$= \frac{\sqrt{3}/2}{1/2} = \sqrt{3}$$

$= 1.73.$ **(Answer).**

Problem 22. White light is passed through a prism of angle 5°. If the refractive indices for red and blue colours are 1.641 and 1.659, respectively, calculate the angle of dispersion between them. (Roorkee; 1979).

Solution. Given: $A = 5°$, $\mu_b = 1.659$, $\mu_r = 1.641$

Angular dispersion $= (\mu_b - \mu_r) A$

$= (1.659 - 1.641) \times 5°$

$= 0.018 \times 5$

$= 0.09°$ **(Answer).**

Problem 23. The minimum deviation suffered by red, yellow and violet beams passing through an equilateral prism are 38.2°, 39.8° and 39.4°, respectively. Determine the dispersive power of the prism material.

Solution. For an equilateral prism, the angle of prism is 60°. Thus, we are given that,

$A = 60°$, $(\delta_m)_r = 38.2°$, $(\delta_m)_y = 38.8°$, $(\delta_m)_v = 39.4°$

$$\therefore \quad \mu_r = \frac{\sin[(A+\delta_m)/2]}{\sin(A/2)} = \frac{\sin[(60°+38.2°)/2]}{\sin(60°/2)}$$

$$= \frac{\sin 49.1°}{\sin 30°} = \frac{0.7558}{(1/2)} = 1.5116$$

$$\mu_y = \frac{\sin[(60°+38.8°)/2]}{\sin 30°} = \frac{\sin(49.4°)}{(1/2)}$$

$$= \frac{0.7593}{(1/2)} = 1.5186$$

$$\mu_v = \frac{\sin[(60°+39.4°)/2]}{\sin 30°} = \frac{\sin(49.7°)}{(1/2)}$$

$$= \frac{0.7627}{(1/2)} = 1.5254$$

∴ The dispersive power of the prism material is given by,

$$\omega = \frac{(\mu_v - \mu_r)}{(\mu_y - 1)} = \frac{(1.5254 - 1.5116)}{(1.5186 - 1)}$$

$$= \frac{0{,}0138}{0.5186} = 0.0266$$ **(Answer).**

Problem 24. A glass sphere of radius 20 cm has a small bubble lying in it along a diameter, at a distance of 10 cm from the centre. If the bubble is viewed along the diameter from the surface nearer to it, how far from the surface will it appear to be ? ($\mu_{glass} = 1.5$).

Solution. As seen in the diagram, the bubble O forms the image I, when seen through the concave surface MQN, along the diameter PQ [note that CM and CQ (= R) are normal at M and Q, respectively; any light ray travelling along them returns along the same path]. μ_1 and μ_2 are the refractive indices of the dense (glass) and rare (air) medium, respectively.

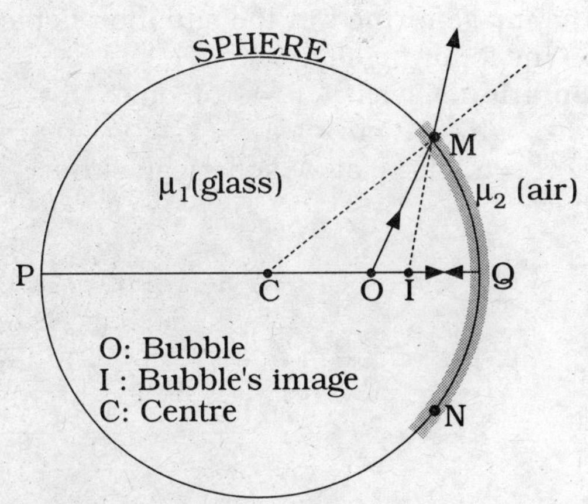

Given: $u = -10$ cm, $R = -10$ cm, $\mu_1 = 1.5$, $\mu_2 = 1$

$$-\frac{\mu_1}{u} + \frac{\mu_2}{v} = \frac{\mu_2 - \mu_1}{R}$$

$$\Rightarrow \quad -\frac{1.5}{(-10)} + \frac{1}{v} = \frac{1-1.5}{(-20)} = \frac{0.5}{20}$$

$$\Rightarrow \quad \frac{1}{v} = \frac{0.5}{20} - \frac{1.5}{10} = -\frac{2.5}{20}$$

$$\therefore \qquad v = -\frac{20}{2.5} = -8 \text{ cm}$$

Thus, the bubble will appear to lie at a distance of 8 cm from the pole Q of the concave surface. **(Answer).**

Problem 25. An air bubble of radius 2 mm is lying inside water. A parallel beam of light travelling in water is incident upon it. Find the position of image formed by refraction at the first surface of the bubble, as well as the position of final image ($\mu_{water} = 4/3$).

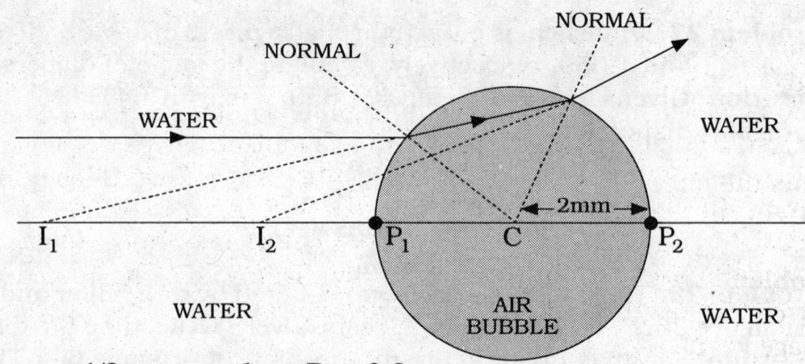

Solution. Given: $\mu_{air} = 1$; $\mu_{water} = 4/3$; $R = 2$ mm $= 0.2$ cm

For image formation at the first surface (convex) of the bubble, we have,

$$u_1 = -\infty; \quad v_1 = ?; \quad \mu_1 = 4/3; \quad \mu_2 = 1; \quad R = 0.2 \text{ cm}$$

Applying the formula for image formation by a curved surface,

$$-\frac{\mu_1}{u_1} + \frac{\mu_2}{v_1} = \frac{\mu_2 - \mu_1}{R}$$

$$\Rightarrow \qquad -\frac{(4/3)}{(-\infty)} + \frac{1}{v_1} = \frac{(1) - (4/3)}{(0.2)} = \frac{-1/3}{0.2} = -\frac{1}{0.6}$$

$$\Rightarrow \qquad v_1 = -0.6 \text{ cm} = 6 \text{ mm}$$

Thus, the image formed by the Ist surface of the bubble is found to be located 6 mm in front of the bubble (at I_1 in the figure; $I_1P_1 = 6$ mm). **(Answer).**

This image acts as the object for the 2nd surface (concave). For image formation by this surface, we have,

$$u_2 = -(I_1P_1 + P_1P_2) = -(6 + 4) \text{ mm} = -1 \text{ cm}; \; v_2 = ?,$$

$$\mu_1 = 1; \quad \mu_2 = 4/3; \quad R = -(0.2) \text{ cm}$$

Again using the relevant formula, we have,

$$-\frac{\mu_1}{u_2} + \frac{\mu_2}{v_2} = \frac{\mu_2 - \mu_1}{R}$$

$$\Rightarrow \qquad -\frac{1}{(-1)} + \frac{(4/3)}{v_2} = \frac{(4/3) - (1)}{(-0.2)} = -\frac{5}{3}$$

$$\Rightarrow \qquad v_2 = -0.5 \text{ cm} = -5 \text{ mm}$$

Thus, the final image I_2 is found to be formed 5 mm in front of the 2nd surface, i.e. 1 mm in front of the Ist surface ($I_2P_1 = 1$ mm). **(Answer).**

Problem 26. Find the position and size of the image formed in the situation depicted in the following figure:

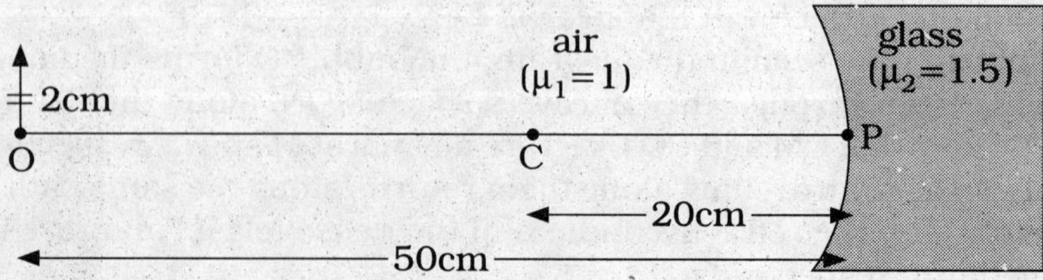

Solution. Given: $u = -50$ cm, $R = -20$ cm, $h_1 = 2$ cm, $\mu_1 = 1$, $\mu_2 = 1.5$

For refraction at a spherical surface, we have,

$$-\frac{\mu_1}{u} + \frac{\mu_2}{v} = \frac{\mu_2 - \mu_1}{R}$$

$$\Rightarrow \qquad \frac{\mu_2}{v} = \frac{\mu_2 - \mu_1}{R} + \frac{\mu_1}{u} = \frac{1.5 - 1}{(-20)} + \frac{1}{(-50)}$$

$$\Rightarrow \qquad \frac{1.5}{v} = -\frac{1}{40} - \frac{1}{50} = -\frac{9}{200}$$

$$\Rightarrow \quad v = -\frac{300}{9} = -33.3 \text{ cm}$$

Magnification, $$m = \frac{h_2}{h_1} = \frac{\mu_1 v}{\mu_2 u}$$

$$\Rightarrow \quad h_2 = h_1 \times \frac{\mu_1 v}{\mu_2 u} = 2 \times \frac{1 \times (-300/9)}{1.5 \times (-50)} = \frac{2 \times (300/9)}{(3/2) \times 50}$$

$$= \frac{2 \times 300 \times 2}{9 \times 3 \times 50} = \frac{8}{9}$$

$$= 0.88 \text{ cm}$$

Thus the image is formed at a distance of 33.3 cm from the pole P, on the same side as the object. Its size is 0.88 cm. **(Answer).**

Problem 27. One end of a cylindrical glass rod of radius 6 cm is rounded in the shape of a hemisphere. If a 1 cm high object is placed in front of this end at a distance of 40 cm, find the position and height of its image formed by the rod ($\mu_{glass} = 1.5$).

Solution. Given: $u = -40$ cm, $R = 6$ cm, $\mu_1 = 1$, $\mu_2 = 1.5$, $h_1 = 1$ cm.

For refraction at a spherical surface, we have,

$$-\frac{\mu_1}{u} + \frac{\mu_2}{v} = \frac{\mu_2 - \mu_1}{R}$$

$$\Rightarrow \quad -\frac{1}{(-40)} + \frac{1.5}{v} = \frac{1.5 - 1}{6} = \frac{1}{12}$$

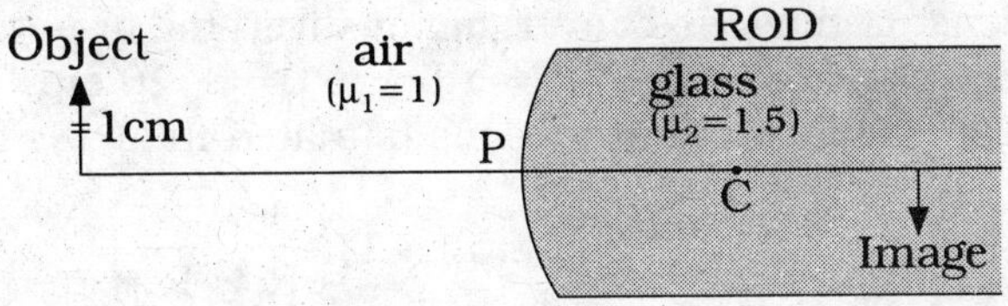

$$\Rightarrow \quad \frac{3}{2v} = \frac{1}{12} - \frac{1}{40} = \frac{7}{120}$$

$$\Rightarrow \quad v = \frac{120 \times 3}{2 \times 7} = \frac{180}{7}$$

$$= 25.7 \text{ cm}$$

Magnification, $$m = \frac{h_2}{h_1} = \frac{\mu_1 v}{\mu_2 u} = \frac{(1) \times (25.7)}{(1.5) \times (-40)} = -\frac{25.7}{60}$$

$$\Rightarrow \quad h_2 = h_1 \times \left(-\frac{25.7}{60}\right) = (1) \times \left(-\frac{25.7}{60}\right)$$

$$= -0.43 \text{ cm}$$

Thus the image is formed inside the rod, at a distance of 25.7 cm from the pole P. Its height is 0.43 cm. **(Answer).**

Problem 28. Two thin converging lenses of focal length 8 cm and 12 cm, respectively, are separated by 20 cm from each other. Locate the position of the image formed by the two lenses together of an object placed at a distance of 40 cm in front of the first lens.

Solution. For the first lens, $u = -40$ cm, $f = 8$ cm

Using the thin lens formula, $$\frac{1}{v} - \frac{1}{u} = \frac{1}{f}$$

$$\Rightarrow \quad \frac{1}{v} = \frac{1}{f} + \frac{1}{u} = \frac{1}{(8)} + \frac{1}{(-40)}$$

$$\Rightarrow \quad = \frac{1}{8} - \frac{1}{40} = \frac{4}{40} = \frac{1}{10}$$

$$\therefore \quad v = 10 \text{ cm}$$

This image acts as an object for the second lens. The +ve sign of v means that the image is formed on the other side of the first lens. Since the separation between the two lenses is 20 cm, it is located (20 – 10) = 10 cm in front of the second lens. Thus, for the second lens,

$u = -10$ cm, $f = 12$ cm

using the thin lens formula for the second lens,

$$\frac{1}{v} = \frac{1}{f} + \frac{1}{u} = \frac{1}{(12)} + \frac{1}{(-10)} = \frac{1}{12} - \frac{1}{10}$$

$$= -\frac{1}{60}$$

$$\therefore \quad v = -60 \text{ cm}$$

Thus the final image is formed at a distance of 60 cm from the second lens, on the same side as the object. It is a virtual image. **(Answer).**

Problem 29. A thin convex lens of focal length 30 cm is in contact with a thin concave lens of focal length 20 cm. Calculate the focal length of the combination. Is the combination equivalent to a converging lens or a diverging lens ?

Solution. Given: $f_1 = +30$ cm; $f_2 = -20$ cm

The focal length, F, of the combination is given by,

$$\frac{1}{F} = \frac{1}{f_1} + \frac{1}{f_2} = \frac{1}{30} - \frac{1}{20}$$

$$= -\frac{1}{60}$$

$$\therefore \quad F = -60 \text{ cm}$$

Thus the focal length of the combination is found to be – 60 cm. The –ve sign means that the combination behaves as a diverging lens. **(Answer).**

Problem 30. A thin lens made of glass ($\mu = 1.5$) has one of its surfaces convex and the other concave. The radii of curvature of the two surfaces are 20 cm and 40 cm, respectively. Determine the focal length of the lens. Is the lens converging or diverging in nature ?

Solution. Given: $\mu = 1.5$ $R_1 = +20$ cm, $R_2 = +40$ cm

The focal length of the lens (situated in air) is given by the lens maker's formula as,

$$\frac{1}{f} = (\mu - 1)\left(\frac{1}{R_1} - \frac{1}{R_2}\right)$$

$$= (1.5 - 1)\left[\frac{1}{20} - \frac{1}{(40)}\right]$$

$$= 0.5 \times \frac{1}{40} = \frac{1}{80}$$

$$\therefore \quad f = 80 \text{ cm}$$

Thus the focal length of the lens is found to be 80 cm. The +ve sign shows that the lens is converging in nature. **(Answer).**

Problem 31. A sharp point image is produced on a screen by a converging beam of light. A lens is inserted in the path of the beam, at 20 cm in front of the screen. Now, to retain the sharpness of the image it becomes necessary to move the screen 10 cm away from the lens. Determine the focal length and nature of the lens.

Solution. Since the insertion of the lens makes it necessary to move the screen away from the lens, it implies that the lens diverges the beam falling upon it, ie. the lens is diverging in nature (e.g. a double concave lens). **(Answer).**

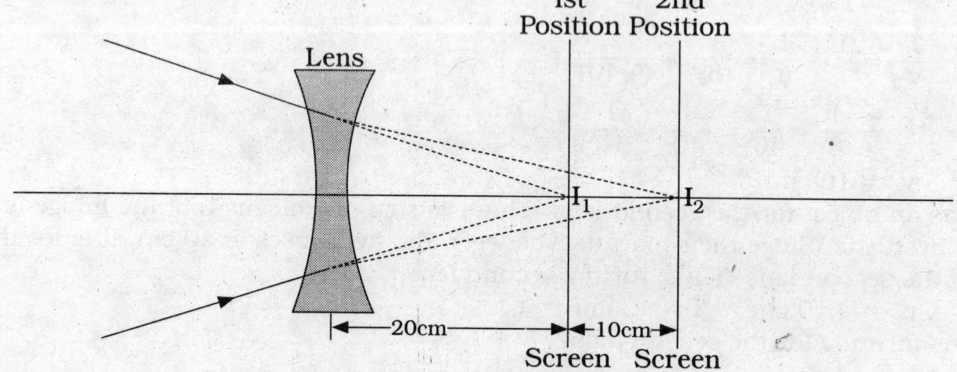

The initial point of convergence of the light beam is I_1. It acts as a virtual object for the lens, which produces the final image at I_2. So we have,

$u = +20$ cm; $v = +(20 + 10) = +30$ cm

using the lens formula, we have,

$$\frac{1}{f} = \frac{1}{v} - \frac{1}{u} = \frac{1}{30} - \frac{1}{20} = -\frac{1}{60}$$

$$\Rightarrow \quad f = -60 \text{ cm}$$

Thus the focal length of the lens is found to be 60 cm (the –ve sign of f confirms the diverging nature of the lens). **(Answer).**

Problem 32. The following diagram shows formation of image I of an object O, by a combination of two thin lenses L_1 and L_2. If the focal length of L_1 is 20 cm, find the focal length of L_2.

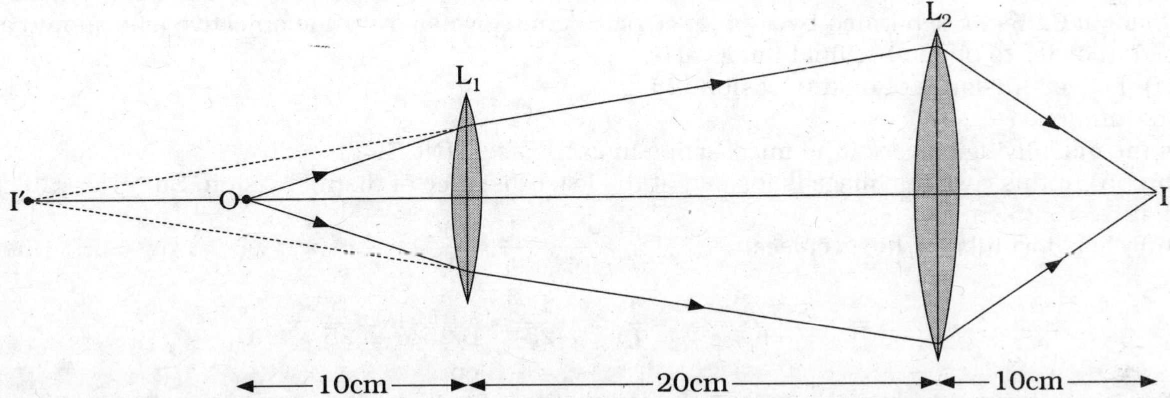

Solution. The lens L_1 forms virtual image of O at I′ ($\because u < f$), which acts as object for the lens L_2, which finally forms the image at I.

For lens L_1: $u = -10$ cm, $f = 20$ cm

Using thin lens formula, $\frac{1}{v} - \frac{1}{u} = \frac{1}{f}$

$$\Rightarrow \quad \frac{1}{v} = \frac{1}{f} + \frac{1}{u} = \frac{1}{20} + \frac{1}{(-10)}$$

$$= \frac{1}{20} - \frac{1}{10} = -\frac{1}{20}$$

$$\therefore \quad v = -20 \text{ cm}$$

Thus I′ (which is a virtual image) is situated at 20 cm in front of L_1. Its distance from L_2 is $(20 + 20) = 40$ cm.

$\therefore$ For lens L_2: $u = -40$ cm, $v = 10$ cm (given)

$$\frac{1}{f} = \frac{1}{v} - \frac{1}{u} = \frac{1}{(10)} - \frac{1}{(-40)}$$

$$= \frac{1}{10} + \frac{1}{40} = \frac{5}{40} = \frac{1}{8}$$

$$\therefore \quad f = 8 \text{ cm}$$

Thus the focal length of L_2 is found to be 8 cm. **(Answer).**

Problem 33. A figure divided into squares of size 1 mm^2 each is viewed at a distance of 9 cm, through a magnifying glass (a converging lens of focal length 10 cm) held close to the eye.

(i) What is the magnification produced by the lens ? How much is the area of each square in the virtual image ?

(ii) What is the angular magnification (magnifying power) of the lens ? (NCERT).

Solution. Given: $u = -9$ cm, $f = 10$ cm, object size (area of one square) = 1 mm^2

$$\frac{1}{v} - \frac{1}{u} = \frac{1}{f}$$

$$\Rightarrow \quad \frac{1}{v} = \frac{1}{f} + \frac{1}{u} = \frac{1}{(10)} + \frac{1}{(-9)} = \frac{1}{10} - \frac{1}{9} = -\frac{1}{90}$$

$$\therefore \quad v = -90 \text{ cm}$$

(i) Linear magnification, $m = -\frac{v}{u} = -\frac{(-90)}{(-9)} = -10$ **(Answer).**

$\therefore$ Area of one square in the image $= (-10)^2 \times 1\ mm^2$

$$= 100\ mm^2 \quad \textbf{(Answer).}$$

(ii) Angular magnification (magnifying power) $= \frac{D}{f} = \frac{25}{10}$

$$= 2.5 \quad \textbf{(Answer).}$$

Problem 34. A compound microscope consists of an objective lens of focal length 2.0 cm and an eyepiece of focal length 6.25 cm, separated by a distance of 15 cm. How far from the objective lens should an object be placed in order to obtain the final image at

(a) the least distance of distinct vision (25 cm),

(b) infinity ?

What is the magnifying power of the microscope in each case ? (NCERT).

Solution. (a) In this case the image is formed at the least distance of distinct vision. So, $v_e = -25$ cm. Also, given that $f_e = 6.25$ cm

Using thin lens formula for the eyepiece: $\frac{1}{v_e} - \frac{1}{u_e} = \frac{1}{f_e}$

$$\Rightarrow \quad \frac{1}{u_e} = \frac{1}{v_e} - \frac{1}{f_e} = \frac{1}{(-25)} - \frac{1}{6.25} = -\frac{5}{25} = -\frac{1}{5}$$

$$\therefore \quad v_e = -5 \text{ cm}$$

Since the image formed by the objective acts as the object for the eyepiece and the distance between the objective and the eyepiece is 15 cm, it follows that the image distance for the objective is $(15 - 5) = 10$ cm. Thus, for the image formation by the objective, $v_o = 10$ cm, $f_o = 2$ cm (given).

Using thin lens formula for the objective: $\frac{1}{v_o} - \frac{1}{u_o} = \frac{1}{f_o}$

$$\Rightarrow \quad \frac{1}{u_o} = \frac{1}{v_o} - \frac{1}{f_o} = \frac{1}{10} - \frac{1}{2} = -\frac{4}{10} = -\frac{2}{5}$$

$$\therefore \quad u_o = -2.5 \text{ cm} \quad \textbf{(Answer).}$$

$\therefore$ Magnifying power $= -\frac{v_o}{u_o}\left(1 + \frac{D}{f_e}\right) = -\frac{10}{(-2.5)}\left(1 + \frac{25}{6.25}\right)$

$$= 20 \quad \textbf{(Answer).}$$

(b) If $v_e = \infty$, we have from the lens formula,

$$-u_e = f_e = 6.25 \text{ cm}$$

$$\therefore \quad v_o = (15 - 6.25) = 8.75 \text{ cm}$$

Using thin lens formula for the objective: $\frac{1}{v_o} - \frac{1}{u_o} = \frac{1}{f_o}$

$$\Rightarrow \quad \frac{1}{u_o} = \frac{1}{v_o} - \frac{1}{f_o} = \frac{1}{8.75} - \frac{1}{2.0} = -\frac{6.75}{17.5}$$

$$\therefore \quad u_o = \frac{17.5}{6.75} = 2.59 \text{ cm} \quad \textbf{(Answer).}$$

$\therefore$ Magnifying power $= \frac{v_o}{u_o}\left(1 + \frac{D}{f_e}\right)$

$$= -\frac{8.75}{(-2.59)}\left(1 + \frac{25}{6.25}\right) = \frac{8.75}{2.59} \times 5$$

$$= 16.9 \quad \textbf{(Answer).}$$

Problem 35. A small telescope has an objective lens of focal length 140 cm and an eyepiece of focal length 5.0 cm. What is the magnifying power of the telescope for viewing distant objects, when,

(a) the telescope is in normal adjustment (i.e. when the final image is at infinity),

(b) the final image is formed at least distance of distinct vision (25 cm) ? (NCERT).

Solution. Given: $f_0 = 140$ cm, $f_e = 5$ cm

(a) For normal adjustment, the magnifying power is given by,

$$m = -\frac{f_0}{f_e} = -\frac{140}{5}$$

$$= -28 \quad \textbf{(Answer).}$$

(b) When the final image is formed at the least distance of distinct vision, the magnifying power is given by,

$$m = -\frac{f_0}{f_e}\left(1+\frac{f_e}{D}\right) = -\frac{140}{5}\left(1+\frac{5}{25}\right)$$

$$= -\left(28\times\frac{6}{5}\right)$$

$$= -33.6 \quad \textbf{(Answer).}$$

Problem 36. (a) A giant refracting telescope at an observatory has an objective lens of focal length 15 m. If an eyepiece of focal length 1.0 cm is used, what is the angular magnification of the telescope ?

(b) If this telescope is used to view the moon, what is the diameter of the image of the moon formed by the objective lens ? The diameter of the moon is 3.48×10^6 m and the radius of the lunar orbit is 3.8×10^8 m. (NCERT).

Solution. Given: $f_0 = 15$ m, $f_e = 1.0$ cm $= 0.01$ m

(a) For **normal adjustment** (i.e. when the final image is formed at ∞, the angular magnification (or magnifying power) is given by,

$$m = -\frac{f_0}{f_e} = -\frac{(15)}{(0.01)}$$

$$= -1500 \quad \textbf{(Answer).}$$

For the case of **image formation at the least distance of distinct vision,** the ang. magnification is given by ($D = 25$ cm $= 0.25$ m),

$$m = -\frac{f_0}{f_e}\left(1+\frac{f_e}{D}\right) = -1500\left(1+\frac{0.01}{0.25}\right)$$

$$= -(1500\times1.04) = -1560 \quad \textbf{(Answer).}$$

(b) The radius of the lunar orbit represents the distance of the observer on the earth to the moon. The angle subtended by the moon at the position of the observer is (for small θ),

3.8×10⁸ m
MOON
3.48×10⁶ m
O θ
(Observer on earth)

$$\theta = \frac{3.48\times10^6}{3.8\times10^8}\text{ radian}$$

θ is also the angle subtended by the moon's image (of size d) at the focus of the objective. So we have,

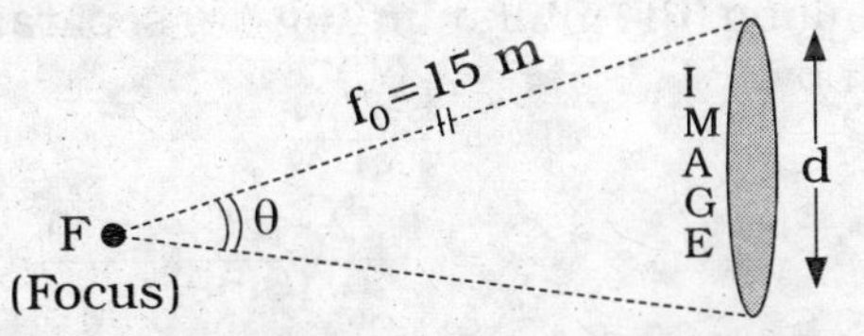

$$\theta = \frac{d}{15} = \frac{3.48\times10^6}{3.8\times10^8}$$

$$\Rightarrow \quad d = \frac{15\times3.48\times10^6}{3.8\times10^8}$$

$$= 0.137 \text{ m}$$

Thus, the diameter of the moon's image is found to be 0.137 m. **(Answer).**

Problem 37. An amateur astronomer wishes to estimate roughly the size of the sun using his crude telescope, consisting of an objective lens of focal length 200 cm and an eyepiece of focal length 10 cm. By adjusting the distance of the eyepiece from the objective, he obtains an image of the sun on a screen 40 cm behind the eyepiece. The diameter of the image is measured to be 6.0 cm. What is the estimate of the sun's size, given that the average earth-sun distance is 1.5×10^{11} m ? (NCERT).

Solution. Given: $f_0 = 200$ cm, $f_e = 10$ cm, $v_e = 40$ cm; earth-sun distance $= 1.5\times10^{11}$ m; diameter of sun's image formed by eyepiece = 6 cm

We use thin lens formula for image formation by the eyepiece,

$$\frac{1}{v_e} - \frac{1}{u_e} = \frac{1}{f_e}$$

$$\Rightarrow \quad \frac{1}{u_e} = \frac{1}{v_e} - \frac{1}{f_e} = \frac{1}{(40)} - \frac{1}{(10)} = -\frac{3}{40}$$

$$\therefore \quad u_e = -\frac{40}{3} \text{ cm}$$

∴ Magnification of the eyepiece is given by,

$$m_e = -\frac{v_e}{u_0} = -\frac{(40)}{(-40/3)} = 3$$

∴ Size of the object for the eyepiece $(h_1) = \dfrac{\text{Image size } (h_2)}{m_e} = \dfrac{6 \text{ cm}}{3}$

$$= 2 \text{ cm}$$

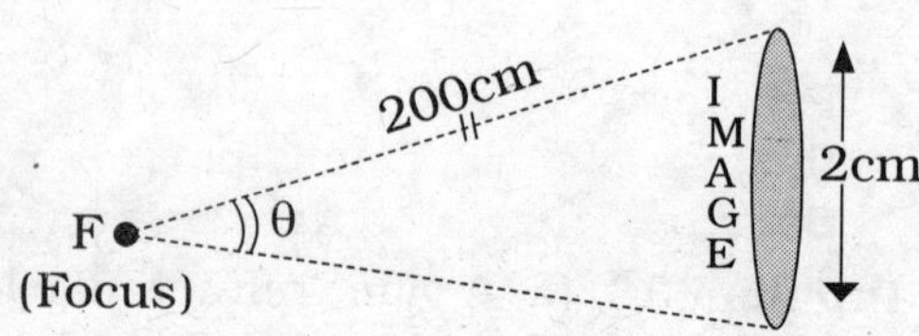

The image formed by the objective acts as the object for the eyepiece. It follows that the diameter of sun's image formed by the objective = 2 cm. In a telescope, the objective forms image at the its focus. ∴ The angle subtended by the sun's image at the objective is given by (for small θ),

$$\theta = \frac{\text{Image diameter}}{f_0}$$

$$= \frac{2 \text{ cm}}{200 \text{ cm}} = 10^{-2} \text{ radian}$$

Let the diameter of sun be d_{sun}. Then, θ is also the angle subtended by the sun's diameter at the position of the objective,

$$\theta = \frac{d_{sun}}{\text{Earth-sun distance}} = \frac{d_{sun} \text{ (meter)}}{(1.5 \times 10^{11} \text{ m})} \text{ radian}$$

Thus,

$$\frac{d_{sun} \text{ (meter)}}{1.5 \times 10^{11} \text{ m}} = 10^{-2}$$

$$\Rightarrow \quad d_{sun} = (1.5 \times 10^{11}) \times 10^{-2}$$

$$= 1.5 \times 10^{9} \text{ m}$$ **(Answer).**

Problem 38. Two closely situated stars, with a separation of 10^{11} m and distant 10^4 light year from earth, are viewed through a telescope on earth. Determine if the telescope will be able to see separate images of the stars. The diameter of the objective of the telescope is 59 cm. (1 light year = 9.46×10^{15} m; mean wavelength of white light = 590 nm).

Solution. Given: Diameter of objective, D = 59 cm = 0.59 m; λ = 590 nm = 59×10^{-8} m;

Earth-stars distance = 10^4 light year = 9.46×10^{19} m; Separation between the two stars = 10^{11} m.

By definition, the minimum angular separation between two objects that can be distinctly seen by a telescope is given by,

$$d\theta = \frac{1.22\,\lambda}{D}$$

$$= \frac{1.22 \times (59 \times 10^{-8})}{(0.59)}$$

$$= 1.22 \times 10^{-6} \approx 10^{-6} \text{ radian}$$

Star 1
10^4 light year
10^{11}m
O
dθ′
(Observer)
Star 2

The angle subtended by the stars at the position of the observer on earth is found as,

$$d\theta' = \frac{\text{Separation between the stars}}{\text{Earth-stars distance}} = \frac{(10^{11})}{(9.46 \times 10^{19})}$$

$$\approx 10^{-9} \text{ radian}$$

Thus, it is found that dθ′ is nearly 10^3 times, i.e. 1000 times, smaller than dθ, the minimum resolvable angular separation. Hence, the telescope will fail to resolve the images of the stars, i.e. it will not be able to see the two stars separately (the images of the two stars will merge into a broad single image). **(Answer).**

Problem 39. The radius of curvature of the concave mirror of a reflecting type telescope is 140 cm. If the magnification produced by the telescope is ×28, Determine the focal length of its eyepiece.

Solution. Given: R = 140 cm, m = 28

The focal length of the objective, $f_0 = \frac{R}{2} = \frac{140}{2} = 70$ cm.

The magnification of a reflecting type telescope is given by,

$$m = \frac{f_0}{f_e}$$

$$\Rightarrow \quad f_e = \frac{f_0}{m} = \frac{70}{28}$$

$$= 2.5 \text{ cm}$$

Thus, the focal length of the eyepiece is found to be 2.5 cm. **(Answer).**

Problem 40. A person with myopia of the eye cannot see clearly farther than 40 cm. Calculate the power of the lens that will enable him to clearly see distant objects.

Solution. It is given that the far point of the person is 40 cm away from his eye at which he can see clearly. The image formed by the corrective lens acts as the object for the eye lens. So the corrective lens should form the image of a distant object ($u = -\infty$) at a distance of nearly 40 cm from it, i.e. v = (–) 40 cm. Using the lens formula, we have,

$$\frac{1}{v} - \frac{1}{u} = \frac{1}{f}$$

$$\Rightarrow \quad \frac{1}{(-40)} - \frac{1}{-(\infty)} = \frac{1}{f}$$

$$\Rightarrow \quad f = -40 \text{ cm} = -0.4 \text{ m}$$

$$\therefore \quad P = \frac{1}{f} = -\frac{1}{0.4} = -2.5 \text{ D}$$

Thus, the required power of the corrective lens is – 2.5 D. The –ve sign of P (and f) shows that it has to be a concave (diverging) lens. **(Answer).**

Problem 41. A hypermetropic lady is unable to clearly see objects located closer than 125 cm from her eye. What has to be the power of the lens that enables her to clearly see objects located at 25 cm from the eye ?

Solution. It is given that the near point of the lady is 125 cm away from her eye, at which she can clearly see an object. The image formed by the corrective lens acts as the object for the eye lens. Therefore the corrective lens should form the image of an object located nearly 25 cm away from it at a distance nearly 125 cm from it, i.e. u = (–) 25 cm and v = (–) 125 cm.

Using the lens formula, we get,

$$\frac{1}{v} - \frac{1}{u} = \frac{1}{f}$$

$$\Rightarrow \quad \frac{1}{(-125)} - \frac{1}{(-25)} = \frac{1}{f}$$

$$\Rightarrow \quad \frac{1}{f} = \frac{1}{25} - \frac{1}{125} = \frac{4}{125}$$

$$\Rightarrow \quad f = \frac{125}{4} \text{ cm} = \frac{5}{16} \text{ m}$$

$$\therefore \quad P = \frac{1}{f} = \frac{16}{5} = (+)\, 3.2 \text{ D}$$

Thus the required power of the lens is found to be 3.2 D. The +ve sign of P (and f) shows that it has to be a convex (converging) lens **(Answer).**

Example 42. An old person suffering from myopia as well as presbyopia has his far point at 125 cm and his near point at 50 cm. He uses two pairs of spectacles, one each for viewing distant objects and for reading purposes, respectively. Find the power of the spectacle lenses in each case. Also find his far point with spectacles in the second case and the near point with spectacles in the first case. (Least distance of distinct vision = 25 cm).

Solution. (i) Power of lenses (Far point: 125 cm):

The focal length of the corrective lens has to be such that it forms image of a distant object ($u = -\infty$) at about 125 cm from the eye, where it is clearly seen (i.e. $v = -125$ cm).
Using the lens formula, we have,

$$\frac{1}{v} - \frac{1}{u} = \frac{1}{f}$$

$$\Rightarrow \quad \frac{1}{f} = \frac{1}{(-125)} - \frac{1}{(-\infty)} = -\frac{1}{125}$$

$$\Rightarrow \quad f = -125 \text{ cm} = -1.25 \text{ m}$$

$$\therefore \quad P = \frac{1}{f} = \frac{1}{(-1.25)} = -0.8 \text{ D} \quad \textbf{(Answer).}$$

The –ve sign means that the lenses are diverging in nature

(ii) Power of lenses (near point: 50 cm):

The focal length of the corrective lens has to be such that it forms image of an object located at least distance of distinct vision ($u = -25$ cm) at about 50 cm, where it is clearly seen by the unaided eye (i.e. $v = -50$ cm). Again, using the lens formula, we get,

$$\frac{1}{v} - \frac{1}{u} = \frac{1}{f}$$

$$\Rightarrow \quad \frac{1}{f} = \frac{1}{(-50)} - \frac{1}{(-25)} = \frac{1}{25} - \frac{1}{50} = \frac{1}{50}$$

$$\Rightarrow \quad f = 50 \text{ cm} = 0.5 \text{ m}$$

$$\therefore \quad P = \frac{1}{f} = \frac{1}{0.5} = (+)\,2.0 \text{ D} \quad \textbf{(Answer).}$$

The +ve sign means that the lenses are converging in nature.

(iii) Far point with spectacles in second case ($f = +25$ cm):

Let the far point with spectacles be located at a distance d in front of the eye. Then, an object located at this point ($u = -d$ cm) is imaged at a distance of 125 cm from the eye ($v = -125$ cm), where it can be clearly seen (this image then acts as the object for the eye). Again, using the lens formula, we get,

$$\frac{1}{v} - \frac{1}{u} = \frac{1}{f}$$

$$\Rightarrow \quad \frac{1}{(-125)} - \frac{1}{(-d)} = \frac{1}{50}$$

$$\Rightarrow \quad \frac{1}{d} = \frac{1}{50} + \frac{1}{125} = \frac{7}{250}$$

$$\Rightarrow \quad d = \frac{250}{7} = 35.7 \text{ cm} \quad \textbf{(Answer).}$$

(iv) Near point with spectacles in first case ($f = -125$ cm):

Let the near point with spectacles be locates at a distance d in front of the eye. Then, an object located at this point ($u = -d$ cm) is imaged at a distance of 50 cm from the eye ($v = -50$ cm), where it can be clearly seen. Again, using the lens formula, we have,

$$\frac{1}{v} - \frac{1}{u} = \frac{1}{f}$$

$$\Rightarrow \quad \frac{1}{(-50)} - \frac{1}{(-d)} = \frac{1}{(-125)}$$

$$\Rightarrow \quad \frac{1}{d} = \frac{1}{50} - \frac{1}{125} = \frac{3}{250}$$

$$\Rightarrow \quad d = \frac{250}{3} = 83.3 \text{ cm} \quad \textbf{(Answer).}$$

Problem 43. A lady wears lenses of power – 5.0 D in both the eyes. Are her eyes myopic or hypermetropic ? Calculate the position of her far point with unaided eyes.

Solution. The –ve sign of the power of lenses means that the lady wears diverging lenses. It follows that her eyes are myopic. **(Answer).**

The focal length of the lenses is,

$$f = \frac{1}{P} = -\frac{1}{5}\text{ m} = -\frac{100}{5}\text{ cm} = -20\text{ cm}$$

The far point is the farthest position at which an object is clearly seen by the eye. Suppose it is located d cm in front of the eye. The focal length of the corrective lens is such that it forms the image of a distant object ($u = -\infty$) at the far point (i.e. $v = -d$ cm). Using the lens formula, we get,

$$\frac{1}{v} - \frac{1}{u} = \frac{1}{f}$$

$$\Rightarrow \quad \frac{1}{(-d)} - \frac{1}{(-\infty)} = \frac{1}{(-20)}$$

$$\Rightarrow \quad \frac{1}{d} = \frac{1}{20}$$

$$\Rightarrow \quad d = 20\text{ cm}$$

Thus, with unaided eye, the far point is located at a distance of 20 cm from the eye. **(Answer).**

Problem 44. With unaided eyes, an individual cannot read clearly at a distance nearer than 75 cm. Are his eyes myopic or hypermetropic ? If he puts on lenses of power 2.5 D, what will be his new near point ?
Solution. The fact that the person's near point is farther (at 75 cm) than the near point for the normal eye, viz. 25 cm, means that his eyes are hypermetropic. **(Answer).**

The focal length of the lenses put on by the person is,

$$f = \frac{1}{P} = \frac{1}{2.5}\text{ m} = -\frac{100}{2.5}\text{ cm} = 40\text{ cm}$$

Suppose the new near point is located at a distance of d cm from the eye. The focal length of the lenses is such that the image of an object located at this point ($u = -d$ cm) is formed at 75 cm from the eye, where it is clearly seen (i.e. $v = -75$ cm). (This image then acts as the object for the eye). Using the lens formula, we have,

$$\frac{1}{v} - \frac{1}{u} = \frac{1}{f}$$

$$\Rightarrow \quad \frac{1}{(-75)} - \frac{1}{(-d)} = \frac{1}{40}$$

$$\Rightarrow \quad \frac{1}{d} = \frac{1}{40} + \frac{1}{75} = \frac{23}{600}$$

$$\Rightarrow \quad d = \frac{600}{23} = 26.1\text{ cm}$$

Thus the new near point is located 26.1 cm away from the eye. **(Answer).**

Problem 45. An old man has his near point and far point at 50 cm and 200 cm, respectively. Determine the power of lenses that will enable him to read clearly at 25 cm from his eyes. When he wears these lenses, at what maximum distance can he see objects clearly ?
Solution. The power of the lenses has to be such that an object placed at 25 cm ($u = -25$ cm) is imaged at 50 cm, where it is clearly seen (i.e. $v = -50$ cm). Using the lens formula, we get,

$$\frac{1}{v} - \frac{1}{u} = \frac{1}{f}$$

$$\Rightarrow \quad \frac{1}{f} = \frac{1}{(-50)} - \frac{1}{(-25)} = \frac{1}{25} - \frac{1}{50} = \frac{1}{50}$$

$$\Rightarrow \quad f = 50\text{ cm} = 0.5\text{ m}$$

$$\therefore \quad P = \frac{1}{f} = \frac{1}{0.5} = (+)\,2.0\text{ D}$$ **(Answer).**

The far point for the unaided eye is 200 cm. With the aided eye, let the new far point (i.e. the maximum distance at which the eye can clearly see an object) be located at a distance of d cm from the eye. The lenses will form the image of an object placed at d cm (i.e. $u = -d$ cm) at a distance of 200 cm, where it is clearly seen (i.e. $v = -200$ cm). Again using the lens formula, we have,

$$\frac{1}{v} - \frac{1}{u} = \frac{1}{f}$$

$$\Rightarrow \quad \frac{1}{(-200)} - \frac{1}{(-d)} = \frac{1}{50}$$

$$\Rightarrow \quad \frac{1}{d} = \frac{1}{50} + \frac{1}{200} = \frac{5}{200} = \frac{1}{40}$$

$$\Rightarrow \quad d = 40 \text{ cm}$$

Thus the required power of the lenses is (+) 2.0 D and the new far point is located 40 cm away from the eye. **(Answer).**

Very Short Answer Questions

[Including HOTS Questions]

(1 mark each)

1. How is the principal axis of a spherical mirror drawn ?

2. What is the value of angle of incidence for a ray falling normally on a mirror ?

3. Can the image formed by a convex mirror be obtained upon a screen ? Give reason for your answer.

4. Why is a convex mirror used as a driver's mirror ?

5. Can a virtual image be photographed ?

6. Under what condition a concave mirror forms a virtual image ?

7. What is the focal length of a plane mirror ?

8. What distance is the focal length of a spherical mirror ? How is it related to the radius of curvature of the mirror ?

9. What is meant by the linear aperture of a spherical mirror ?

10. State laws of refraction.

11. When light passes from one medium to another, which of the following remains unchanged: frequency wavelength, speed, intensity ?

12. What type of lens is an air bubble: converging or diverging ?

13. When a convex lens is placed in a particular medium, the light rays incident upon the lens pass undeviated. How are the refractive indices of the medium and the material of the lens related to each other ?

14. If a convex lens forms a virtual image of a point object lying upon the principal axis, then what is the position of the object with respect to the focal point of the lens ?

15. If a point object is placed at the focus of a concave lens, where will its image be formed, with respect to the position of the focus ?

16. What is the value of the critical angle for a medium of refractive index $\sqrt{2}$, when light passes from the medium into air ?

17. With light passing from air into another medium, for the same angle of incidence the angles of refraction for the media P, Q and R are found to be 40°, 30° and 20°, respectively. In which of the three media is the velocity of light maximum ?

18. Is the refractive index of diamond higher or lower than that of common glass ?

19. Is the value of critical angle for a medium dependent on the colour of incident light ?

20. On what phenomenon is the working of optical fibres based ?

21. Is the focal length of a lens for red light higher or lower than that for blue light ?

22. An equiconvex lens of focal length 15 cm is cut end to end into two equal halves. What is the focal length of each half ?

23. On being immersed in a transparent liquid a lens is found to become invisible. How does it happen ?

24. Will the focal length of a convex lens change if it is placed in water instead of air ? If so, will it increase or decrease ?

25. Trace the path of a monochromatic ray through a prism. Mark the angles of incidence (i), emergence (e) and deviation (δ).

26. What is meant by the angle of minimum deviation ?

27. What is meant by dispersion of light ?

28. What is the reason that dispersion of light is caused by refraction, not by reflection ?

29. If a glass prism is placed in water instead of air, how will the angle of prism be affected ?

30. Express the refractive index of a medium in terms of wavelength of light.

31. State the essential condition for observing a rainbow.

32. Define magnifying power of a microscope.

33. Define magnifying power of a telescope.

34. If the eye sees the final image at infinity, what is the magnifying power of a simple microscope ?

35. State whether the final image formed by the following is real or virtual: (i) compound microscope and (ii) refracting type astronomical telescope.

36. Write one major advantage of having the aperture of the objective lens or the mirror of a telescope as large as possible.

37. Define resolving power of a telescope.

38. Does the resolving power of a telescope increase or decrease, upon increasing the aperture of the objective ?

39. Under normal adjustment, what is the distance between the objective and the eyepiece of a telescope ?

40. What is the focal length of a lens having power as +2 dioptre ?

41. Is the nature and size of image seen in a mirror affected by the size of the mirror ?

42. Can an observer on moon see a rainbow ? If not, why ?

43. Why the astronauts in space see the sky as black ?

44. Why is the use of red traffic lights and street light sodium lamps helpful in foggy weather ?

45. What are paraxial rays ?

46. What are peripheral (or marginal) rays ?

47. How is the eye able to image objects located at different distances from it ?

48. How the eye adjusts itself to varying intensities of light ?

49. Why the pupil of the eye always appears black in colour ?

50. What is yellow spot of the eye ?

51. What is blind spot of the eye ?

52. What is meant by accommodation of the eye ?

53. What is least distance of distinct vision ?

54. What are near point and far point of an eye ?

Short Answer Questions

[Including HOTS Questions]

(2 or 3 marks each)

1. How a plane mirror, a concave mirror and a convex mirror can be distinguished from each other, simply by looking into them ?

2. Each of the following types of mirror are known to produce virtual images: plane mirror, concave mirror, convex mirror. How the conditions of formation of such image differ amongst them ?

3. A concave mirror of focal length 15.6 cm is immersed in water (μ_{water} = 1.3). What will be the value of its focal length as seen from air ?

4. Show with the help of a ray diagram the formation of virtual image of a small object placed in front of a concave mirror.

5. Show with the help of a ray diagram the formation of image of a linear object placed in front of a convex mirror.

6. What is meant by the power of a lens ? What is the focal length of a lens of power –1 dioptre ?

7. A convex lens of glass (refractive index = 1.5) behaves as a converging lens in air and as a diverging lens in carbon disulphide (refractive index = 1.65). Explain why.

8. A convex lens made of glass (R.I. = 1.50) behaves as a converging lens when dipped in water (R.I. = 1.33) and as a diverging lens when dipped in carbon disulphide (R.I. = 1.65). Explain why.

9. Explain why the focal length of a lens has different values for red light and blue light. For a convex lens, for which of the two lights is it smaller ?

10. Explain how the sun becomes visible on earth about 2 minutes before the actual sunrise.

11. What are the essential conditions for total internal reflection of light ? Derive the relation between refractive index of a medium and critical angle.

12. Show with the help of appropriate ray diagrams how prisms of different shapes are used to (i) deviate light through 90°, (ii) deviate light through 180°, and (iii) invert an image without deviation.

13. What is a light pipe ? How light pipes help to examine inner parts of a human body ?

14. Explain the brillian sparkling of cut diamonds, when exposed to light.

15. Explain the phenomenon of mirage, witnessed in a desert.

16. A biconvex lens made of glass of refractive index μ_2 is placed in a medium of refractive index μ_1. If a parallel beam of light is incident upon it, trace the path of rays for (i) $\mu_2 > \mu_1$, (ii) $\mu_2 = \mu_1$, and (iii) $\mu_2 < \mu_1$.

17. Derive the thin lens formula, $\frac{1}{v} - \frac{1}{u} = \frac{1}{f}$, for a biconvex lens.

18. Derive the relation between the resultant focal length of a combination of two thin lenses in contact and the individual focal lengths of the lenses.

19. What is meant by dispersion of light ? Draw a ray diagram to show dispersion through a prism. Explain the underlying cause for dispersion.

20. Explain the formation of a rainbow.

21. Explain the blue colour of sky.

22. Explain the red colour of sun at sunrise and sunset.

23. Draw a labelled diagram to illustrate image formation by a compound microscope.

24. Draw a labelled diagram to show final image formation at near point by an astronomical telescope.

25. Draw a labelled diagram to show image formation by a Newtonian reflecting type telescope.

26. State the relative advantages and disadvantages of refracting type and reflecting type telescopes.

27. What is presbyopia of human eye ? Describe its causes and method of correction.

28. What is astigmatism of eye ? Describe its common cause and method of correction.

Long Answer Questions

(5 marks each)

1. Describe with the help of suitable diagrammes the principle and working of optical fibres. What are light pipes ? Describe their use for medical examination of inner parts of a human body.

2. For a monochromatic ray of light passing through a prism, derive the relation between the refractive index of the prism material, angle of prism, and angle of deviation. Draw a graph to show the variation of angle of deviation with angle of incidence and mark the position of minimum deviation.

3. Derive the lens maker's formula,

$$\frac{1}{f} = (\mu - 1)\left(\frac{1}{R_1} - \frac{1}{R_2}\right),$$

where the symbols have their usual meanings.

4. Describe with the help of a labelled ray diagram the image formation by a compound microscope. Derive an expression for the magnifying power of the microscope. How can the magnifying power be enhanced ?

5. Describe with the help of a labelled ray diagram the image formation by a refracting type astronomical telescope. Derive an expression for its magnifying power when the final image is formed at least distance of distinct vision and hence derive the expression when the image is formed at infinity. How can the magnifying power be enhanced in either case ?

6. Explain how the disadvantages associated with a refracting type astronomical telescope are overcome in a reflecting type telescope.

   Describe with the help of a labelled ray diagram the construction and working of a Newtonian reflecting type telescope.

7. Describe, with the help of suitable ray diagrams, what is meant by myopia and hypermetropia of human eye and their common causes and methods of correction.

8. Describe, with the help of a labelled diagram, the construction and working of human eye.

Numerical Exercises

1. A concave mirror forms a real image of an object, located at a distance p from the focus, at a distance q from the focus. Calculate the focal length of the mirror in terms of p and q. **[$f = \sqrt{pq}$]**

2. A concave mirror forms a three times magnified image of an object when the object distance is reduced to 9 cm from 15 cm. Find the focal length of the mirror. [Hint: Use $m = \frac{f}{f - u}$; put u = – 9 cm and – 15 cm, respectively, according to the sign convention]. **[f = – 6 cm]**

3. Find the position of image of an object placed 0.05 m in front of a concave mirror of focal length 0.1 m. **[10 cm behind the mirror]**

4. When an object is placed 60 cm in front of a concave mirror, the magnification produced has value 1/2. If a virtual image of double the size of the object is to be formed by the mirror, where should the object be located ? [Hint: In the first case, use m = – (1/2) for real image, to calculate f]. **[10 cm from the mirror]**

5. The size of image formed by a convex mirror of focal length 30 cm is one-quarter of the size of the object. Find the position of the image. **[22.5 cm behind the mirror]**

6. A square-shaped wire of side length 3 cm is placed 25 m in front of a concave mirror of focal length 10 cm, with the axis of the mirror perpendicular to the square and passing through the centre of the square. Calculate the area enclosed by the image of the square. **[4 cm^2]**

7. A dentist is examining a tooth of a patient with a small concave mirror of radius of curvature 30 mm. How far from the tooth the mirror should be placed to obtain a five times larger image of the tooth ? **[1.2 cm]**

8. If the refractive indices of common glass and diamond be 1.5 and 2.4, respectively, how much faster will light travel in glass than in diamond ? **[1.6 times faster]**

9. An oil film is spread on the surface of water. If a ray of light is incident at 30° on the oil-water boundary, determine the angle of refraction in water (μ_{oil} = 1.2; μ_{water} = 1.33). **[27°]**

10. At what angle with the surface will a fish lying just under the surface of water in a pond see the setting sun (μ_{water} = 1.33) ? **[41.3° with the water surface]**

11. A ray of light travelling in water is incident at 20° to the normal on a water-glass interface. Calculate the angle of refraction in glass (μ_{water} = 1.33; μ_{glass} = 1.50). **[17.7°]**

12. A rectangular vessel of depth 80 cm is half filled with water and the rest half with an oil. Calculate the apparent depth of the vessel, when viewed normally from the top (μ_{water} = 1.33; μ_{oil} = 1.40). **[58.6 cm]**

13. A metal tank contains two immiscible liquids L_1 and L_2 of heights 4 cm and 3 cm, respectively. The bottom of the tank is lined with glass of thickness 2 cm. Determine the position of a red mark made on the lower side of the glass slab, when viewed normally from the top (μ_{glass} = 1.5; μ_{L1} = 1.3; μ_{L2} = 1.4). **[2.45 cm from the bottom]**

14. A tank filled with water has a small lamp resting on the bottom, at a depth of 1.6 m. Calculate the area of the water surface out of which the light from the lamp will emerge (μ_{water} = 1.33). **[10.4 m^2]**

15. A concave spherical glass surface of radius of curvature 20 cm has a point source of light placed in front of it in air at a distance of 10 cm. Find the position and nature of the image of the source formed by the glass surface (μ_{glass} = 1.5). **[12 cm in front; virtual]**

16. A small needle is located 0.2 m in front of a convex lens of focal length 0.15 m. What is the position and nature of its image ? **[60 cm from lens on the other side of the object; Real]**

17. An equiconvex lens made of glass has focal length 0.5 m. Determine the radius of curvature of its each face (μ_{glass} = 1.5). **[50 cm each]**

18. Calculate the refractive index of the material of a convex lens of focal length 0.27 m, with radii of curvature of its two faces as 0.2 m and 0.6 m, respectively. **[μ = 1.56]**

19. Calculate the focal length of a combination two thin lenses of focal lengths 9 cm and 18 cm respectively, placed in contact with each other. **[6 cm]**

20. The refractive index of the material of a prism of angle 60° is 1.62 for sodium light. What is its angle of minimum deviation ? (CBSE; 1991). **[48.2°]**

21. A screen is placed 90 cm from an object. The image of the object on the screen is formed by a convex lens at two different locations separated by 20 cm. Calculate the focal length of the lens. (NCERT). **[21.4 cm]**

22. A ray of light is incident on an equilateral glass prism, with the angle of incidence equal to the angle of emergence and the angle of emergence equal to 0.75 times the angle of prism. Determine the angle of deviation of the ray. [Hint: $A + \delta = i + e$]. **[30°]**

23. The refractive indices for blue, yellow, and red colours for flint glass are 1.664, 1.654, and 1.644, respectively. Calculate the angular dispersion between blue and red colours produced by a flint glass prism of angle 60°. Also find the dispersive power of the prism. **[1.2°; 0.031]**

24. A compound microscope forms final image at the least distance of distinct vision (25 cm), with a magnification of ×60. If the focal length of its eyepiece is 5 cm, determine the magnification produced by its objective. **[×1]**

25. The objective and eyepiece of a compound microscope have focal lengths 4 cm and 6 cm, respectively. An object is placed 6 cm in front of it. Find the magnification produced by the microscope (least distance of distinct vision = 25 cm). **[m = 10.3]**

26. An astronomical telescope has magnifying power as ×9. Its objective and eyepiece are separated by a distance of 40 cm. Calculate their focal lengths. **[36 cm & 4 cm]**

27. The radius of curvature of the mirror (objective) of reflecting type telescope is 100 cm and the focal length of its eyepiece is 2 cm. Calculate the magnifying power of the telescope. **[×25]**

28. A myopic person can see objects clearly only when they are within 2 m of his sight. Determine the power of the lens that will enable him to see distant objects. **[– 0.5 D]**

29. A hypermetropic person can only see those objects clearly which lie within 50 cm of his eyes. Find the power of the lens that will enable him to read at the least distance of distinct vision (25 cm). Is the lens converging or diverging in nature ? **[2.0 D; Converging]**

30. A lady with myopia can distinctly read at 15 cm away from her eyes. Find the focal length of the lens required by her to distinctly read at a distance of 25 cm. **[f = – 37.5 cm]**

31. A person with defective vision can clearly see objects lying within 80 cm of his eyes and not beyond it. Find the power of lenses that will enable him to see distant objects. **[– 1.25 D]**

32. A hypermetromic man cannot distinctly see objects at distances nearer than 90 cm from his eyes. Determine the power of the lenses required to clearly read 30 cm away from the eyes. **[2.2 D]**

33. Find the power of the lens required to change the reading distance of a man from 80 cm to 20 cm. **[3.75 D]**

CHAPTER 6 (B)

WAVE OPTICS; HUYGENS' CONSTRUCTION; INTERFERENCE; DIFFRACTION; POLARIZATION

6B. 01. Nature of Light

The nature of light radiation, viz. whether it consists of particles or waves, remained a subject of controversy for nearly one and a half century. The question was finally settled only in the beginning of the last century, when it was accepted that light has a dual nature. Sometimes it behaves as waves and sometimes as particles, depending upon the phenomenon under consideration. However, the long process of trying to resolve the controversy has been extremely rewarding. It has very richly contributed to our knowledge of the foundations of physics. In fact, to a large extent, the long history of the resolution of the controversy has been the history of the evolution of physics. The whole process has had numerous significant spin-offs, like the development of quantum mechanics, which is an altogether new field of knowledge of physics, that has enabled us to understand several unexplained physical phenomena.

In the second half of the seventeenth century, two different views of the nature of light were put forward by the contemporaries Christian Huygens (Dutch physicist; 1629-1695) and the great scientist Sir Isaac Newton (British physicist; 1642-1727). According to Huygens, the light propagated as a wave motion (wave theory of light). Newton proposed that the light radiation consisted of tiny particles, called corpuscles (Corpuscular theory of light*). The then known significant optical phenomena of rectilinear propagation of light (light travelling along a straight path) and reflection/refraction of light could be explained on the basis of both the theories, but in different ways. In the wave theory, the rectilinear propagation and the reflection/refraction were accounted for on the basis of Huygens' construction of wavefronts, that we shall describe in the next sections. However, the assumptions made for explaining the rectilinear propagation were not satisfactory. Newton's thinking was guided by the then prevailing laws of mechanics. Accordingly, in the corpuscular theory the optical phenomena were largely treated as mechanical processes. The corpuscles were believed to travel in straight line unless acted upon by external forces. So the rectilinear propagation of light, as commonly observed by the formation of shadows, could be explained straightaway as a natural process. In fact, it remained as the strongest evidence in support of the particle nature of light. The reflection of light was interpreted on the basis of elastic collisions of the corpuscles with the surface on which they were incident. For explaining the phenomenon of refraction, it was postulated that the particles of the denser medium strongly attracted the incident light corpuscles, resulting in a bending of their path at the interface.

Because of the dominance of Newton in the scientific community at that time, the Huygens' wave theory did not receive the attention it really deserved, for more than a century. In the beginning of the nineteenth century, decisive experiments on the interference and diffraction of light were performed by Thomas Young (British physicist; 1773-1829) and Augustin Jean Fresnel (French physicist; 1788-1827), respectively, which could only be explained by considering light as a wave motion. We shall describe these phenomena in the later sections in this chapter. Thus, interest in the wave theory of light was revived. Later, about the middle of the nineteenth century, Maxwell developed the theory of electromagnetic waves, as already described in the last chapter, which established the wave theory of light on a firm footing and heralded the end of the corpuscular theory. Before the Maxwell's work, the light waves were believed to be mechanical waves (like sound waves), which required the universal existence of an extremely dilute and highly elastic hypothetical medium called 'ether', for their propagation. Otherwise, it could not be possible to account for their propagation through free space, i.e. vacuum. Since an electromagnetic wave consists of oscillations of electric and magnetic fields, it is essentially a non-mechanical wave, requiring no material for its propagation. Thus, the need to postulate the presence of a medium like ether was also eliminated through the Maxwell's work.

*The corpuscular theory was originally proposed by the French scientist and philosopher, Rene Descartes (1596-1650).

However, later towards the beginning of the twentieth century, the conviction of the scientists in the wave nature of light was rudely shaken when the phenomenon of photoelectric effect was discovered. The phenomenon consists of emission of electrons from a metal surface, when light is incident upon it. Very significantly, the kinetic energy of the emitted electrons does not depend on the light intensity. It only depends on the frequency of the incident light. We shall study the phenomenon in detail in the next chapter. It was explained by the great scientist Albert Einstein (German physicist, later settled in USA; 1879-1955) in 1905 on the particle nature of light. If could not be interpreted on the basis of the wave theory. Thus once again the wave vs. particle controversy was revived. Ultimately, the issue was settled by accepting the fact that the light radiation has dual nature. Later, this wave-particle duality has been further confirmed by the development of the notion of matter waves, that we shall study in the next chapter. We shall learn that even material particles like electrons and neutrons, and indeed all matter, have a wave character associated with them.

6B.02. Wavefronts; Huygens' Principle; Huygens' Construction

A **wavefront** may be **defined** as "a surface of constant phase" **or** "the locus of points in the same phase". (Remember that **phase** represents the state of vibration of a particle).

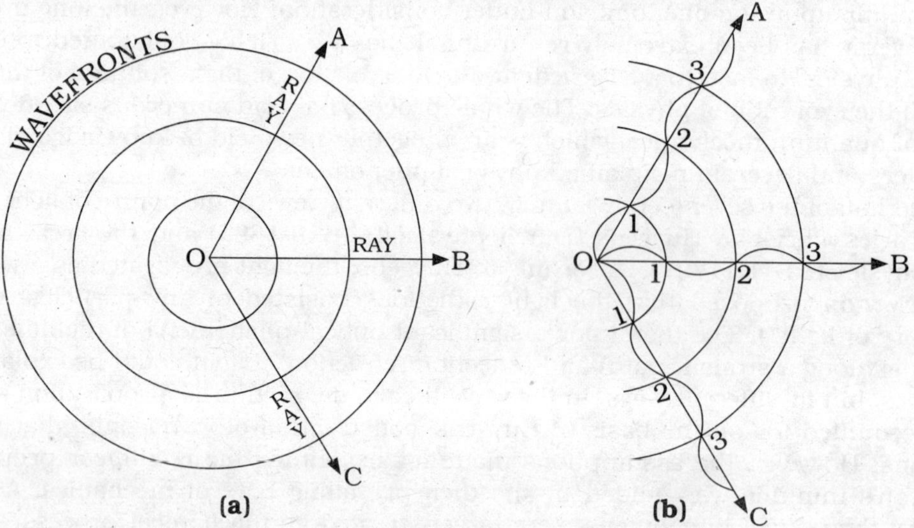

Fig. 6.47

The Fig. 6.47 (a) shows a point source of light situated at O. The light rays emitted by it all around have a spherical symmetry. Three such rays OA, OB, OC have been drawn in the figure. The cross-section of a sphere in the plane of the paper is a circle. Three such circles corresponding to different time intervals are seen in the figure. They all represent **spherical wavefronts.** As a wavefront advances outwards, its radius of curvature increases and correspondingly its curvature decreases At distances far away from the source the curvature is so small that a small part of the wavefront may be treated as a **plane wavefront,** as shown in Fig. 6.48.

The light rays lie along the radii of the spherical wavefronts and along the normals to the plane wavefronts. They indicate the direction of travel of the wavefronts. The fact that the concentric spheres drawn in the Fig. 6.47 (a) represent wavefronts, as defined by us, is more clearly illustrated in Fig. 6.47 (b), which shows light waves at a given instant of time along OA, OB, OC. The positions 1, 2, 3 along the three waves denote three positions of different but constant phase. The respective surfaces of the three spheres are the locii of positions 1, 2, 3 and so they represent wavefronts of three different phases. The following points should be noted:

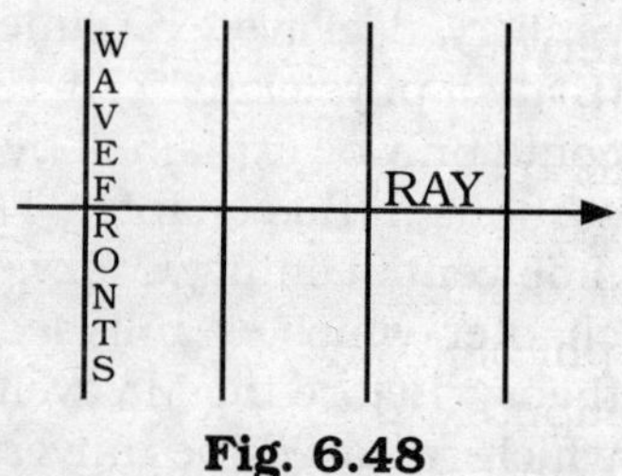

Fig. 6.48

(i) A wavefront is always perpendicular to the light rays.

(ii) The time taken by light to travel from one wavefront to another is the same for all the rays.

If instead of a point source we have a linear source of light, like light passing through a slit, **cylindrical wavefronts** are formed. Once again, at long distances from the source, small portions of theses wavefronts effectively behave as plane wavefronts.

Huygens' principle: The Huygens' wave theory is based on the following so-called **Huygens' principle:**

"Each point on a given wavefront acts as a source of secondary spherical wavelets. The new position of the wavefront after a time interval is the geometrical envelope of these wavelets".

Huygens' construction: The future position of a given wavefront is conveniently found by the following so-called **Huygens' construction,** which is based on the aforesaid Huygens' principle:

(i) To know the new position of a wavefront after a time interval Δt, draw spherical wavelets of radius $c\Delta t$ from the points on the wavefront (c = velocity of light).

(ii) Draw a surface tangentially touching the wavelets in the forward direction. This surface is the new wavefront.

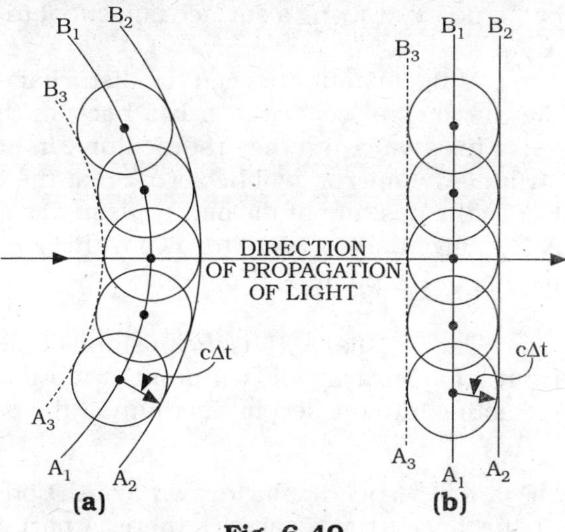

Fig. 6.49

The construction is illustrated in Figs. 6.49 (a) and (b) for a spherical wavefront and a plane wavefront, respectively. In a homogeneous medium, i.e. a medium with same optical properties everywhere, the light originating from a source only moves forward. Therefore, the geometrical envelope of the secondary wavelets is only drawn in the forward direction. However, if there is a change of medium on the way, the light may be reflected at the interface of the two media. Then the secondary wavelets on the back side also need to be considered. In the Figs. 6.49 (a) and (b), A_1B_1 is the initial position of the wavefront and A_2B_2 is its new position after a time interval Δt. A_3B_3 represents the hypothetical position of the wavefront in the backward direction.

Now we proceed to explain the laws of reflection and refraction of light on the basis of the Huygens' wave theory. For the sake of simplicity, only plane wavefronts incident at plane interfaces between two media will be considered.

6B. 03. Reflection of a Plane Wave at a Plane Surface

The two laws of reflection of light are:

(i) The angle of incidence of a light ray is equal to the angle of reflection.

(ii) The incident ray, the reflected ray and the normal at the point of incidence lie in the same plane.

We shall establish the laws on the basis of the Huygens' wave theory. Let a plane wavefront A_1B_1 be incident upon the surface of a plane mirror MN, as shown in Fig. 6.50. Three incident rays 1, 2, 3 perpendicular to the wavefront are also drawn. Suppose at the instant $t = 0$ of the ray 1 and the point A_1 of the wavefront just touch the mirror's surface. According to the Huygens' Principle, all points on the wavefront A_1B_1 are secondary sources of disturbance, spreading out spherical wavelets around them. Suppose the wavelet from the point B_1 takes a time Δt to reach the mirror's surface. Then (c = velocity of light),

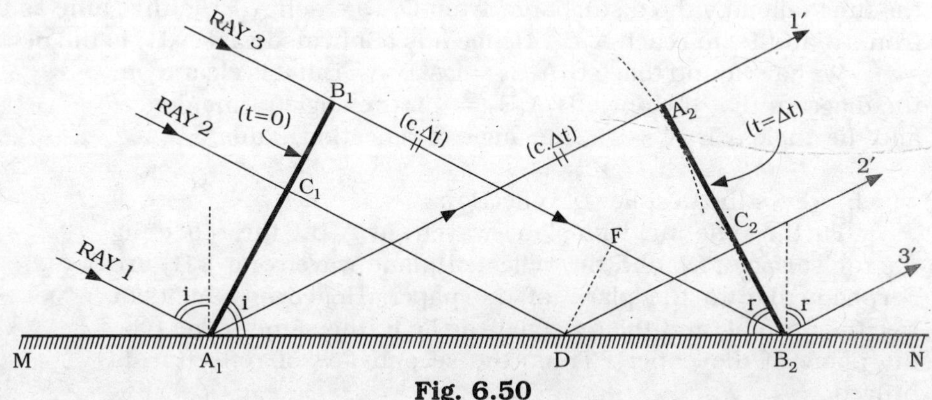

Fig. 6.50

$$B_1B_2 = c.\Delta t$$

Since the mirror's surface is opaque, the disturbance from the point A_1 cannot travel in the forward direction. So it has to travel back in the first medium. To know its direction of travel, we draw an arc of radius $c.\Delta t$, with A_1 as the centre. Obviously, this arc, drawn by dashes in the figure, is a part of the spherical wavelet from A_1 at time Δt. According to the Huygens' construction, the position of the new wavefront is to be found by drawing a surface tangential to such arcs, corresponding to the various points on the wavefront A_1B_1.

At the instant Δt, when the disturbance from B_1 just reaches the surface at B_2, it also cannot penetrate the mirror and so has to travel back in the first medium. The corresponding spherical wavelet will just begin to expand into the first medium. In other words, the distance travelled by it in the first medium, after striking the mirror, will be zero. So at the time Δt this secondary wavelet will just be a point located at B_2. Hence the position of the new (reflected) wavefront at time Δt is found by drawing from the point B_2 a plane A_2B_2 tangential to the arc (drawn with A_1 as the centre). A_2B_2 is the new (reflected) wavefront. It just touches the arc at the point A_2. Note that,

$$A_1A_2 = c.\Delta t = B_1B_2$$

The fact that A_2B_2 is the position of the new wavefront can be confirmed by proving that the disturbance from an arbitrarily chosen point C_1 on the original wavefront also reaches A_2B_2 at the instant t=Δt, after refracting into the second medium at the point D on the interface. The total distance travelled by it is

$$C_1D + DC_2 = B_1F + DC_2$$

DF is the perpendicular drawn from D on B_1B_2. D becomes a secondary source of disturbance when the disturbance from C_1 arrives there. A part of the spherical wavelet at C_2 has been drawn with dashes, with D as the centre.

In the triangles $A_1B_1B_2$ and $A_1A_2B_2$, we have,

$$A_1A_2 = B_1B_2 \ (=c.\Delta t); \angle B_1 = \angle A_2 \ (= 90°); A_1B_2 \text{ is common.}$$

So the two triangles are congruent. Therefore,

$$\angle B_1A_1B_2 = \angle A_2B_2A_1$$

Now, in the Δ^S FDB_2 and C_2DB_2, we have,

$$\angle FDB_2 \ (= \angle B_1A_1B_2) = \angle C_2B_2D \ (= \angle A_2B_2A_1);$$

$$\angle F = 90° = \angle C_2;$$

$$DB_2 \text{ is common.}$$

So the two triangles are congruent. Therefore,

$$DC_2 = FB_2$$

$$\Rightarrow B_1F + DC_2 = B_1F + FB_2$$

$$= B_1B_2$$

$$= c.\Delta t$$

But $(B_1F + DC_2)$ is the total distance travelled by the disturbance in reaching from C_1 to C_2. It follows that the time taken by the disturbance from C_1 to reach A_2B_2 is the same as the time taken by the disturbances from A_1 and B_1 to reach A_2B_2. Hence it is confirmed that A_2B_2 is the new reflected wavefront.

We have found that $\angle B_1A_1B_2 = \angle A_2B_2A_1$. But it is clear from the diagram that the angle $B_1A_1B_2 = \angle i$, the angle of incidence, and the angle $A_2B_2A_1 = \angle r$, the angle of reflection. Thus,

$$\angle i = \angle r,$$

which proves the first law of reflection

Further, the incident plane wavefront A_1B_1, the reflecting mirror surface MN, and the reflected plane wavefront A_2B_2 are perpendicular to the plane of the paper. Hence any incident ray, the normal, and the reflected ray lie in the same plane (viz. the plane of the paper). Thus, the second law of reflection is proved.

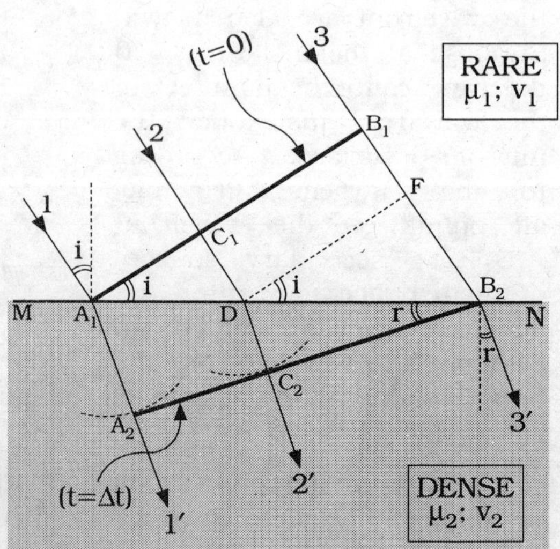

Fig. 6.51

6B. 04. Refraction of a Plane Wave at a Plane Surface

The two laws of refraction of light are (vide Section 6A.01, Chapter 6A):

(i) The incident ray, the refracted ray, and the normal to interface at the point of incidence lie in the same plane.

(ii) The ratio of the sine of angle of incidence to the sine of angle of refraction is constant for a pair of media (Snell's law):

$$\frac{\sin i}{\sin r} = \frac{\mu_2}{\mu_1} = \mu_{21}$$

or $\mu_1 \sin i = \mu_2 \sin r$ (Snell's law; general form)

We shall establish these laws on the basis of the Huygens' wave theory.

The Fig. 6.51 shows a plane wavefront A_1B_1 incident on the plane interface between two media of refractive index μ_1 and μ_2, respectively, with $\mu_2 > \mu_1$. (The student may imagine the first medium as air and the second as a block of glass). Let the velocity of light in the two media be v_1 and v_2 respectively. Three incident rays perpendicular to the wavefront and marked as 1, 2, 3 are also shown in the figure. Suppose at the instant t = 0 the ray 1 and the point A_1 of the wavefront just touch the interface. According to the Huygens' principle, all points on the wavefront A_1B_1 are secondary sources of disturbance, spreading out spherical wavelets around them. Suppose the wavelet from the point B_1 takes a time Δt to reach the interface at B_2. Then,

$$B_1B_2 = v_1.\Delta t$$

During this time Δt, the secondary wavelet from A_1 travels in the second (dense) medium with velocity v_2. We draw an arc with radius $v_2\,\Delta t$, with A_1 as the centre, as shown in the figure (dashed line). Obviously, this arc is a part of the spherical wavelet from A_1 at time Δt. According to the Huygens' construction, the position of the new (refracted) wavefront is to be found by drawing a surface tangential to such arcs, corresponding to the various points on the incident wavefront A_1B_1.

At the instant Δt, the disturbance from the point B_1 on the incident wavefront just reaches the point B_2 on the interface. So B_2 also becomes a (secondary) source of disturbance. The corresponding spherical wavelet from B_2 will just begin to expand into the second medium. It means that the distance travelled by it in the second medium, after striking the interface, will be zero. So at the time Δt this secondary wavelet will just be a point located at B_2. Hence the position of the new (refracted) wavefront at time Δt is found by drawing from the point B_2 a plane A_2B_2 tangential to the arc (drawn with A_1 as the centre). A_2B_2 is the new (refracted) wavefront. It just touches the arc at the point A_2. Note that,

$$A_1A_2 = v_2.\Delta t$$

The fact that A_2B_2 is the position of the new wavefront can be confirmed by proving that the disturbance from an arbitrarily chosen point C_1 on the original wavefront also reaches A_2B_2 at the instant t = Δt, after refracting into the second medium at the point D on the interface. The total distance travelled by it is

$$C_1D + DC_2 = B_1F + DC_2$$

DF is the perpendicular drawn from D on B_1B_2. D becomes a secondary source of disturbance when the disturbance from C_1 arrives there. A part of the spherical wavelet at C_2 has been drawn with dashes, with D as the centre.

From the figure it is seen that the angle of incidence i and the angle of refraction r are equated as,

$$\angle B_1A_1B_2 = \angle FDB_2 = \angle i$$

$$\angle A_1B_2A_2 \text{ (or } \angle DB_2C_2) = \angle r$$

Now,

$$\sin i = \frac{B_1B_2}{A_1B_2} = \frac{v_1.\Delta t}{A_1B_2}$$

$$\sin r = \frac{A_1A_2}{A_1B_2} = \frac{v_2.\Delta t}{A_1B_2}$$

Dividing, we get,

$$\frac{\sin i}{\sin r} = \frac{v_1}{v_2}$$

Next,

$$\frac{FB_2}{DC_2} = \frac{FB_2/DB_2}{DC_2/DB_2} = \frac{\sin i}{\sin r} = \frac{v_1}{v_2}$$

$$\Rightarrow \quad \frac{FB_2}{DC_2} = \frac{v_1.\Delta t}{v_2.\Delta t}$$

The above equation means that in the time Δt when the disturbance from F reaches the new wavefront at B_2 with a velocity v_1, the disturbance from D reaches the point C_2 on the wavefront with a velocity v_2. Thus, it is confirmed that A_2B_2 is the new refracted wavefront.

The incident plane wavefront A_1B_1, the plane interface MN, and the refracted plane wavefront A_2B_2 are perpendicular to the plane of the paper. Hence any incident ray, the normal, and the reflected ray lie in the same plane (viz. the plane of the paper). Thus the first law of refraction is proved.

Next, the refractive index μ for a medium is defined as the ratio of the velocity of light in vacuum to the velocity of light in the medium ($\mu = c/v$). So we have,

$$\mu_1 = c/v_1$$

$$\mu_2 = c/v_2$$

$$\Rightarrow \quad \frac{v_1}{v_2} = \frac{\mu_2}{\mu_1}$$

$$\therefore \quad \frac{\sin i}{\sin r} = \frac{v_1}{v_2} = \frac{\mu_2}{\mu_1}$$

or

$$\mu_1 \sin i = \mu_2 \sin r,$$

which is the Snell's law in general form. Thus the second law of refraction is proved.

6B. 05. Interference of Waves; Principle of Superposition of Waves

In the coming sections, we shall study three important optical phenomena based on the wave nature of light, viz. interference, diffraction and polarization. By now, you are well familiar with waves and wave motion. Still, it will be useful to revise the knowledge of some salient features in this regard, relevant to our study.

A wave represents a disturbance in space. The disturbance has a regularity about it. In a material medium, e.g. air, water, glass, etc., we can have two types of waves, viz. longitudinal and transverse. A longitudinal wave consists of compressions and rarefactions of the medium about their mean positions; the disturbance occurs along the direction of propagation of the wave, e.g. in a sound wave. A transverse wave comprises oscillations of the medium about their mean positions; the disturbance occurs perpendicular to the direction of propagation of the wave, e.g. in a water wave. The electromagnetic waves, e.g. light waves and radio waves, do not require a material medium for their propagation. They are transverse waves, consisting of oscillations of electric and magnetic field, perpendicular to the direction of propagation. So, unlike the sound waves or water waves, they can travel in free space, i.e. vacuum. All waves carry energy with them. An important point to note about the waves travelling in a material medium is that in spite of carrying energy with them in the direction of propagation, there is no transport of material in the direction of propagation. For instance, as a water wave advances in an ocean, the water is not carried forward along with it. Only the water mass moves up and down about the surface.

When two or more waves mutually superimpose, i.e. they pass simultaneously through the same region (e.g. water waves produced by a simultaneous close movement of several boats in a river), the resultant wave in the region of crossing is given by the **principle of superposition,** which states:

"When two or move waves simultaneously pass through a given region, the resultant disturbance in the region is the vector sum of the individual disturbances due to the various waves."

A most remarkable feature is that as the waves emerge after the overlap, they look exactly the same and keep on propagating in the same direction as before, as if nothing had happened! In other words, irrespective of their individual shapes, the waves freely cross each other without any modification of their shapes. This is a unique property of the waves, which is of great practical value in life. It is in sharp contrast to a meeting together of the particles. If two or more particles happen to cross each other, they collide and change their directions of motion. Their energy values also change. But when two waves overlap, there is a redistribution of energy only within the region of overlap. Beyond that the waves move on with the same distribution of energy in them as before the overlap.

"The phenomenon of combination of two or more waves in a given region of space" is called **interference.** The concerned waves are said to **interfere** with each other. Prima facie, such a mixing of the waves appears to be a complex matter. But the principle of superposition enables us to look at it in a simple manner. Each wave makes its own contribution to the resultant wave, irrespective of the presence of the other waves. The resultant wave is simply determined by vectorially summing up the different individual contributions. Its magnitude and nature depends on to what extent the waves are in phase, i.e. in step, with each other. We illustrate it by the following simple example.

Suppose two sinusoidal waves of the same amplitude and wavelength but with phase difference ϕ are produced along a stretched string to travel in the same direction x,

$$y_1 = y_m \sin(kx - \omega t) \quad \text{..... (6.50)}$$

$$y_2 = y_m \sin(kx - \omega t + \phi) \quad \text{..... (6.51)}$$

It can be worked out that the resultant wave, found as the sum $y = y_1 + y_2$, is given by,

$$y = [2y_m \cos(\phi/2)] \sin[kx - \omega t + (\phi/2)] \quad \text{..... (6.52)}$$

It is seen that this is also a sinusoidal wave travelling along the same direction x, but with a different phase constant ($\phi/2$) and a different amplitude,

$$\text{Resultant amplitude} = 2y_m \cos(\phi/2) \quad \text{..... (6.53)}$$

Three plots of the resultant wave, representing the displacement as a function of the distance x, are shown in Fig. 6.52. Two of these are of particular interest:

(i) Fully constructive interference: From the Eq. 6.53, the resultant amplitude has the maximum value when $\phi = 0$, i.e. the phase difference between the two interfering waves is zero or, say, the two waves are completely in phase with each other. It is twice the amplitude of the either interfering wave. In this condition, the interference is said to be **fully constructive interference** (Fig. 6.52 a).

(ii) Fully destructive interference: For $\phi = 180°$ or π radian, the two interfering waves are completely out of phase. The resultant amplitude is zero. It is said to be the condition of **fully destructive interference** (Fig. 6.52 b). In spite of two waves existing along it, now the string looks stationary!

The phase difference ϕ is related to the path difference ΔL between the two waves as,

$$\phi = \left(\frac{2\pi}{\lambda}\right)\Delta L \quad \text{..... (6.54)}$$

So the conditions of interference can also be expressed in terms of path difference between the two interfering waves. Further, since a sine function periodically repeats itself every 2π radians, the conditions can be generally expressed as follows:

$$\left.\begin{array}{l}\text{(i) } \phi = 2n\pi \ (n = 0, \pm1, \pm2, \ldots\ldots\ldots) \\ \Rightarrow \Delta L = n\lambda\end{array}\right\} \text{FULLY CONSTRUCTIVE INTERFERENCE} \quad \text{..... (6.55)}$$

$$\left.\begin{array}{l}\text{(ii) } \phi = (2n+1)\pi \\ \Rightarrow \Delta L = \left(n + \frac{1}{2}\right)\lambda\end{array}\right\} \text{FULLY DESTRUCTIVE INTERFERENCE} \quad \text{..... (6.56)}$$

In the Fig. 6.52 (a), $\phi = 0$ and the resultant amplitude has the maximum value $2y_m$. In the Fig. 6.52 (b), $\phi = 180°$ and the resultant amplitude is zero. For an intermediate value of ϕ, the resultant amplitude has a value between zero and $2y_m$ (Fig. 6.52 c). Accordingly, such a condition of interference, which is partly constructive and partly destructive, is called the condition of **intermediate interference** of the waves.

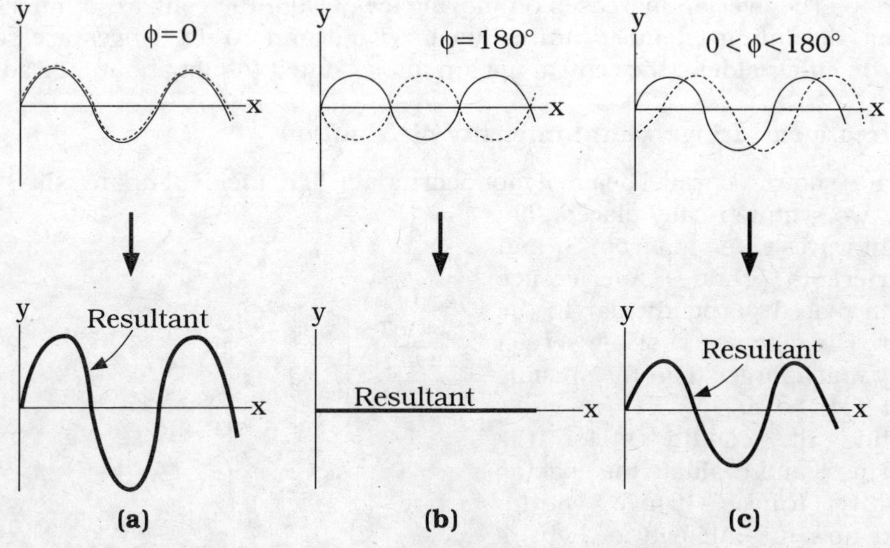

Fig. 6.52

6B. 06. Interference of Light; Young's Double Slit Experiment

In 1801, a very conclusive experiment was performed by the British physicist Thomas Young (1773-1829) in favour of the wave theory of light, involving interference of light waves.

(i) Qualitative treatment

Monochromatic light (i.e. light of single wavelength) falls on a slit S made in a sheet P and then upon two slits S_1 and S_2, made in a symmetrically placed sheet Q (Fig. 6.53 a). The sheets P and Q are perpendicular to the plane of the paper. The spherical wavefronts emerging from S, S_1, S_2 are shown as parts of circle in the plane of the paper. The solid circles correspond to the peaks (crests) and the dashed circles to the valleys (troughs) of the light waves. The waves overlap and mutually interfere in the region beyond S_1,S_2. Constructive interference takes place where a peak meets a peak or a valley meets a valley. The corresponding positions are marked with light small black circles in the figure. Destructive interference occurs where a peak meets a valley (dark circles). The resultant distribution of intensity of light, i.e. the resultant wave, in the region of crossing of the two separate wave patterns is determined by the principle of superposition. If a screen R perpendicular to the plane of the paper is placed in the region, a series of alternate **bright and dark bands** are seen to have formed. They are called **interference bands or fringes.** Since they correspond to positions of constructive and destructive interference, they are also referred to as **interference maxima and minima.** The entire sequence of the bright and dark fringes is termed as an interference pattern. An actual photograph of the central part of such a pattern is shown in Fig. 6.53 (b). (Young's original experiment was performed with pinholes and white light. However, the interference pattern is much more well defined and the maxima are brighter when the slits and monochromatic light are employed). With white light, coloured fringes are formed. As will he seen in the following treatment, the component wavelengths produce closely spaced fringes of different colours. Therefore the maxima and minima are ill-defined, as they partially overlap. The overlap increases on moving away from the centre, so much so that after a few fringes the maxima and minima tend to mutually cancel out and so no fringes are seen thereafter. The number of fringes on either side of the central maximum is limited to just about 5 or so).

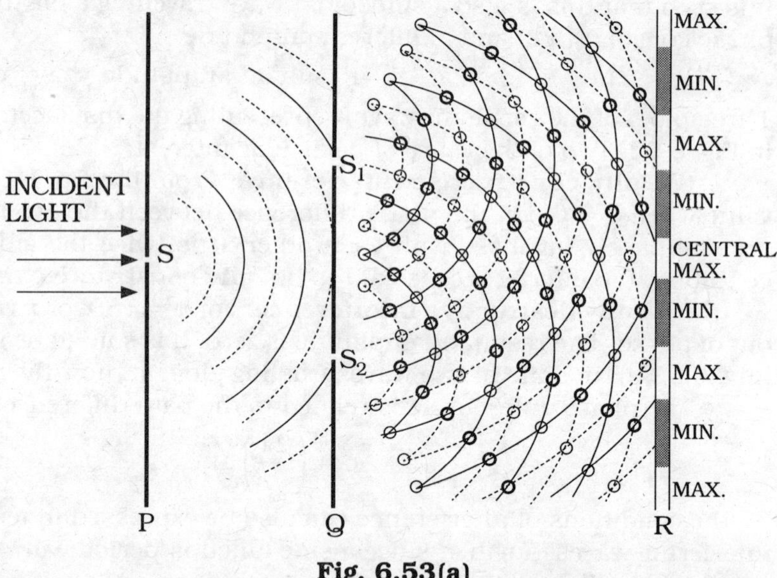

Fig. 6.53(a)

Fig. 6.53(b)

(ii) Quantitative treatment; fringe width; intensity distribution

The Fig. 6.54(a) shows a parallel beam of monochromatic light incident upon a slit S. The light emerging from S falls upon two symmetrically placed slits S_1 and S_2. The light waves emerging from S_1 and S_2 mutually interfere to produce interference fringes on a screen placed perpendicular to the plane of the paper. The distance D of the screen from S_1, S_2 is very much larger than the spacing d between S_1 and S_2 (D>>d).

We shall find the conditions for the formation of the fringes and evaluate the spacing between consecutive fringes (fringe width). We shall also find how the intensity of light is distributed among the fringes.

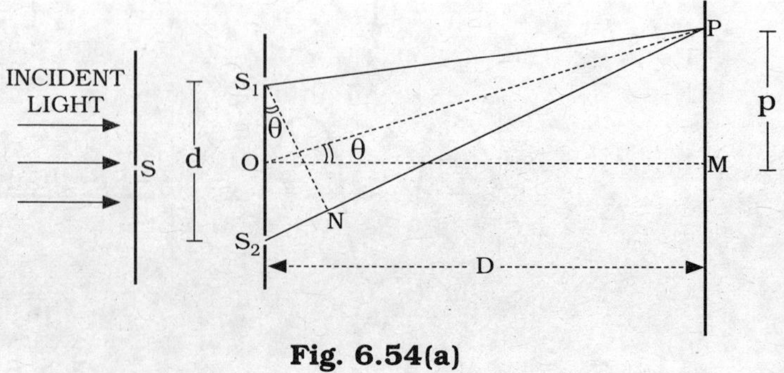

Fig. 6.54(a)

Conditions of formation of the fringes: S_1 and S_2 are two coherent sources of light (to be discussed in the next section). The light waves just emerging from them are in phase. By the time they reach an arbitrary point P on the screen, they develop a phase difference ϕ between them, because they travel different distances S_1P and S_2P. Let A_1 and A_2 be the amplitudes of the waves. Then the waves can be represented as,

$$y_1 = A_1 \sin(kx - \omega t) \quad \text{..... (6.57)}$$

$$y_2 = A_2 \sin(kx - \omega t + \phi) \quad \text{..... (6.58)}$$

Applying the principle of superposition, the resultant wave is given by,

$$\begin{aligned} y &= y_1 + y_2 = A_1 \sin(kx - \omega t) + A_2 \sin(kx - \omega t + \phi) \\ &= A_1 \sin(kx - \omega t) + A_2 \sin(kx - \omega t) \cos\phi + A_2 \cos(kx - \omega t) \sin\phi \\ &= \sin(kx - \omega t)[A_1 + A_2 \cos\phi] + \cos(kx - \omega t)[A_2 \sin\phi] \end{aligned}$$

Let,

$$(A_1 + A_2 \cos\phi) = A\cos\delta \quad \text{..... (i)}$$

$$A_2 \sin\phi = A\sin\delta \quad \text{..... (ii)}$$

Then,

$$y = A[\sin(kx - \omega t)\cos\delta + \cos(kx - \omega t)\sin\delta]$$

$$\Rightarrow \quad y = A\sin(kx - \omega t + \delta) \quad \text{..... (6.59)}$$

Thus, the resultant wave is also a sinusoidal wave. Its amplitude is A and it has a phase difference of δ w.r.t. the first wave. The values of A and δ are found from Eqs. (i) and (ii):

$$\begin{aligned} A^2\sin^2\delta + A^2\cos^2\delta &= A^2 = (A_1 + A_2\cos\phi)^2 + (A_2\sin\phi)^2 \\ &= A_1^2 + A_2^2 + 2A_1A_2\cos\phi \end{aligned}$$

$$\Rightarrow \quad A = \sqrt{A_1^2 + A_2^2 + 2A_1A_2\cos\phi} \quad \text{..... (6.60)}$$

Further,

$$\frac{A\sin\delta}{A\cos\delta} = \frac{A_2\sin\phi}{(A_1 + A_2\cos\phi)}$$

$$\Rightarrow \quad \tan\delta = \frac{A_2\sin\phi}{(A_1 + A_2\cos\phi)} \quad \text{..... (6.61)}$$

From Eq. 6.60 for the amplitude of the resultant wave, the resultant intensity [(amplitude)^2] of light is maximum for $\phi = 0, 2\pi, 4\pi,$.......... and minimum for $\phi = \pi, 3\pi, 5\pi,$........ Accordingly, the conditions for fully constructive and destructive interference, i.e. for the formation of bright fringes (maxima) and dark fringes (minima), are,

$$\left.\begin{aligned} \phi &= 2n\pi \quad \text{(MAXIMA)} \\ \phi &= (2n+1)\pi \quad \text{(MINIMA)} \end{aligned}\right\} \quad \text{..... (6.62a)}$$

where n is an integer ($n = 0, \pm 1, \pm 2, \pm 3,$...................).

Since the phase difference ϕ is related to the path difference ΔL as (vide Eq. 6.46),

$$\phi = \left(\frac{2\pi}{\lambda}\right)\Delta L,$$

the said conditions can also be expressed in terms of path difference as,

$$\left.\begin{aligned} \Delta L &= n\lambda \quad \text{(MAXIMA)} \\ \Delta L &= \left(n + \frac{1}{2}\right)\lambda \quad \text{(MINIMA)} \end{aligned}\right\} \quad \text{..... (6.62b)}$$

At the point M in Fig. 6.54 (a) the path difference $\Delta L = S_1P - S_2P = 0$. It corresponds to $n = 0$ in the Eq. 6.62 (b) and so represents the position of the central maximum of the fringes.

Fringe width; determination of wavelength of light (λ). The fringe width is the spacing between two consecutive bright fringes (maxima) or two consecutive dark fringes (minima) on the screen.

Since $D \gg d$, S_1P, S_2P and OP can be deemed as nearly parallel to each other, as shown in Fig. 6.54 (b). Under this approximation,

$$\angle POM = \theta \approx \angle S_2S_1N,$$

and the path difference,

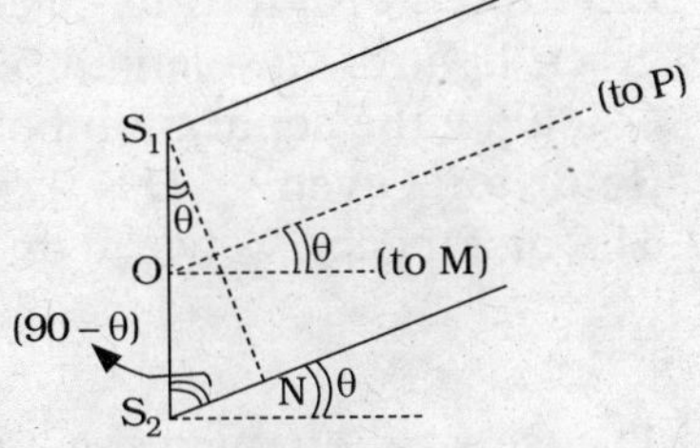

Fig. 6.54 (b)

$$(S_2P - S_1P) \approx S_2N = \Delta L = d \sin\theta$$

For small θ - values, $\sin\theta \approx \theta \approx \tan\theta$. Therefore,

$$\Delta L \approx d \tan\theta = d.\frac{p}{D}$$

From the Eq. 6.62 (b), the maxima (bright fringes) and the minima (dark fringes) correspond to $\Delta L = n\lambda$ and $\Delta L = \left(n+\frac{1}{2}\right)\lambda$, respectively, i.e. they correspond to,

$$p = n\frac{D\lambda}{d} \quad \text{(MAXIMA)}$$

and

$$p = \left(n+\frac{1}{2}\right)\frac{D\lambda}{d} \quad \text{(MINIMA)} \qquad \text{..... (6.63)}$$

The fringe width (β) is found as the difference between two consecutive values of p, corresponding to, say, n and (n+1),

$$\beta = \frac{D\lambda}{d} \qquad \text{..... (6.64 a)}$$

The same value of β is obtained by choosing any other consecutive values of n, say (n+5) and (n+6). Thus it is found that the bright and dark fringes are equally broad and the fringes are equally spaced. Accordingly, the distance of an nth fringe, on either side, from the central maximum is given as,

$$x_n = \frac{n.D\lambda}{d} \qquad \text{..... (6.64 b)}$$

By measuring the values of D (distance of the screen from the centre O of the slits S_1, S_2), d (separation between S_1 and S_2) and β (spacing between two consecutive bright or dark fringes), the wavelength (λ) of the light used to illuminate the slits can be calculated from the Eqs. 6.64. Further, with reference to the Eqs. 6.64, the following conclusions are drawn:

(a) If the distance D of the screen from the slits S_1,S_2 is increased, the fringe width β is increased, i.e. the fringes become widely spaced ($\beta \propto D$). The effect is opposite if D is decreased.

(b) If the separation d between the slits S_1,S_2 increases, the fringe width decreases, i.e. the fringes become closely spaced ($\beta \propto 1/d$). If d is excessively increased, i.e. S_1 and S_2 are far apart, the fringes become too close to be seen and the screen looks uniformly illuminated.

On the other hand, if $d \approx 0$, i.e. S_1 and S_2 are infinitely close to each other, the fringes will be so far apart that except for the central maximum, none will be seen on the screen.

(c) The fringe width increases with increase in the wavelength of the illuminating light ($\beta \propto \lambda$). Thus $(\beta)_{red} > (\beta)_{blue}$. (At the end of the earlier qualitative treatment of the formation of fringes, we had discussed the effect of using white light in place of monochromatic light).

(d) If the medium in which the experiment is being carried out is changed, e.g. the whole apparatus is immersed in water, the fringe width will change. The frequency and the phase difference φ do not change, but the wavelength (λ) changes. If water is substituted for air, λ will decrease, because the velocity of light in water is less than in air. It will become λ/μ_{water}. Accordingly, the fringe width will reduce by the factor $(1/\mu_{water})$.

(e) The source slit S should be very narrow, otherwise the visibility of the fringes is reduced. A broad slit means several narrow slits put close together, each of which produces its own interference pattern, whose overlap on the screen reduces visibility. Ideally, the width of S should be of the order of the wavelength (λ) of light, i.e. ~10^3 nm or 10^{-3} mm. For the same reason, the slits S_1 and S_2 should be very narrow.

Example 1. An interference pattern is produced on a screen by illuminating two slits, 0.11 mm apart, with green light of wavelength 5500 Å. The screen is placed at a distance of 1.5 m from the plane of the slits. Calculate the separation between the 3rd bright fringes formed on the two sides of the central maximum.

Solution. Given: D = 1.5 m, d = 0.11 mm = 0.11×10^{-3} m, λ = 5500 Å = 55×10^{-8} m, n = 2.

The distance x_n between an nth fringe, on either side, from the central maximum is given by,

$$x_n = \frac{n.D\lambda}{d}$$

The separation between two nth fringes, situated on the two sides of the central maximum, is $2x_n$. Therefore,

$$\text{Required separation} = 2x_n = \frac{2n.D\lambda}{d} = \frac{2\times(3)\times(1.5)\times(55\times10^{-8})}{(0.11\times10^{-3})}$$
$$= 45\times10^{-3}\text{ m} = 45\text{ mm} \quad \textbf{(Answer).}$$

Intensity distribution in the fringe pattern: If the slits S_1 and S_2 are identical, then in the Eq. 6.60 for the resultant amplitude,

$$A_1 = A_2 = A_0 \text{ (say)}$$
$$\therefore \quad A = \sqrt{A_0^2 + A_0^2 + 2A_0^2\cos\phi}$$
$$\text{or,} \quad A^2 = 2A_0^2(1 + \cos\phi)$$

The intensity is proportional to the square of the amplitude [$I \propto$ (amplitude)^2]. So the resultant intensity is,

$$I = 2I_0(1 + \cos\phi)$$
$$\Rightarrow \quad I = 4I_0\cos^2\frac{\phi}{2} \qquad \text{..... (6.65)}$$

I_0 is the intensity due to any one of the slits S_1 and S_2. The Eq. 6.65 gives the resultant intensity as a function of the phase difference ϕ. The centres of the bright fringes (maxima) correspond to $\phi = 2n\pi$ and those of the dark fringes (minima) to $\phi = (2n+1)\pi$ (vide Eq. 6.62 a). The corresponding values of the resultant intensity are $I = 4I_0$ and $I = 0$, respectively. For the intermediate values of ϕ, I lies between $4I_0$ and 0 ($4I_0 > I > 0$). A plot of I as a function of ϕ is shown in Fig. 6.55.

The average value of a $\cos^2$ function, viz. $\cos^2\phi$, is 1/2. Therefore the average intensity of light in the interference pattern is,

$$\frac{1}{2}\cdot 4I_0 = 2I_0 \quad \text{(Fig. 6.55)}$$

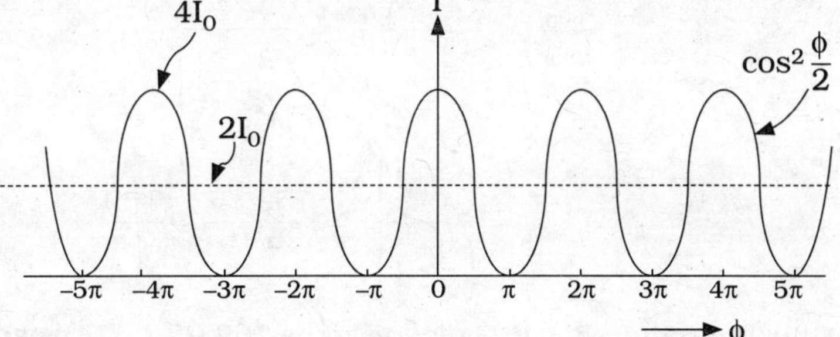

Fig. 6.55

Note that this value is equal to the sum of intensities (I_0 each) due to the slits S_1 and S_2. Thus, the principle of conservation of energy remains valid, as expected. **There is only a redistribution of light energy in the interference pattern.**

In the above discussion, we had assumed that the intensity of light emerging from the two slits S_1 and S_2 is the same. So in the Eq. 6.52 for the resultant amplitude, $A_1 = A_2$. If this is not the case (as will happen, for example, if S_1 and S_2 are not placed symmetrically w.r.t. the source slit S in the Fig. 6.54 a, then $A_1 \neq A_2$. We proceed to find that how the intensity of light is then relatively distributed among the maxima and minima in the interference pattern.

Maxima The maxima correspond to $\phi = 2n\pi$. Therefore the corresponding resultant amplitude (Eq. 6.60) is ($\cos\phi = +1$),

$$A_{max} = \sqrt{A_1^2 + A_2^2 + 2A_1A_2.1}$$
$$= \sqrt{(A_1 + A_2)^2} = (A_1 + A_2)$$

The corresponding intensity can be written as ($I \propto$ (amp.)^2],

$$I_{max} = K(A_1 + A_2)^2, \qquad \text{.... (iii)}$$

where K is the constant of proportionality.

Minima The minima correspond to $\phi = (2n+1)\pi$. Therefore the corresponding value of A is ($\cos\phi = -1$),

$$A_{min} = \sqrt{A_1^2 + A_2^2 + 2A_1A_2.(-1)}$$
$$= \sqrt{(A_1 - A_2)^2} = (A_1 - A_2)$$

The corresponding intensity is,

$$I_{min} = K(A_1 - A_2)^2 \qquad \text{..... (iv)}$$

The Eqs. (iii) and (iv) yield,

$$\frac{I_{max}}{I_{min}} = \left[\frac{(A_1 + A_2)}{(A_1 - A_2)}\right]^2 \qquad \text{..... (6.66)}$$

Example 2. Two coherent sources of light, with intensities in the ratio 81:16, produce interference pattern on a screen. Find the ratio of intensities in the maxima and minima of the pattern.

Solution. Given, $\frac{I_1}{I_2} = \frac{81}{16}$

If A_1 and A_2 are the amplitudes of the light waves from the two sources, then,

$$\frac{I_1}{I_2} = \frac{A_1^2}{A_2^2}$$

$$\Rightarrow \quad \frac{A_1}{A_2} = \sqrt{\frac{I_1}{I_2}} = \sqrt{\frac{81}{16}} = \frac{9}{4}$$

The ratio of intensities of the maxima and minima in the interference pattern is given by,

$$\frac{I_{max}}{I_{min}} = \frac{(A_1+A_2)^2}{(A_1-A_2)^2} = \frac{\left(\frac{A_1}{A_2}+1\right)^2}{\left(\frac{A_1}{A_2}-1\right)^2}$$

$$= \frac{\left(\frac{9}{4}+1\right)^2}{\left(\frac{9}{4}-1\right)^2} = \frac{169/16}{25/16} = \frac{169}{25}$$

Thus the required ratio is found to be 169:25. **(Answer).**

6B. 07. Coherent Sources; Sustained Interference Pattern

The **coherent sources** are "those sources which produce waves of the same frequency (or wavelength), with a constant phase difference between them". They can be obtained in one of the following two manners:

(i) Division of wavefront. Light from a point source (or a narrow slit) falls on a pair of pinholes (or narrow slits), thus dividing a single wavefront into two parts. The pair acts as two coherent sources of light. This method was employed in the Young's experiment, described in the last section.

(ii) Division of amplitude. A light beam is divided into two or more beams by partial reflection. Thus, the amplitude of a beam is divided into two or more parts. The resulting beams act as coherent sources of light.

If a thin reflecting film (e.g. a silver film) is coated on a surface and light falls upon it, a single ray of light is divided into two rays by reflection at the upper and lower surfaces of the film, as shown in Fig. 6.56. A ray AB is divided into two rays 1 and 2 by reflection at B and C, respectively, at a partially reflecting film of a very small thickness t (just a few wavelengths, i.e. $\sim 10^{-3}$ mm).

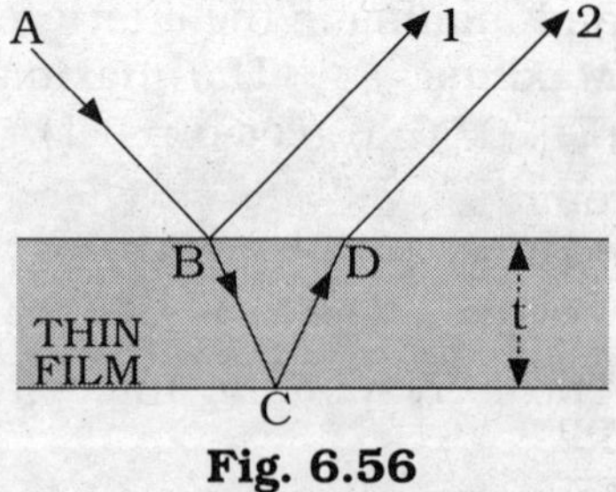

Fig. 6.56

Two independent conventional type of light sources, e.g. candles, electric lamps, etc., are not coherent sources. In such a source, the light comes from a very large number of independent atoms. Each atom emits light for just about 10^{-9} second, i.e. it emits a pulse lasting just about a nanosecond (i.e. a billionth of a second). Then it becomes inactive for some time, before emitting light once again. Thus, different groups of atoms (billions of them) emit light at different times. So if in the Young's experiment the slits S_1 and S_2 are illuminated separately by two such sources, the waves emerging from S_1 and S_2 will constantly be changing their phases. They will have a constant phase difference between them for just about 10^{-9} second (called coherence time). Hence the interference pattern will be changing every nanosecond. The human eye can follow intensity changes of frequency not more than about 20 per second (a cinema reel moves with about this speed). So it observes a uniform intensity of light on the screen, instead of any fringes.

Thus, we find that the beams obtained from two independent conventional sources of light do produce an interference pattern, but it lasts for too small a period of time to be observable, i.e. the pattern is not sustained over a reasonable period. To produce a **sustained interference pattern,** it is essential to have coherent sources of light.

However, two independent **lasers** can be used as coherent sources of light. In a laser, the atoms emit light in a cooperative manner. Hence no phase changes of the type occurring in the conventional sources of light exist and therefore the light emitted by them is coherent.

Additional knowledge

(i). The conventional light sources pose two more problems, viz. (a) they have relatively small (so-called) coherence length and (b) the emitted light is not strictly monochromatic.

During the active time (the so-called coherence time) of about 10^{-9} second of emission of light by a group of atoms in a conventional source, a wavetrain measuring only upto a few metres is produced. It is called coherence length of the light produced by the source. This does not pose any problem for the formation of an interference pattern on the screen, because when one wavetrain is substituted by the next wavetrain the phase is randomly changed, but that change is simultaneously commumicated to S_1 and S_2 and so the phase difference ϕ still remains the same. But if the path difference between the beams from S_1 and S_2 in reaching the screen is large, the second wavetrain may arrive well after the first wavetrain has passed. Then, there will be no overlap and so no interference will take place. Hence the path difference has to be kept small for observing a well defined fringe pattern on the screen. For the same reason, the width of the source slit S has to be extremely small, so that it does not allow simultaneous passage of many wavetrains through it.

The coherence time during which a light pulse is produced is $\sim 10^{-9}$ second. The optical frequencies are $\sim 10^{15}$ per second. Therefore, even this short pulse consists of about 10^6 oscillations (i.e. a million oscillations). Hence it is nearly monochromatic. However, problem appears because the commonly available conventional light sources often produce more than one close wavelenghts. For example, a sodium lamp commonly employed in laboratories emits two close wavelengths 589.0 nm and 589.6 nm, i.e. there is a wavelength spread of 0.6 nm. Other factors, e.g. the so-called pressure broademing, also lead to wavelength spread. This lack of monochromatism reduces the visibility of the interference fringes.

The modern laser sources of light are far superior in all respects. The light emitted by them is highly monochromatic and the coherence length can measure upto a few kilometres. Besides, a laser beam is extremely thin and spreads negligibly during its propagation. So it can be focussed to a width approximating to the wavelength of light. The light intensity is also high.

(ii) Colour of thin films. One often sees colours in soap bubbles, on the oily surface of small pools of water, etc. They are formed by interference between the two coherent beams arising from reflection from the upper and lower surfaces of the film. The overlapping maxima and minima due to the various wavelengths present in the incident light produce the observed colours.

(iii) Antireflectin coating. Photography camera lenses are coated with a suitable reflecting film of thickness $\lambda/4$. Then the incident light produces two coherent interfering beams by reflection from the upper and lower film surfaces, with phase difference $\lambda/2$. So they destructively interefere, resulting in zero reflected intensity and correspondingly inreased transmitted intensity.

The same coating is used for glass panels fitted on doors and windows. The people inside the house can see the things outside. but not vice versa. The panels are semi-transparent, but look black from outside.

Conditions for observing sustained interference pattern with good visibility in the Young's experiment

(i) The essential condition for obtaining a sustained interference pattern is that the two sources of light, which produce the interfering waves, should be coherent sources.

(ii) The source slit S and the secondary coherent sources S_1 and S_2 should be extremely narrow.

(iii) The separation between S_1 and S_2 should be small.

(iv) The source slit S should not be too close to the plane of S_1, S_2.

(v) The amplitudes of the light waves reaching the screen from S_1 and S_2 should preferably be equal.

6B. 08. Diffraction; of Light

If a wave is obstructed in its path by an obstacle, it tends to overcome the obstacle by bending around it. In general, **diffraction** is "the phenomenon of bending of waves round the corners of the obstacles lying in their path". All waves, e.g. water waves, sound waves, radio waves, light waves, etc., show diffraction. Greater is the wavelength, more is the bending.

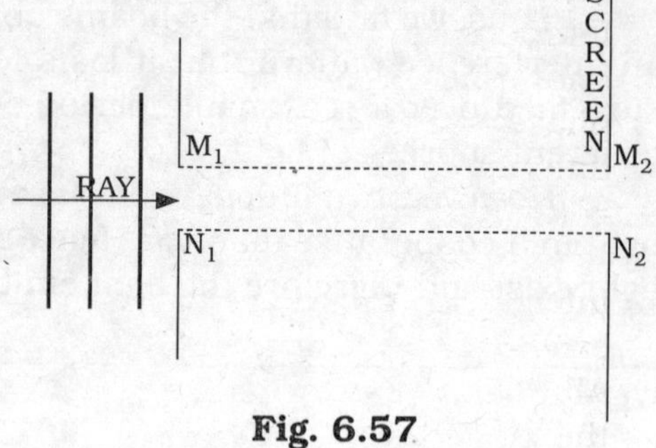

Fig. 6.57

Because of their small wavelength (~ 500 nm, i.e. 5×10^{-4} nm), normally the light waves diffract through inappreciably small angles. So normally the diffraction of light remains unnoticed. It becomes pronounced only when the light passes through narrow apertures (e.g. a pinhole or a narrow slit) or by the side of sharp edges (e.g. the edge of a knife or a razor blade). The Fig. 6.57 shows a plane light wave incident upon a slit M_1N_1 cut into a metal sheet. If the size of the slit is relatively large, measuring a few millimeters or more, it will cast a sharp shadow M_2N_2 on the screen. If the size of the slit is reduced to a small fraction of a millimeter, the edges M_2, N_2 of the shadow will no longer look sharp. Light will be found to have spread out around them, both within the open section M_2N_2 and inside the shadow. The smaller is the size of the slit, greater is the spread. We shall study it in detail in the next section.

The **diffraction of light** is **defined** as "the spreading out of a light wave when it passes through a narrow opening or by the side of a sharp edge".

The phenomenon of diffraction of light can be divided into two categories:

(a) Fresnel* diffraction

(b) Fraunhofer* diffraction

In the Fresnel diffraction, the light source and the screen are situated at finite distances from the diffracting aperture. In the fraunhofer diffraction, they are at infinite distances from the aperture. The infinite distances are easily achieved by (i) placing the light source at the focal point of a converging lens and (ii) placing the screen in the focal plane of another such lens, on the other side of the aperture (Fig. 6.58). The first lens renders the incident light beam parallel, thus effectively removing the source to infinity. The second lens enables the screen to receive a parallel beam of light, thus effectively making it infinitely distant from the slit.

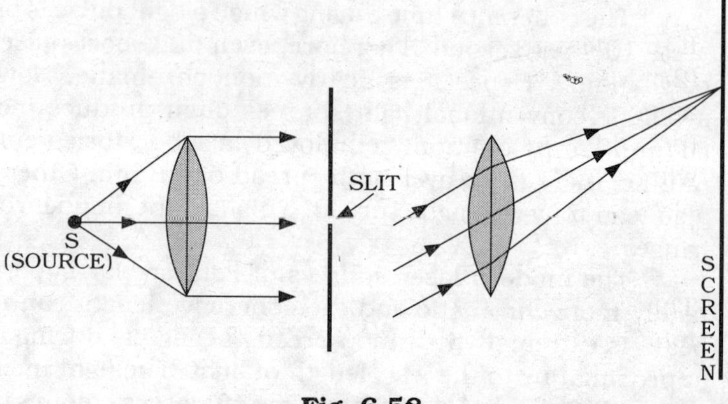

Fig. 6.58

The Fraunhofer diffraction is easier to deal with. We shall describe an example of this category of diffraction in the coming section.

6B. 09. Diffraction by a Single Slit

The Fig. 6.59 (a) shows a parallel beam of monochromatic light normally incident on a narrow slit of width b (b ≈ a few λ's). Therefore the associated incident wavefronts are plane wavefronts. The light beams diffracted by the slit in different directions are collected by a converging lens, which focusses them at different points on a screen placed in the focal plane of the lens. D is the distance between the slit and the screen. [Note that the figure is not drawn to scale, otherwise for a slit of dimension ~10^{-3} mm the lens and the screen will measure a few tens of metre! For the sake of clarity, the slit has been shown disproportionately long].

Accordingly to the Huygens' construction, when the incident wavefront reaches the slit, each point of the slit becomes a source of secondary (spherical) wavelets. After emerging from the slit, these wavelets mutually interfere to produce a diffraction pattern on the screen. We proceed to find the positions of fully constructive

*It is a rare coincidence that two contemperary scientists, Joseph von Fraunhofer (German physicist; 1787-1826) and Augustin Jean Fresnel (French physicist; 1788-1827), worked on the same phenomenon in two different countries and lived an equally short span of life of 39 years each.

interference (maxima) and fully destructive interference (minima) on the screen. As we know, these positions depend upon the mutual phase difference, and therefore the mutual path difference, between the waves [phase difference = $(2\pi/\lambda)$ path difference].

We first determine the condition of interference at the central point P_0, found by extending the central incident ray to the screen. All the rays starting from different point of the slit reach P_0 after travelling equal optical path*. Hence they are in phase and interfere constructively. We refer to P_0 as the position of **central maximum.**

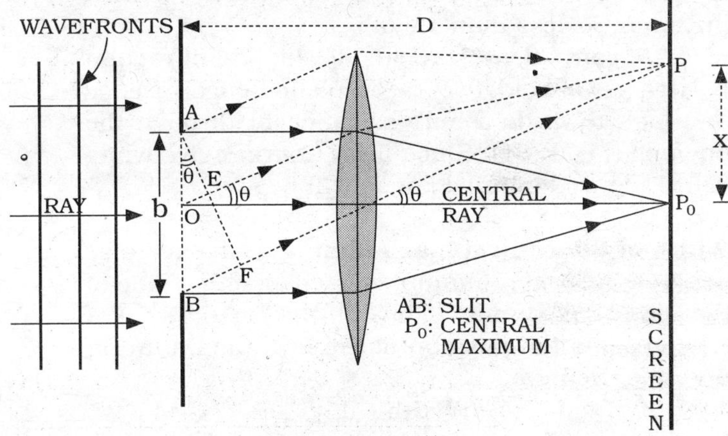

Fig. 6.59 (a)

Next, we find the condition of interference at an arbitrary point P, where the parallel rays emerging from the slit at an angle θ to the central ray OP_0 are focussed. We draw a perpendicular AEF from the uppermost point A of the slit to the rays. AEF represents the plane wavefront associated with the rays (remember that a ray is in the direction of the normal to the corresponding wavefront and determines the direction of propagation of the wavefront). The optical paths of all the rays starting from this wavefront and reaching P are essentially equal. Therefore the path difference between the rays starting from the uppermost point A and the middle point O of the slit is (Fig. 6.59 b)

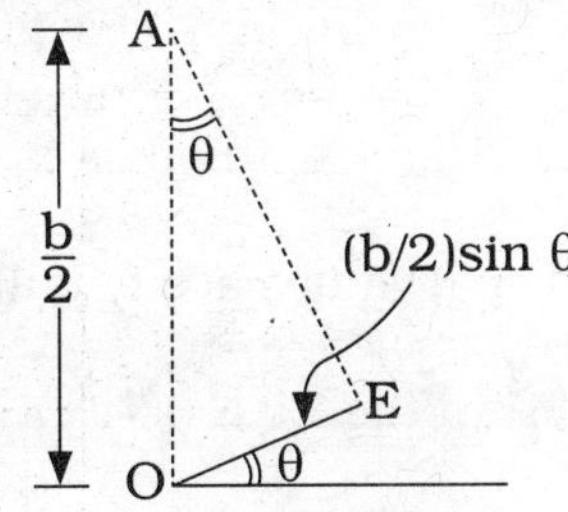

Fig. 6.59 (b)

$$(b/2)\sin\theta$$

If this path difference be equal to $(\lambda/2)$, the corresponding phase difference will be $(2\pi/\lambda).(\lambda/2) = \pi$. Then the two rays will destructively interfere with each other. It is easily seen that a ray from any other point in the upper half of the slit will similarly (destructively) interfere with the ray from a point (b/2) below it in the slit. Thus the whole slit can be divided into pairs of such points, the rays from which will interfere destructively at P and so produce darkness there. Then P will be the position of Ist minimum of interference, given by,

$$(b/2)\sin\theta = (\lambda/2)$$

$$\Rightarrow \quad b\sin\theta = \lambda \text{ (Ist Min.)} \qquad \text{..... (6.67)}$$

The same will happen if $b\sin\theta = 2\lambda, 3\lambda, 4\lambda,$................ Thus, in general, the positions of minimum of intensity are given by (including points on the lower half of the screen),

$$b\sin\theta = n\lambda \text{ (Minima)}, \qquad \text{..... (6.68)}$$

where $n = \pm1, \pm2, \pm3,$................

From considerations of symmetry, the maxima will lie nearly midway between the minima, given by,

$$b\sin\theta = (2n+1)\frac{\lambda}{2}, \qquad \text{..... (6.69)}$$

because then the phase difference has the value $2n\pi$ (0, 2π, 4π,................) and so the diffracted rays interfere constructively.

The **intensity distribution** in the diffraction pattern is shown in Fig. 6.60, for a typical case. The intensity of the maxima falls off very sharply on either side of the central maximum. The expression for intensity at the arbitrary point P on the screen can be derived as,

$$I = I_0\frac{\sin^2\beta}{\beta}, \qquad \text{..... (6.70)}$$

where,

$$\beta = \left(\frac{\pi}{\lambda}\right) b\sin\theta \qquad \text{..... (6.71)}$$

I_0 is the intensity of the central maximum at P_0, corresponding to $\theta = 0$ (so $\beta = 0$). For the other maxima, it is calculated that the intensity of the Ist maximum on either side of P_0 is less than 5% of I_0, that of the IInd

* If a ray travels a distance d in a medium of refractive index μ, the optical path is equal to μd. When light rays starting from a point on one side of a lens meet at another point on the other side of the lens, the optical path travelled by all of them is same, although their geometrical path lengths are different. If a ray travels a longer length in air, it travels a shorter length within the lens (glass) and vice versa. Hence the total optical path travelled by all the rays is same.

maximum is less than 2% of I_0, and that of the IIIrd maximum is less than 1% of I_0.

A convenient way to viewing a diffraction pattern is to hold two razor blades edge to edge and look towards a bulb or tubelight through the opening (Fig. 6.61). Since these sources give white light, coloured parallel bands will be seen.

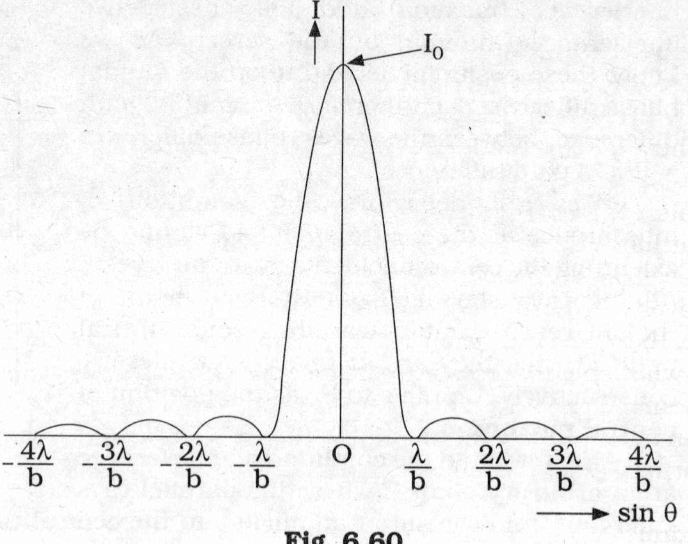

Fig. 6.60

Width of the central maximum: It is "the distance between the first minimum located on either side of the central maximum". Thus, in the Fig. 6.59 (a), if P represents the position of the first minimum on one side, we have,

$$(\text{Width})_{\text{central max.}} = 2x$$

From Eq. 6.67,the position of the Ist minimum is given by,

$$b \sin\theta = \lambda$$

$$\Rightarrow \quad \sin\theta = (\lambda/b) \qquad \text{..... (6.72 a)}$$

In practice, the lens is placed close to the slit. Therefore,

$$\sin\theta \approx (x/D)$$

The θ values are usually small ($x \ll D$). So, $\sin\theta \approx \theta$. Thus we have

$$\sin\theta \approx \theta = (x/D) \qquad \text{..... (6.72 b)}$$

From the Eqs. 6.72 (a) and 6.72 (b),

$$\frac{x}{D} = \frac{\lambda}{b} \quad \text{or} \quad x = \frac{D\lambda}{b}$$

θ is half the angular width and correspondingly x is half the linear width of the central maximum. The total linear width (2x) is given by,

$$W = (\text{width})_{\text{central max.}} = 2x = \frac{2D\lambda}{b} \qquad \text{..... (6.73)}$$

Since D is usually a large value, measuring a metre or more, and the lens is placed very close to the slit, we have $D \approx f$, the focal length of the lens. So we also have,

$$W = (\text{width})_{\text{central max.}} = \frac{2f\lambda}{b} \qquad \text{..... (6.74)}$$

From the Eq. 6.73, the following points should be noted:

(a) Like the Young's double slit experiment, the single slit diffraction pattern consists of bright and dark bands (however, the conditions of formation of the maxima and minima are reversed; compare the Eqs. 6.68 and 6.69 for path difference with the Eq. 6.62 (b) for the case of Young's experiment). But the various maxima are not equally bright and the central maximum is very much brighter than the other maxima. Also, now, while the higher maxima are nearly equally spaced, the central maximum is nearly twice as wide as the higher maxima (Fig. 6.60).

(b) Since practically all the light in the diffraction pattern is concentrated in the central maximum, the size of the image of the single slit as seen on the screen or directly seen by the eye, is determined by the width of the central maximum.

Fig. 6.61

(c) The width of the central maximum is inversely proportional to b, the width of the single slit ($W \propto 1/b$). Since λ is a very small value ($\sim 5\times10^{-4}$ mm), the value of W will not be appreciable unless b is very small, in spite of D being large (~ 1m, i.e. about 10^3 mm). Ideally, **b ≈ a few λ's. This is an essential condition for diffraction of light to be observable.** It also explains why diffraction of light is not ordinarily observable, viz. because normally the places through which the light passes are too wide and λ_{light} is extremely small.

(d) The width of the central maximum is directly proportional to λ ($W \propto \lambda$). The width of an mth maximum is nearly m times this value. Thus, the width of the maxima is greater for red light ($\lambda_{red} \sim 700$ nm) than for the blue light ($\lambda \sim 400$ nm),

$$(W)_{red} > (W)_{blue}$$

Hence if **white light** is used to illuminate the slit, the diffraction bands due to various λ's will overlap only partially. Hence the diffraction pattern will look coloured.

(e) The position of the first minimum is given by b sin θ = λ. So the angular spread of light in the central maximum is,

$$\theta \approx \sin\theta = (\lambda/b)$$

This expression roughly holds for the **sound waves,** too. [The sound waves are longitudinal in nature, but like the light waves they also have wavefronts and rays (which give the direction of propagation of the wavefronts)]. Since $\lambda_{sound} \sim 1$m and $\lambda_{light} \sim 10^{-6}$ m, for a given value of b the angular spread of the sound waves is orders of magnitude higher than the spread of the light waves. In fact they can bend through 180°, i.e. they can completely reverse their direction of propagation even when passing through wide spaces. That is what one observes in daily life, viz. sound waves completely bending round the edges of a window in one room to enter into the adjoining room through a neighbouring window. Thus a person in the adjoining room can hear persons talking in the first room. For the same reason, the sound waves can easily bend round the corners of an obstacle lying in their path.

Example 3. A diffraction pattern is produced by illuminating a single slit of width 0.2 mm with green light of wavelength 5460 Å. Calculate the total angular spread for the first order maximum in the pattern.

Solution. Given: $b = 0.2 \text{ mm} = 0.2\times10^{-3}$ m, $\lambda = 5460$ Å $= 54.6\times10^{-8}$ m

The formation of the maxima is given by the condition,

$$b \sin\theta = (2n+1)\frac{\lambda}{2}$$

∴ For the first order maximum, we have (n = 1),

$$b \sin\theta = (2\times1+1)\frac{\lambda}{2} = \frac{3\lambda}{2}$$

For small θ, sinθ ≈ θ. Therefore, we have,

$$b.\theta = \frac{3\lambda}{2}$$

$$\Rightarrow \qquad \theta = \frac{3\lambda}{2b} = \frac{3\times(54.6\times10^{-8}}{2\times(0.2\times10^{-3})}$$

$$= 4.09\times10^{-3} \text{ radian}$$

This value of θ represents the angular spread between the central maximum and one of the two first order maxima. The total angular spread for the first order maximum is double of this value.

$$\therefore \qquad \text{Total angular spread} = 2\theta = 2\times(4.09\times10^{-3})$$

$$= 8.18\times10^{-3} \text{ radian} \qquad \textbf{(Answer).}$$

Example 4. A diffraction pattern is produced on a screen, when light of wavelength 6200 Å falls on a 0.3 mm wide slit, placed at a distance of 1.2 m in front of the screen. Determine the separation between the two dark bands situated on the two sides of the central bright band in the pattern.

Solution. Given: D = 1.2 m, $b = 0.3 \text{ mm} = 3\times10^{-4}$ m, $\lambda = 6200$ Å $= 62\times10^{-8}$ m

By definition, the distance between the two dark bands situated on the two sides of the central bright maximum, represents the width 2x of the central maximum, given by,

$$2x = \frac{2D\lambda}{b} = \frac{2\times(1.2)\times(62\times10^{-8})}{3\times10^{-4}}$$

$$= 4.96\times10^{-3} \text{ m}$$

Thus the required separation is found to be 4.96×10^{-3} m. **(Answer).**

Example 5. A 0.25 mm wide slit produces a diffraction pattern on a screen, placed 2.5 m away from the lens. The first minima are found to be situated at 6 mm on the two sides of the central maximum. Determine the wavelength of the light employed for illuminating the slit.

Solution. Given: $b = 0.25 \text{ mm} = 0.25\times10^{-3}$ m; D = 2.5 m; distance between first minimum and central maximum, $x = 6 \text{ mm} = 6\times10^{-3}$ m

We know, $$x = \frac{D\lambda}{b}$$

$$\therefore \qquad \lambda = \frac{bx}{D} = \frac{(0.25\times10^{-3})\times(6\times10^{-3})}{2.5} = 6\times10^{-7} \text{ m} = 6000 \text{ Å} \qquad \textbf{(Answer).}$$

6B.10. Comparison between the Phenomena of Interference and Diffraction of Light

It may be clearly understood that both the phenomena, viz. interference and diffraction of light, are based on two common principles, viz. (i) the Huygens' principle of formation of secondary (spherical) wavelets and (ii) the principle of superposition of waves. (Note that in the Young's double slit experiment, if one of the slits is covered, the single slit diffraction pattern is observed on the screen).

However, if one compares the Young's double slit interference pattern with the single slit diffraction pattern, the following points of difference are noted:

Serial No.	Young's double slit interference pattern	Single slit diffraction pattern
1.	Formed by superposition of two sets of wavelets originating from two distinct coherent sources.	Formed by superposition of a large number of wavelets originating from different poin's of the same wavefront.
2.	The maxima have the same intensity	The maxima have unequal intensities. Bulk of the intensity is concentrated in the central maximum.
3.	The maxima are equally spaced.	The central maximum is nearly twice as spread as the other maxima.

6B. 11. Resolving Power of Telescopes and Microscopes; Rayleigh's Criterion

If light is incident upon a narrow circular aperture, the diffraction pattern consists of a series of concentric alternate bright and dark circular bands, as seen in Fig. 6.62. It is known as Airy's pattern. Like the single slit pattern, the central disc (called Airy's disc) is brightest, carrying nearly 85% of the total diffracted intensity of light. The intensity distribution through a diametral section of the pattern looks almost the same as for a single slit pattern, shown earlier in Fig. 6.60. The position of the first minimum, which also determines the angular radius of the central disc, is given by

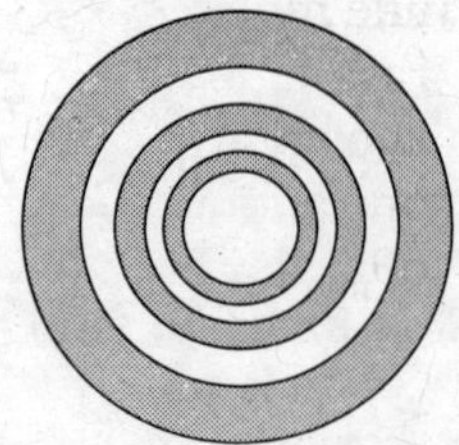

Fig. 6.62

$$\theta_{min} \approx \sin\theta_{min} = \frac{1.22\,\lambda}{d}, \qquad ..... (6.75)$$

where d is the diameter of the circular aperture. This expression is similar to that for a single slit, viz. $\sin\theta = (\lambda/b)$, except for the multiplying factor 1.22.

The diffraction of light by a circular aperture finds direct application to the image formation by optical instruments, which mostly use converging lenses for image formation. When light from a point source is incident on a lens, the wavefront is limited by the periphery of the lens, which acts like a circular aperture. So the light passing through the lens is diffracted. The very important result that follows is that the image of a point object is never a point image but an extended diffraction pattern. However, we know that bulk of the diffracted light intensity is concentrated in the central disc. Therefore, for all practical purposes, the central disc may be considered as representing the extended image of the point object (note that the lens also performs the function of forming the image of the object, by converging the light to its focal plane). The size of the image is given by the Eq. 6.75. Clearly, it depends on the size of the lens.

An important consequence of such an image formation is that if two point objects lie close together, they may not be seen as separate because of overlapping of their respective images. This is clearly seen in part (a) of Fig. 6.63, when we say that the images of the two objects are not resolved. "If the separation between the two objects is such that the central maximum of the diffraction pattern of one coincides with the first minimum of the diffraction pattern of the other, the images are just resolved" [as seen in part (b) of the Fig. 6.63]. This condition of resolution of images is known as **Rayleigh's criterion.** For any further separation between the objects, the images are clearly seen as separated [part (c) of the Fig. 6.63]. Since the position of the first minimum is

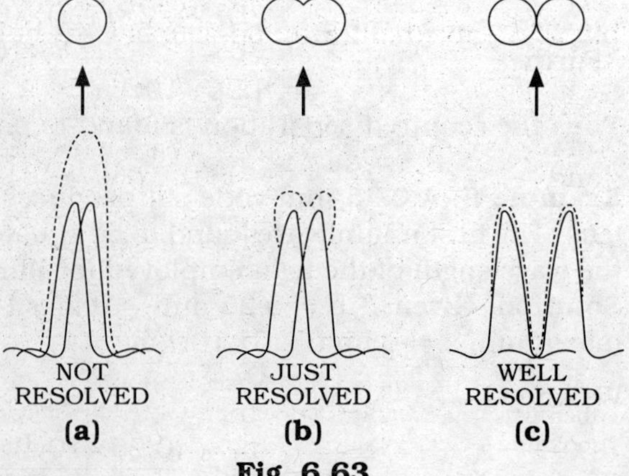

Fig. 6.63

given by Eq. 6.75, the Rayleigh's criterion implies that two closely situated objects are just resolved when their angular separation $d\theta$ is,

$$d\theta_{min} = \frac{1.22\lambda}{d} \quad \text{(RAYLEIGH'S CRITERION)} \qquad .....(6.76)$$

(The angular separation means the angle subtended by the two point objects at the centre of the lens. It is clear from Fig. 6.64, which shows image formation for two very distant point objects, like stars).

In the last chapter (Section 6A.16), the **resolving power of an optical instrument** was **defined** as the "ability of the instrument to form distinct images of two objects lying close to each other". Quantitatively, it is the reciprocal of the **limit of resolution** of the instrument,

$$\text{R.P.} = \frac{1}{\text{Limit of resolution}} \qquad .....(6.77)$$

The **limit of resolution** is **defined** as the "minimum (angular or linear) separation between two objects, to be seen distinctly by the optical instrument". For instance, the (angular) limit of resolution of the human eye* is nearly 1 minute, i.e. two objects subtending an angle of less than 1 minute at the eye cannot be seen separated by the eye. In the following, we shall briefly discuss the resolving power of telescopes and microscopes.

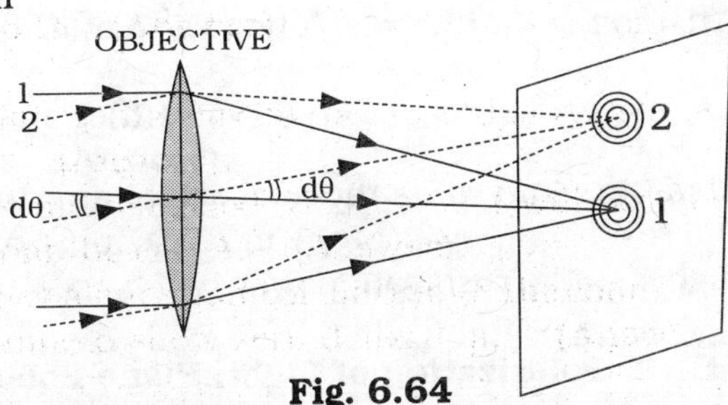

Fig. 6.64

Telescopes: A telescope is generally used for viewing distant objects like moon, planets, stars, etc. So closely situated objects, like two stars 1 and 2, act as point objects and the light rays emanating from them and reaching the earth in slightly different directions form parallel sets of rays, as shown in Fig. 6.64. Their diffraction images 1 and 2 formed by the objective lens are shown well separated from one another. Thus, they are said to have been well resolved by the instrument. In case they are just resolved and $d\theta_{min}$ is the corresponding angular separation between them, the resolving power of the telescope is given by,

$$(\text{R.P.})_{telescope} = \frac{1}{d\theta_{min}} = \frac{D}{1.22\lambda} \qquad .....(6.78)$$

D is the diameter (or aperture) of the objective. So larger is the value of D, higher is the resolving power of the telescope. Note that a high resolution telescope not only helps to see closely lying heavenly objects, but also to reveal details of the objects, e.g. the details of the surface of moon.

Microscopes: A microscope is employed for viewing near objects. The Fig. 6.65 shows two point objects A_1 and B_1 lying close to the objective lens of diameter D. Their angular separation is $d\theta$. They form diffraction images centred at A_2 and B_2 in the focal plane of the lens. u and v are the object distance and the image distance, respectively. The cone of the incident rays from A_1 forms half angle θ at the periphery of the lens.

For a microscope, it is convenient to express the resolving power in terms of the linear separation A_1B_1 between the two point objects. When the objects are just resolved, their angular separation has the minimum value $d\theta_{min}$. From the geometry of the figure and Eq. 6.76 we have,

$$d\theta_{min} = \frac{1.22\,\lambda}{D} = \frac{A_2\,B_2}{v}$$

$$\Rightarrow \quad A_2B_2 = \frac{1.22\,\lambda}{D}.v$$

Further,

$$\sin\theta = \frac{D/2}{u} \text{ or } u = \frac{D}{2\sin\theta}$$

and

$$\frac{A_1B_1}{A_2B_2} = \frac{u}{v}$$

$$\therefore \quad A_1B_1 = \frac{u}{v}A_2B_2 = \frac{D}{2\sin\theta.v}.\frac{1.22\,\lambda}{D}.v$$

$$= \frac{1.22\,\lambda}{2\sin\theta}$$

Fig. 6.65

*An idea of the limited resolution of the human eye can be simply obtained by marking two close dots on a piece of paper, which is then fixed on a wall or a door and viewed from gradually increasing distances. After a particular distance, the two dots will tend to look like a single spot.

In this treatment, A_1 and B_1 have been assumed to be two independent self-luminous objects, so that their intensities can be arithmetically added up. But in practice, they are usually illuminated by a common source. Hence they have some constant phase relationship between them. As a result, an appropriate expression for A_1B_1 is obtained by omitting the factor 1.22 from the numerator,

$$A_1B_1 = \frac{\lambda}{2\sin\theta}$$

A_1B_1 can be reduced, and so the resolving power enhanced, by replacing air with a medium (usually an oil) of refractive index μ. Then, the minimum resolvable linear separation between the objects becomes,

$$A_1B_1 = \frac{\lambda}{2\mu\sin\theta}$$

A_1B_1 is the limit of resolution for a microscope. Therefore,

$$(R.P.)_{microscope} = \frac{2(\mu\sin\theta)}{\lambda} \quad .....(6.79)$$

($\mu\sin\theta$) is known as the **numerical aperture (n.a.)** of the microscope. The higher is the n.a., the higher is the resolving power of the microscope. So, as for a telescope, the bigger is the size of the objective lens, the higher is the resolving power.

6B.12. Polarization of Light; Plane Polarized Light

In the last chapter, we had learnt that the light radiation is a kind of electromagnetic waves, consisting of oscillations of electric and magnetic fields, such that the electric field, the magnetic field, and the direction of propagation of the wave are mutually perpendicular.

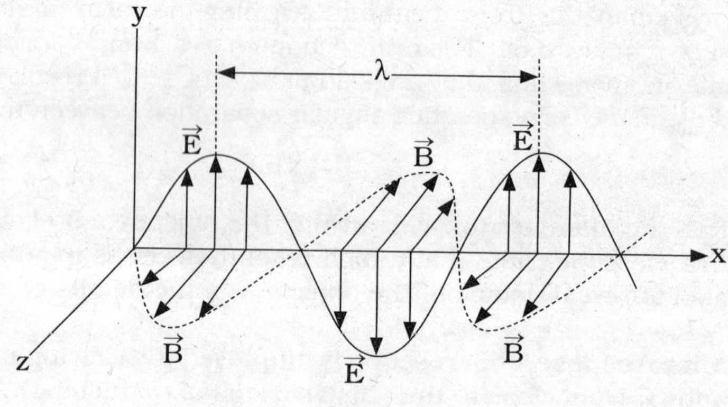

Fig. 6.66

Such a wave is shown in Fig. 6.66 (it is the same as Fig. 5.1 of Chapter 5). The electric vector $\vec{E}$ oscillates along y-direction, the magnetic vector $\vec{B}$ oscillates along z-direction, and the wave propagates along x-direction. Let us focus our attention on the electric vector $\vec{E}$. In the figure, it is shown oscillating in the xy-plane, which is the plane of the paper. But if it oscillates in another plane containing the x-axis, say a plane inclined at 30° to the plane of the paper, the condition of an electromagnetic wave will still be satisfied (the magnetic vector $\vec{B}$ will also correspondingly turn through 30°). In fact an infinite number of such planes, inclined at various angles to the plane of the paper, are available for wave propagation. Our choice of restricting the oscillations to the xy-plane was purely arbitrary. The Fig. 6.67(a) shows some of the other possible random choices for the oscillation of the electric vector. The x-axis, which now is the direction of propagation of the wave, is now directed out of and ⊥ to the plane of the paper.

The light obtained from common sources like sun, electric lamp, candle, etc., consists of such random oscillations, because their atoms emit light independent of each other. It is called **unpolarized light** or **randomly polarized light,** as the electric (and magnetic) field has no definite polarity of oscillation. The Fig. 6.67(a) is one way of representing such light. All the possible oscillations of the electric vector lie in the yz-plane, which is the plane of the paper. However, this representation obviously looks disordered. A simple ordered representation is obtained by resolving each electric vector

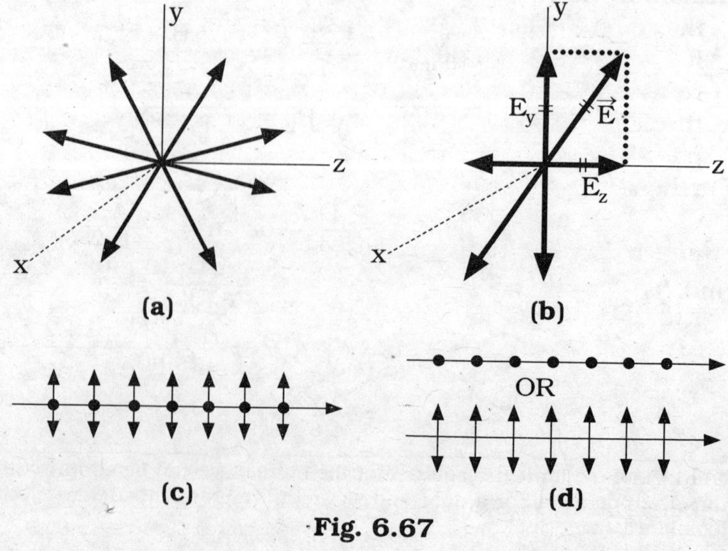

Fig. 6.67

$\vec{E}$ into its y- and z-components (E_y and E_z) along y and z directions, respectively, as shown in Fig. 6.67(b).

When unpolarized light is suitably passed through a polarizing device, about which we shall learn in the next sections, all oscillations except in one particular direction are prevented. "The phenomenon of restriction of oscillations of electric (and magnetic) vector to a particular direction" is called **polarization of light.** Since the oscillations have a linear restriction, the light so obtained is called **linearly polarized light.** It is also called **plane polarized light,** because the oscillations are restricted to a particular plane, e.g. the xy-plane in Fig. 6.66. The plane to which the oscillations (**or** vibrations) of the electric field are confined is known as the **plane of oscillation (or vibration).** It contains the electric vector and the direction of propagation of the wave. The plane ⊥ to it is termed as the **plane of polarization.** Thus, for the polarized wave shown in Fig. 6.66, the xy-plane is the plane of oscillation and the ⊥ xz-plane is the plane of polarization.

The oscillations of the electric vector in and ⊥ to the plane of the paper are customarily denoted by double arrows (↕) and dots (.), respectively. Accordingly, unpolarized light can also be represented as shown in Fig. 6.67(c) and the polarized light in one of the two ways shown in Fig. 6.67(d). In both the cases, the e.m. wave is taken to travel from left to right hand side.

If the incident unpolarized light is not fully polarized by the polarizing device, we get **partially polarized light.**

The polarization of light illustrates the transverse character of the light waves. The phenomena of interference and diffraction only identify the wave nature of light, without specifying whether the waves are longitudinal or transverse.

6B. 13. Production of Polarized Light; Polarizers

The unpolarized light can be polarized through more than one pocesses, which will be briefly described in the following. A device, whether natural or artificial, that produces polarized light is termed as a **polarizer.**

(i) Polarization by reflection; Brewster's law

The Fig. 6.68 shows a beam of unpolarized light incident upon a glass surface. Its electric vector $\vec{E}$ is resolved into two components of equal magnitude. One lies in the plane of incidence, which is the plane of the paper. We call it parallel component, $E_{\parallel}$, shown by double arrows. The other is perpendicular to the plane of incidence, marked by dots. We call it normal component, $E_{\perp}$. The oscillations of the parallel component are in the direction normal to the incident ray, as required by the electromagnetic nature of light.

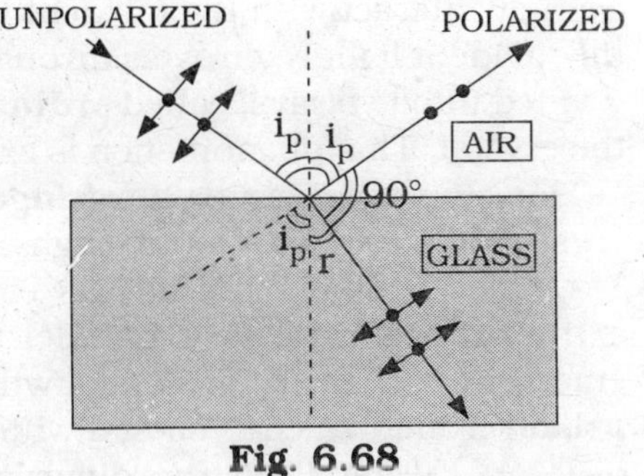

Fig. 6.68

At the air-glass interface, a part of the incident beam is reflected and the rest is refracted. When the angle of incidence is such that the angle between the reflected beam and the refracted beam is 90°, i.e. the reflected beam is at right angles to the refracted beam, the reflected light is found to be fully polarized perpendicular to the plane of incidence. It is so because in the process of refraction the perpendicular component $E_{\perp}$ is not affected, but now in the reflected beam $E_{\parallel}$ has no component with respect to the refracted beam (in general, if θ is the angle between the reflected and refracted beams, this component will be $E_{\parallel}\cos\theta$, which becomes zero for θ = 90°). So the parallel component disappears in the reflected beam. It is fully passed on to the refracted beam, which also retains the unreflected part of perpendicular component $E_{\perp}$.

The angle of incidence i_p for which the reflected light is found to be fully polarized is called **Brewster's angle or polarising angle.** It is related to the refractive index as follows. From the geometry of Fig. 6.68,

$$i_p + r = 90° \quad \text{or} \quad r = (90° - i_p)$$

r is the angle of refraction. By definition, the refractive index μ is given by,

$$\mu = \sin i/\sin r$$
$$= \sin i_p/\sin (90° - i_p)$$
$$= \sin i_p/\cos i_p$$

Or $$\mu = \tan i_p \qquad .....(6.80)$$

In general, if light refracts from a medium of refractive index μ_1 into a medium of refractive index μ_2, it can be similarly shown that,

$$\mu_{21} = \tan i_p \qquad .....(6.81)$$

The Eqs. 6.80 and 6.81 express the so-called **Brewster's law,** named after the scientist who investigated and established it experimentally.

The values of the Brewster's angle i_p for refraction from air to glass ($\mu = 1.5$) and air to water ($\mu = 1.33$) are approximately 57° and 53°, respectively.

For the angles of incidence other than i_p, the reflected light is partially polarized. Note that the refracted light has a reduced component of $E_\perp$ in it and so it is also partially polarized, but parallel to the plane of incidence. In fact, if repeated reflections take place, the finally refracted beam will also be almost fully polarized, parallel to the plane of incidence.

Example 6. Linearly polarized light is produced by reflection from a glass plate ($\mu_{glass} = 1.52$). Find the angle of refraction of light in the glass.

Solution. Given: $\mu = 1.52$

According to the Brewster's law,

$$\mu = \tan i_p$$

$$\therefore \quad i_p = \tan^{-1}(\mu) = \tan^{-1}(1.52) = 56.7°$$

∴ The angle of refraction is given by,

$$r = 90° - i_p = 90° - 56.7° = 33.3° \quad \textbf{(Answer).}$$

(ii) Polarization by refraction through crystals

(a) Semi-natural polarizers; Nicol prism: The crystals are characterized by a regular three-dimensional arrangement of atoms inside them. In an ordinary solid, the distribution of the atoms is largely random. In a crystal, the atoms are neatly arranged along rows at equal spacings (~1nm and less) in three dimensions. So they interact with the incident light in a cooperative manner (vide Section 6A.06; Chapter 6A), such that the incident light ray travels through the crystal with different velocities in two different directions. Thus it is divided into two rays, called **ordinary ray and extraordinary ray,** with different values of refractive index for the crystal. This phenomenon is exhibited by most crystals, which are termed as biaxial crystals. It is known as **double refraction** or **birefringence** (the crystals not displaying the phenomenon are known as uniaxial crystals). An object seen through such a biaxial crystal, e.g. a printed line in this book, looks double.

The ordinary ray has its electric vector $\vec{E}$ perpendicular to a principal section of the crystal. The extraordinary ray has $\vec{E}$ parallel to the principal section. So if one of these rays can be eliminated, the emergent light from the crystal will be polarized. This is successfully achieved by making a so-called **Nicol prism,** from a calcite* crystal which looks like glass. The refractive indices and μ_o and μ_e of calcite for the ordinary ray and he extraordinary ray are nearly 1.7 and 1.5, respectively. A suitably shaped piece of a calcite crystal is cut into two halves along its diagonal section: The two halves are rejoined with a transparent glue called Canada balsam. Its refractive index lies between μ_o and μ_e. The rejoined calcite crystal is known as Nicol prism. When a light ray (unpolarized light) enters the prism, it is divided into an ordinary and an extraordinary ray. When the two rays meet the layer of Canada balsam, the ordinary ray passes from a dense to a rare medium. Also, because of the suitable shape of the crystal, it has the angle of incidence greater than the critical angle. Therefore, it is totally internally reflected and so eliminated, leaving the extraordinary ray alone to pass through the prism. Hence the emergent light is plane polarized.

(b) Natural polarizers; Dichroic crystals. The Nicol prism is a semi-natural polarizer. Some biaxial crystals possess the natural property of selectively absorbing one of the two components $E_\perp$ and $E_\parallel$, when unpolarized light passes through them. They are called **dichroic crystals** and the property of selective absorption of one of the $\vec{E}$ components is known as **dichroism.** The unabsorbed component passes through the crystal with little loss of intensity. Hence the emergent light is almost fully polarized. Tourmaline** is one such dichroic crystal. Thus it is termed as a **natural polarizer.**

*Chemically, calcite is calcium carbonate ($CaCO_3$). Interestingly, the ordinary chalk used for writing on a blackboard is also calcium carbonate, but in noncrystalline form. The calcite is a naturally occurring crystal.

**Tourmaline is a naturally occuring crystal. It is a boron aluminosilicate, with the complex chemical composition (Na, Ca)(Li, Al)$_3$ Al_6 $(OH)_4$ $(BO_3)_3$ Si_6 O_{18}.

(c) Artificial polarizers; Polaroids. A polaroid is a thin plastic sheet containing a very thin stretched transparent film, which has parallel rows of long chain molecules of a dichroic material spread upon it. The film is made of a hydrocarbon material, like polyvinyl alcohol. The molecules are actually constituents of ultramicroscopic crystals of a dichroic material, like iodine. They absorb one of the $\vec{E}$ components (parallel to the rows of the molecules) of the unpolarized light incident upon the polaroid and allow the other perpendicular component to pass through. Hence the emergent light is fully polarized in a direction perpendicular to the rows of the molecules, which we call as the **polarising direction.** Since one of the $\vec{E}$ components is completely absorbed inside the polaroid, the intensity of the emergent polarised light is half of the intensity of the incident unpolarized light.

The polaroid sheets are available in circular, rectangular or square shapes. They look like pieces of an X-ray film. Their commercial production is both cheap and convenient. They also provide a convenient large area of polarization. Hence they have emerged as most popular polarizing devices.

(iii) Polarization by scattering of light

If unpolarized light is incident upon a gas or a vapour, then a beam scattered at right angles to the incident beam is found to be fully polarized. If the unpolarized light is incident on a gas molecule in the z-direction, as shown in Fig. 6.69, its $\vec{E}$ vector can be resolved in two components along the perpendicular x- and y-directions, respectively. By virtue of the nature of the electromagnetic waves, the component travelling along the y-direction will have oscillations along x alone and the other travelling along the x-direction will have oscillations along y alone. Thus a viewer along y finds the light fully polarized along x and another along x finds the light fully polarized along y. For a direction of view other than x or y, the scattered light is partially polarized.

Fig. 6.69

The skylight is always polarized. The degree of polarization depends on the direction of sight of the viewer on earth with respect to the position of sun. The sunlight is unpolarized. As explained in the last chapter in Section 6A.06, the blue colour of the sky is due to the scattering of sunlight by the atoms and molecules present in the earth's atmosphere.

The production of polarized light by scattering is another aspect of the important phenomenon of scattering of light, which we had briefly described in general terms in the Section 6A.06. Another very significant aspect of the scattering of light, known as **RAMAN EFFECT,** was discovered by the reputed Indian physicist Sir C.V. Raman and his coworkers, for which he was awarded Nobel prize in physics for 1930. In the Raman effect, the scattering by the molecules of a medium results in obtaining slightly different wavelengths in the scattered light, in addition to the main λ (cf. Rayleigh scattering, in which there is no change in λ). The study of these extra wavelengths provides valuable information on the molecular structure of the medium.

6B.14. Analysis (Detection) of Polarized Light; Law of Malus

A combination of two polaroids (or any other polarizers) can help us to know if a given light beam is polarized or not and also that to what extent it is polarized.

As we know, the unpolarized light can be resolved into two components $E_{\parallel}$ and $E_{\perp}$, which have the oscillations of the electric vector perpendicular to each other. When this light falls on a polaroid P_1, it allows only the component $E_{\parallel}$ to pass through, because only this component has its oscillations parallel to the polarising direction of P_1 (Fig.

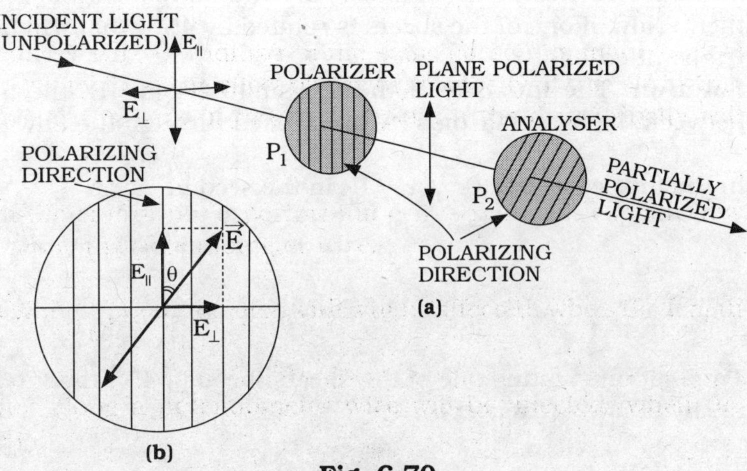

Fig. 6.70

6.70a). Hence the light emerging from P_1 is polarized. The other component $E_\perp$ is absorbed. Thus, half of the original intensity I_o of light is lost and so the transmitted intensity is,

$$I = (1/2)\, I_o$$

If P_1 is rotated about the direction of the incident light beam, there is no change in the intensity of light. That is how we can detect that the incident light is unpolarized. To the naked eye, all states of polarization appear to be the same.

Next, let the plane polarized light emerging from P_1 fall on another polarizer P_2, as shown in the figure. Now, if P_2 is rotated about the incident beam, the intensity of the light emerging from P_2 is found to change. It remains same as the original intensity if the polarising directions of P_1 and P_2 are parallel, but starts decreasing as P_2 is rotated, until it becomes zero when the two directions are at right angles to each other. Such a variation in the intensity of incident light beam when the polarizer is rotated about the beam readily enables us to know that the incident light is polarized. That is how we can detect both unpolarized and polarized light. The positions in which the polarizers P_1 and P_2 have their polarizing directions parallel and perpendicular to each other are termed as **parallel** and **crossed** positions, respectively, of the two polarizers. P_1 and P_2 are called **polarizer** and **analyzer,** respectively.

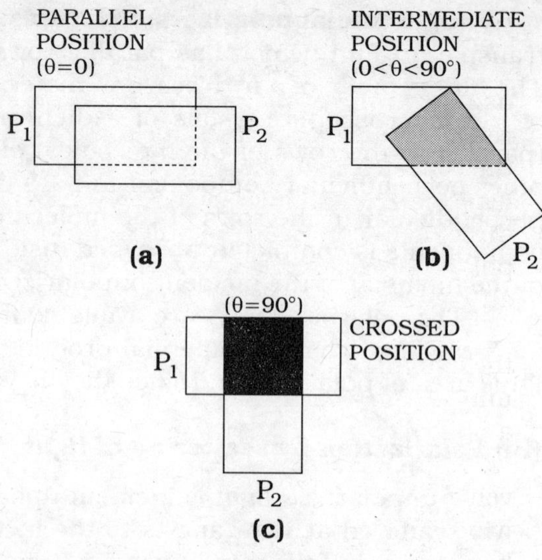

Fig. 6.71

The qualitative variation in the intensity of light for mutual rotation of two rectangular polaroids is shown in Fig. 6.71. θ is the angle between the polarizing directions of the two polarizers.

How the light intensity varies quantitatively can be easily calculated. The Fig. 6.70(b) shows the position of the electric vector $\vec{E}$ of the polarized light when it is just incident on the analyser P_2. We resolve it as $E_\parallel$ and $E_\perp$ along and perpendicular to the polarizing direction, respectively. Only the component $E_\parallel$ is allowed to pass through by P_2,

$$E_\parallel = E \cos\theta$$

θ is the angle between $E_\parallel$ and $\vec{E}$. Since light intensity is proportional to the square of amplitude, the intensities I and I_o of the transmitted and incident light will be proportional to $(E_\parallel)^2$ and $(E)^2$ respectively. Therefore, the variation of the transmitted intensity is given by,

$$I = I_o \cos^2\theta \qquad .....(6.82)$$

This is known as the **law of Malus.** Note that, as expected, $I = I_o$ for $\theta = 0$ and $I = 0$ for $\theta = 90°$ (cf. Fig. 6.71). Thus the law of Malus enables to estimate the quantity of polarized light transmitted through an analyzer. Note that during a full rotation of 360° of the analyzer, there will be two positions each of transmission of maximum and minimum (zero) intensity of the polarized light, corresponding to $\theta = 0, 2\pi$ and $\theta = \pi/2, 3\pi/2$, respectively (Figs. 6.71a and 6.71c, respectively). For intermediate positions, intermediate values of intensity are transmitted (Fig. 6.71b).

Example 7. Two polaroid sheets are mutually positioned to transmit maximum intensity of light through them. Now, if one of the sheets is rotated by 45°, what will be the percentage reduction in the intensity of the transmitted light ?

Solution. The intensity of the transmitted light (I), the intensity of the incident light (I_o) and the angle between the two polaroids (θ) are related through the Law of Malus,

$$I = I_o \cos^2\theta$$

In the first case, $\theta = 0$. $\therefore I = I_o$. In the second case,

$$I = I_o (\cos 45°)^2 = I_o\left(\frac{1}{\sqrt{2}}\right)^2$$

$$= \frac{1}{2}(I_o)$$

Thus, upon rotating one of the sheets through 45°, the intensity of the transmitted light is reduced by half, i.e. by 50%. **(Answer).**

6B.15. Some Common Uses of Polarized Light and Polaroids

(i) Sunglasses. The light reflected from glass, water, snow, glazed surfaces, etc., produces glare (or dazzle) in the eyes, because the region of reflection shines brightly when light falls upon it. We know that reflected light is polarized, partially or fully. The reflecting surfaces are mostly horizontal. Accordingly, the reflected light is polarized horizontally. If such surfaces are viewed with **polarizing sunglasses,** whose lenses are actually polaroids with their polarizing direction vertical, the glare is appreciably reduced. The direct light coming from other objects in view is unpolarized and so its intensity is hardly affected. Thus the polarizing glasses are superior to the conventional sunglasses fitted with dark lenses, that greatly reduce all kinds of light entering them.

(ii) Photography cameras. The atmospheric air scatters all around the light falling upon it. When the photograph of an object is taken with a camera, this scattered light also enters the lens, producing unnecessary background darkening in the photograph. We know that scattered light is polarized. By fitting a polaroid in front of the lens, the scattered light is appreciably cut off.

(iii) Motor headlights. The strong light falling in a driver's eyes from the headlights of he motor vehicles moving in opposite direction poses a great risk in night. The risk can be reduced by fitting the windshields of the vehicles with polaroids having their polarizing direction, say, horizontal and fitting the headlights with polaroids having polarizing directions vertical. The windshield polaroids do not affect the passage of unpolarized light from other objects.

(iv) Watches, calculators, laptops, etc.. We are familiar with the liquid crystal display (LCD) of letters and numbers in watches, calculators, portable computers (laptops), etc. The liquid crystals have long molecules whose orientation can be changed by a controlled variation of the applied electric field. The variation is coupled with a suitable modification of the polarized light produced by a polaroid (polarizer) such that its polarizing direction is perpendicular to the polarizing direction of another polaroid (analyzer), which cuts off the light. The controlled dark regions so produced are used to form various letters and numbers, for the desired display.

Additional knowledge

(i) In addition to the linearly (**or** plane) polarized light, we also have circularly and elliptically polarized light, in which the tip of the electric vector moves along a circle and an ellipse, respectively. However, they both can be resolved into linear components.

(ii) The iodine atoms in the parallel long chains of molecules in a polaroid provide high electrical conductivity. Therefore the electric field parallel to the chains is absorbed in setting up electric currents in the chains, while the electric field $\perp$ to the chains passes through. Thus, the transmitted light is found polarized.

(iii) The human eye cannot distinguish between unpolarized and polarized light but the eyes of a bee do. The skylight results from the scattering of sunlight by the atmospheric air and so it is polarized. Long careful experiments have revealed that the bees not only distinguish between the polarized and unpolarized light, they can also determine the direction of polarization. They successfully use this natural ability for moving at and returning safely to their hives.

(iv) As early as the eighth century, the sea traders (also pirates!) from Scandinavian countries successfully used polarized light for navigation through North Sea, where the sun remains below horizon even duing daytime, because of its high latitude. They discovered cordierite* crystals (of course, without knowing their polarizing property), which changed colour on looking through them towards the sky and rotating them about the line of sight (sky light is polarized). Thus, they could locate the hidden sun and so decide their direction of travel.

*Cordierite is a naturally occuring aluminosilicate of magnesium, with chemical formula as Mg_2 [$(Al, Si)_9$ O_{18}].

Solved Conceptual Problems

Problem 1. Name the optical phenomena which conclusively support the wave nature of light.
Solution. Three optical phenomena: Interference, diffraction, and polarisation of light.

Problem 2. State the shape of the wavefronts in the following cases: (i) Light emitted by a point source, (ii) light emerging from a convex lens, with a point source of light placed at its focus, and (iii) light arriving at an observer on earth from a star.
Solution. (i) Concentric diverging spherical wavefronts.
(ii) Plane wavefronts (parallel rays emerge from the lens, which are normal to the wavefronts).
(iii) Plane wavefronts (The star acts as a point source of light, emitting spherical wavefronts. But because of the enormous distance of the star from the earth, the small part of a wavefront intercepted by the observer on earth is effectively plane).

Problem 3. When monochromatic light falls on the interface between two media (e.g. air-glass), the frequency of the reflected and refracted light remains the same as that of the incident light. Explain how.
Solution. Both the reflection and the refraction result from the interaction of the incident light with the atoms at the surface of separation (e.g. glass). These atoms can be regarded as tiny oscillators. When monochromatic light falls on the surface, these atoms are forced to oscillate with the same frequency as that of the incident light. The frequency of light emitted by them has the same frequency as the frequency of their oscillations. Hence, both the reflected and the refracted light have the same frequency as that of the incident light.

Problem 4. When light passes from a rare to a dense medium, its velocity is reduced. Does it mean that the energy of the light waves is also reduced ? Explain.
Solution. The energy of the waves directly depends upon the intensity of light, which is proportional to the square of the wave amplitude [$I \propto (\text{amp.})^2$]. The amplitude remains unchanged upon passage of the light into a different medium. Hence the energy carried by the waves remains unchanged.

Problem 5. In the interference pattern obtained in the Young's double slit experiment, show that the maximum light intensity is 4 times the light intensity received from each slit. Hence show that the law of conservation of energy is not violated; the interference of the waves only results in a redistribution of energy in the pattern.
Solution. We know that the intensity, $I \propto (\text{amplitude})^2$. Let the wave amplitude of the light from each slit be a. Then the light intensity due to each slit is,

$$I_{slit} \propto a^2 \quad \text{or} \quad I_{slit} = ka^2 \text{ (k: proportionality constant)}$$

For destructive interference (positions of minima), the resultant amplitude = a – a = 0.

$$\therefore \text{Minimum intensity in the pattern} = 0$$

For constructive interference (positions of maxima), the resultant amplitude = (a + a) = 2a.

$$\therefore \text{Maximum intensity} = k(2a)^2 = 4(ka^2)$$

Thus, it is proved that the max. intensity in the pattern is 4 times I_{slit}.

The average intensity in the pattern is given by,

$$I_{average} = \frac{0+4(ka^2)}{2} = 2(ka^2) = ka^2 + ka^2$$

$$= \text{Sum of intensities due to each slit}$$

Thus, we find that the total intensity in the pattern is the same as that obtained from the two slits together. There is only a redistribution of energy in the pattern. The law of conservation of energy is followed.

Problem 6. In the Young's double slit experiment, how the interference pattern is affected in the following cases, when

(i) a white light source is employed in place of the monochromatic source,
(ii) the monochromatic light source has shorter wavelength,
(iii) the separation between the slits is increased,
(iv) the distance of the screen from the slits is increased,

(v) wide slits are used instead of narrow slits,

(vi) the original monochromatic light source is removed and independent conventional monochromatic point sources are placed at the positions of the slits ?

Solution. (i) The fringe width (β) and the distance (x_n) of the nth fringe from the central maximum in the interference pattern are given by,

$$\beta = \frac{D\lambda}{d}$$

$$x_n = \frac{n.D\lambda}{d}$$

Each wavelength in the white light gives it own interference pattern. The resultant pattern consists of an overlap of almost an infinite number of such pattern. Since both β and x_n are different for different λ's, the net result is that the resultant pattern has a central white light fringe ($n = 0; \therefore x_n = 0$ for all λ), with a few coloured fringes lying on either side.

(ii) $\beta \propto \lambda$. ∴ fringe width reduces when λ is shorter.

(iii) $\beta \propto (1/d)$. ∴ fringe width reduces upon increasing the slit separation, d.

(iv) $\beta \propto D$. ∴ fringe width increases upon increasing D, i.e. upon moving the screen away from the slits.

(v) A wide slit can be regarded as consisting of a large number of parallel narrow slits, each of which produces its own interference pattern, slightly shifted from a similar pattern produced by its neighbouring slit. The net effect of the overlap of the patterns on the screen is that the visibility of the fringes is reduced, because maxima and minima of intensity tend to overlap.

(vi) Such independent light sources are incoherent sources of light, with no constant phase relationship between them. ∴ no interference fringes are seen on the screen.

Problem 7. The pictures on the television screens are often found to be slightly disturbed when an aircraft flying low passes over a place. Explain how.

Solution. The incoming direct T.V. signal is partially reflected by the body of the aircraft. Then, the interference between the direct signal and the reflected signal produces the observed disturbance on the T.V. screens. If the aircraft flies high, the reflected signal becomes too weak by the time it reaches a T.V. location.

Problem 8. Explain why the interference pattern in the Young's double slit experiment becomes undetectable, when the two slits are (i) far apart and (ii) infinitely close to each other.

Solution. (i) The fringe width (β) and the distance (x_n) of the nth fringe from the central maximum are given by,

$$\beta = \frac{D\lambda}{d}$$

$$x_n = \frac{n.D\lambda}{d}$$

Thus, both β and x_n are inversely proportional to the separation (d) between the slits. ∴ when d becomes too large, both β and x_n become too small for the interference pattern to be detectable on the screen.

(ii) When d becomes extremely small, the fringe width β becomes so large that a single fringe may occupy the whole screen. Besides, the fringe separation x_n also becomes too large. Thus, once again no pattern is detectable on the screen.

Problem 9. When one looks through a bird's feather or a muslin cloth towards sun, a number of coloured spectra are seen. Explain how.

Solution. The fine slits between the individual hairs of the feather or the threads of the muslin cloth act as a coarse diffraction grating. When white light of the sun passes through them, each colour produces its own diffraction pattern. The overlap of the patterns results in the observed coloured spectra.

Problem 10. How the size and intensity of the central maximum in a single slit diffraction pattern will be affected, if the width of the slit is doubled ?

Solution. When the slit width (b) is doubled, the angular half-width (θ) of the central maximum becomes half ($\theta = \lambda/b$).

The intensity is a measure of the light per unit area. When θ will be reduced to 1/2 of its original value, the area of the central maximum will be reduced to 1/4 of its original value. Hence its intensity will become 4 times.

Problem 11. What are the common principles behind the phenomena of interference and diffraction of light ? In the Young's double slit experiment, if one of the slits is covered, will a diffraction pattern be observed on the screen ?

Solution. The common principles are: (i) Huygen's principle of formation of secondary (spherical) wavelets and (ii) the principle of superposition of waves.

If one of the slits is covered in the Young's experiment, the single slit diffraction pattern is observed on the screen.

Problem 12. Two polaroids are placed in a crossed position (i.e. with their polarising directions at 90° to each other), so that the intensity of the transmitted light is zero. If now another polaroid is introduced between them with its polarising direction at 45° to either of the polarising direction of the original polaroids, will the transmitted intensity still be zero ? If not, what will be its value ?

Solution. The transmitted intensity will not be zero. If the original intensity is I_0 and the angle between the polarising directions of two polaroids is θ, the transmitted intensity is given by the law of Malus as,

$$I = I_0 \cos^2 \theta$$

In our case, the transmitted intensity for one of the original polaroids and the new polaroid will be,

$$I_1 = I_0 \cos^2 45° = I_0\left(\frac{1}{\sqrt{2}}\right)^2 = \frac{1}{2} I_0$$

I_1 is the original intensity for the combination of new polaroid and the second original polaroid. Therefore the final transmitted intensity will be,

$$I_2 = \left(\frac{1}{2}I_0\right)\cos^2 45° = \left(\frac{1}{2}I_0\right)\left(\frac{1}{\sqrt{2}}\right)^2 = \frac{1}{4} I_0$$

Thus, the finally transmitted intensity is (1/4)th of the original intensity.

Solved Numerical Problems

Problem 1. In Young's double slit experiment, the slits are separated by 0.28 mm and the screen is placed 1.4 m away. The distance between the central maximum and the fourth bright fringe is measured to be 1.2 cm. Determine the wavelength of light used in the experiment. (NCERT)

Solution. Given: $D = 1.4$ m, $d = 0.28$ mm $= 0.28\times10^{-3}$ m, $n = 4$, $x = 1.2$ cm $= 12\times10^{-2}$ m

We know that the distance of the nth bright fringe from the central maximum is given by,

$$x = \frac{n.\,D\lambda}{d}$$

$$\therefore \quad \lambda = \frac{x.\,d}{n.\,D} = \frac{(1.2\times10^{-2})\times(0.28\times10^{-3})}{(4)\times(1.4)}$$

$$= 6\times10^{-7}\text{ m} = 6000\text{ Å} \quad \textbf{(Answer).}$$

Problem 2. In Young's double slit experiment, while using a source of light of wavelength 5000 Å, the fringe width is 0.6 cm. If the distance between the screen and slit is reduced to half, what should be the wavelength of the source to get fringes 0.003 m wide ? (PSEB; 1991).

Solution. In the first case,

$$\lambda_1 = 5000\text{ Å} = 5\times10^{-7}\text{ m}, \quad \beta_1 = 0.6\text{ cm} = 6\times10^{-3}\text{ m}, \quad D_1 = D\text{ (say)}$$

In the second case

$$\lambda_2 = ?, \quad \beta_2 = 0.003\text{ m} = 3\times10^{-3}\text{ m}, \quad D_2 = (D/2)$$

We have, $\beta_1 = \dfrac{\lambda_1 D_1}{d}$ and $\beta_2 = \dfrac{\lambda_2 D_2}{d}$

$$\Rightarrow \quad \frac{\beta_1}{\beta_2} = \frac{\lambda_1 D_1}{\lambda_2 D_2}$$

$$\therefore \quad \lambda_2 = \frac{\lambda_1 D_1 \beta_2}{\beta_1 D_2} = \frac{(5\times10^{-7})\times(D)\times(3\times10^{-3})}{(6\times10^{-3})\times(D/2)}$$

$$= 5\times10^{-7}\ \text{m} = 5000\ \text{Å} \quad \textbf{(Answer).}$$

Problem 3. A double slit is illuminated by light of wavelength 6000 Å. The slits are 0.1 cm apart and the screen is placed 1 m away. Calculate (i) angular position of 10th maximum in radian and (ii) separation of two adjacent minima. (HBSE; 1996).

Solution. Given: λ = 6000 Å = 6×10^{-7} m, $d = 0.1$ cm = 10^{-3} m, $D = 1$ m, $n = 10$

(i) The distance of the nth maximum from the central maximum is given by,

$$x = n\frac{D\lambda}{d}$$

The angular position of the 10th maximum is given by

$$\sin\theta \approx \theta = \frac{x}{D} \quad \text{or} \quad x = D.\theta$$

$$\therefore \quad D.\theta = \frac{n\,D\lambda}{d}$$

$$\Rightarrow \quad \theta = \frac{n\,\lambda}{d} = \frac{(10)\times(6\times10^{-7})}{(10^{-3})}$$

$$= 6\times10^{-3} = 0.006 \text{ radian} \quad \textbf{(Answer).}$$

(ii) The separation between two adjacent minima is equal to the width of the central maximum, which is given by,

$$\beta = \frac{\lambda D}{d} = \frac{(6\times10^{-7})\times(1)}{(10^{-3})}$$

$$= 6\times10^{-4}\ \text{m (or 0.6 mm)} \quad \textbf{(Answer).}$$

Problem 4. In an interference pattern, the ratio of the intensities of maxima and minima is measured to be 25:9. Find the ratio of (i) amplitudes and (ii) intensities of the two coherent light sources producing the pattern.

Solution. (i) Given, $\frac{I_{max}}{I_{min}} = \frac{25}{9}$

Let a and b be the required amplitudes. Then, since Intensity $\propto$ (Amplitude)^2, we have,

$$\frac{(a+b)^2}{(a-b)^2} = \frac{25}{9}$$

$$\Rightarrow \quad \frac{(a+b)}{(a-b)} = \frac{5}{3} \quad \Rightarrow \quad 5a - 5b = 3a + 3b$$

$$\Rightarrow \quad (a/b) = (4/1)$$

Thus, the ratio of amplitudes of the two sources is found to be 4:1. **(Answer).**

(ii) The ratio of the intensities is given by,

$$\frac{I_1}{I_2} = \frac{a^2}{b^2} = \left(\frac{4}{1}\right)^2 = \frac{16}{1}$$

Thus, the ratio of the intensities is found to be 16:1. **(Answer).**

Problem 5. In a two slit experiment with monochromatic light, fringes are obtained on a screen placed at some distance from the plane of the slits. If the screen is moved by 5×10^{-2} m towards the slits, the change in fringe width is 3×10^{-5} m. If the distance between the slits is 10^{-3} m, Calculate the wavelength of the light used. (Roorkee; 1992).

Solution. Given; $d = 10^{-3}$ m; change in screen's distance $(\Delta D) = 5\times10^{-2}$ m; change in fringe width $(\Delta\beta) = 3\times10^{-5}$ m

We know,

$$\beta = \frac{D\lambda}{d} = \left(\frac{\lambda}{d}\right)D$$

$$\therefore \quad \Delta\beta = \left(\frac{\lambda}{d}\right)\Delta D$$

$$\Rightarrow \qquad \lambda = \frac{(\Delta\beta)\times D}{(\Delta D)} = \frac{(3\times10^{-5})\times(10^{-3})}{(5\times10^{-2})}$$

$$= (6\times10^{-7}\text{ m} = 6000\text{ Å} \quad \textbf{(Answer).}$$

Problem 6. Employing sodium light (λ = 5890 Å), interference fringes of angular width 0.5° are produced in a double slit experiment. If the whole set-up is immered in water (μ = 4/3), what will be the angular fringe separation ?

Solution. The angular fringe width is simply related to the linear fringe width β (which is the distance between two successive maxima or minima) as,

$$\sin\theta \approx \theta = \frac{\beta}{D}$$

But, we know, $\beta = \frac{D\lambda}{d}$ or $\frac{\beta}{D} = \frac{\lambda}{d}$

$$\therefore \qquad \theta = \frac{\lambda}{d}$$

After immersion in water, the wavelength will change, say to λ'.
Then, the new angular width will be,

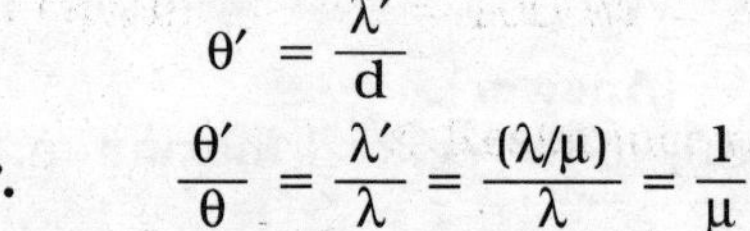

$$\theta' = \frac{\lambda'}{d}$$

$$\therefore \qquad \frac{\theta'}{\theta} = \frac{\lambda'}{\lambda} = \frac{(\lambda/\mu)}{\lambda} = \frac{1}{\mu}$$

(In water, the new wavelength λ' will become λ/μ).

$$\Rightarrow \qquad \theta' = \frac{\theta}{\mu} = \frac{(0.5)}{(4/3)} = 0.375°$$

Thus, the new angular fringe width will be 0.375 Å. **(Answer).**

Problem 7. A beam of light of wavelength 600 nm from a distant source falls on a single slit 1.00 mm wide and the resulting diffraction pattern is observed on a screen 2 m away. What is the distance between the first dark fringe on either side of the central maximum ? (IIT; 1994).

Solution. Given ; D = 2 m, b = 1 mm = 10^{-3} m, λ = 600 nm = 6×10^{-7} m

The distance between the fist dark fringe on either side of the central maximum is equal to the width of the central maximum, which is given by,

$$W = \frac{2D\lambda}{b} = \frac{2\times(2)\times(6\times10^{-7})}{10^{-3}}$$

$$= 24\times10^{-4}\text{ m}$$

$$= 2.4\text{ mm} \quad \textbf{(Answer).}$$

Problem 8. A slit of width b is illuminated by red light of wavelength 6500 Å. For what value of b will the

(i) first minimum fall at an angle of diffraction of 30° ?

(ii) first maximum fall at an angle of diffraction of 30° ? (CBSE; 1998. HBSE; 1990).

Solution. Given: λ = 6500 Å = 65×10^{-8} m, θ = 30° = (π/6) radian

(i) The angular spread of the first minimum is given by,

$$\theta = \frac{\lambda}{b}$$

$$\therefore \qquad b = \frac{\lambda}{\theta} = \frac{(65\times10^{-8})}{(\pi/6)} = \frac{6\times65\times10^{-8}}{3.14}$$

$$= 124.2\times10^{-8}\text{ m}$$

$$= 0.00124\text{ mm (or 1.24 μm)} \quad \textbf{(Answer).}$$

(ii) The angular spread of the first maximum is given by,

$$\theta = \frac{(3\lambda/2)}{b} = \frac{3\lambda}{2b}$$

$$\therefore \quad b = \frac{3\lambda}{2\theta} = \frac{3\times(65\times10^{-8})}{2\times(\pi/6)} = \frac{3\times65\times6\times10^{-8}}{2\times3.14}$$

$$= 186.3\times10^{-8} \text{ m}$$

$$= 0.00186 \text{ mm (or } 1.86\ \mu\text{m)} \quad \textbf{(Answer).}$$

Problem 9. In a single slit diffraction pattern, first minimum for red light (λ = 750 nm) is found to coincide with first maximum for another wavelength λ'. Determine the value of λ'.

Solution. Given; λ = 750 nm

The position of the first minimum for the wavelength λ is given by,

$$\theta = \frac{\lambda}{b}$$

The position of the first maximum (given by the same value of θ as the first minimum) for the wavelength λ' is given by,

$$\theta = \frac{(3\lambda'/2)}{b} = \frac{3\lambda'}{2b}$$

Thus,

$$\frac{\lambda}{b} = \frac{3\lambda'}{2b}$$

$$\therefore \quad \lambda' = \frac{2\lambda}{3} = \frac{2\times750}{3}$$

$$= 500 \text{ nm} \quad \textbf{(Answer).}$$

Problem 10. Unpolarized light of intensity I_o is incident on a combination of two polaroids. If we want the intensity of the transmitted polarized light to be reduced to $I_o/4$, what should be the angle between the axes of the polaroids ?

Solution. The original intensity is I_o. The transmitted intensity I is $I_o/4$.

$$\therefore \quad \frac{I}{I_o} = \frac{(I_o/4)}{I_o} = \frac{1}{4}$$

According to the Law of Malus, if θ is the angle between the axes of the polaroids, we have,

$$I = I_o \cos^2\theta$$

$$\Rightarrow \quad \cos^2\theta = \frac{I}{I_o} = \frac{1}{4}$$

$$\Rightarrow \quad \cos\theta = \sqrt{1/4} = \frac{1}{2}$$

$$\therefore \quad \theta = \cos^{-1}\frac{1}{2} = 60° \quad \textbf{(Answer).}$$

Problem 11. If a light ray is incident on a glass plate at the polarising angle, determine the angle of refraction of the ray (μ_{glass} = 1.54).

Solution. Given: μ = 1.54

According to the Brewster's law, the polarising angle is given by,

$$\mu = \tan i_p$$

$$\therefore \quad i_p = \tan^{-1}\mu = \tan^{-1}(1.54)$$

$$= 57°$$

$$\therefore \quad \text{Angle of refraction, } r = 90° - i_p = 90° - 57°$$

$$= 33° \quad \textbf{(Answer).}$$

Problem 12. An observer standing on the beach of a calm ocean and receiving the sunlight from the water surface finds the light to be linearly polarised. Calculate at what angle above the horizon is the sun (μ_{water} = 1.33).

Solution. Given: $\mu = 1.33$

According to the Brewster's law, the polarising angle is given by,

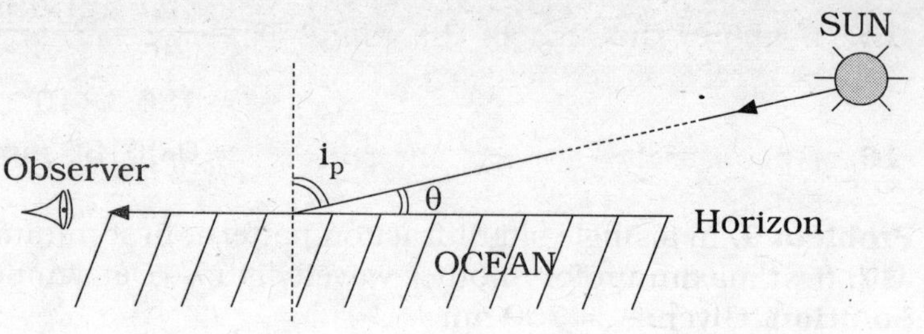

$$\mu = \tan i_p$$

$$\therefore \quad i_p = \tan^{-1}\mu = \tan^{-1}(1.33)$$

$$= 53.06°$$

The horizon coincides with the water surface. Therefore the angle of sun above the horizon is,

$$\theta = 90° - i_p = 90° - 53.06°$$

$$= 36.94° \quad \textbf{(Answer).}$$

Very Short Answer Questions

[Including HOTS Questions]

(1 mark each)

1. Define a wavefront.
2. State Huygens' principle.
3. How is the phase difference in a wave related to the path difference ?
4. Name two phenomena which confirm the wave nature of light,.
5. State the shape of the wavefront when (i) light is emitted by a point source and (ii) when light arrives at an observer on earth from a very distant object.
6. When a light wave enters from air into water, does its energy change ? Does its frequency change ?
7. Write the essential condition to be satisfied for observing interference fringes, when light waves from two sources arrive at a place.
8. Write the relations for (a) constructive interference and (b) destructive interference between light waves from two coherent sources.
9. What is meant by coherent sources of light ?
10. In the Young's double slit experiment, how is the fringe width affected when (a) the screen is moved nearer to the slits and (b) the separation between the slits is reduced.
11. In the Young's double slit experiment, if the main light source is replaced by a source of shorter wavelength, will the fringe width (a) increase, (b) decrease, or (c) remain unchanged ?
12. In the Young's double slit experiment, what is the ratio of fringe widths for bright and dark fringes ?
13. State the principle of superposition of waves.
14. In a single slit diffraction pattern, how will the size of the central maximum be affected, if the slit width is reduced by half ?

15. If one of the slits in the Young's interference experiment is covered, will a diffraction pattern be observed on the screen ?

16. For observing a single slit diffraction pattern, ideally what should be the approximate width of the slit ?

17. Is the width of central maximum in a single slit diffraction experiment greater for red light or blue light ?

18. What is plane polarized light ?

19. What is Brewster's angle ? Does it depend on the colour of light ?

20. A polarizer and an analyzer are so mutually positioned that no light is transmitted through the combination. What is the angle between their axes ? [Hint: 90°].

21. If the angle between the axes of a polarizer and an analyzer is 30°, what percentage of incident light is transmitted through the combination ? [Hint: 75°].

22. If light passes from air into a medium of refractive index $\sqrt{3}$, what is the value of polarizing angle ? [Hint: 60°]

23. Name the optical phenomenon which clearly illustrates the transverse nature of light waves.

Short Answer Questions

[Including HOTS Questions]

(2 or 3 marks each)

1. Two electric bulbs are identical in all respects, e.g. wattage, size, construction, etc. Can they act as coherent sources of light ? Give reasons for your answer.

2. In the Young's double slit experiment, explain how the interference pattern is effected when (a) the monochromatic source of light is replaced by a white light source and (b) wide slits are used, instead of narrow slits.

3. Explain how the phenomenon of interference does not violate the law of conservation of energy.

4. Distinguish between the phenomena of interference and diffraction of light.

5. Draw intensity distribution curves for
(a) Young's double slit experiment
(b) Diffraction at a single slit

6. If one looks towards sun or a high intensity lamp through a bird's feather or muslin cloth, coloured spectra are observed. Explain how.

7. Explain why a single slit diffraction pattern is not visible unless the width of the slit is extremely small.

8. Explain how do the size and intensity of the central maximum in a single slit diffraction pattern change upon changing the slit width.

9. Explain how it is far easier to detect diffraction of sound waves than the diffraction of light waves.

10. Distinguish between unpolarized and polarized light. Illustrate with suitable line diagrams.

11. Explain how polarization of light occurs by reflection. Hence deduce the Brewster's law.

12. Explain how scattering leads to polarization of light.

13. What is a polaroid ? Explain its action for producing plane polarized light. How can it be employed for distinguishing between unpolarized light and plane polarized light.

14. Explain how on viewing the sky through a polaroid the intensity of transmitted light varies on rotating the polaroid.

15. Briefly describe, with necessary explanation, two common uses of polaroids.

Long Answer Questions

(5 marks each)

1. Derive the laws of reflection of light on the basis of Huygens' wave theory.

2. Derive the laws of refraction of light on the basis of Huygens' wave theory.

3. Derive the conditions for formation of bright and dark fringes in the Young's double slit experiment. Hence derive expressions for (a) fringe width and (b) resultant intensity in the interference pattern, as a function of phase difference.

4. Explain on the basis of Huygens' construction the formation of single slit diffraction pattern. Draw the intensity distribution curve for the pattern.

Numerical Exercises

1. In a Young's double slit experiment, light of wavelength 6500Å is incident upon the two slits, placed 0.2 mm apart. The fringes are observed on a screen located 2 m away from the slits. Find (a) fringe width and (b) change in the fringe width, when the screen is moved an additional 0.5 m away from the slits. **[(a) 6.5 mm (b) 1.6 mm]**

2. Interference fringes are produced by two coherent sources of light, having intensity ratio as 64:1. Calculate the intensity ratio of maxima and minima in the fringe system. **[81:49]**

3. Employing a source of light of wavelength 5000 Å, the fringe width in a Young's double slit experiment is found to be 0.6 cm. Determine the wavelength of light to be used for obtaining fringes 0.36 cm wide, after the slit to screen distance has been cut down to half. **[6000 Å]**

4. Interference fringes are produced in a Young's double slit experiment with incident light of wavelength 6200 Å. Find the path difference for (a) the 10th bright fringe and (b) the dark fringe between the 3rd and 4th bright fringes. **[(a) 62×10^{-5} cm (b) 21.7×10^{-5} cm]**

5. Employing light of wavelength 6000 Å in a Young's double slit experiment, 52 interference fringes are observed in the field of view. Determine how many fringes will be visible within the same field of view, if the light source is replaced by another one, emitting light of wavelength 5200 Å. **[n = 60]**

6. A fine slit illuminated by light of wavelength 5000 Å produces a diffraction pattern on a screen placed 2 metre away. The first minima in the pattern are found to lie 5 mm on either side of the central maximum. Find the slit width. **[0.1 mm]**

7. A diffraction pattern is produced by illuminating a slit of adjustable width by light of wavelength 6000 Å, such that the first minimum lies at an angle of diffraction of 30°. Calculate the slit width. Also calculate the slit width for producing another diffraction pattern in which the first maximum lies at an angle of diffraction of 30°. **[12×10^{-5} cm; 18×10^{-5} cm]**

8. Monochromatic light of wavelength 5×10^{-7} m falls on a slit 0.03 cm wide, to produce a diffraction pattern on a screen placed 1.5 m from the slit. Find the width of the central maximum in the pattern. Further, find the same width when the whole experimental set-up is immersed in water (μ_{water} = 4/3). **[0.5 cm; 0.38 cm]**

9. Two fine slits, 1 mm apart, are illuminated by light of wavelength 500 nm and form an interference pattern of fringes on a screen placed 1 m away from the slits. When one of the slits is covered, a single slit diffraction pattern is obtained on the screen. Determine the width of each slit required to obtain 10 interference maxima within the diffraction pattern's central maximum. [Hint: Equate the width of 10 interference maxima $\left(=\dfrac{10\lambda D}{d}\right)$ to the width of diffraction central maximum $\left(=\dfrac{2\lambda D}{b}\right)$, to get $b = \dfrac{d}{5} = \dfrac{0.1}{5}$ cm]. **[0.2 mm]**

10. Determine the Brewster's angle for an air-water interface, on which yellow light is incident (μ_{water} for yellow light = 1.33). **[53°]**

11. The polarizing angle for a medium surrounded by air is measured to be 60°. Find the refractive index of the medium. **[1.73]**

12. If a glass plate is to be used as a polarizer, calculate what should be the value of the polarizing angle. Also calculate the corresponding angle of refraction (μ_{glass} = 1.5). **[56.3°; 33.7°]**

13. Two nicol prisms are mutually positioned to transmit maximum amount of incident light through the combination. If now one of them is rotated first by 30° and next by another 30°, calculate by what percentage the transmitted intensity is reduced in the two cases. **[75% & 25%]**

14. The critical angle for a transparent medium situated in air has the value 60°. When a light beam is incident upon the medium, find the angle of incidence for which the reflected and refracted rays are mutually perpendicular. **[49.1°]**

UNIT 7
DUAL NATURE OF RADIATION AND MATTER

CHAPTER 7
Dual Nature of Radiation and Matter

CHAPTER 7
DUAL NATURE OF RADIATION AND MATTER

Dual Nature of Radiation & Matter

In the introductory part of the last chapter (Section 6B.01), we had given a brief historical account asto how over a period covering nearly two and a half centuries, the dual nature of light radiation came to be known and established. The crucial part played by the discovery of photoelectric effect was emphasized. Further, it was pointed that the whole exercise also led to the development of important concepts of quantum mechanics and matter waves.

In the present chapter, we shall learn about photoelectric effect and matter waves in some detail and also have a glimpse of quantum physics.

7.01. Photon; Quantum Physics

If the light radiation had a particle nature, what exactly this particle is ? A definite answer to this question was first provided by Einstein* in 1905, who identified the particle as a packet (or quantum) of energy that we now call as a photon (derived from the Greek word 'photos', meaning 'light'), while providing explanation for photoelectric effect. This proposal also led to laying the foundations of quantum physics (which is also known as quantum mechanics or quantum theory or wave mechanics).

A **photon** is "A packet (or quantum) of energy associated with a light wave of frequency ν ($E = h\nu$; h: Planck Constant). It represents the minimum energy that a light wave can have."

In quantum physics, the minimum amount that a physical quantity like energy or momentum can possess is called the **quantum** of that quantity. Any other amount of the quantity has to be an integral multiple of the quantum, e.g. energy can only have the following discrete values,

$$E = h\nu,\ 2h\nu,\ 3h\nu, \ldots\ldots$$

Or

$$E = n(h\nu)\ (n = 1,2,3, \ldots\ldots) \qquad \ldots\ldots (7.1)$$

In this sense, the physical quantity is said to be **quantized.** The photon is a quantum of light energy.

A quantum is roughly similar to the minimum of currency of a country. For instance, in India the smallest unit of currency is 1 paisa. The large unit is 1 rupee (= 100 paisa). No payment can be physically made which is less than 1 paisa (e.g. 0.4 paisa) or which involves a fraction of a paisa (e.g. 865.7 paisa = Rupees 8.657). Thus, loosely speaking, 1 paisa is the 'quantum' of currency in India. The difference is that while this minimum of currency has been fixed by the government of India, the minimum of a physical quantity like energy is fixed by the laws of nature!

Important properties of photons. Some important properties of the photons are enumerated in the following:

(i) A photon always travels with the velocity of light in vacuum ($c = 3.0\times10^8\ ms^{-1}$).

(ii) A photon has a definite energy (E) and a definite momentum (p), given by

$$E = h\nu = \frac{hc}{\lambda} \qquad \ldots\ldots (7.2a)$$

$$p = \frac{h}{\lambda} = \frac{E}{c} \qquad \ldots\ldots (7.2b)$$

h is a universal constant (like c, the velocity of light), called Planck constant, with value,

$$h = 6.63\times10^{-34}\ J.s = 4.14\times10^{-15}\ eV.s$$

[Energy $= mc^2 = h\nu$ or $m = h\nu/c^2$; $\therefore$ Momentum $= mc = E/c$ or h/λ].

(iii) The photons are electrically neutral (cf. electrons, which are negatively charged particles). Hence they are not affected by electric and magnetic fields.

*We have already written about Einstein in the last chapter (at the end of Section 6B.01). A full identification of photon as a particle was actually made a few years later by him, when the photon was found to possess not just energy but also a linear momentum. He was awarded Nobel Prize in physics for 1921 for his work on photoelectric effect.

(iv) When light interacts with matter, the photons collide with material particles, e.g. electrons. Then both **photon absorption** and **photon emission** may occur, i.e. some photons may be absorbed and some new photons may be emitted. Hence the total number of photons may not be conserved in the collisions. However, the total energy and the total momentum of the photons is always conserved during the collisions.

(v) When a photon travels in a medium other than vacuum, its velocity and wavelength change, but such that the frequency remains the same. ($v = \nu\lambda$; $\therefore \nu = v/\lambda$).

(vi) If the intensity of a source of monochromatic light (i.e. fixed λ) is increased, the number of photons emitted by the source increases, but the energy of each photon remains unchanged.

(vii) The rest mass of a photon is zero (cf. electron, which has the rest mass $m_e = 9.11\times10^{-31}$ kg). [However, this fact has little physical significance, because there is no frame of reference relative to which the photon can be at rest. The photon moves with the velocity c in vacuum relative to all possible frames of reference].

Example 1. Find the energy and linear momentum of a photon of wavelength 5000 Å.

Solution. Given: $\lambda = 5000\text{ Å} = 5000\times10^{-10}\text{ m} = 5\times10^{-7}\text{ m}$

The energy of a photon is given by,

$$E = \frac{hc}{\lambda} = \frac{(6.6\times10^{-34})\times(3\times10^{8})}{5\times10^{-7}}$$

$$= 3.96\times10^{-19}\text{ J} \quad \textbf{(Answer).}$$

The linear momentum of the photon is given by,

$$p = \frac{h}{\lambda} = \frac{(6.6\times10^{-34})}{(5\times10^{-7})}$$

$$= 1.32\times10^{-27}\text{ kg.m.s}^{-1} \quad \textbf{(Answer).}$$

Example 2. A photon has energy 63 eV. Calculate its frequency.

Solution. Given: $E = 63\text{ eV} = 63\times(1.6\times10^{-19})\text{ J} = 10.08\times10^{-18}\text{ J}$

We know that the energy is related to the frequency ν as,

$$E = h\nu$$

$$\therefore \quad \nu = \frac{h}{E} = \frac{(10.08\times10^{-18})}{(6.6\times10^{-34})}$$

$$= 1.53\times10^{16}\text{ Hz} \quad \textbf{(Answer).}$$

7.02. Photoelectric Effect

The **photoelectric effect** is "the phenomenon of emission of electrons from a metal surface, when irradiated by light of suitable frequency (visible light or ultraviolet light)". The electrons so emitted are called **photoelectrons** and the electric current produced by them is called **photoelectric current.**

[Essentially, the phenomenon involves conversion of light energy into electrical energy. It was first observed by Hertz in 1887, before the discovery of electron, as a change in the electric charge on the metal, when electrons were emitted from its surface. Before that, **cathode rays** had been discovered by William Crookes in 1870, as a coloured glow discharge, produced in an evacuated discharge tube between two electrodes, maintained at a potential difference of a few hundred volt. They were identified as consisting of fast moving negatively charged particles. In 1897, J.J. Thompson confirmed this and measured the velocity and e/m ratio of these particles, which were then termed as electrons. Thus the discovery of electron was made by him (for which he was awarded Nobel Prize in physics for 1906). The particles emitted in the photoelectric effect were found to be the same in nature as the electrons. Hence they were called photoelectrons. Around the same period, the particles emitted by heating certain metals were also found to be electrons (thermionic emission). Now we know that electrons can also be pulled out of a metal by applying a very strong field (~10^5 kV per meter) across it (field emission)].

An experimental set-up for studying the photoelectric effect is shown in Fig. 7.1. Two metal plates C and A are sealed inside an evacuated glass tube, fitted with a quartz window. (Quartz allows the passage of ultraviolet light as well as the visible light, whereas glass absorbs the ultraviolet). The plates C and A are connected to a suitable voltage source. The voltage can be varied with a rheostat, as well as its polarity can be reversed with the help of a commutator (rheostat and commutator not shown in the figure). When the incident light falls on the plate C, photoelectrons are emitted, which are collected at the plate A. Usually A is positive with respect to C, but it can also be made negative w.r.t. to C. C is called **cathode** (or **emitter)** and A is called **anode** (or **collector**). The photocurrent produced by the flow of the photoelectrons can be read in the ammeter (its direction is opposite to the flow of electrons). The applied voltage can be read in the voltmeter.

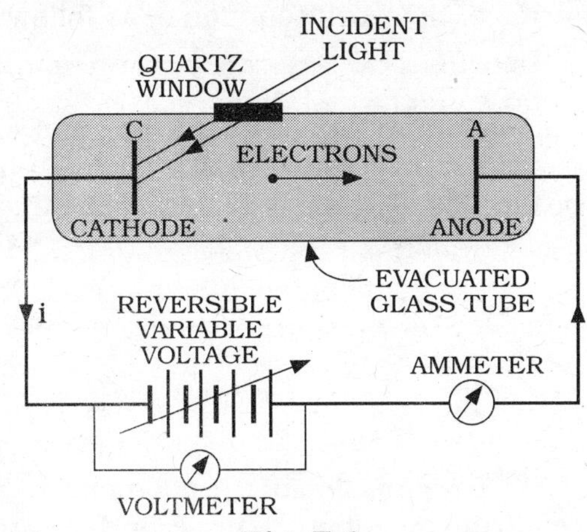

Fig. 7.1

Einstein's photoelectric equation; Work function. We now examine the physical mechanism of the emission of the photoelectrons. We know that a metal contains an enormous number of free electrons, which freely wander inside the metal. They result from their detachment from the outermost orbits of the atoms, thus leaving behind an ocean of positive ions in the metal. Normally, a free electron cannot leave the metal because of the attractive force of the +ve ions. But if it can gain sufficient energy from outside, it can eject from the metal. The **work function** (ϕ) of a metal is **defined** as "the minimum amount of energy required to remove an electron from the metal". Its value is of the order of a few electron volts for the metals, e.g. nearly 4.5 eV for copper.

If monochromatic light of frequency ν (wavelength: λ) falls on the surface of a metal, then the photons with energy $E = h\nu$ each collide with the free electrons of the metal. Suppose $E > \phi$. After colliding with a photon, a free electron may travel in any arbitrary direction. If during the collisions a photon transfers its entire energy E to a free electron and the electron does not suffer any further collisions with the atoms of the metal, the electron will come out with an energy $(E–\phi)$. But if it suffers any such collisions, it may still come out, but with energy less than $(E–\phi)$, which in the limit can be zero. Thus the kinetic energy of an ejected photoelectron can be any value between the minimum O and the maximum $(E–\phi)$. We can write the maximum possible value of the kinetic energy of the photoelectron as $(E = h\nu)$,

$$E_{max} = h\nu - \phi \qquad \text{..... (7.2a)}$$

(Einstein's Photoelectric Equation)

Eq. 7.2(a) is the celebrated **Einstein's photoelectric equation.** It is an expression of the law of conservation of energy for the case of a single photon absorption by a metal of work function ϕ. It can also be expressed in terms of wavelength λ as $(c = \nu\lambda)$,

$$E_{max} = \frac{hc}{\lambda} - \phi \qquad \text{..... (7.2b)}$$

Note that the **value of E_{max} depends only on the value of ν, the frequency of the incident light. It does not depend on the intensity of light.**

The Einstein's equation successfully explains all the important observed features of the photoelectric effect, which we now proceed to consider one by one in the following.

Threshold (or cut-off) frequency (ν_0) [and threshold (or cut-off) wavelength (λ_0)]. The Einstein's Eqs. 7.2 (a),(b) give the relation between E (photon energy), ν/λ (photon frequency/wavelength), ϕ (work function), and E_{max} (maximum possible energy of a photon).

According to the experiment, no photocurrent i is produced in the circuit shown in Fig. 7.1, unless the frequency of the incident light is equal to or greater than a minimum value ν_0 ($\nu \geq \nu_0$, or $\lambda \leq \lambda_0$). $\boldsymbol{\nu_0}$ and λ_0 are called **threshold (or cut-off) frequency** and **threshold (or cut-off) wavelength,** respectively. ν_0 can be **defined** as "the minimum photon frequency required for the emission of photoelectrons from a metal surface". Similarly for λ_0 (maximum wavelength required).

The existence of ν_0 (or λ_0) is readily explained by the Einstein equation, which also gives its value in terms of the work function ϕ, as follows. If the photon frequency has a value ν_0 such that,

$$\nu_0 = \frac{\phi}{h} \quad \text{or} \quad \phi = h\nu_0, \qquad \text{..... (7.3a)}$$

the maximum kinetic energy E_{max} is zero. It means that in this case a photoelectron will just be able to come out of the metal surface. If $\nu<\nu_0$, the photon energy $h\nu<\phi$ $(=h\nu_0)$, the work function, and so no electron can be emitted. The same holds for the threshold wavelength λ_0,

$$\lambda_0 = \frac{hc}{\phi} \text{ or } \phi = \frac{hc}{\lambda_0} \qquad \text{..... (7.3 b)}$$

Using Eqs. 7.3 (a,b), the Einstein equations 7.2 (a,b) can also be written as $(\phi = h\nu_0)$,

$$E_{max} = h(\nu-\nu_0) \qquad \text{..... (7.4a)}$$

or

$$E_{max} = hc\left(\frac{1}{\lambda}-\frac{1}{\lambda_0}\right) \qquad \text{..... (7.4b)}$$

The emission of the photoelectrons, when the threshold frequency is reached, is almost instantaneous, requiring just about 10^{-9} second, i.e. only about a billionth of a second,

Effect of anode potential (V) on photoelectric current (i); Stopping (or cut-off) potential. The effect of variation of anode potential (V) on the photoelectric current (i) is shown in Fig. 7.2 for three different intensities (I) of the incident light $(I_3>I_2>I_1)$. The curves look similar. As a typical case, let us consider the curve abcd for the light intensity I_3. First consider the part bcd, for +ve values of V. Even when V=0, a number of photoelectrons emitted at the cathode reach the anode, producing a photocurrent corresponding to the point b in the figure. Not all the emitted electrons are able to reach the anode, because the cloud of electrons between the cathode and the anode forms a negative **space charge,** which tends to repel the electrons. As V increases, more electrons are able to reach the anode and so i gradually increases. Finally, the position d is reached when all the emitted photoelectrons reach the anode and so i attains its highest value, called **saturation current.** Any further increase of V does not increase i.

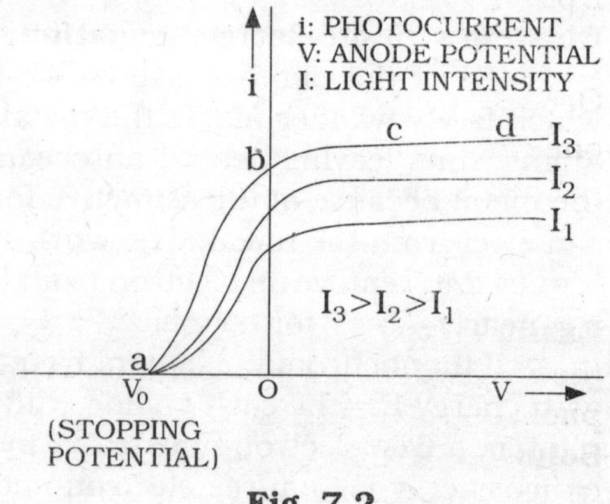

Fig. 7.2

Next, consider the part ba of the curve, corresponding to –ve values of V, i.e. when the anode is made –ve with respect to the cathode. Now the anode repels the electrons moving towards it. The photocurrent rapidly decreases as V is increased. Finally, it becomes zero for a sharply defined value $\mathbf{V_0}$, called **stopping (or cut-off) potential.** It may be **defined** as "the minimum negative potential to be applied across a photoelectric tube in order to make the photoelectric current zero". Note that, like the threshold frequency (ν_0), **$\mathbf{V_0}$ does not depend on the intensity of the incident light. It only depends on the frequency of the light,** as examined in the following part.

From the Fig. 7.2, it is seen that the photocurrent (i) increases with increase in I, the intensity of the incident light. [Note that the frequency (and wavelength) is same in all the cases]. The reason is when I increases (from I_1 to I_2 to I_3), progressively larger number of photons fall on the metal surface, producing progressively larger number of photoelectrons. The photoelectric current (i) is directly proportional to the light intensity (I), as shown in Fig. 7.3, i.e.

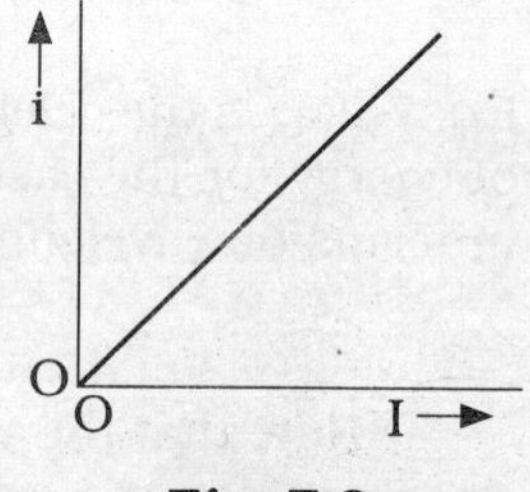

Fig. 7.3

$$i \propto I$$

Effect of frequency (ν) on stopping potential (V_0); Determination of h (planck Constant), ϕ (work function), and ν_0 (threshold frequency). As already stated, the kinetic energy of the photoelectrons emitted from the cathode varies between 0 and E_{max}. Since the photoelectric current becomes zero when the stopping potential V_0 is applied, it means that even the emitted electrons of maximum energy E_{max} fail to reach the anode. They are turned back just before reaching the anode, when their potential energy is eV_0 and the kinetic energy ($1/2$ mv^2) is zero. So their total energy is,

$$E_{max} = eV_0$$

Substituting in the Einstein's Equation 7.2a, we have,

$$E_{max} = h\nu - \phi$$

$$\Rightarrow \quad eV_0 = h\nu - \phi$$

$$\Rightarrow \quad V_0 = \left(\frac{h}{e}\right)\nu - \frac{\phi}{e} \qquad \text{..... (7.5 a)}$$

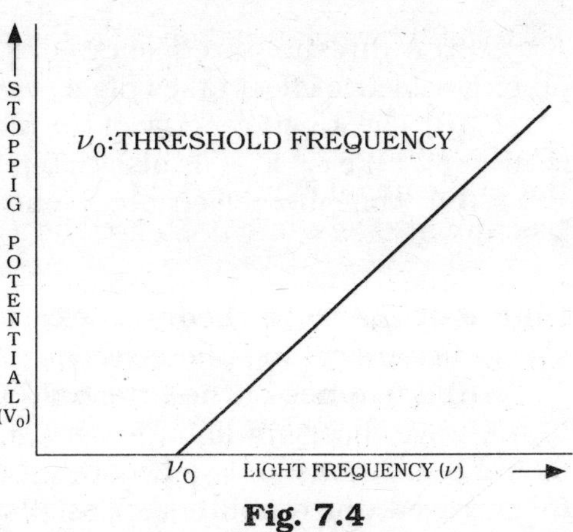

Fig. 7.4

This equation represents the variation of the stopping potential (V_0) with frequency (ν). It has the form y = mx + c. Hence the variation is linear, as shown in Fig. 7.4. The intercept made by the straight line on the x-axis is,

$$\text{Intercept} = \frac{\phi}{e} \qquad \text{..... (7.6 a)}$$

and the slope of the line is,

$$\text{Slope} = \frac{h}{e} \qquad \text{..... (7.6 b)}$$

Thus, from the measurement of the intercept and the slope of the line, one can calculate the value of ϕ (work function of the metal of the cathode) and h (Planck constant), from the Eqs. 7.6a and 7.6b, respectively. Further, since (vide Eq. 7.3a),

$$\phi = h\nu_0,$$

Or

$$\nu_0 = \frac{\phi}{h}, \qquad \text{..... (7.6 c)}$$

one can calculate the value of ν_0, the threshold frequency, from the knowledge of the values of ϕ and h.

The Eq. 7.5(a) can also be written in terms of wavelength as ($c = \nu\lambda$),

$$V_0 = \frac{hc}{e}\left(\frac{1}{\lambda}\right) - \frac{\phi}{e} \qquad \text{..... (7.5 b)}$$

Example 3. Light of wavelength 4500 Å is incident on the surface of a material. If the work function of the material is 2 eV, determine (i) energy of the incident photons in electron-volt, (ii) kinetic energy of the photoelectrons liberated from the material, and (iii) stopping potential.

Solution. Given: $\lambda = 4500\ \text{Å} = 45\times10^{-8}\text{m}$; $\phi = 2\ \text{eV} = 2\times(1.6\times10^{-19})\ \text{J} = 3.2\times10^{-19}\ \text{J}$

(i) The energy of a photon is given by,

$$E_{ph} = \frac{hc}{\lambda} = \frac{(6.6\times10^{-34})\times(3\times10^{8})}{(45\times10^{-8})}$$

$$= 4.4\times10^{-19}\ \text{J} = \frac{4.4\times10^{-19}}{(1.6\times10^{-19})}\ \text{eV}$$

$= 2.75$ eV **(Answer).**

(ii) The maximum k.e. of a photoelectron is given by,

$$E_{max} = h\nu - \phi = [(4.4\times10^{-19}) - (3.2\times10^{-19})]\text{J}$$

$= 1.2\times10^{-19}$ J **(Answer).**

(iii) The stopping potential V_0 is given by,

$$E_{max} = eV_0$$

$$\Rightarrow \quad V_0 = \frac{E_{max}}{e} = \frac{(1.2\times10^{-19})}{(1.6\times10^{-19})}$$

$= 0.75$ volt **(Answer).**

Example 4. A photoemissive surface has a threshold frequency of 5×10^{14} Hz. What will be the minimum energy of incident photons in electron-volt, required for causing photoemission from the surface ?

Solution. Given: $\nu_0 = 5\times10^{14}$ Hz

The minimum photon energy required for photoemission is given by,

$$E_{min} = h\nu_0 = (6.6\times10^{-34})\times(5\times10^{14})$$

$$= 33\times10^{-20}\ \text{J} = \frac{33\times10^{-20}}{(1.6\times10^{-19})}\ \text{eV}$$

$= 2.06$ eV **(Answer).**

Summary of the salient features of photoelectric effect

(i) The emission of photoelectrons, when light of a suitable frequency is incident upon a metal surface, is almost an instantaneous process.

(ii) A threshold frequency ν_0 exists for a given metal. If the incident light has frequency less than ν_0, no photoelectric effect takes place, even if the light has a high intensity.

(iii) The kinetic energy of the photoelectrons varies from the minimum zero to a maximum value E_{max} ($= h\nu - \phi$). Like ν_0, E_{max} is also independent of the intensity of light.

(iv) The photoelectric current (i) increases with increase in the intensity (I) of the incident light ($i \propto I$).

Failure of the wave theory to explain photoelectric effect. The wave theory of light fails to explain the salient features of the photoelectric effect, as detailed below.

(i) Existence of the threshold frequency (ν_0). As explained in Chapter 6A (Section 6A.04), according to the wave theory of light, when light is incident on a material the electrons of the material are set into forced oscillations and so become secondary sources of light of the same frequency. If the intensity of light is increased, the amplitude of oscillation goes up, and so the energy of the electrons increases. Thus, by enhancing the light intensity the electron energy may be increased enough to exceed the work function, whatever may be the frequency of the light. Then, emission of photoelectrons should take place. But that is contrary to observation. However strong is the light source, the photoemission does not take place unless the light's frequency $\nu \geq \nu_0$. On the other hand, the observation is straightway understood in terms of the photon energy, which is fully determined by the frequency ($E = h\nu$). The photoemission occurs as soon as the photon energy ($h\nu_0$) exceeds ϕ, the work function, even if the light intensity is very weak.

(ii) Non-dependence of E_{max} (maximum energy of the photoelectrons) on light intensity. As argued above, on the basis of the wave theory, higher light intensity should mean higher energy of the photoelectrons. Therefore, the value of E_{max} should continuously increase if the light intensity is steadily increased, which is contrary to the observation. But the observation is immediately understood in terms of photons, as increasing the light intensity only increases the number of photons, not the photon energy, which only depends upon the frequency of light ($E = h\nu$).

(iii) Almost instantaneous emission of photoelectrons, even with very weak light intensity. When light is incident on a photosensitive metal, the electrons of the metal continuously absorb the incident energy. However, the calculation shows that if the light source is weak, it may take several hours before an electron absorbs enough energy to be ejected out of the metal surface. But the reality is that even with very weak light, the photoemission is almost instantaneous. Again, the phenomenon finds ready explanation in terms of photons, viz. that the emission depends only on the frequency of light, not its intensity.

7.03. Photoelectric Cell (or Photocell or Phototube or Electric Eye)

A **photoelectric cell (or photocell or phototube or electric eye)** is a device for converting a light signal into an electric signal. We have 3 types of photocells:

(i) Photoemissive cell. It is directly based on the emission of electrons from the surface of a photosensitive material by the action of light falling upon it.

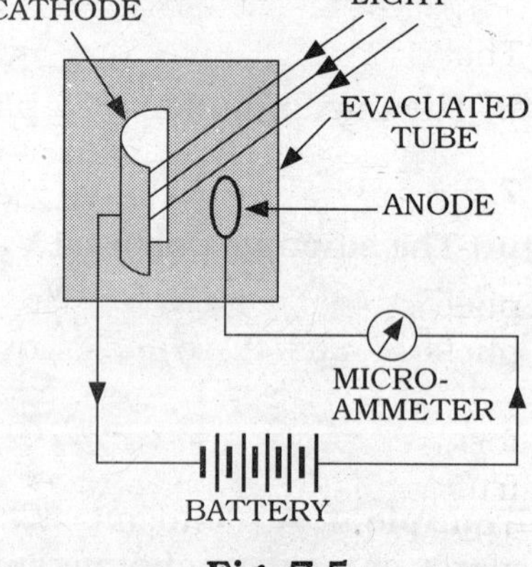

Fig. 7.5

It essentially consists of a small evacuated glass (or quartz) tube, containing a semi-cylindrical cathode and an anode (Fig. 7.5). The cathode (emitter) is usually an alloy on which a photosensitive thin film of antimony-caesium (or any other photosensitive material) is deposited. The anode (collector) is a wire loop. The cathode and anode are connected to a battery. When light falls on the cathode, photoelectrons are emitted, which are collected by the anode. The resulting electric current has a small value (~a few tens of microampere), which can be read by a sensitive microammeter placed in the circuit. In a variation of this kind of photocell, the photosensitive material is directly deposited on the wall of the glass tube.

(ii) Photoconductive cell. It is based on the increase in electrical conductivity in a circuit when light falls on a photosensitive semiconducting material (you will study semiconductors in Chapter 9).

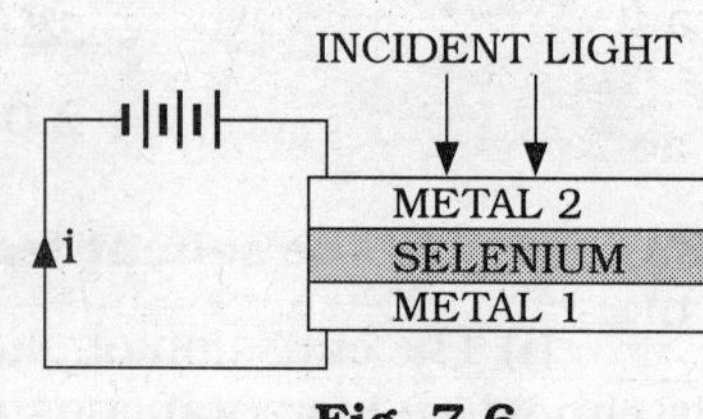

Fig. 7.6

A thin layer of the semiconducting material, e.g. selenium, is sandwiched between two metallic layers, to form a compact small slab (Fig. 7.6). The upper metallic layer is very thin. So it is partially transparent to light. A potential difference is applied across the upper and the lower metallic layers, which makes a small electric current to flow in the circuit. When light is incident on the photosensitive material, the current in the circuit is enhanced, because the conductivity of the semiconductor is increased, as the electrons in the valence band receive energy from the incident light. The changes in current follow the changes in the intensity of light.

Because of their small size, compactness, simplicity of construction, and relatively large value of electric current, this type of photocells are widely used for various applications.

(iii) Photovoltaic cell. It is based on the generation of emf when two dissimilar materials are put in contact.

The two materials may, for instance, be layers of a metal and a semiconductor or layers of two semiconductors of opposite polarity. A forward voltage appears when their junction is illuminated by light and so a small current flows in the circuit. No external voltage is applied across the junction. The action of light is same as for a photoconductive cell.

Applications of photocells

The photocells find three important kinds of application:

(i) Measurement of light intensity. e.g. in photometry, light meters for photographic cameras, etc.

(ii) Faithful reproduction of rapid variations in light intensity. e.g. a beam of light modulated by the sound track of a cinefilm (or by the scanning process in a television tube) falls on a photocell, which converts it into variations of electric current. These variations, in turn, are used to reproduce the sound when the cinefilm is being projected on the screen.

(iii) For detecting changes in light intensity and using the resulting electric signal for controlled operation of a system. If the light signal in a photocell is momentarily interrupted, it causes a corresponding momentary break and remake of the electric circuit of the photocell, which can be utilised for controlled operation of a system. The same can be done if the light intensity exceeds or falls below a specified value. In this sense, the photocell works as an 'electric eye'. Thus it can be used for a variety of purposes, such as,

(a) Burglar alarm and fire alarm (by making to ring an electric bell or sounding a siren), etc.

(b) Automatic opening and shutting of a door when a person enters or leaves a room, automatic switching on and switching off the streetlights at dusk and dawn, respectively, etc.

(c) As a counting device for persons entering an auditorium one by one, for objects on a conveyor belt in a factory, for currency notes in a packet, etc.

The electric current produced in a photocell is usually too small to operate the control system. Therefore it is suitably amplified through appropriate electrical devices before being fed into the control system.

7.04. Wave Nature of Particles; Matter Waves; de Broglie Relation

We have seen that the light radiation has a dual character. Sometimes it behaves as waves, e.g. in the phenomena of interference, diffraction and polarization, and sometimes as particles (photons), e.g. in the phenomenon of photoelectric effect.

The nature has two basic physical constituents, matter and radiation, both of which exist in numerous forms. And we know from observation that the nature prefers to be symmetrical. For example, all flowers/fruits on a tree look alike. They also smell/taste alike. Guided by these two fundamental facts of nature, in 1924 de Broglie* suggested that if radiation could have a dual character (wave and particle), then all kinds of particles (i.e. matter in general) should also have a wave character associated with them. Thus the concept of **matter waves** was born. The matter waves may be **defined** as "the waves of different wavelengths associated with various kinds of matter in general and with various kinds of particles in particular" (e.g. electrons, neutrons, various atoms and molecules, etc.). The existence of these waves was actually demonstrated just a few years later in 1929 by Davisson and Germer, which will be described in the next section.

de Broglie Relation

In the Section 7.01, we had stated that as a material particle a photon possesses both energy and momentum, whose values are given by Eqs. 7.2a,b [$E = h\nu$ and $p = \lambda/h$]. Thus a photon has a wave of

*Louis Victor de Broglie; French physicist (1892-1987). He was awarded Nobel Prize in physics for 1929 for this work.

wavelength λ associated with it. According to the **de Broglie's hypothesis,** all the material particles in nature should similarly have waves associated with them. The wavelength associated with such a particle is given by,

$$\lambda = \frac{h}{p} = \frac{h}{mv} \quad \text{(de Broglie wavelength)} \qquad \text{..... (7.7)}$$

m is the mass of the particle and v is its velocity. λ is known as the **de Broglie wavelength** of the matter wave associated with the particle and the Eq. 7.7 is known as the **de Broglie relation.**

Derivation of de Broglie relation. We know that the energy associated with a photon is given by

$$E = h\nu \qquad \text{..... (i)}$$

According to the de Broglie's hypothesis, this equation holds equally well for any other material particle. Further, according to the famous Einstein's mass-energy relationship (to be studied in the next chapter), the energy of the particle is also given by,

$$E = mc^2 \qquad \text{..... (ii)}$$

From the Eqs. (i) and (ii) we have,

$$h\nu = mc^2 \quad \text{or} \quad m = \frac{h\nu}{c^2}$$

Now, the momentum p is given by,

$$p = \text{mass}\times\text{velocity}$$

$$= m\times c = \frac{h\nu}{c^2}.c$$

$$= \frac{h\nu}{c}$$

But, $\nu = c/\lambda$. Therefore,

$$p = \frac{h}{\lambda}$$

$$\Rightarrow \quad \lambda = \frac{h}{p}$$

$$\Rightarrow \quad \lambda = \frac{h}{p} = \frac{h}{mv},$$

which is the de Broglie relation (Eq. 7.7).

de Broglie wavelength of an electron. The electrons as carriers of electricity have a special importance for us as material particles. In fact, the de Broglie's hypothesis was proposed in context of the electrons. The wavelength λ_{el} associated with an electron can be simply derived as follows. Suppose an electron is accelerated from rest through a potential difference V and acquires a velocity v. Then,

$$\text{Gain in the electron's kinetic energy} = \frac{1}{2}mv^2$$

$$\text{Gain in the electron's potential energy} = -eV$$

(gain in p.e. is equal to the work done on the electron).

According to the principle of conservation of energy,

$$\frac{1}{2}mv^2 + (-eV) = 0$$

$$\Rightarrow \quad \frac{1}{2}mv^2 = eV$$

$$\Rightarrow \quad v = \sqrt{\frac{2eV}{m}}$$

$$\therefore \quad \lambda_{el} = \frac{h}{mv} = \frac{h}{m.\sqrt{2eV/m}}$$

$$\Rightarrow \quad \lambda_{el} = \frac{h}{\sqrt{2meV}} \qquad \text{..... (7.8 a)}$$

If we substitute the standard values of h,m and e, we get,

$$\lambda_{el} = \frac{1.227}{\sqrt{V}} \text{ nm} \qquad \text{..... (7.8 b)}$$

$$(1\text{nm} = 10^{-9} \text{ metre})$$

Example 5. An electron beam is accelerated through a potential difference of 50 V. What is the de broglie wavelength of an electron in the beam?

Solution. Given: V = 50 V

The required de Broglie wavelength is given by,

$$\lambda = \frac{h}{\sqrt{2meV}} = \frac{6.6\times10^{-34}}{\sqrt{2\times(9.1\times10^{-31})\times(1.6\times10^{-19})\times50}}$$

$$= \frac{6.6\times10^{-34}}{3.82\times10^{-24}} = 1.73\times10^{-10} \text{ m}$$

$$= 1.73 \text{ Å}$$ **(Answer).**

7.05. Davisson and Germer Experiment; Verification of Wave nature of Electrons; Electron Diffraction

The wave character of the electrons was first verified by C.J. Davisson and L.H. Germer in 1927 through a historical experiment that we shall now briefly describe.

We have known that when the light radiation is incident upon a material, it interacts with the electrons of the material, setting them into forced oscillations. Thus the electrons become secondary sources of light of the same frequency as the incident light. The light radiation does not interact with the material on an atomic level, because the wavelength of light (~ 500 nm) is far greater than the spacing between the atoms (~ 0.1 nm). But X-rays have wavelength ~ 0.1 nm and so they can effectively interact with the atoms. The most suitable materials for such an interaction are crystals. The atoms within a crystal are regularly arranged in three dimensions. Therefore, they are expected to produce observable diffraction effects when exposed to a beam of X-rays. [Such an observation was first qualitatively made by the German Physicist Max von Laue in 1912 in a Nobel Prize winning work. Later, the father and son combination of the British physicists W.H. Bragg and W.L. Bragg worked out the theory of the diffraction and successfully employed the phenomenon for determining crystal structures (i.e. finding actual positions of the atoms in a crystal) in another Nobel Prize winning work].

If the atoms in a crystal are regularly arranged, then the crystal can be considered as consisting of several sets of parallel atomic planes of different atomic spacings and having different orientations. One such set is shown in Fig. 7.7, with interplanar spacing as d. A beam of X-rays is incident upon the planes at a glancing angle θ, now known as the Bragg angle. Since the X-rays are penetrating electromagnetic radiation, they penetrate through hundreds of the parallel atomic planes. The theory shows that the diffraction effect is such that at each plane they undergo a mirror-like reflection. Thus, in this case the diffraction is equivalent to reflection. The reflected beams have constructive interference if the following condition is satisfied (any two successive reflected beams like 1 and 2 have their phase difference 2d sin θ as an integral multiple of the X-ray wavelength λ),

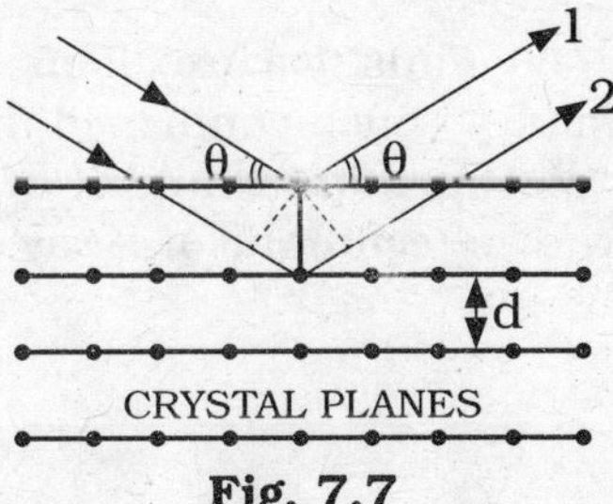

Fig. 7.7

$$2d \sin\theta = n\lambda \quad \text{(Bragg's law)} \qquad \text{..... (7.9 a)}$$

For the strongest first order diffraction (n = 1),

$$2d \sin\theta = \lambda \qquad \text{..... (7.9 b)}$$

The Davisson and Germer experiment is based on the above equations. The experimental set-up is shown in Fig. 7.8 The electrons are emitted from an electron gun, which consists of a metallic filament heated by a low voltage supply. They are accelerated towards a nickel crystal by applying a suitable positive potential from a high voltage supply. They are also collimated into a fine beam by making them to pass through a cylinder with fine holes on the two sides. After falling upon the crystal they penetrate a small thickness of the crystal and get diffracted in different directions by the different sets of parallel atomic planes of the crystal. The diffracted electron beams are received by a collector, which is movable along a circular scale. It is connected to a sensitive ammeter, to record the current (i) produced by the diffracted electrons. By rotating the collector along the circular scale, the intensity of the diffracted beams corresponding to various Bragg angles θ in Eq. 7.9b can be known. The whole apparatus, except the voltage supplies, is kept in an evacuated chamber. (For the sake of clarity, the diffracting planes have been extended outside the crystal in Fig. 7.8).

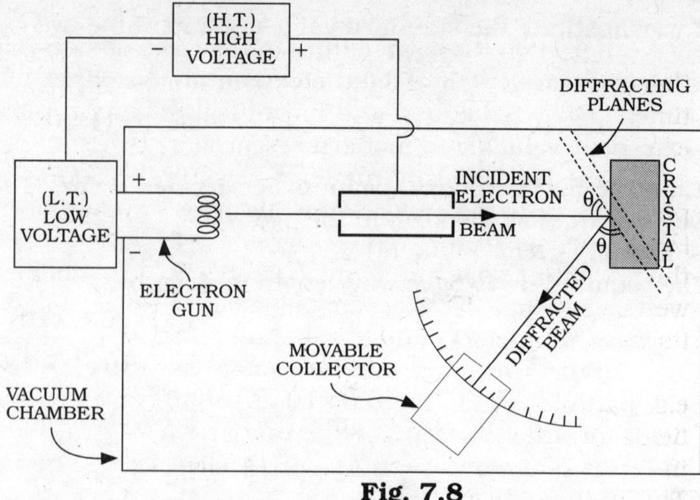

Fig. 7.8

The Eq. 7.9b means that if electrons of an appropriate wavelength λ are incident upon a set of parallel planes of interplanar spacing d at a glancing angle θ, a strong diffracted beam (equivalent to a reflected beam) will be produced with the same value of the glancing angle θ. In the Davisson and Germer experiment, plots of the current i as a function θ of were made for values of accelerating potential V ranging from 44V to 68V. The variation of V meant variation of the electron wavelength λ_{el} [vide Eq. 7.8; $\lambda_{el} = h/\sqrt{2meV} = (1.227/\sqrt{V})$ nm]. A sharp and strong peak in the i vs θ curves was obtained for V = 54 volt and θ = 65°. Thus, the experimental value of λ_{el} is obtained from the Eq. 7.9b as (d is a known value; d = 0.091 nm),

$$\lambda_{el} = 2d \sin\theta$$
$$= 2\times 0.091 \times \sin 65°$$
$$= 0.165 \text{ nm}$$

According to the de Broglie hypothesis, the theoretical value of λ_{el} is given by Eq. 7.8b as,

$$\lambda_{el} = \frac{1.227}{\sqrt{V}}$$
$$= \frac{1.227}{\sqrt{54}}$$
$$= 0.166 \text{ nm}$$

Thus, an excellent agreement was found to exist between the experimental and the theoretical values, which confirmed the wave character of the electrons. It further provided the scientists with the new experimental technique of **electron diffraction** for studying crystal structures, in addition to the already known technique of X-ray diffraction.

Additional knowledge

(i) Whether light (or any other radiation) will behave as waves or particles in a given situation depends on the nature of the phenomenon involved in the situation. In the phenomena of image formation, interference, diffraction and polarization, light exhibits the wave character, but in the phenomenon of photoelectric effect it shows particle character. Until now, no single situation has been encountered in which the light radiation (or any other radiation) behavs both as waves and particles.

(ii) Apart from the electrons, even the bigger fundamental particles protons and neutrons have been found to exhibit wave character. In fact, **neutron diffraction** has emerged as another useful technique for studying crystal structures. The phenomenon of interference has also been found to be exhibited by them, as also by atoms and molecules, e.g. by iodine (I_2) molecules, which are abont 5×10^5 times massive than electrons, and by the molecules of the new interesting complex compounds fullerenes (C_{60} and C_{70}; they are new synthetic forms of carbon, containing 60 and 70 atoms, respectively, per molecule).

Based on the wave nature of the electrons, which have wavelength several thousand times shorter than the wavelength of light, **electron microscopes** have been constructed. They have several thousand times higher resolution and magnification, as compared to the optical microscopes, which has proved extremely valuable in modern research work.

(iii) Theoretically, matter waves may be said to be associated with all the objects in the nature. But physically it is meaningless for all macroscopic objects, because the wavelength associated with them ($\lambda = h/mv$) is extremely small. For example, even for a tiny mustard seed weighing just abont 1 gram and thrown with a speed of about 10 ms^{-1}, the value of λ turns out to be merely $\approx 10^{-28}$ m (or 10^{-19}nm). So we can say that the wavelength associated with a macroscopic object tends to zero, or in other words, its wave character tends to vanish.

(iv) A travelling wave is usually described in terms of the oscillatory motion of a physical entity, e.g. particles of the medium for a sound wave or a wave on a string or the electric field (or magnetic field) for a light wave. For the matter waves, it is somewhat complicated. A matter wave is described in terms of a complex function ψ, called **wave function.** It is a function of space and time [$\psi(x,y,z,t)$]. The probability of finding a particle at a given position in the wave is proportional to ψ^2. In other words, the position of a particle in the wave cannot be located exactly but only approximately. In this sense, a matter wave (including light waves) is said to be a **probability wave.** This concept agrees with the famous Heisenberg's uncertainy principle (Heisenbery, 1927), according to which the position of a particle can never be fixed with certainty. It is quantitatively expressed as,

$$\Delta x.\ \Delta p \geq \hbar$$

Δx and Δp are uncertainties in fixing the position and momentum of the particle, respectively, and $\hbar = h/2\pi$ ($h = 6.6\times10^{-34}$ Js). If we absolutely fix the position of the particle, then $\Delta x = 0$. But then Δp becomes infinite, which is not possible. Hence Δx has to be a finite value, which means that there is always an uncertainty in fixing the position of the particle. A matter wave is described by the so-called Schrödinger's equation (Schrödinger, 1926), which essentially involves the wave function ψ.

The concept of a photon as a quantum of energy, the concept of matter wavs and probability waves, the Schrödinger's equation, the Heisenberg's uncertainty principle, etc. formed the foundation of the new branch of physics that we now know as quantum mechanics (or quantum theory or quantum physics or wave mechanics). All this knowledge evolved in solving the problem of dual nature of light. However, we still do not know fully abont the dual nature of radiation as also about the quantum mechanics. But we can draw a rough boundary line where the classical Newtonian mechanics fails and the new quantum mechanics takes over. The classical mechanics works satisfactorily for particles of linear size greater than 10^{-3}mm. For smaller particles like electrons, neutrons, atoms, molecules, etc., only quantum mechanics can describe their dual nature.

Solved Conceptual Problems

Problem 1. Explain the mechanism of photoelectric emission from a metal surface, including the role played by the work function of the metal in it.

Solution. A metal consists of a very large number of free electrons. When monochromatic light (frequency ν) falls on its surface, the photons of energy $E = h\nu$ each collide with them. If during the collisions a photon transfers its entire energy E to a free electron and the electron does not suffer any further collisions with the atoms of the metal, the electron will come out with an energy ($E-\phi$), where ϕ is the work function of the metal (it represents the minimum amount of energy required to remove an electron from the metal). But if it suffers any such collisions, it may still come out, but with energy less than ($E-\phi$), which in the limit can be zero ($E=\phi$). Thus the kinetic energy of an ejected photoelectron can be any value between the minimum zero and the maximum ($E-\phi$), leading to the famous Einstein's photoelectric equation ($E=h\nu$),

$$E_{max} = h\nu - \phi$$

Problem 2. Why the photoemission from a surface only depends on the frequency of the radiation and not on the intensity of the radiation ?

Solution. The process of the photoelectric emission is governed by the Einstein's photoelectric equation,

$$E_{max} = h\nu - \phi,$$

where $E = h\nu$ is the energy of the incident photon and E_{max} is the energy of the ejected photoelectron. ϕ is the work function of the material on which the light is incident. It is seen that the value of E_{max} only depends on the value of ν, the frequency of the incident light, and not at all on the intensity of light.

Problem 3. Why cannot the photoelectric effect be explained on the basis of wave nature of light ?

Solution. (i) According to the wave theory, when light is incident upon a material, its free electrons are set into free oscillations and so become secondary sources of light of the same frequency. If the intensity of light be increased, the amplitude of oscillation and hence the energy of the electrons should increase. Thus, by sufficiently enhancing the light intensity, the electron energy may be made to exceed the work function. Then emission of electrons should take place. But that does not happen. Howsoever strong may be the light source, the photoemission does not occur unless the light's frequency is greater than ν_0, the threshold frequency.

(ii) If the intensity of incident light is continuously increased, the energy of the ejected photoelectrons should also continuously increase. But again it is contrary to observation (maximum energy of the photoelectrons is found to be governed by the Einstein's photoelectric equation, $E_{max} = h\nu - \phi$).

(iii) Calculation shows that if the source of incident light is very weak, it may take several hours for the ejection of even one photoelectron. But the fact is that the emission of photoelectrons is instantaneous.

Problem 4. In photoelectric effect, does each incident photon cause emission of an electron from the metal ? Give reason for your answer.

Solution. No. One reason is that when an incident photon collides with a free electron in the metal, it may not transfer its entire energy (or sufficient energy) to the electron. Secondly, even if such a transfer takes place, the electron may suffer further collisions with the atoms of the metal and so lose energy. Hence it may not be able to come out of the metal.

Problem 5. Explain how a threshold frequency exists for photoemission. How is it independent of light intensity ?

Solution. The threshold frequency (ν_0) is the minimum photon frequency required for the emission of photoelectrons from a material. Its existence is explained by the Einstein's photoelectric equation,

$$E_{max} = h\nu - \phi$$

If the photon frequency has a value ν_0 such that $\phi = h\nu_0$, then the maximum kinetic energy E_{max} of the emitted photoelectron is zero. It means that the photoelectron will just be able to come out of the material. If $\nu < \nu_0$, the photon energy $h\nu < \phi$, the work function. Hence no photoelectron will be emitted.

Any increase in the intensity of incident light only means an increase in the amplitude of oscillations of electric and magnetic field in the light. The frequency of oscillations remains unchanged. Hence the threshold frequency is independent of light intensity.

Problem 6. If it is easier to remove an electron from a metal M_1 than from another metal M_2, explain which one of the two metals has higher threshold frequency for photoemission.

Solution. From the given data, it follows that the work function ϕ_1 for the metal M_1 is less than the work function ϕ_2 for the metal M_2. Since the work function ϕ is related to the threshold frequency ν_0 as,

$$\phi = h\nu_0,$$

it means that the threshold frequency $(\nu_0)_2$ for the metal M_2 is higher than the threshold frequency $(\nu_0)_1$ for the metal M_1.

Problem 7. The work function for potassium is 2.25 eV. Will photoelectric effect occur for it if sodium light of wavelength nearly 589 nm is incident upon it ?

Solution. The threshold wavelength for potassium will be given by,

$$\lambda_0 = \frac{hc}{\phi} = \frac{(6.6\times10^{-34})\times(3\times10^{8})}{(2.25\times1.6\times10^{-19})\ \text{J}}$$

$$= 5.5\times10^{-7}\ \text{m} = 550\ \text{nm}$$

λ_0 is the maximum wavelength for which photoelectric effect can occur. Light of wavelength greater than λ_0 cannot cause photoemission. Hence sodium light of wavelength 589 nm cannot cause photoelectric effect in potassium.

Problem 8. If the intensity of light falling upon a metal surface is enhanced, explain how will it affect the magnitude of the resulting photocurrent and the stopping potential.
Solution. An increase in the intensity of incident light will increase the emission of photoelectrons. Hence it will result in a rise in the photocurrent.

The stopping potential (V_0) is given by,

$$V_0 = (h/e)\nu - (\phi/e)$$

Thus, it only depends on the frequency ν and not at all on the light intensity. Hence any increase in the light intensity will not affect the value of stopping potential.

Problem 9. Is photoelectric effect restricted to metals alone ?
Solution. No. Any material having work function less than the energy of the incident photons will show photoelectric effect.

Solved Numerical Problems

Problem 1. If the wavelength of a spectral line is 5400 Å, what are its frequency and energy values ?
Solution. Given: $\lambda = 5400 \text{ Å} = 5400\times10^{-10}\text{ m} = 54\times10^{-8}\text{ m}$
The frequency of the spectral line is given by ($c = \nu\lambda$),

$$\nu = \frac{c}{\lambda} = \frac{(3\times10^{8})}{(54\times10^{-8})}$$

$$= 5.56\times10^{14}\text{ Hz} \quad \textbf{(Answer).}$$

The corresponding energy is given by,

$$E = h\nu = (6.6\times10^{-34})\times(5.56\times10^{14})$$

$$= 3.67\times10^{-19}\text{ J} \quad \textbf{(Answer).}$$

Problem 2. Monochromatic light of wavelength 5000 Å is emitted by a 30 W source. What is the number of photons emitted in one second ?
Solution. Given: $\lambda = 5000 \text{ Å} = 5\times10^{-7}\text{ m}$; Power = 30 W
The energy of each photon is given by,

$$E_{ph} = \frac{hc}{\lambda} = \frac{(6.6\times10^{-34})\times(3\times10^{8})}{(5\times10^{-7})}$$

$$= 3.96\times10^{-19}\text{ J}$$

The number of photons emitted in 1 second is given by,

$$N_{ph} = \frac{\text{Total energy emitted in 1s}}{\text{Energy of 1 photon}}$$

$$= \frac{(30\text{ W}\times1\text{s})}{(3.96\times10^{-19}\text{ J})}$$

$$= 7.58\times10^{19} \quad \textbf{(Answer).}$$

Problem 3. Sodium has a work function of 2.3 eV. Will it show photoelectric effect for red light of wavelength 7000 Å ?
Solution. Given: $\phi = 2.3$ eV; $\lambda = 7000 \text{ Å} = 7\times10^{-7}$ m
The energy of a photon of red light is,

$$E = \frac{hc}{\lambda} = \frac{(6.6\times10^{-34})\times(3\times10^{8})}{(7\times10^{-7})}$$

$$= 2.83\times10^{-19}\text{ J} = \frac{2.83\times10^{-19}}{(1.6\times10^{-19})}\text{ eV}$$

$$= 1.77\text{ eV}$$

Since this energy is less than the work function of sodium, it follows that red light cannot cause photoemission for sodium. **(Answer).**

Problem 4. The threshold frequency of a metal is f_0. When light of frequency $2f_0$ is incident on the metal plate, the maximum velocity of the electrons emitted is v_1. When the frequency of the incident radiation is increased to $5f_0$, the maximum velocity of the electrons emitted is v_2. Find the ratio v_1 to v_2. (CBSE; 2004).
Solution. According to the Einstein's photoelectric equation, the maximum kinetic energy of an emitted photoelectron is given by,

$$E_{max} = \frac{1}{2}mv^2 = h\nu - \phi$$

The work function ϕ is related to the threshold frequency ν_0 as,

$$\phi = h\nu_0$$

$$\therefore \quad E_{max} = \frac{1}{2}mv^2 = h\nu - h\nu_0$$

In our question, $\nu_0 = f_0$. For light of frequency $2f_0$, we have,

$$\frac{1}{2}mv_1^2 = h.(2f_0) - h.(f_0) = hf_0$$

For light of frequency $5f_0$, we have,

$$\frac{1}{2}mv_2^2 = h.(5f_0) - h.(f_0) = 4hf_0$$

Dividing, we get,

$$\frac{(1/2)\,mv_1^2}{(1/2)\,mv_2^2} = \frac{hf_0}{4\,hf_0} = \frac{1}{4}$$

$$\therefore \quad \frac{v_1}{v_2} = \frac{1}{2} \quad \text{or} \quad v_1 : v_2 = 1 : 2$$ **(Answer).**

Problem 5. A neutron has kinetic energy of 0.05 eV. What is the de Broglie wavelength associated with it ? ($m_n = 1.67\times10^{-27}$ kg).
Solution. Given: $E = 0.05 \text{ eV} = 0.05\times(1.6\times10^{-19}) \text{ J} = 8\times10^{-21}$ J; $m_n = 1.67\times10^{-27}$ kg

We have,

$$E = \frac{1}{2}mv^2 \quad \text{or} \quad v = \sqrt{\frac{2E}{m}}$$

$$\Rightarrow \quad mv = \sqrt{2mE}$$

The de Broglie wavelength is given by,

$$\lambda = \frac{h}{mv} = \frac{h}{\sqrt{2mE}}$$

$$= \frac{6.6\times10^{-34}}{\sqrt{2\times(1.67\times10^{-27})\times(8\times10^{-21})}}$$

$$= \frac{6.6\times10^{-34}}{5.17\times10^{-24}} = 1.28\times10^{-10} \text{ m}$$

$$= 1.28 \text{ Å}$$ **(Answer).**

Problem 6. An α-particle and a proton are accelerated from rest through the same potential difference V. Find the ratio of de Broglie wavelengths associated with them. (CBSE; 2005).
Solution. The k.e. ($=1/2\ mv^2$) gained by a particle of charge q when accelerated through a p.d. of V volt is given by,

$$E = \frac{1}{2}mv^2 = q.V$$

$$\Rightarrow \quad v = \sqrt{\frac{2qV}{m}}$$

The de Broglie wavelength associated with a particle is given by,

$$\lambda = \frac{h}{mv} = \frac{h}{m\sqrt{(2qV)/m}} = \frac{h}{\sqrt{2mqV}}$$

Accordingly, we have,

$$\lambda_\alpha = \frac{h}{\sqrt{2m_\alpha q_\alpha V}}$$

$$\lambda_p = \frac{h}{\sqrt{2m_p q_p V}}$$

$$\therefore \quad \frac{\lambda_\alpha}{\lambda_p} = \sqrt{\frac{m_p q_p}{m_\alpha q_\alpha}}$$

The charge on a proton is e. The charge on an α-particle is 2e and the mass of an α-particle is 4 times the mass of a proton.

$$\therefore \quad \frac{\lambda_\alpha}{\lambda_p} = \sqrt{\frac{m_p.e}{4m_p.2e}} = \frac{1}{\sqrt{8}} = \frac{1}{2\sqrt{2}}$$

Thus, $\lambda_\alpha : \lambda_p = 1 : 2\sqrt{2}$ **(Answer).**

Problem 7. An electromagnetic wave of wavelength λ is incident on a photosensitive surface of negligible work function. If the photoelectrons from this surface have the de Broglie wavelength λ_1, prove that $\lambda = \left(\frac{2mc}{h}\right)\lambda_1^2$. (CBSE; 2008).

Solution. Given: Wavelength of incident photons = λ; de Broglie wavelength of the emitted photoelectrons = λ_1

The energy of each incident photon is given by,

$$E = \frac{hc}{\lambda}$$

Since the photosensitive surface has a negligible value of work function (i.e. $\phi \approx 0$) in the Einstein's photoelectric equation, we have,

$$E_{max} = \frac{hc}{\lambda} - \phi \approx E - 0 \approx E$$

i.e. The energy E_{max} of a photoelectron is almost the same as the energy of an incident photon. Now the de Broglie wavelength associated with a photoelectron is given by,

$$\lambda_1 = \frac{h}{mv} = \frac{h}{p} \quad \text{(p: momentum)}$$

But momentum is given by,

$$p = \sqrt{2mE} = \sqrt{2m \times \frac{hc}{\lambda}}$$

$$\therefore \quad \lambda_1 = \frac{h}{p} = \frac{h}{\sqrt{2m \times (hc/\lambda)}}$$

$$\Rightarrow \quad \lambda_1^2 = \frac{h^2\lambda}{2mhc} = \frac{h\lambda}{2mc}$$

$$\Rightarrow \quad \lambda = \left(\frac{2mc}{h}\right)\lambda_1^2 \quad \textbf{(Proved).}$$

Very Short Answer Questions

[Including HOTS Questions]

(1 mark each)

1. What is meant by the work function of a metal ?

2. What is a photon ? What is its rest mass ?

3. Name a physical phenomenon which supports the particle nature of light.

4. Which photon carries greater energy, one belonging to blue light or the one belonging to red light ?

5. How will the energy of the photons of an electromagnetic radiation be affected, if the wavelength of the radiation is increased by 50% ?

6. What is the linear momentum of a photon of frequency ν ?

7. What is meant by threshold wavelength ? Does it depend on the nature of the photoelectric material exposed to light ?

8. Two metals A and B have work functions 2eV and 5eV, respectively. Which metal has lower threshold wavelength ? (CBSE; 2004).

9. Write Einstein's photoelectric equation.

10. Which factors determine the stopping potential for photoelectrons ?

11. Electrons are emitted from a photosensitive surface when it is illuminated by green light, but electron emission does not take place by yellow light. Will the electrons be emitted when the surface is illuminated by (i) red light and (ii) blue light ? (CBSE; 2005).

12. Out of microwaves, ultraviolet rays and infrared rays, which radiation will be most effective for emission of electrons from a metallic surface ? (CBSE; 1994).

13. If the intensity of radiation incident on a metal surface is reduced to half, how will it affect the kinetic energy of the electrons emitted ?

14. Red light, however bright it is, cannot produce the emission of electrons from a clean zinc surface. But even weak ultraviolet light can do so. Why ? (CBSE; 2004).

15. Show graphically how the stopping potential for a given photosensitive surface varies with the frequency of the incident radiation.

16. What is meant by de Broglie wavelength ?

17. The de Broglie wavelength associated with an electron accelerated through a potential difference V is λ. What will be its wavelength when the accelerating potential is increased to 4V ? (CBSE; 2006).

18. Write de Broglie relation. Are matter waves electromagnetic in nature ?

19. An electron and a proton have the same de Broglie wavelength associated with them. How are their kinetic energies related to each other ? (CBSE; 2008).

20. Why are de Broglie waves associated with a moving football not detectable ?

21. With what purpose was the famous Davisson-Germer experiment with electrons performed ? (CBSE; 2006).

22. Name any two phenomena which together illustrate the dual nature of radiation.

23. The cathode rays cast a sharp shadow of an object placed in their path. What do we conclude from this observation ?

Short Answer Questions

[Including HOTS Questions]

(2 or 3 marks each)

1. Briefly explain the working of a photoelectric cell. Give its two uses.

2. Derive a relation between stopping potential and the frequency of incident radiation for a metal surface.

3. Show graphically the variation of stopping potential with the frequency of incident radiation for a metal surface. How can the value of Planck constant (h) be determined from this graph ?

4. Derive the de Broglie relation for matter waves.

5. Describe the Davisson and Germer experiment, to establish the wave nature of electrons. Draw a labelled diagram of the apparatus used. (CBSE; 2003).

6. Show graphically the variation of photocurrent (i) with anode potential (V) for three different values of intensity (I) of incident light. How does it lead to the concept of stopping potential ?

7. State any four important properties of photons.

Numerical Exercises

1. 2.2×10^{-3} W of power is spent by a laser source in emitting monochromatic light of frequency 5×10^{14} Hz. Calculate the energy of a photon in the light produced and the number of photons produced per second by the source. [Hint: N = Power ÷ Energy of a photon]. **[3.3×10^{-19} J; 6.7×10^{15}]**

2. Ultraviolet light of wavelength 2271 Å from a 100 W mercury source radiates a photocell made of molybdenum metal. If the stopping potential is 1.3 V, estimate the work function of the metal. How would the photocell respond to high intensity (10^5 W.m^{-2}) red light of wavelength 6328 Å produced by a He–Ne laser ? (CBSE; 2005).
[$\phi = 6.64\times10^{-19}$ J; $\nu_0 = 1.01\times10^{15}$ Hz; $\nu_{red} = 4.74\times10^{10}$ Hz; no photoemission, since $\nu_{red} \ll \nu_0$]

3. The stopping potential for a photosensitive material is 0.5 V. If the work function of the material is 2.2 eV, determine the threshold frequency for the material and the wavelength of the incident light.
[$\nu_0 = 4.85\times10^{14}$ Hz; λ = 240 nm]

4. A silver surface is radiated by monochromatic ultraviolet light of wavelength 2000 Å. If the threshold wavelength for silver is 2640 Å, find the maximum energy of the emitted photoelectrons. [Hint: $E_{max} = hc\left(\frac{1}{\lambda}-\frac{1}{\lambda_0}\right)$]. **[1.5 eV]**

5. If photoelectrons are to be emitted from a potassium surface with a speed of 6×10^{-6} ms^{-1}, what frequency of incident radiation must be used ? Threshold frequency of potassium is 4.22×10^{14} Hz. (CBSE sample paper; 1997). **[6.7×10^{14} Hz]**

6. Calculate the values of de Broglie wavelength associated with an electron and a ball weighing 200 g, moving with speeds of 6×10^6 ms^{-1} and 20 ms^{-1}, respectively. **[1.2×10^{-10} m and 1.7×10^{-34} m]**

7. If the value of kinetic energy happens to be the same for a proton, an electron and an alpha-particle, find which of the three will have shortest value of de Broglie wavelength associated with it. **[α-particle]**

8. The ratio of the de Broglie wavelength associated with a charged particle to that associated with an electron is 1.813×10^{-4}. If the particle travels three times faster than the electron, find its mass. Also identify the particle. **[1.67×10^{-27} kg; proton]**

9. The sunlight falling on the earth's surface has an energy flux of 1.4×10^3 $W.m^2$. If the average wavelength of the photons in this light is 5500 Å, find the number of photons reaching $1m^2$ of the earth's surface in one second. [Hint: No. of photons per second per m^2 = Energy flux ÷ Energy of one photon]. **[3.9×10^{21}]**

UNIT 8
ATOMS & NUCLEI

CHAPTER 8 (A)
Atoms; Composition & Properties of Nucleus

CHAPTER 8 (B)
Radioactivity; Nuclear Fission & Fusion

CHAPTER 8 (A)
ATOMS; COMPOSITION & PROPERTIES OF NUCLEUS

In the earlier classes, you have broadly known about the structure of atom. In this unit. We shall describe it in greater detail. Besides, we shall specifically deal with the atomic nucleus, the core of an atom, and the important phenomena associated with it, viz. radioactivity, nuclear fission, and nuclear fusion.

The concept of an atom was first arrived at around 1810 by the famous British Chemist John Dalton, mainly from experiments on the expansion of gases by heat. The atoms were proposed to be indivisible. All matter in the universe is believed to be made of atoms of various kinds of elements. An atom is the smallest unit of an element, which carries all the properties of the element. The existence of negatively charged electrons in an atom was convincingly demonstrated by discharge tube and other experiments (the discovery of electron in 1897 is attributed to J.J. Thomson). Since an atom had to be electrically neutral, it followed that an atom must also contain an amount of positive charge equal to the negative charge of the electrons. Thus a search began for the +ve charge and its distribution inside an atom.

8A.01. Thomson's and Lenard's Models of Atom

In 1898, J.J. Thomson proposed that an atom was a +vely charged soft sphere, with the electrons embeded in it like seeds in a watermelon or fruits in a cake (Fig. 8.1 a). The +ve charge was supposed to be uniformly distributed over the volume of the sphere. This model could successfully explain a number of physical phenomena known at that time, e.g. ejection of the –vely charged electrons alone upon heating certain metals. Later, in 1903, based on the observation that cathode rays (which consist of electrons) passed through small thicknesses of materials almost unbent, Lenard concluded that an atom must have a large amount of empty space in it. Accordingly, he proposed that like the electrons the +ve charge was also distributed inside the atom in the form of tiny particles. However, this model met with the problem that why upon heating a metal the +ve charge was not ejected alongwith the –ve charge.

8A.02. Rutherford's Model of Atom; α-Particles Scattering Experiments by Geiger and Marsden; Discovery of atomic Nucleus

In 1911, the British physicist Ernest Rutherford proposed the right model of the atom. The model assumes that the +ve charge is concentrated at the centre of the atom, called **nucleus**. The nucleus carries almost all the mass of the atom. The –vely charged electrons revolve around the nucleus in different circular orbits, like the planets revolving around the sun. The +ve charge on the nucleus is exactly equal to the total –ve charge on the electrons, so that the atom as a whole is electrically neutral. The space between the nucleus and the electrons is empty. Thus, an atom mostly consists of empty space. The Fig. 8.1 (b) depicts the Rutherford's model for an atom with two electrons.

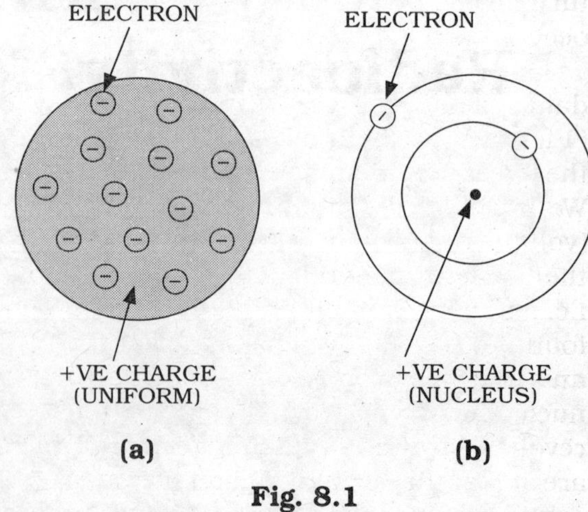

Fig. 8.1

The Rutherford's model was based on experiments involving bombardment of matter with α-particles. The experiments were carried out in 1911 by Rutherford's collaborators, Geiger and Marsden (the latter was a mere 20-year old student, studying for his bachelor's degree). The experimental set-up is shown in Fig. 8.2. A radioactive source, S, of α-particles is enclosed in a lead cavity (you will read about radioactivity in the next chapter; lead is chosen because it strongly absorbs radioactive radiations). A collimated beam of α-particles comes out from the narrow opening of the cavity. It is further collimated by passing through narrow openings between lead bricks and falls on a very thin gold foil (thickness ~10^{-7} m). The gold atoms scatter the beam in various directions. The scattered α-particles are detected by a rotatable detector, which consists of a zinc sulphide (ZnS) screen and a microscope. ZnS is a fluorescent material. When the α-particles strike it, bright green flashes of light (scintillations) are produced. The scintillations produced at different angles can be counted through the microscope.

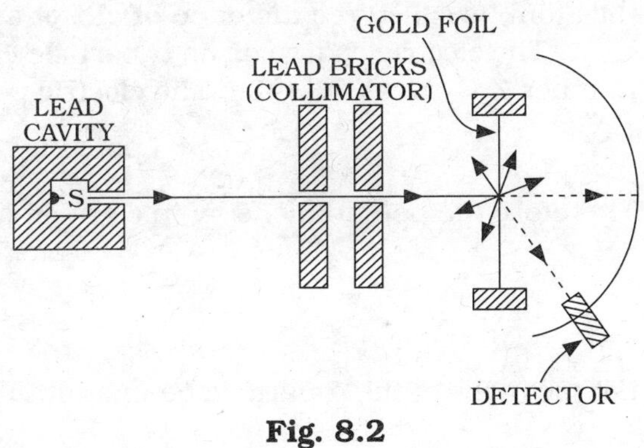

Fig. 8.2

An α-particle is a helium atom that has lost both of its electrons. So it has 2 units of +ve charge and has almost the same mass as that of a helium atom. After striking the gold foil, the α-particles were found to be scattered through different angles θ, as shown in Fig. 8.3. (The figure is not drawn to scale. For the sake of clarity, the size of the nucleus has been shown very much larger than its actual size). Most of them passed undeviated or with a small deviation (θ<π/2; Fig. a, c, d). But a few deflected by large angles (θ>π/2; Figs. b, e), i.e. they were turned back. Their proportion was only about 1 in 8000. A plot of the number of particles (N) deflected through different angles θ is shown in Fig. 8.4. In our considerations here, we assume that because of the extremely small thickness of the gold foil, on an average an α-particle meets only one gold atom in its path.

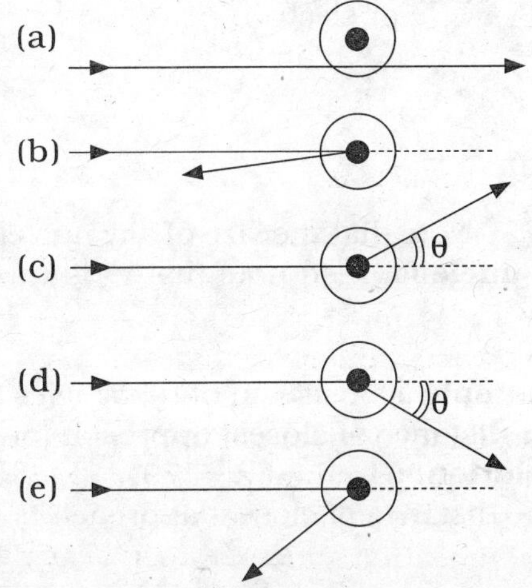

Fig. 8.3

The result that some α-particles, although in a small number, were deflected back by the atoms of the gold foil, was most unexpected. According to the Thomson model, the incident α-particles could only meet resistance from the electrons present in a gold atom. But an α-particle is 7,350 times heavier than an electron. Therefore it could easily brush aside a tiny electron coming in its way. The same would be the case for any similar tiny +ve charges, proposed in the Lenard model. Thus, all the α-particles must pass undeviated through the foil. But it is contrary to observation. It follows that the α-particle must have struck a very heavy particle in its path. The mass of the particle must have been almost equal to the mass of the gold atom. Further, since a very high majority of the α-particles were found to pass undeflected through the gold foil (only 1 in 8000 turned back), it also follows that most of the space in an atom is empty. This in turn means that the said heavy particle (later called as 'nucleus') must occupy a very small volume.

A quantitative analysis could be made from the scattering data. The linear size of the 'nucleus' was found to be nearly 10 fermi (1 fermi = 1 femtometre = 10^{-15} metre). This value is nearly 10^{-5} times the linear size (diametre) of the atom in which the nucleus is contained. We know that volume is proportional to the cube of the linear size [volume of a sphere = $(4/3)\pi r^3$; r:radius]. It means that the volume of the 'nucleus' is about $(10^{-5})^3 = 10^{-15}$ times the volume of the atom, i.e. a tiny fraction of the volume of the atom. Further, the nucleus was found to have a +ve charge. Accordingly, the **Rutherford's model of an atom** was proposed as consisting of a top heavy positively charged nucleus, carrying virtually all the mass of the atom, with electrons revolving around it, like planets revolving around the sun. The electrons had to be mobile, because if they are at rest they would fall into the nucleus due to the strong Coulomb attraction. Their movement in circular orbits provides the centripetal force necessary to balance the Coulomb attraction.

Fig. 8.4

Distance of closest approach (r_0); Size of nucleus. Suppose an α-particle is approaching an atom. Both the α-particle and the nucleus of the atom are +vely charged. Therefore they repel each other. The Coulomb force of repulsion fast increases as the α-particle comes near to the atom. Finally, the repulsive force becomes so strong that the particle cannot approach the atom any further. Then it moves away from the atom. "The distance of the nucleus of an atom from the point beyond which a moving α-particle cannot further approach the atom" is called the **distance of closet approach** of the α-particle. It is usually denoted as r_0.

The electric charge on an α-particle is equal to (+) 2e. The charge on the nucleus of an atom of atomic number Z is equal to (+)Ze. The electric potential due to the nucleus at a distance r_0 is,

$$\text{potential} = \frac{Ze}{(4\pi\epsilon_0)r_0}$$

Therefore the potential energy (p.e.) of an α-particle at the distance of closest approach (r_0) is,

$$\text{p.e.} = \text{potential} \times \text{charge}$$

$$= \frac{Ze}{(4\pi\epsilon_0)r_0} \times 2e = \frac{2Ze^2}{(4\pi\epsilon_0)r_0}$$

Let the mass of the α-particle be m and its velocity be v. Then its kinetic energy (k.e.) is,

$$\text{k.e.} = \frac{1}{2}mv^2$$

As the α-particle approaches the nucleus, its velocity and therefore its k.e. decreases and its p.e. increases. At the distance r_0 from the nucleus, its entire k.e. is converted into p.e. Therefore at the distance of closest approach (r_0) we have,

$$\frac{1}{2}mv^2 = \frac{2Ze^2}{(4\pi\epsilon_0)r_0}$$

$$\Rightarrow \qquad r_0 = \frac{4Ze^2}{(4\pi\epsilon_0)mv^2} \qquad \text{..... (8.1)}$$

Note that the size of any nucleus must necessarily be less than r_0. Thus r_0 gives an estimate of the **size of nucleus.** Also note that r_0 is inversely proportional to the kinetic energy of the α-particle [k.e. = (1/2) mv^2].

Example 1. A stream of α-particles of kinetic energy 5 MeV are incident head-on upon a gold foil. Calculate the distance of closest approach for the particles (Z_{gold} = 79).

Solution. Given: Z = 79; k.e. = 5 MeV = 5×10^6 eV = $5 \times 10^6 \times (1.6 \times 10^{-19})$ J

The distance of closest approach is given by,

$$r_0 = \frac{1}{4\pi\epsilon_0} \cdot \frac{2Ze^2}{(\text{k.e.})}$$

$$= (9 \times 10^9) \cdot \frac{2 \times (79) \times (1.6 \times 10^{-19})^2}{[5 \times 10^6 \times (1.6 \times 10^{-19})]}$$

$$= 45.5 \times 10^{-15} \text{ m (or 45.5 fm)}$$ **(Answer).**

Impact parameter: The **impact parameter** (b) is the "perpendicular distance between the velocity vector of an α-particle and the central line of the nucleus, when the particle is far away from the nucleus" (Fig. 8.5).

The path followed by an α-particle approaching the nucleus depends both on the distance of closest approach (r_0) and the impact parameter (b). The force experienced by the α-particle due to the nucleus varies as 1/(distance)^2, i.e. it has an inverse square law character. Hence for large values of b the α-particle passes almost undeviated and for small b's it suffers large scattering. For a head-on collision, b≈0. Then the particle rebounds, like a ball thrown against a wall.

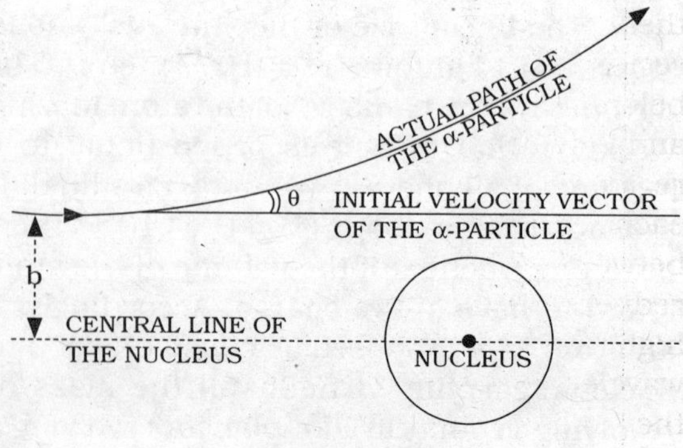

Fig. 8.5

The relation between b and the angle of scattering θ was analytically found by Rutherford as,

$$b = \frac{Ze^2}{(4\pi\epsilon_0)[(1/2)mv^2]} \cot(\theta/2) \qquad \text{..... (8.2)}$$

(1/2) mv^2 is the k.e. of the α-particle. It is seen that larger is the value of b (i.e. more distant is the α-particle from the nucleus), smaller is the value of θ, the angle of scattering . Further, if the velocity is high, i.e. the α-particle has a high k.e., b can be small.

Example 2. An alpha-particle of energy 6 MeV is scattered through an angle of 20°, while approaching the nucleus of a gold atom. Find the impact parameter for the particle ($Z_{gold} = 79$).

Solution. Given: $Z = 79$; $\theta = 20°$; k.e. = 6 MeV=6×10^6 eV = $6\times10^6\times(1.6\times10^{-19})$ J

The impact parameter is given by,

$$b = \frac{1}{4\pi\epsilon_0}\cdot\frac{Ze^2\cot(\theta/2)}{\text{k.e.}} = \frac{1}{4\pi\epsilon_0}\cdot\frac{Ze^2\cot 10°}{\text{k.e.}}$$

$$= \frac{(9\times10^9)\times(79)\times(1.6\times10^{-19})^2\times(5.672)}{[6\times10^6\times(1.6\times10^{-19})]}$$

$$= 107.5\times10^{-15}\text{ m (or 107.5 fm)}$$ **(Answer).**

8A.03. Atomic Spectra; Emission and Absorption Spectra

Although the Rutherford's model of atom was basically sound, it faced some serious shortcomings. Before we discuss about them, it is necessary that the student acquires some necessary elementary knowledge of the nature and formation of atomic spectra.

The student has earlier learnt how a light spectrum (VIBGYOR) is formed by the dispersion of light through a prism (Chapter 6A; Section 6A.05). If the incident light from the source is suitably guided through a very fine slit before falling upon the prism and the emergent light is focussed by a telescope on a screen (as done in a spectrometer; vide Section 6A.06), each wavelength present in the light source forms a line image of the slit at a well-defined position on the screen.

If the light so[illegible] happens to be a white light source (e.g. sunlight, an electric lamp, etc.), consisting of innumerable wavelengths, the spectrum obtained upon the screen consists of a set of extremely close parallel bright lines of different colours. Because of their closeness, they cannot be seen separately from each other. They merge together in a continuous manner. "Such a spectrum, which consists of a continuous range of wavelengths with no sharp boundaries between them", is called a **continuous spectrum.** It looks like a rainbow spectrum, with continuous variation of colour (VIBGYOR).

When a material is heated to a high temperature, it emits electromagnetic radiation, consisting of different wavelengths. The radiation may consist of only wavelengths in the visible region (VIBGYOR) or also in the infrared and/or ultraviolet region. The strongly heated filament of an electric lamp or a white hot iron piece emit all the wavelengths in the visible region and so produce a continuous spectrum in the visible region (VIBGYOR). But generally the light emitted by an element does not consist of a continuous range of wavelength, but only some selected wavelengths (the light may be produced by enclosing the element in a sealed tube at low pressure and vapourising it by passing an electric current through it; gaseous elements like hydrogen, nitrogen, etc. need no vapourisation). Such "a spectrum consisting of lines located at specified positions and of specified intensities" is called a **line spectrum**. Each element has its characteristic spectrum, which acts like a finger print of the element and thus helps to unambiguously identify the element. As an example, the line spectrum of hydrogen is diagrammatically shown in Fig. 8.6. It consists of wavelengths in the visible region (approximately 400 nm to 750 nm), as well as in the ultraviolet and infrared regions. Note that the lines belong to three separate groups, known as Paschen series (infrared region), Balmer series (visible region), and Lyman series (ultraviolet region). Further note that in each series the separation between the spectral lines reduces on moving from the higher wavelength to lower wavelength side. Towards the end of the series the lines become so close that they merge into each other (shaded regions), as in a continuous spectrum.

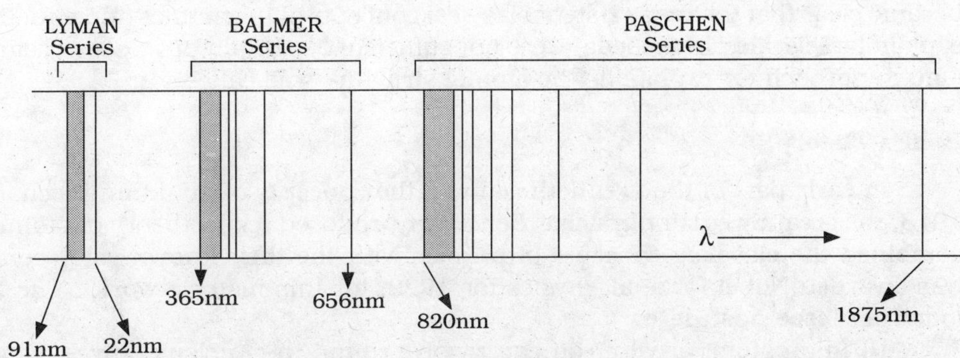

Fig. 8.6

The wavelengths to which the observed lines correspond are represented by the following equation,

$$\frac{1}{\lambda} = R\left(\frac{1}{n^2} - \frac{1}{m^2}\right) \qquad \text{..... (8.3)}$$

n and m are integers, with m>n; R = 1.09737×10^7 m^{-1} (it is called Rydberg constant). The Lyman series is obtained by putting n=1 and m=2,3,4,.........., the Balmer series by putting n=2 and varying m from 3 onwards, and the Paschen series by putting n=3 and varying m from 4 onwards.

Emission and Absorption Spectra

"The spectrum produced by light coming directly from a source" is called **emission spectrum** of the source. Thus the hydrogen line spectrum illustrated in Fig. 8.6 is an emission spectrum. Same for the continuous spectra produced by sunlight, an electric lamp, etc. An emission spectrum consists of bright lines superimposed upon a dark background (the student should not be misled by Fig. 8.6, which shows just the opposite; it is so because it is a print made from the original, which reverses the colour shade).

Conversely, if light from a white light source passes through a gas or the vapour of an element, the spectrum produced has dark lines superimposed upon a bright background. The dark lines precisely occupy the same positions as the bright lines do in the emission spectrum of the element. It is so because these white light wavelengths (or frequencies) which correspond to the emission lines are used up, or say 'absorbed', in causing excitations in the atoms of the element, when the white light passes through the vapour (or gas) of the element. "A spectrum produced when white light passes through the vapour (or gas) of an element" is called an **absorption spectrum.**

8A.04. Bohr's Theory; Energy Levels; Hydrogen Spectrum

As mentioned earlier, gradually serious shortcomings came to be noticed in the Rutherford's model of atom. The fundamental idea of electrons revolving round the nucleus, like a sun-planet system, was found to be sound, but two serious problems were seen. In the sun-planet system, the necessary centripetal force for the revolution of the planet is provided by the gravitational force between the sun and the planet. In the atom, this force for the revolution of an electron is provided by the Coulomb force of attraction between the +vely charged nucleus and the –vely charged electron. Now, an object moving in a circle, and so constantly experiencing a force, is constantly under acceleration (force = mass×acceleration). According to the electromagnetic theory, an accelerated charged particle emits radiation in the form of electromagnetic waves. Thus it should continuously lose energy, with the result that it should continuously spiral inwards and finally fall into the nucleus (Fig. 8.7). It follows that such an atom cannot be stable. But that is contrary to observation; the atoms are found to be perfectly stable. Secondly, the e.m. theory also tells us that the frequency of the emitted e.m. waves is the same as the frequency of revolution of the electron. If the electron spirals inwards, its angular velocity and so its frequency of revolution will constantly change. Therefore, in turn, the frequency of the waves will also change continuously. It means that a continuous spectrum of wavelengths will be emitted. But that is again in disagreement with observation. The element hydrogen was found to have a line spectrum, consisting of discrete wavelengths (Fig. 8.6). It became clear that the then existent classical concepts of mechanics and e.m. theory, which could successfully explain large scale phenomena, were not sufficient to explain processes occuring at atomic scale. Therefore, a new approach for explaining the atomic structure was called for.

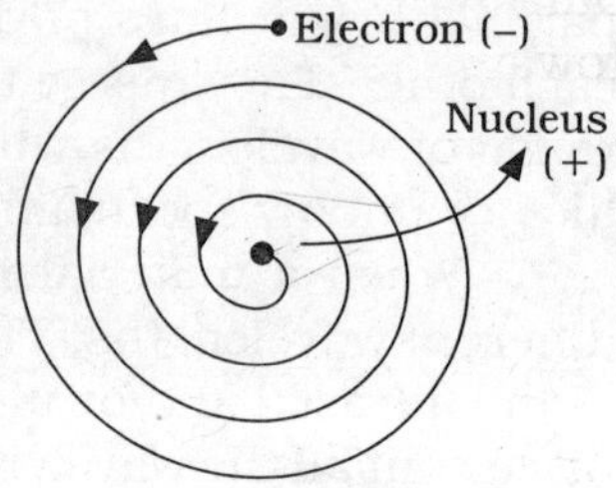

Fig. 8.7

Bohr's Theory

In early part of the twentieth century, the concepts of quantum mechanics had started developing. In 1913, the great Swedish physicist Niels Bohr proposed a new theory of atomic structure, which intelligently combined the classical concepts of physics with the then known early concepts of quantum physics (he was awarded Nobel Prize in Physics for 1922, for this historic work). The **Bohr's theory** is based on the following **three postulates:**

(i) In an atom, an electron can revolve round the nucleus only in some fixed stable orbits. In these orbits, the electron does not radiate energy (contrary to the prediction of e.m. theory). The orbits are called **stationary orbits.** They are assumed to be circular orbits.

(ii) In these stationary orbits, the angular momentum (L) of the electron about the nucleus is an integral multiple of $h/2\pi$ (h: Planck constant; $h = 6.6 \times 10^{-34}$ J.s). (Thus, this postulate defines each stationary orbit, by associating a definite value of energy with it).

$$L = \frac{nh}{2\pi} \ (n = 1,2,3,\ldots\ldots) \qquad \ldots\ldots (8.4)$$

(It means that the angular momentum of the electron can only have some discrete 'quantized' values, not all possible values. Accordingly, this postulate is known as 'Bohr's quantization rule'). 'n' is called **principal quantum number** of the orbit.

(iii) The electron can jump from a stationary orbit of higher energy E_2 to another stationary orbit of lower energy E_1. In doing so, a photon of energy $(E_2 - E_1)$ is emitted. The frequency (and wavelength) of the emitted photon is given by (energy = $h\nu$ and $\nu\lambda = c$),

$$h\nu = \frac{hc}{\lambda} = (E_2 - E_1) \qquad \ldots\ldots (8.5)$$

Energy of Electron in an Orbit; Frequency of Revolution and Orbital Period

We shall now use the Bohr's postulates to find the energy of an electron in different stationary orbits. Suppose the nucleus has Z protons, and so it has a +ve charge Ze, and an electron (of mass m and –ve charge e) revolves around it with a constant velocity v in a circular orbit of radius r. The magnitude of the Coulombian force of attraction experienced by the electron is given by,

$$F = \frac{(Ze).(e)}{4\pi\epsilon_0 r^2} = \frac{Ze^2}{4\pi\epsilon_0 r^2}$$

This force is balanced by the centripetal force (= mass×acceleration) experienced by the electron. From our knowledge of mechanics, we know that the magnitude of the acceleration of the electron is v^2/r, directed towards the centre of the orbit. Therefore, we have,

$$\frac{Ze^2}{4\pi\epsilon_0 r^2} = \frac{mv^2}{r} \qquad \ldots\ldots [A]$$

$$\Rightarrow \quad r = \frac{Ze^2}{4\pi\epsilon_0 mv^2} \qquad \ldots\ldots (8.6\ a)$$

and

$$v = \sqrt{\frac{Ze^2}{4\pi\epsilon_0 mr}} \qquad \ldots\ldots (8.6\ b)$$

The angular momentum of the electron is mvr. From Eq. 8.4 (Bohr's quantization rule), we have,

$$mvr = \frac{nh}{2\pi} \qquad \ldots\ldots [B]$$

Substituting for r from Eq. 8.6a, we get,

$$v = \frac{Ze^2}{2\epsilon_0 nh} \qquad \ldots\ldots (8.6\ c)$$

Each value of n represents an allowed orbit for the electron (n = 1,2,3,..............). The **kinetic energy** of the electron in the nth orbit is given by,

$$K = \frac{1}{2}mv^2 = \frac{mZ^2e^4}{8\epsilon_0^2 n^2h^2} \qquad \ldots\ldots (8.7\ a)$$

The **potential energy** of the electron is given by [p.e. = potential×charge = $(Ze/4\pi\epsilon_0 r)\times(-e)$; the potential of the electron is due to the nuclear charge Ze],

$$V = -\frac{Ze^2}{4\pi\epsilon_0 r} = -\frac{mZ^2e^4}{4\epsilon_0^2 n^2h^2} \qquad \ldots\ldots (8.8)$$

Therefore, the **total energy** of the electron in the nth orbit is given by,

$$E_n = K + V$$

$$\Rightarrow \quad E_n = -\frac{mZ^2e^4}{8\epsilon_0^2 n^2h^2} \qquad \ldots\ldots (8.9\ a)$$

Using Eq. 8.6a, the k.e. can also be written as,

$$K = \frac{1}{2}mv^2 = \frac{Ze^2}{8\pi\epsilon_0 r} \qquad \ldots\ldots (8.7\ b)$$

Accordingly, the total energy can also be written as (using Eq. 8.8),

$$E_n = K + V = \frac{Ze^2}{8\pi\epsilon_0 r} - \frac{Ze^2}{4\pi\epsilon_0 r} = -\frac{Ze^2}{8\pi\epsilon_0 r} \quad \text{..... (8.9 b)}$$

The **frequency of revolution** of an electron in its orbit is given by (ν is no. of revolutions per second and the distance covered in one revolution is $2\pi r$),

$$\nu = \frac{v}{2\pi r} \quad \text{..... (8.6 d)}$$

Correspondingly, the **time period of revolution (orbital period)** is given by,

$$T = \frac{1}{\nu} = \frac{2\pi r}{v} \quad \text{..... (8.6 e)}$$

Bohr Radius; Radii of Allowed Orbits

Substituting for v from Eq. 8.6c in Eq. 8.6a, and now writing r_n in place of r, we get,

$$r_n = \frac{\epsilon_0 n^2 h^2}{\pi m Z e^2} \quad \text{..... (8.10 a)}$$

The allowed orbits for the electron are given by various values of n (=1,2,3,............). Thus, it is found that the allowed orbits have their radii proportional to n^2. The radius of the first orbit (n = 1) is given by,

$$r_1 = \frac{\epsilon_0 h^2}{\pi m Z e^2}$$

For the simplest atom, hydrogen, Z = 1. Substituting in the above equation and putting the values of various constants (h,m, etc.), we get,

r_1 (hydrogen) = a_0 (say) = 0.053 nm (or 53 pm)

[pm denotes picometre; 1 pm = 10^{-12} m]. "The length represented by the radius of the first Bohr orbit of a hydrogen atom" is called **Bohr radius**. It is denoted by the symbol a_0 (= 53 pm or 0.053 nm).

From the Eq. (8.10a), it is seen that the radii of the various allowed orbits for a hydrogen atom are a_0, $4a_0$, $9a_0$, $16a_0$,............. In general, the radius of the nth orbit for a hydrogen atom is given by,

$$r_n = n^2 a_0 \text{ (hydrogen atom)} \quad \text{..... (8.10 b)}$$

For a hydrogen-like ion, having Z proton in the nucleus and one electron revolving round the nucleus (e.g. He^+, Li^{++}, etc.), we have,

$$r_n = \frac{n^2 a_0}{Z} \text{ (hydrogen-like ion)} \quad \text{..... (8.10 c)}$$

Energy Levels (Hydrogen Atom); Ground and Excited States

We have seen that the total energy of the electron in the nth orbit is given by (Eq. 8.9a),

$$E_n = -\frac{mZ^2 e^4}{8\epsilon_0 n^2 h^2} \quad \text{..... (8.9 a)}$$

The Bohr's theory was essentially developed for the simplest atom, viz. a hydrogen atom, which has just one proton in the nucleus and just one revolving electron. It is also applicable to hydrogen-like ions (He^+, Li^{++}, Be^{+++}, etc.), which also have just one revolving electron, but Z (>1) protons in the nucleus. So, now onwards, our treatment of the Bohr's theory will be restricted to the **hydrogen atom** (Z = 1). From the Eq. 8.9a, it is seen that the energy of the electron is inversely proportional to n^2,

$$E_n \propto \frac{1}{n^2}$$

The energy of the electron in the Ist orbit (n=1) of a hydrogen atom (Z = 1) is (after substituting the values of the constants involved in Eq. 8.9a),

$$E_1 = -\frac{m e^4}{8\epsilon_0 h^2} = -13.6 \text{ eV}$$

For the higher orbits, the energy values are given by,

$$E_n = \frac{E_1}{n^2}$$

Thus, for the 2nd, 3rd, 4th, etc., orbits the energy values are found as,

$$E_2 = -\frac{13.6}{(2)^2} = -\frac{13.6}{4} = -3.4 \text{ eV}$$

$$E_3 = -\frac{13.6}{9} = -1.5 \text{ eV}$$

$$E_4 = -\frac{13.6}{16} = -0.85 \text{ eV, and so on.}$$

The –ve sign for energy arises from the electrostatic (Coulombian) attraction of the electron to the nucleus, which binds the electron to the nucleus. It means that if the electron has to be removed from the atom, necessary energy will have to be supplied from outside, e.g. by heating, through electric discharge, etc.

Note that because of the –ve sign the innermost orbit (corresponding to E_1) has lowest energy, viz. –13.6 eV. The higher orbits have progressively higher values, viz. – 3.4 eV, – 1.5 eV, etc., but at the same time it must be understood that the electron in the first orbit is most strongly bound to the nucleus as compared to the electrons in higher orbits, because highest magnitude of energy (13.6 eV) is needed to remove it from the atom.

When an atom has lowest energy, it is said to be in its **ground state.** The states of the atom with higher energies are called **excited states** of the atom. Thus, a hydrogen atom has energy –13.6 eV in its ground state and energies –3.4 eV, –1.5 eV, etc., in the first, second, etc., excited states.

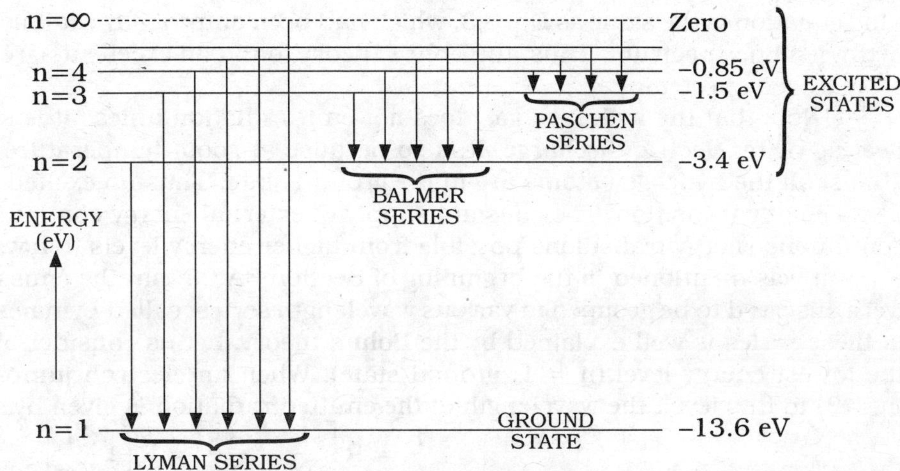

Fig. 8.8

The energy values corresponding to the various stationary orbits can also be represented as **energy levels,** drawn as parallel horizontal lines on a suitable energy scale. The collection of such energy levels is called an **energy level diagram**. The energy level diagram for a hydrogen atom is illustrated in Fig. 8.8. The energy levels for the ground state (n = 1) and four excited states (n = 2,3,4, and ∞), with their corresponding energy values, have been indicated in the diagram. The energy values when n→∞, i.e. when the nucleus and the electron are widely separated, are very nearly zero; for n=∞, the energy becomes exactly zero. The groups of parallel vertical lines, marked as Lyman series, etc., in the diagram are explained as follows.

Hydrogen Spectrum; Spectral Series

We have seen that the higher is the orbit, the higher is the energy of the electron. If the electron jumps from a higher mth orbit to a lower nth orbit (m>n), its energy reduces from E_m to E_n. The difference of energy is emitted as a photon of electromagnetic radiation. The wavelength of the radiation is given by (energy = $h\nu$ and $\nu\lambda = c$),

$$h\nu = \frac{hc}{\lambda} = (E_m - E_n)$$

$$\Rightarrow \quad \frac{1}{\lambda} = \frac{(E_m - E_n)}{hc}$$

$$= \frac{mZ^2e^4}{8\epsilon_0^2 h^3 c}\left(\frac{1}{n^2} - \frac{1}{m^2}\right) \quad \text{(from Eq. 8.9)}$$

Or

$$\frac{1}{\lambda} = RZ^2\left(\frac{1}{n^2} - \frac{1}{m^2}\right) \qquad \text{.....(8.11)}$$

where,

$$R = \frac{me^4}{8\epsilon_0^2 h^3 c} \qquad \text{..... (8.12)}$$

[R is known as **Rydberg constant**. On substituting the values of the quantities m,e,h, etc., its value is found to be nearly 1.1×10^7 m^{-1}. The energy of the electron in the nth state can be expressed in terms of R as,

$$E_n = -\frac{RhcZ^2}{n^2} \quad \text{..... (8.13 a)}$$

Note that,

$$Rhc = \frac{me^4}{8\epsilon_0 h^2} = E_1 = 13.6 \text{ eV} \quad \text{..... (8.14)}$$

The value Rhc (= 13.6 eV) is known as one **rydberg** of energy.

Thus, the Eq. (8.13a) can also be written as,

$$E_n = -\frac{13.6\ Z^2}{n^2} \text{ eV} \quad \text{..... (8.13 b)}$$

The Eq. 8.11 gives the possible wavelengths of the radiation emitted, for different possible values of m and n. For a hydrogen atom (Z = 1), it reduces to,

$$\frac{1}{\lambda} = R\left(\frac{1}{n^2} - \frac{1}{m^2}\right)$$

This equation is the same as Eq. 8.3, which had been empirically established for the experimentally observed hydrogen line spectrum. Thus, the Bohr's theory found an excellent agreement with experiment, for the case of hydrogen spectrum.

[Note that the hydrogen gas does not emit radiation unless it is supplied with enough energy, e.g. by heating or by electric discharge. It is so because at room temperature and relatively lower temperatures, almost all the hydrogen atoms are in the ground state. Thus no excited states exist and therefore there can be no energy transition. The constant supply of external energy continuously creates excited states, making continuous energy transitions possible from higher energy levels to lower energy levels.].

It was mentioned in the beginning of Section 8A.03 that the emission lines in the hydrogen spectrum were observed to be grouped in various wavelength series, called Lyman series, Balmer series, etc. The origin of these series is well explained by the Bohr's theory. Let us consider all the possible energy transitions to the lowest energy level (n = 1; ground state). When an electron jumps from the next higher energy level (n = 2) to this level, the wavelength of the emitted radiation is given by,

$$\frac{1}{\lambda} = R\left[\frac{1}{(1)^2} - \frac{1}{(2)^2}\right] = R\left(1 - \frac{1}{4}\right) = \frac{3R}{4}$$

Or $\lambda = 121.6$ nm

If the electron jumps from the highest level (n = ∞) to the level n = 1, the emitted wavelength will be given by,

$$\frac{1}{\lambda} = R\left(\frac{1}{1} - \frac{1}{(\infty)^2}\right) = R\,(1 - 0) = R$$

Or $\lambda = 91.2$ nm

Thus, all the possible energy transitions to the energy level n = 1 lead to the emission of a series of discrete wavelengths within the range 121.6 nm to 19.2 nm (ultraviolet region). This spectral series is termed as **Lyman Series**.

In a similar way, the possible energy transitions to the energy level n = 2 give rise to the emission of radiation in the wavelength range 656.3 nm to 365.0 nm (visible region). This wavelength range is known as **Balmer series.** The transitions to the energy leve! n = 3 similarly lead to the wavelength series ranging from 1875 nm to 822 nm (infrared region), known as **Paschen series.** Similarly one gets more series of wavelengths in the optical region, for possible transitions to higher energy levels, known as Brackett series and Pound series (not shown in Fig. 8.8). Thus, the existence of the various observed spectral series in the line spectrum of hydrogen is satisfactorily accounted for by the Bohr's theory.

Example 3. Determine the radius of the second Bohr orbit of a hydrogen atom. Also determine the energy of the electron in this orbit.

Solution. Given: n = 2; Z = 1 (for hydrogen)

The radius of the nth orbit is given by,

$$r_n = \frac{Ze^2}{4\pi\epsilon_0\, mv^2} = \frac{e^2}{4\pi\epsilon_0\, mv^2}$$

The value of v is found as,

$$v = \frac{Ze^2}{2\epsilon_0\, nh} = \frac{(1)\times(1.6\times10^{-19})^2}{2\times(8.85\times10^{-12})\times(2)\times(6.6\times10^{-34})}$$

$$= 11\times10^5 \text{ ms}^{-1}$$

$$\therefore \quad r_2 = \frac{1}{4\pi\epsilon_0}\cdot\frac{e^2}{mv^2} = (9\times10^9)\times\frac{(1.6\times10^{-19})^2}{(9.1\times10^{-31})\times(11\times10^5)}$$

$$= 21\times10^{-11}\text{ m} = 0.21\text{nm (or 2.1 Å)} \quad \textbf{(Answer).}$$

[Alternatively, we could simply use the formula,

$$r_n = \frac{n^2a_0}{Z} \text{ (}a_0\text{ = Bohr radius = 53 pm = 0.053 nm)}$$

$$= \frac{(2)^2\times(0.053)}{(1)}\text{ nm} = 0.21\text{ nm]}$$

The energy in the second orbit is given by,

$$E_n = -\frac{Rhc\,Z^2}{n^2}$$

$$= -\frac{13.6\,Z^2}{n^2}\text{ eV } (\because Rhc = 13.6\text{ eV})$$

$$= -\frac{13.6\times(1)^2}{(2)^2}\text{ eV} = -3.4\text{ eV} \quad \textbf{(Answer).}$$

Example 4. Determine the values of maximum and minimum wavelengths in the spectral Lyman series of hydrogen atom (Rydberg constant, $R = 1.1\times10^7\text{ m}^{-1}$).

Solution. For a hydrogen atom, the possible wavelengths emitted are given by,

$$\frac{1}{\lambda} = \frac{1}{R}\left(\frac{1}{n^2}-\frac{1}{m^2}\right)$$

For Lyman series (n = 1), we have,

$$\frac{1}{\lambda} = \frac{1}{R}\left(\frac{1}{(1)^2}-\frac{1}{m^2}\right) = \frac{1}{R}\left(1-\frac{1}{m^2}\right)$$

The maximum wavelength corresponds to m = 2,

$$\frac{1}{\lambda_{max}} = (1.1\times10^7)\times\left(1-\frac{1}{2}\right) = \frac{1.1\times10^7}{2}$$

$$\Rightarrow \quad \lambda_{max} = \frac{2}{1.1\times10^7} = 1.818\times10^{-7}\text{ m}$$

$$= 181.8\text{ nm (or 1818 Å)} \quad \textbf{(Answer).}$$

The minimum wavelength corresponds to $m = \infty$,

$$\frac{1}{\lambda_{min}} = (1.1\times10^7)\times\left(1-\frac{1}{\infty}\right) = 1.1\times10^7$$

$$\Rightarrow \quad \lambda_{min} = \frac{1}{1.1\times10^7} = 0.9090\times10^{-7}\text{ m}$$

$$= 90.9\text{ nm (or 909 Å)} \quad \textbf{(Answer).}$$

Ionization Potential; Ionization Energy

We have known that in a hydrogen atom the energy of the electron in the ground state (around room temperature) is –13.6 eV. If the atom is supplied with an energy of +13.6 eV (e.g. by heating), the total energy of the electron becomes zero. A zero energy corresponds to the excited state $n = \infty$, which effectively means that the electron-nucleus distance is a large value. In other words, the electron is removed from the atom. Then the atom is said to be **ionized**. It no longer remains electrically neutral. It acquires a unit +ve charge. The same condition holds for the atoms of other elements.

"The minimum energy required to ionize an atom" is termed as the **ionization energy** of the atom. "The potential difference through which an electron should be accelerated to acquire the necessary ionization energy" is termed as the **ionization potential** of the atom. The values of ionization energy and ionization potential for a hydrogen atom are 13.6 eV and 13.6 V, respectively.

Additional knowledge

In spite of its brilliant success in explaininng the hydrogen spectrum, the Bohr's theory failed to explain the spectra of atoms and ions having more than one electron. It could not account for even the helium spectrum. As progressively greater accuracy was achieved in the measuremtnt of wavelengths, some discrepancies started becoming noticeable even for the hydrogen spectrum. Such difficulties could be overcome only later, after fuller development of quantum mechanics. In quantum mechanics, the basic idea of the electron being localized in a stationary orbit is replaced by the concept of a one-electron wave function, called 'orbital', which expresses the probability of finding the electron at a given position in space around the nucleus. The quantum mechanical 'orbital' has no resemblance with the 'orbit' of the Bohr's theory. (Another physical consideration for the failure of Bohr's theory in explaining the spectra of many electron atoms is that an electron can interact as well with another electron as with the +ve nucleus, since only electric charges are involved. Hence complications increase as the number of electrons in the system goes up).

Nevertheless, the early remarkable success of the Bohr's theory proved to be a milestone in the development of basic concepts of physics. It provided support to the existing early concepts of quantum theory and so resulted in great encouragement to further development of quantum mechanics (the celebrated Schrödinger wave equation, which provided a concrete mathematical basis to quantum mechanics, came up 13 years later in 1926).

8A.05. Composition and Properties of Nucleus

Composition (Nucleons; Nuclides). A nucleus consists of +vely charged **protons** and electrically neutral **neutrons.** The number of protons contained in a nucleus is called the **atomic number** (or **proton number**) of the nucleus. It is represented by the **symbol Z**. The number of neutrons in the nucleus is called **neutron number (N).** The total number of protons and neutrons in a nucleus is called the **mass number (A)** of the nucleus. We have,

$$A = Z + N \qquad \text{..... (8.15)}$$

The protons and neutrons are collectively called **nucleons**. Thus, the mass number A represents the total number of nucleons in the nucleus. The neutron number is equal to (A – Z).

"A collection of nuclei with the same value of Z and N" is called a **nuclide.** For instance, all the nuclei of sodium have Z = 11 and A = 23, i.e. each of them consists of 11 protons and 12 neutrons. So collectively they form the nuclide of the element sodium. It is represented as $^{23}_{11}Na$. The subscript represents the atomic number Z and the superscript represents the mass number A. Thus, in general, a nuclide is denoted as $^{A}_{Z}X$.

Example 5. Calculate the number of electrons, protons, and neutrons contained in 24 gm of $^{12}_{6}C$ (Avogadro number = 6×10^{23} mol^{-1}).

Solution. The symbol $^{12}_{6}C$ implies that one carbon atom consists of 6 electrons, 6 protons, and (12–6) = 6 neutrons.

The gm-atomic weight of carbon ($^{12}_{6}C$) is 12 gm. So 24 gm of $^{12}_{6}C$ is equivalent to 2 gm-atomic weight of carbon. Since the number of atoms in one gm-atomic weight is 6×10^{23} (Avogadro number), the required numbers of particles are,

$$\text{No. of electrons} = 2\times6\times(6\times10^{23}) = 72\times10^{23} \quad \textbf{(Answer).}$$

$$\text{No. of protons} = 2\times6\times(6\times10^{23}) = 72\times10^{23} \quad \textbf{(Answer).}$$

$$\text{No. of neutrons} = 2\times6\times(6\times10^{23}) = 72\times10^{23} \quad \textbf{(Answer).}$$

Example 6. Determine the number of neutrons and protons contained in one atom each of $^{22}_{11}Na$ and $^{59}_{27}Co$.

Solution. (i) $^{22}_{11}Na$: Here Z = 11, A = 22

$\therefore$ No. of protons = Z = 11

No. of neutrons = (A – Z) = 22 – 11 = 11

(ii) ${}^{59}_{27}$Co: Here Z = 27, A = 59

∴ No. of protons = 27 **(Answer).**

No. of neutrons = 59 – 27 = 32 **(Answer).**

Mass on Atomic and Subatomic Scale [kg; MeV/c²; a.m.u. or u]. A proton has a mass nearly 1840 times the mass of an electron. A neutron is slightly heavier than a proton. Their respective masses, expressed in kilogram, are,

$$\text{Mass of a proton } (m_p) = 1.6726231 \times 10^{-27} \text{ kg}$$

$$\text{Mass of a neutron } (m_n) = 1.6749286 \times 10^{-27} \text{ kg}$$

According to the Einstein's mass-energy relationship, $E = mc^2$, the atomic and subatomic mass can also be expressed in terms of energy. Thus we have,

$$m_p = 938.27231 \text{ MeV/c}^2$$

$$m_n = 939.56563 \text{ MeV/c}^2$$

The mass of an electron can similarly be expressed as,

$$m_e = 9.1093897 \times 10^{-31} \text{ kg} = 510.99 \text{ keV/c}^2.$$

The energy corresponding to the mass of a particle when the particle is at rest is called the **rest mass energy** of the particle. Thus, the above values of m_p, m_n, m_e represent the respective rest mass energies of proton, neutron and electron.

As we see, the atomic and subatomic masses expressed in kg are extremely small values. Hence they are difficult to handle. Therefore, for the sake of convenience, another unit known as **atomic mass unit (a.m.u.)** has been chosen. It is used both in nuclear physics and atomic physics and so it is also called **unified atomic mass unit,** simply expressed as **u**. The relation between u and kg is,

$$1\text{u} = 1.6605402 \times 10^{-27} \text{ kg}$$

As some typical examples, the atomic masses of Na (Sodium), iron (Fe), gold (Au) and uranium (U), expressed in the atomic mass unit u, are nearly 23, 56, 197 and 238 u, respectively. [By international agreement, the carbon-12 atom ${}^{12}_{6}$C has been assigned a mass of 12 atomic mass units (u). Thus, we have,

$$1\text{u} = \frac{\text{mass of } {}^{12}_{6}\text{C atom}}{12} = \frac{1.9926482 \times 10^{-26}}{12} \text{ kg}$$

Accordingly, the atomic masses of all the elements are basically expressed in terms of the atomic mass of ${}^{12}_{6}$C as the standard value].

Example 7. Determine equivalent energy for an electron and a proton ($m_e = 9.1 \times 10^{-31}$ kg; $m_p = 1.67 \times 10^{-27}$ kg).

Solution. The mass-energy relationship is given by,

$$E = mc^2$$

$$\therefore \quad E_e = m_e c^2 = (9.1 \times 10^{-31}) \times (3 \times 10^8)^2 \text{ J}$$

$$= \frac{9.1 \times 9 \times 10^{-15}}{(1.6 \times 10^{-19})} \text{ eV}$$

$$= 0.512 \times 10^6 \text{ eV (or 0.512 MeV)}$$ **(Answer).**

$$E_p = m_p c^2 = \frac{(1.67 \times 10^{-27}) \times (3 \times 10^8)^2}{(1.6 \times 10^{-19})} \text{ eV}$$

$$= 939.4 \times 10^6 \text{ eV} = 939.4 \text{ MeV}$$ **(Answer).**

Isotopes, Isobars and Isotones. The atoms (or nuclides) having the same atomic number Z but different mass number A are called **isotopes**. For example, uranium has two isotopes, ${}^{235}_{92}$U and ${}^{238}_{92}$U. In some cases the number can be very large, e.g. the element gold has 32 isotopes, with A = 173 to 204. The isotopes of an element have the same number of electrons (= number of protons). Therefore, in spite of different physical properties, they have same chemical properties.

The atoms (or nuclides) having the same mass number A but different atomic number Z are called **isobars**, e.g. ${}^{54}_{24}$Cr and ${}^{54}_{26}$Fe.

The atoms (or nuclides) having the same neutron number N but different atomic number Z are called **isotones**, e.g. ${}^{37}_{17}$Cl and ${}^{39}_{19}$K.

Nuclear size; Radius of nucleus. We assume a nucleus as a sphere of radius r (only for convenience; strictly speaking, it is not true). Its volume is proportional to the mass number,

$$\frac{4}{3}\pi r^3 \ \alpha\ A$$

Or $$r \ \alpha\ A^{1/3}$$

Or $$r = r_0 A^{1/3} \qquad \text{..... (8.16)}$$

r_0 is a constant. Its approximate value is,

$$r_0 \approx 1.2\times10^{-15} \text{ m (or 1.2 fm)}$$

The values of the nuclear radii for various elements lie in the approximate range $10^{-15} - 10^{-14}$ m (or 1–10 fm).

Example 8. Find the nuclear radius of an oxygen atom (r_0 = 1.2 fm).
Solution. For oxygen, mass number, A = 16. Also, given that r_0 = 1.2 fm = 1.2×10^{-15} m.
The nuclear radius of an atom is given by,

$$r = r_0 A^{1/3}$$

$$\therefore \quad r_{oxygen} = (1.2\times10^{-15})\times(16)^{1/3}$$

$$= 1.2\times2.52\times10^{-15}$$

$$= 3.02\times10^{-15} \text{ m (or 3.02 fm)} \quad \textbf{(Answer).}$$

Nuclear density. The masses of a proton and a neutron are very nearly equal. As stated earlier, the protons and neutrons together are called nucleons. Let m_{av} be the average mass of nucleons. The total number of nucleons in a nucleus is A, the mass number. So we have,

$$\text{Total mass of the nucleus} = m_{av} A$$

$$\text{Volume of the nucleus} = \frac{4}{3}\pi r^3$$

$$= \frac{4}{3}\pi r_0^3 A \ (\because r = r_0 A^{1/3})$$

Then, the density of the nucleus is given by (density = mass/volume),

$$\rho_{nucleus} = \frac{m_{av} A}{(4/3)\pi r_0^3 A}$$

Or $$\rho_{nucleus} = \frac{3m_{av}}{4\pi\, r_0^3} \qquad \text{..... (8.17)}$$

m_{av} and r_0 are standard constant values. Thus, we find that the **nuclear density** is a constant quantity. This is an interesting and significant result. Further, since $r_0 \approx 10^{-15}$, $r_0^3 \approx 10^{-45}$. Hence the nuclear density turns out to be an extremely high values, viz. of the order of 10^{17} kg/m^3 (cf. density of iron $\approx 10^4$ kg/m^3 and density of water = 10^3 kg/m^3).

Example 9. Determine nuclear density of an atom of gold (mass no., A, for gold = 197; r_0 = 1.2 fm)
Solution. Given: r_0 = 1.2 fm = 1.2×10^{-15} m
The nuclear radius of an atom is given by,

$$r = r_0 (A)^{1/3}$$

$\therefore$ Volume of the nucleus is given by,

$$V = \frac{4}{3}\pi r^3 = \frac{4}{3}\pi (r_0)^3 A$$

$$= \frac{4\times3.14\times(1.2\times10^{-15})^3}{3}\times197$$

$$= 7.23\times10^{-45}\times197$$

The gold nucleus has 197 nucleons (A = 197). Their total mass is [$197\times(1.67\times10^{-27})$] kg.

The nuclear density of gold is given by,

$$\rho_{gold} = \frac{\text{Mass}}{\text{Volume}} = \frac{(197\times1.67\times10^{-27})}{(7.23\times10^{-45}\times197)}$$

$$= 2.31\times10^7 \text{ kg m}^{-3} \quad \textbf{(Answer).}$$

(Note that the density can be obtained even if the value of A is not given, because A cancels out in the numerator and the denominator. It only confirms that the nuclear density is a constant value for all elements).

Nuclear spin. Like the electrons (vide Chapter 3B, Section 3B.08), the protons and neutrons (together called nucleons) also possess orbital angular momenta. Besides, like the electrons, they also possess internal spin angular momentum, with the spin quantum number m_s as +(1/2) or –(1/2). The total angular momentum, comprising the resultant of all the spin and angular momenta of the nucleons, is called the **nuclear spin** of the nucleus.

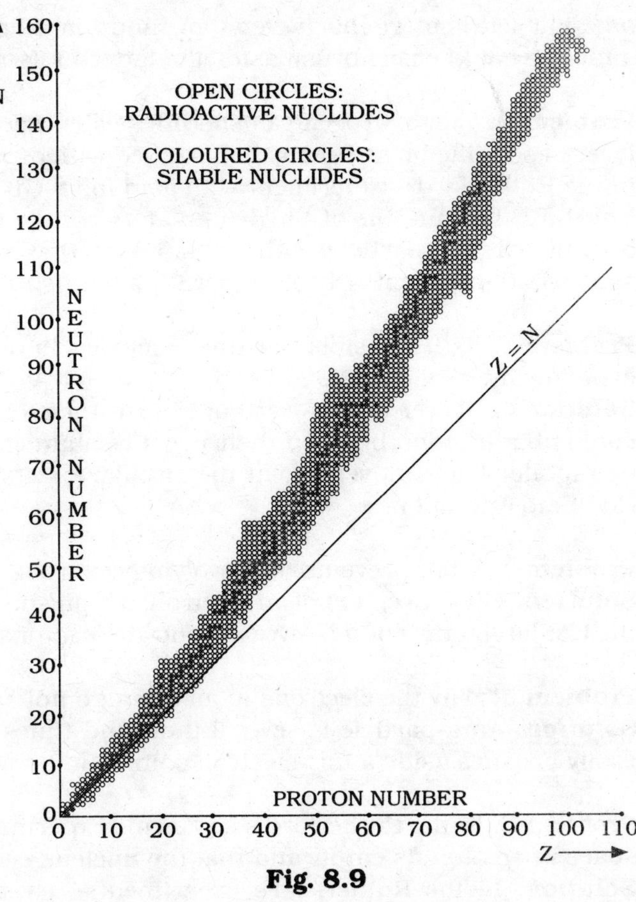

Fig. 8.9

8A.06. Nuclear Stability; Nuclidic Chart

As already stated, the isotopes of an element represent nuclides with the same atomic number Z but different mass numbers A. All the isotopes of a given element may not be stable. In fact, usually only a few of them are stable. The rest are unstable, with different degrees of instability. An unstable nuclide emits some kind of radiation (α,β or γ-radiation) and transforms into a stable or a less unstable nuclide. So far, more than 2000 nuclides have been identified, of which only about 260 have been found stable. The unstable nuclides are called **radioactive nuclides.**

Useful information on the behaviour of the nuclides is obtained through the **nuclidic chart** shown in Fig. 8.9. It is a plot of the neutron number N against the proton number (or atomic number) Z, for the known nuclides. The open circles represent the stable nuclides and the coloured circles represent the unstable nuclides. Together they form a nearly straight zigzag band, inclined at nearly 55° to the Z axis. For comparison, the nearby straight line Z = N is inclined at 45° to Z or N axis. The stable nuclides lie somewhat in the middle of the band. The unstable nuclides lie above and below them. The light stable nuclides (small Z-values) tend to lie close to the line Z = N (i.e. the ratio N/Z = 1), i.e. they have nearly the same number of protons and neutrons. The heavier stable nuclides tend to have more neutrons than protons. The ratio N/Z increases for them, becoming nearly 1.5 for the heaviest ones (e.g. bismuth). It means that the stability of a nuclide depends on the relative numbers of protons and neutrons contained in it. The heaviest stable nuclide is bismuth $\left({}^{209}_{83}\text{Bi}\right)$.

Solved Conceptual Problems
[Including HOTS Questions]

Problem 1. Why the Thomson's and Lenard's models of atom did not succeed, in spite of their initial successes ?

Solution. In the Thomson's model, a –ve charge was supposed to be uniformly distributed over the volume of the atom (taken to be spherical), with the electrons embeded in it here and there. The Lenard's model extended it by proposing that a +ve charge was also similarly distributed, like the electrons.

The model met with the problem that why upon heating a metal the +ve charge was not ejected alongwith the –vely charged electrons. Further, the model failed to explain the observed back deflection of some α-particle by gold atoms in the experiments carried out by Rutherford's collaborators, Geiger and Marsden. If the model was right, all the α-particles should have moved undeviated through the foil, because an α-particle is more than seven thousand times heavier than an electron or a similar +vely charged particle and so it could easily brush aside the latter in its passage through the foil.

Problem 2. The hydrogen gas solidifies when cooled to a temperature below 14 K (i.e. –269 °C). If a thin layer of solidified hydrogen is bombarded with α-particles, will the particles be scattered in the same way as in the Rutherford's experiment with gold foil ? Give reasons for your answer.
Solution. The nucleus of a hydrogen atom contains just one proton, of mass 1.67×10^{-27} kg. In comparison, a bombarding α-particle, with mass 6.64×10^{-27} kg, is four times heavier. It collides with a hydrogen atom somewhat like a football colliding with a tennis ball. Consequently, it undergoes insignificant scattering.

Problem 3. Is it possible that the electrons in an atom could remain stationary in their orbits ? If not, explain why.
Solution. If the electrons were at rest in their orbits, they would fall into the nucleus due to the strong coulomb attraction between their negative charge and the protons' positive charge. Therefore, they have to be mobile. Their movement in the circular orbits provides the necessary centripetal force to balance the Coulomb attraction.

Problem 4. What prevents the revolving electrons in an atom from falling into the +vely charged nucleus ?
Solution. The movement of an electron at high speed in its orbit provides a centripetal force, which balances the Coulomb attraction between it and the protons in the nucleus.

Problem 5. Why the electrons in an atom do not scatter the incident α-particles, just as the nucleus does ?
Solution. An α-particle is several thousand times (nearly 7,350 times) heavier than an electron. Hence it easily brushes aside a tiny electron coming in its way. It is like a football colliding with a mustard seed.

Problem 6. In the Rutherford's scattering experiments, what consideration led to conclude that bulk of the space in an atom is empty and that the nucleus occupies a very small volume ?
Solution. In the Rutherford's experiments, a very high majority of the α-particles were found to pass undeflected through the gold foil. Only 1 in 8000 was found to turn back. Hence it led to the conclusion that most of the space in an atom is empty. It also means that the nucleus occupies a very small volume in the atom.

Problem 7. What considerations led to the evolution of Bohr's model of an atom, in spite of the existing basically sound Rutherford's model, suggesting revolution of electrons in circular orbits around the nucleus ?
Solution. According to the electromagnetic theory, an accelerated charged particle emits radiation in the form of electromagnetic waves. Thus, it should continuously lose energy and so continuously spiral inwards and finally fall into the nucleus. It would mean that an atom cannot be stable, but the reality is that the atoms are perfectly stable. Secondly, the e.m. theory also tells us that the frequency of the emitted e.m. waves is the same as the frequency of revolution of the electron. If the electron spirals inwards, its angular velocity and hence its frequency of revolution will also change continuously. It means that a continuous spectrum of wavelengths will be emitted. But that is again in disagreement with observation. The elements give line spectra, not continuous spectra. Hence a modification of the Rutherford's model became necessary.

Problem 8. In the Bohr's model, the energies of the orbiting electrons are negative values. Explain what does it signify.
Solution. The –ve sign for energy arises from the electrostatic (Coulombian) attraction of the electron to the nucleus, which binds the electrons to the nucleus. In other words, the energy arises from an attractive force, which means that if the electron has to be removed from the atom, necessary energy will have to be supplied from outside. e.g. by heating, electric discharge, etc. The closer is the orbit to the nucleus, the higher is the energy of the electron in the orbit, e.g. –13.6 eV, –3.4 eV, –1.5 eV,.................. for first, second, third,......... orbits in a hydrogen atom (disregarding –ve sign). Thus, highest energy is required to remove an electron from the innermost orbit.

Problem 9. Explain how the density of the nucleus of an atom is far greater than the density of the atom.
Solution. The density is mass/volume. An atom consists of protons, neutrons and electrons, while the

nucleus contains only protons and neutrons. But since an electron is more than a thousand times lighter than a proton or neutron, the masses of an atom and its nucleus are practically the same. But their radii and therefore their volumes [$(4/3)\ \pi r^3$] are vastly different [on an average, the nuclear radius ($\sim 10^{-14}$ m) is nearly ten thousand times shorter than the atomic radius ($\sim 10^{-10}$ m)]. Hence the density of the nucleus is far greater than the density of the atom.

Problem 10. How does the nuclear density compare with the atomic density and with the density of water or a common material like iron ?
Solution. The nuclear density is an extremely high value, viz. of the order of 10^{17} kg m^{-3}, which is about $10^{13} - 10^{14}$ times higher than the atomic densities (of the order of $10^3 - 10^4$ kg m^{-3}). The density of water is 10^3 kg m^{-3} and that of a common material like iron is about 8×10^3 kg m^{-3}. (Note that atomic densities are of the same order as the densities of materials, viz. $10^3 - 10^4$ kg m^{-3}, a result which is naturally expected).

Solved Numerical Problems

Problem 1. The distance of closet approach for a head-on collision of an α-particle with the nucleus of a gold atom is found to be 41.3 fermi. Find the energy of the α-particle.
Solution. Given: $r_0 = 41.3$ fermi $= 41.3\times10^{-15}$ m; $Z = 79$.
The energy of the α-particle is given by,

$$E = \frac{2k\,Ze^2}{r_0} = \frac{2\times(9\times10^9)\times(79)\times(1.6\times10^{-19})^2}{(41.3\times10^{-15})}$$

$$= 8.814\times10^{-13}\ \text{J} = \frac{8.814\times10^{-13}}{1.6\times10^{-13}}\ \text{MeV}$$

$$= 5.51\ \text{MeV} \quad \textbf{(Answer).}$$

Problem 2. In a collision of an α-particle of energy 10 MeV with the nucleus of a copper atom (Z = 29), the particle is found to retrace its path. Determine the size of the nucleus.
Solution. Given: $E = 10$ MeV $= 10\times(1.6\times10^{-13})$ J $= 16\times10^{-13}$ J; $Z = 29$.
We know that the size of nucleus is approximately equal to the distance of closet approach, which is given by,

$$r_0 = \left(\frac{1}{4\pi\epsilon_0}\right)\cdot\frac{(Ze)\times(2e)}{E} = (9\times10^9)\cdot\frac{(29)\times(1.6\times10^{-19})\times(2\times1.6\times10^{-19})}{(16\times10^{-13})}$$

$$= 8.35\times10^{-15}\ \text{m (or 8.35 fermi)} \quad \textbf{(Answer).}$$

Problem 3. An α-particle of energy 6 MeV is scattered through an angle of 18°, while passing through a thin copper foil. Calculate the impact parameter for the particle (Z for copper = 29).
Solution. Given: $E = 6$ MeV $= 6\times(1.6\times10^{-13})$ J $= 9.6\times10^{-13}$ J; $\theta = 18°$; $Z = 29$
The impact parameter is given by,

$$b = \frac{Ze^2.\ \cot(\theta/2)}{(4\pi\epsilon_0)\ [(1/2)mv^2]} = \left(\frac{1}{4\pi\epsilon_0}\right)\cdot\frac{Ze^2.\ \cot(9°)}{E} \quad \left(\because E = \frac{1}{2}mv^2\right)$$

$$= (9\times10^9)\cdot\frac{(29)\times(1.6\times10^{-19})^2\times(6.31)}{(9.6\times10^{-13})}$$

$$= 4.39\times10^{-14}\ \text{m} \quad \textbf{(Answer).}$$

Problem 4. Determine the impact parameter for an α-particle of energy 10 MeV, which is scattered through 90° when approaching a gold nucleus (Z = 79 for gold).
Solution. Given: $E = 10$ MeV $= 10\times(1.6\times10^{-13})$ J $= 1.6\times10^{-12}$ J; $\theta = 90°$; $Z = 79$
We know that the impact parameter is given by,

$$b = \frac{Ze^2.\ \cot(\theta/2)}{(4\pi\epsilon_0)\ [(1/2)mv^2]} = \frac{1}{4\pi\epsilon_0}\cdot\frac{Ze^2.\ \cot(45°)}{E} \quad \left(E = \frac{1}{2}mv^2\right)$$

$$= \frac{(9\times10^9)\times(79)\times(1.6\times10^{-19})^2\times(1)}{(1.6\times10^{-12})}$$

$$= 1.14\times10^{-14}\ \text{m} \quad \textbf{(Answer).}$$

Problem 5. The Bohr's nucleus-electron model of atom is analogous to the sun-earth system. Find which of the two has relatively greater empty space. Take radii of electron's orbit and the nucleus as 10^{-10} m and 10^{-15} m, respectively, and the radii of earth's orbit and the sun as 1.5×10^{11} m and 7×10^{8} m, respectively.
Solution. The ratio of the radii of electron's orbit and the nucleus is,

$$R_1 = \frac{r_{el}}{r_n} = \frac{10^{-10}}{10^{-15}} = 10^5$$

The corresponding ratio for the sun-earth system is,

$$R_1 = \frac{r_{earth}}{r_{sun}} = \frac{1.5\times10^{11}}{7\times10^8} \approx 2\times10^2$$

Thus, R_1 is found to be greater than R_2 in the proportion,

$$\frac{R_1}{R_2} \approx \frac{10^5}{2\times10^2} \approx 500$$

It means that in relative terms the distance of the electron from the nucleus is nearly 500 times greater than the distance of earth from the sun. It follows that an atom has far greater empty space than the sun-earth system (in fact, the same holds for the entire solar system). **(Answer).**

Problem 6. In an α-sattering experiment with a gold foil, an 8.1 MeV α-particle comes momentarily to rest and retraces its path, while approaching a gold atom. Calculate its distance of closet approach (atomic no. of gold = 79).
Solution. Given: $E = 8.1 \text{ MeV} = 8.1\times(1.6\times10^{-13}) \text{ J} = 12.96\times10^{-13} \text{ J}$; $Z = 79$
The distance of closest approach is given by,

$$r_0 = \frac{4Ze^2}{(4\pi\epsilon_0)\, mv^2} = \frac{(1/2)\times4Ze^2}{(4\pi\epsilon_0)\, [(1/2)mv^2]}$$

$$= \frac{1}{4\pi\epsilon_0}\times\frac{2Ze^2}{E}$$

$$= \frac{(9\times10^9)\times2\times79\times(1.6\times10^{-19})^2}{(12.96\times10^{-13})}$$

$$= 28.1\times10^{-15} \text{ m (or 28.1 fm)}$$ **(Answer).**

Problem 7. A hydrogen atom has an electron of energy –13.6 eV revolving around the nucleus. What will be the frequency of light emitted by it, in accordance with the electromagnetic theory ?
Solution. Given: Total energy of the electron = $-13.6 \text{ eV} = -13.6\times(1.6\times10^{-19}) \text{ J} = -2.2\times10^{-18}$ J
The expression for total energy of an electron revolving in nth orbit is given by,

$$E_n = E \text{ (say)} = -\frac{e^2}{8\pi\epsilon_0 r} = -2.2\times10^{-18}$$

$$\Rightarrow \quad r = \frac{-e^2}{8\pi\epsilon_0 E} = \frac{1}{4\pi\epsilon_0}\cdot\frac{-e^2}{2E}$$

$$= (9\times10^9)\times\frac{-(1.6\times10^{-19})^2}{2\times(-2.2\times10^{-18})} = 5.3\times10^{-11} \text{ m}$$

The velocity of the electron is given by,

$$v = \sqrt{\frac{Ze^2}{(4\pi\epsilon_0).mr}} = \frac{e}{\sqrt{4\pi\epsilon_0 .mr}} \quad (Z = 1 \text{ for hydrogen})$$

$$= \sqrt{\frac{1}{4\pi\epsilon_0}}\times\frac{e}{\sqrt{mr}} = \sqrt{9\times10^9}\times\frac{(1.6\times10^{-19})}{\sqrt{(9.1\times10^{-31})\times(5.3\times10^{-11})}}$$

$$= 2.2\times10^6 \text{ ms}^{-1}$$

The frequency of revolution of the electron is given by,

$$\nu = \frac{v}{2\pi r} = \frac{(2.2\times10^6)}{(2\times3.14)\times(5.3\times10^{-11})}$$

$$= 6.6\times10^{15} \text{ Hz}$$

The classical e.m. theory suggests that the frequency of e.m. waves emitted by a revolving electron is the same as the frequency of revolution of the electron around the nucleus. Accordingly, the frequency of light emitted in our case is also 6.6×10^{15} Hz. **(Answer).**

Problem 8. Calculate the energy needed to transfer the electron in a hydrogen atom from its ground state (energy: –13.6 eV) to the first excited state.

Solution. For hydrogen atom, the energy of the electron in nth orbit is given by,

$$E_n = -\frac{13.6}{n^2}\ \text{eV}$$

For ground state, n = 1. The corresponding energy is,

$$E_1 = -\frac{13.6}{(1)^2} = -13.6\ \text{eV}$$

For the first excited state, n = 2. The corresponding energy is,

$$E_2 = -\frac{13.6}{(2)^2} = -3.4\ \text{eV}$$

∴ The energy needed for excitatin from ground state to first excited state is,

$$E_2 - E_1 = -3.4 - (-13.6)$$
$$= 10.2\ \text{eV} \quad \textbf{(Answer).}$$

Problem 9. The energy levels of an atom of an element are shown in the following diagram. Which one of the level transition will result in the emission of photons of wavelength 620 nm ? Support your answer with mathematical calculations. (CBSE; 1997).

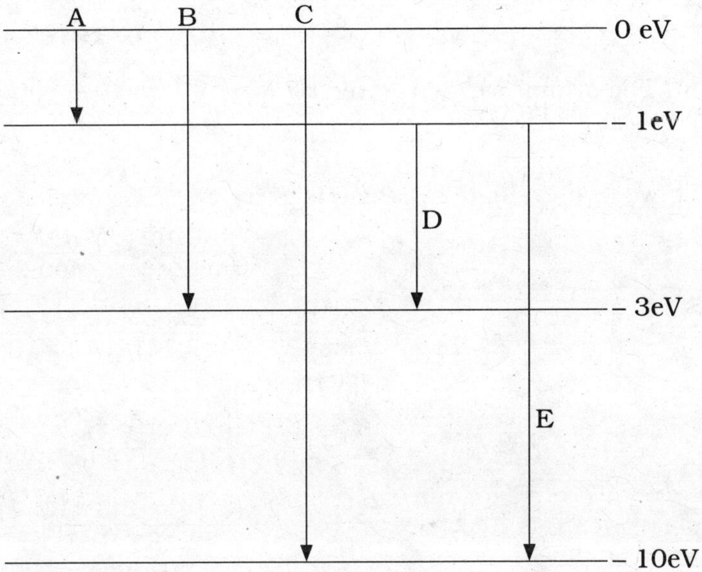

A B C
0 eV
– 1eV
D
– 3eV
E
– 10eV

Solution. Given: $\lambda_{photon} = 620\ \text{nm} = 620\times10^{-9}\ \text{m}$

$$\text{Energy of photon} = h\nu = \frac{hc}{\lambda}$$

$$= \frac{(6.6\times10^{-34})\times(3\times10^{8})}{(620\times10^{-9})}\ \text{J}$$

$$= \frac{(6.6\times10^{-34})\times(3\times10^{8})}{(620\times10^{-9})\times(1.6\times10^{-19})}\ \text{eV}$$

$$= 1.996\ \text{eV} \approx 2\ \text{eV}$$

It follows that the energy transition D from –1eV to –3eV, i.e. [–1– (–3)] = 2eV, will result in the emission of photons of wavelength 620 nm. **(Answer).**

Problem 10. The absorption of a photon by a hydrogen atom excites the atom from ground state to fourth excited state. Calculate the frequency and wavelength of the photon.

Solution. Given: $n_1 = 1$; $n_2 = 4$

The total energy of the electron in the hydrogen atom is given by,

$$E_n = -\frac{13.6\ Z^2}{n^2}\ \text{eV}$$

$$= -\frac{13.6}{n^2}\ \text{eV} \quad (\because Z = 1 \text{ for hydrogen})$$

The energy of the photon absorbed is given by,

$$E = E_4 - E_1 = -13.6\left[-\frac{1}{(4)^2} - \left(-\frac{1}{(1)^2}\right)\right]$$

$$= -13.6\times\frac{15}{16}\ \text{eV} = -\frac{13.6\times15}{16}\times(1.6\times10^{-19})\ \text{J}$$

Since E = hν, the frequency of the photon is given by (now we ignore –ve sign, as it is of no consequence to us here),

$$\nu = \frac{E}{h} = \frac{13.6\times15\times1.6\times10^{-19}}{16\times(6.63\times10^{-34})}$$

$$= 3.06\times10^{15}\ \text{Hz} \quad \textbf{(Answer).}$$

Since νλ = c, the wavelength of the photon is given by,

$$\lambda = \frac{c}{\nu} = \frac{3\times10^{8}}{3.06\times10^{15}}$$

$$= 0.9804\times10^{-7}\ \text{m (or 98.04 nm or 980.4 Å)} \quad \textbf{(Answer).}$$

Problem 11. For a hydrogen atom, determine the velocity and orbital period of the electron in the ground state and the first two excited states.

Solution. Given that, n = 1,2,3. We have to find the corresponding values of velocities (v_1,v_2,v_3) and orbital periods (T_1,T_2,T_3).

(i) The velocity of the electron in the nth orbit is given by,

$$v_n = \frac{Ze^2}{2\epsilon_0 nh} = \frac{e^2}{2\epsilon_0 nh} \quad (Z = 1 \text{ for hydrogen atom})$$

$$\therefore \quad v_1 = \frac{e^2}{2\epsilon_0.1.h} = \frac{(1.6\times10^{-19})^2}{2\times(8.85\times10^{-12})\times(6.6\times10^{-34})}$$

$$= 2.19\times10^6 \text{ ms}^{-1} \quad \textbf{(Answer).}$$

$$v_2 = \frac{e^2}{2\epsilon_0.2.h} = \frac{1}{2}v_1 = 1.095\times10^6 \text{ ms}^{-1} \quad \textbf{(Answer).}$$

$$v_3 = \frac{e^2}{2\epsilon_0.3.h} = \frac{1}{3}v_1 = 0.73\times10^6 \text{ ms}^{-1} \quad \textbf{(Answer).}$$

(ii) The orbital period is given by,

$$T_n = \frac{2\pi r_n}{v_n}$$

So we first evaluate r_n, which is given by,

$$r_n = \frac{\epsilon_0 n^2 h^2}{\pi m Ze^2} = \frac{\epsilon_0 n^2 h^2}{\pi me^2} \quad (Z = 1)$$

$$\therefore \quad r_1 = \frac{\epsilon_0\times(1)^2\times h^2}{\pi me^2} = \frac{(8.85\times10^{-12})\times(1)\times(6.63\times10^{-34})^2}{(3.14)\times(9.1\times10^{-31})\times(1.6\times10^{-19})^2}$$

$$= 5.3\times10^{-11} \text{ m}$$

$$r_2 = n^2 r_1 = (2)^2\times(5.3\times10^{-11}) = 21.2\times10^{-11} \text{ m}$$

$$r_3 = (3)^2\times(5.3\times10^{-11}) = 47.7\times10^{-11} \text{ m}$$

$$\therefore \quad T_1 = \frac{2\pi r_1}{v_1} = \frac{2\times(3.14)\times(5.3\times10^{-11})}{(2.19\times10^6)}$$

$$= 0.152\times10^{-15} \text{ s} \quad \textbf{(Answer).}$$

$$T_2 = \frac{2\pi r_2}{v_2} = \frac{2\times(3.14)\times(21.2\times10^{-11})}{(1.095\times10^6)}$$

$$= 1.22\times10^{-15} \text{ s} \quad \textbf{(Answer).}$$

$$T_3 = \frac{2\pi r_3}{v_3} = \frac{2\times(3.14)\times(47.7\times10^{-11})}{(0.73\times10^6)}$$

$$= 4.10\times10^{-15} \text{ s} \quad \textbf{(Answer).}$$

Problem 12. The energy needed to raise a hydrogen-like ion from ground state to second excited state is found to be 108.8 eV. Calculate the energy required to remove the electron from the ion.

Solution. The energy of the electron in the nth state, expressed in terms of Rydberg constant (R), is given by,

$$E_n = -\frac{RhcZ^2}{n^2} = -\frac{13.6\ Z^2}{n^2} \text{ eV} \quad (\because Rhc = 13.6 \text{ eV})$$

∴ The energy needed for raising the ion from ground state (n = 1) to second excited state (n = 3) is given by,

$$13.6\ Z^2\left[\frac{1}{(1)^2} - \frac{1}{(3)^2}\right] = 108.8$$

$$\Rightarrow \quad Z^2 = \frac{108.8\times9}{13.6\times8} = 9$$

$$\Rightarrow \quad Z = 3$$

The value Z = 3 suggests that the ion under consideration is an Li^{++} ion. The energy required to remove its electron is equal to its energy in the ground state (n = 1), which is given by,

$$E_1 = -\frac{13.6\,Z^2}{(1)^2} = -(13.6\times 9)$$

$$= -122.4 \text{ eV}$$

Thus, the energy needed to remove the electron from the ion is 122.4 eV. **(Answer).**

Problem 13. If the electron in an Li^{++} ion has to be excited to the second energy level, from the first energy level, determine the wavelength of the radiation needed for it (Z for Li = 3).

Solution. The energy of the electron in the nth energy level,, expressed in terms of Rydberg constant (R), is given by,

$$E_n = -\frac{Rhc\,Z^2}{n^2} = -\frac{13.6\,Z^2}{n^2} \quad (\because Rhc = 13.6 \text{ eV})$$

Accordingly, the energies of the electron in the first level (n = 1) and the second level (n = 2) are,

$$E_1 = -\frac{13.6\times(3)^2}{(1)^2} = -(13.6\times 9) \text{ eV}$$

$$E_2 = -\frac{13.6\times(3)^2}{(2)^2} = -\left(13.6\times\frac{9}{4}\right)$$

∴ The energy needed for the required excitation is,

$$E = E_2 - E_1 = -13.6\left(\frac{9}{4}-9\right) = 13.6\times\frac{27}{4}$$

$$= 91.8 \text{ eV} = (91.8\times 1.6\times 10^{-19}) \text{ J} = 146.9\times 10^{-19} \text{ J}$$

Since $E = h\nu = h(c/\lambda)$, we have,

$$\lambda = \frac{hc}{E} = \frac{(6.6\times 10^{-34})\times(3\times 10^8)}{(146.9\times 9\times 10^{-19})} \text{ m}$$

$$= 13.48 \text{ nm (or 134.8 Å)}$$ **(Answer).**

Problem 14. The removal of the electron of a hydrogen-like ion from its ground state requires supply of 4 rydberg of energy. If this electron undergoes an energy transition from second excited state to the ground state, calculate the wavelength of the radiation emitted. Also determine the radius of the second orbit of the electron (Bohr radius, a_0 = 53 pm).

Solution. 1 rydberg = 13.6 eV. Thus, the energy needed to remove the electron from the ion is (4×13.6) eV. We know that it is equal to the energy of the electron in ground state, which means that the energy of the electron in the ground state (n = 1) is,

$$E_1 = -(4\times 13.6) \text{ eV} = -54.40 \text{ eV}$$

The energy of the electron in the second excited state (n = 3) is ($E_n = E_1/n^2$),

$$E_3 = \frac{E_1}{(3)^2} = \frac{-54.40}{9} = -6.04 \text{ eV}$$

∴ The change in energy for the given energy transition is,

$$E = E_3 - E_1 = -6.04 - (-54.40)$$

$$= 48.36 \text{ eV} = 48.36\times(1.6\times 10^{-19}) \text{ J} = 77.34\times 10^{-19} \text{ J}$$

Hence the wavelength of the emitted radiation is ($E = h\nu = hc/\lambda$)

$$\lambda = \frac{hc}{E} = \frac{(6.63\times 10^{-34})\times(3\times 10^8)}{(77.34\times 10^{-19})} \text{ m}$$

$$= 25.72 \text{ nm (or 257.2 Å)}$$ **(Answer).**

Next, The ground state energy of a hydrogen-like ion is given by,

$$E_1 = Z^2\,(13.6 \text{ eV})$$

Since E_1 = (4×13.6) eV in our case, it follows that $Z^2 = 4$. ∴ Z = 2 (He^+ ion).

The radius of the nth orbit for a hydrogen-like ion is given by,

$$r_n = \frac{n^2 a_0}{Z} \quad (a_0 = \text{Bohr radius} = 53 \text{ pm})$$

∴ Radius of the second orbit (n = 2) is,

$$r_2 = \frac{(2)^2 \times 53}{2} = 106 \text{ pm} \quad \textbf{(Answer).}$$

Problem 15. Calculate the velocity of electron in the third orbit of an Li^{++} ion (Z = 3). How does it compare with the velocity of light ?
Solution. Given: Z = 3; n = 3
The velocity is given by,

$$v = \frac{Ze^2}{2\epsilon_0 nh}$$

$$= \frac{(3)\times(1.6\times10^{-19})^2}{2\times(8.85\times10^{-12})\times(3)\times(6.6\times10^{-34})}$$

$$= 2.19\times10^6 \text{ ms}^{-1} \quad \textbf{(Answer).}$$

The ratio of the velocity of light to v is,

$$\frac{c}{v} = \frac{3\times10^8}{2.19\times10^6} = 137$$

Thus, the velocity of light is 137 times the calculated velocity, v. **(Answer).**

Problem 16. Prove that the velocity of electron in the first orbit of a He^+ ion is nearly 110 times less than the velocity of light.
Solution. Given: n = 1; Z = 2
The velocity is given by,

$$v = \frac{Ze^2}{2\epsilon_0 nh}$$

$$= \frac{(2)\times(1.6\times10^{-19})^2}{2\times(8.85\times10^{-12})\times(1)\times(6.6\times10^{-34})}$$

$$= 2.74\times10^6 \text{ ms}^{-1}$$

The ratio of the velocity of light to v is,

$$\frac{c}{v} = \frac{3\times10^8}{2.74\times10^6} = 109.5 \simeq 110$$

Thus, the calculated velocity is found to be nearly 110 times less than the velocity of light. **(Answer).**

Problem 17. Determine which state of a He^+ ion has the same energy as the ground state energy of a hydrogen atom. Also determine the ratio of orbital radii for the two states.
Solution. For a hydrogen atom and hydrogen-like ions, the energy of the electron is given by,

$$E_n = -\frac{13.6\, Z^2}{n^2}$$

Thus,

$$E_n \propto \frac{Z^2}{n^2}$$

Equating the energies of the two given states, we have,

$$\left(\frac{Z^2}{n^2}\right)_{helium} = \left(\frac{Z^2}{n^2}\right)_{hydrogen}$$

$$\Rightarrow \quad \frac{(2)^2}{n^2} = \frac{(1)^2}{(1)^2} = 1$$

$$\Rightarrow \quad n^2 = 4 \quad \text{or} \quad n = 2$$

Thus, the first excited state (n = 2) of a He^+ ion has the same energy as the ground state energy of a hydrogen atom. **(Answer).**

The orbital radius for a hydrogen atom and hydrogen-like ions is given by,

$$r_n = \frac{n^2 a_0}{Z}$$

Thus,

$$r_n \propto \frac{n^2}{Z}$$

∴ The ratio of the two orbital radii in question is found as,

$$\frac{r_0\,(He)}{r_1\,(H)} = \frac{(n^2/Z)_{He}}{(n^2/Z)_H} = \frac{(2)^2/(2)}{(1)^2/(1)} = 2 \quad \textbf{(Answer)}.$$

Problem 18. Show that the density of nuclear matter is nearly 2×10^{14} times the density of water ($m_{average} = 1.67\times10^{-27}$ kg; $r_0 = 1.2$ fm)

Solution. Given: Average nucleon mass, $m_{av} = 1.67\times10^{-27}$ kg; $r_0 = 1.2$ fm $= 1.2\times10^{-15}$ m

The nuclear density is given by,

$$\rho_{nucleus} = \frac{3m_{av}}{4\pi r_0^3} = \frac{3\times(1.67\times10^{-27})}{4\times3.14\times(1.2\times10^{-15})^3}$$

$$= 2.3\times10^{17}\ \text{kg/m}^3$$

The density of water is 10^3 kg/m³. Thus, it is proved that $\rho_{nucleus} \approx 10^{14}\,(\rho_{water})$. **(Answer).**

Very Short Answer Questions

[Including HOTS Questions]

(1 mark each)

1. What is meant by distance of closest approach ?
2. What is meant by impact parameter ?
3. What had been the main shortcoming of Rutherford's model of atom ?
4. What is the impact parameter for an α-particle scattered through 180° ?
5. In the Rutherford's experiments on scattering, for what value of angle of deflection is the number of deflected α-particles least ?
6. Name the hydrogen spectrum series that lies in the visible region of e.m. spectrum.
7. Name the hydrogen spectrum series that lie in (a) ultraviolet region and (b) infrared region of e.m. spectrum.
8. Name the hydrogen spectrum series with minimum value of wavelength.
9. In the Bohr's model of an atom, what is the energy of an electron when $n = \infty$?
10. What is the radius of the first orbit of a hydrogen atom ? What is it called ?
11. What is the energy of the electron in the first orbit of a hydrogen atom ?
12. What is the value of ionization potential of a hydrogen atom ?
13. The orbital radius of a hydrogen atom in ground state is 53 pm. If the radius is raised to 477 pm by exciting the atom, what is the principal quantum number of the excited state ?
14. What is a line spectrum ?
15. What is a nuclide ?
16. What is an a.m.u. ? Give its energy equivalent in MeV.

17. Give the number of electrons and protons in a singly ionised $^{236}_{92}U$ atom.

18. If a helium atom loses 2 electrons, what particle will it transform into ?

19. Are the nuclei $^{7}_{3}X$ and $^{4}_{3}Y$ isotopes of the same element ? Which of them is favoured to be more stable ?

20. The mass numbers of two isotopes X and Y are 14 and 16, respectively, with the number of electrons in X as 7. What is the number of neutrons in Y ?

21. Out of the four nuclei $^{14}_{6}C$, $^{13}_{7}N$, $^{14}_{7}N$, $^{16}_{8}O$, which ones form pairs of isobars and isotones, respectively ?

22. What is the relation between mass number A and radius r of a nucleus, assumed to be a sphere ?

23. How is the nuclear density taken to be the same value for all nuclei ?

24. Out of the four nuclei $^{13}_{6}C$, $^{14}_{7}N$, $^{30}_{15}P$, $^{31}_{15}P$, which ones form pairs of isotopes and isotones ?

25. How is the radius of a nucleus related to its mass number ?

26. How a helium atom is different from an α-particle ?

Short Answer Questions

[Including HOTS Questions]

(2 or 3 marks each)

1. For the Rutherford's α-particle scattering experiments, derive an expression for the distance of closest approach, r_0 ? If the value of r_0 is x for a bombarding α-particle of kinetic energy E, what will be its value for an α-particle of energy 3E ?

2. In the Rutherford's α-particle scattering experiments, what is understood by the impact parameter, b ? For a given value of b, how does the angle of deflection vary with increase in the k.e. of the α-particles ?

3. Explain how the nuclei and not the electrons in the atoms are responsible for the scattering of α-particles in the Rutherford's experiments.

4. What is the significance of negative value of energy of an electron in a Bohr's orbit ?

5. Derive expressions for kinetic energy and potential energy of an electron in a Bohr's orbit.

6. What is meant by ionization energy and ionization potential of an atom ?

7. What is an energy level diagram ? Draw the energy level diagram for a hydrogen atom.

8. What are emission and absorption spectra ?

9. What are isotopes, isobars and isotones ?

10. What is the effect on neutron to proton ratio in a nucleus when an electron is emitted ?

11. Calculate the ratio of density of two nuclei having mass numbers as 27 and 125.

12. Derive an expression for nuclear density, assuming a nucleus to be spherical.

Long Answer Questions

(5 marks each)

1. Describe with the help of a labelled diagram the Geiger-Marsden experiment on scattering of α-particles. Also discuss how the size of nucleus is estimated in the experiment.

2. State the postulates of Bohr's theory of atom. Hence derive expressions for (a) radii of orbits and (b) kinetic energy, potential energy, and total energy of the electron in a hydrogen atom.

3. Describe with the help of a labelled energy level diagram the origin of various spectral series for a hydrogen atom.

Numerical Exercises

1. An α-particle of energy 10 MeV is scattered through 10°, when approaching a gold nucleus. Find the impact parameter for the particle ($Z_{gold} = 79$). **]1.3×10^{-13} m]**

2. In an α-particles scattering experiment, the number of particles scattered at 120° in one minute is found to be 20. Determine the number of particles scattered in 1 minute at an angle of 90°. [Hint: It can be derived that the number of particles scattered at an angle θ is inversely proportional to $\sin^4(\theta/2)$). **[45]**

3. Two energy levels in an atom are separated by an energy difference of 4.6 eV. Calculate the frequency of radiation emitted when an electron jumps from the higher level to the lower level. **[11.1×10^{14} Hz]**

4. In the following energy level diagram, find which of the four transitions A,B,C,D will result in the emission of a photon of wavelength 275 nm. **[Transition B]**

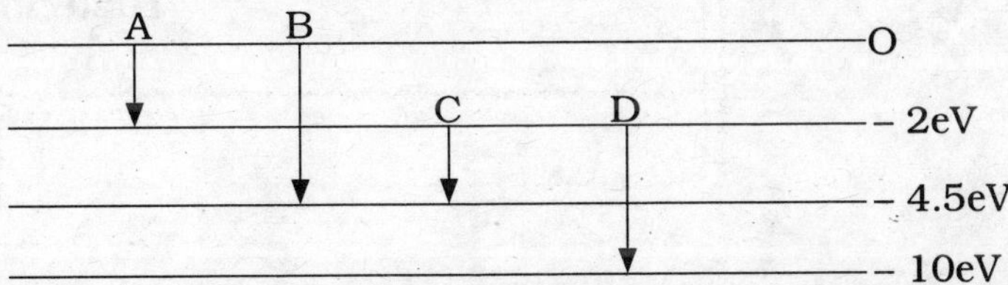

5. Calculate the kinetic energy and potential energy of the electron in the ground state of a hydrogen atom, with total energy as –13.6 eV. **[13.6 eV and –27.2 eV]**

6. A hydrogen atom having potential energy as –13.6 eV in the quantum state n= 1 absorbs energy and goes up to state n = 3. Find its p.e. in the state n = 3 and the amount of energy absorbed by it. Also calculate the wavelength of the emitted radiation if the atom directly goes back to the state n = 1. **[–1.51 eV; 12.09 eV; 102.4 nm]**

7. Determine the energy needed to elevate an electron from ground state to the second excited state of a hydrogen atom. **[12.1 eV]**

8. Find the angular momentum of the electron in an excited state of a hydrogen atom with energy – 3.4 eV. **[2.11×10^{-34} J.s]**

9. An electron is revolving around the nucleus in an orbit of radius 0.5×10^{-10} m in a hydrogen atom. Calculate at what speed should it revolve to counterbalance its Coulombian force of attraction to the nucleus. **[2.25×10^{6} ms^{-1}]**

10. A hydrogen atom in the ground state absorbs a photon to go up to the excited state n =3. Determine the wavelength of the photon, if the ionization energy of the atom is 13.6 eV (Given: Rydberg constant, $R = 1.1\times10^{7}$ m^{-1}). **[102.3 nm]**

11. The wavelength corresponding to the first line of the Lyman series of a hydrogen-like atom is 121.6 nm. Find the wavelength corresponding to the 3rd line of the Paschen series of the atom. [Hint: Using the formula, $\frac{1}{\lambda} = R\left(\frac{1}{n^2} - \frac{1}{m^2}\right)$, for the Lyman series, first find the value of R. Use it to find λ for the Paschen series]. **[1094.4 nm]**

12. A hydrogen atom absorbs energy in its ground state to get excited to the state with n = 4. Calculate the amount of energy absorbed. **[12.95 eV]**

13. Calculate the ratio of the minimum wavelength of the Lyman series to the minimum wavelength of the Balmer series in the hydrogen spectrum. **[1:4]**

14. Find the wavelength of light corresponding to the photon of smallest energy in the Lyman series of hydrogen spectrum. **[121.9 nm]**

15. Calculate the wavelengths of the first four members of the Lyman series in the hydrogen spectrum (Given: Rydberg constant, $R = 1.1\times10^{7}$ m^{-1}). **[121.2 nm; 102.3 nm; 97.0 nm; 94.7 nm]**

16. If the separation of a hydrogen atom into a proton and an electron needs 13.6 eV of energy, what would have been the orbital radius and velocity of the electron inside the atom ? **[5.24×10^{-11} m; 2.2×10^{6} ms^{-1}]**

17. A hydrogen atom in the ground state absorbs energy to get excited to the state n= 3. If it returns to the ground state, what will be the wavelengths present in the radiation emitted ? [Hint: The atom may return directly to the ground state (n = 1) or via the state n = 2. ∴ the λ's present will correspond to the transitions for n-values 3 →1, 3 →2, and 2 →1). **[102.6 nm; 121.8 nm; 653.7 nm]**

CHAPTER 8 (B)
RADIOACTIVITY; NUCLEAR FISSION & FUSION

8B. 01. Radioactivity (or radioactive decay); Alpha (α), Beta (β) & Gamma (γ)-radiations

"The process in which an unstable nuclide spontaneously emits a radiation and transforms into another nuclide" is called **radioactivity** or **radioactive decay** or **radioactive disintegration**. As already stated, the nuclide transforms into a stable or a less unstable nuclide. The phenomenon is not affected by the external conditions of temperature, pressure, chemical combination, etc. It means that the electrons in the atom do not play any part in the phenomenon and so the radioactivity is exclusively a property of the nuclei. The process is spontaneous, in the sense that no external energy needs to be supplied to the radioactive material for the emission of radiation.

The radioactivity was discovered by the French physicist Henry Becquerel in 1896. He observed that uranium salts blackened a nearby photographic plate even when it was wrapped in black paper and separated by a metal (silver) plate, thus implying that the salts emitted some invisible penetrating radiation. Later, the French physicists Marie Curie and her husband Pierre Curie extensively studied the phenomenon and in 1898 isolated the highly radioactive elements radium (10^6 times more active than uranium) and polonium. Marie Curie, Pierre Curie and Becquerel jointly received the Nobel Prize for physics in 1903 for the discovery of radioactivity (later, in 1911, Madam Curie earned the rare distinction of winning another Nobel Prize, for chemistry).

The radioactivity of a material results from the emission of 3 kinds of radiation, known as **α, β, and γ-radiations (or rays)**. Since they all emanate from the nuclei, collectively they are called **nuclear radiation**. They are readily differentiated from one another in terms of their ability to penetrate matter. The α-rays are least penetrating and the γ-rays are most penetrating. A ray consists of a collection of a large number of particles. It is not necessary that a radioactive material emits all the three types of radiations. Some important characteristics of the α, β, and γ-radiations will be briefly described in the following.

α - rays

(i) An α-particle is a doubly charged helium nucleus $\left({}^{4}_{2}\text{He}\right)$, carrying two units of +ve charge (it is a helium atom that has lost both of its orbital electrons, to become a doubly charged He^{++} ion).

(ii) An α-particle has large mass (4 times that of a hydrogen atom and >7000 times the mass of an electron). Hence in spite of a high velocity its penetrating power is very small. The α-rays can be stopped just by a sheet of paper, or by a thin aluminium foil (≈ 0.02 mm thick), or by a few centimeters of air.

(iii) In addition to its large mass, an α-particle also moves with a high velocity ($\approx 10^7$ ms^{-1}). So it possesses a large momentum. Hence the α-rays have a high ionizing power.

(iv) Because of their +ve charge, the α-rays are deflected by electric and magnetic fields.

β - rays

(i) A β-particle is either an electron (charge –e) or a positron (charge +e). [Except for the difference in an charge, a positron is identical with an electron, with mass 9.1×10^{-31} kg. Like the electron, it is a fundamental particle of nature. We may say that it is a 'positive electron'. (At one time the word 'negatron' was suggested for electron, but not accepted)].

(ii) The β-rays travel with a very high velocity, ranging from nearly 33% to 99% of the velocity of light. Because of such high velocity and light mass, they possess a high penetrating power. They can easily penetrate a few millimeters thickness of aluminum and can travel upto several meters in air.

(iii) Because of the low mass of β-particles, the ionizing power of the β–rays is very small as compared to the α-rays, viz. about 1/100.

(iv) Because of their electric charge, like the α-rays the β-rays are also deflected by electric and magnetic fields.

γ- Rays

(i) The γ-rays are electromagnetic radiation of very high frequency (higher than the frequency of X-rays). They consist of highly energetic photons ($E = h\nu$), which are electrically neutral. The rest mass of a γ-ray photon is zero.

(ii) Being an electromagnetic radiation, the γ-rays travel with the velocity of light in air or vacuum. Further, also because they have a high energy and they are electrically neutral, they have a very high penetrating power. They can penetrate a few centimeters thickness of even a heavy metal like lead.

(iii) Because of their high energy, the γ-rays can knock out electrons from the surface of a metal on which they are incident.

(iv) The ionizing power of the γ-rays is much smaller than those of the α-rays and the β-rays.

(v) Being electrically neutral, the γ-rays are not deflected by electric and magnetic fields.

Artificial and Induced Radioactivity

The relatively light nuclei are not radioactive, or we can say that they do not show 'natural radioactivity'. In this sense they are said to be stable nuclei. However, when certain elements are bombarded with energetic particles, they also disintegrate, i.e. they become radioactive. "The phenomenon of disintegration of the normally stable nuclei by bombardment with energetic particles" is called **artificial radioactivity**. Usually, the radioactivity lasts only as long as the bombardment continues. But sometimes it continues even after the bombardment stops. This "persistence of disintegration even after the bombardment of the element with suitable particles has stopped" is called **induced radioactivity**. A familiar example is the disintegration of boron by bombardment with α-particles, (the respective sums of A and Z, before and after the disintegration, should be the same),

$$^{10}_{5}B + ^{4}_{2}He \rightarrow ^{14}_{7}N \rightarrow ^{13}_{7}N + ^{1}_{0}n$$

$$^{13}_{7}N \rightarrow ^{13}_{6}C + ^{0}_{+1}e$$

($^{1}_{0}n$: neutron; $^{0}_{+1}e$: positron). Usually, the source of the bombarding particles is a naturally radioactive material like radium, but accelerated charged particles like protons are also employed.

Example 1. A radioactive nucleus undergoes a series of decays according to the sequence given below. If the mass number and atomic number of A_3 are 172 and 69, respectively, what are the mass number and atomic number of A ? (CBSE; 1994).

$$A \xrightarrow{\beta} A_1 \xrightarrow{\alpha} A_2 \xrightarrow{\alpha} A_3$$

Solution. Let A′ and Z′ be the mass number and atomic number, respectively, of the nucleus A. Then, the given sequence can be expressed as, (assuming that the β-particle is an electron),

$$^{A'}_{Z'}A \longrightarrow ^{0}_{-1}e + ^{A'}_{Z'+1}A_1 \longrightarrow ^{4}_{2}He + ^{A'-4}_{Z'-1}A_2 \longrightarrow ^{4}_{2}He + ^{A'-8}_{Z'-3}A_3$$

We are given that,

$$Z' - 3 = 69 \quad \text{and} \quad A' - 8 = 172$$

$$\therefore \quad Z' = (69+3) = 72 \quad \text{and} \quad A' = (172+8) = 180 \quad \textbf{(Answer)}.$$

8B.02. Law of Radioactive Decay; Half-Life and Mean Life; a.m.u. (or u)

We have known that in the process of radioactive decay (or disintegration) a radioactive nuclide transforms into another nuclide by the emission of nuclear radiation (α, β, or γ-rays). The transformation can be extremely fast (~10^{-15} second) to extremely slow (even upto the age of universe, viz. 10^{10} year). The main feature is that it is a continuous process, with some nuclei transforming themselves all the time. All the nuclei have the same probability of transformation. But it is not possible to identify that which nuclei will transform before the rest. In this sense, the process of radioactive decay is a purely statistical process. We cannot predict when some particular nuclei will decay. We only know that they **shall** decay, sooner or later.

Although it is not possible to fix which particular nuclei will decay at a given moment, we can always work out the rate of decay by estimating how much of the material undergoes decay within a given period of time. Since all the nuclei have the same probability of decay, we know that greater is the amount of radioactive material present at a given time, the greater will be its rate of decay. Thus, if the material contains a total number of N nuclei, its rate of decay ($-$ dN/dt) is proportional to N,

$$-\left(\frac{dN}{dt}\right) \propto N$$

$$\Rightarrow \quad -\left(\frac{dN}{dt}\right) = \lambda N \qquad \text{..... (8.18)}$$

The minus sign arises because the amount of the material reduces with time. The constant of proportionality λ is called **decay constant** (or **disintegration constant**). It has a characteristic value for every radionuclide, i.e. its value differs from one radionuclide to another.

To find the number of nuclei N present at a given time, we evaluate the value of N as a function of time. From Eq. 8.18 we have,

$$\frac{dN}{N} = -\lambda\, dt$$

Let N_0 be the number of radioactive nuclei initially present at time t = 0. Integrating on both the sides we get,

$$\int_{N_0}^{N} \frac{dN}{N} = -\lambda \int_{o}^{t} dt$$

$$\Rightarrow \quad \log_e N - \log_e N_0 = -\lambda t$$

$$\Rightarrow \quad \log_e \frac{N}{N_0} = -\lambda t$$

We take exponential of both sides (an exponential function e^x is the antifunction of the natural logarithm $\log_e x$),

$$\frac{N}{N_0} = e^{-\lambda t}$$

$$\Rightarrow \quad \boxed{\begin{array}{c} N = N_0\, e^{-\lambda t} \\ \text{(Radioactive decay)} \end{array}} \qquad \text{..... (8.19)}$$

We can also use Eq. 8.19 to find the **rate of decay,** R_t, at any given time t,

$$R_t = -\frac{dN}{dt} = \lambda N_0\, e^{-\lambda t}$$

The initial rate of decay (t = 0) is ($e^{\circ} = 1$),

$$R_0 = \lambda N_0 \qquad \text{..... (8.20)}$$

Thus we can write,

$$\boxed{\begin{array}{c} R_t = R_0\, e^{-\lambda t} \\ \text{(Rate of radioactive decay)} \end{array}} \qquad \text{..... (8.21 a)}$$

The Eq. 8.18 can also be rewritten in the form (vide Eq. 8.18),

$$\boxed{\begin{array}{c} R_t = \lambda N \\ \text{(Rate of decay; another form)} \end{array}} \qquad \text{..... (8.21 b)}$$

The decay rate R [number of decays (or disintegrations per second] of a sample of radionuclides is called the **activity** (short form of 'radioactivity') of the sample. Its SI unit is **becquerel (Bq),**

1 becquerel (or 1 Bq) = 1 decay per second

However, a bigger unit **curie (Ci)** is more popularly used,

1 curie (Ci) = 3.7×10^{10} decay per second = 3.7×10^{10} Bq

The 'activity' is usually represented by the symbol A. Since A is essentially same as R_t, from Eq. 8.21 (b) we have,

$$A = \lambda N \qquad \text{..... (8.21 c)}$$

Further, in Eq. 8.21(a) now we can write A_0 for R_0. Then we may write,

$$A = A_0\, e^{-\lambda t} \qquad \text{..... (8.21 d)}$$

Thus, note that both N and A decrease exponentially with time.

Half-life. Two common measures are employed to express the life of a radionuclide, i.e. to know how long will the radionuclide last before it is completely disintegrated. One measure is the **half-life.** It is **defined**

as "the time at which one-half of the radioactive material is disintegrated". It is denoted by $\mathbf{T_{1/2}}$, where T is the total lifetime of the material. During the time $T_{1/2}$, both N and R are reduced to one-half of their initial values. Thus, in Eq. 8.19, $N = N_0/2$ at time $t = T_{1/2}$,

$$\frac{N_0}{2} = N_0\, e^{-\lambda T_{1/2}} \quad \text{or} \quad e^{-\lambda T_{1/2}} = \frac{1}{2}$$

Or
$$-\lambda T^{1/2} = (\log_e 1 - \log_e 2) = (0 - \log_e 2) = -\log_e 2$$

Or
$$T_{1/2} = \frac{\log_e 2}{\lambda} = \frac{0.693}{\lambda} \quad \text{..... (8.22)}$$

In one half-life, the number of atoms reduces to half, i.e.

$$N = \frac{N_0}{2} = N_0\left(\frac{1}{2}\right) \quad \text{[in 1 half-life]}$$

In two half-lives, N will reduce to,

$$N = \frac{1}{2}\left[N_0\left(\frac{1}{2}\right)\right] = N_0\left(\frac{1}{2}\right)^2, \quad \text{[in 2 half-lives]}$$

and so on. Thus in n half-lives, N will reduce to,

$$N = N_0\left(\frac{1}{2}\right)^n \quad \text{..... (8.23 a)}$$

If T is the total time for n half-lives, $T = n \times T_{1/2}$ or $n = T/T_{1/2}$. So eq. 8.23 (a) can also be written as,

$$N = N_0\left(\frac{1}{2}\right)^{T/T_{1/2}} \quad \text{..... (8.23 b)}$$

Mean life (or Average life). Another measure of the life of a radioactive material is **mean life** (or **average** life), **defined** as "the time at which the amount of the radioactive material is reduced to e^{-1} of its initial value". [It is also defined as "the total lifetime of all the atoms of the radioactive material, divided by the total number N of the atoms initially present in the material". When worked out mathematically, it amounts to be the same as the first definition]. It is denoted by $\boldsymbol{\tau}$. During the time τ, both N and R are reduced to e^{-1} of their initial values. Thus from the Eq. 8.19 we get,

$$(e^{-1} N_0) = N_0\, e^{-\lambda\tau}$$

$$\Rightarrow \quad e^{-1} = e^{-\lambda\tau}$$

$$\Rightarrow \quad -1 = -\lambda\tau$$

$$\Rightarrow \quad \tau = \frac{1}{\lambda} \quad \text{..... (8.24)}$$

Using the Eqs. 8.2 and 8.24 we get the relation between the half-life $T_{1/2}$ and the mean life τ as,

$$T_{1/2} = \frac{\log_e 2}{\lambda} = (\log_e 2)\tau = 0.693\,\tau \quad \text{..... (8.25)}$$

Atomic mass unit (a.m.u. or u). Since atomic masses are extremely small, they are usually measured in smaller units called a.m.u. or simply u, instead of in kilogram, which is too large a unit for them,

$$1 \text{ a.m.u. (or 1u)} = 1.66 \times 10^{-27} \text{ kg}$$

The corresponding energy values ($E = mc^2$) are usually measured in electron-volt (eV),

$$1\text{eV} = 1.60 \times 10^{-19} \text{ J}$$

In the relation $E = mc^2$, the multiplying constant c^2 has the value,

$$c^2 = 9.315 \times 10^8 \text{ eV/u} = 931.5 \text{ MeV/u}.$$

Example 2. During the radioactive decay of radon, how much of it would be left after 19 days of decay, if its initial amount was 20 mg ($T_{1/2}$ for radon = 3.8 days) ?

Solution. Given : $T_{1/2} = 3.8$ days; $t = 19$ days; $N_0 = 20$ mg

The number of half-lives in 19 days is given by,

$$n = \frac{T}{T_{1/2}} = \frac{19}{3.8} = 5$$

∴ The amount left after the decay is given by,

$$N = N_0\left(\frac{1}{2}\right)^n = 20\left(\frac{1}{2}\right)^5$$

$$= 0.625 \text{ mg} \quad \textbf{(Answer).}$$

Example 3. A radioactive element with half-life 40 second consists of 10^8 radioactive nuclei. Determine how many of these nuclei would be left after 20 second of decay.

Solution. Given: $N_0 = 10^8$; $T_{1/2} = 40$ s; $t = 20$ s

The number of half-lives in t second is given by,

$$n = \frac{t}{T_{1/2}} = \frac{20}{40} = \frac{1}{2}$$

∴ The number of nuclei left after 20 s, i.e. after 1/2 half-lives, is given by

$$N = N_0\left(\frac{1}{2}\right)^n = N_0\left(\frac{1}{2}\right)^{1/2}$$

$$= 10^8 \times 0.707$$

$$= 7\times10^7 \quad \textbf{(Answer).}$$

8B. 03. Explanation of α, β and γ-Decay

We have known that the radioactive decay occurs by the emission of one or more of the α, β and γ-radiations. In the last section, we had established the laws of radioactive decay (vide Eqs. 8.22 to 8.25). Now we shall provide a brief explanation of the processes of α, β and γ-decay.

The emission of any kind of nuclear radiation changes the original nucleus into a new nucleus. The original nucleus is termed as the **parent nucleus** and the new nucleus as the **daughter nucleus.** All the processes of radioactive decay follow the rule that the sum of the atomic numbers Z (representing the nuclear charge), as well as the sum of the mass numbers A, before and after the decay, should be the same.

α-Decay

In the **α-decay,** the parent unstable nucleus of an element X emits an α-particle $\left({}_2^4He\right)$, which reduces both its atomic (or proton) number Z and the neutron number N by 2 each. The mass number A reduces by 4. Because of the change in the atomic number, the element X changes to a new element Y as,

$${}_Z^AX \longrightarrow {}_{Z-2}^{A-4}Y + {}_2^4He + Q \qquad ..... (8.26)$$

Q is the energy released during the process. As an example, a nucleus of polonium decays by α-emission to the nucleus of lead as follows,

$${}_{84}^{210}Po \longrightarrow {}_{82}^{206}Pb + {}_2^4He + Q$$

The α-decay has been found to occur in all heavy nuclei with the mass number A>210. When a nucleus starts becoming too heavy, i.e. when the number of protons and neutrons contained in it becomes too large, the Coulomb repulsive force between the protons tends to dominate the attractive forces within the nucleus (the nuclear forces will be discussed in the next section). This makes the nucleus unstable. The emission of an α-particle reduces its mass number and thus reduces its instability.

An energy Q is released in the process of radioactive decay, because the rest mass energy of the parent nucleus X is greater than the combined rest mass energy of the daughter nucleus Y and the α-particle ${}_2^4He$. "The difference between the rest mass energy (U_i) of the initial parent nucleus and the combined rest mass energy (U_f) of the final products of radioactive decay is called the **Q-value** or **disintegration energy** of the process of decay",

$$Q = U_i - U_f \qquad ..... (8.27)$$

This definition not only applies to α-decay, but to all the nuclear processes. The disintegration energy Q is shared by the daughter nucleus and the α-particle. Using the Einstein's mass-energy relation $E = mc^2$, the Eq. 8.27 may also be written as

$$Q = (m_X - m_Y - m_{He})\,c^2 \qquad ..... (8.28)$$

The release of disintegration energy explains why a radioactive decay occurs **spontaneously,** i.e. without the supply of any external energy. The excess energy present inside the parent nucleus provides the driving force for the emission of an α-particle. Further, we now have another reason why a heavy nucleus should undergo radioactive decay in general (and α-decay in particular), viz. because its rest mass energy is greater than the total rest mass energy of the decay products.

Velocity and kinetic energy of the α-particles. We can find the values of velocity and k.e. of an α-particle as follows. We consider a parent nucleus at rest, which is undergoing α-decay. Its linear momentum (mass×velocity) is zero. Let v_Y and v_α be the velocities of the daughter nucleus and the α-particle, respectively. According to the principle of conservation of linear momentum,

$$0 = m_\gamma v_\gamma + m_\alpha v_\alpha \text{ or } m_\gamma v_\gamma = -m_\alpha v_\alpha$$

Or
$$v_\gamma = \frac{m_\alpha v_\alpha}{m_\gamma}$$

(We neglect the –ve sign, as it is of no importance to us here. We are only concerned with the magnitude of v_γ).

The disintegration energy Q released in the process of α-decay is shared as kinetic energy by the daughter nucleus Y and the α-particle. Therefore,

$$\frac{1}{2} m_\alpha v_\alpha^2 + \frac{1}{2} m_\gamma v_\gamma^2 = Q$$

Substituting for v_γ, we get,

$$\frac{1}{2} m_\alpha v_\alpha^2 + \frac{1}{2} m_\gamma \left(\frac{m_\alpha v_\alpha}{m_\gamma}\right)^2 = Q$$

$$\Rightarrow \quad \frac{1}{2} m_\alpha m_\gamma v_\alpha^2 + \frac{1}{2} m_\alpha^2 v_\alpha^2 = m_\gamma Q \text{ (multiplying by m on both sides)}$$

$$\Rightarrow \quad \frac{1}{2} m_\alpha v_\alpha^2 (m_\gamma + m_\alpha) = m_\gamma Q$$

We denote the k.e. $\frac{1}{2} m_\alpha v_\alpha^2$ of the α-particle by $[E_k]_\alpha$. Then,

$$\frac{1}{2} m_\alpha v_\alpha^2 = [E_k]_\alpha = \frac{m_\gamma Q}{m_\gamma + m_\alpha} \quad \text{..... (8.29)}$$

Approximate values of the rest mass m_α and m_γ are,

$$m_\alpha = 4 \text{ a.m.u.}$$
$$m_\gamma = (A - 4) \text{ a.m.u.}$$

So the Eq. 8.29 can also be written as,

$$\frac{1}{2} m_\alpha v_\alpha^2 = [E_k]_\alpha = \frac{(A-4)\,Q}{A} \quad \text{..... (8.30)}$$

If the value of $[E_k]_\alpha$ is known, then the value of the velocity v_α can be calculated.

In the Eq. 8.29, m_γ is a fixed value for a given daughter nucleus. m_α is also a fixed value. Therefore, for a given daughter nucleus, all the emitted α-particles have a well defined value of the k.e. $[E_k]_\alpha$ and so also a well defined value of the velocity v_α.

β-Decay

The process of **β-Decay** consists of the emission of either an electron or a positron (vide Section 8B.01), when a parent nucleus changes into a daughter nucleus. The two processes are known as β^--decay and β^+-decay, respectively. They are accompanied with a simultaneous emission of an antineutrino or a neutrino, respectively.

$$^{A}_{Z}X \longrightarrow {}^{A}_{Z+1}Y + \underset{\text{(electron)}}{{}^{0}_{-1}e} + \underset{\text{(antineutrino)}}{\bar{\nu}} + Q \ (\boldsymbol{\beta^-}\textbf{-decay}) \quad \text{..... (8.31 a)}$$

$$^{A}_{Z}X \longrightarrow {}^{A}_{Z-1}Y + \underset{\text{(positron)}}{{}^{0}_{+1}e} + \underset{\text{(neutrino)}}{\nu} + Q \ (\boldsymbol{\beta^+}\textbf{-decay}) \quad \text{..... (8.32 a)}$$

The neutrino and antineutrino are particles with zero rest mass, zero charge (i.e. neutral), and spin ± (1/2). The antineutrino is the antiparticle of the neutrino, just as a positron is the antiparticle of an electron. We give below one example each of the β^--decay and β^+-decay (thorium to protactinium and sodium to neon),

$$^{234}_{90}Th \longrightarrow {}^{234}_{91}Pa + {}^{0}_{-1}e + \bar{\nu} + Q \ (\beta^-\text{-decay})$$

$$^{22}_{11}Na \longrightarrow {}^{22}_{10}Ne + {}^{0}_{+1}e + \nu + Q \ (\beta^+\text{-decay})$$

Being electrically neutral particles, the neutrinos interact very little with matter. They can penetrate matter, even earth, without getting absorbed. This makes their detection very difficult and had been the the reason for their late discovery.

In the β^--decay process, a neutron is converted into a proton inside the nucleus, with simultaneous creation of an electron and an antineutrino. In the β^+-decay process, a proton is converted into a neutron, with simultaneous creation of a positron and a neutrino. If a nucleus has more neutrons than needed for its stability, a neutron tends to transform into a proton to reduce the instability. Similarly, if a nucleus has more protons than necessary, a proton tends to transform into a neutron.

$$\underset{\text{(neutron)}}{n} \longrightarrow \underset{\text{(proton)}}{p} + \underset{\text{(electron)}}{e^-} + \underset{\text{(antineutrino)}}{\bar{\nu}} \quad (\beta^-\text{-decay}) \qquad \text{..... (8.31 b)}$$

$$\underset{\text{(proton)}}{p} \longrightarrow \underset{\text{(neutron)}}{n} + \underset{\text{(positron)}}{e^+} + \underset{\text{(neutrino)}}{\nu} \quad (\beta^+\text{-decay}) \qquad \text{..... (8.32 b)}$$

The disintegration energy Q is shared between the end products of the β-decay. However, because of its comparatively very large mass, the daughter nucleus Y shares very little energy. The energy is practically shared between the β-particle and the neutrino (or antineutrino). Depending upon what fraction of Q is taken away by the neutrino (or antineutrino), the kinetic energy of the β-particle can be any value between 0 and Q (similarly for the neutrino or the antineutrino). This is in sharp contrast to the process of α-decay, in which for a given daughter nucleus all the α-particles have the same value of k.e.

Note that during the process of β-decay the mass number A of the nuclide does not change. Just either a neutron converts into a proton or vice-versa, in accordance with the Eqs. 8.31b and 8.32b.

γ-Decay

The **γ-decay** consists of the emission of γ-ray photons from the parent nuclei of a radioactive material. Since a photon has neither a charge nor a rest mass, the daughter nucleus has the same atomic number Z and the same mass number A as the parent nucleus. In general, the process may be represented as,

$${}^{A}_{Z}X \longrightarrow {}^{A}_{Z}Y + \gamma \qquad \text{..... (8.33)}$$

We already know about the atomic (or electronic) energy levels, corresponding to the revolution of electrons in various orbits around the nucleus. The transition of electron between these levels results in the emission of electromagnetic radiation in the form of visible, ultraviolet and infrared radiations. Similarly, various energy states, known as nuclear energy levels, exist inside an atomic nucleus. The energy of a nucleus, like that of an atom, is quantized, i.e. the nucleus can exist only in discrete quantum states, each with a well defined energy. The energy of the nuclear states is very high, of the order of million electron volt (MeV). The transition of the nucleus from a higher energy state to a lower energy state results in the emission of energy in the form of a γ-ray photon, which is another form of electromagnetic radiation. Since the energies of the nuclear states are of the order of several MeV, the differences of energy are also of the order of several MeV. Hence the energies of the γ-ray photons are very high (of the order of several MeV).

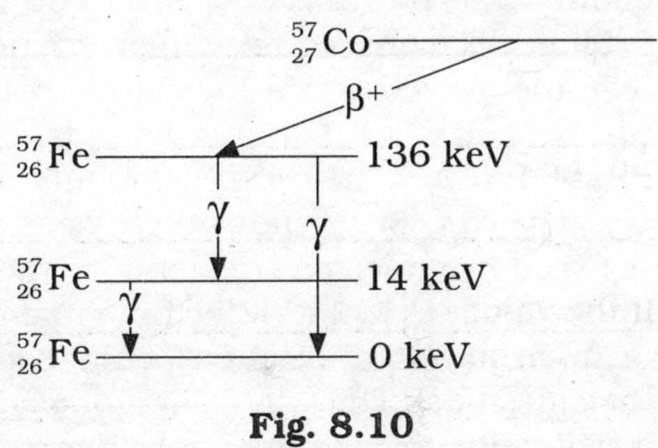

Fig. 8.10

A nucleus cannot be excited to such a high energy state by the usual method of supplying energy to the radioactive material, e.g. by heating. However, the emission of an α-particle or a β-particle leaves the nucleus in an excited state. The nucleus finally comes to its ground state by emitting γ-ray photons in one or more steps. An example is shown in Fig. 8.10. A parent nucleus ${}^{57}_{27}Co$ decays to the daughter nucleus ${}^{57}_{26}Fe$ by β+-decay. The nucleus ${}^{57}_{26}Fe$ is in excited state, with energy 136 keV above the ground state. It can reach its ground state either (i) by emitting a photon of energy 136 keV or (ii) by emitting a photon of energy 122 keV to reach a lower excited state of energy 14 keV and then emitting another photon of energy 14 keV. Thus, if we have a radioactive ${}^{57}_{27}Co$ source, it will be found emitting β+-particles (positrons) and photons of energy 136 keV, 122 keV and 14 keV.

The radioactivity is a sign of the instability of nuclei, as will be explained at the end of Section 8B.05, after we have discussed about forces and binding energy.

8B.04. Nuclear Forces

In the last chapter we have known that an atomic nucleus consists of nucleons, viz. the positively charged protons and the electrically neutral neutrons. Its size is extremely small (≈ a few fm; 1 fm = 10^{-15} m) and hence its density is extremely high (≈ 10^{17} kg/m³). It means that the +vely charged protons are situated extremely close to each other. Then, their strong repulsive Coulombian force should make the nucleus fall apart. Since it does not happen, it follows that some extraordinarily strong attractive forces exist inside the nucleus. "The extremely strong attractive forces that keep the nucleons bound together inside the nucleus" are called **nuclear forces**. They exist alongwith the Coulombian forces (e.g. the repulsive forces between the protons and the attractive forces between the +vely charged protons and the –vely charged orbital electrons), but they are far stronger than them. Their magnitude is such that whereas the work required to separate an orbital electron from an atom is of the order of electron volts (eV), the work required to separate two nuclear protons is of the order of million electron volts (MeV). The nuclear forces are charge-independent. They exist between (a) protons and protons, (b) neutrons and neutrons, and (c) protons and neutrons.

Additional knowledge

Carbon dating. In a living matter containing carbon, the stable carbon isotope ${}^{12}_{6}C$ coexists with a small proportion of the radioactive isotope ${}^{14}_{6}C$, having decay rate of nearly 900 per hour per gram. The decay rate is maintained by the interaction of the living matter with atmosphere. After the living organism is dead, the decay rate starts falling. Then at any subsequent moment of time, which may extend from a few years to about 35,000 years, from the knowledge of half-life (5730 years) of ${}^{14}_{6}C$ and the measured rate of decay at that moment, approximate age of the organism can be calculated. This process is known as **carbon dating**. It is extensively employed to estimate the age of archaeological and fossil remains, rocks, etc.

The influence of the nuclear forces is practically confined to the nucleus itself. The forces do not affect the orbital electrons or any other particle outside the nucleus (e.g. the α-particles in the Rutherford's experiment; Chapter 8A, Section 8A.02). In other words, they are extremely **short-range** forces. Their range of influence exists barely a few fm. The Fig. 8.11 shows how sensitively the nuclear force between two nucleons depends on the separation between the nucleons. It is a plot of the potential energy (MeV) of the pair of nucleons as a function of the separation r (fm) between the nucleons. The –ve values of the p.e. correspond to the attractive nature of the nuclear force (since then energy has to be supplied to separate the two nucleons) and the +ve values to the repulsive nature of the nuclear force. The p.e. is maximum, and so the attractive force is also maximum, when the separation is equal to r_0, which has a value nearly equal to 1 fm (Chapter 8A; Eq. 8.16). The nature of the variation is better understood if the curve is followed along decreasing r. As the value of r decreases, the attractive force gradually increases, until the the value r_0 is reached, after which it starts reducing very fast, so much so that it soon acquires a strongly repulsive character.

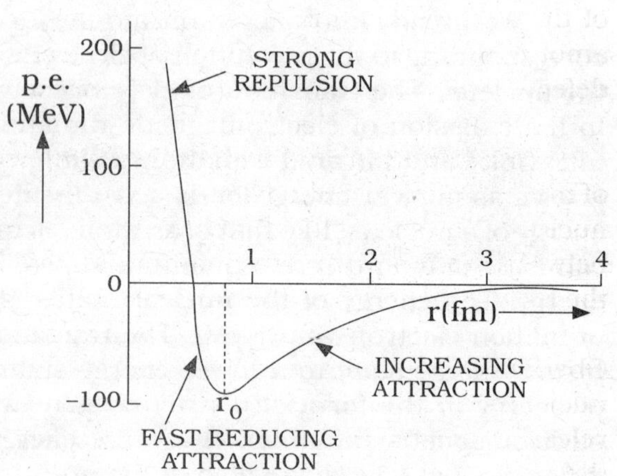

Fig. 8.11

Thus, it is noted that the nuclear forces are characterized by the following **properties**:

(i) The nuclear forces are short-range forces. The **nuclear range**, i.e. the range of their effectiveness, is about a few fm, out of which the maximum effectiveness is limited to just nearly 1 fm.

(ii) The nuclear forces are independent of electric charge. The nuclear force between two neighbouring neutrons is the same as that between two neighbouring protons or between a proton and its neighbouring neutron.

(iii) The nuclear forces are the strongest known forces in nature. They are much stronger than the electrostatic (or electromagnetic) Coulombian forces between the nucleons and far far stronger than the gravitational forces between the nucleons.

(iv) The nuclear forces are non-central forces. They do not solely depend upon the distance between the nucleons. They are also spin-dependent. They depend upon the direction of spin of the nucleons (m_s = + 1/2 or –1/2).

Additional knowledge

The nuclear forces are complex in nature. They are not fully understood as yet. Efforts to acquire a full understanding are still in progress.

Earlier, the nuclear forces were believed to result from a constant exchange of the fundamental particles called **mesons**, between the nucleons (there are 3 types of mesons; the mass of a meson is 270 times the mass of an electron). However, lately neither the protons and neutrons nor the mesons are regarded as fundamental particles, i.e. as ultimate constituents of matter. They all are believed to have an underlying structure of sub-units. The protons and neutrons are thought to be made up of the fundamental particles called **quarks** [there are 6 types of quarks, with different masses, two types of charge (+2/3 and –1/3), and the same spin (1/2)]. The mesons are considered to be quark-antiquark pairs. A strong force is believed to exist inside the nucleons, which binds the quarks together to form neutrons and protons. The nuclear force is now believed to be a spillover effect or by-product of this force. Thus, now the nuclear force is not regarded as a primary fundamental force of the nature, but as a secondary force.

8B.05 Mass Defect; Nuclear Binding Energy; Nuclear Reactions

Prima facie, the mass of a nucleus is expected to be equal to the sum of masses of the nucleons contained in the nucleus. But in reality it is not so. "The difference between the sum of the masses of the nucleons forming a nucleus and the actual mass of the nucleus" is called **mass defect (Δm).** The mass M of the nucleus is found to be less than the total mass Σm of its constituent nucleons. It means that some amount of the mass of the nucleons is lost in the process of formation of the nucleus. Hence the name 'mass defect'.

$$\Delta m = \Sigma m - M \qquad \text{..... (8.34 a)}$$

According to the Einstein's well known relationship, $E = mc^2$, expressing the principle of equivalencce of mass and energy, a mass m has energy mc^2 associated with it. It follows that the rest mass energy Mc^2 of a nucleus is less than the total rest mass energy Σmc^2 of its constituent protons and neutrons. "The difference between the rest mass energy of nucleus and the total rest mass energy of its constituent nucleons" is called the **binding energy of the nucleus**, $(\Delta E)_{be}$,

$$(\Delta E)_{be} = \Sigma mc^2 - Mc^2 = \Delta m.c^2 \qquad \text{..... (8.34 b)}$$

Obviously, the binding energy is the energy equivalent of the mass defect. The mass defect develops when the nucleons 'bind' together to form a nucleus. It means that an amount of energy equal to the binding energy is released (and therefore lost) during the formation of the nucleus. It further means that if we want to reverse the process, i.e. if we want to separate the nucleons from one another, an amount of energy equal to the binding energy will have to be supplied to the nucleus. This provides us **another definition of the binding energy,** viz. "the binding energy is the amount of energy required to be supplied to a nucleus in order to separate its constituent nucleons from one another".

Thus, the binding energy of a nucleus is a measure of how strongly the nucleons are bound together inside the nucleus. Or, it can be regarded as a measure of the strength of the nuclear forces that bind the nucleons together inside the nucleus (thus, it also determines the stability of the nucleus). Hence the name 'binding' energy. Note that it is merely a difference between two mass energy values, that of the nucleus and the nucleons. It is not an energy 'residing' in the nucleus.

Denoting the proton and neutron masses as m_p and m_n, respectively, the Eq. 8.34(b) can be written in terms of the proton number Z and the mass number A as,

$$(\Delta E)_{be} = [Zm_p + (A - Z)m_n - M]\, c^2 \qquad \text{..... (8.35)}$$

The equation can also be written in terms of atomic masses, instead of the nuclear masses, as,

$$(\Delta E)_{be} = \left[Zm\left({}^1_1H\right) + (A - Z)m_n - m\left({}^A_ZX\right)\right] c^2 \qquad \text{..... (8.36)}$$

Here $m\left({}^1_1H\right)$ is the mass of a hydrogen atom and $m\left({}^A_ZX\right)$ is the atomic mass for an element X. Note that the masses of the electrons mutually cancel in this equation, as both the atom ${}^A_ZX$ and the Z hydrogen atoms have Z electrons each.

A more useful measure of the binding energy is the **binding energy per nucleon** $(\Delta E)_{ben}$. It is "the ratio of the binding energy $(\Delta E)_{be}$ of a nucleus to the number A of the nucleons in the nucleus",

$$(\Delta E)_{ben} = \frac{(\Delta E)_{be}}{A} \qquad ..... (8.37)$$

The higher is the value of $(\Delta E)_{ben}$, the higher is the **stability** of the nucleus.

Binding Energy Curve

The physical significance of the binding energy is better appreciated by studying its variation among the nuclei of the various elements (or, say, nuclides). The Fig. 8.12 is a plot of the binding energy per nucleon, $(\Delta E)_{ben}$, as a function of the mass number A, for a large number of nuclei (the positions of a few representative nuclei are marked with black dots). It provides a great deal of information. Some obvious facts are as follows:

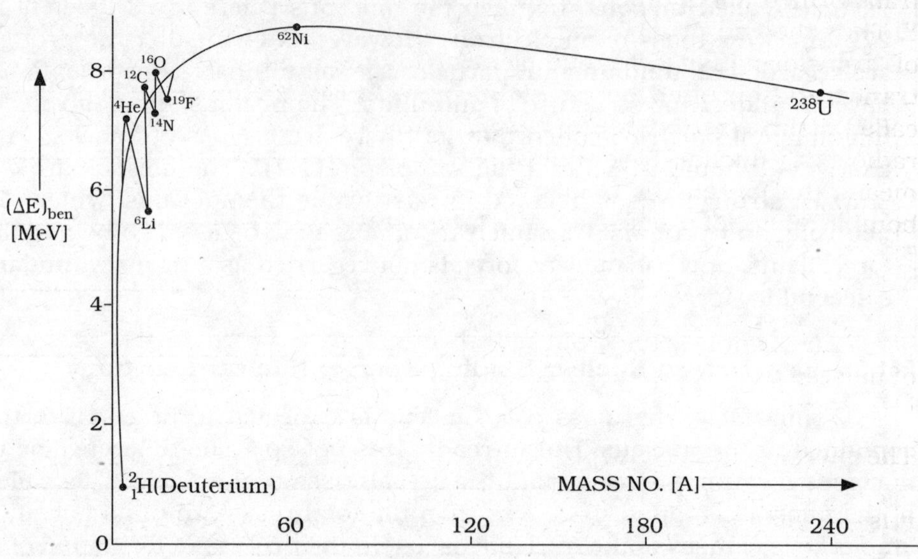

Fig. 8.12

(i) The binding energy per nucleon, $(\Delta E)_{ben}$, is small for light nuclei, e.g. for deuterium (an isotope of hydrogen), ${}^{2}_{1}H$, its value is just 1.1 MeV.

(ii) With increase in the mass number A, $(\Delta E)_{ben}$ increases. For the values 2 to 20 for A, sharply defined peaks are observed for ${}^{4}_{2}He$, ${}^{12}_{6}C$, and ${}^{16}_{8}O$. It means that these nuclei are relatively more stable than their neighbouring nuclei lying along the continuous curve passing between the dots.

(iii) $(\Delta E)_{ben}$ has a maximum value around 8.7 MeV for the range A $\approx$ 50 to 80. The highest value is nearly 8.8 MeV, for ${}^{62}_{28}Ni$.

(iv) As the mass number A increases, i.e. for heavier nuclei, $(\Delta E)_{ben}$ decreases. For one of the heaviest natural element uranium, ${}^{238}_{92}U$, the value drops to nearly 7.5 MeV.

(v) It follows that the nuclei in the intermediate region, A $\approx$ 50 to 80, are most stable.

The above considerations are important for the processes of **nuclear fission** and **nuclear fusion**, that we shall describe in the coming sections. In the nuclear fission, to attain greater stability, a heavy nucleus like uranium splits into two nuclei of lesser total mass. The difference of mass appears in the form of enormous energy. In the nuclear fusion, for the same purpose two light nuclei like deuterium $({}^{2}_{1}H)$ fuse together to form a single heavy nucleus of lesser mass. Once again an enormous amount of energy is released in the process.

Example 4. Calculate the binding energy per nucleon for ${}^{56}_{26}Fe$ (atomic mass of ${}^{56}_{26}Fe$ = 55.934932; mass of proton = 1.007825 a.m.u.; mass of neutron = 1.008665 a.m.u.).

Solution. Number of protons = 26

$\therefore$ Number of neutrons = (56 – 26) = 30

$\therefore$ Mass defect, $\Delta m = Zm_p + (A - Z)m_n - M$

$= (26 \times 1.007825 + 30 \times 1.008665 - 55.934932)$ a.m.u.

$= 0.528461$ a.m.u.

$\therefore$ Total binding energy, $(\Delta E)_{be} = \Delta m.c^2$

$= 0.528461 \times 931.5$ MeV

$= 492.26$ MeV

$\therefore$ Binding energy per nucleon, $(\Delta E)_{ben} = \dfrac{492.26}{56}$

$= 8.79$ MeV **(Answer).**

Nuclear Reactions. The process of nuclear fission, that we shall describe next, involves the so-called nuclear reactions. We shall very briefly discuss them, without going into details.

A **nuclear reaction** represents "the transformation of a stable nucleus into another nucleus by bombardment with some high energy particles".

The bombardment causes artificial radioactivity (cf. Section 8B.01). The parent nucleus X is transformed into a different radioactive nucleus R, which disintegrates to produce a new daughter nucleus Y. Thus the end product element (or nuclide) is different from the parent element (or nuclide). This kind of "transformation from one element to another by producing artificial radioactivity" is called **artificial transmutation** of elements. The process is accompanied either with the release or absorption of the so-called **disintegration energy Q** (cf. Section 8B.03; Eqs. 8.27 and 8.28). If Q is a +ve value, the intermediate radioactive nuclide R spontaneously disintegrates, with release of an amount of energy Q. If Q is –ve, it means that an energy Q has to be supplied to initiate the disintegration. A nuclear reaction caused by bombardment of α-particles may be generally expressed as,

$$\underset{\text{(Parent nucleus)}}{{}^{A}_{Z}X} + \underset{\text{(Alpha particle)}}{{}^{4}_{2}He} \longrightarrow \underset{\text{(Radioactive nucleus)}}{{}^{A+4}_{Z+2}R} \longrightarrow \underset{\text{(Daughter nucleus)}}{{}^{A+3}_{Z+1}Y} + \underset{\text{(Proton)}}{{}^{1}_{1}H} + \underset{\text{(Energy)}}{Q} \quad \text{..... (8.38)}$$

An example is the artificial transmutation of potassium into calcium, via radioactive selenium [this kind of nuclear reaction is known as alpha-proton **(α, p) reaction** (α: bombarding particle; p: product particle)],

$${}^{39}_{19}K + {}^{4}_{2}He \longrightarrow {}^{43}_{21}Se \longrightarrow {}^{42}_{20}Ca + {}^{1}_{1}H + Q$$

The energy Q is shared between the daughter nucleus and the product particle (p).

If the Q-value is +ve, the nuclear reaction is said to be **exothermic or exoergic** (energy is released). If it is –ve, the reaction is termed as **endothermic or endoergic** (energy needs to be supplied).

In addition to obeying the principles of conservation of energy, linear momentum and angular momentum, a nuclear reaction is based on the following rules:

(a) Conservation of mass number

(b) Conservation of electric charge

In the aforementioned nuclear reactions, it may be verified that the total charge and the numbers Z and A on the two sides of an equation are the same.

Nuclear reactions vs. chemical reactions. A **chemical reaction** represents a process in which atoms or molecules interact, e.g.

$$2H_2 + O_2 \rightarrow 2H_2O$$

The chemical reactions also obey the conservation of energy and mass. However, the amount of mass that is transferred into some form of energy, or vice versa, is so small a fraction of the total mass involved that it is impossible to measure it by any physical means available to us. So it goes unnoticed. But in a nuclear reaction, which involves interaction of nuclei and fundamental particles, usually the energy released is about a million times higher. Hence the corresponding change in mass ($E = mc^2$) is easily measurable. Otherwise, a chemical reaction is fundamentally similar to a nuclear reaction. The equations representing the two also look similar (that is why both include the word 'reaction').

Role of neutrons in radioactivity. A free neutron, i.e. a neutron outside the nucleus, is unstable, with a half-life of nearly 15 minutes. It decays into a proton in the same way as a neutron inside a nucleus ($n \rightarrow p + e^- + \bar{\nu}$; Eq. 8.31b), as it is slightly heavier than a proton. For a free proton, on the other hand, such a decay is not possible, as a free proton is a stable particle. It can convert into a neutron only inside the nucleus, by virtue of the existence of extraordinary nuclear forces there (vide Eq. 8.32b). Inside the nucleus, both proton decay and neutron decay have equal chance of occurrence.

The radioactivity is a sign of the instability of nuclei. For stability (cf. the light nuclei), the neutron-proton ratio is 1:1. The ratio increases for heavier nuclei, becoming about 3:2 for very heavy nuclei like uranium. It is so because more neutrons are required to overcome the strong repulsion between neighbouring protons. Being electrically neutral particles, their presence amidst the protons helps to reduce the repulsion between neighbouring protons, as the proton-proton distance is enhanced. Then instability sets in with the neutrons tending to decay (Eq. 8.31b). Similar happens (Eq. 8.32b) if the proton to neutron ratio exceeds 1:1.

Use of neutrons as projectiles; Thermal neutrons. Out of the various high energy particles that can cause a nuclear reaction (and therefore artificial transmutation), the neutrons are particularly suitable. Unlike the photons or α-particles, they experience no repulsing Coulomb force when approaching the nuclei. Even low energy neutrons, of k.e. just about 0.04 eV, are found effective (they are known as **thermal neutrons**, moving slowly in thermal equilibrium with the surrounding matter at room temperature).

8A.06. Nuclear Fission

The **nuclear fission** is "the process of splitting of a heavy nucleus into two or more lighter nuclei, alongwith release of a high amount of energy".

In the Section 8B.05, in the binding energy curve (Fig. 8.12), which is a plot of binding energy of nuclides against their mass number A, we have seen that the binding energy for the middle-mass nuclides ($A \approx 50$ to 80) is higher than the binding energy for the high-mass nuclides. So the former are more stable than the latter. Therefore, in order to achieve greater stability, a heavy nuclide tends to split into two or more lighter nuclides. The total rest mass of the lighter nuclides is less than the rest mass of the single heavy nuclide. This difference in the rest mass appears in the form of a large amount of energy, which gets released during the process of splitting.

A specially suited material for nuclear fission is the uranium isotope $^{236}_{92}U$. However, it is an unstable nuclide and so it does not occur freely in nature. It is produced by bombarding the naturally occuring stable uranium isotope $^{235}_{92}U$ with thermal neutrons. Then it undergoes fission into two middle-mass nuclides $^{140}_{54}Xe$ and $^{94}_{38}Sr$, in roughly equal amounts. Alongwith, two or three neutrons are also emitted. The relevant nuclear reaction is written as,

$$^{235}_{92}U + {}^{1}_{0}n \longrightarrow {}^{236}_{92}U \longrightarrow {}^{140}_{54}Xe + {}^{94}_{38}Sr + 2\,{}^{1}_{0}n + Q \qquad \text{..... (8.39)}$$

Q is the difference between the binding energy of the parent nucleus $^{236}_{92}U$ and the combined binding energy of the daughter nuclei $^{140}_{54}Xe$ and $^{94}_{38}Sr$. From the knowledge of binding energy per nucleon separately for these nuclides, the value of Q is easily calculated as 200 MeV. Thus an energy of 200 MeV per atom of $^{236}_{92}U$ is released during this fission process.

Both the daughter nuclides $^{140}_{54}Xe$ and $^{94}_{38}Sr$ are unstable. They undergo successive β-decay, until a stable end product is reached in each case. The decay chains are,

$$^{140}_{54}Xe \xrightarrow{\beta^-} {}^{140}_{55}Cs \xrightarrow{\beta^-} {}^{140}_{56}Ba \xrightarrow{\beta^-} {}^{140}_{57}La \xrightarrow{\beta^-} {}^{140}_{58}Ce$$

$$^{94}_{38}Sr \xrightarrow{\beta^-} {}^{94}_{39}Y \longrightarrow {}^{94}_{40}Zr$$

The half-life for the various β-decays varies from 14 seconds to 13 days.

[The splitting of the heavy $^{236}_{92}U$ nucleus into the two middle-mass nuclei $^{140}_{54}Xe$ and $^{94}_{38}Sr$ is only one possibility. Several other combinations of the middle-mass nuclei are possible. For instance, two more possibilities are,

$$^{236}_{92}U \longrightarrow {}^{137}_{53}I + {}^{97}_{39}Y + 2\,{}^{1}_{0}n$$

$$^{236}_{92}U \longrightarrow {}^{140}_{56}Ba + {}^{94}_{36}Kr + 2\,{}^{1}_{0}n]$$

Chain reaction

The **chain reaction** is "a collective nuclear reaction which repeatedly produces the particles that start the reaction". Once a chain reaction begins, it continues automatically.

The chain reaction is a very interesting and important feature of nuclear fission. We have seen that the fission of a $^{235}_{92}U$ nucleus by the capture of a thermal neutron leads to the release of 2 or 3 additional neutrons. If an average is taken, about 2.5 additional neutrons per nucleus are produced. Thus more neutrons are produced in the fission than are consumed. Each such neutron can trigger another fission. This leads to a chain reaction of rapidly increasing nuclear fission in the material.

Nuclear reactor/Nuclear (or atom) bomb

If uncontrolled, in a chain reaction the number of fissions in the fissionable material like $^{235}_{92}U$ increases extremely rapidly, with simultaneous releases of a vast amount of energy (mostly heat). The result is an extremely violent explosion, causing a widespread destruction, Thus, the material becomes a so-called **nuclear (or atom) bomb**. Larger is the amount of the material, greater is the destruction.

When the chain reaction is controlled to slow down sufficiently, a large amount of energy can be continuously derived from the fissionable material, which is turn can be utilised to generate a vast amount of electric power for use in our daily lives or otherwise. Thus, the nuclear energy can be harnessed for peaceful purposes. That is what we achieve in a **nuclear reactor**. However, the whole process involves a number of problems. We shall briefly describe a few important ones, alone with the **working** of a nuclear reactor. The Fig. 8.13 (a) shows a schematic representation of main components of a reactor, viz. reactor core, moderator, control rods, steam generator, and turbines. The chain reaction occuring inside the core is schematically depicted in Fig. 8.13 (b).

(i) Surface leakage of neutrons. The neutrons being electrically neutral particles, some of them produced in the chain reaction easily leak out of the surface of the reactor core containing the fissionable material. So they do not further participate in the chain reaction and thus get lost. Since the leakage is essentially a surface effect, whereas the neutrons are produced throughout the volume of the material, the loss is minimised by reducing the surface-to-volume ratio of the reactor core.

(ii) Fast neutrons; use of moderator. The neutrons produced in the fission have kinetic energy $\approx$ 2 MeV. They are called **fast neutrons**. The ${}^{235}_{92}U$ nuclei have a good probability of capturing thermal neutrons ($\approx$ 0.04 eV energy), but a poor probability of capturing the fast neutrons. Therefore, the latter are slowed down by mixing the uranium fuel with a so-called **moderator**. Some moderator materials are water (H_2O), heavy water [D_2O; D: deuterium $\left({}^{2}_{1}H\right)$], graphite, and beryllium. The moderator has to serve two purposes: (i) to slow down the fast neutrons by elastic collision with the atoms of the material of the moderator and (ii) not to absorb the neutrons strongly, so that enough remain available to participate in the chain reaction.

Heavy water (D_2O) is preferred to ordinary water (H_2O) as moderator [deuterium $\left({}^{2}_{1}H\right)$ has nearly double the atomic mass of hydrogen $\left({}^{1}_{1}H\right)$; hence the name 'heavy water' for D_2O]. The reason is that ordinary water has a strong absorption for neutrons; a neutron has a strong tendency to combine with the proton of a hydrogen atom ${}^{1}_{1}H$. Hence sufficient number of neutrons do not remain available to sustain the chain reaction.

(iii) Neutron capture by ${}^{238}_{92}U$; Control rods. Usually the nuclear fuel in a reactor is natural uranium, in the shape of cylindrical rods. The proportion of the fissionable ${}^{235}_{92}U$ in the natural uranium is only 0.7%. The rest 99.3% is the isopote ${}^{238}_{92}U$. Although ${}^{238}_{92}U$ is nonfissionable, it is a very good absorber of the neutrons with energy in the range 1–100 eV. When a fast neutron slows down to the desired 0.04 ev from the initial 2 MeV, it passes through this 1–100 eV energy range. Then it has a strong chance of being captured by a

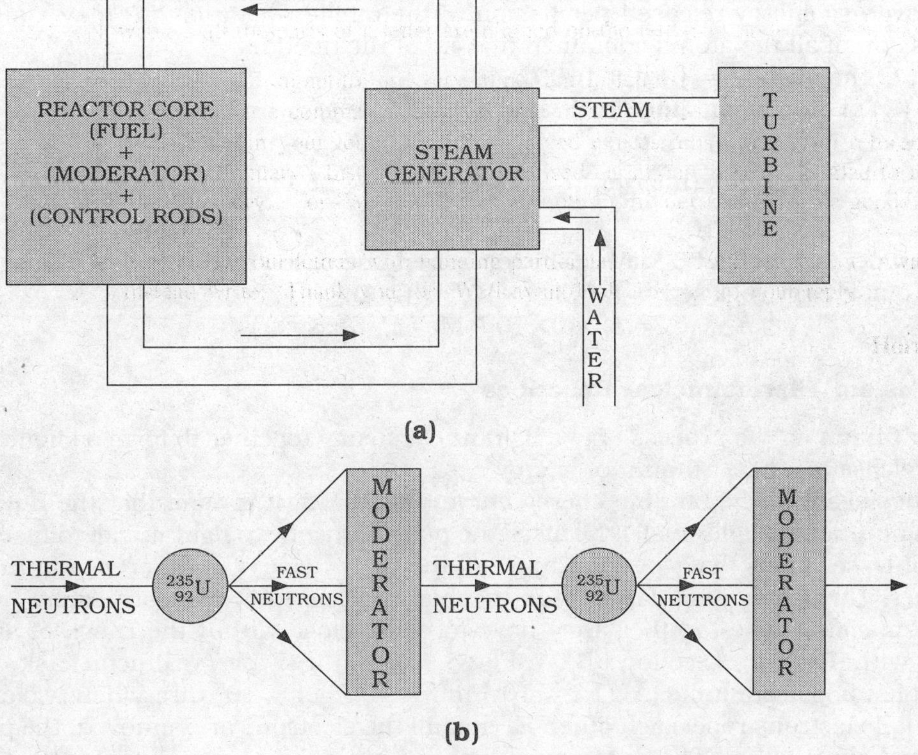

Fig. 8.13

$^{238}_{92}U$ nucleus and thus getting lost for the required fission chain. The situation is remedied by adjusting the spacing between the uranium rods, such that a fast neutron produced in one rod slows down to thermal energies before reaching another rod.

If the rate of production of the fast neutrons is high, it can be brought down by inserting the so-called **control rods** to the required depth in the reactor. These rods are made of cadmium, which is a very efficient neutron absorber. Thus, a steady rate of nuclear fission can be maintained inside the reactor. Under ideal condition, at any instant of time as many neutrons are available for further fission as are captured by the $^{235}_{92}U$ nuclei. Then the reactor is said to be **critical**. Otherwise, it is either 'overcritical' or 'subcritical'.

The uranium fuel in a typical nuclear reactor is in the form of uranium oxide pellets, which are inserted into long metal tubes. These **fuel rods** form the core of the reactor. The moderator is usually in fluid form (usually water or heavy water), surrounding the fuel rods. A suitable circulating **coolant** (e.g. water or air) transfers the heat generated in the reactor core to a steam generator. The high pressure steam so generated operates turbines, which drive electric generators. Thus a huge amount of electric power can be produced.

Additional knowledge

(i) The energy output from a nuclear reactor is comparatively very much higher than that from a conventional source, like coal, oil and hydroelectricity. A rough estimate shows that complete fission of 1 kg of $^{235}_{92}U$ will produce energy in the form of heat at a rate of 1000 MW per day. If this energy is converted into electrical energy at an efficiency of 30%, about 3×10^5 kW per day of electrical energy will be produced. This output of energy is equivalent to that from a large power plant, which consumes nearly 2,500 tons of coal per day.

(ii) The fact that on an average more than two neutrons are produced in a fission opens the possibility that more nuclear fuel may be produced in a reactor than is consumed. This process is called **breeding** and the nuclear reactors using it are known as **breeder reactors**. They also use plutonium (Pu) or thorium (Th) as the nuclear fuel, in addition to uranium.

Since our earth has a limited store of natural uranium, the breeder reactors hold high potential for future.

Example 5. The average energy released per fission of the nuclide $^{239}_{94}Pu$ is 200 MeV. Calculate the energy released by the fission of all the atoms contained in 1 kg of the nuclide.

Solution. Amount of the nuclide = 1 kg = 1000 g

The total number (N) of atoms contained in 1 kg of the nuclide is given by,

$$N = \frac{\text{Avogadro's No.}}{\text{Mass No. (in gram)}}\times 1000\text{ g} = \frac{6.023\times10^{23}\times10^{3}}{239}$$

$$= 25.201\times10^{23}$$

The energy released per fission is 200 MeV. ∴ The total energy released by the fission of 1 kg of the nuclide is,

$$\text{Total energy released} = (25.201\times10^{23})\times(200)$$
$$= 5.040\times10^{26}\text{ MeV} \quad \textbf{(Answer)}.$$

8B.07. Nuclear Fusion; Thermonuclear Reactions

The **nuclear fusion** is "the process of two light nuclei fusing together to form a single nucleus of lesser mass, alongwith release of a high amount of energy".

Referring once again to the binding energy curve (Fig. 8.12), it is seen that the binding energy rises rapidly with mass number for light nuclei. It raises the possibility of two light nuclei to fuse together to form a single nucleus of lesser mass than the total mass of the two nuclei. The difference in mass appears as a vast amount of energy, which is much larger than the energy released in nuclear fission.

For the two nuclei to fuse together, they have to come close within the range of strongly attractive nuclear force, i.e. within ≈ 1 fm (Section 8B.04; Fig. 8.11). But it is a very difficult task. The radius of an atom is typically a few angstrom units (Å) (1Å = 10^{-1} nm = 10^{-10} m = 10^5 fm). When two atoms are pushed together, their orbital electrons repel each other. Even if all the electrons are removed, the positively charged nuclei strongly repel one another. The only way to fuse the atoms together is to supply sufficient energy to them by providing a great thermal motion to them. For this purpose, the material is heated to an extremely high temperature, which removes all the electrons from the atoms and also enables the atoms to fuse together (the enormous heat renders the material to a gaseous form, if it is not already a gas to start with, like hydrogen).

The process of fusion of hydrogen atoms needs very high temperatures ($\approx 10^7$ K, i.e. about ten million degree). The fusion of heavier nuclei needs even higher temperatures, $\approx 10^8$ K. Since such extremely high temperatures are required, the related nuclear reactions are called **thermonuclear reactions**. Increasingly higher temperatures are needed for the fusion of heavier elements, because Z increases and so the Coulomb repulsion between the positively charged nuclei also increases.

A very large number of cyclic thermoreactions are possible for the fusion of various elements (or nuclides). Two of them are particularly notable, as they are believed to be responsible for energy generation in sun and other similar stars: (i) proton-proton cycle and (ii) carbon-nitrogen (or proton-carbon) cycle. Besides the creation of new nuclei, they also involve combination with or emission of electrons (e^-), positrons (e^+), neutrinos (ν) and gamma-rays (γ).

(i) Proton-proton (p, p) cycle. It is represented by the following set of reactions:

$$ {}^1_1H + {}^1_1H \to {}^2_1H + e^+ + \nu + 0.42 \text{ MeV} (\times 2) \quad \text{..... (8.40 a)} $$

$$ e^+ + e^- \to \gamma + \gamma + 1.02 \text{ MeV} (\times 2) \quad \text{..... (8.40 b)} $$

$$ {}^2_1H + {}^1_1H \to {}^3_2He + \gamma + 5.49 \text{ MeV} (\times 2) \quad \text{..... (8.40 c)} $$

$$ {}^3_2He + {}^3_2He \to {}^4_2He + {}^1_1H + {}^1_1H + 12.86 \text{ MeV} \quad \text{..... (8.40 d)} $$

$$ 4\,{}^1_1H + 2e^- \to {}^4_2He + 2\nu + 6\gamma + 26.7 \text{ MeV} \quad \text{..... (8.40 e)} $$

Two protons $\left({}^1_1H\right)$ collide to form a deuteron $\left({}^2_1H\right)$, with simultaneous creation of a positron (e^+) and a neutrino (ν). The positron combines with an electron (e^-). The two particles annihilate together to produce two gamma-ray photons (γ). The deuteron collides with another proton to form a ${}^3_2He$ nucleus. Two such ${}^3_2He$ nuclei combine to form an α-particle $\left({}^4_2He\right)$ and two protons. Note that for this reaction to occur, the first three reactions have to occur two times each. Different amounts of energy are released in these four reactions. The overall process is represented by the last Eq. 8.40 (e), with release of a total amount of 26.7 MeV of energy. We can say that hydrogen 'burns' as a fuel to produce helium and a vast amount of energy. Temperatures $\approx 10^7$ K are generated. The (p, p) cycle is believed to be mainly responsible for the generation of energy in sun.

The (p, p) reactions occur extremely slowly, which is fortunate because then the sun can have a long life. Otherwise, it would explode.

(ii) Carbon-nitrogen (or proton-carbon) cycle. It is believed to be mainly operative in hotter stars, with temperature $\approx 10^8$ K (i.e. 100 million degree). It consists of the following set of reactions:

$$ {}^{12}_6C + {}^1_1H \to {}^{13}_7N + \gamma + 1.93 \text{ MeV} \quad \text{..... (8.41 a)} $$

$$ {}^{13}_7N \to {}^{13}_6C + e^+ + \nu + 1.20 \text{ MeV} \quad \text{..... (8.41 b)} $$

$$ {}^{13}_6C + {}^1_1H \to {}^{14}_7N + \gamma + 7.6 \text{ MeV} \quad \text{..... (8.41 c)} $$

$$ {}^{14}_7N + {}^1_1H \to {}^{15}_8O + \gamma + 7.39 \text{ MeV} \quad \text{..... (8.41 d)} $$

$$ {}^{15}_8O \to {}^{15}_7N + e^+ + \nu + 1.71 \text{ MeV} \quad \text{..... (8.41 e)} $$

$$ {}^{15}_7N + {}^1_1H \to {}^{12}_6C + {}^4_2He + 4.99 \text{ MeV} \quad \text{..... (8.41 f)} $$

$$ 4\,{}^1_1H + {}^{12}_6C \to {}^4_2He + {}^{12}_6C + 2e^+ + 2\nu + 3\gamma + 24.8 \text{ MeV} \quad \text{..... (8.41 g)} $$

The overall effect is again the fusion of 4 hydrogen nuclei into a helium nucleus. The carbon and nitrogen are not used up in the process. They merely act as catalysts.

Example 6. In a star, 3 α-particles combine to form a ${}^{12}_6C$ nucleus. Find the energy released in this process (atomic mass of ${}^4_2He = 4.002603$ u; atomic mass of ${}^{12}_6C = 12$).

Solution. The difference in the mass of 3 α-particles and the mass of ${}^{12}_6C$ nucleus is given as,

$$\text{Mass difference} = (3\times\text{mass of } {}_2^4\text{He} - \text{mass of } {}_6^{12}\text{C})$$
$$= (3\times4.002603)\text{u} - 12\text{u}$$
$$= 0.007809 \text{ u}$$

∴ The energy released in the process is ($E = mc^2$),

$$\text{Energy released} = (0.007809 \text{ u})\, c^2$$
$$= (0.007809 \text{ u})\times(931 \text{ MeV/u})$$
$$= 7.27 \text{ MeV} \quad \textbf{(Answer).}$$

Hydrogen bomb. An uncontrolled thermonuclear fusion leads to a very violent explosion of the material. The hydrogen bomb is such a fusion device. The energy yield per unit mass of the material can be almost an order of magnitude higher than in an atom bomb based on nuclear fission. The high temperature required for the nuclear fusion may be obtained by a prior nuclear fission explosion.

8B.08. controlled Thermonuclear Reactions (laboratory Nuclear Fusion)

As for the case of nuclear fission, a controlled nuclear fusion can be used to produce vast amounts of energy for peaceful purposes. However, the (p, p) cycle described in the last section is not suitable for a terrestrial fusion reactor, because it is extremely slow. The following reactions are suitable. The first two are deuteron-deuteron (D-D) reactions and the third is deuteron-triton (D–T) reaction:

$${}_1^2\text{H} + {}_1^2\text{H} \rightarrow {}_2^3\text{He} + \text{n} + 3.3 \text{ MeV (D–D)} \quad \text{..... (8.42 a)}$$

$${}_1^2\text{H} + {}_1^2\text{H} \rightarrow {}_1^3\text{H} + {}_1^1\text{H} + 4.0 \text{ MeV (D–D)} \quad \text{..... (8.42 b)}$$

$${}_1^2\text{H} + {}_1^3\text{H} \rightarrow {}_2^4\text{H} + \text{n} + 17.6 \text{ MeV (D–T)} \quad \text{..... (8.43)}$$

Deuterium occurs in nature with an abundance of about 1 part in 7000 of hydrogen. Thus, it is available in an almost unlimited quantity is sea water, from which it can be cheaply separted. Thus, a controlled use of thermonuclear reactions might provide a practically infinite source of energy, with a fuel of negligible cost. However, the process involves the following serious problems:

(i) A very high temperature. An extremely high temperature is needed, so that the thermal motion of the interacting particles is strong enough to overcome the Coulomb barrier (repulsion) which tends to keep the particles apart.

(ii) Plasma confinement period. At the high temperatures needed for fusion, the orbital electrons are totally removed from the atoms. Hence a completely ionised gas, called **plasma**, is obtained. It consists of electrons and atomic nuclei (protons, deuterons, tritons, etc.) in rapid random motion. No available material can be used to contain such a superhot plasma (in the sun it is contained by a tremendous gravitational pressure). A further major problem is to confine the hot plasma for a period long enough to fuse sufficient amount of nuclear fuel. Major efforts all over the globe are in progress to solve these problems, based on two approaches: (a) magnetic field confinement and (b) inertial confinement.

(a) Magnetic field confinement; Tokamak. Since the plasma consists of electrically charged particles, it can be controlled through a suitable magnetic field. A **tokamak** is a fusion reactor based on this principle. The plasma is suitably held in a toroidal (ring-shaped) region by a specially designed magnetic field, so that it does not touch the walls of the confinement chamber. A confinement of the plasma for short duratinos (≈ a few seconds) has been achieved, allowing successful nuclear fusion.

The required heating is achieved by passing a high frequency current through the plasma and by bombarding the plasma with a highly accelerated beam of particles.

(b) Intertial confinement. In this method powerful laser beams are used to confine the plasma. The beams are directed from several directions on tiny pellets conatining deuteriun and tritium,. They serve both to vapourize the pellets into a plasma and to compress the plasma gas from all directions. The α-particles produced during the fusion are also forced to remain inside the plasma. The loss of their kinetic energy in the plasma enables a further rise in the temperature of the plasma.

Additional knowledge

(i) We know that the sun has an enormous amount of energy, which is the source of life on earth. No life would exist without it. The sun has been emitting electromagnetic energy (including the familiar sunshine) for billions of years, at the rate of 3.9×10^{26} W [1 watt (W) = 1 joule per second (Js^{-1})]. Estimates show that neither any chemical reactions nor a conversion of gravitational energy into heat energy can provide such a long lasting huge source of energy. The source has to be the nuclear fusion, i.e. thermonuclear reactions. The nuclear fission of heavy elements is also ruled out, because the occurence of these elements in the sun is found to be much small. The light elements dominate, with hydrogen and helium forming nearly 90% of the sun's matter, with nearly equal distribution by weight.

As already described, the source of the sun's energy is believed to be the (p, p) cycle of thermonuclear reactions. Such reactions, in which effectively hydrogen 'burns' to produce helium, have been in existence for about 5 billion years (5×10^9 y). They are likely to go on for another about 5 billiion years. By then the sun's core will be largely helium, which will begin to cool, making the sun to start collapsing under it own gravity. Then the core temperature will rise, causing the outer envelope to expand and thus turning the sun into a 'red giant' [see next point (ii)]. If the temperature touches $\approx 10^8$ K, energy may be produced again by fusion, but this time by 'burning' helium into carbon,

$$ {}^4_2He + {}^4_2He + {}^4_2He \rightarrow {}^{12}_6C + \gamma \qquad \text{..... (8.44)} $$

The process can continue, with other heavier elements being formed by different fusion reactions. However, the Coulomb repulsion becomes progressively more obstructive to fusion as the atomic number (Z) increases. Hence increasingly higher temperatures are needed for the heavier elements to fuse. But, obviously, the elements heavier than those in the peak region of the binding energy curve (e.g. iron and nickel) cannot be produced by further fusion, because for them the binding energy per nucleon is maximum. However, they can be produced by a different process of neutron absorption followed by β-decay, during extremely violent explosions of stars (stellar explosions). A star undergoing such an explosion is called a 'supernova', which suddenly becomes very much brighter. It looks like a huge bright ball, instead of point-like. The matter within the star is thrown out in space all around the star to very long distances. New stars are believed to form out of this debris (which consists of a large variety of elements), by condensation under the influence of the gravitational force. Our own solar system has possibly condensed out of such interstellar matter. Thus, all the elements on the earth, including those in our bodies, were possibly manufactured within the stars that now do not exist. In this sense, we may be regarded as 'children of the stars' !

(ii) The stars can be classified according to their luminosity (or brightness), density and temperature. The sun is one among a class of stars known as main sequence stars. They have core temperatures $\approx 10^7 - 10^8$ K and densities between 0.1 to 10 times that of the sun. Besides this main sequence are the stars called white dwarfs, red giants, variables, novas, and supernovas. The white dwarfs are faint, i.e. they have small luminosities, but are extremely dense, with densities possibly 10^5 times greater than the sun. On the other hand, the red giants have very low densities but high luminosities. The variable stars show both periodic or irregulr variations in luminosity and temperature. The novas and supernovas show sudden large inrease in luminosity. The different types of stars have different mechanisms of energy production.

(iii) We have known that the nuclear fusion produces greater energy than the nuclear fission per unit mass of the nuclear fuel. Besides, unlike the process of fission, the fusion produces no hazardous radioactive waste material, the management and disposal of which poses a serious problem. The fusion reactions are clean. Further, the nuclear fuel deuterium required for the fusion process is found in water, of which almost an unlimited supply is available in the oceans on earth. Thus the nuclear fusion is regarded as the future source of power on the earth. Hence vigorous efforts are being made all over the world, including India, to harness this rich and promising source of energy. Since the process involves serious technical problems and enormous expenditure, a joint international programme involving a large number of countries, including India, has been undertaken in France. It is known as Internatinoal Thermonuclear Experimental Reactor (ITER). In view of the enormity of the task, it is planned for just not a few years but a few decades. Powerful compact superconducting magnets are proposed to be employed for plasma confinement.

Solved Conceptual Problems

[Including HOTS Questions]

Problem 1. In what sense is the radioactivity said to be a spontaneous process ?
Solution. The process is spontaneous in the sense that no external energy needs to be supplied to the radioactive material for the emission of radiation.

Problem 2. Explain how radioactivity is connected with the instability of nuclei.
Solution. For the stability of nuclei, the neutron-proton ratio has to be nearly 1:1, as is the case for light nuclei in the binding energy vs. mass number curve. The ratio increases for heavier nuclei, becoming about 3:2 or more for very heavy nuclei. It is so because then more neutrons are required to overcome the strong Coulombian repulsion between neighbouring protons. Being electrically neutral particles, their presence amidst the protons helps to reduce the repulsion between adjacent protons, because the proton-proton distance is enhanced. Then instability sets in, with the neutrons tending to decay as,

$$n \rightarrow p + e^- + \bar{\nu} \quad (\beta^- \text{ - decay})$$

Similar happens if proton to neutron ratio exceeds 1:1,

$$p \rightarrow n + e^+ + \nu \quad (\beta^+ \text{ - decay})$$

Thus, radioactivity is to be regarded as a sign of instability of nuclei.

Problem 3. What is the physical significance of the term 'binding energy' of a nucleus ?
Solution. The binding energy is the energy equivalent of mass defect. The mass defect (Δm) develops when the nucleons 'bind' together to form a nucleus, such that the mass M of the nucleus is less than the total mass Σm of its constituent nucleons,

$$\Delta m = \Sigma m - M.$$

It physically means that an amount of energy equal to the 'binding' energy is released in the process of formation of the nucleus. It further means that if the whole process is to be reversed, i.e. if the nucleons have to be separated from one another, an amount of energy equal to the binding energy has to be supplied to the nucleus. To sum up, the binding energy of a nucleus is indicative of how strongly the nucleons are bound together in the nucleus.

Problem 4. Differentiate between a β-particle and an electron.
Solution. A β-particle can be either an electron or a positron, depending upon whether the decay process is β^- - decay, or β^+ - decay,

$$n \rightarrow p + \underset{\text{(electron)}}{e^-} + \bar{\nu} \quad (\beta^- \text{ - decay}) \; (\bar{\nu} : \text{antineutrino})$$

$$p \rightarrow n + \underset{\text{(positron)}}{e^+} + \nu \quad (\beta^+ \text{ - decay}) \; (\nu : \text{neutrino})$$

Thus, there is no difference between an electron and a β-particle emitted from a nucleus through β^--decay process.

Problem 5. An atomic nucleus does not contain any electrons or positrons. Then how does it emit β-rays ?
Solution. The processes of β^--decay (conversion of a neutron into a proton, an electron, and an antineutrino) and β^+-decay (conversion of a proton into a neutron, a positron, and a neutrino) are responsible for the emission of electrons and positrons, respectively, from the nuclei,

$$n \rightarrow p + \underset{\text{(electron)}}{e^-} + \bar{\nu} \quad (\beta^- \text{ - decay})$$

$$p \rightarrow n + \underset{\text{(positron)}}{e^+} + \nu \quad (\beta^+ \text{ - decay})$$

Problem 6. A neutron is known to be an unstable particle, with half-life of about 15 minutes. Then why do not the neutrons within a nucleus decay into protons ? Also, why is a proton not unstable like a neutron ?
Solution. A free neutron, i.e. a neutron outside the nucleus, is unstable and decays into a proton (plus an electron and an antineutrino), because it is slightly heavier than a proton. For a free proton, on the other hand, such a decay is not possible, because it is lighter than a neutron. Thus, unlike a free neutron, a free proton is a stable particle.

However, a proton inside the nucleus (so not free) can decay like a neutron, by virtue of the existence of strong extraordinary nuclear forces in the nucleus. Inside the nucleus, both neutron decay and proton decay have equal chances of occurrence.

Problem 7. Differentiate between a neutrino and a photon.
Solution. Both a photon and a neutron have charge and rest mass equal to zero. But the neutrino has spin 1/2, against zero spin of a photon.

The photon is an electromagnetic radiation, representing one quantum of energy ($E = h\nu$), whereas the neutrino is a fundamental particle and so a constituent of matter. The photon has no antiparticle, but the neutrino has antineutrino as its antiparticle.

Problem 8. How thermal neutrons are particularly suitable as projectiles for initiating a nuclear reaction ?
Solution. The neutrons are particularly suitable, because they are electrically neutral. Hence, unlike the charged particles, like protons and α-particles, they experience no repulsive or attractive Coulomb force when approaching a nucleus. Even low energy neutrons, of kinetic energy just about 0.04 eV (called thermal neutrons), are found effective.

Problem 9. What is a chain reaction ? How is it significant in the nuclear fission of $^{235}_{92}U$?
Solution. A chain reaction is a collective nuclear reaction, which repeatedly produces the particles that start the reaction.

The fission of a $^{235}_{92}U$ nucleus is triggered by the capture of a thermal neutron, which leads to the release of 2 or 3 additional neutrons. On an average, 2.5 additional neutrons per nucleus are produced. Thus, more neutrons are produced in the fission process than are consumed. Each such neutron can trigger another fission. This leads to a chain reaction of rapidly increasing nuclear fissions in the material.

Problem 10. Both a nuclear reactor and an atom bomb have energy generated by the process of nuclear fission. Then in what way the two are different ?
Solution. The chain reaction in an atom bomb is uncontrolled. Hence the number of fissions increases extremely rapidly, with simultaneous release of a vast amount of energy (mostly heat), leading to an extremely violent explosion and consequent widespread destruction.

On the other hand, in a nuclear reactor the chain reaction is suitably controlled, which enables a continuous release of energy, that can be utilised for a continuous generation of power.

Problem 11. What is the role of a moderator in a nuclear reactor ?
Solution. The neutrons produced by fission inside the nuclear reactor have high energies $\approx$ 2 MeV (fast neutrons). The $^{235}_{92}U$ nuclei have a high probability of capturing slow (thermal) neutrons ($\approx$ 0.04 eV), but a poor probability of capturing the fast neutrons. Therefore the latter are slowed down by mixing the uranium fuel with a so-called **moderator**, like heavy water, graphite, etc. It has two purposes: (i) to slow down the fast neutrons by elastic collisions with the atoms of the material of the moderator and (ii) not to absorb the neutrons strongly, so that enough remain available to participate in the chain reaction.

Problem 12. What is the function of control rods in a nuclear reactor ?
Solution. If during the process of nuclear fission inside a reactor the rate of production of the fast neutrons is high, it can be brought down by inserting the so-called **control rods** to the required depth in the reactor. These rods are made of cadmium, which is a very efficient absorber of neutrons.

Problem 13. Under what condition a nuclear reactor is said to be critical ?
Solution. During the process of nuclear fission inside a reactor, under ideal condition at any instant of time as many neutrons are available for further fission as are captured by the $^{235}_{92}U$ nuclei. Then the reactor is said to be **critical**. Otherwise, it is either 'overcritical' or 'subcritical'.

Problem 14. What is the essential difference between nuclear fission and nuclear fusion ?
Solution. Both nuclear fission and nuclear fusion are based on the principle of conversion of mass into energy ($E = mc^2$). The difference is that in nuclear fission a heavy nucleus is split into two or more lighter nuclei with total rest mass less than the rest mass of the single heavy nucleus, whereas in nuclear fusion two light nuclei fuse together to form a single nucleus of lesser mass. In both the cases, the difference of mass manifests as release of a vast amount of equivalent energy.

Problem 15. Why is heavy water preferred to ordinary water as moderator in a nuclear reactor ?
Solution. The ordinary water (H_2O) has a strong absorption for neutrons (a neutron strongly tends to combine with the proton of a hydrogen atom ^1_1H, present in H_2O, to form a deuterium atom ^2_1H). Hence in passing through ordinary water as moderator, sufficient number of neutrons do not remain available to continue the chain reaction. No such problem exists with heavy water (D_2O). Therefore its use as moderator is preferred to H_2O.

Problem 16. What is a thermonuclear reaction and its role in nuclear fusion ?
Solution. For the process of nuclear fusion to take place, two light nuclei have to fuse together. For this purpose they have to come close together within the range of strongly attractive nuclear forces, i.e. within ~1fm. But it is not easily possible because of strong Coulombian repulsion between the +vely charged nuclei. it becomes possible only at extremely high temperatures of ~10^7 K and more. The related nuclear reactions occurring at such high temperatures are called **thermonuclear reactions**, e.g. in sun and other stars.

Solved Numerical Problems

Problem 1. The radioactive decay of a $^{238}_{92}U$ nucleus results in the emission of (a) an α-particle and (b) a β-particle. How the neutron to proton ratio is affected in the two cases ?
Solution. The decay process is given by the following sequence,

$$^{238}_{92}U \xrightarrow[(\alpha-\text{decay})]{^4_2He} {}^{234}_{90}Th \xrightarrow[(\beta-\text{decay})]{^0_{-1}e} {}^{234}_{91}Pa$$

(a) The neutron to proton ratios for the α-decay are,

$$\text{(i)} \quad \frac{(238-92)}{92} = \frac{146}{92} = 1.587 \quad \text{(Before decay)}$$

$$\text{(ii)} \quad \frac{(234-90)}{90} = \frac{144}{90} = 1.600 \quad \text{(After decay)}$$

Since 1.6 > 1.587, it follows that the ratio increases after decay. **(Answer).**

(b) The ratios for the β-decay are,

$$\text{(i)} \quad \frac{(234-90)}{90} = 1.600 \quad \text{(Before decay)}$$

$$\text{(ii)} \quad \frac{(234-91)}{91} = \frac{143}{91} = 1.589 \quad \text{(After decay)}$$

As 1.589 < 1.600, it follows that the ratio reduces after decay. **(Answer).**

Problem 2. The sequence of decay of a radioactive nucleus is,

$$D \xrightarrow{\alpha} D_1 \xrightarrow{\beta} D_2 \xrightarrow{\alpha} D_3 \xrightarrow{\alpha} D_4$$

If nucleon number and atomic number of D_2 are 176 and 71, respectively, what are their values for D and D_4 ? (CBSE; 1998).
Solution. The mass number and charge number for an α-particle are 4 and 2, respectively. Therefore, for the product D_4 we have,

$$A = 176 - 8 = 168 \quad \textbf{(Answer).}$$
$$Z = 71 - 4 = 67 \quad \textbf{(Answer).}$$

For a β-particle, the charge number is –1 and its mass number is 0. ∴ For the element D we have,

$$A = 176 + 0 + 4 = 180 \quad \textbf{(Answer).}$$
$$Z = 71 - 1 + 2 = 72 \quad \textbf{(Answer).}$$

Problem 3. After a certain lapse of time, the fraction of radioactive polonium undecayed is found to be 12.5% of the initial quantity. What is the duration of this time lapse, if the half-life of polonium is 139 days ? (CBSE; 1998).

Solution. Given: $T_{1/2} = 139$ days;

$$\frac{N}{N_0} = 12.5\% = \frac{12.5}{100} = \frac{1}{8}$$

We know that, $$N = N_0\left(\frac{1}{2}\right)^n$$

Here, n is the number of half-lives in time t, during which N reduces to $N_0\,(1/2)^n$. Thus, we have

$$\frac{N}{N_0} = \left(\frac{1}{2}\right)^n \qquad \text{..... (i)}$$

We are given that,

$$\frac{N}{N_0} = \frac{1}{8} = \left(\frac{1}{2}\right)^3 \qquad \text{..... (ii)}$$

The comparison of Eqs. (i) and (ii) gives,

$$n = 3$$

Thus, it is found that during the time lapse t, 3 half-lives are spent. Therefore the required time lapse is,

$$t = 3\times T_{1/2} = 3\times(139 \text{ days})$$
$$= 417 \text{ days} \quad \textbf{(Answer).}$$

Problem 4. A polonium source of α-particles of strength 10 mci is required for an experiment. Calculate the quantity of ${}^{210}_{84}\text{Po}$ needed to provide such a source (half-life of polonium = 139 days).

Solution. Given: $T_{1/2} = 139$ days $= 139\times24\times60\times60$ s $= 1.20\times10^7$ s; $R_t = 10\times(3.7\times10^7) = 3.7\times10^8$ dps (R_t is the rate of decay, i.e. the no. of decays per second).

We know that the half-life is given by,

$$T_{1/2} = \frac{0.693}{\lambda} = \frac{0.693}{(R_t/N)} \quad (\because R_t = \lambda N)$$

Or $$N = \frac{R_t\times T_{1/2}}{0.693}$$

$$= \frac{(3.7\times10^8)\times(1.20\times10^7)}{0.693}$$

$$= 6.41\times10^{15} \text{ atoms}$$

Thus (6.41×10^{15}) atoms of polonium are needed to provide a source of strength 10 mci. The mass number of polonium is 210 and the value of Avogadro's number is 6.022×10^{23}, which means that (6.002×10^{23}) atoms of polonium are contained in 210 g quantity of polonium. ∴ the quantity of polonium containing (6.41×10^{15}) atoms is given by,

$$\text{Quantity of } {}^{210}_{84}\text{Po} = \frac{(210)\times(6.41\times10^{15})}{(6.022\times10^{23})} \text{ g}$$

$$= 2.24\times10^{-6} \text{ g} \quad \textbf{(Answer).}$$

Problem 5. If 2 g of radium emits 7.34×10^{10} α-particles per second, calculate the half-life of radium in years (atomic weight of radium = 226).

Solution. Given: $\frac{dN}{dt} = 7.34\times10^{10}\ \text{s}^{-1}$

The number of atoms in 2 g of radium is given by,

$$N = \frac{2\times(\text{Avogadro's No.})}{\text{At. wt.}}$$

$$= \frac{2\times(6.022\times10^{23})}{226}$$

We know, $$\frac{dN}{dt} = \lambda N$$

$$\Rightarrow \quad \lambda = \frac{(dN/dt)}{N}$$

$$= \frac{(7.34\times10^{10})\times226}{2\times6.022\times10^{23}}$$

$$\therefore \quad T_{1/2} = \frac{0.693}{\lambda} = \frac{0.693\times2\times6.022\times10^{23}}{7.34\times10^{10}\times226} \text{ s}$$

$$= 50315\times10^{6} \text{ s} = \frac{50315\times10^{6}}{(24\times60\times60\times365)} \text{ year}$$

$$= 1595.4 \text{ year} \quad \textbf{(Answer).}$$

Problem 6. The activity of a radioactive substance is found to reduce to 1/32 of its initial value in 30 year. Determine the half-life of the substance.

Solution. Given: Total period of decay, $T = 30$ year

$$\frac{N}{N_0} = \frac{1}{32} = \left(\frac{1}{2}\right)^5$$

We use the relation,

$$\frac{N}{N_0} = \left(\frac{1}{2}\right)^{T/T_{1/2}}$$

$$\Rightarrow \quad \left(\frac{1}{2}\right)^5 = \left(\frac{1}{2}\right)^{T/T_{1/2}}$$

$$\Rightarrow \quad \frac{T}{T_{1/2}} = 5$$

$$\Rightarrow \quad T_{1/2} = \frac{T}{5} = \frac{30}{5} = 6 \text{ year} \quad \textbf{(Answer).}$$

Problem 7. The atomic masses of proton, neutron, and $^{209}_{83}$Bi are 1.007825 a.m.u., 1.008665 a.m.u., and 208.980388 a.m.u., respectively. Find the binding energy per nucleon for $^{209}_{83}$Bi (1 a.m.u. = 931.5 MeV).

Solution.

No. of protons = 83

No. of neutrons = 209 – 83 = 126

Total mass of protons (M_p) = 83×1.007825 = 83.649475 a.m.u.

Total mass of neutrons (M_n) = 126×1.008665 = 127.091790 a.m.u.

∴ Total mass of nucleons = $M_p + M_n$ = 210.741265 a.m.u.

Mass of $^{209}_{83}$Bi nucleus = 208.980388 a.m.u.

∴ Mass defect, Δm = 210.741265 – 208.980388 = 1.760877 a.m.u.

∴ Binding energy = (1.760877)×931.5 MeV = 1640.26 MeV

$$\therefore \quad \text{Binding energy per nucleon} = \frac{1640.26}{209} = 7.85 \text{ MeV} \quad \textbf{(Answer).}$$

Problem 8. Calculate the binding energy per nucleon of $^{35}_{17}$Cl nucleus. (Given: mass of $^{35}_{17}$Cl = 34.980000 u; mass of proton = 1.007825 u; mass of neutron = 1.008665 u; 1 a.m.u. (1u) = 931 MeV).

Solution.

No. of protons = 17

No. of neutrons = 35 – 17 = 18

Total mass of protons (M_p) = 17×1.007825 = 17.133025 u

Total mass of neutrons (M_n) = 18×1.008665 = 18.15597 u

$\therefore$ Total mass of nucleons $= M_p + M_n = 35.288995$

$\therefore$ Mass defect $= 35.288995 - 34.980000$

$= 0.308995$ u

$\therefore$ Binding energy $= 0.308995 \times 931$ MeV

$= 287.67$ MeV

$\therefore$ Binding energy per nucleon $= \dfrac{287.67}{35}$

$= 8.22$ MeV **(Answer).**

Problem 9. Three α-particles coalesce together to form a carbon $\left({}^{12}_{6}C\right)$ nucleus, with simultaneous release of 7.27 MeV of energy. Determine the mass of an α-particle.

Solution. If 3 α-particles $\left({}^{4}_{2}He\right)$ can be made to fuse together, they will form a carbon $\left({}^{12}_{6}C\right)$ nucleus,

$${}^{4}_{2}He + {}^{4}_{2}He + {}^{4}_{2}He \rightarrow {}^{12}_{6}C$$

If m_α and $m\left({}^{12}_{6}C\right)$ are the masses of an α-particle and a carbon atom, respectively, the mass defect is given by,

$$\Delta m = [3m_\alpha - m\left({}^{12}_{6}C\right)] \text{ a.m.u.}$$
$$= (3m_\alpha - 12) \text{ a.m.u.}$$

($\because$ The atomic mass of ${}^{12}_{6}C$ is known to be exactly 12 a.m.u.). Δm is equivalent to the energy released in the process, which is given to be equal to 7.27 MeV,

$$\text{Energy released} = 7.27 \text{ MeV} \triangleq \frac{7.27}{931.5} \text{ a.m.u.}$$
$$= 0.0078 \text{ a.m.u.}$$

$$\therefore \quad (3m_\alpha - 12) = 0.0078 \text{ a.m.u.}$$

$$\Rightarrow \quad m_\alpha = \frac{12.0078}{3} = 4.003 \text{ a.m.u.}$$ **(Answer).**

Problem 10. A copper coin of mas 6 gram totally consists of ${}^{63}_{29}Cu$ atoms. Determine the amount of energy needed to separate its constituent protons and neutrons from each other (mass of proton = 1.00783 a.m.u.; mass of neutron = 1.00867 a.m.u.; mass of ${}^{63}_{29}Cu$ atom = 62.92960 a.m.u.).

Solution. Given: Mass of coin = 6g

$$\text{No. of } {}^{63}_{29}Cu \text{ atoms present in 6g} = 6 \times \frac{\text{Avogadro's No.}}{\text{At. wt. of } {}^{63}_{29}Cu}$$
$$= \frac{6 \times (6.02 \times 10^{23})}{(63)}$$
$$= 5.73 \times 10^{22}$$

No. of protons $= 29$

No. of neutrons $= 63 - 29 = 34$

$\therefore$ Mass of protons $(M_p) = 29 \times 1.00783 = 29.2271$ a.m.u.

Mass of neutrons $(M_n) = 34 \times 1.00867 = 34.2948$ a.m.u.

$\therefore$ Total mass of nucleons $= 29.2271 + 34.2948$

$= 63.5219$ a.m.u.

$\therefore$ Mass defect $= 63.5219 - 62.9296$

$= 0.5923$ a.m.u.

$\therefore$ Binding energy per atom $= 0.5923 \times 931.5$ MeV

$= 551.73$ MeV

Hence the energy needed to separate all the consitituent nucleons is given by,

$$\text{Energy} = (\text{Total no. of atoms}) \times (\text{B.E. per atom})$$
$$= (5.73 \times 10^{22}) \times (551.73)$$
$$\approx 3.16 \times 10^{25} \text{ MeV}$$ **(Answer).**

Problem 11. In one hour the radioactivity of a material is found to reduce to 250 s^{-1}, from its initial value of 300 s^{-1}. Determine the half-life of the material.

Solution. Given: $A_0 = 300\ s^{-1}$; $A = 250\ s^{-1}$; time of decay, $T = 60$ minute

We use the formula,

$$A = A_0\, e^{-\lambda t}$$

$$\Rightarrow \quad 250 = 300\, e^{-\lambda \times 60}$$

$$\Rightarrow \quad \lambda = \frac{\ln\left(\frac{300}{250}\right)}{60} = \frac{\ln(1.2)}{60}$$

The half-life is given by,

$$T_{1/2} = \frac{\ln(2)}{\lambda} = \frac{\ln(2)}{\ln(1.2)} \times 60$$

$$= \frac{0.693}{0.182} \times 60$$

$$= 228.5 \text{ minute} \quad \textbf{(Answer).}$$

Problem 12. Express the radioactive β^--emission decay of $^{23}_{10}Ne$. Work out the maximum kinetic energy of the electrons emitted during the process (masses of $^{23}_{10}Ne$ and $^{23}_{11}Na$ are 2.994466 a.m.u. and 22.989770 a.m.u., respectively; 1 a.m.u. = 931.5 MeV).

Solution. The said β^--decay is represented by the equation,

$$^{23}_{10}Ne \longrightarrow {}^{23}_{11}Na + \underset{\text{(electron)}}{e^-} + \underset{\text{(antineutrino)}}{\bar{\nu}} + Q$$

Q is the shared kinetic energy between the emitted electron and antineutrino. The k.e. of the electrons has maximum value when the electrons alone share the entire energy. It is given by (neglecting the mass of electron, which is comparatively very small),

$$(E_{el})_{max} = \left[m\left({}^{23}_{10}Ne\right) - m\left({}^{23}_{11}Na\right)\right] c^2$$

$$= (22.994466 - 22.989770) \times (931.5) \text{ MeV}$$

$$= 0.004696 \times 931.5 \text{ MeV}$$

$$= 4.37 \text{ MeV} \quad \textbf{(Answer).}$$

Problem 13. Two elements with mass numbers as 250 and 125 have binding energies as 7.5 MeV and 8.3 MeV, respectively. If the heavier nucleus with A = 250 splits into two nuclei with close values of A, determine the amount of energy released in the process.

Solution. Let the numbers of protons and neutrons in the heavier nucleus (A = 250) be Z and N, respectively, and let its binding energy be B. Its rest mass energy is given as,

$$E = Mc^2 = (Zm_p + Nm_n)\, c^2 - B$$

$$= (Zm_p + Nm_n)\, c^2 - 250 \times (7.5 \text{ MeV})$$

Let the rest mass energies of the two split fragments 1 and 2 be E_1 and E_2, respectively. Let the numbers of protons and neutrons in the two fragments be Z_1, N_1 and Z_2, N_2, respectively. Then, we have,

$$E_1 = M_1c^2 = (Z_1m_p + N_1m_n)\, c^2 - (Z_1 + N_1) \times (8.3 \text{ MeV})$$

$$E_2 = M_2c^2 = (Z_2m_p + N_2m_n)\, c^2 - (Z_2 + N_2) \times (8.3 \text{ MeV})$$

The energy released in the process of splitting is given by,

$$(E)_{split} = E - (E_1 + E_2)$$

$$= [(Z - Z_1 - Z_2)\, m_pc^2 + (N - N_1 - N_2)\, m_nc^2$$

$$+ (Z_1 + Z_2 + N_1 + N_2) \times (8.3) \text{ MeV} - (250 \times 7.5) \text{ MeV}]$$

But,

$$Z_1 + Z_2 = Z$$

$$N_1 + N_2 = N$$

$$Z_1 + Z_2 + N_1 + N_2 = Z + N = 250$$

$$\therefore \quad (E)_{split} = [0 + 0 + 250 \times (8.3 - 7.5) \text{ MeV}]$$

$$= 250 \times 0.8 = 200 \text{ MeV} \quad \textbf{(Answer).}$$

Problem 14. In the three fusion reactions given below, determine the respective Q-values:

$$\text{(i) } {}_1^2\text{H} + {}_1^3\text{H} \rightarrow {}_2^4\text{He} + \text{n}$$

$$\text{(ii) } {}_1^2\text{H} + {}_1^2\text{H} \rightarrow {}_2^3\text{He} + \text{n}$$

$$\text{(iii) } {}_1^2\text{H} + {}_1^2\text{H} \rightarrow {}_1^3\text{H} + {}_1^1\text{H}$$

(Given: m_n = 1.008665 a.m.u.; m_p = 1.007825 a.m.u.; $m\left({}_1^2\text{H}\right)$ = 2.014102 a.m.u.; $m\left({}_1^3\text{H}\right)$ = 3.016049 a.m.u.; $m\left({}_2^3\text{He}\right)$ = 3.016029 a.m.u.; $m\left({}_2^4\text{He}\right)$ = 4.002603 a.m.u.; 1 a.m.u. = 931 MeV).

Solution. (i) Mass of Reactants $= m\left({}_1^2\text{H}\right) + m\left({}_1^3\text{H}\right)$

$= 2.014102 + 3.016049$

$= 5.030151$ a.m.u.

Mass of Products $= m\left({}_2^4\text{He}\right) + m_n$

$= 4.002603 + 1.008665$

$= 5.011268$ a.m.u.

∴ Mass defect $= 5.030151 - 5.011268$

$= 0.018883$ a.m.u.

∴ Q-value $= 0.018883 \times 931 = 17.58$ MeV **(Answer).**

(ii) Mass of Reactants $= 2 \times 2.014102$

$= 4.028204$ a.m.u.

Mass of Products $= 3.016029 + 1.008665$

$= 4.024694$ a.m.u.

∴ Mass defect $= 0.003510$ a.m.u.

∴ Q-value $= 0.00351 \times 931 = 3.27$ MeV **(Answer).**

(iii) Mass of Reactants $= 2 \times 2.014102$

$= 4.028204$ a.m.u.

Mass of Products $= 3.016049 + 1.007825$

$= 4.023874$ a.m.u.

∴ Mass defect $= 0.004330$ a.m.u.

∴ Q-value $= 0.00433 \times 931 = 4.03$ MeV **(Answer).**

Problem 15. A nuclear reactor of 800 MW capacity generates energy totally by the fission of ${}_{92}^{235}\text{U}$. It works 3/4th of the time in a day and uses up one-fourth of its fuel in 2 year. Determine the amount of fuel contained in the reactor to start with. Take the average energy generated by the fission of one atom of ${}_{92}^{235}\text{U}$ as 200 MeV (Avogadro's No. = 6.0×10^{23}; 1 eV = 1.6×10^{-19} J).

Solution. 800 MW (Power) = (800×10^6) W = 8×10^8 Js^{-1}

[Power = Energy ÷ Time; ∴ 1 W = 1 Js^{-1}];

2 year = $(2 \times 365 \times 24 \times 60 \times 60)$s = 63072×10^3 s;

200 MeV = (200×10^6) eV = $(200 \times 10^6) \times (1.6 \times 10^{-19})$ J = 32×10^{-12} J.

$$\text{Number of atoms per gram of } {}_{92}^{235}\text{U} = \frac{6 \times 10^{23}}{235}$$

∴ Energy generated per gram is given by,

$$\text{Energy per gram} = \frac{(6 \times 10^{23})}{235} \times (32 \times 10^{-12} \text{ J})$$

$$= 817 \times 10^8 \text{ J}$$

This energy is generated per second. So it is equivalent to $(817 \times 10^8 \text{ Js}^{-1}) = 817 \times 10^8$ W of power.

Thus, to generate 817×10^8 W of power, 1g of fuel is consumed. ∴ To generate 800 MW power for 2 year, with 3/4 (= 0.75) utilisation, the required amount of fuel will be,

$$(\text{Amount of fuel})_{2\text{ year}} = \frac{(63072 \times 10^3) \times (0.75) \times (8 \times 10^8)}{(817 \times 10^8)} \text{ g}$$

$$= 463.2 \times 10^3 \text{ g} = 463.2 \text{ kg}$$

This amount of fuel is 1/4th of the total amount consumed by the reactor. ∴ The initial amount is,

Initial amount $= 4 \times (463.2$ kg$)$

$= 1852.8$ kg **(Answer).**

Very Short Answer Questions

[Including HOTS Questions]

(1 mark each)

1. What is meant by radioactive decay constant ?

2. Arrange α, β and γ-radiations in decreasing order of penetrating power.

3. Which part of the atom is responsible for radioactivity ?

4. How is an electron different from an α-particle ?

5. For which of the three radiations α, β, γ an electric field has no effect on its passage in free space ?

6. How is mass quantitatively taken as equivalent to energy ?

7. Calculate the energy equivalent to 10 mg of mass.

8. Give the relationship between half-life and decay constant.

9. If the half-life of tritium is 12.5 year, what fraction of it will be let after 25 year have passed ?

10. How α-particles possess greater ionizing power than γ-rays, in spite of the latter being far more energetic ?

11. Name two elementary particles, which are believed to possess infinite lifetime.

12. Give relationship between (a) half-life and average life and (b) average life and decay constant, of a radioactive material.

13. On the binding energy per nucleon vs. mass number graph, name 3 nuclei lying on peaks of the curve.

14. When a nuclide $^{235}_{92}X$ converts into another nuclide $^{219}_{88}X$, how many α and β particles are given out per atom ?

15. When we say that binding energy per nucleon for $^{209}_{83}Bi$ is 7.8 MeV, what does it mean in physical sense ?

16. Out of the four particles electron, α-particle, proton and neutron, which one is unstable ?

17. What is the approximate half-life of a free neutron ?

18. The sequence of decay of a radioactive nucleus is,

$$D \xrightarrow{\alpha} D_1 \xrightarrow{\beta} D_2 \xrightarrow{\alpha} D_3 \xrightarrow{\alpha} D_4$$

If nucleon number and atomic number of D_2 are 176 and 71, respectively, what are their values for D and D_4 ? (CBSE; 1998)

19. In the process of β-decay, an electron or a positron is emitted from a nucleus. Does it result in the affected atom acquiring an opposite charge ?

20. Does the emission of γ-rays from a nucleus result in loss of mass of the nucleus ?

21. Are β-rays different from cathode rays ? If so, how ?

22. How do the protons coexist in the tiny nucleus, in spite of a strong Coulombian repulsion between them ?

23. The binding energies of a deuteron nucleus $\left({}_{1}^{2}H\right)$ and an α-particle $\left({}_{2}^{4}He\right)$ are 1.25 MeV per nucleon and 7.2 MeV per nucleon, respectively. Which of the two is favoured to be more stable ?

24. Write two properties of nuclear forces.

25. Is the ratio of number of neutrons to the number of protons in heavy nuclei >1 or <1 or equal to 1 ?

26. Out of the following three nuclei, which is more favoured to undergo nuclear fission: ${}_{14}^{30}X, {}_{3}^{6}Y, {}_{40}^{130}Z$?

27. What are thermal neutrons ?

28. Why neutrons are considered particularly suitable for nuclear fission ?

29. What processes are believed to be the source of enormous energy of stars,including our sun ?

30. Complete the following equation:

$$ {}_{92}^{235}U + {}_{0}^{1}n = $$

31. Mention one point each of similarity and dissimilarity between the processes of nuclear fission and nuclear fusion.

32. What is a thermonuclear reaction ?

33. What is meant by mass defect of a nucleus ?

34. What is a chain reaction ?

Short Answer Questions

[Including HOTS Questions]

(2 or 3 marks each)

1. Out of the 3 radiations α, β and γ, which one

(a) is similar to X-rays
(b) is similar to cathode rays
(c) has highest speed
(d) gets readily absorbed by matter

2. Explain why the heavy nuclei generally contain more neutrons than protons.

3. Discuss if the neutron to proton ratio for a nucleus should decrease or increase during (i) α-decay and (ii) β-decay.

4. Prove that at a given instant of time the rate of radioactive decay, R_t, of an element is related to the number of radioactive nuclei, N, present as $R = \lambda N$.

5. Briefly explain the process of nuclear fusion and the main difficulty to bring it in operation.

6. Briefly explain the function of (a) moderator and (b) control rods, in a nuclear reactor.

7. For a radioactive decay process, derive the relation $N = N_0e^{-\lambda t}$, where the symbols have usual meaning.

8. What is meant by decay constant and half-life ? Derive the relation between them.

9. Explain how a radioactive material emits β-rays, in spite of its constituent nuclei not containing any β-particles.

10. Distinguish between the processes of nuclear fission and nuclear fusion. How is the latter considered advantageous over the former ?

11. Explain why the neutrons in a nucleus do not decay, in spite of a neutron known to be an unstable particle.

12. Energy is known to be liberated if three helium nuclei can be made to combine to form a carbon nucleus. Explain why cannot the helium nuclei combine spontaneously and thus minimise the total energy.

13. Explain why heavy water is preferred over ordinary water for use in a nuclear reactor.

Long Answer Questions

(5 marks each)

1. What is meant by radioactivity ? Present a comparative description of the properties of α, β, and γ-radiations, with stress on their points of mutual similarities and dissimilarities.

2. Draw the curve of variation of binding energy per nucleon with mass number of elements. State the important conclusions drawn from it. Explain how does it basically account for the release of energy in the processes of nuclear fission and nuclear fusion.

3. Explain how the process of nuclear fission leads to release of enormous amount of energy. Describe with the help of a schematic diagram the working of a nuclear reactor, clearly detailing that how the serious problems involved are successfully overcome in the reactor.

Numerical Exercises

1. Complete the following decay sequences:

(i) ${}^{19}_{8}O \longrightarrow {}^{19}_{9}F +$

(ii) ${}^{25}_{13}Al \longrightarrow {}^{25}_{12}Mg +$

(iii) ${}^{226}_{88}Ra \longrightarrow \alpha +$ **[(i) $e^- + \bar{\nu}$; (ii) $e^+ + \nu$; (iii) ${}^{222}_{86}Rn$]**

2. Express the following processes in the form of equations:

(a) β^--decay of ${}^{210}_{83}Bi$ (b) β^--decay of ${}^{32}_{15}P$

(c) β^+-decay of ${}^{11}_{6}C$ (d) α-decay of ${}^{226}_{88}Ra$

(e) α-decay of ${}^{242}_{94}Pu$

[(a) ${}^{210}_{83}Bi \rightarrow {}^{210}_{84}Po + e^- + \bar{\nu}$ (b) ${}^{32}_{15}P \rightarrow {}^{32}_{16}S + e^- + \bar{\nu}$ (c) ${}^{11}_{6}C \rightarrow {}^{11}_{5}B + e^+ + \nu$ (d) ${}^{226}_{88}Ra \rightarrow {}^{222}_{86}Rn + {}^{4}_{2}He$ (e) ${}^{242}_{94}Pu \rightarrow {}^{238}_{92}U + {}^{4}_{2}He$]

3. At a given time, the number of active nuclei present in a radioactive material are found to be 6.0×10^{18}. Calculate the number of nuclei left in the same active state after two half-lives have passed. **[2.25×10^{16}]**

4. Find (i) the decay constant λ, (ii) average life τ, and (iii) radioactivity, for 3.0 mg of gold $\left({}^{198}_{79}\text{Au}\right)$. (The half-life and atomic weight of ${}^{198}_{79}$Au are 64.8 hour and 198 g.mol^{-1}, respectively). [Hint: 3.7×10^{10} decay per second = 1 Ci]. **[2.9×10^{-6} s^{-1}; 93.6 h; 713.5 Ci]**

5. For a radioactive substance, the decay rate is observed to be reduced to 2×10^6 s^{-1} from the initial value of 8×10^6 s^{-1}, in 40 hour. Calculate the decay rate after another 160 hour have passed. **[7.8×10^3]**

6. A free proton undergoes β-decay to transform into a proton. Determine the decay constant λ and the energy released in the process ($T_{1/2}$ of neutron = 15 minutes]. **[7.7×10^{-4} s^{-1}; 783 keV]**

7. The activity of a radioactive sample with half-life as 10 hour is initially observed to be 1 Ci. Calculate the number of disintegrations taking place in the 10th hour of its decay. [Hint: First find the activity and hence the no. of disintegrations after 9 h of decay and then the same after 10 h of decay. Then the required number is found as $[103.02 - 96.10)\times10^{13}]$. **[$6.92\times10^{13}$]**

8. Taking the masses of a nitrogen nucleus, a hydrogen atom $\left({}^1_1\text{H}\right)$ and a neutron as 14.00307 a.m.u., 1.00783 a.m.u. and 1.00867 a.m.u., respectively, calculate the binding energy of the nitrogen nucleus (1 a.m.u. = 931.5 MeV). **[104.77 MeV]**

9. Calculate the binding energy per nucleon for ${}^{197}_{79}$Au (atomic mass of ${}^{197}_{79}$Au = 196.96 a.m.u.; 1 a.m.u. = 931.5 MeV). **[7.94 MeV]**

10. Determine the binding energy per nucleon for a deuteron nucleus, having mass as 2.013554 a.m.u. (1 a.m.u. = 931.5 MeV). **[1.367 MeV]**

11. Calculate the maximum kinetic energy of the electron emitted in the following β^--decay process,

$${}^{198}\text{Au} \rightarrow {}^{198}\text{Hg}^{*} + \beta^- + \bar{\nu}$$

^{198}Hg is a mercury nucleus in excited state, with energy 1.088 MeV above the ground state (atomic masses of ^{198}Hg and ^{198}Au are 197.966760 a.m.u. and 197.968233 a.m.u., respectively; 1 a.m.u. = 931.5 MeV). **[0.283 MeV]**

12. A β-decay process is represented as,

$${}^{176}\text{Lu} \rightarrow {}^{176}\text{Hf} + e + \bar{\nu}$$

Determine the maximum energy that the emitted β-particle can possess (atomic masses of ^{176}Lu and ^{176}Hf are 175.942694 a.m.u. and 175.941420 a.m.u., respectively; 1 a.m.u. = 931.5 MeV). **[1.186 MeV]**

13. If one neutron each has to be separated from the nuclei ${}^{41}_{20}$Ca and ${}^{27}_{13}$Al, calculate the values of energies required for this purpose (the masses of a neutron and the nuclei of ${}^{41}_{20}$Ca, ${}^{40}_{20}$Ca, ${}^{27}_{13}$Al, ${}^{26}_{13}$Al are 1.008665 a.m.u., 40.962278 a.m.u. 39.962591 a.m.u., 26.981541 a.m.u., and 25.986895 a.m.u., respectively; 1 a.m.u. = 931.5 MeV). **[8.36 MeV; 13.06 MeV]**

14. Determine the energy needed to remove one neutron from the nucleus of ${}^{13}_{6}$C. The binding energy per nucleon for ${}^{13}_{6}$C and ${}^{12}_{6}$C are 7.469 MeV and 7.678 MeV, respectively. **[4.961 MeV]**

15. A radioactive decay process is represented as,

$${}^{238}\text{Pu} \rightarrow {}^{234}\text{U} + \alpha$$

Calculate the kinetic energy of the α-particle emitted. The atomic masses of ^{238}Pu, ^{234}U, and ${}^{4}_{2}$He are 238.04955 a.m.u., 234.04095 a.m.u., and 4.002603 a.m.u., respectively (1 a.m.u. = 931.5 MeV). **[5.58 MeV]**

16. A radioactive decay is observed to occur as follows,

$$^{228}\text{Th} \rightarrow {}^{224}\text{Ra} + \alpha$$
$$^{224}\text{Ra} \rightarrow {}^{224}\text{Ra} + \gamma \text{ (217 keV)}$$

Find the kinetic energy of the α-particle emitted in the process. The atomic masses of ^{228}Th, ^{224}Ra, and $^{4}_{2}$He are 228.028726 a.m.u., 224.020196 a.m.u., and 4.002603 a.m.u., respectively (1 a.m.u = 931.5 MeV). **[5.304 MeV]**

17. Calculate the energy required to split an α-particle into its constituents (masses of an α-particle, a proton, and a neutron are 4.0039 a.m.u., 1.0078 a.m.u., and 1.0087 a.m.u., respectively; 1 a.m.u. = 931.5 MeV). **[27.09 MeV]**

18. Two deuterons fuse together to form a helium $\left(^{4}_{2}\text{He}\right)$ nucleus. Calculate the energy released in the process. The binding energy per nucleon for a helium nucleus and a deuteron $\left(^{2}_{1}\text{H}\right)$ nucleus are 7.0 MeV and 1.1 MeV, respectively. **[23.6 MeV]**

19. If 4 protons fuse together to form a helium $\left(^{4}_{2}\text{He}\right)$ nucleus and 2 positrons (as believed to be occurring in sun), find the energy released in the process (take masses of a proton, a positron, and a $^{4}_{2}$He nucleus as 1.007825 a.m.u., 0.000549 a.m.u. and 4.002603 a.m.u. respectively; 1 a.m.u. = 931.5 MeV). **[25.70 MeV]**

20. Determine the amount of energy released by nuclear fission of all the atoms contained in 750 g of plutonium $\left(^{239}_{94}\text{Pu}\right)$, if the average energy released per fission is 180 MeV. **[3.40×10^{26} MeV]**

21. Different amounts of energy are generated by nuclear fission of 500 g of $^{235}_{92}$U in a nuclear reactor and by nuclear fusion of 500g of hydrogen in a star. Calculate and compare the two respective amounts of energy. [Hint: In the said fusion, four $^{1}_{1}$H nuclei combine to form a $^{4}_{2}$He nucleus, with release of an average amount of about 26 MeV of energy per event. In the fission, the average energy released per atom of $^{235}_{92}$U is 200 MeV).
[2.55×10^{26} MeV; 19.5×10^{26} MeV; the latter is nearly 8 times of the former]

UNIT 9
SOLIDS & SEMICONDUCTOR DEVICES

CHAPTER 9 (A)
Energy Bands is Solids Semiconductors

CHAPTER 9 (B)
Semiconductor Devices

CHAPTER 9 (A)
ENERGY BANDS IN SOLIDS; SEMICONDUCTORS

In the present and the next chapter we shall essentially be involved with the conduction of electricity in solids. We shall concern ourselves only with the crystalline solids.

9A.01. Crystalline and Noncrystalline (or amorphous) Solids

We know that solids consists of atoms and molecules (a molecule itself is made up of a group of atoms). A question naturally arises as to how the atoms in a solid are packed together. Are they randomly packed like the particles of sand contained in a bag or they are orderly packed, e.g. linearly, circularly, spirally, etc. ?

Both the possibilities exist. In a well developed crystal, e.g. in a crystal of sugar or rocksalt or in a gem crystal like diamond, throughout the volume of the crystal the atoms are linearly arranged along the three dimensions X, Y, Z. They are periodically spaced along each dimension, although the spacings may be different along the three dimensions. Usually the interatomic spacings are a few angstrom units (Å) (1Å = 0.1 nm or 10^{-10} m). A framework of such intersecting lines in 3 dimensions is called a **space lattice**. When atoms are placed at the points of intersection, as in an actual crystal, the whole arrangement is called a **crystal structure**. (Unfortunately, in many books the 'crystal structure' is referred to as 'crystal lattice', which is wrong. The lattice is purely a geometrical arrangement of lines (or points), whereas the crystal structure is a real entity). A part of such a structure is shown in Fig. 9.1. Each black dot represents an atom of the material of the crystal (if the dots are removed, the arrangement will be called space lattice).

Thus, we find that a crystal structure is built up by a translation of atoms at regular intervals in 3 dimensions. Another useful way to visualise the creation of a crystal structure is by a regular repetition of the unit cell of the structure in three dimensions. A **unit cell** is "The fundamental building block of a crystal structure". The unit cell of the crystal structure in Fig. 9.1 is shown by solid lines. It is like a brick in a building. The building is constructed by stacking the bricks in various directions. A crystal is built up by stacking identical unit cells in three dimensions in space.

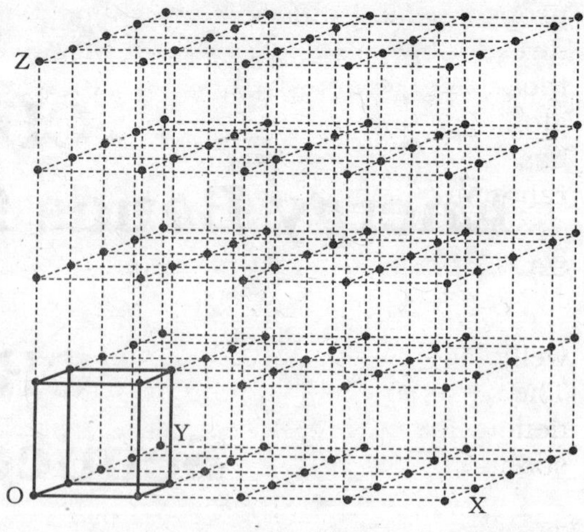

Fig. 9.1

The unit cell dimensions along X, Y, Z are denoted as a, b, c, respectively. A total of 14 types of space lattices are possible, depending upon the values of the cell dimensions a, b, c and the interaxial angles α, β, γ. The simplest type of the space lattice is a simple cubic (s.c.) lattice. A crystal structure based on this lattice has the unit cell with $a = b = c$ and $\alpha = \beta = \gamma = 90°$. It is shown in Fig. 9.2a, with atoms located at the corners of the cell alone. If the unit cell also has an atom at the centre of the cell, the corresponding structure is called body-centred cubic (b.c.c.) structure (Fig. 9.2b). If instead of the centre it has atoms at the centres of all the 6 faces (up-down, left-right, front-back) of the cell, it is termed as face-centred cubic (f.c.c.) structure (Fig. 9.2c). In addition, structures having other lattices are possible, with centring on either A, B, or C faces (⊥ to X, Y, or Z axis, respectively) alone. A C-centred

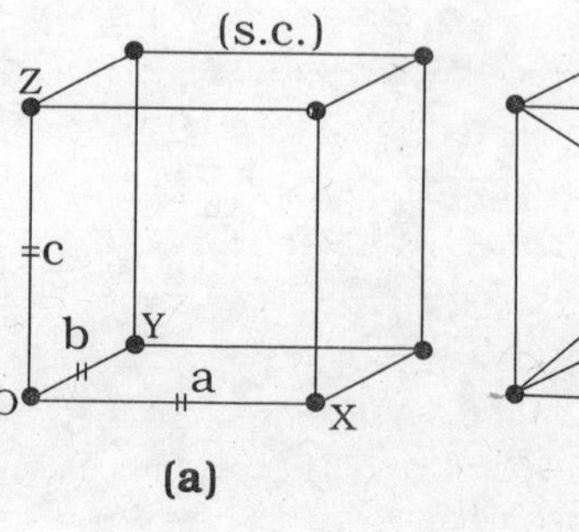

(a)

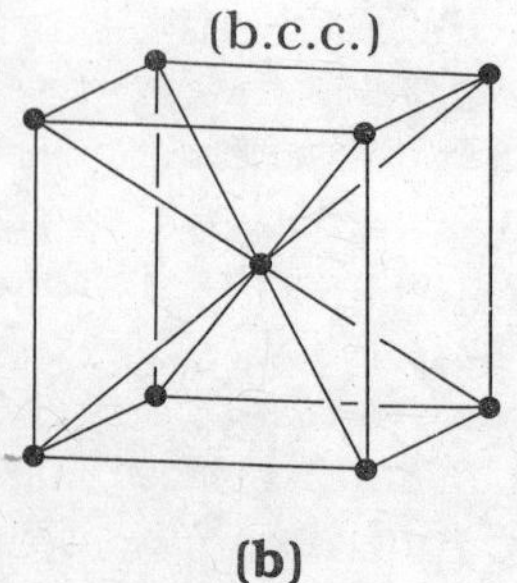

(b)

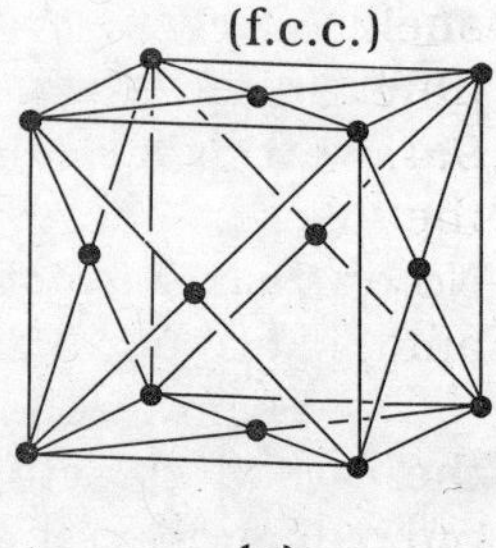

(c)

Fig. 9.2

unit cell is shown in Fig. 9.3. The crystal structure of diamond is based on an f.c.c. lattice, with its carbon atoms located in the unit cell as shown in Fig. 9.2c.

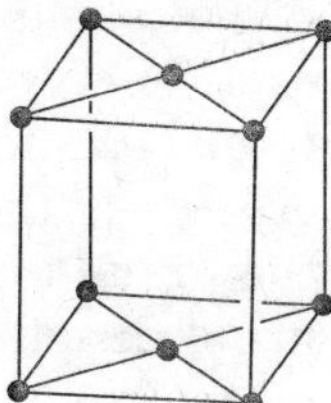

Fig. 9.3

A **crystalline solid** is one which possesses the said periodic 3-dimensionnal arrangement of atoms. If the structure extends throughout the volume of the solid without any break, the solid is called a **single crystal**. The familiar examples are the naturally grown gem crystals like diamond, sapphire, emerald, topaz, etc. Other solids can also be grown in single crystal form in the laboratory by suitable crystal growth techniques. A solid which completely lacks in a regular arrangement of the atoms inside it is called a **noncrystalline or amorphous** solid. Examples are wood, rubber, glass, etc. If the regularity of the atomic arrangement is discontinuous inside the solid, it is called a **polycrystalline solid.** All metals and alloys, e.g. a copper wire or a gold ring, are usually polycrystalline is character. They consist of a vast number of microsize grains packed together. Each grain is a tiny single crystal [from the knowledge of the Avogadro number ($N_A = 6.02\times10^{23}$ mol^{-1}), it can be estimated that even such a tiny grain consists of billions of atoms]. By special laboratory techniques, they can be grown as single crystals, too. X-ray diffraction (also electron diffraction and neutron diffraction) has revealed that even the so-called noncrystalline solids are not absolutely lacking in an orderly arrangement of the atoms inside them; the order only extends over a short range, e.g. a few tens to a few hundred atomic spacings. Similarly, it is found that the so-called single crystals are also not perfect. They have a small percentage of breaks in the structure, because of the presence of various kinds of defects (or imperfections) in the structure. They may be point defects, line defects, planar defects, impurities, etc. (these defects, although small in content, play very important and interesting role in determining the properties of the solids). (The observed lack of perfection in single crystals and the existence of a little crystalline character in the noncrystalline solids are supported by theoretical considerations. However, they are beyond the scope of this book).

Isotropic and anisotropic solids: The solids in which the value of a physical property, e.g. conductivity, is same in different directions are called **isotropic solids**. The solids having different values of a physical property along different directions are called **anisotropic solids**. The value of a property in a certain direction depends upon the arrangement of atoms along that direction. Since the crystalline solids, except those belonging to the most symmetrical cubic system ($a=b=c$; $\alpha=\beta=\gamma=90°$), e.g. diamond, have different atomic arrangements along different directions, they are anisotropic in nature. Only the crystalline solids based on cubic system are isotropic in nature. On the other hand, the amorphous solids have their atoms randomly distributed in all directions. Hence they all are isotropic in nature.

9A.02. Energy Bands in Solids

According to the Bohr's theory, in an isolated atom the electrons revolve in shells and subshells with well defined energy values. In other words, the electrons occupy discrete energy levels in an isolated atom. The shells are represented by the serial numbers 1, 2, 3,.................... and the subshells of each shell are denoted by the letters s, p, d, f,................ Thus, the electronic configuration in an isolated atom of the metal sodium (Z = 11) is represented as,

$$(1s)^2\ (2s)^2\ (2p)^6\ (3s)^1$$

The energy levels 1s, 2s, 2p are completely filled. The level 3s has just one electron, although it can accommodate two electrons, i.e. it is partially filled. All the energy levels above 3s are empty, e.g the next allowed energy level 3p,which can accommodate six electrons, is empty. All the electrons in a particular shell or subshell, e.g. the 6 electrons in the 2p shell, have the same energy but they are distinguished by different quantum states. The electronic configuration is schematically shown in Fig. 9.4, with each electron represented by a black dot. Note that the electrons only possess the discrete energy values represented by the shells and the subshells 1s, 2s, 2p, 3s. No other energy values, viz. between 1s and 2s, between 2s and 2p, and between 2p, and 3s, are permissible.

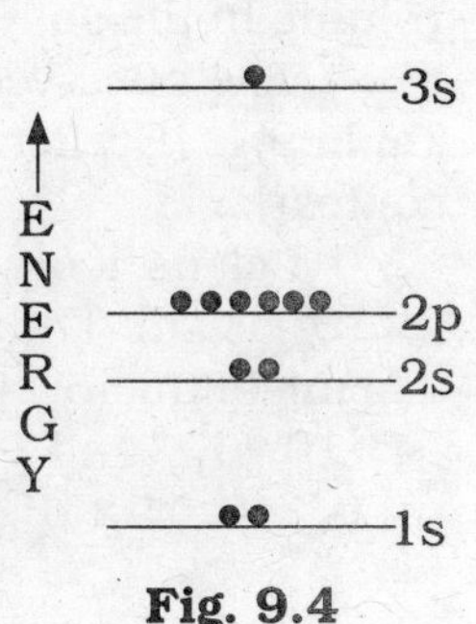

Fig. 9.4

We now consider what happens to the energy levels when we start building up the crystal structure of sodium, by assembling rows of identical Na atoms around our isolated sodium atom, in three dimensions. As the first step, let us see what happens when the first Na atom is brought near to it. Then the electron systems of the two atoms mutually interact (the mutual interaction is negligibly weak when the two atoms are far apart from each other). The interaction makes each energy level of the original atom to split into two energy levels, as shown in Fig. 9.5 (a). The nearer the two atoms are, the greater is the splitting. As more Na atoms are brought around the initial isolated atom, the crystal structure of sodium is built up. If the structure finally consists of N atoms, then each energy level of an isolated Na atom splits into N levels. For instance, if 10 atoms have been brought around, each energy level will split into 10 levels. This is shematically shown in Fig. 9.5(b). In an actual crystal, the value of N is extremely large, being $\sim 10^{24}$. Therefore the split energy levels are extremely close to each other. There energy values lie both above and below the initial energy level.

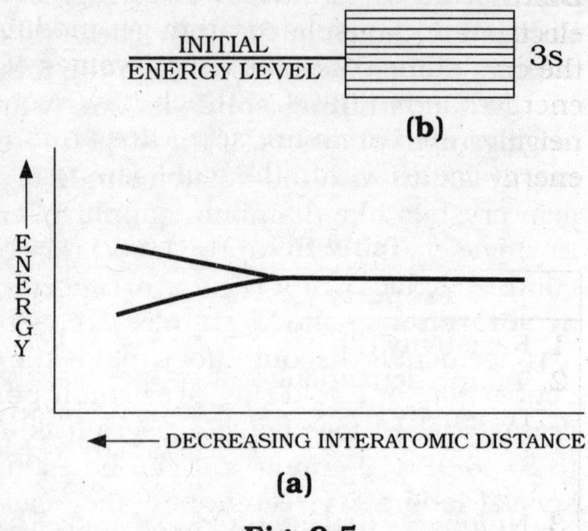

Fig. 9.5

Thus, we find that when a crystalline solid is built up by assembling the required atoms, the initial energy levels of the individual atoms transform into **energy bands**, as shown in Fig. 9.6 (cf. Fig. 9.4). Each energy band consists of very closely spaced energy levels. Any two neighbouring bands are separated by a gap called **energy gap or band gap (E_g)**. The energy values within the band gaps are **forbidden** for the electrons in the solid, i.e. the electrons in the solid cannot possess this range of energies. The bands of lower energy are narrower than those of higher energy. It is so because the electrons of the higher shells interact more strongly than those belonging to the inner shell, as they are less influenced by the positive charges of the nuclei. The Fig. 9.6 has been drawn for a general case, not particularly for sodium. As we shall see shortly, in particular cases one of more bands may only be partially filled or neighboring bands may overlap with each other.

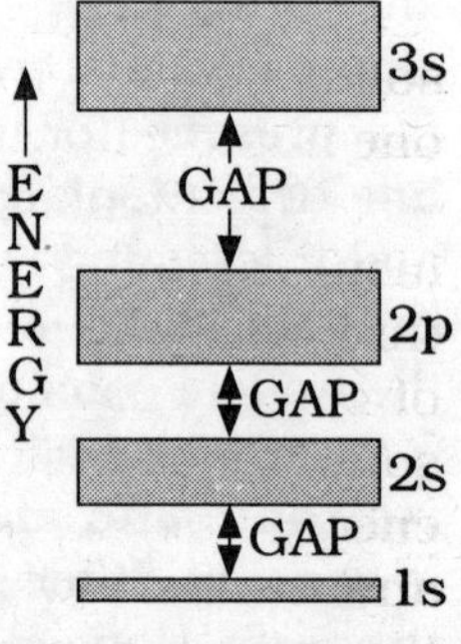

Fig. 9.6

9A.03. Difference between Metals, Insulators, and Semiconductors

Electrically, i.e. from the viewpoint of electric conduction, the metals, insulators, and semiconductors (all solids) can be distinguished in term of some physical properties as well as on the basis of band theory of solids, i.e. in terms of energy bands in solids. While the measurement of the physical properties provides quantitative estimate of the difference, the band theory provides the basic answer, viz. why the properties should differ for the said three kinds of solids,. We shall discuss both the aspects.

Distinction on the basis of physical properties. The metals, insulators, and semiconductors can be differentiated by the measurement of one or more of the following physical properties:

(i) **Resistivity, ρ** (or conductivity, σ). vide Chapter 2A; Section 2A.03; Eq. 2.9,

$$\rho = \frac{A}{\ell} R$$

(ii) **Temperature coefficient of resistivity, α.** vide Chapter 2A; Section 2A.10; Eqs. 2.21 and 2.23,

$$\alpha = \frac{\rho_T - \rho_0}{\rho_0 (T - T_0)} R \quad ; \quad \alpha = \frac{R_T - R_0}{R_0 (T - T_0)}$$

(iii) **Number density of charge carriers, n.** The density, i.e. the number of charge carriers per unit volume in the solid, can be found by a suitable electrical method.

The observed variation in these physical properties among the three types of solids is provided in Table 9.1. [Earlier, in Chapter 2A (Section 2A.08), we had partially examined such variation in a different context].

Let us look into some comparative values as examples. The resistivity (ρ) of the metal copper is 2×10^{-18} Ω.m. The resistivity of diamond, an insulator, is about 10^{24} times this value, whereas that of silicon, a semiconductor, is about 10^{11}times. The temperature coefficient of resistivity (α) is ~0 for diamond and $\sim 4\times10^{-3}$ K^{-1} and ~-70×10^{-3} K^{-1} for copper and silicon, respectively. The number density of charge carriers (n) is ~ 0, 9×10^{28} m^{-3}, 1×10^{16} m^{-3} for diamond, copper and silicon, respectively.

Distinction on the basis of energy bands in solids. We have seen that the discrete energy levels of the electrons in an isolated atom get modified when such atoms come together to form a crystalline solid. In the crystalline solid, the energy values of the electrons are confined within bands of different widths, called energy bands (Fig. 9.6). Each band consists of a very large number of closely spaced energy levels. The neighboring bands are separated from each other through gaps, called energy gap or band gap (E_g). The energy values within the band gaps are forbidden for the electrons.

Table 9.1. Observed variation in the electrical properties of various kind of solids

Property	Metals	Insulators	Semiconductors
1. Resistivity (ρ)	Very low	Very high	Intermediate
2. Temp. dependence of resistivity	ρ increases with temperature. Temp coeff. of resistivity (α) is small and +ve.	Negligible effect. $\alpha \approx 0$.	ρ decreases with temperature. α is high and –ve.
3. Number density of charge carriers (n)	High	Negligible	Low

The nature of the band structure and the value of the band gap (E_g) decide whether a given crystalline solid is a metal (or a conductor, in general), a semiconductor, or an insulator. We shall discuss them one by one in the following.

(a) **Insulators; Valence band and Conduction band; Band gap.** The band structure for a typical insulator is shown in Fig. 9.7 (a). The top energy band is unoccupied (or, say, empty or vacant) and the lower three bands are completely filled. If an electric field is applied across the solid or the solid is heated, a flow of current through the solid will require that the kinetic energy of the electrons increases. It means that some electrons must move to higher energy levels. This is not possible in any of the lower 3 bands, because they all are completely filled. Another possibility is that an electron from the top filled band crosses the energy gap E_g to occupy an energy level in the top vacant band. But in an insulator the band gap E_g is so large ($\approx$ 5eV or more) that normally the energy imparted by an applied electric field or thermal energy supplied by heating the solid to an electron is insufficient for this purpose. Hence normally no electric conduction is possible in an insulator.

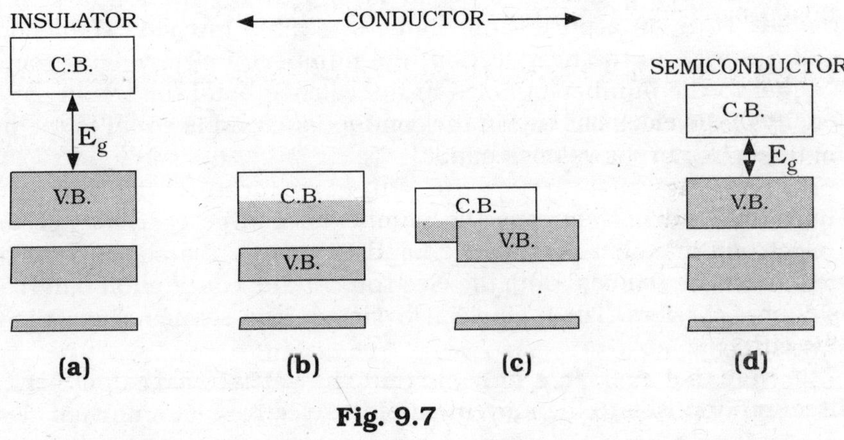

Fig. 9.7

The highest completely filled energy band at absolute zero (0 K or –273 °C) is called **valence band**. Any unfilled energy band above it is called **conduction band** (it may be completely unfilled, i.e. empty, or partially filled). The **band gap** represents the difference in energy between the top of the valence band and the bottom of the conduction band.

Any rise in the temperature of the solid has very little effect on its band structure.

(b) **Metals (or, in general, conductors); Fermi level; Fermi energy.** The energy band structure for a metal (or in general, a conductor) is such that either (i) the highest energy band is partially filled (it is called conduction band; Fig. 9.7b) or (ii) the valence band partly overlaps with the conduction band (Fig. 9.7c). In either case, electric conduction is possible when an electric field is applied across the solid, because within the band itself vacant energy levels are available to the electrons for moving into. Note that no band gap (E_g) exists for a conductor.

"The highest occupied energy level in the conduction band at absolute zero (T = 0 K)" is called **Fermi level** and the energy corresponding to it is called **Fermi energy (E_F).** In expressing the value of E_F, we assume the energy at the bottom of the conduction band to be zero (E = 0), e.g. $(E_F)_{copper}$ = 7 eV. Note that the physical significance of the Fermi level is that all energy levels below it are filled and all those above it are empty.

(c) **Semiconductors.** The band structure for a typical semiconductor is shown in Fig. 9.7d. It is similar to that for an insulator, but with the important difference that the energy gap (E_g) is comparatively quite small, e.g. for the well known semiconductors silicon and germanium the values at room temperature are,

$$(E_g)_{Si} = 1.1 \text{ eV} \quad ; \quad (E_g)_{Ge} = 0.7 \text{ eV}$$

Note that these values are much smaller than the band gap for the insulator diamond, viz. $(E_g)_{diamond} = 5.5$ eV. In general, the value of the band gap for a semiconductor is just a few electron-volt ($E_g \sim 1$eV).

9A.04. Charge Carriers in Semiconductors; Free Electrons and Holes

As mentioned earlier, at absolute zero temperature (0 K) the valence band of a semiconductor is completely filled and the conduction band is completely empty. As the temperature is raised, the electrons gain energy due to thermal agitation. The energy gain is not equal for all the electrons. If an electron gains enough energy, it can overcome the band gap and move over from the valence band to the conduction band. Thus, a vacancy is created in the valence band, which is called a **hole**. In other words, a **hole** is "a vacancy created in the valence band when an electron crosses the band gap and moves over to the conduction band".

When an electric field is applied across the semiconductor, the electrons begin to move in the direction opposite to the field. Then a neighbouring electron can occupy the hole, leading to the formation of another hole at the place from where the electron had left. The process can continue as a chain movement of the hole in the direction opposite to the motion of the electrons. Since the electrons are negatively charged, effectively it is equivalent to the movement of a positive charge in the direction of the field. **Thus, we can treat a hole as an imaginary particle with charge +e.** The electric current due to the movement of the holes is called **hole current**, just as the normal electric current due to the electrons is called **electron current.**

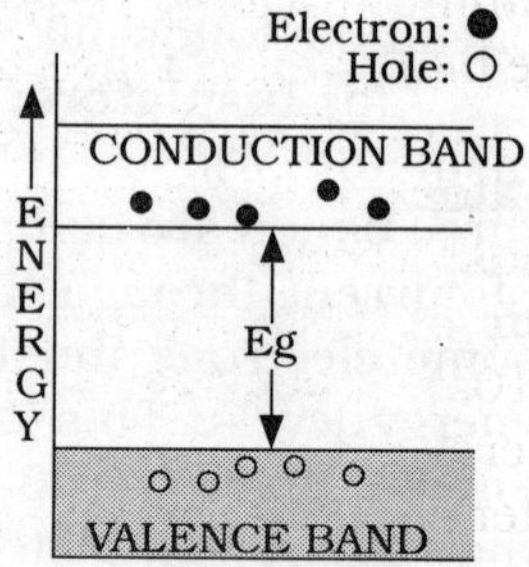

Fig. 9.8

Whenever a valence electron is shifted to the conduction band, a hole is created. Thus, in a pure semiconductor (called intrinsic semiconductor, which we shall describe in the next section) the number of electrons in the conduction band is equal to the number of holes in the valence band, or we an say that the number density of the electrons (n_e) in the conduction band is equal to the number density of the holes (n_h) in the valence band,

$$n_e = n_h \text{ [pure (or intrinsic) semiconductor]}$$

The number density means the number of charge carriers per unit volume. The Fig. 9.8 shows that if 5 electrons have left the valence band to occupy the conduction band, then correspondingly the valence band will have 5 holes. Both the electrons (in the conduction band) and the holes (in the valence band) serve as charge carriers. The total electric current in a semiconductor is the sum of the electron current and the hole current.

Note that **even in a hole current the actual charge carriers are the electrons.** Only they move in a direction opposite to the movement of the electrons in a normal electron current.

9A.05. Intrinsic & Extrinsic Semiconductors; Doping; n-Type & p-Type Semiconductors; Donors & Acceptors

The electrical conductivity of a pure semiconductor can be considerably enhanced by introducing a tiny amount of a suitable impurity into it. This process is known as **doping**. We shall shortly discuss more about it in the following. A pure semiconductor is known as an **intrinsic semiconductor**. A semiconductor containing a tiny amount of a suitable impurity is known as an **extrinsic semiconductor.** There can be two types of extrinsic semiconductors: (i) n-type and (ii) p-type. The **n-type** of semiconductors have an excess of the **n**egatively charged electrons in them. On the other hand, the **p-type** have an excess of the **p**ositively charged holes in them. We shall now discuss about them in some detail.

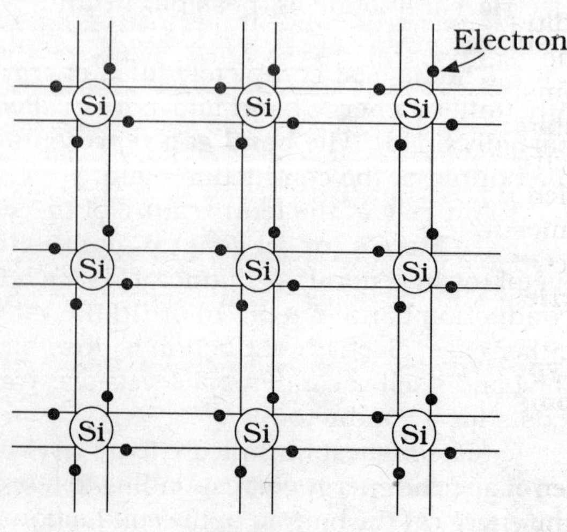

Fig. 9.9

n-Type Extrinsic Semiconductors; Donor Atoms

We shall illustrate the formation of extrinsic semi-conductors through a representative example of the well known semiconductor silicon (Si). It has atomic number Z = 14 and its electronic configuration is,

$$(1s)^2\ (2s)^2\ (2p)^6\ (3s)^2\ (3p)^6$$

The crystal structure of silicon is such that each Si atom is linked with 4 neighboring Si atoms through covalent bonds (Fig. 9.9), i.e. it is tetravalent (valency = 4). In each covalent bond, two electrons are shared by the two neighboring atoms. The energies of all such electrons are such that they belong to the valence band of silicon. If sufficient additional energy becomes available, e.g. by heating the semiconductor, such an electron may shift to the conduction band (vide Fig. 9.8). The minimum energy required for this purpose is the band gap energy E_g. For silicon, E_g = 1.1 eV, at room temperature.

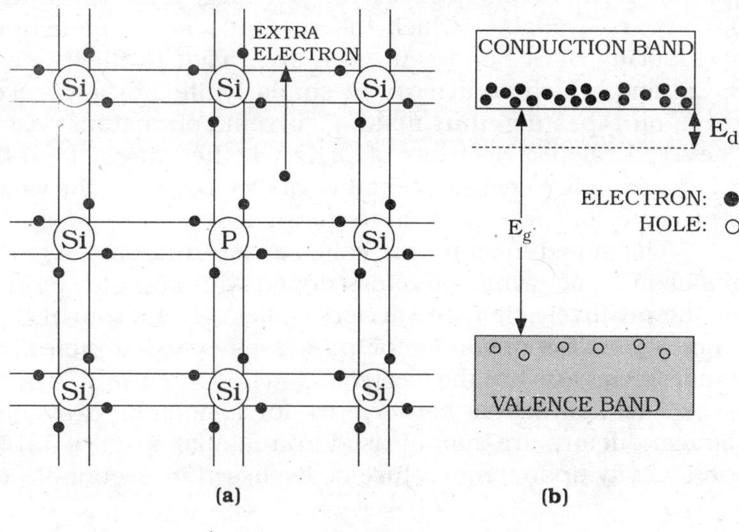

Fig. 9.10

Now, suppose a silicon crystal is doped with phosphorus (P), which has Z = 15. Then, some of the Si atoms are displaced and get substituted by the P atoms, i.e. their positions are occupied by the P atoms. Phosphorus is pentavalent (valency = 5), i.e. each P atom has 5 valence electrons. A doped P atom shares four of these electrons with the neighboring four Si atoms. The fifth 'extra' electron becomes free (Fig. 9.10a). Its energy is higher than the energy of a valence electron locked in a covalent bond. The average energy of these 'extra' electrons is such that they occupy energy levels within the band gap, at a small energy interval E_d below the bottom of the conduction band (Fig. 9.10b). For the doping of phosphorus into silicon, $E_d \approx 0.05$ eV. Since $E_d \ll E_g$, the energy needed to shift these 'extra' electrons to the conduction band is much smaller than the corresponding energy needed in an intrinsic semiconductor to shift electrons from the valence band to the conduction band.

The energy required to shift the 'extra' electron obtained from the phosphorus atoms to the conduction band is so small that at room temperature and above, almost all such electrons pass on to the conduction band. Although only a tiny amount of the phosphorus is added to silicon, so many additional charge carriers (the free electrons) become available that the conductivity goes up about 10^6 times, i.e. about one million times. That is how the electric conduction in a semiconductor is highly sensitive to the introduction of a suitable impurity. Each P atom contributes one free electron to the structure.

The doped phosphorus atoms are called **donor** atoms, as they readily donate additional electrons to the conduction band. "The semiconductors doped with donor atoms" are called **n-type semiconductors**. **n** stands for the **n**egative charge carriers, viz. electrons. The impurities like phosphorus are called **donor impurities**. Because an n-type semiconductor has excess of electrons, the electrons contained in it are called **majority carriers** and the holes the **minority carriers.**

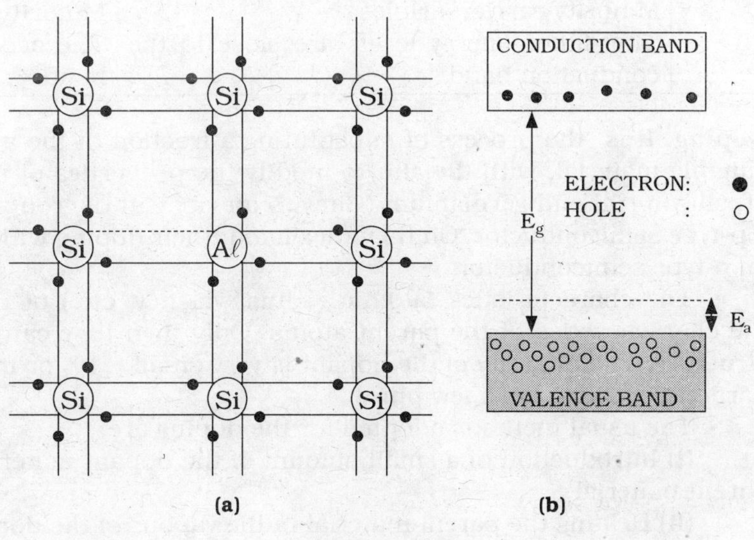

Fig. 9.11

p-Type Extrinsic Semiconductors; Acceptor Atoms

Let us examine the situation when a silicon crystal is doped with aluminium (Al), which is trivalent (valency = 3), with Z = 13. Now, a doped Al atom can covalently bond with just 3 Si atoms, giving rise to a 'missing' electron, i.e. a hole, in one of the Al - Si bond (Fig. 9.11 a; the missing bond is shown by broken lines). If an extra energy of about 0.07 eV is available, which is really so at room temperature and above, a valence electron from a neighboring Si-Si bond can easily free itself to fill this hole. At the same time a hole is created in this Si-Si bond, which in turn can be similarly filled by a valence electron from another Si-Si bond. The process keeps on repeating, thus making current conduction possible in the structure by the movement of holes. The freed valence electrons of slightly higher energy ($\approx$ 0.07 eV) occupy energy levels within the band gap E_g, at an average energy interval E_a above the top of the valence band (Fig. 9.11b). Each substituted Al atom contributes one free electron to the structure.

The doped Al atoms are called **acceptor** atoms, as they readily accept electrons from the valence band of silicon. "The semiconductors doped with acceptor atoms" are called **p-type semiconductors**. **p** stands for the **p**ositively charged carriers (holes). The impurities like aluminium are called **acceptor impurities**. Since a p-type semiconductor possesses excess of holes, the majority carriers in it are the holes and the minority carriers are the electrons.

The distinction between (i) intrinsic and extrinsic semiconductors and (ii) n-type and p-type semiconductors are summarised in a tabular form in Tables 9.2 and 9.3, respectively (the dependence of conductivity upon temperature is discussed in Section 9A.07).

Table 9.2. Distinction between intrinsic and extrinsic semiconductors.

S.No.	Intrinsic Semiconductors	Extrinsic Semiconductors
1.	Pure material; contains no impurities	Prepared by doping a suitable impurity into an intrinsic semiconductor
2.	The number density of free electrons in the conduction band is equal to that of the holes in the valence band ($n_e = n_h$)	$n_e \gg n_h$ in n-type semiconductors and $n_h \gg n_e$ in p-type semiconductors
3.	Low electrical conductivity	High electrical conductivity
4.	The electrical conductivity only depends upon temperature	Electrical conductivity depends both on the temperature and the nature and quantity of dopant.

Table 9.3. Distinction between n-type and p-type semiconductors

S.No.	n-Type Semiconductors	p-Type Semiconductors
1.	Prepared by doping a donor impurity into an intrinsic semiconductor	Prepared by doping an acceptor impurity into an intrinsic semiconductor
2.	The donor atoms provide extra electrons in the structure	The acceptor atoms provide extra holes in the structure
3.	Majority carriers: Electrons Minority carriers: Holes	Majority carriers: Holes Minority carriers: Electrons
4.	The donor energy levels are close to the conduction band	The acceptor energy levels are close to the valence band

Doping. It is "the process of substituting a fraction of the atoms of a material with the atoms of another suitable material, with the aim to modify the properties of the parent material". For example, the doping of silicon (Si) and germanium (Ge)with an element of group 13 of the periodic table (B, Al, etc.) produces a p-type semiconductor. On the other hand, their doping with an element of group 15 (P, As, etc.) produces an n-type semiconductor.

The characteristics (atomic radius, valency, etc.) of the dopant atom have to be compatible with the characteristics of the parent atoms. Only then they can successfully replace the parent atoms in the structure. The quantity of the dopant is very small ~1% or many times much less than that, e.g. just a few parts per million, i.e. a few ppm).

The usual methods adopted for the doping are:

(i) Introduction of a small amount of the dopant material during the process of crystallization of the parent material.

(ii) Heating the parent material in the vapour of the dopant.

(iii) Ion-Implantation: It consists of bombardment of the parent material with ions of the dopant.

Doping and charge neutrality. It is important to note that when electrons leave their parent atoms (the donor atoms in an n-type semiconductor) or attach with new atoms (the acceptor atoms in a p-type semiconductor), the respective atoms become ions (donor atoms: +ve ions; acceptor atoms: –ve ions). Thus, in a p-type semiconductor the total number of –ve ions is equal to the total number of +ve holes created in the solid. Hence, although $n_h \gg n_e$ is such a semiconductor, the solid as a whole remains electrically neutral. Similarly for an n-type semiconductor ($n_e \gg n_h$).

The ions remain locked up at their positions in the structure and so do not participate in electric conduction. The current conduction takes place either by the movement of free electrons (in n-type semiconductors) or by the movement of holes (in p-type semiconductors).

9A.06. Electrical Conductivity of Semiconductors

Consider the flow of current in a rectangular semiconductor, across which an electric field has been applied (Fig. 9.12). The current arises from the motion of electrons in the conduction band and from the motion of holes in the valence band (note that conventionally the direction of flow of the electron current is taken as against the direction of the electric field). We shall use following representations,

- i, i_e, i_h : Total current, electron current and hole current, respectively
- n_e, n_h : Number density of electrons and holes, respectively
- v_e, v_h : Drift velocity of electrons and holes, respectively
- E : Electric field
- A : Area of cross-section of the semiconductor
- ρ, σ : Resistivity and conductivity, respectively ($\sigma = 1/\rho$)
- μ_e, μ_h : Mobility of electrons and holes, respectively ($\mu = v/E$)

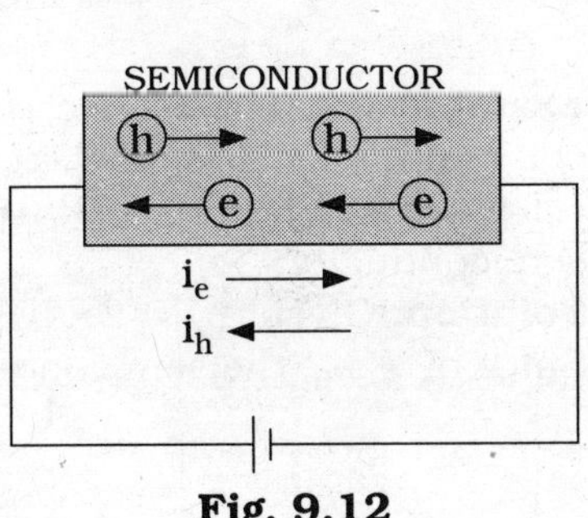

Fig. 9.12

The total current is given by,

$$\begin{aligned} i &= i_e + i_h \\ &= n_e e A v_e + n_h e A v_h \text{ (vide Eq. 2.11, Chapter 2A)} \\ &= eA\,(n_e v_e + n_h v_h) \end{aligned}$$

The current density $J = i/A$ (current per unit area) is given by,

$$J = e\,(n_e v_e + n_h v_h)$$

According to the Ohm's law (vide Eq. 2.10, Ch. 2A)

$$J = E/\rho = E\sigma$$

Hence the electrical conductivity (σ) is given by,

$$\sigma = e\left[n_e \frac{v_e}{E} + n_h \frac{v_h}{E}\right]$$

$$\boxed{\sigma = e\,(n_e\mu_e + n_h\mu_h) \quad \text{(All semiconductors)}} \qquad \text{..... (9.1)}$$

In an intrinsic semiconductor,

$$n_e = n_h = n_i \text{ (say)}$$

Therefore, the conductivity of an intrinsic semiconductor (σ_i) is given by,

Or

$$\boxed{\sigma_i = e n_i\,(\mu_e + \mu_h) \quad \text{(Intrinsic semiconductors)}} \qquad \text{..... (9.2)}$$

Example 1. A pure semiconductor of size 5mm×5mm×1mm possesses 9×10^{20} conduction electrons per cubic metre. Determine the total number of charge carriers (electrons and holes) contained in it.

Solution. Given: $n_e = (9 \times 10^{20})\ \text{m}^{-3}$; volume = 5mm×5mm×1mm = $(5\times10^{-3})\text{m}\times(5\times10^{-3})\text{m}\times(10^{-3})\text{m} = 25\times10^{-9}\ \text{m}^3$

The total number of conduction electrons in the sample is given by,

$$\begin{aligned} N_e &= n_e \times \text{volume} = (9\times10^{20})\times(25\times10^{-9}) \\ &= 22.5\times10^{12} \end{aligned}$$

In an intrinsic (or, say, pure) semiconductor, the number of conduction electrons and holes is equal. Therefore, the total number of holes is,

$$N_h = N_e = 22.5\times10^{12}$$

∴ The total number of charge carriers is,

$$N = N_e + N_h = (22.5+22.5)10^{12}$$
$$= 45\times10^{12} \quad \textbf{(Answer).}$$

Example 2. A p-type sample of silicon is prepared by doping an average of one indium atom per 8×10^6 silicon atoms. If the number density of the silicon atoms in the sample is 5×10^{28} atoms per cubic metre, determine the number of acceptor atoms in one cubic centimeter of the sample.

Solution. Given: No. density of Si atoms $= 5\times10^{28}\ m^{-3}$

The doping of 1 indium atom produces 1 acceptor atom in the sample. Since 1 indium atom is doped per 8×10^6 Si atoms and there are 5×10^{28} Si atoms per cubic metre, the number density (i.e. no. per unit volume) of the acceptor atoms is,

$$(\text{No. Density})_{acceptor} = \frac{5\times10^{28}}{8\times10^6}\ \text{atoms m}^{-3}$$
$$= 625\times10^{19}\ \text{atoms m}^{-3}$$
$$= (625\times10^{19})\times10^{-6}\ \text{atoms cm}^{-3}$$
$$= 625\times10^{13}\ \text{atoms cm}^{-3} \quad \textbf{(Answer).}$$

Example 3. An extrinsic semiconductor has number densities of electrons and holes in it as 2×10^{13} cm^{-3} and 1.5×10^{12} cm^{-3}, respectively. Is it an n-type or a p-type semiconductor ? If the mobilities of the electrons and holes are 3×10^4 cm^2 $(V.s)^{-1}$ and 2×10^2 cm^2 $(V.s)^{-1}$, respectively, calculate the resistivity of the semiconductor.

Solution. Given: $n_e = 2\times10^{13}\ cm^{-3}$; $n_h = 1.5\times10^{12}\ cm^{-3}$; $\mu_e = 3\times10^4\ cm^2\ (V.s)^{-1}$; $\mu_h = 2\times10^2\ cm^2\ (V.s)^{-1}$

Since $n_e > n_h$, the semiconductor is obviously of n-type. Its resistivity is given by,

$$\rho = \frac{1}{\sigma} = \frac{1}{e\,(n_e\mu_e + n_h\mu_h)}$$
$$= \frac{1}{(1.6\times10^{-19})[(2\times10^{13})\times(3\times10^4) + (1.5\times10^{12})\times(2\times10^2)]}$$
$$= \frac{1}{(1.6\times10^{-19})\,[(6\times10^{17}) + (3\times10^4)]} = \frac{1}{9.61\times10^{-2}}$$
$$= 10.41\ \Omega - \text{cm} \quad \textbf{(Answer).}$$

Example 4. In a p-type semiconductor, the highest acceptor level lies 60 meV above the top of the valence band. Find the maximum wavelength of the radiation which, when incident upon the semiconductor, can create a hole in the valence band.

Solution. Given: $E = 60\ meV = 60\times10^{-3}\ eV = (60\times10^{-3})\times(1.6\times10^{-19})\ J = 96\times10^{-22}\ J$

The supply of energy E to the semiconductor by the incident radiation enables an electron in the valence band to occupy an acceptor level (lying slightly above the top of the valence band, within the energy gap), thus creating a hole in the band. ∴ The maximum wavelength of the radiation is given by ($E = h\nu = hc/\lambda$),

$$\lambda_{max} = \frac{hc}{E} = \frac{(6.6\times10^{-34})\times(3\times10^8)}{(96\times10^{-22})}$$
$$= 206\times10^{-7}\ m\ (\text{or } 206\times10^2\ nm) \quad \textbf{(Answer).}$$

(Note that λ_{max} corresponds to the minimum frequency, and therefore to the minimum energy, of the radiation).

9A.07. Effect of Temperature on Band Gap and Electrical Conductivity of Semiconductors; Ohm's Law

Electrical conductivity. From Eq. 9.1, the electrical conductivity of a semiconductor is given by,

$$\sigma = e\,(n_e\mu_e + n_h\mu_h)$$

As the temperature increases, the number density of the charge carriers increases rapidly. Therefore σ tends to increase with rise in temperature. However, the mobility of the charge carriers decreases with temperature as,

$$T^{-m}$$

(For silicon, m ≈ 2.6, for both electrons and holes). But the increase in n_e and n_h is so rapid that it far compensates for any decrease in σ due to reduction in the values of μ_e and μ_h. Thus, the overall effect of the rise in temperature is that σ increases.

This behaviour of semiconductors is in sharp contract to the observed reduction in the conductivity of metals with rise in temperature, because the drift velocity of the charge carriers (electrons) decreases due to their mutual collisions, while their number remains unchanged (vide Chapter 2A; Section 2A.08).

Band gap: The band gap (E_g) of a semiconductor slowly decreases with rise in temperature. For silicon, E_g varies with T as,

$$E_g(T) = 1.21 - 3.6\times10^{-4}\,T \qquad \text{..... (9.3)}$$

Thus, at room temperature (T ≈ 300 K), the value of E_g is reduced from 1.21 eV (the value at absolute zero) to nearly 1.1 eV.

Ohm's law. According to the Ohm's law,

$$J = \sigma E$$

The law holds for the conductors (e.g. metals), for which the conductivity σ is a constant value, so the current density steadily increases with increase in the applied field E. But for the semiconductors σ is not a constant value. The number of charge carriers, both in the valence band and in the conduction band, does not increase uniformly with increase in E. Hence the Ohm's law is not obeyed by semiconductors.

Solved Conceptual Problems

[Including HOTS Questions]

Problem 1. How a crystalline solid differs from an amorphous solid ?
Solution. In a crystalline solid, the atoms are periodically arranged in three dimensions. In an amorphous solid, the atoms are randomly distributed throughout the volume of the solid.

Problem 2. Most crystalline solids are anisotropic and all amorphous solids are isotropic in nature. Explain how.
Solution. The value of a physical property in a certain direction depends upon the arrangement of atoms along that direction. The crystalline solids, except those belonging to the cubic system ($a=b=c$; $\alpha=\beta=\gamma=90°$), have different atomic arrangements along different directions. Hence most of them are anisotropic in nature. On the other hand, the amorphous solids have their atoms randomly distributed in all directions. Hence they all are isotropic in nature.

Problem 3. Explain how energy bands are formed in a solid.
Solution. In an isolated atom, the electrons occupy discrete energy levels. When a large number (say, N of atoms come together to form a solid, each energy level is split into N sub-levels. Since N is very large ($\sim10^{24}$) the sub-levels are extremely close to each other, lying both above and below the initial single energy level. Thus each initial single energy level transforms into an energy band, consisting of extremely close sub-levels of energy. Any two neighbouring bands are separated by a gap, called energy gap or band gap (E_g).

Problem 4. How the flow of current in a semiconductor through free electrons differs from that through holes ?
Solution. The total electric current in a semiconductor is the sum of electron current and hole current. The electron current results directly from the movement of free electrons, which are material particles with definite mass and charge. A hole is an empty space (a vacancy) and therefore it cannot move on its own like an electron. But its movement can be indirectly visualised as follows. When an electric field is applied across the semiconductor, the free electrons begin to move in the direction opposite to the field. Then, a neighbouring electron can occupy the hole, leading to the creation of another hole at its original position. The process can continue as a chain movement of the hole in the direction opposite to the motion of the

electrons. Since the electrons are negatively charged, effectively it is equivalent to the motion of a positive charge in the direction of the field. Thus, a hole can be imagined as a +vely charged particle, with charge +e. Conventionally, the direction of electron current is taken as opposite to the direction of movement of electrons in the solid. Since a hole moves opposite to the direction of movement of free electrons, the hole current has the same direction as the electron current (note that even in a hole current the actual charge carriers are the electrons; only they move in a direction opposite to the motion of electrons responsible for the electron current).

Problem 5. Explain how charge neutrality is maintained in a p-type semiconductor, in spite of the number of holes being far greater than the number of free electrons.
Solution. When a p-type semiconductor is created by doping the pure semiconductor, electrons attach with the new atoms (the acceptor atoms). Then, these atoms become –ve ions. At the same time, +vely charged holes are created at the positions of the attaching electrons. The total number of –ve ions is essentially equal to the total number of newly created +ve holes. Hence, although $n_h \gg n_e$ in a p-type semiconductor, the semiconductor on the whole remains electrically neutral.

Problem 6. Does electric conduction in semiconductors obey Ohm's law ? Justify your answer.
Solution. According to the Ohm's law, the current density is related to the electric field as,

$$J = \sigma E$$

The law holds for conductors (e.g. metals), for which the conductivity σ is a constant value, so the current density steadily increases with increase in the applied field E. But, for the semiconductors σ is not a constant value. The number of charge carriers, both in the valence band and in the conduction band, does not increase uniformly with increase in E. Hence the Ohm's law is not obeyed by the semiconductors.

Problem 7. Unlike a conductor, the electrical conductivity of a semiconductor rises with increase in temperature. Explain why.
Solution. The electrical conductivity of a semiconductor is given by,

$$\sigma = e\,(n_e\mu_e + n_h\mu_h)$$

As the temperature increases, the number density of the charge carriers and hence the value of σ increases rapidly. Although the mobility of the carriers reduces with rise in temperature, but this reduction is far less than the rapid increase of n_e and n_h. Hence the overall effect of the increases in temperature is that the conductivity of the semiconductor rises.

Solved Numerical Problems

Problem 1. The concentrations of electrons and holes in a pure semiconductor have initially the same value of 3×10^{10} m^{-3}. Then the semiconductor is suitably doped to raise the hole concentration to 5×10^{12} m^{-3}. Determine the electron concentration in the doped semiconductor.
Solution. Given: $n_i = n_e = n_h = 3\times10^{10}$ m^{-3} (in pure semiconductor); $n_h = 5\times10^{12}$ m^{-3} (in doped semiconductor)
The doping turns the initial intrinsic semiconductor into an extrinsic semiconductor, in which the electron and hole concentrations are given by,

$$n_e \times n_h = n_i^2$$

$$\Rightarrow \qquad n_e = \frac{n_i^2}{n_h} = \frac{(3\times10^{10})^2}{(5\times10^{12})}$$

$$= 1.8\times10^8 \quad \textbf{(Answer)}$$

Problem 2. A silicon wafer is simultaneously doped with arsenic and indium atoms of concentrations 2×10^{16} and 3×10^{14} per cubic metre, respectively. The number of silicon atoms in the wafer is 5×10^{28} per cubic metre. If $n_i = 1.5\times10^{16}$ m^{-3}, determine the number of electrons and holes in the doped wafer.
Solution. Given: $N_d = 2\times10^{16}$ m^{-3}; $N_a = 3\times10^{14}$ m^{-3}; $n_i = 1.5\times10^{16}$ m^{-3} (the As and In atoms act as donors and acceptors, respectively); (the given value of $N_{si} = 5\times10^{28}$ m^{-3} is immaterial here).

We know, $n_e \times n_h = n_i^2$ ..... (i)

Further, for electrical charge neutrality, we should have,

$$N_d = N_a = n_e - n_h \quad \text{..... (ii)}$$

Substituting for n_h from Eq. (ii) into Eq. (i) and solving, we get the equation,

$$n_e^2 - (N_d - N_a)n_e - n_i^2 = 0,$$

which gives,

$$n_e = \frac{1}{2}\left[(N_d - N_a) + \sqrt{(N_d - N_a)^2 + 4n_i^2}\right]$$

$$= \frac{1}{2}\left[(2\times10^{16} - 3\times10^{14}) + \sqrt{(2\times10^{16} - 3\times10^{14})^2 + 4\times(1.5\times10^{16})^2}\right]$$

$$= \frac{1}{2}\left[(1.97\times10^{16})\times\sqrt{12.88\times10^{32}}\right]$$

$$= \frac{1}{2}[(1.97+3.59)\times10^{16}] = \frac{1}{2}(5.56\times10^{16})$$

$$= 2.78\times10^{16}\ \text{m}^{-3}$$

$$\therefore \quad n_h = \frac{n_i^2}{n_e} = \frac{(1.5\times10^6)^2}{(2.78\times10^{16})}$$

$$= 0.81\times10^6\ \text{m}^{-3} \quad \textbf{(Answer).}$$

Problem 3. A hole-electron pair in a semiconductor recombine such that the energy released in the process is converted into electromagnetic radiation, with maximum wavelength as 710 nm. Find the band gap of the semiconductor.

Solution. Given: $\lambda = 710\ \text{nm} = 710\times10^{-9}\ \text{m}$

The band gap energy (E_g) corresponds with the energy associated with the maximum wavelength emitted (λ) [$E = h\nu = hc/\lambda$],

$$E_g = \frac{hc}{\lambda} = \frac{(6.6\times10^{-34})\times(3\times10^8)}{(710\times10^{-9})}\ \text{J}$$

$$= 0.279\times10^{-18}\ \text{J} = (0.279\times10^{18})\times(1.6\times10^{-19})\ \text{eV}$$

$$= 1.74\ \text{eV} \quad \textbf{(Answer).}$$

Problem 4. The concentrations of electrons and holes in a semiconductor are 4×10^{13} cm^{-3} and 5×10^{14} cm^{-3}, respectively, and their respective mobilities are 20,000 cm^2 (V.s)$^{-1}$ and 120 cm^2 (V.s)$^{-1}$. Determine the resistivity of the semiconductor.

Solution. Given: $n_e = 4\times10^{13}\ \text{cm}^{-3} = 4\times10^{19}\ \text{m}^{-3}$; $n_h = 5\times10^{14}\ \text{cm}^{-3} = 5\times10^{20}\ \text{m}^{-3}$; $\mu_e = 20000\ \text{cm}^2$ (V.s)$^{-1}$ = 2 m^2 (V.s)$^{-1}$; $\mu_h = 120\ \text{cm}^2\ (\text{V.s})^{-1} = 0.012\ \text{m}^2\ (\text{V.s})^{-1}$

The resistivity (ρ) is reciprocal of conductivity, which is given by,

$$\sigma = e\,(n_e\mu_e + n_h\mu_h)$$

$$= (1.6\times10^{-19})\times[(4\times10^{19})\times(2) + (5\times10^{20})\times(0.012)]$$

$$= (1.6\times10^{-19})\times(8.6\times10^{19})$$

$$= 13.76\ (\Omega.\text{m})^{-1}$$

$$\therefore \quad \rho = \frac{1}{\sigma} = \frac{1}{13.76}$$

$$= 0.073\ \Omega.\text{m} \quad \textbf{(Answer).}$$

Problem 5. A 0.2 m long sample of pure silicon of rectangular shape, with area of cross-section as 2×10^{-4} m^2, is connected to a battery of emf 3V. Calculate the current flowing through it. At room temperature, the electron and hole mobilities in it are 0.135 m^2 V^{-1} s^{-1} and 0.048 m^2 V^{-1} s^{-1}, respectively, and the intrinsic charge carrier concentration is 1.5×10^{16} m^{-3}.

Solution. Given: $\ell = 0.2$ m; $A = 2\times10^{-4}$ m^2; V = 3 volt; $\mu_e = 0.135$ m^2 V^{-1} s^{-1}; $\mu_h = 0.048$ m^2 V^{-1} s^{-1}; $n_i = 1.5\times10^{16}$ m^{-3}

The current due to the motion of electrons and holes in the semiconductor is given by,

$$i = eA\,(n_e v_e + n_h v_h)$$
$$= eA\,(n_e.\,E\mu_e + n_h.E\mu_h) \quad \text{(E: Electric field)}$$
$$= eA\,(n_e\mu_e + n_h\mu_h)\times\frac{V}{\ell} \quad \left(E = \frac{V}{\ell}\right)$$
$$= eAn_i\,(\mu_e + \mu_h)\,\frac{V}{\ell} \quad (n_e = n_h = n_i \text{ for an intrinsic semiconductor})$$
$$= (1.6\times10^{-19})\times(2\times10^{-4})\times(1.5\times10^{16})\,[(0.135) + (0.048)]\,\frac{(3)}{(0.2)}$$
$$= (4.8\times10^{-7})\times[0.183]\times(5)$$
$$= 13.2\times10^{-7}\text{ A} \quad \textbf{(Answer)}.$$

Problem 6. Determine the magnitude of electric field that can impart an average 3 eV of energy to a conduction electron in a specimen of copper. The mean free path of conduction electrons in copper may be taken as nearly 3×10^{-8} m.

Solution. Given: Mean free path, $\lambda = 3\times10^{-8}$ m; electron energy = 3 eV

The force on an electron in an electric field E is equal to eE. The electron receives energy in the field by the work done when it moves in the field through a distance d (= λ, in our case). Since work done = force×distance, we have,

$$eE.\,\lambda = 3\text{ eV}$$
$$\Rightarrow \quad E.\,\lambda = 3V$$
$$\Rightarrow \quad E = \frac{3V}{\lambda} = \frac{3V}{(3\times10^{-8})\text{ m}}$$
$$= 10^{8}\text{ V m}^{-1} \quad \textbf{(Answer)}.$$

Very Short Answer Questions

[Including HOTS Questions]

(1 mark each)

1. What is a unit cell ?
2. Name two each of crystalline and amorphous solids.
3. What are valence band and conduction band ?
4. Why an energy gap is deemed as a forbidden gap ?
5. What is the order of energy gap in (i) an insulator, (ii) a semiconductor, and (iii) a conductor ? [Hint: > 5eV; ~1eV; zero).
6. What is meant by Fermi level and Fermi energy ?
7. What is the conductivity of a semiconductor at absolute zero of temperature ? [Hint: Zero].
8. Name two materials for which the conductivity rises with increase in temperature.
9. What are intrinsic and extrinsic semiconductors ?
10. What are the charge carriers in an intrinsic semiconductor ? What is the ratio of their numbers ?
11. What is meant by doping of a semiconductor ?

12. What are (i) donor atoms and (ii) acceptor atoms ?

13. In a semiconductor, where are the donor levels and acceptor levels located ?

14. Does a p-type semiconductor contain any free electrons ?

15. What is meant by the mobility of a charge carrier ? What is its unit ?

16. Do the semiconductors obey Ohm's law ?

17. How a hole participates in current conduction ?

18. What is the charge on a hole ?

Short Answer Questions

[Including HOTS Questions]

(2 or 3 marks each)

1. Distinguish between intrinsic semiconductors and extrinsic semiconductors.

2. Distinguish between n-type and p-type semiconductors.

3. Explain how the resistivity of semiconductors varies with temperature.

4. Draw line diagrams showing the energy band structure for insulators, semiconductors, and conductors.

5. Derive an expression for the electrical conductivity of a semiconductor in terms of mobility of charge carriers.

6. Explain how electric conduction occurs in (i) n-type semiconductors and (ii) p-type semiconductors.

Long Answer Questions

(5 marks each)

1. Explain the formation of energy bands in solids. Distinguish between insulators, semiconductors, and conductors on the basis of energy bands. Explain how does the electric conduction in an intrinsic semiconductor differ from that in an extrinsic semiconductor.

2. Describe how the process of doping leads to the conversion of an intrinsic semiconductor into (i) an n-type semiconductor and (ii) a p-type semiconductor. Draw energy band diagrams for these semiconductors.

Numerical Exercises

1. A sample of pure silicon is doped with pentavalent As of 1 ppm concentration. If it has 5×10^{28} atoms per cubic metre and number density of intrinsic carriers $n_i = 7\times10^5$ per cubic metre, determine the numbers of electrons and holes contained in it. [Hint: $n_e \approx N_d$ (no. density of donor atoms), because the number of thermally produced electrons ($n_i \sim 10^{16}$ m^{-3}) is negligible to the number ($\approx 10^{22}$ m^{-3}) produced by doping. Further, $n_e\, n_h = n_i^2$). **[$n_e = 5\times10^{22}$ m^{-3}; $n_h = 9.8\times10^8$ m^{-3}]**

2. In a pure semiconductor, the electron and hole concentrations have the same value of 7×10^{15} m^{-3}. If on doping it with indium the hole concentration rises to 3×10^{22} m^{-3}, what is concentration of electrons in it ? **[$n_e = 16.33\times10^8$ m^{-3}]**

3. The number density of conduction electrons in an intrinsic semiconductor is 3×10^{20} per cubic metre. Calculate the number of holes contained in a rectangular piece of size 5mm×5mm×2mm of this semiconductor. **[15×10^{14}]**

4. A sample of pure germanium has to be converted into an n-type semiconductor of conductivity $5\Omega^{-1}$ cm^{-1}. If the electron mobility in the n-type sample is 4000 cm^2 (V.s)$^{-1}$, find the number density of the donor atoms required for the necessary conversion. Assume that the contribution of holes to the conductivity is negligible. **[$N_d = 7.8\times10^{14}$]**

5. When an electromagnetic radiation of wavelength <2520 nm is incident upon a semiconductor, its electrical conductivity is found to rise. Determine its energy gap. **[$E_g = 0.49$ eV]**

6. A pure semiconductor with number density of charge carriers as 2.5×10^{13} per cubic centimeter is doped with donor atoms of number density 2×10^8 atoms per cubic centimeter. If the electron mobility is 3800 cm^2 V^{-1} s^{-1}, calculate the conductivity of the doped sample. Also calculate the ratio of conductivities of the doped sample and the undoped sample.
[$12.2\times10^2\ \Omega^{-1}$ cm^{-1}; $\sigma_d : \sigma_i = n_d : n_i = 8\times10^4 : 1$]

7. An electron in the conduction band of a semiconductor combines with a hole in its valence band, resulting in the release of energy in the form of an electromagnetic radiation. If the band gap of the semiconductor is 1.5 eV, find the maximum wavelength that can be emitted in the process. **[$\lambda = 825$ nm]**

CHAPTER 9 (B)
SEMICONDUCTOR DEVICES

The semiconductor devices have revolutionised the world of electronics. Now they are basic integral parts of items of common use in our daily life, like radio, television, taperecorder, computer, washing machine, etc. We shall describe some important ones in this chapter.

9A.01. The p-n Junction

The p-n junctions, in one form or the other, are fundamental components of all semiconductor devices.

A p-n junction is formed by suitably joining a p-type semiconductor with an n-type semiconductor (In practice, usually it consists of wafers of p-type and n-type fused together or grown on each other). For simplicity, we assume the two semiconductors to be bar-shaped (Fig. 9.13a). The plane separating the p-type and n-type regions is the junction plane. We know that the majority carriers in the n-type region are the free electrons and the minority carriers are the holes. The reverse is the case in the p-type region. We shall separately consider the motions of majority and minority carriers across the junction plane.

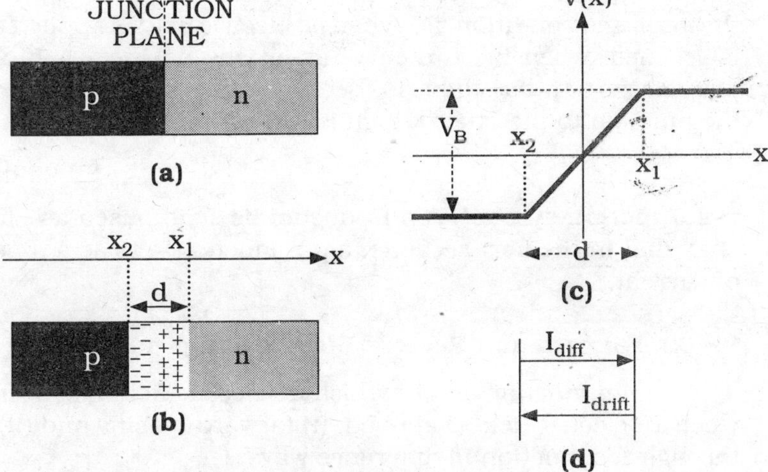

Fig. 9.13

Majority carriers; Diffusion current: Let us begin when the two semiconductors are just brought into mutual contact. The majority free electrons from the n-region tend to diffuse across the junction plane into the p-region, which contains very few free electrons (diffusion is the natural slow process of mixing of particles of two fluids or two solids, due to the particles' kinetic motion, when the two are brought together). They diffuse from right to left, in Fig. 9.13(a). Similarly, +vely charge holes diffuse from left to right, from the p-region to the n-region. The two diffusions together give rise to a **diffusion current** (**I_{diff}**) in the same direction, viz. from left to right in the figure (remenmber that, conventionally, the electron current is taken to flow in the direction opposite to the direction of movement of the electrons).

As already described earlier in the last chapter (Section 9A.05), the crystal structure in the n-region has +vely charged donor ions firmly fixed in position. The electrical neutrality of the structure is preserved by one-to-one correspondence of these ions with the conduction band free electrons present in the structure. When some free electrons near the junction plane diffuse into the p-region, the electrical neutrality is disturbed. Then, the +ve charges of the donor ions near the junction plane are 'exposed', forming a +vely charged space on the right hand side of the junction plane. When the diffusing free electrons arrive in the p-region, they readily combine with the acceptor ions (lacking in one electron each), thus forming a –vely charged space on the left hand side of the junction plane. The majority carriers (holes), diffusing from the p-region to the n-region across the junction plane, have exactly the same effect. The net result is that a **space charge** is built up on the two sides of the junction plane, half positive and half negative (Fig. 9.13b). It largely consists of fixed charges. "The region consisting of +vely charged donor ions and –vely charged acceptor ions on the two sides of the junction plane, which is relatively free of mobile charge carriers", is called **depletion region** (or **depletion zone** or **depletion layer**). The width d of the depletion region is a few microns (1 micron = 1μm = 10^{-6} m). It is practically depleted of mobile charge carriers. Hence the name 'depletion region'.

After crossing the p-n boundary region, the electrons and holes may collide and recombine (or, say, annihilate each other), since they have opposite charges. Thus, such collisions further help to 'expose' the +ve and –ve charges of the donor and acceptor ions, respectively.

The creation of the space charge produces a potential difference V_B across the depletion region (Fig. 9.13c), which strongly obstructs further diffusion of the charges across the junction plane. Therefore $\mathbf{V_B}$ is called **potential barrier**. The electrons slowly diffusing from right to left find themselves moving towards a –ve potential and so tend to turn back. Similarly, the +vely charged holes diffusing towards right move towards a +ve potential and so also turn back. (For silicon, V_B = 0.7 V).

Minority carriers; Drift current. The minority carriers in the n-region are the +vely charged holes. They naturally find it very favourable to move towards a lower (–ve) potential. Hence, unlike the majority carriers (electrons), they fast 'sweep' across the junction plane and move into the p-region. Similarly, the minority carriers (electrons) in the p-region move fast across the junction plane into the n-region. These movements of the two kinds of minority carriers together produce a **drift current ($\mathbf{I_{drift}}$)** in a direction opposite to the diffusion current, I_{diff} (Fig. 9.13d).

Thus, we have two opposite currents I_{diff} and I_{drift} simultaneously flowing across the junction plane. In the unbiased condition, i.e. when no electric field is applied across the p-n junction, ultimately an equilibrium is reached, when the two currents exactly balance each other. Then the net current through the junction plane is zero (it has to be so, because otherwise it will mean a limitless transfer of charge from one end of the junction to the other, which is impossible). the magnitudes of the two opposing currents are equal,

$$I_{diff} = I_{drift} \text{ (at equilibrium)}$$

p-n Junction symbol. A p-n junction is symbolised as —▷|— (p | n). Actually it is the symbol for a diode. As we shall find in the next sections, a p-n junction acts as a diode. The arrow points in the direction of free flow of current.

9B.02. Forward and Reverse Biasing of a p-n Junction; Dynamic Resistance; p-n junction as a Diode

A p-n junction across which no electric field has been applied is said to be **unbiased**. Two cases arise when an electric field is applied, (i) forward biasing and (ii) reverse biasing. They modify the flow of current through the junction in difference ways.

(i) Forward biasing. A p-n junction is said to be forward-biased when the p-region is maintained at a higher potential with respect to the n-region. e.g. by connecting the p-region to the +ve terminal and the n-region to the –ve terminal of a battery. Two examples are illustrated in Figs. 9.14 (a) and (b).

As compared to the unbiased p-n junction, now the n-region becomes more negative and the p-region becomes more positive. Hence the height of the potential barrier V_B is reduced (cf. Fig. 9.13c). Therefore, more of the majority carriers are able to surmount the barrier, which leads to a substantial increase in the diffusion current (I_{diff}). There is another reason for increase in the forward current I_F. The width d of the depletion region is automatically reduced, with the reduction in V_B (Figs. 9.13b and 9.13c). However, the drift current (I_{drift}) is not affected, because the motion of the minority carriers (electrons in the p-region and holes in the n-region) is little disturbed by the change in V_B. **The net result is that a large forward current ($\mathbf{I_F}$) flows in the circuit** (Fig. 9.15a).

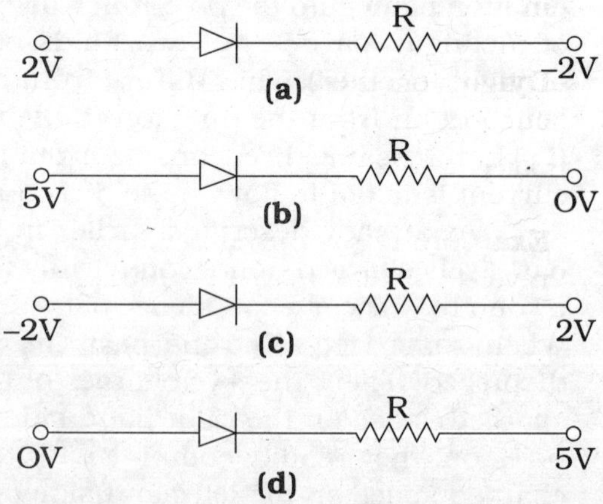

2V R –2V (a)

5V R OV (b)

–2V R 2V (c)

OV R 5V (d)

Fig. 9.14

A qualitative plot of the forward current (I_F) against the forward bias voltage (V_F) is shown in the upper right side quadrant of Fig. 9.16. It is known as **forward I-V characteristic** of the p-n junction. Until the value of V_F is less than the barrier potential V_B, little current I_F flows in the circuit. This cut-of value of V_F is known as **knee voltage**. Beyond it, the current I_F increases very rapidly (exponentially), and soon becomes nearly constant.

(ii) Reverse biasing. The p-n junction is said to be reverse-biased when the p-region is maintained at a lower potential than the n-region, e.g. by connecting the p-region to the –ve terminal and the n-region to the +ve terminal of a battery. The Figs. 9.14 (c) and (d) illustrate two examples.

Now the majority carriers in both the p-region and the n-region tend to move away from the junction. Therefore the potential barrier increases substantially. Also, the width of the depletion region is much

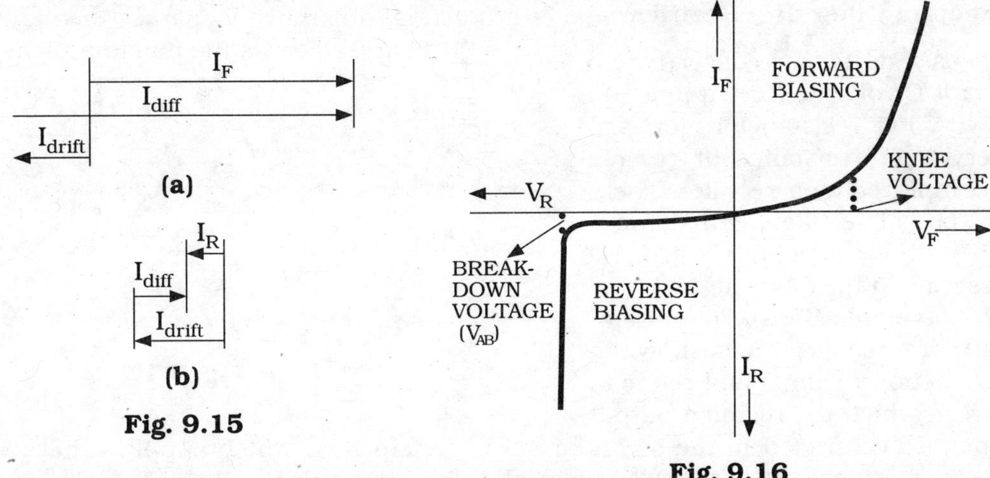

Fig. 9.15

Fig. 9.16

enhanced. Hence the diffusion current (I_{diff}) is substantially reduced, while the drift current (I_{drift}) remains unchanged. **The net result is that now only a small reverse current (I_R)** flows in the circuit (Fig. 9.15b). I_R is also called leakage current. Now the p-n junction offers a high resistance to the flow of current in the circuit.

A plot of I_R vs. the reverse bias voltage V_R is shown in the lower left quadrant of Fig. 9.16. It is known as the **reverse I-V characteristic** of the p-n junction. As V_R increases, I_R remains a nearly constant small value, until a breakdown value is reached, when I_R shoots up abruptly. This value (V_{RB}) of V_R is known as **breakdown voltage or Zener voltage (after the discoverer, C. Zener).** The breakdown occurs because at sufficiently high values of V_R the covalent bonds in the semiconductor are broken, which releases a large number of charge carriers in the circuit. Then, the voltage remains constant and the junction offers a very low resistance to the flow of current. Other possibilities also exist. More about it will be described in Section 9B. 04 on Zener diode.

An electronic device with only two electrodes is called a **diode**. Thus, a p-n junction is essentially a diode. An ideal diode permits a free flow of electric current in one direction and not at all in the opposite direction. A p-n junction is close to an ideal diode, since it allows a very small current (a few microamperes) in reverse bias.

Dynamic resistance. The ratio of the change in applied voltage (ΔV) to the corresponding change in the current (ΔI) in the circuit" is called **dynamic resistance (R)** of the p-n junction,

$$R = \frac{\Delta V}{\Delta I}$$

Example 1. In the following diagrams, indicate which of the diodes are forward biased and which are reverse biased. (CBSE; 1995).

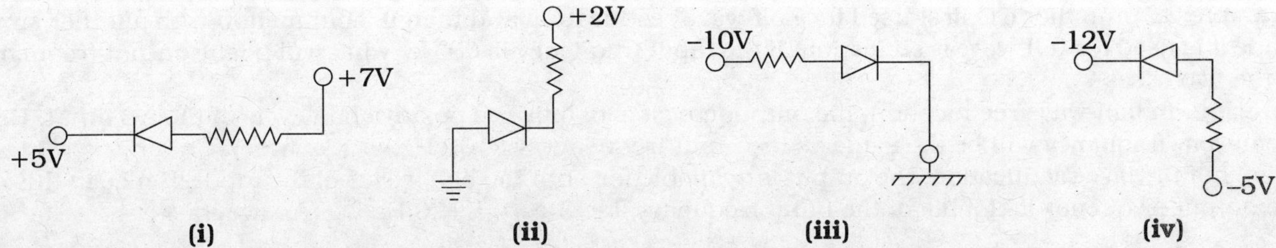

Solution. A p-n diode is forward biased when the potential at the p-terminal is greater than the potential at the n-terminal. The reverse is the case for reverse biasing. Thus, we have,

(i) $V_p > V_n$. ∴ The diode is forward biased. **(Answer).**

(ii) $V_p = 0$. ∴ $V_p < V_n$. ∴ The diode is reverse biased. **(Answer).**

(iii) $V_p = -10V$; $V_n = 0$. ∴ $V_p < V_n$. ∴ The diode is reverse biased. **(Answer).**

(iv) $V_p = -5V$; $V_n = -12V$. ∴ $V_p > V_n$. ∴ The diode is forward biased. **(Answer).**

9B.03. p-n Junction Diode as a Rectifier

A rectifier is a device for converting an AC voltage into a DC voltage. Since a junction diode offers a very low resistance in forward bias and a very high resistance in reverse bias, it can be employed as a rectifier. There are two main types of rectifier: (i) half wave rectifier and (ii) full wave rectifier, which we shall briefly describe in the following.

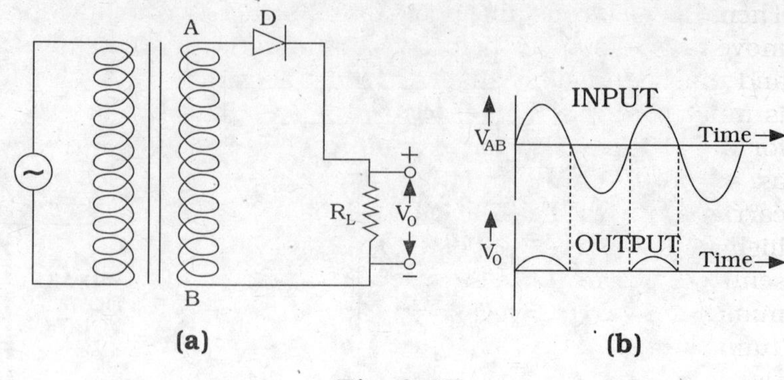

Fig. 9.17

(i) Half wave rectifier. The circuit diagram for half wave rectification is shown in Fig. 9.17(a). A junction diode D and the load resistor R_L (across which the rectified output voltage develops) are connected to the secondary of a transformer, which supplies the AC voltage. When the voltage at A is +ve, the diode is forward biased and so it conducts. When it is –ve at A, the diode is forward biased and so it conducts. When it is –ve at A, the diode is reverse biased and there is practically no conduction. Hence during the **+ve half cycle** current flows in the circuit and so we get an output voltage V_0 across R_L. But during the **–ve half cycle** there is a negligibly small current in the circuit and so the output voltage is almost nil. The process keeps on repeating in the subsequent cycles. Thus, the output is restricted to only one direction (although still varying). In this sense it is regarded as rectified. But since the output is obtained for only a half cycle, the circuit is regarded as a **half wave rectifier.**

(ii) Full wave rectifier. The circuit now consists of two identical diodes D_1 and D_2. One end of the load resistor R_L is now connected to the centre tap C of the transformer.

During the +ve half cycle, when the voltage at A is +ve, the voltage at B is –ve and at C is zero. Hence the diode D_1 is forward biased and so it conducts. But the diode D_2 is reverse biased and so it does not conduct. During the –ve half cycle the condition is reversed, i.e. D_2 conducts and D_1 does not conduct. Thus, unidirectional output voltage is obtained during both the half cycles, i.e. during the full cycle. Therefore, the circuit is known as **full wave rectifier** circuit. Note that during both the half cycles of rectification, the current flows from E to C through the load resistor.

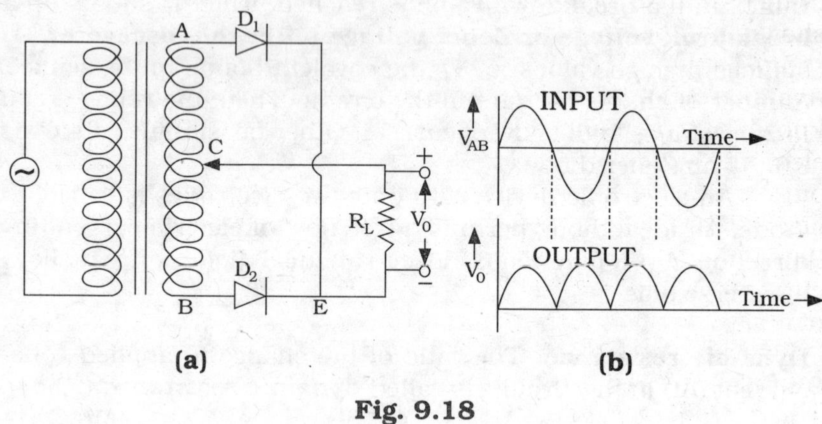

Fig. 9.18

Example 2. A p-n diode is first used for half-wave rectification and then in conjunction with another similar diode it is used for full-wave rectification. If the input frequency is 60 Hz, what will be the output frequencies in the two cases ?

Solution. In half-wave rectification, the output consists of only the positive half cycle of the AC input. Hence the output frequency will be the same as the input frequency, viz. 60 Hz. **(Answer).**

In full-wave rectification, the output is available for both the half cycles of the input. Hence in this case the output frequency is double of the input frequency, i.e. $2 \times 60 = 120$ Hz. **(Answer).**

9B.04. Zener Diode; Voltage Regulation

In the Section 9B.02, we had found that when the reverse bias voltage V_R of a junction diode equals (or exceeds) the value V_{RB}, known as breakdown voltage or Zener voltage, the current I_R shoots up abruptly (Figs. 9.16 and 9.19a). This phenomenon enables to use the diode as a voltage regulator, such that the voltage across the diode remains constant in spite of a change in current flowing through the diode. Under this condition, the junction diode is called a **Zener diode**.

The p-and n-regions of the diode are heavily doped with acceptor and donor impurities, respectively.

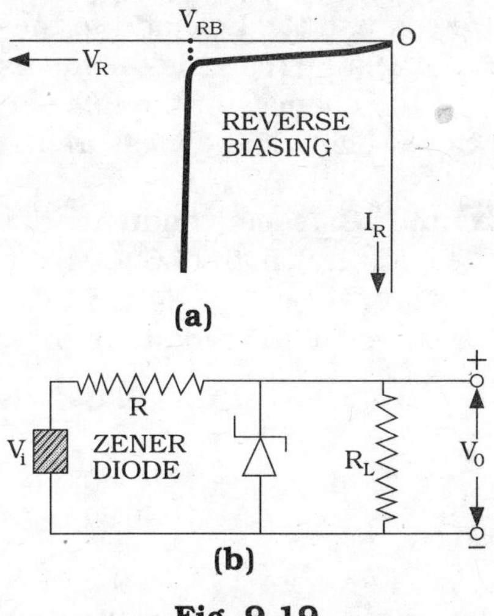

Fig. 9.19

Then, the electrons and holes diffusing across the junction plane cannot move large distances before recombining with other holes and electrons and thus getting annihilated. Hence the width (d) of the depletion region is much reduced. This leads to the creation of a high electric field, e.g. for d $\approx 10^{-7}$m, even for a small input voltage of 5V the field is as large as $\approx 5\times10^{7}$ Vm^{-1}. Under the action of such a high field the minority carriers in both the n- and p-regions are greatly accelerated and acquire high kinetic energies. Their collisions with the valence electrons in the semiconductor breaks the valence bonds, resulting in the release of a large number of valence electrons in the structure. These electrons can easily 'tunnel' through the thin depletion layer to enter the n-region, which gives rise to a high reverse current (I_R) in the circuit, as seen in Fig. 9.19 (a). The corresponding input voltage V_{RB} is the breakdown voltage or the Zener voltage and we refer to the junction diode as a Zener diode. Note the symbol used for such a diode in Fig. 9.19(b).

After the breakdown occurs, the potential drop across the diode does not increase even if the input voltage is increased and I_R increases. Thus, a Zener diode can be used to obtain not just a constant output voltage but a **regulated constant voltage**. The voltage output of a rectifier described in the last section gives a constant output voltage only when the load resistance R_L (and therefore the current flowing through it) of the output circuit remains constant, i.e. it works as an unregulated DC power supply. The Zener diode provides a regulated DC power supply, because the output voltage (V_0) does not change even when the load resistance (and so the current) in the output circuit changes. A typical circuit for Zener diode for use as a regulated DC power supply is shown in Fig. 9.19(b).

Avalanche diode. The Zener-like breakdown in a reverse biased diode can occur through another process, known as **avalanche breakdown**. If a high reverse bias voltage (upto about 200 V) is applied, the existing electrons and holes are highly accelerated. They acquire high enough kinetic energies to be able to knock out outer orbit electrons from the atoms with which they collide. These newly created electron-hole pairs also get accelerated in the field, to produce far more electrons and holes, through further collisions with the atoms Thus an 'avalanche' of charge carriers results, leading to a similar abrupt increase in the reverse current, as for a Zener diode. So once again the diode can be used to obtain a regulated DC power supply, of a higher magnitude of output voltage.

9B.05. Photonic p-n Junction Devices; Photodiode; Solar Cells; Light-Emitting Diode (LED)

We have seen that electric conduction takes place across a p-n junction by the application of an electric field, i.e. electrically. Under suitable conditions, the same can also happen optically, i.e. by the action of light photons. The reverse can also happen, i.e. photons can be produced when electric conduction occurs across the p-n junction (the word photon is derived from the Greek word 'photo', meaning 'light'). Such phenomena have led to the development of a number of important devices, known as **photonic or optoelectronic devices**, since they combine optics and electronics together. We shall briefly discuss a few of them, which include photo-detectors (to detect optical signals, e.g. a photodiode), photovoltaic devices (to convert light into electricity, e.g. a solar cell), and electro-optical devices (to convert electricity into light, e.g. a LED).

Photodiodes

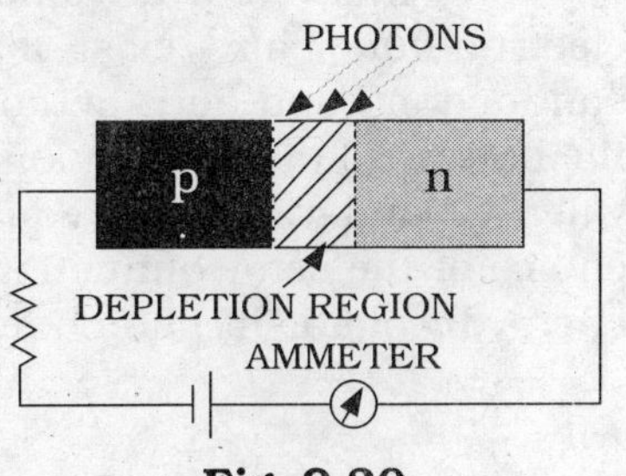

Fig. 9.20

A photodiode in the simplest form is a reverse biased p-n junction diode (Fig. 9.20). As explained in Section 9B.02, in this condition the depletion region has a large width and the junction offers a high resistance to the flow of current in the circuit. Only a small leakage current I_R flows in the circuit.

Suppose the diode is illuminated with light of frequency ν, such that the photon energy is greater than the band gap of the semiconductor ($h\nu > E_g$). Such a photon will excite an electron in the valence band to move to the conduction band, leaving a hole in the valence band. In other words, an electron-hole pair will be generated. As already described, both the electron and the hole add to the conductivity of the semiconductor. Hence the reverse current I_R increases (Fig. 9.19a). Greater is the intensity of light, the more is the increase in I_R. Thus by measuring the changes in I_R, the changes in the intensity of light can be detected or, by a suitable calibration, the intensity of the incident light can be measured.

Example 3. A semiconductor of band gap 2.5 eV is used for making a p-n photodiode. Find if this diode can detect a wavelength of 6200 nm.

Solution. Given: $E_g = 2.5$ eV; $\lambda = 6200$ nm $= 62 \times 10^{-7}$ m

The energy of the radiation (of wavelength 62×10^{-7} m) to be detected is given by ($E = h\nu$; $\nu\lambda = c$),

$$E = \frac{hc}{\lambda} = \frac{(6.6 \times 10^{-34}) \times (3 \times 10^{8})}{(62 \times 10^{-7})}$$

$$= 0.32 \times 10^{-19} \text{ J} = \frac{0.32 \times 10^{-19}}{1.6 \times 10^{-19}} \text{ eV}$$

$$= 0.2 \text{ eV}$$

This energy is less than the value of E_g (= 2.5 eV), the energy gap of the semiconductor. Hence the diode cannot detect the said radiation. **(Answer).**

Solar Cells

In principle, a solar cell works in the same manner as a junction photodiode. But no external bias is applied and the surface area of the cell is large, so that it can collect enough solar radiation. The light energy of the radiation is converted into electrical energy.

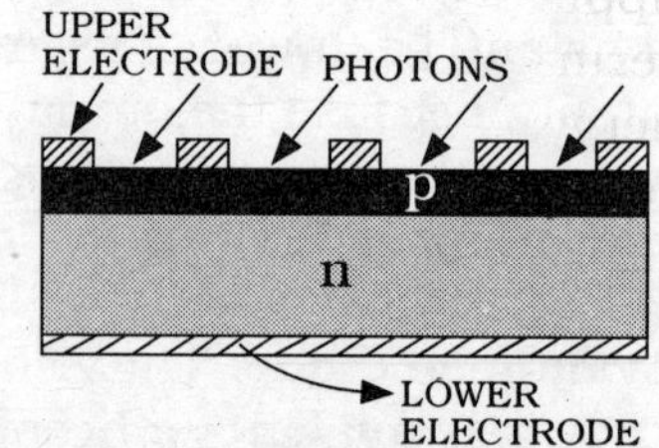

Fig. 9.21

The simplest p-n junction solar cell is illustrated in Fig. 9.21. A thin p-layer of the semiconducting material is deposited over a large and relatively thick layer of n-type material. The n-layer is in contact with a thin metallic layer, which acts as one of the electrodes of the solar cell. The other metallic electrode is in contact with the p-layer, but it is finger-shaped to allow for the incident light radiation from the sun to be absorbed by the layer (in the gaps between the 'fingers').

As for a photodiode, when photons of energy greater than the band gap of the material ($h\nu > E_g$) are incident upon the junction, electron-hole pairs are produced. In this case, a very large number of electron-hole pairs are generated because of the large area of the junction. Because of the space charge effect existing at the junction plane, the free electrons move towards the n-side and the holes towards the p-side. The result is that the n-side acquires a –ve potential and the p-side a +ve potential. Thus, a **photo-voltage** is developed across the junction. Electric current can be drawn in a circuit by connecting it across the junction, i.e. the junction acts as a voltaic cell. Since it works by the action of solar radiation, it is called a solar cell. The small thickness of the p-layer enables the free electrons (minority carriers) to quickly diffuse towards the junction.

Light-Emitting Diode (LED)

A light-emitting diode (LED) is a forward biased p-n junction, which emits light when current passes through the junction. The basic construction is shown in Fig. 9.22. The n- and p-regions are connected in the circuit through metallic contacts.

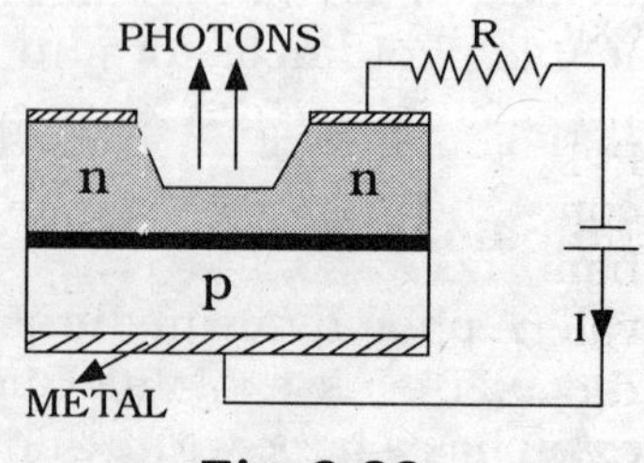

Fig. 9.22

We know that at temperatures greater than absolute zero (T > O K), electron-hole pairs exist in a semiconductor. The electrons reside in the conduction band and the holes in the valence band. Now, if an electron from the bottom of the conduction band falls to combine with a hole at the top of the valence band, an energy equal to the band gap (E_g) is released. With a proper choice of the semiconducting material, the value of E_g may correspond to the energy $h\nu$ of a light photon in the visible region (VIBGYOR),

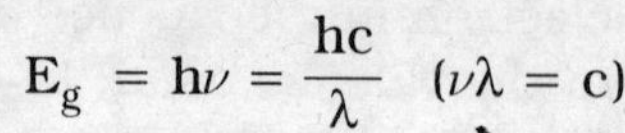

$$E_g = h\nu = \frac{hc}{\lambda} \quad (\nu\lambda = c)$$

Or $$\lambda = \frac{hc}{E_g}$$

Thus, light of wavelength λ is emitted by the junction diode. The colour of the light depends on the value of λ, and therefore on the value of E_g. The lowest permissible value of E_g is ≈ 1.8 eV, which corresponds to red light of wavelength ≈ 6880 Å (or 688 nm). Semiconducting materials with higher E_g values are required to obtain lights of other colours, like orange and green. Gallium doped with various proportions of arsenic and phosphorus (Ga As_{1-x} P_x) has been found suitable to provide semiconductors with different E_g values. For $E_g < 1.8$ V, e.g. for $E_g \approx 1.5$ eV, the diode emits invisible infrared light. The common semiconductors silicon ($E_g \approx 1.1$ eV) and germanium ($E_g \approx 0.7$ eV) are not suitable, because of their low values of E_g. They emit radiation in far infrared region, which is not useful for application purposes.

The LED's are rugged in construction and have a long life. They operate at low voltages and consume little power. Because of spontaneous emission of photons, their action is extremely fast. The light emitted by them is nearly monochromatic. These advantages have led to a widespread use of the LED's in our daily life. They are commonly used as indicator lights. The coloured bright numbers seen glowing at petrol pumps, microwave ovens, telephones, traffic digital signals, digital watches, calculators, etc., are usually light-emitting diodes. The LED's emitting invisible infrared light control burglar alarms, elevators (lifts) in the buildings, etc. and operate television sets by remote control (the light signal from the remote control device is detected by photodiodes in the TV set).

It should be noted that for the creation of an electron-hole pair, some kind of energy has to be supplied from outside, e.g. light energy (called photoexcitation, as in a photodiode), heat energy (called thermal excitation), etc. It follows that when an electron and hole recombine, a corresponding amount of energy is released in the form of light (radiative recombination, as in an LED), heat, etc. (non-radiative recombination).

Additional knowledge

(i) It may be questioned that why a photodiode works in reverse bias when the current is very much higher in the condition of forward bias.

When light falls on a photodiode, equal numbers of electrons and holes are created. So the number of both types of carriers is equally increased ($\Delta n = \Delta p$). In the forward bias, $n \gg p$. Therefore the fractional change $\Delta n/n$ in the majority carriers is far less than the corresponding change $\Delta p/p$ in the minority carriers. Since the total current is almost wholly contributed by the majority carriers, the fractional change in it becomes too small to detect and measure. Hence the reverse bias is preferred.

(ii) What is the main consideration for the choice of semiconducting materials for making solar cells, satisfying the condition $h\nu > E_g$? The light intensity is not the same for all frequencies (or wavelengths) in the solar radiation. It increases with increase in the value of $h\nu$, until a maximum value of nearly 1.5 eV is reached, after which it starts decreasing. Therefore, the semiconducting materials with $E_g \approx 1.5$ eV or lower have higher conversion efficiency. Gallium arsenide (Ga As) with $E_g \approx 1.53$ eV is well suited. It also has a higher absorption coefficient. Silicon ($E_g \approx 1.1$ eV) is also useful.

(iii) Only the simplest forms of photonic devices have been described in the last section. The advanced forms consist of more than two layers, as well as more kinds of layer, e.g. metals, metal oxides, etc.

(iv) Diode laser: A laser (light amplification by stimulated emission of radiation) is based on population inversion, i.e. higher energy levels are more populated than the lower energy levels. Such a condition can also be obtained in a forward biased p-n junction, in which electrons in the conduction band occupy higher energy levels. The transition from the higher to the lower energy levels (in the valence band) results in the emission of very nearly monochromatic light. Thus, a handy low power diode laser is obtained. Such tiny (grain-sized) lasers are built into compact disc (CD) players. They are extensively used in optical fibre based communication systems.

9B.06. Junction Transistors

A junction transistor is a two junctions, three terminals device. Two types are available:

(i) p-n-p transistor. It consists of a thin layer of n-type material sandwiched between two layers of p-type material (Fig. 9.23a). Its symbol is shown in Fig. 9.23b.

(ii) n-p-n transistor. It consists of a thin layer of p-type material sandwiched between two layers of n-type material (Fig. 9.23c). It is symbolised as shown in Fig. 9.23d.

Each transistor is made up of the following three regions:

(i) Emitter. It is heavily doped.

(ii) Base. This is the middle layer. It is very lightly doped. It is very thin [thickness $\approx 1\mu m$ (1 micron)].

(iii) Collector. It is moderately doped. The collector-base contact area is larger than the emitted-base contact area. Thus in reality a transistor looks as in Fig. 9.23(e).

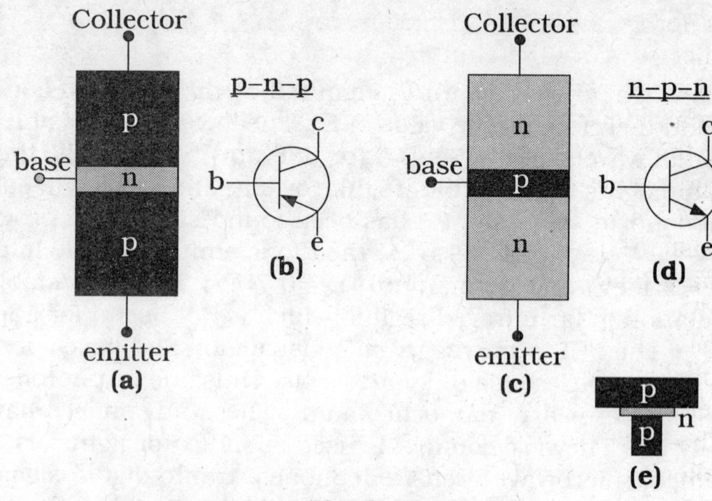

Fig. 9.23

For making external connections with an electric circuit, terminal wires are drawn from the emitter, base, and collector. Thus, a transistor forms a 3-terminal device, with 2 p-n junctions. Normally, the **emitter-base junction** is always forward-biased and the **collector-base junction** is reverse - biased. In the symbolic representations of the transistors, shown in Figs. 9.23 (b) and (d), the arrow mark along the emitter line indicates the direction of flow of current through the emitter-base junction. An n-p-n transistor has a large number of free electrons in the emitter and a small number of holes in the base. When forward-biased, the electrons diffuse from the emitter to the base and the holes from the base to the emitter. Thus, in this case the direction of the current is from the base to the emitter, as indicated by the arrow mark in Fig. 9.23(d). Opposite is the case for a p-n-p transistor (Fig. 9.23b), in which the majority charge carriers responsible for the flow of current are holes, instead of electrons (now the direction of current is the same as the direction of movement of holes).

Biasing of a Transistor; Configurations (or Modes) of a Transistor

For operating a transistor, suitable potential differences have to be applied across the two junctions. It is known as the **biasing** of the transistor.

A transistor can be operated in three possible **configurations (or modes)**. By the configuration of a transistor we mean how the transistor is connected in an electric circuit (the dictionary meaning of 'configuration' is 'the arrangement of parts or components in a particular form'). The possible configurations are:

(a) Common emitter (or grounded emitter)
(b) Common base (or grounded base)
(c) Common collector (or grounded collector)

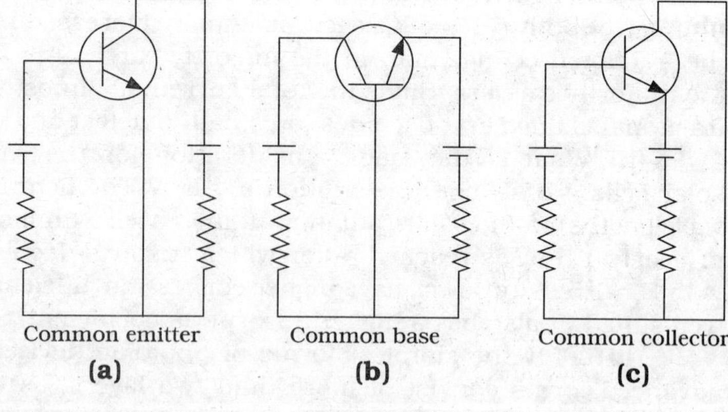

Fig. 9.24

In the common emitter configuration, the emitter is at zero potential and the other two terminals are given suitable potentials (Fig. 9.24a). Similarly for the other two configurations (Figs. 9.24b and 9.24c), i.e. the base and the collector, respectively, are at zero potential.

In any electric circuit, there are four essential terminals, viz. two for the input side and two for the output side. But, since a transistor has only three terminals, one of them has to be common to the input and the output side, as we find in all the aforesaid three possible configurations.

In all the configurations, the emitter-base junction is forward-biased and the collector-base junction is reverse-biased.

Working of a Transistor

Consider an n-p-n transistor with suitable biasing (Fig. 9.25a). Since the emitter-base junction J_1

is forward-biased, electrons are injected from the emitter into the base. Because of the extremely small thickness of the base, most of these electrons cross over into the region of the collector. This process is further helped by the existence of a reverse bias at the collector-base junction J_2, which imparts a +ve potential to the collector. The electrons pass through the batteries V_c and V_e, before returning to the emitter. The current $\mathbf{I_e}$ (conventionally taken in the direction opposite to the movement of the electrons) in the emitter-base circuit is called **emitter current.** The current $\mathbf{I_C}$ in the collector-base circuit is called **collector current.**

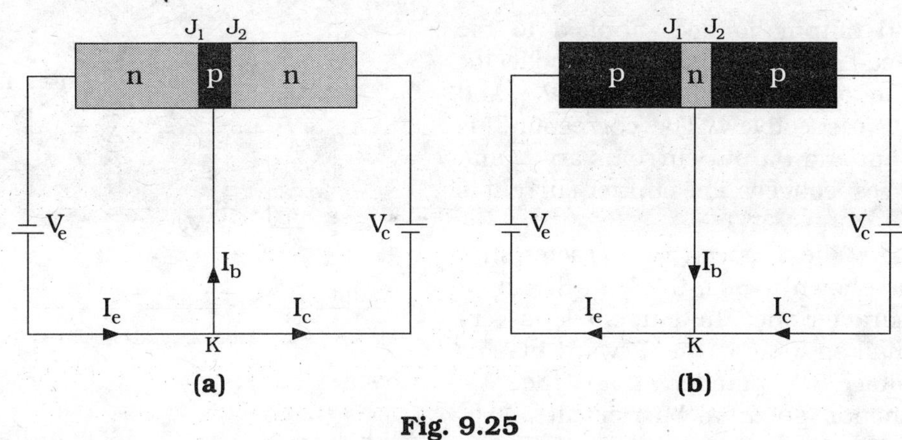

Fig. 9.25

The holes move in the direction opposite to that of the electrons and so produce current in the same direction. But because the base is lightly doped, the hole concentration is relatively very low. Hence the current in an n-p-n transistor is largely due to electrons. The collector current I_c is nearly equal to the emitter current I_e, because practically all the electrons from the emitter pass on to the collector. I_c is slightly smaller than I_e, as some electrons are able to pass directly from the base to the battery V_e, creating a small **base current** $\mathbf{I_b}$ (it is just about 1% to 5% of I_e). Using Kirchhoff's law (at the point K in the circuit), we have

$$I_e = I_b + I_c \qquad .....(9.4)$$

Similar considerations apply to a p-n-p transistor (Fig. 9.25b), in which holes are the main charge carriers. It is easily seen that the Eq. 9.4 holds for this case, too.

Example 4. In an electric circuit containing an n-p-n transistor, 95% of the electrons from the emitter reach the collector. If the collector current is 10 mA, find the emitter current and the base current.

Solution. Given: $I_c = 10 \text{ mA} = 10^{-2} \text{ A}$

It is given that 95% of the electrons from the emitter reach the collector, i.e. the collector current (I_c) is 95% of the emitter current (I_e),

$$I_c = \frac{95}{100} \times I_e = 0.95\, I_e$$

$$\Rightarrow \quad I_c = \frac{I_c}{0.95} = \frac{(10^{-2})}{(0.95)} \text{ A}$$

$$= \frac{(10^{-2}) \times 10^3}{0.95} \text{ mA}$$

$$= 10.53 \text{ mA} \quad \textbf{(Answer).}$$

$$\therefore \quad I_b = I_e - I_c = [(10.53) - (10)] \text{ mA}$$

$$= 0.53 \text{ mA} \quad \textbf{(Answer).}$$

9B.07. Characteristics of a Transistor

Any variation in voltage on the input or output side of a transistor causes a change in the input or output current in the circuit. The plot of variation of the input current with input voltage is known as the **input characteristic** of the transistor. The plot of variation of the output current with output voltage is known as the **output characteristic.** The plot of variation of the output current with the input current is known as the **transfer characteristic.** We shall discuss the characteristics in (i) common emitter configuration and (ii) common base configuration.

(i) Common emitter configuration. The circuit characteristics of an n-p-n transistor in common emitter configuration is shown in Fig. 9.26. The input section contains the base-emitter junction and the voltage source E_b. The output section contains the base-collector junction and the voltage source E_c. The input and output voltages applied to the base-emitter and base-collector junctions, respectively, are V_{be} and V_{bc}, respectively. The corresponding input and output currents are I_b and I_c, respectively. The emitter current is I_e.

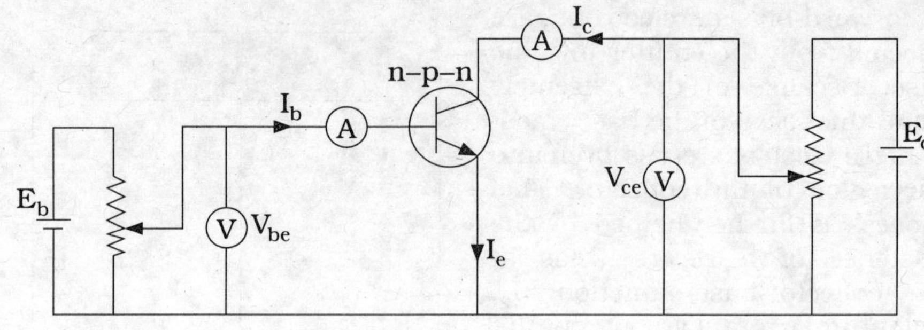

Fig. 9.26

The respective characteristics are shown in Fig. 9.27. In the input characteristic, the current I_b is very small so long as the forward biasing voltage is smaller than the junction potential barrier, after which it increases rapidly. The base current I_b is very small ($\approx$a few μA) even for large V_{be} values, because majority of the electrons moving across the junction pass on to the collector. In the output characteristic, the current I_c is small for small V_{ce} values, since the collector-base junction then becomes reverse biased on account of the base being at a higher +ve potential. After that it rapidly increases with increase in V_{ce}, becoming roughly constant when the junction becomes forward-biased. In the transfer characteristic, the collector current I_c increases rapidly with increase in the base current I_b, even in the forward bias.

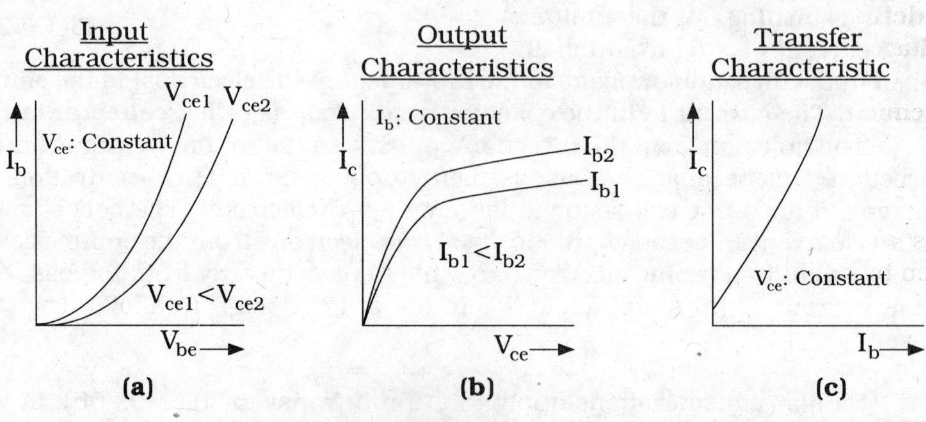

Fig. 9.27

(ii) Common base configuration. The circuit diagram for this case is shown in Fig. 9.28. The corresponding characteristics are shown in Fig. 9.29.

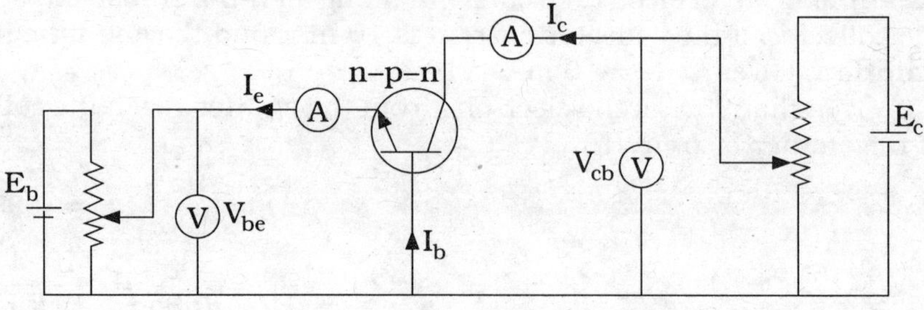

Fig. 9.28

The characteristics for a p-n-p transistor are the same as for an n-p-n transistor. However, the circuit diagrams are partly different (in the Fig. 9.26 and 9.28, the transistor symbol has to be changed and the battery polarities and the current directions have to be reversed).

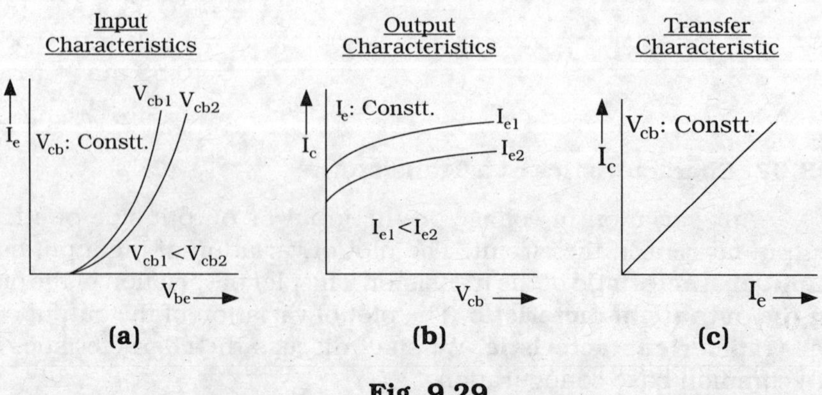

Fig. 9.29

Input Resistance (R_i), Output Resistance (R_0), and Current Gain (α, β) of a Transistor

(i) Input resistance (R_i). It is the ratio of change in base-emitter voltage (dV_{be}) to the resulting change in base current (dI_b) at a constant collector-emitter voltage (V_{ce}),

$$R_i = \left(\frac{dV_{be}}{dI_b}\right)_{V_{ce}} \qquad ..... (9.5)$$

The values of R_1 are ≈a few hundred ohms.

(ii) Output resistance (R_0). It is the ratio of change in collector-emitter voltage (dV_{ce}) to the change in collector current (dI_c) at a constant base current I_b,

$$R_0 = \left(\frac{dV_{ce}}{dI_c}\right)_{I_b} \qquad ..... (9.6)$$

The values refer to the nearly flat portion of the characteristic. Obviously, the R_0-vlaues are very high (≈ 100 kΩ), since I_c changes little with rise in V_{ce}.

(iii) Current gain (α, β). It is also known as **current amplification factor** or **AC current gain.** It is **defined** as the ratio of change in output current to the change in input current, at a constant output voltage.

For the **common base configuration,** it is the ratio of change in collector current (dI_c) to the change in emitter current (dI_e), at a constant collector voltage (V_c),

$$\alpha = \left(\frac{dI_c}{dI_e}\right)_{V_c} \qquad ..... (9.7\ a)$$

α is <1, but very close to 1. If the current values are measured from the starting moment, i.e. when $I_c = I_e = 0$, then by definition, we have,

$$\alpha = \left(\frac{I_c}{I_e}\right)_{V_c} \qquad ..... (9.7\ b)$$

For the **common emitter configuration,** it is the ratio of change in collector current (dI_c) to the change in base current (dI_b), at a constant collector voltage (V_c),

$$\beta = \left(\frac{dI_c}{dI_b}\right)_{V_c} \qquad ..... (9.8\ a)$$

β is >1. Again, if the current values are measured from the starting moment, we have,

$$\beta = \left(\frac{I_c}{I_b}\right)_{V_c} \qquad ..... (9.8\ b)$$

We can readily establish a **relation between α and β.** From the Eq. 9.4,

$$I_e = I_b + I_c$$

$$\therefore \quad dI_e = dI_b + dI_c$$

Dividing by dI_c on both the sides,

$$\frac{dI_e}{dI_c} = \frac{dI_b}{dI_c} + \frac{dI_c}{dI_c}$$

$$\Rightarrow \quad \frac{I}{\alpha} = \frac{1}{\beta} + 1$$

$$\Rightarrow \quad \alpha = \frac{\beta}{1+\beta} \qquad ..... (9.9)$$

$$\text{and} \quad \beta = \frac{\alpha}{1-\alpha} \qquad ..... (9.10)$$

Since I_b ≈ 1 – 5% of I_e, α is nearly 0.95 to 0.99 and β is nearly 20 to 100.

9B.08. Transistor as an Amplifir (Common Emitter Configuration)

An amplifier is a device to enlarge the amplitude of an input AC voltage (or current or power; the Latin word 'amplus' means 'large').

A transistor can be used as an amplifier. We shall discuss the use of an n-p-n transistor in common emitter configuration for this purpose. The necessary circuit diagram is shown in Fig. 9.30. As usual, the base-emitter junction is forward-biased and the collector -base junction in reverse-biased. The input signal to be amplified is connected in series with the biasing battery in the base circuit. A load resistor of large resistance R_O is connected in the collector circuit. The amplified output voltage is taken across this resistor. The capacitors C_1 and C_2 in the input and output circuits, respectively, serve to shield the biasing DC voltages (supplied by the batteries) from the input and output AC signals, respectively.

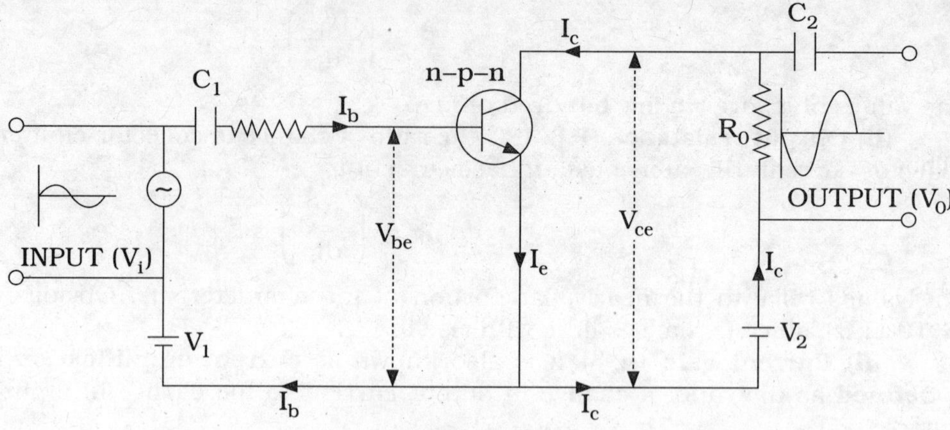

Fig. 9.30

The biasing voltage V_{be} for the base-emitter junction is provided by the battery V_1. The potential difference V_{ce} between the collector and the emitter is maintained by the battery V_2. When the additional input AC signal is applied across the base and the emitter, the biasing voltage V_{be} is modified. As a result the collector current I_c changes, which in turn results in changes in the output current, voltage, and power. We now consider the various parameters related to these changes.

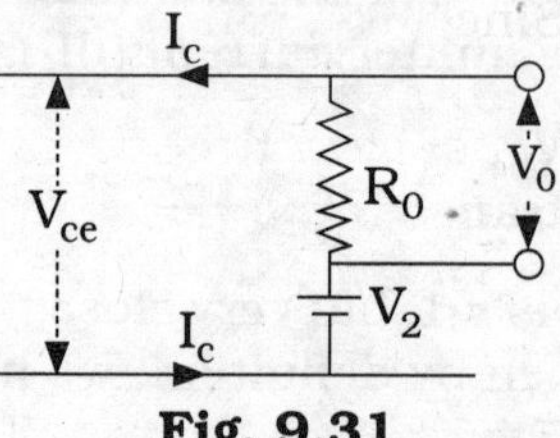

Fig. 9.31

(i) Current gain. As discussed in the last section (Eq. 9.8), the AC current gain β for the common emitter configuration is given by,

$$\beta = \frac{dI_c}{dI_b} \qquad \text{..... (9.8 a)}$$

Again, if I_c and I_b are measured from the starting moment, when $I_c = I_b = 0$, we have,

$$\beta = \frac{I_c}{I_b} \qquad \text{..... (9.8 b)}$$

(ii) Voltage gain (A_v). The voltage gain (or voltage amplification) is defined as the ratio of change in the output voltage V_O to the change in the input voltage (V_i),

$$A_v = \frac{dV_o}{dV_i} \qquad \text{..... (9.11 a)}$$

If the voltage values are measured from the starting moment (when $V_i = V_O = 0$), we have,

$$A_v = \frac{V_o}{V_i} \qquad \text{..... (9.11 b)}$$

Using Kirchhoff's law for the output circuit (Fig. 9.31), we have,

$$V_2 = V_{ce} + V_O = V_{ce} + I_cR_O$$

$$\Rightarrow \quad V_{ce} = V_2 - I_cR_O \qquad \text{..... (9.12)}$$

Differentiating, we get (the battery voltage V_2 is a constant value; so $dV_2 = 0$),

$$dV_{ce} = -R_OdI_c$$

Further,

$$V_O = I_cR_O$$

$$\therefore \quad dV_O = R_OdI_c$$

The input resistance R_i of the transistor is given by (vide Eq. 9.5),

$$R_i = \frac{dV_{be}}{dI_b}$$

$$\Rightarrow \quad dV_{be} = R_i\, dI_b$$

But any change dV_{be} in V_{be} is produced by a change in the input voltage V_i (∵ battery voltage V_1 is constant). Therefore,

$$dV_i = dV_{be} = R_i\, dI_b$$

$$\therefore \quad A = \frac{dV_O}{dV_i}$$

$$= \frac{R_O dI_c}{R_i dI_b}$$

Therefore, from Eq. 9.8,

$$A = \left(\frac{R_O}{R_i}\right)\beta = \left(\frac{R_O}{R_{be}}\right)\beta \qquad \text{..... (9.13)}$$

β is $\approx$ 50 and usually $R_O \gg R_i$. Therefore the voltage gain is very high.

(iii) Power gain (A_p). The power gain is given by,

Power gain = Voltage gain × Current gain

$$\therefore \quad A_p = \left(\frac{R_O}{R_{be}}\right)\beta^2 \qquad \text{..... (9.14)}$$

Since $\beta < 1$, the power gain is slightly less than the voltage gain.

(iv) Transconductance (or Transfer conductance). For a large amplification, a small change in V_{be} should produce a large change in the collector current I_c. This property is expressed by the quantity **transconductance (g_m)**,

$$g_m = \frac{dI_c}{dV_{be}} \qquad \text{..... (9.15)}$$

Example 5. An n-p-n transistor amplifier in common emitter mode yields an AC current gain of 19. Find the changes in base current and collector current, when the emitter current changes by 1 mA.

Solution. Given: $\beta = 19$; $dI_e = 1$ mA

The current gain for a common emitter amplifier is given by,

$$\beta = \frac{dI_c}{dI_b} = \frac{d(I_e - I_b)}{dI_b} = \frac{dI_e - dI_b}{dI_b}$$

$$\Rightarrow \quad dI_b(1+\beta) = dI_e$$

$$\Rightarrow \quad dI_b = \frac{dI_e}{1+\beta} = \frac{1}{1+19} = \frac{1}{20}$$

$$= 0.05 \text{ mA}$$

$$\therefore \quad dI_c = dI_e - dI_b = 1.00 - 0.05$$

$$= 0.95 \text{ mA}$$

Thus, the corresponding changes in the base current and the collector current are found to be 0.05 mA and 0.95 mA, respectively. **(Answer).**

Example 6. An n-p-n transistor in common emitter configuration amplifies a signal, with AC current gain equal to 63. If its internal resistance is 240 Ω and the output resistance is 4800 Ω, calculate the voltage gain and power gain.

Solution. Given: $\beta = 72$; $R_i = 240\ \Omega$; $R_O = 4800\ \Omega$

The voltage gain is given by,

$$A_v = \left(\frac{R_O}{R_i}\right)\beta = \frac{(4800)}{(240)} \times 72$$

$$= 1440 \quad \textbf{(Answer).}$$

The power gain is given by,

$$A_p = \left(\frac{R_O}{R_i}\right)\beta^2 = \frac{(4800)}{(240)} \times (72)^2$$

$$= 1{,}03{,}680 \quad \textbf{(Answer).}$$

Example 7. In a transistor, the base current is changed by 30 μA. This results in a change of 0.03 V in base to emitter voltage and a change of 3 mA in the collector current. (i) Find (a) current gain 'β_{ac}' and (b) transconductance 'g_m'. (ii) If this transistor is used as an amplifier with the load resistance 7.5 kΩ, calculate voltage gain of amplifier. (AI; 2001).

Solution. Given: $dI_b = 30\ \mu A = 30\times10^{-6}$ A; $dV_{be} = dV_i = 0.03$ V; dI_c = 3mA = 3×10^{-3} A; R_0 = 7.5 kΩ = 7500 Ω
The AC current gain of an amplifier is given by.

$$\beta_{ac} = \frac{dI_c}{dI_b} = \frac{(3\times10^{-3})}{(30\times10^{-6})}$$

$$= 100 \quad \textbf{(Answer).}$$

The input resistance,

$$R_i = \frac{dV_i}{dI_b} = \frac{(0.03)}{(30\times10^{-6})}$$

$$= 1000\ \Omega$$

Therefore, the voltage gain (A_v) and transconductance (g_m) are,

$$A_v = \frac{R_0}{R_i}\times\beta_{ac} = \frac{(7500)}{(1000)}\times 100 = 750 \quad \textbf{(Answer).}$$

$$g_m = \frac{dI_c}{dV_i} = \frac{\beta_{ac}}{R_i} = \frac{(100)}{(1000)} = 0.1\ \Omega^{-1} \quad \textbf{(Answer).}$$

Phase relationship between input voltage and output voltage (for common emitter configuration). During the +ve half of the input AC signal, the voltage V_{be} increases, which in turn results in increase in the collector current I_c. Hence according to the Eq. 9.12,

$$V_{ce} = V_2 - I_c R_0,$$

the collector voltage V_{ce} decreases ($\because$ V_2 and R_0 are constant values). It means that the +ve end of V_{ce} becomes less positive, i.e. it becomes negative with respect to the initial value. Thus, when the input AC signal goes through a +ve half cycle, the output voltage goes through a –ve half cycle. Similarly, during the–ve half cycle of the input signal the output voltage goes through a +ve cycle.

Thus, for a common emitter amplifier circuit, the input voltage and the output voltage are in opposite phase, i.e. they are **180° (or π) out of phase** with each other, as depicted in Fig. 9.30.

For **common base configuration,** it can be worked out that the output voltage and the input voltage are in phase with each other. However, the current gain for a common emitter amplifier is more than that for a common base amplifier. Hence usually the former is preferred over the latter.

9B.09. Transistor as an Oscillator

An oscillator is a device to produce an amplified AC signal without using an input AC signal, i.e. by using DC voltages alone in the input and the output circuits,. This is achieved by (i) employing an LC network to produce oscillatory voltage in the output circuit and (ii) feeding back a part of the output into the input circuit.

The simplest oscillator circuit is shown in Fig. 9.32 (a). It is similar to the amplifier circuit (Fig. 9.30), with the difference that the resistors in both the input circuit (base – emitter circuit) and the output circuit (collector–emitter circuit) are replaced by the inductance coils L_1 and L_2, respectively, and an LC-network (called tank circuit) is incorporated in the output circuit. The coils L_1 and L_2 are wound around the same core. So they are **inductively coupled** through their mutual inductance, i.e. any electromagnetic oscillations in L_2 induce similar oscillations in L_1 and vice versa. When current starts flowing in the output circuit, electromagnetic oscillations of angular frequency $\omega_0 = 1/\sqrt{LC}$ or linear frequency

$$\nu_0 = \frac{1}{2\pi\sqrt{LC}}$$

are set up in the tank circuit (vide Chapter 4B, Section 4B.06).

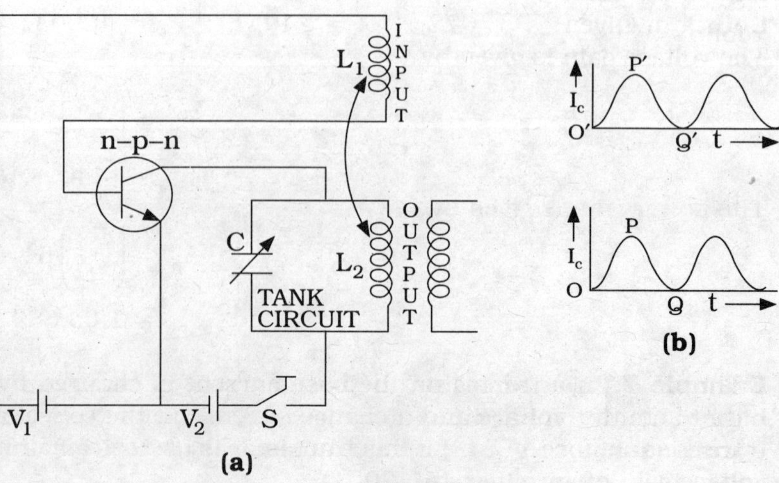

(a) (b)

Fig. 9.32

A switch S is included in the circuit. As soon as it is closed, the biasing voltages V_1 and V_2 become available in the circuit and the currents I_c, I_e, I_b start building up. The build-up of the collector current I_c is shown in the lower half of Fig. 9.32 (b). It gradually rises from O (zero displacement) to P (full displacement). This induces a similar oscillatory current I_c in the base-emitter circuit, making it to rise from O′ to P′ (upper half of Fig. 9.32b). Thus, we have a **positive feedback** from the output circuit to the input circuit. I_c has maximum value at P, when it cannot rise further. Then the transistor is said to be **saturated.** The magnetic field around L_2 ceases to grow, i.e. it momentarily becomes static. Therefore, there is no further feedback to L_1 and the current I_e begins to fall. Thus, beyond P the magnetic field around L_2 starts decaying, which makes I_c and therefore I_e, to fall. When the position Q is reached, the transistor is said to be cut-off and both the currents I_c and i_e cease flowing. Then the process starts repeating itself.

The input current I_e is amplified in the circuit, as explained earlier in the last section. Thus, now we obtain an amplified AC voltage, without any input AC signal.

9B.10. Logical Variables and Functions; Binary Numbers (vs. Decimal Numbers); Analogue Signals vs. Digital Signals

The students are already familiar with the elements of **Boolean algebra** (named after the English mathematician George Boole). However, for the sake of convenience and ready reference, we shall recapitulate the main features.

Usually, a given situation offers scope for a large number of possibilities, e.g. in a class of 50 students if the teacher wants to talk to one student, he has 50 choices available to him. But many times we have a situation with just two possibilities, e.g. a statement being TRUE or FALSE, the answer to a question being NO or YES, an electric lamp being ON or OFF, etc. The two possibilities can be mathematically represented by two numbers, 0 and 1. They are referred to as **binary numbers (or digits or bits)** (derived from the Latin word 'bini', meaning 'two together'). They are sufficient to represent any variable with only two possibilities.

Logical variables and functions; Truth tables. As an example, consider the simple circuit shown in Fig. 9.33. It contains a lamp, a power source and two switches S_1 and S_2. The power source is constantly in operation. Whether the lamp is ON or Off depends on the state (ON or OFF) of the switches S_1 and S_2. Thus we have three variables here:

A: State of the switch S_1
B: State of the switch S_2
Y: State of the lamp

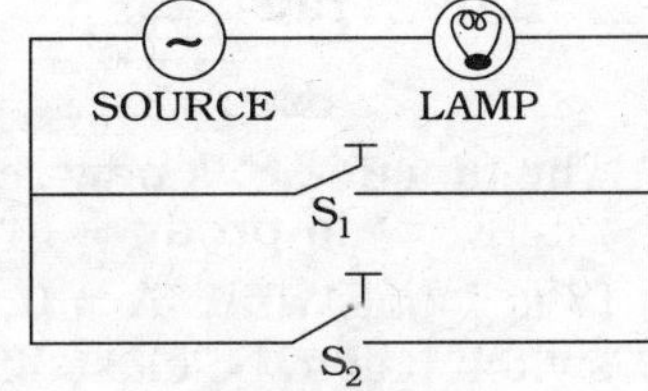

Fig. 9.33

Let,

(i) A = 0 for S_1 open and A = 1 for S_1 closed
(ii) B = 0 for S_2 open and B = 1 for S_2 closed
(iii) Y = 0 for the lamp off and Y = 1 for the lamp on

The state of S_1 does not depend on the state of S_2 or the lamp. Similarly for S_2. But the state of the lamp depends on the states of S_1 and S_2, as given in the following table.

Table 9.1

S_1	S_2	Lamp	A	B	Y
open	open	off	0	0	0
open	closed	on	0	1	1
closed	open	on	1	0	1
closed	closed	on	1	1	1

The table shows that A and B are independent variables, but Y is a **function** of A and B. The table specifies the value of the function for all possible combinations of the variables A and B. Such a table is called the **truth table** for the concerned function. The **function is represented** as,

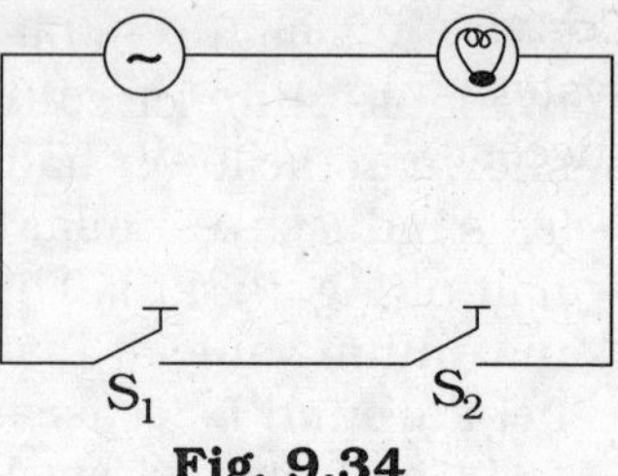

Fig. 9.34

$$Y = A \text{ OR } B \quad \text{or} \quad Y = A + B$$

[Y = 0 if both A and B are zero. Y = 1 only if either or both of A and b are 1. Hence the word OR].

Another example of a function of two independent variables is presented in Fig. 9.34. The circuit is now such that the lamp lights only when both the switches S_1 and S_2 are closed. The truth table for this case is the following (Table 9.2).

Table 9.2

S_1	S_2	Lamp	A	B	Y
open	open	off	0	0	0
open	closed	off	0	1	0
closed	open	off	1	0	0
closed	closed	on	1	1	1

The function Y, as specified in the Table 9.2, is now represented as,

$$Y = B \text{ AND } B \quad \text{or} \quad Y = A.B \text{ (or AB)}$$

[Y = 0 if either or both of A and B are zero. Y = 1 only if both A and B are 1. Hence the word AND].

We can have a function of a single variable, too, as represented in Fig. 9.35. Now the circuit has just one switch S. If S is open the lamp is on. If S is closed, the lamp is off (because then the current prefers to flow through the near zero resistance path containing S). Let the states of the switch and the lamp be represented by A and Y, respectively. Now, obviously Y is a function of A. The corresponding truth table is given below (Table 9.3).

Table 9.3

S	Lamp	A	Y
open	on	0	1
closed	off	1	0

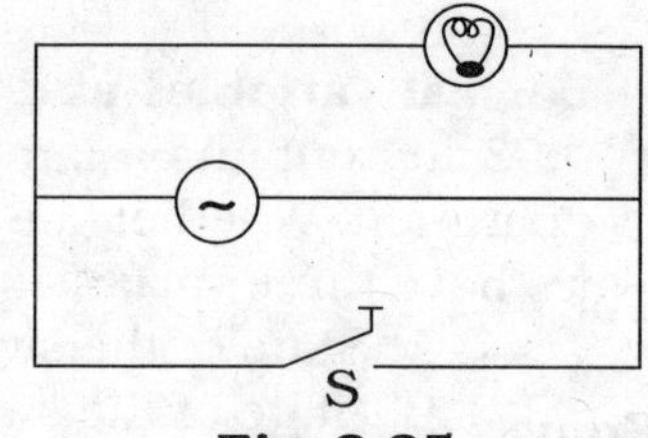

Fig. 9.35

The function Y is now represented as,

$$Y = \text{NOT } A \quad \text{or} \quad Y = \overline{A}$$

[Y is 1 only when A = 0, i.e. when the switch S is not closed. Hence the word NOT. (Note that normally the switch has to be closed for lighting a lamp in an electric circuit)].

A **logical variable** is a variable which can have only two possible values. A **logical function** is a function of logical variables. The **symbols AND, OR, NOT denote operations on logic variables.** The first two involve two variables and the third involves just one variable. These three are basic operations. By a repetitive and mixed use of them a vast number of other logical functions can be generated, e.g. Y = [(A OR B) AND B].

Binary Number and Decimal Number Systems

The student is well familiar with the **decimal number system,** which uses ten digits, viz. 0, 1, 2, 3, 4, 5, 6, 7, 8, 9. In any number each digit has a well defined positional value, determined as a multiple of 10 [units: 10^0; tens: 10^1; hundreds: 10^2; thousands: 10^3, and so on]. For instance, the number 5936 is expressed as,

$$\begin{aligned} 5936 &= 5\times10^3 + 9\times10^2 + 3\times10^1 + 6\times10^0 \\ &= 5000 + 900 + 30 + 6 \end{aligned}$$

It is clearly seen that **the base of the decimal number system is 10.**

The **binary number system** employs only two digits, viz. 0 and 1. **The base of the binary system is 2.** The decimal system can be converted into the binary system and vice verse, as follows.

Conversion from decimal to binary system. The procedure is that keep on dividing the given decimal system number by 2, until the quotient becomes zero. The binary denotation is obtained by writing the successive remainders in reverse order. We give a few examples.

(i) **Decimal number $(2)_{10}$**

2	2	Remainder
2	1	0
	0	1

The **binary denotation is $(10)_2$.** Thus, $(2)_{10} \triangleq (10)_2$

(ii) **Decimal number $(5)_{10}$**

2	5	Remainder
2	2	1
2	1	0
	0	1

The **binary denotation is $(101)_2$.** Thus, $(5)_{10} \triangleq (101)_2$

(iii) **Decimal number $(25)_{10}$**

2	25	Remainder
2	12	1
2	6	0
2	3	0
2	1	1
	0	1

The **binary denotation is $(11001)_2$.** Thus, $(25)_{10} \triangleq (11001)_2$.

Conversion from binary to decimal system. The procedure is to multiply the last digit of the decimal system number by 2° (i.e. 1), the next digit by 2^1, and so on, until the first digit is reached. Then all these products of multiplication are added up. As an example, we find the decimal equivalent of the **binary number $(101)_2$.**

$$
\begin{aligned}
1 \times 2^{\circ} &= 1 \\
0 \times 2^1 &= 0 \\
1 \times 2^2 &= 4 \\
\hline
&\quad 5
\end{aligned}
$$

The **decimal equivalent is 5.** Thus, $(101)_2 \triangleq (5)_{10}$.

Binary equivalent of a fractional decimal number. The procedure is to keep on multiplying the given number and the successive fractional numbers by 2 and to record every time the 'carry' in the integer position (i.e. the digit before the decimal point), until the required equivalent binary digits or zero are obtained. The 'carries' expressed in the direct order give the decimal binary equivalent. As an example, we find the binary equivalent of the **fractional decimal number $(0.75)_{10}$.**

$$0.75 \times 2 = 1.50, \text{ with carry } 1$$
$$0.50 \times 2 = 1.00, \text{ with carry } 1$$

The binary decimal equivalent is $(0.11)_2$. Thus, $(0.75)_{10} \triangleq (0.11)_2$.

We give another example, involving both an integer and a fractional decimal number, e.g. we find binary equivalent of the **decimal number $(5.75)_{10}$.** Here we separately find the binary equivalent of 5 and 0.75 (we have already done so in the previous examples):

$$(5)_{10} \bumpeq (101)_2$$
$$(0.75)_{10} \bumpeq (0.11)_2$$

Thus, the **binary decimal equivalent** is $(101.11)_2$. Or,

$$(5.75)_{10} \bumpeq (101.11)_2$$

Decimal equivalent of a fractional decimal binary number. The procedure is to multiply the successive digits with 2^{-1}, 2^{-2}, 2^{-3} and so on, and add up the products of multiplication. As an example, we evaluate the decimal equivalent of the **fractional decimal binary number $(0.011)_2$.**

$$(0.011)_2 \bumpeq 0 \times 2^{-1} + 1 \times 2^{-2} + 1 \times 2^{-3}$$
$$\bumpeq 0 + \frac{1}{4} + \frac{1}{8}$$
$$\bumpeq \frac{3}{8} \bumpeq (0.375)_{10}$$

Analogue (or Analog) and Digital Signals

An **analogue (or analog) voltage (or current) signal** is one in which the voltage (or current) continuously varies with time. The voltage waveform shown in Figs. 9.36 (a) is an example of such a signal. The electronic circuits for amplifiers, oscillators, etc. produce such signals. Accordingly, they are called analogue circuits.

A digital voltage (or current) signal is one in which the voltage (or current) has only two discrete values, one low and the other high. An example is presented in Fig. 9.36 (b). The electronic circuits that produce such signals are known as digital circuits.

The two discrete low and high voltage (or current) values produced by an electronic digital circuit can be represented by the binary number 0 and 1, respectively. This has led to the development of the extremely useful **digital electronics,** that finds wide application in our daily life, e.g. in calculators, computers, televisions, digital cameras, washing machines, etc.

Note that the digits 0 and1 do not mean voltages of zero volt and one volt, respectively. They simply indicate a relatively low voltage (e.g. 0.5 V) and a relatively high voltage (e.g. 3V), respectively, as seen in Fig. 9.36 (b).

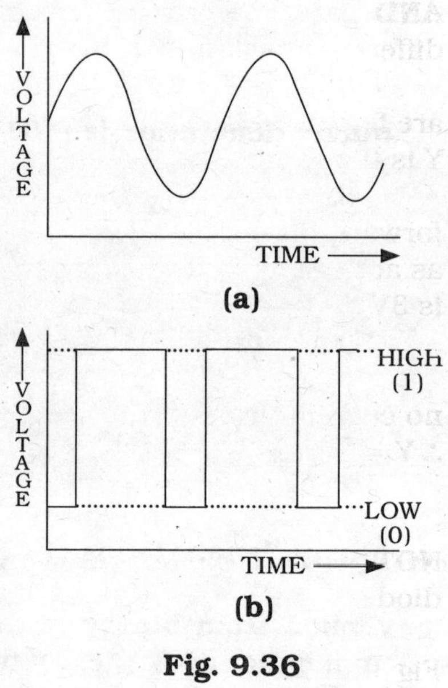

Fig. 9.36

9B.11. Logic Gates

The logic gates are the fundamental components of digital electronics. A **logic gate** is an electronic digital circuit, used to evaluate a logical function, by processing the binary 0 and 1 voltage signals [0: low voltage; 1: high voltage]. The circuit has one or more input signals and one or more output signals. There is a logical relationship between the input signals and the output signals. Each input terminal represents an independent variable. The voltage (low or high) appearing at an output terminal determines the value of the function. A given circuit can be used to evaluate more than one functions.

OR, AND, and NOT gates and their realisation. The logic gates used to evaluate OR, AND, and NOT functions are known as **OR, AND, and NOT gates respectively.** The symbols used for them are shown in Fig. 9.37. Each of the first two have two input signals (A and B) and one output signal (Y). The third has one each of the input and output signal.

Now we shall describe how the OR and AND gates can be realised with the help of two p-n junction diodes and the NOT gate with the help of an n-p-n transistor.

We assume that A = 0 represents zero potential at A and A = 1 represents a finite +ve potential (e.g. 3V) at A. The same

Fig. 9.37

holds for B and Y, i.e. B = 0 and Y = 0 mean 0 V at B or Y, and B = 1 and Y = 1 mean +3V at B or Y.

OR (Y=A+B)

Fig. 9.38

OR gate. The circuit employed for an OR gate is shown in Fig. 9.38. We evaluate Y under different possible conditions.

(a) A = 0; B = 0: It means that both A and B are at zero potential and there is no potential difference anywhere in the circuit. ∴ Y = 0.

(b) A =1; B = 0: The potential is 3V at A and 0 at B. The diode D_1 is forward-biased and so it offers practically no resistance to the flow of current (as if A and Y were directly connected with a wire). So the potential at Y is the same as at A, viz. 3V. Hence Y = 1.

(c) A = 0; B = 1: It leads to the same condition as in (b). Hence, again Y = 1.

(d) A = 1; B = 1: Both D_1 and D_2 are forward-biased. So A, B and Y are at the seme potential, viz. 3V. ∴ Y = 1.

These results match with the truth table drawn earlier for this case (Table 9.1),

AND gate. The circuit for this case is shown in Fig. 9.39. We again examine the different possibilities.

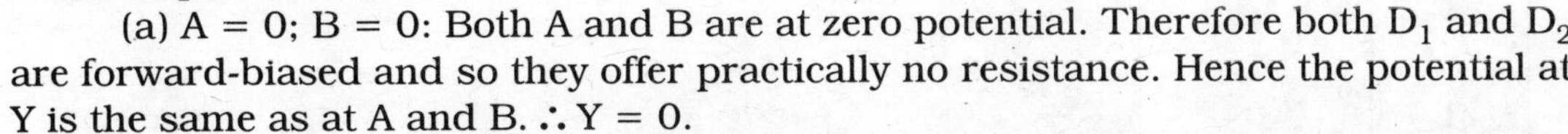

(a) A = 0; B = 0: Both A and B are at zero potential. Therefore both D_1 and D_2 are forward-biased and so they offer practically no resistance. Hence the potential at Y is the same as at A and B. ∴ Y = 0.

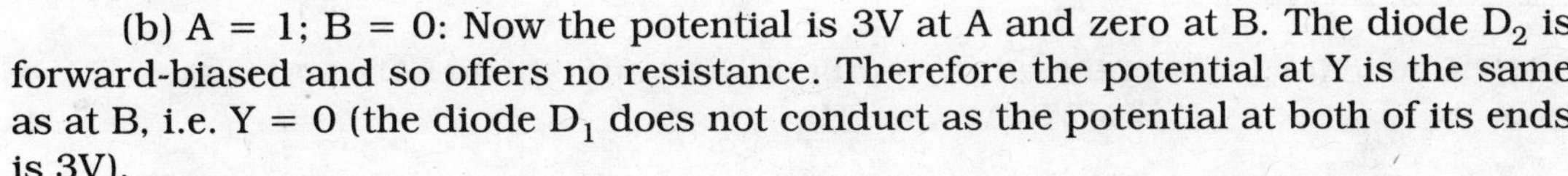

(b) A = 1; B = 0: Now the potential is 3V at A and zero at B. The diode D_2 is forward-biased and so offers no resistance. Therefore the potential at Y is the same as at B, i.e. Y = 0 (the diode D_1 does not conduct as the potential at both of its ends is 3V).

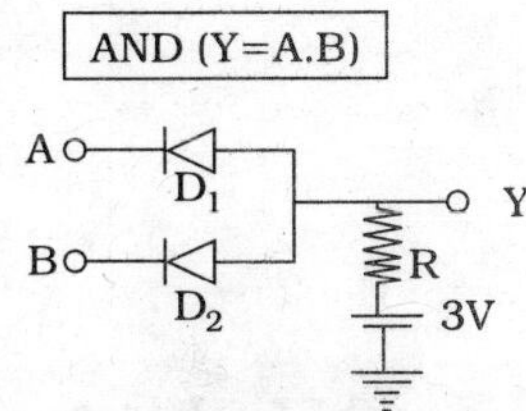

Fig. 9.39

(c) A = 0; B = 1: It leads to the same condition as in (b). Hence, again Y = 0.

(d) A = 1; B = 1: Now the potential at both A and B is 3V. So none of the diodes conducts and therefore no current flows through R. Hence the voltage drop across R (= iR) is zero and so the potential at Y is 3V. ∴ Y = 1.

These results are in agreement with the corresponding truth table (Table 9.2).

NOT gate. The realisation of a NOT gate requires a transistor. It cannot be realised with the help of diodes.

The required circuit consisting of an n-p-n transistor is shown in Fig. 9.40. Now there is only one input. So only two possibilities exist.

(a) A = 0: The emitter-base junction is not biased. So no current flows through it. Hence no current flows in the collector circuit, too (also, the collector-base junction is reverse-biased) and the potential drop across R_c is zero. Therefore the potential at Y is the same as the +ve potential of the battery, viz. 3V. ∴ Y = 1.

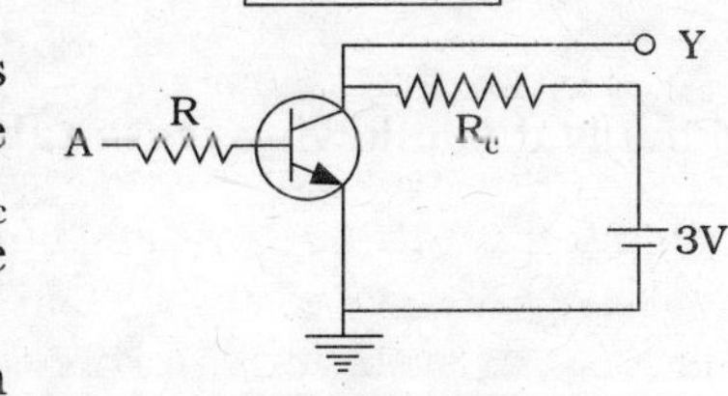

Fig. 9.40

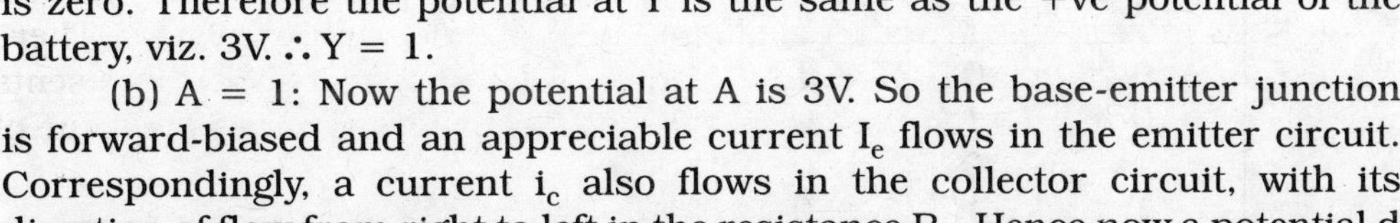

(b) A = 1: Now the potential at A is 3V. So the base-emitter junction is forward-biased and an appreciable current I_e flows in the emitter circuit. Correspondingly, a current i_c also flows in the collector circuit, with its direction of flow from right to left in the resistance R_c. Hence now a potential difference (= I_cR_c) exists across R_c, such that the potential at its left end (and therefore at Y) is zero. ∴ Y = 0.

Once again, the results are consistent with the relevant truth table (Table 9.3).

NAND, NOR and XOR Gates

More logic gates can be produced by a suitable combination of the basic logic gates OR, AND and NOT. Three such commonly used gates are NAND, NOR, and XOR, Their symbols are shown in Fig. 9.41.

(a) NAND gate. The **NAND function** (Y) is obtained by a combination of the operations NOT and AND on the independent variables A and B,

$$Y = \text{NOT (A AND B)}$$

It is expressed as,

$$Y = A \text{ NAND } B \quad \text{or} \quad Y = \overline{A.B} \quad \text{or} \quad Y = \overline{AB}$$

It is evaluated in two steps, first by the operation of AND and then by the operation of NOT, as follows.

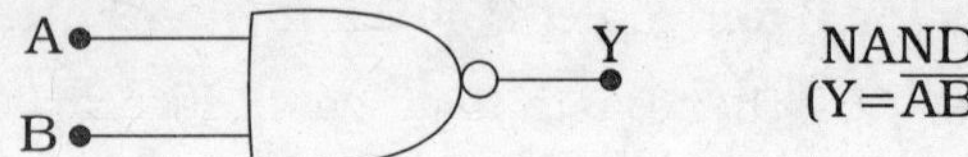

A	B	A AND B (AB)	NOT (A AND B) ($\overline{AB}$)
0	0	0	1
0	1	0	1
1	0	0	1
1	1	1	0

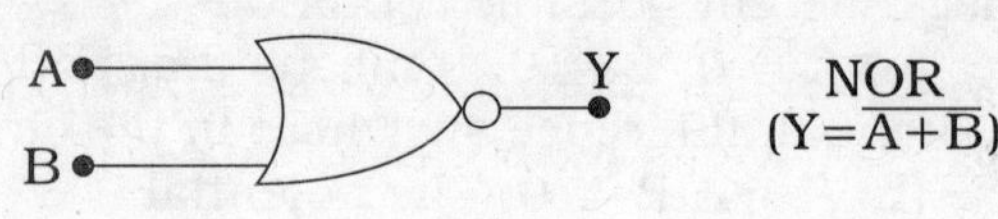

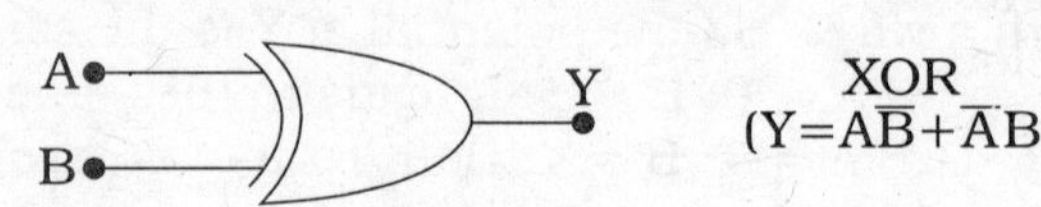

Fig. 9.41

Thus we finally get the following **truth table** for the NAND function (Table 9.4).

Table 9.4

A	B	Y = $\overline{AB}$
0	0	1
0	1	1
1	0	1
1	1	0

(b) NOR gate. The NOR function (Y) is obtained by combining the operations NOT and OR,

$$Y = \text{NOT (A OR B)}$$

It is expressed as,

$$Y = \text{A NOR B} \quad \text{or} \quad Y = \overline{A+B}$$

Again a 2-step operation is performed as below.

A	B	A OR B (A + B)	NOT (A OR B) ($\overline{A+B}$)
0	0	0	1
0	1	1	0
1	0	1	0
1	1	1	0

Finally the following **truth table** is obtained (Table 9.5).

Table 9.5

A	B	Y = $\overline{A+B}$
0	0	1
0	1	0
1	0	0
1	1	0

(c) XOR gate. An XOR function (Y) is obtained by a combination of the AND, OR, and NOT operations as follows,

$$Y = \text{A XOR B} = A\overline{B} + \overline{A}B$$

The relevant **truth table** is the following (Table 9.6).

Table 9.6

A	B	Y = $A\overline{B} + \overline{A}B$
0	0	0
0	1	1
1	0	1
1	1	0

It is seen that Y = 1 if one of the variables A and B is zero and the other is 1. If both the variables are either O or 1, then Y = 0.

Note. All other basic logic gates like AND, OR, NOT can be obtained by a repeated operation of a NAND (or NOR) gate. All digital circuits, howsoever complicated, can be built up by a suitable combination of the various basic gates. Hence a NAND gate is referred to as a **universal gate** or a **digital building block.**

Example 8. The output of a 2 input AND gate is fed to a NOT gate. Draw the logic circuit of the combination of gates. Give its logic symbol and write down its truth table. Identify the new logic gate formed [CBSE & AI; 1997].

Solution. The logic circuit of the combination and the logic symbol of the equivalent logic gate are shown below.

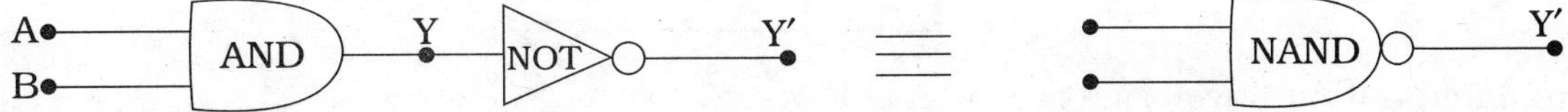

The equivalent gate is a NAND gate . Its truth table is:

A	B	Y	Y′
0	0	0	1
0	1	0	1
1	0	0	1
1	1	1	0

Use of Logic Gates for Modifying Waveforms

We have seen that the logic gates can perform various mathematical logical operations, as listed in truth tables 9.1 to 9.6. They can also be used to modify a given set of waveforms. Different gates modify the waveforms in different ways. The Fig. 9.42 shown the modification of a set of waveforms (due to the variables A,B) by three types of logic gates, viz. OR, AND, and XOR. How the modification is obtained for each gate can be understood with the help of the truth tables for the three gates and the Table 9.7 below, along with the Fig. 9.42.

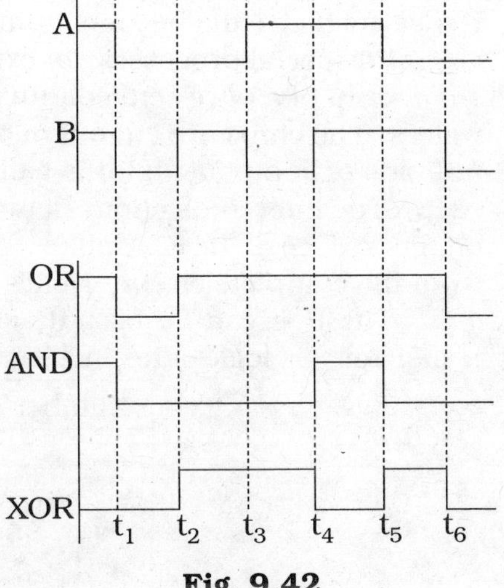

Fig. 9.42

Truth Table (OR)

A	B	Y
0	0	0
0	1	1
1	0	1
1	1	1

Truth Table (AND)

A	B	Y
0	0	0
0	1	0
1	0	0
1	1	1

Truth Table (XOR)

A	B	Y
0	0	0
0	1	1
1	0	1
1	1	0

Table 9.7

t-values	A, B	Y		
		OR	AND	XOR
$\leq t_1$	1, 1	1	1	0
t_1 to t_2	0, 0	0	0	0
t_2 to t_3	0, 1	1	0	1
t_3 to t_4	1, 0	1	0	1
t_4 to t_5	1, 1	1	1	0
t_5 to t_6	0, 1	1	0	1
$> t_6$	0, 0	0	0	0

Example 9. In the figure below, circuit symbol of a logic gate and input waveform are shown. (i) Name the logic gate, (ii) write its truth table and (iii) give the output wavelength. (CBSE; 2003).

Solution. (i) The given logic gate is a NOT gate

(ii) Its truth table is

Input A	Output Y = $\overline{A}$
0	1
1	0

(iii) The input and output waveforms are given below:

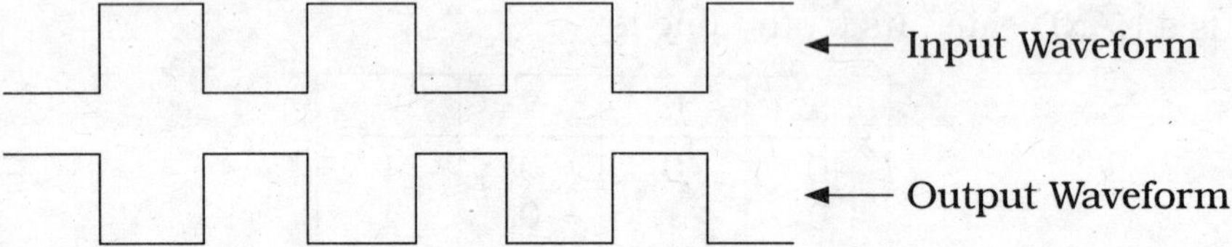

9B.12. Integrated Circuit (IC or Chip or Microchip or Microprocessor Chip)

An electronic circuit can contain a very large number of components like resistors, capacitors, diodes, transistors, etc. (upto several millions of them). In a conventional circuit they are interconnected by wires. Hence such circuits become unmanageably bulky and large-sized. They are also not much shock-proof.

By special processes, an extremely large number of electronic components can be assembled on a very small **chip** of a pure semiconductor material. The material is usually single crystal silicon, dissected in thin wafers. The chips are cut out of a wafer. A chip may measure less than 1mm×1mm and yet contain several millions of components! It is called an **integrated circuit** (or **IC** or **chip** or **microchip** or **microprocessor chip**). The interconnections between the components are made by means of a pattern of conducting material on the surface of the IC. Unlike a conventional electronic circuit, the individual components are not separable from the complete circuit. Hence the name 'integrated circuit'.

The IC's can be broadly classified into 4 categories. The classification is based on the number of transistors or logic gates included in the circuit.

Number of logic gates	Classification
≤ 10	Small Scale Integration (SSI)
≤ 100	Medium Scale Integration (MSI)
≤ 1000	Large Scale Integration (LSI)
> 1000	Very Large Scale Integration (VLSI)

The manufacture of an IC is a very complex and highly technical process. It is also lengthy and repetitive. Some basic steps are the following.

(a) Deposition of n-type or p-type semiconducting layer on the chip.

(b) To produce an insulating layer of SiO_2 by oxidation, for use to separate one region of the silicon chip from another.

(c) Selective etching of different regions of the chip, so that different components can be incorporated in different regions.

(d) Diffusion of various required impurities in various regions.

(e) Deposition of metal films to interconnect the various components of the circuit (called metallisation).

[Because of the tiny size of the chip, incorporating millions of electronic components, even a single speck of dust can ruin the chip. Therefore the whole manufacturing unit requires an **absolutely dust-free atmosphere.** To provide such an atmosphere is also a highly technical and specialized job. The workers in such **clean rooms** have to wear protective clothes. Before moving into the rooms they have to undergo air showers and wipedown stations for removing cosmetics.].

As compared to the conventional electronic circuits the IC's offer several major advantages:

(i) Very much smaller	(ii) Lighter
(iii) Faster	(iv) Use less power
(v) Cheaper	(vi) Much more shock-proof
(vii) Last longer	(viii) Can withstand extreme temperatures

However, even if one component becomes defective, the entire IC has to be replaced. Further, the IC's cannot operate at high power values.

The use of the IC's has led to the era of the so-called **microelectronics.** The IC's are now essential components of most electronic circuits, forming parts of devices of common use like calculators, computers, televisions, taperecorders, washing machines, etc.

Solved Conceptual Problems

[Including HOTS Questions]

Problem 1. Explain why a p-n junction has high resistance when reverse-biased and a negligibly low resistance when forward-biased.

Solution. The resistance is the ratio of voltage to current ($R = V/i$). When the junction is forward-biased, a little increase in the voltage across it results in a large increase in the current flowing across it. Hence R is very low.

In reverse bias, even a large rise in the voltage results in a small increase in the current. Hence R is high.

Problem 2. The current flowing across a p-n junction in forward bias (~mA) is nearly three orders of magnitude higher than the current in reverse bias (~μA). Yet a photodiode is operated in reverse bias, Explain why.

Solution. In an n-type semiconductor, the number density of the majority carriers (free electrons) is very much larger than the number density of the minority carriers (holes), i.e.,

$$n_e \gg n_h$$

When the semiconductor in a photodiode is illuminated, an equal number of additional holes and electrons are created. If their number densities are Δn_h and Δn_e, respectively, the new number densities n_e' and n_h' will be given by,

$$n_e' = n_e + \Delta n_e$$

$$n_h' = nh + \Delta n_h$$

$$(\Delta n_e = \Delta n_h)$$

Since $\Delta n_e = \Delta n_h$ and $n_e \gg n_h$, the fractional changes in the number of majority carriers ($\Delta n_e/n_e$) is far less than the corresponding change in the number of minority carries ($\Delta n_h/n_h$). Since in reverse bias the current is dominated by minority carriers, it means that any changes in current (consequent upon changes in the intensity of illuminating light) are far easily measurable than in forward bias. Hence for measuring light intensities, it is preferable to use a photodiode in reverse bias.

Problem 3. In an electric circuit, an ideal p-n junction diode has an 80 Ω resistor connected with it. What will be the total resistance in the circuit when the junction is (a) forward-biased and (b) reverse-biased ?

Solution. (a) When the junction is forward-biased, its resistance is nearly zero. Hence the total resistance in the circuit will be 80Ω.

(b) When the junction is reverse-biased, its resistance becomes nearly infinite. Hence the total resistance in the circuit will be almost infinite.

Problem 4. Explain why is it essential to have the base in a transistor thin and lightly doped. What will happen if it is doped as much as the emitter ?

Solution. The base is sandwiched between the emitter and the collector. Thus, the conduction of the majority carriers between the emitter and the collector takes place through the base. Some of them combine with charge carriers in the base (electron-hole recombination), to produce a base current. A thin and lightly doped base region contains only a small number of majority charge carriers. It helps to reduce the recombination substantially. Hence the base current is weak and the collector current is nearly as high as the emitter current, resulting in high voltage gain and power gain for the transistor.

The emitter is heavily doped. If the base is also doped equally heavy, the electron-hole recombination will be enormous, resulting in a strong base current and a weak collector current. Thus, the very purpose of the transistor will be lost.

Problem 5. Can a p-n-p transistor be produced by putting two p-n junctions in mutual contact ? If not, explain why.
Solution. By joining two p-n junctions at the n-region, the base region will become very thick, which will greatly reduce the collector current and so defeat the purpose of the transistor.

Problem 6. Normally, in a transistor the emitter-base junction is always forward-biased and the collector-base junction is reverse-biased. What will happen if (i) both the junctions are forward-biased, (ii) both the junctions are reverse-biased, and (ii) the biasing is interchanged ?
Solution. (i) In normal functioning, the forward-biased emitter sends the majority charge carriers to the base region, from where they are collected by the reverse-biased collector. Thus a continuous free flow of the charge carriers and therefore a free flow of current in the circuit is ensured. If the collector-base junction is also forward-biased, the charge carriers will cease to be collected, thus virtually stopping the flow of current.

(ii) The reverse biasing of the emitter-base junction will disable the charge carriers to reach the base from the emitter. Hence once again there will be virtually no flow of current in the circuit.

(iii) In this case the emitter will work as the collector and vice versa. The transistor will work. But since the collector is moderately doped and the emitter is heavily doped, the current in the circuit will be appreciably reduced.

Problem 7. Why is a NAND gate (or a NOR gate) called a universal gate or a digital building block ?
Solution. All other basic logic gates like AND, OR, NOT can be obtained by a repeated operation of a NAND (or a NOR) gate. All digital circuits, howsoever complicated, can be built up by a suitable combination of various basic gates. Hence a NAND gate is referred to as a universal gate or a digital building block.

Solved Numerical Problems

Problem 1. The band gap of a semiconductor is equal to the energy of a photon of mercury green light of wavelength 546 nm. What will be the minimum energy needed to create an electron-hole pair in the semiconductor ?
Solution. Given: $\lambda = 546 \text{ nm} = 546 \times 10^{-9}$ m
The energy of a photon is given by ($E = h\nu$; $\nu\lambda = c$),

$$E = \frac{hc}{\lambda} = \frac{(1242 \text{ eV–nm})}{(546 \text{ nm})}$$
$$= 2.27 \text{ eV}$$

As stated in the question, this E-value is the band gap of the semiconductor. Hence this is also the minimum value of energy needed to create an electron-hole pair, by making an electron to move from the valence band to the conduction band. **(Answer).**

Problem 2. An electric circuit consists of a p-n diode connected to a resistor of 80 Ω and an emf source of 3.5 V. If the potential barrier created in the diode is 0.4 V, what is the current in the circuit ?

Solution. Given: $E = 3.5$ V; $R = 80\ \Omega$; $V_B = 0.4$ V

The voltage drop across the diode is 0.4 V. Therefore,

$$\text{Effective voltage in the circuit} = (3.5 - 0.4)\text{V} = 3.1 \text{ V}$$

$$\therefore \text{The current in the circuit} = \frac{\text{Voltage}}{\text{Resistance}} = \frac{(3.1)}{(80)} = 0.039 \text{ A (or 39 mA)}$$ **(Answer).**

Problem 3. A p-n junction has a potential barrier of 0.5 volt created across it, with the depletion region having a width of 5×10^{-7} m. (i) Find the electric field in the region. (ii) If an electron enters the junction from the n-side with a velocity of 5×10^{5} ms^{-1}, with what velocity will it enter the p-side ?

Solution. Given: $V_B = 0.5$ V; $d = 5\times10^{-7}$ m; velocity of electron at entry to the depletion layer from n-side, $v_1 = 5\times10^{5}$ ms^{-1}

(i) The electric field in the depletion region is given by,

$$E = \frac{V_B}{d} = \frac{(0.5)}{(5\times10^{-7})} = 10^{6} \text{ Vm}^{-1}$$ **(Answer).**

(ii) Let v_2 be the velocity of the electron when it leaves the depletion region and enters the p-side. According to the law of conservation of energy, the kinetic energy and the potential energy of the electron will be related as,

$$(\text{k.e.})_{\text{entry to depletion layer}} = (\text{k.e.})_{\text{leaving the layer}} + (\text{p.e.})_{\text{gained}}$$

$$\Rightarrow \quad \frac{1}{2}mv_1^2 = \frac{1}{2}mv_2^2 + eV_B$$

$$\Rightarrow \quad \frac{1}{2}\times(9.1\times10^{-31})\times(5\times10^{5})^2 = \frac{1}{2}\times(9.1\times10^{-31})v_2^2 + (1.6\times10^{-19})\times(0.5)$$

$$\Rightarrow \quad 113.8\times10^{-21} = 4.55\times10^{-31}\,v_2^2 + 8\times10^{-20}$$

$$\Rightarrow \quad v_2^2 = \frac{3.38\times10^{-20}}{4.55\times10^{-31}} = 7.43\times10^{10}$$

$$\Rightarrow \quad v_2 = 2.73\times10^{5} \text{ ms}^{-1}$$ **(Answer).**

Problem 4. A transistor operating in common emitter mode has AC current gain of 60. If the input resistance in the circuit is 1 kΩ, the collector resistance is 4 kΩ, and the input voltage is 0.02 V, find the output voltage.

Solution. Given: $\beta = 60$; $R_i = 1\text{k}\Omega$; $R_0 = 4\text{k}\Omega$; $V_i = 0.02$ V

The voltage gain is given by,

$$A_v = \left(\frac{R_0}{R_i}\right)\beta = \frac{(4\text{ k}\Omega)}{(1\text{ k}\Omega)}\times(60) = 240$$

The voltage gain is defined as the ratio of change in output voltage to the change in input voltage,

$$A_v = \frac{dV_0}{dV_i}$$

Taking the changes starting from zero value, we have

$$A_v = \frac{V_0}{V_i}$$

$$\Rightarrow \quad V_0 = A_v V_i = (240)\times(0.02) = 4.8 \text{ volt}$$ **(Answer).**

Problem 5. A CE-transistor amplifier with the collector resistance of 4 kΩ and the base resistance of 800 Ω is employed to amplify an audio signal. If the voltage drop across the collector resistance is 3 V and the current amplification factor (current gain) of the transistor is 80, find the input signal voltage and base current.

Solution. Given: $R_0 = 4\ k\Omega = 4\times10^3\ \Omega$; $R_i = 800\ \Omega$; $V_0 = 3$ V; $\beta = 80$

The current amplification factor (β) for a CE - amplifier is defined as the ratio of output current to the input current,

$$\beta = \frac{I_c}{I_b}$$

$$\Rightarrow \quad I_b = \frac{I_c}{\beta} = \frac{V_0/R_0}{\beta} = \frac{(3)/(4\times10^3)}{80}\ \text{A}$$

$$= \frac{3}{4\times80\times10^3}\times10^6\ \mu\text{A}$$

$$= 9.38\ \mu\text{A}$$

The voltage amplification factor (A_v) for a CE-amplifier is given by,

$$A_v = \frac{R_0}{R_i}\times\beta = \frac{(4\times10^3)}{(800)}\times(80)$$

$$= 400$$

But A_v is also defined as the ratio of output voltage to the input voltage,

$$A_v = \frac{V_0}{V_i}$$

$$\Rightarrow \quad V_i = \frac{V_0}{A_v} = \frac{(3)}{(400)}\ \text{V} = \frac{(3)}{400}\times10^3\ \text{mV}$$

$$= 7.5\ \text{mV}$$

Thus, the input signal voltage and the base current are found to be 7.5 mV and 9.38 μA, respectively. **(Answer).**

Problem 6. A CE-transistor amplifier has a load resistance of 4 kΩ. Addition of a signal of 10 mV to the base-emitter voltage brings about a change of 1 mA in the collector current and a change of 15 μA in the base current. Determine (i) the current amplification factor, (ii) the input resistance, (ii) the transconductance, and (iv) the voltage gain in the circuit.

Solution. Given; $R_0 = 4\ k\Omega = 4\times10^3\ \Omega$; $dV_i = 10\ \text{mV} = 10^{-2}$ V; $dI_b = 15\ \mu\text{A} = 15\times10^{-6}$ A; $dI_c = 1$ mA $= 10^{-3}$ A

(i) The current amplification factor (AC current gain) is given by,

$$\beta = \frac{dI_c}{dI_b} = \frac{(10^{-3})}{(15\times10^{-6})}$$

$$= 66.7 \quad \textbf{(Answer).}$$

(ii) Input resistance,

$$R_i = \frac{dV_i}{dI_b} = \frac{(10^{-2})}{(15\times10^{-6})}$$

$$= 667\ \Omega \quad \textbf{(Answer).}$$

(iii) Transconductance,

$$g_m = \frac{dI_c}{dV_i} = \frac{(10^{-3})}{(10^{-2})}$$

$$= 0.1\ \Omega^{-1} \quad \textbf{(Answer).}$$

(iv) Voltage gain,

$$A_v = \left(\frac{R_0}{R_i}\right)\times\beta = \frac{(4\times10^3)}{667}\times(66.7)$$

$$= 400 \quad \textbf{(Answer).}$$

Problem 7. An n-p-n transistor is employed in an amplifier circuit operating in common base configuration. When the emitter current changes by 10 mA, the collector current changes by 9.8 mA. What should be the change in base current to bring about the same change of 9.8 mA in the collector current ? Also calculate the AC current gain.

Solution. Given: $dI_e = 10$ mA; $dI_c = 9.8$ mA

We know,

$$I_e = I_b + I_c$$

$$\Rightarrow \quad dI_e = dI_b + dI_c$$

$$\Rightarrow \quad dI_b = dI_e - dI_c = [(10) - (9.8)] \text{ mA}$$

$$= 0.2 \text{ mA}$$

Thus, the base current has to change by 0.2 mA to cause a change of 9.8 mA in the collector current. **(Answer).**

The AC current gain is given by,

$$\alpha = \frac{dI_c}{dI_e} = \frac{(9.8)}{(10)}$$

$$= 0.98 \quad \textbf{(Answer).}$$

Problem 8. The output of an AND gate is connected to both the inputs of a NAND gate. Draw the logic circuit of this combination of gates and write its truth table. (CBSE; 2000).

Solution. The required logic circuit is given below:

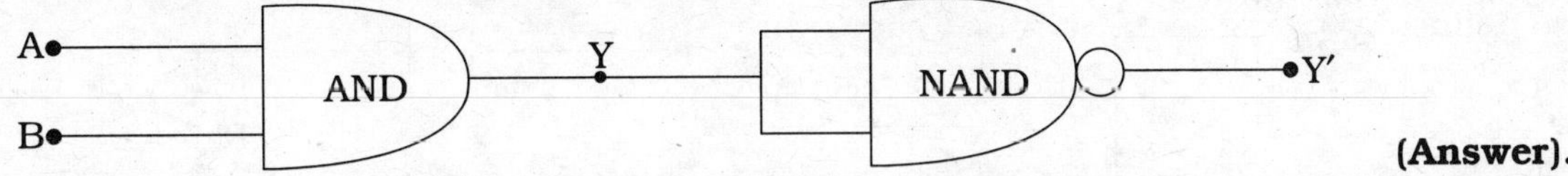

(Answer).

When the two inputs of a NAND gate are connected together, it becomes a NOT gate. The resulting combination of these two gates will be a NAND gate. Its truth table is the following:

A	B	Y	Y'
0	0	0	1
0	1	0	1
1	0	0	1
1	1	1	0

(Answer).

Problem 9. In the figure below, circuit symbol of a logic gate and two input waveforms 'A' and 'B' are shown. (i) Name the logic gate, (ii) write its truth table, and (iii) give the output waveform. (CBSE; 2002).

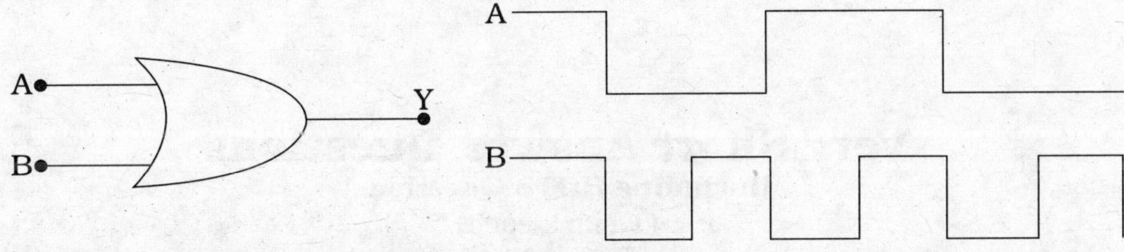

Solution. (i) The given logic gate is an OR gate. **(Answer).**

(ii) Its truth table is given below:

A	B	Y
0	0	0
0	1	1
1	0	1
1	1	1

(Answer).

(iii) The output waveform is given below:

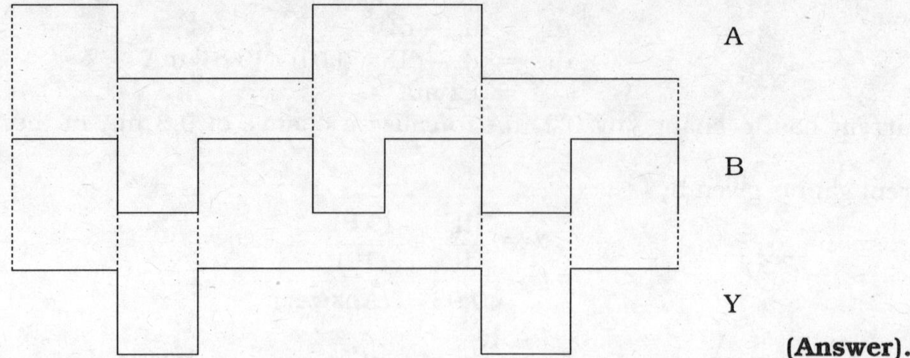

(Answer).

Problem 10. Draw and explain the output waveform across the load resistor R, if the input waveform is as shown in the figure. (CBSE; 2006).

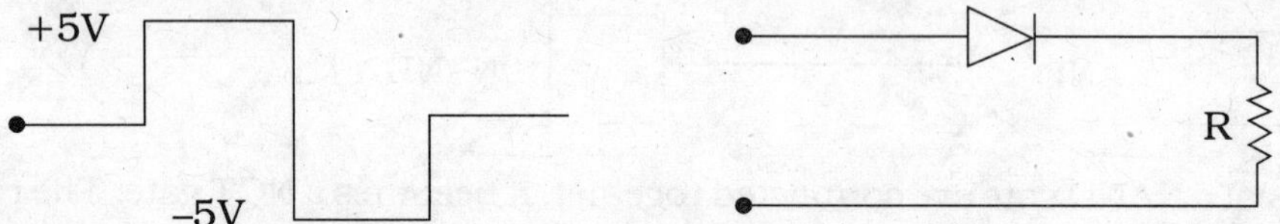

Solution. When in the AC signal the value of the input voltage is +5 V, the p-n diode in the circuit becomes forward-biased, resulting in an output voltage of +5 V across R. When the input voltage is –5 V, the diode gets reverse-biased. Then no output is obtained across R. Thus the output waveform is as given below:

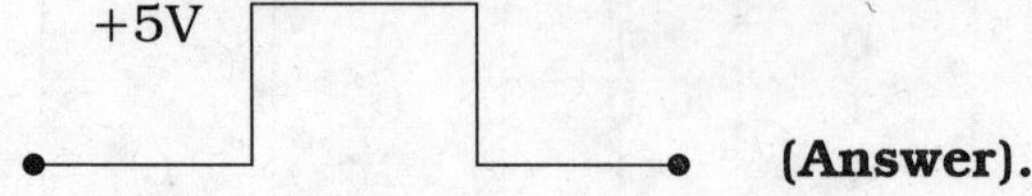

(Answer).

Very Short Answer Questions

[Including HOTS Questions]

(1 mark each)

1. In an unbiassed p-n junction, does the diffusion current flow from p-side to n-side or from n-side to p-side ?

2. In an unbiased p-n junction, does the drift current flow from p-side to n-side or from n-side to p-side ?

3. In an unbiased p-n junction, is the diffusion current greater than or less than or equal to the drift current ?

4. Draw line diagram of a p-n junction in reverse bias.

5. Which type of biasing imparts a very high resistance to a semiconductor diode ?

6. What is an ideal junction diode ?

7. Is a p-n junction an ideal diode ?

8. How is the width of depletion layer in a p-n junction affected when it is (a) forward-biased and (b) reverse-biased ?

9. In the figure below, state if the p-n junction is forward-biased or reverse-biased.

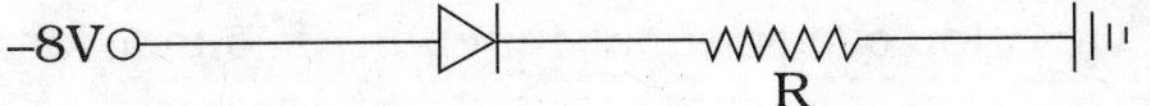

10. What is meant by Zener breakdown ?

11. In a circuit containing a p-n-p transistor, what are the current carriers (i) through the transistor and (ii) outside the transistor ? [Hint: (i) Holes (ii) Electrons].

12. In a circuit containing an n-p-n transistor, what are the current carriers (i) through the transistor and (ii) outside the transistor ? [Hint: Electrons, in both cases].

13. What is meant by the voltage gain of an amplifier ?

14. What is meant by the transconductance of a transistor ?

15. What is meant by the AC current gain for a transistor amplifier in common emitter mode ?

16. What is the phase relationship between the output and input voltages for a transistor amplifier operated in (i) common emitter mode and (ii) common base mode ?

17. What is the advantage of a common emitter amplifier over a common base amplifier ?

18. What is a logic gate ?

19. What are the three basic logic gates ?

20. What is an AND gate ? Write its Boolean expression.

21. Give the truth table for an AND gate.

22. Give the truth table for the combination of gates shown below:

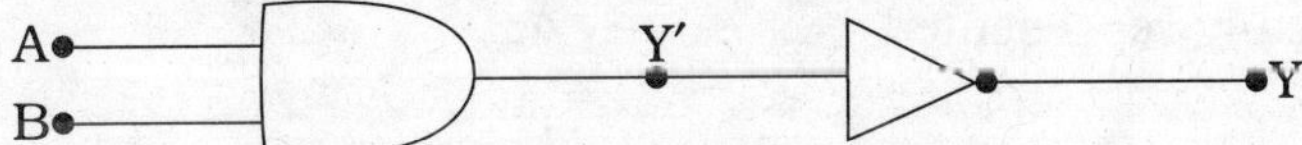

23. Give the values of input A and B for the Boolean expression $(\overline{A+B}).(\overline{A.B}) = 1$.

24. Convert the binary number 1100101 into decimal system.

25. Express the number 39 in binary system.

26. What is meant by an analogue symbol ? Represent it graphically.

27. What is meant by a digital signal. Represent it graphically.

28. What is meant by rectification of an output voltage ?

Short Answer Questions

[Including HOTS Questions]

(2 or 3 marks each)

1. Represent graphically the variation of current as function of voltage for a p-n junction, for both forward bias and reverse bias.

2. Explain the action of a Zener diode. Give its symbolic representation.

3. Which logic gate does the following diagram represent ? Give its truth table:

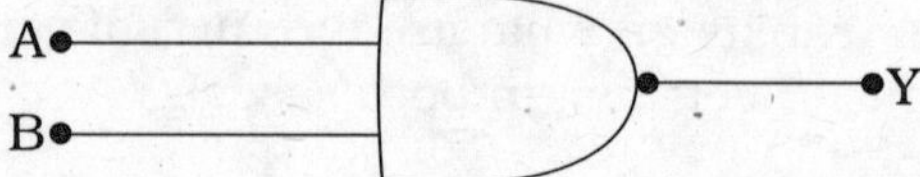

4. In the following diagram, identify the logic gates I and II. Write the truth table for the final output Y, for all possible combinations of the inputs A and B.

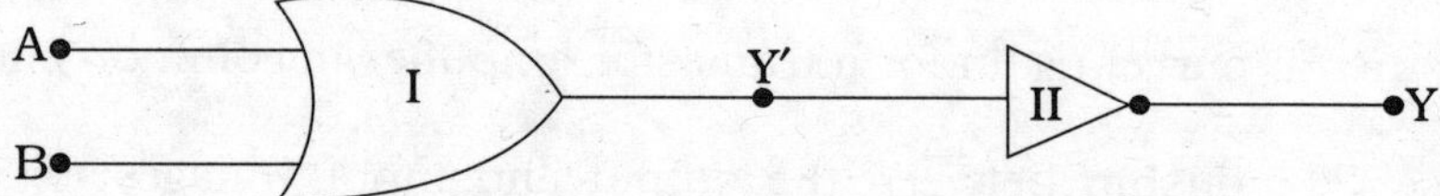

5. Explain the formation of depletion region and potential barrier for a p-n junction.

6. Explain the principle and working of a solar cell.

7. Explain the action of a Zener diode. Sketch it symbolically.

8. Describe with the help of a labelled diagram the use of a p-n junction as a half-wave rectifier.

9. Describe with the help of a labelled diagram how full-wave rectification is achieved with the use of junction diodes.

10. What is an n-p-n transistor ? How does it differ from a p-n-p transistor ? Draw relevant diagrams. How are they symbolically represented ?

11. Explain with the help of a circuit diagram the construction and working of an n-p-n transistor. Show that $I_e = I_b + I_c$.

12. Give the logic symbol, Boolean expression, and truth table for an OR gate. Explain with relevant diagrams how it is realised in practice.
[The same question can be separately framed one each for AND gate, NAND gate, and NOR gate].

Long Answer Questions

(5 marks each)

1. What is a p-n junction ? Explain its working, drawing suitable diagrams and characteristic curves, when it is (a) unbiased, (b) forward-biased and (c) reverse-biased.

2. Explain with the help of a circuit diagram how characteristic curves for an n-p-n transistor are obtained. Draw the input and output characteristics and interpret them.

3. Explain with the help of a circuit diagram the use of an n-p-n transistor as an amplifier, for common emitter configuration. Derive an expression for its voltage gain.

4. Describe with the help of a circuit diagram the use of a transistor as an oscillator.

Numerical Exercises

1. For a transistor operating in common emitter configuration, it is found that if V_{ce} changes by 0.1 V, the collector current changes through 0.002 mA. What is the output resistance in the circuit ? **[50 kΩ]**

2. In an electric circuit containing an n-p-n transistor, a change of 6.35 mA in the collector current is observed, when the emitter current is changed by 6.4 mA. What amount of change in the base current will be required to produce the same change of 6.35 mA in the collector current ? **[51 μA]**

3. For a silicon transistor operating in common emitter mode, the current gain is found to be 18, when the emitter current happens to be 6 mA. Determine the base current and the collector current in the circuit. **[0.32 mA; 5.68 mA]**

4. Two transistor amplifiers of voltage gain 8 and 16, respectively, are connected together in series in an electric circuit. The first amplifier is given an input AC signal of 0.02 V. Find the value of the final output voltage. **[2.56 V]**

5. A transistor is operated in common emitter mode, with a load resistance of 1 kΩ in the collector circuit. The collector is supplied a voltage of 10 V and the voltage drop across the load resistor is 0.6 V. If the current gain, $\alpha = 0.95$, calculate the collector-emitter voltage and base current. **[9.6 V; 0.03 mA]**

6. The input resistance of a silicon transistor is 665 Ω. Its base current is changed by 15 μA, which results in a change of collector current by 2 mA. This transistor is used as a common emitter amplifier with a load resistance of 5 kΩ. Calculate,
   (i) Current gain 'β_{ac}', (ii) transconductance 'g_m',
   (ii) voltage gain 'A_v' of the amplifier. (AI; 2001) **[(i) 133, (ii) 0.2 Ω^{-1}, (iii) 1002]**

7. In the logic circuit given below, identify the gates P and Q. Write the truth table for finding the output at Y for the various possible values of A and B.

A	B	Y
0	0	0
0	1	1
1	0	1
1	1	1

[P is NAND gate and Q is OR gate]

8. In the logic circuit given below, identify the gates P and Q. Write the truth table for finding the output at Y for the various possible value of A and B.

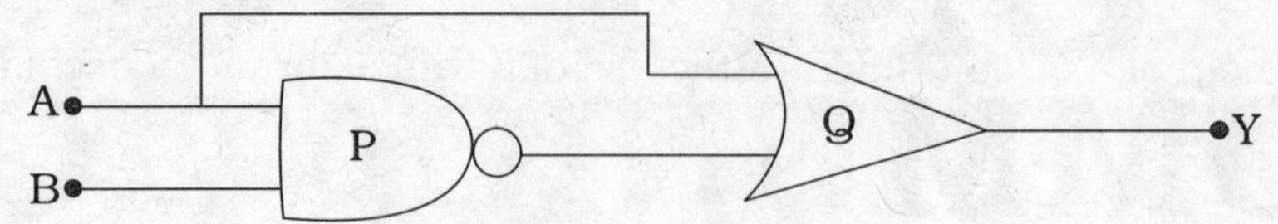

A	B	Y
0	0	1
0	1	1
1	0	1
1	1	1

[P is NAND gate and Q is OR gate]

9. In the given figure, circuit symbol of a logic gate and two input waveforms A and B are shown. (i) Name the logic gate, (ii) write its truth table, (iii) give the output waveform. (CBSE; 2002).

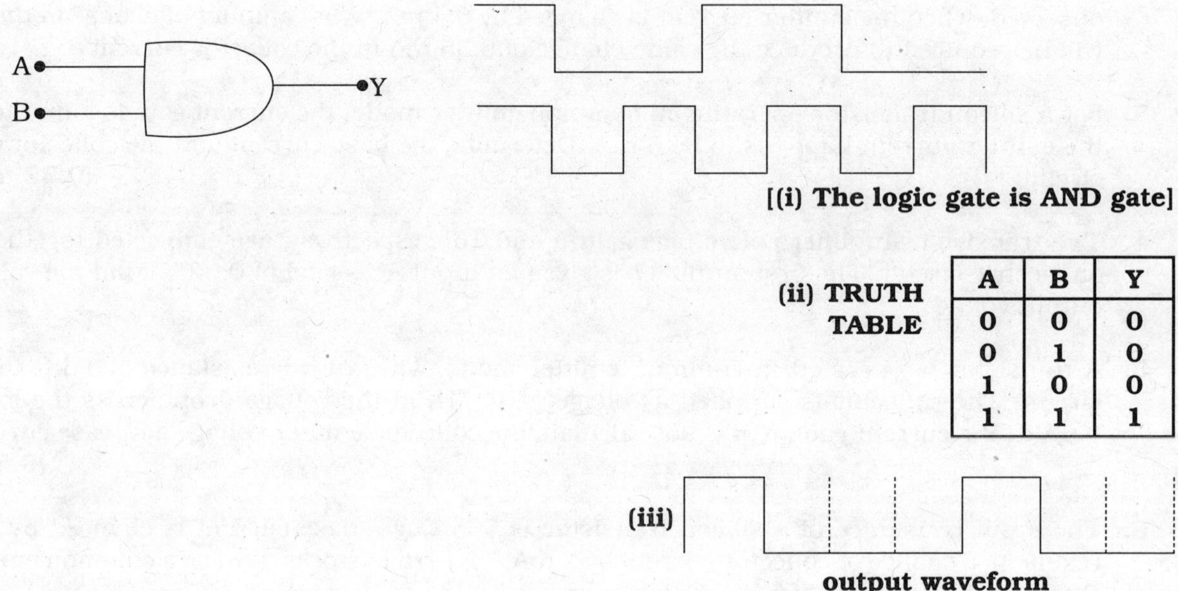

[(i) The logic gate is AND gate]

(ii) TRUTH TABLE

A	B	Y
0	0	0
0	1	0
1	0	0
1	1	1

(iii)

output waveform

10. In the logic circuits drawn below, show that the circuit (i) and (ii) are equivalent to an OR gate and AND gate, respectively:

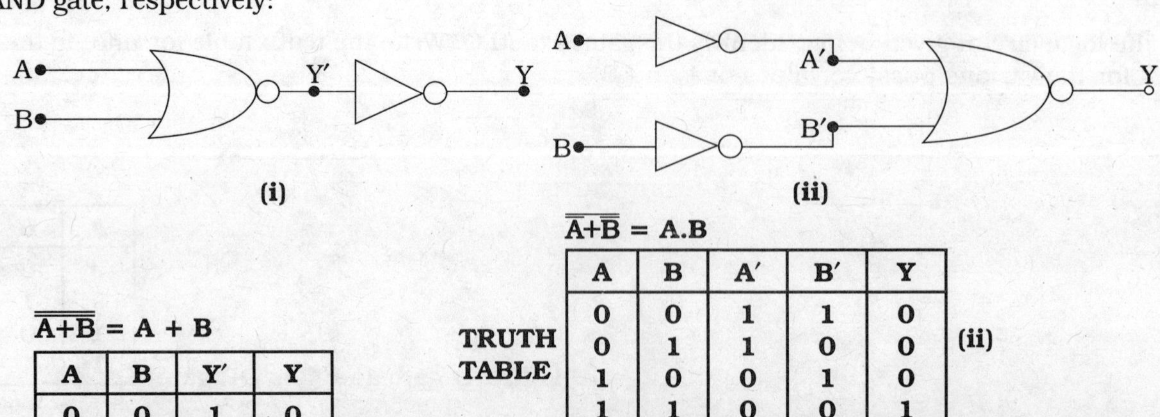

$\overline{\overline{A+B}} = A + B$

TRUTH TABLE (i)

A	B	Y′	Y
0	0	1	0
0	1	0	1
1	0	0	1
1	1	0	1

$\overline{\overline{A}+\overline{B}} = A.B$

TRUTH TABLE (ii)

A	B	A′	B′	Y
0	0	1	1	0
0	1	1	0	0
1	0	0	1	0
1	1	0	0	1

[The truth tables (i) and (ii), derived for the given circuits (i) and (ii), clearly correspond with Or gate and AND gate, respectively. Hence proved]

UNIT 10
COMMUNICATION SYSTEMS

CHAPTER 10
Communication Systems

CHAPTER 10
COMMUNICATION SYSTEMS

10.01. Introduction

Communication means conveying or exchange of information. It is an integral part of our daily lives. The simplest effective way to communicate is by talking, in which sound waves emanating from one person travel in air to be received by another person. But it is restricted to small distances. In the ancient times, communication over longer distances was achieved through a chain of runners or drummers, carrier pigeons, etc. But these modes were both slow and unsuitable for very long distances. Gradually, they have been superceded by electrical communication systems, in which the information at one end is converted into electrical signals, which are then transmitted to the other end, where they are converted back to the original piece of information. Not only the signals can travel over extremely long distances (even to heavenly bodies like planets, stars, and galaxies), they also travel extremely fast (at the speed of light). The results are obvious to us, in the form of devices like radio, television, e-mail, fax, internet, etc., which have not only made our lives convenient and comfortable but have also brought about enormous economy of time, money and resources.

Since we shall be concerned with the propagation of electromagnetic waves in this chapter, the student should first revise his knowledge of the e.m. waves, described earlier in Chapter 5. It will enable the reader to fully understand the various aspects of communication systems to be described in the following.

Telecommunication means communication over long distances (the Greek word 'tele' means 'far').

10.02. Essential components of a Communication System; Block Diagram

(i) Message. The **message** is the information to be communicated.

The message may be in one or more of the following forms: (a) audio (e.g. a spoken word, a song, or any other sound), (b) visual or video (e.g. a dance), (c) written (e.g. a letter).

The term 'signal' may also be used in place of 'message'. However, it is preferable to use it for an 'electrical signal' into which the 'message' is converted by a transducer, that we are going to describe next.

(ii) Input transducer. A **transducer** is a device which converts a message into an electrical signal. For example, a microphone is a transducer which converts the human voice or any other sound into electrical signals (consisting of variations of electric voltage or current), or a photoelectric cell that converts light signals into electrical signals. The **input transducer** converts the input message into an electrical signal, which we may refer to as **input signal** or **baseband signal.**

(iii) Transmitter. The electrical signals obtained from the transducer cannot be directly transmitted over long distances. A **transmitter** suitably modifies the input signal from the transducer, so that it can be

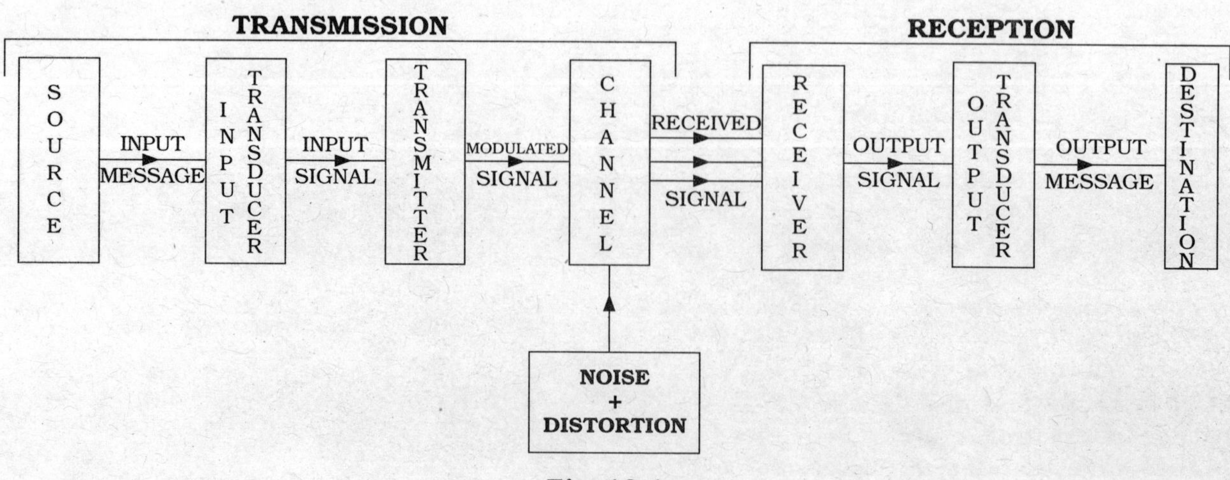

Fig. 10.1

efficiently transmitted over long distances. It consists of one or more subcomponents (e.g. coder, quantizer, modulator, etc). An essential subcomponent is a **modulator,** which superimposes the input signal (message) over a high frequency carrier wave. This process is called **modulation,** which we shall describe in the next section.

(iv) Channel. The **channel** is a medium through which the modulated signal is transmitted, e.g. a cable, an optical fibre, a waveguide, free space, etc.

During its passage through the channel, the transmitted signal undergoes **distortion,** which results from noise, attenuation (i.e. weakening), or other influences. The noise means a disturbance to the signal. It arises from unwanted random electrical signals (sometimes we exclude noise from distortion, thus treating it as a separate factor). The sources of noise are both external (e.g. interference from signals carried by nearby channels) and internal (e.g. thermal motion of electrons in conductors).

(v) Receiver. The **receiver** processes the signals received from the channel, by reversing the signal modifications made at the transmitter and the channel. (An essential part of the process is that the signals are **demodulated,** i.e. they are separated from the high frequency carrier waves).

(vi) Output transducer. The output transducer converts the demodulated electrical signals from the receiver to their original form. viz. the message.

(vii) Destination. The message finally reaches its destination, which may be one or more. (In the case of radio waves, which propagate all around through free space, the destinations are infinite. The waves can be received anywhere, as done is our homes by radio sets and televisions).

A block diagram of a communication system is shown in Fig. 10.1. It consists of two parts, viz. **transmission** and **reception.** (In the simple case of conversation between two persons, the source/transmitter is the vocal cord of the first person, the atmospheric air is the transmission channel, and the ears of the second person are the receiver/destination. No transducers are involved).

Additional knowledge

Noise. It consists of signals which are random and unpredictable. Some external sources are automobile engine iignition, faulty contacts in electric switches, fluorescent lamps, lightning, electrical storms, solar and intergalactic radiation, etc. Some internal sources are diffusion or recombination of charge carriers in electronic devices, random emission of radiation, etc.

The noise is a fundamental factor that affects both the quality and the rate of communication.

10.03. Modulation; Demodulation

The input signal produced on the transmission side is usually not suited for direct transmission over long distances. Therefore it needs to be suitably modified. "The process of suitably modifying an input signal so that it can be transmitted over long distances" is called **modulation.** The device having the electrical circuit to produce modulation is called **modulator.** The reasons for the unsuitability of an unmodified input signal for communication purposes, i.e. need for modulation, will be described in Section 10.06, after the role of channel bandwidth in communication has been discussed in Section 10.05.

The **process of modulation** consists of **superimposing** (or **mounting**) the input signal on a high frequency wave, called **carrier wave** or **carrier signal.** The carrier signal can be either a **continuous wave** or a **pulse.** Accordingly, we have two main types of modulation:

(a) Continuous wave modulation

(b) Pulse Modulation

(a) Continuous wave modulation. The carrier continuous wave is a sinusoidal wave of high frequency, of the general

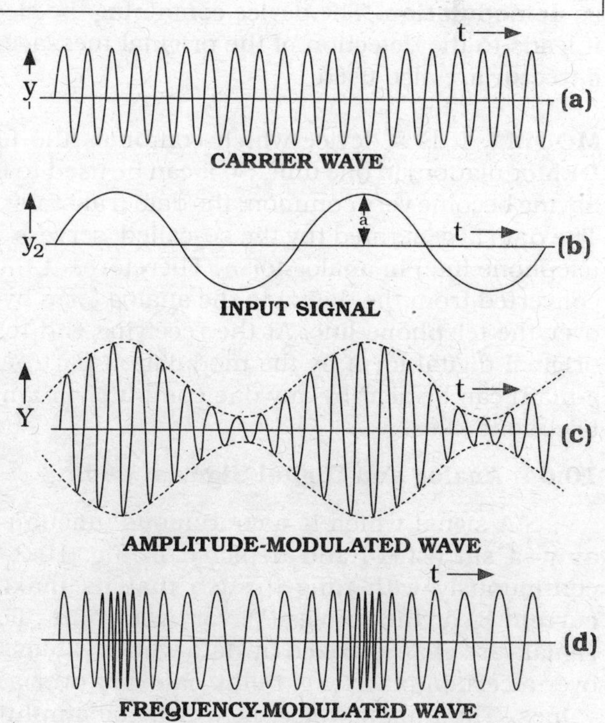

Fig. 10.2

form y=a sin (ωt+φ) (Fig. 10.2a). Any one of its three characteristics [amplitude (a), frequency (ω) or phase (φ)] can be varied in proportion to the input signal. Accordingly. we have the following three types of modulation in this case:

(i) Amplitude modulation (AM)
(ii) Frequency modulation (FM)
(iii) Phase modulation (PM)

The Fig. 10.2 illustrates a carrier wave, an input signal and the AM/FM waves.

In Section 10.07, we shall describe the production and detection of an amplitude-modulated wave.

(b) Pulse modulation. The variable parameters for a pulse are pulse amplitude, pulse width and pulse interval. Accordingly, we first have the following three types of modulation available in this case:

(i) Pulse amplitude modulation (PAM).
(ii) Pulse duration modulation (PDM) or pulse width modulation (PWM).
(iii) Pulse position modulation (PPM).

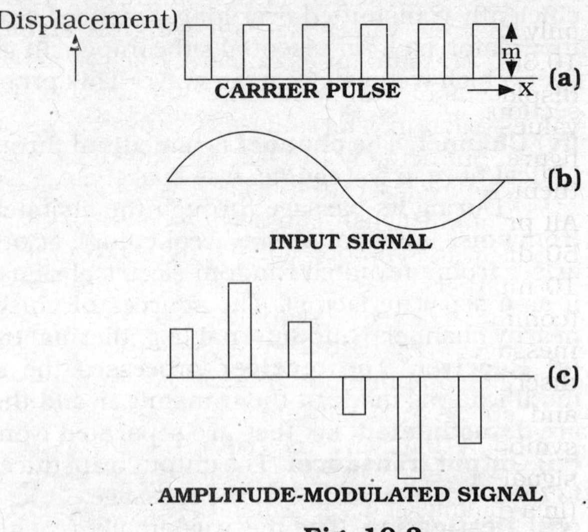

Fig. 10.3

Further, the input signal can first be **digitized** (or **discretized**) in binary code and then used to modulate the carrier signal. This will be explained in Section 10.04. It leads to the following additional kind of modulation:

(iv) Pulse code modulation (PCM)

This modulation is of great practical importance, because virtually all modern digital communication is binary. It affords simplicity, less distortion and ease of detection.

The pulse amplitude modulation (PAM) is illustrated in Fig. 10.3. The pulse amplitude varies with the input signal. In PDM and PPM, the pulse duration and the pulse position (i.e. time of rise or fall of the pulse), respectively, vary with the modulating signal.

When the modulated signal reaches the receiver's side, the original input signal is separated from the carrier wave. This "reverse process of recovery of the original input signal form the carrier signal" is known as **demodulation.** The device containing the electrical circuit for this purpose is called **demodulator.** Since it leads to the detection of the original message at the receiving end, alternatively the terms **detection** and **detector** are also used.

MODEM. It is a device which combines the functions of modulation and demodulation (**MO**dulation + **DEM**odulation) in one unit. So it can be used to transmit as well as to receive data. With the use of computers having become very common, the data transmission from one machine to another has become very necessary. The data is generated (by the so-called 'servers' and computers) in digital form, but it is transmitted (over a telephone line) in analog form. Therefore, at the generation end (e.g. a personal computer) the data is first converted from the digital to the analog form by the demodulation part of the modem and then transmitted over the telephone line. At the receiving end (e.g. another personal computer), it is converted back to the original digital form by the modulation part of the modem of the second computer. In this way, data (e.g. e-mail) can be sent by any one computer in a network to any other computer in the network, via ordinary telephone lines.

10.04. Analog and Digital Signals; Coding

"A signal which is a continuous function of time" is called **analog signal.** A sine wave, represented as y=a sin (ωt+φ) and depicted in Fig. 10.2 (b), represents such a signal. The displacement y varies continuously with time t, such that its maximum value is a (amplitude). In an electrical signal, the current (i) or the voltage (V) vary with time. Their values represent the strength of the signal. The analog signals are characterized by data whose values vary over a continuous range. For instance, the temperature over a certain place, e.g. Delhi, can vary over a wide range, covering almost an infinite number of possible values. The human voice, in which the amplitude can vary over a continuous range, also belongs to the same category.

"A signal which is discontinuous in time, taking only discrete values" is called **digital signal.** The Fig. 10.3(a) presents an example of such a signal; here the displacement can take only one of the two possible values 0 and m. The values are shown equispaced in this figure, but actually the spacing can differ widely among them. A physical example is the printed text on this page. All printed English language is constructed from about 50 discrete symbols, consisting of 26 alphabet letters, 10 numbers (0 to 9), a space, and punctuation marks (comma, colon, fullstop, etc). A Morse key telegraph message is a digital message made up of possible discrete combinations of just two symbols, viz. a mark and space. It is a **binary** message, involving just two symbols. Generally, the two possible values in a digital signal are represented by the **binary numbers** 0 and 1 (in a digital electrical signal, they correspond to the minimum and maximum values, respectively, of current or voltage). They are referred to as **bits.** A group of such bits is termed as a **binary word** or simply as a **byte.** We can have almost an infinite code combinations of the said two bits. For instance, a byte composed of 2 bits can have 4 possible code combinations, viz. 00, 01, 10, 11. In general, the total number (N) of possible code combinations is given by,

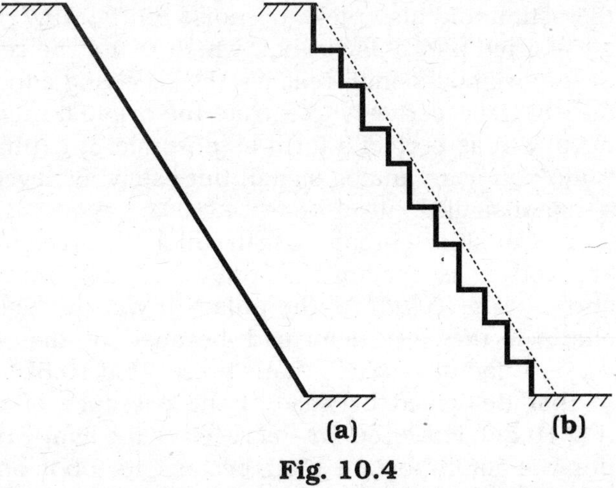

Fig. 10.4

$$N = 2^x$$

where x is the number of bits in a byte. Thus a 10 bits byte has the possible number of code combinations as $2^{10} = 1024$. A digital signal consists of a sequence of such code combinations. The process of expressing a digital signal in this way is called **coding.** (However, it may be borne in mind that an analog signal can also be coded).

Most of the present day communications are made using digital signals. The output of a personal computer (PC) is in digital form.

In a physical sense, the difference between the analog and digital forms is the same as that between a ramp and a staircase, shown in Figs. 10.4 (a) and (b), respectively. Either can be used for climbing up or coming down, but in one case the walk is continuous (smooth) and in the other case it is stepwise.

Additional knowledge

Analog to digital conversion. An analog signal can be converted into a digital signal. The conversion is not exact, but it is very approxiamatly so. The detailed theoretical treatment, based on the so-called 'sampling theorem' is complex and so beyond the scope of a school student. But we can understand it in a simple qualitative way. The analog signal, e.g the sine wave shown in Fig. 10.2 (b), is supposed to be divided into closely spaced 'samples' (looking as parallel vertical lines within the wave). Each 'sample' corresponds to a certain value of the displacement y [$y = a \sin(\omega t + \phi)$)]. In this sense it can be imagined as a waveform of amplitude equal to a particular y-value. Then it is 'quantized', i.e. its amplitude is approximated to an integral multiple of a certain minimum value. Thus, the change from one sample to the neighbouring sample is by a fixed amount. In this way the information is now digitized; each quantized sample can be coded into a sequence of binary numbers. (The envelope of the samples still very approximately looks like the original analog sine wave).

The situation is very similar (but not exactly the same) to that shown in Fig. 10.4 for a ramp and a staircase. If the steps in the staircase are very small in size, they lie very close to each other. Then the staircase very nearly looks like the smooth ramp. At the flat part of the step the value of the signal does not change, but it changes abruptly at the vertical part, i.e. there is a quantum jump of value equal to the height of the step, called step-size.

High Resistance of Digital Signals to Noise and Distortion; Use of Repeater Stations

We have seen that when a message is communicated over a channel, it undergoes some amount of

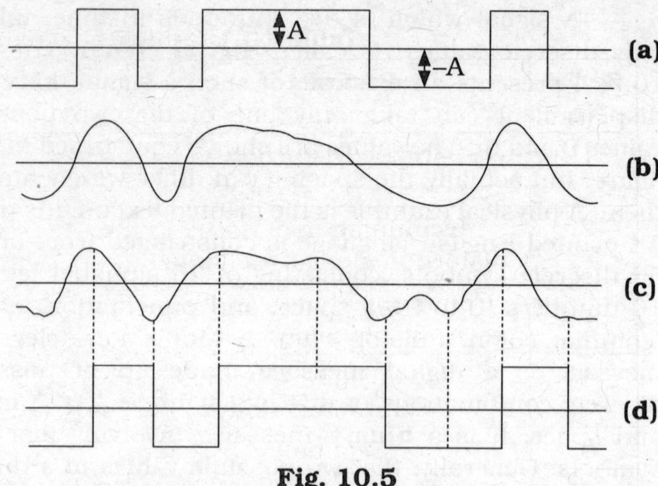

Fig. 10.5

distortion and also picks up noise on the way. Hence a distorted and noisy signal, which is not the replica of the original signal, reaches the receiving end. The task of the receiver is to extract the original message from it in as perfect a form as possible. It is difficult to do so for an analog signal, but easily achieved for a digital signal.

Consider a binary case in which the two symbols are coded as rectangular pulses of amplitudes A and –A (Fig.10.5a). As the pulses travel through the channel, they get deformed because of distortion due to factors other than noise (Fig.10.5b) and further deformed because of the presence of noise (Fig.10.5c). However, for the receiver the shape of the pulse is not important at all, because its job is only to select between the two possible pulses of amplitudes A and –A, respectively. Therefore, unless the said distortion and noise are excessive (then the pulses may be deformed beyond recognition), it can readily decide whether the received pulse is positive or negative. Thus, if the distortion and noise are within limits, the original data can be recovered correctly. The Fig. 10.5d shows the original signal recovered from the deformed signal (Fig. 10.5c), with slight delay.

In comparison, for an analog message the shape of the waveform is important. Even a small deformation due to distortion and noise seriously affects the detection of the received signal. Thus, while analog signals are highly sensitive to the presence of distortion and noise, the digital signals offer high resistance to their presence.

Use of repeater stations. For the digital signals, the said distortion and noise are further efficiently controlled by placing a number of repeater stations in the communication path. Each repeater station detects the incoming digital signal and transmits the so detected fresh 'clean' signal to the next station (the 'cleaning' is automatically achieved, as explained in the foregoing). The distance between two successive repeaters is arranged to be short enough to keep the noise and distortion within a reasonable limit. Such periodic cleaning at the repeating stations prevents any accumulation of noise/distortion along the channel. Hence the digital signals can be transmitted over long distances with high accuracy. Further, the strength of the signal is well maintained all along, by arranging for repeated renewal at each station.

For the analog signals, on the other hand, the said accumulation of noise distortion along the channel cannot be avoided. Further, as the signal advances along the channel, the problem worsens because of continuous natural attenuation (i.e. weakening) of the signal. In this way, while the signal continuously weakens with increasing distance, the noise/distortion grow stronger. Finally, the latter may grow so large that the message is mutilated beyond recognition. (Any amplification of the signal cannot help, because it equally amplifies the distortion/noise alongwith). Consequently, analog signals cannot be communicated over long distances unless they are made very powerful. Hence long distance communication of analog signals needs high power transmitters, which are expensive and therefore increase the cost of communication.

A comparison between analog and digital transmission is presented in Table 10.1. The reasons for the last three factors (numbered 4,5,6) will become clear in the next section. It is seen that the analog transmission is advantageous in all respects, except the channel bandwidth. However, in view of the now available optical fibre and satellite communication, which offer almost infinite bandwidth, the increased bandwidth required for digital signals is not really a problem. Thus, the digital transmission has a decisive overall superiority over the analog transmission. Hence it is increasingly preferred all over the world.

Table 10.1 Comparison between analog & digital transmission

Serial No.	Parameter	ANALOG Transmission	DIGITAL Transmission
1.	Noise/Distortion	Seriously affected by noise and distortion	Highly resistant to noise and distortion
2.	Accumulation of noise/ distortion along the channel	Not avoidable	Avoidable
3.	Communication over long distances	Possible only with high power transmitters	Possible with transmitters of much less power
4.	Quality and rate of communication	Relatively low	Relatively much higher
5.	Economy	Expensive	Relatively cheaper
6.	Available channel bandwidth	Relatively small	Large

10.05. Signal Bandwidth; Channel Bandwidth; Signal-to-Noise Ratio; Quality and Speed of Communication

(i) Signal Bandwidth. It is "the range of frequencies occupied by the signal". In general, no signal has just single frequency. The frequency spread of some common types of signal are listed in Table 10.2. Thus, for example, the speech signal roughly occupies the frequency range 100 Hz to 3 kHz.

Table 10.2. Range of frequencies occupied by some common signals

Serial No.	Signal	Frequency range
1.	Human Speech	100 Hz to 3 kHz
2.	Radio Broadcast	300 kHz to 3 MHz
3.	VHF (Very High Frequency) TV Broadcast	30 to 300 MHz
4.	UHF (Ultra High Frequency) TV Broadcast; includes a part of microwaves	300 MHz to 3 GHz

(ii) Channel Bandwidth (B). It is "the range of frequencies that the communication channel can transmit reliably". By a reliable transmission we mean that the quality and rate of communication remain reasonably good. Thus, for instance, if a channel can transmit with reasonable reliability a signal with frequency range covering 3 MHz to 8 MHz, the channel bandwidth (B) is 5 MHz.

(iii) Signal-to-Noise Ratio (SNR). It is simply **defined** as the "ratio of signal power (S) to noise power (N)". We have already seen that as a signal advances along the communication channel, it gets distorted and also noise is accumulated on the way, i.e. the value of N continuously rises. At the same time, the signal gets continuously attenuated (weakened), i.e. the value of S steadily falls. The net result is that the SNR continuously falls along the channel. Any amplification of the signal is of no help, because it amplifies the noise, too, in the same proportion. In fact, it leads to a further lowering of the SNR, because the noise of the amplifier is added to the noise level.

We shall now examine how the signal power (and therefore the SNR) and the channel bandwidth B vitally affect the quality and speed of communication.

(iv) Quality and speed of communication. The quality and speed of communication depend both on the channel bandwidth B and the signal power S. The detailed theoretical treatment is complex and so beyond the scope of the student. We shall restrict ourselves to an elementary qualitative treatment.

We first consider the role played by the channel bandwidth B. We can increase the speed of communication of a signal by reducing the time period of the signal, because then the time taken by the signal to travel a given distance in the channel is reduced. For example, if the period is reduced to half, the speed will be doubled. Since the time period of a waveform is inversely proportional to its frequency ($\nu \propto 1/T$), it implies that the frequency (and therefore the frequency range of the components of a signal) of the signal will be doubled. This, in turn, will mean that the bandwidth B of the channel will have to be doubled, because only then the channel will be able to transmit the signal without distortion. If the speed of communication has to be raised n times, B will also have to be increased n times. Thus we find that the rate of communication of a signal is directly proportional to B.

The role played by the signal power S is twofold. First, S directly determines the quality of communication, because an increase in the power of the signal naturally lowers the effect of disturbance caused by the channel noise N, i.e. the value of SNR goes up. Another associated benefit of the larger SNR value is that the signal can be transmitted over a longer distance. The second very important role of the signal power S arises from the fact that it is found to be related to the bandwidth B. This makes it possible to exchange B with S, i.e. B may be reduced to increase S and vice versa. Since S determines the value of SNR, it follows that B and SNR are exchangeable. However, the relationship between B and SNR is not simple. It is exponential in nature. A relatively small increase in the bandwidth B results in a large reduction in the signal power S. But a large increase in S yields only a small reduction in the bandwidth. Therefore, in practice, the exchangeability between B and S is mostly employed to increase B (for reducing the transmission power) and not the other way round. These results are of great practical significance in the performance of communication systems. A large value of S requires powerful transmitters, which are expensive to set up and maintain. Therefore an increase in S is avoided, unless found essential. The balance between B and S depends upon the requirement of the communication system under consideration. For instance, a great deal of power is available in a usual telephone system, but its bandwidth is small (nearly 3 kHz). In contrast, in space vehicles only limited power is available, but the availability of bandwidth is almost infinite (because a very wide range of frequencies can be transmitted reliably; vide Section 10.9). The communication system has to be designed accordingly.

10.06. Need for Modulation

As already mentioned in the beginning of Section 10.02, usually the input signal is not directly suited for long distance communication and therefore it needs to be suitably modulated. We shall now briefly describe some important reasons for the unsuitability of the input signal for direct transmission.

(i) Signal power. The audio signals are low frequency signals, viz. nearly 100 Hz to 3 kHz (Table 10.2). Hence they carry small energy ($E = h\nu$). A better parameter for understanding this aspect is the power radiated by an antenna (or aerial). In Section 5.05 (Chapter 5), we had described how an antenna connected to an LCR circuit acts as a source of electromagnetic waves. Theoretical treatment shows that the power radiated from a linear antenna of length ℓ is inversely proportional to the square of wavelength,

$$(\text{Power radiated})_{\text{antenna}} \quad \alpha \quad \left(\frac{\ell}{\lambda^2}\right)$$

Since the frequency $\nu \;\alpha\; (1/\lambda)$, it follows that for the same antenna length the power radiated by small frequency signals is relatively very low. Since long distance transmission needs high signal power, it is essential to use carrier waves of high frequency. In other words, it is essential to modulate the signal.

(ii) Size of the antenna. As already mentioned, the e.m. waves produced by an LCR circuit are radiated all around through an antenna. For even a barely minimum efficiency of radiation of electromagnetic energy by the antenna, it is essential that the size (length) of the antenna be at least about one-tenth of the wavelength of the carrier waves, i.e. nearly $\lambda/10$ in length (the student can better appreciate it by imagining the antenna to be in a horizontal position, along the direction of propagation of an e.m. wave, in Fig. 5.4, Section 5.05, Chapter 5). For many input signals, the wavelengths are too large, requiring antennas of impossible sizes. For example, for speech signals with frequencies between 100 Hz and 3 kHz, the corresponding wavelengths ($\lambda = c/\nu$) lie between 100 km and 3000 km, requiring minimum antenna lengths of 1 km to 30 km, which are practically impossible to produce. When carrier waves of high frequency are used, the necessary antenna size becomes much shorter. For instance, for $\nu = 3$ MHz, we have $\lambda = 100$ m, necessitating an antenna size of nearly 10 m, which is practically feasible to produce. Thus, once again we find it essential to employ high frequency carrier waves for the input signal, i.e. modulation is essential.

Desirable length of a dipole antenna. In Section 5.05 (Ch. 5), it was explained how the transmitting antenna is equivalent to an electric dipole, with its dipole moment varying along its length. These variations result in combined oscillations of electric field and magnetic field, which constitute the radiated e.m. waves.

As mentioned earlier, a workable efficiency of radiation of electromagnetic energy can be achieved with an antenna length of about $\lambda/10$. However, it is clear from Fig. 5.4 (Ch. 5) that for a full efficiency of radiation of energy, the length of the antenna should be equal to $\lambda/2$ (then the antenna can sense the time variation of the signal for full one cycle),

$$\ell = \lambda/2 \quad \text{(Dipole antenna)}$$

What applies to the transmitting (or , say, radiating) antenna, also holds for the receiving antenna.

(iii) Mutual interference of signals. Imagine several radio stations broadcasting their programmes together directly, i.e. without modulation of their input signals. The broadcasted signals would interfere with each other, because they all will have nearly the same bandwidth (about 100 Hz to 3 kHz). The only way to avoid such interference would be to allow only one radio station to broadcast in a given time interval. The same would hold for TV stations. This will not only be highly inconvenient (in fact nearly impossible, on a global scale) but also wasteful, because the channel bandwidths (~ a few MHz) available for the communication of the programmes are 3 to 4 orders of magnitude higher than the signal bandwidth.

The problem is readily solved by modulation of the signals. The signals from different stations modulate carrier waves occupying different ranges of frequency and each station is allotted one such frequency range for transmission. The channel bandwidth allowed to each station may be the same, e.g. 10 kHz, i.e. 0.01 MHz. Thus, if the stations are to broadcast around 2 MHz, the allotted frequency ranges may be 2.00 ± 0.005 MHz, 2.01 ± 0.005 MH_3, 2.02 ± 0.005 MHz, and so on. Then, the signals will not interfere with each other even when they are broadcasted simultaneously. At the receiving end, a signal can be easily separated from the mixed lot by using a tuner (a tunable band pass filter circuit). Thus, the available channel bandwidth is shared by numerous input signals without any mutual interference. The high frequency of the carrier waves easily makes such sharing possible.

Example 1. What should be the length of the dipole antenna for a carrier wave of frequency 3×10^8 Hz ? (CBSE; 2005).

Solution. Given: $\nu = 3\times10^8$ Hz

The required length of the antenna for a proper reception of the e.m. waves is given by, ($\nu\lambda = c$),

$$\text{Length} = \frac{\lambda}{2} = \frac{c}{2\nu} = \frac{(3\times10^8)}{2\times(3\times10^8)}$$

$$= 0.5 \text{ m (or 50 cm)} \quad \textbf{(Answer).}$$

10.07. Analytical Treatment of Amplitude Modulation

We assume that both the carrier wave and the input signal have sinusoidal form, as shown in Figs. 10.2a and 10.2b, respectively. Let their amplitudes be A_c and A_s, respectively, and their angular frequencies be ω_c and ω_s, respectively. Then the two may be represented as,

$$y_1 = A_c \sin \omega_c t$$

$$y_2 = A_s \sin \omega_s t$$

It may be safely assumed that the modulated wave, which is a combination of the high frequency carrier wave and the low frequency audio signal, will also have sinusoidal form (Fig. 10.2c). Its frequency will be the same as that of the carrier wave, but its amplitude will be a sum of the amplitude A_c of the carrier wave and the instantaneous displacement y_2 of the input signal. Thus, it may be represented as,

$$Y = (A_c + A_s \sin \omega_s t) \sin \omega_c t \quad \text{..... (10.1)}$$

$$= A_c \left(1 + \frac{A_s}{A_c} \sin \omega_s t\right) \sin \omega_c t$$

$$\Rightarrow \quad Y = A_c \sin \omega_c t + \mu_m A_c \sin \omega_c t \sin \omega_s t \quad \text{..... (10.2 a)}$$

$\mu_m = (A_s/A_c)$ is called **modulation index.** It is "the ratio of the amplitude of the input signal (A_s) to the amplitude of the carrier wave (A_c)". To minimise distortion, the value of μ_m is generally arranged to be ≤ 1.

Using the relation, $\sin \alpha \sin \beta = \frac{1}{2} [\cos (\alpha - \beta) - \cos (\alpha + \beta)]$, we get,

$$Y = A_c \sin \omega_c t + \frac{\mu_m A_c}{2} [\cos (\omega_c - \omega_s)t - \cos (\omega_c + \omega_s)t] \quad \text{..... (10.2 b)}$$

The Eq. 10.2 b shows that the modulated wave consists of the carrier wave of frequency ω_c plus two sinusoidal waves at frequencies $(\omega_c - \omega_s)$ and $(\omega_c + \omega_s)$, respectively, which are known as **side bands.** In Fig. 10.2c, they are represented by the upper envelope and the lower envelope (shown by broken lines), respectively, of the modulated wave. The displacement Y in Eq. 10.2b may be considered as a sum of three displacements: Y_1 (= A_c sin $\omega_c t$), Y_2 [= $(\mu_m A_c/2)$ cos $(\omega_c - \omega_s)t$], Y_3 [= $(\mu_m A_c/2)$ cos $(\omega_c+\omega_s)t$]. Since amplitude is maximum value of the displacement, it means that the values of amplitude for $\omega = \omega_c$, $(\omega_c - \omega_s)$ and $(\omega_c+\omega_s)$ are A_c, $\mu_m A_c/2$ and $\mu_m A_c/2$, respectively. A plot of the variation of amplitude of the modulated signal as a function of the angular frequency ω is shown in Fir. 10.6.

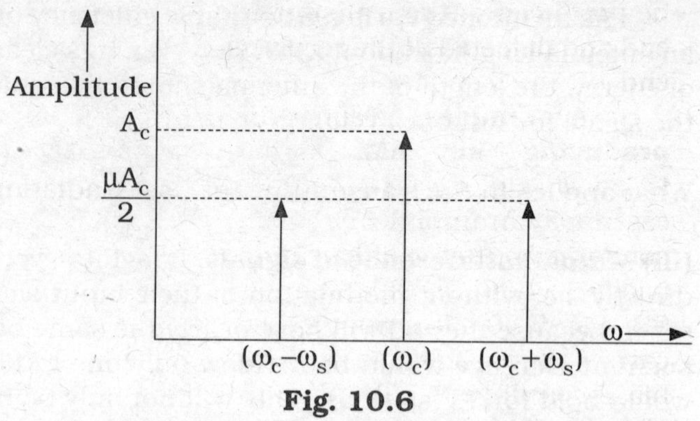

Fig. 10.6

In order to avoid mutual interference of frequencies (of carrier waves) used for broadcasting programmes from various radio stations, it is necessary that the frequencies are adequately separated from each other. Otherwise, the side bands will start overlapping, causing mutual interference of the transmitted signals, resulting in distortion of the signals.

The modulation index (μ_m) can also be expressed in terms of the maximum value (M_1) and the minimum value (M_2) of the amplitude of the modulated wave. In Eq. 10.1, they correspond to the maximum and minimum values of $(A_c + A_s \sin \omega_s t)$. Since the maximum and minimum values of sin $\omega_s t$ are +1 and –1, respectively, we have,

$$M_1 = A_c + A_s$$
$$M_2 = A_c - A_s$$
$$\Rightarrow \quad A_c = \frac{M_1+M_2}{2}; \quad A_s = \frac{M_1 - M_2}{2}$$
$$\therefore \quad \mu_m = \frac{A_s}{A_c} = \frac{M_1 - M_2}{M_1 + M_2} \qquad \text{..... (10.3)}$$

Example 2. A carrier wave of frequency 0.8 MHz and peak voltage of 12 V is modulated by an input signal of frequency 20 kHz and peak voltage of 9 V. Find the value of modulation index and the frequency locations of side bands.

Solution. Given: A_c = 12 V; A_s = 9 V; ν_c = 0.8 MHz = 800 kHz; ν_s = 20 kHz

The modulation index is given by,

$$\mu_m = \frac{A_s}{A_c} = \frac{(9)}{(12)} = 0.75 \quad \textbf{(Answer).}$$

The frequency locations of the side bands are,

(i) $(\nu_c - \nu_s)$ = (800) – (20) = 780 kHz **(Answer).**

(ii) $(\nu_c + \nu_s)$ = (800) + (20) = 820 kHz **(Answer).**

Thus, the side bands are found to be located at the ν–values 780 kHz and 820 kHz, respectively. **(Answer).**

Production and Detection of an Amplitude Modulated Wave

In the Section 10.03, we had mentioned that the process of modulation comprises the superimposition of the input signal on a high frequency carrier wave and that different kinds of modulation are possible. Now we shall briefly describe how an amplitude modulated (AM) signal can

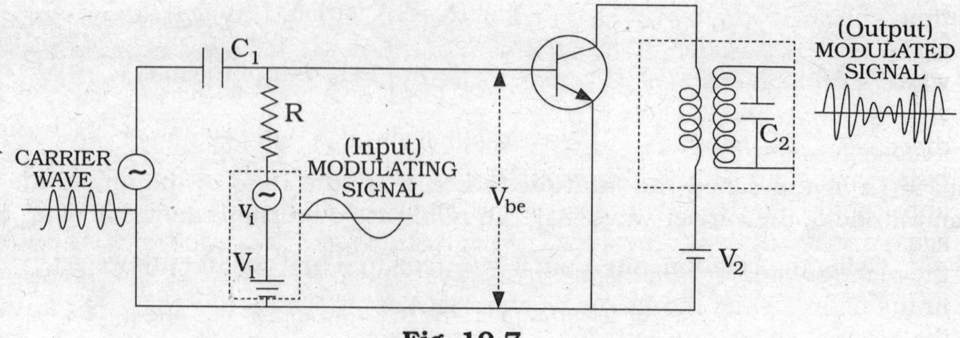

Fig. 10.7

be produced at the transmitter end and detected at the receiver end.

A simple circuit for **producing an AM signal** is shown in Fig. 10.7. It is essentially an amplifier circuit in common emitter configuration (vide Section 9B.08; Chapter 9B; Fig. 9.30). It amplifies the carrier wave signal. The base biasing voltage V_{be} is the sum of a constant voltage V_1 and the input sinusoidal signal voltage, v_i. The input signal modulates the carrier wave signal. As the base biasing voltage V_{be} varies, the amplified modulated signal voltage varies accordingly. That is how we get a time varying high frequency modulated signal as the output voltage (vide Figs.10.2 a, b, c).

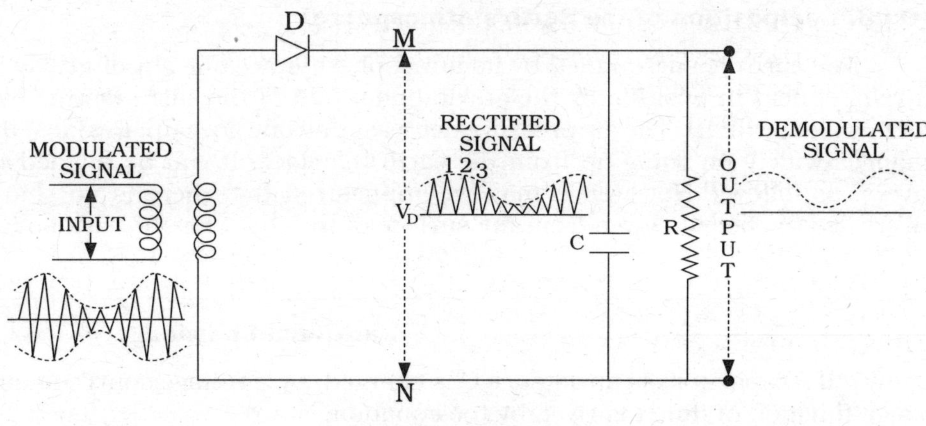

Fig. 10.8

Modulation index. It is the "ratio of the amplitude of the modulating input signal (A_m) to the amplitude of the carrier wave (A_c)",

$$m = \frac{A_m}{A_c}$$

The value of m is generally kept as less than 1 ($m \leq 1$). Otherwise it leads to a rise in distortion of the output signal.

The circuit employed for the **demodulation** (or **detection**) of the modulated signal, in order to get back the original signal, is shown in Fig. 10.8. The **input modulated signal** passes through a diode D, which allows only the positive half cycle of the signal to go through it, The **rectified signal** so obtained feeds an RC filter, which prevents the high frequency pulse-shaped component from passing through,. Thus, finally, the output demodulated signal consists of the enveloping original signal alone (compare with Fig. 10.2 a, b, c). The filter action of the RC-network can be qualitatively understood in relation to charging and discharging of the capacitor C, as follows.

The product RC has the dimensions of time. It is called the **time constant (τ)** of the capacitor C (or, of the circuit).

$$\tau = RC \qquad \text{..... (10.4)}$$

When a capacitor starts discharging, then within the time τ its charge falls to 37% of the initial value. For a typical case, the value of τ is a few millisecond. As seen in Fig. 10.8, the rectified signal voltage consists of the carrier wave pulses, like those marked as 1, 2, 3,.................., contained within an envelope (shown by stripes). As a pulse arrives, the capacitor starts charging. It charges upto the peak voltage of the pulse, after which it starts discharging to the minimum value (represented by the horizontal line drawn in the figure). Then the next pulse arrives and the cycle repeats itself. The value of RC is so chosen that,

$$\frac{1}{\nu_c} (= t_c) \ll \tau \qquad \text{..... (10.5)}$$

ν_c is the frequency of the carrier wave. Correspondingly, t_c is the time period of the wave, i.e. the time within which the next pulse arrives at a certain position in the path of the wave. Since the carrier wave has a high frequency, t_c is very much smaller than τ. Hence the voltage across the RC network changes very little when the capacitor discharges. Because one pulse is very rapidly replaced by the next pulse, sufficient time is not available for the capacitor to discharge appreciably. As the next pulse arrives, the capacitor starts charging again and the cycle is repeated. The net result is that the output voltage across the RC-network (which forms the demodulated original signal) roughly follows the outline of the envelope of the rectified signal, as shown in the figure. Thus the RC-network effectively filters out the high frequency carrier wave and permits only the original superimposed signal to pass through.

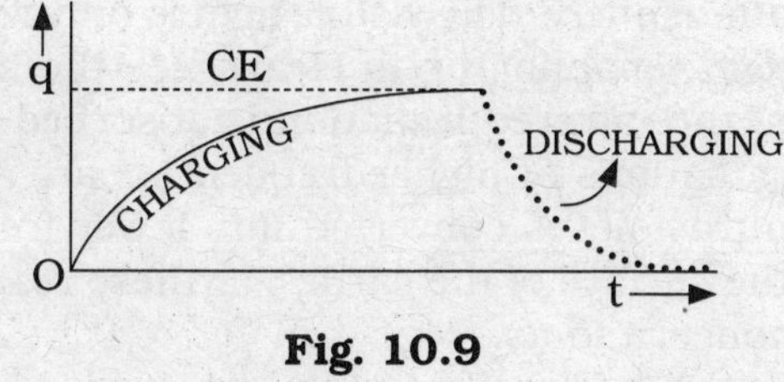

Fig. 10.9

10.08. Composition of the Earth's Atmosphere

The earth is surrounded by its atmosphere upto a height of nearly 500 km. It is a spherical gaseous envelope, held in position by the gravitational pull of the earth (as the earth rotates, the atmosphere also rotates alongwith it). The air pressure decreases as one goes up. Besides, the composition of the atmosphere changes widely on going up from the earth's surface. It can be divided into a number of thick spherical layers, as follows. However, it may be emphasized that there is no sharp division between two adjacent layers. So the heights given from the surface of the earth for the various layers are only a rough guide.

Additional knowledge

If a capacitor of capacitance C is charged, by a voltage source of emf E, its instantaneous charge q as a function of time t is given by the equation,

$$q = CE\,(1 - e^{-t/RC}) = CE\,(1 - e^{-t/\tau}) \qquad \text{..... (10.6)}$$

It is graphically represented in Fig. 10.9. The capacitor becomes fully charged to the value CE, as $t \to \infty$. Within the time τ (= RC), the charge on the capacitor becomes,

$$(q)_\tau = CE\,(1 - e^{-1}) = 0.63\ CE$$

For a capacitor with initial charge as q_0, the discharging through a resistance R is given by,

$$q = q_0\,(e^{-t/RC}) = q_0\,(e^{-t/\tau}) \qquad \text{..... (10.7)}$$

Within the time τ (= RC), the charge on the capacitor reduces to,

$$(q)_\tau = q_0\,e^{-1} = 0.37\ q_0$$

The discharging part is shown in the figure by dotted line.

(i) Troposphere (0 - 12 km). It is closest to the earth and affects the world's weather. Most of the water droplets, water vapour, and ice particles forming clouds are contained in this layer. The air density at its top is about one-tenth (10^{-1}) of the value near the earth's surface.

(ii) Stratosphere (12 - 50 km). The lower part of this layer (about 15 - 30 km) contains ozone, which is formed by the dissociation of molecular oxygen by the ultraviolet radiation from the sun. The density of air at the top of the stratosphere falls to just about one-thousandth (10^{-3}) of the value at the earth's surface.

(iii) Mesosphere (50 - 80 km). It is characterized by a rapid fall of temperature (down to about −100 °C) with height.

(iv) Thermosphere/Ionosphere (80 km upwards). This is the uppermost part of the atmosphere. However, the air density is very low, tending to fall to zero as one goes up. Here the short wave high energy radiation from the sun ionizes the gas atoms, creating electrons and positive ions. So this region has a high concentration of charged particles. That is why it is called **ionosphere.** However, different parts of the region have different concentrations of charge, named as Appleton layer, Heaviside layer, etc. As we shall see later, this region has a special role in communication, as it can reflect back the radio waves. The rest of the atmosphere largely consists of electrically neutral molecules.

The temperature in this region is higher than that in the mesosphere. Hence it is called **thermosphere.**

In free space, the radiation directed towards the earth comes from numerous sources in the cosmos. The main source of the e.m. radiation entering the earth's atmosphere is the sun. The radiation from the sun consists of e.m. waves of different frequencies (or wavelengths). Most of it is absorbed by the atmosphere. The visible light is only weakly absorbed. So the atmosphere is nearly transparent to the visible light, enabling us to clearly see the heavenly objects like the sun, moon, stars, and sky through it. Most of the infrared radiation is absorbed by the atmosphere and used up to heat it. The ultraviolet and other radiations of higher frequencies are absorbed by the ozone layer, which thus protects us from them. The ultraviolet is converted into lower frequency infrared radiation, which usefully heats the atmosphere and the surface of the earth. For these reasons, the suspected depletion of the ozone layer is a matter of serious concern to us.

10.09. Propagation of e.m. Waves in Atmosphere

The atmosphere plays an important role in all modern forms of telecommunication, involving radio broadcasts, television, mobile phones, computers, radar systems, etc. These means of communication require propagation of radio waves, including microwaves, of different frequencies through the atmosphere. In the following, we consider the propagation in increasing order of the linear frequency (ν) of the waves (or decreasing order of the wavelength (λ) of the waves). Remember, once again, that the given values of ν and λ are only a rough guide, not exact values.

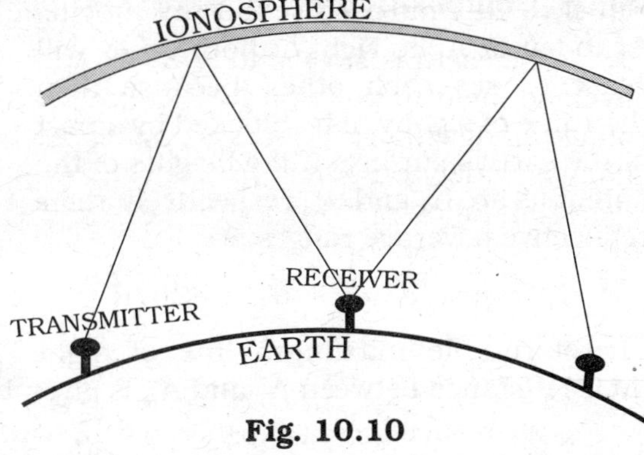

Fig. 10.10

(i) Ground (or Surface) wave propagation (ν: upto 3 MHz; $\lambda > 100$ m). Even at such relatively low frequencies, the radio waves from an antenna on the earth can travel fairly long distances along the surface of the earth. For instance, waves from a medium wave radio station with frequency range ~ 0.5 to 1.5 MHz can travel upto about 200 km. Beyond it, the continuous absorption of the waves as they travel weakens them appreciably. For frequencies > 2 MHz, the absorptions rises rapidly with frequency and so their transmission becomes very weak.

(ii) Sky wave (or Ionospheric) propagation (ν: 3 – 30 MHz; λ: 100 – 10 m). The waves of these frequencies moving upwards are reflected by the ionosphere and sent back to the earth at another place (remember that the waves radiated from an antenna get distributed spherically in all directions). In this way they effectively travel over long distances (Fig. 10.10). In fact, they can be made to travel round the earth by repeated transmission and ionospheric reflection. The receiver both receives and transmits back the waves.

(iii) Space wave propagation (ν: 30 MHz to 3 GHz; λ: 10 m to 10 cm). The waves of frequencies higher than about 30 MHz penetrate the ionosphere, which therefore cannot reflect them back to the earth. Hence they cannot be propagated as sky waves. The reception of these waves by the receiver antenna is only possible if (a) it directly intercepts the signal communicated by the transmitter antenna [called **line of sight communication (LOS)**] or (b) the transmitted signal is sent to a geostationary satellite, which directs it back to the earth (called **satellite communication**). These two modes of propagation of e.m. waves together are called **space wave propagation.**

Because of the earth's curvature, the signal from transmitting antenna can be received only within a limited distance d upon the earth (coverage range), which is related to the height (h) of the antenna and the radius of curvature (R) of the earth as,

$$d = \sqrt{2hR} \qquad \text{..... (10.8)}$$

Fig. 10.11

[The e.m. waves transmitted from the antenna A cannot be seen beyond the tangent points Q_1 and Q_2 (Fig. 10.11). They can be received on the earth's surface only within a radius $AQ_1 = AQ_2 = d$ (called 'radio horizon' of the transmitting antenna; it represents the coverage range of the antenna). In the figure, R is radius of earth and h is the height of the antenna. R = 6400 km (approx.) and h is usually a few hundred metre. Thus $R \gg h$ (note that the Fig. 10.11 is not drawn to scale). In the right-angled $\Delta\, AQ_1O$, we have,

$$AQ_1^2 + OQ_1^2 = AO^2 = (OP + PA)^2$$

$$\Rightarrow \quad d^2 + R^2 = (R + h)^2$$

$$\Rightarrow \quad d^2 = h^2 + 2hR = 2hR\left(\frac{h}{2R} + 1\right)$$

Since $R \gg h$, $\frac{h}{2R} \ll 1$ $\therefore$ on the right hand side, h/2R can be neglected, which gives,

$$d^2 = 2hR \quad \text{or} \quad d = \sqrt{2hR}.$$

The area covered by the transmitted signal is given by,

$$\text{Coverage Area} = \pi d^2 \qquad \text{..... (10.9)}$$

What would be the **maximum line of sight distance (d_M) between a transmitting antenna and a receiving antenna** (Fig. 10.12) ? In the figure, the two antennas A_T and A_R are just able to see each other. If d_M is reduced, they will continue to see each other. But if d_M is reduced, they will continue to see each other.

But if d_M is increased (either by moving A_T to left or A_R to right or both), they will cease to see each other, because then their line of sight will be blocked by a part of the earth's surface. If the heights of the antennas are h_T and h_R, respectively, their respective coverage ranges are,

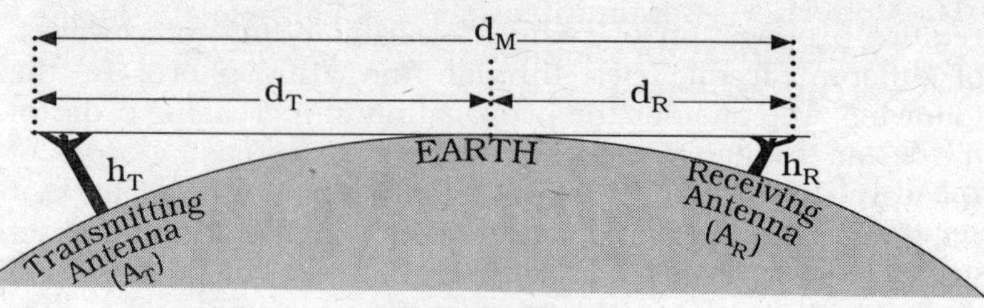

Fig. 10.12

$$d_T = \sqrt{2h_T R} \quad ; \quad d_R = \sqrt{2h_R R}$$

Therefore, the maximum line of sight (LOS) distance between A_T and A_R is given by,

$$d_M = \sqrt{2h_T R} + \sqrt{2h_R R} \qquad \text{..... (10.10)}$$

The television signals having frequencies in the approximate range 30-300 MHz [known as VHF (very high frequency) range] are transmitted and received in one of the aforesaid ways, viz. LOS communication and satellite communication. Because of the limited distance consideration, the TV towers are built as tall as possible, so that they can cover a wide region of transmission around them. The value of d is typically about 40 km. But the TV transmission via satellites can cover the entire surface of earth, as described in the next section].

Microwave propagation (ν: 1 GHz to 500 GHz; λ: 30cm to 1mm). This is another category of space wave propagation. As stated earlier, the microwave signals can be transmitted like a light beam, because by virtue of their small wavelengths these waves neither spread out nor bend round the corners of an obstacle. They are transmitted from one place on the earth to another via geostationary satellites, as described in the next section.

Example 3. Determine the height of a transmitting antenna, required to telecast TV programme over a coverage range of 32 km. Also find how much population the telecast will cover, if the average population density in the area around the TV tower is 500 persons per km^2. (Radius of earth = 6400 km).

Solution. Given: d = 32 km (or 32×10^3 m); R = 6400 km (or 64×10^5 m); Population density = 500 km^{-2}

The height (h) of the antenna is related to the coverage range (d) as,

$$d = \sqrt{2hR}$$

$$\Rightarrow \quad h = \frac{d^2}{2R} = \frac{(32\times10^3)^2}{2\times(64\times10^5)}$$

$$= 80 \text{ m} \quad \textbf{(Answer).}$$

$$\text{Population covered} = (\text{Coverage area})\times(\text{Population density})$$

$$= (\pi d^2)\times(500 \text{ km}^{-2})$$

$$= \pi\times(32 \text{ km})^2\times(500 \text{ km}^{-2})$$

$$= 16.1\times10^5 \text{ (or 16.1 lakh).} \quad \textbf{(Answer).}$$

Example 4. A TV music programme transmitted from a 50 m high antenna is received by another antenna of height 32 m. Calculate the maximum distance between the two antennas for LOS communication between them. (Radius of earth = 6400 km).

Solution. Given: h_T = 50 m; h_R = 32 m; R = 6400 km = 64×10^5 m

The required maximum distance for LOS mode of communication is given by,

$$d_M = \sqrt{2h_T R} + \sqrt{2h_R R}$$

$$= \sqrt{2\times(50)\times(64\times10^5)} + \sqrt{2\times(32)\times(64\times10^5)}$$

$$= 80\times10^2\times\sqrt{10} + 64\times10^2\times\sqrt{10}$$

$$= (144\times10^2)\times3.16$$

$$= 45504 \text{ m (or 45.5 km)} \quad \textbf{(Answer).}$$

10.10. Modern Communication Channels; Space Communication via Satellites; Line Communication via Optical Fibres

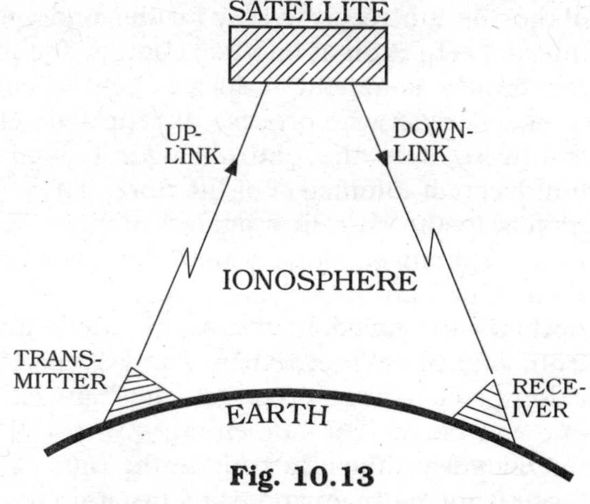

Fig. 10.13

In this section we shall very briefly describe two modern forms of space and line communication, which have resulted in faster and vastly improved quality of communication and lower costs. They have virtually revolutionised the communication systems.

As explained earlier in Sections 10.05 and 10.06, a large channel bandwidth permits simultaneous transmission of a large number of signals without mutual interference. Besides, it provides both fast and high quality communication. But such enormously large bandwidths are possible only at very high frequencies. Two modern communication channels are now available for communication at such high frequencies. One of them, viz. **satellite communication,** is a form of space communication and the other, **optical communication,** is a form of line communication.

(i) Satellite communication. This form of communication employs very high radio frequencies in the UHF and microwave region (vide Table 10.2, Section10.05). These radio waves can cross the ionosphere and reach a satellite located thousands of kilometer high in the sky. The satellite carries a device called **transponder,** which can receive the signal transmitted from a ground station on earth as well as transmit it back to earth (at a different frequency, to avoid any confusion). In this way, the signal from a ground station on earth is first **up-linked** with the satellite and then **down-linked** with another ground station on earth (Fig. 10.13).

Mostly geostationary satellites are used for such communications. As the student is aware, a geostationary satellite is one which appears stationary all the time with respect to the earth. It occupies a geosynchronous circular orbit in the equatorial plane of the earth. It is located very high above the earth's surface (~ 35, 000 km). Because of such a large height, a satellite can communicate over nearly one-third of the earth's surface. Thus, three geostationary satellites placed nearly 120° apart in the equatorial plane can provide communication all over the globe. [Two more orbits of lesser use for satellites are the so-called polar circular orbit (passing over or very near to the poles) and highly elliptical orbit. These satellites are located realtively much nearer to the earth and provide communication in limited regions].

(ii) Optical communication. It is the modern form of line communication. The conventional way of line communication, which also continues to be in use but progressively lesser (e.g. in telephonic and telegraphic communications), employs either a pair of twisted wires or coaxial cables as the communication channel. It has serious disadvantages at high frequencies, e.g. interference with other electromagnetic signals, creation of secondary sources of radiation, strong dielectric loss, etc. (Note that the space communication is free from such defects, because input signals travel in free space, without any point-to-point contact between the transmitter and the receiver). In the optical communication, the communication channel consists of the so-called **optical fibres,** to be described next. They offer many advantages as compared to the conventional line communication, which we shall list at the end of this section.

As the name implies, the optical communication employs light waves as the carrier waves, i.e. the carrier waves have optical frequencies. The optical frequencies in the visible region are extremely high (~ 10^{15} Hz) as compared to the radio waves and microwaves frequencies (~ 10^5 Hz to 10^{11} Hz). They are even higher in the ultraviolet region. Hence a light wave can carry several thousands time or more information than the radio waves and microwaves. The flow of information traffic in the present day world is already very high and it is likely to grow further. Only the light waves may be able to cope with it.

Optical Fibres. An optical fibre is an extremely thin thread-like structure. The thickness is ~ 50 μm (about 0.05 mm). It can guide a light beam entering it over long distances, based on the phenomenon of **total internal reflection.** The fibre consists of a cylindrical central core of glass cladded by a dielectric material of slightly lower refractive index. If the refractive indices of glass and the cladding material are μ_1 and μ_2, respectively, then $\mu_1 > \mu_2$. A small part of an optical fibre is shown in Fig. 10.14. For a light ray entering the fibre, if the angle of incidence θ at the glass-cladding interface is greater than the critical angle θ_c [$= \sin^{-1} (\mu_2/\mu_1)$], the ray undergoes total internal reflection at the interface. Because of the cylindrical symmetry of the fibre, the reflected ray further undergoes total internal reflection at another point of the interface. Since there is almost no loss of light intensity in a total reflection, the process of repeated reflections continues over the entire length (which can be hundreds of kilometer) of the fibre. Thus, the light travels in the core in a guided manner, with little loss of intensity, while remaining confined to the core. Thus, an optical fibre works as an excellent **optical waveguide.** It effectively guides a light beam from one place to another. The Fig. 10.15 shows a long rolled fibre, in which the light entering at one end on the left side emerges with little loss of intensity from the other end at the right. The small loss of intensity during transmission results from the little scattering and absorption of light within the glass core.

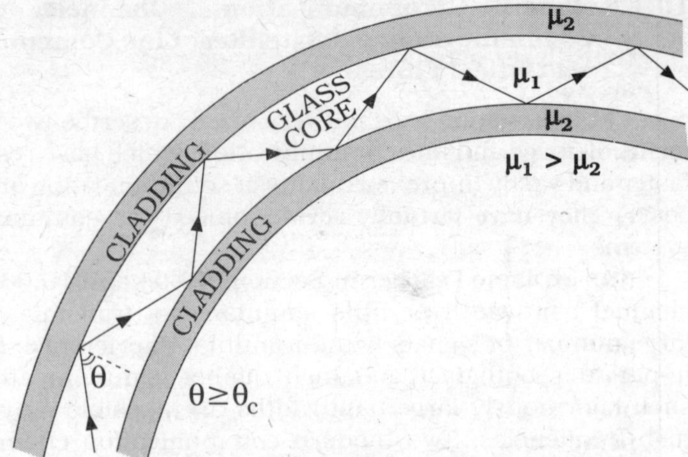

Fig. 10.14

A large number of optical fibres can be put together to form a **bundle** of fibres. A bundle may consist of hundreds of thousands of fibres. It can be rolled up like a coil of electric wires or cable.

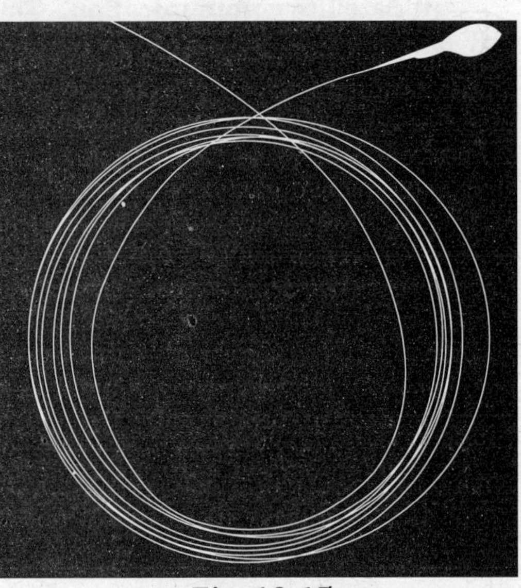

Fig. 10.15

Light source. For optical communication, a small but high intensity source of monochromatic light is needed as **transmitter.** A semiconductor light emitting diode **(LED)** or a miniature semiconductor laser (called **laser diode**) are suitable sources. The latter is superior but expensive. The varying voltage or current from the input signal **modulates** the light waves from the source. At the receiving end, a miniaturised semiconductor photodetector, e.g. a photo diode, acts as the **demodulator** (or **detector**).

Advantages of optical communication over the conventional two-wire or cable communication

1. It provides extremely large channel bandwidth, resulting in (a) an extremely high information carrying capacity (hundreds of thousands of signals per kilometer), (b) very fast communication of information, and (c) high quality of communication (low noise and distortion).

2. The transmission power losses are comparatively very low, being less than 5% per kilometer. Hence the distance between successive repeater stations can be as large as nearly 250 km (cf. a few kilometers spacing for the conventional line communication).

3. The cost of material (glass) is much lower than the conventionally used copper. Besides, while the supply of copper on the earth is limited, the supply of glass (which is basically silica) is almost unlimited.

4. The dielectric material of the cladding makes the optical communication immune to interference from other signals.

5. Unlike the space communication, the transmitted information remains confined to the fibre. Hence it cannot be tampered with easily. Thus, secret information, pertaining to defence, banking, etc., can be safely transmitted.

Additional knowledge

(a) The idea of using light waves for communication is not new. Way back in 1880, Graham Bell demonstrated it in a remarkable experiment, with a suitable illuminated diaphragm as the transmitter and a selenium cell plus a telephone as the receiver. The difficulties lay with a suitable light source (until laser arrived) and the scattering/absorption of light in passage through air (until optical fibres arrived).

(b) Plastics are not suitable for making optical fibres, because they have high scattering/absorption of light passing through them.

(c) In addition to long distance communication, now the optical fibres are being extensively employed for local networking of telephones, TV's, computers, and robots in offices and cities.

(d) Besides communication, the optical fibres have other important uses. For instance, optical fibre medical probes called **endoscopes** enable doctors to examine inner parts of human body.

(e) The rods and cones that constitute the retina of humani eye have a similar structure as that of an optical fibre. They consist of minute cylindrical rods of a dielectric material surrounded by another dielectric of slightly lower refractive index [core diameters measure just a few microns (μm)]. The light absorbed in these 'optical guides' produces electrical signals, which ae transmitted to the brain through nerves.

(f) The branch of physics dealing with optical fibres has come to be known as **fibre optics.**

Additional knowledge

The fast progress made in the communication systems in the last few decades has virtually revolutionised our lives, giving a new turn to the way we live. Some popular remarkable advancements are mentioned in the following.

(i) Internet. It is an **inter**national **net**work of computer installations, enabling mutual sharing of all kinds of information by billions of users all over the world. It is accessible to anybody through a MODEM link (which involves the use of telephose lines) to one's computer. It was initially developed in the 1970's for military purposes and steadily grew to become available for public use in 1990's. In combination with TV, telephone, and conventional publishing, a new term **cyberspace** has been coined to represent a world of informatin in digital form. It signifies a notional environment, provided by the modern communication systems.

The internet provides a variety of services, some of which are given below.

(a) World Wide Web (www). It is an extensive information system, designed for multimedia use. Computer installations store specific informations, which can be accessed by others on the so-called **websites.** Individuals or organizatinos make available information about their activities and services on their respective websites to the users, mostly free of cost but sometimes paid.

Search engines like Yahoo, Netscape, Google, etc. help the users to access the sites. A so-called **hypertext** provides an automatic linkage bettween the various sets of information (documents).

(b) e-mail (electronic mail). It is a system of sending typed messages by electronic means from one computer user to another, through ISP's (Internet Service Providers). Since the information travels with the velocity of light, the delivery of the 'mail' is almost instantaneous, unlike the conventional postal system.

(c) Chat. It is an exchange of information between friends or people with common interests through typed messages, thus electronically simulating verbal conversation. The messages are received almost instantaneously.

(ii) Mobile phone (or cell phone). It is a very convenient portable pocket telephone, based on a cellular radio system. The system employs local low-power radio transmitters, covering small areas (cells), measuring a few square kilometers or less, depending on the numbre of customers. The transmitters receive and transmit calls in association with the telecommunication network. When a cell phone user crosses from one cell to another, the transfer is so smooth that it goes unnoticed by the user. The frequencies employed are in the UHF range (~1 GHz). Recently, satellite telephony has started coming in use, which does away with the need of cells.

(iii) FAX. It is short form of **facsimile,** which means 'an exact copy'. A fax machine scans the contents of a document (e.g a hand-written or typed page, a photograph, etc). and converts them into electronic signals. The signals are transmitted via telephone line to another FAX machine at the receiver's end, where they are reconverted as a replica (i.e. an exact copy) of the original document.

Solved Conceptual Problems
[Including HOTS Questions]

Problem 1. Briefly mention the main reasons which make it essential to modulate a signal for communication.

Solution. The modulation of a signal means superimposing the signal on a carrier wave of high frequency. It is required for the following reasons.

(a) The power radiated from an antenna is directly proportional to the length (ℓ) of the antenna and to the square of the frequency ν of the wave,

$$(\text{Power radiated})_{\text{antenna}} \;\alpha\; (\ell\nu^2)$$

The audio signals are low frequency signals (~ 100 Hz to 3 kHz). Hence they carry small energy ($E = h\nu$). But long distance transmission needs high signal power, which is possible only with the carrier waves of high frequencies, which are 3 to 4 orders of magnitude higher than the signal frequencies.

(b) For an effective radiation (and reception) of the e.m. waves, the length of the antenna should be at least $\lambda/10$ (ideally it should be $\lambda/2$). Since the speech (audio) signals have relatively low frequencies (~100 Hz to 3 kHz) and therefore very long λ's (~100 km to 3000 km), they need minimum antenna lengths of ~1 km to 30 km, which are practically impossible to make. But the antenna lengths needed for high frequency carriers waves are nearly 10 m or less, which are possible to produce.

(c) A direct broadcast of the audio signals together from hundreds of radio and TV stations around the world will make them to mutually interfere, since they will have nearly the same bandwidth (~100 Hz to 3 kHz). With high frequency carrier waves, which have frequencies 3-4 orders of magnitude higher (~a few MHz), the channel bandwidth is so very high that it can be easily shared between the stations, by allotting a specific frequency range to each station. Hence there can be no mutual interference of the signals, even when all the stations broadcast together.

Problem 2. What are the causes for distortion of a signal during its communication ? What is meant by noise ? Mention some of its possible sources.

Solution. The causes for distortion are noise, attenuation and other influences.

The noise means a disturbance to the signal. It arises from unwanted random electrical signals. The possible source can be both external (e.g. interference from signals carried by nearby channels, faulty electrical contacts, automobile engine ignition, etc.) and internal (e.g. thermal motion of electrons in conductors, random emission of radiation, etc.)

Problem 3. Briefly explain how the quality and speed of communication of a signal is vitally affected by the channel bandwidth and the signal power.

Solution. (i) Channel bandwidth: For increasing the speed of communication of a signal the frequency of the signal (and therefore the frequency of the components of the signal) needs to be increased. This, in turn, requires the channel bandwidth to be increased proportionately, because only the channel will be able to transmit the signal without distortion. Thus, the speed of communication of a signal is directly proportional to the channel bandwidth.

(ii) Signal power: An increase in the power of the signal reduces the effect of channel noise. Thus, the signal power (S) directly determines the quality of communication. (Another side benefit is that the signal can be transmitted over a longer distance, in spite of a continuous loss in its strength as it travels).

Another advantage arises from the fact that the channel bandwidth (B) is found to be related to the signal power (S). Thus, B and S are exchangeable to some extent. This helps in designing communication systems according to their specific requirements.

Problem 4. Explain why is it necessary to employ short waves for long distance radio broadcasting.

Solution. The ground wave propagation of radio waves along the surface of earth is limited to just about 200 km. Besides, for frequencies >2 MHz, the absorption of the waves rises rapidly with frequency and so their transmission becomes very weak.

The short waves (in an approximate frequency range 3-30 MHz) moving upwards are reflected by the ionosphere and sent back to earth. Therefore they can be made to travel long distances round the earth by repeated transmission and ionospheric reflection (sky wave propagation).

Problem 5. Why ground wave and sky wave propagations are not suitable for transmission of TV signals ? What modes of communication are employed for them ? Why TV transmission towers are made tall ?
Solution. The TV signals cover an approximate frequency range of 30 MHz to 3 GHz. The ground wave propagation (along the surface of earth) is not suitable for them, because waves of frequency >2 MHz are rapidly absorbed in transmission. The sky wave propagation also does not suit because such energetic waves ($E = h\nu$) penetrate the ionosphere, which therefore cannot reflect them back to earth.

The reception of these waves (by the receiver antenna) is only possible through space wave propagation which consists of two possible modes: (i) line of sight (LOS) communication, in which the receiver antenna directly intercepts the signal communicated by the transmitting antenna and (ii) satellite communication, in which the transmitted signal is sent to a geostationary satellite, where it is received as well as sent back to earth, through a so-called transponder (a satellite, placed ~ 35000 km above earth, can communicate over nearly one-third of the earth's surface).

Because of the earth's curvature, the signal from a transmitting antenna can be received only within a limited distance (coverage range) d, which is related to the height (h) of the antenna and the radius of curvature (R) of the earth as,

$$d = \sqrt{2hR} \quad \text{(coverage range)}$$

To have a large coverage range (d), it is necessary that h is a large value, i.e. the TV tower should be as high as possible.

Solved Numerical Problems

Problem 1. A T.V. tower has a height of 400 m at a given place. Calculate its coverage range, if the radius of the earth is 6400 km. (CBSE; 2004).
Solution. Given: $h = 400$ m; $R = 6400 \text{ km} = 64\times10^5$m
∴ The required coverage range is,

$$\begin{aligned} d &= \sqrt{2hR} = \sqrt{2\times(400)\times(64\times10^5)} \\ &= 16\times10^3\times\sqrt{20} = 16\times10^3\times4.47 \\ &= 71.5\times10^3 \text{ m (or 71.5 km)} \quad \textbf{(Answer)}. \end{aligned}$$

Problem 2. A transmitting antenna at the top of a tower has a height of 36 m and the height of the receiving antenna is 49 m. What is the maximum distance between them, for satisfactory communication in the LOS mode ? (Radius of earth = 6400 km).
Solution. Given: $h_T = 36$ m; $h_R = 49$ m; $R = 6400 \text{ km} = 64\times10^5$ m
The required maximum distance is,

$$\begin{aligned} d_M &= \sqrt{2h_T R} + \sqrt{2h_R R} \\ &= \sqrt{2\times(36)\times(64\times10^5)} + \sqrt{2\times(49)\times(64\times10^5)} \\ &= (48\times10^2)\times\sqrt{20} + (56\times10^2)\times\sqrt{20} \\ &= (104\times10^2)\times4.47 = 446.88\times10^2 \text{ m} \\ &= 46.5 \text{ km} \quad \textbf{(Answer)}. \end{aligned}$$

Problem 3. If the coverage range of a TV tower of height 200 m has to be increased 1.5 times, by what amount its height will need to be increased ? What will be the coverage area after increased height ? (Radius of earth = 6400 km).
Solution. Given: $h = 200$ m; $R = 6400 \text{ km} = 64\times10^5$ m
Let the increased height be h′ and the increased coverage area be d′. Then,

$$d = \sqrt{2hR} \quad ; \quad d' = 1.5d = \sqrt{2h'R}$$

Dividing, we get, $\sqrt{\frac{h'}{h}} = 1.5$

$$\Rightarrow \quad h' = (1.5)^2 \times h = 2.25 \times (200)$$
$$= 450 \text{ m}$$

$\therefore$ Increase in height $= (450 - 200)$ m
$= 250$ m **(Answer).**

The new coverage area, with increased height h′, will be,

$$\text{New coverage area} = \pi(d')^2 = \pi\,(1.5\,d)^2$$
$$= \pi\,(1.5 \times \sqrt{2hR})^2 = \pi \times (1.5)^2 \times 2hR$$
$$= 3.14 \times 2.25 \times 2 \times (200) \times (64 \times 10^5)$$
$$= 1808.6 \times 10^7 \text{ m}^2 = 1808.6 \times 10^7 \times 10^{-6} \text{ km}^2$$
$$= 18086 \text{ km}^2$$ **(Answer).**

Problem 4. The values of maximum and minimum amplitudes of an amplitude-modulated wave are found to be 16 V and 4 V, respectively. (i) What is the magnitude of modulation index ? (ii) What will be the magnitude if the minimum amplitude is reduced to zero ?

Solution. Given: $M_1 = 16$ V; $M_2 = 4$ V

(i) The modulation index is given by,

$$\mu_m = \frac{M_1 - M_2}{M_1 + M_2} = \frac{(16) - (4)}{(16) + (4)}$$
$$= 0.6$$ **(Answer).**

(ii) For $M_2 = 0$, we have,

$$\mu_m = \frac{M_1 - 0}{M_1 + 0} = \frac{M_1}{M_1} = 1$$

Thus, when $M_2 = 0$, the value of the modulation index is always equal to 1, irrespective of the value of M_1. **(Answer).**

Very Short Answer Questions

[Including HOTS Questions]

(1 mark each)

1. What is meant by modulation of a signal ?

2. What is meant by demodulation of a signal ?

3. What is a MODEM ?

4. Distinguish between an analog signal and a digital signal.

5. Diagrammatically represent an analog signal and a digital signal.

6. What is meant by signal bandwidth ?

7. What is meant by channel bandwidth ?

8. What is meant by signal - to - noise (SNR) ratio ?

9. What is modulation index ?

10. What are side bands in a modulated signal ?

11. How is the effective power radiated by a transmitting dipole antenna related to the antenna length and the frequency of e.m. waves ?

12. Which layer of the earth's atmosphere provides us protection against the ultraviolet radiation in the sunlight ?

13. In the absence of the earth's atmosphere, will the average temperature on earth be more or less than the existing value ? [Hint: Less, because greenhouse effect will disappear and so the infrared radiation from earth will be radiated out].

14. Name the various layers comprising the earth's atmosphere.

15. Why is ionosphere so named ?

16. Which layer of the earth's atmosphere is relevant to long distance radio broadcasts on earth ?

17. Which mode of communication will be suitable for sending a signal of frequency 100 kHz over a distance of 50 km ?

18. What is a geostationary satellite ?

19. How are the microwaves specially useful in satellite communication ? [Hint: By virtue of their small λ's they neither spread out much nor bend round corners much].

20. On which phenomenon is the long distance propagation of light through optical fibres based ?

21. What is the function of cladding in an optical fibre ?

Short Answer Questions

[Including HOTS Questions]

(2 or 3 marks each)

1. What do the terms message, transducer, modulator, channel, and receiver mean ?

2. Draw a labelled block diagram of a communication system.

3. Briefly explain the function of a MODEM.

4. Briefly explain the causes that make it essential to modulate a signal for communication.

5. Give an analytical treatment of amplitude modulation of a signal, explaining how side bands are created. Graphically represent the variation of amplitude with angular frequency (ω).

6. Briefly describe the transmission of signals through satellite communication.

7. Enumerate the advantages of optical communication over the conventional cable communication.

8. Describe with the help of a labelled diagram the construction and working of an optical fibre.

Long Answer Questions

(5 marks each)

1. Describe, with necessary diagrams, why digital transmission is preferred over analog transmission. Also, present the advantages of the former over the latter in a tabular form.

2. Describe, with necessary labelled circuit diagrams, the processes of production and detection of an amplitude-modulated wave.

Numerical Exercises

1. What should be the length of a dipole antenna for a carrier wave of frequency 6×10^8 Hz ? **[25 cm]**

2. A T.V. tower has a height of 500 m at a given place. Calculate its coverage range, if the radius of the earth is 6400 km. (CBSE; 2004). **[80 km]**

3. The average population density around a TV tower is 1000 km^{-2}. If the height of the tower is 120 m, how much population will be covered by the telecast from the tower ? (Radius of earth = 6400 km) [Hint: Population coverage = Population density$\times(\pi d^2)$ = Pop. density$\times\pi\times 2hR$ = 1000 km$^{-2}\times 3.14\times 2\times 0.12$ km$\times$6400 km]. **[48.2 lakh]**

4. An input signal modulates a carrier wave of peak voltage 15 V. If the modulation index is 60%, what is the peak voltage of this modulating signal? [Hint: μ_m = 60% = 0.6]. **[9 volt]**

Appendices

APPENDIX A

THE GREEK ALPHABET AND THE PREFIXES USED FOR VARIOUS MEASUREMENTS

The Greek Alphabet

LETTER		SYMBOL	
ENGLISH	**HINDI**	**SMALL**	**CAPITAL**
Alpha	ऐल्फा	α	A
Beta	बीटा	β	B
Gamma	गामा	γ	Γ
Delta	डेल्टा	δ	Δ
Epsilon	ऐप्साइलौन	ϵ	E
Zeta	ज़ीटा	ζ	Z
Eta	ईटा	η	H
Theta	थीटा	θ	Θ
Iota	आयोटा	ι	I
Kappa	कप्पा	κ	K
Lambda	लैम्बडा	λ	Λ
Mu	म्यू	μ	M
Nu	न्यू	ν	N
Xi	ज़ाई	ξ	Ξ
Omicron	औमिक्रौन	o	O
Pi	पाई	π	Π
Rho	रो	ρ	P
Sigma	सिगमा	σ	Σ
Tau	टाउ	τ	T
Upsilon	अप्साइलन	υ	Y
Phi	फाई	ϕ, φ	Φ
Chi	काई	χ	X
Psi	साई	ψ	Ψ
Omega	ओमिगा	ω	Ω

The Prefixes used for Various measurements [e.g. nanometer (nm)]

PREFIX		SYMBOL	FACTOR DENOTED
ENGLISH	**HINDI**		
yocto	यौक्टो	y	10^{-24}
zepto	ज़ैप्टो	z	10^{-21}
atto	ऐट्टो	a	10^{-18}
femto	फैम्टो	f	10^{-15}
pico	पीको	p	10^{-12}
nano	नैनो	n	10^{-9}
micro	माइक्रो	μ	10^{-6}
milli	मिली	m	10^{-3}
centi	सैन्टी	c	10^{-2}
deci	डैसी	d	10^{-1}
deka	डैका	da	10^{1}
hecto	हैक्टो	h	10^{2}
kilo	किलो	k	10^{3}
mega	मैगा	M	10^{6}
giga	गीगा	G	10^{9}
tera	टैरा	T	10^{12}
peta	पैटा	P	10^{15}
exa	ऐक्सा	E	10^{18}
zetta	ज़ैट्टा	Z	10^{21}
yotta	यौट्टा	Y	10^{24}

APPENDIX B

THE INTERNATIONAL SYSTEM OF UNITS (SI)

(i) Base Units

QUANTITY		UNIT		DIMENSION
NAME	SYMBOL	NAME	SYMBOL	
Length	ℓ,L	metre	m	L
Mass	m,M	kilogram	kg	M
Time	t	second	s	T
Electric current	i	ampere	A	I
Temperature	T	kelvin	K	K
Amount of substance	m,M	mole*1	mol	M
Luminous Intensity	I	candela*2	cd	–

(ii) Supplementary Units

QUANTITY	UNIT NAME	SYMBOL
(Plane) Angle	radian	rad
Solid Angle	steradian	sr

*1 Equal to the quantity containing as many elementary units as there are atoms in 0.012 kg of carbon –12.

*2 This unit has not been used in the present book.

(iii) Derived Units of Well known physical Quantities

QUANTITY	UNIT NAME and/or OTHER IDENTITY	SYMBOL	DIMENSIONS
Area	m^2	**A**	L^2
Volume	m^3	**V**	L^3
Density	kg/m^3	**ρ**	M/L^3
Velocity	m/s	**v,u**	L/T
Acceleration	m/s^2	**a**	L/T^2
Force	newton (N)	**F**	ML/T^2
Pressure	pascal (Pa; N/m^2)	**P,p**	M/LT^2
Momentum	kg.m/s	**p**	ML/T
Work	joule (J; N.m)	**W**	ML^2/T^2
Energy	joule (J; N.m)	**E,U,K**	ML^2/T^2
Power	watt (W; J/s)	**P**	ML^2/T^3
Gravitational constant	$N.m^2/kg^2$	**G**	L^3/MT^2
Angular velocity	radian/s	**ω**	T^{-1}
Angular acceleration	radian/s^2	**α**	T^{-2}
Angular momentum	$kg.m^2/s$	**ℓ**	ML^2/T
Torque	N.m	**τ**	ML^2/T^2
Moment of inertia	$kg.m^2$	**I**	ML^2
Period	s	**T**	T
Frequency	hertz (Hz; s^{-1})	**ν**	T^{-1}
Angular frequency	radian/s	**ω**	T^{-1}
Wavelength	m	**λ**	L
Young's modulus	N/m^2	**Y**	M/LT^2
Bulk modulus	N/m^2	**B**	M/LT^2
Shear modulus	N/m^2	**η**	M/LT^2
Surface tension	N/m	**S**	M/T^2
Coefficient of viscosity	$N.s/m^2$	**η**	M/LT
Light intensity	W/m^2	**I**	M/T^3
Heat (quantity)	joule (J; N.m)	**Q**	ML^2/T^2
Specific heat	J/kg.K	**c**	L^2/T^2k
Stefan's constant	$W/m^2.K^4$	**σ**	M/T^3K^4
Thermal conductivity	W/m.K	**K**	ML/T^3K
Charge (quantity)	coulomb (C; A.s)	**q,Q**	IT
Potential (or voltage)	volt (V; J/C)	**V**	ML^2/IT^3
Current density	A/m^2	**J**	I/L^2
Electrical conductivity	1/Ω.m (mho/m)	**σ**	I^2T^3/ML^3
Dielectric constant	—	**K**	—
Electric dipole moment	C.m	**p**	LIT
Electric field	V/m (N/C)	**E**	ML/IT^3
Electric flux	V.m	**ϕ**	ML^3/IT^3
Electric resistance	ohm (Ω; V/A)	**R**	ML^2/I^2T^3
Inductance	henry (H; V. s/A)	**L**	ML^2/I^2T^2
Capacitance	farad (F; A.s/V)	**C**	I^2T^4/ML^2
Permittivity constant	$C^2/N.m^2$ (F/m)	ϵ_0	I^2T^4/ML^3
Permeability constant	N/A^2	μ_0	ML/I^2T^2
Electromotive force	volt (V)	**E**	ML^2/IT^3
Magnetic dipole moment	N.m/T	p_m	IL^2
Magnetic field	tesla (T; Wb/m^2)	**B**	M/IT^2
Magnetic flux	weber (Wb; V.s)	ϕ_B	ML^2/IT^2

APPENDIX C

SOME USEFUL CONVERSION FACTORS

Mass
1 kg = 1000 g = 6.02×10^{26} a.m.u. (or u)
1 a.m.u. (or u) = 1.66×10^{-27} kg
1 lb (pound) = 16 0z (ounce) = 453.6 g
1 ton = 2000 lb = 907.2 kg
1 tonne (metric ton) = 1000 kg

Length
1m = 10^3 mm = 10^2 cm = 10^{-3} km
1nm = 10^{-9} m= 10Å (angström); 1Å = 10^{-10} m
1yard = 3ft (feet); 1ft (foot) = 12 inch = 30.48 cm; 1 inch = 2.54 cm
1mile = 1.61 km = 1760 yard; 1 nautical mile = 1.15 mile

Area
1m^2 (square metre) = 1.196 sq. yard
1 acre = 4840 sq. yard; 1 hectare = 10^4m^2 = 2.47 acre

Volume
1L (litre) = 10^3 c.c. (cubic centimetre)
1 U.S. gallon = 3.79 L; 1 British gallon = 4.55 L
1m^3 (cubic metre) = 1000 litre = 264 U.S. gallon = 220 British gallon

Time
1h (hour) = 60 min (minute); 1 min = 60 s (second)
1 year = 365 $\frac{1}{4}$ day; 1 day = 24 h

Density
1 kg/m^3 = 10^{-3} g/cm^3

Angle
1 rev (ravolution) = 360°
π rad (radian) = 180° = 1/2 rev; 1 rad = $\frac{180}{\pi}$ = 57.3°

Force
1 N (newton) = 10^5 dyne = 0.225 lb
1 lb = 4.45 N

Pressure
1 Pa (pascal) = 1 N/m^2 = 10 dyne/cm^2 = 1.45×10^{-4} lb/in^2
1 atm (atmosphere) = 1.01 Pa = 76 cm-Hg = 14.7 lb/in^2

Energy; Work; Heat
1 cal (calorie) = 4.19 J (joule); 1 eV (electron-volt) = 1.6×10^{-19} J
1 kW.h (kilowatt -hour) = 3.6×10^6 J
1 J = 10^7 erg = 0.24 cal = 2.78 kW.h = 6.24×10^{18} eV

Power
1 hp (horsepower) = 746 W (watt) = 550 ft – lb/s (foot – pound per second)

Magnetic flux; Magnetic field
1 Wb (weber) = 10^8 maxwell
1 T (tesla) = 1 Wb/m^2 = 10^4 G (gauss); 1 G = 10^{-4} T

APPENDIX D (a)

SOME USEFUL MATHEMATICAL SIGNS

SIGN	MEANING
$\pm$	Plus or Minus
$=$	Equal to
$\approx$	Approximately equal to
$\sim$	Of the order of magnitude of
$\neq$	Not equal to
$\equiv$	Identical to
$>$	Greater than
$\gg$	Much greater than
$<$	Less than
$\ll$	Much less than
$\geq$	Greater than or equal to
$\leq$	Less than or equal to
$\propto$	Proportional to
Σ	Summation; The sum of
$\bar{x}$	Average value of x

APPENDIX D (b)

SOME USEFUL MATHEMATICAL RELATIONS

Algebra

1. $(a \pm b)^2 = a^2 + b^2 \pm 2ab$

2. $(a + b)^3 = a^3 + b^3 + 3ab\,(a+b)$

3. $(a^2 - b^2) = (a + b)\,(a - b)$

4. Roots of a quadratic equation:

Equation: $ax^2 + bx + c = 0$

Roots : $x = \dfrac{-b \pm \sqrt{b^2 - 4ac}}{2a}$

Geometry

(a) Area and Volume

(i) Triangle (base: c; height: h): Area $= \dfrac{1}{2}\,ch$

(ii) Circle (radius:r): Area $= \pi r^2$: Circumference $= 2\pi r$

(iii) Sphere (radius: r): Surface Area $= 4\pi r^2$; Volume $= \dfrac{4}{3}\pi r^3$

(iv) Right circular cylinder (radius: r; height: h):

Surface Area $= 2\pi r\,(r+h)$; Volume $= \pi r^2 h$

(b) Triangles

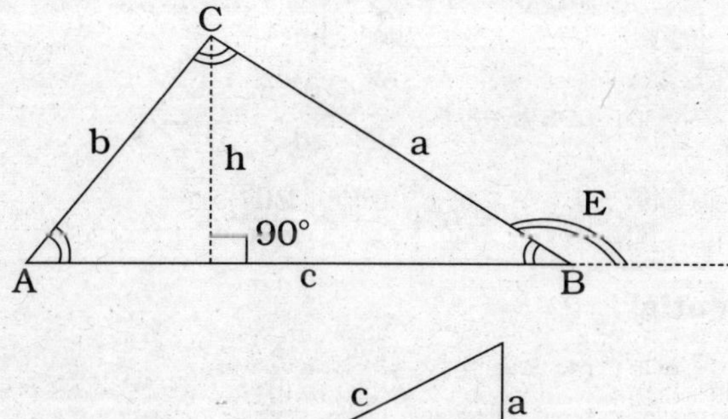

For the given triangle, we have:

(i) $\angle A + \angle B + \angle C = 180°$

(ii) Exterior angle, $\angle E - \angle A + \angle C$

(iii) $\dfrac{\sin A}{a} = \dfrac{\sin B}{b} = \dfrac{\sin C}{c}$

(iv) $c^2 = a^2 + b^2 - 2ab\cos C$

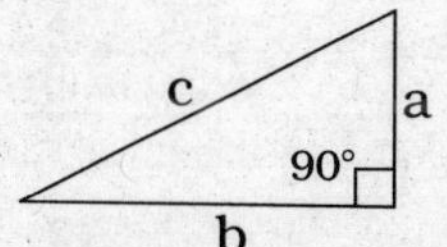

(c) Pythagorean relation (for a right triangle): $a^2 + b^2 = c^2$

Trigonometry

1. $\sin(2n\pi \pm \theta) = \pm \sin\theta$ (n = 0, 1, 2, 3,....................)

2. $\cos(2n\pi \pm \theta) = \cos\theta$

3. $\sin(90 - \theta) = \cos\theta$

4. $\cos(90 - \theta) = \sin\theta$

5. $\sin(90 + \theta) = \cos\theta$

6. $\cos(90 + \theta) = -\sin\theta$

7. $(\sin^2\theta + \cos^2\theta) = 1$

8. $(\sec^2\theta - \tan^2\theta) = 1$

9. $(\text{cosec}^2\theta - \cot^2\theta) = 1$

10. $\dfrac{\sin\theta}{\cos\theta} = \tan\theta$

11. $\sin 2\theta = 2\sin\theta\cos\theta$

12. $\cos 2\theta = \cos^2\theta - \sin^2\theta = 1 - 2\sin^2\theta = 2\cos^2\theta - 1$

13. $\sin(\alpha \pm \beta) = \sin\alpha\cos\beta \pm \cos\alpha\sin\beta$

14. $\cos(\alpha \pm \beta) = \cos\alpha\cos\beta \mp \sin\alpha\sin\beta$

15. $\tan(\alpha \pm \beta) = \dfrac{\tan\alpha \pm \tan\beta}{1 \pm \tan\alpha\tan\beta}$

16. $\sin\alpha \pm \sin\beta = 2\sin\frac{1}{2}(\alpha \pm \beta)\cos\frac{1}{2}(\alpha \mp \beta)$

17. $(\cos\alpha + \cos\beta) = 2\cos\frac{1}{2}(\alpha + \beta)\cos\frac{1}{2}(\alpha - \beta)$

18. $(\cos\alpha - \cos\beta) = -2\sin\frac{1}{2}(\alpha + \beta)\sin\frac{1}{2}(\alpha - \beta)$

Expansions

(i) Trigonometric (θ in radian)

$$\sin\theta = \theta - \frac{\theta^3}{\lfloor 3} + \frac{\theta^5}{\lfloor 5} - \ldots\ldots\ldots\ldots$$

$$\cos\theta = 1 - \frac{\theta^2}{\lfloor 2} + \frac{\theta^4}{\lfloor 4} - \ldots\ldots\ldots\ldots$$

$$\tan\theta = \theta + \frac{\theta^3}{3} + \frac{2\theta^5}{15} + \ldots\ldots\ldots\ldots$$

(ii) Exponential

$$e^x = 1 + x + \frac{x^2}{\lfloor 2} + \frac{x^3}{\lfloor 3} + \ldots\ldots\ldots\ldots$$

(iii) Logarithmic ($x < 1$)

$$\log(1 + x) = x - \frac{x^2}{2} + \frac{x^3}{3} - \ldots\ldots\ldots$$

(iv) Binomial theorem ($x^2 < 1$)

$$(1 + x)^n = 1 + \frac{nx}{\lfloor 1} + \frac{n(n-1)x^2}{\lfloor 2} + \ldots\ldots\ldots\ldots$$

Vector Products

(i) For unit vectors $\hat{i}, \hat{j}, \hat{k}$ (along x, y, z directions, respectively):

$$\hat{i}.\hat{i} = \hat{j}.\hat{j} = \hat{k}.\hat{k} = 1$$

$$\hat{i}.\hat{j} = \hat{j}.\hat{k} = \hat{k}.\hat{i} = 0$$

$$\hat{i}\times\hat{i} = \hat{j}\times\hat{j} = \hat{k}\times\hat{k} = 0$$

$$\hat{i}\times\hat{j} = \hat{k},\quad \hat{j}\times\hat{k} = \hat{i},\quad \hat{k}\times\hat{i} = \hat{j}$$

(ii) A vector $\vec{a}$ with components a_x, a_y, a_z (along the x, y, z directions, respectively) can be expressed as:

$$\vec{a} = a_x\,\hat{i} + a_y\,\hat{j} + a_z\,\hat{k}$$

(iii) For any arbitrary vectors $\vec{a}$, $\vec{b}$, $\vec{c}$

(a) $$\vec{a}\times(\vec{b}+\vec{c}) = (\vec{a}\times\vec{b}) + (\vec{a}\times\vec{c})$$

(b) With θ as the smaller angle between two vectors $\vec{a}$ and $\vec{b}$:

$$\vec{a}.\vec{b} = \vec{b}.\vec{a} = a_xb_x + a_yb_y + a_zb_z = ab\cos\theta$$

$$|\vec{a}\times\vec{b}| = ab\sin\theta$$

(c) $$(\vec{a}\times\vec{b}) = -(\vec{b}\times\vec{a}) = \begin{vmatrix}\hat{i} & \hat{j} & \hat{k}\\ a_x & a_y & a_z\\ b_x & b_y & b_z\end{vmatrix}$$

$$= \hat{i}\begin{vmatrix}a_y & a_z\\ b_y & b_z\end{vmatrix} - \hat{j}\begin{vmatrix}a_x & a_z\\ b_x & b_z\end{vmatrix} + \hat{k}\begin{vmatrix}a_x & a_y\\ b_x & b_y\end{vmatrix}$$

$$= (a_y b_z - b_y a_z)\hat{i} + (a_zb_x - b_za_x)\,\hat{j} + (a_x b_y - b_x a_y)\,\hat{k}$$

(d) $$\vec{a}.(\vec{b}\times\vec{c}) = \vec{b}.(\vec{c}\times\vec{a}) = \vec{c}.(\vec{a}\times\vec{b})$$

(e) $$\vec{a}\times(\vec{b}\times\vec{c}) = (\vec{a}.\vec{c})\,\vec{b} - (\vec{a}.\vec{b})\,\vec{c}$$

Differentials

The differentiation is made with respect to x. A and B are functions of x.

1. $\frac{d}{dx}(x) = 1$

2. $\frac{d}{dx}(kA) = k\frac{dA}{dx}$ (k:a constant)

3. $\frac{d}{dx}(A+B) = \frac{dA}{dx} + \frac{dB}{dx}$

4. $\frac{d}{dx}(x^m) = mx^{m-1}$ (m: a constant)

5. $\frac{d}{dx}(\log x) = \frac{1}{x}$

6. $\frac{d}{dx}(AB) = A\frac{dB}{dx} + B\frac{dA}{dx}$

7. $\frac{d}{dx}(e^x) = e^x$

8. $\frac{d}{dx}(e^A) = e^A\frac{dA}{dx}$

9. $\frac{d}{dx}(\sin x) = \cos x$

10. $\frac{d}{dx}(\cos x) = -\sin x$

11. $\frac{d}{dx}(\tan x) = \sec^2 x$

12. $\frac{d}{dx}(\cot x) = -\operatorname{cosec}^2$

13. $\frac{d}{dx}(\sec x) = \tan x \sec x$

14. $\frac{d}{dx}(\operatorname{cosec} x) = -\cot x \operatorname{cosec} x$

15. $\frac{d}{dx}(\sin A) = \cos A\frac{dA}{dx}$

16. $\frac{d}{dx}(\cos A) = -\sin A\frac{dA}{dx}$

Integrals

The integration is made with respect to x. A and B are functiions of x. (An arbitrary constant of integration has to be added to each indefinite integral).

1. $\int dx \, (= 1 \int 1.dx) = x$

2. $\int kB \, dx = k \int B \, dx$ (k: A constant)

3. $\int (A+B) \, dx = \int Adx + \int Bdx$

4. $\int x^m \, dx = \frac{x^{m+1}}{(m+1)}$ (m: a constant; ≠ 1)

5. $\int A \frac{dB}{dx} dx = AB - \int B \frac{dA}{dx} dx$

6. $\int \frac{1}{x} dx = \log x$

7. $\int \sin x \, dx = -\cos x$

8. $\int \cos x \, dx = \sin x$

9. $\int \tan x \, dx = \log (\sec x)$

10. $\int \sin^2 x \, dx = \frac{1}{2} x - \frac{1}{4} \sin 2x$

11. $\int e^{-kx} dx = -\frac{1}{a} e^{-kx}$

12. $\int \frac{dx}{\sqrt{x^2 + a^2}} = \log\left(x + \sqrt{x^2 + a^2}\right)$

APPENDIX E

SOME FUNAMENTAL PHYSICAL CONSTANTS

CONSTANT	SYMBOL	VALUE
Velocity of light (in free space)	**c**	3.00×10^{8} ms^{-1}
Charge on electron	**e**	1.60×10^{-19} C
Avogadro constant	$\mathbf{N_A}$	6.02×10^{23} mol^{-1}
Planck constant	**h**	6.63×10^{-34} J.s
Gravitational constant	**G**	6.67×10^{-11} N.m^{2}.kg^{-2}
Boltzmann constant	**k**	1.38×10^{-23} JK^{-1}
Stefan-Boltzmann constant	σ	5.67×10^{-8} Wm^{-2} K^{4}
Universal gas constant	**R**	8.31 J mol^{-1} K
Permittivity constant	ϵ_0	8.85×10^{-12} Fm^{-1}
Permeability constant	μ_0	1.26×10^{-6} Hm^{-1}
Mass of electron	$\mathbf{m_e}$	9.11×10^{-31} kg
Mass of proton	$\mathbf{m_p}$	1.67×10^{-27} kg
Mass of neutron	$\mathbf{m_n}$	1.68×10^{-27} kg
Electron charge-to-mass ratio	$\mathbf{e/m_e}$	1.76×10^{11} C kg^{-1}
Ratio of proton mass to electron mass	$\mathbf{m_p/m_e}$	1840
Bohr radius	$\mathbf{r_B}$	5.29×10^{-11} m
Rydberg constant	**R**	1.10×10^{7} m^{-1}
Electron magnetic moment	μ_e	9.28×10^{-24} JT^{-1}
Proton magnetic moment	μ_p	1.41×10^{-26} JT^{-1}
Bohr magneton	μ_B	9.27×10^{-24} JT^{-1}
Nuclear magneton	μ_N	5.05×10^{-27} JT^{-1}

APPENDIX F

SOME IMPORTANT APTRONOMICAL VALUES

Distance from Earth

Moon (mean distance): 3.82×10^8 m

Sun (mean distance): 1.50×10^{11} m

Nearest Star (Proxima Centauri): 4.04×10^{16} m

Edge of the observable universe: $\sim 10^{26}$ m

Comparative Values for Sun, Earth, Moon

ENTITY	SUN	EARTH	MOON
Mass (kg)	1.99×10^{30}	5.98×10^{24}	7.36×10^{22}
Mean radius (m)	6.96×10^8	6.37×10^6	1.74×10^6
Mean density (kg m^{-3})	1410	5520	3340
Free-fall acceleration Near surface (ms^{-2})	274	9.81	1.67
Escape velocity (km s^{-1})	618	11.2	2.38
Period of rotation*1	37 day at poles; 26 day at equator*2	23 h 56 min	27.3 day

*1 Measured with respect to distant stars.

*2 The sun does not rotate as a single rigid body, as it consists of inhomogeneous gaseous matter.

APPENDIX G

IMPORTANT PROPERTIES OF SOME SELECTED ELEMENTS*

ELEMENT	SYMBOL	ATOMIC NO. (Z)	MOLAR MASS (g/mol)	DENSITY (g/cm^3)	MELTING PT. (°C)	BOILING PT. (°C)
Aluminum	**Al**	13	26.98	2.70	660	2450
Antimony	**Sb**	51	121.75	6.69	630.5	1380
Argon	**Ar**	18	39.95	0.0017	–189.4	–185.8
Arsenic	**As**	33	74.92	5.78	817 (28 atm)	613
Barium	**Ba**	56	137.34	3.59	729	1640
Beryllium	**Be**	4	9.01	1.85	1287	2770
Bismuth	**Bi**	83	208.98	9.75	271.4	1560
Boron	**B**	5	10.81	2.34	2030	—
Bromine	**Br**	35	79.91	3.12 (liquid)	–7.2	58
Cadmium	**Cd**	48	112.40	8.65	321	765
Calcium	**Ca**	20	40.08	1.55	838	1440
Carbon	**C**	6	12.01	2.26	3727	4830
Cesium	**Cs**	55	132.91	1.87	28.4	690
Chlorine	**Cl**	17	35.45	0.0032 (0 °C)	–101	–34.7
Chromium	**Cr**	24	52.00	7.19	1857	2665
Cobalt	**Co**	27	58.93	8.85	1495	2900
Copper	**Cu**	29	63.54	8.96	1083	2595
Fluorine	**F**	9	19.00	0.0017 (0 °C)	–219.6	–188.2
Gallium	**Ga**	31	69.72	5.91	29.8	2237
Germanium	**Ge**	32	72.59	5.32	937.3	2830
Gold	**Au**	79	196.97	19.32	1064	2970
Helium	**He**	2	4.00	0.00016	–269.7	–268.9
Hydrogen	**H**	1	1.01	0.00008	–259.2	–252.7
Indium	**In**	49	114.82	7.31	156.6	2000
Iodine	**I**	53	126.90	4.93	113.7	183
Iridium	**Ir**	77	192.2	22.5	2447	(5300)
Iron	**Fe**	26	55.85	7.87	1536	3000
Krypton	**Kr**	36	83.80	0.0035	–157.4	–152
Lanthanum	**La**	57	138.91	6.19	920	3470
Lead	**Pb**	82	207.19	11.35	327.45	1725
Lithium	**Li**	3	6.94	0.53	180.6	1300
Magnesium	**Mg**	12	24.31	1.74	650	1107
Manganese	**Mn**	25	54.94	7.44	1244	2150
Mercury	**Hg**	80	200.59	13.55	–38.9	357

ELEMENT	SYMBOL	ATOMIC NO. (Z)	MOLAR MASS (g/mol)	DENSITY (g/cm³)	MELTING PT. (°C)	BOILING PT. (°C)
Molybdenum	**Mo**	42	95.94	10.22	2617	5560
Neon	**Ne**	10	20.18	0.0008	–248.6	–246.0
Nickel	**Ni**	28	58.71	8.90	1453	2730
Niobium	**Nb**	41	92.91	8.57	2468	4927
Nitrogen	**N**	7	14.01	0.0012	–210	–195.8
Oxygen	**O**	8	16.00	0.0013	–218.8	–183.0
Palladium	**Pd**	46	106.4	12.02	1552	3980
Phosphorus	**P**	15	30.97	1.83	44.3	280
Platinum	**Pt**	78	195.09	21.45	1769	4530
Plutonium	**Pu**	94	(244)	19.8	640	3235
Polonium	**Po**	84	(210)	9.32	254	—
Potassium	**K**	19	39.10	0.86	63.2	760
Radium	**Ra**	88	(226)	5.0	700	—
Radon	**Rn**	86	(222)	0.0100 (0 °C)	(–71)	–61.8
Rhodium	**Rh**	45	102.91	12.41	1963	4500
Rubidium	**Rb**	37	85.47	1.53	39.5	688
Selenium	**Se**	34	78.96	4.79	221	685
Silicon	**Si**	14	28.09	2.33	1412	2680
Silver	**Ag**	47	107.87	10.49	960.8	2210
Sodium	**Na**	11	22.99	0.97	97.9	892
Strontium	**Sr**	38	87.62	2.54	768	1380
Sulphur	**S**	16	32.06	2.07	119.0	444.6
Tantalum	**Ta**	73	180.95	16.6	3014	5425
Tellurium	**Te**	52	127.60	6.24	449.5	990
Thallium	**Tl**	81	204.37	11.85	304	1457
Thorium	**Th**	90	(232)	11.72	1755	(3850)
Tin	**Sn**	50	118.69	7.30	231.9	2270
Titanium	**Ti**	22	47.90	4.54	1670	3260
Tungsten	**W**	74	183.85	19.3	3380	5930
Uranium	**U**	92	(238)	18.95	1132	3818
Vanadium	**V**	23	50.94	6.11	1902	3400
Xenon	**Xe**	54	131.30	0.0055	–111.8	–108
Yttrium	**Y**	39	88.91	4.47	1526	3030
Zinc	**Zn**	30	65.37	7.13	419.6	906
Zirconium	**Zr**	40	91.22	6.51	1852	3580

*All the listed values are for 1 atmosphere pressure, unless indicated otherwise. The density values are for 20 °C, unless indicated otherwise. For radioactive elements, the bracketed value of the molar mass is the mass number of the longest-lived isotope of the element.

Subject Index

SUBJECT INDEX

A

B

D

E

F

G

H

I

J

K

L

M

N

Q

R

S

T

U

V

W

X

Y

Z